XXVII IUPAP Conference on Computational Physics (CCP 2015)

AA006627

Journal of Physics: Conference Series Volume 759

Guwahati, India
2 – 5 December 2015

Editors:

Sitangshu Bikas Santra
Purusattam Ray

ISBN: 978-1-5108-3226-8
ISSN: 1742-6588

Printed from e-media with permission by:

Curran Associates, Inc.
57 Morehouse Lane
Red Hook, NY 12571

Some format issues inherent in the e-media version may also appear in this print version.

Printed by Curran Associates, Inc. (2016)

For permission requests, please contact the Institute of Physics
at the address below.

Institute of Physics
Dirac House, Temple Back
Bristol BS1 6BE UK

Phone: 44 1 17 929 7481
Fax: 44 1 17 920 0979

techtracking@iop.org

Additional copies of this publication are available from:

Curran Associates, Inc.
57 Morehouse Lane
Red Hook, NY 12571 USA
Phone: 845-758-0400
Fax: 845-758-2633
Email: curran@proceedings.com
Web: www.proceedings.com

Table of contents

Volume 759

XXVII IUPAP Conference on Computational Physics (CCP2015)
2–5 December 2015, Guwahati, India

Accepted papers received: 22 September 2016
Published online: 11 November 2016

Computational Biological Physics

Material and Nano Sciences

Atomic, Molecular and Optical Physics

Astrophysics, Plasma, Gravitation and Cosmology

High energy, Nuclear and Particle Physics

Fluid Dynamics, Oceanography, Geophysics and Climate modeling

Software and Hardware development

Journal of Physics: Conference Series **759** (2016) 011001 doi:10.1088/1742-6596/759/1/011001

XXVII IUPAP Conference on Computational Physics

Sitangshu Bikas Santra[1] **and Purusattam Ray**[2]

[1]Department of Physics, Indian Institute of Technology Guwahati, Guwahati-781 039, India
[2]Theoretical Physics, The Institute of Mathematical Sciences, Chennai - 600 113, India

E-mail: [1]`santra@iitg.ernet.in`, [2]`ray@imsc.res.in`

Editorial

The 27th IUPAP Conference on Computational Physics, CCP2015, was held in the heritage city Guwahati, in the eastern part of India, next to the mighty river Brahmaputra, during December 2-5, 2015. The Conference on Computational Physics is organized annually under the auspices of Commission 20 (C20) of the IUPAP (International Union of Pure and Applied Physics). This is the first time it has been held in India. Almost 300 participants from 25 countries convened at the auditorium and lecture halls at the Indian Institute of Technology Guwahati for four days. Thirteen plenary speakers, fifty six invited speakers, three presnters from the computer industries and two hundred and eight contributory participants coverd a broad range of topics in computational physics and related areas. Thirty eight women participated in CCP2015 and seven of them presented invited talks. This volume of Journal of Physics: Conference Series contains the proceedings of the scientific contributions presented at the Conference.

The main purpose of the meeting was to discuss the progress, opportunities and challenges of common interest to physicists engaged in computational research. Computational physics has taken giant leaps during the lat few years, not only because of the enormous increases in computer power but especially because of the development of new methods and algorithms. Computational physics now represents a third leg of research alongside analytical theory and experiments. A meeting such as CCP, must have sufficient depth in different areas and at the same time should be broad and accessible. The topics covered in this conference were: Materials/Condensed Matter Theory and Nanoscience, Strongly Correlated Systems and Quantum Phase Transitions, Quantum Chemistry and Atomic Physics, Quantum Chromodynamics, Astrophysics, Plasma Physics, Nuclear and High Energy Physics, Complex Systems: Chaos and Statistical Physics, Macroscopic Transport and Mesoscopic Methods, Biological Physics and Soft Materials, Supercomputing and Computational Physics Teaching, Computational Physics and Sustainable Energy.

As organizers and editors of these Proceedings, we are very pleased with the number and the quality of the papers provided by the participants. The papers cover a good cross-section of what was presented at the meeting. We are sure that these will represent the state of the computational physics today. The remainder of this Preface contains lists detailing the organizational structure of CCP2015, endorsers and sponsors of the meeting, plenary and invited talks, and a presentation of the 2015 IUPAP C20 Young Scientist Prize.

Finally, we would like to express our sincere thanks to our sponsors and endorsers: C20 Commission of International Union of Pure and Applied Physics (IUPAP); Division of Computational Physics, American Physical Society (APS); European Physicsl Society (EPS);

Association of Asia Pacific Physical Societies (AAAPS), Intel Software; Nvidia; Fujitsu; Netweb Technologies; The Institute of Mathematical Sciences Chennai, and finally Indian Institute of Technology Guwahati. We are grateful to the organizing committee, the International Advisory Board, the local organizing committee and participants who helped in making CCP2015 a success.

Conference Organization

Chairs
Sitangshu Bikas Santra, IIT Guwahati
Purusattam Ray, IMSc Chennai

Vice Chairs
Bipul Bhuyan, IIT Guwahati
Ronojoy Adhikari, IMSc Chennai

Secretary
Sourav Chattopadhyay, IIT Guwahati

Local organizing committee
Jasjeet Singh Bagla, IISER Mohali
Saurabh Basu, IIT Guwahati
Swati Bhattacharya, IIT Guwahati
Jaydeep Bhattacharya, NISER Bhubaneswar
Biplab Bose, IIT Guwahati
Sayan Chakraborty, IIT Guwahati
Kedar S. Damle, TIFR Mumbai
Santabrata Das, IIT Guwahati
G. P. Das, IACS Kolkata
A. K. Dasmahapatra, IIT Guwahati
Tarak Nath Dey, IIT Guwahati
S. Digal, IMSc Chennai
Subhradip Ghosh, IIT Guwahati
Smarajit Karmakar, TIFR, Hyderabad
Sanjay Kumar, BHU Varanasi
Gagan Kumar, IIT Guwahati
P. B. Sunil Kumar, IIT Chennai
Prabal Kumar Maiti, IISc Bangalore
Pinaki Majumdar, HRI Allahbad
P. K. Padmanabhan, IIT Guwahati
A. N. Panda, IIT Guwahati
Poulose Poulose, IIT Guwahati
Manabendra Sarma, IIT Guwahati
Amarendra K. Sarma, IIT Guwahati
Anjan Ananda Sen, JMI New Delhi
Tarun Souradeep, IUCAA Pune
L. Sriramkumar, IIT Chennai
Gaurav Trivedi, IIT Guwahati

S. Vemparala, IMSc Chennai

International advisory board

Joan Adler, Technion, Israel
James F Amundson, Fermilab USA
Norbert Attig, Juelich Supercomputing Centre, Germany
Amanda Barnard, CSIRO, Australia
Eddie A. Baron, University of Oklahoma, USA
George Batrouni, Universite de Nice-Sophia Antipolis, France
Rajeev Bhatla, Banaras Hindu University, India
Bimalendu B. Bhattacharya, SNBNCBS, India
Kurt Binder, Johannes Gutenberg-University in Mainz, Germany
Gautam Biswas, IIT Guwahati, India
Indrani Bose, Bose Institute, India
Nithaya Chetty, University of Pretoria, RSA
Giovanni Ciccotti, University of Rome La Sapienza, Italy
R. Dickman, NIST for Complex Systems, Brazil
O. Eriksson, Uppsala University, Sweden
Daan Frenkel, University of Cambridge, UK
J. Gubernatis, Los Alamos National Laboratory, USA
Sourendu Gupta, TIFR Mumbai, India
Satoshi Hamaguchi, Osaka University, Japan
Alex Hansen, NTNU, Norway
Takanori Hara, KEK, Japan
Morten Hjorth-Jensen, Michigan State University, USA, and University of Oslo, Norway
Werner Krauth, Ecole Normale Suprieure, France
David P. Landau, University of Georgia, USA
R. H. Landau, Oregon State University, USA
H. Q. Lin, Beijing Computational Science Research Center, China
Priya Mahadevan, SNBNCBS, India
J. Marro, University of Granada, Spain
Giuseppina Orlandini, University of Trento, Italy
Giorgio Parisi, Sapienza University of Rome, Italy
S. Ramasesha, IISc Bangalore, India
Claudio Rebbi, Boston University, USA
Mathieu Salanne, Universit Pierre et Marie Curie, Paris, France
Anders Sandvik, Boston University, USA
Srikanth Sastry, JNCASR, India
Abhijit Sen, Institute for Plasma Research, India
Lev Shchur, Landau Institute of Theoretical Physics, Russia
Rajiv Singh, University of California, USA
Berend Smit, University of California, USA
Balasubramanian Sundaram, JNCASR, India
Hideaki Takabe, Osaka University, Japan
David Wales, University of Cambridge, UK
J. S. Wang, National University of Singapore, Singapore
Marek S. Wartak, Wilfrid Laurier University, Canada
S. R. White, University of California, USA
Julia Yeomans, University of Oxford, UK

IUPAP Young Scientist Award

The 2014 IUPAP Young Scientist Prize for Computational Physics was awarded during CCP2015 at IIT Guwahati by the C20 Commission to Professor Mathieu Salanne of Université Pierre et Marie Cure, Paris, France, as Dr Salanne could not attend the CCP2014, in recognition of his development of appropriate methods to allow realistic atomistic simulation of molten salts and ionic liquids in situations of relevance to electrochemistry.

The 2015 Young Scientist Prize awarded to Dr. Wei-Min Wang at Laser-plasma physics of the Institute of Physics, Chinese Academy of Sciences, China for his significant achievements in computational plasma physics with applications to advanced schemes of inertial confined fusion and novel laser-plasma based particle accelerators and radiation sources. However, the prize was not presented to Dr. Wang at the CCP2015 because of his inability to attend CCP2015.

For further information on the Prize and the nomination process, please see information online at `http://phycomp.technion.ac.il/~C20/prizes.html`

Endorsers and Sponsors of CCP2015

The organizations who sponsor or endorse CCP2015 are listed below:

- The International Union of Pure and Applied Physics (IUPAP)
- The American Physical Society (APS)
- The European Physical Society (EPS)
- The Association of Asia Pacific Physical Societies (AAAPS)
- Intel Software
- Nvidia
- Fujitsu
- Netweb Technologies
- The Institute of Mathematical Sciences, Chennai, India
- Indian Institute of Technology Guwahati, India

We would like to take this opportunity to thank all the sponsors for their generous support, which was critical to the success of CCP2015.

Plenary Presentations

A total of 13 plenary talks were presented over the four days of the meeting. The title of the presentation with presenters affiliation are listed below:

- Annette Zippelius, University of Goettingen, Germany, *Dense Granular Flow of Frictional Particles*
- Barry I. Schneider, National Institute of Standards and Technology, USA, *How Novel Algorithms and Access to High Performance Computing Platforms are Enabling Scientific Progress in Atomic and Molecular Physics*
- Carlo Massimo Casciola, University of Rome La Sapienza, Italy, *Energy fluxes, scale energy and turbulent separation*
- Gabriel Kotliar, Rutgers University, USA, *Understanding Strongly Correlated Materials thru Theory Algorithms and High Performance Computers*

- Hans J. Herrmann, ETH Zurich, Switzerland, *Packing of wires in cavities and growing surfaces*
- Luca Baiotti, Osaka University, Japan, *Numerical relativity for binary neutron-star mergers*
- Norbert Attig, Juelich Supercomputing Centre, Germany, *Impacts of Current Hardware and Software Developments on Simulation Sciences*
- Richard Martin, Illinois State University, USA, *Undergraduate Computational Physics Education: Uneven History and Promising Future*
- Shyam S Rai, Indian Institute of Science Education and Research Pune, India, *Imaging Earths Interior through Noise Interferometry*
- Steve Gottlieb, Indiana University, USA, *Lattice Field Theory Applications in High Energy and Nuclear Physics*
- Turab Lookman, Los Alamos National Laboratory, USA, *Information-driven approach to materials discovery and design*
- William Goddard, California Institute of Technology, USA, *In silico predictions of 3D structures and binding sites for GPCRligand complexes*
- Yutaka Ishikawa, RIKEN Advanced Institute for Computational Science, Japan, *An Overview of the Next Flagship Supercomputer in Japan*

Industry Sessions

- Austin Cherian, INTEL Software
- Ashok Chaudhary, FUJITSU
- Gaurav Mishra, NVIDIA
- Rama Kishan V Malladi, INTEL Software

Parallel and Poster Sessions

The parallel sessions were organized as eleven different tracks with a mix of invited (30 minute) and contributed (15 minute) talks, and one session on General Computational Physics with eight contributed talks. In total, 56 invited talks and almost 100 contributed presentations were given. Poster sessions with almost 100 presenters were held during the conference. The invited talks are listed below:

Statistical Physics, Complex system and Non-linear Dynamics

- Jian-Sheng Wang, National University of Singapore, Singapore, *Computational methods for quantum thermal transport in nanostructures*
- Mathis Plapp, Ecole Polytechnique, Palaiseau, France, *Phase-field models for the growth and evolution of complex structures*
- Pinaki Chaudhuri, IMSc Chennai, India, *Onset of flow in soft glasses: linking percolation to flow*
- Sitabhra Sinha, IMSc Chennai, India, *Chaos, Complexity and Computation: Patterns in oscillatory media*
- Smarajit Karmakar, TIFR Hyderabad, India, *Short-time β-relaxation in glass-forming liquids is cooperative in nature*

- Wolfhard Janke, Institut fr Theoretische Physik, Germany, *Finite-Size Scaling at the Droplet Condensation-Evaporation Transition*

Soft materials and Polymer Physics

- Michael Bachmann, University of Georgia, *Statistical Analysis of Transitions in Finite Polymer Systems*
- Rajarshi Chakrabarti, IIT Bombay, India, *Single polymer chain with internal friction*
- Sumesh P Thampi, IIT Madras, India, *Intrinsic free energy in active nematics*
- Takahiro Hatano, University of Tokyo, Japan, *Critical slowing down at jamming transition*
- Tanja Schilling, University of Luxembourg, Luxembourg, *Percolation in colloidal model systems*

Computational Biological Physics

- Birgit Strodel, Structural Biochemistry Research Centre Juelich, Germany, *Thermodynamics and kinetics of amyloid aggregation from atomistic simulations*
- Durba Sengupta, NCL Pune, India, *Association at the membrane: Do lipophobic effects mediate membrane protein assembly?*
- Prabal Kumar Maiti, IISc Bangalore, India, *Effect of Dendrimer in HIV-1 inhibition*
- Sanjoy Bandyopadhyay, IIT Kharagpur, India, *Correlated conformational features and hydration dynamics of protein-DNA complexes*

Quantum many body and strongly correlated systems

- Federica Agostini, Max-Planck-Institut Halle, Germany, *Electron-Nuclear Correlation: A Time-Dependent Perspective*
- Hong-Gang Luo, Lanzhou University, China, *Phase diagram of one-dimensional t-J model with long-range dipolar interactions*
- Kedar S. Damle, TIFR Mumbai, India, *Sign-free Quantum Monte Carlo simulation of certain frustrated quantum magnets*
- Mark Jarrell, Louisiana State University, USA, *Typical Medium Dynamical Cluster Simulations of Interacting and Strongly Disordered Systems*
- Silke Biermann, École Polytechnique, France, *Screened exchange dynamical mean field theory*
- Wei Ku, Brookhaven National Lab, USA, *Spin/Orbital Correlation and Itinerancy-Enhanced Quantum Fluctuation in Iron-Based Superconductors*

Material and Nano Sciences

- A Marco Saitta, UPMC, France, *Reaction networks and free-energy landscapes via a novel topological approach: ab initio Miller-like experiments and beyond*
- G. P. Das, IACS Kolkata, India, *Substrate induced modulation of physical and chemical properties of quasi two-dimensional nanostructures*
- Priya Mahadevan, SNBNCBS Kolkata, India, *The dilute magnetic semiconductors: A playground for the ZaanenSawatzky-Allen phase diagram*
- Rajeev Ahuja, Uppsala University, Sweden, *Materials for Li & Na Batteries : A Computational Materials Science Point of View*
- Stefano Sanvito, Trinity College Dublin, Ireland, *The magnetic genome project*
- Yuan-Ping Feng, NUS, Singapore, *2D Materials and Devices for Spintronics: First-principles Studies*

Atomic, Molecular and Optical Physics

- Dibakar Roy Chowdhury, Mahindra Ecole Centrale Hyderabad, India, *Near field coupled Metamaterials for Terahertz Wavelengths*
- E. Krishnakumar, TIFR Mumbai, India, *Dynamics of Electron - Molecule Resonances: A Challenge to Computational Physics*
- Gagan Kumar, IIT Guwahati, India, *Terahertz guided wave devices using plasmonic metamaterials*
- Gautam Vemuri, Indiana University, USA, *Realization of a PT-symmetric dimmer in coupled semiconductor lasers*
- Harishankar Ramachandran, IIT Madras, India, *Modelling Dielectric Structures with Bends using the Beam Propagation Method*
- Harshawardhan Wanare, IIT Kanpur, India, *Interplay between light and magnetic field in inducing anisotropy in atomic systems*
- Satrajit Adhikari, IACS Kolkata, India, *Beyond Born-Oppenheimer Theories: Diabatic PESs for Spectroscopic & Scattering Processes*

Astrophysics, Plasma, Gravitation and Cosmology

- Adam Amara, ETH Zurich, Switzerland, *Information from Cosmology Experiments*
- Ganesh Rajaraman, IPR Gandhinagar, India, *Strongly correlated classical plasmas under external forcing and dissipation - examples using molecular dynamics approach*
- Parameswaran Ajith, ICTS Bangalore, India, *Towards gravitational-wave astronomy*
- Pankaj Joshi, TIFR Mumbai, India, *Gravitational Collapse, Black Holes and Space-time Singularities in Einstein Gravity*
- Tarun Souradeep, IUCAA Pune, India, *Bayesian Cosmological inference beyond statistical isotropy*

High energy, Nuclear and Particle Physics

- Andreas Juettner, University of Southampton, UK, *Flavour physics and lattice QCD*
- Nilmani Mathur, TIFR Mumbai, India, *Heavy quark physics using large scale simulations*
- Pushan Majumdar, IACS Kolkata, India, *Lattice QCD simulations using the OpenACC platform*
- Satyaki Bhattacharya, Saha Institute of Nuclear Physics, India, *Multivariate techniques in analysis of the LHC data*
- Yoshinobu Kuramashi, University of Tsukuba, Japan, *Tensor Network scheme for lattice gauge theories*

Fluid Dynamics, Oceanography, Geophysics and Climate modeling

- Gautam Biswas, IIT Guwahati, India, *Simulation of Multi-Fluid System: Understanding Drop Impact*
- L. S. Luo, Old Dominion University, USA, *Multiscale Modeling and Simulation of Gaseous Microflows*
- Prasad Perlekar, TIFR Hyderabad, India, *Two and three-dimensional binary-fluid turbulence*
- Rajeev Bhatla, Banaras Hindu University, India, *Simulation of different phases of Indian summer monsoon using Regional Climate Model*
- Ronojoy Adhikari, IMSc Chennai, India, *Computation of many-body hydrodynamic interactions in active colloidal suspensions*

- Tapan K Sengupta, IIT Kanpur, India, *Non-equilibrium thermodynamics of Rayleigh-Taylor instability*
- Tapan Mukerji, Stanford University, USA, *Computational Rock Physics: Imaging and Process Modeling at the Pore Scale*

Computational Physics Education

- Ananda Dasgupta, IISER Kolkata, India, *Open Source Computation in Physics Education*
- KPN Murthy, University of Hyderabad, India, *Non-Boltzmann Ensembles and Monte Carlo Simulation*
- Scott A Lathrop, University of Illinois, USA, *Expanding Access to Computational Science Education*

Software and Hardware development

- Prabhu Ramachandran, IIT Bombay, India, *Extensible, reusable, and reproducible computing: a case study of PySPH*
- Scott A. Klasky, ORNL, USA, *SIRIUS: Science-driven Data Management for Multi-tiered Storag*

Peer review statement

All papers published in this volume of *Journal of Physics: Conference Series* have been peer reviewed through processes administered by the proceedings Editors. Reviews were conducted by expert referees to the professional and scientific standards expected of a proceedings journal published by IOP Publishing.

Liquid Fluidized Beds of Granular Particles

Andrea Fiege
Georg-August-Universität Göttingen, Institut für Theoretische Physik,
Friedrich-Hund-Platz 1, 37077 Göttingen

Annette Zippelius
Georg-August-Universität Göttingen, Institut für Theoretische Physik,
Friedrich-Hund-Platz 1, 37077 Göttingen

E-mail: annette@theorie.physik.uni-goettingen.de

Abstract.
We suggest a simple model for the dynamics of granular particles in suspension which is suitable for an event driven algorithm, allowing to simulate $N = \mathcal{O}(10^6)$ particles or more. As a first application we consider a dense granular packing which is fluidized by an upward stream of liquid, i.e. a fluidized bed. In the stationary state, when all forces balance, we always observe a well defined interface whose width is approximately independent of packing fraction. We also study the dynamics of expansion and sedimentation after fast and slow changes in flow rate and determine the timescales to reach a stationary state.

1. Introduction

Fluidized beds are ubiquitous in chemical and pharmaceutical processes, including tablet coating, catalytic cracking, coal combustion and incineration [1]. They consist of a collection of solid particles immersed in a fluid which can be a gas or a liquid. The fluid is flowing upwards through the particles and fluidizes the packing. Despite the simple setup, fluidized beds show a variety of complex flow regimes. Gas-fluidized beds usually are unstable and exhibit bubbling instabilities [2; 3], however intervals of stable fluidization have also been observed [4; 5] Liquid-fluidized beds are more stable and a wider range of flowrates gives rise to homogeneous fluidization [6], in particular for low Reynolds number flow. Instabilities have been observed in the form of voidage waves. These waves can be one-dimensional in narrow beds [7] or develop transverse structures in wider beds [8], that have been conjectured to cause bubbles [9].

Despite their seemingly simple ingredients, fluidized beds are far from being well understood. As fluidized beds consist of particles and a surrounding fluid, both, interparticle- and particle-fluid-interactions must be accounted for in an appropriate model. Modeling the fluidized bed by multiphase continua has been proven successful to describe the onset and propagation of instabilities [10; 11]. Stronger simplifications were adopted by Johri and Glasser [12; 13], who showed that a suspension with nonuniform concentration can behave like a continuum with nonuniform density subject to a density-dependent force. Although this is a major simplification, the model is able to predict wavelike instabilities in narrow fluidized beds.

Homogeneous fluidized beds have proven useful to study granular matter near the jamming transition [14; 15]. Fluidization allows to adjust the volume fraction, so that e.g. the onset of mechanical stability in random loose packings can be studied [16]. It is this regime of approximately homogeneous density in a stable fluidized bed that we focus on here. We propose an event driven algorithm to simulate three-dimensional fluidized beds. Our simulation is based on a recently developed event-driven algorithm for hard particles, experiencing drag by a surrounding fluid [17; 18]. Following Johri and Glasser [12; 13], we assume a *density dependent* drag force and incorporate the density dependence into the viscosity of the fluidizing liquid. The paper is organized as follows: we first introduce the model (sec. 2) and give details of the simulation (sec. 3). We then show (sec. 4) that indeed an approximately homogeneous profile is established for all investigated packing fractions $0.3 \leq \bar{\phi} \leq 0.55$. We determine the width of the interface and discuss expansion and sedimentation after a sudden change in density. Finally, we summarize our results and suggest possible extensions (sec. 5).

2. Model

We consider a monodisperse collection of N spheres of mass m, radius R, volume $V = \frac{4}{3}\pi R^3$ and density $\rho_s = m/V$. The particles are inserted in a fluid with density ρ_f and viscosity η. The fluid is flowing from below to the top of the system with velocity $v_{ex} > 0$. The particles are subject to friction with the surrounding fluid, a gravitational force, g, as well as bouyancy and inelastic collisions, so that the Langevin equation for the velocity of particle i in z-direction reads

$$\dot{v}_{i,z} = -\gamma(v_{i,z} - v_{ex}) - g + \frac{F_b}{m} + \left.\frac{dv_{i,z}}{dt}\right|_{coll} . \tag{1}$$

The friction coefficient in the drag force, γ, is allowed to depend on packing fraction, $\gamma = \gamma_0 \cdot f(\phi)$, via the function $f(\phi)$, which will be specified below. The three dimensional packing fraction is given by $\phi = \frac{4}{3}\pi R^3 n_0$, where n_0 denotes the number density. The bare friction coefficient is given by the Stokes value: $\gamma_0 = \frac{6\eta\pi R}{m}$. We only consider viscous drag and neglect long-range hydrodynamic interactions between the granular particles. This approximation is expected to hold for experiments on fluidized granular beds, where the typical Reynolds number is larger than one [14–16]. Our study is furthermore restricted to dense suspensions, where the hydrodynamic interactions are screened [19; 20]. The bouyancy force F_b is equal to the fluid mass that is displaced by the particle, $F_b = \rho_f g V$.

The particles are modeled as hard particles which collide inelastically. The degree of inelasticity is characterized by a coefficient of incomplete restitution ε. Upon collision of particles i and j, their relative velocity $\mathbf{g} = \mathbf{v}_i - \mathbf{v}_j$ changes to the postcollisional relative velocity $\mathbf{g}'$ according to: $(\mathbf{g} \cdot \mathbf{n})' = -\varepsilon(\mathbf{g} \cdot \mathbf{n})$, where $\mathbf{n} = (\mathbf{r}_i - \mathbf{r}_j)/|\mathbf{r}_i - \mathbf{r}_j|$. For the sake of simplicity, we assume a constant coefficient of restitution ε here.

The density dependence of $\gamma(\phi)$ is determined experimentally [14] by measuring the sedimentation velocity, U, of the particles in a resting fluid. A single particle in suspension moves with a constant velocity, denoted by U_0, such that gravity and bouyancy balance the drag force: $\frac{\gamma_0}{m}U_0 = (\rho_f - \rho_s)Vg = \Delta\rho V g$. With increasing ϕ, the settling velocity U decreases. Batchelor [21] derived the first-order correction: $\frac{U}{U_0} = 1 - 6.55\phi + \mathcal{O}(\phi^2)$. Higher order terms were derived in [22–25], using a virial expansion. The above expansion is helpful as long as dilute packings are considered; in the case of dense fluids one must resort to empirical relations. Here we assume that the relation of Richardson and Zaki [26] captures the settling behavior of the suspension appropriately as found in recent experiments [27; 28] and simulations by [29] and [30]. Hence we take $U/U_0 = (1 - \phi)^n$, where n depends on the Reynolds number. To obtain a relation between the density dependent drag $\gamma(\phi) = f(\phi)\gamma_0$ and U/U_0, we again use force balance, but now for a dense suspension: $\frac{\gamma}{m}U = \frac{\gamma_0 f(\phi)}{m}U = \Delta\rho V g$. Together with the

Richardson-Zaki relation, this implies for the density dependent drag

$$\frac{\gamma(\phi)}{\gamma_0} = \frac{U_0}{U(\phi)} = (1-\phi)^{-n}. \tag{2}$$

In steady state, the average particle velocity vanishes, $\sum_i v_{i,z} = 0$, and the external flow field, v_{ex}, controls the average packing fraction, $\bar{\phi}$ according to

$$\dot{v}_{i,z} = 0 = \gamma(\bar{\phi})\bar{v}_{ex} - g\frac{\Delta\rho}{\rho_s}. \tag{3}$$

We now expand Eq.(1) around the homogeneous steady state and keep only linear deviations: $\phi(\mathbf{r}_i) = \bar{\phi} + \delta\phi_i$ and $\gamma = \gamma(\bar{\phi}) + \gamma'(\bar{\phi})\delta\phi_i$. We also allow for fluctuations in the external flow field $v_{ex} \to v_{ex} + \xi_i$. Linearizing Eq. (1) around the steady state of Eq. (3) we find

$$\dot{v}_{i,z} = -\gamma(\bar{\phi})v_{i,z} + \gamma(\bar{\phi})\xi_{i,z} + \left.\frac{dv_{i,z}}{dt}\right|_{coll} + \frac{\gamma'(\bar{\phi})}{\gamma(\bar{\phi})}\frac{\Delta\rho}{\rho_s}g\delta\phi_i \tag{4}$$

The driving force towards the steady state in Eq.(4) is the density, or more precisely its deviation from the average value. If locally $\delta\phi_i < 0$, particles in this region will be accelerated downward, if $\delta\phi_i > 0$, particles will be accelerated upward. The equations of motion (4) can be simulated with an event driven algorithm, as detailed in the next paragraph.

3. Event driven simulation

To implement Eq. (4) in an event driven algorithm, we model the driving force, resulting from the inbalance of gravity and buoyancy, by discrete *kicks*. In other words, the particles are not accelerated continuously, but instead are kicked with a given frequency. These kicks are treated as events in the simulation, with the kick frequency, f_{dr}, comparable to the collision frequency, f_{coll}. Similarly, the noise $\{\xi_{i,\alpha}\}$ is modelled by *random* kicks with average zero and variance ξ_0^2. In the steady state, energy losses due to drag forces and inelastic collisions balance energy input due to driving, according to:

$$\frac{3}{4}f_{coll}(1-\varepsilon^2)T_G + 2\gamma(\bar{\phi})T_G = f_{dr}m\gamma^2(\bar{\phi})\xi_0^2. \tag{5}$$

Here we have introduced the granular temperature $T_G = 2/3E_{kin}$ in terms of the average kinetic energy (for details see [17]).

In experiments on fluidized beds of granular particles, the thermal heat bath provided by the surrounding fluid is small in the sense that the gravitational energy of a grain is much larger than the thermal energy. Nevertheless the grains show random motion with typical velocities of the order of a few mm/s. Energy is fed into the system by pushing a liquid through the fluidized bed. This energy input does not only compensate the frictional losses due to drag but in addition provides the energy input to sustain the random motion of the particles in the steady state.

We are interested in the fluidization and sedimentation of the system. The density dependent contribution to the particle motion in x- and y-direction is therefore negligible as also found by [31]. To compute the local packing fraction ϕ_i, we count the number of particles in a layer of defined thickness Δ around each particle at the time of the kick, which determines the deviation of the local density from its global value, $\delta\phi_i = \phi_i - \bar{\phi}$. For most of the simulations we choose $\Delta = 10R$, but have checked other values (see below). Finally, we use the Richardson-Zaki relation to rewrite Eq. (4)

$$\dot{v}_{i,\alpha} = -\gamma(\bar{\phi})(v_{i,\alpha} - \xi_{i,\alpha}) + \left.\frac{dv_{i,\alpha}}{dt}\right|_{coll} + \delta_{z,\alpha}\frac{ng\Delta\rho}{\rho_s}\frac{\delta\phi_i}{1-\bar{\phi}}. \tag{6}$$

We regard our ansatz as a simplification for more complicated simulations as in [32] and [33], where 2400 and 5000 particles, respectively, were simulated. Both methods are combinations of molecular dynamics, capturing the motion of single particles, and computational fluid dynamics, accounting for the interstitial fluid. The main advantage of our simulation ansatz is that the event-driven simulation can handle $N = \mathcal{O}(10^6)$ and more particles.

We use systems of $N = 10000$ and $N = 200000$ monodisperse particles in a box with hard walls, i.e. colliding particles are reflected elastically. At high densities, we have to circumvent the inelastic collapse, which is done in the same way as in [34]. Alternatively, one could use a velocity dependent coeffcient of restitution [35–37]. The range of packing fractions is chosen to be $0.3 \leq \bar{\phi} \leq 0.55$ and $n = 4.65$ as discussed in [28]. We choose the parameters as $\varepsilon = 0.7$, and $\frac{\rho_s - \rho_f}{\rho_s} = 0.6$, to mimick a typical experiment with glass spheres. The drag coefficient is determined from Stokes law $\gamma_0 = 6\pi\eta R/m$, with the viscosity of water $\eta \sim 0.01\text{g}/(\text{cm s})$. For a typical example of an experimental realisation, we choose $R = 300\mu m$, implying a mass $m = \rho_s \cdot V = 2.8274 \cdot 10^{-7}$ kg. The granular temperature is taken to be $T_G = 10^{-9}\text{kg m}^2/\text{s}^2$, which fixes the driving amplitude according to Eq.(5). In the results section, we will use these specific values, but want to point out that the same data also apply to other systems: Since we model the particles as hard spheres, there is no inherent length scale associated with the interaction and the particle diameter basically sets the unit of length. In other words the same data can also be interpreted in terms of differently sized particles with correspondingly scaled m and T_G.

4. Results

Our primary interest is the fluidization and sedimentation in a granular suspension, driven by an upward flow. As a first step we compute the density profile for a given average volume fraction, corresponding to a given flow rate (2). We characterize the resulting interface and check for effects due to the finite resolution for the density $\delta\phi_i$. Subsequently the dynamics of fluidization and sedimentation is simulated by changing the average volume fraction and monitoring the following expansion or compactification.

Interface formation and packing fraction profile: As a first step, we investigate the formation of an interface for a given average packing fraction $\bar{\phi}$. To that end we prepare the system initially in a homogeneous state by equilibrating the system without gravity in a closed box, corresponding to the desired volume fraction. Subsequently the simulation box is enlarged in the z-direction and gravity and density dependent drag force are switched on.

We monitor the particles' position to compute the density profile

$$\phi(z) = \frac{\bar{\phi}}{N} \sum_i \delta(z - z_i) \tag{7}$$

In Fig. 1 this profile is shown at different times, starting from a sharp profile. We observe the formation of an interface of finite width with a steady state reached after around 1000 collisions per particle. At the bottom $z \approx R$, the packing fraction is locally increased, because the particles tend to arrange at a height $z \approx R$ as this configuration allows for the most efficient packing.

A simple model can account for the observed profiles and allows us to quantify the width of the interface. We consider a bed, extended to negative z, so that the probability to find particle i at position z in the homogeneous state is given by: $p_{hom}(z) = p_0\theta(h - z)$. At the interface, several processes disturb the particle position. We model these by a Gaussian with zero mean and variance σ^2. Hence the probability to find the particle at position z is given by a convolution

$$p(z) = (p_{hom} \star p_\sigma)(z) = \frac{p_0}{2}\left(1 - \text{erf}\left(\frac{z - h}{\sqrt{2}\sigma}\right)\right) . \tag{8}$$

Journal of Physics: Conference Series **759** (2016) 012001 doi:10.1088/1742-6596/759/1/012001

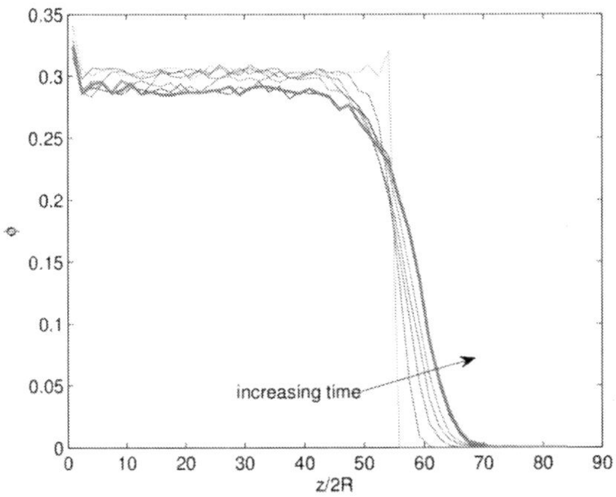

Figure 1. Interface formation for a fluidized bed with steady state packing fraction $\bar{\phi} = 0.3$. Time step in between two subsequent density profiles $\sim$ 20, 50, 100, 200, 400, 400 collisions per particle; note that the last two curves are alomost on top of each other.

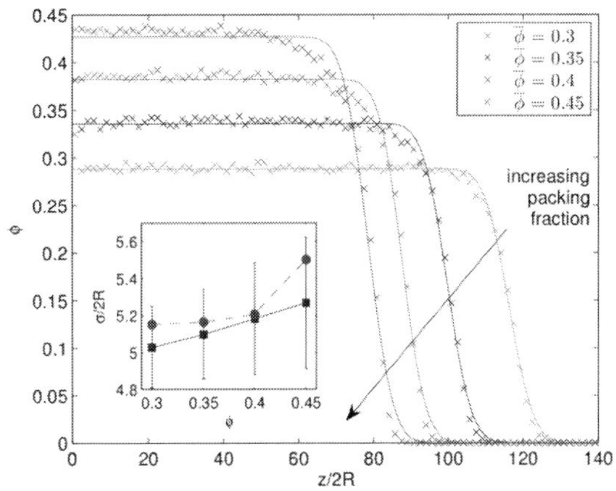

Figure 2. Packing fraction profiles for fluidized states of different flow rates, corresponding to $\bar{\phi} = 0.3$, 0.35, 0.4 and 0.45. Data points (crosses) are fitted (solid lines) to Eq. 9; inset: width of the interface as a function of $\bar{\phi}$ for two different values of the resolution in z-direction, Δ.

The packing fraction is proportional to this probability and the proportionality constant can be fixed by requiring $\phi = \bar{\phi}$ well inside the bulk

$$\phi(z) = \frac{\bar{\phi}}{2}\left(\text{erfc}\left(\frac{z-h}{\sqrt{2}\sigma}\right)\right). \tag{9}$$

Hence we have two free parameters, we call h the height of the bed and σ is a measure for the sharpness of the interface: The smaller σ the sharper the interface.

We fit the packing fraction profile to Eq.(9) for fluidized states of several packing fractions. As discussed above, at the bottom the packing fraction is locally increased. Therefore, we exclude these data points from the following procedure. The data and the resulting fits are plotted in Fig. 2. Our modeled packing fraction profile in Eq. (9) matches the measured data very well for moderately dense packings, but deviates for higher densities in the region where the bulk value starts to drop down. Nevertheless we can use the fit procedure to determine the width of the interface, which is shown in the inset of Fig. 2 as a function of packing fraction. The width of the interface σ is in the range of 10 particle radii and is approximately independent of ϕ.

To calculate the local packing fraction we have used a slice of thickness $\Delta = 10R$. We do not expect that the resolution in the local density has strong effects in the bulk of the sample. However the interface and in particular its width might depend on Δ. To check this point we simulate the same fluidized bed but with a different value of $\Delta = 6R$. The width of the interface σ is compared for the two resolutions in the inset of Fig. 2. On average, σ is slightly smaller for $\Delta = 6R$. However the effect is small and we do not expect that a further refinement of the resolution for the density will change these results.

Sedimentation and Expansion So far we have shown that our simple model allows for a steady state with a well defined interface whose width has been characterized. In this section we study the relaxational dynamics of the fluidization and sedimentation process which in experiment is achieved by changing the flow rate. The initial and final flow rate give rise to different steady packing fractions. In our simple model sedimentation and compactification is modeled by a sudden (instantaneous) change in density. Specifically, we compact a fluidized bed with $N = 10^4$ particles from $\bar{\phi} = 0.3$ to a different packing fraction $\bar{\phi} = 0.55$. The initial state is chosen as the steady state for $\bar{\phi} = 0.3$ and then at $t = 0$ the average volume fraction is set to $\bar{\phi} = 0.55$ in Eq.(6). To illustrate the sedimentation process, we depict the compactifying bed at six different times during the sedimentation in Fig. 3. Already from this figure, especially the second and third frame, one can see that the sedimentation process is not homogeneous.

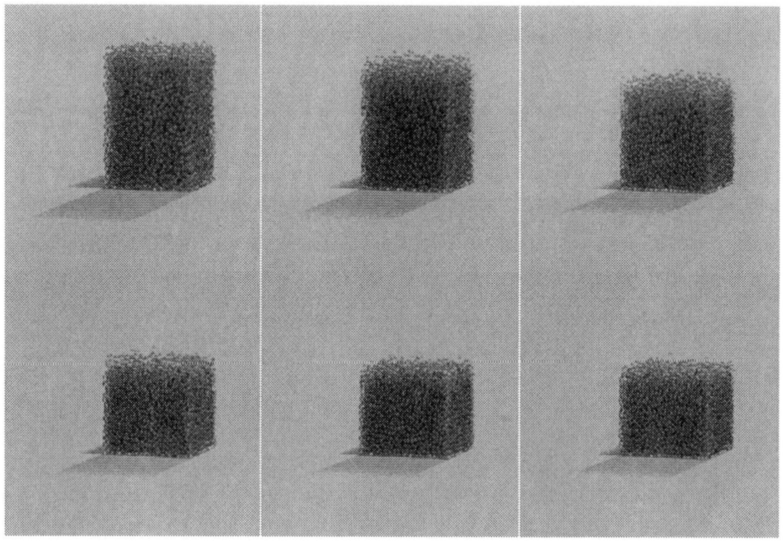

Figure 3. Sedimentation process from $\bar{\phi} = 0.3$ (upper left) to $\bar{\phi} = 0.55$ (lower right), at times $t = 0, 0.1, 0.2, 0.3, 0.4, 0.5s$.

To quantify this process, we plot in Fig.4 the density profile for several intermediate time steps. The compactification starts from the bottom of the fluidized bed, working its way up to the top. The packing fraction starts to increase to the final value $\bar{\phi} = 0.55$ at the bottom, subsequently the layers above are compacting and after about 500 collisions per particle at $t = 0.5s$ the new steady state is attained.

Similarly, we monitor the reverse process, i.e. expansion of the bed. The initial state is chosen as the steady state for $\bar{\phi} = 0.55$ and then at $t = 0$ the average volume fraction is set to $\bar{\phi} = 0.3$ in Eq.(6). The density profile is shown in Fig.5 for several timesteps. The expansion process is inhomogeneous as well. The bed rapidly expands at the top with a simultaneous inhomogeneous dilution in the compactified region. Subsequently the upper part of the system continues to expand until the packing fraction has ultimately flattened throughout the system. The expansion process is faster than the sedimentation. After about 90 collisions per particle at time $t = 0.2s$ the bed has attained the expanded state. The above expansion corresponds to a drop of almost 50% in density and is by no means quasistatic. To study the latter, we consider a much smaller drop in density, from $\bar{\phi} = 0.35 \to 0.3$. The profiles are shown in Fig. 6 and observed to be approximately monotonic as a function of height within the bed.

We want to extract a timescale for the expansion and study its dependence on average packing fraction and coefficient of restitution. To that end we define the average deviation from the initial profile

$$L^2(t) = \int_0^\infty dz |\phi(z,t) - \phi(z,0)|^2 \tag{10}$$

and compute its temporal evolution. As seen in Fig. 7, there is a rapid increase for short times

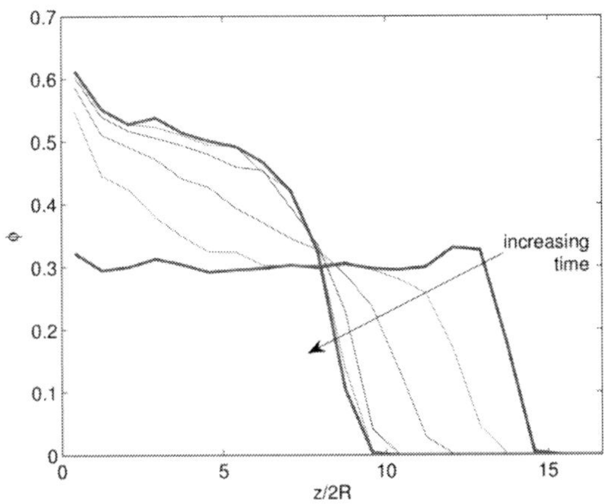

Figure 4. Packing fraction profiles for sedimentation from $\bar{\phi} = 0.3$ to $\bar{\phi} = 0.55$, time step $\Delta t = 0.1s$. Same data as in Fig. 3.

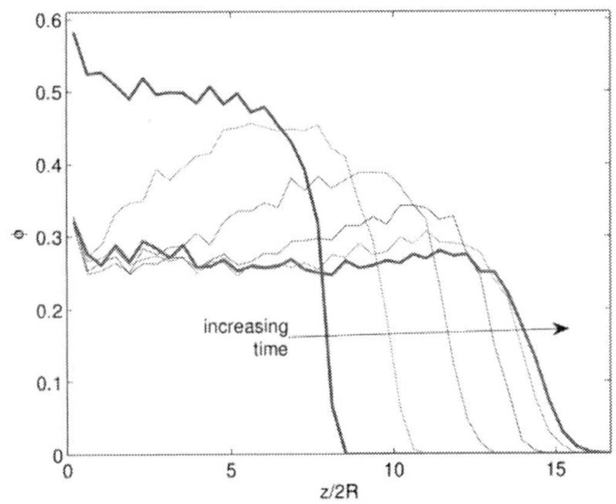

Figure 5. Expansion from $\bar{\phi} = 0.55$ to $\bar{\phi} = 0.3$, time step $\Delta t = 0.04s$.

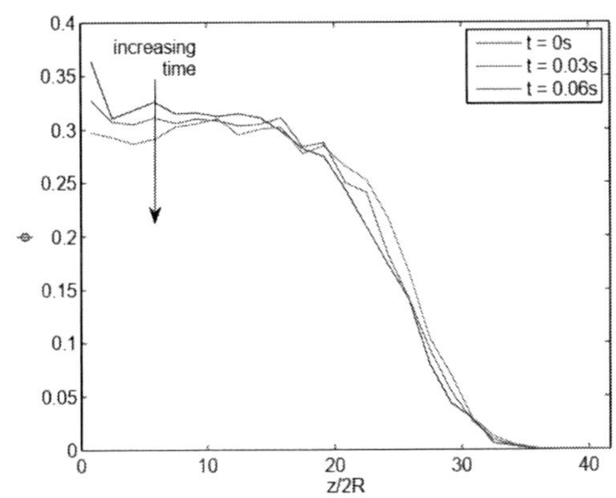

Figure 6. Expansion from $\bar{\phi} = 0.35$ to $\bar{\phi} = 0.3$.

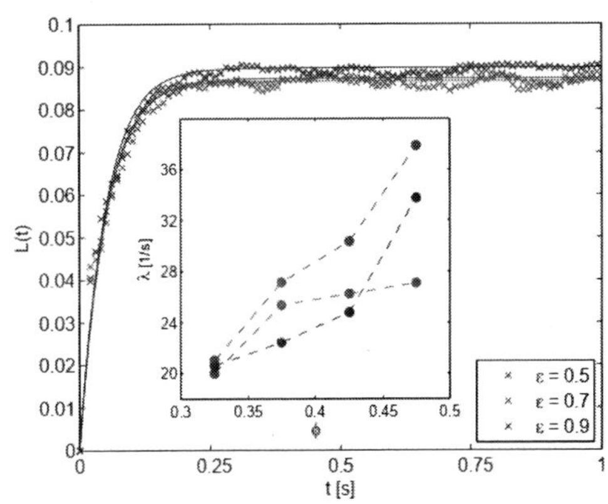

Figure 7. Approach to the steady state, as measured by $L(t)$ for the expansion from $\bar{\phi} = 0.35$ to $\bar{\phi} = 0.3$; inset: $\lambda(\bar{\phi})$ for three different coefficients of restitution, ε.

followed by a plateau at longer times. The data are well fitted by an exponential increase according to: $L(t) = const.(1 - e^{-\lambda t})$. In the inset of Fig. 7 we show $\lambda(\bar{\phi})$ as a function of average packing fraction for small changes of packing fraction, i.e. the first point corresponds to $\bar{\phi} = 0.35$ to $\bar{\phi} = 0.3$, the second point to $\bar{\phi} = 0.4$ to $\bar{\phi} = 0.35$ and so on. The timescale of expansion increases with increasing volume fraction and is slightly larger for the more inelastic systems.

5. Conclusion and Outlook

We have shown how to simulate a fluidized bed with an event-driven method that is capable of simulating large numbers of particles. The algorithm is based on an expansion around the homogeneous fluidized state. We found stable homogeneous fluidized beds, whose packing fraction can be adjusted by changing the flow rate. A stable bed is characterized by an interface at the top of the system, whose width is of the order of 5-10 particle radii. A stable bed also forms after changing the flow rate to a lower value, which increases the packing fraction (sedimentation), and by increasing the flow rate, which decreases the packing fraction (expansion). In general, the processes of sedimentation and expansion are non homogeneous, i.e. the packing fraction in the dense region does not deform homogeneously but rather changes from bottom to top, ultimately flattening to the previously observed steady state profiles.

We plan to further investigate fluidized beds using our algorithm. An obvious first extension is the temperature profile. Furthermore, we can easily compute velocity distributions and mean square displacements, which are also accessible to experiment by introducing markers. So far, we have focused on approximately homogeneous states. However, it is known [38] that suspensions of sedimenting particles exhibit a rich spectrum of instabilities which should be accessible to our simulations. In fact, we have already seen striped phases in our simulations for very small noise levels, but postpone a sytematic study to future work.

Acknowledgments
We thank W. T. Kranz, M. Schroeter and S. Ulrich for useful discussions. We furthermore acknowledge support from DFG by FOR 1394.

References

[1] Kunii D and Levenspiel O 1991 *Fluidization engineering* vol 2 (Butterworth-Heinemann Boston)
[2] Davidson J F and Harrison D 1971 *Fluidization* (Academic Press)
[3] Valverde J, Quintanilla M, Castellanos A and Mills P 2003 *Physical Review E* **67** 016303
[4] Geldart D 1986 *Gas fluidization technology* (John Wiley and Sons Inc., New York, NY)
[5] Tsinontides S C and Jackson R 1993 *Journal of Fluid Mechanics* **255** 237–274
[6] M H J, Thomas S, Guazzelli E, Homsy G M and Anselmet M C 1990 *J. Multiphase Flow* **16** 171–185
[7] Anderson T B and Jackson R 1969 *Industrial & Engineering Chemistry Fundamentals* **8** 137–144
[8] El-Kaissy M M and Homsy G M 1976 *International Journal of Multiphase Flow* **2** 379–395
[9] Batchelor G K 1991 *Of fluid mechanics and related matters. Proc. Symp. Honoring John W. Miles on his 70th birthday*
[10] Anzerais F M and Jackson R 1988 *J. Fluid Mech.* **195** 437–462
[11] Jackson R 2000 *The dynamics of fluidized particles* (Cambridge University Press)
[12] Johri J and Glasser B J 2002 *AIChE Journal* **48** 1645–1664
[13] Johri J and Glasser B J 2004 *Computers & Chemical Engineering* **28** 2677–2693
[14] Schröter M, Goldman D I and Swinney H L 2005 *Physical Review E* **71**(3) 030301
[15] Metayer J F and et al 2011 *Europhys. Lett.* **93** 64003
[16] Jerkins M, Schröter M and Swinney H L 2008 *Phys. Rev. Lett.* **101** 018301
[17] Fiege A, Grob M and Zippelius A 2012 *Granular Matter* **14**(2) 247–252 ISSN 1434-5021
[18] Fiege A, Vollmayr-Lee B and Zippelius A 2013 *Phys. Rev. E* **88**(2) 022138
[19] Segre P, Liu F, Umbanhowar P and Weitz D 2001 *Nature* **409**(2) 594
[20] Koch D L and Shaqfeh E 1991 *Journal of Fluid Mechanics* **224** 275–303
[21] Batchelor G K 1972 *Journal of Fluid Mechanics* **52** 245–268
[22] Cichocki B, Ekiel-Jeżewska M L, Szymczak P and Wajnryb E 2002 *The Journal of Chemical Physics* **117** 1231

[23] Cichocki B, Ekiel-Jeżewska M L and Wajnryb E 2003 *The Journal of Chemical Physics* **119** 606–619

[24] Batchelor G K and Green T J 1972 *J. Fluid Mechanics* **56** 401–427

[25] Batchelor G K 1977 *J. Fluid Mechanics* **83** 97–117

[26] Richardson J F and Zaki W N 1954 *Transactions of the Institution of Chemical Engineers* **32** 35–53

[27] Duru P and Guazzelli E 2002 *Journal of Fluid Mechanics* **470** 359–382

[28] Pätzold W M 2012 *Dynamik von angetriebenen Suspensionen* Master's thesis Georg-August-Universität Göttingen

[29] Abbas M, Clément E, Simonin O and Maxey M R 2006 *Physics of Fluids* **18** 121504

[30] Yin X and Koch D L 2007 *Physics of Fluids* **19** 093302

[31] Nguyen N Q and Ladd A J C 2005 *Journal of Fluid Mechanics* **525** 73–104

[32] Xu B H and Yu A B 1997 *Chemical Engineering Science* **52** 2785–2809

[33] Hoomans B P B, Kuipers J A M and Van Swaaij W P M 2000 *Powder Technology* **109** 41–48

[34] Fiege A, Aspelmeier T and Zippelius A 2009 *Physical Review Letters* **102** 98001

[35] Zenit R and Hunt M L 1999 *Journal of Fluids Engineering* **121** 179–184

[36] Gondret P, Lance M and Petit L 2002 *Physics of Fluids* **14** 643

[37] Müller P, Hecker M, Sack A and Pöschel T 2013 *Physical Review Letters* **111** 254301

[38] Guazelli E and Hinch J 2011 *Annu. Rev. Fluid Mech.* **43** 97

How Novel Algorithms and Access to High Performance Computing Platforms are Enabling Scientific Progress in Atomic and Molecular Physics

Barry I. Schneider
Applied and Computational Mathematics Division
National Institute of Standards and Technology∗
Gaithersburg, MD 20899-8910, USA
E-mail: bis@nist.gov

Abstract. Over the past 40 years there has been remarkable progress in the quantitative treatment of complex many-body problems in atomic and molecular physics (AMP). This has happened as a consequence of the development of new and powerful numerical methods, translating these algorithms into practical software and the associated evolution of powerful computing platforms ranging from desktops to high performance computational instruments capable of massively parallel computation. We are taking the opportunity afforded by this CCP2015 to review computational progress in scattering theory and the interaction of strong electromagnetic fields with atomic and molecular systems from the early 1960's until the present time to show how these advances have revealed a remarkable array of interesting and in many cases unexpected features. The article is by no means complete and certainly reflects the views and experiences of the author.

1. Introduction

The time independent and time dependent many-body Schrödinger equations show a remarkable array of features which often only emerge subsequent to laborious and time consuming calculations. Progress in the field has been marked by the development of novel theoretical and numerical methods, backed by solid algorithms, software development and the continuing evolution of computing platforms ranging from powerful desktops to supercomputers. In this article, I outline historical progress in the treatment of atomic and molecular continua in scattering processes and how these ideas have been extended to the time-dependent interaction of short, intense laser pulses with such targets.

1.1. Early History: The 1960's and 1970's

I begin with a short history of early developments in electron-atom/molecule collisions.

(i) Computers advance sufficiently that quite high quality calculations could be performed on the scattering of electrons and photoionization of simple atoms such as H, He and Ne.

(ii) Significant experimental advances; A number of experimental groups in AMO physics develop instruments able to extract high quality total and differential electron scattering cross sections and photoabsorbtion cross sections (See Figure 1 [1, 3])

∗Official contribution of the National Institute of Standards and Technology; not subject to copyright in the US

Journal of Physics: Conference Series **759** (2016) 012002 doi:10.1088/1742-6596/759/1/012002

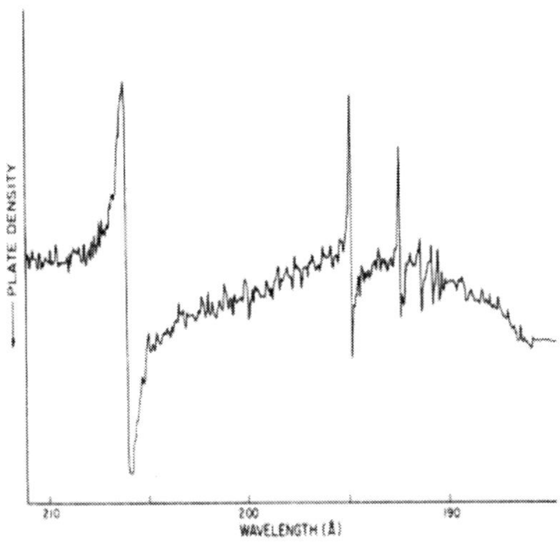

Densitometer trace of the absorption spectrum of He. The absorption is enhanced on the low-frequency side and reduced on the high-frequency side. See [1]

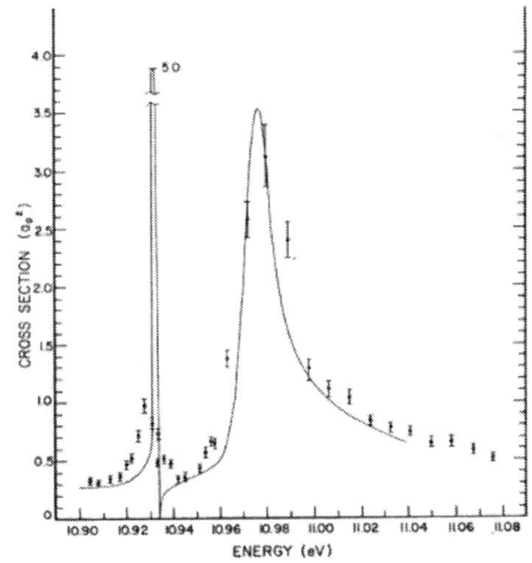

Photoabsorption of H^-. Comparison of theory (solid curve) [See [2]] and experiment (with error bars) [See [3]]. Data from experiment normalized to theory at 10.90 eV.

Figure 1: Some simple examples of resonances in atoms

(iii) Computation (See 3.1 for discussion of close-coupling (CC)) and experiment discover interesting and in some cases unexpected features such as Feshbach and shape resonances in an otherwise smooth background continuum [4, 5]. Such long-lived, quasi-bound states were seen in nuclear scattering earlier.

(iv) Extension of CC to a number of first row atoms

(v) An intuitive but useful approach called the Stabilization method [6] developed to treat resonances (3.2)

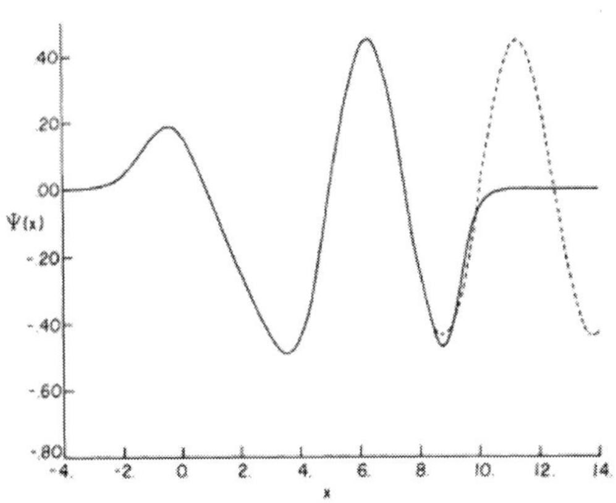

Figure 2: ((L) Comparison of exact and stabilization wavefunction in nonresonant region for simple model potential [6]. (R) Mentor and close friend Howard S. Taylor [9/17/1935-4/17/2015] circa 1970's

(vi) Recognition that better numerical approaches were needed: Early developments include the R-and-J-matrix methods, (3.3, 3.4) Kohn and Schwinger variational methods (4.1), and complex rotation/scaling (3.5)

(vii) Extensions to small diatomic molecules started. Effects of nuclear motion are discussed.

Some comments. The CC approach primarily used well established numerical techniques, although special care was required to deal with long range interactions. The use of matrix and variational techniques really enabled much more complex and sophisticated calculations. These matrix based methods have proven their mettle and are even today the methods of choice for most problems.

1.2. Later Developments; 1980-present

From the 1980's to the mid 2000's there were a number of important computational advancements.

(i) R-matrix and Kohn variational methods become reliable in predicting the cross sections of even moderately complex polyatomic molecules

(ii) New and more efficient discretization techniques such as the Finite Element Discrete Variable Representation were developed (FEDVR) 4.2

(iii) Scattering methods were extended to the time-dependent domain where new experimental facilities such as free electron lasers are becoming available to study the interactions of short, intense laser radiation with atomic and molecular targets. Now possible to "see the electrons dance" and even control their motion using optical techniques. Looking directly at electron correlation effects becomes a reality computationally and experimentally

2. Personal View

In my opinion, the most effective approaches to treat problems in atomic and molecular systems involving a continuous spectrum, have been based on techniques which try to avoid the details of the asymptotic form of the wavefunction. There are a few options. One may divide either physical or function space into two (or perhaps more) regions. In the innermost region, one needs to deal with the full interacting set of nuclei and electrons but does not have to worry about the details of asymptopia. This inner region is then quite similar to a bound state problem and it becomes unsurprising that the most successful computational approaches to scattering borrow liberally from those developed by the atomic structure and quantum chemistry communities for discrete states. In the other regions, the Hamilitonian often "simplifies" in some way that makes it computationally easier to solve. In real space, one might be able to set the interaction to zero or to some problem where an analytic solution is available 3.3. In function space, one could assume the expansion coefficients of the asymptotic functions are known outside the interaction region 3.4. How the different regions are joined, and how the scattering information is extracted becomes the central difference of all the various approaches. There is also the possibility of avoiding the asymptotic region entirely by analytically continuing from the complex plane, rotating the scattering coordinates into the complex plane, or relating a discrete basis set expansion to a numerical quadrature. Among the first class of methods are the R-and-J matrix method, and the Kohn and Schwinger variational principles. The latter category is exemplified by either full or exterior complex scaling and the equivalent quadrature technique. Lets us look briefly at a few of these.

3. The Early Days; 1960-1980

3.1. Close Coupling

The close-coupling method [5] proceeds in the most straightforward manner; expand the scattering wavefunction in a set of target states times unknown one-electron scattering functions, insert into the Schrödinger equation, project onto the target states to obtain a set of coupled integro-differential equations and solve those equations by numerical integration. The close coupling or expansion in target states approach became routinely employed over the 1960's-1970's to both neutral and ionized

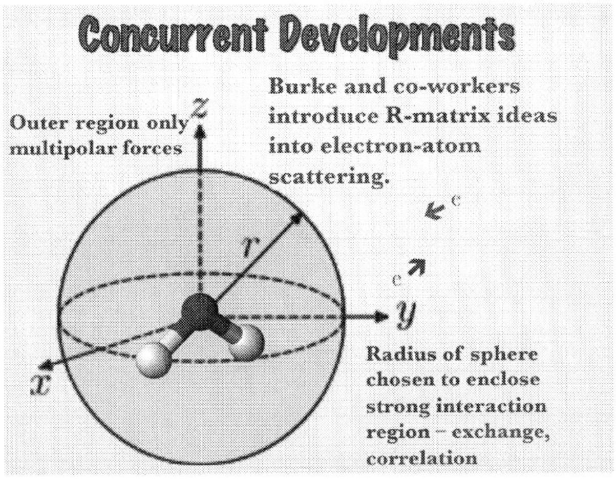

Figure 3: Partitioning Space in the R-Matrix Method into an Inner and Outer Region

R-Matrix Equations

$$\Psi_k = \sum_{c=1}^{n}\sum_{i=1}^{m} \psi_c u^0_{ci,k} a_{ci,k} + \sum_{q=1}^{M} \chi_q b_{q,k} \quad (1a)$$

$$\left[H_{N+1}(\mathbf{r};\mathbf{R}) + L_{N+1} - E(\mathbf{R})\right]\Psi(\mathbf{r};\mathbf{R}) = L_{N+1}\Psi(\mathbf{r};\mathbf{R}) \quad (1b)$$

$$\left[H_{N+1} + L_{N+1} - E_k(R)\right]\Psi_k = 0 \quad (1c)$$

$$L_{N+1} = \sum_{i=1}^{N+1} \frac{1}{2}\delta(r_i - a)\left(\frac{\partial}{\partial r_i} - b\right) \quad (1d)$$

$$u_c(a) = (\psi_c|\Psi)_{r=a} = \sum_{d=1}^{n} \mathscr{R}_{c,d}\left[\frac{\partial u_d(r)}{\partial r} - b u_d(r)\right]_{r=a} \quad (2a)$$

$$\mathscr{R}_{c,d} = \frac{1}{2}\sum_k \frac{u_{c,k}(a) u_{d,k}(a)}{[E_k(R) - E(R)]} \quad (2b)$$

$$u_{c,k}(a) = \sum_i a_{ci,k} u^0_{ci,k}(a) \quad (2c)$$

species. The computed cross sections were used in astrophysical and fusion modeling. Computing photoionization cross sections became a simple matter of calculating the dipole matrix element between the electron ion wavefunction and the initial bound state, recognizing the dipole selections rules, a nice benefit once the scattering calculation has been performed. The method was extended to include pseudostates to account for the polarizability of the true target states and to represent the ionization continuum. However, it also became evident that the direct solution of the integrodifferential equations of scattering theory, was not the most efficient way to attack the problem numerically and other approaches were initiated.

3.2. Stabilization Method

The "Stabilization Method", was developed by A. Hazi and H. Taylor [6] (See Figure 2) at USC. The success of the approach had a significant influence on much of the work that followed. Taylor correctly and intuitively argued that a discrete basis set expansion would converge, inside the range of the basis, to the true scattering function within an unknown normalization constant. By diagonalizing the Hamiltonian in a larger and larger discrete basis, certain eigenvalues "stabilized" with basis set size. Those eigenvalues that stabilized, were the resonances, and they stabilized simply due to the fact that resonant wavefunctions tend to be localized with rather small asymptotic tails. It was also noticed that the basis could describe the background continuum, but again only to an unknown normalization factor. There are a number of approaches which utilized the stabilized solution to find estimates of the width. Some are motivated by the connection between the stabilized eigenfunction and the variational method proposed by Harris [7]. Another method used the diagonalization to find an approximation to the background continuum function and then computed the width via the Fermi Golden Rule and an approximation to the density of states suggested by Miller [8].

3.3. The R-Matrix Method

The R-matrix method was developed in nuclear physics by Wigner and Eisenbud [9]. Phil Burke [10] and his colleagues in the UK, turned the R-matrix method into one of the most successful computational approaches to treat the scattering of electrons from atoms. A few years later, Schneider [11, 12] generalized the method to treat electron scattering from molecules and in a significant development [13, 14] showed how the theory could be used effectively to treat nuclear motion beyond the usual adiabatic approximation. The central R-matrix idea, including the relevant equations, is shown in Figure 3. One encloses the strong interaction region in a sphere, imposes a boundary condition on the surface of the

Journal of Physics: Conference Series **759** (2016) 012002 doi:10.1088/1742-6596/759/1/012002

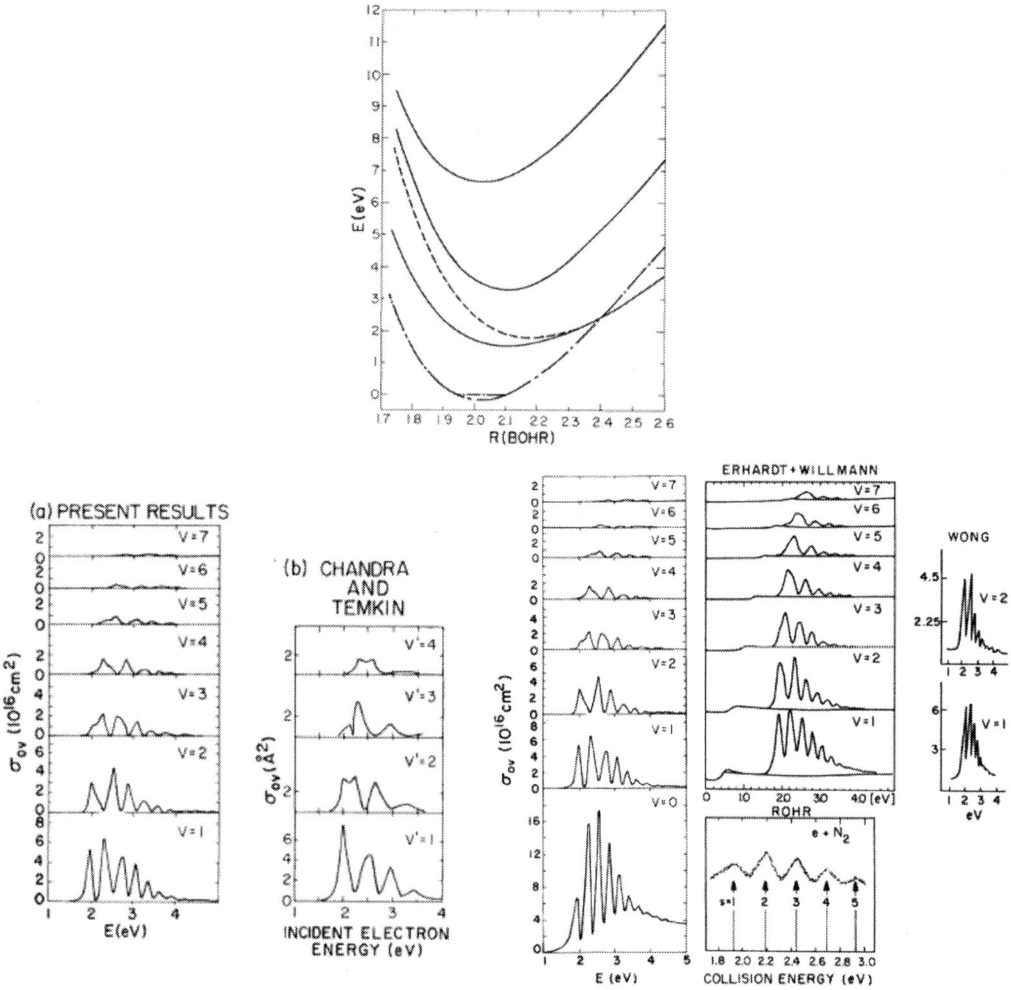

Figure 4: Upper Panel: N_2^- R-matrix Born-Oppenheimer Potential Curves. Lower Panels: Comparison of Vibrational Excitation Cross Sections. (L) R-Matrix [14] vs Vibrational Close-Coupling [15] and (R) Theory [14] vs Experiment. [16, 17].

sphere to get a discrete eigenvalue problem, solves that problem for its eigenvalues and eigenfunctions and uses that information to extract the scattering matrix by matching to asymptotically known functions. In contrast to nuclei, where one may assume that the potential is zero outside the interaction region, the atomic scattering problem needs to treat the long-range multipolar interaction of the electron and target more precisely. In the atomic R-matrix method this is done using solutions to coupled long range *differential* equations using these multipolar potentials. One may also take the R-matrix on the inner sphere and directly propagate it to larger distances using the coupled differential equations. The external region is devoid of exchange and electron correlation and this makes for a much simplified treatment. The functions used to expand Ψ_k are completely arbitrary. In practice, they are chosen such that the χ_q vanishes at $r = a$. The channel functions, ψ_c, describe all the internal coordinates of the residual target and $u^0_{ci,k}$ is a radial basis function which is non vanishing at $r = a$. The $u^0_{ci,k}$ contain all the information needed for extracting the scattering information in channel c. The unknown coefficients a and b are determined by inserting the expansion of the wavefunction (1a) into (1c), projecting onto the basis states, thus reducing the equation to the diagonalization of a large matrix. Once this diagonalization has been performed, the R-matrix and therefore the scattering information may be calculated at any energy by matching at $r = a$, a very efficient process. At this point we have fixed the nuclei in space. This restriction may be relaxed [13, 14] to include nuclear motion. One approach which works away from threshold and away from resonances, is to take the fixed nuclei scattering amplitudes at a series of geometries, multiply by a ro-vibrational wavefunction and integrate over the nuclear coordinates. Near a resonance this does not work since the electron is trapped in a quasi-bound state and the molecular cloud wants to adjust to the change in the field but they do so adiabatically. So, the essence of the Born-Oppenheimer approximation is retained. The compound state vibrational energies determine the positions of the resonances and the overlap of the compound state and neutral vibrational eigenfunctions (Franck-Condon factors) are the strengths. The lower left panel in Figure 4 compares the vibrational excitation cross sections from [15] and [14]. Note that [15] fails to reproduce the details of the compound levels. The lower right panel compares [14] with experiment [16, 17]. The calculations agree with experiment quite well and in addition, allowed the absolute normalization of the cross sections to be determined for the first time. Finally, I must mention a new and extremely robust R-matrix approach, using a B-spline, non-orthogonal basis, that has been developed and applied to electron atom collisions by Zatsarinny [18]

3.4. The J-Matrix Method

The J-Matrix method was developed by Reinhardt, Heller, Yamani and Fishman [19, 20, 21]. Applications to both single channel and multichannel problems may be found in the literature. [22, 2]. The basic idea is straightforward. For many simple potentials, for example, the free particle or coulomb potential, there are tri-diagonal, analytic expressions available for matrix elements of the appropriate one-particle Hamiltonian in a known L^2 basis. There are also analytic expressions for the coefficients of the regular and irregular solutions to these one particle operators which are exploited in the final solution of the scattering problem. The fundamental idea is quite similar to the R-matrix theory. The matching in space to the unknown linear combination of regular and irregular solutions to the asymptotic Hamiltonian is replaced by the analytically known expansion coefficients when the potential has assumed its asymptotic form. The coefficient joining the two solutions provides the scattering information. All the unknown coefficients are determined by projection onto the complete basis set using the known tridiagonal form of the asymptotic Hamiltonian in the basis.

3.5. Complex Scaling

Simon [23, 24] extended and made the work of Balslev and Combes [25] on the spectral theory of Schrödinger operators known to atomic physicists. These authors introduced the original concept of complex scaling (let $r \to r\exp(i\theta)$), an approach that was quickly exploited by a number of researchers, particularly the quantum chemistry community, as a practical tool to compute the positions and widths of scattering resonances. An unfortunate problem was encountered early on in that the complex scaling made the computations of the integrals involving the now, highly oscillatory inner orbitals, quite difficult. Another issue, raised by McCurdy and Rescigno [26] was that complete complex scaling would not work for molecules in a practical sense (within the BO approximation) and suggested the use of a mixed set of real and complex basis functions as a substitute. Simon [27] put the idea of using exterior complex scaling (ECS) on a firm mathematical foundation and this method remains one of most robust approaches even today. It has been advantageously used to produce the first truly accurate calculation of impact ionization of H [28] (Figures 5 and 6)

3.6. Equivalent Quadrature

The idea of relating the diagonalization of a Hamiltonian in an L^2 basis to a numerical quadrature was put forward by Heller, Reinhardt and Yamani [29]. This relationship yields a numerical approximation to the Greens's function which properly treats the $i\varepsilon$ limit required in the scattering problem. The concept was interesting from a historical perspective but never took off as a consequence of the difficulty of having to relate the L^2 basis to a quadrature. This can be done only for a few basis sets. I mention it here for completeness.

Journal of Physics: Conference Series **759** (2016) 012002 doi:10.1088/1742-6596/759/1/012002

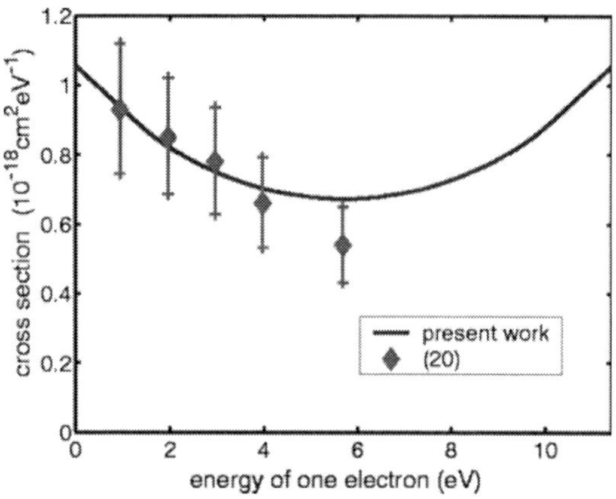

Figure 5: Single Differential ionization cross section for e$^-$ + H collisions at 25eV incident energy. At 5.7eV there is equal energy sharing as expected.

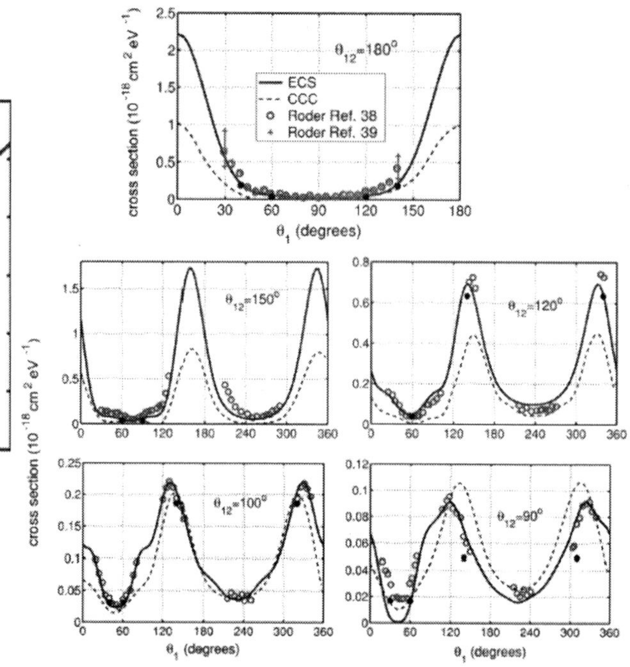

Figure 6: Coplanar triple differential cross section for 17.6-eV incident e$^-$ + H collisions. ECS(solid), Convergent Close Coupling (broken) and experiment(dots). Experimental cross sections are absolute. See [28]

4. The Modern Era;1980-Today

4.1. The Kohn Variational Methods

Arguably, the Kohn [30, 31, 32, 33, 34] (KVP) and the Schwinger [35, 36](SVP) variational methods, have, along with the R-matrix method, been the most productive of the methods developed for the electron scattering problem. The Schwinger variational method has the advantage that it does not require the trial function to have the correct asymptotic behavior. A practical disadvantage is that it is based on an integral equation, and requires matrix elements of the Green's function. Here I will concentrate on the Kohn variational method. The Kohn stationary principle is stated in Equation 3 and the trial wavefunction in Equation 4. Depending on the choice of the long range functions, R and I, O can be the T, S or K matrix. The trial wavefunction is inserted into the Schrödinger equation and then projected onto the basis to get a large set of linear equations for d and O. The first two summations in (4) contain only L^2 functions and are chosen to account for all of the details of electron correlation and exchange as well as the inner parts of the scattering functions. R and I are regular at $r = 0$, are chosen to be orthogonal to ψ_q and to go asymptotically to the regular and irregular solutions of an appropriately chosen model problem, such as a free wave or a coulomb function. The terms on the right hand side of the linear equations involve matrix elements containing the known functions, $\Phi_\Gamma R_\Gamma$. If we define the projection operators as in Equation 5 then one may formally eliminate the terms involving the Q-space functions in terms of an energy dependent, non-local, optical potential, where P now contains only the functions, $\Xi_\Gamma \psi_q$, and $\Phi_\Gamma I_\Gamma$. Formally solving the linear equations for the trial value of O and then substituting its value back into the Kohn expression, gives the final stationary value of $[\mathbf{O}]$, Equation 6. $\mathbf{M}$ consists of matrix elements of the optical potential between the functions $\Xi_\Gamma \psi_q$, $\Phi_\Gamma R_\Gamma$ and $\Phi_\Gamma I_\Gamma$. Importantly, forcing R and I to be orthogonal to ψ_q eliminates any difficult free-free and bound-free matrix elements from the calculation. Only overlap integrals of these functions are required. If the KVP is used with principal value boundary conditions, the $\mathbf{M_{qq}}$ can be zero leading to what are known as the Kohn anomalies. While there are various ways to get around this problem, it turns out that when outgoing wave boundary conditions are imposed, the $\mathbf{M_{qq}}$ matrix is complex symmetric and at real energies there are no anomalies. This approach termed the Complex Kohn Variational Method is therefore preferred albeit one must deal with complex rather than real matrices. Two examples of the application of the complex Kohn method to polyatomic molecules are shown in Figures 7 and 8 [37, 38].

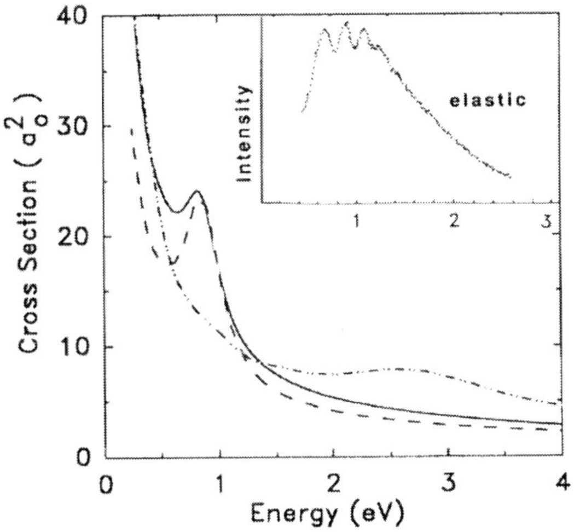

Figure 7: e + CH_2O exhibiting a π^* antibonding shape resonance. [37]

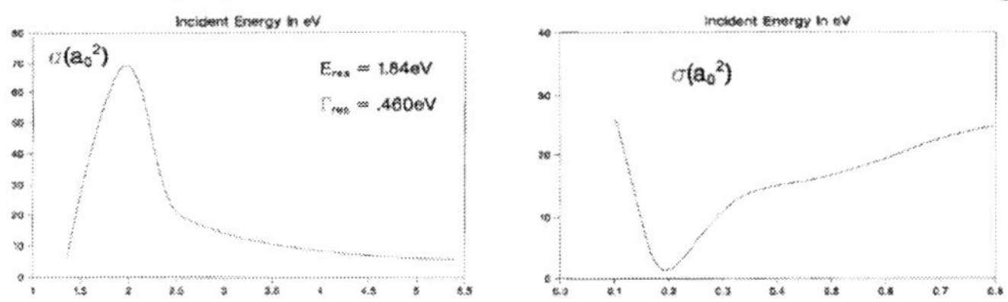

Figure 8: e + C_2H_4 scattering. (L) Exhibiting a π^* antibonding shape resonance and (R) a ^2A_g Ramsauer-Townsend minimum. [38]

Kohn Variational Principle

$$[O_{\Gamma,\Gamma'}] = O_{\Gamma,\Gamma'} - 2\langle\Psi^\Gamma|H-E|\Psi^{\Gamma'}\rangle \tag{3}$$

Trial Wavefunction

$$|\Psi^\Gamma> = \sum_{\Gamma',q} d^\Gamma_{\Gamma' q}|\Xi_{\Gamma'}(\mathbf{X_N})\psi_q(X_{N+1})> + \sum_K d^\Gamma_K|\Theta_K(\mathbf{X_{N+1}})> + \sum_{\Gamma'}|\Phi_{\Gamma'}(\mathbf{X_N},\Omega_{N+1}> [R^\Gamma_{\Gamma'}\delta_{\Gamma\Gamma'} + O_{\Gamma,\Gamma'}I^\Gamma_{\Gamma'}] \tag{4}$$

Projection Operators

$$Q = \sum_\mu |\Theta_k><\Theta_k|$$

$$P = I - Q \tag{5}$$

The Effective Matrix Hamiltonian

$$\mathbf{H_{eff}} = \mathbf{H_{PP}} + (\mathbf{H}-\mathbf{E})_{\mathbf{PQ}}(\mathbf{E}-\mathbf{H})^{-1}_{\mathbf{QQ}}(\mathbf{H}-\mathbf{E})_{\mathbf{QP}}$$

The Stationary Value of $[\mathbf{O}]$

$$[\mathbf{O}] = -2(\mathbf{M}_{00} - \mathbf{M}_{0q}\mathbf{M}^{-1}_{qq}\mathbf{M}_{q0}) \tag{6}$$

4.2. Time-Dependent (TD) Processes

In the past ten years there has been a significant effort devoted to computational studies of the interaction of intense, short pulse (attosecond) electromagnetic fields with atoms and molecules. The computational effort that is required is substantial, even for relatively small atomic and molecular targets. As a consequence, it is not surprising that the use of supercomputers and parallel processing techniques have become the norm. In the TD approach, one works with wavepackets [having an energy spread], needs to [often] propagate for long distances and times and then extracts the scattering information by projection onto asymptotic states. The plus side is that there is no need to impose boundary conditions on the wavefunction, which can be difficult to accomplish in ionizing collisions. Space limitations prevent only a brief look at this fascinating and growing field of physics. Arguably, the most effective approach to the space discretization in one and two-electron atomic and molecular systems is the FEDVR [39]. The advantage of this approach in practical calculations is that it leads to a sparse and structured Hamiltonian matrix which is amenable to efficient parallelization. The FEDVR discretization procedure is briefly shown in Equation (7). When coupled to the Short Iterative Lanczos (SIL) process as outlined in Equations (8), (9) and (10), one obtains a powerful computational approach to the TD Schrödinger equation. As can be seen in Figure 10, the method has been successfully applied to the double ionization of the He atom from the energy region where only direct double ionization is possible up to the sequential threshold. The scaling of the method on a variety of parallel processors is shown in Figure 11. One should note that after an initial erratic beginning, the curve flattens indicating that each core is now performing $1/n$ of the total work. As one can see, only very sophisticated numerical methods are able to reproduce

Summary of FEDVR Discretization

(i) Within each finite element:

$$u_i(x) = w_i \sum_n \phi_n(x)\phi_n(x_i) \quad (7a)$$

$$u_i(x_j) = \delta_{i,j} \quad (7b)$$

$$\langle u_i|x|u_j\rangle = x_i\delta_{i,j} \quad (7c)$$

$$\langle u_i|F(x)|u_j\rangle = F(x_i)\delta_{i,j} \quad (7d)$$

where $\phi_i(x)$ are members of a classical set of orthogonal functions and x_i and w_i are the corresponding Gauss quadrature points and weights.

(ii) Connection between elements
The elements are connected by ensuring that the functions at the element boundaries are continuous.

The structure of the Hamiltonian in one dimension is,

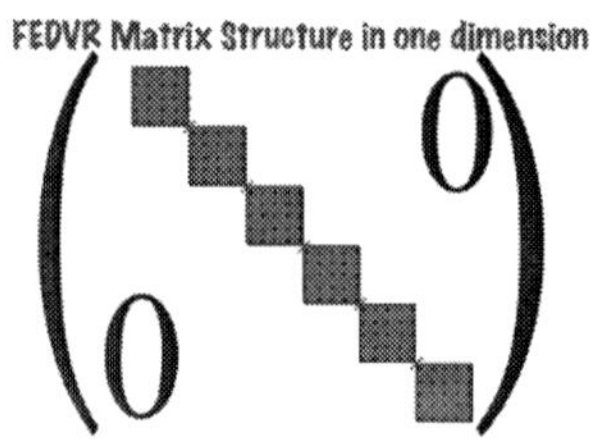

Figure 9: FEDVR Hamiltonian in 1D

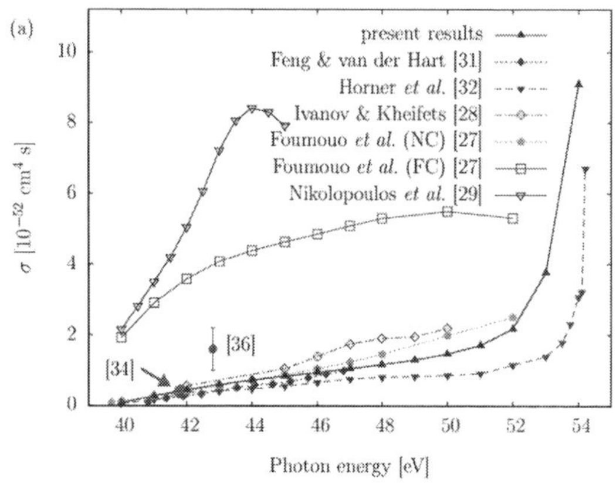

Figure 10: Comparison of a number of theoretical calculation of the direct double photoionization of He [See [40]]. Also shown are two experimental measurements.

Summary of TD Propagation

Short Time Lanczos (SIL) propagator

$$U(t+\delta t) \approx \exp[-\mathrm{i}\mathbf{H}(t)\delta t] \quad (8)$$

The Three Term Lanczos Recursion

$$\beta_{n+1}|n+1\rangle = |q\rangle = [\mathbf{H}(t) - \alpha_n]|n\rangle - \beta_n|n-1\rangle \quad (9a)$$

$$\alpha_n = \langle n|\mathbf{H}(t)|n\rangle \quad (9b)$$

$$\beta_{n+1} = \langle q|q\rangle \quad (9c)$$

The Exponential Propagator in the Diagonal Basis

$$\langle a|\exp[-\mathrm{i}\mathbf{H}(t)\delta t]|C\rangle = \sum_n \langle A|n\rangle \exp[-\mathrm{i}\lambda_n\delta t]\langle n|B\rangle$$

Table 1: Performance of the He FEDVR code on various platforms.

Name	CPU's	GHz	cores/CPU	CPU's/cores per node
Coyote	Opteron	2.60	1	2/2
Lobo	Opteron	2.20	4	4/16
Ranger	Opteron	2.30	4	4/16
Lonestar	Xeon	2.66	2	2/4
Abe	Xeon	2.33	4	2/8
Queen Bee	Xeon	2.33	4	2/8
Kraken	Opteron	2.33	4	2/8

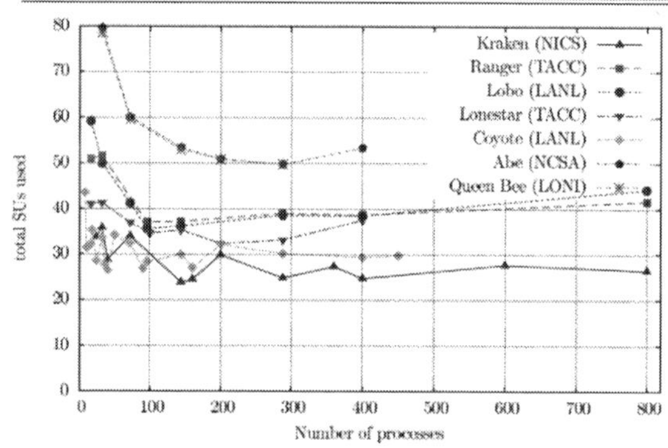

Figure 11: Scaling of FEDVR/SIL on Supercomputers. Plotted are total service units (≈ core hours) on the x-axis. and processor number on the y-axis. Strong scaling would be characterized by a perfectly flat curve.

the experimental results. The large rise of the cross section approaching the sequential threshold has been called virtual sequential double ionization in analogy to what is seen from the pole structure due to a resonance. There are two important points to emphasize. The first is that for many problems only very refined physical models, requiring intensive computation, can be relied upon to be predictive. The second is that it is critical to develop algorithms and software that scale well with processor count. A nice summary of what has been happening in the area of attosecond physics may be found in [41]

5. Conclusion

In this article I have tried to give the reader a flavor of many of the exciting theoretical and computational developments in atomic and molecular physics since it became possible to perform large scale calculations on computers. The early pioneers has to contend with slow processors and tiny memories by todays standards. Luckily, their processors and memories were quite sophisticated. It is a tribute to them that even with substantial digital obstacles they were able to produce many new and interesting results. Today, we are looking towards computers that will operate at the exascale level. The challenges to use that computer power effectively are formidable. If we can meet these challenges by developing the algorithms and software to use these new devices, I am sure the rewards will be significant and we will the truly be able to control atoms and molecules at their most fundamental level.

Acknowledgements

The author is deeply indebted to Prof. William P. Reinhardt for a critical reading of the manuscript and suggestions which greatly improved both the readability and understanding of the material. Prof. Klaus Bartschat did a critical reading of the manuscript and offered some valuable suggestions.

References

[1] Madden, R P and Codling, K 1963 *Phys. Rev. Letts.* **10** 516
[2] Broad, J T and Reinhardt, W P 1976 *Phys. Rev.* **A14** 2159
[3] Bryant, H C et. al. 1977 *Phys. Rev Letts.* **38** 228
[4] Fano, U 1961 *Phys. Rev.* **124** 1866
[5] Burke, P G and Schey, H 1962 *Phys. Rev.* **126** 147
[6] Hazi, A U and Taylor, H S.1970 *Phys. Rev.* **A1** 1109
[7] Harris, F E 1967 *Phys. Rev. Letts.* **19** 173
[8] Miller, W H 1966 *Phys. Rev.* **152** 70
[9] Wigner, E P and Eisenbud, L 1947 *Phys. Rev.* **72** 29
[10] Burke, P G, Hibbert, A and and Robb, W D 1971 *J. Phys. B: At. Mol. Phys.* **4** 153
[11] Schneider, B I 1975 *Chem. Phys. Letts.* **31** 237
[12] Schneider, B I 1975 *Phys. Rev.* **A11**, 1957
[13] Schneider, B I, LeDourneuf, M and Burke, P G 1979 *J. Phys. B.* **12** 1365
[14] Schneider, B I, LeDourneuf, M and Lan, V K 1979 *Phys. Rev. Letts.* **43**, 1926
[15] Chandra, N and Temkin, A 1976 *Phys. Rev.* **A13** 188
[16] Erhardt, H and Willman, K. 1967 *Z. Phys.* **204** 462
[17] Wong, S F Unpublished
[18] Zatsarinny, O 2006 *Comp. Phys. Comm.* **174** 273
[19] Heller, E J and Yamani, H A 1974 *Phys. Rev.* **A9** 1201
[20] Heller, E J and Yamani, H A 1974 *Phys. Rev.* **A9** 1209
[21] Yamani, H A and Fishman L 1975 *J. Math. Phys.* **16** 410
[22] Broad, J T Reinhardt, W P 1976 *J. Phys. B* **9** 1491
[23] Simon, B 1972 *Commun. Math. Phys.* **27** 1
[24] Simon, B 1973 *Ann. Math* **97** 247
[25] Balslev, E and Combes, J M 1971 *Commun. Math. Phys* **22** 280
[26] McCurdy, C W and Rescigno,T N 1978 *Phys. Rev. Letts* **41** 1364
[27] Simon, B 1973 *Phys. Lett.* **71 A** 211
[28] Baertschy, M, Rescigno, T N, Isaacs, W A , Li, X and McCurdy, C W 2001 *Phy. Rev.* **A63** 022712
[29] Heller, E J, Reinhardt, W P and Yamani, H A.1973 *J. Comp.Phys.* **13 3** 536
[30] Kohn W 1948 *Phy. Rev.* **74** 1763
[31] Nesbet, R K 1980 *Variational Methods in Electron-Atom Scattering Theory* (New York:Plenum)
[32] Rescigno, T N and Schneider, B I 1988 *Phys. Rev.* **A37** 1044
[33] Schneider, B I and Rescigno, T N 1988 *Phys. Rev.* **A37** 3749
[34] Rescigno, T N, McCurdy, C W, Orel, A E. and Lengsfield B H 1995 *The Complex Kohn Variational Method* in *Computational Methods for Electron-Molecule Collisions* Editor Gianturco F. A. (USA:Springer)
[35] Watson, D and McKoy, V 1979 *Phys. Rev.* **A21** 112
[36] Lucchese, R R and McKoy, V 1979 *J. Phys B* **12** L421

[37] Rescigno, T N, McCurdy, C W and Schneider, B I 1989 *Phys. Rev. Letts.* **63** 248

[38] Schneider, B I, Rescigno, T N, Lengsfield, B H and McCurdy, C W 1991 *Phys. Rev. Lett.* **66** 2728

[39] Rescigno, T N and McCurdy, C W 2000 *Phys. Rev.* **A62** 032706; Schneider, B I and Collins, L A 2005 *J. Non-Cryst. Solids*, **351** 1351

[40] Feist, J, Nagele S, Pazourek, R, Persson E, Schneider, B I, Collins L A and Burgdörfer, J *Phys. Rev.* **A77** 043420; Guan X, Bartschat, K and Schneider, B I *ibid* 043421

[41] Bartschat, K, Guan, X, Noble, C J, Schneider, B I and Zatsarinny, O. 2011 *Quantum Dynamic Imaging* 13; Schneider, B I, Feist, J, Nagele, S, Pazourek, R, Hu, S, Collins, L A and Burgdörfer, J *ibid* 149 Edited by Bandrauk, A D and Ivanov, M (New York:Springer-Verlag) http://www.springer.com/978-1-4419-9490-5

Journal of Physics: Conference Series **759** (2016) 012003 doi:10.1088/1742-6596/759/1/012003

Energy fluxes in turbulent separated flows

J.-P. Mollicone, F. Battista, P. Gualtieri and C.M. Casciola

Department of Mechanical and Aerospace Engineering, Sapienza University of Rome

E-mail: carlomassimo.casciola@uniroma1.it

Abstract. Turbulent separation in channel flow containing a curved wall is studied using a generalised form of Kolmogorov equation. The equation successfully accounts for inhomogeneous effects in both the physical and separation spaces. We investigate the scale-by-scale energy dynamics in turbulent separated flow induced by a curved wall. The scale and spatial fluxes are highly dependent on the shear layer dynamics and the recirculation bubble forming behind the lower curved wall. The intense energy produced in the shear layer is transferred to the recirculation region, sustaining the turbulent velocity fluctuations. The energy dynamics radically changes depending on the physical position inside the domain, resembling planar turbulent channel dynamics downstream.

1. Introduction

Most of the statistical theory of turbulence concerns the mechanisms that sustain turbulent fluctuations against dissipation, namely production and fluxes of turbulent kinetic energy in the physical space [1] and across the different scales of motion [2]. In the classical picture, the energy flux and production are single point statistical observables unable to distinguish between the different scales of motion. On the other hand, the spectrum of turbulent kinetic energy and the spectral balance allow to address each scale of motion separately [3]. As a main drawback, the spectral description does not permit to localise the energy transfer process in space, due to projection on non-local Fourier modes. Physical and spectral views were already reconciled in the early stages of turbulent research, when Kolmogorov devised his description of turbulence in terms of structure functions. The implications of his extremely elegant and beautiful theory for inhomogeneous and complex flows was not recognised until recently through generalisations to shear dominated anisotropic flows.

The scale-by-scale approach, originally introduced by Kolmogorov to develop his theory in homogeneous and isotropic turbulence [5], has been progressively extended to anisotropic and inhomogeneous conditions, for example, from shear dominated flows, see [6, 7], to channel flow, see [8], [9] and [10]. Casciola et al. (2003) address the link between the intermittency and anisotropy in homogeneous shear flows [6]. The generalised Kolmogorov equation is then employed to distinguish the shear-dominated scales by the small isotropic scales dominated by dissipation. Numerical data shows that the dissipation scales are independent on the mean shear, so the intermittency correction is universal. Danaila et al. (2001) provide a generalised form of the Kolmogorov equation adding additional terms to account for the "large-scale turbulent diffusion acting from the walls through to the centreline of the channel" [8]. The aim is to apply the Kolmogorov theory to inhomogeneous turbulent flows with moderate Reynolds number. Hot-wire measurements allow the prediction of large scale effects on the dissipation rate at the

Journal of Physics: Conference Series **759** (2016) 012003 doi:10.1088/1742-6596/759/1/012003

channel centre. The scale-by-scale budget in a planar turbulent channel flow is numerically studied by means the generalised form of Kolmogorov equation, see [9] and [10]. The DNS data sets are analysed to investigate how the inhomogeneity produced by the wall affects transfer of energy in physical and scale space at different distances from the wall. Such understanding is fundamental to tune innovative techniques to control boundary layer transition to turbulence or to design innovative turbulence models, see the discussion in [11], since it is able to capture the correct dynamics of the fluctuations for inhomogeneous flows.

The present work investigates, through Direct Numerical Simulation (DNS), the scale-by-scale energy production, transport and dissipation in a channel with a lower curved wall which is a relatively more complex geometry with respect to previous studies in homogeneous or shear flows. The results section shows how the generalised Kolmogorov equation is a viable tool to characterize the turbulence dynamics in anisotropic and strongly inhomogeneous flows. The energy can be transferred in the physical space from production regions (mainly the shear layer behind the bump) to regions where it is dissipated or alternatively, the energy produced in some specific region of the flow domain gives rise to an energy cascade in the separation spaces, depending on the distance from the wall.

2. Computational Approach

Direct numerical simulations are used to solve the incompressible Navier-Stokes equations,

$$\frac{\partial u_i}{\partial t} + u_j \frac{\partial u_i}{\partial x_j} = -\frac{\partial p}{\partial x_i} + \frac{1}{Re}\frac{\partial^2 u_i}{\partial x_j^2} + f_i \qquad \frac{\partial u_i}{\partial x_i} = 0, \tag{1}$$

where u_i is the ith velocity component, p is the hydrodynamic pressure and f_i is the body force sustaining the flow inside the channel. $Re = h_0 U_b/\nu$ is the bulk Reynolds number, where h_0 is half the nominal channel height, U_b is the bulk velocity and ν is the kinematic viscosity. All length scales are made dimensionless with the nominal channel half-height, time with h_0/U_b and pressures with ρU_b^2. When needed, the adoption of wall units will be explicitly highlighted by the customary superscript $+$, and the index $i = 1, 2, 3$ will be replaced with the corresponding label, x, y or z. Angular brackets denote the temporal and z-direction averaging. The setup was simulated for enough time to reach steady state conditions before time-uncorrelated data sets were extracted for statistical analysis. The mesh contains approximately 120 million grid points. Large computational power was required for the simulation and therefore the simulation was run on the FERMI BlueGene/Q supercomputer located in Bologna, Italy. The computational fluid dynamics solver used is Nek5000 [12], which is based on the spectral element method (SEM), see [13].

Figure 1 shows a sketch of the simulated geometry. The curved wall inside the channel on the lower wall is commonly referred to as a "bump". The domain dimensions are $(L_x \times L_y \times L_z) = (26 \times 2 \times 2\pi) \times h_0$ where x, y, and z directions are the stream-wise, wall-normal and span-wise

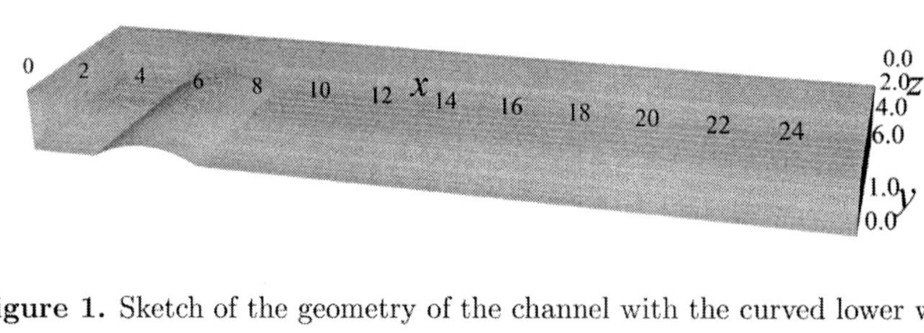

Figure 1. Sketch of the geometry of the channel with the curved lower wall.

directions respectively. Flow is from left to right in the x direction with periodic boundary conditions in both the streamwise, x, and spanwise, z, directions. No slip boundary conditions are enforced at the top and bottom walls. The simulation has a bulk Reynolds number equal to $Re = 2500$ and the maximum friction Reynolds number equal to $Re_\tau = u_\tau h_0/\nu = 300$.

3. The Generalised Kolmogorov Equation

The Kolmogorov equation follows directly from the equations of motion and describes all the dynamical effects occurring at each scale of the turbulent flow, see [14] and [15]. The generalised form of the Kolmogorov equation, in order to account for inhomogeneous conditions, is derived by [4], see also [16] and [17] for a different approach, and reads

$$\begin{aligned}&\frac{\partial\langle\delta u^2\delta u_j\rangle}{\partial r_j}+\frac{\partial\langle\delta u^2\delta U_j\rangle}{\partial r_j}+\frac{\partial\langle\delta u^2 u_j^*\rangle}{\partial X_{cj}}+\frac{\partial\langle\delta u^2 U_j^*\rangle}{\partial X_{cj}}+\\&2\langle\delta u_i\delta u_j\rangle\frac{\partial\delta U_i}{\partial r_j}+2\langle u_j^*\delta u_i\rangle\frac{\partial\delta U_i}{\partial X_{cj}}=\\&-4\,\varepsilon^*+2\nu\frac{\partial^2\langle\delta u^2\rangle}{\partial r_j\partial r_j}+\frac{\nu}{2}\frac{\partial^2\langle\delta u^2\rangle}{\partial {X_{cj}}^2}+\frac{2}{\rho}\frac{\partial\langle\delta p\delta u_i\rangle}{\partial X_{ci}}\,.\end{aligned}\tag{2}$$

Hereafter, any quantity denoted by an apostrophe indicates that is taken at x_i' which is a position separated by the vector $r_i = x_i' - x_i$ with respect to x_i. The coordinate of the mid-point between points x_i' and x_i is defined by $X_{c_i} = (x_i + x_i')/2$. Lower-case letters refer to fluctuations whilst upper-case letters refer to averaged quantities. The asterisk denotes the mid-point average, e.g. $u_i^* = (u_i' + u_i)/2$, whilst the δ denotes an increment, e.g. $\delta U_i = U_i' - U_i$.

Since the present domain has only one homogeneous direction, the span-wise one, the generalised Kolmogorov equation is defined in a five dimensions space where the independent coordinates are the two physical X_c and Y_c and the three separation r_x, r_y and r_z coordinates. In the simplified case of homogeneous turbulent flow, equation (2) reduces to the von Kármán equation, defined in the three dimensional separation space,

$$\frac{\partial\langle\delta u^2\delta u_i\rangle}{\partial r_i}=-4\,\varepsilon^*+2\nu\frac{\partial\langle\delta u^2\rangle}{\partial r_i\partial r_i}\,.\tag{3}$$

In order to highlight the physical description of each single term of equation (2), the generalised Kolmogorov equation can be recast as follow,

$$\left(\frac{\partial\Phi_{r,j}}{\partial r_j}+\frac{\partial\Phi_{c,j}}{\partial X_{cj}}\right)=-4\,\varepsilon^*-(\Pi_r+\Pi_c)\,,\tag{4}$$

where ε is the turbulent kinetic energy dissipation, Π_r and Π_c are the energy production in the separation and physical spaces respectively,

$$\Pi_r=2\langle\delta u_i\delta u_j\rangle\frac{\partial\delta U_i}{\partial r_j}\tag{5}$$

$$\Pi_c=2\langle u_j^*\delta u_i\rangle\frac{\partial\delta U_i}{\partial X_{cj}}\,,\tag{6}$$

and $\Phi_{r,j}$ and $\Phi_{c,j}$ are the energy flux in the separation and physical spaces respectively,

$$\Phi_{r,j}=\langle\delta u^2\delta u_j\rangle+\langle\delta u^2\delta U_j\rangle+2\nu\frac{\partial\langle\delta u^2\rangle}{\partial r_j}\tag{7}$$

$$\Phi_{c,j}=\langle\delta u^2 u_j^*\rangle+\langle\delta u^2 U_j^*\rangle+\frac{\nu}{2}\frac{\partial\langle\delta u^2\rangle}{\partial X_{cj}}+\frac{2}{\rho}\frac{\partial\langle\delta p\delta u_j\rangle}{\partial X_{cj}}\,.\tag{8}$$

Equation (4) provides a way to distinguish between the transfer of energy from production to dissipation in both physical and separation spaces as opposed to only observing one-point statistics, as, for example, in the mean flow kinetic energy and the turbulent kinetic energy equations, which provide information concerning the transport, production and dissipation only in physical space.

4. Results

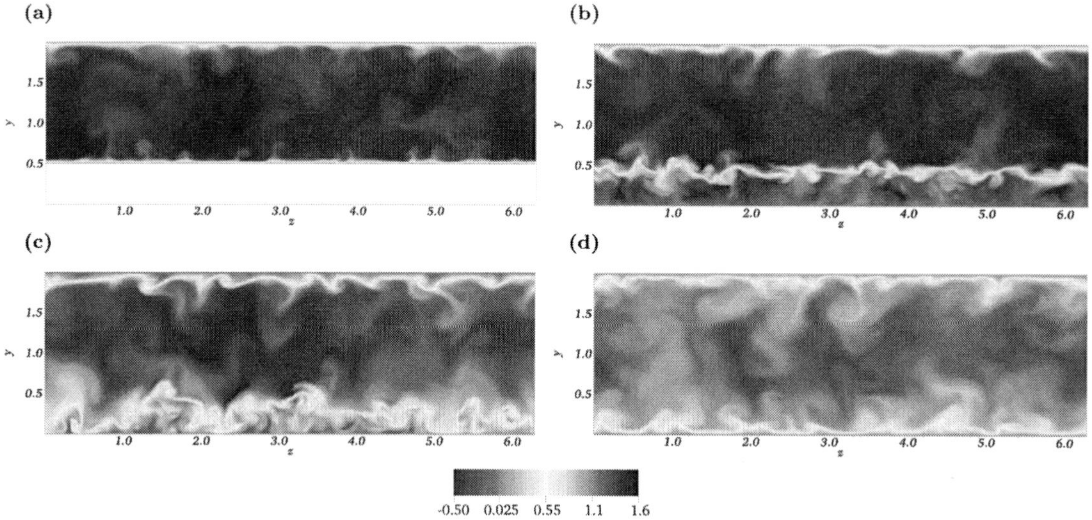

Figure 2. Instantaneous streamwise velocity contour plots in y-z planes at the tip of the bump in panel (a), end of the bump in panel (b), towards the end of the separation bubble in panel (c) and at $x = 24$ in panel (d).

Figure 2 shows instantaneous streamwise velocity contour plots in y-z planes at various x locations inside the domain. Panel (a) shows the flow at the tip of the bump, where the boundary layer becomes thinner towards the lower wall due to the intense flow acceleration induced by the decrease in channel section. At the end of the bump, panel (b), the separation induced by the abrupt change in channel section increase is evident and an intense shear layer detaches from the bump separating the recirculation bubble from the bulk flow. The shear layer is corrugated by large scale structures. Effects on the top wall are seen by the thickening of the boundary layer which is more evident further downstream towards the end of the separation bubble, panel (c), where the flow reattaches to the bottom wall. The flow stabilises downstream at $x = 24$, panel (d), where it resembles a classical planar turbulent channel flow.

Figure 3 shows a vorticity iso-surface which highlights the turbulent structures just behind the bump. Small three-dimensional structures that fill the recirculation region in the bottom half of the channel are generated by the shear layer. These structures tend to disappear downstream and elongated structures, similar to the planar channel streaky structures, appear close to the walls. Figure 4 shows the contour plots of mean streamwise velocity. The separation bubble behind the bump is evident and a smaller stagnation region in front of the bump is also present. Due to the diminished channel section at the bump, the flow accelerates. The contour plot confirms the existence of a strong shear layer between the bulk flow and the separation bubble. As anticipated, the bump also affects the flow close to the top wall where boundary layer thickening occurs.

Figure 3. Instantaneous vorticity surfaces at constant magnitude.

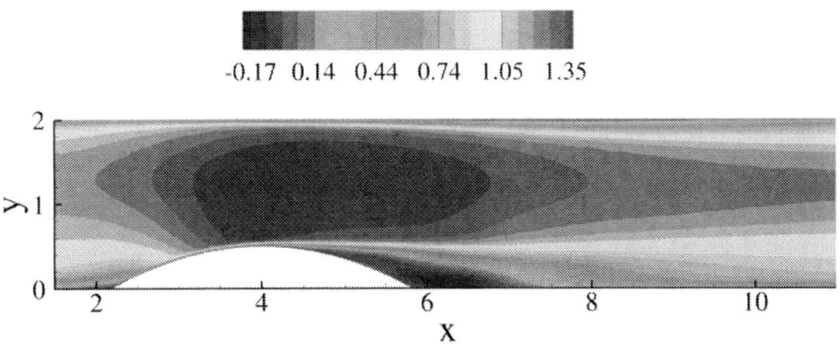

Figure 4. Contour plot of the streamwise mean velocity. Only a small portion of the flow domain near the bump is shown.

Energy fluxes, production and dissipation are now discussed following the background provided in section 3. We recall that the proper phase space involves five dimensions, namely three separations and two center point positions. The results discussed in the following concern the two-dimensional sub-spaces (Y_c, r_z) at fixed streamwise stations X_c with $r_x = r_y = 0$, see figure 5. Figure 6 reports the energy fluxes, Φ_r and Φ_c, as vectors and the right-hand side of equation (4) as background contour plots. All the terms in equation (2) are normalised with the dissipation $-4\varepsilon^*$ and, for better visualisation, the vectors are shown in arbitrary units. Proceeding downstream from panel (a) to panel (f) in figure 6, the energy dynamics shows a substantial dependence on the streamwise position X_c. Three main regions can be distinguished: the zone before the bump in panel (a), the zone after the bump where the flow is separated in panels (b)-(d) and the well reattached flow region downstream in panel (f).

In panel (a), the intense localised production at small scales, $r_z^+ \approx 50$, and close to the wall, $y^+ \approx 25$, is associated with the small recirculation just ahead of the bump. In this plane, the wall-normal component of the spatial flux, $\Phi_{c,y} = \langle \delta u^2 u_y^* \rangle + \langle \delta u^2 U_y^* \rangle + \frac{\nu}{2}\partial\langle \delta u^2 \rangle/\partial X_{cy} + \frac{2}{\rho}\partial\langle \delta p \delta u_y \rangle/\partial X_{cy}$, is dominated by the contribution arising from the mean flow and overwhelms the r_z-component of the flux, $\Phi_{r,r_z} = \langle \delta u^2 \delta u_z \rangle + 2\nu\partial\langle \delta u^2 \rangle/\partial r_z$. The physical interpretation is that energy generated near the wall is moved towards the center of the channel. We observe that the fluxes apparently do not absorb all the energy produced by the localised source, suggesting

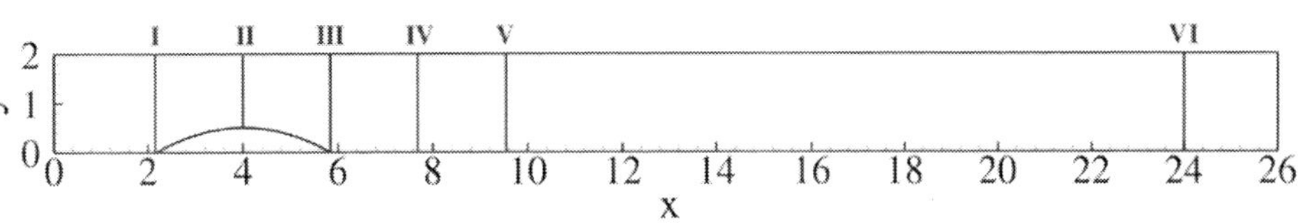

Figure 5. Sketch of the stream-wise stations selected for the statistical analysis.

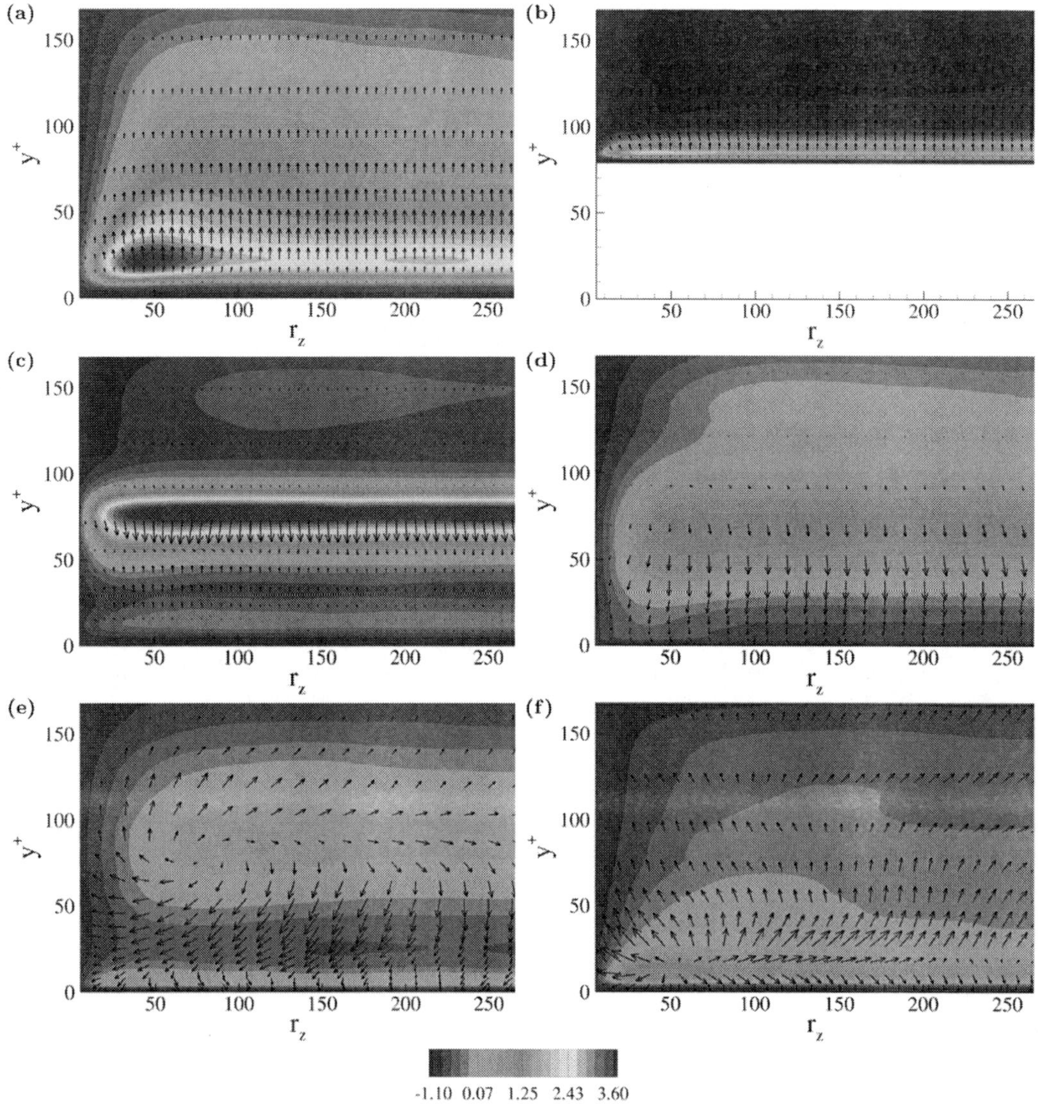

Figure 6. Contour plots of $(\Pi - 4\epsilon^*)/(4\epsilon^*)$ with vectors representing Φ in the reduced space (Y_c, r_z), i.e. in the plane $r_x = r_y = 0$, $X_c = const$. Panels (a) to (f) represent stations I to VI respectively, see Figure 5. Only the lower half of the channel, where the bump is present, is shown.

that the excess energy is captured by the other components of the flux acting in directions orthogonal to the plane.

At the tip of the bump, panel (b), production occurs across all the scales, the qualitative behaviour of the fluxes is basically unchanged, and again, the energy produced at a given spanwise separation is moved by the spatial flux towards the center of the channel.

The most interesting behaviour occurs downstream, after the bump, where the interaction of the main bulk flow and the recirculation bubble produces an intense shear layer. Energy production is intense over a large range of scale separations, for $r_z^+ > 20$, and is spatially located at $y^+ \simeq 75$ which corresponds to the position of the shear layer away from the wall. Apart from the intense energy production due to the shear layer, a second, small intensity local peak (light blue) of $\Pi - 4\epsilon^*$ can be observed close to the bottom wall deep inside the recirculating region. These two regions of production (the shear layer and the second minor peak) feed energy to the recirculating region. The absence of a significant r_z-component of the flux suggests that the fluid structures do not change their spanwise size.

Downstream, the most intense energy production peak gradually decreases while the second peak close to the bottom wall becomes more prominent and stabilises to turbulent channel-like behaviour in panel (f). The energy behaviour in panel (f) is in accordance with what already observed in the literature [10]. The downstream evolution from panels (d) to panels (e) shows the change between fluxes directed towards the wall inside the recirculation bubble below the shear layer and fluxes directed towards the channel center downstream of the bubble. In particular, in panel (e), the fluxes take a clockwise orientation as the r_z-component rises in intensity, indicating a transfer of energy to smaller spanwise scales below the nominal position of the shear layer and to larger scales above.

5. Conclusions

Turbulent separation was addressed by a Direct Numerical Simulation of turbulent channel flow with a lower curved wall. The presence of the bump produces a stable, fully turbulent recirculation bubble which is separated from the core flow by a strong shear layer.

This configuration is sufficiently general to reproduce the general features of separated flows, yet it is simple enough to allow to explore the potential of novel advanced statistical tools for turbulence dynamics. In particular, new partial results were discussed concerning the generalised Kolmogorov equation which, up to now, was only used to deal with simpler configurations where no separation occurs. The Kolmogorov equation allows to study the energy dynamics not only in the physical space but also in the separation space, i.e. it extends the scale-by-scale budget originally used by Kolmogorov to characterise the homogeneous and isotropic turbulence.

The channel section variation induced by the lower curved wall greatly affects the energy transfer across different scales and positions. The most intense peak of energy production is found in the shear layer where strong velocity fluctuations and intense mean velocity gradients occur. Production and dissipation of energy are not uniform throughout both physical and separation spaces. The scale-by-scale analysis is found able to account for the inhomogeneous effects coming from the recirculation region. It highlights how the turbulent fluctuations behind the bump are sustained by spatial energy fluxes arising from the shear layer to feed the recirculation bubble. Although the presented results provide a partial view, the analysis can be extended to cover all the five components of scale and spatial fluxes and their relationships with production and dissipation in turbulent flow separation.

Acknowledgments

The research leading to these results has received funding from the European Research Council under the ERC Grant Agreement no. 339446. PRACE, under grant no. 2014112647, has awarded access to resource FERMI based in Bologna, Italy.

References

[1] Townsend A 1976 *Cambridge UP, Cambridge*
[2] Frisch U 1995 *Turbulence: the legacy of AN Kolmogorov* (Cambridge university press)
[3] Monin A and Yaglom A 1971 *Vol. I and II MIT Press, Cambridge*
[4] Hill R J 2002 *Journal of Fluid Mechanics* **468** 317–326
[5] Kolmogorov A N 1941 *Dokl. Akad. Nauk SSSR* vol 30 pp 299–303
[6] Casciola C M, Gualtieri P, Benzi R and Piva R 2003 *Journal of Fluid Mechanics* **476**
[7] Casciola C, Gualtieri P, Jacob B and Piva R 2005 *Physical review letters* **95** 024503
[8] Danaila L, Anselmet F, Zhou T and Antonia R 2001 *Journal of Fluid Mechanics* **430** 87–109
[9] Marati N, Casciola C M and Piva R 2004 *Journal of Fluid Mechanics*

[10] Cimarelli A, Angelis E D and Casciola C M 2013 *Journal of Fluid Mechanics* **715**
[11] Thiesset F, Antonia R A, Danaila L and Djenidi L 2013 *Physical Review E* **88** 011003
[12] Fischer P, Lottes J W and Kerkemeier S G Nek5000 - Open source spectral element CFD solver. Argonne National Laboratory, Mathematics and Computer Science Division, Argonne, IL, see http://nek5000.mcs.anl.gov (2008).
[13] Patera A T 1984 *Journal of Computational Physics* **54** 468–488
[14] Thiesset F, Danaila L and Antonia R A 2014 *Fluid-Structure-Sound Interactions and Control*
[15] Cambon C and Gracia B J 2013 *Journal of Turbulence*
[16] Yakhot V 2001 *Physical Review E* **63** 026307
[17] Kurien S and Sreenivasan K R 2001 *Physical Review E* **64** 056302

Modeling of neutron-star mergers: a review while awaiting gravitational-wave detection

Luca Baiotti

Graduate School of Science, Osaka University, Toyonaka, 560-0043, Japan

E-mail:

Abstract.
I review some recent results on simulations of mergers of binary neutron stars, highlighting some of the several significant advances published in the literature in the years 2013 to 2015.

1. Introduction

We are living in an exciting time for research on binary neutron-star (BNS) mergers. In fact, we are in a very dynamical phase of research, in which many accomplishments have been achieved (especially in the years 2013-2015), while many more need to be achieved to describe such fascinating physical phenomena. The first detection of gravitational waves [1] adds excitement to the field. Such first detection concerned a binary black-hole coalescence, but measurements of BNS mergers are also expected in the next months.

The details of the genesis of BNS systems are still unclear, but the general picture seems well accepted [2]. The more massive star of the binary explodes as a supernova once it has evolved off the main sequence and through its giant phase. The remnant of such an explosion becomes the first neutron star of the binary, with a potential recoil. Subsequently, the secondary star evolves off the main sequence and explodes as a supernova becoming the second neutron star.

Many astronomical observations revealed that BNS indeed exist [3]. The inspiral and merger of two neutron stars in binary orbit is the inevitable fate of close-binary evolution, since their angular momentum is dissipated through the emission of gravitational radiation. Here is a brief description of their evolution. As the inspiral progressively speeds up and the two neutron stars approach each other, tidal waves produced by the tidal interaction become visible at their surface. At the merger, the two stars collide with a rather large impact parameter. A *vortex sheet* (or *shear interface*) develops, where the tangential component of the velocity exhibits a discontinuity. This condition is known to be unstable to very small perturbations and it can develop a *Kelvin–Helmholtz instability*, which curls the interface forming a series of vortices at all wavelengths [4, 5]. Even if this instability is purely hydrodynamical and it is likely to be important only for binaries with very similar masses, it can have strong consequences if the stars have magnetic fields. It has been shown in most simulations [6, 7, 8, 9, 10, 11] that, in the presence of an initially poloidal magnetic field, this instability may lead to an exponential growth of the toroidal component. Such a growth is the result of the exponentially rapid formation of vortices that curl magnetic field lines that were initially purely poloidal. The exponential growth caused by the Kelvin–Helmholtz instability leads to an overall amplification of the magnetic field of up to a few orders of magnitude [10, 11].

After the merger, unless the progenitor stars have very small masses or for some particular equations of state (EoSs), the compact object newly formed is expected to collapse to a black hole, either promptly or after a certain amount of time. If the collapse is not prompt, the merged object is expected to be temporarily a hypermassive neutron star (HMNS), namely a neutron star whose rest mass exceeds the maximum rest mass of nonspinning or uniformly spinning neutron stars and is sustained against collapse by differential rotation. The merged object oscillates violently, with a dominant $m = 2$ non-axisymmetric character [12]. Dissipative effects like viscosity, magnetic fields or gravitational-wave emission bring then the star towards a configuration which is unstable to gravitational collapse. The time interval between merger and collapse can be roughly estimated to be of the order of $1-100$ s or more. The large uncertainty in determining this is due particularly to the difficulty of performing accurate numerical simulations over these very long time-scales. During this time, if the merged object has a sufficiently high ratio of rotational to gravitational binding energy, it could also become dynamically unstable to nonlinear instabilities leading to a barmode deformation [13, 14, 15]. Both magnetic fields [7, 10, 11] and radiative processes [16, 17, 18, 9, 19] can modify significantly the lifetime of the HMNS and hence change the properties of the black hole and of the surrounding torus. In most cases when a black hole is formed some amount of matter remains outside of it. This leads to the formation of an accretion torus that may be rather dense ($10^{12} - 10^{13}$ g cm^{-3}) and extended horizontally for tens of kilometers and vertically for a few tens of kilometers.

BNS mergers are expected (i) to be significant sources of gravitational radiation, not only during the inspiral, but also during and after the merger, (ii) to be possible progenitors for short-gamma-ray bursts (GRBs), whose short rise times suggest that their central sources have to be highly relativistic objects, and (iii) to be the possible sources of other electromagnetic (e.g. macronovae [20, 21]) and neutrino emission.

The typical scenario leading to GRBs assumes that a system composed of a rotating black hole and a surrounding massive torus is formed after the merger. If the disc had a mass $\gtrsim 0.01$–$0.1\,M_{\odot}$, it could provide the large amount of energy observed in SGRBs, either through neutrino processes or by extracting the rotational energy of the black hole via magnetic fields [22, 23, 8]. Such a configuration is expected to form in most BNS mergers.

As far as gravitational-wave emission is concerned, BNS systems are expected to produce signals of amplitude large enough to be relevant for Earth-based gravitational-wave detectors such as LIGO [24], Virgo [25], or Kagra [26]. They are also expected to be sufficiently frequent sources: Advanced interferometric detectors are in fact expected to observe them at a rate of $\sim$ 0.4–400 events per year, with a most realistic rate of $\sim$ 40 per year [27]. Among other results, observations of gravitational waves, photons, and neutrinos emitted during and after BNS mergers will give strong indications about the EoS of matter at nuclear densities, which cannot be probed in laboratories on Earth and is not fully understood at the theoretical level [28].

During the inspiral phase, when the two stars are still well separated, sophisticated post-Newtonian expansions offer a good approximate description of the dynamics, but only full numerical simulations can describe the highly nonlinear regime of the merger and post-merger phases. The final goal of numerical computations may be a simulation that includes the solution of the Einstein equations, the relativistic hydrodynamic and (resistive) magnetohydrodynamics (MHD) equations, EoSs based on microphysical calculations, neutrino and photon radiation transport, nuclear-reaction networks. On top of all this, codes should possibly implement high-order, high-accuracy numerical methods and run fast enough to allow parameter-space exploration. This is a goal still very far in the future. But when we achieve all this, the reward will be great, namely an accurate interpretative and predictive description of the processes related to BNS systems and involving and causing short GRBs, jet formation and emission, particle formation and emission, other electromagnetic emission like that of macronovae, heavy-element abundance. As far as the study and interpretation of gravitational-wave observations is concerned, we may be able to obtain satisfactory results even without the full machinery introduced above. In fact, the interesting gravitational radiation is probably emitted before neutrino, radiation transport, and nuclear reactions become important. The evolution of the merged object, instead, depends on a large number of details, such as the mass-ratio of the initial stars, the EoS, the neutrino physics and precise energy transport, the dynamical relevance of the magnetic field, etc.

Several groups are working on BNS simulations with their own independent codes. Most of the codes solve the full Einstein Equations without approximations (except for the truncation error and the errors associated to numerical methods), but there are two notable exceptions, namely the code of Rosswog [29, 30, 31, 32] and the code of Janka/Bauswein [33, 34, 35]. These two smoothed-particle-hydrodynamics (SPH) codes make theoretical approximations to the evolution of gravity / spacetime but are among the most advanced in treating microphysical processes.

The first successful fully general-relativistic simulations of BNS mergers were presented at the end of the past century in [36] and huge advances have been made since. Several groups have developed numerical codes for general-relativistic simulations. All of these codes share the following capabilities:

- Implementation and use of high-order finite-differencing techniques[1] for the solution of the Einstein equations
- Implementation and use of high-resolution shock-capturing (HRSC) methods for the solution of the GRHD/GRMHD equations
- Implementation and use of adaptive mesh-refinement (AMR) techniques that provide higher resolution around the orbiting stars and the merged object
- Use of consistent initial data representing a system of BNS in quasi-circular orbits[2]
- Accurate evolution of matter and spacetime (including long-term evolutions of the formed black holes and accretion discs), thanks to the numerical methods mentioned above
- Analysis of the properties of the black holes produced in the merger, *e.g.* through the calculation of trapped surfaces
- Extraction of gravitational waveforms (usually through computation of the Weyl tensor)

Intense work is ongoing and first satisfactory results are being obtained on linking future gravitational-wave observations to physical properties of the emitting system (e.g. relating the main frequency of postmerger oscillations to the neutron-star masses) [41, 42, 43, 44, 45, 46, 47, 48] and on heavy-element production and macronovae [49, 31, 50, 32, 51, 52, 53, 54, 55, 56, 57, 58, 59, 19, 60]. There are then still issues that are very open, like the accurate dynamics of magnetic fields after the merger [8, 10, 11] (and before the merger for resistive MHD [61, 62, 63, 64, 65]), neutrino treatment [16, 17, 66, 18, 67, 68, 9, 69, 70, 71, 19] and photon radiation transport [70].

Most of the general-relativistic codes [36, 72, 73, 74, 10, 75, 76, 77, 78, 79, 61, 80, 81, 82, 83, 62, 84, 85, 86, 87] solve the field equations in the Baumgarte-Shapiro-Shibata-Nakamura formulation [88, 89]. The remainder [90, 91, 92, 93] use the generalized-harmonic formulation [94, 95]. Most groups use finite-difference methods for the metric evolution and

[1] One group is developing a code based on pseudo-spectral techniques, but no publication on BNS mergers has been produced yet, except the preliminary studies of [37, 38]. It is named the *SpEC-hydro* code and employs a dual-grid method in which the metric fields are solved using pseudospectral methods on a multidomain spectral grid, and the matter fields are solved using high-resolution shock-capturing finite volume schemes on a Cartesian finite-difference grid.

[2] Some studies focus on purpose on different types of initial data [39, 40, 31, 32].

conservative, HRSC schemes for the hydrodynamics evolution. In all cases, these unigrid algorithms are then interfaced with some sort of AMR.

Some groups have implemented the MHD equations in full general relativity; since these codes all make use of conservative HRSC methods, their main differences are in how they meet the challenge of preserving the div·$\mathbf{B} = 0$ constraint, especially in AMR[3]. Besides MHD, the other major advances in the physical modelling for numerical relativity codes have been in the arena of microphysics. While the polytropic or ideal-fluid EoSs have been the community standard for quite some time, most codes now allow for a nuclear theory-based EoSs and/or use various parameterized, piecewise polytropic EoSs inspired by the range of plausible nuclear EoSs [96]. Some groups have also started to account for neutrino transport via simplified leakage schemes [16, 17, 66, 18, 67, 68, 9, 69] and are also studying formulations for more accurate treatments [70, 71, 19], which show much promise for numerical relativity simulations with neutrino physics beyond the leakage approximation.

2. Status of models, techniques, and results for general-relativistic simulations

General relativistic hydrodynamics simulations of BNS started being performed more than 15 years ago [72, 36, 97, 98]. Even if nowadays many state-of-the-art codes are able to solve more physical equations (for magnetic fields, neutrinos, radiation, etc.), simulations involving only general-relativistic hydrodynamics (*pure GRHD*) are still the benchmark for any new code and the necessary testbed for more advanced codes. Furthermore, in many cases, results obtained with pure GRHD - notably inspiral gravitational waveforms - are a sufficiently good description of BNS systems. During the merger and after a single object forms, possibly surrounded by a massive disc, it is necessary to consider additional physical components in order to have a satisfactory description.

One particularly important aspect of BNS simulations is the EoS. While detectable differences between simulations that employed different EoSs already appear during the inspiral [96, 99], it is really the post-merger phase that is markedly different. Hence, an accurate description of the post-merger evolution is essential to extract information concerning the neutron-star interior structure. These particularly interesting parts of the waveforms that are more likely to yield fundamental clues to the physics beyond nuclear density are at very high frequencies and thus probably only marginally detectable by detectors like advanced LIGO. Third-generation detectors, such has ET [100], may provide the first realistic opportunity to use gravitational waves to decipher the stellar structure and EoS [101].

First attempts to single out the effects of the EoS on the inspiral and merger were done in Refs. [102, 103, 74, 104]. The authors there focus mostly on the dynamics of equal-mass binaries, as these are thought to be the most common [105] and are easier and faster to compute, since symmetries of the configuration can be exploited to save computational resources. In Baiotti et al. [104], an investigation was made using two analytic EoSs: a *cold*, *i.e.* polytropic, EoS, or a *hot*, *i.e.* ideal-fluid (also called *simple fluid* or *gamma-law*), EoS. Although the polytropic EoS is isentropic and thus unrealistic for describing the post-merger evolution, it provides an adequate description of the inspiral phase, during which the neutron stars are expected to interact only gravitationally [106]. The most marked differences between *cold* and *hot* EoSs are in the post-merger phase.

The study of the effect of realistic EoSs in GRHD simulations has been subsequently especially brought forward by the Shibata group[4], starting from Ref. [107], where they employ the Akmal-Pandharipande-Ravenhall (APR) EoS [108] augmented, after the onset of merger, by a *hot* part through an ideal-fluid EoS. In another work of the same group [109], the authors studied the

[3] Straightforward finite-difference evolutions of the magnetic field would generically lead to magnetic monopoles and, hence, unphysical behavior.

[4] Other groups have implemented the ability to use tabulated EoSs on their codes, see [68, 9]

dependence of the dynamical behavior of BNS mergers on the EoS of the supernuclear-density matter with piecewise-polytropic EoSs [110]. Table I in [109] usefully summarizes the piecewise polytropic parameters of several realistic EoSs.

In a work published soon after, the same group showed results of simulations performed by incorporating both nucleonic and hyperonic finite-temperature EoS [17, 18]. Reported results are of interest, even if the hyperonic EoS used here is limited by the fact that it only takes Λ hyperons into account and that it cannot produce a stable neutron star with realistic masses[5]. It was found that also for the hyperonic EoS, a HMNS is first formed after the merger and subsequently collapses to a black hole. The radius of such a HMNS decreases in time because of the increase of the mass fraction of hyperons and the consequent decrease in supporting pressure. Such a shrinking is noticeably larger than the one simply due to angular-momentum loss through gravitational-wave emission that is present also in nucleonic EoSs. These dynamics are clearly visible in the gravitational-wave signal, whose characteristic peak frequency has an increase of 20%-30% during the HMNS evolution. By contrast, for the nucleonic EOS the peak gravitational-wave frequency in the HMNS phase is approximately constant. The authors stress that their results raise a warning about using the peak frequency of the gravitational-wave spectrum to extract information of the neutron-star matter, because it may evolve and so make the relation of the peak frequency with the HMNS structure ambiguous. Finally, the authors found that the torus mass for the hyperonic EoS is smaller than that for the nucleonic EoS. This makes hyperonic EoSs less favorite for the description of short gamma-ray bursts.

The most recent works of the Shibata group on EoSs are Ref. [44, 99], where they use their AMR SACRA code [74] to perform a large number of simulations with a variety of EoSs and mass ranges. On the basis of their results they derive a fitting formula for the quasiperiodic gravitational waveforms, which they say may be used for the analysis of future observed gravitational-wave signals. They choose two types of EoSs: piecewise polytropic EoSs [110] and the tabulated Shen EoS [114]. In both cases, they add approximate finite-temperature effects to the cold EoSs through an additional ideal-fluid term.

They put in evidence that the universal features of gravitational waves emitted by HMNSs include, in order: (i) a peak in frequency and amplitude soon after the merger starts, (ii) a decrease in amplitude during the merger and an new increase when the HMNS forms, (iii) a damped oscillation of the frequency during the HMNS phase lasting for several oscillation periods and eventually settling to an approximately constant value (although a long-term secular change associated with the change of the state of HMNSs is always present), (iv) a final decrease in the amplitude during the HMNS phase, either monotonical or with modulations. Based on this, they find an optimal 13-parameter fitting function. Using the fitted functions it may be possible to constrain the neutron-star radius with errors of about 1 km.

3. Outlook

In this article, because of limited space, I have been able to review only one topic about BNS merger simulations, namely the influence of the EoS. The phenomenology and physics of BNS mergers, however, are much richer. Simulations of BNS merger and post-merger phases will soon be connected to observations of gravitational waves and will allow us to interpret them and therefore to uncover the mysteries of the physical and astrophysical properties of relativistic stars.

References

[1] Abbott B P, Abbott R, Abbott T D, Abernathy M R, Acernese F, Ackley K, Adams C, Adams T, Addesso P, Adhikari R X and et al 2016 *Physical Review Letters* **116** 061102 (*Preprint* 1602.03837)

[5] Hyperonic EoSs, which are generally soft, are currently disfavored by the observation of a $2M_\odot$ star [111, 112]. However, the existence of exotic phases in neutron stars remains unconstrained [113].

[2] Kalogera V, Belczynski K, Kim C, O'Shaughnessy R and Willems B 2007 *Physics Reports* **442** 75–108 (*Preprint* arXiv:astro-ph/0612144)
[3] Kramer M, Lyne A, Burgay M, Possenti A, Manchester R, Camilo F, McLaughlin M, Lorimer D, D'Amico N, Joshi B, Reynolds J and Freire P 2004 The double pulsar – a new testbed for relativistic gravity *Binary Pulsars* ed Rasio and Stairs (Chicago: PSAP)
[4] Chandrasekhar S 1981 *Hydrodynamic and hydromagnetic stability* (New York, USA: Dover Edition)
[5] Bodo G, Massaglia S, Ferrari A and Trussoni E 1994 *Astron. Astrophys.* **283** 655–676
[6] Price D J and Rosswog S 2006 *Science* **312** 719–722 (*Preprint* astro-ph/0603845)
[7] Giacomazzo B, Rezzolla L and Baiotti L 2011 *Phys. Rev. D* **83** 044014
[8] Rezzolla L, Giacomazzo B, Baiotti L, Granot J, Kouveliotou C and Aloy M A 2011 *Astrophys. J. Letters* **732** L6 (*Preprint* 1101.4298)
[9] Neilsen D W, Liebling S L, Anderson M, Lehner L, O'Connor E and Palenzuela C 2014 *Phys. Rev. D* **89** 104029 (*Preprint* 1403.3680)
[10] Kiuchi K, Kyutoku K, Sekiguchi Y, Shibata M and Wada T 2014 *Phys. Rev. D* **90** 041502 (*Preprint* 1407.2660)
[11] Kiuchi K, Cerdá-Durán P, Kyutoku K, Sekiguchi Y and Shibata M 2015 *Phys. Rev. D* **92** 124034 (*Preprint* 1509.09205)
[12] Stergioulas N, Bauswein A, Zagkouris K and Janka H T 2011 *Mon. Not. R. Astron. Soc.* **418** 427–436 (*Preprint* 1105.0368)
[13] Shibata M, Baumgarte T W and Shapiro S L 2000 *Astrophys. J.* **542** 453–463 (*Preprint* astro-ph/0005378)
[14] Baiotti L, de Pietri R, Manca G M and Rezzolla L 2007 *Phys. Rev. D* **75** 044023 (*Preprint* astro-ph/0609473)
[15] Franci L, De Pietri R, Dionysopoulou K and Rezzolla L 2013 *Journal of Physics Conference Series* **470** 012008 (*Preprint* 1309.6549)
[16] Sekiguchi Y, Kiuchi K, Kyutoku K and Shibata M 2011 *Physical Review Letters* **107** 051102 (*Preprint* 1105.2125)
[17] Sekiguchi Y, Kiuchi K, Kyutoku K and Shibata M 2011 *Phys. Rev. Lett.* **107** 211101 (*Preprint* 1110.4442)
[18] Sekiguchi Y, Kiuchi K, Kyutoku K and Shibata M 2012 *Progress of Theoretical and Experimental Physics* **2012** 01A304 (*Preprint* 1206.5927)
[19] Sekiguchi Y, Kiuchi K, Kyutoku K and Shibata M 2015 *Phys. Rev. D* **91** 064059 (*Preprint* 1502.06660)
[20] Li L X and Paczyński B 1998 *Astrophys. J.* **507** L59–L62 (*Preprint* astro-ph/9807272)
[21] Metzger B D, Martínez-Pinedo G, Darbha S, Quataert E, Arcones A, Kasen D, Thomas R, Nugent P, Panov I V and Zinner N T 2010 *Mon. Not. R. Astron. Soc.* **406** 2650–2662 (*Preprint* 1001.5029)
[22] Paczynski B 1986 *Astrophys. J. Lett.* **308** L43–L46
[23] Eichler D, Livio M, Piran T and Schramm D N 1989 *Nature* **340** 126–128
[24] Abadie J, Abbott B P, Abbott R, Abernathy M, Adams C, Adhikari R, Ajith P, Allen B, Allen G, Amador Ceron E and et al 2010 *Nuclear Instruments and Methods in Physics Research A* **624** 223–240 (*Preprint* 1007.3973)
[25] Accadia T *et al.* 2011 *Class. Quantum Grav.* **28** 114002
[26] Kuroda K and LCGT Collaboration 2010 *Class. Quantum Grav.* **27** 084004
[27] Abadie J, Abbott B P, Abbott R, Abernathy M, Accadia T, Acernese F, Adams C, Adhikari R, Ajith P, Allen B and et al 2010 *Class. Quantum Grav.* **27** 173001 (*Preprint* 1003.2480)
[28] Lattimer J M and Prakash M 2007 *Phys. Rep.* **442** 109–165 (*Preprint* arXiv:astro-ph/0612440)
[29] Rosswog S, Davies M B, Thielemann F K and Piran T 2000 *Astron. Astrophys.* **360** 171–184 (*Preprint* astro-ph/0005550)
[30] Rosswog S 2010 *Journal of Computational Physics* **229** 8591–8612 (*Preprint* 0907.4890)
[31] Rosswog S, Piran T and Nakar E 2013 *Mon. Not. R. Astron. Soc.* **430** 2585–2604 (*Preprint* 1204.6240)
[32] Piran T, Nakar E and Rosswog S 2013 *Mon. Not. R. Astron. Soc.* **430** 2121–2136 (*Preprint* 1204.6242)
[33] Oechslin R, Rosswog S and Thielemann F K 2002 *Phys. Rev. D* **65** 103005 (*Preprint* gr-qc/0111005)
[34] Oechslin R, Janka H T and Marek A 2007 *Astron. Astrophys.* **467** 395–409 (*Preprint* astro-ph/0611047)
[35] Bauswein A, Janka H and Oechslin R 2010 *Phys. Rev. D* **82** 084043 (*Preprint* 1006.3315)
[36] Shibata M and Uryū K 2000 *Phys. Rev. D* **61** 064001 (*Preprint* gr-qc/9911058)
[37] Kaplan J D 2014 *Where Tori Fear to Tread: Hypermassive Neutron Star Remnants and Absolute Event Horizons or Topics in Computational General Relativity* Ph.D. thesis California Institute of Technology
[38] Foucart F, Haas R, Duez M D, O'Connor E, Ott C D, Roberts L, Kidder L E, Lippuner J, Pfeiffer H P and Scheel M A 2015 *ArXiv e-prints* (*Preprint* 1510.06398)
[39] Gold R, Bernuzzi S, Thierfelder M, Brügmann B and Pretorius F 2012 *Phys. Rev. D* **86** 121501 (*Preprint* 1109.5128)
[40] East W E and Pretorius F 2012 *Astrophys. J.* **760** L4 (*Preprint* 1208.5279)

[41] Bauswein A and Janka H T 2012 *Phys. Rev. Lett.* **108** 011101 (*Preprint* 1106.1616)
[42] Bauswein A and Janka H T 2012 *Phys. Rev. Lett.* **108** 011101 (*Preprint* 1106.1616)
[43] Bauswein A, Janka H T, Hebeler K and Schwenk A 2012 *Phys. Rev. D* **86** 063001 (*Preprint* 1204.1888)
[44] Hotokezaka K, Kiuchi K, Kyutoku K, Muranushi T, Sekiguchi Y i, Shibata M and Taniguchi K 2013 *Phys. Rev. D* **88** 044026 (*Preprint* 1307.5888)
[45] Bauswein A, Stergioulas N and Janka H T 2014 *Phys. Rev. D* **90** 023002 (*Preprint* 1403.5301)
[46] Takami K, Rezzolla L and Baiotti L 2014 *Phys. Rev. Lett.* **113** 091104 (*Preprint* 1403.5672)
[47] Clark J, Bauswein A, Cadonati L, Janka H T, Pankow C and Stergioulas N 2014 *Phys. Rev. D* **90** 062004 (*Preprint* 1406.5444)
[48] Takami K, Rezzolla L and Baiotti L 2015 *Phys. Rev. D* **91** 064001 (*Preprint* 1412.3240)
[49] Rosswog S 2013 *Nature* **500** 535–536
[50] Rosswog S 2013 *Royal Society of London Philosophical Transactions Series A* **371** 20272 (*Preprint* 1210.6549)
[51] Bauswein A, Goriely S and Janka H T 2013 *Astrophys. J.* **773** 78 (*Preprint* 1302.6530)
[52] Tanaka M and Hotokezaka K 2013 *Astrophys. J.* **775** 113 (*Preprint* 1306.3742)
[53] Hotokezaka K, Kyutoku K, Tanaka M, Kiuchi K, Sekiguchi Y, Shibata M and Wanajo S 2013 *Astrophys. J.* **778** L16 (*Preprint* 1310.1623)
[54] Metzger B D and Fernández R 2014 *Mon. Not. R. Astron. Soc.* **441** 3444–3453 (*Preprint* 1402.4803)
[55] Rosswog S, Korobkin O, Arcones A, Thielemann F K and Piran T 2014 *Mon. Not. R. Astron. Soc.* **439** 744–756 (*Preprint* 1307.2939)
[56] Grossman D, Korobkin O, Rosswog S and Piran T 2014 *Mon. Not. R. Astron. Soc.* **439** 757–770 (*Preprint* 1307.2943)
[57] Perego A, Rosswog S, Cabezón R M, Korobkin O, Käppeli R, Arcones A and Liebendörfer M 2014 *Mon. Not. R. Astron. Soc.* **443** 3134–3156 (*Preprint* 1405.6730)
[58] Wanajo S, Sekiguchi Y, Nishimura N, Kiuchi K, Kyutoku K and Shibata M 2014 *Astrophys. J.* **789** L39 (*Preprint* 1402.7317)
[59] Just O, Bauswein A, Pulpillo R A, Goriely S and Janka H T 2015 *Mon. Not. R. Astron. Soc.* **448** 541–567 (*Preprint* 1406.2687)
[60] Goriely S, Bauswein A, Just O, Pllumbi E and Janka H T 2015 *Mon. Not. R. Astron. Soc.* **452** 3894–3904 (*Preprint* 1504.04377)
[61] Dionysopoulou K, Alic D, Palenzuela C, Rezzolla L and Giacomazzo B 2013 *Phys. Rev. D* **88** 044020 (*Preprint* 1208.3487)
[62] Palenzuela C 2013 *Mon. Not. R. Astron. Soc.* **431** 1853–1865 (*Preprint* 1212.0130)
[63] Palenzuela C, Lehner L, Ponce M, Liebling S L, Anderson M, Neilsen D and Motl P 2013 *Phys. Rev. Lett.* **111** 061105 (*Preprint* 1301.7074)
[64] Palenzuela C, Lehner L, Liebling S L, Ponce M, Anderson M, Neilsen D and Motl P 2013 *Phys. Rev. D* **88** 043011 (*Preprint* 1307.7372)
[65] Dionysopoulou K, Alic D and Rezzolla L 2015 *Phys. Rev. D* **92** 084064 (*Preprint* 1502.02021)
[66] Kiuchi K, Sekiguchi Y, Kyutoku K and Shibata M 2012 *Class. Quantum Grav.* **29** 124003 (*Preprint* 1206.0509)
[67] Kiuchi K, Sekiguchi Y, Kyutoku K and Shibata M 2012 Gravitational Waves and Neutrino Emissions from the Merger of Binary Neutron Stars *Numerical Modeling of Space Plasma Slows (ASTRONUM 2011)* (*Astronomical Society of the Pacific Conference Series* vol 459) ed Pogorelov N V, Font J A, Audit E and Zank G P p 85
[68] Galeazzi F, Kastaun W, Rezzolla L and Font J A 2013 *Phys. Rev. D* **88** 064009 (*Preprint* 1306.4953)
[69] Palenzuela C, Liebling S L, Neilsen D, Lehner L, Caballero O L, O'Connor E and Anderson M 2015 *Phys. Rev. D* **92** 044045 (*Preprint* 1505.01607)
[70] Shibata M, Kiuchi K, Sekiguchi Y and Suwa Y 2011 *Progress of Theoretical Physics* **125** 1255–1287 (*Preprint* 1104.3937)
[71] Radice D, Abdikamalov E, Rezzolla L and Ott C D 2013 *Journal of Computational Physics* **242** 648–669 (*Preprint* 1209.1634)
[72] Shibata M 1999 *Phys. Rev. D* **60** 104052 (*Preprint* gr-qc/9908027)
[73] Shibata M 1999 *Progress of Theoretical Physics* **101** 1199–1233 (*Preprint* gr-qc/9905058)
[74] Yamamoto T, Shibata M and Taniguchi K 2008 *Phys. Rev. D* **78** 064054 (*Preprint* 0806.4007)
[75] Baiotti L, Hawke I, Montero P and Rezzolla L 2003 A new three-dimensional general-relativistic hydrodynamics code *Computational Astrophysics in Italy: Methods and Tools* vol 1 ed Capuzzo-Dolcetta R (Trieste: MSAIt) p 210
[76] Baiotti L 2004 *Numerical relativity simulations of non-vacuum spacetimes in three dimensions* Ph.D. thesis SISSA, International School for advanced studies

[77] Baiotti L, Hawke I, Montero P J, Löffler F, Rezzolla L, Stergioulas N, Font J A and Seidel E 2005 *Phys. Rev. D* **71** 024035 (*Preprint* gr-qc/0403029)
[78] Giacomazzo B and Rezzolla L 2007 *Classical and Quantum Gravity* **24** 235 (*Preprint* gr-qc/0701109)
[79] Kellerman T, Baiotti L, Giacomazzo B and Rezzolla L 2008 *Class. Quantum Grav.* **25** 225007 (*Preprint* 0811.0938)
[80] Kastaun W, Galeazzi F, Alic D, Rezzolla L and Font J A 2013 *Phys. Rev. D* **88** 021501 (*Preprint* 1301.7348)
[81] Baiotti L, Shibata M and Yamamoto T 2010 *Phys. Rev. D* **82** 064015 (*Preprint* 1007.1754)
[82] Anderson M, Hirschmann E W, Lehner L, Liebling S L, Motl P M, Neilsen D, Palenzuela C and Tohline J E 2008 *Phys. Rev. D* **77** 024006 (*Preprint* 0708.2720)
[83] Anderson M, Hirschmann E W, Lehner L, Liebling S L, Motl P M, Neilsen D, Palenzuela C and Tohline J E 2008 *Phys. Rev. Lett.* **100** 191101 (*Preprint* 0801.4387)
[84] Löffler F, Faber J, Bentivegna E, Bode T, Diener P, Haas R, Hinder I, Mundim B C, Ott C D, Schnetter E, Allen G, Campanelli M and Laguna P 2012 *Class. Quantum Grav.* **29** 115001 (*Preprint* 1111.3344)
[85] Zilhão M and Löffler F 2013 *International Journal of Modern Physics A* **28** 1340014 (*Preprint* 1305.5299)
[86] Moesta P, Mundim B, Faber J, Noble S, Bode T, Haas R, Loeffler F, Ott C, Reisswig C and Schnetter E 2013 General relativistic magneto-hydrodynamics with the Einstein Toolkit *APS Meeting Abstracts* p 10001
[87] Thierfelder M, Bernuzzi S and Brügmann B 2011 *Phys. Rev. D* **84** 044012 (*Preprint* 1104.4751)
[88] Shibata M and Nakamura T 1995 *Phys. Rev. D* **52** 5428–5444
[89] Baumgarte T W and Shapiro S L 1999 *Phys. Rev. D* **59** 024007 (*Preprint* gr-qc/9810065)
[90] Duez M D *et al.* 2008 *Phys. Rev. D* **78** 104015 (*Preprint* 0809.0002)
[91] Anderson M, Hirschmann E W, Liebling S L and Neilsen D 2006 *Classical and Quantum Gravity* **23** 6503–6524 (*Preprint* gr-qc/0605102)
[92] Lindblom L, Scheel M A, Kidder L E, Owen R and Rinne O 2006 *Classical and Quantum Gravity* **23** 447 (*Preprint* gr-qc/0512093)
[93] East W E, Pretorius F and Stephens B C 2012 *Phys. Rev. D* **85** 124010 (*Preprint* 1112.3094)
[94] Garfinkle D 2002 *Phys. Rev. D* **65** 044029 (*Preprint* gr-qc/0110013)
[95] Pretorius F 2005 *Classical and Quantum Gravity* **22** 425–451 (*Preprint* gr-qc/0407110)
[96] Read J S, Baiotti L, Creighton J D E, Friedman J L, Giacomazzo B, Kyutoku K, Markakis C, Rezzolla L, Shibata M and Taniguchi K 2013 *Phys. Rev. D* **88** 044042 (*Preprint* 1306.4065)
[97] Shibata M and Uryū K 2002 *Progress of Theoretical Physics* **107** 265–303 (*Preprint* gr-qc/0203037)
[98] Shibata M, Taniguchi K and Uryū K 2003 *Phys. Rev. D* **68** 084020 (*Preprint* gr-qc/0310030)
[99] Hotokezaka K, Kyutoku K, Okawa H and Shibata M 2015 *Phys. Rev. D* **91** 064060 (*Preprint* 1502.03457)
[100] Punturo M *et al.* 2010 *Class. Quantum Grav.* **27** 084007
[101] Andersson N, Ferrari V, Jones D I, Kokkotas K D, Krishnan B, Read J S, Rezzolla L and Zink B 2011 *General Relativity and Gravitation* **43** 409–436 (*Preprint* 0912.0384)
[102] Shibata M, Taniguchi K and Uryū K 2005 *Phys. Rev. D* **71** 084021 (*Preprint* gr-qc/0503119)
[103] Shibata M and Taniguchi K 2006 *Phys. Rev. D* **73** 064027 (*Preprint* astro-ph/0603145)
[104] Baiotti L, Giacomazzo B and Rezzolla L 2008 *Phys. Rev. D* **78** 084033 (*Preprint* 0804.0594)
[105] Osłowski S, Bulik T, Gondek-Rosińska D and Belczyński K 2011 *Mon. Not. R. Astron. Soc.* **413** 461–479 (*Preprint* 0903.3538)
[106] Giacomazzo B, Rezzolla L and Baiotti L 2009 *Mon. Not. R. Astron. Soc.* **399** L164–L168
[107] Kiuchi K, Sekiguchi Y, Shibata M and Taniguchi K 2009 *Phys. Rev. D* **80** 064037 (*Preprint* 0904.4551)
[108] Akmal A, Pandharipande V R and Ravenhall D G 1998 *Phys. Rev. C* **58** 1804–1828 (*Preprint* arXiv:hep-ph/9804388)
[109] Hotokezaka K, Kyutoku K, Okawa H, Shibata M and Kiuchi K 2011 *Phys. Rev. D* **83** 124008 (*Preprint* 1105.4370)
[110] Read J S, Markakis C, Shibata M, Uryū K, Creighton J D E and Friedman J L 2009 *Phys. Rev. D* **79** 124033 (*Preprint* 0901.3258)
[111] Antoniadis J, Freire P C C, Wex N, Tauris T M, Lynch R S, van Kerkwijk M H, Kramer M, Bassa C, Dhillon V S, Driebe T, Hessels J W T, Kaspi V M, Kondratiev V I, Langer N, Marsh T R, McLaughlin M A, Pennucci T T, Ransom S M, Stairs I H, van Leeuwen J, Verbiest J P W and Whelan D G 2013 *Science* **340** 1233232 (*Preprint* 1304.6875)
[112] Demorest P B, Pennucci T, Ransom S M, Roberts M S E and Hessels J W T 2010 *Nature* **467** 1081–1083 (*Preprint* 1010.5788)
[113] Bhowmick B, Bhattacharya M, Bhattacharyya A and Gangopadhyay G 2014 *Phys.Rev.* **C89** 065806 (*Preprint* 1403.0341)
[114] Shen H, Toki H, Oyamatsu K and Sumiyoshi K 1998 *Progress of Theoretical Physics* **100** 1013–1031 (*Preprint* nucl-th/9806095)

Journal of Physics: Conference Series **759** (2016) 012005 doi:10.1088/1742-6596/759/1/012005

Undergraduate computational physics education: uneven history and promising future

Richard F Martin

Physics Department, Illinois State University, Normal, IL 61790-4560 USA

rfm@ilstu.edu

Abstract. While some physics educators have included computing in courses or have developed specialized courses for over 50 years, computational physics education has only slowly made inroads into the broader physics education community. Even now, when computation is arguably more important than ever in physics research and applications, it is rare that a physics department offers more than a single course in the topic to its undergraduate students. There have been several times over the years when interest in a more global approach to computational physics education has surged, only to subside without attaining the goal that computing finally take an essential role in the education of undergraduate physicists. In this presentation I will review some of the history of computational physics education, briefly discuss our experience with the program at Illinois State University, and suggest some direction for the future.

1. Introduction

Physics and computing have been associates since the beginning of the computer era. From Atanasoff's development [1] of the first electronic digital computer in the late 1930's, the Manhattan Project's use of computing as a calculational aid propelling Metropolis [2] into algorithm design, and ground-breaking simulations [3] of Fermi, Pasta, and Ulam, physicists have been at the forefront of computational science. The close relationship between physics and computation has grown spectacularly since the mid 20th century so that computational physics is arguably on par with theory and experiment as a research methodology in physics.

The same cannot be said of computational physics education (CPE), at least at the undergraduate level in the United States. By this I mean teaching young physicists in their first four years of post-secondary education the concepts, methods, and experience of approaching physics problems with computational tools, just as we have done for more than a century with theoretical and laboratory experimental physics. Examples of CPE goals might include, among other ideas, teaching students to: Model physical systems for computational solution, understand computational algorithms & techniques, have familiarity with numerical analysis, computer architecture, simple complexity theory, implement models/algorithms as efficient code, use appropriate diagnostics and analyze simulation data, communicate results via data visualization and beyond.

I do not include in CPE learning aids in physics, such as computer-aided instruction systems, animated modules designed to teach specific physics concepts, online courseware, etc. While these can be quite useful, I choose to concentrate on developing physics majors who are in control of the technology, models, algorithms, and code themselves, and not just consumers of someone else's computational skills.

A great deal of energy in the physics education community over the past 20 years has been directed toward physics education research (PER), producing a wealth of information on how students learn physics and suggesting new approaches to effective teaching. Almost no PER investigations have focused on educating students in computational physics. I suggest that the time is now for redirecting some of that "educational energy" toward rethinking physics curricula with an eye toward computation, and toward the development of innovative new ways of guiding physics students toward expertise in computation physics.

In that light, I offer the following brief history of CPE (Section 2), a discussion of three "third generation" degree programs in CPE (Section 3), a short look at the current state of affairs (Section 4), and a brief final analysis. I limit my discussion mainly to programs in the USA as I cannot claim expertise beyond North America, yet I hope these ideas find some resonance with physicists beyond my home country and I welcome new information on such programs.

2. Where have we been? A brief history of CPE

Physicists are a creative bunch and it's not surprising that there were "early adopters" in the classroom of the new methodology they had used in research. Alfred Bork was one such pioneer who influenced a generation of CPE advocates. In his 1963 article [4] entitled *A physics independent study course with computer*, Bork presented several reasons why physics faculty and students might want to have some familiarity with the computer (all quotes from the article):

A. The modern computer makes workable problems which were for all practical purposes impossible only a few years ago.
B. Using numerical analysis permits the consideration of material beyond the students' analytic ability.
C. The "feel" of numbers promotes understanding of analytic relations.

The first two represent a statement of the power of computation in both physics and education. Keep in mind that this was a time of mainframe computers programmed with cards. In Bork's course at the University of Alaska, an IBM 1620 computer was used. Students programmed in FORTRAN with punch cards and a "typewriter" for I/O. Ditto masters were used in the typewriter so output could be reproduced for the whole class. Students had no prior training in programming and received only the briefest introduction in this class. Despite these limitations it is clear that Bork understood the consequences of the new technology. Indeed, the course he developed in 1962 does not look that different than a first course in computational physics might today. Point (C) above indicates that Bork appreciated the improved insight and enhanced physical intuition obtainable by simulating physical systems.

Bork's article is characteristic of what one might call the "first generation" of CPE: a devoted advocate developing course modules or, rarely at that time, special course designed to give students their first experience in solving physics problems computationally. It would be interesting to know the future of the students in his class in 1962; how many of them would go on to be computational research physicists, or to further develop CPE as future faculty?

If not Bork's students, there were others who followed in his shoes. The 1970's and 1980's saw an explosion of single course development. Like Bork, these were frequently a single faculty member enthusiast, often experienced with computational research, who saw the writing on the wall: computing is here to stay and we owe it to our students to prepare them for that future. This "second generation" of CPE also began to formalize the content of those courses in several textbooks [5]. These computational physics texts (and CP courses) tend to differ significantly from a traditional numerical analysis text, by being more physics and project-oriented. While still explaining the algorithms involved, CPE texts tend to begin with a physics problem, explaining how to select a solution technique that fits the problem. In discussing CPE courses with physics colleagues, a fair generalization might be that more time is spent on code testing and verification based on physical constraints, like conservation laws and checking simulation behavior against known limiting cases and ones' physical intuition, rather than focusing on analytic error estimation, for example.

Around the late 1980's and through the 1990's a "third generation" became notable: movement toward developing degree sequences in computational physics. The 1988 *Computers in Physics Instruction* was an early conference devoted explicitly to CPE and stimulated many of its participants to implement their new ideas as courses and programs. In the USA, this momentum was reinforced by a demographic issue, namely a significant enrolment dip nationally in undergraduate physics programs. One of the proposed remedies emanating from community discussions and conferences such as the 1996 AAPT *Physics at the Crossroads* conference and the 1999 APS/AAPT/NSF *Physics Revitalization Conference: Building Undergraduate Physics Programs for the 21st Century* [6], included providing more breadth to undergraduate physics programs. Degree options and sequences in CPE fit that recommendation nicely. In that time period new degree programs in computational physics were initiated, CP was integrated into regular physics major courses, concentrations/options in CP were introduced, and interdisciplinary degrees, such as dual degrees in physics and computer science, were created. The next section explores three such programs in more detail.

The first decade of this century saw new collaborative momentum, new programs and a continuing trend toward a wider comprehension that future physicists need to have experience in computational physics. An explosion of conferences and journal special issues serve to both chronicle and stimulate this new energy:

- 2004 APS March Meeting, session: *Computers in Physics Education*
- 2005 AAPT, "cracker barrel": *Building a physics computing community for undergraduate education*
- 2006 AAPT, session: *Computation in physics courses*
- 2006 Sept. Computers in Science E: Focus topic issue on CPE
- 2007 AAPT topical conference: *Computational Physics for Upper Level Courses*
- 2007 fall AJP theme issue: *Computation and Computer-Based Instruction*
- 2008 Gordon Conference: *Computation and Computer-Based Instruction*.

The last two decades' CPE developments promised to produce lasting reform of physics curricula. A few short case studies may help shed some light on the reality of the dream.

3. Three comprehensive programs

3.1. Integration or specialization?

Lawrence University, in Appleton, Wisconsin, is a small, liberal-arts oriented university, with a Carnegie Classification of "Baccalaureate Colleges: Arts & Sciences Focus". The physics department has five faculty and offers a B.A. degree in physics and a "3-2" joint engineering degree with a partner engineering university. Their innovation, beginning in 1985, was to integrate computational methodology into their regular physics courses [8], a stated goal being *"to embed the use of general purpose graphical, symbolic, and numerical computational tools throughout our curriculum"*. Their program had two explicitly computational courses, Computational Mechanics and Computational Physics, with many CP methods integrated into physics major courses. The program's primary architect, David Cook, disseminated the program well, including talks, articles, and a self-published customizable textbook, available through the author's website. Prof. Cook is now retired and remains active in the AAPT. The CP integration program is now absent from the Lawrence U. physics department website and from the course catalog, although the Computational Mechanics course remains required for all majors and the Computational Physics course is now integrated with an Advanced Mechanics course, the combination being a senior elective class.

Oregon State University is a large university in Corvallis, Oregon with a Carnegie Classification of "Doctoral Universities: Highest Research Activity". The physics department, with 17 faculty, offers B.S. and B.A. degrees in physics, as well as M.S. and PhD graduate degrees. In 2001, a full B.S. in Computation Physics was initiated. Under the leadership of Rubin Landau, this program consisted of five CP courses (Scientific Computing I and II, Computational Physics Lab, Simulation I and II), two numerical math courses, and two related computer science courses. In addition, a senior research thesis was required of all students, and CP students did a computational project. This program was one of the most comprehensive CP degrees, and was well disseminated. A CP textbook [11] is in third edition, and online courseware is available through the author's web page. Landau has gone further in CPE pedagogy, arguing that CPE methods are better for learning physics [12]. Prof Landau is now retired, remaining active in CPE. The Oregon State B.S. in Computational Physics had been eliminated due to insufficient enrolment. However, CP remains as one of 8 "options" on the regular physics degree, with two dedicated courses.

Illinois State University is a mid-size university in Normal, Illinois. While the university carries a Carnegie Classification of "Doctoral Universities: Higher Research Activity", the physics department is undergraduate only. With 10 faculty, the department offers four B.S. degrees: physics, computational physics, physics teacher education, and engineering physics (a "3-2" joint program). Beginning around 1990, a team of three faculty, George Skadron, Robert Young, and myself initiated a program of integrating selected CP methods into physics major courses. Student feedback encouraged the development of a full CP degree track, which began in 1998. Students take 3-5 CPE courses and two computer science courses. This program was not as well disseminated as the other two, but an early article [10] describes the initiative. Of the originators, two are retired. The program continues under the leadership of younger, computationally-oriented faculty.

3.2. A closer look at the Illinois State program

I am naturally most familiar with the Illinois State University CP degree and will discuss it more depth as an example of a longer-lived program. While the CP bachelor's degree began in 1998, it's roots go back to "second generation" in the late 1970's when Robert Young first offered a single course using computational methods entitled Computers in Physics. In the late 1980's we started integrating CP into physics courses, guided by a "skills matrix" which mapped out which CP topics, including basic numerical, simulation, and visualization methods, would naturally be used in which courses. For example, both the over-relaxation method for Laplace's equation and 3D surface plotting were in the intermediate E&M "matrix cell". In a short time our students were graduating with significantly more modeling and programming experience than their peers at other institutions. Graduates who went on to graduate school and those who entered the technical workplace both reported how their computing experience served them well; several said they wish they could have done more. With concerns about physics enrolment at the time, we decided to develop a degree program. NSF funded our "Undergraduate Computation Science Lab" in the mid-1990's, a project to develop the B.S. degree and to enhance computational undergraduate research. The "Computer Physics" bachelor's degree became official in 1998 and its first graduate, preparing ahead for the requirements, appeared a year later. This name, incidentally, was changed from "Computational Physics" early on after surveying high school physics students who thought "computational" meant "mathematical". "Computer Physics" cleared up any ambiguity.

The degree sequence requirements include all core physics courses that regular physics majors take: Three semesters of introductory physics, and courses in intermediate mechanics, E&M, quantum, mathematical physics, and laboratory physics. This is complemented by the same five math courses that physics majors take, one chemistry course, and two advanced physics courses (*Thermal Physics* and either senior *E&M* or *Quantum Mechanics*). Beyond these common classes, CP majors take dedicated CP courses *Methods of Computational Physics*, *Advanced Computational Physics*, and a capstone project course *Computational Research in Physics*. Supplemented by computationally-oriented electives (one required), such as *Nonlinear Dynamics*, and one or two courses from the

Information Technology department (*Programming for Scientists* if the student has no programming experience and *Hardware and Software Concepts* – an introductory computer architecture overview). With additional computing integrated into regular physics courses, the program gives students a strong mix of physics and computation. For those seeking research participation, several faculty offer computational research topics, and the majority of CP majors do get involved. Indeed, both of the department's finalists for the APS LeRoy Apker Award were working on computational problems.

I've highlighted courses taken by CP majors, but what physics material was "sacrificed" in favor of CP courses? This question is a common one from colleagues who are skeptical of CPE, and it deserves a straight answer. We dropped the following requirements: one senior lab course, one senior physics course, one senior elective, and 1 semester of chemistry. Table 1 summarizes the trade-offs. Both sequences require 21 courses, and are within 1 semester hour total.

Table 1. Comparison of courses required for the CP degree and the Physics degree. Numbers of courses included in the category are in parentheses.

Computer Physics	Physics
Core physics (8)	Core physics (8)
Math (5)	Math (5)
Chemistry (1)	Chemistry (2)
Computer science (1)	Advanced physics (3)
Advanced physics (2)	Advanced Lab (1)
Methods of CP (1)	Senior electives (2)
Advanced CP (1)	
Senior elective (1)	
Capstone CP project (1)	

The basic methods course Methods of Computational Science is frequently team-taught by one physics and one chemistry faculty, motivating the computational methods by problems in those fields. Sample topics include: discretization and numerical errors; numerical integration for dynamic bridge stability; ODE solving for molecular vibration, nonlinear oscillations, and chemical kinetics; Fourier methods for NMR spectroscopy, and Monte Carlo for neutron shielding.

The Advanced Computational Physics course is a projects course taught by three faculty, each one presenting a five week project. Students are responsible for writing the code, analyzing the results, and producing a written report. A variety of projects have been covered over the years, including neural networks for time series prediction or pattern recognition, molecular dynamics simulation of microclusters, an original classical ensemble method for calculating time-dependent quantum wave functions, Monte Carlo simulation of light scattering off a turbid medium, and finite element simulation of heat transfer in nonuniform solids. The projects are chosen to display the power of simulation physics and to open the student's eyes to computational thinking. One graduate returning several years post-graduation told stories of tech industry job interviews requiring immediate problem analysis and coding up a solution. Initially panicking at one of these, he realized that a neural network might solve the problem he was assigned. Based on his one experience with neural nets in this course, and his own creativity, he succeeded in the interview. It doesn't take many cases like that to convince you that CPE is essential for today's physics graduates.

Finally, the capstone research project is a chance for students to really innovate. They have more-or-less complete freedom to choose a topic. All they need is a faculty mentor to agree to work with them. Given that the student is responsible to do the modelling, coding, and analysis, and present a talk to the department or to an appropriate research venue. Topics are often physics-based, such as particle-in-cell plasma simulation with parallelized code, galaxy classification in Hubble Space Telescope deep-field images using image processing and pattern recognition, slab geometry atomic scattering using matrix methods. Non-physics projects encompass many other student interests. One student was curious about gerrymandered US congressional districts and decided to compute their correlation dimension to see if they're complex enough to have a fractal boundary. A student athlete designed a project to analyze some common weight lifts. She was able to get experimental data from the kinesiology department, developed a simple model of two basic lifts, and analyzed the data relative to her model. Another project entailed a two dimensional rule-based (cellular automaton) model of traffic flow and was able to reproduce the effect of a simple yield growing into a major traffic jam. While not physics, per se, all of these projects are precisely the kind of problem we'd like to think physicists could solve, in particular, computational physics students were able to make progress on each of these.

The original goals of the program were to fill a perceived need reported by graduates, to provide a recruiting tool for attracting new physics majors, and as a retention tool to offer flexibility to current students. We did a program assessment in in 2004 that included a survey of graduates. This was continued informally, with more formal studies in 2008 and 2014. The survey's typical response rate is about 35% although the first one was closer to 50%. Survey questions covered course and program satisfaction, perceived value in career, and open ended general questions.

Respondents reported high satisfaction with CP courses with an average of 4.1 on a 5 point scale over the three dedicated CP courses. Satisfaction was somewhat lower for the computer science courses. The introductory computer architecture course was rated at 2.5, with student comments indicating the course is not very challenging. Our students, however, do not take all the prerequisites that would allow them entry into a more advance architecture class. The question is under discussion in the physics department. Regarding course utility in respondents' current position, the CP classes were rated somewhat lower at an average score of 3.3 on the 5 point scale. Comments indicate that students have learned much more post-graduation that is relevant to their current position so this decrease is understandable. Next, graduates were asked to rate their satisfaction with the methodologies they were taught in the program. Not surprisingly, computational methods were ranked highest at 4.2 out of 5, with research a close second at 4.0, and theory and modeling tied at 3.8. Interestingly, experimental techniques were rated lowest at 3.3. Perhaps this is telling us that facility with programming and developing numerical models is a rather different skill than building and using experimental instrumentation. Finally CP graduates ranked their overall satisfaction with the degree at 4.5 out of 5.

Some basic student data provides further assessment details. Graduation data (Figure 1) show that the fraction of CP graduates grew rapidly from inception in 1998 to about 20% of the department's total graduates (which averages 18-20 per year).

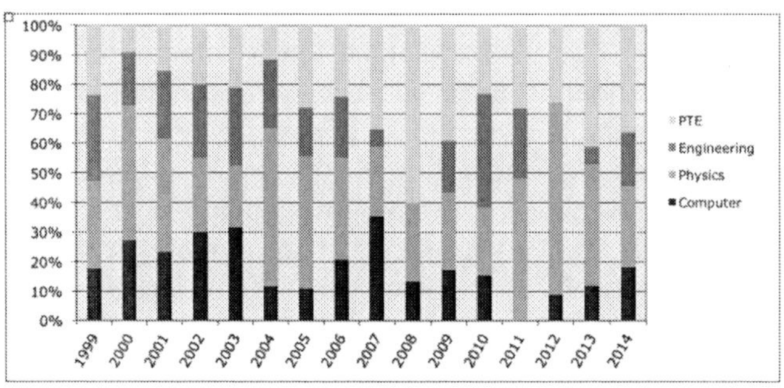

Figure 1. Annual percentage of total graduates in each of four physics degree programs at Illinois State University. The black bar is the computational physics degree.

Fluctuations are large due to the small numbers, but 20% means roughly 4 CP students graduate per year, not insignificant number for a department this size. Comparing the number of incoming freshman physics majors over all four degree sequences, to the number of graduates, gives another useful bit of information (Figure 2): the program is not meeting our original expectations that it would attract majors from high schools.

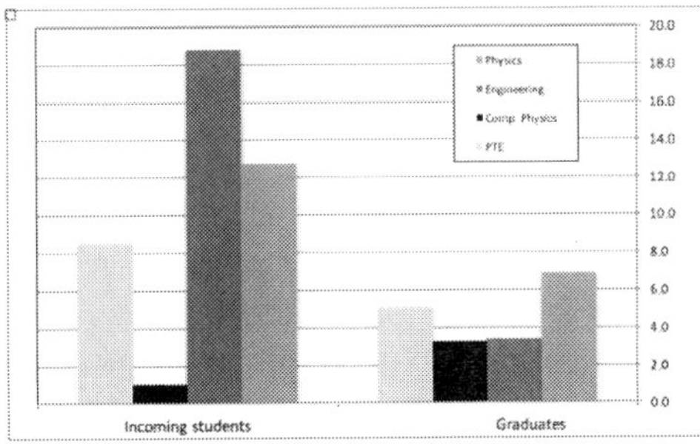

Figure 2. Number of incoming freshmen physics majors (left) and number of physics graduates (right), 2000-2014, shown for each degree sequence. The black bar is the CP degree.

In fact, Figure 2 implies that the majority of CP graduates are internal transfers from on of our other programs. It is actually the only one of the four degrees that has more graduates than incoming freshmen. The other three programs show noticeable attrition, especially the engineering physics degree, simply indicating that 18 year olds don't always make the right choice of future career. This argues that the CP degree program is more of a retention tool than a recruiting tool. We have anecdotal evidence that those who switch to CP from another physics program would have left physics had the CP program not been there.

Graduates of the Computer Physics program distribute themselves in similar percentages as physics majors to graduate school(~40%) and directly to employment (~60%). Of those taking jobs directly after graduation, 75-80% go into a technical field (35% computer related, 40% engineering, 25% other) with nearly 12% of those attempting to organize start-up companies. Graduate school bound graduates are less likely than their physic degree counterparts to study physics. About half go on in physics, while 40% choose to switch fields into electrical engineering or computer science, and 10% other fields.

4. Why isn't this easy?

4.1. Lessons from the recent past

We saw in Section 3 some indication of why computational physics programs do not exist in every physics department. First, the fact that such programs to date have required one or more devotees to champion the cause, implies that there is not universal acceptance of CPE in physics departments. Indeed, discussions at topical conferences indicate a relatively high rate of recalcitrance toward CPE among physics faculty. The fact that the Oregon State and Lawrence programs did not survive their creators' retirement may be symptomatic of this issue. Perhaps due to the long tradition of theory vs. experiment, the fact is that many of our colleagues find computation a "lesser" approach to physics and are unwilling to replace theoretical and/or experimental training either in specific courses or in curricula. In a web and email survey that included input from 252 physics departments a decade ago, Robert Fuller [12] found that even though a significant fraction of respondents agreed that computation is important in physics, when it came to actually teaching CP in physics departments, fewer than 20% included any graded CP assignments in their courses. Another issue affecting faculty perception of the field is that CP can change rapidly and physics faculty are not used to this. The discipline changes remarkably slowly on the timescale of a faculty career. Newton's laws are not going to disappear even though their domain of validity has been significantly restricted. Computation is different. When I started programming the idea that writing code to be efficient on a multi-core parallel architecture machine was not on my radar screen. The point is that there are reasons why some non-computational colleagues are not friends of CPE, and one may need to deal with this in creating a new CP program.

A second point is implicit in the decision to terminate the CP bachelor's degree at Oregon State: low numbers. The Illinois State program only graduates 3-4 students per year. Viewed as an independent program, that would be difficult to justify. The administration at Illinois State is able to look more globally and treat the CP sequence as one component of the overall physics department educational output, which also includes three other degrees and a large service course and general education load. Administrative support is also important in how curricular reform is valued in tenure and promotion decisions. If one does not get rewarded, the hard work of creating something new may not seem worth the effort. CPE requires resources, both human and hardware. If colleagues and administrators are open to faculty searches for computationalists, and if funds are available for computing needs, the cause of improving computation education are made that much easier.

Finally, it is worth accounting for student attitudes toward CPE. Surveying high school students in physics classes, we found that such students do not have computational physics on their radar screens. They simply don't know what it is and why it might lead to exciting research or an employable degree. This is consistent with the ISU finding that most CP majors come through internal transfers, and it argues that spending a lot of time and funds on recruitment at high schools for CPE majors may not be a high priority. Some students already majoring in physics actively dislike computing when they encounter it embedded in a physics course. Others find it difficult to distinguish computational physics from computer science, seeing them as more-or-less the same field, but perceiving a computer science degree as easier and more employable. Being aware of these attitudes can help faculty advise students more effectively.

4.2. The current state of CPE

The number of conferences and journal special issues in the past decade or so indicate an increasing energy in the CPE community. A web search for physics departments that have any sort of CP in the curriculum shows a growth in the number of such programs. Seventeen programs advertise full degree programs as well as CP concentrations (also called options, tracks, specializations, etc.). Table 2 summarizes some findings. The number of physics courses "lost" was estimated by comparing the CP degree course requirements with those of the physics degree. The last row gives information on the basic Carnegie Classification of the institutions, which gives a sense of their primary mission (R/VH indicates a very high level of research, R/H is a high level, and M is for institutions that grant Masters degrees as their highest degree).

Table 2.. Characteristics of US computational physics programs. Column 2 is for programs with more than 2 CP courses, Column 3 programs have fewer than 2. Source: web search November 2015.

Characteristic	> 2 CP courses	< 2 CP courses
Degree type	4 B.S., 7 Concentration	2 B.S., 5 Concentration
Mean # CP courses	2.4	0.9
Mean # CS/math courses	2.7	4.0
"Lost" physics courses	4.2	3.3
Carnegie class	4 R/VH, 2 R/H, 3 M	2 R/VH, 2 R/H, 3 M

There is a clear distinction between the institutions that offer two or more CP courses and those that offer 0 or 1 course. The latter group requires more computer science and/or computer math courses, with little actual computational physics. In that respect a physics major and a computer science minor, which could be done at may universities, may be similar experience for students. Of the two institutions in this group advertising B.S. degrees in CP one has no CP courses and no indication of CP integration, while the other does not have a separate physics degree, so this degree in "Physics and Computational Science" is the only program with any physics courses. The institutions offering two or more CP courses read more like a program concentrating on CP. Several have some CP integrated into other physics courses, and some have senior theses which can also be computational. Of the 18 total institutions with a web presence in computation physics education, only 6 are classified as primarily research universities (R/VH), indicating that curricular reform is just as likely at less research-oriented universities.

5. Conclusions

Developing a curriculum including computational physics in a meaningful way is not straightforward. I have extracted some relevant information out of a short history of computational physics education at the undergraduate level, out of direct experience with one program, and out of a search for current programs. I summarize below what may be worth thinking about if one is to create such a program, and what can be expected22 along the way.

1. As with most new projects, a dedicated team of proponents is usually necessary to get started. Then a broader team of supporters can help keep the program viable as it matures.
2. Recalcitrant faculty may need to be brought along by educating them on the value of CP for physics students both in future research and in future employment, as well CP's place in the discipline now and that perhaps some experiment and theory can realistically be replaced with computational physics.
3. Administrative support is important in clarifying the value the institution places on curriculum development, and in taking a broad view of the standard measures used to allocate resources.
4. CPE community discussions (conferences, etc) can be very motivational and allow like minded people to communicate what works and what doesn't.
5. Do not expect CP curricula to enhance recruitment of students to CP out of high schools, but it can work as a retention tool, offering those not satisfied with traditional physics a very viable option.
6. Do expect that student satisfaction can be quite high for those who find a home in computational physics.
7. 18 current programs advertising CP curricula were discovered via web search. Of those, none are programs that integrate computation into existing physics courses, 11 are physics-based CP programs, with two or more CP courses, and 7 seem more like joint physics-computer science-math programs, with one or no CP courses. Fewer than 50% of the institutions with CP programs are classified as primarily research universities.

It was pointed out that one difference between CPE and teaching traditional physics is that CP changes more rapidly. This conference presented new directions in computational physics research, many of which will be appropriated for teaching as the next generation of CP faculty develop courses and curricula. Already we are seeing techniques from high performance computing and "big data" finding their way into undergraduate courses. Can machine learning and computational model building be far behind? With so many recent conferences and special journal issues devoted to CPE, as well as many new CP programs being created and tested, it is a time of energy and innovation in the field. If momentum continues, I suspect that the value of applying computational thinking to teaching physics will begin to be better appreciated. The time would be right for physics education researchers to include computational pedagogy in their purview. I, for one, look forward to CPE making a major impact on how physics is taught and learned.

Acknowledgments

The author would like to acknowledge all those who played a role in the development and maintenance of the Illinois State University computational physics program, especially Bob Young, George Skadron, Hiroshi Matsuoka, Jean Standard, Rainer Grobe, Charles Su, and Brian Clark. NSF support at a critical time in its initial years was crucial and support from the Research Corporation for computational research with undergraduates was invaluable.

6. References

[1] Mackintosh, A R 1982 The first electronic computer *Physics Today* **40** 25-32

[2] Anderson, H L 1986 Metropolis, Monte Carlo, and the MANIAC *Los Alamos Science* **14** 96-107

[3] Weissert, T P 1997 *The Genesis of Simulation in Dynamics: Pursuing the Fermi–Pasta–Ulam Problem* (New York: Springer)

[4] Bork, A M 1963 A physics independent study course with computer *American Journal of Physics* **31** 364-368

[5] Koonin, S E 1986 *Computational Physics* (Addison-Wesley); Gould H and Tobochnik J 1988 *An Introduction to Computer Simulation Methods: Applications to Physical Systems* (Reading, MA: Addison-Wesley); Schmidt E, Spitz G, and Losch W 1988 *Theoretical Physics on the Personal Computer* (Berlin: Springer-Verlag)

[6] Kirkpatrick, L D 1999 Guest editorial: Revitalizing Physics, *The Physics Teacher* **37** 71

[7] Redish, E F and J S Risley 1990 *Computers in Physics Instruction. Proceedings of a Conference* (Reading: Addison-Wesley)

[8] Cook, D M 2008 Computation in undergraduate physics: The Lawrence approach *American Journal of Physics* **76** 321-326

[9] Cook, D M 2001 *Computation and Problem Solving in Undergraduate Physics* (AAPT, http://www.compadre.org/psrc/items/detail.cfm?ID=5593)

[10] Martin, R.F., Jr., G. Skadron, and R.D. Young 1991 Computers, Physics and the Undergraduate Experience, *Computers in Physics*, **5**, 302-310 (available as free content on the AIP/Scitation website http://scitation.aip.org/content/aip/journal/cip/5/3)

[11] Landau R H, Peaz M J, and Bordeianu C 2015 *Computational Physics, problem solving with Python* (Berlin: Wiley-VCH)

[12] Landau, R H 2006 Computational physics: A better model for physics education?", *Computing in Science & Engineering*, **8(5)**, 22-30

Journal of Physics: Conference Series **759** (2016) 012006 doi:10.1088/1742-6596/759/1/012006

Seismic interferometry and ambient noise tomography: theoretical background and application in south India

Ritima Das* and S S Rai

Indian Institute of Science Education & Research- Pune, Dr. Homi Bhabha Road, Pune-411008. India

*E-mail: ritima08@gmail.com

Abstract. Seismic interferometry can be used to extract useful information about Earth's subsurface from the ambient noise wave field. It is an important new tool for exploring seismically quiescent areas. The method involves extraction of empirical Green's function from the background ambient vibrations of the Earth, followed by computation of group or phase velocity and tomographic imaging. Here we provide a review of seismic interferometry and ambient noise tomography (ANT) and present an example of the method in south India.

1. Introduction

The traditional methods in seismology for imaging the earth's interior use time and amplitude of earthquake waves recorded over seismographs. The methods therefore could be best used only in seismologically active areas. Elsewhere, it has limited resolution. Over the last decades, a new method known as seismic or wave field interferometry has revolutionized passive seismology imaging. The method is based on ambient seismic noise caused mainly by wind, ocean waves, rock fracturing and anthropogenic activity that constantly travel through the Earth. This complex wave field contains information about the Earth's subsurface. The feasibility of the method has been established experimentally (Weaver and Lobkis, 2001; Loarose et al., 2005) and theoretically (Snieder, 2004; Wapenaar, 2004). The methodology allow us to decode the information contained in the ambient noise wave field to create an artificial seismogram that can then be used to image the subsurface of the Earth following traditional seismological tomographic or imaging methods.

Surface wave empirical Green functions (EGFs) can be determined from cross-correlations between long time sequences of ambient seismic noise observed at different stations. Shapiro and Campillo (2004) demonstrated that EGFs estimated from ambient noise possess dispersion characteristics similar to earthquake derived measurements. The dispersion measured from surface wave EGFs have been inverted to produce the corresponding tomography maps in several regions (Shapiro et al., 2005; Moschetti et al., 2007; Kao et al., 2013, Borah et al., 2014,Guo et al., 2015). Most of these studies obtained Rayleigh wave group velocity maps in the period range from 5 s to 70 s.

In this paper we review the theory of seismic interferometry as applied to ambient noise data. We then construct the virtual seismogram using noise propagating across south India and 3-D tomogrphic images are generated thereof.

2. Seismic Interferometry: Theoretical Background

Interferometry is a general interference phenomenon between pairs of signals and is used to gain useful information about the medium. Compared to many other tomographic imaging tools with long history, application of ambient noise to Earth imaging is recent (Wapenaar, 2004; Campilo and Paul, 2003; Wapenaar et al, 2011). The basic theory behind seismic interferometry is that Green's function between two seismic stations can be estimated by cross correlating long time series of ambient noise recoded at those stations. This Green's function may be thought of as the seismogram recorded at one location due to an impulsive or instantaneous source of energy at the other.

Suppose two receivers at positions r1 and r2 are surrounded by energy sources located on an arbitrary surroundings boundary S (Figure 1). The wave field emanating from each source propagates into medium in the interior of S and is recorded at both receivers. The signals recorded at the two receivers are then cross-correlated. When the cross-correlation of all the sources are added together, the energy that travels along the path will add constructively and the energy that doesn't travel along the path will add destructively. Thus the Green's function obtained between r1 and r2 will be as if one of the receivers had actually been a source (Figure 1) (Wapenaar, 2004). This case is actually observed where each source is fired sequentially and impulsively. For the case of random noise, a surface S exists such that it joins up all the noise sources and since noise sources may all fire at the same or at overlapping time, their recorded signals at two receivers are already summed together which actually takes place naturally. Snieder (2004) showed that the seismic sources located around the extensions of the inter-receiver path contribute most to the interferometric Green's function construction and thus whole boundary of source is not necessary in order to approximate the inter-receiver Green's function.

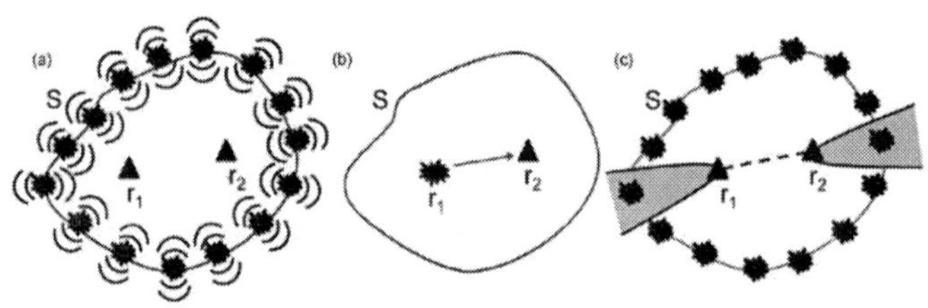

Figure 1: (a) Two receivers (triangles) are surrounded by a boundary S of sources each of which sends a wavefield into the interior and exterior of S. (b) the seismic interferometry method turns one of the receivers (r1) into virtual source from which a real seismogram is obtained. (c) sources located within the grey regions contribute the most to the Green's function computation. (Nicolson et al., 2012).

2.1. Conceptual Framework for Correlation and Green's Function

Recent works by Wapenaar & Fokkema (2005), and a host of others have confirmed both theoretically and practically the 3D elastodynamic generalization of Claerbout's (1967) 1-D postulate that the autocorrelation of noise generated at depth at a surface station yields the reflection profile of the earth under that station and using two different surface points yields the impulse response function between these points, as is indicated by the cartoon in Figure 2.

The Green's function of a medium between two points A and B represents the record obtained at A if an impulsive source is applied at B. In a completely random wavefield, the cross-correlation of signals recorded between two points converges to the complete Green's function of the medium, including all reflection, scattering and propagation modes (Weaver & Lobkis, 2001). If the sources of the ambient noise are evenly distributed and recorded by sensors at two points $A(\varphi_A)$, $B(\varphi_B)$, then time derivative of the correlation is the exact Green's function of the medium, as if the source was at A or B (Weaver and Lobkis, 2001; Wapennaar, 2004):

$$\frac{d}{dt}C_{AB}(\tau) = \frac{d}{dt}\int \varphi_A(t)\varphi_B(t+\tau)dt \propto G^+(A,B,\tau) - G^-(A,B,-\tau) \tag{1.0}$$

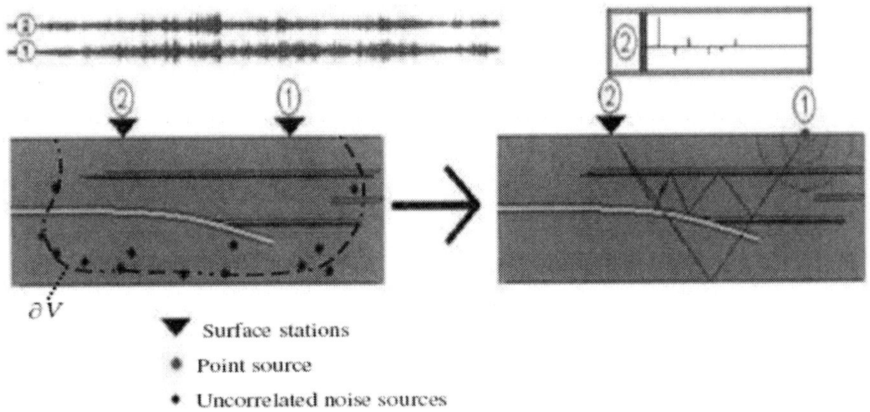

Figure 2: Schematic diagram showing the cross-correlation of noise at two surface stations.

It has been established that normalized cross-spectral density $C_{A,B}(\omega)$ at frequency ω between two receivers A and B separated by a distance r is

$$C_{A,B}(\omega) = \frac{\sin(\omega r/c)}{\omega r/c} \tag{1.1}$$

In time domain (taking Fourier transform) the normalized correlation function is

$$C_{A,B}(t) = \frac{1}{2\pi}\int_{-\infty}^{\infty} C_{A,B}(\omega)\, e^{i\omega t}\, d\omega$$

Therefore

$$\frac{d}{dt}C_{A,B}(t) = \frac{1}{4\pi r/c}\left[\delta(t + r/c) - \delta(t - r/c)\right] \tag{1.2}$$

The above equation (1.2) is our required exact Green's function.

The surface wave parts of inter-receiver Green's functions appear particularly clearly in seismograms constructed from seismic interferometry, because strong sources of seismic noise are in general restricted to locations within or on the Earth's crust. Surface wave travel along the interfaces between different layers; within the Earth, they propagate particularly strongly within the crust and upper-mantle. Seismic surface waves can be divided into Rayleigh waves which have longitudinal and vertical motion, and Love waves which have transverse horizontal vibration. Both these types of surface waves are observable on cross-correlations of ambient seismic noise.

3. Ambient Noise Tomography

The ambient noise data processing is best described by Bensen et al. (2007). The procedure is divided into four main phases (1) single station data preparation, (2) cross-correlation and temporal stacking, (3) measurement of dispersion curves and (4) quality control, including error analysis and selection of the acceptable measurements. Single station data preparation includes removal of instrument response, mean and trend; band-pass filter and cut to a length of one day for each individual station. Earthquake signals and instrument irregularities are then to remove by temporal normalization. After temporal normalization, the signals are whitened in frequency. Before whitening, ambient noise is most energetic in the microseism bends below 20 sec period. Frequency whitening is carried out to broaden the period band of the dispersion measurement. Next is to perform cross-correlation between each pair of station for each available day. In order to enhance the signal to noise ratio (SNR), stacking of available daily cross-correlation for each component and station pair is done. This stacked cross-correlation can be converted to Empirical Green's function by introducing an additive phase factor (Lin et al., 2008). Next step is to calculate the dispersions. One particularly useful property of surface wave is that they are dispersive: the longer period waves within a packet of surface wave energy have a longer wavelength and hence penetrate deeper into Earth. Since different frequencies are sensitive to properties at different depths, study of surface wave dispersion allows us to infer information about how seismic velocity varies with depth in the Earth (e.g. Dziewonski et al., 1969). Typically, periods below about 20 s are mainly sensitive to crustal structure and properties, and above 20 s are also sensitive to properties of the upper mantle. Inverting surface wave velocities at different periods, measured for many paths within a given region, to obtain models of the Earth's velocity structure with depth is known as surface wave tomography. Since seismic interferometry does not depend on the location of impulsive sources such as earthquakes, rather only the location of the receivers, the resolution of ambient noise tomography in relatively aseismic regions can be greater than that achieved by local surface wave tomography using earthquakes.

3.1. Ambient noise tomography in South India

We applied ambient noise tomography to image the subsurface of southern India. Seismograms recorded at 57 broadband sites (Figure 3a) have been analyzed for the study. These stations operated for approximately three years data (from February 2009 to April 2012).

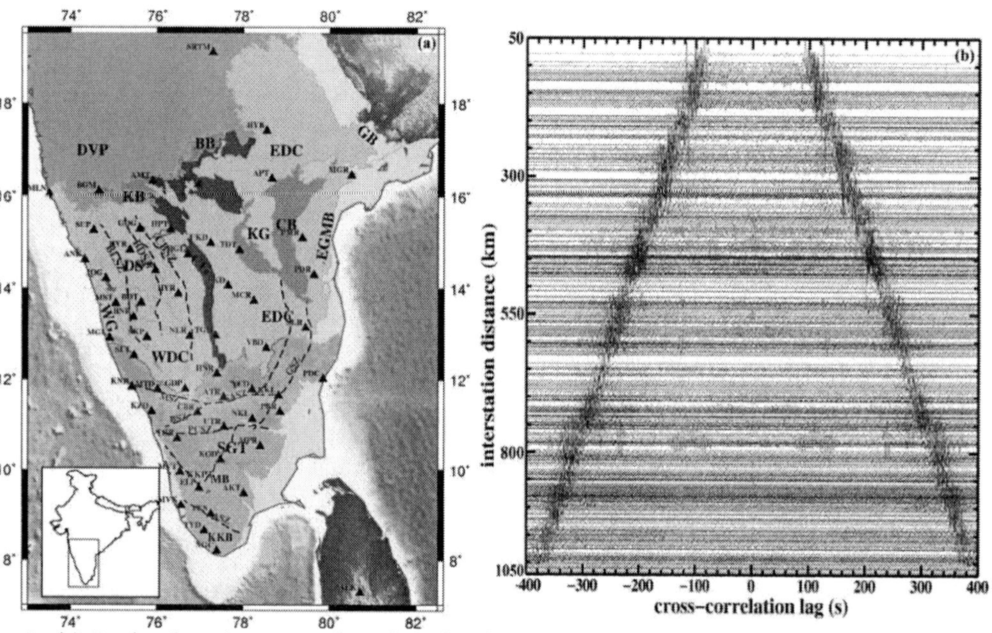

Figure 3: (a) Station location map of south India, (b) Cross-Correlation gather for the station SUP. All waveform are bandpass between 5 and 10-s period.

We have followed the data preparation steps as discussed in the previous section. For each station the data has been resampled to 10 samples per second and split into one day segments, followed by the removal of mean, trend and instrument response. The resulting waveform is then tapered, bandpass filtered between periods 1-60 s followed by time domain normalization and spectral whitening. After all of these preprocessing steps were applied, cross-correlation between each of the resulting day segment was computed. The resulting cross-correlations are two sided functions, where the signal at positive and negative lag times (i.e. causal and acausal components) represents energy travelling in opposite directions between pair of stations. The cross-correlation functions (CCFs) were calculated for the daily waveforms of each pair of station and stacked to increase the signal to noise ratio (SNR). Cross-correlation gather for a station SUP is shown in Figure 3b. We have applied frequency time analysis (FTAN) multiple-filter technique (Dziewonski et al. 1969) with phase match filtering (Levshin and Ritzwoller, 2001) to measure the group velocity dispersion curves. After calculating the dispersion for all EGF's, we selected data with the signal to noise ratio (SNR) greater than 10 and the interstation spacing more than three wavelengths at a given period (Bensen et al 2008). SNR is defined as the ratio of peak amplitude in the signal window to root mean square (rms) noise in the trailing noise window (Bensen et al., 2007). We used 2514 interstation paths with a distance ranging from 44.7 km to 1359 km. The maximum number of dispersion curve measurement is 1350 for 10 s period. For the shortest periods (T < 4 s) the number of measurements decreases significantly as waveforms with poor signal to noise ratio have been discarded.

Before performing a tomographic inversion with surface wave travel time data it is important to understand how well the subsurface structure could be resolved by the geometry of stations and virtual sources. This is achieved by generating synthetic data for each of a series of checkerboard models which represent an imaginary Earth's group velocity structures, calculating a solution model using the synthetic data and comparing the solution velocity models with the synthetic Earth models. The resolution of the surface wave tomography depends primarily on path coverage (Figure 4) and their azimuth distribution, has been studied using checkerboard test where, input models are a square cells of side 0.7°. Velocity perturbations of ±8% are assigned to each cell with average velocity shown at the bottom left corner at each period in Figure 4.

After computing the group velocity dispersion measurement a tomographic inversion is performed for periods between 2 and 10 s (Figure 5) using the non-linear 2-D tomographic inversion technique developed by Rawlinson and Sambridge (2003). The method combines the Fast Marching Method (FMM) (Rawlinson and Sambridge, 2004a, b) for calculation of forward problem and gradient method based on subspace technique (Kennett et al., 1998) for inversion, where the minimization is carried out simultaneously along several search directions that together span a subspace of the model space. The method has been successfully used for ambient noise tomography of several regions (Young 2011, Saygin & Kennet 2012, Ren et al. 2013 etc). A detailed explanation of the methodology is presented in Rawlinson and Sambridge (2004a) and Saygin and Kennett (2012).

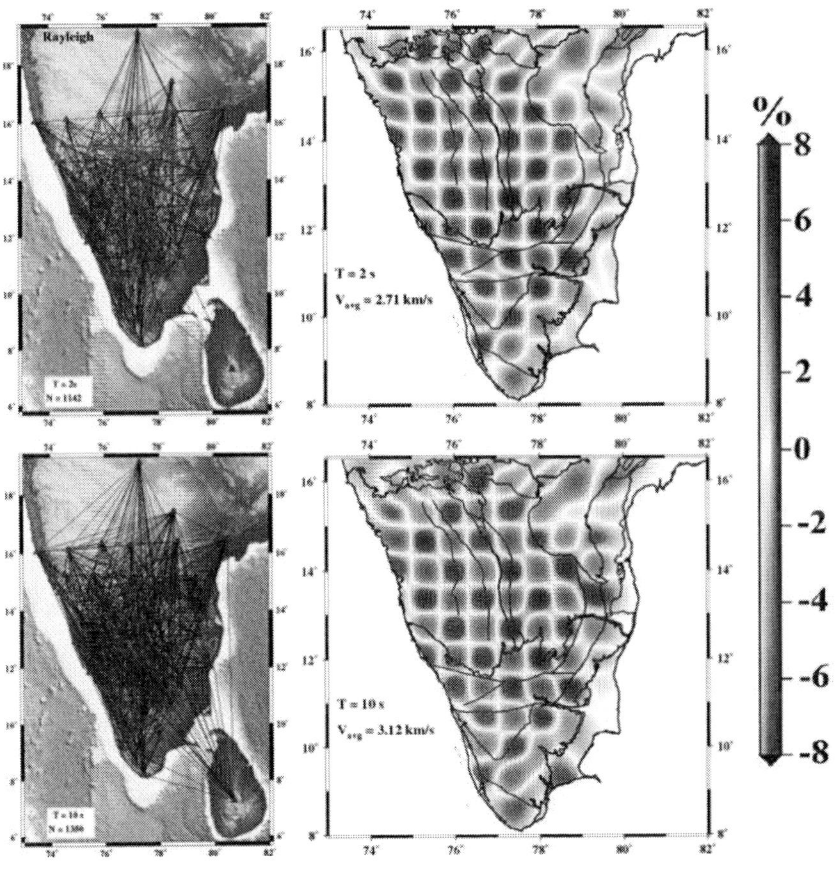

Figure 4: Ray path coverage and checkerboard resolution test at 2 s and 10 s period.

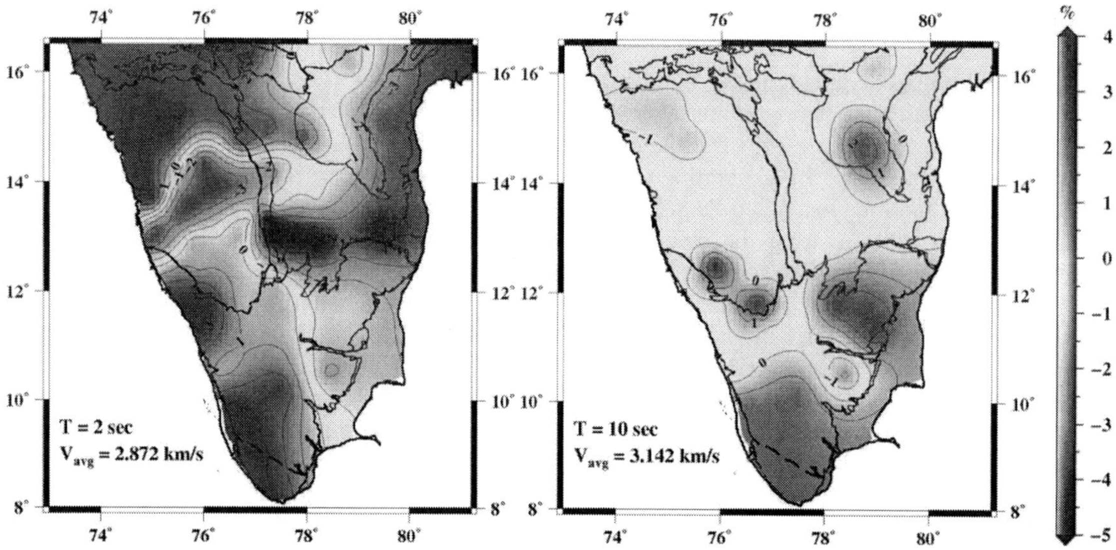

Figure 5: Group velocity perturbation (%) maps for selected periods of 2 and 10 s. Velocity perturbations are relative to the average group velocity plotted at the bottom left corner of each plot.

4. Discussions and Conclusions

In this paper we described the background of seismic interferometry and the ambient noise tomography method. We have shown that seismic interferometry can be used to compute inter-receiver surface wave using ambient noise data recorded in southern India. We have presented the first surface wave group velocity map of south India from ambient seismic noise and they contain useful information of surface geology. The group velocity can be converted to shear velocity for 3-D variations in subsurface structures with the depth. Further work will aim to use Love as well as Rayleigh waves to achieve better lithological constraints along south India.

5. Acknowledgement

We greatly appreciate help of Rishikesh Meena, B N V Prasad, Utpal Saikia, Somasish Bose, and K Nagaraju during the field deployment of seismographs. We are also grateful to Nick Rawlinson for providing us the FMST code (http://rses.anu.edu.au/~nick). Most of the data processing was made using Seismic Analysis Code (SAC) and figures were made using Generic Mapping Tools (GMT) (Wessel and Smith, 1998).

References

[1] Bensen, G D, Ritzwoller M H, Barmin M P, Levshin A L, Lin F, Moschetti M P, Shapiro N M, Yang Y, 2007, Processing seismic ambient noise data to obtain reliable broad band surface wave dispersion measurements, *Geophys. J. Int.*, **169(3)**, 1239–1260.

[2] Bensen, G D, Ritzwoller M H and Shapiro N M, 2008 Broadband ambient noise surface wave tomography across the United States, *J. Geophys. Res.*, **113(B5)**.

[3] Borah, K, Rai S S, Prakasam K S, Gupta S, Priestley K and Gaur V K, 2014 Seismic imaging of crust beneath the Dharwar Craton, India, from ambient noise and teleseismic receiver function modelling, *Geophys. J. Int.*, **197(2)**, 748–767.

[4] Campillo, M, Paul A, 2003 Long-range correlations in the diffuse seismic coda. *Science* **299**, 547–549.

[5] Claerbout, J, F, 1967, Synthesis of a layered medium from its acoustic transmission response, *Geophysics*, **33**, 264.

[6] Dziewonski, A, Bloch S, and Landisman M, 1969 A technique for the analysis of

transient seismic signals, *Bull. Seismol. Soc. Am.*, ***59*(1)**, 427–444.

[7] Guo, Z, Chen Y J, Ning J, Feng Y, Grand S P, Niu F, Kawaktsu H, Tanaka S, Obayashi M, Ni J, 2015 High resolution 3-D crustal structure beneath NE China from joint inversion of ambient noise and receiver functions using NECESSArray data, *Earth Planet. Sci. Lett.*, **416**, 1-11.

[8] Kao, H, Behr Y, Currie, C A, Hyndman R, Townend J, Lin F C, Ritzwoller M H, Shan S, He J, 2013 Ambient seismic noise tomography of Canada and adjacent regions: part I. crustal structures, *J. Geophys. Res.*, **118**, 5865-5887.

[9] Kennett, B, L, N, Seismic wave propagation and seismic tomography, Research School Of Earth Sciences,Institute of Advanced Studies, The Australian National University, Canbera, 1998.

[10] Larose E., Derode A, Clorennec D, Margerin L, and Campillo M, 2005 Passive retrieval of Rayleigh waves in disordered elastic media, *Phys. Rev. E,* **72(4)** 46,607.

[11] Levshin, A L, and Ritzwoller M H 2001 Automated detection, extraction, and Measurement of Regional Surface Waves, *Pure Appl. Geophys.*, **158(8),** 1531–1545.

[12] Lin, F., M.P. Moschetti, and M.H. Ritzwoller (2008). Surface wave tomography of the western United States from ambient seismic noise: Rayleigh and Love wave phase velocity maps, *Geophys. J. Int.*, **173(1)**, 281–298.

[13] Moschetti M P, Ritzwoller M H and Shapiro N M, 2007 Surface wave tomography of the western United States from ambient seismic noise: Rayleigh wave group velocity maps, *Geochem. Geophys. Geosys.,* **8**, Q08010.

[14] Nicolson, H, Curtis A, Baptie B, Galetti E, 2012 Seismic interferometry and ambient noise tomography in British Isles, *Proceeding of the Geologists Association*, **123**, 74-86.

[15] Rawlinson, N, Sambridge M, 2003 Seismic traveltime tomography of the crust and lithospare, *Advances in Geophysics*, pp. 81–198, Elsevier {BV}.

[16] Rawlinson, N, Sambridge M, 2004a Multiple reflection and transmission phases in complex layered media using a multistage fast marching method, *Geophysics*, **69(5)**, 1338.

[17] Rawlinson, N, Sambridge M, 2004b Wave front evolution in strongly heterogeneous layered media using the fast marching method, *Geophys. J. Int.*, **156(3)**, 631–647.

[18] Ren, Y, Grecu B, Stuart G, Houseman G, Hegedus E, 2013 Crustal structure of the Carpathian Pannonian region from ambient noise tomography, *Geophys. J. Int.*, **195(2)**, 1351-1369

[19] Shapiro N, Campillo M, 2004 Emergence of broadband Rayleigh waves from correlations of the ambient seismic noise. *Geophysical Research Letters* **31 (7)**, 1615-1619.

[20] Shapiro N, Campillo M, Stehly L, Ritzwoller M H, 2005 High-resolution surfacewave tomography from ambient seismic noise. *Science* **307**, 1615.

[21] Snieder R, 2004 Extracting the Green's function from the correlation of coda waves: a derivation based on stationary phase. *Physical Review* **E69** 046610.1- 046610.8.

[22] Wapenaar K, 2004 Retrieving the elastodynamic Green's function of an arbitrary homogeneous medium by cross correlation. *Physical Review* **E 69**, 046610.

[23] Wapenaar, K, Fokkema J, 2005 Seismic interferometry, time-reversal and reciprocity, 67th EAGE Conference & Exhibition.

[24] Wapenaar, K, Ruigrok E, Neut V D, Draganov J D, 2011 Improved surfacewave retrieval from ambient seismic noise by multi-dimensional deconvolution. *Geophysical Research Letters* **38**, L01313.

[25] Weaver R, Lobkis O, 2001 Ultrasonics without a source: thermal fluctuation correlations at MHz frequencies. *Physical Review Letters* **87 (13)**, 134301.

[26] Wessel, P. & Smith, W.H.F., 1998. New, improved version of the Generic Mapping Tools released, EOS, *Trans. Am. geophys. Un.,* **79,** 579.

[27] Young, M K. Rawlinson N, Arroucau P, Reading A M, and Tkalčič H, 2011 High frequency ambient noise tomography of southeast Australia: New constraints on Tasmania's tectonic past, *Geophys. Res. Lett.*, **38(13).**

Lattice field theory applications in high energy physics

Steven Gottlieb
Department of Physics, Indiana University, Bloomington, IN 47405, USA
E-mail: sg@indiana.edu

Abstract. Lattice gauge theory was formulated by Kenneth Wilson in 1974. In the ensuing decades, improvements in actions, algorithms, and computers have enabled tremendous progress in QCD, to the point where lattice calculations can yield sub-percent level precision for some quantities. Beyond QCD, lattice methods are being used to explore possible beyond the standard model (BSM) theories of dynamical symmetry breaking and supersymmetry. We survey progress in extracting information about the parameters of the standard model by confronting lattice calculations with experimental results and searching for evidence of BSM effects.

1. Introduction

In 1974, Kenneth Wilson invented lattice Quantum Chromodynamics (QCD), a non-perturbative approach to Nature's strong force [1]. Wilson's formulation was based on using elements of the Lie group $SU(3)$, rather than elements of the Lie algebra, which is used in the continuum formulation of the theory. This approach allowed Wilson to exactly preserve gauge invariance, which was not possible when formulating the theory in terms of finite difference operators applied to elements of the Lie algebra. Gauge invariance is familiar to us from electromagnetism, but in QCD it is much richer, as it is based on the non-Abelian group $SU(3)$, not the Abelian group $U(1)$. Gauge symmetry determines three of Nature's forces: electromagnetic, strong, and weak.

In the years since Wilson's initial paper, which discussed quark confinement based on a strong coupling expansion, there have been monumental advances in algorithms, formulations of the theory, and computer power. In the early days, it was necessary to neglect the contribution of quantum fluctuations related to quark-antiquark production and annihilation in the vacuum. This is called the quenched approximation and cannot be systematically improved. However, it is now possible to include the quantum fluctuations of the four lightest quarks: up, down, strange, and charm. The still heavier bottom and top quarks have negligible effect at current precision. We can even make the up and down quarks as light as in Nature, which has traditionally been a difficult computing challenge. Work on lattice QCD has progressed to the stage that a number of interesting quantities can be calculated to sub-percent level, and there have been predictions of particle masses and decay properties, not just postdictions.

Lattice QCD is now extensively used in theoretical studies of elementary particle and nuclear physics. The proton-neutron mass difference, the spectrum of excited baryons, parton moment distributions, and properties of light nuclei have all been studied with varying degrees of precision. The properties of QCD at non-zero temperature have been studied to understand the quark-gluon plasma produced in heavy-ion collisions. In particle theory, the quark masses for up, down, strange, charm, and bottom, i.e., all except top, have been calculated. So, has the strong coupling α_s and many weak decays that are needed to determine the Cabibbo-Kobayashi-Maskawa (CKM) quark mixing matrix. Lattice field theory has also been used to study theories with dynamical symmetry breaking as an alternative to the spontaneous symmetry breaking of the simple Higgs boson theory. As exciting as it was to discover the Higgs boson at the LHC, knowing the mass of the Higgs boson does not solve the mysteries of the Standard Model, and we would dearly love to find evidence of Beyond the Standard Model physics. This might come from seeing new particles at the LHC, but it could just as easily come from observing anomalies in high precision experiments, where the anomalies come from small interactions caused by new (virtual) particles that are too heavy to be produced at the LHC. Lattice QCD has an important role to play here in determining the elements of the CKM matrix. Another area of significant recent progress is in the formulation of lattice theories with supersymmetry.

Unfortunately, I cannot cover all of these topics, so I shall point the interested reader to the annual International Symposium on Lattice Field Theory. The most recent one was held in July, 2015 in Kobe, Japan [2]. There were over a dozen plenary talks relevant to nuclear and particle physics. In this talk, I have relied heavily either on results from my own collaborations, MILC and Fermilab Lattice-MILC, (referred to as FNAL/MILC below), or state-of-the-art summaries prepared by the Flavor Lattice Averaging Group (FLAG) [3] an international group of scientists who critically review work on a large number of quantities in lattice QCD and prepare averages for ease of use by a wider (non-lattice) community. The last FLAG review appeared in 2013 [3] and a new one will appear in early 2016. I am pleased to be a member of FLAG.

2. The Standard Model and lattice QCD

The Standard Model is a theory of quarks, leptons, gauge bosons, and the Higgs boson. It describes only three of the known forces, as gravity is not included. The model is described by its symmetries and the matter content. The symmetry is $SU(3) \times SU(2) \times U(1)$. The group $SU(3)$ is the symmetry of QCD and $SU(2) \times U(1)$ is that of the electroweak interactions. Spin-1 particles are the force carriers. They are called gluons (for QCD), and the photon and weak bosons (specifically, $W^{\pm}$ and Z) for the electromagnetism and the weak force, respectively. The Higgs boson has no intrinsic spin. The quarks and leptons are spin-1/2 matter particles. The quarks interact with all the force carriers. The charged and neutral leptons don't interact with the gluons, but they do interact with the weak force carriers. The charged leptons interact with the photon, but the neutral ones (neutrinos) do not. One of the reasons we think there is physics beyond the the Standard Model is that the model has many undetermined parameters. These parameters must be determined from experiment (with various inputs from theory). In a more fundamental theory, there might be relations between the parameters, so they would not seem as arbitrary as they do now.

For each of the three symmetry groups there is a coupling constant. For $SU(3)$, it is called g_s. For $SU(2) \times U(1)$, the two couplings are g and g'. There are six quark masses. There are three masses for the charged leptons. Now that we know neutrinos have mass, there are also three neutrino masses. The Cabibbo-Kobayashi-Maskawa quark mixing matrix (detailed in the next section) is complex and unitary. An arbitrary 3×3 complex matrix would have 18 real parameters; however, because of unitarity and our ability to chose some phases of the quark fields, there are only four independent parameters that determine the CKM matrix. These are commonly described as three angles and a complex phase factor that determines CP violation. The combination of discrete symmetries charge conjugation C, and parity is denoted by CP. There is a similar matrix for the neutrinos called PMNS for Pontecorvo, Maki, Nakagaw and Sakata. However, since the neutrinos do not interact strongly, we will have no more to say about that. Lattice QCD input is important for determination of ten parameters of the Standard Model: $\alpha_s = g_s^2/(4\pi)$, the four parameters that determine the CKM matrix, and m_u, m_d, m_s,

m_c, and m_b. The sixth quark, the top quark, decays weakly before it can form bound states, so we do not need lattice QCD to study its mass.

Lattice QCD provides a nonperturbative treatment of the quantum field theory that describes the strong interaction. Because the coupling is strong, many phenomena cannot be calculated perturbatively. Quantum field theories require regularization and renormalization. The lattice technique provides one such regularization. However, numerical errors must be carefully controlled. Errors come from the non-zero lattice spacing (continuum limit), finite volume (infinite volume limit), and unphysical light quark masses (chiral extrapolation). In parentheses, we have the limit or operation that must be done to control the systematic error from each effect. In addition, there are statistical errors. Groups are increasingly able to work with up and down quark masses very close to their physical value, which greatly reduces errors from the chiral extrapolation that were seen in earlier calculations. There are at least five popular ways to deal with the quarks in lattice QCD. They go by the names: Wilson/Clover, staggered, domain wall, twisted mass, and overlap. Each method has different systematic errors at nonzero lattice spacing, so it is useful to use different methods and compare the final results after all errors are controlled. The number of dynamical flavors used also varies by collaboration. The most phenomenologically relevant calculations use dynamical up, down, and strange quarks, or those plus charm. These are denoted $N_f = 2+1$ or $2+1+1$, respectively, because the up and down quarks are usually treated as if their masses were identical. (Their average mass is used.)

3. CKM matrix

It has been observed for many years that the Universe contains much more matter than antimatter. This is known as the baryon asymmetry. Kobayashi and Maskawa won the Nobel prize for their realization that with three (or more) generations we can have CP violation, which might explain the baryon asymmetry of the Universe. However, we now know that the CP violation in the strong interaction is probably too weak for this purpose, and it may be the CP violation appearing in the PMNS matrix for neutrino mixing that accounts for the baryon asymmetry. Here is the CKM mixing matrix (bold notation) augmented with some of the decay or mixing processes that can be used to determine each matrix element:

$$\begin{pmatrix} \mathbf{V_{ud}} & \mathbf{V_{us}} & \mathbf{V_{ub}} \\ \pi \to l\nu & K \to l\nu & B \to l\nu \\ & K \to \pi l\nu & B \to \pi l\nu \\ \mathbf{V_{cd}} & \mathbf{V_{cs}} & \mathbf{V_{cb}} \\ D \to l\nu & D_s \to l\nu & B_c \to l\nu \\ D \to \pi l\nu & D \to K l\nu & B \to D^{(*)} l\nu \\ \mathbf{V_{td}} & \mathbf{V_{ts}} & \mathbf{V_{tb}} \\ B_d \leftrightarrow \bar{B}_d & B_s \leftrightarrow \bar{B}_s & \end{pmatrix} . \tag{1}$$

In the second and fifth rows, we have meson decays called leptonic, because only a charged lepton and a neutrino appear in the final state. The third and six rows contain decays called semi-leptonic because there is also a meson in the final state. The last row contains two meson mixing processes that determine the CKM matrix elements just above them. Since the CKM matrix is unitary, each row and each column is a complex unit vector. Also, each row (column) is orthogonal to the other rows (columns) leading to the so-called unitarity triangle in the complex plane. Violations of unitarity are evidence of non-standard-model physics. Further, if two different processes are used to determine an element of the matrix and they do not agree, that is evidence for new BSM physics, which we would dearly love to find. We will examine both these tests of the SM.

If we could do experiments on free quarks, it would be easy to determine mixing; however, confinement means we need to deal with bound states. Thus, LQCD input for decay constants and form factors is needed to determine elements of the CKM matrix. For example, the branching fraction for the leptonic decay of a $D_{(s)}$ meson is given by

$$\mathcal{B}(D_{(s)} \to \ell\nu_\ell) = \frac{G_F^2 |V_{cq}|^2 \tau_{D_{(s)}}}{8\pi} f_{D_{(s)}}^2 m_\ell^2 m_{D_{(s)}} \left(1 - \frac{m_\ell^2}{m_{D_{(s)}}^2}\right)^2 \qquad (2)$$

where the unknowns are $|V_{cq}|$, the absolute value of the CKM matrix element with $q = d$ or $q = s$ for the D or D_s meson, respectively, and $f_{D_{(s)}}$ is the corresponding decay constant, which needs to be calculated in LQCD. The other quantities, such as masses, lifetimes and the Fermi constant are easily found from experiment.

3.1. Light quarks and the first row

We will begin our discussion with results for mesons that contain only the three lightest quarks up, down, and strange. The ground states are called pions and kaons. In Fig. 1(L), we see the FLAG summary of LQCD results stretching back over a decade by various groups. Results in red are deemed to have insufficient control of all systematic errors. Results with a solid green symbol are included in the FLAG estimate (black point with error bar and vertical gray bands). Points with light green have been superseded or not been refereed, hence not included in the FLAG estimate. Blue points come from the Particle Data Group (PDG) [4] or a non-LQCD method. Calculations with different numbers of dynamical quarks are considered separately. In some cases, f_π has been used to set the lattice spacing (or scale), so only f_K is shown. We see excellent agreement with the values from the PDG. The ratio f_K/f_π can be calculated accurately and used to determine $|V_{us}/V_{ud}|$ from precise measurements of the ratio of pion and kaon decay rates which show that $\left|\frac{V_{us}}{V_{ud}}\right| \frac{f_{K^\pm}}{f_{\pi^\pm}} = 0.2758(5)$. Figure 1(R) shows results for the decay constant ratio.

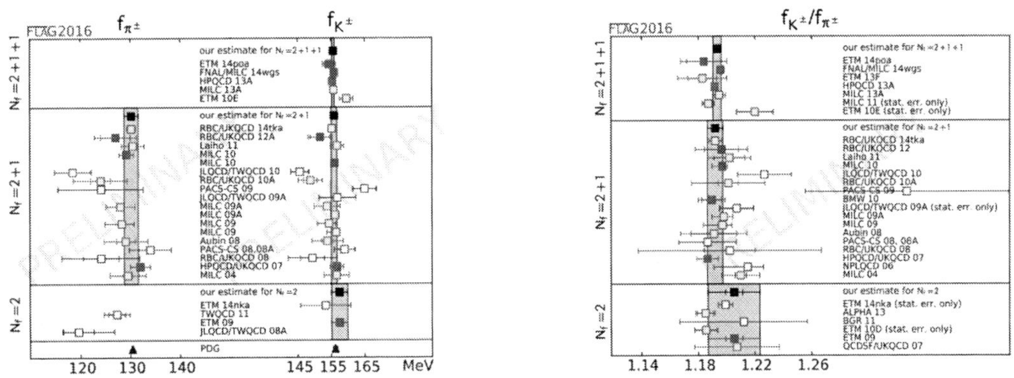

Figure 1. (L) FLAG [3] summary of LQCD results for the charged pion and kaon decay constants. (R) The ratio of charged kaon to charged pion decay constants. The meaning of the colors, symbols and bands is explained in the text.

Let's turn to the semileptonic kaon decay. Semileptonic decays have three-body final states, so there is one kinematic variable, usually denoted q^2, which is the square of the momentum transfer to the leptons. From 4-momentum conservation, $p_K = p_\pi + q_l + q_\nu$ and $q = q_l + q_\nu$, where we have used p for hadron momenta and q for lepton momenta, with the subscript denoting the particle. To extract $|V_{us}|$, we just need the form factor at zero momentum transfer, i.e., $f_+(0)$ as experiment tells us that $|V_{us}|f_+(0) = 0.2163(5)$. This can be combined with the FNAL/MILC

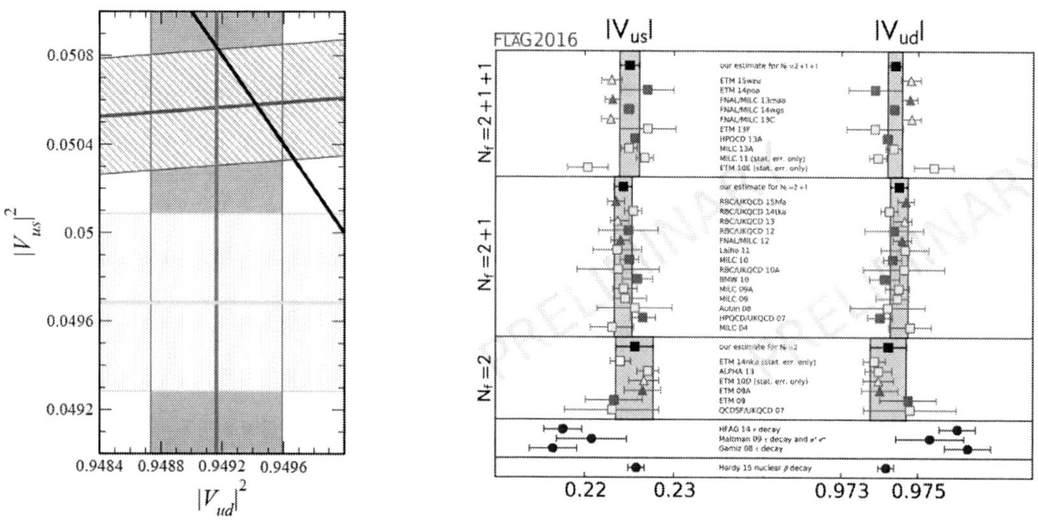

Figure 2. (L) Unitarity test for the first row of the CKM matrix from Ref. [6]. Black line shows unitarity constraint. Colored bands explained in text. (R) Subset of FLAG [3] summary for $|V_{us}|$ and $|V_{ud}|$. Most red points eliminated for readability.

$N_f = 2+1+1$ result [5] $f_+(0) = 0.9704(24)(220$ to determine an error band for $|V_{us}|$. First row unitary states $|V_{ud}|^2 + |V_{us}|^2 + |V_{ub}|^2 = 1$. However, as we will see below $|V_{ub}| \approx 4 \times 10^{-3}$, so the last term can be neglected as the errors on the first two terms are a few times 10^{-4}. Thus, the unitarity constraint will be a straight line in the $|V_{ud}|^2$ - $|V_{us}|^2$ plane. In Fig. 2(L), we show the unitary constraint as a black line, along with the angled error band from leptonic pion and kaon decay, the horizontal error band from kaon semileptonic decay, and a vertical error band from nuclear β-decay that is independent of LQCD calculations. We see that there is some tension between the two types of decay studied in LQCD, but that unitarity, leptonic decays, and β-decay are in good agreement.

A summary of many determinations of $|V_{ud}|$ and $|V_{us}|$, based on either leptonic or semileptonic decays, has been prepared by FLAG. This is shown in Fig. 2(R) where squares indicate leptonic decays and triangles indicate semileptonic. The blue points, as usual, are from non-lattice calculations. Careful inspection of the $N_f = 2+1+1$ 2014 results from MILC and FNAL/MILC shows the tension we have seen above between leptonic and semileptonic decays. Other calculations do not yet have the precision to confirm the discrepancy or rule it out. The errors on the FLAG estimates for $N_f = 2$ and $2+1$ are larger than for $2+1+1$ and they do not have a tension with unitarity. The FLAG estimate for $2+1+1$ does show some tension with unitarity. First row unitarity is not the first place we would expect to find evidence for BSM flavor physics, so it will be interesting to improve these calculations, particularly to calculate the kaon semileptonic form factor over its entire kinematic range. There is also the interesting tension between the results near the bottom of the figure, based on τ decays, and those for pion and kaon decay.

3.2. Charm decays and the second row

Leptonic and semileptonic decays of the D and D_s mesons have been studied on the lattice, but not as extensively as for the pion and kaon. It has been about a decade since decay constant predictions of FNAL/MILC were tested at CLEO-c [7]. Initial errors were about 10%, but current errors from FNAL/MILC are only 0.6%. A great deal of the improvement is due to the use of highly improved staggered quarks (HISQ) that were developed by the HPQCD collaboration. Figure 3(L) compares the 2014 FNAL/MILC results (labeled "This work") with prior calculations, mostly by the European Twisted Mass collaboration and HPQCD. The FNAL/MILC results are $f_{D^+} = 212.6(0.4)(^{+1.0}_{-1.2})$ MeV, $f_{D_s} = 249.0(0.3)(^{+1.1}_{-1.5})$ MeV, and $f_{D_s}/f_{D^+} = 1.1712(10)(^{+29}_{-32})$ for the decay constant ratio, for which there is some cancellation of the systematic errors. For references to the other work and the HISQ action see Ref. [6].

To make use of these decay constants, we rely on the work of Rosner and Stone [8] to summarize the experimental results. They find $f_D|V_{cd}| = 46.06(1.11)$MeV, and $f_{D_s}|V_{cs}| = 250.66(4.48)$MeV. The experimental errors are 1.8–2.4%.

Combining the experimental results and the FNAL/MILC decay constants gives $|V_{cd}| = 0.217(1)(5)(1)$, and $|V_{cs}| = 1.010(5)(18)(6)$, where the errors are lattice, experiment and structure-dependent electromagnetic, respectively. Thus, the experimental errors are currently dominant. In Fig. 3R, we see evidence for an $\approx 1.8\sigma$ tension with unitarity for the two leptonic charm decays. The black line is the unitarity constraint. The horizontal blue band is for D_s decay and the vertical green band is for D^+ decay. Once again, the third element of the row V_{cb} is too small to make a difference at the current level of precision. The semileptonic form factors for $D_{(s)}$ mesons are much less studied than for light quarks; however, there should be some updates in the coming year. We refer the reader to FLAG [3] for details.

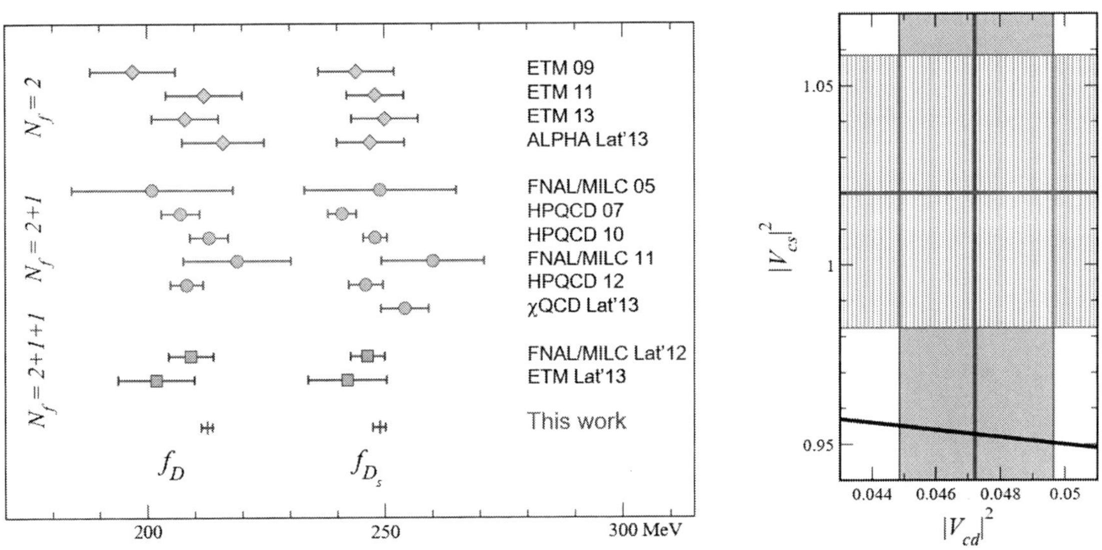

Figure 3. (L) Summary of results for the charm meson decay constant prepared by FNAL/MILC [6]. "This work" refers to the calculation presented there. (R) Unitarity test for the second row of the CKM matrix from Ref. [6].

3.3. B meson decays

The b quark is the heaviest one that forms bound states that can be studied with LQCD. Both leptonic and semileptonic decays of B and B_s mesons have been studied. In addition to the usual decays that produce a charged lepton and a neutrino, there are a number of rare decays that involve the so-called flavor changing neutral current (FCNC). In the SM, the FCNC vanishes at the tree level, so small quantum loop effects from new physics may be visible. That makes the rare decays a promising topic to study. In fact, there have been some recent results from the LHCb experiment at the Large Hadron Collider that show tensions with the SM predictions.

Journal of Physics: Conference Series **759** (2016) 012007 doi:10.1088/1742-6596/759/1/012007

These rare decays also present an alternative way to determine $|V_{td}|$ and $|V_{ts}|$ that can be compared with the meson mixing processes indicated in our CKM matrix.

FLAG has summarized results for decay constants f_B and f_{B_s} [3]. The errors on these decay constants are about 2% for $N_f = 2+1$ and $2+1+1$. For $N_f = 2+1$, $f_B = 190.5(4.2)$MeV and $f_{B_s} = 227.7(4.5)$MeV. Unfortunately only $B \to \tau\nu$ has been observed so far and the error is about 20%. So, in this case the LQCD calculation is ahead of the measurement. This allows a determination of $|V_{ub}|$, but it is not competitive with the value from semileptonic decay.

The semileptonic decays $B \to \pi\ell\nu$ and $B_s \to K\ell\nu$ have been studied on the lattice. The former has been observed at BaBar and Belle, but the latter has not been observed. Another way to determine $|V_{ub}|$ is from inclusive decays. There is a long standing tension between that determination and the one from $B \to \pi\ell\nu$. The central value of $|V_{ub}|$ based on the SM analysis of the leptonic decay is between that from the semileptonic exclusive decay and the inclusive method. However, its error bar, limited by experiment, is too large to help clarify the situation. Belle II will improve the $B \to \tau\nu$ measurement, which should really help resolve these issues. Figure 4(L) shows the FLAG summary from 2013. There are new results for $B \to \pi\ell\nu$ from FNAL/MILC and RBC/UKQCD. There is also a new determination of $|V_{ub}|$ based on decay of the Λ_B baryon recently seen at LHCb. These determinations are compared in Fig. 4(R) [9]. FLAG computed values of $|V_{ub}|$ based on the BaBar and Belle experiments. They found $|V_{ub}| = 3.37(21) \times 10^{-3}$ and $3.47(22) \times 10^{-3}$, respectively, based on 2+1 flavor LQCD. The new FNAL/MILC result which uses results from both BaBar and Belle is $3.72(16) \times 10^{-3}$ which reduces, but does not eliminate the tension between exclusive and inclusive decays.

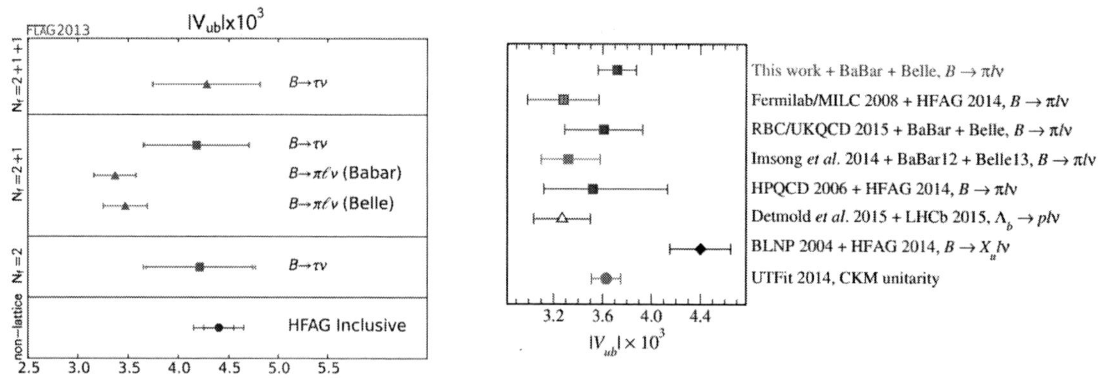

Figure 4. (L) FLAG [3] summary for $|V_{ub}|$ including results for $N_f = 2$, 2+1, and 2+1+1, as well as the inclusive result. (R) FNAL/MILC [9] summary of results for $|V_{ub}|$, including determination from Λ_b decay. See Ref. [9] for original sources.

The last entry in the second row of the CKM matrix, V_{cb}, can be studied in the exclusive decays $B \to D^*\ell\nu$ and $B \to D\ell\nu$. It can also be determined in inclusive decays where the decay products must include charm quarks. FNAL/MILC has recently compared $|V_{cb}|$ based on both determinations and there is again some tension between inclusive and exclusive results. As can be seen in Fig. 5(L), the errors from the decay to D^* are somewhat smaller than that from decay to D. With current errors, those decays agree with each other reasonably well and the real tension is between $B \to D^*\ell\nu$ and the inclusive value of $|V_{cb}|$.

Turning to rare B decays, FNAL/MILC has recently calculated the form factors needed for both SM and BSM decays through a FCNC [10, 11]. As mentioned above, this is a promising place to look for new physics. There is some tension between the SM prediction and recent LHCb measurements of $B^+ \to \pi^+\mu^+\mu^-$ and $B^+ \to K^+\mu^+\mu^-$. The LHCb measurement is smaller than the SM prediction in three of four fairly wide bins of q^2, the square of the momentum transfer to the muons. Figure 5(R) shows the comparison for $B^+ \to K^+\mu^+\mu^-$ where the difference is more pronounced. In Ref. [12], both processes are shown. For all four bins for the two processes, the tension is 1.7σ.

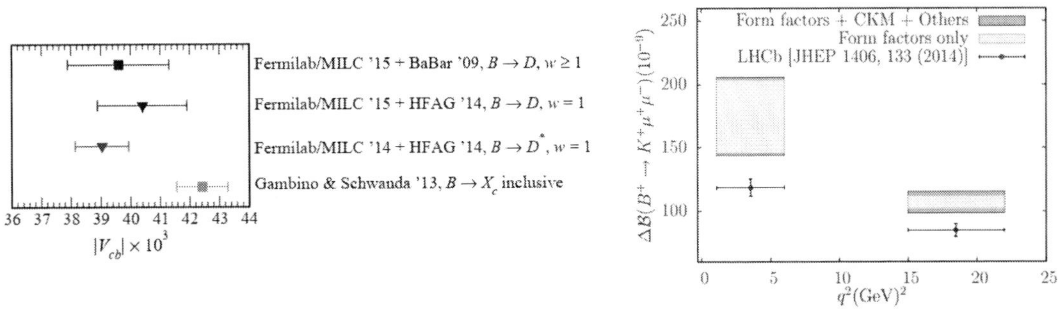

Figure 5. (L) FNAL/MILC [13] summary of results for $|V_{cb}|$, including determination from inclusive B decays to D or D^*, and inclusive decays. (R) Branching fraction for $B^+ \to K^+\mu^+\mu^-$ showing tension between SM prediction [12] and recent LHCb measurement.

4. Conclusions

Much progress has been made in lattice QCD, and more generally in lattice field theory. We concentrate here on quantities needed for the study of the CKM matrix. Calculational precision is now high enough that we can begin to look for evidence of BSM physics. We have seen a number of tensions between 1.5 and 2 standard deviations related to the CKM matrix. It will be interesting to see if reduced errors from both theory and experiment result in stronger hints (or perhaps significant evidence) of BSM physics. In the oral presentation, quark masses and α_s were also discussed. (See Ref. [3] for details.)

Acknowledgments

I thank all my colleagues in the MILC (especially C.B.) and Fermilab Lattice Collaborations, as well as all the current and past members of the Flavor Lattice Averaging Group. It has been a real pleasure working with them to produce new results and to make it easy for others to follow the progress in LQCD. This work was supported by the US Department of Energy grant DE-SC0010120. I also thank Profs. Santra and Ray for organizing such a wonderful conference, inviting me to attend, providing great hospitality, and helping to support my visit.

References

[1] Wilson K G 1974 *Phys. Rev.* **D10** 2445–2459
[2] *Lattice 2015* URL `http://www.aics.riken.jp/sympo/lattice2015/`
[3] Aoki S *et al.* 2014 *Eur. Phys. J.* **C74** 2890 (*Preprint* 1310.8555)
[4] Olive K A *et al.* (Particle Data Group) 2014 *Chin. Phys.* **C38** 090001
[5] Bazavov A *et al.* 2014 *Phys. Rev. Lett.* **112** 112001 (*Preprint* 1312.1228)
[6] Bazavov A *et al.* (Fermilab Lattice, MILC) 2014 *Phys. Rev.* **D90** 074509 (*Preprint* 1407.3772)
[7] Artuso M *et al.* (CLEO) 2005 *Phys. Rev. Lett.* **95** 251801 (*Preprint* hep-ex/0508057)
[8] Rosner J L and Stone S 2013 (*Preprint* 1309.1924)
[9] Bailey J A *et al.* (Fermilab Lattice, MILC) 2015 *Phys. Rev.* **D92** 014024 (*Preprint* 1503.07839)
[10] Bailey J A *et al.* (Fermilab Lattice, MILC) 2015 *Phys. Rev. Lett.* **115** 152002 (*Preprint* 1507.01618)
[11] Bailey J A *et al.* 2016 *Phys. Rev.* **D93** 025026 (*Preprint* 1509.06235)
[12] Du D, *et al.* 2016 *Phys. Rev.* **D93** 034005 (*Preprint* 1510.02349)
[13] Bailey J A *et al.* (MILC) 2015 *Phys. Rev.* **D92** 034506 (*Preprint* 1503.07237)

An Overview on Short and Long Time Relaxations in Glass-forming Supercooled Liquids

Smarajit Karmakar

E-mail: smarajit@tifrh.res.in
Centre for Interdisciplinary Sciences, Tata Institute of Fundamental Research, 21 Brundavan Colony, Narsingi, Hyderabad, 500075, India

Abstract. Density fluctuations in supercooled liquids near the glass transition relax in multiple steps. The short time relaxation is known as β-relaxation and the final long time relaxation is called α-relaxation. It is believed that the long time α-relaxation is a cooperative phenomena associated with a growing length scales, whereas the short-time β-relaxation is often attributed to spatially local processes involving the rattling motion of a particle in the transient cage formed by its neighbors. Using molecular dynamics simulations of few model glass-forming liquids, we show that the β-relaxation is also cooperative in nature and the length scale extracted from the detailed finite-size scaling analysis of β-relaxation is found to be the same as that of the length scale that describes the spatial heterogeneity of local dynamics in the long-time α-relaxation regime. These results provide a clear connection between short-time dynamics and long-time structural relaxation in glass-forming liquids.

1. Introduction

Dramatic increase in viscosity or relaxation time with decreasing temperature in glass forming liquids remain as an unsolved problem even after decades of research [1, 2, 3]. The increase in viscosity is so dramatic that with few tens of degrees of change in temperature, viscosity can change as much as 13 orders of magnitude and the temperature at which viscosity reaches 10^{13} Poise is termed as calorimetric glass transition temperature, T_g. The density-density correlation functions at low temperatures for these glass forming liquids show multiple relaxation steps. The short time relaxation is termed as β-relaxation and the final longer time scale relaxation is called α-relaxation as depicted in Fig.1. This long time α-relaxation gets sluggish with decreasing temperature and it is generally believed that this process is cooperative in nature. On the other hand the short time relaxation process is often assumed to be trivially associated with motion of some part of a complex molecule in glass forming liquids, e.g. rotation of side groups, leading to short time scale relaxation. All short time relaxation mechanisms, however, need not related to such trivial intra-molecular motions and can happen even for rigid molecules. A detailed understanding of these relaxation processes is important as it is believed that the short time relaxation is the main relaxation process below calorimetric glass transition temperature and is probably one of the main relaxation processes responsible for aging and rejuvenation in glasses.

Many different approaches to understand the slowing down of long time α-relaxation invoke the existence of correlation lengths that grow [1, 2, 4, 5, 6, 7, 8, 9, 10, 11] as the temperature is decreased. It is already established that there are at least two different length scales in the system : a dynamic length scale associated with the heterogeneity in local dynamics and a static length scale associated subtle amorphous order. Recent theoretical and experimental studies on dynamic heterogeneity in supercooled glass forming liquids [12, 13, 14, 15] provided clear

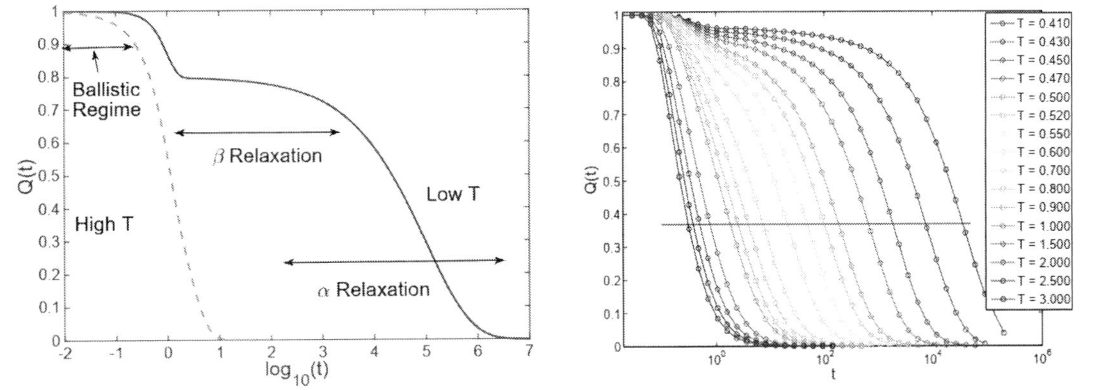

Figure 1. Left panel: Sketch of two point density-density correlation function $Q(t)$. Right panel: Calculated $Q(t)$ using molecular dynamics simulations for a binary glass forming liquids for different temperatures as indicated in the legend. Notice the appearance of two step relaxation with decreasing temperature.

evidence for the existence of this dynamical length scale [16, 17, 18, 19, 20, 10, 25]. These spatial correlations have been studied through a four-point susceptibility [21, 22, 23, 24] defined as

$$\chi_4(t) = \frac{1}{N}[\langle Q^2(t)\rangle - \langle Q(t)\rangle^2], \quad \text{where} \quad Q(t) = \sum_{i=1}^{N} w(|\mathbf{r}_i(0) - \mathbf{r}_i(t)|), . \tag{1}$$

where $\mathbf{r}_i(t)$ is the position of particle i at time t, N is the total number of particles, $w(r) = 1$ if $r \leq a_0$ and zero otherwise, and a_0 is a short-distance cutoff chosen to be close to the distance at which the root-mean-square displacement (RMSD) of the particles as a function of time exhibits a plateau. The window function $w(r)$ is used to coarse grain the small-amplitude vibrational motion of a particle in the "cage" formed by its neighboring particles. The peak value of $\chi_4(t)$ is defined as $\chi_4^P \equiv \chi_4(t = \tau_4)$, where τ_4 is the time at which $\chi_4(t)$ attains its maximum value (see left panel of Fig.2) and $\tau_4 \simeq \tau_\alpha$ at all temperatures. Typical time-temperature dependence of two point correlation function $Q(t)$ is shown in the right panel of Fig.1. The α-relaxation time, τ_α is defined at the time where this correlation function decays to $1/e$ to its initial value as $Q(t = \tau_\alpha) = 1/e$.

Predictions obtained from inhomogeneous mode coupling theory (IMCT) [5, 16, 17] suggest an initial power-law growth of the cooperativity length scale in time up to the β-relaxation time followed by a saturation region that continues up to a time scale comparable to the α-relaxation time. However, it is still not understood how the dynamics cross over from its short-time behaviour to that at long times. There are many ideas and observations that seem to suggest a close relation between dynamical features observed at short and long time scales [26, 27, 28, 29, 30, 31, 32, 33]. Thus, a proper understanding of relaxation processes in both β and α regimes and their underlying relation (if any) is extremely important in the overall understanding of glassy dynamics and its rich phenomenology.

In Ref.[36] the method of finite size scaling (FSS) [34] is used to extract the growing dynamic correlation length in a model glass forming liquid. This method has been used extensively in past for obtaining accurate numerical results for critical properties of a system near a conventional continuous phase transition [34]. Ref.[36] clearly demonstrated that the method of FSS is very useful [21, 35, 36] to study glass transition and also showed the importance of FSS compared to other conventional methods. In the left panel of Fig.2 we have plotted $\chi_4(t)$ as a function of time for different temperatures. $\chi_4(t)$ shows a peak at time close to the α-relaxation time and both the peak height and peak position increase with decreasing temperature. The peak height measures the typical size of the cooperatively rearranging region (CRR) in the system. This peak value of $\chi_4(t)$, χ_4^P has been found to show finite size effect and the effect becomes

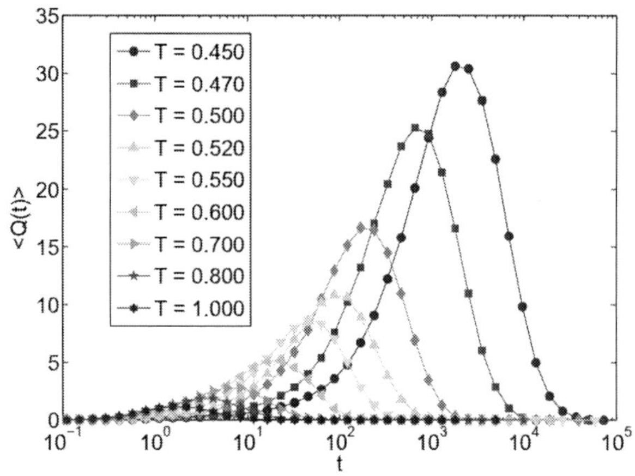

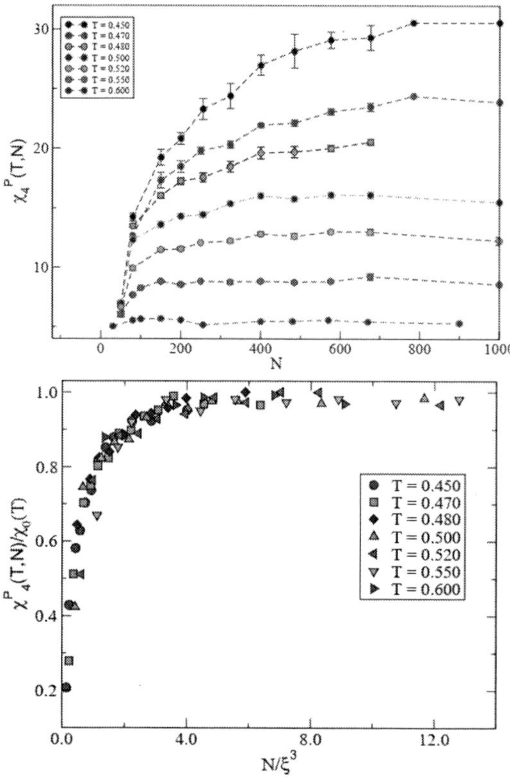

Figure 2. Left panel: $\chi_4(t)$ as a function of time for different temperature for 3dKALJ model. Top right panel: System size dependence of peak value of $\chi_4(t)$, χ_4^P for different temperatures. Notice the stronger dependence on system size with decreasing temperature. Bottom right panel: Scaling collapse of same data to extract the dynamic heterogeneity length scale. The right top and bottom panel figures are reproduced from Ref[36].

much more stronger as one decreases temperature. In the right top panel we have shown the system size dependence of χ_4^P for different temperatures for Kob-Andersen model in three dimensions (referred here as3dKALJ) [45] and one can clearly see that at higher temperature, the dependence is weak and it grows with decreasing temperature. In the right bottom panel, we have shown the finite size scaling done to extract the length scale associated with this cooperativity. It was subsequently shown that the heterogeneity length obtained in this way is in good agreement with that obtained by analyzing the four-point structure factor [37, 38].

In this article, we describe the results of FSS studies of a short time scale τ_β and long time scale τ_α for two generic model glass-formers. The simulation details are provided in Sec.[2]. We compare the estimated length scales with those obtained from FSS analysis of χ_4^P, which quantifies dynamic heterogeneity at the α relaxation time scale, and find that the two length scales agree well for all the model systems we studied whereas the length scale associated with the long time scale are completely different. We also find a universal relationship between β time scale and α relaxation time over the whole range of timescale accessible in simulations. These results are described in Sec.[3]. Sec.[4] contains a discussion of our results and a summary.

2. Method and Simulation Details

We report studies done on two model liquids in this work. They are: (i) the three-dimensional Kob-Andersen binary Lennard-Jones mixture (referred here as 3dKALJ) [45], (ii) a three-dimensional system characterized by a repulsive inverse power-law potential (3dR10) [47] whose range of interaction is smaller compared to that in the first model. For the first model system the interaction range covers the second neighbouring shell whereas for the second case it is truncated within the first shell. These models have been studied extensively and found to be very good glass formers. The interaction potential of 3dKALJ is given by

$$V_{\alpha\beta}(r) = 4\epsilon_{\alpha\beta}[(\frac{\sigma_{\alpha\beta}}{r})^{12} - (\frac{\sigma_{\alpha\beta}}{r})^{6}] \qquad (2)$$

where $\alpha, \beta \in \{A, B\}$ and $\epsilon_{AA} = 1.0$, $\epsilon_{AB} = 1.5$, $\epsilon_{BB} = 0.5$, $\sigma_{AA} = 1.0$, $\sigma_{AB} = 0.80$, $\sigma_{BB} = 0.88$. The interaction potential was cut off at $2.50\sigma_{\alpha\beta}$ and the number density of particle is $\rho = 1.20$. We have done simulations in 3D in a cubic simulation box with periodic boundary conditions. Simulation was done in canonical ensemble (NVT) with Newtonian dynamics. We have used modified leap-frog algorithm to do the constant temperature simulation with Berendsen thermostat. Length, energy and time scales are measured in units of σ_{AA}, ϵ_{AA} and $\sqrt{\sigma_{AA}^2/\epsilon_{AA}}$. We have done simulation in the temperature range $T \in [0.90, 0.45]$. We have equilibrated the system for $\sim 10^8 - 10^9$ MD steps depending on the system size and the production runs are at least $5 - 6$ times bigger than the equilibration runs. We have used integration time steps $dt = 0.005$. The system sizes studied are in the range $N \in [150, 28160]$. The second model is purely repulsive and the potential is given by [47]

$$V_{\alpha\beta}(r) = \epsilon_{\alpha\beta}\left(\frac{\sigma_{\alpha\beta}}{r}\right)^n \tag{3}$$

where $\alpha, \beta \in \{A, B\}$ and $\epsilon_{\alpha\beta} = 1.0$, for all α and β. $\sigma_{AA} = 1.0$, $\sigma_{AB} = 1.18$, $\sigma_{BB} = 1.40$, $n = 10$. The interaction potential was cut off at $1.38\sigma_{\alpha\beta}$. For this model, system sizes studied are in range $N \in [64, 4000]$ at number density $\rho = 0.81$ for 9 temperatures in the range $T \in [0.50, 1.000]$. All other simulations details are same as the previous model. Results for another model system (3dIPL model) [46] are not reported here and details of the model and some of the results can be found in [48].

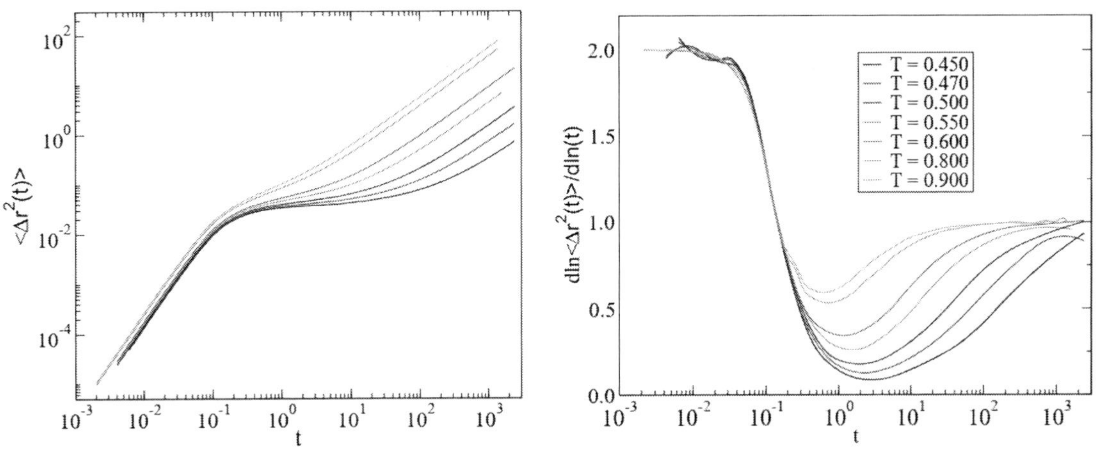

Figure 3. Mean square displacement (MSD) (left panel) and the derivative (right panel) of the logarithm of MSD with respect to the logarithm of time t, shown as a function of t. The minimum of the derivative defines τ_β. (see text for details). The figure is reproduced from Ref[48].

In [41], τ_β was defined as the time at which an inflection occurs in a log-log plot of the mean squared displacement *vs.* time. We use the same definition to evaluate τ_β as a function of system sizes and temperature. In left panel of Fig. 3, we show the mean squared displacement(MSD) as a function of time in a log-log plot for different temperatures. Notice the appearance of plateau in the intermediate time regime with decreasing temperature. The derivative is plotted in the right panel of Fig.3. Note that as expected the slope at small time is 2, indicating the ballistic motion of particles and then a dip appears at intermediate time and the slope tends to 1 at long time. Thus the observed clear dip in the derivative and its variation with temperature indicate that this estimation of τ_β can be done unambiguously and without much uncertainty. We use this definition of τ_β in the present work.

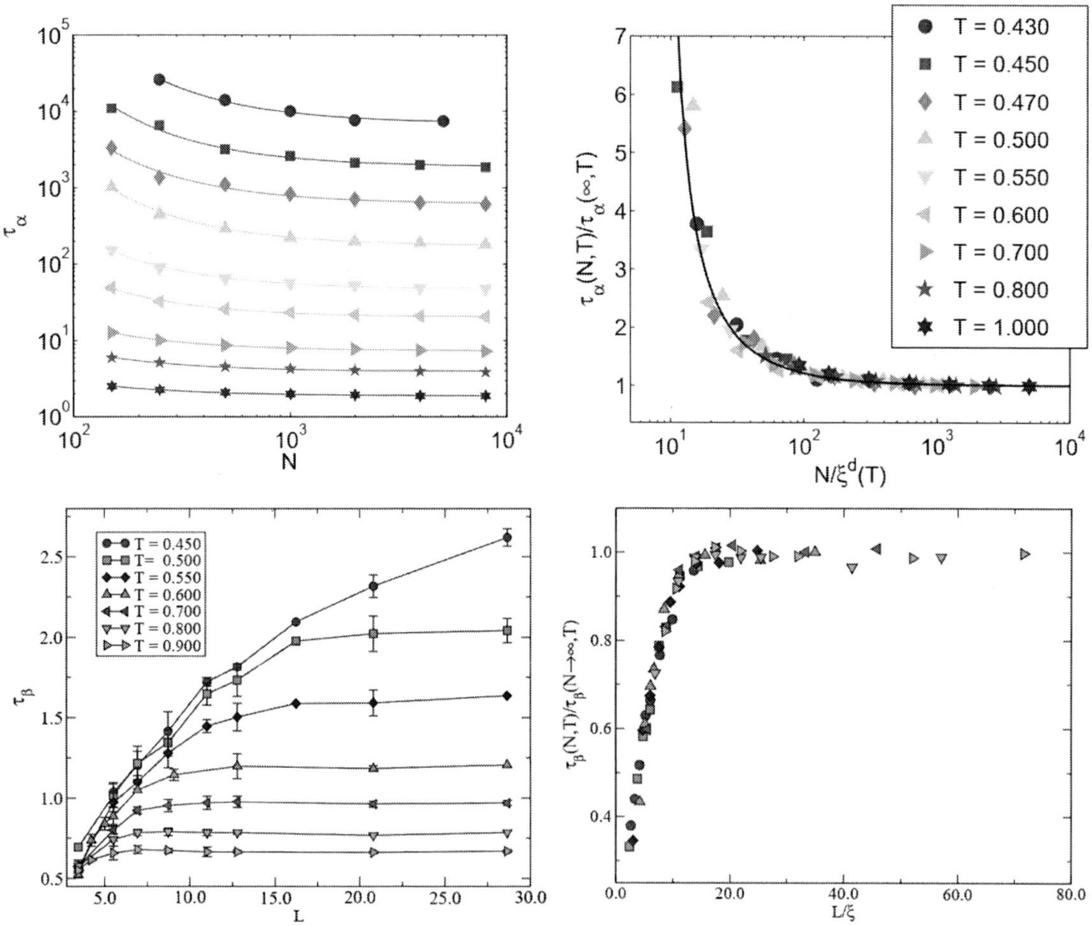

Figure 4. Top left panel: τ_α for the 3dKALJ model, shown as a function of system size for different temperatures in the interval $T \in [0.45, 0.90]$. τ_α decreases with increasing system size and saturates at a value which increases with decreasing temperature and the right panel shows the scaling collapse of the same data. Bottom panel: τ_β increases with increasing system size in stark contrast with the behaviour seen for τ_α. The right panel shows the data collapse. The collapse in both top and bottom panels are observed to be very good. Top panel is reproduced from Ref[3] and bottom panel from Ref[48].

3. Results

In Fig. 4, we show the system-size dependence of τ_α (top panel) and τ_β (bottom panel) for different temperatures for the 3dKALJ model. Notice the dramatic difference in the system size dependence of these two time scales. τ_α decreases with increasing system size [36] whereas τ_β shows a completely opposite behaviour [48]. This was interpreted as indicating a mixing of activated and non-activated mechanisms of structural relaxation (see also [42, 43]). It has been suggested [43] that behaviour described by IMCT, without the influence of activated dynamics, should be expected instead in the short-time dynamics, in a temperature window above the mode coupling transition temperature. Indeed the system size dependence of τ_β is observed to be what one would expect for normal time scale determined by an underlying length scale in the system. For high temperatures the dependence is weak and the asymptotic value is reached for small system sizes, but this characteristic size becomes increasingly large with decreasing temperature. We note that the range over which τ_β varies with system size and temperature is modest, unlike the behaviour observed for τ_α.

We next scale the system size N at each temperature by an empirically determined correlation volume $\xi^d(T)$ such that the values of $\tau_\alpha(N,T)/\tau_\alpha(N \to \infty, T)$ for all N and

T [$\tau_\alpha(N \to \infty, T)$ is the large-N asymptotic value of $\tau_\alpha(N,T)$] collapse onto a master curve when plotted versus $N/\xi^d(T)$. d is the dimensionality of the system ($d = 3$ here). The data collapse obtained this way, shown in top right panel of Fig. 4, is very good. This allows one to determine the length scale at the α-regime. Similar analysis for τ_β is shown in bottom right panel of Fig.4. Again the data collapse is found to be very good. This gives us the confidence on the reliability of the extracted length scales.

We then compare the length scale estimated in this way with the heterogeneity length estimated from system size dependence of χ_4^P [36]. This comparison is shown in the Fig. 5. In order to make the comparison, we scale the present estimate so that the two length scales match at $T = 0.6$. Length scales obtained by FSS are known only up to a multiplicative factor, and hence this procedure does not introduce any additional arbitrariness. We find that the temperature dependence of the two length scales agrees very well. IMCT predicts that ξ grows as a power law in time up to τ_β and stays at the value τ_β up to τ_α. Our results therefore confirm this expectation. Independently of IMCT predictions, the agreement is remarkable in pointing to an intimate connection between dynamics at short times (β regime) and long times (α regime). This result suggests that the

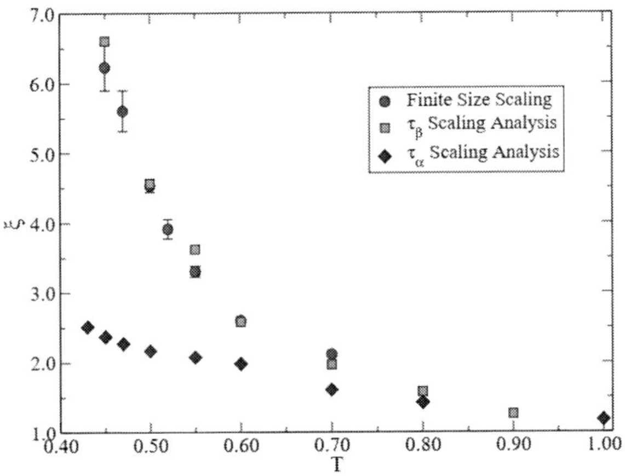

Figure 5. Different length scales are plotted as a function of temperature. Notice that dynamic length scale and length scale obtained form the short time relaxation is same and the length scale obtained from long time relaxation is completely different.

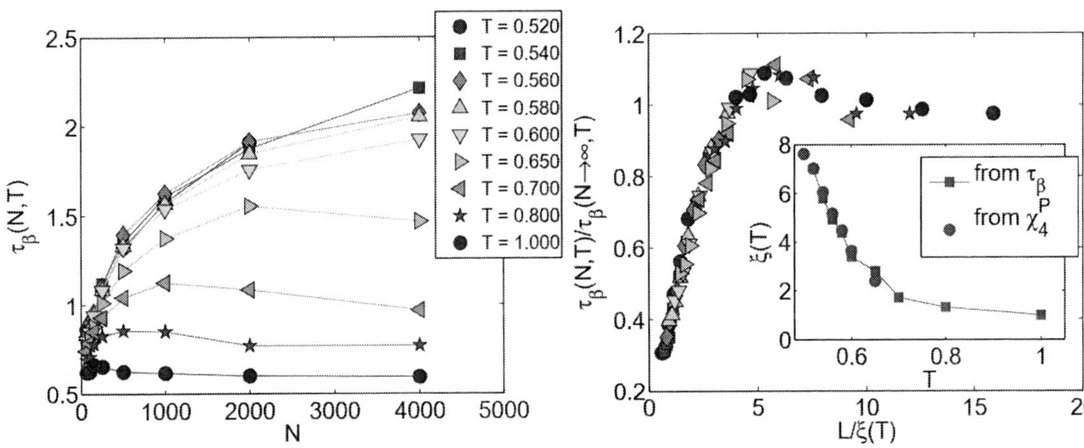

Figure 6. Left Panel: System size dependence of τ_β for different temperature for 3dR10 model. Right panel: Data collapse of τ_β to obtain the length scale associated with the cooperativity in β- relaxation. Inset: Comparison of the correlation length extracted from the system-size dependence of τ_β with the dynamical length scale. The two length scales agree with each other to a good accuracy. Taken from Ref.[48].

heterogeneity present in the dynamics in the α regime has been built up already in the β regime, and therefore in principle, essential information about the α relaxation can be obtained by studying short-time dynamics. Our result therefore lends support to many investigations and ideas that aim to relate dynamical behaviour at short times to long-time structural relaxation[26, 27, 28, 29, 30, 31, 32, 33]. It is important to note here that the length scale obtained from the finite size scaling analysis of α-relaxation time is very different from these two other length scales. The growth of this length scale is much milder with temperature and many studies suggests that this length scale is same as the static length scale (see Refs. [2, 3] for in-depth discussion on the inter relation between different length scales in glassy systems). Thus although the current results suggest a strong connection between dynamics at short and long time scales, some puzzles still remain to be answered to understand the intricate connection between relaxation processes at short and long time scales.

To see whether the results obtained for the 3dKALJ model are generic, we have done similar analysis for the 3dR10 model. In Fig.6, we have shown, in the left panel we show the system-size dependence of the short time scale τ_β for different temperatures and in the right panel the corresponding finite size scaling analysis done to obtain the length scale. The scaling collapse observed in this case again is very good. We find that the temperature dependence of the length scale obtained from the FSS of τ_β matches quite well with the dynamic heterogeneity length scale, as shown in the inset of the right panel of Fig.6. The results obtained for the 3dIPL model are similar and are not reported here (see Ref.[48] for details).

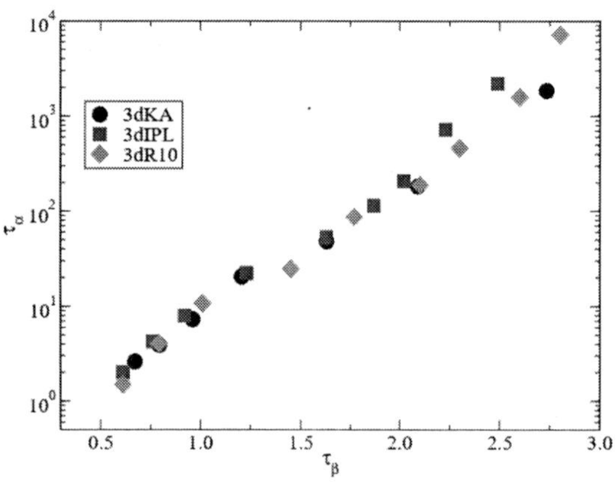

Figure 7. Long time α-relaxation timescale is plotted as a function of the short time relaxation timescale τ_β for different model systems studied. It seems to suggest a possible universal relationship between these two time scales.

As long time dynamics seem to be connected to short time dynamics, as suggested by the development of dynamic heterogeneity within the β-relaxation time scale and maintaining that at least up to α-relaxation timescale, one might wonder whether short timescale and long time scale in these systems are related to each other by an universal relationship. In Fig. 7 we have plotted logarithm of the long time α-relaxation timescale, τ_α as a function of τ_β for different model system studied. Over four orders of magnitude in α-relaxation timescale, it seems to follow a universal relationship. Although without a proper microscopic understanding, the existence of such a universal relationship can not be established with certainty from the limited range of the data presented here, but the current results do suggest a possible interesting connection worth further investigation.

4. Discussion

In this work, we have studied the system-size dependence of the β and α relaxation time for two model liquids in three dimensions. For β relaxation, we find that the time scale initially increases with increasing system size and saturates for large values of the system size, exhibiting behaviour expected for the size dependence of a quantity that depends on a length scale. This is unlike the system-size dependence of the α relaxation time that is found to decrease with increasing system size [36]. The length scale extracted from FSS of β relaxation time scale matches very well with the length scale extracted from the FSS of the four-point susceptibility $\chi_4^P(T)$ measured at the α-relaxation time scale, whereas the α relaxation time itself seems to be controlled by a completely different length scale. The behaviour observed in β-relaxation is in qualitative agreement with the predictions of inhomogeneous mode coupling theory and suggests a possible deep connection between short and long time dynamics in glass forming liquids [29, 30, 31, 32, 33, 48] with the caveat that the α-relaxation time itself seems to be controlled by a completely different length scale. Finally the observed universal relationship between short and long time relaxation timescales is worth future investigation.

5. Acknowledgements

The author would like to thank Chandan Dasgupta and Srikanth Sastry for many useful and illuminating discussions on the theory of glass transition. Saurish Chakrabarty is acknowledged for many useful discussions and critical reading of this manuscript.

References

[1] L. Berthier and G. Biroli Rev. Mod. Phys. **83**, 587 (2011).
[2] S. Karmakar, C. Dasgupta and S. Sastry, Annu. Rev. Condens. Matter Phys. **5**, 255 (2014).
[3] S. Karmakar, C. Dasgupta and S. Sastry, Rep. on Prog. in Phys. **79** (1), 016601 (2015).
[4] G. Adam and J. H. Gibbs, J. Chem. Phys. **43**, 139 (1965).
[5] G. Biroli, J.-P. Bouchaud, K. Miyazaki, and D. R. Reichman, Phys. Rev. Lett. **97**, 195701 (2006).
[6] T. R. Kirkpatrick, D. Thirumalai, and P. G. Wolynes, Phys. Rev. A **40**, 1045 (1989).
[7] V. Lubchenko and P. G. Wolynes, Annu. Rev. Phys. Chem. **58**, 235 (2007).
[8] F. Ritort and P. Sollich, Adv. Phys. **52**, 219 (2003).
[9] D. Chandler *et al.*, *Phys. Rev. E* **74**, 051501 (2006).
[10] L. Berthier, Phys. Rev. E. **69**, 020201(R) (2004).
[11] S. Whitelam, L. Berthier and J. P. Garrahan, Phys. Rev. Lett. **92**, 185705 (2004).
[12] M. D. Ediger, Annu. Rev. Phys. Chem. **51**, 99 (2000).
[13] R. Yamamoto and A. Onuki, J. Phys. Soc. Jpn. **66**, 2545 (1997).
[14] M. M. Hurley and P. Harrowell, Phys. Rev. E **52**, 1694 (1995).
[15] C. Donati, *et al.*, Phys. Rev. Lett. **80**, 2338 (1998).
[16] L. Berthier, *et al.*, J. Chem. Phys. **126**, 184503 (2007).
[17] L. Berthier, *et al.*, J. Chem. Phys. **126**, 184504 (2007).
[18] L. Berthier, *et al.*, Science **310**, 1797 (2005).
[19] C. Dalle-Ferrier, *et al.* Phys. Rev. E **76**, 041510 (2007).
[20] S. Capaccioli, G. Ruocco, and F. Zamponi, J. Phys. Chem. B **112**, 10652 (2008).
[21] C. Dasgupta, *et al.*, Europhys. Lett. **15**, 307 (1991).
[22] T.R. Kirkpatrick and D. Thirumalai, Phys. Rev. A **37**, 4439 (1988).
[23] S. Franz and G. Parisi, J. Phys. Condens. Matter **12**, 6335 (2000).
[24] C. Donati, *et al.*, J. Non-Cryst. Sol. **307-310**, 215 (2002).
[25] M. Campellone, G. Parisi and P. Ranieri, *Phys. Rev. B* **59**, 1036 (1999).
[26] R. W. Hall and P. G. Wolynes *J. Chem. Phys.* 86 2943 (1987).
[27] U. Buchenau and R. Zorn, Europhys. Lett. 18, 523 (1992).
[28] J. C. Dyre, *Rev. Mod. Phys.* **78** 953 (2006).
[29] A. Widmer-Cooper, H. Perry, P. Harrowell and D. R. Reichman, *Nature Physics* **4**, 711 (2008).
[30] C. Brito and M. Wyart, Europhys. Lett., **76** 149 (2006).
[31] F. W. Starr, S. Sastry, J. F. Douglas, and S. C. Glotzer, Phys. Rev. Lett. **89**, 125501 (2002).
[32] L. Larini, A. Ottochian, C. De Michele and D. Leporini, *Nature Physics* **4**, 42 (2008).
[33] D. S. Simmons and J. F. Douglas, Soft Matter, **7**, 11010 (2011).
[34] Finite size scaling and numerical simulations in statistical systems, V. Privman (ed) (World Scientific, Singapore, 1990).
[35] L. Berthier, Phys. Rev. Lett. **91**, 055701 (2003).
[36] S. Karmakar, C. Dasgupta, and S. Sastry, Proc. Nat. Acad. Sci (USA) **106**, 3675 (2009).
[37] S. Karmakar, C. Dasgupta, and S. Sastry, *Phys. Rev. Lett.* **105** 015701 (2010).
[38] S. Karmakar, C. Dasgupta, and S. Sastry, *Phys. Rev. Lett.* **105** 019801 (2010).
[39] E. Flenner and G. Szamel, Phys. Rev. Lett. **105** 217801 (2010).
[40] G. Szamel and E. Flenner, Phys. Rev. E **81**, 031507 (2010)
[41] R.S.L. Stein and H.C. Andersen, Phys. Rev. Lett. **101**, 267802 (2008).
[42] S. M. Bhattacharyya, B. Bagchi, P. G. Wolynes *Proc Natl Acad Sci USA* **105** 16077 (2008).
[43] G. Biroli and J.-P. Bouchaud, The Random First-Order Transition Theory of Glasses: A Critical Assessment, in *Structural Glasses and Supercooled Liquids: Theory, Experiment, and Applications (eds P. G. Wolynes and V. Lubchenko), John Wiley & Sons, Inc., Hoboken, NJ, USA.* (2012).
[44] T. Haxton, and A.J. Liu, Europhysics Lett. **90(6)**, 66004, (2010).
[45] W. Kob and H. C. Andersen, Phys. Rev. E **51**, 4626 (1995).
[46] U.R. Pedersen, T.B. Schrder, and J.C. Dyre, Phys. Rev. Lett. **105**, 157801 (2010).
[47] S. Karmakar, E. Lerner, I. Procaccia and and Jacques Zylberg Phys. Rev. E **82** 031301 (2010).
[48] S. Karmakar, C. Dasgupta and S. Sastry, Phys. Rev. Lett. **116**, 085701 (2016).

Numerical test of finite-size scaling predictions for the droplet condensation-evaporation transition

Andreas Nußbaumer[1], **Johannes Zierenberg**[2], **Elmar Bittner**[3] and **Wolfhard Janke**[2]

[1] Institut für Physik, Johannes Gutenberg Universität Mainz, Staudinger Weg 7, D-55128 Mainz, Germany

[2] Institut für Theoretische Physik, Universität Leipzig, Postfach 100 920, D-04009 Leipzig, Germany

[3] Institut für Theoretische Physik, Universität Heidelberg, Philosophenweg 16, D-69120 Heidelberg, Germany

E-mail: `wolfhard.janke@itp.uni-leipzig.de`

Abstract. We numerically study the finite-size droplet condensation-evaporation transition in two dimensions. We consider and compare two orthogonal approaches, namely at fixed temperature and at fixed density, making use of parallel multicanonical simulations. The equivalence between Ising model and lattice gas allows us to compare to analytical predictions. We recover the known background density (at fixed temperature) and transition temperature (at fixed density) in the thermodynamic limit and compare our finite-size deviations to the predicted leading-order finite-size corrections.

1. Introduction

Droplet formation is an essential process in nature with a variety of analogues in biological systems and material science. Of course, this generally involves non-equilibrium processes, e.g., the formation of nucleation prerequisites from local fluctuations while the surrounding gas acts as a density bath. Here, we consider instead a canonical setup in a finite system of size V with fixed temperature T and particle number N. If the gas is supersaturated, a variation of N or T results in the formation of equilibrium droplets [1, 2, 3, 4, 5]. In simple terms, the particle excess is subdivided to form a single macroscopic droplet in equilibrium with the surrounding vapor plus remaining excess. The resulting theory has been supported by numerous computational studies at fixed temperature, including the two-dimensional lattice gas [6, 7, 8] and three-dimensional Lennard-Jones gas [9, 10, 11]. The orthogonal approach at fixed density has received less attention [12], but recently enabled us to come closer to the asymptotic scaling regime for two- and three-dimensional lattice gas and three-dimensional Lennard-Jones gas [13]. In the following, we will compare the leading-order scaling corrections for a two-dimensional lattice gas at fixed temperature and fixed density from analytical predictions with numerical results.

2. Model

We consider a lattice gas in $d = 2$ dimensions. Excluded volume is modeled as lattice sites being either occupied by exactly one particle or empty, $n_i = \{0, 1\}$. Short-range interaction is included by nearest-neighbor interaction ($\langle i, j \rangle$). The Hamiltonian is then

$$\mathcal{H} = -\sum_{\langle i,j \rangle} n_i n_j, \tag{1}$$

which is equivalent to an Ising model at fixed magnetization with $T^{\rm Is} = 4T$ and coupling constant $J = 1$ [14]. This originates in the identification of the spin state $s_i = 2n_i - 1$, which allows one to rewrite the Ising Hamiltonian

$$\mathcal{H}^{\rm Is} = -\sum_{\langle i,j \rangle} s_i s_j = 4\mathcal{H} - d(V - 4N), \tag{2}$$

considering that, on a simple hypercubic lattice, the sum over nearest neighbors yields d contributions and $N = \sum n_i$. The equivalence is then established by equating the Boltzmann factors $\exp(-\beta\mathcal{H})$, where $\beta = (k_{\rm B}T)^{-1}$. Evaluating $\beta^{\rm Is}\mathcal{H}^{\rm Is} = \beta\mathcal{H}$, where the constant shift is neglected due to physical unimportance, results in a simple rescaling of the temperature. When making use of the Ising equivalence all energy-related observables have to be rescaled correspondingly. The canonical Ising model is then equivalent to a grand-canonical lattice gas.

Exploiting the equivalence to the 2D Ising model, the critical point is located at $\beta_c = 4\beta_c^{\rm Is} = 2\ln\left(1 + \sqrt{2}\right) \approx 1.763$ or $k_{\rm B}T_c \approx 0.567$. For temperatures below this point, a spontaneous magnetization m_0 is observed and described by the Onsager-Yang solution [15, 16]:

$$m_0(\beta^{\rm Is}) = \left[1 - \sinh^{-4}(2\beta^{\rm Is})\right]^{1/8}. \tag{3}$$

This is directly related to the (grand-canonical) equilibrium background density $\rho_0 = (1 - m_0)/2$. The magnetic susceptibility is connected with the isothermal compressibility $\chi = \beta\kappa = \hat{\kappa}$ [8] and may be evaluated from sufficiently long series expansions (see, e.g., Ref. [17] and references therein), where

$$\chi(\beta^{\rm Is}) = \beta^{\rm Is} \sum_{i=0}^{n} c_i u^{2i} \quad \text{with} \quad u = \frac{1}{2\sinh(2\beta^{\rm Is})}, \tag{4}$$

and $c = \{0, 0, 4, 16, 104, 416, 2224, 8896, 43840, 175296, 825648, 3300480, 15101920, ...\}$,[1] here considered up to the $300^{\rm th}$ term. The equilibrium shape of a 2D Ising droplet is described by the Wulff plot (or shape), given by [18]

$$W = \frac{4}{(\beta^{\rm Is})^2} \int_0^{\beta^{\rm Is}\sigma_0} dx \, \cosh^{-1}\left[\frac{\cosh^2(2\beta^{\rm Is})}{\sinh(2\beta^{\rm Is})} - \cosh(x)\right], \tag{5}$$

where $\sigma_0 = 2 + \ln[\tanh(\beta^{\rm Is})]/\beta^{\rm Is}$ and $\cosh^{-1}$ is referring to the inverse hyperbolic cosine. This will be relevant for the surface free energy of a (Wulff shaped) droplet of unit volume $\tau_{\rm W}^{\rm Is} = 2\sqrt{W}$. Being energy-related, the interface tension gets converted as $\tau_{\rm W} = \tau_{\rm W}^{\rm Is}/4$.

3. Method

In order to obtain numerical data at fixed temperature and at fixed density, we employed two different kinds of simulations. In both cases the underlying algorithm is the multicanonical method [19, 20, 21, 22]. The condensation transition is a first-order phase transition for which this method is well suited because it potentially allows one to overcome barriers in the free energy. The principle idea is to replace the Boltzmann weight $\exp(-\beta E)$ by an a priori unknown weight function $W(E)$, which is iteratively adapted in order to yield a flat histogram over a

[1] The coefficients were obtained from `http://www.ms.unimelb.edu.au/~iwan/ising/Ising_ser.html` [17].

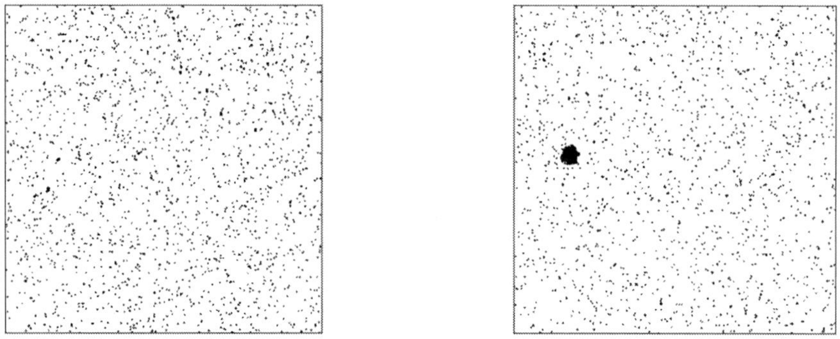

Figure 1. Snapshots of a two-dimensional lattice gas at fixed density above (left) and below (right) the condensation-evaporation temperature, showing a homogeneous gas phase and a droplet in equilibrium with surrounding vapor, respectively. Both plots show $N = 2500$ particles on a lattice of linear size $L = 500$, i.e., $\rho = 0.01$.

desired energy range. That way, energy states which are suppressed at a first-order phase transition are artificially enhanced and the simulation may transit back and forth between the involved coexisting phases. In the end, the simulation data has to be reweighted to estimates of expectation values at any temperature for which the probability distribution is covered by the flat histogram. The "optimal" weight function is reached if a simulation run produces a flat histogram, i.e., for a fixed c it holds $h(E_{\min})/h(E_{\max}) > c$. In part, we further make use of a parallel implementation [23, 24], which exploits the fact that in each iteration the histogram is an estimate of the probability distribution belonging to the current weight function. This allows one to distribute the sampling to independent Markov chains and to obtain a joint estimate as a simple sum of individual histograms. The procedure scales very well for the problem at hand [8].

The involved Monte Carlo updates include particle displacements to free nearest neighbors and random jumps to free sites. This corresponds to local and global Kawasaki updates. Errors are obtained using the Jackknife method [25] and standard error propagation. For further details of the employed methods we refer to Refs. [6, 7, 8, 13].

4. Theory of leading-order correction

A natural approach to particle condensation/evaporation is the consideration of a grand-canonical ensemble at fixed temperature, where the system relaxes to an equilibrium background contribution $N_0 = \rho_0 V$. Below the corresponding critical temperature the system is dominated either by particles (fluid branch) or by void space (gas branch) and $\rho_0 \neq 1/2$. Let us now consider the dilute gas branch well below the critical temperature and fix $N > N_0$. This results in the canonical ensemble of a supersaturated gas with particle excess $\delta N = N - N_0$. Initially, this excess goes into the gas phase while for sufficiently large excess droplet formation occurs.

In equilibrium droplet formation, the probability for intermediate-sized droplets was shown to vanish [3] and the scenario reduces to a homogeneous gas phase and an inhomogeneous phase of a droplet in equilibrium with surrounding vapor. This may be considered as the interplay of entropy maximization by fluctuations in the gas phase and energy minimization by forming a droplet [3, 4], see also Fig. 1. At fixed temperature it is possible to consider (fixed) thermal fluctuations and relate them to infinite-size temperature-dependent quantities, like the isothermal compressibility $\hat{\kappa}$ and the normalized surface free energy τ_W, see Sec. 2. The free energy may then be approximated by a contribution $F_{\rm fluc}$ from the fluctuation of particle excess δN and a contribution F_{drop} from the single macroscopic droplet of size V_{D}:

$$F_{\text{fluc}} = \frac{(\delta N)^2}{2\hat{\kappa}V} \qquad \text{and} \qquad F_{\text{drop}} = \tau_{\text{W}}(V_{\text{D}})^{\frac{d-1}{d}}. \tag{6}$$

These contributions are idealized with possible sources of corrections in both the Gaussian approximation and the droplet shape for finite systems.

In the two-phase scenario, the particle excess may be decomposed into the excess inside the droplet δN_{D} and the excess in the fluctuating phase δN_{F}, i.e., $\delta N = N - N_0 = \delta N_{\text{D}} + \delta N_{\text{F}}$. Linking the droplet size to the particle excess inside the droplet, one expects $\delta N_{\text{D}} = (\rho_{\text{L}} - \rho_0)V_{\text{D}}$, where ρ_{L} and ρ_0 are the background liquid and gas density, respectively. Then, one may introduce a scalar fraction

$$\lambda = \delta N_{\text{D}}/\delta N, \tag{7}$$

such that $\delta N_{\text{D}} = \lambda\delta N$ and $\delta N_{\text{F}} = (1-\lambda)\delta N$. The total free energy $F = F_{\text{drop}} + F_{\text{fluc}}$ becomes

$$F = \tau_{\text{W}}\left(\frac{\lambda\delta N}{\rho_{\text{L}} - \rho_0}\right)^{\frac{d-1}{d}} + \frac{(1-\lambda)^2(\delta N)^2}{2\hat{\kappa}V} = \tau_W\left(\frac{\delta N}{\rho_{\text{L}} - \rho_0}\right)^{\frac{d-1}{d}}\left(\lambda^{\frac{d-1}{d}} + \Delta(1-\lambda)^2\right), \tag{8}$$

with a dimensionless "density" parameter

$$\Delta = \frac{(\rho_{\text{L}} - \rho_0)^{\frac{d-1}{d}}}{2\hat{\kappa}\tau_W}\frac{(\delta N)^{\frac{d+1}{d}}}{V} = \frac{(\rho_{\text{L}} - \rho_0)^{\frac{d-1}{d}}}{2\hat{\kappa}\tau_{\text{W}}}(\rho - \rho_0)^{\frac{d+1}{d}}\, V^{\frac{1}{d}}. \tag{9}$$

At fixed temperature $\rho_{\text{L}}, \rho_0, \hat{\kappa}, \tau_W$ are constants and Δ may be interpreted as an unusual density. In principle, all constants may be estimated and for the present case, equivalent to the 2D Ising model, the parameters are even known exactly or with very high precision.

This leading-order formulation allows one to obtain the fraction of particles inside the largest droplet λ as a function of Δ in the limit of large systems, by minimizing Eq. (8) with respect to λ. It turns out (for details see Refs. [3, 4]) that there exists a constant Δ_{c} below which no droplet forms ($\lambda = 0$) and above which a single macroscopic droplet exists with non-trivial $\lambda > \lambda_{\text{c}}$:

$$\Delta_{\text{c}} = \frac{1}{d}\left(\frac{d+1}{2}\right)^{\frac{d+1}{d}} \overset{\text{2D}}{=} 0.9186... \qquad \text{and} \qquad \lambda_{\text{c}} = \frac{2}{d+1} \overset{\text{2D}}{=} 2/3. \tag{10}$$

The result $\lambda(\Delta)$ describes the expectation value of the equilibrium droplet size in the limit of large systems without any free parameter. As mentioned before, this already includes the leading-order finite-size corrections for idealized assumptions. In fact, Eq. (9) may be rewritten at each finite-size transition density ρ_{c}, where $\Delta(\rho_{\text{c}}) = \Delta_{\text{c}}$. For a lattice gas model with particle-hole symmetry, $\rho_{\text{L}} = 1 - \rho_0$, this yields to leading order

$$\rho_{\text{c}} = \rho_0 + \left(\frac{2\hat{\kappa}\tau_{\text{W}}\Delta_{\text{c}}}{(1-2\rho_0)^{\frac{d-1}{d}}}\right)^{\frac{d}{d+1}} V^{-\frac{1}{d+1}}, \qquad \text{or} \qquad m_{\text{c}} = m_0 - 2m_0\left(\frac{\chi\tau_{\text{W}}^{\text{Is}}\Delta_{\text{c}}}{2m_0^2}\right)^{\frac{d}{d+1}} V^{-\frac{1}{d+1}}, \tag{11}$$

where the second formulation is in terms of the Ising model at fixed magnetization. This is the notation of Biskup et al. [3, 4], but is in quantitative agreement with the (independent) result of Neuhaus and Hager [2]. For the 2D Ising model, they find the same leading scaling behavior $\Delta m(L) = A_{\text{cond}}L^{-2/3}$, with the amplitude $A_{\text{cond}} = 0.23697...$ for $\beta^{\text{Is}} = 0.7$ and Eq. (11) yields $A = 0.236965...$ for the corresponding constants.[2]

[2] For $\beta^{\text{Is}} = 0.7$ we obtain $m_0 \simeq 0.99016$, $\chi \simeq 0.019310$ and $\tau_{\text{W}}^{\text{Is}} \simeq 4.5758$.

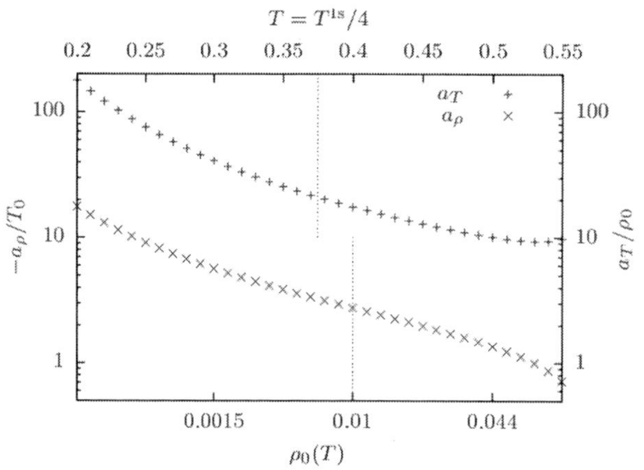

Figure 2. Leading-order normalized finite-size correction amplitude at fixed temperature T and fixed density ρ. The densities $\rho_0(T)$ on the lower x-axis are calculated from the Onsager-Yang solution Eq. (3). The data points are obtained by numerically evaluating Eq. (14) and Eq. (15). Dashed lines indicate the parameters considered below.

In the following, we will numerically review the leading-order finite-size scaling predictions. We consider directly the scaling of the factual transition density (or magnetization) at fixed temperature [6, 7] and compare to the scaling of the transition temperature at fixed density [13]. Albeit the possibility of logarithmic corrections, we consider empirical higher-order corrections as powers of the leading term if necessary. Our fit ansatz for the leading-order correction is

$$\rho_c = \rho_0 + a_T V^{-1/3} + \mathcal{O}\left(V^{-2/3}\right) \qquad \text{at fixed } T, \tag{12}$$

$$T_c = T_0 + a_\rho V^{-1/3} + \mathcal{O}\left(V^{-2/3}\right) \qquad \text{at fixed } \rho. \tag{13}$$

By comparing to Eq. (11), a_T can be related to Δ_c. Similarly, a relation for a_ρ follows from a Taylor expansion around T_0 of a reformulated Eq. (9), namely $\Delta^{2/3}V^{-1/3} = f(\rho, T)$. From the equivalence to the Ising model, we identify $f(\rho, T)$ with $f(m, T^{\mathrm{Is}}) = (m_0(T^{\mathrm{Is}}) - m)[m_0(T^{\mathrm{Is}})/2]^{1/3}[\chi(T^{\mathrm{Is}})\tau_{\mathrm{W}}^{\mathrm{Is}}(T^{\mathrm{Is}})]^{-2/3}$ (see Ref. [13] for details), which leads to

$$\Delta_c = a_T^{3/2} 2 m_0^{1/2} / \chi\tau_{\mathrm{W}}^{\mathrm{Is}} \qquad \text{at fixed } T, \tag{14}$$

$$\Delta_c = \left[a_\rho 4 f'(m, T_0^{\mathrm{Is}})\right]^{3/2} \qquad \text{at fixed } \rho. \tag{15}$$

This may be evaluated numerically considering Eqs. (3)–(5) and (10) to yield estimates of the leading-order correction. Figure 2 shows results for a range of temperatures and according densities, related by the Onsager-Yang solution Eq. (3). The relative leading-order corrections at fixed temperature are almost an order of magnitude larger than at fixed density. However, in the practical finite-size scaling analysis higher-order corrections are of more significance, which cannot be estimated by the current theory.

5. Results

We will start the discussion with our findings at fixed temperature $k_{\mathrm{B}}T = 0.375$. Most results are analyses of Monte Carlo time series data obtained by refined methods analogous to Refs. [6, 7]. For each system size, we set up individual simulations at selected densities around the expected transition region, see Fig. 3 (left). For densities below the finite-size transition point $\Delta < \Delta_c$

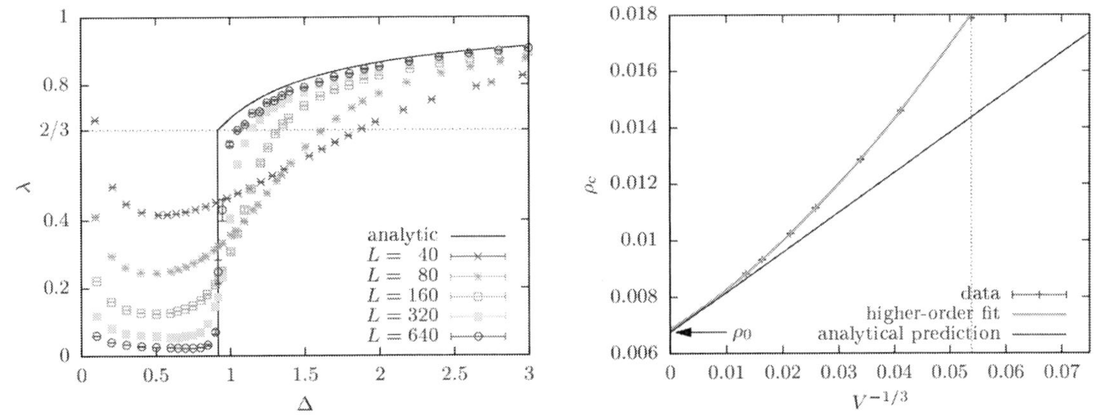

Figure 3. Droplet formation at fixed temperature $k_BT = 0.375$ [6, 7]. (left) Fraction of excess versus rescaled dimensionless density Δ. (right) Finite-size scaling of the transition density ρ_c. The black line shows the predicted leading-order scaling and the blue line is a higher-order fit to the data. ρ_0 is the analytically known thermodynamic limit. The dotted vertical line indicates the end of the fitting range ($L = 80$).

the system is in the gaseous phase and for $\Delta > \Delta_c$ a single macroscopic droplet forms in accordance with theory. In fact, the results are obtained in the equivalent formulation of an Ising model at fixed magnetization, but here discussed in the generic formulation of a lattice gas. In order to ensure ergodic sampling, we augmented the Kawasaki Monte Carlo update with a multicanonical scheme to locally sample a flat histogram including transition states and ensuring a good sampling of the energy probability distribution up to suppressions of 30 orders of magnitude. The fraction of excess in the largest droplet, Eq. (7), is obtained by a two-step process. Firstly, all clusters in the system are determined, where we define a cluster as alike sites connected via the nearest-neighbor property. The largest cluster is the background (or gaseous phase), while the second largest cluster forms the droplet we are interested in. Then, the volume of the droplet is identified as all sites confined by the boundary, including the enclosed holes. The fraction λ is then the ratio of the droplet volume and the expected equilibrium volume of full excess $V_\delta = (N - \rho_0 L^2)/(1 - 2\rho_0)$. The result of this procedure is shown in Fig. 3 (left). We also show the limiting case that can be found as the solution of Eq. (8) as solid black line. Every data point is the average of 10^6 Monte Carlo sweeps (L^2 updates). At the infinite-size transition density, the analytical prediction is $\lambda_c = 2/3$, see Eq. (10). Hence, we estimate the finite-size transition point as the density ρ_c for which $\lambda(\rho_c, L) = 2/3$. Technically, we do a linear interpolation of the data points in Fig. 3 (left) and search for the intersection.

The resulting scaling of the transition density is shown in Fig. 3 (right) in dependence on $V^{-1/3}$ as expected from Eq. (12). The thermodynamic limit is given by the Onsager-Yang solution, $\rho_0 \simeq 0.00675$. From the analytical prediction in Eq. (14), we expect for the leading-order correction at fixed temperature $k_BT = 0.375$ that $\Delta_c \simeq 0.919 \approx (6.6833\, a_T)^{3/2}$ and thus $a_T \approx 0.141$,[3] shown in the figure as a black line. The data does not allow for a qualitatively satisfying leading-order fit, but the largest system size is already close to the predicted finite-size deviation. Considering the next higher empirical correction, i.e., $\rho_c = \rho_0 + a_T V^{-1/3} + b_T V^{-2/3}$, allows for a decent fit which yields for $L > 80$ the result $\rho_0 = 0.00676(5)$ and $a_T = 0.135(3)$ with goodness-of-fit parameter $Q \approx 0.02$. The estimated limit is in good agreement with the

[3] For $k_BT = k_BT^{\rm Is}/4 = 0.375$ we obtain $m_0 \simeq 0.9865$ or $\rho_0 \simeq 0.00675$, $\chi = \hat{\kappa} \simeq 0.02708$, and $\tau_W^{\rm Is} \simeq 4.2454$ or $\tau_W \simeq 1.0613$.

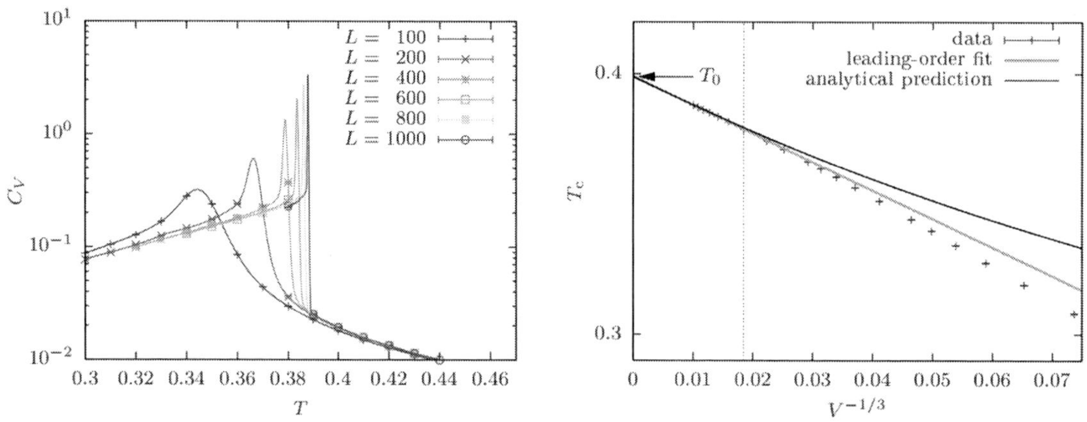

Figure 4. Droplet formation at fixed density $\rho = 10^{-2}$ [13]. (left) Specific heat with exemplary data points that indicate the size of the error. (right) Finite-size scaling of the transition temperature. The black line shows the numerical evaluation of Eq. (9) at fixed ρ and the blue line is a leading-order fit. T_0 is the analytically known thermodynamic limit. The dotted vertical line indicates the end of the fitting range ($L = 400$).

analytical prediction ρ_0 marked by the arrow. Using Eq. (15) with error propagation yields the estimate $\Delta_c = 0.86(3)$, which is below but consistent with the theoretical prediction. We notice that the choice of the intersection in $[0, 2/3]$ strongly influences the size of the higher-order corrections.

We next turn to the orthogonal setup of a fixed density $\rho = 10^{-2}$ with varying temperature [13]. Here, the multicanonical method is more straightforward and for each system size we may perform a single (yet parallel) simulation with up to 128 cores and 1.28×10^6 measurements in the final production run. This yields a full temperature range of expectation values and the transition temperature may be determined precisely as the peak of the specific heat $C_V = k_B\beta^2 \left(\langle E^2\rangle - \langle E\rangle^2\right)/N$, see Fig. 4 (left). For details see Ref. [13], from which we recapture the data in order to compare to the leading-order results at fixed temperature. For $\rho = 10^{-2}$, the numerical evaluation of the Onsager-Yang solution Eq. (3) yields $T_0 \simeq 0.39882$, which is the expected thermodynamic limit. From the analytical prediction Eq. (15), we expect for the leading-order correction that $\Delta_c \simeq 0.919 \approx (-0.85177\ a_\rho)^{3/2}$ and thus $a_\rho \approx -1.109$.

Figure 4 (right) shows the finite-size transition temperature as a function of the expected leading-order scaling correction. In two dimensions, we can numerically evaluate Eq. (9) at fixed density for various system sizes. This gives the (analytical) leading-order finite-size estimate of the transition temperature, shown as a black line. We can see that this predicted leading-order behavior is only approximately reached for very large system sizes of about $L \simeq 500$. For small system sizes, the leading-order solution strongly deviates, which may be expected from the simplified assumptions. A leading-order fit for $L \geq 400$ yields $T_0 = 0.39891(5)$ and $a_\rho = -1.091(4)$ with $Q \approx 0.25$. The limit is in good agreement with the analytical value. Using Eq. (15) with error propagation yields the estimate $\Delta_c = 0.896(5)$. This is close to, but differs slightly from, the predicted value. As for the case of fixed temperature, this is below the theoretical prediction. The fit error is too small in order to be fully consistent, which may be accounted to missing higher-order corrections. We also tried the empirical higher-order corrections, however, it did not improve our estimates. This may be explained by additional corrections, e.g., logarithmic ones, which were not considered. Still, the qualitatively good leading order fit shows that in this setup we are close to the asymptotic scaling regime.

6. Conclusions

We numerically investigated the leading-order finite-size scaling corrections of the two-dimensional droplet condensation-evaporation transition at fixed temperature and density, respectively. In both cases we could recover the analytically known thermodynamic limits, in part by considering empirical higher-order corrections. The leading-order corrections were found to be slightly smaller than but consistent with the analytical predictions [3, 4]. This shows that the established theory is able to qualitatively describe the finite-size deviations; in case of the two-dimensional lattice gas already for system sizes $L \simeq 500$.

Acknowledgments

The project was funded by the European Union and the Free State of Saxony. This work has been partially supported by the DFG (Grant No. JA 483/31-1), the Leipzig Graduate School "BuildMoNa", and the Deutsch-Französische Hochschule DFH-UFA (Grant No. CDFA-02-07). The authors gratefully acknowledge the computing time provided by the John von Neumann Institute for Computing (NIC) on the supercomputer JUROPA at Jülich Supercomputing Centre (JSC).

References

[1] Binder K and Kalos M H 1980 *J. Stat. Phys.* **22** 363
[2] Neuhaus T and Hager J 2003 *J. Stat. Phys.* **113** 47
[3] Biskup M, Chayes L and Kotecký R 2002 *Europhys. Lett.* **60** 32
[4] Biskup M, Chayes L and Kotecký R 2003 *J. Stat. Phys.* **116** 175
[5] Binder K 2003 *Physica A* **319** 99
[6] Nußbaumer A, Bittner E, Neuhaus T and Janke W 2006 *Europhys. Lett.* **75** 716
[7] Nußbaumer A, Bittner E and Janke W 2008 *Phys. Rev. E* **77** 041109
[8] Zierenberg J, Wiedenmann M and Janke W 2014 *J. Phys: Conf. Ser.* **510** 012017
[9] MacDowell L G, Virnau P, Müller M and Binder K 2004 *J. Chem. Phys.* **120** 5293
[10] MacDowell L G, Shen V K and Errington J R 2006 *J. Chem. Phys.* **125** 034705
[11] Schrader M, Virnau P and Binder K 2009 *Phys. Rev. E* **79** 061104
[12] Martinos S, Malakis A and Hadjiagapiou I 2007 *Physica A* **384** 368
[13] Zierenberg J and Janke W 2015 *Phys. Rev. E* **92** 012134
[14] Lee T D and Yang C N 1952 *Phys. Rev.* **87** 410
[15] Onsager L 1949 *Nuovo Cimento (Suppl.)* **6** 261
[16] Yang C N 1952 *Phys. Rev.* **85** 808
[17] Boukraa S, Guttmann A J, Hassani S, Jensen I, Maillard J M, Nickel B and Zenine N 2008 *J. Phys. A: Math. Theor.* **41** 455202
[18] Leung K and Zia R K P 1990 *J. Phys. A: Math. Gen.* **23** 4593
[19] Berg B A and Neuhaus T 1991 *Phys. Lett. B* **267** 249
[20] Berg B A and Neuhaus T 1992 *Phys. Rev. Lett.* **68** 9
[21] Janke W 1992 *Int. J. Mod. Phys. C* **3** 1137
[22] Janke W 1998 *Physica A* **254** 164
[23] Zierenberg J, Marenz M and Janke W 2013 *Comput. Phys. Commun.* **184** 1155
[24] Zierenberg J, Marenz M and Janke W 2014 *Physics Procedia* **53** 55
[25] Efron B 1982 *The Jackknife, the Bootstrap and Other Resampling Plans* (Society for Industrial and Applied Mathematics)

Kinetics of Order-Disorder Transitions in Binary Mixtures: A Monte Carlo Study

Prasenjit Das[1], Tanusri Saha-Dasgupta[2], Sanjay Puri[1]
[1] School of Physical Sciences, Jawaharlal Nehru University, New Delhi 110067, India.
[2] S. N. Bose National Centre for Basic Sciences, Kolkata 700098, India.

E-mail: prasenjit.jnu@gmail.com

Abstract. We study ordering kinetics in symmetric and asymmetric binary mixtures, undergoing an order-disorder transition below the critical temperature. We model the microscopic dynamics via an anti-ferromagnetic Ising model with Kawasaki spin-exchange kinetics. This conserves the composition but the order parameter, (staggered magnetization) is not a conserved quantity. The order-parameter correlation function shows dynamical scaling, and the scaling function is independent of mixture composition. The average domain size exhibits power law growth: $L(t) \sim t^{\alpha}$. The asymptotic growth regime has $\alpha = 1/2$, though there can be prolonged transients with $\alpha < 1/2$ for asymmetric mixtures. Our unambiguous observation of the asymptotic universal regime is facilitated by use of an accelerated Monte Carlo technique.

1. Introduction

Phase ordering processes are of great interest in the fields of material science and metallurgy for the designing of new materials [1, 2, 3]. When a disordered binary mixture A_xB_{1-x} is suddenly quenched below the order-disorder critical temperature, $T_c(x)$, the system evolves toward two degenerate (ABAB and BABA) ordered domains. The mixture is said to be symmetric if $x = 0.50$, otherwise asymmetric, for ordering in bipartite lattices [4, 5]. The kinetics of order-disorder transitions in symmetric mixtures is well-studied by using microscopic and coarse-grained models [6, 7]. At the microscopic level, the order-disorder transition is investigated using the nearest-neighbour (n.n) Ising anti-ferromagnet with Kawasaki spin-exchange kinetics. The average domain size $L(t)$ grows with time as $L(t) \sim t^{1/2}$, corresponding to diffusive growth with non-conserved order parameter (Cahn-Allen theory) [8]. The driving force of this growth process is the local curvature of the interfaces between differently-ordered domains. The domain morphology is studied using the order-parameter correlation function $C(r,t)$, and structure factor $S(k,t)$. These quantities show dynamical scaling with the scaling forms:

$$C(r,t) = g(r/L), \tag{1}$$
$$S(k,t) = L^d f(kL), \tag{2}$$

where d is the spatial dimensionality. Here, $g(x)$ and $f(p)$ are master functions, which are independent of time.

The kinetics of ordering has been studied experimentally for both symmetric and asymmetric compositions [9, 10]. For symmetric mixtures, the average domain size grows very slowly at the early stages, with a crossover to the $t^{1/2}$-law at late times [9]. The first experimental study of ordering kinetics in asymmetric mixtures is due to Shannon *et al.* [11]. They studied ordering kinetics in sputtered films of $Cu_{0.75+x}Au_{0.25-x}$, which has *fcc* structure. Below T_c, for symmetric compositions with $x = 0.00$, domain coarsening is consistent with $t^{1/2}$-law. The domain growth in asymmetric films with $x = 0.04$ is much slower, and shows logarithmic time dependence. The "logarithmic" regime could be a consequence of quenched impurities [12, 13]. It is also possible that it could be a transient regime prior to an asymptotic power law growth.

The first theoretical study of phase ordering in asymmetric binary mixtures is due to Porta *et al.* [14], using the n.n anti-ferromagnetic Ising model with Kawasaki spin exchange dynamics. They studied the effect of composition asymmetry on phase ordering in A_xB_{1-x} mixture, with $x \leq 0.50$ in two dimensional square lattices. The characteristic length scale of domain growth follows power law $L(t) \sim t^{\alpha}$, with exponent $\alpha \sim 0.40 - 0.50$ for x lying in the range 0.40 to 0.50. However, they predicted the growth law $L(t) \sim t^{1/2}$, irrespective of the asymmetry. First principle study of Sanati and Zunger [15] also confirms the above growth law in the late stage of ordering in *fcc* Cu_3Au alloys. In this paper, we numerically study the kinetics of order-disorder transitions in A_xB_{1-x} mixture. For asymmetric mixtures, there is lack of clarity regarding the asymptotic growth regime due to slow transients. We use an accelerated Monte Carlo (MC) algorithm to umambiguously demonstrate the following:

(i) The asymptotic domain growth law is $L(t) \sim t^{1/2}$, regardless of the composition of the mixture.

(ii) The scaling functions, which characterize the evolution morphologies, are universal for different values of x.

This paper is organised as follows. In Sec. 2, we describe our numerical techniques including the accelerated MC algorithm. In Sec. 3, we present detailed numerical results. Finally in Sec. 4, we conclude with a brief summary and discussion.

2. Details of Numerical Simulations

For the microscopic model, we start with the Ising model Hamiltonian:

$$\mathcal{H} = J \sum_{\langle ij \rangle} S_i S_j. \tag{3}$$

In Eq. (3), J is the strength of exchange interaction and is positive for order-disorder transition. The subscript $\langle ij \rangle$ represents n.n interaction and $S_i = +1$ or -1 corresponds to A or B atoms of the binary mixture respectively. We study ordering dynamics of our model given by Eq. (3) with Kawasaki spin exchange kinetics in $d = 2$ [1, 2]. This microscopic kinetics individually conserves numbers of $S_i = \pm 1$. The system size is N^2 with $N = 4096$. We employ periodic boundary conditions in all the directions.

The initial condition of our MC simulation consists of random distribution of A and B with number densities $c_B = 1 - c_A$, mimicking the disordered state before the quench. At $t = 0$, the system was quenched below $T < T_c$, the A and B atoms order on alternate sublattices of the square lattice. The appropriate order parameter is *staggered magnetization* M, which is the difference between two sublattice *magnetizations* of square lattices. It is not conserved under the Kawasaki kinetics and is equivalent to the ferromagnetic Ising model ($J < 0$) with Glauber kinetics [1, 2]. A randomly chosen pair of unlike spins have interchanged according to the above stochastic move, corresponding to a change in configuration from $\{S_i\} \to \{S_i'\}$. The change is accepted with the probability p, given by

$$p = \begin{cases} 1 & : \Delta E \leq 0 \\ \exp(-\beta \Delta E) & : \Delta E > 0 \end{cases} \tag{4}$$

where $\Delta E = \mathcal{H}(\{S_i^{'}\}) - \mathcal{H}(\{S_i\})$ is the energy difference between the final and initial configurations and $\beta = (k_B T)^{-1}$ with $k_B = 1$. Here one Monte Carlo Step (MCS) corresponds to N^2 attempted updates. All the statistical results presented here are obtained as averages over ten independent runs.

In the accelerated approach, we use the algorithm proposed by Marko and Barkema (MB) for phase-separating binary alloys [16]. The MB algorithm accelerates to reach asymptotic growth law by suppressing diffusion along the interfaces and favouring interdomain bulk diffusion. This is because, in the late stage of evolution, a small fraction of time is spent on intradomain transport processes, which generate actual growth law and dynamical scaling. Here, we modify the MB algorithm for the order-disorder transitions. In the modified approach, we keep track of the coordination number of each site i,

$$Q(i) = \sum_{j \in \langle i,j \rangle} \delta(S_i, -S_j). \tag{5}$$

The $Q(i)$'s run from 0 to Z, where Z is the lattice coordination number. Here, $Z = 4$ in two dimensional square lattices. Isolated spins i whose neighbours are same in sign have $Q(i) = 0$, while spins i, whose neighbours have opposite sign have $Q(i) = Z$. The coordination numbers provide sufficient information to compute energy changes due to spin exchanges. The change in energy resulting from the exchange of two nearest neighbour spins i and j of opposite sign is given by

$$\Delta E = 4J\left[Q(i) + Q(j) - (Z+1)\right]. \tag{6}$$

We order all the sites having equal coordination number into lists. Thus, we have $Z+1$ lists and all the sites in a given list have the identical environment. When the system quenches below T_c, ordering starts throughout the system. As a result, the size of the list with $Q(i) = Z$ will be increasing by shrinking size of the lists with $Q(i) < Z$. The steps of the algorithm are same as those proposed by MB [16]. However for the sake of completeness, we will mention the steps. We choose a step from the ensemble of all possible spin exchanges according to how likely it is to occur per unit time, making time step of appropriate duration. One step of our dynamics for a cubic lattice consists of the following parts:

(i) to increment time by

$$\Delta t = \left[\sum_q \left(1 - \frac{q}{Z}\right) N_q e^{-4Jq}\right]^{-1}, \tag{7}$$

where N_q is the number of elements in the list with q neighbours. We call this time as *Monte Carlo time* (MCT).

(ii) to select the list of q neighbours with probability

$$P_q = \Delta t \left(1 - \frac{q}{Z}\right) N_q e^{-4Jq}. \tag{8}$$

(iii) to randomly select a site i from the list of q neighbours.

(iv) to randomly select a neighbour j of site i with $S_i \neq S_j$.

(v) to exchange the spins S_i and S_j according to the probability given by Eq.(4). Adjust Q values of the sites i, j and their neighbours and update the lists.

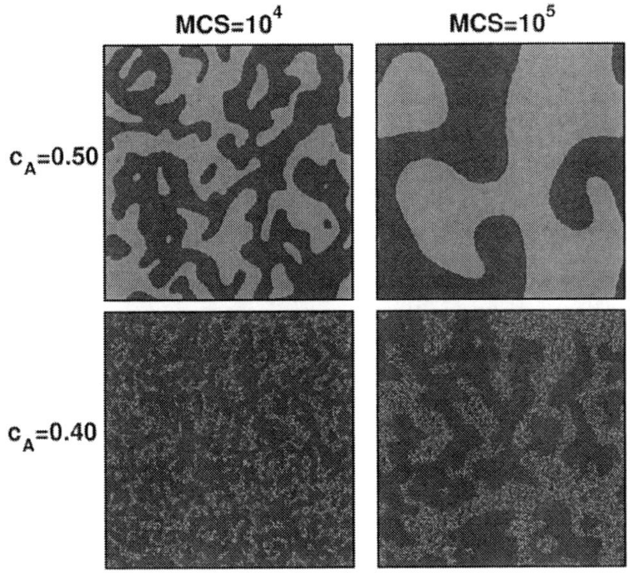

Figure 1. Evolution snapshots of staggered magnetization M, which is the magnetization difference between two sublattices, obtained from $d = 2$ MC simulations of Ising model described by Eq. (3). First row represents the same for mixtures with $c_A = 0.50$ and second row for $c_A = 0.40$ at two different MCS, as mentioned. The lattice size was 4096^2 and periodic boundary conditions were applied in both the directions. The initial condition for each run consisted of random mixtures of A and B, according to the desired ratio. The quenching temperature was $T = 0.9$ for both the cases. Regions with $M > 0$ are marked in red and that of with $M < 0$ are marked by green. From the evolution snapshots of the mixtures with $c_A = 0.40$, it is clear that the excess B-atoms (number of B-atoms, which is greater than A-atoms) were dissolved into the bulk.

3. Numerical Results

In Fig. 1, we show the evolution snapshots of the staggered magnetization field for $c_A = 0.50$ and $c_A = 0.40$ at different MCS, as mentioned. Details are given in the figure caption. For the symmetric mixture with $c_A = c_B = 0.50$, immediately after the quench below T_c, the system evolves toward two degenerate ABAB and BABA ordered states. At the late stage, the domain growth is driven by the removal of the interfaces between two differently ordered domains. But for asymmetric mixture with $c_A = 0.40$, ordering starts throughout the system and the excess B atoms start accumulating along the interface of the ordered regions immediately after quench below T_c. At the late stage of the domain growth, the excess B atoms start migrating into the bulk of the ordered domains by minimizing the energy. It is evident from the evolution snapshots for $c_A = 0.40$ in Fig. 1. For all other asymmetric compositions, similar dynamics is observed.

In order to study the morphology of domain growth, we calculate the correlation function and structure factor of the staggered magnetization field, defined as $\sigma_{ij} = (-1)^{(i+j)} S_{ij}$ for a spin S_{ij} at the site (i, j) in two dimensions. The equal time correlation function $C(\vec{r}, t)$ is defined as follows

$$C\left(\vec{r}, t\right) = \left[\left\langle \sigma(\vec{R}, t)\sigma(\vec{R} + \vec{r}, t)\right\rangle - \left\langle \sigma(\vec{r}, t)\right\rangle \left\langle \sigma(\vec{R} + \vec{r}, t)\right\rangle\right]. \tag{9}$$

Here the angular brackets represent the average over different initial conditions [18, 20]. Similarly, we calculate *structure factor* $S(k, t)$, which is defined as the Fourier transform of

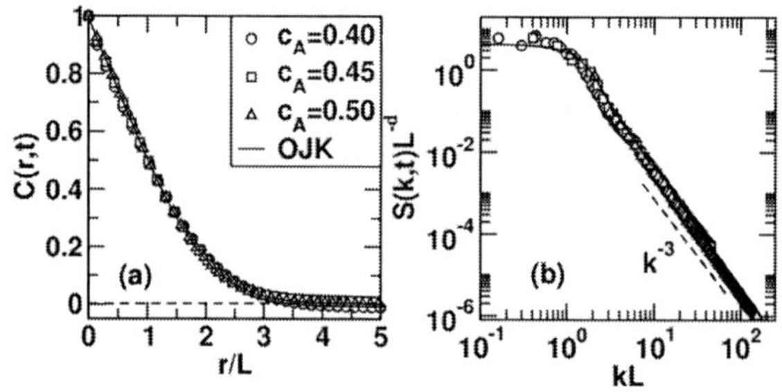

Figure 2. Scaling plot of spherically averaged correlation functions $C(r,t)$ and structure factors $S(k,t)$ of $\sigma(\vec{r},t)$-field at MCS=10^5 for different concentration of A. The system size is $L_x \times L_y = 4096 \times 4096$. Results are averaged over ten independent runs. (a) Plot of $C(r,t)$ at MCS=10^5 for different c_A, as mentioned. We define the length scale L, which is the distance at which $C(r,t)$ falls to half from its maximum (=1 at r=0.0). The solid line shows the OJK function in Eq. (11), collapsed appropriately. (b) Plot of $S(k,t)$ *vs.* kL on log-log scale. The solid line denotes the Fourier transform of OJK function, scaled appropriately. The line labeled with k^{-3} represents the Porod's law with $d=2$ and $n=1$. Rest of the details are same as Fig.1.

the correlation function $C(r,t)$ as follows

$$S\left(\vec{k},t\right) = \int d\vec{r} e^{i\vec{k}.\vec{r}} C(\vec{r},t), \tag{10}$$

at wave vector $\vec{k}$. The correlation function for the ordering in 50%-50% mixture is well studied and solved by Ohta *et al.* (OJK), who showed that length scale grows as $L_\sigma(t) \sim t^{1/2}$ [19]. The functional form of the correlation function is

$$C_{OJK}\left(r,t\right) = \frac{2}{\pi} \sin^{-1}\left(e^{-r^2/L_\sigma^2}\right). \tag{11}$$

In Fig. 2(a), we plot $C(r,t)$ *vs.* r/L at MCS=10^5 for different values of c_A, as mentioned. We also plot the OJK function in Eq. (11). We see that numerical data for all values of c_A are indistinguishable from the OJK function, which confirms that the morphology of ordering in symmetric and asymmetric binary mixtures is same. In Fig. 2(b), we plot $S(k,t)L^{-d}$ *vs.* kL at MCS=10^5 for different values of c_A, mentioned. We also plot the Fourier transform of the OJK function - this also matches with our numerical data. The data collapse confirms the dynamical scaling. In the limit $k \to \infty$, $S(k,t)$ decays as k^{-3}, following the *Porod's law* as $S(k,t) \sim k^{-(d+n)}$ with spatial dimensionality $d=2$ and the number of order parameter components $n=1$ [17]. This implies sharp interfaces formed between two degenerate ordered states, irrespective of the amount of surplus materials present in the system.

We calculate the length scale of domain growth L from the equal-time correlation function of the staggered magnetization field, as described earlier for regular MC. It is defined as the distance at which correlation decays from 1(at r = 0) to 0.5. The variation of average domain size L *vs.* MCT for different c_A's are shown in Fig. 3. Details are given in the figure caption. Line with growth exponent $\alpha = 1/2$ in Fig. 3, corresponds to diffusive growth for non-conserved order parameter. Here the growth exponent is $\alpha = 1/2$ for all the mixtures. This is the well-known *Allen-Cahn Law:* $L(t) \sim t^{1/2}$ [8]. Asymmetric mixtures have less surface tension than symmetric ones. This is because of the saturation value of the order-parameter is less in the bulk of a ordered domain. This can give rise to slow transients. In Fig. 3, the $L(t)$ for $c_A = 0.50$ is smaller than that of for $c_A < 0.50$. This is because we quenched the systems for different c_A's at same temperature and there is conversion factor between MCT and MCS, which is less for asymmetric mixtures than the symmetric mixture.

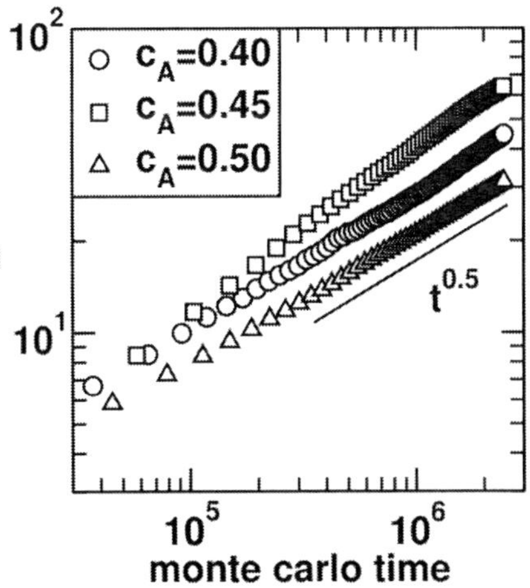

Figure 3. Time-dependence of the characteristic length scales L of M-field for various concentration of A component obtained from accelerated MC simulation. The system size is 2048^2 and quench temperature $T = 0.9$ for all the mixtures. Solid line with exponent $\frac{1}{2}$ is shown, corresponds to growth due to the diffusion process. Clearly, the growth exponent is same irrespective of asymmetry in compositions.

4. Conclusion

Let us conclude this paper with the summary and discussion of our results. We carried out phase ordering kinetics for symmetric and asymmetric binary mixtures via antiferromagnetic Ising model with Kawasaki spin-exchange dynamics. The equal-time correlation functions of staggered magnetization field for symmetric and asymmetric composition show data collapse, indicating the morphological similarity of domain growth. In the limit $k \to \infty$, structure factor tails decay as k^{-3} for both symmetric and asymmetric mixtures, implying sharp interfaces formed between two degenerate phase of staggered magnetization field irrespective of the composition. The universal domain growth law: $L \sim t^{1/2}$ for order-disorder transition in symmetric and asymmetric mixtures is obtained from accelerated MC algorithm. These results confirm the universality in the ordering of symmetric and asymmetric binary mixtures.

5. References

[1] *Kinetics of Phase Transitions*, edited by Puri S. and Wadhawan V. (CRC Press, Boca Raton, 2009).
[2] Dattagupta S. and Puri S., *Dissipative Phenomena in Condensed Matter: Some Applications* (Springer-Verleg Heidelberg 2004).
[3] Bray A.J., Adv. Phys. **43**, 357 (1994).
[4] Kessler M., Dieterich W. and Majhofer A., Phys. Rev. B **67**, 134201 (2003).
[5] Feldman E. P., Stefanovich L. I. and Gumennyk K. V., J. Stat. Phys. **132** 501 (2008).
[6] Sahni P. S., Dee G., Gunton J. D., Phani M., Lebowitz J. L. and Kalos M., Phys. Rev. B **24**, 410 (1981).
[7] Grest G. S., Safran S. A. and Sahni P. S., J. Appl. Phys. **55**, 2432 (1984).
[8] Cahn J. W. and Allen S. M., Acta Metall. **27**, 1085 (1979).
[9] Hashimoto T., Nishimura K. and Takeuchi Y., Phys. Lett. **65**, 250 (1978).
[10] Phani M. K. and Lebowitz J. L., Phys. Rev. Lett. **55**, 366 (1980).
[11] Shannon, Jr. R. F., Harkless C. R. and Nagler S. E., Phys. Rev. B **38**, 9327 (1988); Nagler S. E., Shannon, Jr. R. F., Harkless C. R., Singh M. A. and Nicklow R. M., Phys. Rev. Lett **61**, 918 (1988).
[12] Lippiello E., Mukherjee A., Puri S. and Zannetti M., Euro. Phys. Lett. **90**, 46006 (2010).
[13] Corberi F., Lippiello E., Mukherjee A., Puri S. and Zannetti M., Phys. Rev. E. **88**, 042129 (2013).
[14] Porta M. and Castán T., Phys. Rev. B **54**, 166 (1996).
[15] Sanati M. and Zunger A., J. Phys.: Condens. Matter **19** 086201 (2007).
[16] Marko J. F. and Barkema G. T., Phys. Rev. E. **52**, 2522 (1995).
[17] Porod G., in *Small-Angle X-Ray Scattering*, Glatter O. and Kratky O. (eds.), (Academic Press, New York, 1982, 42).
[18] Oono Y. and Puri S., Mod. Phys. Lett. B **2** 861 (1988).
[19] Ohta T., Jasnow D. and Kawasaki K., Phys. Rev. Lett. **49** 1223 (1982).
[20] Puri S., Bray A. J. and Lebowitz J. L., Phys. Rev. E. **56**, 758 (1997).

Journal of Physics: Conference Series **759** (2016) 012011 doi:10.1088/1742-6596/759/1/012011

Disorder induced hysteresis in diluted kinetic Ising model

Sourav Chattopadhyay* **and S B Santra**

Department of Physics, Indian Institute of Technology Guwahati, Guwahati-781039, Assam, India.

E-mail: c.sourav@iitg.ernet.in*

Abstract. Dynamic phase transition in diluted kinetic Ising model in two dimensions has been studied in presence of external time dependent oscillating magnetic field applying Glauber Monte Carlo spin flip dynamics. A dynamic phase transition from a dynamically ordered to dynamically disordered ferromagnet is identified studying dynamic order parameter against temperature for different concentrations of disorder. For a given temperature, field strength and frequency for which there was no hysteresis, it is observed that disorder is able induce hysteresis in the system. Effect of increasing concentration of disorder on hysteresis loop area has also been studied for different amplitude and time period of the external field.

1. Introduction

Disordered soft magnetic materials have a wide range of industrial applications because of the property of fast magnetization reversal with minimal magnetic loss [1, 2]. Important desirable technical properties for soft magnetic materials include high permeability, low hysteresis loss, large saturation and remnant magnetization and high Curie temperature. Manipulation of chemical composition and optimization of microstructure are required to achieve all such properties in a single soft magnetic material. Though large amount of experimental effort has been devoted in the recent past to develop such materials [3], theoretical understanding of the magnetic properties of these system remain incomplete. Disordered magnetic materials under time independent magnetic field provide a rich static critical phenomena [4]. Dynamic phase transition and hysteresis effect in spin-1/2 Ising model without disorder have been studied using mean-field theory as well as employing Monte Carlo (MC) simulation technique and reported in [5–7]. However dynamic phase transition in disordered magnetic materials are rarely reported. Disordered magnetic material can be modeled by replacing magnetic elements randomly by non-magnetic elements and their magnetic properties under static as well as dynamic conditions can be obtained as function of disorder concentration.

In this paper, dynamic phase transition and hysteresis effect in two dimensional site diluted (or disordered) spin-1/2 Ising system using Glauber Monte Carlo stochastic spin flip dynamics are reported for different frequencies, amplitudes of the external magnetic field and concentrations of disorder.

Journal of Physics: Conference Series **759** (2016) 012011 doi:10.1088/1742-6596/759/1/012011

2. Model

A diluted kinetic Ising model (DKIM) is developed on a two dimensional square lattice of size L. A site is called disordered site if it is not occupied by a magnetic ion. Disordered sites are assigned randomly throughout the system with disorder concentration q by calling a random number r for each lattice site. If $r \leq q$ the site is marked disordered. Otherwise the site is regular. All the regular lattice sites are assigned with a Ising spin $S_i = \pm 1$. Each spin can interact with its nearest neighbour spin, if available, under periodic boundary condition with ferromagnetic interaction strength J. The system is studied under an external time varying oscillating magnetic field $H(t)$ given by

$$H(t) = h_0 \sin(2\pi t/\tau) \tag{1}$$

where h_0 and τ are the amplitude and time period of oscillating magnetic field. The Hamiltonian of the DKIM under external field $H(t)$ is given by,

$$\mathcal{H} = -J \sum_{\langle ij \rangle} c_i c_j S_i S_j - \mu H(t) \sum_i c_i S_i \tag{2}$$

where the occupation index $c_i = 0$ corresponding to a disordered site and $c_i = 1$ represents a site that is occupied with a spin.

In order to equilibrate the system with the heat bath at temperature T, the system is evolved without magnetic field following Glauber Monte Carlo (MC) single spin flip dynamics. The spin flip acceptance ratio W according to the Glauber MC dynamics is given by

$$W(S_i \rightarrow -S_i) = \begin{cases} \frac{e^{-\Delta E/k_B T}}{1+e^{-\Delta E/k_B T}} & \text{if } \Delta \mathrm{E} > 0 \\ 1 & \text{Otherwise} \end{cases} \tag{3}$$

where ΔE is the change in energy due to spin flip. One MC step includes checking and updating of all the spins in the lattice. The equilibrium states for different temperatures T and disorder concentrations q with zero external field are obtained after 10^5 MC time steps of iteration starting from all spin up states. As the system achieves a constant magnetization in equilibrium at a given temperature T for a given q, the external time dependent magnetic field $H(t)$ is switched on. A spin is chosen randomly and updated using Glauber acceptance ratio W in which ΔE is calculated using the full Hamiltonian. A MC step is taken as one unit of time. One cycle of the magnetic field of time period τ then consists of τ MC steps. To equilibrate the system with external magnetic field $H(t)$, the system is evolved further over 256 cycles. Data collection starts only after the dynamical equilibrium in the system is achieved.

3. Results and Discussions

Simulations are performed on a square lattice of size $L = 256$. All the dynamical quantities are averaged over 1024 cycle for 8 different initial disorder configurations for a given q and other macroscopic conditions. Any average mentioned below then corresponds to cycle average as well as configuration average. In the simulation $J = 1$ has been taken as the simplest choice. The temperature is measured in the unit of J/k_B where k_B is the Boltzmann constant. Magnetic field intensity is measured in the unit of J/μ where μ is the magnetic moment of an individual spin. For a given disorder concentration q, the magnetic state of the system is defined by the values of field amplitude h_0, field time period τ and temperature T.

The average instantaneous magnetization $\langle m(t) \rangle$ of the system under a time varying magnetic field $H(t)$ can be estimated as

$$\langle m(t) \rangle = \frac{1}{N_s} \sum_{k=1}^{N_s} \left\{ \frac{1}{N_c} \sum_{j=1}^{N_c} m_{jk}(t) \right\} \quad \text{where} \quad m_{jk}(t) = \frac{1}{L^2} \sum_i c_i S_i, \tag{4}$$

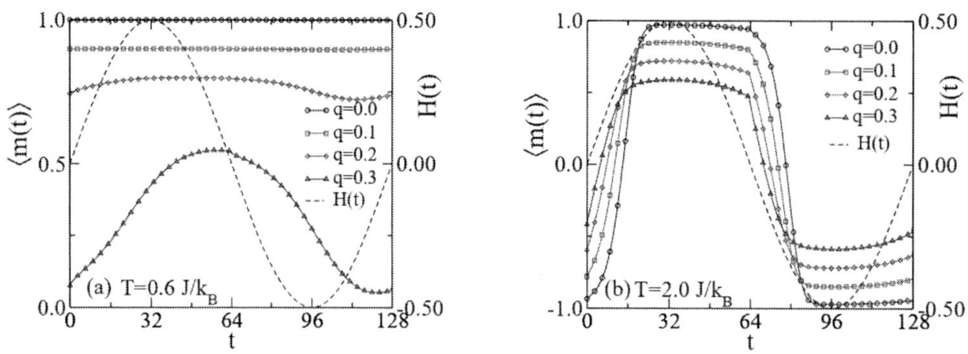

Figure 1. Plot of $\langle m(t)\rangle$ against MC step t for different q for fixed $h_0 = 0.5$ and $\tau = 128$ are shown at (a) $T = 0.6$ and (b) $T = 2.0$ over a complete cycle of $H(t)$. In both the figures variation of $H(t)$ against t is shown in dashed line.

the instantaneous magnetization during jth cycle on kth configuration and $N_s = 8$ is the number of configurations, $N_c = 1024$ are cycles per configuration. Variation of $\langle m(t)\rangle$ over a complete cycle ($\tau = 128$) for different values of q are shown in Fig.1 at two different temperatures (a) $T = 0.6$ and (b) $T = 2.0$. It is found that at $T = 0.6$ and low disorder concentration $q = 0$ or 0.1, $\langle m(t)\rangle$ has almost a constant value and remain unaffected over the cycle of $H(t)$. From $q \geq 0.2$, $\langle m(t)\rangle$ follows $H(t)$ with certain phase lag. Due to slow relaxation, $\langle m(t)\rangle$ lags behind $H(t)$. As q increases $\langle m(t)\rangle$ gets more and more influenced by $H(t)$. At a fixed T, h_0 and τ for any nonzero value of q, some of the spins in the system lose one or more of their neighboring spins. As a result a spin with one or more disordered site in its neighbour becomes weakly coupled. In such a situation thermal energy can randomize the ferromagnetic order easily where it was not possible at zero or little disorder. Hence for higher disorder, $\langle m(t)\rangle$ shows better response to the field. On the other hand, at a higher temperature $T = 2.0$ for any q the thermal energy wins over ferromagnetic coupling and randomize the system. The magnetization $\langle m(t)\rangle$ follows the external field $H(t)$ and oscillate symmetrically around zero with certain phase lags depending on disorder concentration q. With increase in q, the maximum of $\langle m(t)\rangle$ decreases as total number of spins in the system decreased.

Depending on temperature T, the system is found to belong to two distinct dynamical phases, one with a positive finite $\langle m(t)\rangle$ and other with $\langle m(t)\rangle$ oscillating about zero. To study the dynamical phase transition in the system, a dynamical order parameter $\langle Q\rangle$ is defined and studied as a function of temperature T for various values of q. The order parameter Q over a full cycle of the magnetic field is defined as,

$$\langle Q\rangle = \left\langle \frac{1}{\tau}\int_0^\tau m(t)dt \right\rangle \tag{5}$$

where $\langle Q\rangle$ is function of q, T, h_0 and τ. In Fig.2(a), $\langle Q\rangle$ is plotted against T for different values of q taking $h_0 = 0.5$ and $\tau = 128$. For a given q, there exists a transition temperature T_c at which the order parameter vanishes [8]. Though the qualitative behavior of $\langle Q\rangle$ versus T remains same for all qs, the values of the saturation order parameter and the critical temperature T_c are found to decrease as q is increased. The decrease in saturation value of $\langle Q\rangle$ is expected as the magnetization per lattice site depends on total number of magnetic spins present in the system. However, the decrease in T_c is due to the fact that some of the spins nearby the disordered sites becomes weakly coupled with the rest of the spins. A dynamical phase transition is then found to occur in diluted spin systems too. Below T_c it corresponds to dynamically ordered phase (DOP) and above T_c it corresponds to dynamically disordered phase (DDP). The nature of the

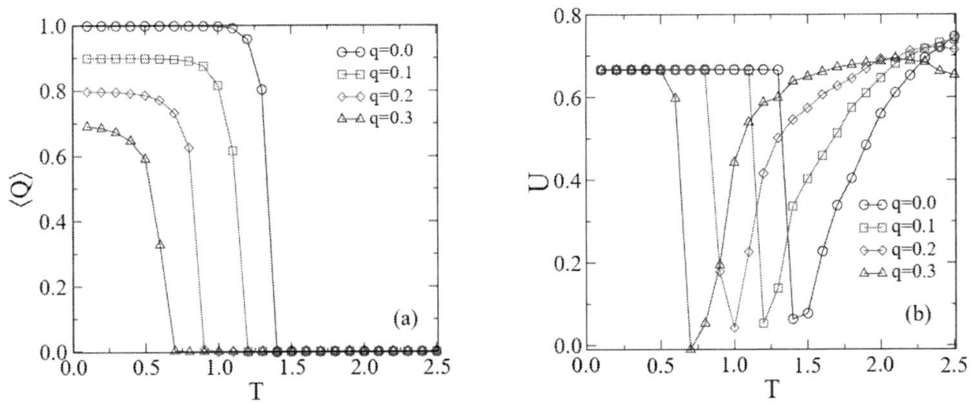

Figure 2. Plot of (a) dynamic order parameter $\langle Q\rangle$ and (b) Binder cumulant U against T for different concentrations of disorder q for $h_0 = 0.5$ and $\tau = 128$.

transition is asserted calculating Binder Cumulant U defined as,

$$U = 1 - \frac{\langle Q^4\rangle}{3\langle Q^2\rangle} \tag{6}$$

where $\langle Q^2\rangle = \langle \int_0^\tau m^2(t)dt/\tau\rangle$ and $\langle Q^4\rangle = \langle \int_0^\tau m^4(t)dt/\tau\rangle$ are the second and the fourth moment of magnetization. In Fig.2(b) Binder Cumulant U is plotted with temperature T. Around the transition temperature T_c, there is a sudden dip in U for all values of qs which is a signature of a first order phase transition in contrast to the fact that U decreases continuously with temperature T and goes to zero in second order phase transition [9, 10].

Since appearance of hysteresis loop is often associated with a first order phase transition, a through study of the variation of hysteresis loop is performed over a wide range of parameter values. Since in DDP, $\langle m(t)\rangle$ varies symmetrically about zero value with $H(t)$, the hysteresis loop is also expected to be symmetric about $\langle m\rangle = 0$ line [11]. On the other hand, in DOP, $m(t)$ does not vary symmetrically about zero value with $H(t)$. Hence, the loop is expected to be asymmetric about $\langle m\rangle = 0$ line [11]. In Fig.3, $\langle m(t)\rangle$ is plotted against $H(t)$ over a complete cycle of $H(t)$ for different parameter values. Keeping $h_0 = 0.5$, $\tau = 128$ and $q = 0.1$ fixed, the variation in the hysteresis loop with T is shown in Fig.3(a). As expected, the nature of the hysteresis loop changes from asymmetric to symmetric about $\langle m\rangle = 0$ line as T passes through T_c. Similar variations in the hysteresis loop with h_0 in Fig.3(b), with q in Fig.3(c) and with τ in Fig.3(d) are shown. Apart from T_c for fixed values of h_0, τ and q, critical values of h_0, τ and q (for fixed values of other parameters) can also be identified monitoring the change in the shape of the hysteresis loop. As T the thermal energy is responsible to destroy the ferromagnetic order. However, as h_0 is increased for a fixed T is increased, the system acquires energy from the external field to overcome the ferromagnetic order. At fixed T and h_0, as the value of τ increases the hysteresis loop area is also increases resulting more and more loss of magnetic energy which eventually allow the thermal energy to randomize the spin orientation. On the other hand, as q is increased, the system becomes more and more ferromagnetically weakened. Hence a transition to DDP is possible increasing q at a fixed T and h_0 or τ.

Beside the change in the shape of the hysteresis loop, the loop area A is also found to vary with system parameters. The hysteresis loop area A is defined as

$$A = -\oint m(t)dH = -\frac{2\pi h_0}{\tau}\int_0^\tau m(t)\cos(2\pi t/\tau)dt, \tag{7}$$

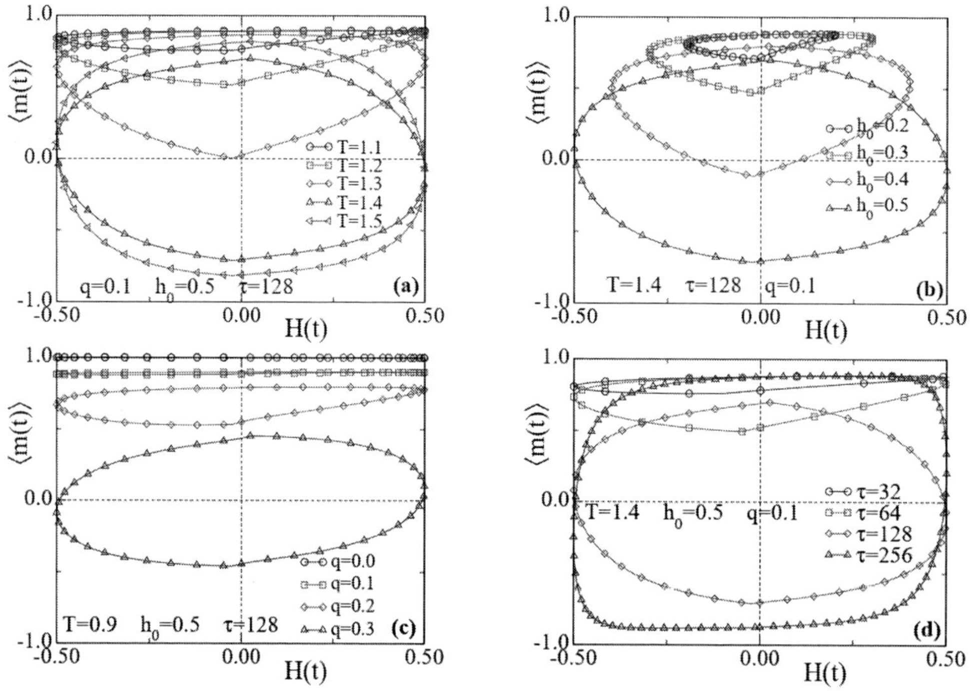

Figure 3. Plot of $\langle m(t)\rangle$ versus $H(t)$ over a complete cycle, (a) for different T at fixed $q = 0.1$, $h_0 = 0.5$ and $\tau = 128$, (b) for different h_0 at fixed $T = 0.9$, $\tau = 128$, $q = 0.1$, (c) for different values of q at fixed $T = 0.9$, $h_0 = 0.5$, $\tau = 128$ (c) for different values of τ at fixed $T = 0.9$, $h_0 = 0.5$, $q = 0.1$.

which is a measure of the loss in magnetic energy over a cycle. In Fig.4(a), A is plotted against T for different values of q. For a given q, A increases from zero to a maximum value and then decreases to a small finite value as T is increased starting from a low value. At low temperature $m(t)$ does not follow $H(t)$ and remains constant at its saturation value leading to no hysteresis loop. As T increases, $m(t)$ starts following $H(t)$ and loop area starts increasing. At the dynamical phase transition temperature T_c, a maximum change in A [11] occurs as shown in Fig.4(b). The loop area increases further with the increase of temperature as the system becomes more susceptible with external field. However, further increase of temperature beyond the temperature corresponding to maximum loop area the system becomes more and more paramagnetic and reduces the loop area. For the present field strength, a complete paramagnetic phase is not achieved in the temperature range used. In presence of disorder, the ferromagnetic order can be overcame at lower temperature and the peak of A shifts towards lower values of T. At the same time, the maximum value of A decreases as q is increased due to scarcity of spins in the system.

4. Conclusion

Dynamical phase transition in a diluted kinetic Ising model is studied for various different values of dilution of spins. Dynamical order parameter Q vanishes at the transition point from DOP to DDP. The critical temperature T_c of the transition found to depend not only on h_0 and τ but also is found to depend strongly on q, amount of spins dilution in the system. Spin dilution makes the system magnetically softer and a dynamical transition occurs at a lower and lower T as dilution increases. Hysteresis loop is found to change its shape from asymmetric to symmetric about zero

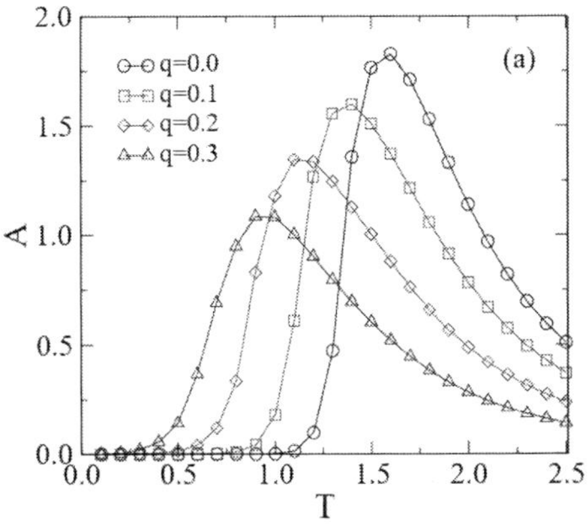

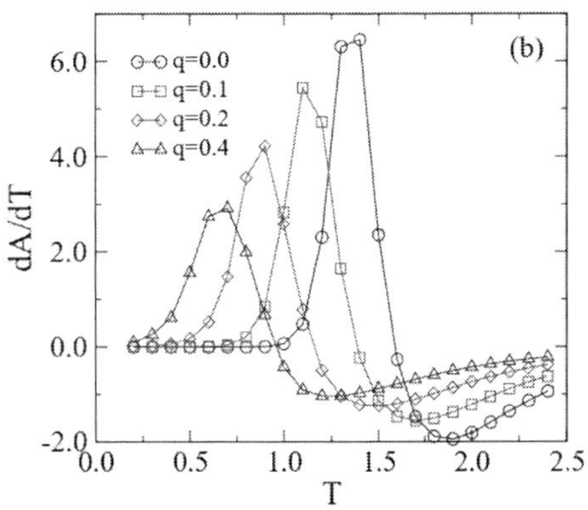

Figure 4. Variation of (a) loop area A and (b) dA/dT, are shown against T for different values q for $h_0 = 0.5$ and $\tau = 128$.

magnetization line around T_c along with vanishing dynamical order parameter. Beside decrease in T_c, dilution is found to have strong effect on magnetic hysteresis. For a given T, h_0 and τ, not only hysteresis loop area is found to increase with q but also the remnant magnetization is found to increase with q.

References

[1] McHenry M E, Willard M A and Laughlin D E 1999 *Prog. Mat. Sci.* **44** 291 – 433

[2] Hasegawa R 2006 *J. Mag. Mag. Mat.* **304** 187 – 191

[3] Hasegawa R 2004 *Mat. Sci. Eng. A* **375-377** 90 – 97

[4] Kaul S N 1985 *J. Mag. Mag. Mat.* **53** 5 – 53

[5] Rao M, Krishnamurthy H R and Pandit R 1990 *Phys. Rev. B* **42**(1) 856–884

[6] Chakrabarti B K and Acharyya M 1999 *Rev. Mod. Phys.* **71**(3) 847–859

[7] Acharyya M 2005 *Int. J. Mod. Phys. C* **16** 1631–1670

[8] Tomé T and de Oliveira M J 1990 *Phys. Rev. A* **41** 4251

[9] Sides S W, Rikvold P A and Novotny M A 1998 *Phys. Rev. Lett.* **81**(4) 834–837

[10] Park H and Pleimling M 2013 *Phys. Rev. E* **87**(3) 032145

[11] Acharyya M 1998 *Phys. Rev. E* **58**(1) 179–186

Search for the Heisenberg spin glass on rewired cubic lattices with antiferromagnetic interaction

Tasrief Surungan

Department of Physics, Hasanuddin University, Makassar 90245, Indonesia

E-mail: tasrief@unhas.ac.id

Abstract. Spin glass (SG) is a typical magnetic system which is mainly characterized by a frozen random spin orientation at low temperatures. Frustration and randomness are considered to be the key ingredients for the existence of SGs. Previously, Bartolozzi *et al.* [Phys. Rev. B73, 224419 (2006)] found that the antiferromagnetic (AF) Ising spins on scale free network (SFN) exhibited SG behavior. This is purely AF system, a new type of SG different from the canonical one which requires the presence of both FM and AF couplings. In this new system, frustration is purely due to a topological factor and its randomness is brought by irregular connectivity. Recently, it was reported that the AF Heisenberg model on SFN exhibited SG behavior [Surungan *et al.*, JPCS, 640, 012005 (2015)/doi:10.1088/1742-6596/640/1/012005]. In order to accommodate the notion of spatial dimension, we further investigated this type of system by studying an AF Heisenberg model on rewired cubic lattices, constructed by adding one extra bond randomly connecting each spin to one of its next-nearest neighbors. We used Replica Exchange algorithm of Monte Carlo Method and calculated the SG order parameter to search for the existence of SG phase.

1. Introduction

Spin glass (SG) is a fascinating subject in condensed matter physics. It is a random magnetic system having a unique type of order called temporally ordered phase [1] rather than spatially ordered phase. This type of order is associated with frozen random spin configuration at low temperature. From the theoretical point of view, SG is a perfect ground test for several complex problems such as NP-hard and optimization algorithm [2, 3]. SGs have found some technological applications, for example, as a material for giant magneto-resistance (GMR)[4]. The phenomenon of SGs was firstly observed in experiment by Cannella and Mydosh in the early 70s [5]. They reported the presence of a cusp instead of a sharp peak in the plot of ac-susceptibility of transition metal impurities hosted by nobel metals ($CuMn$ and $AuFe$). These are magnetic alloy with a small concentration of magnetic impurities (Mn and Fe) randomly substituted into the lattice of non-magnetic hosts (Cu and Au).

Frustration and randomness are considered to be the pivotal ingredients for the existence of SG. Spins are frustrated if they fail to simultaneously satisfy their nearest neighbors for minimizing the interaction energy. The example of this is a plaquette where both ferromagnetic (FM) and antiferromagnetic (AF) interaction exist, or by an AF triangular unit. Spins interacting ferromagnetically (anti-ferromagnetically) will prefer to be aligned (anti-aligned) to minimize the free energy.

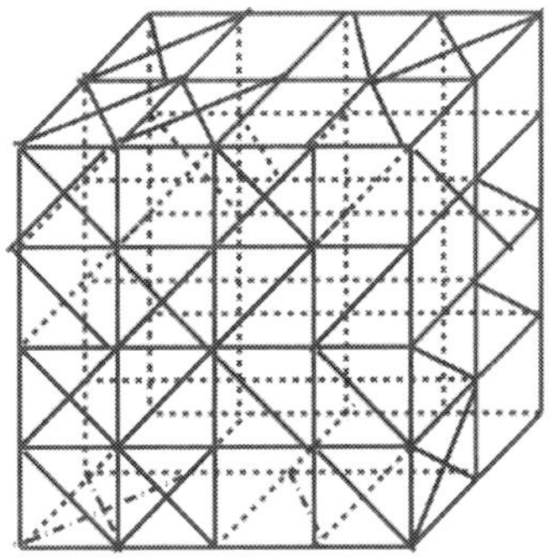

Figure 1. (Color online) An example of a rewired cubic lattice with an extra link randomly connects each spin to one of the next-nearest neighbors.

SGs exhibit many such remarkable phenomena as aging, rejuvenation and memory effects [6], which have been the subject of some recent studies[7, 8]. Due to the randomness and frustration, the ground state configuration is highly degenerate, leads the system to possess finite value ground state entropy, commonly called residual entropy. It is also due to the random frustration that the energy landscape of the system is occupied by a large number of local minima. Accordingly, finding the ground state energy of the systems is extremely difficult and has been a challenge for many optimization algorithms such as genetic and hysteresis algorithms [9, 10].

The so-called non-canonical SG has recently attracted a lot of interest [11, 12, 13, 14]. This is a new type of SGs with purely AF where randomness is in connectivity and the frustration is brought by the topological factor. Its existence was firstly reported on scale free network (SFN) for the Ising model[11]. We have studied this type of model for Heisenberg case and found SG phase transition[13]. As an SFN corresponds to a lattice with very large spatial dimension and due to the importance of dimensionality in phase transition, it is desirable to find the existence of SG phase on finite dimension regular lattices. While our recent study on rewired square lattice found no SG for Heisenberg model [14], we observed SG for the Ising case[15].

We further elaborate this type of models on higher dimensional lattice, the rewired cubic lattice. It is motivated by the notion that larger coordination number should preserve any existing order phase including SGs. The remaining part of the paper is organized as follows: Section II describes the model and method. The result is discussed in Section III. Section IV is devoted to the summary and concluding remarks.

2. Model and Simulation

The SG model is written with the following Hamiltonian

$$H = -\sum_{\langle ij \rangle} J_{ij}\ \vec{s}_i \cdot \vec{s}_j \tag{1}$$

where J_{ij} is the coupling interaction between spins residing on sites i-th and j-th of the lattice. For a purely AF model, $J_{ij} = J < 0$. We consider a rewired cubic lattice with one extra link added to each site. The link connects each spin to one of the next-nearest spins. Therefore, a spin can have in maximum up to 13 links. The summation is performed over all directly connected spins of the rewired lattice. Figure 1 shows one particular realization of the lattices. A randomly rewired cubic lattice is a quasi-regular lattice where each site has different number of neighbors. We define an average number of links (ANOL) for this type of lattice, associated with a coordination number n for a regular lattice. As we add only one extra link to each site of the cubic lattice ($n = 3$), then for the resulted lattice, ANOL = 4. In principle we can generate any lattice with fractional ANOL. As shown in Fig. 1, a large number of triangular units exist, on which spins are frustrated due to AF interaction. Randomly adding extra links is a simple way of transforming a purely AF system into a non-canonical SG model. This type of procedure had been done for example in Ref. [12]. We do not follow this procedure as it breaks totally the translational symmetry of regular lattices, while we want to preserve the notion of spatial dimension.

Monte Carlo simulation is a standard numerical method to study complex systems such as SGs. Due to the presence of local minima in the energy landscape of the system, the performance of the conventional Metropolis algorithm is poor. A random walker can easily get trapped in one of the local minima and suffers from critical slowing down, particularly when approaching critical temperature. Cluster algorithms are commonly used to overcome this phenomenon. However, for SG system, due to the presence of frustration, a cluster algorithm is not applicable. Instead, one implements extended ensemble algorithms, *i.e.*, Replica Exchange (RE) algorithm [16].

The basic idea of RE algorithm is to avoid a random walker being trapped in one local minimum by duplicating the original system into several replicas. Each replica is in contact with a corresponding heat bath and independently treated by a standard Metropolis algorithm. Then, the simulation is run in parallel where replicas of two adjacent temperatures are exchanged. Configurations of pairs in two consecutive trials of exchanges have to be different. For example, if we start with M replicas and a set of invers temperatures β, the first trial of exchange involves pairs of replicas: X_1 and X_2, X_3 and $X_4, \cdots X_{M-1}$ and X_M; then followed by the next trial which involves pairs: X_2 and X_3 and X_4 and $X_5 \cdots X_{M-2}$ and X_{M-1}. This is important to enable any random walker to escape from any local minimum. An exchange can be performed, for example, in every 5 MCSs.

The final result of the simulation is obtained by combining all results from each replica (temperature). The probability distribution of finding the whole M replicas in a state $\{X\} = \{X_1, X_2, \ldots, X_M\}$ is given by,

$$P(\{X,\beta\}) = \prod_{m=1}^{M} \frac{\exp(-\beta_m H(X_m))}{Z(\beta_m)} \tag{2}$$

where $Z(\beta_m)$ is the partition function for the m-*th* replica. We can then define an exchange matrix between replicas, $W(X_m, \beta_m | X_n, \beta_n)$, which is the probability to switch the configuration X_m of β_m with the configuration X_n of β_n. With the requirement of keeping the entire system at equilibrium, we use the detailed balance condition on the transition matrix

$$\begin{aligned} &P(\{X_m,\beta_m\},\ldots,\{X_n,\beta_n\},\ldots)\cdot W(X_m,\beta_m|X_n,\beta_n) \\ &= P(\{X_n,\beta_m\},\ldots,\{X_m,\beta_n\},\ldots)\cdot W(X_n,\beta_m|X_m,\beta_n), \end{aligned} \tag{3}$$

along with Eq. (2), so that we have

$$\frac{W(X_m,\beta_m|X_n,\beta_n)}{W(X_n,\beta_m|X_m,\beta_n)} = \exp(-\Delta), \tag{4}$$

where $\Delta = (\beta_n - \beta_m)(H(X_m) - H(X_n))$. With this constraint, we can choose the matrix coefficients according to the standard Metropolis algorithm which gives the following

$$W(X_m,\beta_m|X_n,\beta_n) = \begin{cases} 1 & \text{if } \Delta < 0, \\ \exp(-\Delta) & \text{if } \Delta > 0. \end{cases} \tag{5}$$

Due to the fact that the acceptance ratio decays exponentially with $(\beta_n - \beta_m)$, we restrict the exchange temperatures next to each other, i.e., the terms $W(X_m, \beta_m | X_{m+1}, \beta_{m+1})$. In performing the simulation, we define one MC step (MCS) as updating once each spin of each

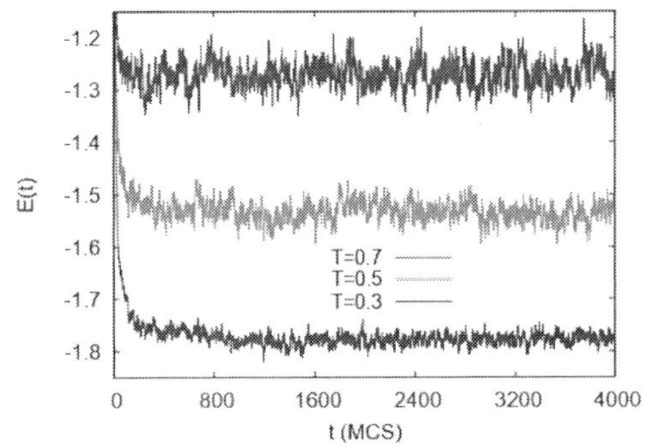

Figure 2. (Color online) Time series of energy at $T = 0.7, 0.5$ and 0.3 for linear size $L = 10$.

replica, either consecutively or randomly. After a number of MCSs, we exchange configurations from pairs of temperatures next to each other based on the probability according to Eq. 5.

For each connectivity realization we start from a random spin configuration. Then, we perform 10^4 MCSs for equilibration before taking a total of 10^4 sample points for the thermal averages. To avoid temporal correlation between samples, we performed several extra MCSs between two sample points. We consider high and low temperatures to search for SG phase transition. The thermal averages obtained for each lattice size then averaged over many lattice connectivities, which is a standard procedure in doing MC simulation for a random system such as SG. In the next section, we present the results of our study.

3. RESULTS AND DISCUSSION

3.1. Energy and the specific heat

We have simulated AF Heisenberg model on rewired cubic lattices with ANOL = 4.0 for several linear sizes $L = 8, 10, 12$, and 16; the number of spins is L^3. The periodic boundary condition is imposed so that each site of the native cubic lattice (before being rewired) has six nearest neighbors. For each system size we took many realizations of connectivities, then average the results over the number of realizations. Each realization corresponds to one particular connectivity distribution randomly generated. For the result to be reliable, we have to take a reasonable number of realizations. In our previous study of Heisenberg SG on SFN, we took 1000 realizations. Here we took smaller number of realizations ($N_r = 144$) due to its degree of randomness is less compared to the previous system.

To check the reliability of our simulation we evaluated the energy time series of the system, from which the average energy and specific heat can be extracted. One can tell whether the system is in equilibrium or not by analyzing the energy time series. In MC simulation, time

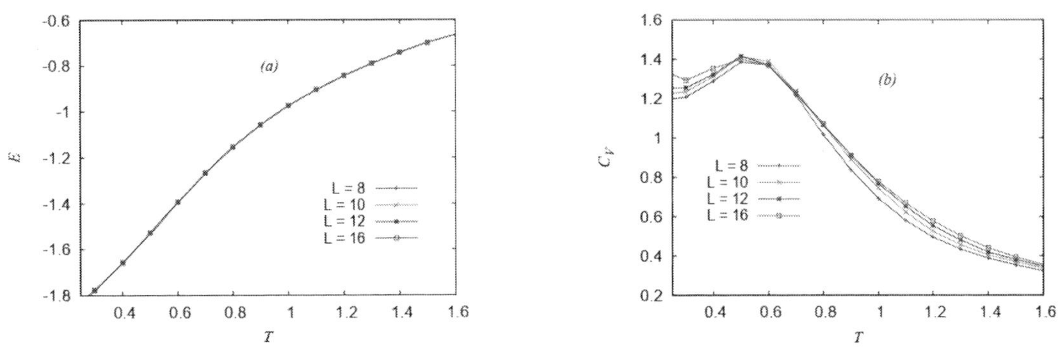

Figure 3. (Color online) (a) Temperature dependence of average energy and (b) specific heat for different sizes.

corresponds to a series of MCSs. We perform M_s MCSs for each temperature and take N_s samples out of M_s. To make sure the system is well equilibrated, we perform enough initial M_i MCSs, usually M_i around 10^4 MCSs, before doing measurement.

The energy time series (ETS) of three different temperatures, *i.e.*, $T = 0.7, 0.5$ and $T = 0.3$, were taken for linear size $L = 10$. As shown in Fig. 2, the system reaches equilibrium after several MCSs, depending on temperature, which is around 2000 for $T = 0.3$. Fluctuation at larger system size is more pronounced compared to smaller system size. We extracted two quantities from ETS, namely the ensemble average of energy, $\langle E\rangle = \frac{1}{N_s}\sum_{N_s} E_i$, and the specific heat defined as follows

$$C_v = \frac{N}{kT^2}\left(\langle E^2\rangle - \langle E\rangle^2\right) \tag{6}$$

where $N = L^3$ and k are respectively the number of spins and Boltzmann constant. These quantities are shown in Fig. 3(a) and (b).

As the specific heat plot has a constant peak, which corresponds to critical exponent $\alpha = 0$, it can not precisely signify the presence, neither the absence of spin glass phase transition in the system. A more convincing evidence for the existence or absence of SG phase is obtained from the SG order parameter which is presented in the next sub-section.

3.2. Spin Glass Order Parameter

The exitence of SG phase is characterized by the follwing order parameter [17]

$$q_{EA} \equiv \left[\langle |\frac{1}{N}\sum_i \vec{s}_i^{(\alpha)} \otimes \vec{s}_i^{(\beta)}|\rangle\right]_{\mathrm{av}}, \tag{7}$$

where $\vec{s}_i^{\alpha}$ and $\vec{s}_i^{\beta}$ are two sets of vector spins from different replicas having the same structure of connectivity. The essence of this parameter is to capture the frozen configuration in SG phase; which is merely introduced for numerical simulation. In real experiment it is almost impossible to obtain two duplicated systems with exactly the same realization of randomness. It is to be noticed that we have additional replicas for the calculation of overlapping parameter, apart from replicas related to the RE method.

The overlapping parameter is much simpler for Ising spins compared to the vector spins which involves tensor product [13, 14]. If system is frozen, their overlapping parameter will give finite value and vanish otherwise. The plot of temperature dependence of q_{EA} for different sizes, $L = 8, 10, 12$, and 16 is shown in Fig. 4. As indicated, the parameter is increasing as temperature decreases. However, the finite values of the parameter at lower temperature decrease as system

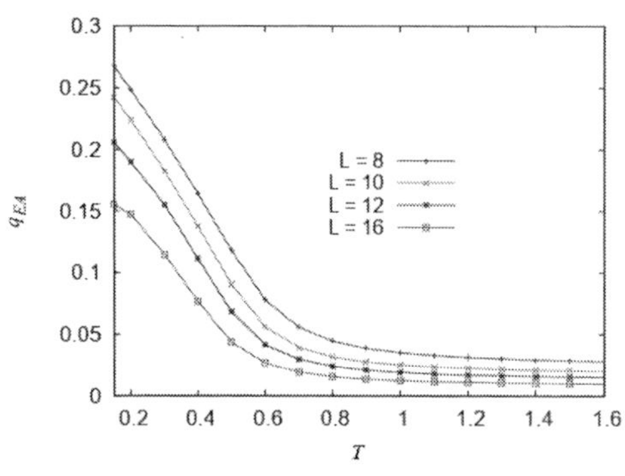

Figure 4. (Color online) Temperature dependence of the overlapping order parameter for various system sizes.

sizes increase. This is the indication of the absence of SG phase on the rewired cubic lattice with ANOL = 4.0 in the thermodynamic limit Therefore, it is desirable to study this type of model with larger value of ANOL, associated with system with larger spatial dimension. This argument is comparable with the case in the canonical SG model where the existence of SG phase on cubic lattice remains a controvercy, while convincingly found in the structure with four spatial dimensions [18].

4. SUMMARY AND CONCLUSION

We have studied antiferromagnetic Heisenberg model on rewired cubic lattices with average number of links (ANOL) or connectivity density 4.0; obtained by adding one extra link to each site of the cubic lattice. Each added link is constrained to connect to one of the next-nearest neighbors of each site. We use Replica Exchange Monte Carlo method which is considered to be very powerful in dealing with complex systems such as SGs. We calculate several physical observables such as energy, specific heat and overlapping SG order parameter.

We observed no finite temperature SG phase transition which is indicated by the decreasing value of the overlapping parameter q_{EA} as the system sizes increase at low temperatures. This result suggested that the AF rewired cubic lattice with ANOL=4 can not host the non-canonical Heisenberg SG phase. With our previous report on the existence of this type of SG on SFN[13], which corresponds to regular lattice with very large spatial dimension, the quest for finding lower critical dimension or ANOL for this system still remains. The study of the system with larger ANOL is required. A systematic study of system with larger fractional ANOL is still in progress. The results will be reported elsewhere.

Acknowledgement

The author wishes to thank Y. Okabe, Md. A. Khan, S. B. Santra and P. Ray for valuable discussions. The computation of this work was performed using parallel computing facility in the Department of Physics Hasanuddin University and the HPC facility of Indonesian Institute of Science. The work is supported by HIKOM (Hibah Kompetensi) research grant 2015 of the Indonesian Ministry of Research, Technology and Higher Education.

References

[1] H. Nishimori, Statistical Physics of Spin Glasses and Information Processing, Oxford Univ. Press, (2001).
[2] F. Barahona, J. Phys. A 15, 3241 (1982).
[3] N. Kawashima J. Phys. Soc. Jpn., Vol. 75, No. 7, 073002, (2006).
[4] B. Idzikowski et al., Europhys. Lett. **45**, 714, (1999)/ doi:10.1209/epl/i1999-00226-8.
[5] V. Cannella and J. A. Mydosh, Phys. Rev. B, 6,4220 (1972).
[6] V. Dupuis et al., Phys. Rev. B 64, 174204 (2001).
[7] M. Hudl, R. Mathieu and P. Nordblad, Sci. Rep. 6, Art. No.: 19964 (2016); doi:10.1038/srep19964.
[8] S. V. Ketov et al., Nature 524, 200203, (2015).
[9] S. V. Isakov, I. N. Zintchenko, T. F. Rnnow and M. Troyer, Comput. Phys. Commun., 192, 265, (2015).
[10] J. H. Holland, Adaptation in Natural and Artificial Systems, Univ. of Michigan Press, Michigan. (1975).
[11] M. Bartolozzi, T. Surungan, D.B. Leinweber and A.G. Williams, Phys. Rev. B73, 224419 (2006).
[12] C. P. Herrero, Phys. Rev. E **77**, 041102, (2008).
[13] T. Surungan, F. P. Zen, and A.G. Williams, JPCS 640, 012005 (2015).
[14] T. Surungan, Bansawang BJ, D. Tahir, AIP Conf. Proc. 1719, 030006 (2016); doi:10.1063/1.4943701.
[15] T. Surungan (To be published).
[16] K. Hukushima and K. Nemoto, J. Phys. Soc. Japan 65, 1863 (1996).
[17] S. F. Edwards and P. W. Anderson, J. Phys. F: Metal Phys. **5**, 965, (1975).
[18] B. Coluzzi, J. Phys. A: Math. Gen. 28 , 747 (1995).

The impact of bonded interactions on the ground-state geometries of a small flexible polymer

Tomas Koci[1], **Kai Qi**[1] **and Michael Bachmann**[1,2,3]
[1] Soft Matter Systems Research Group, Center for Simulational Physics, The University of Georgia, Athens, Georgia 30602, USA
[2] Instituto de Física, Universidade Federal de Mato Grosso, 78060-900 Cuiabá (MT), Brazil
[3] Departamento de Física, Universidade Federal de Minas Gerais, 31270-901 Belo Horizonte (MG), Brazil

E-mail: koci@smsyslab.org, qi@smsyslab.org, bachmann@smsyslab.org
Homepage: http://www.smsyslab.org

Abstract. Bonded interactions in coarse-grained models of elastic polymers are commonly represented by the finitely extensible nonlinear elastic (FENE) potential. In this study, we perform parallel multicanonical Monte Carlo simulations to examine the impact of an additional Lennard-Jones term in the bonded potential on the geometry of ground-state structures of a short polymer. Employing microcanonical inflection point analysis and conformational analysis, we construct a hyper-phase diagram and identify ground-state structures with two distinct geometries.

1. Introduction

The physical and chemical functionality of biopolymers and synthetic variants is intimately related to their conformational structure. Therefore, the systematic study of their dynamic and structural properties is of a fundamental importance and a major topic of current interdisciplinary research. Theoretical approaches are generally not adequate due to the prohibitive complexity of even the simplest molecular systems [1]. On the other hand, computational methods such as the replica-exchange [2], multicanonical [3, 4], and Wang-Landau sampling [5], have proven to be most effective. Despite the vast increase in the availability of computational resources over the past decade, simulations of fully atomistic models remain a major challenge. Hence a clever choice of suitable coarse-grained models with a reduced number of parameters is often a prerequisite to a successful computational study. In this paper, we consider a generic model of the flexible elastic homopolymer and examine the effects of adding an extra Lennard-Jones term to the bonded potential. The unexpected consequences for the ground-state structures highlight the importance of a careful choice of model parameters.

2. Model and Methods

We employ a generic model of a flexible elastic homopolymer to study the effects of the shape and width of the bonded potential on the formation of low-energy structures for a polymer chain with $N = 15$ monomers. Finite-size effects are essential and the 15mer shows a particularly distinct and clear transition behavior, which makes it a perfect candidate for this systematic study.

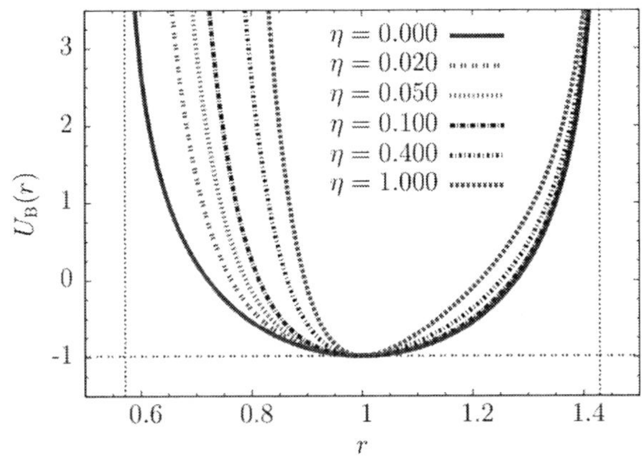

Figure 1. The modified bonded potential $U_B(r)$, modeled by the FENE and LJ interactions. The strength of the Lennard-Jones term is controlled by the bond flexibility parameter $\eta \in [0, 1]$.

Non-bonded interactions are represented by the truncated and shifted Lennard-Jones (LJ) potential

$$U_{\mathrm{LJ}}^{\mathrm{trunc}}(r_{ij}) = \begin{cases} U_{\mathrm{LJ}}(r_{ij}) - U_{\mathrm{LJ}}(r_c), & \text{if } r_{ij} \le r_c, \\ 0, & \text{if } r_{ij} > r_c, \end{cases} \tag{1}$$

where

$$U_{\mathrm{LJ}}(r) = 4\epsilon \left[\left(\frac{\sigma}{r} \right)^{12} - \left(\frac{\sigma}{r} \right)^{6} \right]. \tag{2}$$

The energy scale is set to $\epsilon = 1$ and the van-der-Waals radius to $\sigma = r_0/2^{1/6}$, where $r_0 = 1.0$ is the location of the potential minimum. We select a cut-off radius at $r_c = 2.5\,\sigma$ and introduce a shift $U_{\mathrm{LJ}}(r_c) \approx -0.0163169\,\epsilon$ to avoid discontinuities in the potential. In our model, monomers adjacent in the linear chain interact via a modified bonded potential

$$U_{\mathrm{B}}(r) = -\frac{1}{2} K R^2 \ln \left[1 - \left(\frac{r - r_0}{R} \right)^2 \right] + \eta (U_{\mathrm{LJ}}(r) + \epsilon) - (\epsilon + U_{\mathrm{shift}}), \tag{3}$$

where $R = 3/7$ and $K = 98/5$. In addition to the standard anharmonic FENE (finitely extensible nonlinear elastic) potential [6, 7, 8], the modified potential contains an additional Lennard-Jones term adjusted by a bond flexibility control parameter η. The bonded potential is shifted by $-(\epsilon + U_{\mathrm{shift}})$ in order to match the minimum energy with the non-bonded interactions.

The maximum bond extension is limited by the FENE potential, which diverges as $r \to r_0 \pm R$. Increasing the value of η introduces asymmetry to the bonded potential and the energy cost associated with non-optimal bond lengths is increased. In particular, compressed bonds result in high energy penalties as η becomes large. The effects of different η values on the shape and width of the bonded potential is exhibited in Fig. 1. The total energy of a configuration $\mathbf{X} = (\vec{r}_1, ..., \vec{r}_N)$ with the monomer-monomer distances $r_{ij} = |\vec{r_i} - \vec{r_j}|$ is given by

$$E(\mathbf{X}) = \sum_{i<j}^{N} U_{\mathrm{LJ}}(r_{ij}) + \sum_{i=1}^{N-1} U_{\mathrm{B}}(r_{i\,i+1}). \tag{4}$$

In order to enhance the sampling of low-energy configurations, a parallel version of multicanonical sampling [3, 4, 9] is employed in this simulation. The standard multicanonical runs are performed in K replicas independently with the same initial weight functions but different random seeds. Displacement updates are proposed within a cubic box of edge lengths $d = 0.3r_0$ and accepted according to the probability

$$P(X \to X') = \min[1, W(E(X'))/W(E(X))], \tag{5}$$

where $W(E(X))$ represents the weight function of a given configuration X. After the ith iteration, since the weights are identical in each thread, the energy histograms obtained for each replica can simply be summed up:

$$H^i(E) = \sum_{k=1}^{K} H_k^i(E). \tag{6}$$

The total histograms are combined with the current weights to calculate the weights for the subsequent iteration by utilizing the error-weighted recursive scheme [1, 3, 4].

To construct the hyper-phase diagram, we use the generalized microcanonical inflection-point analysis [1, 10]. This approach has the advantage of uniquely and systematically locating the transition energies and thus is commonly employed to study pseudophase transitions in finite-size systems. By applying the principle of minimal sensitivity [11] to the derivatives of microcanonical entropy $S(E)$, the $(2n+1)$th-order transition (n is a positive integer) is identified from the least sensitive inflection point of the $2n$th-derivative of entropy and the positive valley in the $(2n + 1)$th-derivative curve. For a $2n$th-order transition, the least sensitive inflection point in the $(2n - 1)$th-derivative of entropy together with the negative peak in the $2n$th-order derivative curve are utilized to locate the transition energy. The specialty of the first-order-like transition is that it can be distinguished from the back-bending region in the inverse temperature $\beta(E) \equiv T^{-1}(E) = dS(E)/dE$ and the corresponding positive peak in the $\gamma(E) = d\beta(E)/dE$ curve.

3. Results

3.1. Canonical and microcanonical analysis

First we discuss the results of canonical statistical analysis applied to our generic model. Heat-capacity curves as functions of temperature (Fig. 2(a)) are constructed using the time series of data collected in the multicanonical production run. At $T \approx 0.34$, broad prominent peaks, indicating the freezing transition, are identified for all simulated η values. At this transition, globular structures change to more compact crystalline or amorphous structures. For $\eta = 0$, an additional peak emerges at $T = 0.11$, suggesting the existence of a solid-solid transition. With increasing values of η the peak becomes more pronounced as it gradually shifts towards zero temperature and finally disappears when $\eta \geq 0.1$. However, the order of the individual transitions remains ambiguous and the broad peaks of the freezing transition may envelope several transitions which cannot be resolved by the methods of canonical analysis. Therefore we further examine the system using a more systematic and robust approach. One such method that has proven to reliably signal transitions is the microcanonical inflection-point analysis [10].

The microcanonical results are shown in Fig. 2(b-d) for six different values of η. Careful inspection of the first and second derivatives of $\beta(E)$ in the energy region $E \in [-45.5, -33]$ reveals that the broad peak in the canonical specific heat encloses two distinct transition signals; clear indication that the freezing transition is a two-step process. The first signal located at $E \approx -44$, is a fourth-order transition indicated by the corresponding least sensitive inflection point in $\delta(E)$. The second transition, found at $E \approx -38$, is of third order for $\eta \leq 0.2$, but it is classified as a second-order transition for higher values of the bond flexibility control parameter. In agreement with the canonical results, we have also identified signals corresponding to a solid-solid transition for values of $\eta \leq 0.1$. For $\eta = 0.02$ and 0.05, the negative-valued peaks in $\gamma(E)$ at energies $E = -49.7$ and -50.4 respectively, indicate a second-order transition. The inflection point in $\gamma(E)$ and the corresponding positive valley in $\delta(E) = d\gamma(E)/dE$ at $E = -48.92$ reveal that for $\eta = 0.00$ the solid-solid transition is of third order.

The hyper-phase diagram is constructed on the basis of the signals identified in the microcanonical analysis and plotted in Fig. 3. In the "gas" pseudophase at high energies, the

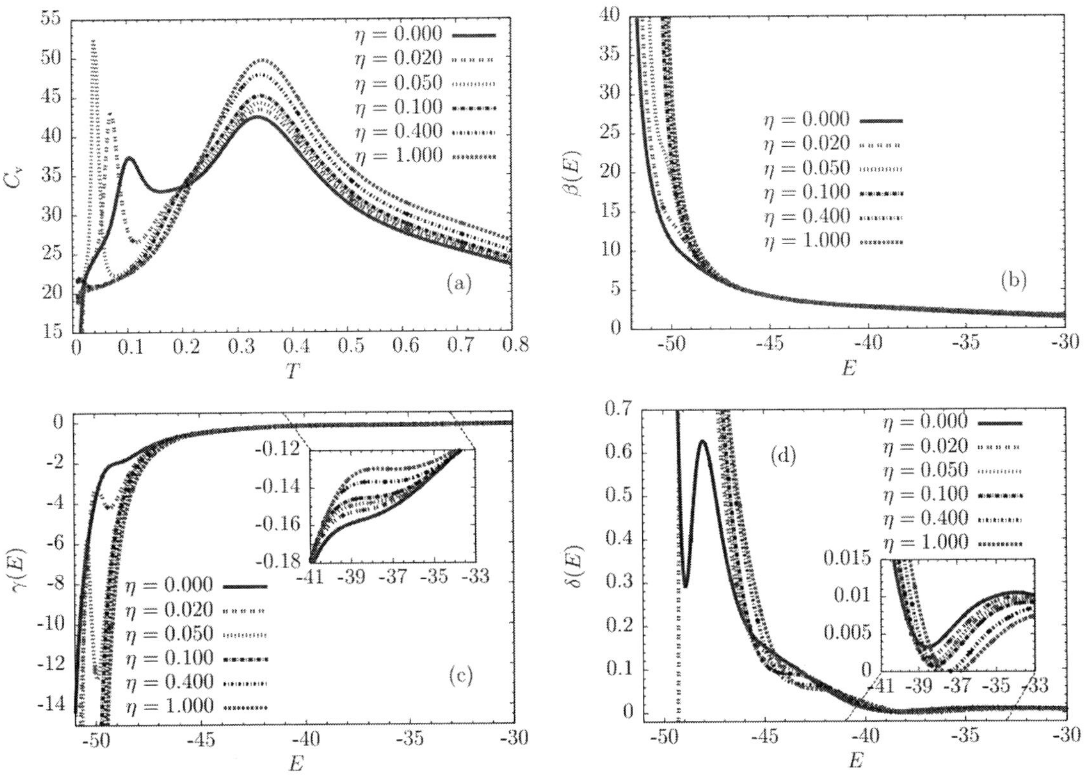

Figure 2. (a) Heat capacity C_V, (b) microcanonical inverse temperature $\beta(E)$, and (c,d) its first and second derivatives $\gamma(E)$ and $\delta(E)$, respectively.

polymer expands in free space and forms random-coil structures. As the energy decreases, the expanded chain undergoes a second-order collapse transition and enters the "liquid" pseudophase consisting mainly of globular structures. Passing the third/second-order transition associated with the nucleation process, the polymer enters the S_{ic} pseudophase in which incomplete icosahedral structures become dominant. Further decrease in energy weakens thermal fluctuations and allows for the formation of a stable surface layer. The transition associated with the surface formation process is of fourth order. Visual inspection of low-energy structures reveals that icosahedral geometries are dominant. However, for $\eta \leq 0.1$, the additional solid-solid transition suggests the existence of low-energy conformations with unexpected geometric properties. In order to examine the low-energy structures systematically, we carry out a careful structural analysis utilizing a suitable set of order parameters.

3.2. Structural analysis

Various order parameters, such as the number of monomer-monomer contacts, radius of gyration, or radial and angular distributions, have proven to provide valuable insight into the thermodynamic and structural properties of polymer systems. Based on the microcanonical results in Fig. 3, we expect the existence of two solid phases when the strength of the bond flexibility parameter is sufficiently small ($\eta \leq 0.1$). We aim to identify the dominant structures in the low-energy phases and to gather additional data supporting the existence of the solid-solid transition line. For this purpose, we employ a set of order parameters exploiting the symmetry

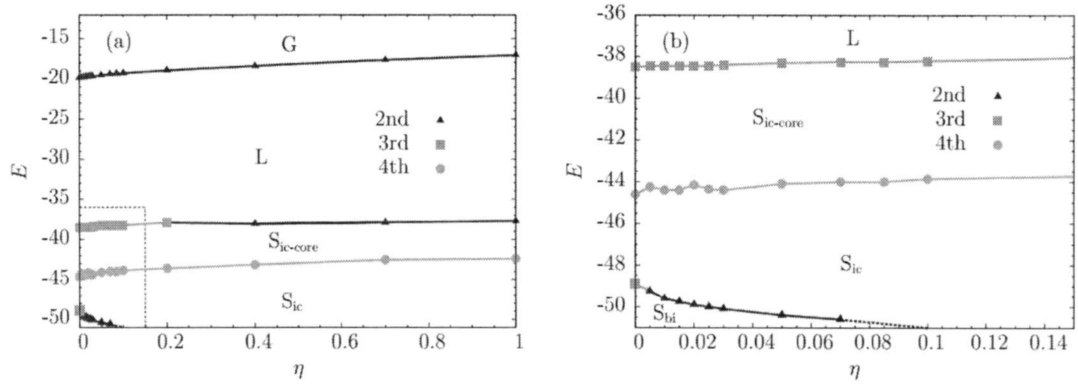

Figure 3. (a) Microcanonical hyper-phase diagram parameterized by energy and the bond flexibility control parameter η. Here, G, L, and S stand for "gas", "liquid", and "solid" structural phases, respectively. The $S_{\text{ic-core}}$ pseudophase consists mostly of incomplete icosahedral structures. S_{ic} and S_{bi} incorporate compact icosahedral and bihexagonal structures, respectively. (b) Enlarged section detailing the low energy region for $\eta < 0.15$. The solid-solid transition line is extended by a dashed empirical line constructed based on the data provided by structural analysis.

properties of real spherical harmonics [12].

We define a polymer core to consist of monomers within a distance $r_{\text{core}} < 1.25\sigma$ of the central monomer, which has been chosen to be nearest to the center of mass. Let $C = \{\vec{r}_1, \ldots.., \vec{r}_M\}$ be the coordinates of a core with M monomers. Various core geometries can be distinguished using the set of rotationally invariant order parameters

$$Q_l = \left[\frac{4\pi}{2l+1}\sum_{m=-l}^{l}|\rho_{l,m}|^2\right]^{1/2}, \tag{7}$$

where

$$\rho_{l,m} = \frac{1}{M}\sum_{i=0}^{M} \mathrm{Y}_{l,m}(\vec{r}_i) \tag{8}$$

is the average of the real spherical harmonics evaluated at the locations of the core monomers. The connection between the real and complex spherical harmonics is given by

$$\mathrm{Y}_{lm} = \begin{cases} \frac{i}{\sqrt{2}}\left[\mathrm{Y}_l^m - (-1)^m \mathrm{Y}_l^{-m}\right] & \text{if} \quad m < 0, \\ \mathrm{Y}_l^m & \text{if} \quad m = 0, \\ \frac{1}{\sqrt{2}}\left[\mathrm{Y}_l^{-m} + (-1)^m \mathrm{Y}_l^m\right] & \text{if} \quad m > 0. \end{cases} \tag{9}$$

Using of the order of 10^6 polymer structures per value of η, we computed Q_l up to $l = 6$ and found that Q_6 can be used most effectively to resolve the geometries of the low-energy conformations. We present the results in the form of intensity plots in Fig. 4. The probability of detecting a structure with a specific value of the order parameter at an energy E is represented by shading; red indicating the maximum probability and black corresponding to zero. In agreement with the microcanonical and canonical results, we detect a single solid phase for $\eta > 0.1$, corresponding to

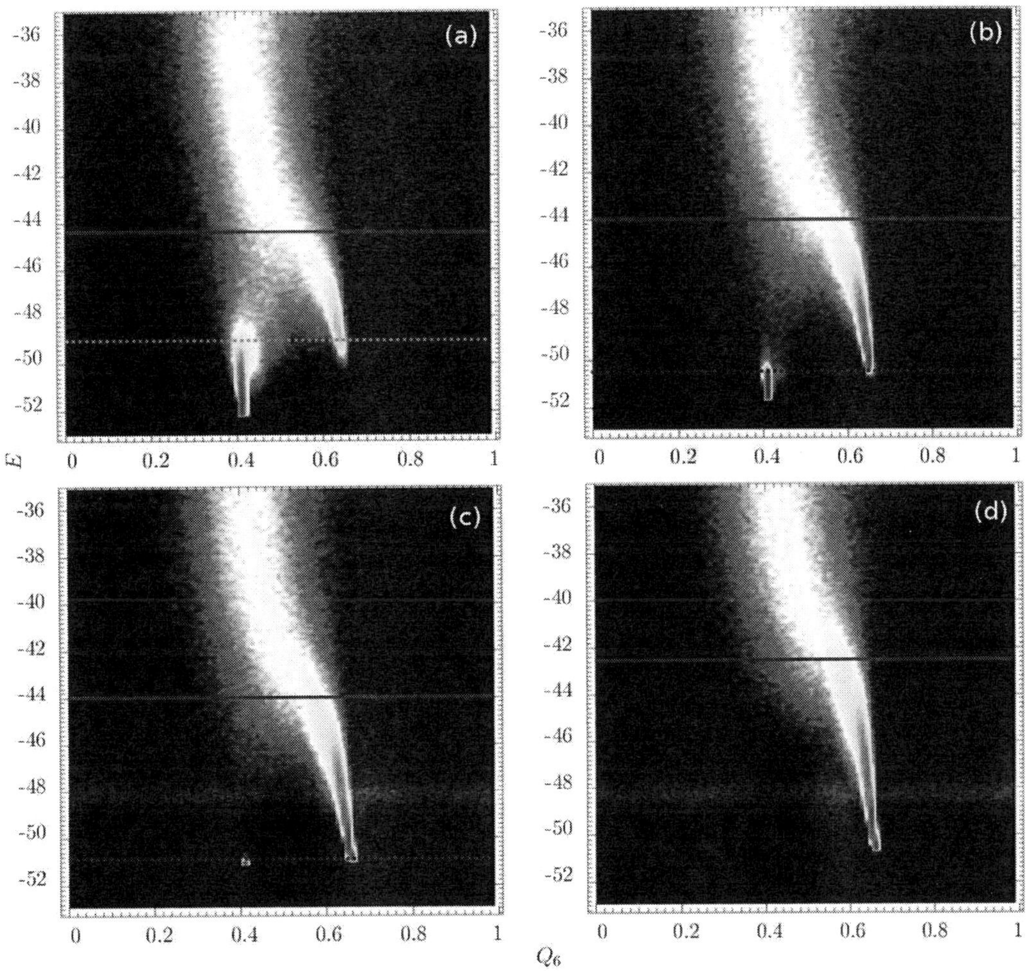

Figure 4. (a,b,c,d) Intensity plots of the Q_6 order parameter for a 15mer with $\eta = 0.00, 0.05, 0.10, 1.0$. The shading indicates the probability of detecting a configuration with a given value of the order parameter, red being the maximum probability and black being the lowest. The freezing and the solid-solid transitions are indicated by solid and dashed horizontal lines respectively. For $\eta \leq 0.1$, the polymer has two distinct solid phases. In addition to the icosahedral phase ($Q_6 \approx 0.65$) the polymer is found in the bihexagonal phase at low energies ($Q_6 \approx 0.41$).

the narrow funnel in Q_6 below the transition line at $E \approx -43$. The dominant structures in this region contain an icosahedral core which is typically found in the ground-state conformations of many short polymer chains. Below $\eta = 0.1$, the 15mer explores an additional solid phase, as indicated by the appearance of a second funnel centered around $Q_6 \approx 0.41$. Structures which populate this phase possess an unusual bihexagonal geometry which has also been recently found in longer chains. The energy at which both phases coexist agrees well with the microcanonical estimates for the solid-solid transition. Representative structures of both solid phases are shown in Fig. 5.

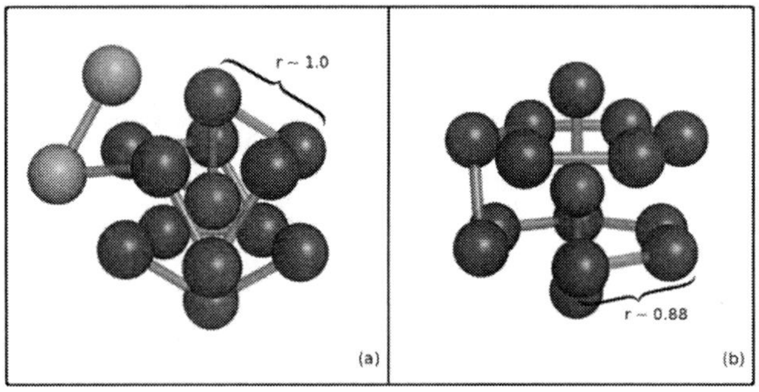

Figure 5. Two distinct low-energy structures of the elastic 15-mer. (a) Compact structure with a stable icosahedral core and two monomers displaced onto the incomplete second layer. (b) The bihexagon is the preferred ground-state geometry for $\eta \leq 0.1$.

The shape of the bonded potential has undoubtedly a strong effect on the geometry of the ground state. Having identified the two dominant structure types, we may ask why the additional LJ term in the bonded potential eventually precludes the existence of the bihexagonal phase. The answer is readily obtained by comparing the average bond lengths for the icosahedral and bihexagonal structures. The bihexagon accommodates all monomers into a single shell allowing for a larger number of non-bonded interactions and consequently lower energy. However, the two six-monomer rings of the bihexagon contain significantly compressed bonds ($r_{\mathrm{bond}} \approx 0.88 r_0$), which become energetically infeasible as η increases. In contrast, we find near-optimal bond lengths in the icosahedron ($r_{\mathrm{bond}} \approx r_0$), hence the "narrowing" of the bonded potential imposes no additional energetic penalty.

4. Summary

By means of multicanonical simulations of a generic model for elastic, flexible polymers, we have investigated the structural behavior of a 15-mer upon changing a model parameter η that controls the shape of the bonded potential. For small values of this parameter, a freezing transition into an icosahedral phase precedes a solid-solid transition into low-energy states with bihexagonal geometry. The non-optimal bond lengths found in bihexagonal conformations cause a large energy penalty due to the "narrowing" of the bonded potential if η is increased. Hence only a single solid phase remains for $\eta > 0.1$, which is icosahedral. The striking consequences of a relatively small modification to the standard model of elastic, flexible homopolymers illustrate the importance of a careful choice of model parameters.

References

[1] Bachmann M 2014 *Thermodynamics and Statistical Mechanics of Macromolecular Systems* (Cambridge: Cambridge University Press)
[2] Hukushima K and Nemoto K 1996 *J. Phys. Soc. Jpn.* **65** 1604
[3] Berg B A and Neuhaus T 1991 *Phys. Lett.* B **267** 249
[4] Janke W 1998 *Physica* A **254** 164
[5] Wang F and Landau D P 2001 *Phys. Rev. Lett.* **86** 2050
[6] Bird R B, Curtiss C F, Armstrong R C and Hassager O 1987 *Dynamics of Polymeric Liquids* 2nd ed. (New York: Wiley)
[7] Kremer K, Grest G S, 1990 *J. Chem. Phys.* **92** 5057
[8] Milchev A, Bhattacharya A, Binder K 2001 *Macromolecules* **34** 1881
[9] Zierenberg J, Marenz M and Janke W 2013 *Comp. Phys. Comm.* **184** 1155
[10] Schnabel S, Seaton D T,Landau D P and Bachmann M 2011 *Phys. Rev.* E **84** 011127
[11] Stevenson P M 1981 *Phys. Rev.* D **23** 2916
[12] Neirotti J P, Calvo F, Freeman D L and Doll J D 2000 *J. Chem. Phys.* **112** 10340

How does the tail length and the separation between the tagged monomers influence the reconfiguration of a chain with internal friction for different solvent-quality?

Nairhita Samanta and Rajarshi Chakrabarti*
Department of Chemistry, Indian Institute of Technology Bombay, Mumbai, Powai 400076
E-mail: rajarshi@chem.iitb.ac.in

Abstract.
Recent experiments involving single proteins have shown the importance of internal friction in protein dynamics. In addition to internal friction, solvent-quality also plays an important role in elucidating the structure and dynamics of proteins. Using a single chain based model where internal friction is introduced via dashpots between the neighbouring monomers and the solvent-quality is taken care of by introducing a mean field Flory type parameter, we monitor the effect of separation of two tagged monomers along the chain and the tail length on the reconfiguration time. We show that irrespective of the degree of compactness, the reconfiguration time remains unchanged beyond a tail length. Our calculations can also reproduce different trends observed in experiments.

1. Introduction

In recent past single molecule based studies have shown not only the existence of internal friction in proteins but also its profound effect on protein dynamics [1, 2]. In particular, measurements of the reconfiguration dynamics of some unfolded proteins such as cold shock protein and spectrin domain R17 [1, 2] reconfirmed the importance of internal friction in proteins. The reconfiguration times shown to have a linear dependence on the solvent viscosity but with a non-zero intercept that attributes to the time scale for internal friction. However it was Khatri and McLeish who first showed that internal friction can be introduced starting with the celebrated Rouse model by placing a dashpot in parallel to the harmonic springs which connect the neighbouring monomers [3]. It is obvious that the amount of internal friction a polymer would experience should depend on its conformation, solvent-quality and the concentration of the denaturant added to the solvent. Keeping these in mind, very recently we have proposed a generalised polymer model which can take care of internal friction, solvent-quality and denaturant concentration [4]. This new model is named "solvent dependent compacted Rouse with internal friction (SDCRIF)" [4] and is an extension of existing models [5, 6, 7]. To take care of the solvent-quality the excluded volume effect is included at the mean field level following the work by Panja and Barkema [8] by introducing a parameter ν analogous to the Flory exponent in the expression for the correlation function of the normal modes of the polymer. In experiments denaturant are usually added to proteins for expanding them from a more compacted state to an unfolded state and as a

(a)

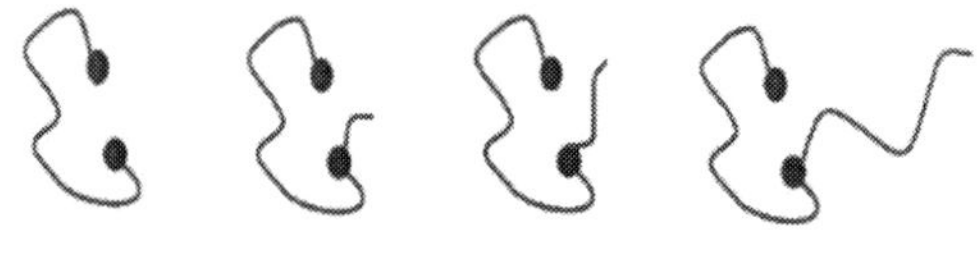

(b)

Figure 1. (a) Scheme showing how the relative separation between the monomers is changed along the chain. (b) Scheme showing how the tail is added to one end of the chain and then the tail length is increased.

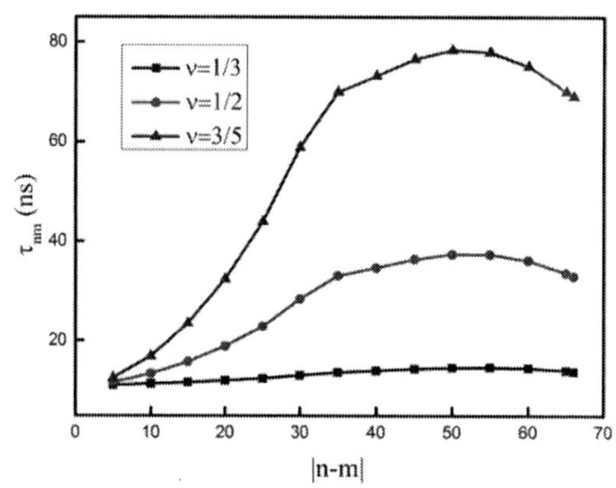

Figure 2. τ_{nm} vs $|n - m|$ for different solvent-quality for $k_c = 0.5k_{c,0}$.

consequence internal friction also decreases with reduced non-bonded interactions [9, 10]. This has been also taken into account in SDCRIF model by introducing a harmonic potential in the system which can control the compactness of the polymer. Further the strength of this potential is considered to be proportional to the internal friction.

In this paper primarily two questions have been addressed. The first one is how the reconfiguration time changes in different solvent condition with the separation of monomers along the chain. This is important since in single molecule FRET based experiments [1], the donor and acceptor need not always be at the two ends of the protein but could also be placed in the interior of the chain (Fig. (1)). The next problem is to study the behaviour of the reconfiguration time by keeping the position of the monomers fixed and introducing additional

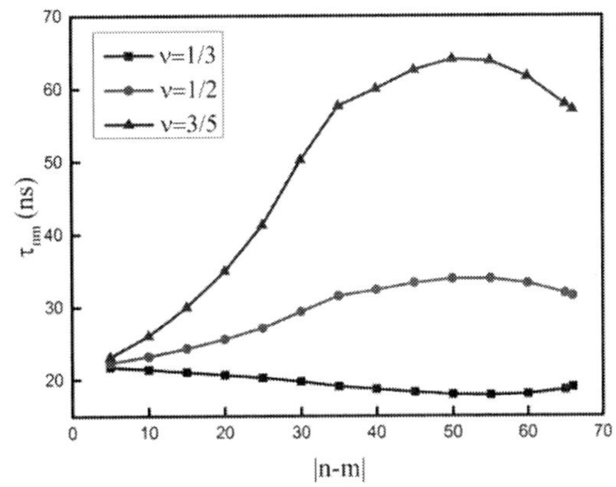

Figure 3. τ_{nm} vs $|n-m|$ for different solvent-quality for $k_c = k_{c,0}$.

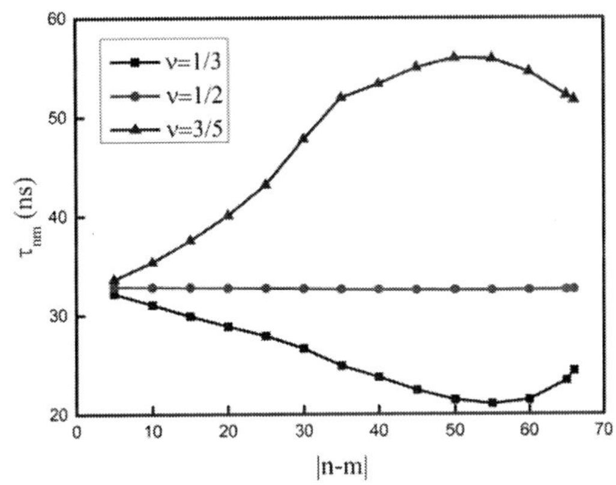

Figure 4. τ_{nm} vs $|n-m|$ for different solvent-quality for $k_c = 1.5k_{c,0}$.

monomers (or tail) to one end of the chain chain as presented in the second scheme of the Fig. (1).

The paper is arranged as follows. The polymer model is discussed in Section 2 and the results are presented in Section 3. The conclusion is made in Section 4.

2. Polymer model

The simplest description of a polymer chain is given by the Rouse model [11, 12]. In this model the polymer is represented as a series of Brownian particles connected by harmonic springs without any hydrodynamics interactions and excluded volume effects. The equation of motion for the nth monomer of a Rouse chain is given by

$$\xi \frac{\partial R_n(t)}{\partial t} = k \frac{\partial^2 R_n(t)}{\partial n^2} + f(n,t) \tag{1}$$

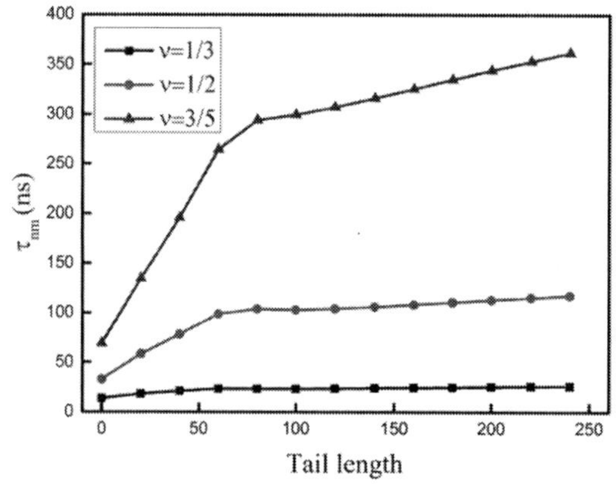

Figure 5. τ_{nm} vs tail length for different solvent-quality for $k_c = 0.5k_{c,0}$.

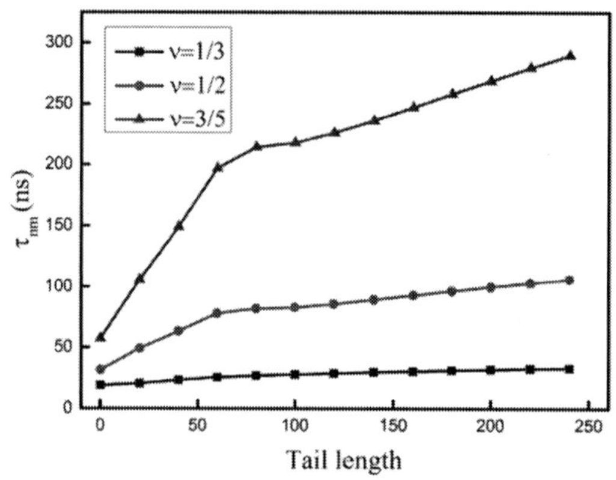

Figure 6. τ_{nm} vs tail length for different solvent-quality for $k_c = k_{c,0}$.

Here, n can take up any value from 0 to N for a polymer having $(N+1)$ monomers connected via springs having force constant $k = \frac{3k_BT}{b^2}$ and ξ is the friction co-efficient. $f(n,t)$ is the random force acting on the nth with first and second moments

$$\langle f(n,t)\rangle = 0, \langle f_\alpha(n,t_1)f_\beta(m,t_2)\rangle = 2\xi k_BT\delta_{\alpha\beta}\delta(n-m)\delta(t_1-t_2) \quad (2)$$

The novelty of our recently proposed model, "solvent dependent compacted Rouse with internal friction (SDCRIF)" lies in the fact that it takes care of the effects of solvent-quality, denaturant concentration in addition to the internal friction to the chain dynamics [4]. In normal mode description, this SDCRIF behaves very similarly to the simple Rouse chain and has the following equation of motion holds

$$\xi_p \frac{dX_p(t)}{dt} = -k_pX_p(t) + f_p(t) \quad (3)$$

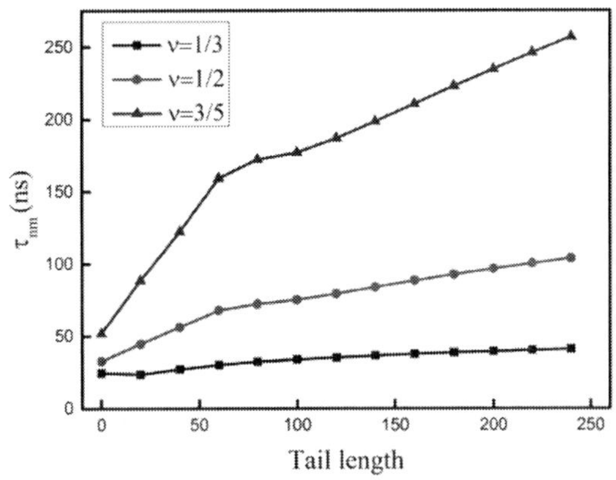

Figure 7. τ_{nm} vs tail length for different solvent-quality for $k_c = 1.5k_{c,0}$.

Where, $R_n(t) = X_0 + 2\sum_{p=1}^{\infty} X_p(t)cos(\frac{p\pi n}{N})$, $\xi_p = 2N\xi + \frac{2\pi^2 p^{2\nu+1}\xi_{int}}{N^{2\nu}}$ and $k_p = 2Nk_c + \frac{6\pi^2 k_B T p^{2\nu+1}}{N^{2\nu}b^2}$. The relaxation time for p^{th} normal mode is given by $\tau_p = \frac{\xi_p}{k_p} = \frac{2N\xi}{k_p} + \tau_{int}$ where, $\tau_{int} = \frac{\xi_{int}}{k+k_c N^{2\nu+1}/p^{2\nu+1}\pi^2}$. Here, ν, k_c and ξ_{int} are the three parameters which take care of solvent-quality, denaturant concentration and internal friction respectively. ν is the the mean field Flory exponent that account for the solvent-quality. Value of ν speaks for the solvent-quality. Typically, $\nu = 3/5$ corresponds to the good solvent, 1/2 for the θ solvent and 1/3 for the poor solvent [13, 14]. On the other hand, the compactness of the polymer is controlled by introducing a harmonic potential to the system of strength $-\frac{\partial}{\partial R_n}\left(\frac{k_c}{2}(R_n - 0)^2\right)$. This is to replicate the denaturation of proteins from their native structures to completely unfolded state. It is obvious that k_c and ξ_{int} should be related to each other since internal friction is higher when the polymer is in more compacted state and vice-versa. As mentioned in our earlier work k_c and ξ_{int} can be connected using an ansatz $k_c = \tilde{k}_c + A\frac{\xi_{int,0}}{\tau_{int}}(c_0 + c_1 n_b + c_2 n_b^2 + ...)$ and $\xi_{int} = \xi_{int,0}(c_0 + c_1 n_b + c_2 n_b^2 + ...)$ [7], with n_b being the number of non-neighbouring monomers interacting with the monomer and $k_{c,0} = \tilde{k}_c + A\frac{\xi_{int,0}}{\tau_{int}}$, which corresponds to a situation when only the two nearest neighbours of a monomer contribute to its internal friction and this corresponds to $c_0 = 1$. Even in the absence of internal friction $\tilde{k}_c$ can vary the compactness of the polymer which makes the ansatz applicable for the polymers with no internal friction.

The time-correlation function for the normal modes is

$$\langle X_{p\alpha}(0)X_{q\beta}(t)\rangle = \frac{k_B T}{k_p}\delta_{pq}\delta_{\alpha\beta}\exp\left(-t/\tau_p\right) \quad (4)$$

The position correlation function of a vector connecting say n^{th} and m^{th} monomer can be determined from the correlation function of the normal modes

$$\phi_{nm}(t) = \langle R_{nm}(0).R_{nm}(t)\rangle = 12\sum_{p=1}^{\infty}\frac{k_B T}{k_p}exp\left(-t/\tau_p\right)\left[cos\left(\frac{p\pi n}{N}\right) - cos\left(\frac{p\pi m}{N}\right)\right]^2 \quad (5)$$

The reconfiguration time is calculated as follows [15]

$$\tau_{nm} = \int_0^\infty dt \tilde{\phi}_{nm}(t) \tag{6}$$

Where, $\tilde{\phi}_{nm}(t) = \frac{\phi_{nm}(t)}{\phi_{nm}(0)}$

3. Results and Discussions

We make use of Eq. (6) to calculate the reconfiguration time (τ_{nm}) between two interior monomers of the polymer and have plotted against the separation between the monomers in Fig. (2). For this we follow the scheme as depicted in Fig. (1), thus by keeping one of the end monomers fixed the other monomer is moved towards the interior of the chain starting at the other end. From the Fig. (2) it can be clearly seen that the reconfiguration time increases with increasing solvent-quality and this is obvious since the polymer goes to a more unfolded state from collapsed state as the solvent-quality changes from poor to good. Another interesting aspect of the figure is the trend of reconfiguration time with the increasing separation between the monomers n and m. The plot shows an initial increase of τ_{nm} with $|n-m|$, but as the two monomers move towards the two opposite ends of the chain, τ_{nm} starts decreasing. In other words τ_{nm} passes through a maxima. This happens presumably because when the distance between two monomers is large, the contributions from the lower normal modes will be predominant. As a consequence the correlation decay of R_{nm} becomes slow which results in slow reconfiguration of the chain. However, with increasing distance the MSD also increases which results in faster correlation decay and consequently smaller τ_{nm}. Because of these two competing factors, reconfiguration time passes through a maxima [6, 7]. In Fig. (3) and Fig. (4) the same plots are shown at higher values of k_c. Surprisingly the trend of reconfiguration time against $|n-m|$ changes with increasing k_c. It is observed that there exists two opposite trends. In one case, τ_{nm} passes over a maxima and in other case it goes through a minima as $|n-m|$ increases. It is observed that k_c and ν both controls this trend. When the polymer is in more compact state, which means higher k_c and lower ν, the reconfiguration time decreases with increasing distance between the monomers, however with increasing solvent-quality the trend reverses. The decreasing nature of τ_{nm} with the increasing distance between the monomers was observed by Nettels *et al* [16] while performing single molecule experiments with cold shock proteins.

Next we have looked into the effect of increasing tail length on reconfiguration time in Fig. (5). We follow the scheme depicted in Fig. (1) to add tail to one of the ends of the chain. In the beginning τ_{N0} i.e. the end-to-end reconfiguration time for a polymer with length $N = 66$ is calculated, which is followed by addition of more monomers to one end of the polymer and calculation of end-to-interior reconfiguration time between the same two monomers i.e. $n = 66$ and $m = 0$. Similar calculations are done for different solvent-quality and shown in the same Fig. (5). It is observed that the reconfiguration time reaches a saturation value after increasing for a while with the increasing tail-length. This is in consistent with the fact that although chain reconfiguration becomes slower for the increment of the total chain-length, when the tail becomes too long it fails to affect the reconfiguration of the other end of the chain. Similar trends for looping times were observed by Doucet *et al* when they performed simulation for in polypeptide chains [17]. For two higher k_c values the reconfiguration time has been studied as a function of tail-length as shown in Fig. (6) and Fig. (7) and in every cases the τ_{nm} for the polymer in poor solvent saturates early in comparison to the one in good solvent for a relatively shorter tail-length. This is expected given the more collapsed state of the polymer in poor solvent makes one end of the polymer less susceptible to the effects coming from the other end. Throughout the paper the value of parameters used are: $\xi = 3.8 \times 10^{-12} kgs^{-1}$ and $T = 300K$.

4. Conclusions

Using a model of single chain with internal friction and a mean field description of the solvent-quality we show how these factors influence the reconfiguration dynamics of a chain. In particular we analyse the effect of adding a tail and increasing the separation between the monomers along the chain to the reconfiguration time. Our calculation shows that the parameters controlling the solvent-quality (ν), denaturant concentration (k_c) and internal friction (ξ_{int}) can reproduce different trends observed in experiments [16].

References

[1] Soranno A, Buchli B, Nettels D, Cheng R R, Müller-Späth S, Pfeil S H, Hoffmann A, Lipman E A, Makarov D E and Schuler B 2012 *Proc. Natl. Acad. Sci.* **109** 17800
[2] Borgia A, Wensley B G, Soranno A, Nettels D, Borgia M B, Hoffmann A, Pfeil S H, Lipman E A, Clarke J and Schuler B 2012 *Nat. Commun.* **3** 1195
[3] Khatri B S and McLeish T C B 2007 *Macromolecules* **40** 6770
[4] Samanta N and Chakrabarti R 2016 *Physica A* **450** 165
[5] Samanta N and Chakrabarti R 2013 *Chem. Phys. Letts.* **582** 71
[6] Samanta N, Ghosh J and Chakrabarti R 2014 *AIP Adv.* **4** 067102
[7] Samanta N and Chakrabarti R 2015 *Physica A* **436** 377
[8] Panja D and Barkema G T 2009 *J. Chem. Phys.* **131** 154903
[9] Yasin U M, Sashi P and Bhuyan A K 2013 *J. Phys. Chem. B* **117** 12059
[10] Sashi P and Bhuyan A K 2015 *Biochemistry* **54** 4453
[11] Doi M and Edwards S F 1988 *The Theory of Polymer Dynamics* (Clarendon Press. Oxford)
[12] Kawakatsu T 2004 *Statistical Physics of Polymers An Introduction* (Springer)
[13] Flory P J 1953 *Principles of Polymer Chemistry* (Cornell University Press)
[14] Flory P J 1969 *Statistical mechanics of chain molecules* (Interscience Publishers)
[15] Makarov D E 2010 *J. Chem. Phys.* **132** 035104
[16] Nettels D, Hoffmann A and Schuler B 2008 *J. Phys. Chem. B* **112** 6137
[17] Doucet D, Roitberg A and Hagen S J 2007 *BioPhys. J* **92** 2281

Collapse transition of short polymers on simple cubic lattice

Asweel Ahmed A Jaleel and S. V. M. Satyanarayana
Department of Physics, Pondicherry University, Puducherry, India
E-mail: svmsatya@gmail.com

Abstract. Denatured proteins and polymers exhibit two types of conformations in solution. Extended coil conformation and compact globule conformation. There is a phase transition associated with these conformation change as a function of temperature or solvent concentration etc. These phase transition is usually studied using Self Avoiding Walk(SAW). We have generated Self avoiding walk using chain growth algorithm. Here we have compared our generated density of states with Exact enumerated data of the same. Also we have presented the Pseudo order parameter method to identify collapse temperature. The results are compared with Partition function zero analysis.

1. Introduction

Proteins are macromolecules of amino acids vital for almost all biological functions. There are about 10^5 proteins in human body[1]. Geometrical shapes of proteins are very important in their functions[2]. Linear chains of amino acids are called primary structure of proteins. Primary structures fold to form secondary structures through weak interactions and they in turn fold to form tertiary and quaternary structures. Secondary, tertiary and quaternary structures are together called native conformations. Native conformations are thermodynamically stable. Understanding the structure of proteins is very important in understanding their functions. Studying proteins using computational models have found significant attention from research community as the memory capacity and speed of computers have increased considerably. Several force fields have been used to simulate all atom conformations of specific proteins. but they are not very much useful in understanding thermodynamics of protein functions. Several measuring equipments are working in a time range that is beyond the reach of such models[3]. Therefore we are forced to remove some degrees of freedom from our model (e.g. intermolecular structures of amino acids, each amino acids are considered as a monomer). This reduction of degrees of freedom to increase relevant length scales is termed as coarse graining. Coarse grained models are further justified by the observation that protein topology has a grater role in their function than their atomistic details and interactions[4]. Coarse graining helps in understanding general problem of protein folding, we cannot make precise or specific predictions. Coarse grained models have been proposed in both continuum off lattice models and discreet lattice models. We are only going to see lattice models.

Most proteins are inside biological cell and are always surrounded by cytoplasm. Polymer physicists have been studying isolated polymers in solution for quiet considerable time. A polymer or protein in solution is modeled via Self Avoiding Walks (SAW) . Random walks which does not cross itself are called Self Avoiding Walks (SAW). Self avoiding condition takes care of excluded volume interaction, i.e. two monomers cannot share same space. In lattice model SAW, polymers are considered as a long chain of rigid beads connected by bonds. Each bead represents a monomer, and they are sitting on a lattice point of a regular lattice (square lattice, triangular lattice etc in 2D and any of Bravais lattices in 3D). The effect of solvent is incorporated by considering Lattice points which are not occupied by polymer beads are occupied by solvent molecules. The monomers which are sitting in nearest neighbor lattice points and are not connected by bonds are having an attractive interaction of Lenard Jones type. This attraction gives rise to energy for each polymer. SAWs with the associated energy definition is called Interacting Self Avoiding Walk. The number of monomers in a polymer is denoted as walk length or chain length in SAW terminology.

From the previous studies it has been observed that isolated polymers in solution has mainly two types of conformations[5]. A random coil conformation and a comparatively ordered globule conformation. The pH, concentration or temperature of solvent can influence the conformational transition[6]. The temperature at which this transition happens is called Theta transition. But since polymers are finite size systems, they are some times termed as collapse transition. Collapse transition is a second order transition. Recent simulations shows a first order kind transition at a lower temperature and is called freezing transition. We will generate an ensemble of ISAWs to study these transitions.

2. Exact Enumeration and Monte Carlo Methods

Generating ensemble of walks is tedious task. We can exactly calculate the number of SAWs of chain length N by exhaustively Counting them using a computer. This becomes increasingly difficult since number of conformations increases exponentially with the increase in walk length. Each conformation has unique Energy. Number of conformations with same energy is called density of that state. Monte Carlo methods are also powerful method to generate ensemble of ISAWs. In chain growth method, growth starts by putting a monomer on some arbitrary lattice site and then next monomer is added to any of randomly chosen nearest neighbor site. This addition of monomers are repeated till we reach the specified number. We will generate an ensemble of walks to study the statistical mechanics of polymer system[7].

In a recent study J. H. Lee et al. has published exact enumeration studies of ISAWs on Simple Cubic Lattice up to chain length 24[8]. They have computed exact partition function zeros and analysis of zeros showed two different loci corresponding to two transitions. Earlier in 2007 T.Vogel et al. has done a Monte Carlo study using nPERMss and discussed collapse transition and freezing transition in detail[9].

3. Simulation and Density of States

In our study we have used chain growth algorithm, the simple RR walk[10]. By using the concept of atmosphere, which is a random variable we can calculate the density of states of any ensemble of walks[11]. Local atmosphere is the number of accessible nearest neighbor sites available for the walk to grow. Random variable 'atmosphere' for a complete walk is the product of all local atmospheres of each step. Atmosphere of a trapped walk is zero.

$$C_N^m = \frac{1}{M}\sum_{l=1}^{M_m} A_l \tag{1}$$

C_N^m is the estimated density of energy $-m$. M is total number of walks including trapped ones in ensemble. M_m is number of successful walks with energy $-m$. A_l is the atmosphere of lth

walk. Comparison of computed density of states with the same from exact enumeration shows they are almost same up to walk length 24 (24 is the maximum available exact enumeration data on Simple Cubic). Fig. 1 shows micro canonical entropy obtained from Simulation and Exact enumeration. Statistical error is less than the marker size. Statistical error is more for minimum energy conformations.

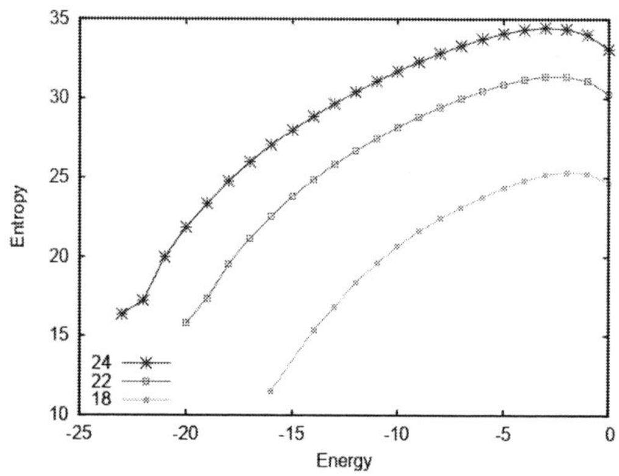

Figure 1. Entropy of walk lengths 18, 22 and 24

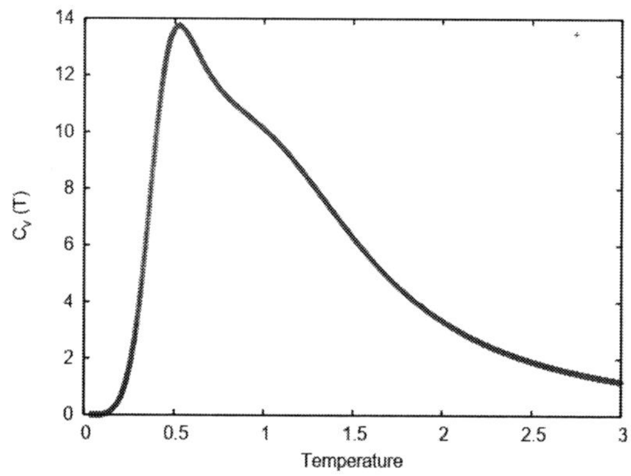

Figure 2. Specific heat capacity of walk length 24.

Looking at the specific heat curve in Fig: 2, we can clearly see a maximum followed by a shallow shoulder. Maximum corresponds to excitation transition, which is a pseudo effect only present in Simple cubic lattice[9]. The shallow shoulder denotes the collapse transition other wise known as coil globule transition. First order freezing transition is not visible in this plot, because of smaller walk length.

4. Partition Function Zeros and Pseudo order parameter

Partition function zeros are celebrated method for studying collapse transition. Characterization of phase transition using the zeros of partition function in the complex fugacity plane is introduced by Yang and Lee [12]. The method is generalized by Fisher [13] for treating the phase transitions in canonical ensemble. This is based on the analysis partition function zeroes on a complex temperature plane. The partition function can be computed from density of states,

$$C_N(\beta) = \sum_{m=0}^{m_N^{max}} C_N^m e^{-m\epsilon\beta} \tag{2}$$

Partition function is thus a polynomial of order m_N^{max} and will have same number of roots which are complex. It has been already observed that the zeros of partition function provide a good estimate of the collapse transition for ISAWs.Phase transition occurs if the roots intersect the positive real axis [14]. For a given finite sized ISAW, one can define first zeros as a pair of roots closest to the positive real axis. Fig 3 shows the first zeros from both exact and Monte Carlo estimate of density of states. First zeros of Monte carlo estimated DOS for even and odd walks are plotted using different color.

Further, density of states was used to find the canonical probability function. Phase transition was studied by plotting the canonical probability function for different inverse temperature.

$$P_N(\beta, m) = \frac{\sum_{m=0}^{m_N^{max}} C_N^m e^{-m\epsilon\beta}}{C_N(\beta)} \tag{3}$$

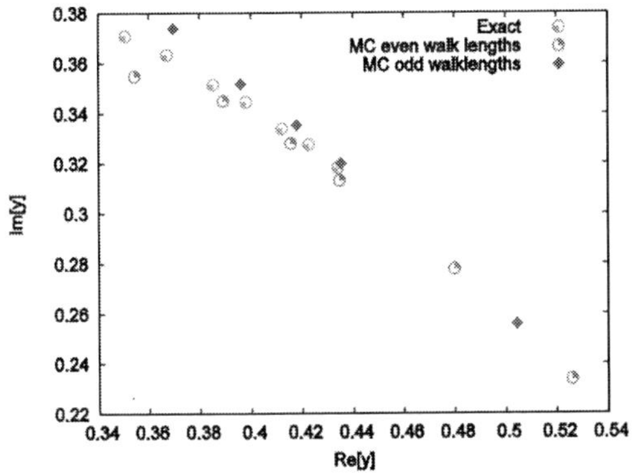

Figure 3. First zeros of Partition function.

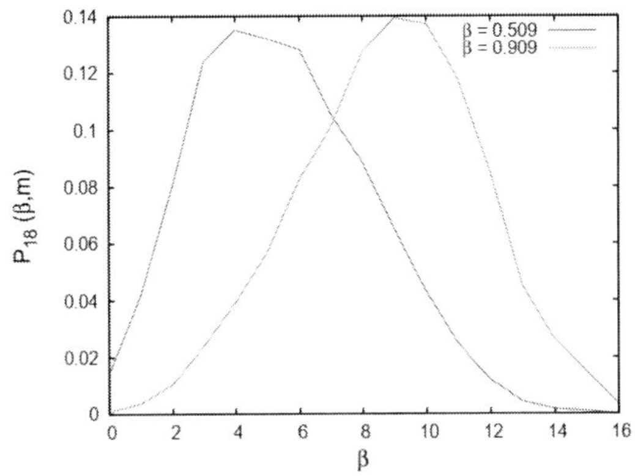

Figure 4. canonical partition function of walk lenth 18 above and below collapse temperature.

For any second order transition, we will be able to see an asymmetric shape with one long tail region for canonical partition function at any inverse temperature. This long tail shifts from side to the other as we vary temperature across collapse transition as in fig 4. This shifting can be used as a pseudo order parameter to find the theta transition[15].

$$O_N(\beta) = \frac{m(min[P_N(\beta, m)])}{m_N^{max}} \tag{4}$$

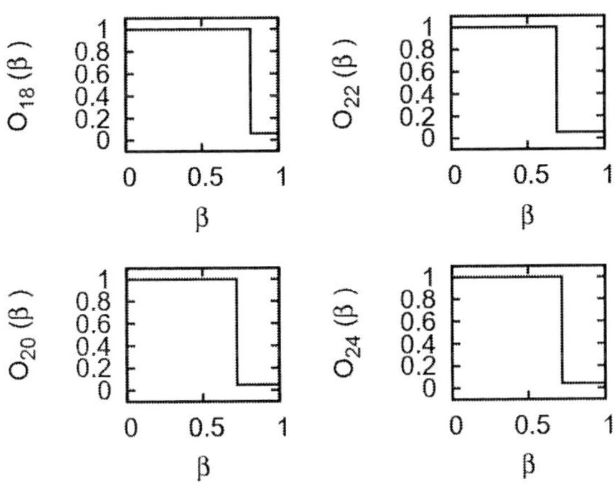

Figure 5. Pseodo order parameter registers a jump from 1 to 0 across collapse temperature

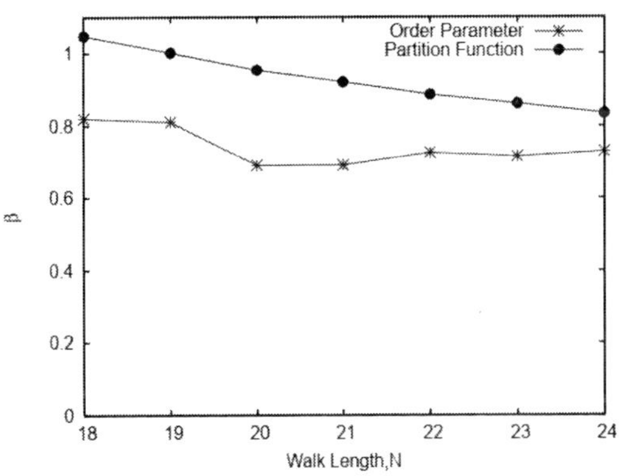

Figure 6. Fluctuation in collapse transition. Results from both PFZ analysis and pseudo order parameter is shown

Fig 5 shows pseodo order parameter defined above. Since this method gives a binary signal, estimation of collapse temperature is unmistakable. In Fig 6 we have plotted inverse collapse temperature (β_c) for walk lengths 18 to 24 from both PFZ method as well as from pseudo order parameter method.

T. Vogel et al. reported theta temperature is 3.72(1) ($\beta = 0.268$) [9]and J.H.Lee repoted same to be 3.76(32) ($\beta = 0.266$) and 3.03(12) ($\beta = 0.330$) with and without logarithmic correction[8].

Collapse transition for short walks are always smaller than that. In our simulation we got first zeros of partition function same as those from exact enumeration. Collapse temperature from pseudo order parameter is not same as those from first zeros of partition function. But our pseudo order parameter method gives collapse temperature slightly closer to theta temperature than partition function zero. This is not because of any artifacts from estimation of density of states. Pseudo order parameter calculated using exact enumeration data also show similar behavior. We still don't have an explanation for that.

5. Conclusion

Simulated density of states using chain growth algorithm are in good agreement with exact enumeration results for interacting Self Avoiding walk on Simple Cubic Lattice. Computed specific heat showed signatures of excitation transition and collapse transition. Collapse transition temperature is estimated using the analysis of partition function zeros and pseudo order parameter method.

Acknowledgment

Asweel Ahmed A Jaleel acknowledge UGC for grant (201112-MUS-KER-4227) under the scheme of Moulana Azad National Fellowship. We would like to thank Jae Hwan Lee for providing exact enumeration data on SC lattice. We would like to thank M Ponmurugan, Department of physics, Central University of Tamilnadu for fruitful discussions.

References

[1] C. Tang 2000 *Physica A* **288** 31
[2] M Bachmann and W Janke 2008 *Lect. Notes Phys.* **736** 203
[3] V.Tozzini 2005 *Current Opinion in Structural Biology* **15** 144
[4] D. Baker 2000 *Nature* **405** 39
[5] P.G. De Gennes 1979 *Scaling concept in physics*,(Cornell Univ. Press,Ithaca)
[6] Guy Ziv, D. Thirumalai, and Gilad Haran 2009
[7] Ponmurugan M, Sridhar V, Narasimhan S L and Murthy K P N. 2011 *Physica A***390** 1258
[8] Jae Hwan Lee,Seung-Yeon Kim , and JulianLee 2012 *Phys. Rev. E* **86** 011802
[9] T. Vogel, M. Bachmann and Wolfhard Janke 2007 *Phys. Rev. E* **76** 061803
[10] M. N. Rosenbluth and A. W. Rosenbluth 1955 *J. Chem. Phys.* **23** 356
[11] A. Rechnitzer and E.J.Janse van Rensburg 2002 *J.Phys.A:Math. Gen.* **35** L605
[12] C. N. Yang ,T. D. Lee 1952 *Phys.Rev.* **87** 404
[13] M. E. Fisher 1965. Lectures in Theoretical Physics (University of Colorodo press, Boulder)
[14] Jae Hwan Lee,Seung-Yeon Kim , and JulianLee 2010 *J. Chem. Phys.* **133** 114106
[15] M. Ponmurugan and S. V. M. Satyanarayana 2012 *J. Stat. Mech.* P06010

Birefringence and temperature-gradient refractive index study of a laterally Fluorinated Terphenyl LC compound.

M. Lyndem[1], D. Bhattacharjee[1], R. Dabrowski[2] and A. Bhattcharjee[1].

[1]Department of Physics, NIT Meghalaya, Shillong, Meghalaya – 793003, India,
[2]Institute of Chemistry, Military University of Technology, 2 Kaliskiego str.,00–908 Warsaw, Poland

Corresponding Author's e-mail address : ayonbh@gmail.com

Abstract. The temperature behaviour of the refractive indices in the nematic range of a laterally fluorinated terphenyl LC compound, viz, 2,3-difluoro-1,4 bis(4',4"-dipropyl phenyl) benzene was determined using the thin prism experiment. Birefringence of this LC compound was calculated from the measured refractive indices. Using the four parameter model, the temperature variation of the refractive indices and birefringence were theoretically fitted. The compound is found to have high value of birefringence. The temperature gradients of refractive indices i.e, $\frac{dn_e}{dT}$ and $\frac{dn_o}{dT}$ were determined, as a result, a cross over temperature was also found. The values of order parameter were obtained in two ways one by direct calculation using the effective geometry parameter α_{eg} .and the other by using the modified Vuks' method. Both show the same trend. It has been observed also α_{eg} changes almost linearly with the birefringence.

1. The Introduction

Liquid crystals (LCs) are a class of materials whose thermodynamic phase lies between that of a conventional liquid and a crystalline solid [1]. As a result of this, they share some of the properties of liquids and some of crystalline solids. One such case is the anisotropy in the optical properties. The refractive indices of LCs and their temperature dependence play a crucial role from the application point of view [2, 3]. An aligned LC behaves like uniaxial crystal and therefore it has two refractive indices viz., ordinary (n_o) and extraordinary (n_e) refractive index. The difference of the two values i.e., $\Delta n = n_e - n_o$ defines the birefringence of the LC. High birefringence nematic LCs are of great interest in display and communication devices because they help in reducing the cell gap which is required for achieving a fast response time [4, 5, 6]. Refractive indices and hence birefringence of LC depends on the wavelength of the incident light, molecular structure and temperature. The temperature gradients of refractive indices ($\frac{dn_e}{dT}$ and $\frac{dn_o}{dT}$) and the crossover temperature, defined as the temperature at which $\frac{dn_o}{dT}$ changes sign from negative to positive, play an important role in the characterization of LC materials in display technology and non-linear optical effects [7,8]. The effective geometry parameter, defined by $\alpha_{eg} = \frac{n_o}{n_e}$, is also very useful in understanding of light propagation in LC and it can be used to determine the order parameter [9-11].

In this paper, we present the study of the variation of the refractive indices with temperature in the nematic range of an LC compound, viz, 2,3-difluoro-1,4 bis(4',4"-dipropyl phenyl) benzene, or in short as C3PP(23F)PC3. Birefringence of this LC compound was calculated from the measured refractive indices. The experimental results were fitted theoretically using the Four Parameter Model [12]. The temperature gradients of refractive indices were determined using the formulae derived by Wu [7] and a crossover temperature was also calculated. The values of order parameter were obtained by two different methods. One is by direct calculation using the effective geometry parameter as reported in literature [10, 13]; and the other is by using modified Vuks' method [17, 18].

2. The Theoretical Framework

The expression for the average refractive index is given as

$$<n> = \frac{n_e + 2n_o}{3} \tag{1}$$

Using the experimental data of n_e and n_o, the average refractive index $<n>$ decreases linearly as the temperature increases and can be expressed as

$$<n> = A - BT \tag{2}$$

The values of A and B are obtained by plotting the temperature (K) and $<n>$ by linear regression. And the results validated by solving system of linear equations given by (2).
The birefringence is also given as [14]

$$\Delta n = \Delta n_0 \left(1 - \frac{T}{T_c}\right)^{\beta} \tag{3}$$

where, Δn_o is the birefringence in the crystalline state (S=1) obtained by extrapolating the birefringence at T = 0K. T_c is the clearing temperature of LC under investigation, and β is a material constant which is obtained by fitting the Δn data. The four-parameter model [12] describes the temperature variation of the LC refractive indices which eventually gives the theoretical value of n_o and n_e as

$$n_e(T) \approx A - BT + \frac{2\Delta n_0}{3}\left(1 - \frac{T}{T_c}\right)^{\beta} \tag{4a}$$

$$n_o(T) \approx A - BT - \frac{\Delta n_0}{3}\left(1 - \frac{T}{T_c}\right)^{\beta} \tag{4b}$$

The temperature gradients of refractive indices are obtained from (4a) and (4b) respectively as:

$$\frac{dn_e}{dT} = -B - \frac{2\beta\Delta n_0}{3T_c}\left(1 - \frac{T}{T_c}\right)^{\beta-1} \tag{5a}$$

$$\frac{dn_o}{dT} = -B + \frac{\beta\Delta n_0}{3T_c}\left(1 - \frac{T}{T_c}\right)^{\beta-1} \tag{5b}$$

The value of T in equation (5b) at which $\frac{dn_o}{dT} = 0$ gives the crossover temperature as

$$T_{CO} = T_c\left[1 - \left\{\frac{3BT_c}{\beta\Delta n_0}\right\}^{\frac{1}{\beta-1}}\right] \tag{6}$$

The order parameter S in terms of effective geometry parameter α_{eg} is given as [13]

$$S = \frac{3 < n > \left(1-\alpha_{eg}\right)}{\left(2\alpha_{eg}+1\right)\Delta n_0} \tag{7}$$

In the modified Vuks' method, the order parameter S is calculated using Vuks' hypothesis [19]:

$$S\left(\frac{\Delta\alpha}{\langle\alpha\rangle}\right) = \frac{n_e^2 - n_o^2}{\langle n^2\rangle - 1} \tag{8}$$

where $\frac{\Delta\alpha}{\langle\alpha\rangle}$ is a scaling factor which can be determined by plotting the term $\frac{n_e^2 - n_o^2}{\langle n^2\rangle - 1}$ as a function of $\ln\left(1-\frac{T}{T_c}\right)$. The graph is a straight line and can be extrapolated to T = 0K at which S=1. Thus, the intercept at T = 0 K gives the value of the scaling factor. Then the required order parameter S was obtained by substituting this value of scaling factor in the equation (8).

3. Experimental

The LC compound (Fig.1.) used in this work was obtained from Military University of Technology, Poland. To determine the refractive indices, the thin prism method (Fig.2) was used [15].

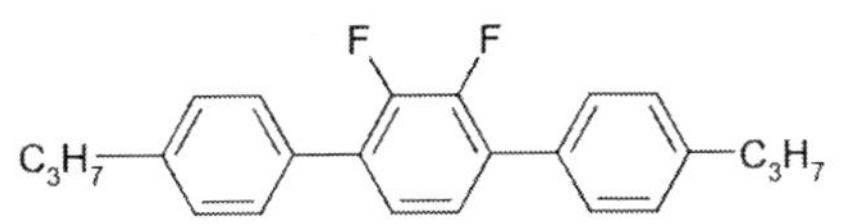

$$Cr \xrightarrow{94.9^0 C} N \xrightarrow{130^0 C} Iso$$

Fig.1. Molecular structure of 2,3-difluoro-1,4 bis(4',4"-dipropyl phenyl) benzene

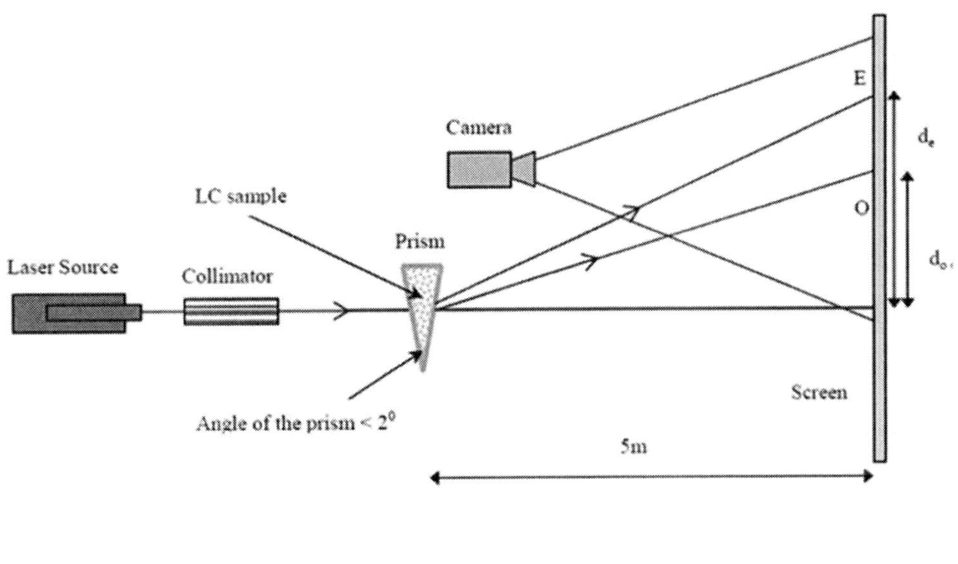

Fig.2. Experimental Setup

The sample was introduced in a thin prism (refracting angle less than 2°). The prism was made of glass slides whose inner surfaces were coated with a 1% polyvinyl alcohol (PVA) solution and

Journal of Physics: Conference Series **759** (2016) 012016 doi:10.1088/1742-6596/759/1/012016

unidirectional rubbing was done for a homogeneous alignment of sample. The prism was enclosed inside an electrically heated hot stage with a circular aperture, whose temperature is controlled manually to ±0.5°C. He–Ne laser (λ= 632.8 nm) beam was focused onto the prism through a collimator. Two refracted spots from the LC–filled prism corresponding to e-ray and o-ray were obtained on a screen placed at a distance of about 5 meters from the prism. The images of the spots were photographed using a high resolution digital camera and those photos were analysed, and the data obtained were used to calculate the required refractive indices.

The ordinary (n_o) and extraordinary (n_e) refractive indices were calculated using the formula [16]:

$$n_e = \frac{1+2d_e}{d} \quad (8a)$$

$$n_o = \frac{1+2d_o}{d} \quad (8b)$$

where 'd_e' and 'd_o' are respectively the distance of extraordinary spot and that of ordinary spot from the direct spot on the screen. The direct spot was obtained on the screen by passing the laser beam through the empty prism without the sample. 'd' is the distance between the position of the direct spot and the spots formed by reflected beams from the two reflecting surfaces of the empty prism when the prism was placed symmetrically with its apex facing the direct beam.

4. Results and Discussions

The measurements of n_e and n_o were done during cooling of the sample from its isotropic phase. The variation of refractive indices with temperature is shown in Figure 3.

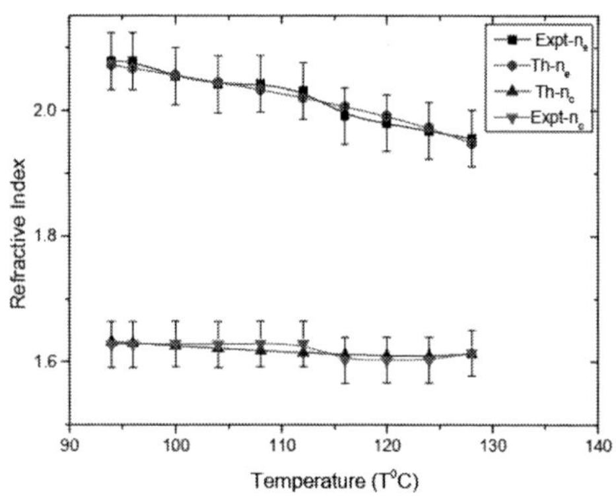

Fig.3. Temperature variation of refractive indices. In the Figures, (Th-stands for theoretical and Expt-stands for experimental.

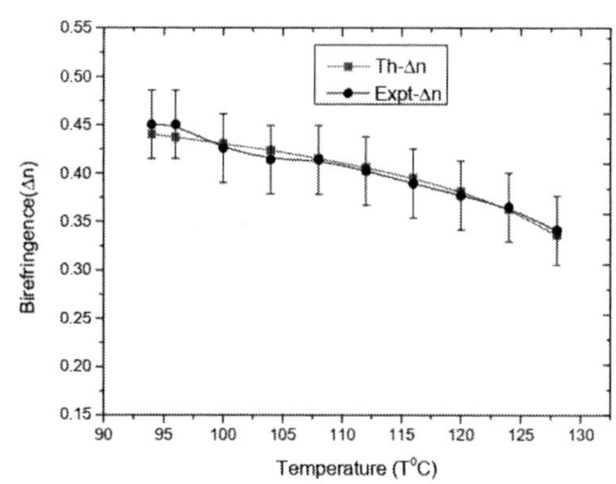

Fig. 4. Temperature variation of birefringence. (Th-stands for theoretical and Expt-stands for experimental)

From this experiment, The nematic–isotropic transition temperatures and the crystal–nematic transition is observed at 133° C and 94° C respectively, which are very close to those observed under DSC recorded by the manufacturer of the compound. As seen in Fig.3, it was found that n_e and n_o follow the general behaviour of temperature dependent refractive indices of NLCs except that n_o does not increase much.

The values of Δn are found to be high and decrease with temperature as shown in Fig.4. The theoretical values fit well with the experimental values in both the cases

The values of ⟨n⟩ as expected also decrease with temperature as shown in Fig.5. The value of A and B are given in the following table:-

Table. Fitting parameters

A	B [K^{-1}]	Δn_o	β
2.36557	0.0016	0.599206	0.13158

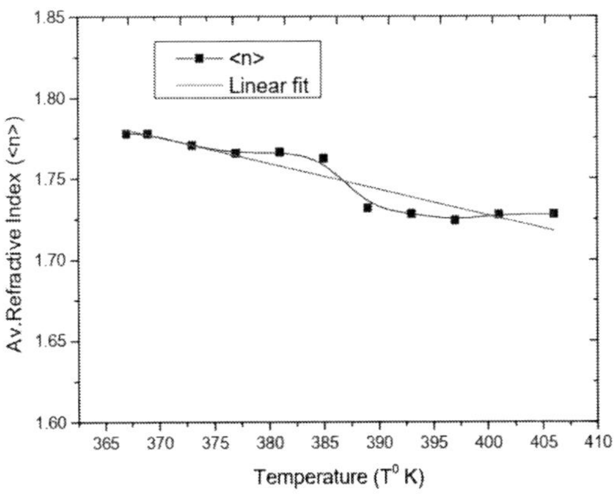

Fig.5. Temperature variation of average refractive index

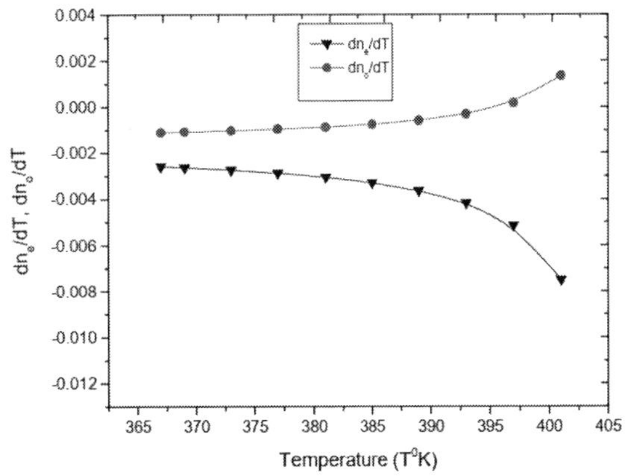

Fig.6. Temperature variation of temperature gradients of n_e and n_o

From Fig. 6, we can see that dn_e/dt remains negative throughout the nematic range. However, there is crossover behaviour of n_o as dn_o/dt changes its sign from negative to positive at a particular temperature known as crossover temperature, T_{CO}. Using Eq (6) it was found that T_{CO} is equal to 395.896° K and from the graph of Fig. 6. , it is found to be 395.908° K. This implies that below T_{CO}, n_o decreases with temperature but above T_{CO}, it increases as temperature increases.

It is observed that α_{eg} decreases linearly with the birefringence as shown in Fig.7. This shows that with increase in the molecular order of the LC compound, α_{eg} decreases. LCs having higher value of the α_{eg} parameter is known to exhibit a lower deflection of light [11]. The order parameter S was determined in two ways one by using Eq (7) and by Eq (8), its temperature dependent is shown in Fig.8.

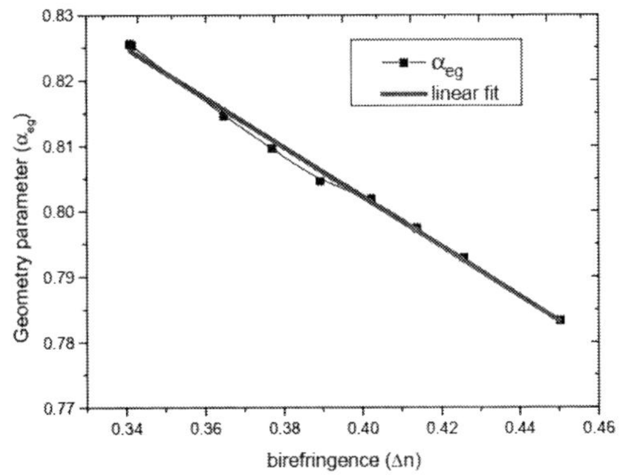

Fig.7. Effective geometry parameter α_{eg} vs birefringence Δn

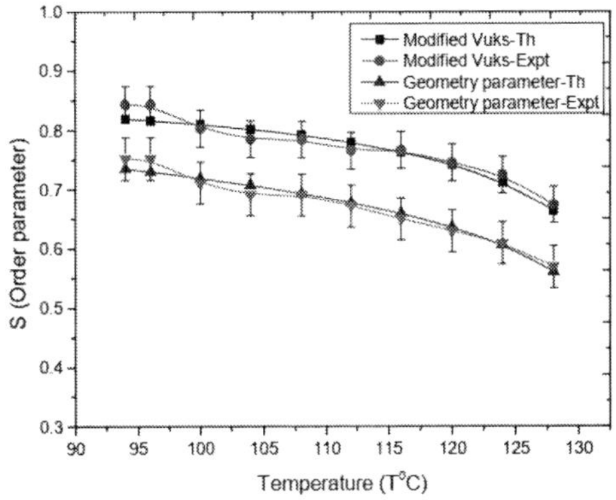

Fig.8. Order parameter S vs Temperature.

The theoretical values fit well with the experimental values in both the cases and the values of S determined from both the methods follow the same trend and they differ from each other by 0.1 approximately.

5. Conclusion

By using the thin prism technique, refractive indices and birefringence of an LC compound: 2,3-difluoro-1,4 bis(4',4"-dipropyl phenyl) benzene were determined and their variation with temperature was found to follow a normal trend. Birefringence of this LC compound was found to be high with values between 0.34~0.45. We have used a Four-parameter model for describing the temperature effect on the refractive indices and birefringence of this LC compound. Excellent agreement between the experimental data and fitted value is obtained. The temperature gradients of refractive indices i.e, dn_e/dt and dn_o/dt shows behaviour which are similar to the ones reported for other compounds [7, 8, 10]. Finally, we had determined the order parameter values without any density study but using only the values of the effective geometry parameter α_{eg} and the modified Vuks' method.

References

[1] de Gennes P G and Prost J 1993 *The Physics of Liquid Crystals 2nd ed.* (Oxford University Press, New York).
[2] Li J and Wu S T 2004 *J. Appl. Phys* **95** 896
[3] Singh A K, Manohar R, Shukla J P and Biradar A M 2006 *Acta Phys. Pol.* A **116** 485
[4] Parish A, Gauza S, Wu S T, Dziaduszek J and Dabrowski R 2008 *Mol. Cryst. Liq. Cryst.* **489** 22–39
[5] Song Q, Gauza S, Xianyu H, Wu S T, Liao Y M, Chang C Y and Hsu C S 2010 *Liq. Cryst.* **37** 139–47
[6] Dabrowski R, Dziaduszek J, Garbat K, Filipowicz M, Urban S, Gauza S and Sasnouski G 2010 *Liq. Cryst.* **37** 1529–37
[7] Li J, Gauza S and Wu S T 2004 *Optics Express.* **12** 2002–10
[8] Kumar A 2013 *Liquid Crystals.* **40** 1422-26
[9] Satiro C and Moraes F 2008 *European Physics Journal E.* **25** 425–29
[10] Devi G S, Prasad P V D, Latha D M and Pisipati V G K M 2013 *Liquid Crystals.* **40** 1274–81
[11] Ozbek H, Ustunel S, Kutlu E and Cetinkaya M C 2014 *Journal of Molecular Liquids.* **199** 275–86
[12] Li J, Gauza S and Wu S T 2004 *J. Appl. Phys.* **96** 19-24
[13] Sastry S S, Kumari T V , Mallika K, Rao B G S, Ha ST and Lakshiminarayana S 2012 *Liquid Crystals.* **39** 295–301
[14] Haller I, 1975 *Prog. Solid State Chem.* **10** 103−18
[15] Thingujam K D, Sarkar S D, Choudhury B and Bhattacharjee A 2012 *Acta Physica Pol A.* **122** 754–57
[16] Yakhmi J V, Kelkar V K, Shukla R P and Manohari C 1979 *Mol. Cryst. Liq. Cryst.* **53** 55
[17] Thingujam K D, Alapati P R, Choudhury B and Bhattacharjee A 2013 *Liquid Crystals.* **40** 810–16
[18] Kumar A 2007 *Acta Phys. Pol. A* **112** 1213–21
[19] Vuks M F 1966 *Opt. Spectros.* **20** 361-364

Miscibility studies of Polyethylene Glycol with Polystyrene in Toluene by Various Physical and Advanced Techniques

R Padmanaban, K Venkatramanan

Department of Physics, Sri Chandrasekharendra Saraswathi Viswa Mahavidyalaya, Enathur, Kanchipuram – 631561, Tamilnadu, India

E-mail: kv.scsvmv@gmail.com

Abstract. Polyethylene glycol (PEG) is a chemical that has an extensive variety of applications in the world of medicine. It is used as a base to manufacture certain medicines, assist in drug delivery, and is also used as an agent in some medical procedures. It is an osmotic laxative. Polyethylene glycol works by retaining water in the stool, resulting in softer stools and more frequent bowel movements. Polyethylene glycol does not affect glucose and electrolytes in the body. PEG refers to a hydrocarbon molecule that can have a variable size, and different sizes can have different physical properties, giving this compound a great deal of flexibility in its application. In the present study, Polyethylene Glycol (PEG) (Molar mass : 1500) is blended with Polystyrene (PS) (Molar mass: 35000) in Toluene. The miscibility nature of the poly blend is analyzed by Ultrasonic velocity, viscosity, density and refractive index techniques at 303K. The compatibility nature of the blend is confirmed by Differential Scanning Calorimetry (DSC) studies.

1. Introduction

Polyethylene Glycols (PEG) have extensive applications in pharmaceutical preparations, cosmetics and in the metal costing process [1]. Polyethylene glycol is edible polymer film used for food packing and directly in foodstuffs [2]. To satisfy the growing needs of new materials with specific properties such as engineering materials, new polymers have been synthesized [3]. However, the mixture of two or more polymers, forming a polymer blend, continues to be an economical method to obtain new polymeric materials [4]. Blending of polymers is one of the simplest methods to obtain a variety of physical and chemical properties from the constituent polymers at a molecular level. The gain in newer properties depends on the degree of compatibility or miscibility of the polymers. There are a number of techniques to study the miscibility of polymer blends such as ultrasonic, viscosity, refractive Index, FTIR, XRD, DSC and SEM [5]. The final properties of a polymeric blend will commonly depend on the properties of its polymeric components, its composition and mainly on the miscibility of the constituent polymers. Several researchers [6-9] have reported on the miscibility studies of PEG. In the present case, PEG (1500) and PS (35000) polymer blend solutions are prepared in toluene and their miscibility nature is analysed using density, viscosity, ultrasonic velocity, and refractive index measurements. By solution casting method, polymer blend films are prepared and their miscibility nature is further analysed using DSC techniques. The compatibility of these polymer blends is discussed and it is concluded that the presence of hydrogen bonding groups in these two polymers is responsible for the polymer-polymer interaction.

2. Material and Methods

In the present study, Polyethylene Glycol (PEG 1500) is blended with Polystyrene (PS 350000) in toluene. High purity spectra grade PEG 1500 and PS 35000 are obtained from KAVIN SCIENTIFIC, Chennai, India. 1% concentration of PEG 1500 and PS 35000 are separately prepared by dissolving 4 gms of each polymer in 400 ml Toluene. Then various compositions of the blend PEG 1000 / PS 35000 in the following ratio (0/100, 20/80, 40/60, 50/50, 60/40, 80/20, 100/0) are prepared by proper mixing at 30° C (accuracy ± 0.5 ° C). Magnetic stirrer [REMI make] is used for this purpose at a rate of 1000 rpm. The relative viscosity of the polymer solutions is studied for all the blend composition ranges at 303K using Brookfield Viscometer (USA) (accuracy ±0.01cP). The temperature is maintained using a thermostat with a thermal stability of ± 0.05 K. The density values are measured using specific gravity method. The mass of the liquid is measured using a K-ROY make electronic balance, with an accuracy of ±0.001gm. Refractive index studies are performed at 303 K using Abbe Refractometer (uncertainty ± 0.5%). The ultrasonic velocity measurements are performed using Mittal make single frequency Ultrasonic interferometer at 2 MHz (F-81 model) (uncertainty ± 0.01m/s). The temperature is maintained at 303 K by circulating water from a thermostat with a thermal stability of ± 0.05 K through the double wall jacket of the cell. The glass transition temperature is studied using Differential Scanning Calorimetry (DSC), NETZSCH Technologies, Chennai.

3. Results and Discussion

Viscometry is a simple and effective technique for monitoring the interactions of polymer blend solutions. From these measurements, relative viscosity of pure polymers and their blends are obtained. Figure 1 shows the variation of PEG 1500 / PS 35000 in toluene at 313 K for different weight fractions of polymer 1 and polymer 2 respectively. From this figure, the variation is linear from 50/50 to 100/0 blend compositions and non linear for remaining blend composition ranges. From this, it may be concluded that the blend system is partially miscible for blend composition ranges from 50/50 to 100/0.

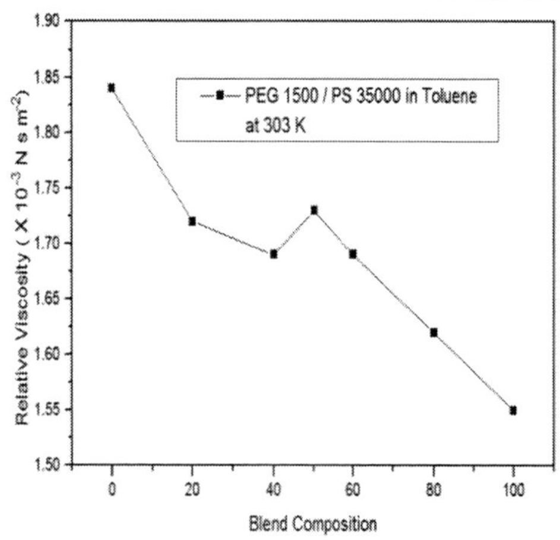

Figure 1. Variation of Relative viscosity against blend composition of PEG 1500

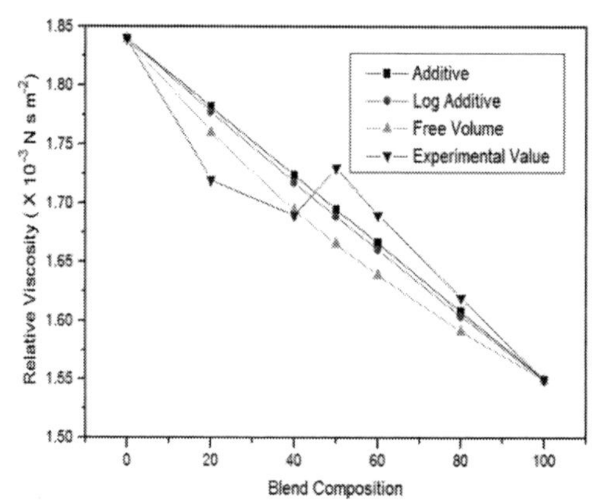

Figure 2. Variation of Relative viscosity against blend composition of PEG 1500 by additive rule

Gupta et al [10] have recommended viscometric method to study polymer-polymer miscibility. The basic idea of using viscosity as a parameter for compatibility determination of polymer blends lies in the fact that in solution the repulsive interaction may cause shrinkage of polymer coils resulting in the viscosity of the polymer mixture that is lower than the value calculated from viscosities of the pure components on the assumption of additive law. Venkatramanan et al [7] have used a few empirical and semi empirical equations for predicting the miscibility of polymer blends based on viscosity viz., the additive rule, log additive rule, and free volume additive rule given by Eq. 1, Eq. 2 and Eq. 3.

$$\eta_b = W_1\eta_1 + W_2\eta_2 \quad (1)$$

$$\log \eta_b = W_1 \log \eta_1 + W_2 \log \eta_2 \quad (2)$$

$$1/\log \eta_b = W_1(1/log\eta_1) + W_2(1/log\eta_2) \quad (3)$$

Where η_b is the viscosity of the blend, η_1, η_2 that of the components and W_1, W_2 the weight fractions of the components.

In the case of completely miscible blends, that is 100% miscible blends, the experimental values will be the same as the values calculated from the additive rule of the mixture. When the blend is less than 100% miscible or partially miscible the values predicted from log additive rule of the mixture will be close to the experimental value. Figure 2 shows the variation of relative viscosity against blend composition of PEG 1500 using additive rule method. Here from 50/50 to 100/0 blend composition range, it is observed that the experimental values are positively deviating. This shows that the blend is miscible. For other blend compositions, the experimental values are negatively deviating which means that the blend may be immiscible. From this, it is concluded that the blend system is partially miscible from 50/50 to 100/0 blend compositions. For further confirmation density studies are performed. It is observed from the density results (Figure 3) that in the composition range from 50/50 to 100/0, the variation is linear showing the miscible nature of the blend and from 0/100 to 50/50 composition, the blend is immiscible due to non-linear variation. Throughout the system, the blend is semi-compatible. To confirm the compatibility nature of the blend further, refractive index studies are carried out. Refractive index is a useful technique to analyze miscibility of polymer blends. In the present case (Figure 4), for the composition range from 0/100 to 50/50, the variation is nonlinear which shows that the blend is immiscible and for other composition ranges, the variation is linear which shows the miscible nature of blends. From this, it may be concluded that the blend system is partially miscible from 50/50 to 100/0 composition range. For further confirmation, ultrasonic velocity was calculated.

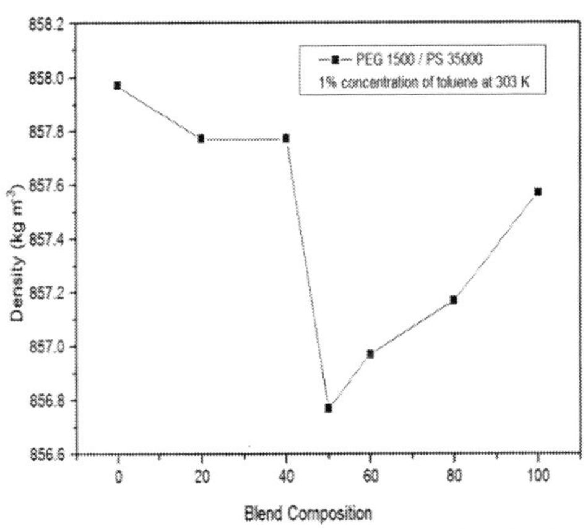

Figure 3. Variation of Density against blend composition of PEG 1500

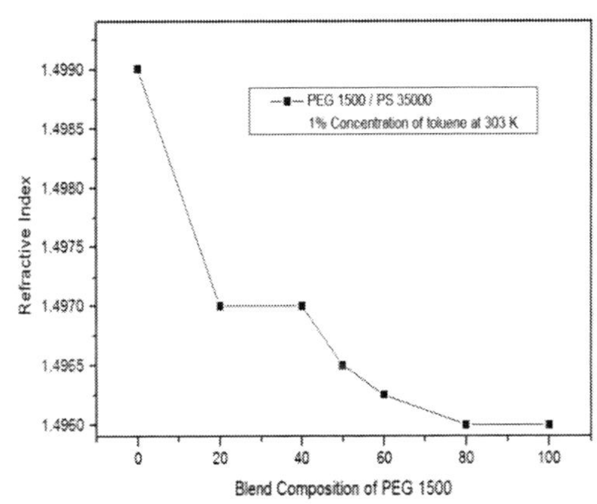

Figure 4. Variation of Refractive Index against blend composition of PEG 1500

Ultrasonic technique plays a vital role in determining the compatibility of the polymer blends. It has been reported by Sidkey et al [11] that (a) if the variation of ultrasonic velocity with blend composition is of S -type, the blend is immiscible, (b) if the ultrasound velocity shows a linear relation with blend composition, then the blend is compatible or miscible and (c) for semi compatible system, the nature of the curve is in between straight line and s- shape. In the present case [PEG 1500 / PS

35000] the nature of the curve is s- shape (Figure 5). It may be concluded that the blend is immiscible in nature. To confirm further, DSC studies were done.

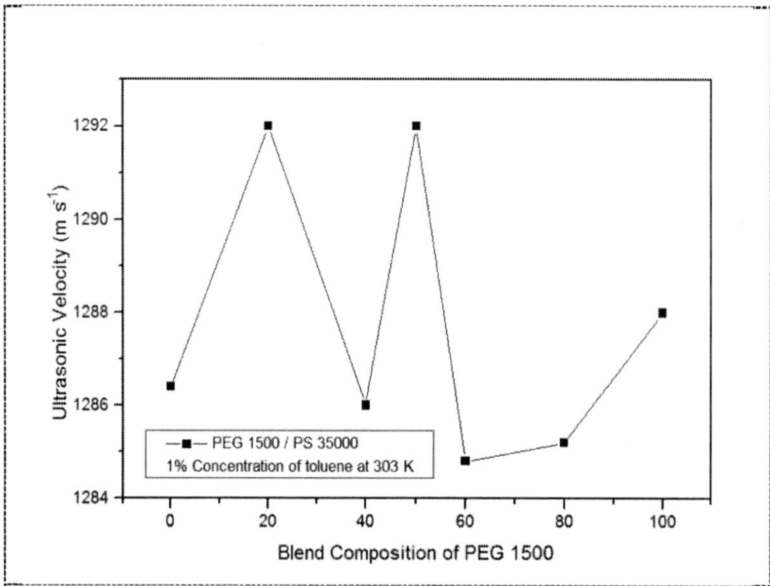

Figure 5. Variation of Ultrasonic Velocity against Blend Composition of PEG 1500

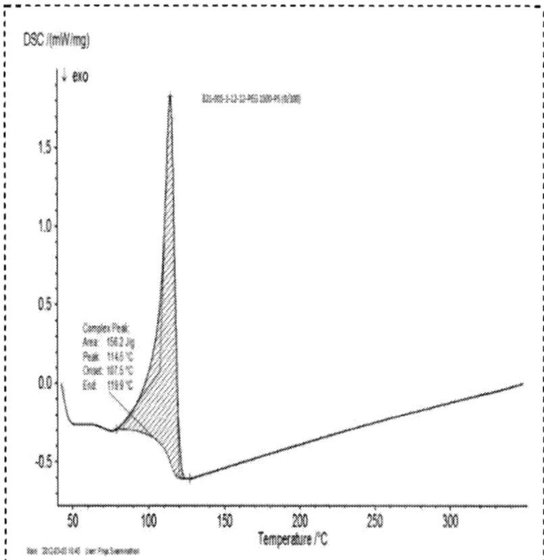

Figure 6. DSC Studies of PEG 1500 / PS 35000 in Toluene at 303 K (0/100 composition)

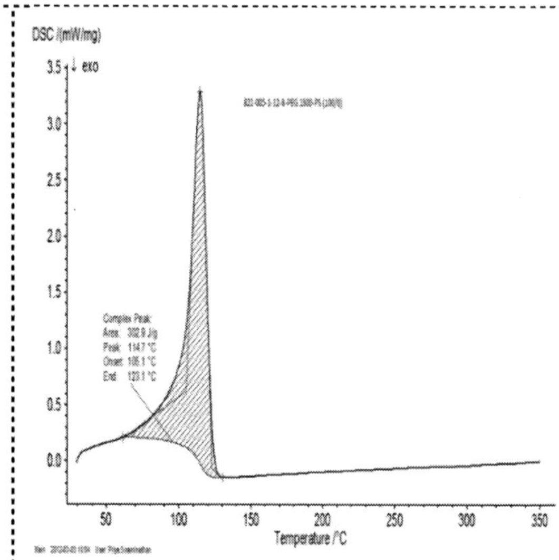

Figure 7. DSC Studies of PEG 1500 / PS 35000 in Toluene at 303 K (100/0 composition)

DSC is a valuable thermal analysis technique which provides with important information on the glass transition of polymer blends. The miscibility and interaction between components are the most important aspects for defining polymer blends. Measuring the glass transition temperature of the desired polymer blend will quickly and easily determine their compatibility. The blend system PEG 1500 / PS 35000 was first cooled from 120°C to -140°C at 5K/min and then measured by heating from -160°C to 600°C at 10 K/min in nitrogen atmosphere at 50ml/min to determine the temperature at which the glass transition occurs. The glass transition range for PS 35000 extends over an area of 156.2 J/g between 107.5°C and 119.9°C with a peak at 114.5°C. For PEG 1500, it extends over an area of 302.9 J/g between 105.1°C and 123.1°C with a peak at 114.7°C. Figure 8 shows the glass transition range which extends over an area of 260.4 J/g for PEG 1500/PS 35000 blend system [50/50] between 111.1°C and 124.2°C has a peak at 118.5°C. The occurrence of just one glass transition observed in the polymer blend system PEG 1500 / PS 35000 (50/50 blend composition) confirms that polymer components are compatible [12-14] from 50/50 to 100/0 blend compositions.

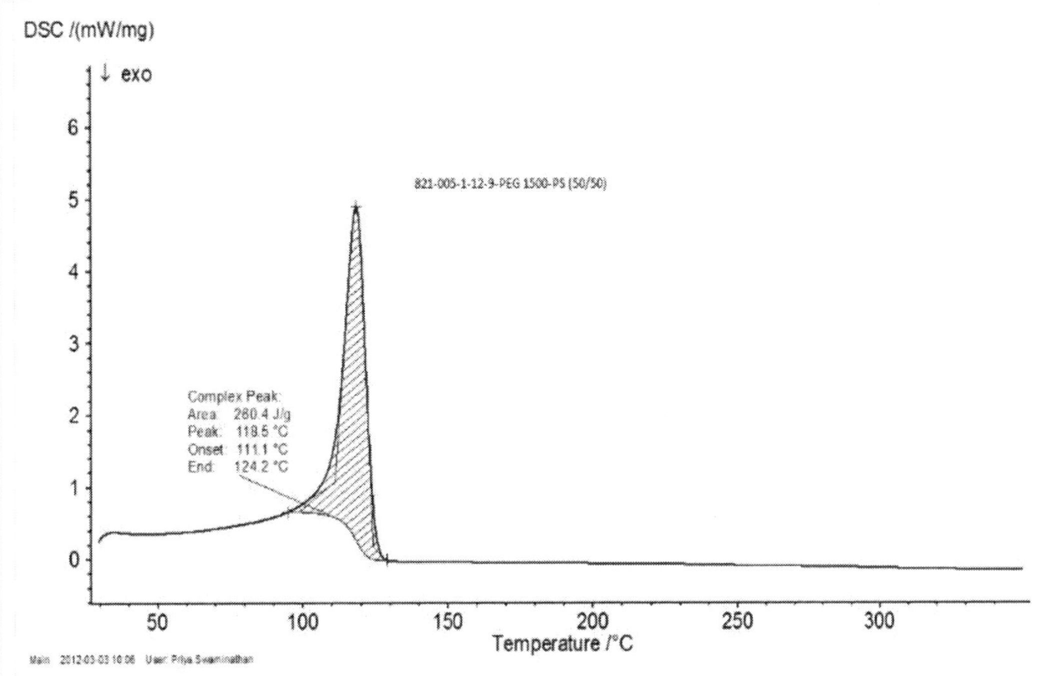

Figure 8. DSC Studies of PEG 1500 / PS 35000 in Toluene at 303 K (50/50 composition)

4. Conclusion

PEG 1500 is blended with PS 35000 in toluene at 303 K and the compatibility nature of the blend is analysed using various physical methods like viscosity, density, refractive index and ultrasonic and DSC techniques. The blend system showed partial miscibility from 50/50 to 100/0 blend compositions. DSC studies confirmed the partial miscibility nature of this blend.

References

[1] Syal V K, Anita Chauhan, and Suvarcha Chauhan 2005 *J. Pure Appl. Ultrason.* **27** 61-69
[2] Sudhir Ramswamy Illiger, Chandralekha Fadnis, Demappa T, Jayaraju J and Keshavayya J 2009 *Carbohydr. Polym.* **75** 484-488
[3] Chowdoji Rao K, Subha M C S, Suresh Reddy C, Kumara Babu P, Sudhaka K, Prabhakar M N, Maruthi Y, and Sajan Kumarji Rao 2013 *Int. J. Basic App. Chem. Sci.* **3** 73-83
[4] Crispim E G, Rubira A F and Muniz E C 1999 *Polym. J.* **40** 5129-35

[5] Sudharsan Reddy K, Prabhakar M N, Kumara Babu P, Venkatesulu G, Sajan Kumarji Rao U, Chowdoji Rao K and Subha M C S 2012 *Int. J. Carbohydr. Chem.* **2012** 1-9
[6] Mruthyunjaya Swamy T M and Siddaramaiah 2007 *J. Appl. Polym. Sci.* **104** 2048-53
[7] Venkatramanan K, Padmanaban R and Kalidass D 2013 *J. Pure Appl. Ultrason.* **35** 105-108
[8] Venkatramanan K and Arumugam V 2008 *AIP Conf. Proc.* **1004** 235-40
[9] Jayaraju J, Basavaraju K C, Keshavayya J and Rai S K 2006 *J. Macromol. Sci., Phys.* **45** 741-51
[10] Ashok Kumar Gupta and Gautam Kumar Sah 2013 *Nanosystems: Physics, Chemistry, Mathematics* **4** 670–79
[11] Sidkey M A, Yehia A A, Abd El Malaka N A, and Gaafar M S 2002, *Mat. Chem. Phys.* **74** 23-32
[12] Vasile Cristian Grigoras and Virgil Barboiu 2008 *Reveu Roumaine de Chimie* **53** 127-31
[13] Marina Vivas de Meftahi, Jean MJ Frechet 1988 *Polymer* **29** 477-82
[14] Everaert, Groeninckx, Koch and Reynaers H 2003 *Polymer* **44** 3491–08

Journal of Physics: Conference Series **759** (2016) 012018 doi:10.1088/1742-6596/759/1/012018

Heterogeneous dynamics in linear polymer melts near glass transition temperature

Jalim Singh and Prasanth P. Jose

School of Basic Sciences, Indian Institute of Technology Mandi, Kamand, Himachal Pradesh 175005, India

E-mail: prasanth@iitmandi.ac.in

Abstract. Equilibrium molecular dynamics simulations of a model linear polymer melt is carried out near its glass transition temperature T_c. We have elucidated the role played by intra and inter chain monomers in formation of secondary structures using radial distribution functions on reduction of temperature towards T_c. Our studies on dynamics using intermediate scattering function $F_s(k,t)$ shows step like relaxation, which is a signature of heterogeneous dynamics of the glassy domains that is widely known in earlier studies of glass transition. Further studies of heterogeneity in the dynamics using four-point correlation function $\chi_4{}^s(t)$ shows the growth of correlated region in this system as temperature is reduced towards T_c.

1. Introduction

Glass transition and glassy dynamics has been a subject of intense research in last several decades: as glassy systems are abundant and glass transition is ubiquitous in nature[1, 2]. Structural glasses are special class of glassy systems which requires structural rearrangement for relaxation near the glass transition temperature T_c. Many systems exhibit structural glass transition, examples include systems of polymers, bio-polymers *etc.*

An important theoretical approach to understand changes in the microscopic dynamics during the glass transition is via studies of the mode coupling theory(MCT)[3–5]. The ideal MCT[6, 7] has explained the temperature dependency of α-relaxation time, viscosity, and diffusion coefficients *etc.*, above glass transition temperature T_c. One of the prominent feature of glass transition is exponential slowing down of the relaxation process; in experiments the glass transition occurs at around viscosity equal to 10^{13} poise(or, relaxation time $\approx$ 100 sec)[8]. Below the crossover temperatures of supercooled liquids, particle motion become increasingly correlated, resulting in formation of cages by neighboring particles. The relaxation of a particle within a cage results in β relaxation and relaxation of the cages classified as α relaxation[9, 10]. The structures of these cages around each particle differ which results in dynamic heterogeneity[11, 12]. Computational studies of the glass transition in various glass forming model liquids are successful in explaining many microscopic features such as microscopic relaxation, dynamic heterogeneity that is concomitant to the glass transition. In the past, dynamic heterogeneity of glass forming liquids has been investigated by many authors [13–22] using two point correlation functions such as, intermediate scattering function, mean-squared displacements, non-Gaussian parameters *etc.* Many investigations have argued that conventional two point correlation functions are not adequate for fully describing the spatially heterogeneous dynamics, and the relation between length and time scales of relaxation in supercooled and glassy liquids[23]. To measure the extent of dynamic heterogeneity, several authors have employed four point correlation functions in computational studies of model glass formers [14, 17, 24–35].

There are large variety of materials show glass transition, including polymers[36]. This work focus on the glass transition in polymer melts, which are well known glass formers and have numerous commercial, industrial, and daily life applications[37]. As polymers have many degrees of freedom, it is computationally expensive to address glass transition in polymer systems. One of the simplest model of a polymer liquid is a system of linear chains. We have performed equilibrium molecular dynamics(MD) simulations of supercooled system of linear polymer melt of a model polymer proposed by Grest and Kremer(GK)[38], which was used for earlier studies of glass transition by Bennemann *et al.*[39]. They studied the divergence of α-relaxation time at temperatures above T_c. Increasingly correlated motion, which was more pronounced well above T_c, has been studied by Bennemann *et al.*[40] in the same model of polymer melts but at higher densities. Spatially correlated dynamics and string like motion of the same system has been studied by Glotzer and coworkers[21, 41]. Starr *et al.*[42] compared string lengths to the length scale of Adam-Gibbs theory[43] and random first order transition theory [44–46] of the glass transition. Extensive MD simulations has been performed by Leporini *et al.*[47–51] to measure the spatial correlated motion of the particles in the supercooled polymer melt. Jump like motions near the T_c are investigated by Helfferich *et al.*[52, 53] in a glassy polymer melt using continuous time random walk approach. It is now interesting to extend earlier studies to further establish the connection between microscopic structure and dynamics of this system especially by calculating the four-point correlation functions. The concept of four point correlation functions has been proposed by Dasgupta *et al.*[54] in their simulation studies. Donati *et al.*[55] showed that the growth of correlation length can be given by the volume integral of four point correlation functions. Four point density correlation functions have been computed to understand increasingly correlated motion and dynamic correlation length in various glass forming liquids[14, 34, 35, 42].

We describe basic theory of correlation functions in Sec.2; the model and simulations detail is presented in Sec.3; finally we will present the results and analysis part of this work in Sec.4.

2. Basic theory of static and dynamic correlations

For homogeneous and isotropic systems, radial distribution function $g(\mathbf{r})$ is only function of separation of a pair of particles, *i.e.*, r and then the $g(r)$ is defined as[56]

$$g(r) = \frac{1}{N\rho}\left\langle \sum_{i=1}^{N} \sum_{j=1,j>i}^{N} \delta(\mathbf{r} - \mathbf{r}_j + \mathbf{r}_i) \right\rangle. \tag{1}$$

The $g(r)$ of polymer melt system can be expressed as $g(r) = g_{intra}(r) + g_{inter}(r)$, where sum for $g_{intra}(r)$ takes contribution from monomers of same chain, and sum for $g_{inter}(r)$ takes contribution from monomers that belong to different chains[57].

The single particle relaxation dynamics of the system can be analyzed using incoherent intermediate scattering function[56], which can be defined as

$$F_s(k,t) = \frac{1}{N}\left\langle \sum_{i=1}^{N} \exp(i\mathbf{k}.[\mathbf{r}_i(t) - \mathbf{r}_i(0)]) \right\rangle, \tag{2}$$

and the structural α-relaxation time, *i.e.*, τ_α can be defined as $\tau_\alpha = \int_0^\infty F_s(k,t)\,\mathrm{d}t$.

The structural dynamic susceptibility ${\chi_4}^s(k,t)$ can be defined from the variance of $F_s(k,t)$[27, 31, 58]

$${\chi_4}^s(k,t) = N[\left\langle F_s(k,t)^2 \right\rangle - \langle F_s(k,t)\rangle^2]. \tag{3}$$

We can extract the length and time scale of dynamic heterogeneity from $\chi_4{}^s(k,t)$ of a model glass-former.

3. Model and simulation details

We simulated coarse-grained bead-spring polymer melt model system for modeling the polymer chains, each linear chain consists of 10 monomers. The non-bonding interactions are modeled by Lennard-Jones(LJ) potential[59] that is given by

$$V_{LJ}(\mathbf{r}_{ij}) = \begin{cases} 4\epsilon\left[\left(\frac{\sigma}{\mathbf{r}_{ij}}\right)^{12} - \left(\frac{\sigma}{\mathbf{r}_{ij}}\right)^{6}\right], & r_{ij} \leq 2 \times 2^{1/6}\sigma \\ 0 & , \quad r_{ij} > 2 \times 2^{1/6}\sigma \end{cases} . \tag{4}$$

The LJ potential is truncated at distance of $r_{cut} = 2 \times 2^{1/6}\sigma$, where σ and ϵ are diameter of a particle and depth of the potential respectively. The bonding interactions are modeled by finitely extensible non-linear elastic(FENE) potential[38] which is given by $V_{FENE}(\mathbf{r}_{ij}) = -0.5k_0R^2 ln[1 - (\mathbf{r}_{ij}{}^2/R^2)]$, where $k_0 = 30\epsilon\sigma^{-2}$ and $R = 1.5\sigma$. We integrated the Newton's equations of motion using Velocity-Verlet algorithm[59] with time step $\delta t = 0.002$. We simulated a linear polymer melt system of $N_c = 1000$ polymer chains. The equilibration time used in the simulation is the time required for relaxation of end-to-end vector to 10% of its initial value at all temperatures[52]. The system is studied at the temperatures T = 2.0, 1.0, 0.90, 0.80, 0.70, 0.60, 0.50, 0.45, and 0.40 at fixed density $\rho = 0.85$ in NVE ensemble. All the parameters reported from our simulation study are in LJ simulation units, *i.e.*, the physical quantities are represented as follows: distance in units of σ; the reduced density, $\rho^* = \rho\sigma^3$; temperature, $T^* = k_BT/\epsilon$; time, $t^* = (\epsilon/m\sigma^2)^{1/2}t$[59].

4. Results and discussions

Fig.1(a) shows the $g(r)$ of the system at various temperatures, which shows a broad peak at $r = 1.0$ at high temperatures. When temperature is reduced, the first peak shows prominent splitting from temperature $T = 1.0$. The first peak at low temperatures in Fig.1(a) corresponds to average bond length($r_{bond} \approx 0.96$) and second peak is due to LJ interaction this is around $r = 1$. When temperature decreases towards T_c, first peak and the subsequent peaks grow that shows formation of the secondary structure in this system. As polymers can form complex structures within same molecule, it is meaningful to look separately in structures formed within each polymer molecules and also the structure formed by surrounding molecules. In the inset of Fig.1(b), $g_{intra}(r)$ shows the structure formed around a bead of a chain by monomers of the same chain. Like $g(r)$, the peaks in $g_{intra}(r)$ grow while reducing the temperature towards T_c. The average radius of gyration of the polymer melt chains is $\langle R_g\rangle \approx 1.44$ in equilibrium, which has weak temperature dependence. The second peak and subsequent peaks grow at low temperatures due to the confinement of a chain via LJ interaction among monomers of the same chain, that results in the globule like structure of chains in these simulations. The peak of the $g_{intra}(r)$ near $r = 2$ also broadens showing multiple secondary structures within a chain similar to binary LJ glass forming systems[60] near T_c. In Fig.1(b), we show $g_{inter}(r)$, which shows the structure formed by the monomers belong to neighboring chain that surrounds a tagged chain. It is clear from this plot that $g_{inter}(r)$ contributes more to the $g(r)$ in comparison with $g_{intra}(r)$ in the formation of secondary structure. In this case also the peaks that is around $r = 2$ broadens to show formation of the secondary structure. Now it is interesting to look into how the relaxation dynamics affected by these multiple secondary structures that is formed at low temperatures.

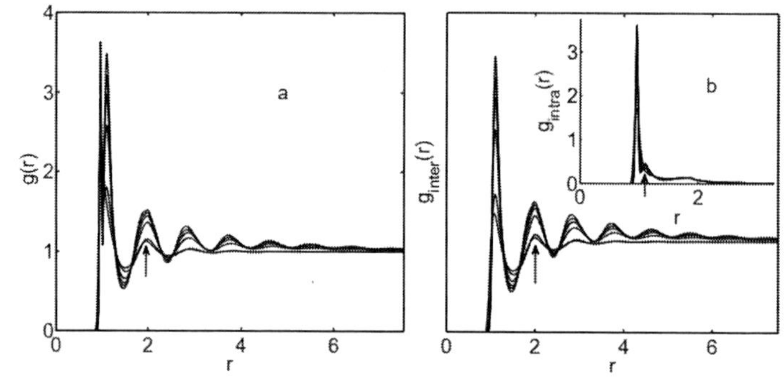

Figure 1: In the direction of arrow, temperature changes as T = 2.0, 1.0, 0.60, 0.50, 0.45, and 0.40. Radial distribution functions: (a) Fig-1(a) shows $g(r)$, the splitting of primary peak is clear at $T = 1.0$ and second peak shows the formation of secondary structure at $T = 0.40$; (b) Fig.1(b) shows $g_{inter}(r)$ and inset of Fig.1(b) shows $g_{intra}(r)$.

We have analyzed single particle relaxations by computing the $F_s(k,t)$, from Eq.2, which is plotted in Fig.2. The wavenumber used in the calculation is $k = 6.99$, this corresponds to the maximum of static structure factor $S(k)$. The $F_s(k,t)$ fits well with exponential function at higher temperature $T = 4.0$, below this temperature, it is fitted by empirical Kohlrausch-Williams-Watts function of the form $f(t)^{KWW} \propto \exp(-(\frac{t}{\tau})^{\beta})$ with value of exponent β varying from 0.87 to 0.61, as temperature decreases for this polymer melt system. At temperatures, above and near T_c, particles shows pronounced two step relaxations, showing β and α regime of the relaxation function in supercooled linear polymer melt system. The onset of plateau in the $F_s(k,t)$ starts at $T = 0.50$, which signifies the heterogeneous relaxation dynamics in the system. We computed τ_α at aforementioned temperatures; the τ_α *vs* T curve is fitted using MCT and Vogel-Fulcher-Tammann relations, *i.e.*, $\tau_\alpha \propto (T - T_c)^{-\gamma}$ and $\tau_\alpha \propto \exp(DT_0/(T - T_0))$. The fitted parameters for the linear polymer melt system, at given thermodynamical conditions, of these two equations are given as follows: MCT temperature, $T_c \approx 0.40$; the MCT exponent, $\gamma \approx 1.65$; the fragility index, $D \approx 2.34$; the dynamic divergence temperature, $T_0 \approx 0.30$. Many parameters we obtained are comparable to the values obtained in the work of Bennemann *et al.*[39] in NVT and NPT ensembles, however, our calculations are for a larger system and in NVE ensemble. The ideal glass transition temperature T_g and extrapolated dynamic divergence temperature T_0 are related through the relation $T_g \approx 1.2T_0$[42]. The approximate T_g for this system is 0.36.

Now we look into the four-point correlation functions to understand the increasingly correlated motion of particles in this linear polymer melt. In Fig.3, we have given self part of four point ${\chi_4}^s(t)$ of monomers of linear polymer chains. The height of the peak and corresponding time grows in ${\chi_4}^s(t)$, while reducing temperature towards the T_c in this system. Interestingly, time corresponding to the peak of ${\chi_4}^s(k,t)$ in our system shows same variation as the α-relaxation time. Berthier *et al.*[61] reported experimentally that the time corresponding to the peak of $\chi_4(t)$ is equivalent to the average relaxation time of the system, *i.e.*, ${t_4}^{max} \approx \tau_\alpha$, where ${t_4}^{max}$ time corresponding to the maximum in ${\chi_4}^s(t)$, in supercooled glycerol and colloidal hard spheres. The pronounced peak of ${\chi_4}^s(k,t)$ reveals the increasingly correlated motion of the particles in supercooled polymer melt, which is due to the spatially heterogeneous dynamics. This increasing height of the peak of ${\chi_4}^s(k,t)$ towards T_c gives the signature of the growing correlation length in the system. This growing correlation length and time quantify the spatio-temporal heterogeneity of the system in the vicinity of T_c.

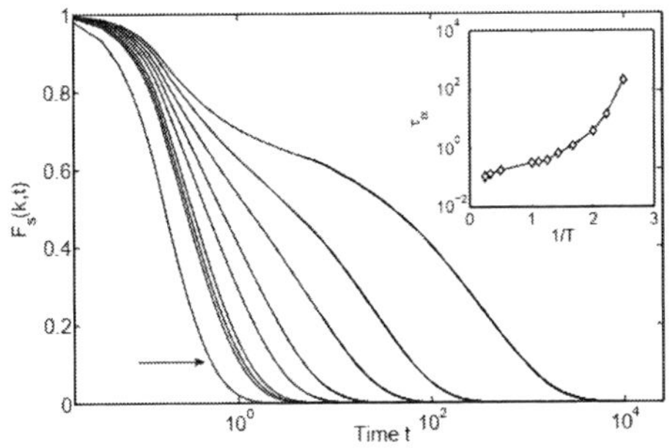

Figure 2: In the direction of arrow, temperatures changes as T = 2.0, 1.0, 0.90, 0.80, 0.70, 0.60, 0.50, 0.45 and 0.40. Incoherent intermediate scattering function of linear polymer melts at wavenumber k=6.99; inset shows τ_α vs $1/T$ plot, in which τ_α increases with decreasing temperature

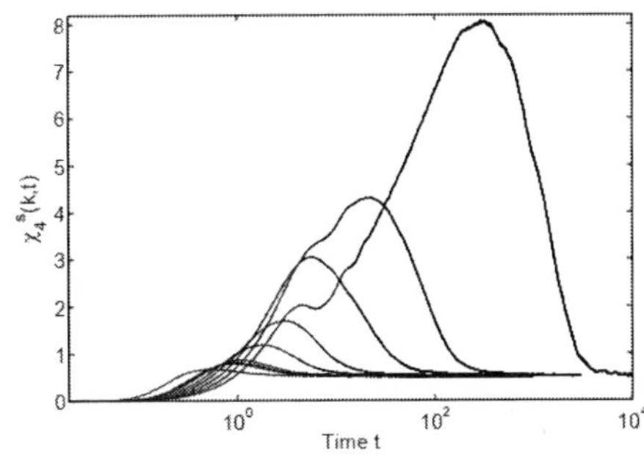

Figure 3: $\chi_4^s(k,t)$ of linear polymer melts at wave vector k = 6.99; multiple peaks appear from T=0.45. Peak height of $\chi_4{}^s(t)$ increases with decreasing temperature, *i.e.*, T = 2.0, 1.0, 0.90, 0.80, 0.70, 0.60, 0.50, 0.45 and 0.40.

5. Conclusion and summary

We have reported results of the long time molecular simulation study of linear polymer melt system of 10^3 particles in micro-canonical ensemble near its glass transition temperature. Our studies looked into the microscopic structure of the slowly relaxing domains near glass transition temperature T_c. Explicit analysis of the radial distribution functions of linear polymer melts shows formation of multiple structures in this system as glass transition is approached. We have computed the structures formed by intra and inter chain monomers using $g_{intra}(r)$ and $g_{inter}(r)$. Analysis of $g_{intra}(r)$ and $g_{inter}(r)$ revealed the role played by these monomers in formation of secondary structures. As reported in earlier studies [39], we also find appearance of plateau region in the $F_s(k,t)$ and divergence of α-relaxation time[39] in linear polymer melts at low temperatures, which is a signature of spatially heterogeneous dynamics in the system. To further characterize spatio-temporal dynamic heterogeneity, in this system of polymer melt, we have computed structural dynamic susceptibility $\chi_4^s(k,t)$. The calculated $\chi_4^s(k,t)$ gives the correlation volume of a system, which diverges at low temperatures, signifies the increasingly correlated motion of the particles in this linear polymer melt system.

6. Acknowledgments

We acknowledge central HPC facility at IIT Mandi for computational support.

References

[1] Debenedetti P G 1996 *Metastable liquids: concepts and applications* (Princeton University Press)
[2] Wolynes P G and Lubchenko V 2012 *Structural Glasses and Supercooled Glasses: Theory, Experiments, and Applications* (John Wiley & Sons, Inc.)
[3] Götze W 2009 *Complex Dynamics of Glass-Forming Liquids: A mode-coupling theory* (Oxford University Press, New York)
[4] Das S P 2004 *Rev. Mod. Phys.* **76** 785
[5] Reichman D R and Charbonneau P 2005 *J. Stat. Mech.* **05** P05013
[6] Götze W 1991 *Liquids, Freezing and the Glass Transition, edited by J. P. Hansen, D. Levesque and J. Zinn-Justin;* (Les Houches. Session LI, 1989,(North-Holland, Amsterdam))
[7] Götze W and Sjögren L 1992 *Rep. prog. Phys.* **55** 241
[8] Angel C A 1995 *Science* **267** 1924

[9] Doliwa B and Heuer A 1998 *Phys. Rev. Lett.* **80** 4915
[10] Buchalla G, Dersch U, Gotzess W and Sjogren L 1988 *J. Phys.: Condens. Matter* **21** 4239
[11] Ediger M D 2000 *Ann. Rev. Phys. Chem.* **51** 99
[12] Berthier L, Biroli G, Bouchaud J, cipelletti L and Saarloos W (eds) 2011 *Dynamical Heterodeneities in Glasses, Colloids and Granular Media* (Oxford University Press, New York)
[13] Kob W, Donati C, Plimpton S J, Poole P H and Glotzer S C 1997 *Phys. Rev. Lett.* **79** 2827
[14] Karmakar S, Dasgupta C and Sastry S 2010 *Phys. Rev. Lett.* **105** 015701
[15] Vollmayr-Lee K and Zippelius A 2005 *Phys. Rev. E* **72** 041507
[16] Cicerone M T, Zhong Q and Tyagi M 2014 *Phys. Rev. Lett.* **113** 117801
[17] Flenner E, Staley H and Szamel G 2014 *Phys. Rev. Lett.* **112** 097801
[18] Mizuno H and Yamamoto R 2011 *Phys. Rev. E* **84** 011506
[19] Donati C, Glotzer S C, Poole P H, Kob W and Plimpton S J 1999 *Phys. Rev. E* **60** 3107
[20] Donati C, Douglas J F, Kob W, Plimpton S J, Poole P H and Glotzer S C 1998 *Phys. Rev. Lett.* **80** 2338
[21] Aichele M, Gebremichael Y, Starr F W, Baschnagel J and Glotzer S C 2003 *J. Chem. Phys.* **119** 5290
[22] Biroli G and Garrahan J P 2013 *J. Chem. Phys.* **138** 12A301
[23] Lačević N, Starr F W, Schrøder T B, Novikov V N and Glotzer S C 2002 *Phys. Rev. E* **66** 030101
[24] Glotzer S C, Novikov V N and Schrøder T B 2000 *J. Chem. Phys.* **112** 509
[25] Lačević N, Starr F W, Schrøder T B and Glotzer S C 2003 *J. Chem. Phys.* **119** 7372
[26] Dauchot O and Marty G 2005 *Phys. Rev. Lett.* **95** 265701
[27] Chandler D, Garrahan J P, Jack R L, Maibaum L and Pain A C 2006 *Phys. Rev. E* **74** 051501
[28] Flenner E and Szamel G 2009 *Phys. Rev. E* **79** 051502
[29] Kim K and Saito S 2009 *Phys. Rev. E* **79** 060501
[30] Flenner E and Szamel G 2013 *J. Chem. Phys.* **138** 12A523
[31] Kim K and Saito S 2013 *J. Chem. Phys.* **138** 12A506
[32] Gupta B S and Das S P 2014 *Phys. Rev. E* **90** 012137
[33] Karmakar S, Dasgupta C and Sastry S 2014 *Ann. Rev. Condens. Matter Phys.* **5** 255
[34] Karmakar S, Dasgupta C and Sastry S 2008 *Proc. Natl. Acad. Sci. USA* **106** 3675
[35] Sengupta S, Karmakar S, Dasgupta C and Sastry S 2012 *Phys. Rev. Lett.* **109** 095705
[36] Colmenero J 2015 *J. Phys.: Condens. Matter* **27** 103101
[37] Doi M and Edwards S F 1986 *The Theory of Polymer Dynamics* (Clarendon Press: Oxford)
[38] Grest G S and Kremer K 1986 *Phys. Rev. A : Rapid Comm.* **33** 3628
[39] Bennemann C, Paul W, Binder K and Dünweg B 1998 *Phys. Rev. E* **57** 843
[40] Bennemann C, Donati C, Baschnagel J and Glotzer S C 1999 *Nature* **399** 246
[41] Gebremichael Y, Schrøder T B, Starr F W and Glotzer S C 2003 *J. Chem. Phys.* **119** 5290
[42] Starr F W, Douglas J F and Sastry S 2013 *J. Chem. Phys.* **138** 12A541
[43] Adam G and Gibbs J H 1965 *J. Chem. Phys.* **43** 139
[44] Kirkpatrick T R, Thirumalai D and Wolynes P G 1989 *Phys. Rev. A* **40** 1045
[45] Kirkpatrick T R and Thirumalai D 2015 *Rev. Mod. Phys.* **87** 183
[46] Bouchaud J P and Biroli G 2004 *J. Chem. Phys.* **121** 7347
[47] Puosi F and Leporini D 2012 *J. Chem. Phys.* **136** 164901
[48] Ottochian A, Michele C D and Leporini D 2009 *J. Chem. Phys.* **131** 224517
[49] Barbieri A, Campani E, Capaccioli S and Leporini D 2004 *J. Chem. Phys.* **120** 437
[50] Ottochian A, Molin D, Barbieri A and Leporini D 2009 *J. Chem. Phys.* **131** 174902
[51] Puosi F and Leporini D 2011 *J. Phys. Chem. B* **115** 14046
[52] Helfferich J, Ziebert F, Frey S, Meyer H, Farago J, Blumen A and Baschnagel J 2014 *Phys. Rev. E* **89** 042603
[53] Helfferich J, Ziebert F, Frey S, Meyer H, Farago J, Blumen A and Baschnagel J 2014 *Phys. Rev. E* **89** 042604
[54] Dasgupta C, Indrani A V, Ramaswamy S and Phani M K 1991 *Europhys. Lett.* **15** 307
[55] Donati C, Glotzer S C and Poole P H 1999 *Phys. Rev. Lett.* **82** 5064
[56] Hansen J P and McDonald I R 2008 *Theory of simple liquids* (Academic press)
[57] Jose P P and Szamel G 2007 *J. Chem. Phys.* **127** 114905
[58] Flenner E and Szamel G 2007 *J. Phys.: Condens. Matter* **19** 205125
[59] Allen M P and Tildesley D J 1991 *Computer simulation of liquids* (Oxford University Press, New York)
[60] Kob W and Anderson H C 1995 *Phys. Rev. E* **51** 4626
[61] Berthier L, Biroli G, Bouchaud J P, Cipelletti L, ElMasri D, L'Hôte D, Ladieu F and Pierno M 2005 *Science* **310** 1797

Microscopic structure and dynamics of glass like domains in a low density binary mixture

Anna Varughese and Prasanth P. Jose
School of Basic sciences, IIT Mandi, Mandi, 175005, H. P., India
E-mail: prasanth@iitmandi.ac.in

Abstract.
We present results from extensive equilibrium molecular dynamic simulations of a binary Lennard Jones 80:20 mixture. Our study focuses on the changes in structural and dynamic properties of this system during its transition from a stable phase to an unstable phase as temperature reduces. This system, phase separate into a mixture of glass like domains and gas at lowest temperatures. We compute coarse-grained density(CGD) to identify density distribution in the system, which shows formation of very dense domains. Studies of bond order parameter shows a mixture of crystal structures in these domains at low temperatures. These results show that the structure of the dense domains formed at low density is akin to glassy domains that is found in studies of glass transition at high density, at low temperatures. Our studies on mean square displacement of particles at various state points shows pronounced sub-diffusive regime and plateau that marks the existence of dynamic arrest due to cage formation known in the studies of glassy and supercooled liquids.

1. Introduction

In simple model of liquids, such as Lennard-Jones (LJ) potential, show universal phases of gas, liquid and solid. It is a well studied fact that a binary mixture of LJ system that consists of particles differ in their radii and interaction strength, at state points of lower temperatures and high density, form a glassy or amorphous state[1, 2, 3]. A computationally well studied example of this system is 80:20 binary mixture used in the studies of Kob and Andersen which is extensively used for computational studies of glass transition [4, 5, 6, 7, 8, 9, 10]. One of the outstanding problem of the glass forming liquids is to understand nature and relaxation properties of metastable states formed at state points that is close to the glass transition temperature T_g [11, 12, 7, 13]. There are many computational studies of glass transition in the binary mixture which are done at high densities [14, 15, 16, 17, 18, 19, 20, 21]; where, it is difficult for the structural rearrangement to occur at low temperatures. Then one of the interesting question to ask is, what is the role of density in the formation of glassy domains or how the cooperative rearrangements that results in glassy dynamics is effected by density of the system. To investigate the role played by the density of the system in the formation of glassy or crystalline phases, let us first look into phase changes of a single component LJ system at low densities.

In many regions of the phase diagram of LJ liquid, see point (d) in Fig.1, at low densities and low temperatures where system cannot exist as a single phase: the system exists as a mixture of domains of vapor and liquid phases. These state points are considered as unstable states in the phase diagram of the system. In between (d) and (h) in Fig.1, where the system phase separate into random shaped gas and liquid domains [22, 23, 24]. In state points where temperature is further lowered, see point (h) in Fig.1, the liquid domains freeze into crystallites and exist as a mixture of crystallites and dilute gas. The crystallites formed can be considered as local potential energy minimized structures. Unlike a single component LJ system which at high densities and low temperatures, freezes into a crystal state, the glass forming binary mixture at these state points forms droplets of glassy domains [25, 24]. In non-equilibrium simulation studies of this binary mixture by Testard *et al.* [25, 24], they showed the temperature dependency of phase separation. Their study in binary mixture showed gel-like structures exist at low densities and temperatures. It is now interesting to study the structure and dynamics of these dense domains

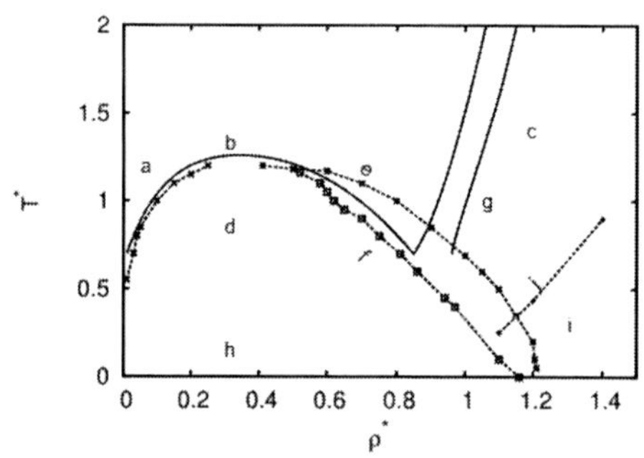

Figure 1. A comparison of phase diagram of simple and binary Lennard-Jones system. Solid line represents simple Lennard-Jones(SLJ) and dotted lines represent binary Lennard-Jones(BLJ) fluid. a is the gas regime for both BLJ and SLJ. b represents the critical point, c is the solid phase of SLJ. For BLJ, d shows the gas+liquid regime, e and f represents binodal and spinodal respectively, g represents liquid, h is a mixture of gas and glass, i represents glass and j represents the glass transition line. Schematic phase diagram is based on Ref.[26, 27, 25].

in the equilibrium state of this binary mixture at low temperatures and low densities. As these systems have excess free volume [28], therefore, the relaxation of these dense domains may differ from the glassy phase formed at high densities. In addition, it is also very interesting to see what are the local structures formed in a binary mixture at low densities. The structural changes of this binary fluid, in the state points of phase separation, while temperature is lowered, result in the structural arrest. We have studied these structural changes using the radial distribution function $g(r)$. A useful way of distinguishing dense domains is by using coarse grain density (CGD) defined in the studies of Testard *et al.* [24]. We have used this method to characterize the domains obtained in our simulations. Our work shows coexistence of dense amorphous domain as well as gas in the low density and low temperature state points of the binary mixture we have studied.

Another feature that characterizes clusters formed is their local structure. In an important work that characterize local structures of LJ liquid by Steinhardt *et al.* [29], they used bond order parameters to identify various possible ordered structures. Bond order parameter has been used in many studies of glass forming liquids to identify the local structures [30, 31, 32]. Studies of bond order parameter of 80:20 binary mixture at state points near glass transition at high densities shows coexisting collection of structures like hcp, fcc, bcc, *etc.*[33]. To identify the existence of local structures in the system we looked into the bond order parameter of nearest neighbors of each molecule at various state points we have studied which also show coexistence of the mixture of structures found at high density.

To further identify the cage relaxation process in the low density binary mixture we have computed mean squared displacement which clearly shows the sub-diffusive regime that get prolonged as temperature reduces. At lowest temperatures we find a plateau region in the mean squared displacements computed, which is a characteristic property of a glass forming system that is found in many studies [4, 34, 35, 36, 37].

This paper is organized as follows: In next Sec. model and computation details are presented. In section III we discuss the results from the simulations. Summary and conclusions are presented in section IV.

2. Model and Computational Details

We simulated 80:20 binary mixture widely used in studies of glass transition. Interaction potential between particles of binary Lennard Jones system is given by [4, 5, 6].

$$V_{\alpha\beta}(r) = 4\epsilon_{\alpha\beta}\left[\left(\frac{\sigma_{\alpha\beta}}{r}\right)^{12} - \left(\frac{\sigma_{\alpha\beta}}{r}\right)^{6}\right] \tag{1}$$

where, $\alpha, \beta \in \{A, B\}, \epsilon_{AA} = 1.0, \sigma_{AA} = 1.0, \epsilon_{AB} = 1.5, \sigma_{AB} = 0.8, \epsilon_{BB} = 0.5$, and $\sigma_{BB} = 0.88$ [4, 5, 6]. The interaction potential was truncated at a cutoff distance of $2.5\sigma_{\alpha\beta}$. All the quantities presented in reduced units of AA interaction, the Lennard Jones well depth ϵ_{AA} is the unit of energy, while the Lennard Jones diameter σ_{AA} is the unit of length, time in units of $t^* = \sqrt{m\sigma_{AA}^2/\epsilon_{AA}}$ and Boltzmann constant is set as $k_B = 1$. The equation of motion are determined by integrating Verlet algorithm with time step of 0.01 [24].

To study the properties of this system, we prepare equilibrated homogeneous system at high temperature $T = 3.0$ and at desired density. Then the system is quenched to state points of desired temperature and equilibrated before the production runs. Equilibration time used in our simulation is time that required for domain growth to saturate this is determined from the average time dependent chord length[24] reaches a plateau; it is nearly $10^3 t^*$ at lowest temperatures. State points that we investigated are at $T = 0.1, 0.2, 0.4, 0.6, 0.8, 1.0, 3.0$ and at density $\rho = 0.6$. Our system has 10^4 particles (8000 A and 2000 B) and the total length of the simulation is 10^6 number of steps.

3. Results and discussion

We have used radial distribution function $g(r)$ to identify the formation of secondary structures while temperature is reduced. The partial radial distribution function for the binary mixture is defined as[38] $g_{\alpha\alpha}(r) = \frac{V}{N_\alpha(N_\alpha - 1)}\langle\sum_{i=1}^{N_\alpha}\sum_{j=1}^{N_\alpha}\delta(\vec{r} - \vec{r}_{ij})\rangle$; $g_{AB}(r) = \frac{V}{N_A N_B}\langle\sum_{i=1}^{N_A}\sum_{j=1}^{N_B}\delta(\vec{r} - \vec{r}_{ij})\rangle$ where, $\alpha \in \{A, B\}$, $\vec{r}_{ij}$ is position vectors of the particles. The radial distribution function of the particles of three distinct pairs, AA, BB, and AB are given by $g_{AA}(r)$, $g_{BB}(r)$, and $g_{AB}(r)$. These radial distribution functions gives how respective species of the particles are distributed in the system. In the simulation of the binary mixture at high densities, these radial distribution functions show splitting of the second peak due to the formation of multiple structures at state points of low temperature[4, 39, 40, 31].

Figs. 2 shows radial distribution functions $g_{AA}(r), g_{AB}(r)$ and $g_{BB}(r)$ at $\rho = 0.6$ and at different temperatures. For better representation, curves have been displaced vertically in Fig. 2. As state points of lower temperatures peaks become sharper in all three radial distribution functions presented. In addition, $g_{AA}(r), g_{AB}(r)$ and $g_{BB}(r)$ show multiple splitting of second peaks at state points of low temperatures which is the signature of different local structures present in the system; this splitting of the secondary peaks similar to what found in the simulation studies of binary mixture at high densities [40, 4]. In contrast results of simulation studies at high density system, in our simulations, splitting of second peak at low density

Journal of Physics: Conference Series **759** (2016) 012019 doi:10.1088/1742-6596/759/1/012019

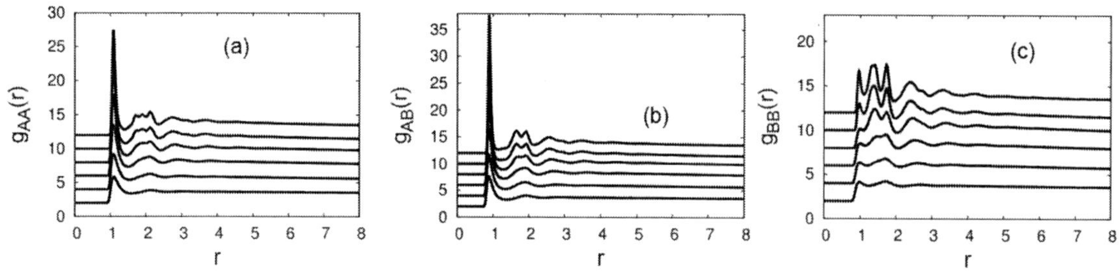

Figure 2. $g_{AA}(r), g_{AB}(r)$ and $g_{BB}(r)$ are shown in (a), (b), (c) respectively. Plots are arranged in the decreasing order of T from bottom to top, at values of $T = 1.0, 0.8, 0.6, 0.4, 0.2, 0.1$.

shows multiple peaks, which indicates the existence of the various structures that differ from each other in this system. Another characteristic of $g_{AA}(r), g_{AB}(r)$ and $g_{BB}(r)$ is , at low temperatures their mean go above value 1 which shows clustering of particles the system, which is the expected behavior of the system in unstable region given in Fig.1. It is now interesting to see the characteristics of these dense domains formed and the corresponding structures.

To identify the nature of high density domains we have used method of CGD proposed by Testard *et al.*[24]. This method can directly measure the average density of a domain and it also this method helps to identify various domains and their interfaces. The methodology can be summarized as follows. First discretize the simulation box into multiple cells of linear size ξ_b, so a total of V/ξ_b^3 sites[24]. Then for each cell we compute discrete density defined as $\rho(\vec{r}) = \frac{3}{4\pi\xi_s^3}\sum_{i=1}^{N}\theta(\xi_s - |(\vec{r} - \vec{r}_i)|)$ where $\theta(x)$ is the Heaviside function, ξ_s is the radius of sphere from center of each cell. The final CGD computed by considering the neighboring cells too and it is given by

$$\bar{\rho}(\vec{r}) = \frac{1}{8}[2\rho(\vec{r}) + \rho(\vec{r} + \xi_b\vec{e}_x) + \rho(\vec{r} + \xi_b\vec{e}_y) + \rho(\vec{r} + \xi_b\vec{e}_z) + \rho(\vec{r} - \xi_b\vec{e}_x) + \rho(\vec{r} - \xi_b\vec{e}_y) + \rho(\vec{r} - \xi_b\vec{e}_z)] \quad (2)$$

Then the probability distribution of CGD is computed.

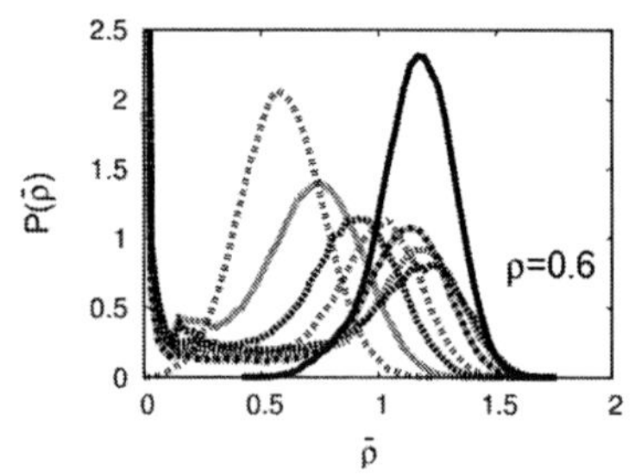

Figure 3. CGD distribution plots are arranged in the decreasing order of $T = 3.0, 1.0, 0.8, 0.6, 0.4, 0.2, 0.1$ from left to right. Solid line represent CGD computed from simulations at ρ=1.2, and T=0.4 in our model.

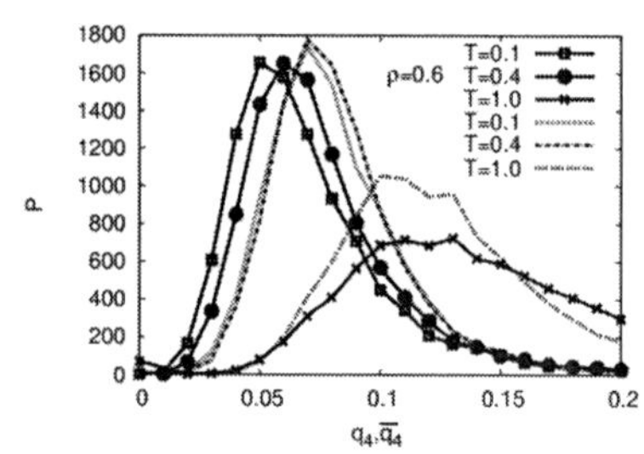

Figure 4. Dotted lines show local bond order distribution q_4 and solid lines show average local bond order distribution $\bar{q}_4$ for various temperatures.

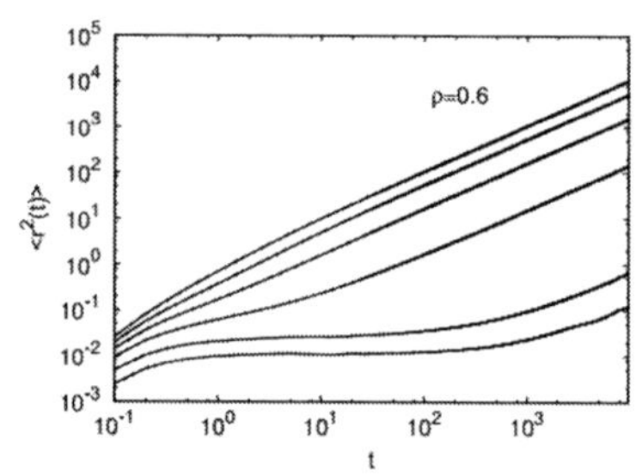

Figure 5. Mean squared displacement. Plots are arranged in the decreasing order of $T = 1.0, 0.8, 0.6, 0.4, 0.2, 0.1$ from top to bottom.

The CGD distribution we get is similar to that obtained by Testard *et al.* [24] at state points $T = 0.1$ and $\rho = 0.6$. Fig. 3 shows CGD at $\rho = 0.6$ for various temperatures. At low temperatures CGD shows two peaks: first belong to regions of gas phase; second belongs to dense phase. Between these phases there are regions of interface which also appears in the CGD. At high temperatures, $T = 3.0$ and density $\rho = 0.6$, CGD shows a homogeneous distribution of densities in the cells that appears as a single peak. We see that at low temperatures, the system prefer to exist in a dense domains whose CGD ranges from 0.8 to 1.4, even though simulation are performed at a reference density of 0.6; in a homogeneous system CGD peaks around reference density. It is an indication that there exist a preferred range of densities in in these dense domains. Many simulations of glass transition in this binary mixture is performed at $\rho = 1.2$, CGD computed at this density near glass transition temperature in our model is given by the solid line in Fig. 3, which show same CGD as that found at low density.

In order to look into various structures formed in these dense domains, we compute the bond order parameter which indicate the existence of crystalline as well as amorphous structures [29, 41]. In the liquid where there is only a transient orientational order, bond order parameters are small. When local structures stabilize as temperature is reduced there is considerable increase in the value of bond-order parameters. The local bond order parameter is defined as[29, 42] $q_l(i) = \sqrt{\frac{4\pi}{2l+1}\sum_{m=-l}^{l}|q_{lm}(i)|^2}$ where, $q_{lm}(i)$ is the local orientational order vectors. The averaged form of local bond order is given by $\bar{q}_l(i) = \sqrt{\frac{4\pi}{2l+1}\sum_{m=-l}^{l}|\bar{q}_{lm}(i)|^2}$ where, $\bar{q}_{lm}(i)$ is the local orientational order vectors averaged over particle i and its surroundings. We computed the average local bond order, see Fig. 4 for the system and compared with standard values q_4=0.08, $\bar{q}_4$=0.03 for bcc, q_4=0.01, $\bar{q}_4$=0.08 for hcp, See Tab.I of Ref. [42]. A similar result has been obtained in the calculation of bond order parameters at high densities in the same system [33]. This shows structural similarities of dense domains obtained in this simulations with density distribution of a glass forming system near its glass transition temperature.

Many studies support the view that glass transition occurs due to the dynamical arrest or cage formation in glass systems. One way to characterize the single particle cage is by studying the mean square displacement $\langle r^2(t)\rangle = \frac{1}{N}\sum_{i=1}^{N}\langle|\vec{r}_i(t) - \vec{r}_i(0)|^2\rangle$ where $\vec{r}_i(t) - \vec{r}_i(0)$ is the particle displacement of ith particle over a time t. The cage formation in the supercooled and glassy system is indicated by the long sub-diffusive regime that appears at lower temperatures [4, 34, 35, 36, 37] near T_g. Fig. 5 shows plot of mean square displacement for $\rho = 0.6$ for various temperatures. At high temperatures $\langle r^2(t)\rangle$ of the particles changes from ballistic region to diffusive region. Below temperature T=0.6, $\langle r^2(t)\rangle$ shows a sub-diffusive region between ballistic and diffusive regions as shown in the Fig. 5. A sub-diffusive region which extends when temperature reduces called plateau region, suggest cage formation in this system [4, 16], that results in slow relaxation, which is found in the studies of many glass forming liquids at high density $\rho = 1.2$ [4, 34, 35, 36, 37] and in colloids [43].

4. Summary and Conclusions

We have done long time molecular dynamic simulation study on the formation of glass like domains at the unstable region of the 80:20 binary mixture which is widely used in the studies of glass transition at high densities[4, 5, 6]. The radial distribution functions $g_{AA}(r)$, $g_{BB}(r)$ and $g_{AB}(r)$ computed have shown the existence of multiple local structures at low temperatures. In comparison to studies of same system at high densities [40, 4], the second peak of the radial distribution functions split into multiple peaks. The local structure of these dense domains are studied using method of CGD given by Testard *et. al.*[24]. We found that uniform phase at high temperature splits into gas and liquid phases at intermediate temperature. At very low temperatures the system show coexistence of dense domains whose density distribution is comparable to that near the state points used in simulation studies of glass transition at high density [4, 5, 6]. Multiple structures that appear in these dense domains are characterized by the bond order parameters [29]. We find the existence of mixture of bcc, hcp structures in this system which is also found at high density[33]. The cage relaxation properties of these dense systems are studied using mean square displacement of particles of the system. The mean squared displacement shows clearly the growth of the sub-diffusive regime which indicate the formation and caging that is found in the computational studies of models of glassy and supercooled liquids[4, 34, 35, 36, 37]. Thus our studies indicate presence high density domains in phase separating low density regime of 80:20 binary mixture, whose structure is comparable to what found high density studies of same system near glass transition. The nature of diffusive motion of particle at these state points is also comparable to that found in high density studies of glass transition in the same system.

PPJ acknowledge DRDO for partial financial support and HPC facility at IIT Mandi for computational support.

References

[1] Bengtzelius U, Gotze W and Sjolander A 1984 *J. Phys.: Condens. Matter* **17** 5915
[2] Wahnstrom G 1991 *Phys. Rev. A.* **44** 3752
[3] Ediger M D, Angell C A and Nagel S R 1996 *J. Phys. Chem.* **100** 13200
[4] Kob W and Andersen H C 1995 *Phys. Rev. E.* **51** 4626
[5] Kob W and Andersen H C 1995 *Phys. Rev. E.* **52** 4134
[6] Kob W and Andersen H C 1994 *Phys. Rev. Lett.* **73** 1376
[7] Debenedetti P G and Stillinger F H 2001 *Nature* **410** 259
[8] Sastry S, Debenedetti P G and Stillinger F H 1998 *Nature* **393** 554
[9] Mukherjee A, Bhattacharyya S M and Bagchi B 2002 *J. Chem. Phys.* **116** 4577
[10] Karmakar S, Dasgupta C and Sastry S 2009 *Proc. Natl. Acad. Sci. USA* **106** 3675
[11] Cavagna A 2009 *Physics Reports.* **476** 51
[12] Angell C A, Ngai K L, McKenna G B, McMillan P F and Martin S W 2000 *J. App. Phys.* **88** 3113
[13] Berthier L and Biroli G 2011 *Rev. Mod. Phys.* **83** 587
[14] Donati C, Glotzer S C, Poole P H, Kob W and Plimpton S J 1999 *Phys. Rev. E.* **60** 3107
[15] Donati C, Douglas J F, Kob W, Plimpton S J, Poole P H and Glotzer S C 1998 *Phys. Rev. Lett.* **80** 2338
[16] Flenner E and Szamel G 2005 *Phys. Rev. E.* **72** 011205
[17] Berthier L and Tarjus G 2009 *Phys. Rev. Lett.* **103** 170601
[18] Pedersen U R, Schroder T B and Dyre J C 2010 *Phys. Rev. Lett.* **105** 157801
[19] Berthier L 2004 *Phys. Rev. E.* **69** 020201
[20] Malins A, Eggers J, Tanaka H and Royall C P 2013 *Faraday Discuss.* **167** 405
[21] Jack R L, Hedges L O, Garrahan J P and Chandler D 2011 *Phys. Rev. Lett.* **107** 275702
[22] Debenedetti P G 1995 *Metastable liquids: concepts and principles* (New Jersey: Princeton university press)
[23] Majumder S and Das S K 2011 *EPL* **95** 46002
[24] Testard V, Berthier L and Kob W 2014 *J. Chem. Phys.* **140** 164502
[25] Testard V, Berthier L and Kob W 2011 *Phys. Rev. Lett.* **106** 125702
[26] Hansen J P and Verlet L 1969 *Phys. Rev.* **184** 151
[27] Johnson J K and Gubbins K E 1992 *Mol. Phys.* **77** 1033
[28] Cohen M H and Grest G S 1979 *Phys. Rev. B.* **20** 1077
[29] Steinhardt P J, Nelson D R and Ronchetti M 1983 *Phys. Rev. B.* **28** 784
[30] Coslovich D and Pastore G 2007 *J. Chem. Phys.* **127** 124504
[31] Vollmayr K, Kob W and Binder K 1996 *J. Chem. Phys.* **105** 4714
[32] Leocmach M and Tanaka H 2012 *Nat. Commun.* **3** 974
[33] Banerjee A, Chakrabarty S and Bhattacharyya S M 2013 *J. Chem. Phys.* **139** 104501
[34] Binder K and Kob W 2011 *Glassy Materials and disordered Solids* (Singapore: World Scientific)
[35] Varnik F, Bocquet L and Barrat J L 2004 *J. Chem. Phys.* **120** 2788
[36] Berthier L and Kob W 2007 *J. Phys.: Condens. Matter* **19** 205130
[37] Schroder T B and Dyre J C 1998 *J. Non-Cryst. Solids* **235** 331
[38] Hernandez-Rojas J and Wales D J 2003 *Phys. Rev. B.* **68** 144202
[39] Marcotte E, Stillinger F H and Torquato S 2013 *J. Chem. Phys.* **138** 12A508
[40] Ingebrigtsen T S, Schroder T B and Dyre J C 2012 *Phys. Rev. X.* **2** 011011
[41] Tanaka H, Russo J, Leocmach M and Kawasaki T 2013 *AIP Conf. Proc.* **1518** 143
[42] Lechner W and Dellago C 2008 *J. Chem. Phys.* **129** 114707
[43] Weeks E R, Crocker J C, Levitt A C, Schofield A and Weitz D A 2000 *Science* **287** 627

Energetics of dendrimer binding to HIV-1 gp120-CD4 complex and mechanismic aspects of its role as an entry-inhibitor

Suman Saurabh, Anil Kumar Sahoo and Prabal K. Maiti

Center for Condensed Matter Theory, Indian Institute of Science, Bangalore 560012, Karnataka, India

E-mail: `maiti@physics.iisc.ernet.in`

Abstract. Experiments and computational studies have established that de-protonated dendrimers (SPL7013 and PAMAM) act as entry-inhibitors of HIV. SPL7013 based Vivagel is currently under clinical development. The dendrimer binds to gp120 in the gp120-CD4 complex, destabilizes it by breaking key contacts between gp120 and CD4 and prevents viral entry into target cells. In this work, we provide molecular details and energetics of the formation of the SPL7013-gp120-CD4 ternary complex and decipher modes of action of the dendrimer in preventing viral entry. It is also known from experiments that the dendrimer binds weakly to gp120 that is not bound to CD4. It binds even more weakly to the CD4-binding region of gp120 and thus cannot directly block gp120-CD4 complexation. In this work, we examine the feasibility of dendrimer binding to the gp120-binding region of CD4 and directly blocking gp120-CD4 complex formation. We find that the process of the dendrimer binding to CD4 can compete with gp120-CD4 binding due to comparable free energy change for the two processes, thus creating a possibility for the dendrimer to directly block gp120-CD4 complexation by binding to the gp120-binding region of CD4.

1. Introduction

The entry of HIV into target CD4$^+$ T cells in humans is initiated by the binding of the viral surface protein gp120 to the target cell receptor protein CD4 [1]. This binding induces structural changes in gp120, which help it bind to the target cell trans-membrane protein CCR5 or CXCR4 leading to further conformational changes resulting in membrane fusion. The disruption of binding between the proteins that facilitate viral entry is one of the many drug-discovery strategies aimed at developing an HIV inhibitor. Small molecule drugs like maraviroc, which blocks binding of gp120 to CCR5 (second stage in the entry process), utilize such a strategy [2]. While the swift resistance mechanism of the virus renders individual drugs incapable of performing their pre-defined function after a brief period of use, the search for a drug with a large genetic barrier to resistance is still on.

Dendrimers are highly branched polymers. They consist of a core, a branching unit and terminal groups [3]. Among many families of dendrimer molecules, SPL7013 and PAMAM dendrimers are experimentally known to exhibit anti-retroviral activity [4]. Time-of-addition experiments indicate that these dendrimers act as entry inhibitors [5]. The multivalent nature of the dendrimer allows it to bind in multiple possible conformations with various bio-molecules.

Unlike small molecule drugs, which bind to the cavities on protein surfaces, the dendrimer, being a large molecule, can bind a protein over a very large contact area. Although there is no specific binding conformation, the huge loss in entropy due to dendrimer binding imparts partial specificity to the binding process. SPL7013 is thought to act via different mechanisms on different viral strains (R5, X4 and R5X4). In a recent computational study of the stability of R5 HIV-1 gp120-CD4 complex in the presence of SPL7013, we have elucidated the potential mechanism of its action against R5 viruses, which was previously unknown [6]. The dendrimer destabilizes the gp120-CD4 complex by docking to R5 gp120, allosterically breaking many key gp120-CD4 contacts and in the process helping preclude viral entry. A similar effect of the PAMAM dendrimer on the gp120-CD4 complex was observed in another computational study [7], where the binding energy of the complex was found to decrease drastically in the presence of the dendrimer.

The HIV-1 gp120 is structurally complex. It has two main domains (inner and outer) connected by a bridging sheet (see Figure 1). The protein also contains fluctuating loops which produce a constantly moving target for the immune system, enabling the virus to evade human immune responses.

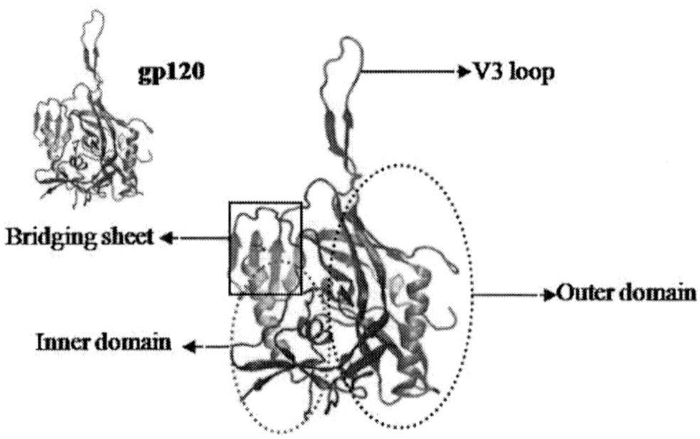

Figure 1. Structure of gp120 (top view), with different regions labeled. The black and blue ellipses highlight the outer and inner domains respectively. The rectangle covers the bridging sheet (β-sheets in green). V3 loop is shown as the extended structure (mauve).

gp120 binds to CD4 with low affinity. The CD4 binding region of gp120 contains many unstructured loops. Thus the huge enthalpy of binding ($\sim$ -62 kcal/mol) [8] is offset by the large loss in entropy ($\sim$ -53 kcal/mol) [8]. The dendrimers further destabilize the complex and prevent the formation of the number of gp120-CD4 complexes necessary for viral entry.

In this work, we study the molecular details of SPL7013-gp120-CD4 ternary complex, using molecular docking, equilibrium simulations and binding energy calculations. We focus on the V3 loop and study various effects of its strong positive charge on dendrimer-gp120 binding. We also aim at verifying the experimental findings regarding the mechanism of action of SPL7013. The dendrimer when docked to gp120-CD4 complex binds to gp120. In our previous study [6] we found that when the dendrimer docked to gp120 in the absence of CD4, there were only a few low ranked docking output conformations from ZDOCK [9] wherein SPL7013 was bound to the CD4-binding region of gp120. We showed that the binding energy of such binding conformations is very low compared to the one where SPL7013 binds to the V3 loop of gp120. This shows that the dendrimer cannot prevent gp120-CD4 binding by blocking the CD4 binding region of gp120. That SPL7013 binds weakly to R5 gp120 compared to the gp120-CD4 complex is known [6]. When, in experiments, R5 HIV-1 was exposed to SPL7013 and then used to infect cells in the absence of the dendrimer, the dendrimer was found to be inefficacious in preventing viral entry. Whereas the dendrimer was very effective with R5 HIV-1 when present in culture with the virus and target cells. This indicates two possible mechanisms for the dendrimer to stop gp120 from binding to CD4. Firstly, the dendrimer can bind to gp120 in complex with CD4 and allosterically hinder their complexation by breaking key contacts. Secondly, before gp120-CD4 complexation occurs, it can bind to the region of CD4 that goes into the CD4-binding cavity of gp120 in the gp120-CD4 complex (the gp120-binding region of CD4) and block their binding. Although the first possibility is well explored and is energetically feasible, the feasibility of the second possibility has not been studied. Here, we study the energetics and consequent feasibility of the second option.

2. Materials and Methods

2.1. Building the SPL7013 dendrimer

The SPL7013 dendrimer was built using Dendrimer Builder Toolkit (DBT) [10] developed some years back in our laboratory. The core, repeating units and terminal residues were designed as per the chemistry and topology of the dendrimer. All the residues with cap(s) were optimized using GAUSSIAN03 [11] with the HF/6-31+G (d,p) basis set. Restrained electrostatic potential (RESP) [12] charges were calculated using Antechamber module [13] of AMBER. During RESP calculation, cap charges were set to zero. The net charge on the core and repeating residues were also set to zero. To have the anionic form, we set the charge on the terminal de-protonated residue to be -2. Post optimization and RESP charge calculation, the caps were removed from the residues using xleap module in AMBER. The de-capped residues thus obtained were used for building the SPL dendrimer of different generations using DBT. The optimized structure of generation 4 (G4) SPL7013 dendrimer was thus obtained.

2.2. MD simulations of individual molecules

We used the AMBER12 [14] software package with GAFF [15] set of parameters for the SPL7013 G4 dendrimer. The dendrimer was solvated with TIP3P water model with a 13 Å hydration shell in all three directions using the xleap module. Additionally, 64 Na^+ ions were added to make the system charge neutral.

For the gp120-CD4 complex, we have used the crystal structure of the YU2 gp120 core complexed with CD4 and a functionally sulfated antibody F12d (PDB id: 2QAD) [16]. The antibody was removed from the crystal structure to get the initial 3-D model of the gp120-CD4 complex. The complex was solvated using TIP3P model of water using xleap module of AMBER12 [14]. 9 Cl^- ions were also added to make the solvated gp120-CD4 system charge neutral. The system contained 275726 atoms with 89310 water molecules and 9 Cl^- ions.

Similarly, to study the CD4 structure, we removed both gp120 and the F12d antibody and obtained the structure of monomeric CD4. The structure was solvated in TIP3P water with a water buffer of 30 Å in the x- and y-directions and 15 Å in the z-direction. The solvated system was then neutralized by adding 4 Cl^- ions using xleap.

We have used ff99SB parameters [17] to describe inter and intra-molecular interactions corresponding to the proteins. Ions were described using the Joung-Cheatham parameter set [18]. All the above three systems namely solvated SPL system, solvated gp120-CD4 complex as well as solvated CD4 system were energy minimized using the following protocol: we first performed 1000 steps of steepest descent minimization. This was followed by another 2000 steps of conjugate gradient minimization. During the minimization, the solute atoms were fixed to their initial co-ordinates using harmonic constraints with a force constant of 500 kcal/mol/$Å^2$ while the water molecules were allowed to reorganize and eliminate unfavorable contacts with the solute. The systems were further subjected to 5000 steps of conjugate gradient minimization, with the harmonic constraints on the solute going from 20 kcal/mol/$Å^2$ to 0 with a reduction of 5 kcal/mol/$Å^2$ every 1000 steps. After minimization the systems were gradually heated from 0 to 300K during a 40 ps long MD simulation. During heating, the solute atoms were restrained to their initial position with a force constant of 20 kcal/mol/$Å^2$, which allowed slow relaxation of the solute. SHAKE method [19] was use for constraining bonds involving hydrogen with a geometrical tolerance of 5 x 10^{-4} Å. This allowed the use of a 2 fs time step. The long range electrostatic interactions were handled with the Particle Mesh Ewald (PME) method. The real space cut-off was set to 9 Å. Finally, the dendrimer was subjected to 82 ns long NPT simulation. While for the protein systems, a 500 ps long NPT run was performed followed by 100 ns and 60 ns long NVT runs for gp120-CD4 complex and monomeric CD4 respectively. The resulting structures of CD4 and the gp120-CD4 complex were employed for further docking and simulation studies.

2.3. Docking of SLP7013 to CD4 and to the gp120-CD4 complex

ZDOCK [9], an automated protein docking server, was used for docking the gp120-CD4 complex and the de-protonated SPL7013 dendrimer. ZDOCK uses a scoring function which is based on sum of scores from electrostatics, surface complementarity and desolvation energy. Desolvation energy is defined as the energy required for breaking two amino acid-water bonds and forming two amino acid-amino acid bonds. The gp120-CD4 complex after 40 ns of MD simulation in the NVT ensemble and the dendrimer after 80 ns of MD simulation in the NPT ensemble (see above) were employed as the receptor and the ligand, respectively, for docking. We also employed several other structures of the dendrimer between 70 and 80 ns of MD simulation to examine the robustness of our findings towards the initial structures employed, and found very similar docking conformations. The input parameters were kept at their default values in ZDOCK. Of the resulting docked structures, we considered the top 8 for further analysis.

We repeated a similar procedure with monomeric CD4 (using structures after 16 ns, 20 ns and 50 ns of MD simulation docked to 45 ns, 60 ns and 80 ns dendrimer structures respectively) as receptor and found that almost all of the docking outputs had the dendrimer bound to the region of CD4 that goes into the CD4 binding site of gp120 in the gp120-CD4 complex. We selected one conformation out of the top ten ZDOCK predictions for each input receptor-ligand pair. The three selected CD4-SPL7013 complexes were further used for equilibrium simulations and energy calculations.

2.4. MD simulation and energetics of docked structures

Even though ZDOCK assigns ranks based on a sophisticated scoring function, in light of the CAPRI test runs, where ZDOCK predicted correct binding poses for only half of the targets used in the study, it is important to check whether ZDOCK ranked the complexes correctly. Accordingly, we performed MD simulations of the top 8 SPL7013-gp120-CD4 docked structures to compute their binding energies and examine their stability.

Each of the docked structures was immersed in a water box (with at least a 30 Å water layer in all the three directions). 64 Na^+ ions and 9 Cl^- ions were added for charge neutrality. The structures were allowed to equilibrate and were subjected to 60 ns or more of MD simulation in an NVT ensemble. The binding energy between gp120-CD4 (receptor) and the dendrimer (ligand) was then calculated using the MMGBSA [20] module of AMBER12. Entropy calculations were also performed using normal mode analysis (see below). From the binding energy calculations, we found that the highest ranked ZDOCK structure was not the most stable. We re-ranked the ZDOCK outputs based on our energy calculations. In the text we will mention the complexes according to their energy ranks (complex 1 to complex 8). We extended the NVT runs for complexes 1, 3, and 8 to a total of 100 ns. The structures of the ternary complexes studied in this work are shown at appropriate places in the text.

For CD4-SPL7013 docking, the SPL7013 docks near the gp120-binding region of CD4 for majority of the ZDOCK generated structures (28 out of 30 structures). We selected three structures with considerably different binding conformations (see Results and Discussion). These systems were solvated in TIP3P water box with a buffer of 33 Å in the x- and y- directions and 13 Å in the z-direction. 64 Na^+ ions and 4 Cl^- ions were added for charge neutralization. The resulting solvated structures were first energy minimized using the same protocol as described above, after which a 500 ps long NPT simulation was performed followed by a 50 ns long NVT simulation.

2.5. Calculation of binding free energy of various complexes

We used the MMGBSA method employed in the MMPBSA.py [20] module of AMBER12 [14] to calculate the binding free energy of various binary and ternary complexes. The free energy is computed as $\Delta G = \Delta E_{bind}$-T$\Delta S_{bind}$, where the binding enthalpy, $\Delta E_{bind} = \Delta E_{ele} + \Delta E_{vdw} + \Delta E_{int} + \Delta E_{sol}$ consists of changes in the electrostatic energy, ΔE_{ele}, van der Waals energy, ΔE_{vdw}, the internal energy from bonded terms, ΔE_{int}, and the solvent contribution, ΔE_{sol}. The latter contribution, $\Delta E_{sol} = \Delta E_{es} + \Delta E_{nes}$, is the sum of the electrostatic energy, ΔE_{es}, calculated using the Generalized Born (GB) method, and the non-electrostatic energy, ΔE_{nes}, given by γSASA+β, ($\gamma = 0.00542$ kcal/Å^2 is the surface tension, $\beta = 0.92$ kcal/mol, and SASA is the solvent-accessible surface area of the molecule [21]). We performed similar calculations to estimate the binding energies of CD4-SPL7013 complexes.

Binding entropy was also calculated for all the complexes. Rotational and translational entropies were calculated by considering the receptor and ligand as rigid rotors. Harmonic approximation was employed. The use of rigid rotor model enables the use of a reduced Hessian (6X6) that needs to be diagonalized. The vibrational term was calculated using normal mode analysis which also uses harmonic approximation for entropy calculation. This method requires the diagonalization of the full fledged Hessian (3N X 3N, N being the number of atoms) for determination of the normal modes.

3. Results and Discussion

3.1. Energetics of dendrimer binding to gp120-CD4 complex

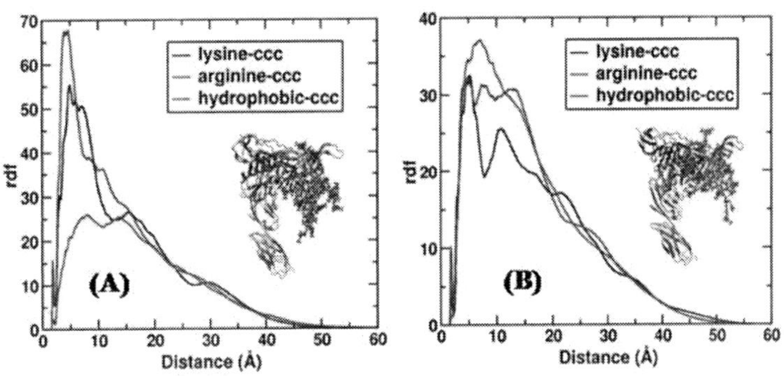

Figure 2. (A) and (B) Distribution of the terminal aromatic groups of SPL dendrimer (ccc) around the charged and hydrophobic residues of two different dendrimer binding conformations corresponding to complex 1 and complex 3 respectively. The distribution has been calculated around the charged and hydrophobic residues that form atomic contacts with the dendrimer. The ternary complexes are shown in the inset.

The SPL7013 dendrimer has naphthalene disulphonic acid based anionic aromatic residue as the terminal group [6]. The dendrimer interacts with gp120 mainly through electrostatic and hydrophobic interactions. The dendrimer binds to gp120 in a manner in which the interactions of the terminal groups are optimally satisfied. We plot (see Figure 2) the radial distribution of the terminal groups (named ccc) around positively charged and hydrophobic residues of gp120 which form contacts with the dendrimer in case of complex 1 and complex 3. We find that the terminal groups of the dendrimer are placed very close to the positively charged residues of gp120 (~5 Å), whereas the average distance from the hydrophobic residues is a bit larger (~7.5 Å) and the corresponding peaks are smaller than those corresponding to the positively charged residues. On closer observation we find that the terminal rings are oriented with their planes parallel to the guanidinium groups of arginine. Such stacking interaction between aromatic rings and the basic residues is well known in literature [22,23] and provides stability over and above the long-range electrostatic interaction. We have also calculated the residue-wise energy contributions to the protein-dendrimer binding energy for complex 1 and complex 3 corresponding to various gp120 residues that form atomic contacts with the dendrimer (see Figure 3). Atomic contacts were defined to exist when an atom of a gp120 residue fell within 3 Å of an atom belonging to the dendrimer. We observe that the ARG and LYS residues have the largest negative contribution to the total binding energy (see Figure 3(A) and (B)). In case of complex 1, ARG-142, LYS-143, ARG-151, LYS-172, ARG-239, ARG-267 and ARG-271 provide large stabilization to the ternary complex. In case of complex 3, LYS-45, ARG-142, LYS-143, ARG-151, ARG-164, ARG-246 and ARG-267 have large contributions to the stability of the ternary complex. We also see that in case of complex 3 GLU-156 has a significant positive contribution to the binding energy owing to its negative charge. All this shows that the electrostatic interaction plays a dominant role in the binding of the dendrimer to the gp120-CD4 complex and the binding poses for the complex are arrived at with a priority to electrostatic considerations.

The V3 loop of gp120 plays a very important role in stabilizing the SPL7013-gp120-CD4 complex. The base of the V3 loop is a region of very high positive charge density and the poly-anionic dendrimer can gain large electrostatic stabilization by choosing to bind around this region. In Figure 3(C), we show the positions of the positively charged residues that have large contributions to the stability of the ternary complexes 1 and 3. We observe that all these residues lie on the V3 loop, near the V3 loop base and the bridging sheet. We performed a simulation with the dendrimer initially placed at a center of mass distance of 50 Å from gp120 in complex with CD4. When the system was time evolved, we found that the dendrimer very quickly approached gp120 and docked on the V3 loop as shown in Figure 4(A). For the SPL7013-gp120-CD4 ternary complexes with initial docking position of SPL7013 away from V3 loop (complex 7 and 8), the dendrimer was found to move towards the V3 loop during the course of the simulation. We calculated the binding energy between gp120-CD4 and the dendrimer for complex 7 at different stages of the simulation and found that as the dendrimer approaches the V3 loop, the binding energy increases significantly (see Figure 4(B)). This highlights the importance of the V3 loop region of gp120 in dendrimer binding and suggests that if the virus attains mutations in this region, it would significantly affect the efficacy of the dendrimer.

3.2. Dendrimer destabilizes gp120-CD4 complex by inducing binding-position dependent conformational changes into gp120

It is known from earlier studies that the dendrimer binds to gp120 in the gp120-CD4 complex and disrupts key contacts between the two proteins by modifying the conformation of gp120 [6]. In this section we show that the manner and degree of conformational change induced in gp120 depends mainly on the position of dendrimer binding, and so does the degree of destabilization of the gp120-CD4 complex. Unlike small drug molecules, which bind to specific binding pockets on the target protein, dendrimer can bind at different positions, engage a large number of gp120 residues in the binding process and thus is expected to have a huge genetic barrier to resistance, which is a key property of a robust drug. In our earlier work, we studied the effect of dendrimer binding on the conformation of gp120 for complex 3 [6]. We found that the dendrimer induces a significant global tilt in gp120 with respect to CD4. In addition to that, it physically forms contacts with the residues on the bridging sheet near the CD4 binding site and through

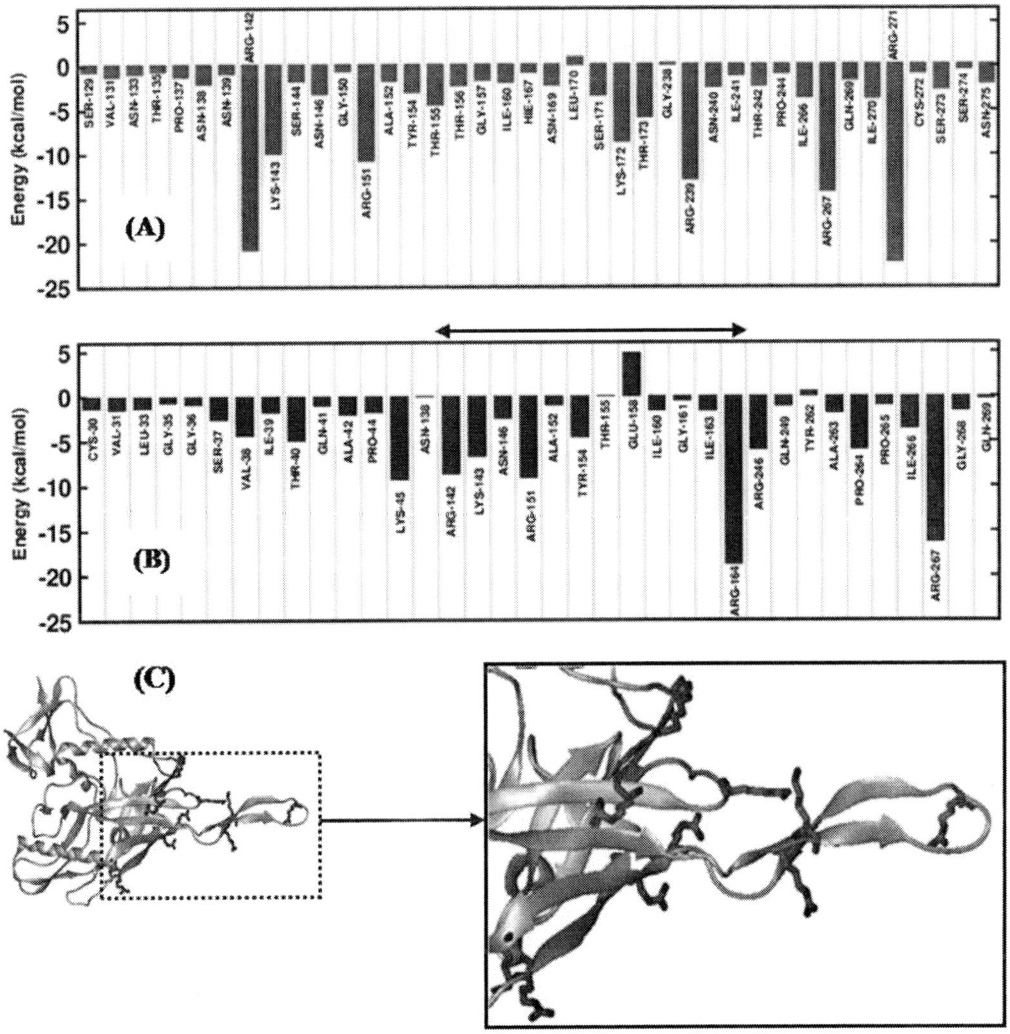

Figure 3. (A) The residue-wise contributions corresponding to various gp120 residues to the dendrimer-protein binding energy for complex 1. (B) The residue-wise contributions corresponding to various gp120 residues to the dendrimer-protein binding energy for complex 3. The residues belonging to V3 loop in case of complex 3 are marked on the x-axis. (C) Locations of positively charged residues of gp120 that have a large energy contribution to the binding of the dendrimer to gp120-CD4 complex.

electrostatic interactions, pulls various gp120 residues out of the contact of CD4. To show that the dendrimer can actually cause the disruption in contacts in multiple ways, we study the effect of the dendrimer on complex 8 with the dendrimer binding exactly on top of the V3 loop, as shown in Figure 5. In our previous study [6] we had found that complex 3 and complex 8 showed significant reduction in gp120-CD4 binding energy as compared to the gp120-CD4 binary complex in the absence of the dendrimer.

First thing to notice is that there is no relative tilt in this case as was seen for complex 3 (see Figure 5). So the tilt was a result of a particular dendrimer binding position. The most

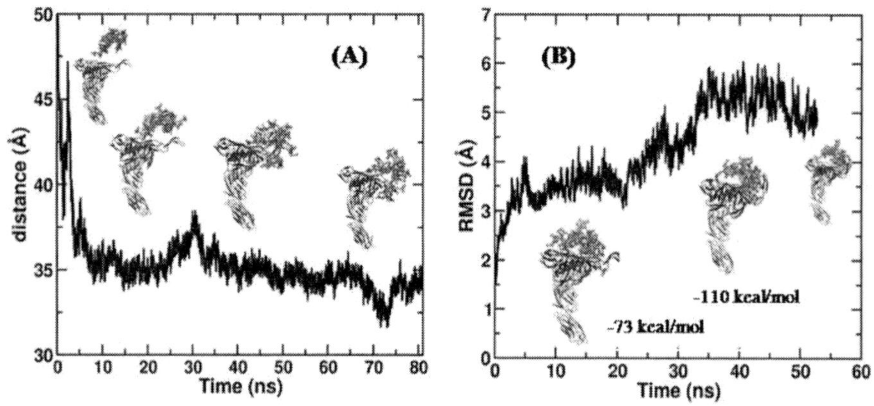

Figure 4. (A) Time evolution of the distance between gp120 and the dendrimer initially placed 50 Å apart along with snapshots at various times. The dendrimer approaches gp120 and attaches to the V3 loop. (B) RMSD of complex 7 with respect to the minimized structure. The figures in the insets show the conformation of complex 7 at various stages of the simulation. One can clearly see how the dendrimer shows displacement towards the V3 loop.

significant difference between the structure without dendrimer and the structure of complex 8 is in the conformation of the V3 loop with respect to the other domains of gp120. It is well known that the V3 loop acts as an electrostatic modulator of gp120 conformation [24] and thus the conformation of gp120 will strongly depend on the conformation of the V3 loop. By freezing

Figure 5. Conformational changes in gp120-CD4 complex on dendrimer binding. The structures corresponding to the average of 400 conformations corresponding to the last 4 ns of MD trajectory of complex 8 (red and yellow) and no dendrimer (blue and mauve) simulation superposed with the CD4 structures aligned. The difference in orientations of the V3 loop is clearly visible. The position of the dendrimer is shown in the beads representation on the red gp120.

the V3 loop in an unusual orientation, the dendrimer would induce conformational changes in other regions of gp120, like the bridging sheet and hence would affect gp120-CD4 binding. So, in case of complex 8 the dendrimer does not disrupt gp120-CD4 binding by physically pulling gp120 residues out of the influence of CD4. It does so by unusually deforming the V3 loop which in-turn induces conformational changes in gp120. The comparison of the mode of action of the dendrimer for complexes 3 and 8 indicate that the dendrimer can act in a number of different ways to disrupt the gp120-CD4 complex, depending on its binding position on gp120. Thus the virus may not be able to escape the entry-inhibitory property of the dendrimer by undergoing mutations that destabilize a particular binding conformation of the dendrimer as there will always be an alternative dendrimer binding conformation that can still destabilize the gp120-CD4 complex.

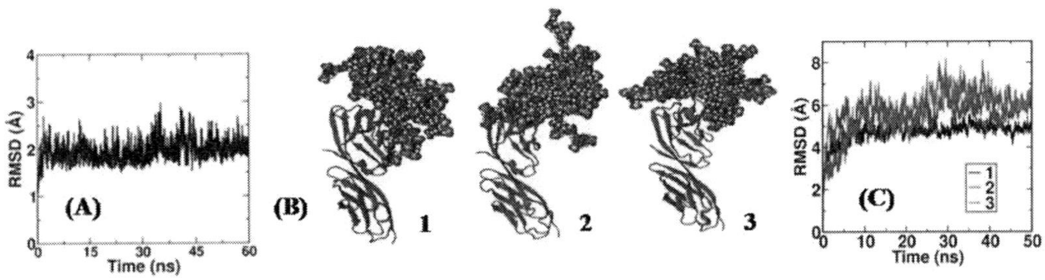

Figure 6. (A) RMSD of CD4 through the 60 ns long MD simulation. The starting conformation of CD4 was obtained by removing gp120 from the gp120-CD4 structure obtained after 40 ns of MD simulation. The RMSD saturates starting from a very early stage in the production run. (B) The starting conformations of the three CD4-SPL7013 docked structures used for MD simulations and energy calculations. (C) RMSD for the three CD4-SPL7013 complexes shown in (B) with respect to their minimized structure.

3.3. Dendrimer binding to CD4

In this section, we study the feasibility of SPL7013 binding to the gp120-binding region of CD4. In an earlier study, we docked the dendrimer to the gp120-CD4 binary complex and found that almost all the docking predictions had the dendrimer bound to gp120, while there were no predictions with the dendrimer bound to CD4 [6]. We performed simulations with gp120-SPL7013 complexes in the absence of CD4 and found that the dendrimer binding to gp120 is much weaker than its binding to the gp120-CD4 complex [6]. The binding energies suggested that the dendrimer could not block gp120-CD4 complexation by sterically blocking the CD4 binding region of gp120. Here, to check the feasibility of SPL7013 binding to CD4 and blocking its binding to gp120, we removed gp120 from the gp120-CD4 complex obtained after 40 ns of MD simulation, and simulated the resulting CD4 molecule for 60 ns using the protocol described in the Materials and Methods section. The RMSD for the CD4 structure seemed to saturate very early during the course of the simulation at a value in the range of 2-2.5 Å (see Figure 6 (A)). We extracted three different CD4 structures along the MD trajectory and docked them with three different dendrimer structures, obtained from the 80 ns long NPT simulation of the dendrimer (see Materials and Methods). Surprisingly, for the three different CD4 structures extracted from different stages of the MD trajectory, docked to three different dendrimer conformations, we found that almost all the top 10 structures predicted by ZDOCK had the dendrimer bound to the gp120-binding region of CD4. To estimate the binding energy between CD4 and SPL7013, with an aim to establish whether the dendrimer can utilize its binding to CD4 for blocking viral entry, we selected three CD4-SPL7013 binary complexes among the ZDOCK output structures. The selected structures are shown in Figure 6(B). The three complexes have considerably different binding conformations. We performed 50 ns long MD simulations of the three binary complexes (RMSDs shown in Figure 6(C)), and performed binding energy calculations using MMPBSA module of AMBER12 [14]. The results from the binding energy calculation are shown in Table 1 along with a comparison with other complexes.From the data in Table 1 we see that the binding energy (BE) for the gp120-CD4 complex (-62 kcal/mol) is much larger than the BE between gp120 and SPL7013 (-41 kcal/mol), with the dendrimer bound to gp120 at its CD4-binding region. The entropy of binding for the gp120-CD4 complex is -53 kcal/mol which gives a binding free energy (ΔF_{avg}) of -9 kcal/mol. As the free dendrimer constitutes a strongly fluctuating system, the entropy loss when the dendrimer binds to gp120 is expected to be huge.

Table 1. Average free energy of binding for various complexes. (All energies are reported in kcal/mol.)

	gp120-CD4	SPL7013-(gp120-CD4) (V3 loop)	gp120-SPL7013 (V3 loop)	gp120-SPL7013 (CD4-binding region)	CD4-SPL7013 (gp120-binding region)
BE	-62±3 [8]	-136±10 [6]	-99±10 [6]	-41±25 [6]	-75±9.
TΔS	-53±3 [8]	-96±9 [6]	–	–	-61±6.
ΔF_{avg}	-9 kcal/mol	-40 kcal/mol	–	–	-14 kcal/mol.

The gp120-SPL7013 complex with SPL7013 bound to the CD4-binding region of gp120 is thus expected to be unstable. So, thermodynamically, gp120 would prefer to bind CD4 than to bind SPL7013 at its CD4-binding region. In contrast, the ΔF_{avg} for CD4-SPL7013 complex (-14 kcal/mol, averaged over the three complexes studied), with the dendrimer binding at the gp120-binding region of CD4 is comparable to that for the gp120-CD4 complex. These two processes can thus compete with each other. Hence, in an experimental mixture of the target cells, virions and dendrimer, the dendrimer and gp120 would compete with each other to bind CD4. Thus, the calculation suggests that the dendrimer can block gp120-CD4 complexation by binding to CD4 at its gp120-binding region and sterically hindering the process. As entropy is expected to play a very important role, a more accurate calculation of the entropy of binding can establish this on a firmer basis as a parallel mechanism of entry-inhibition by the SPL7013 dendrimer.

4. Conclusion

In this work we study the molecular level energetics of the process of the dendrimer binding to the HIV-1 gp120-CD4 complex. Through residue-level energy calculations, we find that the V3 loop region has a major contribution to the binding and the complexation is driven by electrostatics. We also find that the dendrimer utilizes a large number of gp120 residues for binding. Drug evasion by the virus would require it to mutate at multiple sites. Such large scale mutations may induce significant conformational changes in gp120, rendering it incapable of inducing viral entry. In addition to this, the dendrimer can affect gp120-CD4 binding in multiple ways. For different positions at which it binds the gp120-CD4 complex, it disrupts a different set of contacts. Thus mutations that hinder a particular binding pose may not be sufficient for the virus to dodge the dendrimer. We also studied the binding between CD4 and SPL7013 and deduced through end-state energy calculations that the process of the dendrimer binding to CD4 may energetically compete with gp120-CD4 binding, in the process, sterically blocking gp120 from binding CD4. The validity of this finding needs to be confirmed by experiments.

Acknowledgments

This work was funded by the DST Mathematical Biology Initiative at the Indian Institute of Science, Bangalore.

References

[1] Chan D C and Kim P S 1998 *Cell* **93** 681-84
[2] MacArthur R D and Novak R M 2008 *Clin. Infect. Dis.* **47** 236-41
[3] Maiti P K, Cagin T, Wang G and Goddard W A 2004 *Macromolecules* **37** 6236-54
[4] Jiminez J L, Pio M, Mata F J, Gomez R, Munoz E, Leal M and Munoz-Fernandez M 2012 *New J. Chem.* **36** 299-309
[5] Tyssen D *et al.* 2010 *PLos One* **5** e12309

[6] Nandy B , Saurabh S, Sahoo A K, Dixit N M and Maiti P K 2015 *Nanoscale* **7** 18628-41
[7] Nandy B, Bindu D H, Dixit N M and Maiti P K 2013 *J. Chem. Phys.* **139** 024905
[8] Myszka D G, Sweet R W, Hensley P, Brigham-Burke M, Kwong P D, Hendrickson W A, Wyatt R, Sodroski J and Doyle M L 2000 *Proc. Natl. Acad. Sci. USA* **97** 9026-31
[9] Chen R, Li L and Weng Z 2003 *Proteins* **52** 80-87
[10] Chan D C, Kim P S, Maingi V, Jain V, Bharatam P V and Maiti P K 2012 *J. Comput. Chem.* **33** 1997-2011
[11] Frisch M J *et al.* 2004 *Gaussian03 Revision C.02* (Wellington CT: Gaussian Inc.)
[12] Bayly C I, Cieplak P, Cornell W D and Kollman P A 1993 *J. Phys. Chem.* **97** 10269-80
[13] Case D *et al.* 2012 *AMBER12* (San Fransisco: University of California)
[14] Wang J, Wang W, Kollman P A and Case D A 2006 *J. Mol. Graphics Modell.* **25** 247260
[15] Wang J, Wolf R M, Caldwell J W, Kollman P A and Case D A 2004 *J. Comput. Chem.* **25** 1157-74
[16] Huang C C *et al.* 2007 *Science* **317** 1930-34
[17] Hornak V, Abel R, Okur A, Strockbine B, Roitberg A and Simmerling C 2006 *Proteins* **65** 712-25
[18] Joung I S and Cheatham T E III 2008 *J. Phys, Chem.* B **112** 9020-41
[19] Ryckaert J-P, Ciccotti G and Berendsen H J C 1977 *J. Comput. Phys.* **23** 327-41
[20] Miller B R, McGee T D, Swails J M, Homeyer N, Gohlke H and Roitberg A E 2012 *J. of Chem. Theory Comput.* **8** 3314-21
[21] Sitkoff D, Sharp K A and Honig B 1994 *J. Phys. Chem.* **98** 1978-88
[22] Flocco M M and Mowbray S L 1994 *J. Mol. Biol.* **235** 709-17
[23] Mitchell J B, Nandi C L, McDonald I K, Thornton J M and Price S L 1994 *J. Mol. Biol.* **239** 315-31
[24] Yokoyama M, Naganawa S, Yoshimura K, Matsushita S and Sato H 2012 *PLoS One* **7** e37530

When the clock strikes: Modeling the relation between circadian rhythms and cardiac arrhythmias

Pavithraa Seenivasan[1,2], **Shakti N. Menon**[1], **S. Sridhar**[1,3] **and Sitabhra Sinha**[1]

[1] Institute of Mathematical Sciences, C. I. T. Campus, Taramani, Chennai 600 113, India
[2] Molecular Biophysics Unit, Indian Institute of Science, Bangalore 560 012, India
[3] Department of Physics and Astronomy, Ghent University, St. Pietersnieuwstraat 33, 9000 Ghent, Belgium

E-mail: `sitabhra@imsc.res.in`

Abstract. It has recently been observed that the occurrence of sudden cardiac death has a close statistical relationship with the time of day, viz., ventricular fibrillation is most likely to occur between 12am–6am, with 6pm–12am being the next most likely period. Consequently there has been significant interest in understanding how cardiac activity is influenced by the circadian clock, i.e., temporal oscillations in physiological activity with a period close to 24 hours and synchronized with the day-night cycle. Although studies have identified the genetic basis of circadian rhythm at the intracellular level, the mechanisms by which they influence cardiac pathologies are not yet fully understood. Evidence has suggested that diurnal variations in the conductance properties of ion channel proteins that govern the excitation dynamics of cardiac cells may provide the crucial link. In this paper, we investigate the relationship between the circadian rhythm as manifested in modulations of ion channel properties and the susceptibility to cardiac arrhythmias by using a mathematical model that describes the electrical activity in ventricular tissue. We show that changes in the channel conductance that lead to extreme values for the duration of action potentials in cardiac cells can result either in abnormally high-frequency reentrant activity or spontaneous conduction block of excitation waves. Both phenomena increase the likelihood of wavebreaks that are known to initiate potentially life-threatening arrhythmias. Thus, disruptive cardiac excitation dynamics are most likely to occur in time-intervals of the day-night cycle during which the channel properties are closest to these extreme values, providing an intriguing relation between circadian rhythms and cardiac pathologies.

"And now it is almost midnight, the moment when the page of the night turns over into day. Almost midnight, the hour when the figure of Death strikes the golden bell of the clock. And what will happen when the clock strikes ?"

When the Clock Strikes by Tanith Lee (1981)

1. Introduction

Heart disease constitutes a significant public health burden worldwide, being the leading cause of death in most developed countries - for instance, being responsible for about 25% of all deaths in the United States [1, 2]. A large fraction of these can be classified as *sudden cardiac death* occurring as a result of certain types of *cardiac arrhythmia*, i.e., disturbances in the normal

Published under licence by IOP Publishing Ltd

rhythmic activity of the heart that severely impair the normal functioning of the organ. There are many factors that determine the occurrence of cardiac pathologies. While ageing results in a progressive decline in the functioning of cardiovascular systems in general [3], genetics [4] and lifestyle choices [5] can result in a higher predisposition to arrhythmias. Apart from these, the role of circadian rhythms - which refer to temporal variations in physiological processes having a period of approximately 24 hours - in initiating cardiac dysfunction has become the focus of recent clinical interest. Circadian rhythms help in coordinating the behavior of organisms with the terrestrial day-night cycle. Such biological clocks in the body are realized by a set of proteins that enable the generation of rhythmic activation through self-sustained, transcriptional positive and negative feedback loops [6, 7]. Experimentally, it has been observed that the circadian rhythm is not driven by the environment but rather is an intrinsic property of the organism [7]. This is supported by the observation that these rhythms persist even in systems maintained in vitro [8, 9]. It has been postulated that an internal clock mechanism enables the organisms to anticipate temporal changes in their environment, thereby allowing biological processes to occur at a favorable time in the day [10, 11, 12]. Circadian clocks are entrained to the external day-night cycle by environmental factors (*zeitgebers*) that influence the timing of the molecular mechanism [13].

In this paper, we have used computational modeling to investigate the relation between onset of arrhythmias in the heart and the circadian rhythm that have been revealed by recent experimental studies. Over the last few decades, sophisticated models of electrical activity in cardiac cells (that lead to mechanical contraction and hence the pumping action of the heart) have been developed with the aim of reproducing experimentally observed phenomenon. The focus of such modelling efforts has been to capture relevant details from the level of the cell to that of the whole heart. Increased availability of experimental data and vastly improved computational power has enabled the development of these detailed models of cardiac activity in order to study the normal and pathological functioning of the heart. We use such a biologically realistic model to simulate the dynamics of cardiac tissue in order to understand the higher propensity for the onset of arrhythmias at certain times of the day-night cycle. In particular we show that the diurnal variations that cause ion channel conductances to take extreme values can result in significant changes in the repolarization properties of the excitation waves. This can lead to either abnormally rapid high-frequency activity or spontaneous conduction block of propagating excitation - both of which increase the likelihood of wavebreaks that can initiate potentially life-threatening arrhythmia. In the following section we describe the biological phenomena underlying circadian rhythm and its possible relation to onset of arrhythmia. Following this we describe the mathematical model we have used for our computational investigation of excitation dynamics in simulated cardiac tissue. In the subsequent section we describe the results of our investigation linking circadian rhythms and genesis of cardiac arrhythmias. We conclude with a brief summary of our results.

2. Circadian Rhythms

The mammalian circadian clock involves an interaction between three negative loops and one positive loop, executing a series of transcriptional, translational and post-translational events [14]. The *clock* and *bmal1* genes, which form the preliminary units of the clock system, encode the proteins CLOCK and BMAL1. These proteins, on hetero-dimerization, recognize the promoter regions of the genes, *per* and *cry*, and transcribe them in the nucleus [15, 16]. The *per* and *cry* messenger RNAs get translated in the cytosol to form proteins, PER and CRY. These proteins, on hetero-dimerization [17], translocate back into the nucleus and inhibit the activation of CLOCK/BMAL1. This, in turn, inhibits CLOCK/BMAL1-mediated transcription, thereby decreasing the expression of *per* and *cry* genes, ultimately relieving the inhibition on CLOCK/BMAL1 [18, 19, 14]. This mechanism constitutes the dominant negative transcriptional loop of the mammalian circadian clock at the molecular level. On the other hand, the positive transcriptional loop of the circadian clock results from CLOCK/BMAL1 and/or PER2-mediated induction of *bmal1* expression [20].

Depending on the cell type, mammalian circadian clocks are divided into two major classes: the central clock located within the suprachiasmatic nucleus (SCN) of the brain, and peripheral clocks located in all non-SCN cells of the organism [21, 22]. *Zeitgebers*, the factors that reset or entrain circadian clocks, are different for the two classes of clocks. The SCN is reset by light (through electrical signals transmitted along the retino-hypothalamic tract) from the environment, whereas peripheral clocks are reset by various neuro-humoral factors, specific to it [23, 24]. For example, Norepinephrine [9] and Vasoactive intestinal peptide (VIP) [25] have been identified to be the neuro-humoral factors that entrain the cardiomyocyte peripheral clock to the SCN.

2.1. Sudden cardiac death and the circadian clock

The incidence of sudden cardiac death has been observed to exhibit a diurnal variation with peaks occurring in the early morning and late evening hours [26]. The circadian expression of clock genes and clock-controlled genes (genes that are transcribed by the clock genes) in the heart have been recently studied with the aim of understanding this variation [27, 28]. One such clock controlled gene observed to exhibit endogenous circadian rhythmicity in the heart is *klf15* [27]. KChIP2, the regulatory β subunit for the repolarizing transient outward K^+ current was identified to be the transcriptional target of *klf15*. That is, *klf15* transcribes the gene *kcnip2* that encodes the potassium ion channel protein, KChIP2. Experimentally, it has been observed that myocardial repolarization and the expression of ion channels exhibit endogenous circadian rhythmicity. In *klf15*-null mice, the QTc interval (an index of myocardial repolarization) is prolonged and the I_{to} (outward transient K^+ current density) reduced. In contrast, the QTc interval shortens and I_{to} increases in the case of transgenic mice with overexpressed *klf15*. This implies that *klf15*-dependent transcriptional regulation of rhythmic KChIP2 expression plays a key role in the rhythmic variation of myocardial repolarization. A similar observation has been made with respect to the sodium ion channel proteins [28]. The clock gene *bmal1* has been found to directly regulate the circadian oscillation of *scn5α* gene that encodes Nav1.5, the principal voltage-gated Na^+ ion channel. Ventricular myocytes isolated from *bmal1*-deleted mutants have less Na^+ current compared to the ones isolated from *bmal1*-expressed mice. This suggests that the expression of *scn5α* is a significant factor that influences the susceptibility to cardiac arrhythmias. To summarize, the susceptibility to cardiac arrhythmias is increased when the temporal variation (circadian variation) in myocardial repolarization and ion channel expression is impaired. This is reflected in the observation of lengthened/shortened QT interval downstream of a disruption to the circadian genes.

3. The computational model

Electrical activity in cardiac muscle is typically described mathematically by a generic class of models known as excitable systems. An excitable medium is characterized by a stable resting state, a metastable excited state, and a threshold which needs to be exceeded for the system to be excited. A supra-threshold stimulus gives rise to an excitation which is followed by a period of slow recovery (referred to as the refractory period) during which the system cannot be re-excited. This refractory property results in the annihilation of waves when they collide with each other. This implies that a wavefront, on encountering a region that has not yet completely recovered, develops wave breaks that can give rise to spatial patterns such as spiral waves of excitation [29].

In this paper, we have used a modified version of the Luo-Rudy I (LRI) model describing the electrical activity of cardiac myocytes in the guinea pig ventricle [30]. The model is based on the Hodgkin-Huxley formalism [31] developed for describing the action potential of a squid giant axon. The generic form of such models are described by a partial differential equation for the transmembrane potential V:

$$\frac{\partial V}{\partial t} + \frac{I_{\rm ion}}{C} = D\,\nabla^2 V,$$

where C is the membrane capacitance, D is the diffusion constant and $I_{\rm ion}$ is the total ionic current density. In the Luo-Rudy I model, the total ionic current density is the sum of six components:

$$I_{\rm ion} = I_{\rm Na} + I_{\rm K} + I_{\rm K1} + I_{\rm Kp} + I_{\rm Ca} + I_{\rm b}\,,$$

where $I_{\rm Na} = G_{\rm Na}\, m^3\, h\, j\,(V - 54.4)$ is the fast inward Na^+ current, $I_{\rm K} = G_{\rm K}\, x\, x1\,(V + 77.62)$, $I_{\rm K1} = G_{\rm K1}\, K1_{\infty}\,(V + 87.95)$, and $I_{\rm Kp} = 0.0183\, K_{\rm p}\,(V + 87.95)$ are respectively the time-varying, the time-invariant and the plateau K^+ currents $I_{\rm Ca} = G_{\rm Ca}\, d\, f\,(V - E_{\rm Ca})$ is the slow inward Ca^{2+} current where $E_{\rm Ca} = 7.7 - 13.0287\,\ln([{\rm Ca}^{2+}]_i)$ is the reversal potential, dependent on the intracellular ion concentration $[{\rm Ca}^{2+}]$, and $I_{\rm b} = 0.03921\,(V + 59.87)$ is the background current (leakage current). The currents are determined by ion channel gating variables m, h, d, f and x, whose time evolution is governed by ordinary differential equations of the form:

$$\frac{\partial \varepsilon}{\partial t} = \frac{\varepsilon_{\infty} - \varepsilon}{\tau_{\varepsilon}},$$

where $\varepsilon_{\infty} = \frac{\alpha\,\varepsilon}{\alpha\,\varepsilon + \beta\,\varepsilon}$ is the steady state of ε, $\tau_{\varepsilon} = \frac{1}{\alpha\,\varepsilon + \beta\,\varepsilon}$ is the corresponding time constant and $\alpha\,\varepsilon$ and $\beta\,\varepsilon$ are voltage dependent rate constants, all obtained by fitting experimental data. The parameter values used are identical to those in Ref. [30], except $G_{\rm Ca}$ which is chosen to be 0.07 mS cm^{-2} [32].

The model equations are solved using the forward Euler method for time evolution on a one dimensional fiber, discretized over a spatial grid of size L, using a finite difference scheme for the spatial (Laplacian) term. The values of space step Δx and time step Δt used for integrating the equations are 0.0225 cm and 0.01 ms respectively. The diffusion coefficient D is chosen to be 0.01 cm^2/msec.

4. Results

The key experimental results that motivate our modeling study relate to the observation that knocking out *klf15* results in a marked reduction of $I_{\rm to}$ and that an over-expression of *klf15* leads to large increase of $I_{\rm to}$ compared to the normal range of variation seen over the course of a day, both increasing the susceptibility to cardiac arrhythmia. While the former mutation showed a marked increase in the susceptibility of ventricular arrhythmia on being subjected to programmed electrical stimulation of the heart, the latter mutation exhibited spontaneous occurrence of ventricular arrhythmia and resulted in a high mortality rate. In our model, we use the closest analog of $I_{\rm to}$, viz., the time-dependent potassium current $I_{\rm K}$ and study the effect of increasing or decreasing the corresponding channel conductance $G_{\rm K}$.

Fig. 1 shows the role of $G_{\rm K}$ in setting up a stable reentrant circuit. Typically, such circuits can be organized around a zone of inexcitable or partially excitable region of cardiac tissue (anatomical reentry) or even a transiently inactive zone which prevents the passage of excitation wavefront across it (functional reentry). As the wavefront of excitation goes around it, stable reentry can occur if the circuit is long enough such that the tissue recovers by the time the front completes the circuit. Such circuits act as sources for persistent high-frequency stimulation that compete with the normal rhythmic activity initiated by the sinus node. Note that the period of the waves generated around this circuit is governed by the circuit length which is determined by obstacle size (for anatomical reentry) or refractory period of tissue (for functional reentry). Focusing on the region immediately surrounding the circuit, we study the simplified system

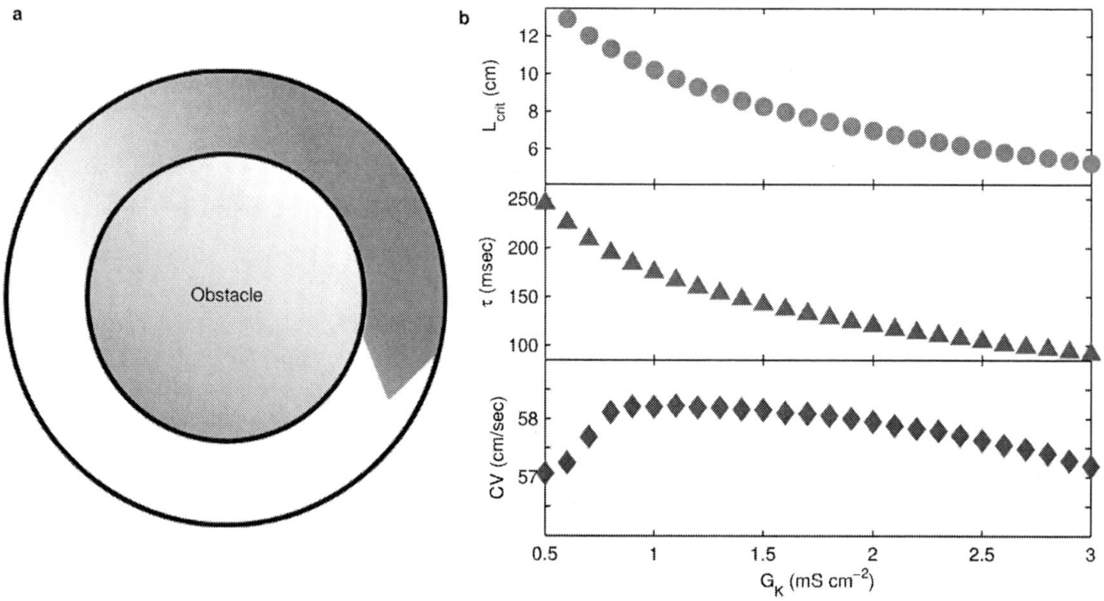

Figure 1. Potential mechanism for genesis of pathological situations in cardiac tissue with a high potassium ion channel conductance G_K. (a) Schematic diagram showing the ring topology of a reentry circuit which describes the motion of the spiral wave around an obstacle. The wavefront of the excitation propagating around the ring is indicated by the arrowhead. The color intensity represents the level of refractoriness with white regions indicating completely recovered tissue. (b) Variation in dynamical properties of reentrant wave propagation as a function of G_K in a ring of critical size, i.e., the minimum perimeter length required for sustained wave activity. The electrical activity of each cell is described by the LRI model. Unidirectional wave propagation is set up using a special initial condition. The critical ring size L_{crit} decreases monotonically on increasing G_K [top], as does the period of reentry τ, i.e., the time required by the wave to complete one circuit [center]. The conduction velocity CV of the wavefront [bottom] remains approximately constant, varying over a very narrow range of values, as G_K is changed. These results show that for high G_K the critical ring size reduces, allowing reentry to occur at higher frequency (i.e., smaller τ), which could potentially lead to wave breakup away from the reentry circuit.

of a one-dimensional reentrant circuit [Fig. 1 (a)]. We observe that increasing G_K results in stable reentry being possible for circuits of smaller lengths, and hence of much higher frequency. As more rapid stimulation of the tissue surrounding the circuit will result in waves that are more likely to result in front breakup as they travel to other regions of the tissue [33, 34, 35], we suggest that the propensity of life-threatening arrhythmia will be enhanced for higher G_K. Note that this is analogous to the situation corresponding to enhanced I_{to} resulting from an over-expression of *klf15*.

Fig. 2 shows an alternative scenario where the role of decreasing G_K in promoting arrhythmia is examined. This comes about through the mechanism of *alternans*, where periodic stimulation of excitable media at a sufficiently high frequency results in an alternating succession of strong and weak responses. In cardiac tissue, this can be observed as variations in the action potential duration (APD), such that long and short APD pulses alternate. Such alternans can appear even at the scale of a single cell as an outcome of the nonlinear restitution property (functional

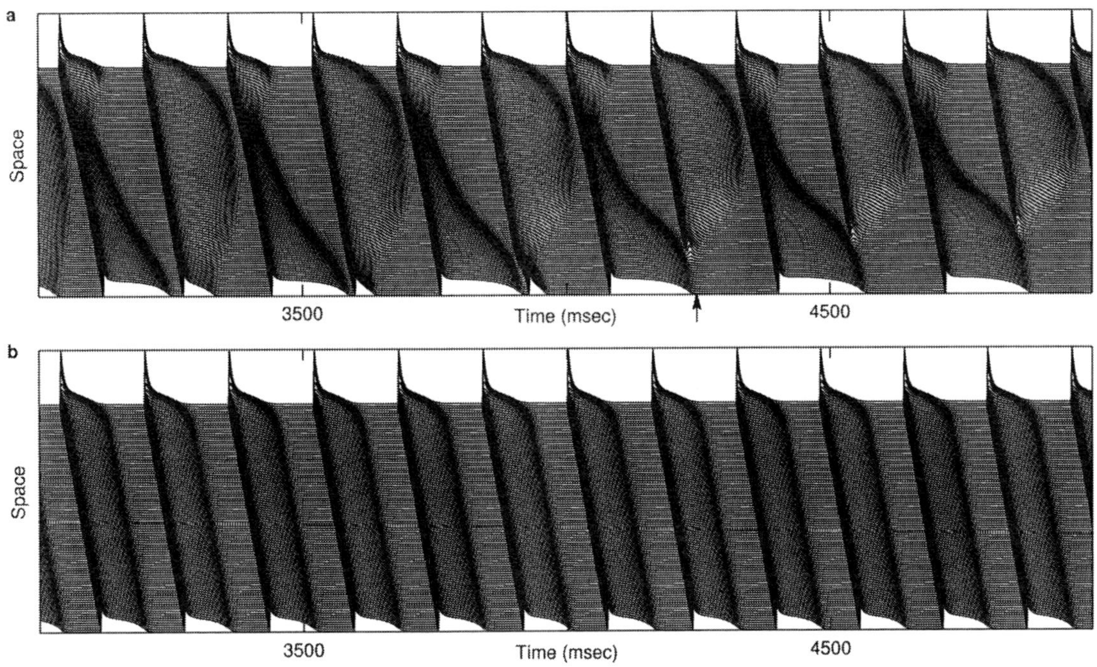

Figure 2. Potential mechanism for genesis of pathological situations in cardiac tissue with a low potassium ion channel conductance G_K. The dynamical response of a one-dimensional chain of cells subjected to external periodic stimulation (pacing) with a period of $T = 160$ ms at the top end is shown for two different values of (a) $G_K = 0.7$ and (b) 0.9 mS cm^{-2}. The electrical activity of each cell is described by the LRI model. For the higher value of G_K (b), we observe a 1:1 periodic response, resulting in a train of identical action potentials. However, for the lower value of G_K (a), pacing at the same frequency results in alternating action potentials varying in duration (alternans). Moreover the pattern of alternans varies across space (discordant). The degree of variation in the action potential duration continues to increase over time, until a wavefront encounters a refractory region behind the preceding wave resulting in a conduction block (see arrow). These results show that for low G_K, periodic activation of cardiac tissue can occasionally cause propagating fronts to be spontaneously blocked, which may then lead to the formation of spiral waves and their breakup.

relation between APD and the interval between successive stimuli), while in combination with conduction velocity dispersion (i.e., dependence of the front propagation speed on the extent of recovery of the medium) an additional spatial variation can be seen in the alternans [36]. Such a variation is manifested in terms of long-short APD alternans in certain regions co-occurring with short-long alternans in other regions, a phenomenon termed as *discordant alternans*.

We study the pacing of a one-dimensional fiber as a simple approximation of cardiac tissue subjected to periodic electrical stimulation. We note that at lower values of G_K the system is much more likely to show enhanced discordant alternans that eventually results in conduction block far from the pacing site. Note that in a higher-dimensional system this will result in wave-break as a block occurs in a section of the wavefront. The broken front can then form a reentrant circuit, resulting in arrhythmia.

5. Discussion

In this paper we have computationally investigated the possible mechanism underlying the recently discovered close statistical relationship between occurrence of sudden cardiac death with the time of day. Essentially this involves understanding how cardiac activity is influenced by the circadian clock, i.e., temporal oscillations in physiological activity with a period close to 24 hours and synchronized with the day-night cycle. As noted in our paper, although studies have identified the genetic basis of circadian rhythm at the intracellular level, the mechanisms by which they influence cardiac pathologies are yet to be fully understood. We focus on evidence suggesting that the key to the circadian-cardiac relation may be provided by the diurnal variations in the conductance properties of ion channel proteins that govern the excitation dynamics of cardiac cells. Thus, we investigate the relationship between the circadian rhythm as manifested in modulations of ion channel properties and the susceptibility to cardiac arrhythmias by using a mathematical model describing the electrical activity in ventricular tissue. We show that changes in the channel conductance that lead to extreme values for the duration of action potentials in cardiac cells can result either in abnormally high-frequency reentrant activity or spontaneous conduction block of excitation waves. Both phenomena increase the likelihood of wavebreaks that are known to initiate potentially life-threatening arrhythmias. Thus, disruptive cardiac excitation dynamics are most likely to occur in time-intervals of the day-night cycle during which the channel properties are closest to these extreme values, providing an intriguing relation between circadian rhythms and cardiac pathologies. For simplicity, here we have examined the spatially homogeneous situation where every element in an excitable medium undergoes similar temporal variation in the parameter controlling repolarization. In real physiological situations, however, there may be considerable heterogeneity in the system - e.g., in terms of the repolarization dynamics of different regions as reflected in potassium channel properties. Spatial variations in the recovery characteristics are known to modulate the propagation speed of excitation wavebacks, which can result in wavebreaks [37, 38]. Furthermore, we have used one-dimensional models that consider rings or chains of cardiac cells. However, even such simple models can capture the essential properties of excitation dynamics that we investigate here. In future studies, we plan to show that the results obtained here under simplifying setting can be reproduced even for heterogeneous systems in two- or three-dimensional spatial geometries.

References

[1] Hennekens C H 1998 *Circulation* **97** 1095.
[2] Heron M 2013 *National Vital Statistics Reports* **62** 1.
[3] Hatch F, Lancaster M K and Jones S A 2011 *Expert Rev. Cardiovasc. Ther.* **1059** 67.
[4] Lara-Pezzi E, Dopazo A and Manzanares M 2015 *Dis. Model. Mech.* **5** 434-43.
[5] Mozaffarian D, Wilson P W and Kannel W B 2008 *Circulation* **117** 3031-8.
[6] Dunlap J C 1999 *Cell* **96** 271-90.
[7] Edery I 2000 *Physiol. Genomics* **3** 59-74.
[8] Balsalobre A, Damiola F and Schibler U 1998 *Cell* **93** 929-37.
[9] Durgan D J, Hotze M A, Tomlin T M, Egbejimi O, Graveleau C, Abel E D, Shaw C A, Bray M S, Hardin P E and Young M E 2005 *Am. J. Physiol. Heart Circ. Physiol.* **289** H1530-41.
[10] Pittendrigh C S 1993 *Annu. Rev. Physiol.* **55** 17-54.
[11] Ouyang Y, Andersson C R, Kondo T, Golden S S and Johnson C H 1998 *Proc. Natl. Acad. Sci. USA* **95** 8660-4.
[12] Paranjpe D A and Sharma V K 2005 *J. Circadian Rhythms* **3** 7.
[13] Sharma V K and Chandrashekaran M K 2005 *Curr. Sci.* **89** 1136-46.
[14] Gachon F, Nagoshi E, Brown S A, Ripperger J A and Schibler U 2004 *Chromosoma* **113** 103-12.
[15] Gekakis N, Staknis D, Nguyen H B, Davis F C, Wilsbacher L D, King D P, Takahashi J S and Weitz C J 1998 *Science* **280** 1564-9.
[16] Hogenesch J B, Gu Y Z, Jain S and Bradfield C A 1998 *Proc. Natl. Acad. Sci. USA* **95** 5474-9.
[17] Kloss B, Price J L, Saez L, Blau J, Rothenfluh A, Wesley C S and Young M W 1998 *Cell* **94** 97-107.

[18] Curtis A M, Seo S B, Westgate E J, Rudic R D, Smyth E M, Chakravarti D, FitzGerald G A and McNamara P 2004 *J. Biol. Chem.* **279** 7091-1097.
[19] Etchegaray J P, Lee C, Wade P A, and Reppert S M 2003 *Nature* **421** 177-82.
[20] Shearman L P *et al* 2000 *Science* **288** 1013-9.
[21] Cermakian N and Sassone-Corsi P 2000 *Nat. Rev. Mol. Cell Biol.* **1** 59-67.
[22] Hirota T and Fukada Y 2004 *Zoolog. Sci.* **21** 359-68.
[23] Berson D M 2003 *Trends Neurosci.* **26** 314-20.
[24] Brown S A, Zumbrunn G, Fleury-Olela F, Preitner N, and Schibler U 2002 *Curr. Biol.* **12** 1574-83.
[25] Schroeder A, Loh D H, Jordan M C, Roos K P and Colwell C S 2011 *Am. J. Physiol. Heart Circ. Physiol.* **300** H241-50.
[26] Muller J E, Ludmer P L, Willich S N, Tofler G H, Aylmer G, Klangos I and Stone P H 1987 *Circulation* **75** 131-8.
[27] Jeyaraj D *et al* 2012 *Nature* **483** 96-9.
[28] Schroder E A, Lefta M, Zhang X, Bartos D C, Feng H Z, Zhao Y, Patwardhan A, Jin J P, Esser K A and Delisle B P 2013 *Am. J. Physiol. Cell Physiol.* **304** C954-65.
[29] Jalife J, Anumonwo J M, Delmar M and Davidenko J M 1999 *Basic Cardiac Electrophysiology for the Clinician* (Armonk: Futura Publishing).
[30] Luo C and Rudy Y 1991 *Circ. Res.* **68** 1501-26.
[31] Hodgkin A L and Huxley A F 1952 *J. Physiol.* **117** 500-44.
[32] Xie F, Qu Z, Garfinkel A and Weiss J N 2001 *Am. J. Physiol. Heart Circ. Physiol.* **280** H535-45.
[33] Panfilov A V and Keener J P 1993 *J. Theor. Biol.* **163**, 439-48.
[34] Fenton F H, Cherry E M, Hastings H M and Evans S J 2002 *Chaos* **12**, 852-92.
[35] Sridhar S, Sinha S and Panfilov A V 2010 *Phys. Rev. E* **82**, 051908.
[36] Sinha S and Sridhar S 2015 *Patterns in Excitable Media: Genesis, Dynamics, and Control* (Boca Raton, FL: CRC Press).
[37] Breuer J and Sinha S 2005 *Pramana* **64**, 553-62.
[38] Sridhar S and Sinha S 2010 *EPL* **92**, 60006.

Temperature dependent conformation studies of Calmodulin Protein using Molecular Dynamics

Sahil Aneja[1], Vivek Kumar Bhartiya[2] and Sunita Negi[3]

[1] Cluster Innovation Centre (CIC), University of Delhi, Delhi, INDIA
[2] Centre for Photonics and Quantum Materials, Skolkovo Institute of Science and Technology, Moscow, Russia
[3]Amity University, Manesar, Gurgaon, Haryana, INDIA

Abstract. Calmodulin (CaM) protein plays a very crucial role in the calcium signaling inside the eukaryotic cell structure [1, 2]. It can also bind to other proteins/targets and facilitate various activities inside the cell [3, 4]. Temperature dependent conformation changes in the CaM protein are studied with extensive molecular dynamics simulations. The quantitative comparison of simulation data with various forms of experimental results probing different aspects of the folding process can facilitate robust assessment of the accuracy of the calculations. It can also provide a detailed structural interpretation for the experimental observations as well as physical interpretation for theory behind different aspects of the experiment. Earlier these kinds of studies have been performed experimentally using fluorescence measurements as in [5]. The calcium bound form of CaM is observed to undergo a reversible conformation change in the range 295-301 K at calcium ion concentration 150 mM. The transition temperature was observed to depend on the calcium ion concentration of the protein. Leap-dynamics approach was used earlier to study the temperature dependent conformation change of CaM [6]. At 290 K, both the N- and C-lobes were stable, at 325 K, the C-lobe unfolds whereas at 360 both the lobes unfold [6]. In this work, we perform molecular dynamics simulations of 100 ns each for the temperatures 325 K and 375 K on the apo form of CaM, 3CLN and 1CFD. A remarkable dependence of the temperature is observed on the overall dynamics of both the forms of the protein as reported in our earlier study [7, 8]. 1CFD shows a much flexible linker as compared to 3CLN whereas the overall dynamics of the lobes mainly N-lobe is observed to be more in later case. Salt bridge formation between the residues 2 (ASP) and 148 (LYS) leads to a more compact form of 1CFD at 325 K. The unfolding of the protein is observed to increase with the increase in the temperature similar to the earlier reported studies [8].

1. Introduction

Calmodulin (CaM) (an abbreviation for calcium-modulated protein) is a calcium binding messenger protein expressed in all eukaryotic cells. CaM is a multifunctional intermediate messenger protein that transduces calcium signals by binding calcium ions and then modifying its interaction with various target proteins. CaM mediates many crucial processes such as metabolism [9], apoptosis, smooth muscle contraction [10], inflammation [11], intracellular movement, short term and long term memory, and the immune response. CaM is expressed in many cell types and can have different subcellular locations, including the cytoplasm, within organelles, or associated with the plasma or organelle membranes. This protein is capable to bind a bulk of the other proteins or small organic compounds very well.

Many of the proteins that CaM binds are unable to bind calcium themselves, and use CaM as a calcium sensor and signal transducer. CaM can also make use of the calcium stores in the endoplasmic reticulum, and the sarcoplasmic reticulum. CaM can undergo post-translational modifications, such as phosphorylation, acetylation, methylation, and proteolytic cleavage, each of which has potential to modulate its actions. CaM is a small, highly conserved protein approximately 148 amino acids long (16706 Daltons). It contains four EF-hand motifs, each of which binds a Ca^{2+} ion. The protein has two approximately symmetrical globular domains (the N- and C- domain), separated by a flexible linker region. Calcium participates in an intracellular signaling system by acting as a diffusible second messenger to the initial stimuli.

Up to four calcium ions are bound by CaM via its four EF hand motifs. EF hands supply an electronegative environment for ion coordination. After calcium binding, hydrophobic methyl groups from methionine residues become exposed on the protein via conformational change. This presents hydrophobic surfaces, which can in turn bind to basic Amphiphilic Helices (BAA helices) on the target protein. These helices contain complementary hydrophobic regions. The flexibility of CaM's hinged region allows the molecule to "wrap around" its target. This property allows it to tightly bind to a wide range of different target proteins compared to the X-ray crystal structure; the C terminal domain solution structure is similar while the EF hands of the N terminal domain are considerably less open. The backbone flexibility within Calmodulin is key to its ability to bind a wide range of targets.

CaM undergoes large conformational changes that would depend upon its interacting partner and the environmental conditions would play a vital role in this interaction. Therefore our work is concerned with, conformational changes, in two different forms of CaM, CFD and 3CLN. CaM can be obtained from the protein data bank (PDB) [12], comprise of 148 amino acid and made of, two distinct domains, by convention, N-lobe (residue 1-64), central linker (residue 65-91) and C-lobe (residue 92-148). The coordinates of residues 1-4 and 148 are not included in the 3CLN PDB file as their coordinate are not reported in the X-ray data, but can be added by following recent studies [13].

Each domain consists of two EF-hands. Helix-loop-helix motifs are building blocks of these segments, with a Ca^{2+} binding site comprising a sequence of 12 residues, out of which the first 10 constitute the loop portion. 1CFD, PDB entry 1CFD, is free of calcium ions, in EF-hand motifs, two in both lobes.

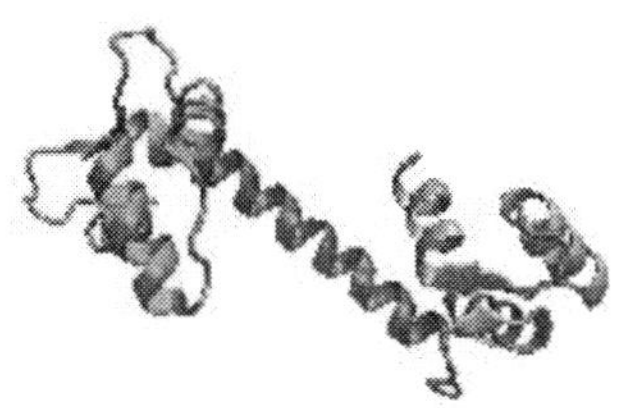

(a)

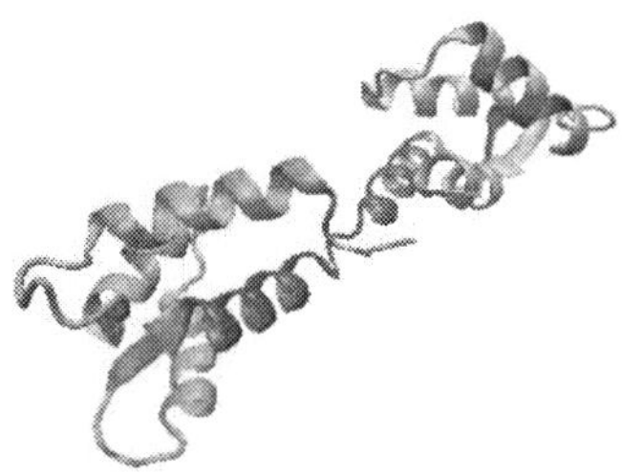

(b)

Figure 1 Initial structures of the simulations (a) 3CLN-Calmodulin protein (b) 1CFD -Calmodulin protein

2. Simulation Details

Various computational techniques are used to prepare, represent and analyze the system. Most importantly, we used NAMD [14] and VMD [15] to study structural time evolution of protein-water system. Simulations were performed for 100 ns each starting the PDB file (coordinates of atoms generated from techniques like X-r^{a}y, NMR etc.) obtained from Protein Data Bank. Protein is immersed in a water box of length 10 Å using the VMD package. Water molecule is assigned as TIP3P model and CharmM27 force field [16] is used for protein and water system energetics. Initial box dimensions are 75 × 76 × 55. These dimensions are not decided by the user, it is calculated once the system is ready by making use of tcl script in VMD. Later cutoff distance and system image ranges are decided by this data.

Particle Mesh Ewald method [17] is used to calculate long range electrostatic interactions. A cutoff distance 12 Å and a switching function at 10 Å is used to accomplish electrostatic calculations and to reduce computational load. Of course, one can increase cross-sectional area hence accuracy in energetics by increasing cutoff distance, but it requires a heavy computational power and time. As prepared protein with water box is not a neutral system. By making use of VMD tools, whole system is neutralized with Na^+ and Cl^- ions. Control in temperature is carried out by Langevin dynamics with a damping coefficient of 5 p/s. Langevin piston is used to control the pressure. Volumetric fluctuations are present to be isotropic. Spatial arrangement of ions is done automatically such that total energy of system remains minimum or ions go in a configuration of minimum energy. It is required to have a stable initial structure for studying the folding dynamics of the protein.

3. Results and Discussion

The changes that take place in the form of conformational changes of the protein are due to the change in the environmental conditions such as change in temperature, ion concentration and pH of the solution etc. Conformational change of CaM for different temperatures is an interesting phenomenon, with the increase of temperature helps the N and C-lobes to come close to each other. This is because the increase in the temperature unfolds the CaM. So unfolding in CaM is the result of some type of non-covalent interaction between lobes.

In this work, we discuss the results of our molecular dynamics simulations for different values of temperatures i.e. 325 K and 375 K for two different Apo forms of Calmodulin, 3CLN and 1CFD when rest of the environmental conditions are kept constant. The concentration of the Ca^{2+} ions is kept at 150 mM and the pH at physiological value of 7.0.

Root mean square deviation (RMSD) and the distance distribution between the residues on the two lobes are used to understand the dynamic conformational changes in the protein structure. We also calculate the energy fluctuations and salt bridge formations in order to satisfy the different conformational changes observed from RMSD and distance distribution, of the protein.

3.1. Root Mean Square Deviation (RMSD)

The root-mean-square deviation (RMSD) is the measure of the average distance between the atoms usually the backbone atoms) of superimposed proteins given as:

$$RMSD = \sqrt{\frac{\sum_{t=0}^{n}\left(x_{1,t} - x_{2,t}\right)^2}{n}}$$

where $x_{1,t}$ and $x_{2,t}$ are the two time series of the protein data.

The RMSD for the two forms of the CaM protein at various temperatures is compared as shown in Figure 2. The RMSD pattern shows a much more flexible linker for the 1CFD as compared to 3CLN. In 3CLN we observed more flexibility of the N-lobe. The protein RMSD of the 3CLN form is consistent with the RMSD of its N-lobe as it shows a similar pattern at 325 K.

The order of the RMSD is observed to remain between 3 Å to 5 Å where at 375 K it raises between 8 Å to 10 Å which shows a significant shift in the RMSD with an increase in temperature. Similar results are observed in the case of 1CFD when we compare our RMSD at 325 K and 375 K, at lower temperature the RMSD is between 5 Å to 10 Å whereas at higher temperature the value of RMSD lies between 10 Å to 15 Å. This increase in the RMSD value would be the result of kinetic energy imparted to the protein residues at higher temperature. The kinetic energy imparted to the protein would affect the stability of the protein which leads to conformation changes on changing the temperature.

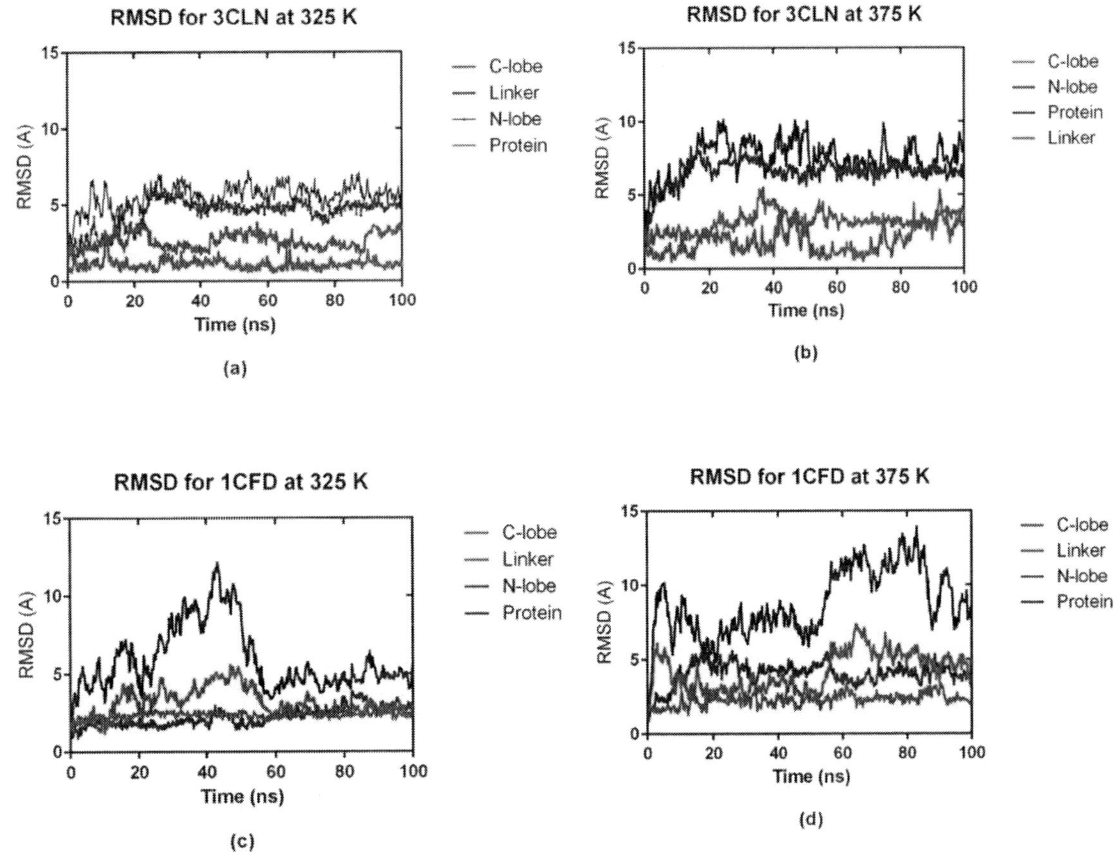

Figure 2 Temporal evolution of the RMSD for (a) 3CLN at 325 K, (b) 3CLN at 375 K, (c) 1CFD at 325 K and (d) 1CFD at 375 K.

3.2. Distance Distribution

The distance distribution is calculated to study the deviation of the protein structure from its mean position. Therefore, in this section we discuss the distance distribution calculated between residues 34 and 110 for the 3CLN and 1CFD proteins at various temperatures i.e. 325 K and 375 K. Figure 3(a) shows the distance distribution of the residues 34 and 110 of the N and C-lobes of the 3CLN protein at 325 K. The protein is observed to have a distribution around 30 Å in 3CLN at 325 K whereas on 375 K the distribution is observed around 45 Å which shows a compact form of protein at lower temperatures. Similarly we observed the compactness of protein in 1CFD but in this case order of compactness is much higher in comparison to $3CLN$, on higher temperature i.e s375 K the distribution lies around 37 Å where at 325 K the distribution is observed to be around 15 Å.

4. Another section of your paper

A thumb rule in physics is the free energy minimization as it becomes necessary to study energetics of a system before studying its properties or dynamics [18]. Because, if a system is not sitting in its lowest hoop on the energy surface, there might exist unbalanced forces, as a consequence these forces may contribute to the dynamics [19]. The initial structural energy minimization is done automatically by NAMD when commands are written in script file. Protein folding is a self-assembly process, guided by the structure of the minimum free energy. Therefore in order to understand the structure of

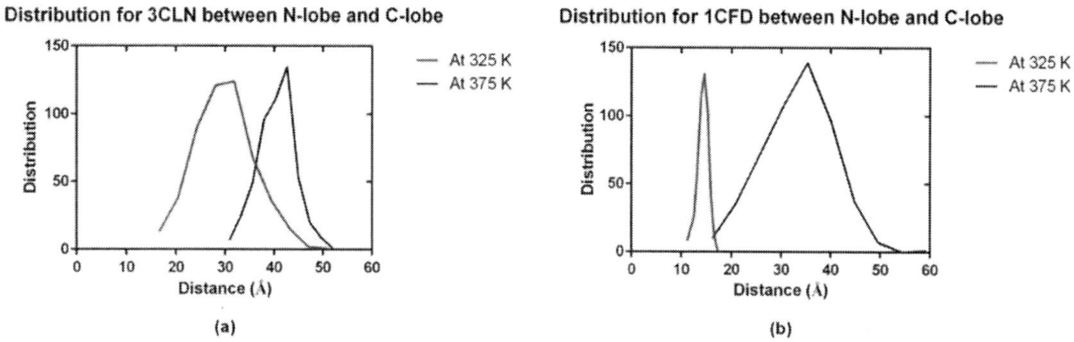

Figure 3: Distance distribution between the residues 34 and 110 of the N and C-lobes of (a) 3CLN at 325 and 375 K (b) 1CFD at 325 and 375 K.

the protein requires knowing of the folding energetics, which also governs the dynamics of protein or conformational changes in protein. The temperature dependent changes in conformation can be related to the energetics of the protein like the linker is more flexible in the case of 1CFD leading to bend compact form during the simulation at 325 K is due to change in temperature governs to break the rigidity between linker.

Here we are concerned with the energetics of various folding events during the time evolution of the system. Energy measurements are therefore a good tool to categorize folding and unfolding events. So in order to study the dynamics of the proteins at various temperatures in detail we perform an energy calculation using the "NAMD energy" plugin within the VMD program. The folding of proteins is due to presence of multiple conformations formed by various contacts between residues of the main and side chains. Figure 4 shows the linker energy term of the 1CFD and Figure 5 shows the N-lobe energy term of the 3CLN at 325 K and 375 K calculated using "namdenergy" plugin in VMD.

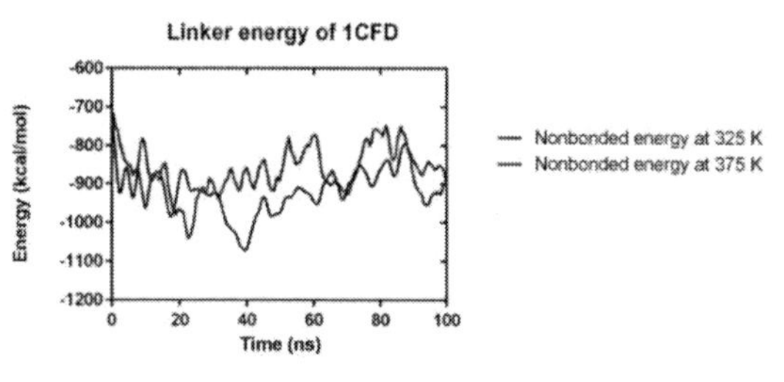

Figure 4: Linker energy calculated for 1CFD at 325 and 375 K using "NAMD Energy" plugin.

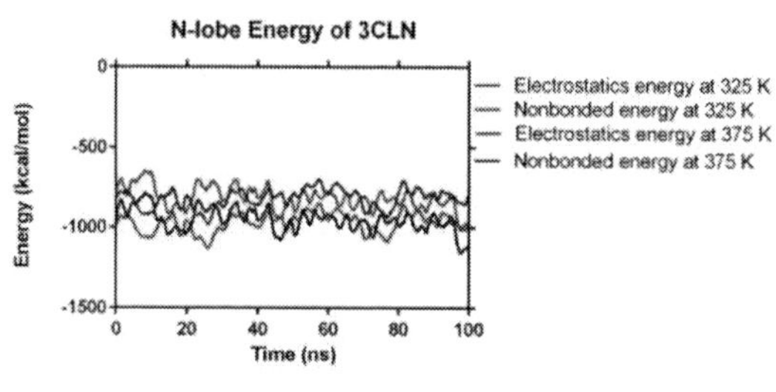

Figure 5: N-lobe energy calculated for 3CLN at 325 and 375 K using "NAMD Energy" plugin.

As discussed in Section 3.1, the flexibility of the linker is observed to be more in the case of 1CFD in comparison to 3CLN. This was confirmed from the RMSD of the two forms which showed a greater flexibility of the N-lobe in the case of 3CLN in comparison with 1CFD. We did the energy calculation using namd which also leads us to same result as shown as in Figure 4 and Figure 5. The Non-bonded energy term for the linker of 1CFD is as in Figure 4 at both the temperatures which represents the flexibility of the linker to be more in this case. On the other hand, the Electrostatics and Non-bonded energy terms for the N-lobe of the 3CLN form are as shown in Figure 5 which verifies the flexibility of N-lobe to be more in this case. The N-lobe energy calculated in the case of 1CFD is observed to change very less showing an insignificant variation in this case. Similarly the linker energy in the case of 3CLN does not vary much from its average value.

5. Salt Bridge Formations

In order to understand the dynamics and the conformational changes occurred with change in temperature, we want to analyze every possible interaction between the residues of the two lobes of the protein. We therefore calculate salt bridges formed between lobes of 1CFD at both temperatures 325 K as well as 375 K. A number of salt bridges are observed to be formed between the residues of the N- and C-lobes of the protein. Three salt bridges are observed to be formed between the residues 2 (ASP) and 148 (LYS), 7 (GLU) and 148 (LYS), and 11 (GLU) and 115 (LYS) of N-lobe and C-lobe of 1CFD at 325 K. Similarly, three salt bridges are observed to be formed between the residues 2 (ASP) and 115 (LYS), 6 (GLU) and 115 (LYS), and 7 (GLU) and 115 (LYS) of N-lobe and C-lobe of 1CFD at 375 K.

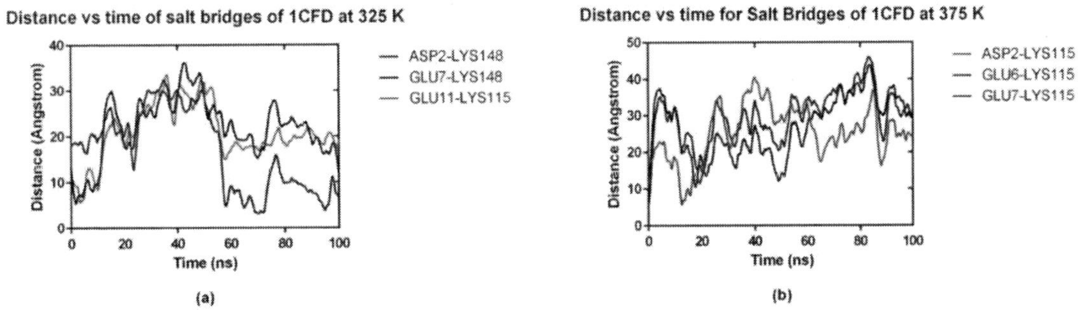

Figure 6: The variation of the salt bridge distance between the residues of N and C-lobes of 1CFD at (a) 325 K and (b) 375 K.

We analyze the distance between different salt bridges during the entire simulation as shown in Figure 6 and observed that in the case of 1CFD at 325 K one of the salt bridges between residues 2 (ASP) and 148 (LYS) distance reduces below 10 Å near 60 ns and remains such till 75 ns. Around 78 ns the residues move apart from each other but again gradually start coming close to each other around 80 ns making the salt bridge. We represent the snapshots of this particular salt bridge as shown in Figure 7, showing the interaction between residues 2 (ASP) and 148 (LYS) at different times of simulation. The salt bridge once formed leads to a more compact form of the protein at 325 K wherever at 375 K the salt bridges are formed but they do not get closer as shown in the Figure 6(b), so the decrease in the temperature shows compactness in protein 1CFD. In this case the linker is also observed to bend at the time of the salt bridge formations thus leading to a more compact structure. This is consistent with our earlier analysis which leads to a compactness of 1CFD at 325 K [8].

6. Conclusions

The functioning of so many proteins is dependent on environmental conditions like in CaM protein (3CLN and 1CFD) calcium ion removal, change in ionic strength and temperature variations leads to change in conformations or flexibility of protein. Extensive MD simulations for 100 ns performed on both Apo form of CaM (3CLN & 1CFD) at different temperatures 325 K and 375 K. All rest of the other environmental factors are maintained constant during these simulations. 1CFD shows a much flexible linker as compared to 3CLN whereas the overall dynamics of the lobes mainly N-lobe is observed to be more in later case. Salt bridge formation between the residues 2 (ASP) and 148 (LYS) leads to a more compact form of 1CFD at 325 K. The unfolding of the protein is observed to increase with the increase in the temperature similar to the earlier reported studies. Thus changing temperature comes out to be the key factor in the flexibility of CaM, RMSD calculations of CaM (3CLN & 1CFD) at different temperatures 325 K & 375 K shows the flexibility of N-lobe and Linker at these

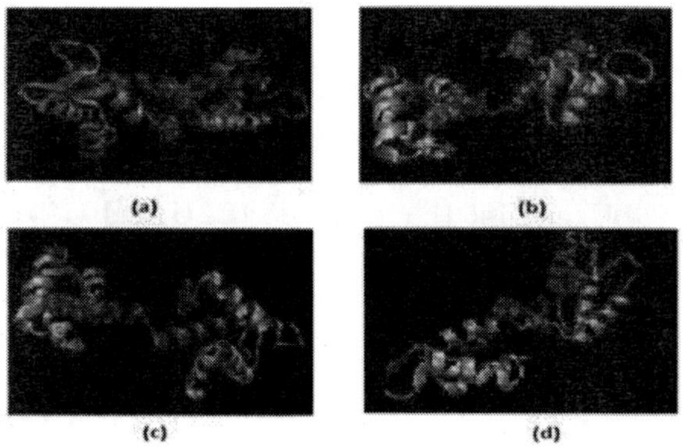

Figure 7: Snapshots showing the salt bridge formation between residue 2 (ASP) and residue 148 (GLU) of the N and C-lobes of 1CFD protein at 325 K at (a) 0 ns (b) 20 ns (c) 40 ns and (d) 58 ns.

temperatures. The flexibility of N-lobe (3CLN) and linker (1CFD) on both the temperatures is consistent with the experimental observations reported earlier. In order to understand this behavior in more detail, we calculate different energy terms corresponding to different parts of the protein. The flexibility of the linker is observed to be more in the case of 1CFD at both the temperatures i.e 325 K and 375 K in comparison to 3CLN which is shown in the form of nonbond energy terms of the linker. In the case of 3CLN, the N-lobe is observed to be more flexible at both the temperature values adding overall flexibility to the protein. The distance-distribution plots are shown at both the temperatures for 3CLN and 1CFD, where 1CFD shows more compactness at lower temperature 325 K. This compactness is observed as a result of salt bridge formations at this temperature in 1CFD whereas no salt bridge formation is observed in the case of 3CLN where the overall dynamics is governed by the N-lobe.

Acknowledgement
This work is funded by Department of Science & Technology (DST), Delhi. Mr. Sahil Aneja is receiving a fellowship from the DST project. We also acknowledge Inter University Accelerator Centre (IUAC), New Delhi for providing us computational resources for doing this research work.

References
[1] H. Frauenfelder, Proceedings of the National Academy of sciences of the USA 99, 2479-2480 (2002).
[2] M. Ikura, JB Ames, Proc. Natl. Acad. Sci. USA 103, 1159-1164 (2006).
[3] JL Fallon, Quiocho, FA Structure 11, 1303-1307 (2003).
[4] M Vandonselaar, RA Hickie, JW Quail, LTJ Delbaere, Nat. Struct. Biol 1, 795-801 (1994).
[5] Preeti Gangola and Harish Pant, Biochemical and Biophysical Research Communications, 111(1): 301-305 (1983).
[6] Jens Kleinjung, Franca Fraternali, Stephen Martin, R. and Peter Bayley, PROTEINS: Structure, Function and Genetics, 50:648–656 (2003).
[7] Sunita Negi, Biophysical Journal, 108, Issue 2, Supplement 1, 46, 108 (2015).
[8] Sunita Negi, The Biophysics Journal, 329703, 9 (2014).
[9] RM Anthony, RJ Dedman, Nature 285, 73 – 77 (1980).
[10] DP Wilson, Cindy Sutherland and MP Walsh, The Journal of Biological Chemistry 277, 2186-2192 (2002).
[11] JM Nakashima, DM Hyde and SN Giri, The American Journal of Pathology 124(3), 528-536 (1986).

[12] YS Babu, JS Sack, TJ Greenhough, CE Bugg, AR Means and WJ Cook, Nature 315, 37-40 (1985).
[13] JL Pascal-Ahuir, EL Mehler and H. Weinstein, Molecular Engineering 1, 231-247 (1991).
[14] L. Kale, R. Skeel, M. Bhandarkar, R Brunner, A Gursoy, N Krawetz, J Phillips, A Shinozaki, K Varadarajan and K Schulten, Journal of Computational Physics 151, 283–312 (1999).
[15] W Humphery, A Dalke and K Schulten, J Mol. Graphics 1, 27-33 (1996).
[16] BR Brooks, RE Bruccoleri, BD Olafson, DJ States, S. Swaminathan and M Karplus, J Comput. Chem 4, 183-187 (1983).
[17] TA Darden, L Perera, L Li and L Pedersen, Structure with Folding and Design 7, 55-60 (1999).
[18] S Negi, Biophysical Journal 108, 46 (2015).
[19] R. L. Baldwin, J. Mol. Biol., 371, (2007).

Journal of Physics: Conference Series **759** (2016) 012023 doi:10.1088/1742-6596/759/1/012023

A Two-Layer Mathematical Modelling of Drug Delivery to Biological Tissues

Koyel Chakravarty and D C Dalal

Department of Mathematics, Indian Institute of Technology Guwahati, Guwahati, Assam, India

E-mail: koyel@iitg.ernet.in

Abstract. Local drug delivery has received much recognition in recent years, yet it is still unpredictable how drug efficacy depends on physicochemical properties and delivery kinetics. The purpose of the current study is to provide a useful mathematical model for drug release from a drug delivery device and consecutive drug transport in biological tissue, thereby aiding the development of new therapeutic drug by a systemic approach. In order to study the complete process, a two-layer spatio-temporal model depicting drug transport between the coupled media is presented. Drug release is described by considering solubilisation dynamics of drug particle, diffusion of the solubilised drug through porous matrix and also some other processes like reversible dissociation / recrystallization, drug particle-receptor binding and internalization phenomena. The model has led to a system of partial differential equations describing the important properties of drug kinetics. This model contributes towards the perception of the roles played by diffusion, mass-transfer, particle binding and internalization parameters.

1. Introduction

Studies related to drug delivery draw much attention to present day researchers for their theoretical and clinical investigations. The controlled drug delivery is a process by means of which a drug is delivered at a pre-determined rate, locally or systemically, for a stipulated period of time. Controlled drug delivery systems may incorporate the maintenance of drug levels within a desired therapeutic range, the need of less number of administrations, optimal use of the required drug and yet possibility of enhanced patient compliance. The quintessential drug delivery should be not only inert, biocompatible, but also at the same time provide patient compliance, capable of attaining high drug loading. However, it should also have preventive measures for the accidental release of drug and at the same time it should be simple to administer and to remove from the body. Controlled release drug delivery involves drug-encapsulating devices from which drug or therapeutic compounds may be released at controlled rates for prolonged span of time.

While many old and new therapeutics are well tolerated, numerous compounds are in requirement of localized advanced drug delivery technologies to reduce toxicity level, enhance therapeutic efficacy and potentially recast bio-distribution. Local drug delivery is the manifestation of drug delivery in which drugs are delivered at a specific site inside the body to a particular diseased organ or tissue. Though the drug delivery, in principle, may be monitored, but the most important hazard is that the design of drug delivery is unclear, which must be used to attain the level of control required for a specific purpose. This is because there exists complex interaction between biology, polymer chemistry and pharmacology [1].

Mathematical modelling of drug delivery and predictability of drug release is a steadily growing field with respect to its importance in academic and industrial areas due to its astronomic future potential. In the light of drug dose to be incorporated in desired drug administration and targeted drug release profile, mathematical prognosis will allow for good estimates of the required composition along with other requirements of the respective drug dosage forms. An extremely challenging aspect is to combine mathematical theories with models quantifying the release and transport of drug in living tissues and cells. Various works are done in the past on drug delivery devices regarding its therapeutic efficiency, optimal design with the aid of either experimental methods or numerical / modelling simulations and sometimes both procedures are used [2, 3, 4, 5]. The investigation in which drug association / dissociation aspect is taken into account with regard to transdermal drug delivery [6] while in another study, drug release process from a microparticle is considered based on solubilisation dynamics of the drug [7]. Very recently, the updated mathematical model considering both the above mentioned aspects is framed and analysed successfully [8].

The study of concern is presented by a phenomenological mathematical model of drug release from a local drug delivery device and its subsequent transport to the biological tissue. The model is comprised of two-phase drug release where the drug undergoes solubilisation, recrystallisation and internalization through a porous membrane. An important aspect of the aforementioned model is appropriate judgement of the model parameters of significance such as diffusion coefficient, solid-liquid transfer rate, mass-transfer, drug association / dissociation rate constants, membrane permeability and internalization rate constant. The numerical simulation provides reliable information on different properties of drug release kinetics.

2. Problem of drug release

To model the local drug delivery device, a two-phase system is considered which is made of: (a) a polymeric matrix that operates as a reservoir where the drug is loaded initially, and (b) the biological tissue where the drug is being transported as target region. The first phase i.e the drug delivery device is framed as a planar slab, encompassed on one side with an impermeable backing and the other side of the device is in contact with layer (b). A rate-controlling membrane protecting the polymeric matrix is present at the interface of the coupled layers.

2.1. Mechanism of drug release

At the beginning, the drug occurs completely in a solid phase embraced within the polymeric matrix (e.g. in crystalline form) (C_0^*) at its maximum concentration. Being in bound state, it cannot be transferred to the tissues directly. Water enters into the polymeric matrix and wets the drug encapsulated inside it, permitting solubilisation of the loaded drug crystals into free state (C_0) which diffuses out of the matrix into the tissue. The rate of transfer of drug from solid state to free state depends not only on solubilisation phenomenon but is also proportional to the difference between C_0^* and C_0. Again, a fraction of solid drug ($\beta_0 C_0^*$) is transformed to its free state which is competent to diffuse. Conversely, through a recrystallisation process, another fraction of free drug ($\delta_0 C_0$) is transferred back to its bound state. Simultaneously, a part of free drug (C_1) diffuses into the tissue. In the similar way, in tissue, a portion of free drug ($k_a C_1$) is metabolised into bound phase (C_1^*), which also unbinds ($k_d C_1^*$) to form free drug. Now, the bound drug is engulfed (internalized) ($k_i C_1^*$) by the cell in the tissue through the process of endocytosis. Endocytosis is an active energy-using transport phenomenon in which molecules (proteins, drugs etc.) are transported into the cell. Thus, bound drug gets transformed into internalized drug particles (C_i). These internalized drug particles, after a span of time, gets degraded by the lysosomes and the drug remnants after degradation ($k_{id}C_i$) is expelled out of the cell into the extracellular fluid. The complete drug transport process is schematically demonstrated in Fig. 1.

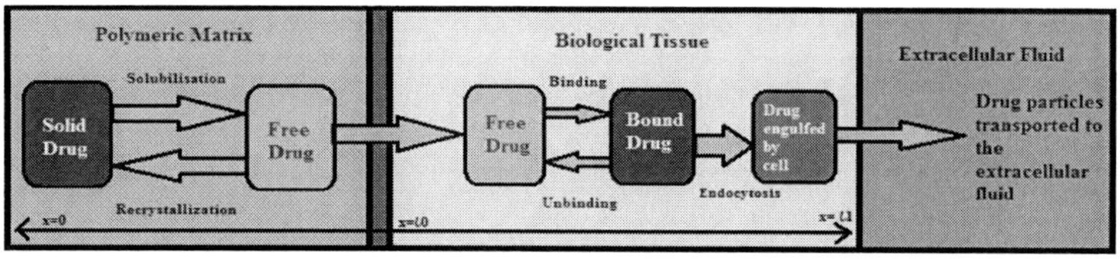

Figure 1. Schematic diagram of drug transport

3. Formulation of the model

Generally, mass transport prevails in the direction normal to the tissue which may be the reason behind the modelling restriction confined to one-dimensional case. In the present study, x-axis is considered to be normal to the layer and aligned with the positive direction outwards.

3.1. Drug dynamics in the polymeric matrix

The governing equations describing the dynamics of drug release in the polymeric matrix phase are

$$\frac{\partial C_0^*}{\partial t} = -\alpha_0(\Phi_0 C_0^* - C_0) - k_m(C_{lim} - C_0) - \beta_0 C_0^* + \delta_0 C_0, \qquad x \in (0, l_0) \tag{3.1}$$

$$\frac{\partial C_0}{\partial t} = D_0 \frac{\partial^2 C_0}{\partial x^2} + \alpha_0(\Phi_0 C_0^* - C_0) + k_m(C_{lim} - C_0) + \beta_0 C_0^* - \delta_0 C_0, \qquad x \in (0, l_0) \tag{3.2}$$

where, $\Phi_0(=\frac{k\varepsilon_0}{1-\varepsilon_0})$ is the ratio of accessible void volume to solid volume, C_0^* denotes available molar concentration of solid drug, C_0 is the available molar concentration of free drug, k stands for partition coefficient, ε_0 denotes porosity, l_0 is the length of the polymeric matrix, k_m is mass transfer coefficient, C_{lim} stands for drug solubilisation limit, β_0 is the dissociation rate constant, δ_0 is the association rate constant, α_0 denotes solid-liquid rate parameter. D_0 is the diffusion coefficient of free drug in the matrix.

3.2. Drug dynamics in the biological tissue

The corresponding equations governing the dynamics of drug in the tissue are

$$\frac{\partial C_1^*}{\partial t} = k_a C_1 - k_d C_1^* - k_i C_1^*, \qquad x \in (l_0, l_1) \tag{3.3}$$

$$\frac{\partial C_1}{\partial t} = D_1 \frac{\partial^2 C_1}{\partial x^2} - k_a C_1 + k_d C_1^*, \qquad x \in (l_0, l_1) \tag{3.4}$$

$$\frac{\partial C_i}{\partial t} = k_i C_1^* - k_{id} C_i, \qquad x \in (l_0, l_1) \tag{3.5}$$

where, C_1 is the available molar concentration of free drug in the tissue, C_1^* is the available molar concentration of solid drug in the tissue, C_i denotes the molar concentration of internalized drug particles, $l_1 - l_0$ is the length of the tissue, k_a depicts the binding rate coefficient, k_d is the dissociation rate coefficient, k_i stands for internalization rate coefficient, k_{id} denotes degradation rate constant in the lysosome, and D_1 is the diffusion coefficient of free drug in the biological tissue.

3.3. Initial, Interface and Boundary Conditions

The initial conditions are as follows:
$C_0^*(x,0) = M$, $C_0(x,0) = 0$, $C_1^*(x,0) = 0$, $C_1(x,0) = 0$, $C_i(x,0) = 0$,where M is the maximum value of loaded solid drug concentration.
A flux continuity must be assigned at the interface, i.e at $x = l_0$, $-D_0\frac{\partial C_0}{\partial x} = -D_1\frac{\partial C_1}{\partial x}$.
At $x = l_0$, a concentration jump may occur: $-D_1\frac{\partial C_1}{\partial x} = P_1(C_0 - C_1)$.
where, P_1 is the overall mass transfer coefficient.
No mass flux can pass to exterior environment due to the presence of impermeable backing and hence no flux condition arises. At $x = 0$, $D_0\frac{\partial C_0}{\partial x} = 0$.
Lastly, at $x = l_1$, $C_1(l_1, t)$ is considered to be finite.

3.4. Model Solutions

For the purpose of reducing the number of model parameters, the entire above-mentioned equations together with all the conditions are made dimensionless. The transformed equations subject to the conditions in dimensionless approach are solved analytically by separation of variables procedure. Thus, the solutions should be read as

$$C_0 = (e^{-m_1 t} - e^{-m_2 t})\cos ax \tag{3.6}$$

$$C_0^* = \frac{(\alpha_0 + k_m + \delta_0)\cos ax}{(m_1 - (\alpha_0\Phi_0 + \beta_0))(m_2 - (\alpha_0\Phi_0 + \beta_0))}\Big[(m_1 - (\alpha_0\Phi_0 + \beta_0))e^{-m_2 t} - (m_2 - (\alpha_0\Phi_0 + \beta_0)) e^{-m_1 t} + (m_2 - m_1)e^{-(\alpha_0\Phi_0+\beta_0)t}\Big] - \frac{k_m C_{lim}}{(\alpha_0\Phi_0 + \beta_0)}\Big[1 - e^{-(\alpha_0\Phi_0+\beta_0)t}\Big] + e^{-(\alpha_0\Phi_0+\beta_0)t} \tag{3.7}$$

$$C_1 = E_2(e^{-n_1 t} - e^{-n_2 t})\cos bx \tag{3.8}$$

$$C_1^* = \frac{E_2 k_a \cos bx}{(k_d + k_i - n_1)(k_d + k_i - n_2)}\Big[(k_d+k_i-n_2)e^{-n_1 t} - (k_d+k_i-n_1)e^{-n_2 t} + (n_2-n_1)e^{-(k_d+k_i)t}\Big] \tag{3.9}$$

$$C_i = \frac{E_2 k_a k_i \cos bx}{(k_d + k_i - n_1)(k_d + k_i - n_2)}\Big[\frac{(k_d+k_i-n_2)}{(k_{id}-n_1)}(e^{-n_1 t} - e^{-k_{id}t}) - \frac{(k_d+k_i-n_1)}{(k_{id}-n_2)}(e^{-n_2 t} - e^{-k_{id}t}) + \frac{(n_2-n_1)}{(k_{id}-k_d-k_i)}(e^{-(k_d+k_i)t} - e^{-k_{id}t})\Big] \tag{3.10}$$

where, a and E_2 are determined from the following relations.

$$\frac{\Pi}{\gamma a}\cot al_0 = 1 + \frac{\Pi}{b}\cot bl_0, \qquad E_2 = \frac{n_1 n_2 \gamma a \sin al_0 (m_2 - m_1)}{m_1 m_2 b \sin bl_0 (n_2 - n_1)},$$

$$m_1 = \frac{A - \sqrt{A^2 + 4B}}{2}, \qquad m_2 = \frac{A + \sqrt{A^2 + 4B}}{2},$$

$$A = \alpha_0(\Phi_0 + 1) + k_m + \beta_0 + \delta_0 - \lambda\gamma, \qquad B = \lambda\gamma(\alpha_0\Phi_0 + \beta_0), \qquad \lambda = -a^2,$$

$$n_1 = \frac{P - \sqrt{P^2 - 4Q}}{2}, \qquad n_2 = \frac{P + \sqrt{P^2 - 4Q}}{2},$$

$$P = k_d + k_i + k_a - \mu, \qquad Q = k_i k_a - \mu(k_d + k_i), \qquad \mu = -b^2, \qquad b = 3.14/2l_1.$$

All the parameters and the variables are expressed in dimensionless form whose expressions are not included here for the sake of brevity.

4. Numerical illustration and discussion

Numerical illustration for the present drug release system is performed by taking various parameter values of the model in order to characterise the pharmacokinetic aspects. The graphical representations of the time-variant concentration profiles of the drug in its different states for both the facets are well illustrated through the Figs. 2 - 6 in order to understand the drug release phenomenon.

The time-variant concentration profiles for solid (loaded) and free drug particles in the polymeric matrix phase for four different axial locations spread over the entire domain are demonstrated in Figs. 2 and 3. The rate of decrease of concentration of loaded drug (C_0^*) becomes higher

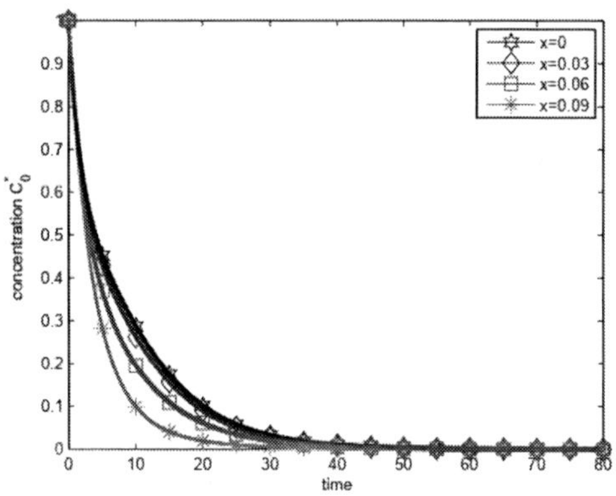

Figure 2. Time variant concentration profile of C_0^* at different locations.

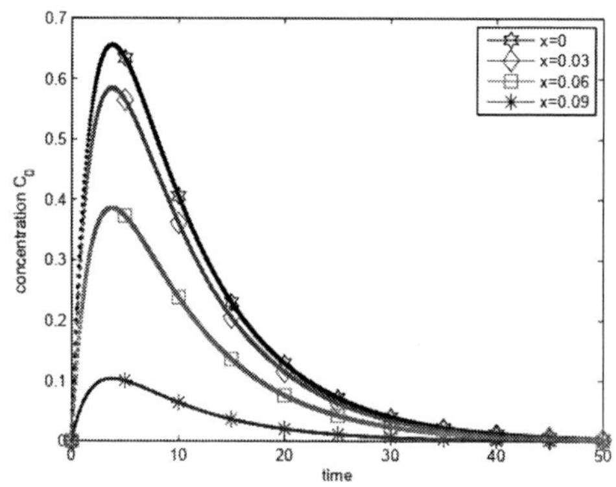

Figure 3. Time variant concentration profile of C_0 at different locations.

and higher as one proceeds away from the commencing region of the polymeric matrix to the interface resulting in early disappearance. When the loaded drug gets exposed to water, a solid-liquid mass transfer is initiated causing drug release from the matrix which is on the process of transformation of free drug C_0. One may note on the other hand that C_0 grows and acquires certain peak in accordance to a specific instant of time followed by a gradual descend for rest of the times.

Figs. 4 - 6 represent the time-variant concentration profiles for free, bound and internalized drug particles respectively in the biological tissue for different axial locations stretched over the entire domain. It is important to observe that C_1 heightens hastily compared to both C_1^* and C_i towards the inception. This observation, as anticipated, reflects in the realm of drug kinetics that free drug (C_1) gets transformed into bound drug (C_1^*) and subsequently the bound drug is metabolised into internalized drug (C_i) after a short passage of time. Ultimately, it is further noted that the internalized drug takes more time to get absorbed completely in the tissue than the characteristics of both bound and free drug particles. In addition to the present findings, one may append that the extended time span distinctly reveals that both loaded and free drug particles in polymeric matrix melt away in a comparatively small span of time than those in the tissue where they need to take time lengthened to get the drug absorbed completely.

One may also explore a variety of cases in order to exhibit the behaviour of the concentration profiles for both the phases under present consideration by varying all the parameter values of significance. The sensitivity

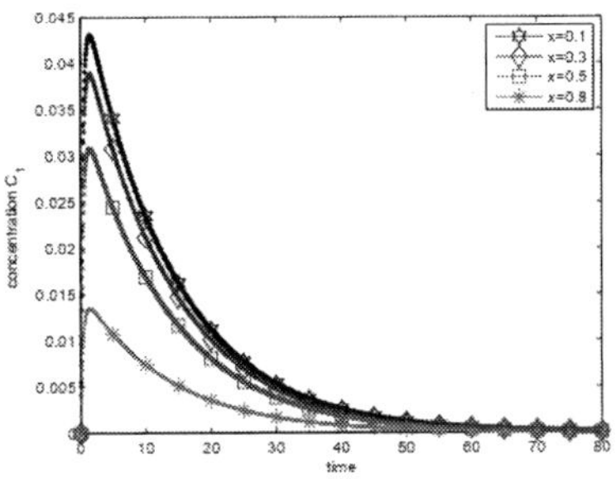

Figure 4. Time variant concentration profile of C_1 at different locations.

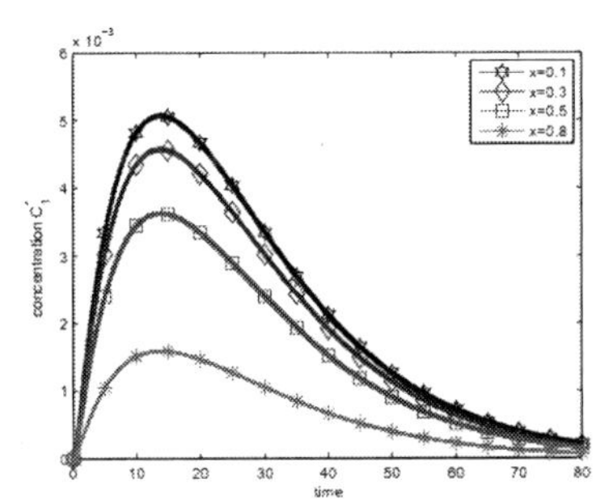

Figure 5. Time variant concentration profile of C_1^* at different locations.

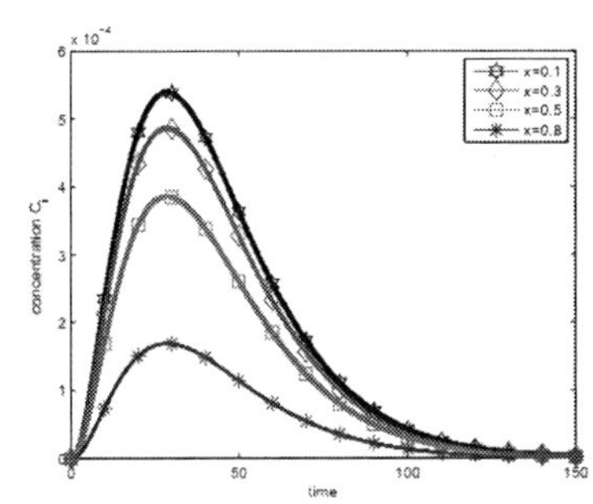

Figure 6. Time variant concentration profile of C_i at different locations.

included in the model formation for future course of studies relevant to this domain. Most of the results exhibited here agree qualitatively well with those of [9] and many others relevant to this realm of research.

5. Conclusions

Mathematical modelling of drug release and drug transport phenomena turns out to be very useful as it helps in prediction of release kinetics which has an important value in the optimization of drug dosage forms. The sensitivity of the model parameters poses challenges to the applicability of drug administration for treatment of patients at large through pharmacotherapy. One may highlight that as both loaded and free drug particles in polymeric matrix melt away in comparatively small span of time than those in the tissue, its influence will certainly persist for a long time before repeated medication occurs and hence care needs to be exercised for maintaining appropriate time-gap before redispensation in order to avoid toxicity by the presence of excess drug.

References

[1] Siepmann J and Siepmann F 2008 *Int. J. Pharm.* B **364(2)** 328
[2] Siepmann J and Gopferich A 2001 *Adv. Drug Delivery Rev.* **48** 229
[3] Crane M, Hurley N J, Crane L, Healy A M, Corrigan O I, Gallagher K M and McCarthy L G 2004 *Simul. Model. Pract. & Theory* **12(2)** 147
[4] Narasimhan B 2001 *Adv. Drug Delivery Rev.* **48** 195
[5] Bozsak F, Chomaz J M and Barakat A I 2014 *Biomech. Model. Mechanobiol.* **13(2)** 327
[6] Pontrelli G and Monte F 2014 *Mathematical Biosciences* **257** 96
[7] Casalini T, Rossi F, Lazzari S, Perale G and Masi M 2014 *Mol. Pharmaceutics* **11** 4036
[8] Chakravarty K and Dalal D C 2016 *Mathematical Biosciences* **272** 24
[9] Zhou Q, Ruan Z R, Yuan H, Jiang B and Xu D H 2006 *World J. Gastroenterol.* **12(17)** 2742

Probing the energy landscape of alanine dipeptide and decalanine using temperature as a tunable parameter in molecular dynamics

A Chatterjee[2] and S Bhattacharya[1]

[1]Department of Physics, Indian Institute of Technology Guwahati, Guwahati, Assam, India 781039
[2]Department of Chemical Engineering, Indian Institute of Technology, Bombay, Maharashtra.

Abstract. We perform several molecular dynamics (MD) calculations of solvated alanine dipeptide and decalanine in vacuum with temperature as a tunable parameter and in the process, generate Markov state models (MSMs) at each temperature. An interesting observation that the kinetic rates appear to obey the Arrhenius rate law allows us to predict the dynamics of alanine dipeptide at 300 K at the microsecond timescales using the nanoseconds long high temperature calculations without actually performing MD simulations at 300 K. We conclude that the energy landscape of alanine dipeptide contains superbasins deeper than k_BT and determine the energy barriers associated with the moves from the Arrhenius rate expression. Similar insights regarding the energy landscape associated with folding/unfolding pathways of a deca-alanine molecule are obtained using kinetic rates calculated at different temperatures.

1. Introduction

A protein molecule can exist in different configurations that determine its biological activity. Consequently, significant efforts have been made over the last three decades to understand the stable configurations, free energy differences between the stable configurations, and the mechanisms for transformations from one configuration to another[1–6]. In recent years, a large number of studies have focussed on the application of Markov State Models to describe the dynamics of solvated biomolecular systems. An MSM enables one to study the dynamics of protein systems in terms of state-to-state transitions thereby providing a systematic approach for understanding the nature and timescales of evolution of the system. Pande and co-workers have pioneered the development of an automated kinetic clustering procedure for building of MSMs by post-analyzing long room temperature MD trajectories[5]. Generating an accurate MSM requires extensive MD calculations especially when the system remains trapped in certain states for extended periods of time, a situation that is frequently encountered when large free energy barriers are present. Several accelerated MD methods have been proposed to overcome the issue of long-lived metastable states leading to poor sampling[7–12].

Our interest lies in methods employing high temperature simulations to overcome energy barriers. Although high temperature simulations are often used for seeding standard simulations and REMD has been widely used for enhanced conformational sampling, very few studies have directly applied high temperature simulations to generate MSMs of biomolecular systems[13]. Establishing the connection between high temperature and low temperature dynamics opens up the possibility of applying temperature accelerated schemes[14,15] to enhance study of rare events in proteins and other biomolecular systems while recovering the correct dynamics at room temperature. We show that high temperature MD calculations can efficiently provide information about the states and the kinetic rates of the solvated alanine dipeptide with a speed-up of almost 500 times over room-temperature MD calculations[16]. In this process, additional insights about the energy landscape are obtained compared to single temperature MD calculations of similar duration.

We employ a swarm of independent MD calculations in parallel at high temperatures to accelerate conformational sampling. The states are determined *on-the-fly* in terms of energy-minimized configurations of the biomolecule using a distance-metric. Once the states have been identified at a high temperature, we perform shorter MD simulations starting from selected states at different temperatures to obtain the rate constants associated with specific moves originating from these states. A consistent picture of the states and kinetic pathways is obtained across different temperatures. A similar method is applied to the decaalanine system in a force spectroscopy setup in vacuum to obtain the energy barriers along a folding/unfolding path.

2. Methodology

2.1 System Preparation

A single molecule of blocked N–acetyl–N'–methyl–L–alanylamide (or alanine dipeptide), was immersed in a rectangular box of 390 pre-equilibrated TIP3P water molecules bringing the total system size to 1192 atoms. The final system was in the form of a cubical cell with dimensions 2.3x2.3x2.3 nm^3. The CHARMM27 force-field[17] was used. We constructed MSMs for the capped deca-alanine (Ala_{10}) with acetylated N-terminus and amidated C-terminus in vacuum, stretched under a force-spectroscopy setup. The C_α atoms at the two ends (residues 1 and 10) are tethered to two anchor points by harmonic restraints with a spring constant of 0.86 $kcal/mol/Å^2$. The initial configuration was obtained from the 104-atom helical model of Ref.[18].

2.2 Simulation Protocols

Prior to the main simulation, the system was minimized using the conjugate gradient method for 1000 steps following which it was equilibrated for 3 ns at 300 K and constant pressure of 1 atm. Subsequent simulations were performed at constant volume and temperature (canonical ensemble) with NAMD[19] with periodic boundary conditions and particle mesh Ewald electrostatics[20] in the canonical ensemble using Langevin thermostat[21] at 600, 450 and 750 K. An integration time-step of 2 fs was used with RATTLE[22] and SETTLE[23] algorithms applied to covalent bonds involving hydrogen in the peptide and water respectively. A swarm of independent MD trajectories are generated that share the state information. The state of the system in each trajectory is monitored on-the-fly. State-matching is performed based on the internal coordinates of the molecule with a tolerance of 1.5 Å following alignment using the Kabsch algorithm. A match is said to occur when respective heavy (non-hydrogen) atoms of the alanine dipeptide molecule are within the tolerance value after the molecule has been aligned to one of the energy-minimized structures via rotation and translation operations. The MD calculation is stopped once the system moves to a new state. The waiting time for each escape is recorded and a new calculation is started from the new state using the same procedure.

3. Results

3.1 Kinetics of alanine dipeptide

A total of 16 states of alanine dipeptide were discovered using the state-identification procedure described above. Of these, five major states (labelled A-E) were identified that accounted for >99% occupation probability of the molecule in the range of temperatures studied. States *A* and *C* are associated with the α_R conformation of the molecule. States *B* and *D* are located in the region commonly associated with the β/PII/$C7_{eq}$ structures of the system. State *E* can be identified with the C_{7ax} conformation. The rate constants were obtained using the maximum likelihood estimate (MLE)[24] *i.e.*,

$k_\alpha = n/t_{MD}$ where n denotes the number of times a move has been observed and t_{MD} is the total MD time elapsed for this move. The Arrhenius plot for the move A-B is shown in Figure 1(b). The rate constants for the move from state A to B from MLE are given by 0.045, 0.102, 0.150 and 0.188 ps^{-1} at 300, 450, 600 and 750 K when plotted against inverse temperature, fall on a straight line in semi-logarithmic coordinates demonstrating the validity of the Arrhenius expression $k_\alpha = \nu_\alpha \exp(-E_\alpha / k_B T)$. The pre-exponential factor and activation barrier for the A-B move were estimated to be 0.53 ps^{-1} and 0.064 eV. An interesting observation is that the pre-exponential factor is not temperature dependent for the range of temperatures studied. The fact that the same pre-exponential factor and activation barrier can be used at different temperatures indicates that the move between two states of alanine dipeptide proceeds via a single mechanism.

The rates estimated at 300 K for the moves from state A using the Arrhenius expression were compared to the ones obtained directly from regular MD using MLE at 300 K to validate our procedure. The estimated rates for the move from state A to state B, C and D are 0.048, 0.0014, 0.0011 ps^{-1}, respectively, while the rate calculated using MLE were 0.045, 0.0015, 0.0011 ps^{-1}, respectively. Although a direct comparison of the kinetic rates with previous studies is not possible due to differences in the system setup and force fields used, we found that the kinetic rates predicted at 300 K from the high temperature simulations are consistent with results from experiments as well as previous simulation studies. The rate for moves from A to B (alpha-helix to beta strand transitions) is estimated to be 0.045 ps^{-1} corresponding to a mean escape time of ~22 ps which is comparable to previous studies[25].

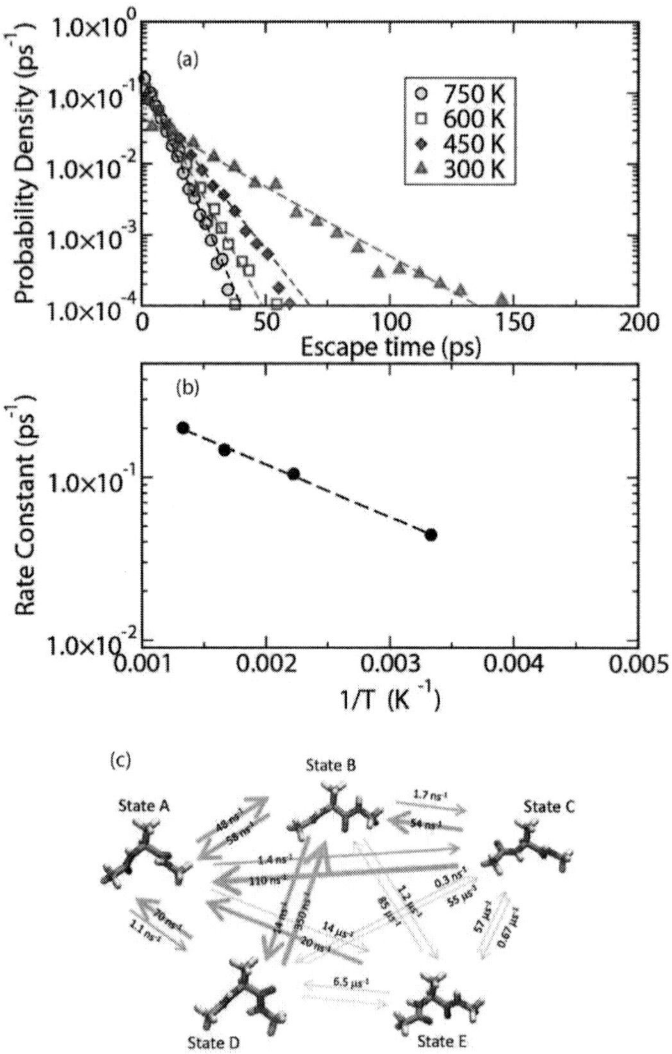

Fig. 1. (a) Probability density for escape (or waiting) times distributions obtained for the move from state A to B at 750, 600, 450 and 300 K. Dashed lines denotes the exponential distribution of escape times using the rate constant calculated by maximum likelihood estimation. (b) Arrhenius plot showing the variation of the rate constants with temperature for the move A-B. The rate constants fall on the dashed line showing the Arrhenius behaviour. (c) Markov State Model (MSM) constructed for solvated alanine dipeptide. The rate constants at 300 K are the obtained from high temperature simulations.

The energy barriers and pre-exponential factors estimated using the Arrhenius expression showed the pre-factors to be temperature independent indicating that each move proceeds via a single mechanism. The largest energy barrier was 8.3 kcal/mol for the move C-E, while the move from C-A had a negligible barrier of 0.054 kcal/mol. The size of the observed energy barriers can be explained by making comparisons to the corresponding barriers for alanine dipeptide in vacuum from previous studies[26,27]. The barriers for the moves *B-E*, *E-B*, and *E-D* are 6.55, 5.7 and 5 kcal/mol, respectively, in the solvated system. The corresponding barriers in vacuum[27] are less than 7, 5 and 5 kcal/mol, respectively, i.e., they are similar to the ones for solvated alanine dipeptide. The location of the states B, D and E in the ϕ-ψ map are slightly altered due to the presence of solvent molecules. The large activation barriers point to the depth of the superbasins being significantly more than k_BT. The barriers involved in these moves are energetic in nature. The origin of these barriers can be attributed to the large energy changes associated with the backbone torsional changes in alanine dipeptide that are also witnessed in vacuum. The presence of solvent molecules introduces ripples in the potential energy landscape

3.2 Kinetics of Decaalanine

We consider a single deca-alanine molecule that is stretched at two ends as shown in Fig. 2a. The C_α atoms (residues 1 and 10) are tethered to two anchor points by harmonic restraints (spring constant of 0.86 kcal/mol/Å^2). The anchors are separated by a distance of 24 Å. The result of the applied force is that the α-helical structure (Fig. 2a, state 1) is no longer the preferred configuration, unlike the behaviour observed in the absence of an applied force. Instead the molecule can stretch dramatically as shown in Fig.2a (state 17).

State-constrained MD calculations were performed at 300, 400 and 500 K starting from the α-helical structure to reveal other states of the system using a procedure analogous to the one for alanine dipeptide. The pathway involving transitions between states 17-6-2-1 can be regarded as the folding pathway. As evident from Table 1, the α-helicity decreases while 3_{10}-helicity increases along this pathway. In addition, the α-helical structure can form a deformed helical state denoted as state 3, which in fact is the dominant state at anchor separation of 24 Å. Unfolding of the deca-alanine entails breaking of the hydrogen bonds, so the question arises how large is the energy change.

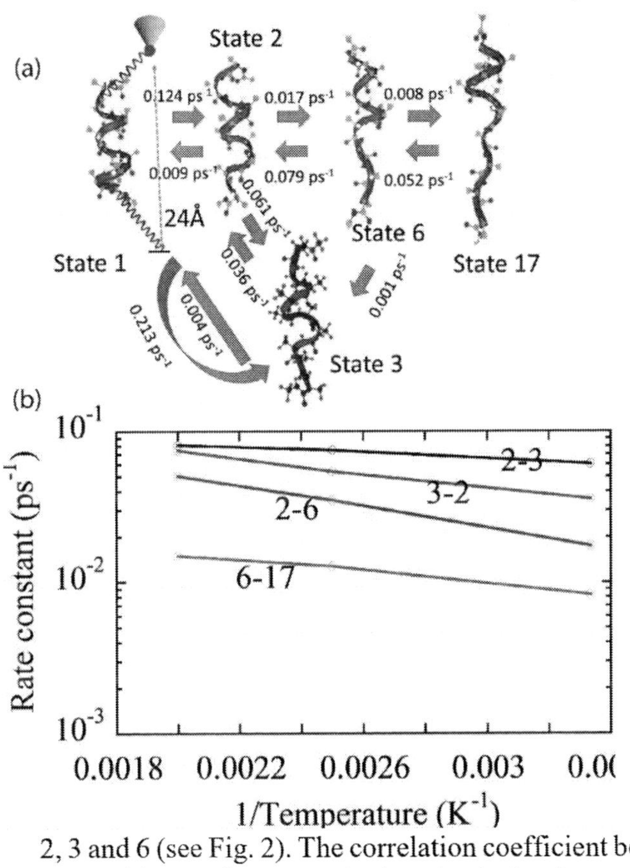

Figure 2. (a) Snapshots showing the states 1,2,3,6 and 17 along the chosen folding/unfolding pathway with the kinetic rates calculated at 300 K with anchor separation 24 Å. (b) The Arrhenius plot showing the variation of the kinetic rates of conversion between the states with inverse temperature in semi-logarithmic scale.

We answer this question by analyzing the temperature dependent kinetics from states 2, 3 and 6 (see Fig. 2). The correlation coefficient between the data and the Arrhenius fit exceeds 0.96. This permits us to calculate the Arrhenius parameters. Surprisingly, the activation barriers involved in the pathways are small. For instance, the barrier for the move from state 2-1, 2-6 and 6-17 are found to be 0.01, 0.069 and 0.039 eV, respectively. These small barriers indicate the folding or unfolding proceeds as the molecules slithers into a compact or elongated form. Minor rearrangements in the process could explain the small barriers. This conclusion is also supported by the α- and 3_{10}-helicity shown in Table 1.

Table 1. Average α-helicity and 3_{10}-helicity of states along a folding pathway.

State	α-Helicity	3_{10}-Helicity
1	0.801	0.08
2	0.201	0.21
6	0.021	0.19
17	0.013	0.15

4. Discussions

We use molecular dynamics (MD) temperature as a probe to explore the overall features of the potential energy landscape of the solvated alanine dipeptide system and the decaalanine system in a force spectroscopy setup in vacuum. In alanine dipeptide, the moves between the states are found to be first order processes and the rate constants were found to obey the Arrhenius expression. Some of the activation barriers extracted from the Arrhenius fits were found to be significantly larger than k_BT showing that the landscape can be characterized in terms of deep energy superbasins. The method was applied to the decaalanine system to obtain the energy barriers along a folding/unfolding pathway in a force spectroscopy setup. The surprisingly small barriers along the pathway chosen in the study shows that the molecule unfolds/folds by small rearrangements, i.e., wriggles into compact or elongated form.

Acknowledgments
SB acknowledges support from the Department of Biotechnology Grant No. BT/PR8311/BID/7/451/2013.

References:

[1] Gnanakaran S, Nymeyer H and Portman J 2003 Peptide folding simulations Kevin Y Sanbonmatsu and Angel E Garcı *Curr. Opin. Struct. Biol.* **13** 168–74
[2] Dodson G G, Lane D P and Verma C S 2008 new windows on mechanisms in biology *EMBO Rep.* **9** 144–50
[3] Oliveira C A F De, Hamelberg D and Mccammon J A 2007 Estimating kinetic rates from accelerated molecular dynamics simulations : Alanine dipeptide in explicit solvent as a case study Estimating kinetic rates from accelerated molecular dynamics simulations : Alanine dipeptide in explicit solvent as a case stu *J. Chem. Phys.* **127** 175105
[4] Hamelberg D, Shen T and Mccammon J A 2005 Relating kinetic rates and local energetic roughness by accelerated molecular- dynamics simulations Relating kinetic rates and local energetic roughness by accelerated molecular-dynamics simulations *J. Chem. Phys.* **122** 241103
[5] Bowman G R, Huang X and Pande V S 2009 Using generalized ensemble simulations and Markov state models to identify conformational states. *Methods* **49** 197–201
[6] Freddolino P L, Harrison C B, Liu Y and Schulten K 2010 Challenges in protein-folding simulations *Nat. Phys.* **6** 751–8
[7] de Oliveira C A F, Hamelberg D and McCammon J A 2007 Estimating kinetic rates from

accelerated molecular dynamics simulations: alanine dipeptide in explicit solvent as a case study. *J. Chem. Phys.* **127** 175105
[8] Wang Y, Harrison C B, Schulten K and McCammon J A 2011 Implementation of Accelerated Molecular Dynamics in NAMD. *Comput. Sci. Discov.* **4** 015002
[9] Voter A F 1997 Hyperdynamics : Accelerated Molecular Dynamics of Infrequent Events *Phys. Rev. Lett.* **78** 3908
[10] Laio A and Gervasio F L 2008 Metadynamics: a method to simulate rare events and reconstruct the free energy in biophysics, chemistry and material science *Reports Prog. Phys.* **71** 126601
[11] Tribello G A, Ceriotti M and Parrinello M 2010 A self-learning algorithm for biased molecular dynamics *Proc. Natl. Acad. Sci.* **107** 17509–14
[12] Sugita Y and Okamoto Y 1999 Replica exchange molecular dynamics method for protein folding simulation. *Chem. Phys. Lett.* **314** 141–51
[13] Buchete N V and Hummer G 2008 Peptide folding kinetics from replica exchange molecular dynamics *Phys. Rev. E - Stat. Nonlinear, Soft Matter Phys.* **77** 1–4
[14] Sorensen M R and Voter A F 2000 Temperature-accelerated dynamics for simulation of infrequent events *J. Chem. Phys.* **112** 9599
[15] Divi S and Chatterjee A 2014 Accelerating rare events while overcoming the low-barrier problem using a temperature program. *J. Chem. Phys.* **140** 184115
[16] Chatterjee A and Bhattacharya S 2015 Uncertainty in a Markov state model with missing states and rates: Application to a room temperature kinetic model obtained using high temperature molecular dynamics. *J. Chem. Phys.* **143** 114109
[17] Mackerell A D, Bashford D, Bellott M, Dunbrack R L, Evanseck J D, Field M J, Fischer S, Gao J, Guo H, Ha S, Kuchnir L, Kuczera K, Lau F T K, Mattos C, Michnick S, Ngo T, Nguyen D T, Prodhom B, Reiher W E, Roux B, Schlenkrich M, Smith J C, Stote R, Straub J, Watanabe M, Wio J, Yin D and Karplus M 1998 All-Atom Empirical Potential for Molecular Modeling and Dynamics Studies of Proteins † *J. Phys. Chem. B* **5647** 3586–616
[18] Park S, Khalili-Araghi F, Tajkhorshid E and Schulten K 2003 Free energy calculation from steered molecular dynamics simulations using Jarzynski's equality *J. Chem. Phys.* **119** 3559–66
[19] Phillips J C, Braun R, Wang W, Gumbart J, Tajkhorshid E, Villa E, Chipot C, Skeel R D and Kalé L 2005 Scalable Molecular Dynamics with NAMD *J. Comput. Chem.* **26** 1781–802
[20] Darden T, York D and Pedersen L 1993 Particle mesh Ewald: An N·log(N) method for Ewald sums in large systems *J. Chem. Phys.* **98** 10089
[21] Allen P and Tildesley D J 1991 *Computer Simulation of Liquids* (Oxford University Press)
[22] Andersen C 1983 Rattle : A " Velocity " Molecular Version of the Shake Dynamics Calculations for *J. Comput. Phys.* **52** 24–34
[23] Miyamoto S and Kollman P a. 1992 Settle: An analytical version of the SHAKE and RATTLE algorithm for rigid water models *J. Comput. Chem.* **13** 952–62
[24] Ensign D E and Pande V S 2009 Bayesian Single-Exponential Kinetics in Single-Molecule Experiments and Simulations *J. Phys. Chem. B* **113** 12410–23
[25] Chekmarev D S, Ishida T and Levy R M 2004 Long-Time Conformational Transitions of Alanine Dipeptide in Aqueous Solution : Continuous and Discrete-State Kinetic Models *J. Phys. Chem. B* **108** 19487–95
[26] Tobias D J and Brooks III C L 1992 Conformational Equilibrium in the Alanine Dipeptide in the Gas Phase and Aqueous Solution: A Comparison of Theoretical Results *J. Phys. Chem.* **96** 3864–70
[27] Ren W, Vanden-Eijnden E, Maragakis P and E W 2005 Transition pathways in complex systems: Application of the finite-temperature string method to the alanine dipeptide. *J. Chem. Phys.* **123** 134109

Journal of Physics: Conference Series **759** (2016) 012025 doi:10.1088/1742-6596/759/1/012025

On the Origin of Photodynamic activity of Perylene Quinone Framework

Dibyajyoti Parida[1], Pattath D. Pancharatna[2] and Musiri M Balakrishnarajan*

[1] School of Chemistry, University of Hyderabad, Hyderabad, India-500046

[2] & [*] Department of Chemistry, Pondicherry University, Pondicherry, India-605014
E-mail: djparida06@gmail.com

Abstract: The basic skeleton of perylenequinone is surprisingly ubiquitous in several naturally occurring pigments, such as Hypocrellins, Cercosporin, etc. to name a few. Several of these molecules and their derivatives are also experimentally characterized as potent candidates for photodynamic therapy and are predicted to be aiding the formation singlet Oxygen. Theoretical calculations that unravel the mystery behind the perylenequinone motif in these bio-molecules. Perylenequinone framework has a unique frontier MOs that aid in facile intersystem crossing of the π-π* excitation. The resulting triplet state remarkably resists phosphorescence that presumably leads to high quantum yield of singlet oxygen production. The excitation assisted change in the nature of conjugation and the attendant out-of-plane distortion of the perylene framework is found to be the general characteristic of all these systems and the substituents at the bay region favourably assist the excited state behavior as shown by time dependent/ independent DFT calculations.

1. Introduction:

Photodynamic therapy(PDT) is an effective noninvasive exogenous clinical procedure[1]. Introduction of Photosensitizers(PS) to living system is sundry. The advantage of the PDT is the indolence of the drugs in the absence of the light source. PS accumulated close the targeted area obliterate target selectively. PDT attributes treatment with minimal drug oriented adverse effects. Delicate surgery and lengthy recuperation periods is avoided along with minimal disfigurement and surgical scar. Introduction to PDT was led by Finsen in the late 1890's[2]. Clinical applications of Hematoporphyrin derivatives in PDT are immense[3] with non-avoidable issues[4].This lead to the search of alternative photosensitizers[5-9]. The basic skeleton of perylenequinone is surprisingly ubiquitous in several naturally occurring pigments[14]; such as Hypocrellins, Cercosporin, etc. Several of these molecules[15] and their derivatives are also experimentally characterized[16] as potent candidates for photodynamic therapy(Figure 1) and are predicted to be aiding the formation singlet Oxygen[17].The perylenequinonoid pigments[16] provide a practical platform from which numerous clinical applications[19] with low toxicity, selective accumulation in targeted cells or tissues retention time absorption[20-21] in the phototherapeutic window between 600 to 900nm[21,34] are achieved. Strong light source absorption, high triplet state quantum yield, slow decay[20-21] low cytotoxicity, easy biological removal are the basis of an effective photosensitizer[18]

2.1.Structural features

Naturally occurring perylenequinonoid pigments consists of planner perylene backbone with four oxygen atoms to the sp^2-hybridised carbons of the perylene with intramolecular as well as intermolecular hydrogen bonding.

2.2 Mechanism of PS:

Several theories have been put to explore the mechanistic approach of the photodynamic therapy. Firstly, irradiated PS form slow decaying excited triplet state via intersystem crossing which transfer the energy to the triplet ground state of oxygen resulting highly reactive singlet oxygen[11-13].

In a second way, irradiated PS with hydrogen(s) at the bay region or any labile Hydrogen(s) on the structure (eg: Hypericin) either transfer hydrogen atom or hydrogen radical to the medium having activity similar to reactive singlet oxygen[10]. Hydrogen atom transfer is solvent independent. Transfer is effected by the bulky substituents at the bay region of the perylenequinones and viscosity of the medium. Experimental Validation is confirmed through hydrogen labeling.

Alternatively, radiation excited PS acidify the medium by means of hydrogen bonding interaction and subsequent removal of protons form the medium[23-25]. Induced P_H change impacts localized anti-viral[20,22,24], apoptosis[1,26-27] and anti-tumour activity[28-29]. Computational investigations on some selected perylenequinone molecules establishes oxygen dependent reactions[32].

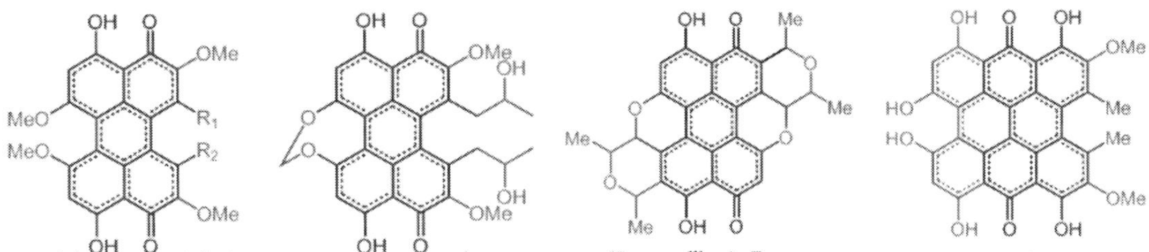

Figure 1. Natural pigments with perylenequinone skeleton. All the structures have a common framework marked in Blue colour. The Red colored region in each structure is the Bay Region.

2.3 Present Work

Current works deals with DFT[30-31] and TDDFT studies on perylene and perylene based systems with most structural resemblance with respect to the naturally occurring perylene quinonoid pigments. The project emphasizes on the qualitative investigation of phototherapeutic potential of some naturally occurring PS having structural similarity.

3. Research Methodology

In Gaussian 09[32], Geometrical optimization and subsequent absorption spectra of all molecular structures are investigated with DFT and TDDFT respectively using B3LYP functional and 6-31G(d)[31] basis set. Molecular Orbit Analysis is done using CACAO[36]; based on EHMO[35].

4. Result and discussion:

Two naphthalene units are molecules connected by sp^2-hybridized carbon-carbon bonds at the 1 and 8 positions. Kekule model for perylene of D_{2h} symmetry reveals that the π-conjugation in the perylene is incomplete due to single bonded carbon–carbon atoms at central region with length 1.476A° as observed on optimized structure(Figure 2 A.) Nonbonding Carbon-Carbon and Hydrogen-Hydrogen interaction at the bay region lie within van der waal's contact. The lowest frequency corresponding out of plane distortion at central carbon-carbon bond occurs around 27cm^{-1}.

Both highest occupied molecular orbital(HOMO) and lowest unoccupied molecular orbital(LUMO) are non-degenerate at the central region. Highest occupied molecular orbital(HOMO) is antibonding and the lowest unoccupied molecular orbital(LUMO) is bonding; favouring electron transition HOMO to LUMO with band gap around 1.8eV. Ground state Perylenequinone dianion is a 1,2,7,8-tetraoxo perylene dianion. Connected oxo group may be either carbonyl linked or hydroxyl linked to perylene. Optimized Singlet perylenequinone dianion reveals forced double bond character at the central region Carbon-Carbon bond with D_{2h} symmetry. The distances of nonbonding carbon-carbon atom and nonbonding hydrogen-hydrogen at the bay region are shorter than that of perylene. Out of plane distortion central carbon-carbon bond region is observed at the lowest frequency of 25cm^{-1}.Both HOMO and LUMO of high electron density are non-degenerate and anti-bonding at the central region with band gap 2.0ev.

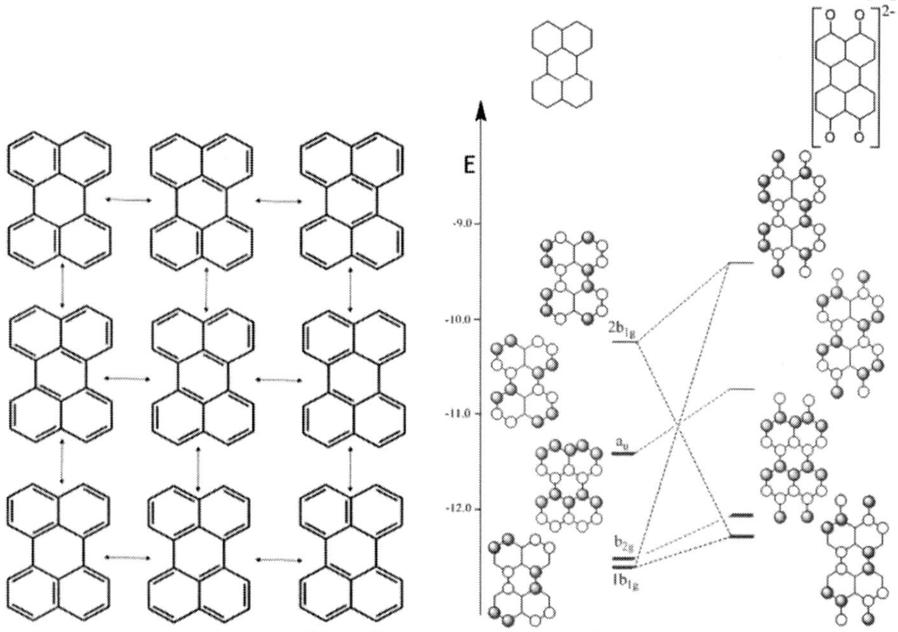

Figure 2 A. *Kekule structure of perylene showing single bond character at the central region* B. *Correlation diagram of the frontier p-MOs of perylene with perylenequinone*

Perylenequinone dianion in triplet state pertaining planner geometry(D_{2h}) is the transition state. The computed minimum for ground state of D_2 symmetry which exhibits out of plane distortion where two naphthalene units have rotation by an angle of 7°.The carbon-carbon bonds neighbouring to oxygen atoms show large elongation as corresponding oxygen atoms are antibonding. The central carbon-carbon bonds are also elongated due to increased electron density. The energy gap between singlet and triplet is 23.3KCalMol^{-1}. The sp mixing results reduced energy gap(Figure 2 B).The geometry of the perylenquinone tetra anion is similar to that of the triplet state of perylenequinone dianion with more pronounced effect due to high electron density as well as out-of-plane distortion by 42cm^{-1} .The rotation angle of the two naphthalene units is 11° which is 4° more than that of triplet dianion. However, there is substantial distortion within individual naphthalene units. Four Chlorine atoms are substituted at the bay region analyzed computationally for Singlet and triplet states. On substitution of four chlorine atoms at the bay region there is enhanced electron density and the steric hindrance. The individual naphthalene units have a waving type distortion. Chlorine substitution increased the distortion. The dihedral angle between the two bay carbons rises to 32° along with a geometry D_2. The computed properties are similar to that of singlet state of tetra-chloro substituted perylene quinone dianion but the extent of effect is larger. The extent of bond compression is more than that of bond elongation in the triplet state. The energy gap between the singlet and triplet is 16.3KCalMol^{-1}.

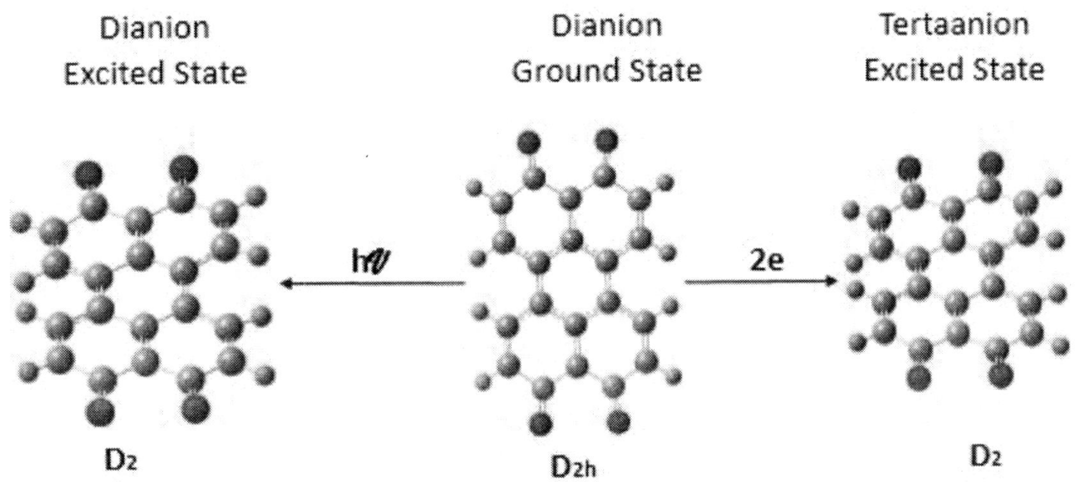

Figure 3. Dianion ground state of perylenequinone of D_{2h} symmetry becomes non-planer D_2 geometry either by radiation or by increased electron density to attain energy minima. The naphthalene units are maneuver out-of-plane rotation .Purple spheres denotes Hydrogen atoms and Red spheres represents the oxygen atoms. Carbon atoms are marked as Grey spheres.

4.3 UV/Visible Spectra

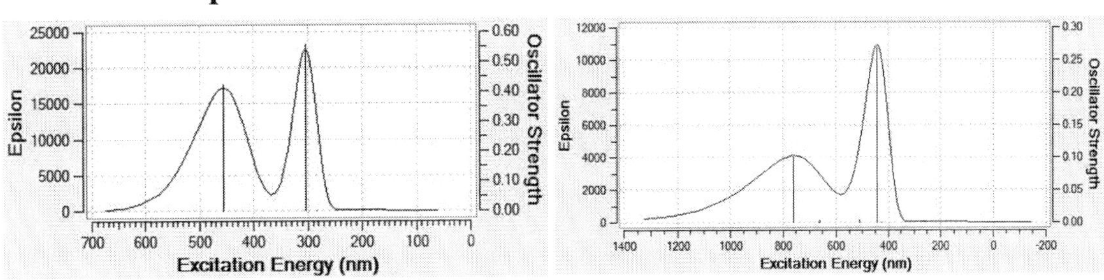

Perylenequinone **Tetrachloroperylenequinone**

Figure 4. UV/Visible spectra of perylenequione based structures

Two peaks in the UV/Visible Spectra are attributed to π-π^* and n-π^* transitions from perylene quinone based structures as per computational data analysis. In case of unsubstituted perylene quinone structures, the π-π^* and n-π^* transitions are strong and both are overlapping to each other marginally. In case of tetra-chlorosubstituted perylenequinone, there is decreased absorption intensities. Both the absorptions are red shifted and fall in the phototherapeutic window. Both transitions are overlapping with each other.

Conclusion:

Forced single bond character at the central carbon-carbon bond plays crucial role in the photo biological activities. Perylenequinone framework has a unique frontier MOs that aid in facile intersystem crossing of the π-π* excitation. The resulting triplet state remarkably resists phosphorescence that presumably leads to high quantum yield of singlet oxygen production. Quinone framework partially removes the forced single bond character implanting rotational restriction. The twisted naphthalene units in the excited state maintains stability. Enhanced non-planarity in the triplet excited state ease energy transfer and subsequent activity of PS. Varying the size of the substituents at the bay region provides finer control to engineer the pigments and synthesis of more efficient photosensitizers.

References

[1] Dougherty T. *et al*, 1998, *J Natl Cancer Inst* , **90**, 889
[2] Bonnett R, 1995, *Chemical Society Reviews*, **24**, 19
[3] McCaughan J, 1987, *Photochem. Photobiol.*, **46**, *903.*
[4] Spikes J, 1986, *Photochem.Photobio.*, **43**, 691.
[5] Aizawa K, Okunaka T, Ohami T, 1987, *Photochem. Photobiol.*, **46**, *789.*
[6] Cincotta L, Foley J, 1987, *Photochem. Photobiol.*, **46**, 751.
[7] Kessel D, 1988, *Photochem. Photobiol.*, **47**, 241.
[8] Kreimer-Bimbaum M, 1989, *Hematol.*, **26**, 157
[9] Pandey R, Majchrzycki D, Smith K, 1989, *SPIE, 1065*, 164.
[10] Konopka K and Goslinski T, 2007, *J Dent Res*, **86(8)**, 694
[11] McLean A, McGarvey D, 1990, *J. Chem. Soc. Faraday Trans*, **86**, 3075
[12] Rodgers M, 1987, *Acs Symp. Ser*, **39**, 58
[13] Kearns D, 1971, *Chemical Rev.*, **71**, 395
[14] Weiss U and Nasini G, 1987, *Prog. Chem. Org. Nat. Prod.*, **52**, 1
[15] Zhenjun D and Lown J, 1990, *Photochem. Photobiol.*, **52**, 609
[16] Zhenjun D and Lown J, 1992, *J. Photochem. Photobiol. A, Chem.*, **64**, 273.
[17] Yang C, Liu Q, 2014, *Analyst*, **139**, 6414
[18]. Reszka K, Hartley J, Lown J, 1990, *Biophys. Chem.*, **35**, 313
[19] Diwu S, Zang C, Lown J, 1992, *J. Photochem. Photobiol. A Chem.*, **66**, 99
[20] Diwu S, Lown J, *J.Photochem.Photobiol. A Chem.*, **1992**, *69*, 191
[21] Miller G, Moore R, Diwu Z, Liu J, Lown J, Begg D, 1995, *J.Photochem.Photobiol.*, **61**, 623
[22] Miskovsky P, 2002, *Current Drug Targets*, **3**, 55.
[23] Fehr M, McCloskey M, Petrich J, 1995, *J. Am. Chem. Soc.*, **117**, 1833.
[24] Sureau S, Miskovsky P, Chinsky P, 1996, *J. Am. Chem. Soc.*, **118**, 9484.
[25] Chaloupka R, Sureau F, Kocisova E, 1998, *Photochem. Photobiol.*,**68**, 44.

Journal of Physics: Conference Series **759** (2016) 012025 doi:10.1088/1742-6596/759/1/012025

[26] Oleinick N, Morris R and Belichenko I, 2002, *Photochem. Photobiol. Sci.*, **1**, 1
[27] Trepel M, Grimmel C, Schabet M, Bremen D, 1997, *Neurol Res*, **19***(5)*, 459
[28] Diwu J, Haugland R, Liu J, Miller G, Moore R, Brown K *et al* , 1996*Free Radical Biology & Medicine*, **20**, 589.
[29] Castano A, Mroz P and Hamblin M, 2006, *Nature Reviews Cancer*, **6**, 535
[30] Becke A,1993, *J. Chem. Phys.*, **98**, 5648
[31] Ditchfield R, Hehre W, Pople J,1971, *J. Chem. Phys.*, **54**, 724
[32] Guedes R, 2007, *Photochem. Photobiol. Sci.*, **6**, 1089
[33] Frisch M *et al*,2003, Gaussian 03, revision B. 03.*Gaussian Inc.,* Pittsburgh*, PA*,
[34] Lown J, 1997, *Can. J Chem.*, **75**, 97
[35] Hoffmann R, 1963, *J. Chem. Phys.*, **39**, 1397
[36] Mealli C, Proserpio D, 1990, *J. Chem.Edu.*, *67*, 399

Journal of Physics: Conference Series 759 (2016) 012026 doi:10.1088/1742-6596/759/1/012026

Allosteric dynamics of SAMHD1 studied by molecular dynamics simulations

K K Patra[1] , A Bhattacharya[2] and S Bhattacharya[1]

[1]Department of Physics, Indian Institute of Technology Guwahati, Guwahati, Assam, India 781039

[2]Department of Biochemistry, MSC 7760, The University of Texas Health Science Center at San Antonio, 7703 Floyd Curl Drive ,San Antonio, TX, USA 78229-3900

swaticb@iitg.ernet.in

Abstract. SAMHD1 is a human cellular enzyme that blocks HIV-1 infection in myeloid cells and non-cycling CD4+T cells. The enzyme is an allosterically regulated triphosphohydrolase that modulates the level of cellular dNTP. The virus restriction is attributed to the lowering of the pool of dNTP in the cell to a point where reverse-transcription is impaired. Mutations in SAMHD1 are also implicated in Aicardi-Goutières syndrome. A mechanistic understanding of the allosteric activation of the enzyme is still elusive. We have performed molecular dynamics simulations to examine the allosteric site dynamics of the protein and to examine the connection between the stability of the tetrameric complex and the Allosite occupancy.

1. Introduction

The SAMHD1 is a cellular enzyme in humans with phosphohydrolase activity, i.e. it converts deoxynucleotide triphosphates (dNTPs) into nucleosides and inorganic triphosphate and in the process depletes the cellular pool of dNTPs thereby preventing viral replication. The role of SAMHD1 in blocking HIV-1 infection in dendritic cells has generated considerable interest in its activity and regulation[1–9].

Previous studies have suggested that the catalytically active protein is a complex allosteric tetramer[9,10]. Tetramerization can occur only in the presence of GTP and dNTPs. The enzyme has two putative allosteric sites and a catalytic site (see Figure 1 b). Allosite 1 selects specifically for GTP or dGTP. Five hydrogen bonds exist between the base edge of the guanine and residues D137, Q142 and R145. GTP can provide one extra hydrogen bond with V117 carbonyl group that is not possible for dGTP. Other nucleotides do not bind to Allosite 1. Allosite 2 selects specifically for dNTPs but not NTPs. The catalytic site can accommodate any dNTP with water mediated non-specific contacts between the dNTP base and the enzyme. X-ray crystallographic structure based analysis revealed that the binding affinity of the various dNTPs to Allosite 2 (dATP>dGTP>dTTP>dCTP) was in the reverse order as that to the catalytic site (dCTP>dGTP/dTTP>dATP). Under physiological conditions, Allosite 1 is occupied by GTP while any of the four dNTPs can bind to Allosite 2 although, under equal concentrations, dATP is most likely to occupy Allosite 2 due to highest affinity. Enzyme assays also indicate a complex matrix of activation cross-talk. The catalysis rate in single dNTP experiments (dA>dT>dC>dG) is exactly the opposite to that in mixed dNTP experiments dG>dC>dT>dA, i.e the rate of phosphohydrolysis at the catsite depends on which dNTP occupies Allosite 2. However, a mechanistic understanding of the allosteric activation has been elusive. As a first step, we examine the question of how the presence or absence of the bound ligand influences the stability of the tetramer.

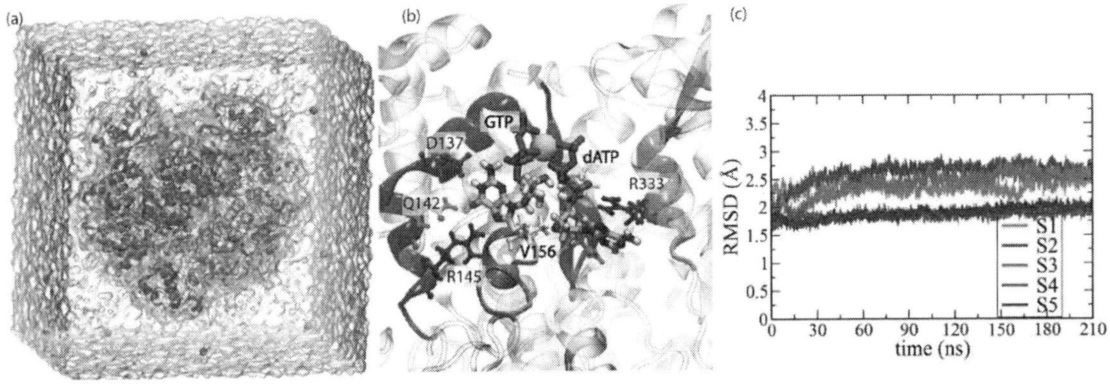

Figure 1.(a) A snapshot showing the SAMHD1 immersed in a box of water. The four subunits of the tetrameric complex are represented by violet, red, green and ochre ribbons. (b) Snapshot showing the allosteric pocket with the GTP and dATP molecues in stick representation. The protein is shown in ribbon representation with the different colors corresponding to different subunits (chain C in violet, B in red and D in green). The residues directly interacting with the GTP and the dATP molecules are shown in stick representation. (c) The RMSD (root mean square deviation) of the protein backbone in the five systems plotted as a function of time.

2. Results.

To examine the dynamics at the allosteric pocket of the SAMHD1 protein, we set up five systems based on the crystal structure (protein databank entry 4TNR) with various occupancies of the Allosites and the catalytic sites summarized in Table 1. The five *wt* systems generated with various occupancies of the Allosites and the catsite were simulated for more than 200 ns each, with the aggregate MD simulation time exceeding 1μs.

System	GTP at Allosite 1	dATP at Allosite 2	dATP at catalytic site	NVT simulation length (ns) 295 K
1	Absent	Absent	Absent	210
2	Absent	Present	Absent	210
3	Present	Absent	Absent	210
4	Present	Present	Absent	210
5	Present	Present	Present (3 out of 4 Cat sites)	210

Table 1. List of the five systems simulated summarizing the occupancy/vacancy at the Allosites and the catalytic sites as well as the length of the MD trajectories generated

An analysis of the root mean square deviation (RMSD) of the protein backbone (see Figure 1 (c)) shows that the two systems with both allosteric sites occupied (systems 4 and 5) having a consistently smaller RMSD compared to the remaining three. We therefore coclude that the binding of the nucleotides at both allosteric sites imparts stability to the tetrameric complex. The removal of either of the nucleotides at the Allosites results in an increased deviation of the protein backbone from the initial crystal structure.

In order to examine the local effects of the Allosite occupation, we calculated the RMSD of short segments of the protein containing the residues in direct interaction with the Allosite bound GTP/dATP molecule (see Figure 2). Four segments were studied: (i) Met115 to Pro130, (ii) Leu131 to Lys148, (iii) Gly151 to Pro158 and (iv) Asp330 to Val340. The first segment contains Val117 and Asn119, which are the residues in direct contact with Allosite 2 dATP. The second segment contains Asp137, Gln142 and Arg145 which interact with the GTP at Allosite 1. The third segment contains the residue Val156 that interact with dATP at Allosite 2. The fourth segment includes residue Arg333 that is has a stacking interaction with the dATP according to the crystal structure. The segments (iii) and (iv) belong to a different subunit compared to the residues of segments (i) and (ii) since the Allosites are at the interface of three different subunits. Hence the plots in Figure 2 are grouped according to the proximity to a given allosteric pocket. As expected, the RMSD of the segments in systems 4 and 5 is small indicating little deviation from the crystal structure. The most prominent deviations are observed in systems 3 and 1. In both these systems, dATP at Allosite 2 is missing. It is interesting to note that the absence of Allosite 2 dATP also causes a prominent increase of the RMSD in segment (ii) that interacts directly with Allosite 1. The results show that the vacancy at Allosite 2 causes local perturbations to the structure. A vacancy at Allosite 1 also causes increased perturbations (see panels (i) and (l) in Figure 2). However, the overall effect is not as large as that caused by Allosite 2 vacancy.

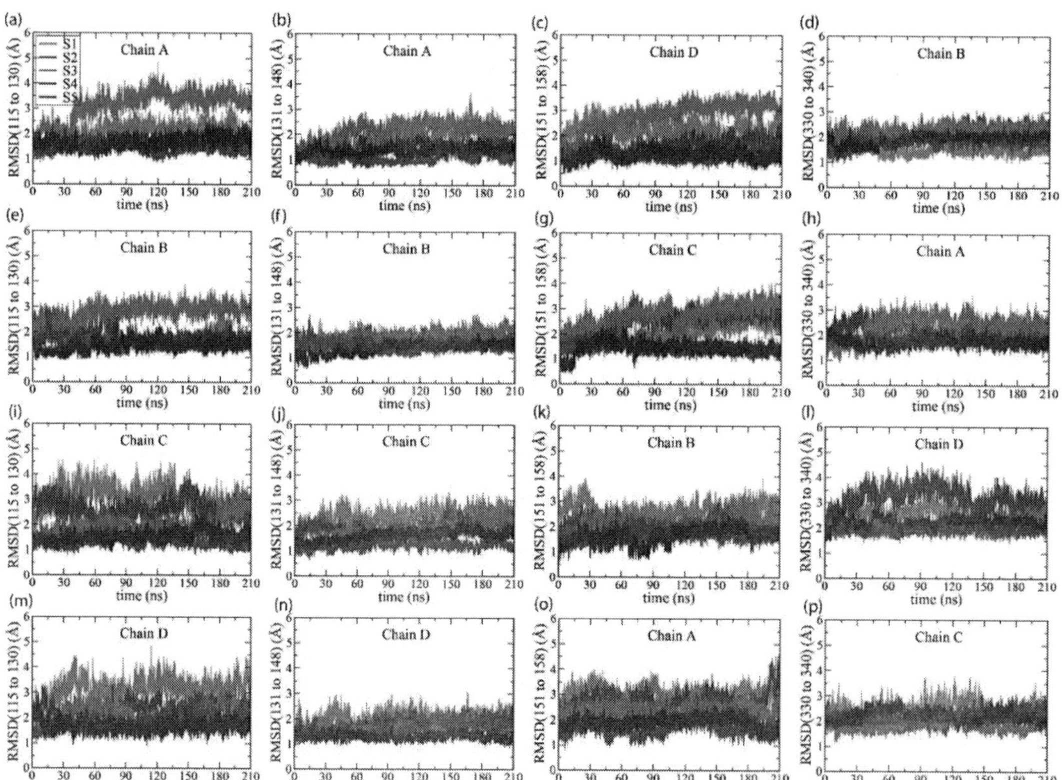

Figure 2. RMSD vs. time plots for the segments : 115 to 130, 131 to 148, 151 to 158 and 330 to 340 for the four chains (A, B, C and D). All the plots in a single row correspond to the protein segments at a given allosteric pocket interacting with a common GTP or dATP. The segments belong to different chains since the allosteric pocket lies at the interface of three chains

3. Discussion

We have studied the allosteric site dynamics of the SAMHD1 enzyme which is an HIV-1 restriction factor. Previous experimental studies have indicated a complex cross-talk between the allosteric sites in the protein, which is not understood as yet. A series of simulations designed to explore the effect of the presence/absence of the preferred Allosite 1 and 2 ligands show that the stability of the tetramer is adversely affected when either or both allosteric sites are vacant. The destabilizing effect is more pronounced when the vacancy is in Allosite 2, i.e. in the dATP-lost form of the protein. The dATP bound to Allosite 2 therefore appears to play a role in pinning the subunits together. A complex network of hydrogen bonds between the dATP/GTP and nearby protein residues are crucial to the stability of the complex. However the exact role of the two Allosite nucleotides in the catalysis remains to be explored.

4. Methods

4.1 Simulation System Setup The crystal structure of SAMHD1 complexed with GTP and dATP obtained from the Protein Data Bank entry 4TNR[11] was used to generate the starting conformation for the all atom MD simulations in an explicit water environment. Three of the four catalytic sites are occupied by dATP while the fourth (in subunit A) is vacant. The pdb structure was used to generate initial structures for five systems as shown in Table 1. In all five systems, the crystallographic waters were retained. The Mg+2 ions coordinated by allosteric site molecules were deleted in System 1. However, systems 4 and 5 contained both allosteric site molecules along with the Mg ions. The missing residues in the loop 278-283 were inserted in the protein structures whereas the missing N terminal and C terminal residues were ignored. The four R206 and N207 residues were mutated back to histidine and aspartate in accord with the sequence of the *wt* SAMHD1 (Uniprot Q9Y3Z3-1). Disulfide bonds were introduced between residues 341 and 350. Each system was solvated in a cubic waterbox with in ~59000 pre-equilibrated TIP3P water molecules. An appropriate number of sodium ions were added to produce a neutral system. Each system consisted of ~210,000 atoms measured 13×12×14 nm.

4.2 General MD Methods. Energy minimizations and molecular dynamics simulations were performed using NAMD 2.9[12,13]. All simulations employed periodic boundary conditions and multiple time-stepping wherein local interactions were calculated every 2 fs and full electrostatic evaluations were performed every two timesteps. The particle mesh Ewald[14] method was adopted for treating long range electrostatic calculations. CHARMM31[15,16] force fields were employed along with the TIP3P water model. During simulations, the SETTLE[17] and RATTLE[18] algorithms were applied to keep the covalent bonds involving hydrogen in water and other molecules rigid. A cut-off of 12 Å and a switching distance of 10 Å were used for treating van der Waals and short-ranged interactions. Conjugate gradient minimization was performed for 3000 steps for each system following which each system was equilibrated in the NPT ensemble using the Nosé-Hoover Langevin piston pressure control at 295 K for at least 5 ns. Finally NVT simulations were performed with the temperature maintained at 295 K using the Langevin thermostat. The coordinates was recorded at 10 ps intervals. Each of the five systems were simulated for about 210 ns bringing the total simulation time to 1.2 μs.

Acknowledgements We are deeply thankful to the National PARAM Supercomputing facility (NPSF), CDAC, for providing us the computational facilities for this study.

References

[1] Behrendt R, Schumann T, Gerbaulet A, Nguyen L a., Schubert N, Alexopoulou D, Berka U, Lienenklaus S, Peschke K, Gibbert K, Wittmann S, Lindemann D, Weiss S, Dahl A, Naumann R, Dittmer U, Kim B, Mueller W, Gramberg T and Roers A 2013 Mouse SAMHD1 has antiretroviral activity and suppresses a spontaneous cell-intrinsic antiviral response *Cell Rep.* **4** 689–96

[2] Hollenbaugh J a, Tao S, Lenzi G M, Ryu S, Kim D-H, Diaz-Griffero F, Schinazi R F and Kim B 2014 dNTP pool modulation dynamics by SAMHD1 protein in monocyte-derived macrophages. *Retrovirology* **11** 63

[3] Hollenbaugh J a, Gee P, Baker J, Daly M B, Amie S M, Tate J, Kasai N, Kanemura Y, Kim D-H, Ward B M, Koyanagi Y and Kim B 2013 Host factor SAMHD1 restricts DNA viruses in non-dividing myeloid cells. *PLoS Pathog.* **9** e1003481

[4] Yan N and Lieberman J 2013 SAMHD1 does it again, in resting T cells *Nat. Med.* **18** 1199–216

[5] Cribier A, Descours B, Valadão A, Laguette N and Benkirane M 2013 Phosphorylation of SAMHD1 by Cyclin A2/CDK1 Regulates Its Restriction Activity toward HIV-1 *Cell Rep.* **3** 1036–43

[6] Baldauf H-M, Pan X, Erikson E, Schmidt S, Daddacha W, Burggraf M, Schenkova K, Ambiel I, Wabnitz G and Gramberg T 2012 SAMHD1 restricts HIV-1 infection in resting CD4+ T cells *Nat. Med.* **18** 1682–9

[7] Beloglazova N, Flick R, Tchigvintsev A, Brown G, Popovic A, Nocek B and Yakunin A F 2013 Nuclease activity of the human SAMHD1 protein implicated in the aicardi-goutieres syndrome and HIV-1 restriction *J. Biol. Chem.* **288** 8101–10

[8] Seamon K J, Sun Z, Shlyakhtenko L S, Lyubchenko Y L and Stivers J T 2015 SAMHD1 is a single-stranded nucleic acid binding protein with no active site-associated nuclease activity *Nucleic Acids Res.* 1–14

[9] Ji X, Wu Y, Yan J, Mehrens J, Yang H, DeLucia M, Hao C, Gronenborn A M, Skowronski J, Ahn J and Xiong Y 2013 Mechanism of allosteric activation of SAMHD1 by dGTP *Nat. Struct. Mol. Biol.* **20** 1304–9

[10] Yan J, Kaur S, DeLucia M, Hao C, Mehrens J, Wang C, Golczak M, Palczewski K, Gronenborn A M, Ahn J and Skowronski J 2013 Tetramerization of SAMHD1 is required for biological activity and inhibition of HIV infection. *J. Biol. Chem.* **288** 10406–17

[11] Ji X, Tang C, Zhao Q, Wang W and Xiong Y 2014 Structural basis of cellular dNTP regulation by SAMHD1 *Proc. Natl. Acad. Sci.* **111** E4305–14

[12] Kal L, Skeel R, Bhandarkar M, Brunner R, Gursoy A, Krawetz N, Phillips J, Shinozaki A, Varadarajan K and Schulten K 1999 NAMD2 : Greater Scalability for Parallel Molecular Dynamics *J. Comput. Phys.* **151** 283–312

[13] Phillips J C, Braun R, Wang W, Gumbart J, Tajkhorshid E, Villa E, Chipot C, Skeel R D and Kalé L 2005 Scalable Molecular Dynamics with NAMD *J. Comput. Chem.* **26** 1781–802

[14] Batcho P F, Case D a. and Schlick T 2001 Optimized particle-mesh Ewald/multiple-time step integration for molecular dynamics simulations *J. Chem. Phys.* **115** 4003–18

[15] Mackerell A D, Bashford D, Bellott M, Dunbrack R L, Evanseck J D, Field M J, Fischer S, Gao J, Guo H, Ha S, Kuchnir L, Kuczera K, Lau F T K, Mattos C, Michnick S, Ngo T, Nguyen D T, Prodhom B, Reiher W E, Roux B, Schlenkrich M, Smith J C, Stote R, Straub J, Watanabe M, Wio J, Yin D and Karplus M 1998 All-Atom Empirical Potential for Molecular Modeling and Dynamics Studies of Proteins † *J. Phys. Chem. B* **5647** 3586–616

[16] Klauda J B, Venable R M, Freites J A, O'Connor J W, Tobias D J, Mondragon-Ramirez C,

Vorobyov I, MacKerell A D and Pastor R W 2010 Update of the CHARMM all-atom additive force field for lipids: validation on six lipid types. *J. Phys. Chem. B* **114** 7830–43

[17] Miyamoto S and Kollman P a. 1992 Settle: An analytical version of the SHAKE and RATTLE algorithm for rigid water models *J. Comput. Chem.* **13** 952–62

[18] Andersen C 1983 Rattle : A " Velocity " Molecular Version of the Shake Dynamics Calculations for *J. Comput. Phys.* **52** 24–34

A Possible Explanation of Oriented Bar Filling-in at the Blind-Spot in the light of Hierarchical Prediction Mechanism

R Raman and S Sarkar[#]
Applied Nuclear Physics Division,
Saha Institute of Nuclear Physics, Kolkata, India
[#] *sandip.sarkar@saha.ac.in*

Abstract. Despite the absence of retinal input, the blind-spot gets filled up with surrounding visual attributes. This phenomenon is called the perceptual filling-in or completion. Various possible neural mechanisms have been proposed but most of these are not entirely consistent with the findings of brain activation and brain organization. In one study, recently, it has been shown that the filling-in of the shifting bar and the anisotropy in the filling-in of misaligned bars could be explained by incorporating a common general principle called hierarchical predictive coding (HPC). In this report, we have extended this proposition to explain the filling-in of oriented bars. We have considered a three level (LGN-V1-V2) HPC model network in which, the blind-spot was emulated by removing the corresponding feed-forward (LGN-V1) connection. We simulated the responses of predictive estimator (PE) neurons at blind-spot while stimulating the network with oriented bar stimuli. Results show that the filling-in is best for aligned bars but it faded away with increasing orientation and moreover, corners were not predicted in any orientation. Qualitatively, these results are consistent with the findings of psychophysical experiments. We discussed this phenomenon in the HPC framework and argue that in the absence feed-forward connections, the best prediction is dominated by learned statistical regularities that include an abundance of bar and similar structures. These results suggest that the filling-in process could be a manifestation of HPC.

1. Introduction

Because of the absence of photoreceptors at optics disc, the retina is unable to send the corresponding signal to the brain and thereby and therefore, some portion of the visual field is hidden. However, we never notice any odd patch due the hidden visual field, known as blind-spot, even in monocular vision, but rather we see the complete scene; filled up in accordance with the surrounding visual attributes [1]. This process is known as perceptual filling-in or simply filling-in. In addition to the blind spot, filling-in also occurs in other visual input deficit conditions, e.g. filling-in at the artificial and natural retinal scotoma [2, 3]. Filling-in also observed in visual illusions such as Neon colour spreading, Craik-O'Brien-Cornsweet illusion, Kanizsa shapes, and etc. and steady fixation condition like Troxler effect (for review see [4].

Psychophysical and physiological studies suggest that the filling-in is an active process, where some neural processes are involved and mainly take place in the early visual cortex [5-7]. Studies show that neural activities, correlated to perceptual experiences, are evoked in the deep layer of primary visual cortex, when filling-in completion occurs [5, 6]. Matsumoto and Komatsu [7] demonstrated that some neuron in BS region in deep layer of primary visual cortex (BS neurons), which possess larger receptive fields that extend beyond the blind spot, exhibits non-linear elevated response when a long moving bar cross over the blind spot and perceptual completion occurs.

Although some attempts have been made to understand the computational mechanism of general filling-in [8-13], little work has been devoted to the study of computational mechanism of filling-in completion at the blind spot and the current knowledge is far from complete. Recently, it has been shown that the filling-in of the shifting bar and the anisotropy in the filling-in of misaligned bars could be predicted by incorporating HPC [14], which has, recently, been argued as a general coding principle of the visual cortex [15-22]. Here in this study, we have expanded the scope of HPC by demonstrating that the some results related to oriented bar filling-in could also be explained using this same general principle.

Functionality of a network in the HPC framework largely depends on a probabilistic hierarchical generative model and the efficient coding of natural images. In this framework, the visual system attempts to infer or estimate the properties of the world from signals coming from receptors [23-25]. This job is hypothesized to be completed by concurrent prediction-correction mechanism along the hierarchy of the visual system. Each higher visual area (say V2) attempt to predict the response at its lower area (say V1) on the basis of the learned statistical regularities, and send that prediction signal to the lower area by feedback connection. Lower area, in response, sends a residual error signal, signifying any mismatch, to the higher area by feed-forward connection. Neural networks, in this paradigm, are constantly attempting to match sensory input with the top-down predictions that are based on the learned statistical regularities of natural scenes. By using its best current predictions as the priors for the levels below and engaging in iterative estimation, the network allows priors and predictions to co-evolve to arrive at the best prediction.

This idea is based on the anatomical architecture of the visual system which is hierarchically organized and reciprocally connected [27]. Recently, several neuronal tuning properties in different visual areas such as the lateral geniculate nucleus (LGN), primary visual cortex (V1) and middle temporal level (MT) have been explained using this framework [19, 20]. For example, Rao [14] suggested that the extra-classical properties of neurons in V1 could be understood in terms of predictive-feedback signal from the secondary visual cortex (V2). We speculated that a similar computational mechanism could also explain the filling-in completion at the blind spot.

In this report, we have investigated the filling-in of oriented bars under a fresh perspective of HPC. We have considered a three level (LGN-V1-V2) HPC model network in which, the blind-spot was emulated by removing the corresponding feed-forward (LGN-V1) connection in early visual area. We simulated the responses of predictive estimator (PE) neurons at blind-spot while stimulating the network with oriented bar stimuli across the blind-spot. Results show that the filling-in is best for aligned bars but it faded away with increasing orientation. Additionally, we also observed that the corners were not predicted in any orientation. These results are consistent with the finding of psychophysical experiment [28].

We discussed this phenomenon in the HPC framework and argue that in the absence feed-forward connections, the best prediction is dominated by learned statistical regularities that include an abundance of bar and similar structures and anisotropy in orientation distribution. These results suggest that the filling-in process could be a manifestation of HPC.

2. Methods

2.1 Hierarchical Predictive coding of natural images

A general computational HPC architecture is illustrated in figure 1. In this framework, on the arrival of an input, predictor estimator modules (PE module) at each visual processing level generate the prediction (or estimate) on the basis of the learned statistical regularities. Each higher area (say V2) then sends these predictions to its immediate lower level (say V1) by feedback connections and in return receives the error signal, by feed-forward connections. An equilibrium state is achieved after several concurrent prediction-

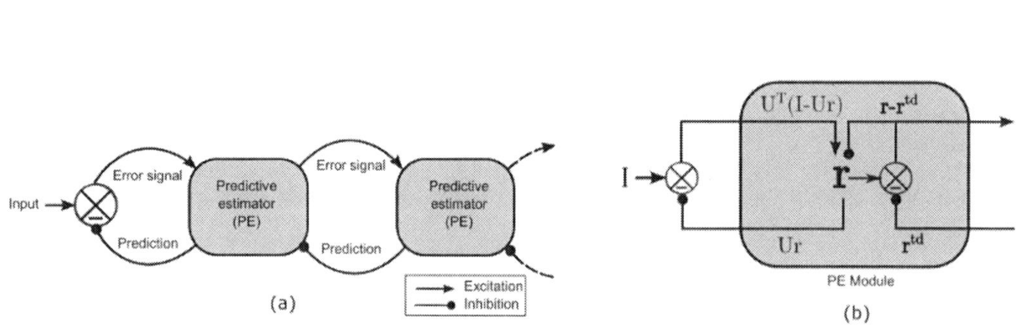

Figure 1. General HPC architecture (adopted from Rao.[15]). a) Illustration of connections between PE modules at different levels, b) Details of computation performed by each PE module.

correction cycles, where the estimate approximately matches the input signal caused by the stimuli. This optimum-estimate is regarded as a representation of the input at that level.

A single PE module (figure 2(b)) consists of, (i) Predictive estimator neurons (PE neurons) which represent the estimate of current input signal $\mathbf{I}$ with response vector $\mathbf{r}$ (state vector), (ii) neurons that carry prediction signal $U\mathbf{r}$ (for the input $\mathbf{I}$) to lower level by feed-back connections, whose synapse encode encoding efficacy matrix U, (iii) neurons that carry feed-forward error signal ($\mathbf{I}-U\mathbf{r}$) from lower level to higher level, whose synapses encoded rows of efficacy matrix U^T, and (iv) error detecting neurons which carry the residual error signal ($\mathbf{r}-\mathbf{r}^{td}$) to the higher level corresponding to the prediction $\mathbf{r}^{td}$ from the higher level.

2.2 Network dynamics and learning rule

In the Bayesian framework (maximum a posteriori (MAP) approach), the dynamics and the learning can be derived by maximizing the posterior probability $P(\mathbf{r},\mathbf{r}^{td},U\,|\,\mathbf{I})$, which is equal to the product of $P(\mathbf{I}\,|\,\mathbf{r},U)$, $P(\mathbf{r}^{td}\,|\,\mathbf{r})$ and prior probabilities $P(\mathbf{r})$ and $P(U)$, with respect to $\mathbf{r}$ and U respectively. By assuming $P(\mathbf{I}\,|\,\mathbf{r},U)$ and $P(\mathbf{r}^{td}\,|\,\mathbf{r})$ as Gaussians of zero mean and variances σ^2 and σ_{td}^2 respectively, the total coding length E can be written as (for details, see [15], [21]),

$$E = \frac{1}{\sigma^2}(\mathbf{I}-U\mathbf{r})^T(\mathbf{I}-U\mathbf{r}) + \frac{1}{\sigma_{td}^2}(\mathbf{r}-\mathbf{r}^{td})^T(\mathbf{r}-\mathbf{r}^{td}) + g(\mathbf{r}) + h(U) \quad (1)$$

here, $g(\mathbf{r})$ and $h(U)$ are the negative log of prior probabilities $P(\mathbf{r})$ and $P(U)$ respectively. Minimizing the coding length E, with respect to $\mathbf{r}$ provides the dynamics of PE module and it is equivalent to recurrently achieving an internal state matching the "sensory driven" input. The visual representation of the predication $U\mathbf{r}$, is regarded as the representation of our perception. Furthermore, the minimization of coding length E, with respect to U provides the learning rule for basis matrix U. A kurtotic prior distribution ($P(r_i) = exp(-\alpha log(1+r_i^2))$) is chosen for the prior probability for embedding sparse representation [25].

3. Simulation

In this study, we have utilised a three level linear HPC network (similar to [14]), where level 1 (equivalent to V1) consists of 9 PE modules. These modules receive input from level 0 (LGN) and send the output to

the sole module at level 2 (equivalent to V2). Therefore, the PE module at level 2 receives input from all the nine level 1 PE modules and sends back the feedback signal to all of them. This architecture is based on the fact that the visual area higher in hierarchy operates on a higher spatial scale. Each PE module at level 1 consists of 64 PE neurons, 144 Prediction carrying neurons, 64 afferent error carrying neurons and 64 error detecting neurons for conveying the residual error to level 2. The layer 2 module consists of 169 PE neurons, 576 prediction carrying neurons and 169 error carrying neurons. Six natural images of size 512×512 pixels were used for training after pre-processing with a low-pass filter of cut-off frequency 200 cycles/image. Variance normalized 1000 batches of 100 image patches of size 30×30 pixel, which were extracted from randomly selected locations from the randomly selected pre-processed images, were given as input to the network. For each batch of image patches, the network was allowed to achieve steady states and the average of these states was used to update the efficacy of neurons initially assigned random value (for details see [25, 26]).
To mimic the blind spot the feed-forward connection in a certain area was removed from the model network, which was pre-trained with usual feed-forward connections. The removal was implemented by setting the efficacy of early feed-forward (level 0 - level 1) neurons to zero. This "pre-training" preserved the experimentally observed binocular nature of these active neurons in the deep layer (5/6) (for the detail see [14]).

4. Results

Initially, the model network was trained with binocular connections. This resulted in the Gabor like receptive field (RF) at level 1 whereas, level 2 learned more abstract RF (e.g. corner, long bar etc.) similar to those reported in earlier studies [14,25,26]. After the completion of the learning process, the network

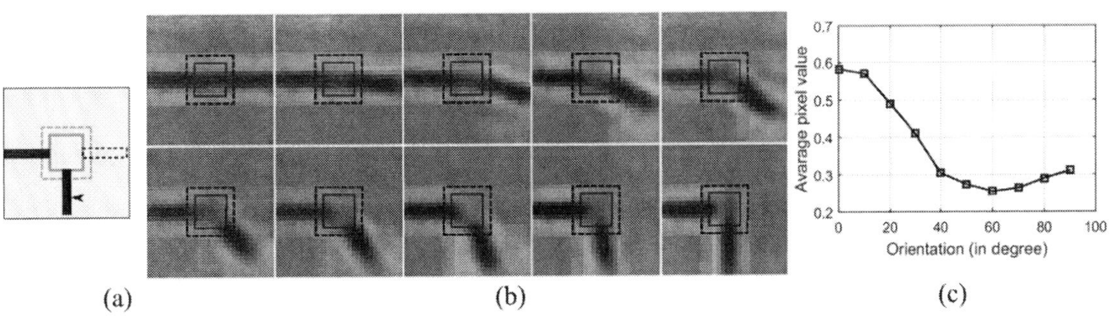

Figure 2. Results of oriented bar filling-in. a) Stimulus used for the study is shown. One part of the stimulus, the horizontal black bar, was fixed. The other part was rotated clockwise from its initial horizontal position (dotted bar) to the vertical position in steps of 10 degree. The central rectangle is BS and the outer rectangle (dotted) is the receptive field. Rectangular regions are shown for illustration only, b) Perceptual images, generated from the response of PE neurons at level 1, corresponding to all ten orientations of stimulus are shown, c) Average response at the centre (2x2) of the BS region in (b) is plotted as a function of orientation angle.

was modified to accommodate the blind spot before stimulated with the bar stimuli shown in figure 2(a). One part of the bar was fixed at a position outside of blind spot. Orientation of the other part was varied from horizontal to vertical position producing ten different input stimuli. We recorded the response PE neurons at level 1 to these stimuli, and generated the corresponding 'perceptual images', which we defined as reconstruction Ur. These 'perceptual images' (figure 2(b)) demonstrate that the filling-in is best when the bars were aligned (horizontal) and with the increasing mismatch in orientation, the filling-in deteriorates. This is illustrated in more detail in figure 2(c), where the average pixel value at the centre of the BS region is plotted as a function of the orientation difference. This plot demonstrates that filling-in deteriorates at a faster rate till 40 degree and for further increase in angle, it still deteriorates but with a slower rate till 60 degree difference in orientation, Thereafter, a little increase is also visible in the plot. The observed decrease in filling-in corroborates the experimental findings reported in [28]. Additionally, in none of these results (figure 2(b)), corners are visible, which leads to the conclusion that corners were never anticipated in the top-down expectation. Regarding the increase in filling-in beyond 60 degree, it is not possible to make any concluding remark and this discussed in detail in the discussion section.

5. Discussion

In this study, we have investigated the computational aspects of oriented bar filling-in at the blind spot. We postulated that this could be understood in the HPC paradigm inside the visual cortex and therefore, conducted simulation studies on the three level HPC model network modified in accordance with the blind-spot architecture. Our studies show that the filling-in completion, which occurs in the case of perfectly aligned bar segments, also happens with a small degree of misalignment, but the extent of filling-in completion deteriorates with increasing misalignment in orientation (figure 2).

These results show good agreement with the physiological and psychophysical results reported earlier [28] and can be discussed in the context of HPC. When an input stimulus was presented in and around the blind spot, higher areas (V2) generated a unified estimate for the input stimuli on the basis of the learned statistical regularities of natural scenes. In nature, these statistical regularities are dominated by the abundance of straight continuous bars and similar structures and therefore, these structures are much more probable, compared to other non-linear continuous structures, in the predictions made by the higher visual area (V2). These predictions remain uncorrected due to the absence of error carrying feed-forward connection in BS region at V1 and therefore, local optimum-estimate is achieved essentially by the top-down expectation.

In case of aligned bars on both sides of the BS region (the first case in figure 2(b)), the visual system received inputs corresponding to the presence bars on both sides. This is interpreted as the presence of a long straight continuous bar because in natural image statistics the presence of a broken bar or two bars side by side is less probable and this is reflected in the response of PE neurons announcing the prediction of a long bar across the BS resulting in the perception of a near perfect filling-in. In other cases, the model system received inputs corresponding to the presence of bars with different orientations. As long as the difference in orientation was small, the model system was able to interpret it as a continuous bar and filling-in occurred. With increasing angles, however, the higher area (V2) faces increased difficulties in interpreting two oriented bars as parts of a long continuous (may be non-linear) bar resulting in the deterioration of filling-in. Beyond a certain threshold, the system fails to predict anything more than what is visible and the filling-in ceases to happen.

To illustrate the varying filling-in with increasing angles, we have plotted average pixel value recorded at the centre of the BS region as a function of angle in figure 2(c). It shows that the value decreases till 60 degree and then increases a little. It may look like that filling-in never fails but varies with increasing angles. The problem lies in the interpretation of a threshold. In similar psychophysical experiments, participants report the occurrence or non-occurrence of filling-in depending on their perceptual experience. However, in our simulation study, there exists no such threshold and in fact, it is very difficult to define an equivalent one without further extension of the current model incorporating much more details properties of the visual cortex. This is precisely why the filling-in never fails (figure 2(c)), though deteriorates, with increasing difference in orientation

In conclusion, we demonstrate that some aspects of oriented bar filling-in could be explained in the computational paradigm of HPC. We show that, in the absence of feed-forward connection due to the blind-spot, the top-down predictions (in accordance with learned statistical regularities of natural scene) mainly determines the nature of filling-in that corroborates well with experimental findings. These results suggest that the filling-in could be a manifestation of a hierarchical predictive coding principle.

References

[1] Ramachandran VS 1992 *Scientific American.* **266(5)** 86–91.
[2] Ramachandran VS, Gregory RL, et al 1991 *Nature.* **350(6320)** 699–702.
[3] Gerrits H, Timmerman G 1969 *Vision Research.* **9(3)** 439–442.
[4] Komatsu H 2006 *Nature reviews neuroscience.* **7(3)** 220–231.
[5] Júnior MF, Rosa M, Gattass R, Rocha-Miranda CE 1992 *PNAS.* **89(18)** 8547–8551.
[6] Komatsu H, Kinoshita M, Murakami I 2000 *The Journal of Neuroscience.* **20(24)** 9310–9319.
[7] Matsumoto M, Komatsu H 2005 *Journal of Neurophysiology.* **93(5)** 2374–2387.
[8] Neumann H, Pessoa L, Hansen T 2001 *Biological Cybernetics.* **85(5)** 355–369.
[9] Neumann H, Sepp W 1999 *Biological Cybernetics.* **81(5-6)** 425–444.
[10] Grossberg S, Mingolla E 1985 *Psychological review.* **92(2)** 173.
[11] Grossberg S, Kelly F 1999 *Vision Research.* **39(22)** 3796–3816.
[12] Grossberg S, Mingolla E, Ross WD 1997 *Trends in neurosciences.* **20(3)** 106–111.
[13] Satoh S, Usui S 2008 *Neural Networks.* **21(9)** 1261–1271.
[14] Raman R, Sarkar S 2015 *arXiv: 1507.01390* (q-bio.NC).
[15] Rao RP, Ballard DH 1999 *Nature neuroscience.* **2(1)** 79–87.
[16] Lee TS, Mumford D 2003 *JOSA A.* **20(7)** 1434–1448.
[17] Friston K 2005 *Phil. Trans. Roy. Soc. London B.* **360(1456)** 815–836.
[18] Hohwy J, Roepstorff A, Friston K 2008 *Cognition.* **108(3)** 687–701.
[19] Jehee JF, Ballard DH 2009 *PLoS computational biology.* **5(5)** e1000373.
[20] Jehee JF, Rothkopf C, Beck JM, Ballard DH 2006 *Journal of Physiology-Paris.* **100(1)** 125–132.
[21] Huang Y, Rao RP 2011 *Wiley Interdisciplinary Reviews: Cognitive Science.* **2(5)** 580–593.
[22] Clark A 2013 *Behavioral and Brain Sciences.* **36(03)** 181–204.
[23] Mumford D 1994 *Large-Scale Theories of the Cortex* (MIT Press). 125–152.
[24] Dayan P, Hinton GE, Neal RM, Zemel RS 1995 *Neural computation.* **7(5)** 889–904.
[25] Olshausen BA, et al 1996 *Nature.* **381(6583)** 607–609.
[26] Olshausen, BA, Field, DJ 1997 *Vision Research.* **37(23)** 3311–3325.
[27] Felleman DJ, Van Essen DC 1991 *Cerebral cortex.* **1(1)** 1–47.
[28] Araragi Y, Nakamizo S 2008 *Vision research.* **48(4)** 618–625.

Spin dependent disorder in a junction device with spin orbit couplings

Sudin Ganguly[a] **and Saurabh Basu**[b]
Indian Institute of Technology Guwahati, Assam-781039, India
E-mail: [a]sudin@iitg.ernet.in, [b]saurabh@iitg.ernet.in

Abstract. Using the multi-probe Landauer-Büttiker formula and Green's function approach, we calculate the longitudinal conductance (LC) and spin Hall conductance (SHC) numerically in a two-dimensional junction system with the Rashba and Dresselhaus spin orbit coupling (SOC) and spin dependent disorder (SDD) in presence of both random onsite and hopping disorder strengths. It has been found that when the strengths of the RSOC and DSOC are same, the SHC vanishes. Further in presence of random onsite or hopping disorder, the SHC is still zero when the strengths of the two types of SOC, that is Rashba and Dressselhaus are the same. This indicates that the cancellation of SHC is robust even in the presence of random disorder. Only with the inclusion of SDD (onsite or hopping), a non-zero SHC is found and it increases as the strength of SDD increases. The physical implication of the existence of a non-zero SHC has been explored in this work. Finally, we have compared the effect of onsite SDD and hopping SDD on both longitudinal and spin Hall conductances.

1. Introduction

The notion of dissipationless spin current [1] has attracted considerable interest recently. The spin-orbit coupling produces a transverse force on a moving electron and this force tends to form a transverse spin current. In the simplest form, a spin current is about the flow of spin-up electrons in one direction, say $+x$, accompanied by the flow of an equal number of spin-down electrons in the opposite direction, $-x$. The total charge current in the x direction is therefore zero, $I_e = e\,(I_\uparrow + I_\downarrow) = 0$. The total spin current is finite: $I_s = e\,(I_\uparrow - I_\downarrow) \neq 0$.

In a two-dimensional electron gas (2DEG) lacking structure inversion symmetry of the confining potential and bulk inversion symmetry, the effective Hamiltonian is given by,

$$H = \frac{\mathbf{p}^2}{2m^*} + \alpha\,(\sigma^x p_y - \sigma^y p_x) + \beta\,(\sigma^x p_x - \sigma^y p_y) \tag{1}$$

where the second term is the Rashba spin-orbit coupling (RSOC) and the third term is the Dresselhaus spin-orbit coupling (DSOC). $\sigma^{x/y}$ is the x/y component of the Pauli matrices, and α and β are the coupling parameters which denote the strengths of RSOC and DSOC respectively.

In this paper, we investigate the effect of a spin dependent random disorder on a four terminal square lattice in presence of both Rashba and Dresselhaus spin orbit interactions. Recently it has become possible to create the spin dependent disorder within a new experimental context, namely that of ultracold fermionic and bosonic atomic gases confined to optical lattices [5, 6]. Few theoretical works have been done on this context [2, 3, 4]. However, till now it is not possible to create experimentally which facilitate studying of such phenomena in mesoscopic physics. Nevertheless, the issue deserves to be studied theoretically.

We organize our paper as follows. The theoretical formalism leading to the expressions for the spin Hall and longitudinal conductances using Landauer Bütiker formula are presented in section II. Section III includes an elaborate discussion on the results obtained for the spin Hall and longitudinal conductances that are helpful in studying the effect of spin dependent disorder on four terminal junction device in presence of spin orbit interactions.

2. Theoretical formulation

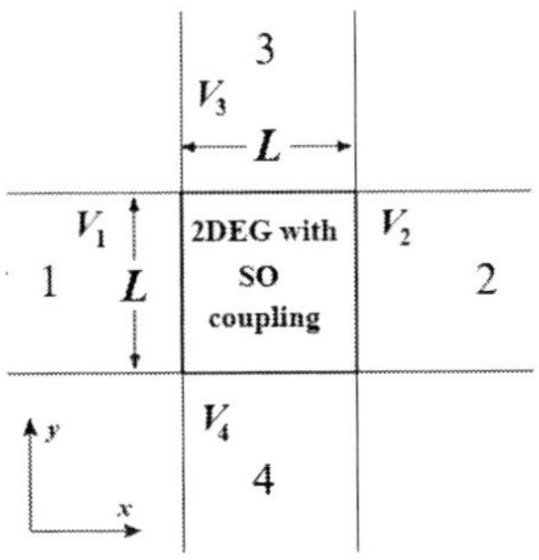

Figure 1. Cross-shaped device with four semi-infinite metallic leads. The spin-orbit coupling exists in the central region only, and the effect of the semi-infinite leads is treated exactly through the self-energy terms.

We choose a four-probe measuring set-up as shown in Fig.1 to observe the spin Hall effect. Here the four ideal semi-infinite leads are attached to the central conducting region which in our case is a square lattice with spin orbit interaction. An unpolarized charge current is allowed to pass through the longitudinal leads along lead-1 and lead-2 (see Fig.1) inducing spin Hall current in the transverse directions that is along lead-3 and lead-4. With the tight binding approximation for nearest neighbour interaction, Eq.(1) reads as,

$$H = \sum_{i,\sigma} \epsilon_{i\sigma} c_{i\sigma}^{\dagger} c_{i\sigma} + \sum_{\langle ij \rangle,\sigma} t_{ij,\sigma} c_{i\sigma}^{\dagger} c_{j\sigma} + V_R \sum_i \left[\left(c_{i\uparrow}^{\dagger} c_{i+\delta_x\downarrow} - c_{i\downarrow}^{\dagger} c_{i+\delta_x\uparrow} \right) - i \left(c_{i\uparrow}^{\dagger} c_{i+\delta_y\downarrow} + c_{i\downarrow}^{\dagger} c_{i+\delta_y\uparrow} \right) \right]$$
$$+ V_D \sum_i \left[(-i) \left(c_{i\uparrow}^{\dagger} c_{i+\delta_x\downarrow} + c_{i\downarrow}^{\dagger} c_{i+\delta_x\uparrow} \right) + \left(c_{i\uparrow}^{\dagger} c_{i+\delta_y\downarrow} - c_{i\downarrow}^{\dagger} c_{i+\delta_y\uparrow} \right) \right] \quad (2)$$

Here $\epsilon_{i\sigma}$ is the random on-site spin dependent potential and $t_{ij,\sigma}$ is the spin dependent hopping strength chosen from a uniform rectangular distribution ($-W$ to W), $V_R = \frac{\hbar\alpha}{a_0}$ and $V_D = \frac{\hbar\beta}{a_0}$ are the Rashba and Dresselhaus coupling strengths, respectively. $\delta_{x/y}$ is the unit vector along x/y direction.

For the four-probe case, where pure spin current is expected to flow through the transverse leads, due to the flow of charge current through the longitudinal leads, the longitudinal and spin Hall conductances are defined as [7],

$$G_L = \frac{I_2^q}{V_1 - V_2} \quad \text{and} \quad G_{SH} = \frac{\hbar}{2e} \frac{I_3^s}{V_1 - V_3} = \frac{\hbar}{2e} \frac{I_3^{\uparrow} - I_3^{\downarrow}}{V_1 - V_3} \quad (3)$$

where I_2^q and I_3^s are the charge and spin current flowing through the lead-2 and lead-3 respectively. V_m is the potential at the m-th lead. $I_3^{\uparrow}$ and $I_3^{\downarrow}$ are the up and down spin currents flowing in lead-3.

The calculation of electric and spin currents is based on the Landauer-Büttiker multi-probe formalism [8]. The charge and spin currents flowing through the lead m with potential V_m, can be written in terms of spin resolved transmission probability as [9]

$$I_m^q = \frac{e^2}{h} \sum_{n\neq m,\sigma,\sigma'} \left(T_{nm}^{\sigma\sigma'} V_m - T_{mn}^{\sigma'\sigma} V_n \right) \tag{4}$$

$$I_m^s = \frac{e^2}{h} \sum_{n\neq m,\sigma'} \left[\left(T_{nm}^{\sigma'\sigma} - T_{nm}^{\sigma'-\sigma} \right) V_m + \left(T_{mn}^{-\sigma\sigma'} - T_{mn}^{\sigma\sigma'} \right) V_n \right] = \frac{e^2}{h} \sum_{n\neq m} \left[T_{nm}^{out} V_m - T_{mn}^{in} V_n \right] \tag{5}$$

where, we have defined two useful quantities as follows,

$$T_{pq}^{in} = T_{pq}^{\uparrow\uparrow} + T_{pq}^{\uparrow\downarrow} - T_{pq}^{\downarrow\uparrow} - T_{pq}^{\downarrow\downarrow} \quad \text{and} \quad T_{pq}^{out} = T_{pq}^{\uparrow\uparrow} + T_{pq}^{\downarrow\uparrow} - T_{pq}^{\uparrow\downarrow} - T_{pq}^{\downarrow\downarrow} \tag{6}$$

Physically, the term $\frac{e^2}{h} \sum_{n\neq m} T_{nm}^{out} V_m$ is the total spin current flowing from the m-th lead with potential V_m to all other n leads, while the term $\frac{e^2}{h} \sum_{n\neq m} T_{mn}^{in} V_n$ is the total spin current flowing into the m-th lead from all other n leads having potential V_n.
The zero temperature conductance, $G_{pq}^{\sigma\sigma'}$ that describes the spin resolved transport measurements, is related to the spin resolved transmission coefficient as [10, 11], $G_{pq}^{\sigma\sigma'} = \frac{e^2}{h} T_{pq}^{\sigma\sigma'}(E)$. The transmission coefficient can be calculated as [12, 13], $T_{pq}^{\sigma\sigma'} = \text{Tr}\left[\Gamma_{\text{p}}^{\sigma} \text{G}_{\text{R}} \Gamma_{\text{q}}^{\sigma'} \text{G}_{\text{A}}\right]$. Γ_p^{σ}'s are the coupling matrices representing the coupling between the central region and the leads, and they are defined by the relation [14], $\Gamma_p^{\sigma} = i\left[\Sigma_p^{\sigma} - (\Sigma_p^{\sigma})^{\dagger}\right]$. Here Σ_p^{σ} is the retarded self-energy associated with the lead p. The self-energy contribution is computed by modeling each terminal as a semi-infinite perfect wire [15].

The retarded Green's function, G_R is computed as $G_R = \left(E - H - \sum_{p=1}^{4} \Sigma_p\right)^{-1}$, where E is the electron Fermi energy and H is the model Hamiltonian for the central conducting region. The advanced Green's function is, $G_A = G_R^{\dagger}$.

Now, following the spin Hall phenomenology, in our set-up since the transverse leads are voltage probes, $I_3^q = I_4^q = 0$. Also, as the currents in various leads depend only on voltage differences among them, we can set one of the voltages to zero without any loss of generality. Here we set $V_2 = 0$. Finally, if we assume that the leads are connected to a geometrically symmetric ordered bridge, so, $\frac{V_3}{V_1} = \frac{V_4}{V_1} = \frac{1}{2}$. Now from Eq.(5), Eq.(3) and Eq.(4) we can write the expression of spin Hall and the longitudinal conductances as,

$$G_{SH} = \frac{e}{8\pi} \left(T_{13}^{out} + T_{43}^{out} + T_{23}^{out} - T_{34}^{in} - T_{31}^{in} \right) \quad \text{and} \quad G_L = \frac{e^2}{h} \left(T_{21} + \frac{1}{2} T_{32} + \frac{1}{2} T_{42} \right) \tag{7}$$

3. Results and Discussion

We have investigated the effect of spin dependent disorder (onsite and hopping) (W) in presence of spin orbit coupling (both Rashba and Dresselhaus terms) on the experimentally measurable quantities such as the longitudinal conductance (G_L) and the spin Hall conductance (G_{SH}).

Before we start computing the physical quantities, we briefly describe the values of different parameters used in our calculation. Throughout our work, we have considered lattice constant, $a = 1$, system size, $L = 20$, onsite term, $\epsilon = \epsilon_L = 0$, hopping term, $t = t_L = t_C = 1$. All the energies are measured in unit of t. Further we choose a unit where $c = h = e = 1$. The longitudinal conductance, G_L is measured in unit of $\frac{e^2}{h}$. The spin Hall conductance, G_{SH} is measured in unit of $\frac{e}{8\pi}$.

Disorder (onsite and hopping) in the system is modeled by a random rectangular distribution between $[-W, +W]$. We consider two different types of disorder:
(i) Spin dependent disorder for the up spin (USDD) with $W_{\uparrow} = W$ and $W_{\downarrow} = 0$,
(ii) Spin dependent disorder for the down spin (DSDD) with $W_{\uparrow} = 0$ and $W_{\downarrow} = W$.

All the results obtained below are averaged over 1000 disorder configurations. For most of our numerical calculations we have used KWANT [16].

There exists a special case at the symmetric point of $V_R = V_D$. The spin Hall conductance (SHC) vanishes at that point [17, 18]. In presence of onsite or hopping spin independent disorder (SID), SHC is still zero in the given symmetric case. However, the main finding of the paper is that when we introduce onsite or hopping spin dependent disorder, we found a non-zero SHC.

Fig.2 shows the variation of spin Hall conductance as a function of energy E, in presence of onsite and hopping SDD when Rashba and Dresselhaus spin orbit interaction have the same strengths ($V_R = V_D$). Here, $W_{O\sigma}$ stands for the onsite disorder with spin σ and $W_{H\sigma}$ is the hopping disorder with spin σ. In Fig.2(a), G_{SH} increases in the negative region with increasing the strength of USDD.

The same thing happens with DSDD, except the value of G_{SH} which acquires non zero values in the positive region as shown in Fig.2(b). Another interesting point is that the hopping SDD is much more efficient to increase SHC in magnitude than onsite SDD.

We can explain this non-zero SHC from the definition of the spin current (Eq.(3)). As we introduce SDD in the up spin electron, $I_3^{\uparrow}$ (up spin current flowing in terminal 3) will decrease and most of $I_3^{\downarrow}$ (down spin current flowing in terminal 3) will reach lead 3, resulting negative G_{SH}. The same argument will render positive G_{SH} in case of DSDD.

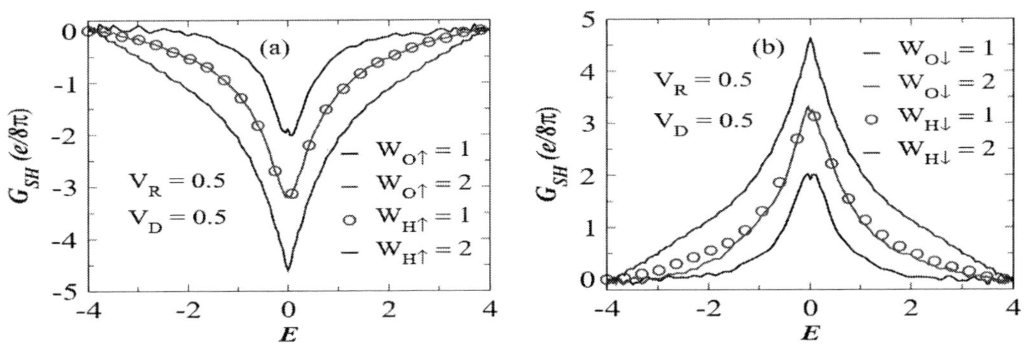

Figure 2. The spin Hall conductance, G_{SH} (in units of $e/8\pi$) is plotted as a function of energy E when both the interactions have the same strength, *i.e.* $V_R = V_D$. (a) Onsite and hopping USDD, (b) onsite and hopping DSDD.

In Fig.3, we have studied the variation of longitudinal conductance, G_L as a function of energy (E) for different spin dependent disorder strengths (W). Fig.3(a) shows G_L vs E plot for onsite and hopping disorder with different USDD strengths. For $W_{O\uparrow} = 2$ and $W_{H\uparrow} = 1$, the value of G_L remains almost same, implying that the hopping disorder is more efficient to decrease the longitudinal conductance. Fig.3(b) shows G_L vs E plot for the DSDD case. Hopping disorder is again found to be more efficient to suppress longitudinal conductance than onsite disorder. Hence we can say that the variation of G_L is independent of SDD, *i.e.*, G_L remains same for both USDD and DSDD irrespective of the onsite and hopping disorder.

We can explain why the longitudinal conductance is independent of spin dependent disorder from (Eq.(3)). G_L is just the ratio between I_3^s (the total charge current flowing in terminal 3) and V_3 (the voltage at terminal 3). Since the expressions of G_L contains the total charge current, no matter what type of SDD is introduced, the total charge current at terminal 3, I_3^q will always be the same. As a result, G_L is independent of SDD.

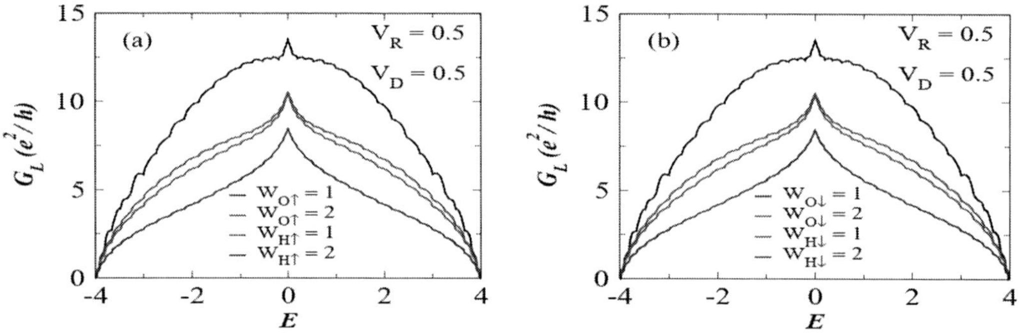

Figure 3. The longitudinal conductance, G_L (in units of e^2/h) is plotted as a function of energy E when both the interactions have the same strength, *i.e.* $V_R = V_D$. (a) Onsite and hopping USDD, (b) onsite and hopping DSDD.

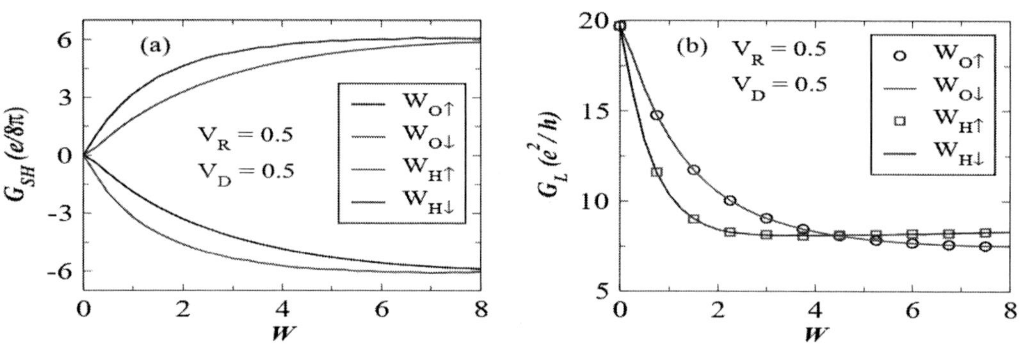

Figure 4. (a) G_{SH} is plotted as a function of disorder strength, W. (b) G_L is plotted as a function of disorder strength.

The variation of G_{SH} as a function of disorder, W at $E = 0$ is shown in Fig.4. Here we set the energy at $E = 0$. From Fig.4(a) we see that for USDD, G_{SH} starts to increase with $W_\uparrow$ and tends to saturate in the negative region. For DSDD, G_{SH} shows a behaviour which looks like a mirror reflection of the corresponding feature for USDD . From Fig.4(a) it is clear that G_{SH} increases more in case of hopping disorder than that of onsite disorder. Fig.4(b) shows the variation of G_L as a function of disorder strength. For both USDD and DSDD, the variation of G_L is exactly the same as we saw in Fig.3. Here G_L decreases with increasing disorder strength and finally tends to saturate in the higher region of W. In case of hopping disorder, G_L decreases more abruptly than the onsite case. G_{SH} becomes negative for USDD and positive for DSDD. In presence of USDD, $I_3^\uparrow$ will be suppressed and for larger value of $W_\uparrow$, $I_3^\uparrow \simeq 0$. As a result G_{SH} tends to saturate in the higher region of $W_\uparrow$.

In order to study the effect of system size on G_{SH} and G_L in presence of SDD, we fixed width of the scattering region to lattice of size $L = 20$ and set energy, $E = 0$. Fig.5(a) shows the variation of G_{SH} as a function of the system size, L. Here G_{SH} is essentially constant except when the width is lower than the length of the scattering region. G_{SH} is almost same for $W_{O\uparrow} = 2$ and $W_{H\uparrow} = 1$. Fig.5(b) shows the variation of G_L as a function of the system size, L. G_L decreases with length and then becomes almost constant.

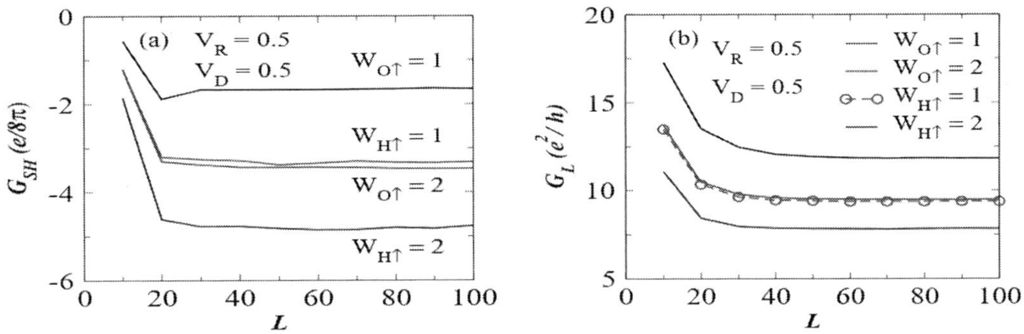

Figure 5. (a) G_{SH} is plotted as a function of system size, L. (b) G_L is plotted as a function of system size.

After a certain value of L, both G_{SH} and G_L remain constant with L. G_{SH} takes appropriate sign depending upon the nature of the SDD as we discussed earlier. Whereas, the variation of G_L with L is found to be independent of the nature of SDD.

4. Summary and Conclusions

In summary, in the present work we have studied the effect of spin dependent disorder effect on a four terminal junction device in presence of Rashba and Dresselhaus spin orbit coupling, specifically when the strengths of the two types of the spin orbit coupling are the same.

Spin dependent disorder can produce a non zero G_{SH} even if the strengths of RSOC and DSOC are the same. Higher the SDD, larger is the G_{SH} (in magnitude). G_{SH} is symmetric about $E = 0$ and takes opposite signs for different types of SDD. Hopping disorder is more efficient to increase G_{SH} than the onsite disorder. G_L is independent of the nature of SDD. With the same disorder strength, G_L decreases more in case of hopping disorder than the onsite case. With increasing the strength of SDD, both G_L and G_{SH} tend to saturate in the higher region of SDD. Further, we have found that essentially G_L and G_{SH} are independent of length (L) of the sample, at least for larger values of L.

References

[1] S. Murakami, N. Nagaosa and S. C. Zhang, Science **301**, 1348 (2003).
[2] R. Nanguneri, M. Jiang, T. Cary, G. G. Batrouni, and R. T. Scalettar Phys. Rev. B **85**, 134506 (2012).
[3] M. Jiang, R. Nanguneri, N. Trivedi, G. G. Batrouni, and R. T. Scalettar, New J. Phys. **15**, 023023 (2013).
[4] K. Makuch, J. Skolimowski, P. B. Chakraborty, K. Byczuk, and D. Vollhardt, New J. Phys. **15**, 045031 (2013).
[5] D. Jaksch, C. Bruder, J. I. Cirac, C. W. Gardiner, and P. Zoller, Phys. Rev. Lett. **81**, 3108 (1998).
[6] I. Bloch, J. Dalibard, and W. Zwerger, Rev. Mod. Phys. **80**, 885 (2005).
[7] M. Dey, S. K. Maiti and S. N. Karmakar, J. Appl. Phys. **112**, 024322 (2012).
[8] M. Büttiker, Phys. Rev. Lett. **57**, 1761 (1986).
[9] T. P. Pareek, Phys. Rev. Lett. **92**, 076601 (2004).
[10] R. Landauer, IBM J. Res. Dev. **1**, 223 (1957).
[11] R. Landauer, Philos. Mag. 21, 683 (1970).
[12] C. Caroli, R. Combescot, P. Nozieres and D. Saint-James, J. Phys C: Solid State Phys. **4**, 916, (1971).
[13] D. S. Fisher and P. A. Lee, Phys. Rev. B **23**, 6851 (1981).
[14] S. Dutta, *Electronic transport in Mesoscopic systems*, University press (Cambridge), (1995).
[15] B. K. Nikolic, Phys. Rev. B **64**, 014203 (2001).
[16] C. W. Groth, M. Wimmer, A. R. Akhmerov, X. Waintal, *Kwant: a software package for quantum transport*, New J. Phys. **16**, 063065 (2014).
[17] J. Li, L. Hu, S. Q. Shen, Phys. Rev. B **71**, 241305(R) (2005).
[18] C. P. Moca and D. C. Marinescu, Phys. Rev. B **72**, 165335 (2005).

First principles study on structural, phase transition and electronic structure of Zinc Sulfide (ZnS) within LDA, GGA and mBJ potential

Kh. Kabita and B. Indrajit Sharma*
Department of Physics, Assam University, Silchar-788011, Assam, India.
*Corresponding Author: indraofficial@rediffmail.com

Abstract. An ab-initio calculation based on full potential linearised augmented plane wave method of density functional theory (DFT) within the localised density approximation (LDA), generalised gradient approximation (GGA) and modified Becke-Johnson (mBJ) for exchange correlation potential is performed to investigate the structural, phase transition and electronic structure of Zinc Sulfide (ZnS). A comparative study is performed for the three approximations and reveal that calculations within the GGA gives better structural stability and phase transition pressure of 17.6 GPa in reasonably close agreement to experimental results. The electronic structure calculation within mBJ predicts higher direct band gap of 3.5eV in comparison to LDA and GGA underestimated value. The trend in the accuracy of energy band gap calculation is found to be LDA<GGA< mBJ-GGA. The electronic band structures are investigated using the total and partial density of states.

1. Introduction

Zinc Sulfide (ZnS) is an important semiconductor which has been a subject of great interest in the recent years due to its technological importance and scientific applications. Its wide and direct band gap makes it ideal for applications in solar cells, laser diodes, high-density optical memory and electronic image display [1,2]. Under ambient pressure conditions it crystallises in zinc blende (ZB) and wurzite (WZ) structure which transforms to a six-fold co-ordinated rock salt (RS) structure at high pressure. The electronic structure plays a significant role in understanding the electrical, optical and transport properties of a material. It helps in better interpretation of the solid state phenomena of a material. The structural and electronic properties of ZnS has been performed by many experimentalist and theorists using different methods but the variation in the theoretical results with experimental data has lead to various disagreements [3,4,5,6]. In order to sort out this discrepancy many methods with different approximation has been used by various theorists.

Recent developments in the computational simulations have resulted in a more systematic and precise study of the structural and electronic properties of a material from the first principle. The theoretical study of the structural and electronic properties of a material are generally performed within the local density approximation (LDA), generalised gradient approximation (GGA) and GW [7,8] but calculations within the LDA and GGA generally underestimates the energy band gap [9]. The best method for studying the energy band gap is the many perturbation theory within the GW approximation which give us very good results in comparison to experimental studies. But this method is computationally very expensive and time consuming. In 2009 Tran and Blah gave an alternative method for the energy band calculation [10] which is computationally cheap, less time consuming and equally effective by adding a potential known as the modified Becke-Johnson (mBJ) potential. The mBJ potential within the Perdew-Burke-Ernzerhof (PBE)-GGA approximation performs energy band calculation similar to GW calculation giving us much improved results close to experimental studies and overcoming the underestimation problem within the LDA and GGA approximation.

2. Computational methods

The first principle study of ZnS calculation is performed using the first principle full potential linearized augmented plane wave (FP-LAPW) [11] method within the local density approximation (LDA), generalized gradient approximation (GGA) of Perdew-Burke-Ernzerhof (PBE-GGA) [12] and modified Becke-Johnson (mBJ) [13] potential within PBE-GGA under the framework of density functional theory (DFT) [14-16]. This method is used as it is one of the most accurate methods in electronic structure calculation of crystals. In this method, the lattice is divided into non-overlapping atomic spheres surrounding each atomic sites and an interstitial region [17]. Inside the muffin tin (MT) region, the potential is a product of radial function and spherical harmonics and expanded up to order $l = 10$. For the interstitial regions that are outside the muffin tin spheres, the potentials are expanded in plane waves. 8000 k-points are used for the integration part which reduces to 256 irreducible k-points inside the Brillouin zone. Convergence is obtained at $R_{MT}K_{max} = 9.0$ where R_{MT} is the atomic sphere radii and K_{max} gives the plane wave cut-off. All calculations are performed with the equilibrium lattice constants which are determined from the plot of the total energy against the unit cell volume by fitting to the Birch-Murnaghan equation of states [18]. The overall calculation is done with WIEN2K [19].

3. Results and discussion

The structural properties plays an important role in the study of solids as it provides detailed information about inter atomic binding forces in a material. The stable structure of ZnS is calculated by calculating the total energy and the corresponding primitive cell volume from series of different values of lattice constant and fitting it to the Birch Murnaghan equation of states. In figure 1 the energy versus volume curve of ZnS for both ZB and RS structure within (a) LDA and (b) GGA methods are given. From the figure we find that in both the calculations ZnS-ZB is more stable than ZnS-RS. A summary of the calculated structural parameters are compared with other experimental and theoretical studies and tabulated in table 1. From the table we find that our present work agrees well with other experimental and theoretical results and hence is used for further calculation of the phase transition and energy band gap.

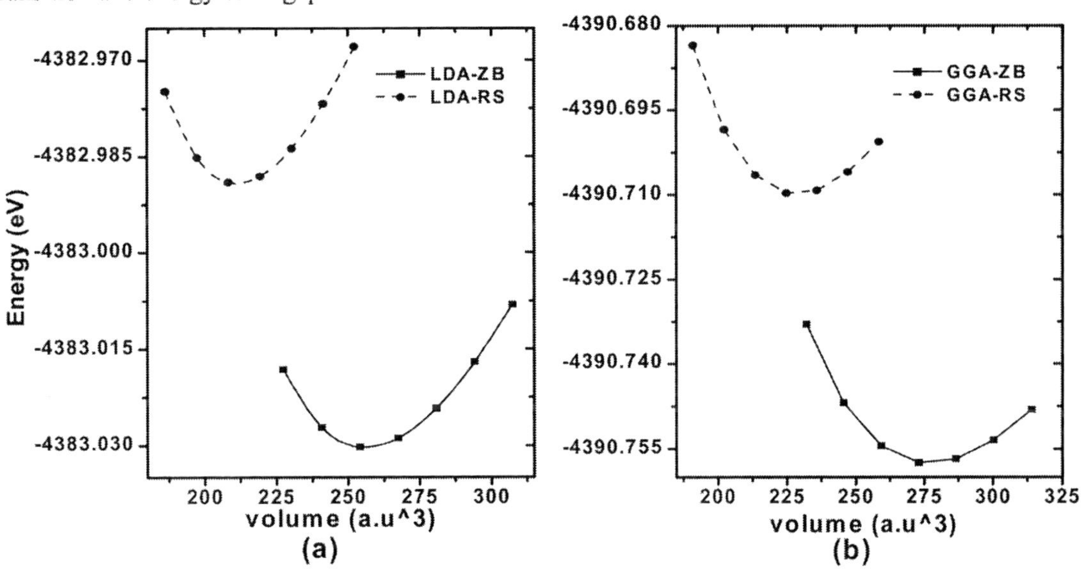

Figure 1. Total energy as a function of volume of ZnS-ZB and ZnS-RS within (a) LDA and (b) GGA.

Table 1. Experimental and calculated ground state structural parameters of ZnS

Structure	Ground state structural parameters	Present work		Experimental studies	Other theoretical calculations
		LDA	GGA		
ZB	a_0 (A^0)	5.336	5.475	5.412[a]	5.328[b], 5.335[c], 5.342[d], 83.8[b], 83.70[c], 89.67[d]
	B_0 (GPa)	88.88	71.51	75[a]	
RS	a_0 (A^0)	5.004	5.135	--	5.066[e]
	B_0 (GPa)	111.75	88.73	--	105.6[e]

[a]Ref [4], [b]Ref [5], [c]Ref [20], [d]Ref [21], [e]Ref [6]

The structural, mechanical and physical properties of a material under pressure can be understood from the phase transition study. Generally the density functional calculation is performed at zero temperature conditions. Therefore the equal enthalpy condition, H = E+PV including the pressure-volume effects is used for the structural phase transition calculations. In figure 2 the calculated enthalpy as a function of pressure is given within both (a) LDA and (b) GGA. From the figure we can see the crossing over of the enthalpy in both LDA and GGA methods indicating a phase transition between the two phases. Within the LDA, the structural transition pressure from the ZnS-ZB to ZnS-RS occurs at 15.9 GPa pressure while GGA shows a structural transition at 17.6 GPa pressure. Thus structural phase transition calculation within the GGA gives us results closer to the experimental study of 18.1GPa pressure [22] and 16.9 GPa pressure[23] than the LDA.

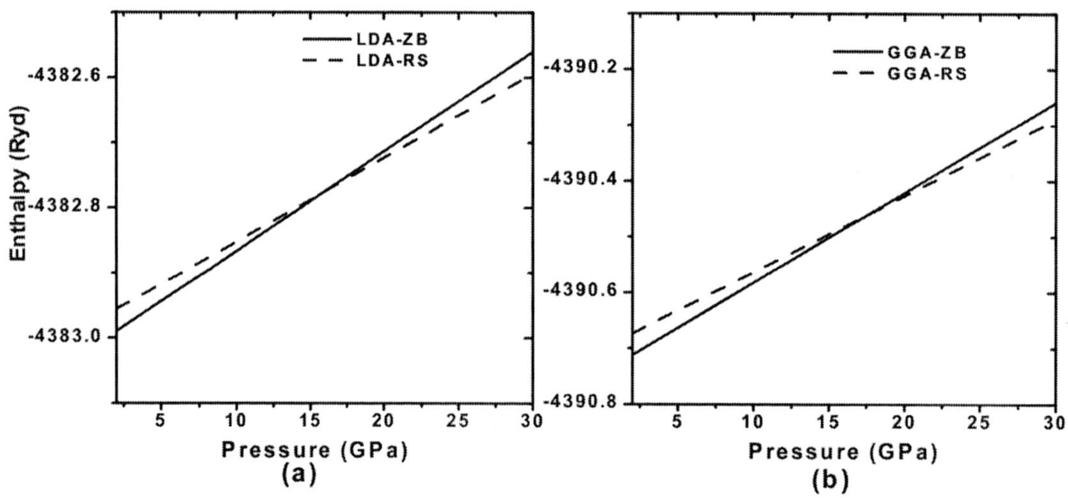

Figure 2. Enthapy as a function of pressure of ZnS-ZB and ZnS-RS within (a) LDA and (b) GGA.

ZnS crystallizes in the ZB structure under ambient pressure and RS structure at high pressure. We have therefore performed the electronic structure calculation at 0 GPa pressure within the LDA, GGA and mBJ-GGA for the ZB phase and shown in figure 3(a), 3(b) and 3(c) respectively. From the figures, we find that in all three cases the valance band maximum and the conduction band minimum are located at the middle of the brillion zone, Γ point indicating a direct band gap. However the energy band gap calculated within the LDA and GGA in figure 3(a) and 3(b) shows a band gap of 1.91 eV and 1.98 eV respectively which is very low as compared to the experimental results of 3.6 eV [24,25,26] while in figure 3(c) the calculation within the mBJ-GGA shows a band gap of 3.5 eV which is very close to the experimental result. The reason for the mere difference in the calculation within the mBJ-GGA and experimental data is that the DFT calculation is performed at absolute temperature while the reported experimental data is at room temperature. One interesting point in the energy band gap calculation is the shifting of the conduction band maximum towards higher energy and wide opening of the band gaps. The order of the energy band gaps of ZnS in the three cases is LDA< GGA< mBJ-GGA. Also the gap between the lowest band and valance band minimum is found decreasing in the order LDA> GGA> mBJ-GGA. Thus the implementation of the mBJ-GGA overcomes the underestimation in the band gap calculation with LDA and GGA and gives us results closer to the experimental studies.

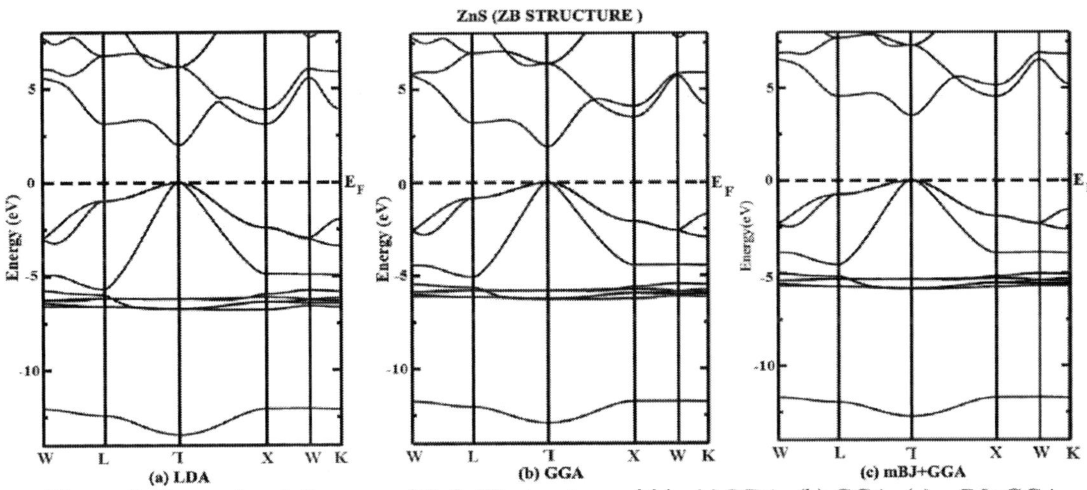

Figure 3. Energy band diagram of ZnS-ZB structure within (a) LDA, (b) GGA, (c) mBJ+GGA

The origin of the band structure is related to the corresponding density of states of the compound. For better understanding of the contribution of the states in the energy band gaps the total and partial density of states of ZnS-ZB within the LDA, GGA and mBJ-GGA at 0 GPa pressure is also studied and shown in figure 4(a), 4(b) and 4(c). From the total DOS study we find that the lowest band is mainly contributed by the s-state (S atom) and its contribution within the LDA, GGA as LDA > GGA> mBJ-GGA. The valance band is found to be mainly contributed by the d-state (Zn atom) whose contribution increases in the order LDA > GGA > mBJ-GGA. The increase in the intensity within the mBJ-GGA calculation leads to better treatment of the band structure than the LDA and GGA methods.

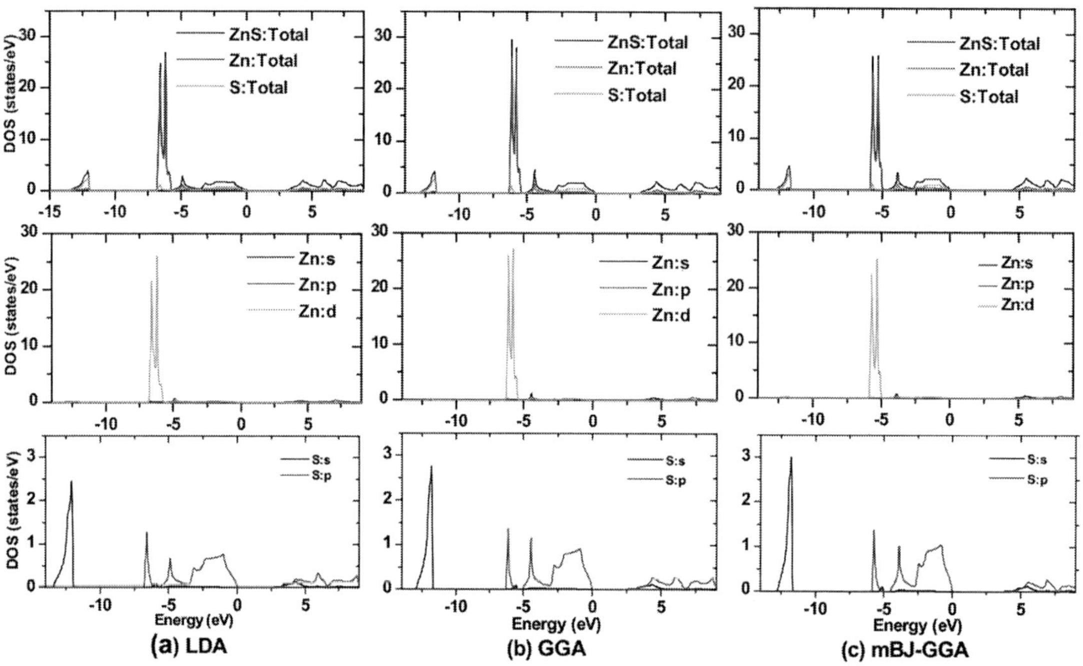

Figure 4. Total and Partial DOS of ZnS within (a) LDA, (b) GGA and (c) mBJ-GGA.

4. Conclusion

The structural parameters of ZnS-ZB and ZnS-RS are studied within the LDA and GGA. The structural phase transition from the ZB to RS phase calculated within the LDA occurs at 15.9 GPa pressure while within the GGA shows a structural phase transition at 17.6 GPa pressure which is closer to the experimental results. The energy band gap within LDA and GGA are 0.91 eV and 0.98 eV respectively while within mBJ-GGA, a band gap of 3.5 eV is shownp. The LDA and GGA method underestimates the energy band gap while mBJ-GGA solves this discrepancy and gives us results very close to the experimental studies. Also the effective results provided by the mBJ-GGA method are found to be a result of the proper treatment of the electronic states. In conclusion, we find that the mBJ-GGA is a more effective tool for electronic structure calculation of ZnS and will validate its applications in optoelectronic devices.

References

[1] E. Bellotti, K. F. Brennan, R. Wang and P. P. Ruden, *J. Appl. Phys.* 82, 2961 (1997).
[2] S. Q. Wang, Appl. *Phys. Lett.* 88, 061902(1)-061902(3) (2006).
[3] A. Mujica, A. Rubio, A. Munoz and R. J. Needs, *Rev. Mod. Phys.* 75, 863 (2003).
[4] Vagelatos N, Wehe D, King J S. *J Chem Phys*, 60(9): 3613 (1974).
[5] Postnikov A V, Pagès O, Hugel J, *Phys Rev B*, 71(11): 115206 (2005).
[6] Sun-Ghil Lee and K.J. Chang, *Phys Rev. B,* 52 (3):1918 (1995).
[7] Wilfried G. Aulbur, Martin Städele, and Andreas Görling, *Phys. Rev. B* 62, 7121 (2000).
[8] M. Shishkin, M. Marsman, And G. Kresse, *Phys. Rev. Lett.* 99 24640 (1-4) (2007)
[9] F Aryasetiawan and O Gunnarsson *Rep. Prog. Phys.* 61, 237 (1998)
[10] F. Tran and P. Blaha, Phys. Rev. Lett 102, 226401 (2009).
[11] Wimmer, E., Krakauer, H., Weinert, M., and Freeman, A. J., *Phys. Rev. B* 24, 864 (1981).
[12] Perdew, J. P., Burke, K., and Ernzerhof, M., *Phys. Rev. Lett.* 77, 3865 (1996)
[13] Becke, A. D., and Johnson, E. R., *J. Chem. Phys.* 124, 221101 (2006)
[14] Hohenberg , P., and Kohn, W., *Phy. Rev.3B*, 136, 864 (1964).

[15] Kohn, W., and Sham, L. J., *Phys. Rev. 4A* 140, 1133 (1965).
[16] Cottenier, S., *"Density Functional Theory and the family of (L)APW-methods: a step-by-step Introduction"* Belgium, Insti-tuut voor Kern-en Stralingsfysica, K.U.Leuven ISBN 90-807215-1-4, (2002).
[17] Andersen, O. K., *Phys. Rev. B*. 12, 3060 (1975).
[18] Birch, F., *Phys. Rev.* 71, 809 (1947).
[19] Blaha, P., K. Schwarz, Madsen, G. K. H., Kvasnicka, D., and Luitz, J., *"An augmented plane wave + local orbitals program for calculating crystal properties",Wien2k. Techn. Universitat* (Wien, Austria) ISBN 3-9501031-1-2. (2001).
[20] Casali R A, Christensen N E, Solid State Commun, 108(10): 793 (1998)
[21] Khenata R, Bouhemadou A, Sahnoun M, et al. Comput Mater Sci, 38(1): 29 (2006).
[22] A. Nazzal and A. Qteish, *Phys. Rev. B* 53 8262 (1996).
[23] Yang Jun, Zhu Feng, Zhang Qain, Wu Ye, Wu Xiang, Qin Shan, Dong Jun-Cai, Chen Dong-Liang, Chin. Phys. Lett. 30 4 046101 (2013)
[24] Kittel, *Introduction to Solid State Physics*, 6th Ed. (New York: John Wiley) 185 (1986).
[25] K. Sooklal, B.S. Cullumn, S.M. Angel, C.J. Murphy, *J. Phys. Chem.* 100, 4551 (1996).
[26] Jiye Zhang, Yonghao Wang, Jun Zhang, Zhang Lin, Feng Huang, and Jiaguo Yu, *ACS Appl. Mater. Interfaces* 5, 1031–1037 (2013).

Thermally assisted tunneling of charge carriers between trapping sites in calcite ($CaCO_3$) mineral

J.M. Kalita* and G. Wary

Department of Physics, Cotton College, Guwahati-781001, India

*E-mail: jitukalita09@gmail.com

Abstract. Thermally Stimulated Luminescence (TSL) mechanism of natural orange calcite mineral has been studied. The glow curve of X-ray irradiate sample has been studied in un-annealed and 673 K annealed samples by a Computerized Glow Curve De-convolution (CGCD) technique. It has been observed that after exposure to X-ray, both the un-annealed and 673 K annealed samples show three distinct peaks at around 323, 360 and 409 K in the glow curves. Study on the effect of post-annealing treatment on the excited sample reveals that the 323 K peak can be thermally bleached by annealing at ~340 K, however it is very significant that the 360 K TSL peak can't be bleached by increasing the post-annealing temperature up to 380 K. CGCD analysis reveals that the activation energies corresponding to the 323, 360 and 409 K peaks are found to be around 0.70, 0.60 and 1.30 eV respectively. The low activation energy peak observed at relatively high temperature and existence of the 360 K peak after post-annealing the sample beyond its peak maximum temperature are due to thermally assisted tunneling of trapped charge from 1.30 to 0.70 eV trapping sites in calcite. This analysis leads to an important conclusion that though theoretically the glow peak observed at relatively high temperature should have high activation energy, but due to charge tunneling phenomenon between the trapping sites, glow peak observed at high temperature may have relatively low activation energy.

1. Introduction

Thermally Stimulated Luminescence (TSL) analysis is an efficient tool to study the presence of trapping sites as well as movement of the charge carriers among the trapping sites. For TSL analysis, at first the sample is exposed under high energy radiation. Due to high energy irradiation the valence electrons excite to the conduction band. If some localized energy state available in between the valence band and conduction band then these charge carriers (electron/hole) may get trapped at the localized energy state. At the stimulation stage the sample is heated at a constant rate. At this stage the trapped electron eject out from the trapping site and recombine with the holes via the conduction band. The recombination is sometime radiative and non-radiative. The radiative recombination caused the emission of light. The plot of the amount of emitted light as a function of temperature of the sample is often called TSL glow curve. The shape and position of a glow peak is characterized by various physical parameters such as activation energy (trap depth), frequency factor, order of kinetics etc. A glow curve may compose of more than one glow peak. Each glow peak in a glow curve corresponds for a unique set of some physical parameters. Therefore proper analysis of the glow curve can give the accurate estimation about the trapping site available in the energy gap which plays a vital role in the whole luminescence phenomenon.

Calcite ($CaCO_3$) is a naturally occurring high sensitive luminescent mineral. Because of its high sensitivity, it becomes an attractive nominee for many applications including geological dating [1], medical dosimetry [2] as well as generic defect studies [3–5]. Townsend et al. [6] studied thermoluminescence (TL) spectra of X-ray irradiated natural calcite and its pure habit, Iceland spar, and observed line structures in the emission spectra. Analysis of the TL spectra showed that Mn impurity was the most recognized recombination site in calcite [3–8]. However, in geological calcite sample some differences in the behavior of the spectra have been observed [9]. Though luminescence properties of calcite mineral had been studied by several researchers, but there are still some extraordinary results that need to be addressed. According to the theory of TSL, the glow peak observed at high temperature should have high activation energy relative to the peak observed at low temperature, but in calcite a glow peak was observed at high temperature that has relatively low activation energy [5]. The luminescence mechanism of calcite is very complicated and it cannot be explained with the help of ordinary luminescence model commonly available in the literature. In the present case luminescence mechanism in calcite has been discussed in detail with the help of TSL analysis.

2. Experimental works

Rarely orange color natural calcite mineral was procured from Geological Survey of Assam, Govt. of Assam (India) which was originated from Umrangso, Karbi Anglong, (India) (latitude: 25°30' N, longitude: 92°44' E). It was crushed by roll crusher and then ground in a ceramic mill to powder form. To study the TSL characteristics, at first the sample was divided into two parts; one part was kept un-annealed and the other part was annealed at 673 K for 30 minutes. For TSL analysis taking 10 mg sample in each aliquot the samples were exposed under X-ray (dose rate: 0.05mGy/s). TSL glow curve of the irradiated sample was recorded by NUCLEONIX-TL 1009I TL reader at 2 K/s heating rate.

3. Results and discussion

Figure1 shows the TSL glow curve of calcite mineral after 60 mGy X-ray irradiation. Three glow peaks P_1, P_2 and P_3 have been observed at around 323, 360 and 408 K in the glow curve. A natural TL (NTL) glow peak (P_0) has been observed at around 472 K in the fresh un-irradiated sample. However after X-ray irradiation no TSL peak has been observed at that temperature. Comparing the glow curves it can be observed that due to the pre-annealing at 673 K, no any dislocation of the peaks is observed. This signifies the thermal stability of the trapping sites that are present in the crystal. The glow curves of both the samples have been analyzed by Computerized Glow Curve De-convolution (CGCD) technique on the basis of Kitis general order equation (equation (1)) [10] and various TL parameters such as activation energy (E), order of kinetics (b) and frequency factor (s) have been estimated.

$$I(T) = I_m b^{b/(b-1)} \exp\left(\frac{E}{kT}\frac{T-T_m}{T_m}\right) \times \left[(b-1)(1-\Delta)\frac{T^2}{T_m^2}\exp\left(\frac{E}{kT}\frac{T-T_m}{T_m}\right) + Z_m\right]^{-b/(b-1)} \quad (1)$$

Here I is TL intensity, T is absolute temperature, I_m is peak maximum intensity, T_m is peak maximum temperature, k is Boltzmann's constant, $\Delta = 2kT/E$, $\Delta_m = 2kT_m/E$ and $Z_m = 1+(b-1)\Delta_m$. In this method, the activation energy, the peak maximum temperature and the order of kinetics are independent variables that are determined by iteration to a best fit. For each glow curve the goodness of fit is testedby the Figure of Merit (FOM) [11]. All de-convoluted glow curves are accepted when their FOM values are lower than 2%. A de-convoluted glow curve of 673 K annealed sample has been shown in Figure 2. The results are reported in Table 1. This analysis reveals that the activation energies corresponding to the 323, 360 and 409 K peaks are found to be around 0.70, 0.60 and 1.30 eV respectively. The activation energy corresponding to the NTL peak is about 1.49 eV. But after X-ray irradiation all the naturally trapped charge carriers move to the other trap sites. As a result no TL peak has been observed.

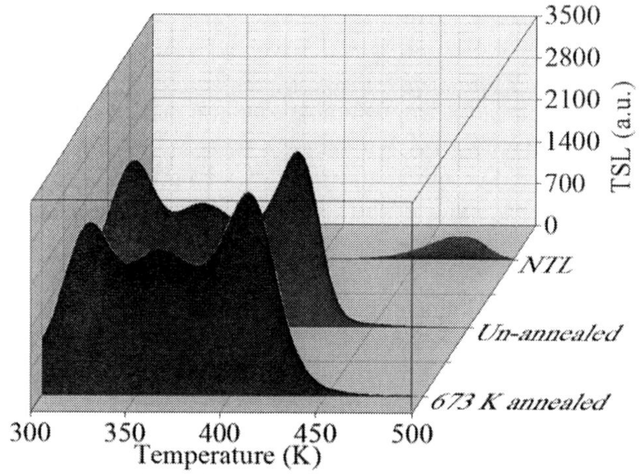

Figure 1. TSL glow curves of natural calcite mineral.

In order to study the trap spectroscopy of calcite in a more detail manner, the un-annealed and 673 K pre-annealed samples are heated at 2 K/s up to 340 K and 380 K temperatures after exposure to 60 mGy X-ray dose. The reason for selecting these two specific temperatures is to thermally bleach the peaks P_1 and P_2 respectively. Figure 3(a,b) shows the glow curves of both the samples after post-annealing at 340 and 380 K temperature for 6 minutes. As expected the 323 K glow peak (P_1) has completely bleached out in both the post-annealed samples; however it is very significant that the P_2 peak observed at ~360 K doesn't disappear after post-annealing at 380 K. In both the cases, a significant shift of the peak P_2 towards high temperature side has been observed but there is no dislocation of the peak P_3 due to the post-annealing treatments. The shift of peak P_2 towards high temperature side is due to the peak overlapping effect as discussed by McKeever [12]. Comparing both the glow curves it can be concluded that the glow curves of 380 K post-annealed samples are exactly identical with the 340 K post-annealed samples. The glow curves have been analyzed by CGCD technique and the results are reported in Table 2. The activation energies of the peaks P_2 and P_3 are found to be 0.61 eV and 1.30 eV respectively. These values are well consistent with the results obtained for pre-annealed samples.

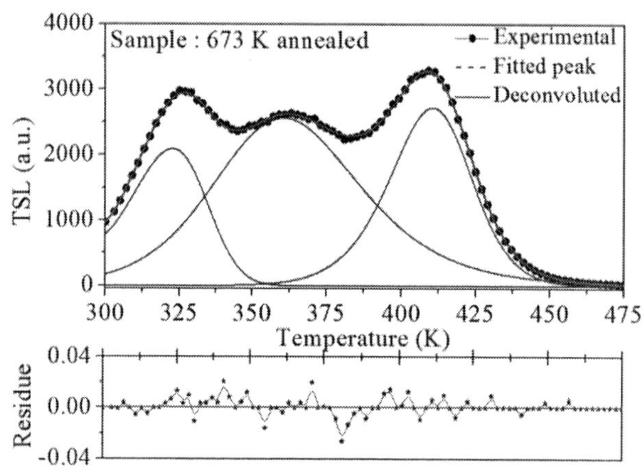

Figure 2. De-convoluted glow curve of 673 K annealed sample along with the residual of the fit.

Table 1. TL parameters evaluated by CGCD analysis.

Sample	Peak	T_m (K)	b	E (eV)	s (s^{-1})	FOM (%)
Un-annealed	P_0	472.0	1.15	1.49	2.54×10^{12}	0.14
	P_1	322.5	1.13	0.70	1.35×10^{11}	0.98
	P_2	360.0	1.90	0.60	2.47×10^{7}	
	P_3	412.4	1.10	1.30	1.36×10^{15}	
Annealed at 673 K	P_1	323.3	1.18	0.71	1.35×10^{11}	1.56
	P_2	360.2	1.93	0.61	1.56×10^{7}	
	P_3	410.4	1.20	1.29	5.26×10^{14}	

This analysis shows that during the excitation the electrons are excited and get trapped at the trapping sites. After excitation when the temperature of the sample increases, at temperature around 323 K, electrons release from the 0.70 eV trapping site and recombine with the holes resulting the P_1 TSL peak. Further, increase in temperature caused tunneling of a few electrons from 1.30 eV trapping site to 0.70 eV trapping site and from 0.70 eV trapping site direct recombination takes place at the recombination center. Due to this center-to-center recombination process, the peak P_2 has been

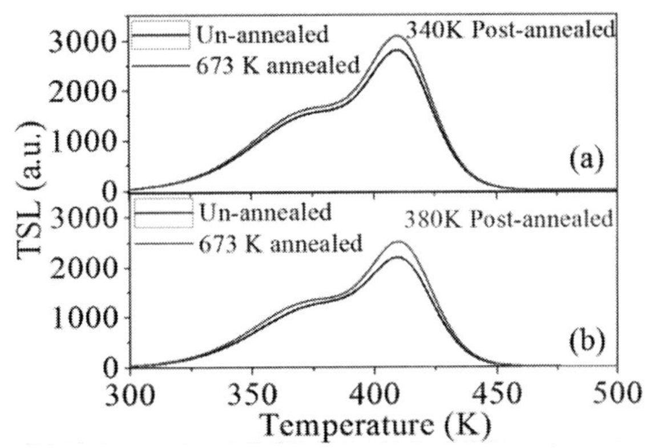

Figure 3. Glow curves of un-annealed and 673 K annealed samples after post-annealing at (a) 340 K and (b) 380 K for 6 minutes.

Table. 2: TL parameters of 340 and 380 K post-annealed samples evaluated by CGCD technique.

Sample	Post-annealing temperature (K)	Peak	T_m (K)	b	E (eV)	s (s^{-1})	FOM (%)
Un-annealed	340	P_2	370.3	1.10	0.60	1.50×10^{7}	1.25
		P_3	411.8	1.15	1.31	9.56×10^{14}	
673 K annealed		P_2	369.4	1.13	0.61	2.15×10^{7}	0.95
		P_3	411.8	1.22	1.31	8.18×10^{14}	
Un-annealed	380	P_2	369.5	1.11	0.61	1.43×10^{7}	1.32
		P_3	412.1	1.12	1.30	9.16×10^{14}	
673 K annealed		P_2	370.1	1.18	0.60	1.95×10^{7}	1.13
		P_3	412.2	1.23	1.30	7.47×10^{14}	

observed at relatively high temperature. However it is significant that, even after this temperature assisted tunneling process, sufficient numbers of electrons left on the 1.30 eV trapping site, and at temperature around 408 K, the residual trapped electrons release from that 1.30 eV trapping site and recombine with the holes that result the P_3 TSL peak. Due to this thermally assisted tunneling process the peak P_2 has not disappear in 380 K post-annealed sample. A plot of peak integrals versus the 340 K and 380 K post-annealed samples has been shown in Figure 4. It has been observed that the peak

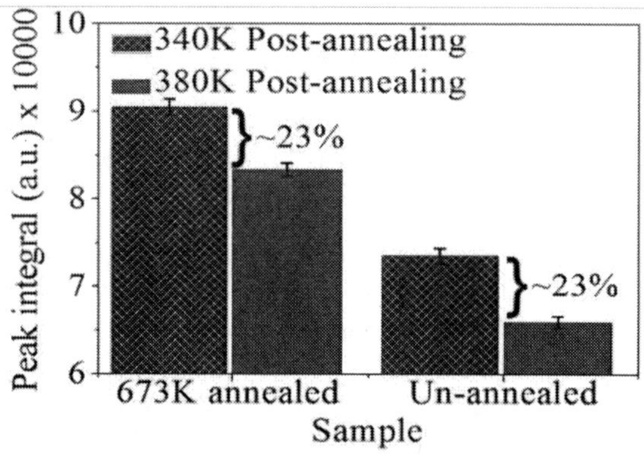

Figure 4. Plot of peak integrals versus the two different types of samples after post annealing at 340 K and 380 K.

integrals of 380 K post-annealed samples reduced to 23% from that of the 340 K post-annealed samples. This report quantitatively signifies that because of annealing up to 380 K for 6 minutes, approximately 23% of trapped electrons transferred from 1.30 eV trapping site to the 0.70 eV trapping site by the tunneling mechanism. This analysis quantitatively give support to the previous conclusion about the thermally assisted tunneling mechanism reported independently in Visocekas et al [13] and Kalita et al [14].

4. Conclusion

Thermally Stimulated Luminescence (TSL) mechanism of natural calcite mineral has been studied in detail. After X-ray irradiation a complex glow curve have been observed. De-convolution analysis reveals that the glow curve composed of three overlapping glow peaks having activation energy around 0.70, 0.60 and 1.30 eV respectively. The post-annealing treatments on the excited sample reveals that the there is thermally assisted tunneling of charge carriers from 1.30 eV to 0.70 eV trapping sites in calcite. Due to this tunneling mechanism a TSL peak having low activation energy has been observed at relatively high temperature. This analysis leads to an important conclusion that though theoretically the glow peak observed at relatively high temperature should have high activation energy, but due to charge tunneling phenomenon between the trapping sites, the glow peak that observed at high temperature may have relatively low activation energy.

Acknowledgment

Authors thank to UGC (University Grand Commission), New Delhi for providing the UGC major research project (Project No.: 42-910/2013 (SR)).

References

[1] Wintle A J 1978 *Can. J. Earth. Sci.* **15** 1977
[2] Gosku H Y, Wieser A, Waibel A, Vogenauer A and Regulla D F, 1989 *Appl. Radiat. Isot.* **40** 905

[3] Medlin W L 1968 in: *Thermoluminescence of Geological Materials*, McDougall D J (Ed.) London, Academic Press
[4] Townsend P D 1968 in: *Thermoluminescence of Geological Materials*, McDougall D J (Ed.) London, Academic Press
[5] Visocekas R 1979 *Euro.PACT.* **23** 258
[6] Townsend P D, Luff B J and Wood R A 1994 *Radiat. Meas.* **23** 433
[7] Down J S, Flower R, Strain J A and Townsend P D 1985 *Nucl. Track. Radiat. Meas.* **10** 581
[8] Orgel L E 1955 *J. Chem. Phys.* **23** 1004
[9] McDougal D J 1968 *Thermoluminescence of Geological Materials*, London, London University Press
[10] Kitis G, Gomez-Ros J M and Tuyn J W N 1998 *J. Phys. D: Appl. Phys.* **31** 2636
[11] Balian H G and Eddy N W 1977 *Nucl. Instrum. Method* **145** 389
[12] McKeever S W S 1985 *Thermoluminescence of Solids*, Cambridge, Cambridge University Press
[13] Visocekas R, Ceva T, Marti C, Lefaucheux F and Robert M C 1976 *Phys. Status Solidi (a)* **35** 315
[14] Kalita J M and Wary G 2014 *Spectrochim. Acta A*. **125** 99

Tunneling conductance through normal metal - superconductor junctions : effects of Rashba spin orbit coupling and magnetic field

Priyadarshini Kapri[1], Sudin Ganguly[2] and Saurabh Basu[3]
Department of Physics, Indian Institute of Technology Guwahati, Guwahati, Assam, India
E-mail: [1]priyadarshini@iitg.ernet.in, [2]sudin@iitg.ernet.in, [3]saurabh@iitg.ernet.in

Abstract. In a system consisting of a metal-(s-wave) superconductor junction, we study the conductance characteristics in presence of Rashba spin orbit coupling (RSOC) and an external magnetic field applied along the plane of the sample. With a selective inclusion of the Rashba coupling either in the metallic or in both we note that there is a distinct effect with regard to the magnitude of the Andreev peak that occurs at a biasing voltage lower than the superconducting gap energy. The height of the peak is sensitive to the RSOC (increases with increase in RSOC) for RSOC to be present only in the metallic region, (same is true when RSOC is present throughout the junction), while the peak height is fairly independent when RSOC is solely present in the superconducting region. The in-plane magnetic field has very interesting effects which show up in the form of having a conductance peak at zero bias, thereby making it possible to realize a Majorana bound state.

1. Introduction

Transport phenomena in a junction device that constitutes a contact between a superconductor (S) and a normal metal (N) with or without the presence of a thin insulating barrier has been a subject of considerable interest both theoretically and experimentally. Very recently, some works have begun to look at the transport behavior of metal superconductor junctions [1, 2, 3]. When a metal coupled to a superconductor, the transport phenomena are dominated by Andreev reflection [4] that occurs at the interface between the metal and superconductor. It is a process in which an electron transforms into a hole which retraces the path of an incoming particle and transfers a charge $2e$ across the interface, avoiding the single particle transmission within the superconducting energy gap. In the usual metal-superconductor junction, the Andreev reflected hole retraces the path of the incident electron.

Due to breaking of inversion symmetry near the normal metal-superconductor interface, a quasiparticle experiences spin orbit coupling (SOC) of the Rashba type [5], even if both the normal and superconducting crystals have bulk inversion symmetries. The effects of such Rashba spin orbit coupling (RSOC) have mostly been overlooked in the previous studies of tunneling conductance across a $N - S$ junction.

Further It has been suggested that Majorana states [6, 7, 8, 9] could exist in metal-superconductor junction due to SOC and in-plane magnetic field which may be realizable from the proximity effect of a ferromagnet [10]. Majorana fermions in nanowires could be used in future application to quantum information processing [11].

In this paper we carry out the study of tunneling conductance through normal metal-superconductor junction in presence of RSOC. We incorporate RSOC selectively in the metallic regime and also in both the metallic and superconducting leads. The effect of RSOC only in the superconducting lead should be trivial, owing to an uniform superconducting order parameter (s-wave) upon which the RSOC should have negligible effects and hence the Andreev reflection should be by and large unaffected.

Further we include an in-plane magnetic field to probe into the presence of zero mode states. In this section we consider various cases, such as, the in-plane magnetic field (B) is present in both normal and superconducting leads and the magnetic field is only in the superconducting lead, RSOC being present everywhere in the system.

2. Conductance of $N - S$ junction

We consider a system consisting of normal metal coupled to a superconductor denoted by Hamiltonian,

$$H = H_N + H_S \tag{1}$$

where H_N, H_S are the Hamiltonians of the normal metallic lead and superconducting lead respectively. The H_N in the tight binding representation in the presence of RSOC and in-plane magnetic field is given by,

$$H_N = \sum_{\langle i,j\rangle,\sigma,\sigma'} c^\dagger_{i\sigma} t^{\sigma\sigma'}_{ij} c_{j\sigma'} - B\sum_{i\sigma\sigma'} (\sigma_x)_{\sigma\sigma'} c^\dagger_{i\sigma} c_{i\sigma'} \tag{2}$$

and H_S is given by,

$$H_S = \sum_{\langle i,j\rangle,\sigma,\sigma'} c^\dagger_{i\sigma} t^{\sigma\sigma'}_{ij} c_{j\sigma'} - B\sum_{i\sigma\sigma'} (\sigma_x)_{\sigma\sigma'} c^\dagger_{i\sigma} c_{i\sigma'} + \sum_i \Delta_i c^\dagger_{i\uparrow} c^\dagger_{i\downarrow} + \sum_i \Delta^*_i c_{i\uparrow} c_{i\downarrow} \tag{3}$$

where $c^\dagger_{i\sigma}$ and $c_{i\sigma}$ are the creation and annihilation operators at the discrete site i, $t_{\sigma\sigma'}$ are the generalized hopping parameters in the σ basis. In $\pm x$ and $\pm y$ directions for square lattice , the hopping parameters are respectively,

$$\begin{pmatrix} -t & t_{so} \\ -t_{so} & -t \end{pmatrix}; \begin{pmatrix} -t & -t_{so} \\ t_{so} & -t \end{pmatrix}; \begin{pmatrix} -t & -it_{so} \\ -it_{so} & -t \end{pmatrix}; \begin{pmatrix} -t & it_{so} \\ it_{so} & -t \end{pmatrix}$$

where t is the nearest neighbor hopping, and $t_{so} = \alpha/2a$ (α: strength of RSOC, a : lattice constant) is the Rashba spin-orbit hopping term. B is strength of in-plane magnetic field, σ_x is Pauli spin matrix, and Δ_i is uniform superconducting order parameter at i th site. This implies $\Delta_i = \Delta_0\Theta(x)$ ($\Theta(x)$: Heaviside function) where the $N - S$ junction is located at $x = 0$, the left being the normal and the superconducting lead is in the right.

The conductance of a $N - S$ junction is given by the theoretical formalism developed by Blonder - Tinkham - Klapwijk (BTK) [12, 13] which essentially solves the Bogoliubov de Gennes equations selfconsistently using appropriate boundary conditions at the interface. We skip the details here and only present the main formula which is relevant for the conductance calculation. The formula for conductance of a $N - S$ junction is given as,

$$G = (1 - c(E) + d(E)) \tag{4}$$

where $c(E) = |C|^2$, C is the amplitude of specular reflection, $d(E) = |D|^2 \frac{k_N^-}{k_N^+}$, and D is amplitude of Andreev reflection. Now k_N^+, the momentum of electrons in the N region is given by,

$$k_N^+ = \sqrt{\frac{2m_N}{\hbar^2}}\sqrt{E_{FN} + \epsilon_k} \tag{5}$$

and k_N^-, the momentum of holes in N side is given by,

$$k_N^- = \sqrt{\frac{2m_N}{\hbar^2}}\sqrt{E_{FN} - \epsilon_K} \tag{6}$$

where N side is characterized by mass, m_N and Fermi energy, E_{FN} and energy dispersion, ϵ_k. The energy dispersion, ϵ_K for a two dimension square lattice is given by,

$$E = -2t(\cos k_x' + \cos k_y') - 2t_{so}\sqrt{\sin^2 k_x' + \sin^2 k_y'} \tag{7}$$

where k_x', k_y' are wavevectors in the first Brillouin zone.

3. Effect of RSOC in tunneling conductance through normal metal s-wave superconductor junction

For the calculation of conductance, G (in units of $\frac{e^2}{h}$), we have taken a square lattice of dimension $L = 10$ (in units of a) with the superconducting order parameter $\Delta = 0.5$ (in units of t) and $\alpha = 0.5, 1.5$ (in units of $2t$ ($t_{so} = t$)). For calculation of conductance we have used Kwant [14]. We have considered RSOC being present in the normal regime (N) and superconducting regime (S) selectively and in both the normal and superconducting leads.

When RSOC is present only in N, we note that there is distinct effect with regard to magnitude of conductance due to Andreev reflection that occurs at a biasing voltage, E (in units of t) less than gap energy, Δ as expected. The height of Andreev peak is sensitive to the strength of RSOC (α) and it starts gaining weight with increasing α (in units of $2t$). The

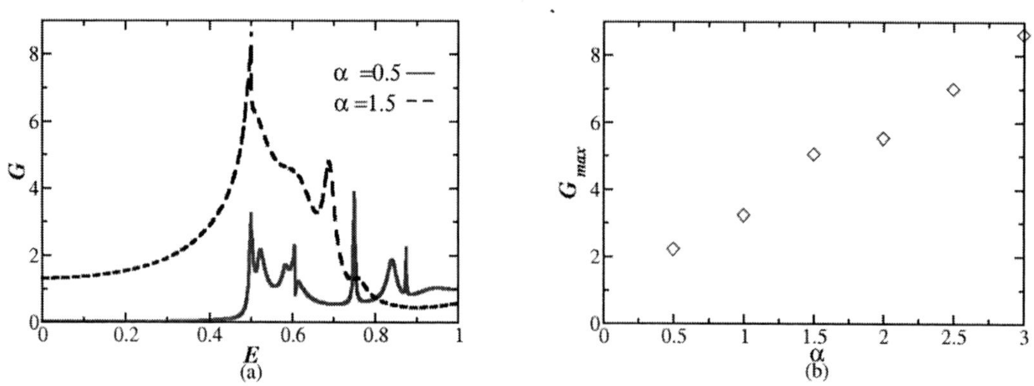

Figure 1. Red (solid) line denotes conductance for $\alpha = 0.5$ where the black (dashed) line denotes conductance for $\alpha = 1.5$. (a) The variation of conductance, G (in units of $\frac{e^2}{h}$) as a function of biasing energy, E (in units of t) for different strengths of RSOC (α) (in units of $2t$) when RSOC is present in N. (b) The variation of conductance peak, G_{max} (in units of $\frac{e^2}{h}$) as the function of strengths of RSOC (α) (in units of $2t$) when RSOC is present in N.

variation of the conductance (G) as a function of energy (E) for different strengths of RSOC is shown in Fig. (1a). Fig. (1a) shows the signature of an Andreev peak at $E \leq \Delta$ and the enhancement of conductance at biasing energy lower the superconducting gap amplitude with increasing α

In Fig. (1b) we have plotted the height of conductance peak (G_{max}) as the function of strengths of RSOC (α). From Fig. (1b) it is well understood that the magnitude of the Andreev peak increases with α. We can understand this results as follows. With increasing strength of α, the momentum of the electrons (Eq. (5)) in the N side decreases while the momentum of the holes (Eq. (6)) in the N side increases, since the energy, ϵ_k decreases (see Eq. (7)). So we can conclude that the term $d(E)$ which is responsible for Andreev reflection in Eq. (4) increases with α. Thus RSOC helps in enhancing Andreev reflection.

Next we consider the presence of RSOC both in the metallic and superconducting regimes. Here also the Andreev reflection which occurs at $E < \Delta$ increases with α. Fig. (2a) and Fig. (2b) shows the variation of conductance (G) as the function of biasing energy (E) for different strengths of RSOC and the variation of conductance peak (G_{max}) as the function of α when RSOC is present in both N and S leads. Similar to Fig. (1b), Fig. (2b) also reveals that the Andreev peak is responsive to the RSOC and it increases with the increasing strength of RSOC. The same argument as given earlier applies here.

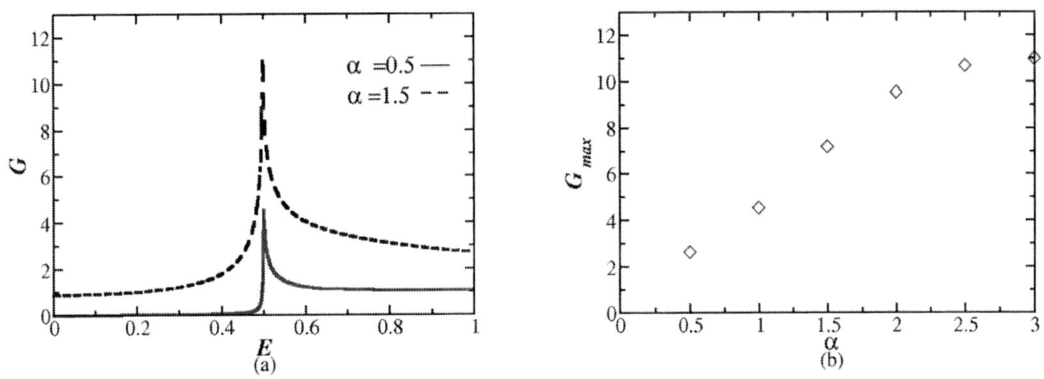

Figure 2. Red (solid) line denotes conductance for $\alpha = 0.5$ where the black (dashed) line denotes conductance for $\alpha = 1.5$. (a) The variation of conductance, G (in units of $\frac{e^2}{h}$) as a function of biasing energy, E (in units of t) for different strengths of RSOC (α) (in units of $2t$) when RSOC is present in both N and S leads. (b) The variation of conductance peak, G_{max} (in units of $\frac{e^2}{h}$) as the function of strengths of RSOC (α) (in units of $2t$) when RSOC is present in both N and S leads.

4. Effect of an in-plane magnetic field

To realize a zero mode state, which are the much talked about Majorana states, we have included an in-plane magnetic field. We have considered two different situations, the magnetic field (B) is present throughout the sample and the magnetic field is present only in the superconducting lead. The RSOC is present everywhere in the sample. Since our focus is to clarify the presence of zero mode states, we have used two representative values for B, namely $B = 1, 3$ in units of t. For both two cases we have found conductance peak at zero biasing energy which points towards the presence of Majorana zero mode states. Fig. (3) shows the variation of conductance as function of biasing energy when in-plane magnetic field and RSOC both are present in throughout the sample

Fig. (4) shows the variation of conductance as function of biasing energy when in-plane magnetic field is present in S. Both figures reveal the distinct presence of a zero mode state.

The presence of the zero mode peak in the conductance data is shown clearly in the inset. Thus the Andreev peak is shifted from $E = \Delta$ to $E = 0$. The above results should be very interesting to the conductance experiments for a $N - S$ junction. The robustness of the zero mode peak can be tested in presence of impurity, disorder or other kind of perturbations.

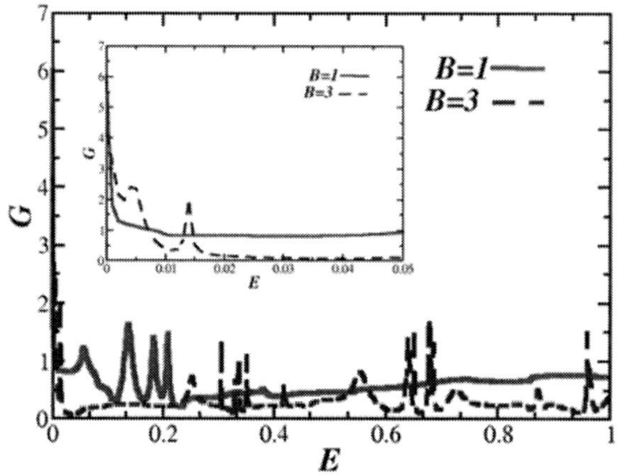

Figure 3. Red (solid) line denotes conductance for $B = 1$ where the blue (dashed) line denotes conductance for $B = 3$. The variation of conductance, G (in units of $\frac{e^2}{h}$) as a function of energy, E (in units of t) for different strengths of in-plane magnetic field (B)(in units of t) when magnetic field and RSOC both are present in both leads.

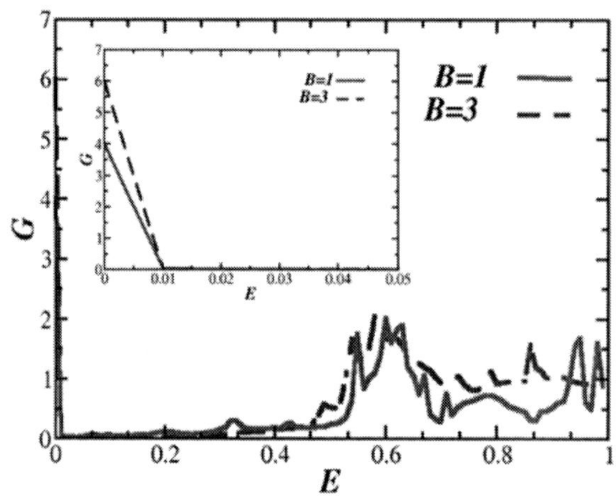

Figure 4. Red (solid) line denotes conductance for $B = 1$ where the blue (dashed) line denotes conductance for $B = 3$. The variation of conductance, G (in units of $\frac{e^2}{h}$) as a function of energy, E (in units of t) for different strengths of in-plane magnetic field, B (in units of t) when magnetic field is present in S lead but RSOC is present in both leads.

Journal of Physics: Conference Series **759** (2016) 012031 doi:10.1088/1742-6596/759/1/012031

5. Conclusions

We conclude saying that the Andreev peak is sensitive to the strength of RSOC. Further, the conductance increases with increasing RSOC at lower biasing energies. Also there are exciting possibilities of realizing zero mode state in presence of in-plane magnetic field in the conduction characteristics of a $N - S$ junction.

6. References

[1] A. R. Akhmerov and C. W. J. Beenakker, *Phys. Rev. B* **75**, 045426 (2007).
[2] Q. Liang, Y. Yu, Q. Wang and J. Dong, *Phys. Rev. Lett.* **101**, 187002 (2008).
[3] S. Bhattacharjee and K. Sengupta, *Phys. Rev. Lett.* **97**, 217001 (2006).
[4] A. F. Andreev and T. M. Rice, *Sov. Phys. JETP* **19**, 1228 (1964).
[5] E. I. Rashba, *Fiz. Tverd. Tela (Leningrad)* **2**, 1224 (1960).
[6] J. Alicea, *Rep. Prog. Phys.* **75**, 076501 (2012).
[7] C. W. J. Beenakker, *Ann. Rev. Cond. Mat. Phys.* **4**, 113 (2013).
[8] T. D. Stanescu and S. Tewari, *J. Phys.: Condens. Matter* **25**, 233201 (2013).
[9] S. R. Elliott and M. Franz, *"Colloquium: Majorana fermions in nuclear, particle and solid-state physics"* (2014).
[10] N. Mohanto an A Taraphder, *Euro. Phys. Lett.* **108**, 6001 (2014)
[11] M. Leijnse and K. Flensberg, *Semicond. Sci. Technol* **27**, 124003 (2012).
[12] G. E. Blonder, M. Tinkham, and T. M. Klapwijk, *Phys. Rev. B* **25**, 4515 (1982).
[13] G. E. Blonder, and M. Tinkham, *Phys. Rev. B* **27**, 112 (1983).
[14] C. W. Groth, M. Wimmer, A. R. Akhmerov, X. Waintal, *New J. Phys.* **16**, 063065 (2014).

Effect of Strong Orbital Magnetic Field on the Exciton Condensation in an Extended Falicov Kimball Model

S Pradhan[1] and A Taraphder[1,2]

[1]Department of Physics, Indian Institute of Technology Kharagpur, W. B. - 721302, India

[2]Center for Theoretical Studies, Indian Institute of Technology Kharagpur, W. B. - 721302, India

E-mail: spradhan@phy.iitkgp.ernet.in

Abstract. A spinless, extended Falicov-Kimball model in the presence of a perpendicular magnetic field is investigated employing a self-consistent mean-field theory in two dimensions. In the presence of the field the excitonic average $\Delta =< {d_i}^{\dagger} f_i >$ is modified: the exciton responds in subtle different ways for different values of the magnetic flux. We examine the effects of Coulomb interaction and hybridization between the localized and itinerant electrons on the excitonic average, for rational values of the applied magnetic field. The excitonic average is found to get enhanced exponentially with the Coulomb interaction while it saturates at large hybridization. The orbital magnetic field suppresses the excitonic average in general, though a strong commensurability effect of the magnetic flux on the behaviour of the excitonic order parameter is observed.

1. Introduction

The Falicov-Kimball model (FKM), involving a dispersive d-electron band, an atomic-like f-electron state and an on-site Coulomb interaction U between them, is perhaps the simplest lattice model to study many-body effects. Since its introduction in 1969 [1], to describe valence or metal-insulator transition in some transition metal oxides, the model has been successful in describing several correlation effects [2], e.g., metal-insulator transition [3], mixed-valence [4], the formation of ionic crystals [5, 6] and orbital [7] and charge-density waves (CDW)[8]. Brandt and Schmidt [9] and Kennedy and Lieb [8] showed that on a bipartite lattice at half-filling ($n_d = n_f = 0.5$), the f-electrons order in the chequerboard pattern. Portengen *et al.* [10, 11] introduced a k-dependent hybridization between d and f-bands and the resulting bound state between d-electrons and f-holes underwent a Bose Einstein condensation (BEC) of $d-f$ 'excitons'. As d and f-states differ by an odd parity, this could lead to an 'electronic ferroelectricity'. In the weak-coupling mean-field theory, the formation of an order parameter and the condensation thereof are concomitant. The Coulomb interaction U gives rise to a nonvanishing $d-f$ coherence $< {d_i}^{\dagger} f_i > \neq 0$ even in the limit of $V \rightarrow 0$ [10]

in the presence of a putative homogeneous ground state under the HF approximation. Czycholl [12] showed later that $< d_i^\dagger f_i > \to 0$ as $V \to 0$, i.e., there is no spontaneous symmetry breaking, consistent (but not contradictory at $T = 0$ though, where Elitzur's theorem does not forbid an order) with the local $U(1)$ gauge symmetry at $V = 0$. For a small non-zero V, the inhomogeneous (CDW) phase is stable, and the EOP is finite. Similar conclusions were then reached for a triangular lattice as well [13]. Indeed the Coulomb interaction between the conduction band electron and the valence band hole causes the formation of an excitonic bound state in some materials.

Recently, there is report of possible transition to an excitonic insulating state in the quasi-1D Ta_2NiSe_5 due to BEC of these bound states [14]. A moot question is what happens to such an excitonic condensate in the presence of orbital field. This kind of gauge-field can be experimentally realised in ultracold particles (fermions and bosons) in optical lattices. Moreover, of late, there are proposals for the realization of FKM in optical lattices with mixtures of light atoms in the correlated, disordered environment of heavy atoms [15, 16]. Therefore, it is quite pertinent to appraise an extended Falicov-Kimball model (EFKM) in optical lattices in the presence of artificial gauge fields. In a spinful situation, strong magnetic field may cause the two Zeeman-split bands to move sufficiently apart, and at low filling only the lower band is relevant, effectively reducing it to a spinless problem. We implement a self-consistent mean-field calculation to study the effect of a perpendicular magnetic field in the spinless model. We first examine the case without a magnetic field and then study the effect of the field on the exciton condensation. Results for commensurate and incommensurate magnetic fluxes are discussed.

2. An extended Falicov Kimball model

We consider an EFKM, a model system with spinless electrons arising from d and f orbitals on every site on a square lattice

$$H = -\sum_{<ij>} (t_{ij} d_i^\dagger d_j + h.c.) + U \sum_i d_i^\dagger d_i f_i^\dagger f_i + E_f \sum_i f_i^\dagger f_i + \sum_i V(d_i^\dagger f_i + h.c.) \quad (1)$$

where $< i,j >$ are nearest-neighbour site indices on a square lattice (lattice constant $a = 1$), $d_i\,(f_i)$ are itinerant (localized) electron annihilation operators at site i. The first term represents the kinetic energy of d-electrons while the second term represents on-site Coulomb interaction between d (density $n_d = \frac{1}{N}\sum_i d_i^\dagger d_i$) and f-electrons (at E_f, with density $n_f = \frac{1}{N}\sum_i f_i^\dagger f_i$; N being the number of sites). The fourth term stands for the hybridization between them. We set the hopping integral t to 1 and all energies are defined in units of t. For $V = 0$, the Hamiltonian commutes with $\hat{n}_{f,i}$, in which case $n_{f,i}$ is a good quantum number. It can be 'solved' numerically by annealing over the f-electron positions. For $V \neq 0$ the local $U(1)$ gauge symmetry is removed and the Hamiltonian is no longer exactly solvable, albeit in the above sense. We, therefore, take recourse to the usual self-consistent mean-field approximation to obtain the excitonic average $\Delta =< d_i^\dagger f_i >$. The inhomogeneous mean-field theory discussed below allows for local excitonic order parameter (EOP) and involves diagonalizing the mean-field Hamiltonian matrix followed by calculation of the EOP. The new EOP's are then fed back in to the Hamiltoanian and the process continues till convergence (up to $10^{-7}\%$) in all the EOP's is reached.

3. Effect of perpendicular magnetic field

For $B \neq 0$, the spinless, mobile fermions 'see' the field via the canonically conjugate momentum. The field then couples to the 'orbital degrees' only. With the choice of Landau gauge $\vec{A}(r) = B(0, x, 0)$ for a uniform magnetic field B perpendicular to the plane of the lattice, the hopping integral does not change in x-direction while along the y-direction it gets a 'Peierls phase' $t_{ij} = -t\exp(\pm ie/\hbar \int_j^i A(\vec{r})d\vec{r}) = -t\exp(\pm 2\pi im\frac{\phi}{\phi_0})$. Here, $\phi = Ba^2$ is the flux per plaquette of a square lattice. This represents the gain of phase by an electron, hopping along a closed path around the plaquette. We consider only rational magnetic flux, i.e., $\phi = \frac{p}{q}\phi_0 = \alpha\phi_0$ with p, q co-prime integers and ϕ_0, the Dirac flux quantum. As is customary, the lattice is discretized by $(x, y) = (ma, nb)$ - each site is then indexed by two integers "(m, n)" along (x, y)-directions (m, n now stand for the x, y coordinates in this discrete geometry) respectively. Peierls phases leave the Hamiltonian invariant only for such translations that are in the magnetic translation group [17]. This group is associated with a magnetic unit cell q times larger than the original unit cell, so as to enfold an integer flux $p\phi_0$. With the choice of a Landau gauge, the hopping in y-direction is associated with a phase which again depends on the x-component of the position vector "m". Therefore, to accommodate a magnetic flux $B = \frac{2\pi}{L}$, the resultant magnetic supercell will have to be a strip of length L [18] (we use maximum $L = 36$). We work in the half-filled limit ($n_f + n_d = 1$), in the particle-hole symmetric case at zero temperature. In the non-interacting limit, the problem reduces to the original Hofstadter model (as the f-electrons do not enter in the problem now) and shows the well-known energy spectrum: a self-similar structure in which widths and gaps open and close depending on the values of the magnetic flux [19, 20].

4. Results and discussions

To check our numerical procedure we start with the extended ($V \neq 0$) FKM in the symmetric case ($E_f = 0$ and $n_d = n_f = 0.5$) without the transverse field. We study the effect of Coulomb interaction on the stability of excitons. It is evident from Fig. 1 that $\Delta = 0$ in the $V \to 0$ as expected on symmetry grounds. For a finite V, there is a nonvanishing Δ that is strongly enhanced as U increases. As we increase hybridization between d and f electrons, for a fixed Coulomb correlation, we find an enhancement in the EOP. In the $V \to 0$ limit we find a CDW ground state, which melts on increasing V. Moreover, we study the effect of Coulomb interaction: the effect of U is more for small-V regime: the EOP is exponentially enhanced with U as expected. These results are in complete agreement with previous results in a wide parameter regime [12, 13].

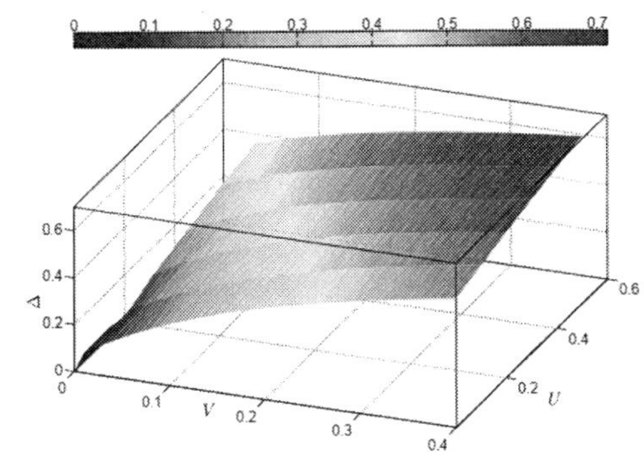

Figure 1. The mean-field behaviour of $\Delta = < {d_i}^\dagger f_i >$ in the UV plane in the zero-field limit.

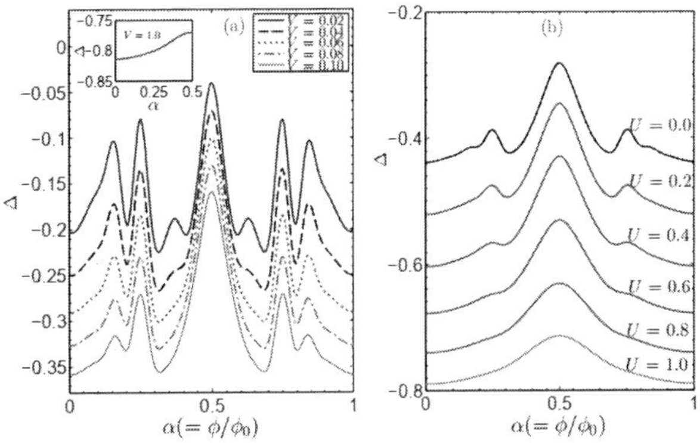

Figure 2.(a) Δ versus α for different V (inset: $V = 0.1$) at constant $U = 0.2$. (b) EOP with α for different U at fixed $V = 0.20$.

The effect of orbital magnetic field on the evolution of excitons is a question we address next. With flux $\alpha \neq 0$ per plaquette, the hopping term is associated with a phase which affects the exciton formation. In Fig. 2, the EOP is suppressed by the field (note the negative sign of the EOP, so it increases downwards) and a prominent peak at $\alpha = 0.25$ and a dip at $\alpha = 1/3$ can be seen for $U = 0$. There is a large peak at $\alpha = 0.5$; signifying the value of α at which Δ is a minimum: same is the case for all the U. The large oscillations of Δ reduce with larger U. As the Hamiltonian is symmetric, between α and $1 \pm \alpha$, Fig. 2 is symmetric about $\alpha = 0.5$ and displays typical Hofstadter characteristics like the non-interacting limit. The variation of Δ with V more or less follows the same pattern as in the absence of magnetic field (see Fig. 3(a)). However, for a fixed U, the general tendency is that the excitonic average reduces with increase in magnetic filed. However, it is also possible to get an enhancement in the excitonic average in a region where hybridization and correlation effects are quite weak (see Fig. 3(b)) for special (here 1/3) values of the flux. This is likely to be related to the extra stabilization in the band energy at special flux values, also observed in the Hofstadter spectrum [19].

In the presence of field, there is serious commensurability effect on the excitonic average. For certain values of $d - f$ correlation and hybridization, the system is essentially in the homogeneous state. From Fig. 4, we see that for some magnetic flux values, the magnetic unit cell is commensurate with the lattice size (in the present study on a 24×24 lattice, $1/24, 2/24, 3/24..12/24$ and so on, represent commensurate fluxes) and the excitonic average is uniform throughout the lattice. With increase in the value of the commensurate flux, magnitude of Δ decreases and the EOP is uniform. On the other hand, the EOP varies in a quite different way when the magnetic flux is incommensurate. In this case, the excitonic average exhibits a one-dimensional modulation and the modulation length changes with the flux (Fig. 4(b), (c)).

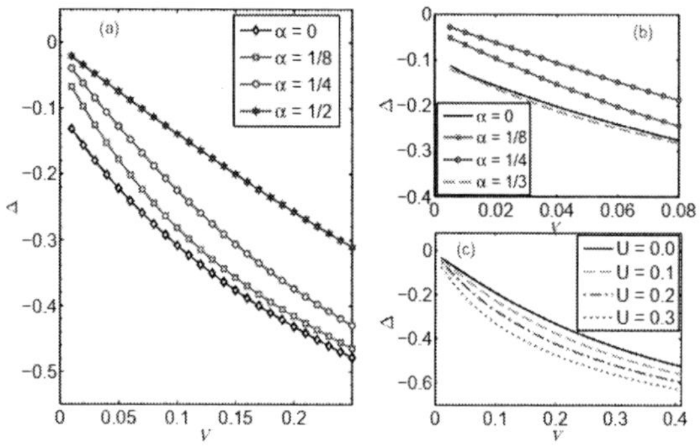

Figure 3. (a) EOP with V for several α at $U = 0.20$. (b) $\alpha = 1/3$ (dashed line). (c) Δ vs V at different U with $\alpha = 0.25$.

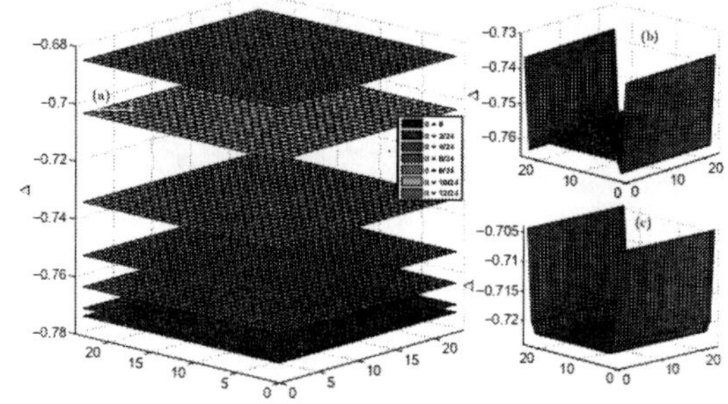

Figure 4. The EOP on a 2D lattice plane for α (a) commensurate (b) 0.1, (c) 0.37. At $U = 1.0$ and $V = 0.2$.

5. Summary and conclusion

We observe the competition between magnetic field and Coulomb correlation in a prototype correlated lattice model. The Coulomb interaction exponentially enhances the excitonic average while the orbital magnetic field has a localizing effect on mobile d-electrons, affecting the excitonic coherence: a drop in the value of excitonic average (hence electronic ferro-electricity) with both commensurate and incommensurate flux. In the small V limit, there is a special value ($\alpha = 1/3$) of the field which gives extra stability to the exciton. These observations open up the possibility of tuning ferro-electricity via applied magnetic field.

6. References

[1] Falicov L M and Kimball J C 1969 *Phys. Rev. Lett.* **22** 997
[2] Freericks J K and Zlatić V 2003 *Rev. Mod. Phys.* **75** 1333
[3] Freericks J K and Falicov L M 1990 *Phys. Rev.* B **5** 2163
[4] Remirez R and Falicov L M 1971 *Phys Rev.* B **3** 2425
[5] Gruber C, Lebowitz J L and Macris N 1993 *Phys. Rev.* B **48** 4312
[6] Lemanski R, Freericks J K and Banach G 2004 *J. Stat. Phys.* **116** 699
[7] Yadav U K, Maitra T, Singh I, Taraphder A 2010 *J. Phys Condens Matter* **22** 295602
[8] Kennedy T and Lieb E H 1986 *Physica* A **138** 320
[9] Brandt U and Schmidt R 1986 *Z. Phys.* B **63** 45
[10] Portengen T, Östreich T, and Sham L J 1986 *Phys. Rev. Lett.* **76** 3384
[11] Portengen T, Östreich T, and Sham L J 1996 *Phys. Rev.* B **54** 17452
[12] Czycholl G 1999 *Phys. Rev.* B **59** 2642
[13] Yadav U K, Maitra T, Singh I and Taraphder A 2011 *Euro Phys Lett.* **93** 47013
[14] Seki K et al 2014 *Phys. Rev.* B **90** 155116
[15] Dalibard J, Gerbier F, Gediminas J and Patrik O 2011 *Rev. Mod. Phys.* **83** 1523
[16] Ates C and Ziegler K 2005 *Phys. Rev.* A **71** 063610
[17] Zak J 1964 *Phys. Rev.* A **136** 776
[18] Pradhan S, Chakraborty M and Taraphder A 2016 Phys. Rev. B **93** 115109
[19] Hofstadter D R 1976 *Phys. Rev.* B **14** 2239
[20] Pradhan S 2016 arXiv: 1606.06009

Study of 4f electron based compound $Sm_{1-x}Gd_xAlO_3$, a thermoelectric (TE) material: FP-LAPW method

Sandeep*[1], D P Rai[2], A Shankar[3], M P Ghimire[4] and R K Thapa[1]

[1] Department of Physics, Mizoram University, Aizawl-796001, Mizoram, India
[2] Department of Physics, Pachhunga University College, Aizawl-796001, India
[3] Department of Physics, University of North Bengal, Darjeeling -734013, India
[4] Condensed Matter Physics Research Center, Butwal, Rupandehi, Nepal

E-mail: *sndp.chettri@gmail.com

Abstract. Gadolinium doping effects on the thermoelectric properties in $Sm_{1-x}Gd_xAlO_3$ (x=0%, 50%, and 100%) were studied using first-principles calculations based thermal transport property measurement. The result indicates that the compound is an intrinsic n-type material. Samarium doping has a positive effect on the overall thermoelectric performance of the $Sm_{1-x}Gd_xAlO_3$ system, with a sharp increase in the figure of merit (ZT) observed when x=0, 25, 50, 75 and 100% up to 800K. Compared to x=0 and 100%, the case of x=50% was found to have a more positive increment in ZT value suggesting that the doping to have a positive effect on the figure of merit in $Sm_{1-x}Gd_xAlO_3$. Furthermore, all the samples show stable thermoelectric compatibility factors over a broad temperature range from 700 to 1000 K, which could have great benefits for their practical applications. It is concluded that the overall thermoelectric performance of the $Sm_{1-x}Gd_xAlO_3$ could be highly enhanced using doping techniques.

1. Introduction

Thermoelectric compounds have drawn renewed interest in the recent past due to their wide applicability in the field of the device and high-temperature components [1]. The perovskites have been known for their potential applications as electronic and thermoelectric devices [2-3] due to the rich exhibition of phase transformation with temperature and doping [4]. $RAlO_3$ (R=Eu) have been studied using mBJ and GGA+U based approximation to compute the electronic structures and were found to have a high density of states at EF supporting its candidature as thermoelectric device fabrication [5]. Thermoelectric power and electrical conductivity of $LaAl_{0.95}Mg_{0.05}O_{3-\delta}$ and $ErAl_{0.95}Mg_{0.05}O_{3-\delta}$ have been studied as a function of oxygen partial pressure [6]. The isobaric heat capacity of $SmAlO_3$ in the range 0-300 K has been measured with adiabatic calorimetry and its dependence on the electronic structures were reported with emphasis on $4f$ state electrons [7]. Thermo-electric efficient materials in pure or doped form have attracted the researchers in the last few decades, because of their striking features like easy synthesis, good conductance, price and stability [8]. Wilfried et al. studied the large Seebeck coefficient and its dependence on effective electron mass and suggested doping and increase of lattice constant [9]. In this present paper we present various doped concentrations of Gd on $SmAlO_3$ and the increase in the thermoelectric efficiency or figure of merit (ZT) as a function of temperature.

2. Methodology

The crystal structure of $Sm_{1-x}Gd_xAlO_3$ for x=0%, x=25%, x=50%, x=75% and x=100% are shown in Fig 1 (a-e).

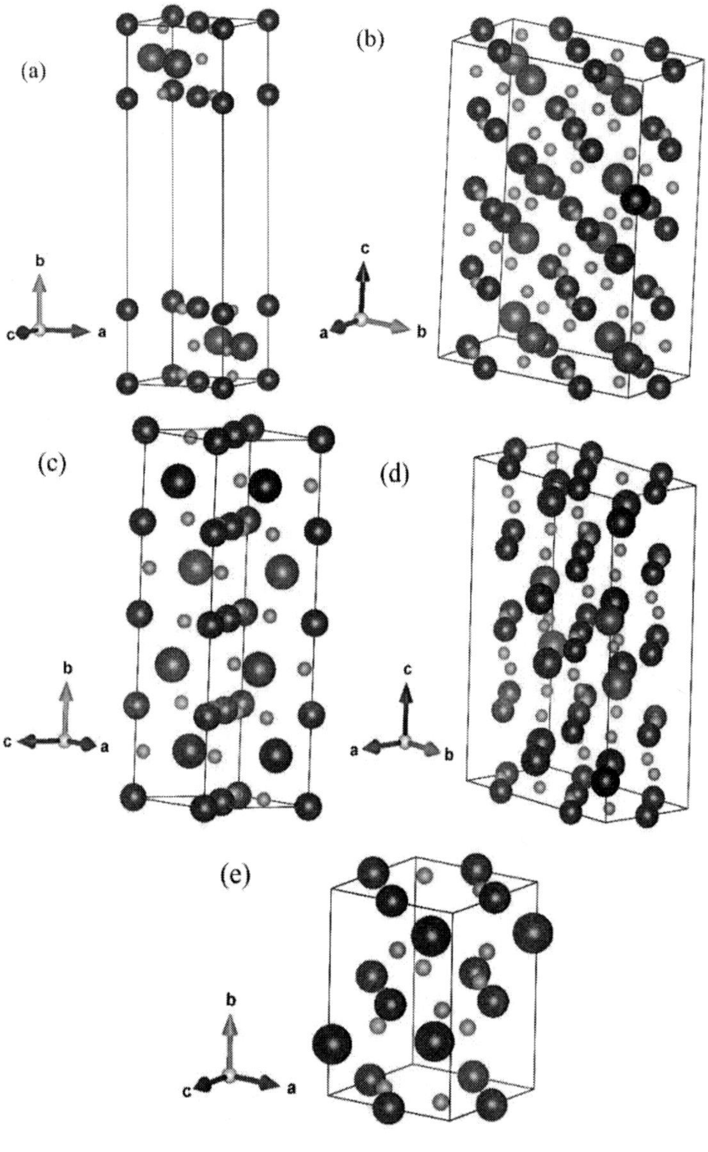

Figure 1. Crystal structure of $Sm_{1-x}Gd_xAlO_3$ for (a) x=0%, (b) x=25%, (c) x=50%, (d) x=75% and (e) x=100% Red=Sm, Black=Gd, Blue= Al, Green=O

Seebeck coefficient (S), thermal conductivity (κ) and electrical conductivities (σ) for an isotropic electronic structure using BoltzTraP code [10]. Eq. (1) interprets electrical conductivity tensors [11]

$$\sigma_{\alpha,\beta} = e^2 \sum_{i,k} \left[\frac{-\partial f_0(T,\varepsilon,\mu)}{\partial \varepsilon} \right] v_\alpha v_\beta \tau_k \qquad (1)$$

where, α, β are the tensor indices, v_α, v_β are the group velocities, *e* is the electron charge and τ_k is the relaxation time. The relaxation time is neglected for isotropic systems as the variation of is not known.

The electrons contribution remain near the chemical potential (μ) in a narrow range of μ-k_BT < ε < μ+k_BT, where k_B is the Boltzmann constant [12]. The transport distribution is written as [13]

$$\Xi_{i,k} = e^2 \sum_{i,k} v_\alpha v_\beta \tau_k \qquad (2)$$

which is the kernel of all transport coefficients. From the rigid band approach, the electrical conductivity, thermal conductivity and Seebeck coefficient can be written as a function of temperature (T) and chemical potential (μ) by integrating the transport distribution [10, 14]

$$\sigma = \frac{1}{\Omega}\int \Xi_{i,k}\left[\frac{-\partial f_o(T,\varepsilon)}{\partial\varepsilon}\right]d\varepsilon \qquad (3)$$

$$\kappa = \frac{1}{e^2}k_B^2 T\int \Xi_{i,k}\left(\frac{\varepsilon-\mu}{k_BT}\right)^2\left[\frac{-\partial f_0(T,\varepsilon)}{\partial\varepsilon}\right]d\varepsilon \qquad (4)$$

$$v_{\alpha\beta} = \frac{k_B}{e\Omega}\int \Xi_{i,k}\left(\frac{\varepsilon-\mu}{k_BT}\right)\left[\frac{-\partial f_0(T,\varepsilon)}{\partial\varepsilon}\right]d\varepsilon \qquad (5)$$

here f_0 is a Fermi-Dirac distribution function and is the volume of a unit cell. The Seebeck coefficient (S) is determined as

$$S_{ij} = E_i(\nabla_j T)^{-1} = (\sigma^{-1})\alpha_i v_{\alpha j} \qquad (6)$$

The thermoelectric efficiency is presented as

$$ZT = \frac{S^2\sigma T}{\kappa} \qquad (7)$$

Thus the efficiency of thermoelectric materials (ZT) can be tuned by optimizing any of the three parameters S, σ and κ at defined temperature T.

3. Results and Discussions

The negative value of S indicates n-type carriers (Fig. 2a). Since S depends on temperature, the sharp decrease of S at lowest temperature is due to the depopulation of phonon modes. Strongly correlated heavy Fermion compounds of Ce and Yb also show a similar type of low-temperature phenomena [15].

The value of S at room temperature was found to be -2.19 ×10-4 μV/K. The figure of merits is shown in Fig 2(b) for five different doping concentrations x=0%, 25%, 50%, 75% and 100% respectively. The ZT values for different doping states were of similar nature. They showed sharp increment with a temperature below 150 K. Compared to x=0% and x=50% doping configuration, the other doping concentrations were found to have a more positive effect of the temperature on them. ZT as high as 0.75 at temperature near 300 K suggests that the presence of doped Gd in $SmAlO_3$ enhances the thermoelectric efficiency of the material and is also indication that a doping of such other concentrations may yield higher ZT values which may be useful for thermoelectric device fabrications. A more thorough and comparative study using *GGA+U* and *mBJ+U* functionals along with experimental results may lead to the realisation of a critical value of the doping concentration.

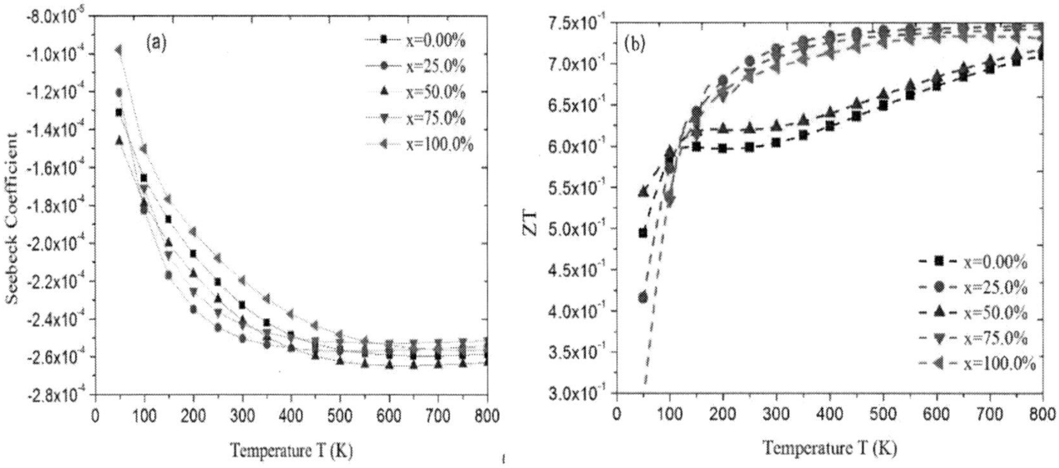

Figure 2. Temperature dependence of (a) Seebeck coefficient (S) and (b) figure of merit (ZT) for $Sm_{1-x}Gd_xAlO_3$ for x=0%, x=25%, x=50%, x=75% and x=100%

4. Acknowledgements

SD acknowledges a research grant from DST-SERB, Dy No. SERB/3586/2013-14 dated on 6.09.2013, New Delhi, India. DPR acknowledges a Start-Up Grant from UGCBSR, FRPS, Lett.No.F.30-52/2014 RKT a research grant from UGC (New Delhi, India).

5. References

[1] Di Salvo F J 1999 *Science* **285** 703
[2] Sandeep, Rai D P, Shankar A, Ghimire M P, Khenata R and Thapa R K 2016 *Mod. Phys. Lett. B* **30** 1650028
[3] Hashimoto H, Kusunose T and Sekino T 2011 *Journal of Ceramic Processing Research* **12** 223
[4] Karvonen L, Tomes P and Weidenka A 2011 *Material Matters* **6** 9299
[5] Sandeep, Rai D P, Shankar A, Ghimire M P, Khenata R and Thapa R K 2015 Phys. Scr. **90** 065803
[6] Ahlgren E O, Ranlov J and Poulsen F W 1995 *J Electrochem Soc.* **142** 4230
[7] Kopan A R, Gorbachuk M P, Lakiza S M and Tishchenko Ya S 2012 *Powder Metallurgy and Metal Ceramics* **51** 209
[8] Hochbaum A I, Chen R, Delgado R D, Liang W, Garnett E C, Najarian M, Majumdar A and Yang P 2008 *Nature* **451** 163
[9] Wunderlich W, Ohta S, Ohta H, Koumoto K 2005 *Proc. Of 24th International Conference On Thermoelectrics, (ICT 2005 IEEE Proceedings)* p. 237
[10] Madsen G H K, Singh D 2006 *Comput. Phys. Commun.* **175** 67
[11] Yamamoto A, Miyazaki H, Takeuchi T 2014 *J. Appl. Phys.* **115** 023708
[12] Takeuchi T 2009 *Materials Transactions* **50** 2359
[13] Mahan G D and Sofo J O 1996 *Proc. Natl. Acad. Sci. USA* **93** 7436
[14] Scheidementel T J, Draxl C A, Thoonhauser T, Badding J V, Sofo J O 2003 *Phys. Rev. B* **68** 1252101
[15] Dilley N R, Bauer E D, Maple M B, Dordevic S, Basov D N, Freibert F, Darling T W, Migliori A, Chakoumakos B C, Sales B C 2000 *Phys. Rev. B* **61** 4608

Three body interaction effects on the phase diagram of spinor bosons

Sk Noor Nabi[1] **and Saurabh Basu**[2]
Department of Physics, Indian Institute of Technology Guwahati, Guwahati, Assam 781039, India
E-mail: [1]sk.noor@iitg.ernet.in, [2]saurabh@iitg.ernet.in

Abstract. We include a three body density interaction in the Bose Hubbard model and study its effects on the phase diagram for spinor ($S = 1$) bosons on an optical lattice via a mean field theory. The Mott insulating (MI) phases are noted to stabilize, in the sense that the MI phases extend to larger values of the system parameters alongwith widening of the particle-hole excitation spectrum as the three body interaction term is included for both the polar (spin dependent interaction being positive) and the ferromagnetic (spin dependent interaction being negative) cases. Another remarkable feature emerges as the phase diagram corresponding to the ferromagnetic case becomes distinct from that of its spinless variant, which in the absence of the three body term is indistinguishable from that of the scalar particles. A strong coupling perturbation theory is employed to provide analytical support to the above results.

1. Introduction

The experimental observation of the superfluid (SF) to Mott insulator (MI) transition with ultracold atoms trapped in optical lattices gives a prominent signature of quantum phase transition [1]. The quantum fluctuations that arises due to Heisenberg's uncertainty principle plays a vital role to study the quantum many body phenomena in light of atomic physics and optics. With the technological revolution, atoms in optical trap preserve their hyperfine spin, F (^{87}Rb, ^{23}Na) and thereby known as spinor Bose gas unlike the magnetic trap which ceases the z component of spin degrees of freedom, F_z making them as scalar or spin-0 Bose gas [2]. Significant progress has been made to study the spinor Bose gas where the SF to MI phase transition gets modified due to the hyperfine spin being present and hence posseses much richer phase diagrams compared with the spin-0 counterpart [2–5].

Following the seminal paper by Fisher *et. al.* [6], the dynamics of cold atoms in optical lattices can be modeled by the Bose Hubbard model (BHM) where the superfluid to Mott insulator phase transition is controlled by a single parameter, namely the ratio of tunneling to onsite interaction energy [7]. Since then, theoretical efforts have been made to study the BHM in its different forms like including the spin-orbit coupling, nearest neighbour extended interaction, three and higher body interaction terms etc. Among them, the study of three body interaction on the BHM for both the scalar [8–11] and spinor [12–14] Bose gas has gained much momentum after its successful experimental observation by Will *et. al.* [15].

In this work, we will study the spinor BHM in presence of the three body repulsive interaction strengths through mean field approach (MFA) [16]. We observe the effect of three body interaction on the Mott insulating phase and thereby change in critical tunneling strength for the SF to MI transition. Finally we performed a perturbation calculation to provide a strong support against the mean field phase diagrams.

2. Spinor Bose Hubbard Hamiltonian

The Bose Hubbard Hamiltonian (BHM) for spin-1 ultracold atoms with three body interaction term is [9, 10, 13, 17],

$$\begin{aligned}\hat{H} = & - t\sum_{<i,j>}\sum_{\sigma}(\hat{a}^{\dagger}_{i\sigma}\hat{a}_{i\sigma} + h.c) - \mu\sum_i \hat{n}_i + \frac{U_0}{2}\sum_i \hat{n}_i(\hat{n}_i - 1) \\ & + \frac{W}{6}\sum_i \hat{n}_i(\hat{n}_i - 1)(\hat{n}_i - 2) + \frac{U_2}{2}\sum_i(\mathbf{S}_i^2 - 2\hat{n}_i)] \end{aligned} \quad (1)$$

where t is the tunneling matrix, $\langle i,j\rangle$ are the nearest neighbor sites, μ is the chemical potential, U_0 describes the spin independent and U_2 is the spin dependent on-site interaction. U_0 and U_2 related to the scattering lengths, a_0 and a_2 as $U_2/U_0 = (a_2-a_0)/(a_0+2a_2)$ corresponding to S=0 and S=2 channels and W is the strength of three body interaction term. The spin dependent interaction, U_2 is antiferromagnetic (AF) for $a_2 > a_0$ and ferromagnetic (F) for $a_2 < a_0$. The total spin at site i is given by, $\mathbf{S}_i = \hat{a}^{\dagger}_{i\sigma}\mathbf{F}_{\sigma\sigma'}\hat{a}_{i\sigma'}$ where $\mathbf{F}_{\sigma\sigma'}$ are the components of spin-1 matrices and σ =+1, 0, -1. The particle number operator, $\hat{n}_i = \sum_{\sigma}\hat{n}_{i\sigma}$, $\hat{n}_{i\sigma} = \hat{a}^{\dagger}_{i\sigma}\hat{a}_{i\sigma}$ where $\hat{a}^{\dagger}_{i\sigma}(\hat{a}_{i\sigma})$ is the boson creation (annihilation) operator at a site i. Applying mean field approximation [3, 16] to decouple the hopping part of the Hamiltonian, the BHM can be written as a sum of single site Hamiltonians as, $H = \sum_i H_i^{MF}$ where

$$\begin{aligned} H_i^{MF} = & \underbrace{-zt(\psi^*_{i\sigma}\hat{a}_{i\sigma} + h.c) + zt\sum_{i\sigma}\psi^2_{i\sigma}}_{\mathrm{V}} \\ - & \underbrace{\mu\hat{n}_i + \frac{U_0}{2}\hat{n}_i(\hat{n}_i - 1) + \frac{U_2}{2}(\mathbf{S}_i^2 - 2\hat{n}_i) + \frac{W}{6}\hat{n}_i(\hat{n}_i - 1)(\hat{n}_i - 2)}_{H^0} \end{aligned} \quad (2)$$

where V and H^0 are the perturbed and unperturbed Hamiltonian and the superfluid order parameter at a site i as $\psi_{i\sigma} \equiv \langle\hat{a}_{i\sigma}\rangle$ and the local density, ρ can be computed using, $\rho = \sum_{\sigma}\langle\psi^{eq}|\hat{n}_{\sigma}|\psi^{eq}\rangle$, where ψ^{eq} is the self consistently computed ground state.

3. Results

At first we consider the atomic limit $t = 0$ to see the effect of three body interaction. In the atomic limit $t = 0$, the Hamiltonian consist only the unperturbed term *i.e* H^0. Since $\mathbf{S}$ and $\hat{n}$ both commute with each other, they have a common eigenstate $|S, S_z, n\rangle$ and the eigenvalue E^0 is given by $E^0 = \frac{U_0}{2}n(n-1) + \frac{U_2}{2}[S(S+1) - 2n] + \frac{W}{6}n(n-1)(n-2) - \mu n$. In the atomic limit, the system consists of only the Mott insulating (MI) phase where there is a gap in the particle hole excitation spectrum. The energy gap, E_g for a particular filling of the MI lobe n corresponds to the upper and lower boundaries of the MI region as $E^0(S-1, n-1) < E^0(S, n) < E^0(S+1, n+1)$ [18].

In the AF case, this inequality leads to the following equations corresponding to the even MI lobes as $(n-1) - \frac{2U_2}{U_0} + \frac{W}{2U_0}(n-1)(n-2) < \mu/U_0 < n + \frac{W}{2U_0}n(n-1)$ and for odd MI lobes as $(n-1) + \frac{W}{2U_0}(n-1)(n-2) < \mu/U_0 < n - \frac{2U_2}{U_0} + \frac{W}{2U_0}n(n-1)$ and also in the ferromagnetic case, $(n-1)(1+\frac{U_2}{U_0}) + \frac{W}{2U_0}(n-1)(n-2) < \mu/U_0 < n(1+\frac{U_2}{U_0}) + \frac{W}{2U_0}n(n-1)$ and is shown in Fig.1. The chemical potential width, μ in both AF and F cases becomes wider due to W/U_0 which suggest the increase in energy gap E_g and hence the MI lobes becomes more stable compared to the

SF phase.

In the AF case, although the chemical potential width of first MI lobe ($n = 1$) remains unaffected but the higher lobes, like even MI phase with $n = 2$ becomes more broader compared to the previous one in absence of W/U_0 [Fig.1]. Interestingly the chemical potential width for the third MI lobe with $n = 3$ remarkably increases and it vanishes beyond $U_2/U_0 = 1$ value. This suggests that without W/U_0, all odd MI lobes ($n = 1, 3...$) vanishes beyond $U_2/U_0 = 0.5$. However for $W/U_0 \neq 0$, the critical value of U_2/U_0 for the disappearance for the first MI lobe ($n = 1$) is still at $U_2/U_0 = 0.5$, but the corresponding value changes to 1 from 0.5 for the next odd (that is third) MI lobes [Fig.1].

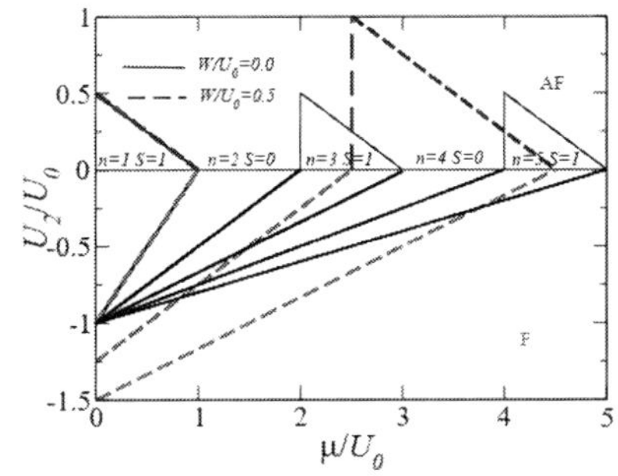

Figure 1. (Color online) The MI phase in atomic limit, $t = 0$ for $W/U_0 = 0.5$.

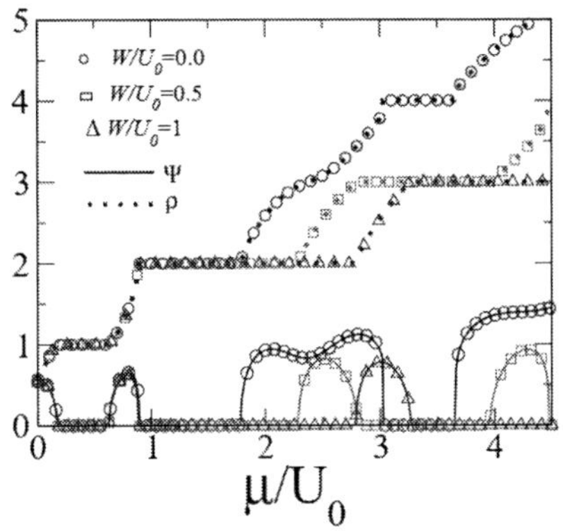

(a)

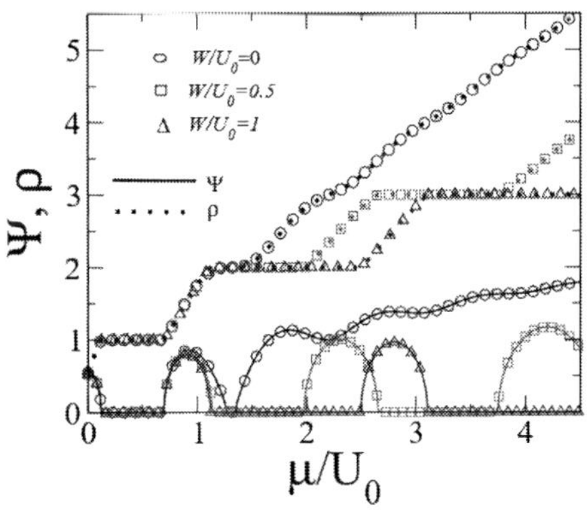

(b)

Figure 2. (Color online) The variation of SF order parameter ψ and occupation density ρ at W/U_0= 0.5,1 for $U_2/U_0 = 0.1$ (AF) in (a) and $U_2/U_0 = -0.2$ (F) in (b).

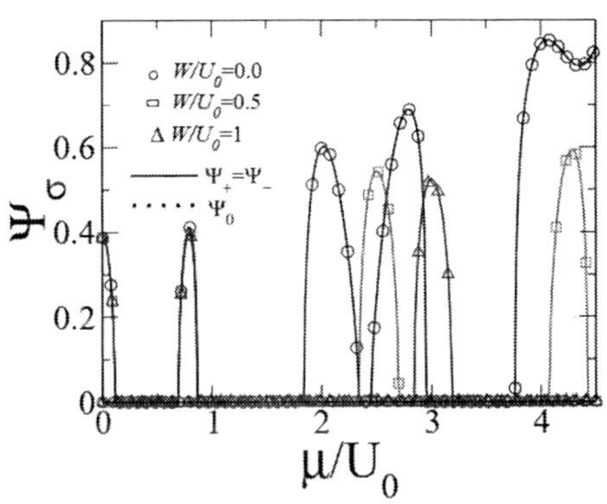

(a)

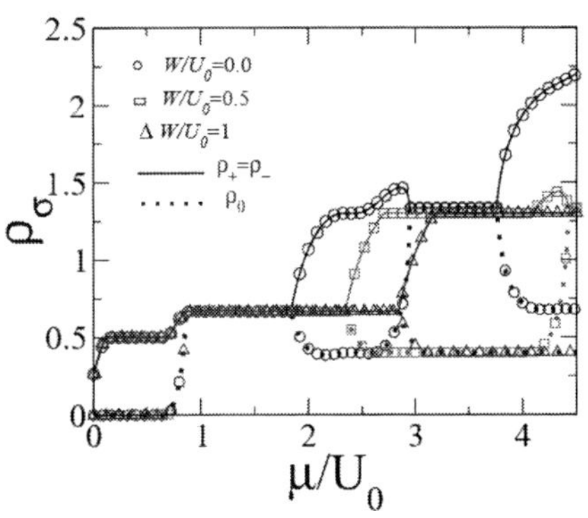

(b)

Figure 3. (Color online) The variation of spinor components ψ_σ (a) and occupation density ρ_σ (b) for $W/U_0 = 0.5$, 1 and $U_2/U_0 = 0.1$ (AF).

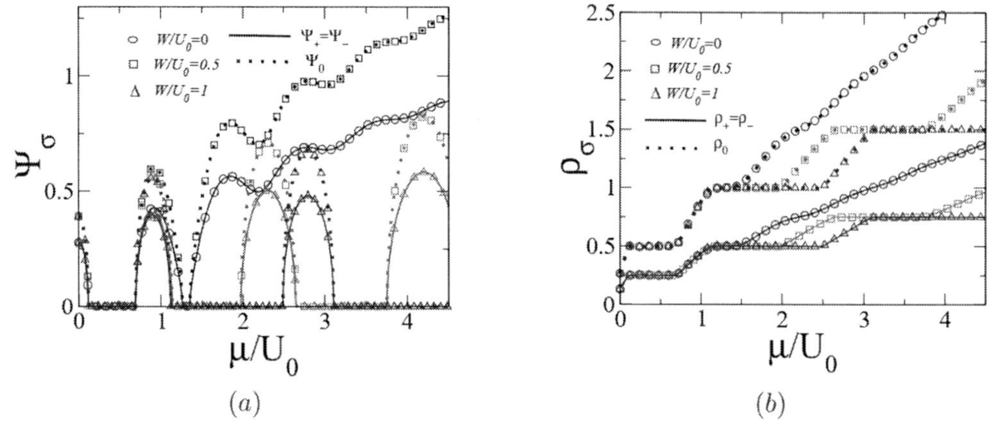

Figure 4. (Color online) The variation of spinor components ψ_σ (a) and occupation density ρ_σ (b) for $W/U_0 = 0.5$, 1 and $U_2/U_0 = -0.2$ (F).

Similarly in ferromagnetic case, the μ width for $n = 1$ MI lobe remains unaltered but the higher order MI lobes are affected by W/U_0 and the system more likely to be in the insulating regime than in the SF regime [Fig.1].

We now turn on the hopping term t and diagonalize the matrix $\langle \hat{n}'_\sigma | H^{MF} | \hat{n}_\sigma \rangle$ to obtain the ground state energy and equilibrium order parameter ψ^{eq}. In the homogeneous case, the Hamiltonian is site independent and we neglect the site index i to compute the SF order parameter ψ and local density ρ to study the MI to SF transition as a function of μ/U_0 in presence of W/U_0 [Fig.2]. The variation of ψ_σ and ρ_σ corresponding to both in antiferromagnetic (AF) and ferromagnetic (F) cases are shown in Figs.3 and 4.

The variation of ψ and ρ show that in the MI phase, ψ is 0 and $\rho =$ integer while in SF phase, ψ shows finite values and $\rho =$ non integer [Figs.2(a),(b)]. With the three body interaction, W/U_0, the MI-SF phase transition for the first MI lobe occurs at same value of μ/U_0 but for second and

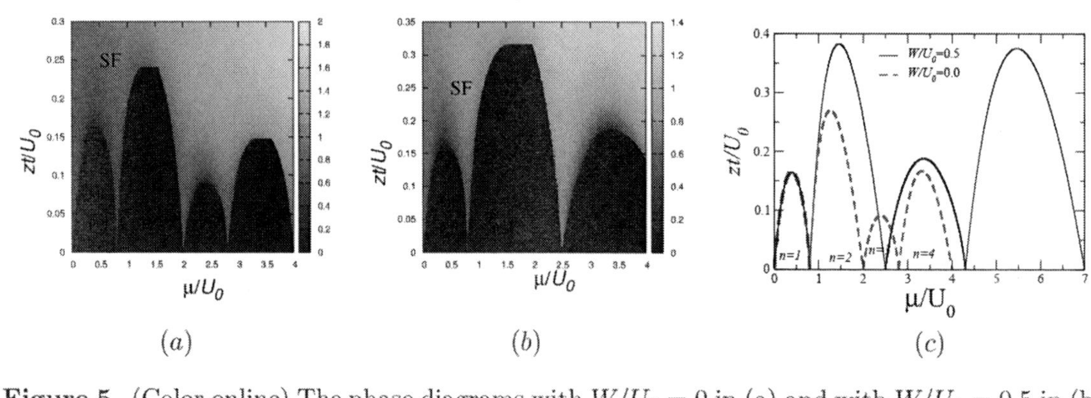

Figure 5. (Color online) The phase diagrams with $W/U_0 = 0$ in (a) and with $W/U_0 = 0.5$ in (b) for $U_2/U_0 = 0.1$ (AF). Phase diagram obtained from perturbation calculation for $W/U_0 = 0.5$ and $U_2/U_0 = 0.1$ in (c).

higher MI lobes, the SF order parameter and occupation density now show transition from zero and integer values to finite and non integer values at higher value of μ/U_0 for both in AF and F cases as seen from Figs.2(a),(b). We have also studied the SF order parameter components ψ_σ and ρ_σ as a function of W/U_0 for the antiferromagnetic [Fig.3(a),(b)] and ferromagnetic cases

[Fig.4(a),(b)]. We verify that $\psi_+ = \psi_- \neq 0$, $\psi_0 = 0$ and $\rho_+ = \rho_-$ in AF case and $\psi_+ = \psi_- \neq 0$, $\psi_0 = \sqrt{2}\psi_+$ in the ferromagnetic case which was studied earlier without W/U_0 in [3] and they show similar kind of behaviour as that of ψ and ρ with W/U_0. Also the critical tunneling strength zt_c/U_0 for MI-SF phase transition now occurs at higher value of zt/U_0 due to the three body interaction strength as this interaction forces the system towards the Mott insulating regime which can also be seen from the perturbation calculation in the following discussion.

In order to get a clear visualization of the MI-SF phase transition and the effect of W/U_0, we compute the phase diagrams based on the SF order parameter ψ [Figs.5 and 6]. The phase diagram without W/U_0 for AF case is in Fig.5(a)] and ferromagnetic case is shown in Fig.6(b). In the AF case, the even MI lobes with occupation densities $\rho = 2, 4...$ get enlarged compared

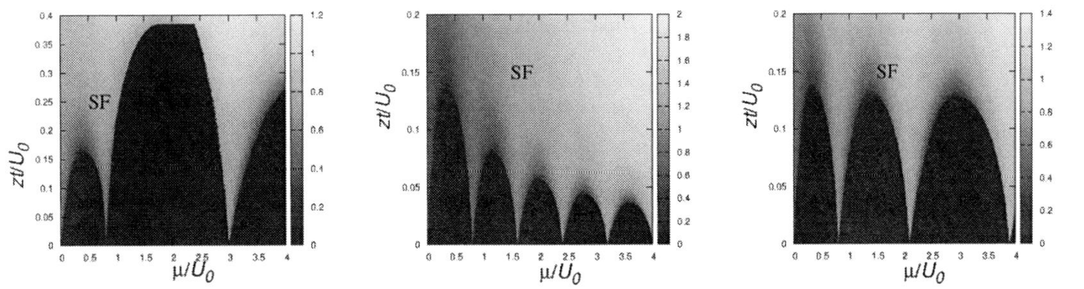

(a) (b) (c)

Figure 6. (Color online) The phase diagram with W/U_0=1 for $U_2/U_0 = 0.1$ (AF) in (a). Phase diagram without W/U_0 in (b) and with W/U_0=0.5 in (c) for $U_2/U_0 = -0.2$ (F).

to odd MI lobes ($\rho = 1, 3..$) due to formation of singlet pairs [3,4] [Fig.5(a)]. At $W/U_0 = 0.5$, all the MI lobes except the first one encroaches towards the SF regime [Figs.5(b) and 6(a),(c)] and the critical tunneling strength zt_c/U_0 for the MI-SF phase transition also increases with W/U_0.

In the AF case, the even MI lobes are found to be more stable than the odd MI lobes [Figs.5(b) and 6(a)] and the later also expanded compared with their previous counterparts. In the ferromagnetic case, the higher order MI lobes also show similar properties as that of W/U_0 in AF [Fig.6(c)].

We now include a second order perturbation calculation to provide a strong theoretical support for the mean field phase diagram presented following the approach developed in ref. [4]. As we have seen in atomic limit $t = 0$ [Fig.1], using the non degenerate perturbation theory, the second order correction for the even MI lobes in the AF case is given by, $E^2 = \sum_{q \neq |0,0,\hat{n}\rangle} \frac{|\langle q|V|0,0,\hat{n}\rangle|^2}{E^0_{0,0,\hat{n}} - E^0_q}$

Now calculating all the non zero elements of $\langle q|V|0,0,\hat{n}\rangle$, the ground state energy can be written in terms of series expansion of ψ as $E^n = E^0 + E^2\psi^2 + O(\psi^4)$. To find the tip of MI lobes, we minimize the ground state energy E^n with respect to ψ which leads to the following equation,

$$zt^{-1} = \frac{(n+3)/3}{nU_0 - \mu + \frac{W}{2}n(n-1)} + \frac{n/3}{\mu - U_0(n-1) + 2U_2 - \frac{W}{2}(n-1)(n-2)} \tag{3}$$

Similarly using degenerate perturbation theory in [4], for the odd MI lobes,

$$\begin{aligned} zt^{-1} &= \frac{(n+2)/3}{\mu - U_0(n-1) - \frac{W}{2}(n-1)(n-2)} + \frac{(n+1)/3}{-\mu + nU_0 - 2U_2 + \frac{W}{2}n(n-1)} \\ &+ \frac{4(n-1)/15}{\mu - U_0(n-1) + 3U_2 - \frac{W}{2}(n-1)(n-2)} + \frac{4(n+4)/15}{-\mu + nU_0 + U_2 + \frac{W}{2}n(n-1)} \end{aligned} \tag{4}$$

If we plot these two equations for $W/U_0 = 0.5$ and $U_2/U_0 = 0.1$, the obtained phase diagram is shown in Fig.5(c) and found that it is in complete agreement with the mean field phase diagram Fig.5(b). If we solve the above equations, which are quadratic in μ shows that the critical hopping strength, t_c/U_0 (by equating μ_+ and μ_-) for MI-SF phase transition increases with W/U_0.

4. Conclusion

We study the effect of three body repulsive interaction on spin-1 BHM through mean field analysis. At first, we first consider the atomic limit that is $t = 0$ to the Mott lobes both in antiferromagnetic (AF) and ferromagnetic (F) cases. All the Mott lobes except the first one gets extended and becomes more stable compared to the SF phase due to the presence of three body interaction strengths. We also compute the SF order parameter and local density to study the MI-SF phase transition with three body interaction strength and obtain the phase diagrams based on the SF order parameter both in AF and ferromagnetic cases. Finally a perturbation calculation is provided to support of the mean field phase diagrams.

Acknowledgments

SB thanks CSIR, India for financial support under the grants F no: 03(1213)/12/EMR-II.

References

[1] Greiner M, Mandal O, Esslinger T, Haensch T W and Bloch I, Nature(London), **415**, 39 (2002).
[2] Stamper-Kurn D M, Andrews M R, Chikkatur A P, Inouye S, Miesner H J, Stenger J, and Ketterle W, Phys. Rev. Lett. **80**, 2027 (1998).
[3] Pai R V, Sheshadri K and Pandit R, Phys. Rev. B, **77**, 014503 (2008).
[4] Tsuchiya S, Kurihara S, and Kimura T, Phys. Rev. A, **70**, 043628 (2004).
[5] Nabi S N and Basu S, J. Phys. B: At. Mol. Opt. Phys., **49**, 125301 (2016).
[6] Fisher M P A, Weichman P B, Grinstein G and Fisher D S, Phys. Rev. B **40**, 546(1989)
[7] Jaksch D, Bruder C, Cirac J I, Gardiner C W, and Zoller P, Phys. Rev. Lett., **81**, 3108 (1998).
[8] Schmidt K P, Dorier J, and Luchli A M , Phys. Rev. Lett., **101**, 150405 (2008).
[9] Chen B, Huang X, Kou S and Zhang Y, Phys. Rev. A, **78**, 043603, (2008).
[10] Sowinski T, Phys. Rev. A, **85**, 065601, (2012).
[11] Ejima S, Lange F, Fehske H, Gebhard F and Mnster K Z, Phys. Rev. A, **88**, 063625 (2013).
[12] Mazza L, Rizzi M, Lewenstein M and Cirac J I, Phys. Rev. A, **82**, 043629 (2010).
[13] Mahmud K W and Tiesinga E, Phys. Rev. A, **88**, 023602, (2013).
[14] Nabi S N and Basu S, ArXiv:1606.02897v2.
[15] Will S, Best T, Schneider U, Hack-ermuller L, Luhmann D and Bloch I, Nature, **465**, 197 (2010).
[16] Sheshadri K, Krishnamurthy H R, Pandit R, and Ramakrishnan T V, Europhys. Lett., **22**, 257 (1993).
[17] Ho T L, Phys. Rev. Lett., **81**, 742 (1998).
[18] Paganelli S., Lacki M., Ahufinger V., Zakrzewski J., and Sanpera A., J Low Temp Phys, **165**, 227 (2011).

Disorder, three body interaction and Bose glass phase in a spinor atomic gas in an optical lattice

Sk Noor Nabi[1] **and Saurabh Basu**[2]
Department of Physics, Indian Institute of Technology Guwahati, Guwahati, Assam 781039, India
E-mail: [1]`sk.noor@iitg.ernet.in`, [2]`saurabh@iitg.ernet.in`

Abstract. We study the effects of disorder on the spin dependent interaction term of a spinor Bose Hubbard model with a three body interaction potential. The signature of the Bose glass (BG) phase is observed by computing the fraction of the lattice sites having finite superfluid (SF) order parameter and non integer occupation densities. We obtain the phase diagram both for the antiferromagnetic (AF) and ferromagnetic (F) cases via a percolation analysis. In the AF case, the BG phase intervenes between the odd-even Mott insulating (MI) lobes (for example, the lobes corresponding to occupation densities, n=1 and (n=2) but not between the even-odd MI lobes. In the ferromagnetic case, the presence of the BG phase is observed between all the MI lobes irrespective of them being even or odd. The BG phase almost destroys the first MI lobe while the MI phase looks more stable than the SF phase both in the AF and F cases due to the presence of the three body interactions.

1. Introduction

Ultracold atoms trapped in optical lattices offer a novel platform to explore the quantum phenomena where quantum fluctuations are responsible for phase transition from one macroscopic phase to another one. In magnetic trapping, ultracold atoms freeze their internal degrees of freedom and thereby behaves like scalar or spin-0 Bose gas while in optical trapping, they retains their spin degrees of freedom and show spinor like behaviour [1].

The properties of cold atoms in optical lattices first theoretically described by the well known Bose Hubbard model (BHM) developed by Jaksch *et. al.* [2] where the phase transition is entirely determined between the competition in tunneling and two body interaction strengths. Following the proposal by Jaksch *et. al.*, Greiner *et. al.* [3] first experimentally realized the superfluid (SF) to Mott insulator (MI) transition of bosonic cold atoms loaded in optical lattices.

Several progress have been made to study the scalar or spinor Bose gas with different extensions of BHM in both homogeneous and inhomogeneous case from theoretical as well as experimental point view. Among them, one possible extension of the BHM include the study of three body interactions [4–7] on the SF to MI transition gained much attention after the successful experimental realization of three and higher body interaction strengths by Will *et. al.* [8].

In this work, we will study the effect of three body repulsive interaction strengths on spinor ultracold Bose gas in presence of disorder through mean field approach (MFA) [9] where an additional glassy type of phase known as Bose glass (BG) phase [10–12, 14] intervene in between SF and MI due to the presence of disorder. Since MFA is quite inadequate to overcome the site dependencies in the Hamiltonian, numerical approach like quantum Monte Carlo (QMC) [10], stochastic mean field theory (SMFT) [11], probabilistic mean field perturbative (PMFA) approach [12, 13] and strong coupling expansion [14] have been developed to study the BHM in presence of disorder.

Following the approach developed in [15], here we use the MFA to study the disordered BHM and define a new quantity which is the fraction of lattice sites having finite SF order parameter to remove the inhomogenities in the Hamiltonian and did a percolation analysis to find the extent of the different phases based on the appearance of SF percolating cluster.

2. Spinor Bose Hubbard Hamiltonian

The Bose Hubbard Hamiltonian (BHM) for spin-1 ultracold atoms in presence of random disorder with three body interaction term is [4–6, 16],

$$\begin{aligned}\hat{H} = & -t\sum_{<i,j>}\sum_{\sigma}(\hat{a}^{\dagger}_{i\sigma}\hat{a}_{i\sigma}+h.c)-\mu\sum_i\hat{n}_i+\frac{U_0}{2}\sum_i\hat{n}_i(\hat{n}_i-1)\\ & +\frac{W}{6}\sum_i\hat{n}_i(\hat{n}_i-1)(\hat{n}_i-2)+\sum_i\frac{1}{2}(U_2+\epsilon_i)(\mathbf{S}_i^2-2\hat{n}_i)]\end{aligned} \quad (1)$$

where t is the tunneling amplitude and μ is the chemical potential. $\hat{a}^{\dagger}_{i\sigma}(\hat{a}_{i\sigma})$ is the boson creation (annihilation) operator at a site i with hyperfine spin state σ =+1, 0, -1 and the particle number operator is $\hat{n}_i = \sum_\sigma \hat{a}^{\dagger}_{i\sigma}\hat{a}_{i\sigma}$. The hyperfine spin operator at site i is given by $\mathbf{S}_i = \hat{a}^{\dagger}_{i\sigma}\mathbf{F}_{\sigma\sigma'}\hat{a}_{i\sigma'}$ where $\mathbf{F}_{\sigma\sigma'}$ are the components of spin-1 matrices. W represents the three body interaction strength and U_0 and U_2 are spin independent and dependent two body interaction strengths respectively. U_0 and U_2 are related to the scattering lengths, a_0 and a_2 as $U_2/U_0 = (a_2 - a_0)/(a_0 + 2a_2)$ and is called antiferromagnetic (AF) interaction if $U_2/U_0 > 0$ and ferromagnetic (F) if $U_2/U_0 \leq 0$. ϵ_i is the on-site disorder is introduced at site i, which is randomly chosen from a box distribution in the interval $[-\Delta, \Delta]$ where Δ is the strength of the disorder. To decouple the hopping part in Eq.(1), we shall use mean field approximation as [17]

$$\hat{a}^{\dagger}_{i\sigma}\hat{a}_{j\sigma} \simeq \langle\hat{a}^{\dagger}_{i\sigma}\rangle\hat{a}_{j\sigma}+\hat{a}^{\dagger}_{i\sigma}\langle\hat{a}_{j\sigma}\rangle-\langle\hat{a}^{\dagger}_{i\sigma}\rangle\langle\hat{a}_{j\sigma}\rangle \quad (2)$$

where $\langle\ \rangle$ denotes the equilibrium value of an operator. Now defining the superfluid order parameter at a site i as, $\psi_{i\sigma} \equiv \langle\hat{a}^{\dagger}_{i\sigma}\rangle \equiv \langle\hat{a}_{i\sigma}\rangle$ and $\psi_i = \sqrt{\psi^2_{i\sigma}}$, the BHM can be written as a sum of single site Hamiltonians as, $H = \sum_i H_i^{MF}$ where

$$\begin{aligned}H_i^{MF} = & -t\sum_{\sigma}(\phi^*_{i\sigma}\hat{a}_{i\sigma}+\phi_{i\sigma}\hat{a}^{\dagger}_{i\sigma})+t\sum_{\sigma}\phi^*_{i\sigma}\psi_{i\sigma}-\mu\hat{n}_i+\frac{U_0}{2}\hat{n}_i(\hat{n}_i-1)\\ & +\frac{W}{6}\hat{n}_i(\hat{n}_i-1)(\hat{n}_i-2)+\frac{1}{2}(U_2+\epsilon_i)(\mathbf{S}_i^2-2\hat{n}_i)\end{aligned} \quad (3)$$

where $\phi_{i\sigma} = \sum_j \psi_{j\sigma}$, j includes all nearest neighbors at the site i of a square lattice with $z = 4$, z being the coordination number. The local density, ρ_i and compressibility, κ_i can also be computed using,

$$\rho_i = \langle\psi_i^{eq}|\hat{n}_{i\sigma}|\psi_i^{eq}\rangle;\ \ \kappa_i^2 = \langle\rho_i^2\rangle - \langle\rho_i\rangle^2 \quad (4)$$

where ψ_i^{eq} is the equilibrium value of ψ_i after self consistent ground state is obtained.

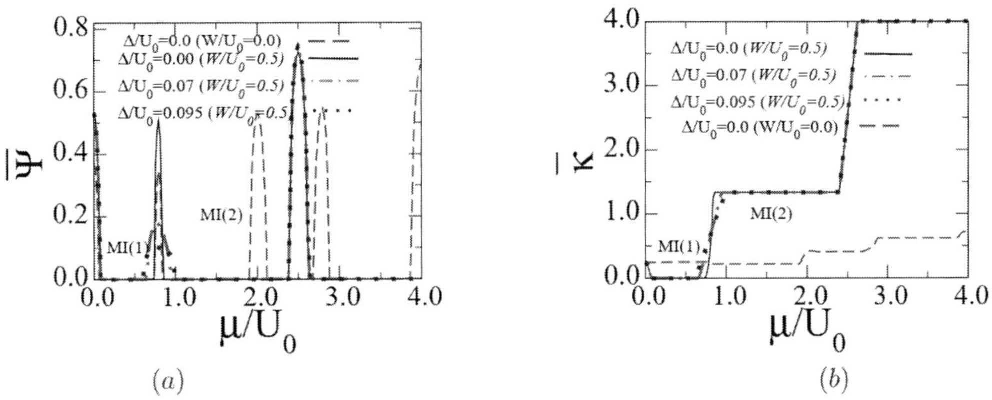

Figure 1. (Color online) The SF order parameter $\bar{\psi}$ (a) and $\bar{\kappa}$ (b) with Δ/U_0 for $U_2/U_0 = 0.1$ (AF) and $W/U_0 = 0.5$.

3. Results

In order to see the effect of disorder and three body interaction term, first we consider the atomic limit *i.e* $t = 0$. In the atomic limit $t = 0$, system will consist only the Mott insulating (MI) phase where there is a gap in the particle hole excitation spectrum. The energy gap, E_g for a particular filling of the MI lobe n corresponds to the upper and lower boundaries of the MI region as $\mu_-(n) < \mu < \mu_+(n)$ [12]. In the AF case, this inequality leads to the following equations corresponding to the even MI lobes as $(n-1) - \frac{2U_2}{U_0} + \frac{W}{2U_0}(n-1)(n-2) < \mu/U_0 < n + \frac{W}{2U_0}n(n-1)$ and for odd MI lobes as $(n-1) + \frac{W}{2U_0}(n-1)(n-2) < \mu/U_0 < n - \frac{2U_2}{U_0} + \frac{W}{2U_0}n(n-1)$ and also in the ferromagnetic case, $(n-1)(1+\frac{U_2}{U_0}) + \frac{W}{2U_0}(n-1)(n-2) < \mu/U_0 < n(1+\frac{U_2}{U_0}) + \frac{W}{2U_0}n(n-1)$ and shown in Fig.3(a). It shows that both in the AF and F cases, the chemical potential width μ increases due to W/U_0 which suggests that the MI lobes becomes more stable compared to the SF phase. Interestingly the critical value of U_2 for the disappearance of the odd MI lobes in the AF case now depends on the three body interaction strengths which was 0.5 without W/U_0.

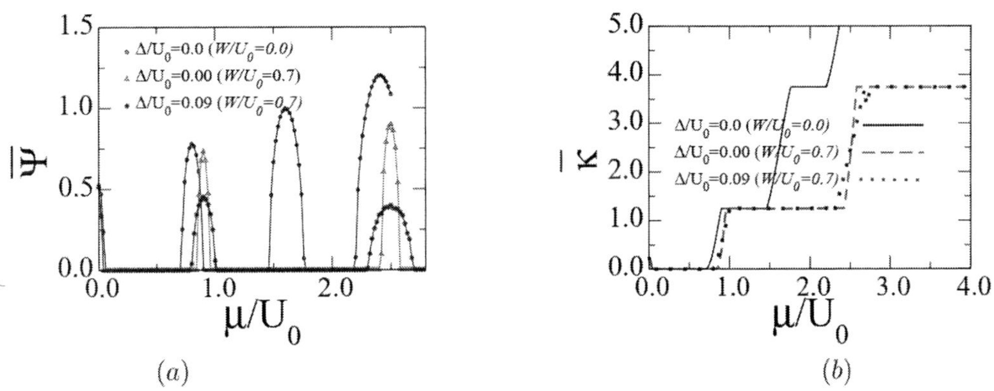

Figure 2. (Color online) The $\bar{\psi}$ (a) and $\bar{\kappa}$ (b) with Δ/U_0 for $U_2/U_0 = -0.2$ (F) and $W/U_0 = 0.7$.

We now turn on the disorder and diagonalize the mean field Hamiltonian on a 2D square lattice of size $L = 30$ self consistently to find the ground state energy and the equilibrium order parameter ψ_i^{eq} and $\bar{\psi} = \frac{1}{L^2}\sum_i \psi_i^{eq}$ and the compressibility $\bar{\kappa} = \frac{1}{L^2}\sum_i \kappa_i$ which are shown in Figs.1 and 2. In a disorder free scenario, the SF order parameter $\bar{\psi}$ and compressibility $\bar{\kappa}$ change from zero to

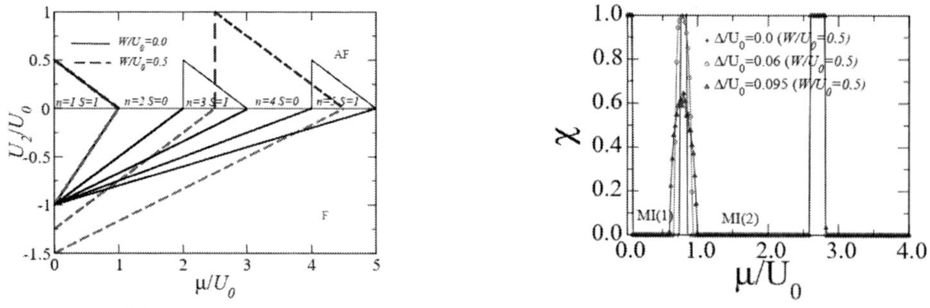

(a) (b)

Figure 3. (Color online) MI phase in atomic limit $t = 0$ for $W/U_0 = 0.5$ (a). The variation of χ with Δ/U_0 in U_2 for $U_2/U_0 = 0.1$ and $W/U_0 = 0.5$ in (b).

finite value indicating direct MI to SF phase transition [Figs.1 and 2]. The transition point for the MI-SF phase increases with W/U_0 since it pushes the system towards the insulating regime. As disorder is introduced, the BG phase appears in between the MI and the SF phase which is characterized by zero SF order parameter but finite compressibility. In the AF case, no BG phase occurs between the even ($\rho = 2, 4..$) and odd ($\rho = 1, 3..$) MI lobes [Figs.1(a),(b)(MI(2))] while for the odd and even MI lobes, the BG phase appears[Figs.1(a),(b)(MI(1))]. In the atomic limit, $t = 0$ for AF case, the chemical potential width μ for the even-odd transition is parallel to U_2 axis while for odd-even transition, it depends upon U_2 and disorder in U_2 will change the chemical potential width and hence the MI lobes [Fig.3 (a)]. In F case, the BG phase intervenes in all the MI lobes since disorder in U_2 will affect the occupation densities [Figs.2 (a),(b)].

From the above discussion, although we have been able to identify the BG phase but in order to get a clear understanding, we have to determine the boundaries of these three phases and hence the phase diagram in presence of disorder. In MFA, all SF parameter varies randomly due to disorder, thus it is quite difficult to determine the location of different phases and hence the phase diagrams based on the average $\bar{\psi}$ and $\bar{\kappa}$.

To remove the inhomegenities in the Hamiltonian, we define a new quantity, called χ as,

$$\chi = \frac{sites\ with\ \psi_i \neq 0\ \ and\ \rho_i \neq integer}{total\ numbers\ of\ sites} \tag{5}$$

The variation of χ for AF case is shown in Fig.3(b). It shows that χ switches from 0 to 1 indicating a direct MI to SF phase transition in clean state while in presence of disorder, there is gradual increase in slope due to the intervening BG phase in between MI and SF phases.

Interestingly in AF case, χ only assumes 0 and 1 values corresponding to even-odd MI transition in both clean and disordered cases leaving the signature of BG phase [Fig.3(b)(MI(2))]. While for the odd-even transition χ shows a gradual increase between 0 and 1, indicating the intervention of BG phase in between MI and SF phases as discussed earlier [Fig.3(b)(MI(1))]. In F, χ takes all finite values signaling the appearance of BG phase for all MI lobes and shows similar properties if disorder is present in μ.

From the variation of χ, it is now quite clear that $\chi = 0$ corresponds to MI and $\chi = 1$ corresponds to SF phase but we have to determine if there is any critical value of χ *i.e* χ_c below which the system is in BG phase above which it belongs to SF phase for $0 < \chi < 1$ region [Fig.3(b)]. To answer the above concern, we will follow the percolation analysis to know if there is any SF percolating cluster exists. We have characterized the three phases based on the appearances of SF percolating cluster [15]. A SF percolating cluster means cluster of sites with non integer densities connected with their nearest neighbors that percolates throughout

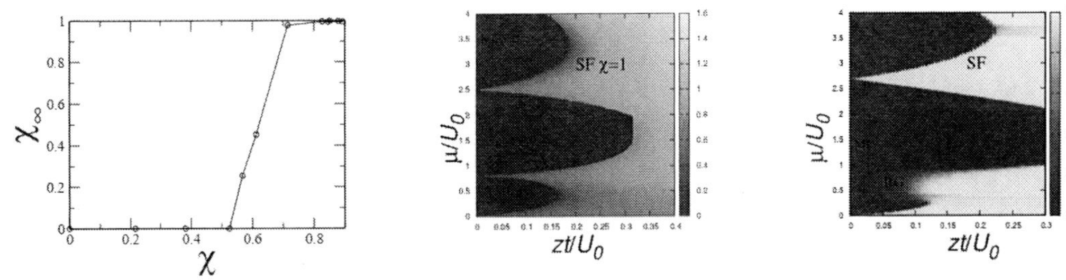

(a) (b) (c)

Figure 4. (Color online) Percolation probability χ_∞ with χ for lattice size $L = 120$ for $U_2/U_0 = 0.1$ with disorder present in $\mu/U_0 = 0.4$ in (a). Phase diagram for $W/U_0 = 0.5$ with $\Delta/U_0 = 0$ at $U_2/U_0 = 0.1$ in (b) and with $\Delta/U_0 = 0.095$ in U_2/U_0 and $W/U_0 = 0.5$ in (c).

the lattice, that is from left to right and/or top to bottom. In MI phase, all sites have integer densities thus no SF percolating cluster. While in SF phase, all sites have non integer densities and thus have at least one SF percolating cluster. But in the BG phase, some sites have non integer densities, while the rest will have integer densities. Hence the BG phase can be identified with having a SF cluster which does not percolate.

Whether a SF cluster will percolate can be well understand from the following quantity, χ_∞ known as percolation probability, defined as the fraction of sites within the SF percolating cluster out of total no of occupied sites [18]. To find a SF percolating cluster, we shall use Hoshen-Kopelman(HK) algorithm to find a SF percolating cluster [19]. It is based on the special application of the union-find algorithm where it aims to assign a label to each cluster. For that we represent our 2D lattice as a matrix and label all occupied sites initially with -1 and unoccupied sites with 0. Our cluster index starts with 1. Corresponding to each occupied site, every time we check the neighbors at the top and left corner of the current site. If both sites are empty, we label a new cluster number that has not used so far. Else, if the site has one occupied neighbor, then we assign same cluster number to the current site. If both neighboring sites are occupied, then we set the smallest number cluster label of the occupied neighbors to use as the label for the current site. To link two clusters, we create a union between both labels and set the site as the lowest of the two labels. When we burn the lattice for a second time, we collect the unions and update the lattice [19]. χ_∞ is analogous to critical percolation probability p_c which is 0.593 for infinite square lattice that is $L = \infty$ and the variation of χ_∞ with χ is shown in Fig.4(a). It shows that χ_∞ is 0 till $\chi < 0.524$ means no SF percolating cluster exists while it is non zero above $\chi \geq 0.524$ where SF percolating cluster exists.

At this stage, we have now all information to determine the location of different phases and thereby the phase diagrams entirely based on χ and χ_∞. In clean state, phase diagram for AF and F cases are obtained by either ψ or χ as shown in Figs.4(b) and 5(b). Although there is no change in the first MI lobe but higher order MI lobes becomes more stable and the transition from MI to SF phase increases due to W/U_0. In the disordered case, we have obtained the phase diagrams by incorporating the $\chi = 0$ for MI phase, for BG phase, $0 < \chi < \chi_c = 0.43$ corresponding to AF and $0 < \chi < \chi_c = 0.53$ corresponding to F cases below which $\chi_\infty = 0$ and $\chi_c \leq \chi \leq 1$ for the SF phase and shown in Figs.4 and 5. It is quite helpful to mention that the colors code in the legend are as follows-blue: MI, red: BG and yellow: SF phases respectively. In AF case, BG phase only appears in between odd-even MI phase and due to inclusion of three body interaction, the intervening BG phase becomes quite thin compared to the MI phase [Fig.4(c)]. While in F case, the BG appears in between all MI lobes shows similar phase diagrams if disorder is present in μ [Fig.5(c)]. We have also considered the disorder in μ

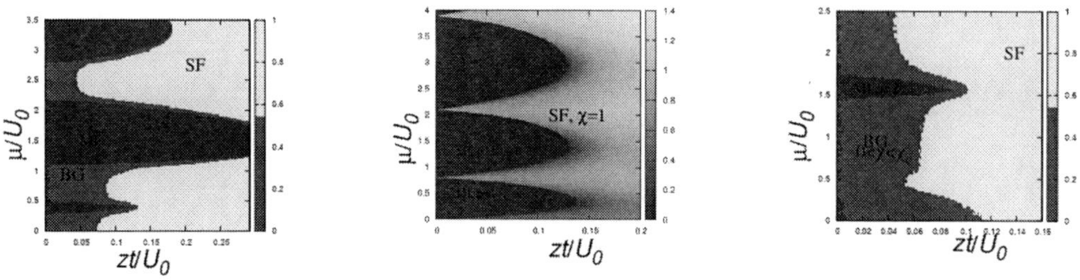

(a) (b) (c)

Figure 5. (Color online) Phase diagram with disorder in AF for $\Delta/U_0 = 0.5$ in μ and $W/U_0 = 0.5$ (a). Phase diagrams in ferromagnetic case without disorder at $U_2/U_0 = -0.2$ (F) and $W/U_0 = 0.7$ (b) and with disorder, $\Delta/U_0 = 0.55$ (c).

in AF case [Fig.5(a)] and interestingly from Fig.5(a), the BG can not destroy the MI lobes since the disorder strength needed to dismiss the respective MI lobes is now lower than the critical value Δ_c which depends on W/U_0 [12]. Similarly in F case, the first MI lobe vanishes due the BG phase as Δ is higher than the critical disorder strength [Fig.5(c)] [12].

4. Conclusion

We study the effect of spin dependent disorder in spinor BHM through percolation analysis in presence of three body interactions. The different phases in presence of disorder are observed by the SF order parameter and compressibility. The boundary for the MI, BG and SF phases are determined with the help of percolation analysis and finally obtain the phase diagrams based on χ. In AF case, the BG appears only for the odd-even MI transition and very thin compared with the MI lobes due to the presence of three body interaction. We have also studied the effect of disorder in ferromagnetic case and found that it has similar properties as that of disorder in the chemical potential μ.

Acknowledgments

SB thanks CSIR, India for financial support under the grants F no: 03(1213)/12/EMR-II.

References

[1] Stamper-Kurn D M *et. al.*, Phys. Rev. Lett. **80**, 2027 (1998).
[2] Jaksch D, Bruder C, Cirac J I, Gardiner C W, and Zoller P, Phys. Rev. Lett., **81**, 3108 (1998).
[3] Greiner M, Mandal O, Esslinger T, Haensch T W and Bloch I, Nature(London), **415**, 39 (2002).
[4] Chen B, Huang X, Kou S and Zhang Y, Phys. Rev. A, **78**, 043603, (2008).
[5] Sowinski T, Phys. Rev. A, **85**, 065601, (2012).
[6] Mahmud K W and Tiesinga E, Phys. Rev. A, **88**, 023602, (2013).
[7] Tsuchiya S, Kurihara S, and Kimura T, Phys. Rev. A, **70**, 043628 (2004).
[8] Will S, Best T, Schneider U, Hack-ermuller L, Luhmann D and Bloch I, Nature, **465**, 197 (2010).
[9] Sheshadri K, Krishnamurthy H R, Pandit R, and Ramakrishnan T V, Europhys. Lett., **22**, 257 (1993).
[10] Lee J W, Cha M C, and Kim D, Phys. Rev. Lett., **87**, 247006 (2001).
[11] Bissbort U and Hofstetter W, Europhys. Lett., **86**, 50007 (2009).
[12] Lacki M, Paganelli S, Ahufinger V, Sanpera A, and Zakrzewski J, Phys. Rev. A, **83**, 013605 (2011).
[13] Paganelli S., Lacki M., Ahufinger V., Zakrzewski J., and Sanpera A., J Low Temp Phys, **165**, 227 (2011).
[14] Freericks J K and Monien H, Phys. Rev. B, **53**, 2691 (1996).
[15] Nabi S N and Basu S, J. Phys. B: At. Mol. Opt. Phys., **49**, 125301 (2016).
[16] Ho T L, Phys. Rev. Lett., **81**, 742 (1998).
[17] Pai R V, Sheshadri K and Pandit R, Phys. Rev. B, **77**, 014503 (2008).
[18] Stauffer D and Aharony A, *Introduction to Percolation Theory* (Taylor and Francis, 1994).
[19] Hoshen J and Kopelman R, Phys. Rev. B, **14**, 3438 (1976).

Condensate characteristics of bosons in a tilted optical lattice

Sunayana Dutta* and Saurabh Basu†

Department of Physics, Indian Institute of Technology Guwahati, Guwahati, Assam, India 781039

E-mail: *sunayanadutta@iitg.ernet.in and †saurabh@iitg.ernet.in

Abstract. In this work we study the properties of interacting bosons in a tilted triple well optical lattice potential. A tilted optical lattice is possible to create in an experiment with charged bosons in presence of an electric field. The ground state energy, occupation densities in each well and first order correlation function of occupation densities as a function of the tilt parameter for $N = 90$ bosons using recently developed Multiconfigurational time-dependent Hartree method for bosons (MCTDHB) are studied. The ground state energy falls off with the tilt parameter for a given value of the barrier height and interparticle interaction. Also the tilt parameter is observed to restrict the fragmentation of the occupation densities in each well and thus promote condensation. Further the correlation between the bosons in different wells enhances with the tilt parameter and at large values the bosons in the first and the third wells are correlated apart from the correlation among the bosons in the same well, while the occupancy in the middle one seems to be uncorrelated with the other two. Such correlation between different wells is not observed in absence of the tilt effect for the same barrier height of the potential and interparticle interaction strength.

1. Introduction

In recent years, the successful experimental realization of Bose-Einstein condensates with the help of optical lattices provided a suitable paradigm to study different exotic quantum phenomena of cold atoms[1]. The Gross-Pitaevskii mean-field theory is one of the efficient tool to study the static and dynamic properties of Bose-Einstein condensates. Though this theory is an excellent tool for explaining weakly correlated bosonic systems, it has some limitations. As this theory deals with only one time-dependent single-particle state (orbital) to explain Bose-Einstein condensates, it can not explain phenomena like fragmentation of condensates or Mott-insulator phase transition which require more than one single-particle state.

In recent years, an essentially exact, numerically efficient and multi-orbital many-body theory is developed which is known as multi-configuration time-dependent Hartree approach (MCTDH). Very recently, a bosonic version of MCTDH, called Multiconfigurational time-dependent Hartree method for bosons (MCTDHB)[2, 3, 4] is developed. It describes the static and dynamical properties of fragmented condensates. So MCTDHB is the generalization of the (one-orbital) Gross-Pitaevskii mean field theory.

In this work, we wish to study a system of weakly correlated bosons in an optical lattice potential and specifically a triple well. S.Sachdev *et.* al[5] have studied a system of utltracold charged bosonic atoms in an optical lattice potential and in presence of a gradient (produced possibly by an applied electric field). Motivated by this paper, we study the static properties of a system of bosonic atoms in a tilted triple optical potential well using MCTDHB.

2. Formalism

Alon *et al.*[3] provided the complete derivation of the MCTHDB equations and also the details of the numerical implementation of MCTDHB. The use of time-adaptive optimized basis states is the main idea of this method. This method is a fully variational approach and there is no restriction for the number of orbitals, geometry, dimensionality, and interparticle interactions of the time-dependent many-boson problem. In this section we discuss in brief about this method.

The starting point of MCTDHB approach is the time-dependent variational principle (TDVP)[6]. The variational principle is based on the action, S for the time development of a wave function for a certain time interval. The Lagrangian of a many-body quantum system with a constant number of particles can be defined as,

$$\mathcal{L}(\psi(t), \psi^*(t)) = \langle \psi(t)|\hat{H} - i\frac{\partial}{\partial t}|\psi(t)\rangle \tag{1}$$

Then the principle of least action (namely, $\delta \mathrm{S}(\psi(t), \psi^*(t)) = 0$) is employed to determine the equations of motion. Here the Hamiltonian of Eq.(1) reads as,

$$\hat{H}(\mathbf{r}_1, \mathbf{r}_2, ..., \mathbf{r}_N) = \sum_{j=1}^{N} \hat{h}(\mathbf{r}_j) + \sum_{k>j=1}^{N} \hat{W}(\mathbf{r}_j - \mathbf{r}_k) \tag{2}$$

where $\mathbf{r}_j$ is the position of the jth boson. The first term $\hat{h}$ is the one-body Hamiltonian and is given as $\hat{h}(\mathbf{r}) = \hat{T}(\mathbf{r}) + \hat{V}(\mathbf{r})$, where $\hat{T}(\mathbf{r})$ is kinetic energy and $\hat{V}(\mathbf{r})$ is the external potential (that is, $V_{trap}(x)$ here). The second term $\hat{W}(\mathbf{r}_j - \mathbf{r}_k) = \lambda_0 \delta(\mathbf{r}_j - \mathbf{r}_k)$ is the two-body interaction term. It is approximated as a δ-function to incorporate the hardcore feature of bosons in the interparticle interaction term. The ground state energy is given by[7, 8],

$$E = N\Big\{ \int \psi^* \hat{h} \psi d\overrightarrow{r} + \frac{\lambda_0(N-1)}{2} \int |\psi|^4 d\overrightarrow{r} \Big\} \tag{3}$$

where we have written the coefficient of the interaction term as, $\lambda = \lambda_0(N-1)$, where λ is the interaction parameter and λ_0 is the interaction strength. A particular value of λ, namely $\lambda = 2$ is chosen here, so that fragmentation of the condensate can be noted for a given barrier height and tilting parameter. This issue is elaborated later.

In second quantized notations, the bosonic field operator which annihilates a particle at position $\mathbf{r}$, corresponding to a set of M orthonormal, time-dependent functions (orbitals) $\{\phi_k(\mathbf{r}, t)\}$ is given as,

$$|\Psi(\mathbf{r})\rangle = \sum_{k=1}^{M} \hat{a}_k \phi_k(\mathbf{r}, t) \tag{4}$$

Here the bosonic annihilation and creation operators obey their usual commutation relations at any given time. The time-dependent permanents with the help of the bosonic creation operators, $\hat{a}_k^\dagger$ can be written as,

$$|\vec{n}; t\rangle = \frac{[a_1^\dagger(t)]^{n_1} [a_2^\dagger(t)]^{n_2} ... [a_M^\dagger(t)]^{n_M}}{\sqrt{n_1! n_2! n_3! ... n_M!}} |0\rangle \tag{5}$$

where $\vec{n} = (n_1, n_2, n_3, ...n_M)$ are the occupations of the orbitals preserving the total number of particles as $n_1 + n_2 + ... + n_M = N$.

The ansatz made for many-boson wave function is,

$$|\Psi(t)\rangle = \sum_{n} C_{\vec{n}}(t)|\vec{n}; t\rangle \tag{6}$$

the summation in Eq.(6) runs over all possible configurations, that means N bosons are distributed over M accessible orbitals. The time-dependent many particle Schrödinger equation which governs the time evolution of N correlated bosons is given as ($\hbar = 1$),

$$\hat{H}\Psi = i\frac{\partial \Psi}{\partial t} \tag{7}$$

Now the main objective is to find out the exact solution of the many-boson time-dependent Schrödinger equation[3, 4], that is, to determine the time evolution of the expansion coefficients $\{C_{\vec{n}}(t)\}$ and the orbitals $\{\phi_k(\mathbf{r}, t)\}$. Now we employ the Lagrangian formalism of the time-dependent variational principle[6] to evaluate the time evolution of $\{C_{\vec{n}}(t)\}$ and $\{\phi_k(\mathbf{r}, t)\}$ which in turn gives the MCTDHB equations of motion. The different steps of our approach are:

(i) Variation with respect to the expansion coefficients $C_{\vec{n}}(t)$:

Here the variation of the functional action with respect to the coefficients is considered and the equation of motion for expansion coefficients is obtained as[3, 4],

$$H(t)C(t) = i\frac{\partial C(t)}{\partial t} \tag{8}$$

This equation of motion is a first-order differential equation and the matrix $H(t)$ is time-dependent as it is evaluated in the time-dependent permanents $|\vec{n}; t\rangle$ and $|\vec{n}'; t\rangle$.

(ii) Variation with respect to the orbitals $\{\phi_k(\mathbf{r}, t)\}$:

Considering variation with respect to the orbitals, the equation of motion for orbitals is obtained as[3, 4],

$$i|\dot{\phi}_j\rangle = \hat{\mathbf{P}}[\ \hat{h}|\phi_j\rangle + \sum_{k,s,q,l=1}^{M} \{\rho(t)\}^{-1}_{jk}\rho_{ksql}\hat{W}_{sl}|\phi_q\rangle\ ] \tag{9}$$

where

$$\hat{\mathbf{P}} = 1 - \sum_{j'=1}^{M} |\phi_{j'}\rangle\langle\phi_{j'}| \tag{10}$$

here $\rho(t)_{jk}$ is the one body reduced density matrix (RDM) and ρ_{ksql} is the two body reduced density matrix (RDM).

2.1. RDM and Natural occupations

The pth order reduced density matrix (RDM) is defined as[4],

$$\hat{\rho}^{(p)} = \frac{N!}{(N-p)!}Tr_{N-p}[|\Psi(t)\rangle\langle\Psi(t)|] \tag{11}$$

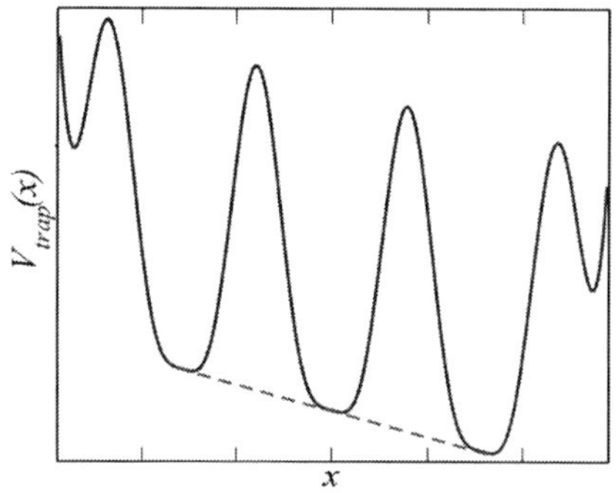

Figure 1. This figure shows the trapping potential which is a tilted triple well.

Here $Tr_{N-p}[.]$ is the partial trace runs over $(N-p)$ particles and diagonal $\rho^{(p)}$ corresponds to p-particle probability distribution at time t. The pth order RDM, $\rho^{(p)}$ expanded in its eigenfunctions is given as,

$$\rho^{(p)}(\mathbf{r}_1...,\mathbf{r}_p|\mathbf{r}_1^{'}...,\mathbf{r}_p^{'};t) = \sum_i n_i^{(p)}(t)\alpha_i^{(p)}(\mathbf{r}_1...,\mathbf{r}_p;t)\alpha_i^{*(p)}(\mathbf{r}_1^{'}...,\mathbf{r}_p^{'};t) \quad (12)$$

where $n_i^{(p)}(t)$ is the ith eigenvalue of the pth order RDM and $\alpha_i^{(p)}$ is the eigenfunction in real space. For $p=1$, $\alpha_i^{(1)}$ is known as natural orbitals and $n_i^{(1)}(t)$ is known as natural occupations.

The pth order correlation function of an N boson system at a time t is given as,

$$g^{(p)}(\mathbf{r}_1^{'},...,\mathbf{r}_p^{'},\mathbf{r}_1,...,\mathbf{r}_p;t) = \frac{\rho^{(p)}(\mathbf{r}_1,...,\mathbf{r}_p|\mathbf{r}_1^{'},...,\mathbf{r}_p^{'};t)}{\sqrt{\prod_{i=1}^{p}\rho^{(1)}(\mathbf{r}_i|\mathbf{r}_i;t)\rho^{(1)}(\mathbf{r}_i^{'}|\mathbf{r}_i^{'};t)}} \quad (13)$$

The diagonal $g^{(p)}(\mathbf{r}_1^{'},...,\mathbf{r}_p^{'},\mathbf{r}_1,...,\mathbf{r}_p;t)$ of the pth order correlation function in the real space is a measure of the pth order coherence.

Eqs.(12) and (13) are the working formulae for obtaining the occupation densities and first-order correlation function that will yield the coherence properties of bosons in a tilted triple well respectively.

3. Physical Observables and Results

We consider a representative system of $N=90$ bosons in a tilted triple well potential of the form as follows (Fig.(1)),

$$V_{trap}(x) \;=\; -\alpha x + V_0(\sin(2x))^4 \quad (14)$$

This trapping potential has sufficiently large barrier height, that is, $V_0 = 250$ in our case and it is maintained for the entire proceedings. Here α is the tilting parameter. The tilting term αx is similar to 'potential energy' of a charged particle in an electric field. In our case α ranges from 0 to 5. A brief justification of using such values for α is included later. As mentioned earlier, we have chosen $\lambda = 2$, such that the fragmentation of the condensate is observed in absence of the tilting effect $\alpha = 0$ for a barrier height, $V_0 = 250$. Non zero values of α are benchmarked with respect to this. Further, we have considered three basis states ($M=3$). Our main aim is to study the static properties of weakly correlated bosons in a tilted triple well. For this purpose we have used an open source version of the MCTDHB algorithm[3].

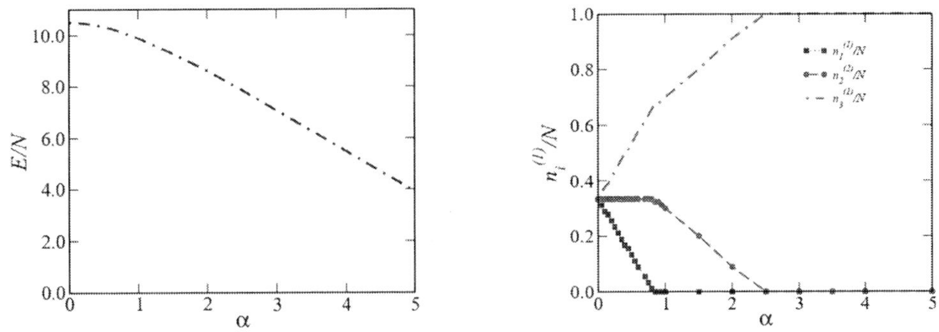

(*a*) (*b*)

Figure 2. (*a*) This figure shows the energy per particle, E/N as a function of α and (*b*) shows occupation density, $n_i^{(1)}/N$ as a function of α. Here $n_1^{(1)}$, $n_2^{(1)}$ and $n_3^{(1)}$ are the eigenvalues of the first-order reduced density matrix.

3.1. Bose-Einstein Condensation and Fragmentation

In order to make the terminologies that have been used in this paper clear, we define a few relevant quantities such as condensation and fragmentation of the condensates. Bose-Einstein condensation can be defined with the help of natural orbital occupation numbers $n_i^{(1)}$[4, 9]. An identical bosonic system is said to be in a condensed state if the largest eigenvalue of the first-order RDM is of the order of the number of particles in the system, that is[4, 9],

$$n_1^{(1)} \;=\; \mathcal{O}(N) \tag{15}$$

So a bosonic state with N-bosons is said to be maximally coherent when $n_1^{(1)} = N$ and $|g^{(1)}| = 1$. Bosonic states with largest natural occupation number close to N are known as depleted condensates. When two natural orbital occupation number is of the order of the total number of particles, that is, $n_1^{(1)}, n_2^{(1)} = \mathcal{O}(N)$, then one get a two-fold fragmentation of the condensate. Again for, $n_1^{(1)}, n_2^{(1)}, n_3^{(1)} = \mathcal{O}(N)$ a three-fold fragmented condensate emerges and so on.

3.2. Ground state energy and occupation density

We study the ground energy per particle, E/N and occupation density, $n_i^{(1)}/N$ as function of the tilting parameter, α for constant values of interparticle interaction, λ_0 and barrier height, V_0. From Fig.2(a), it is observed that E/N decreases with α. The decrease is slow for smaller values of α. From $\alpha \simeq 0.8$ onward, a rapid fall of E/N is observed.

Fig.2(b) shows the behavior of the occupation density, $n_i^{(1)}/N$ of natural orbitals of the many body wave function as a function of titling parameter. Without titling (that is with $\alpha = 0$), there is a three-fold fragmentation of the condensate. With the inclusion of the tilting angle, the three-fold fragmented state disappears. There is a gradual decrease in the occupation density of the first natural orbital, $n_1^{(1)}/N$ and correspondingly a gradual increase in the occupation density of the third natural orbital, $n_3^{(1)}/N$ till $\alpha \simeq 0.8$. Whereas the occupation density of the second natural orbital, $n_2^{(1)}/N$ remains constant till $\alpha \simeq 0.8$. However, beyond $\alpha \simeq 0.8$, a rapid change in the occupation density is observed. For higher values of α ($\alpha = 5$), the third natural orbital becomes fully condensed and correspondingly the occupation densities in the second and first natural orbitals become negligibly small. So 'tilting' leads to a transition from a fragmented to a condensed state for the bosons.

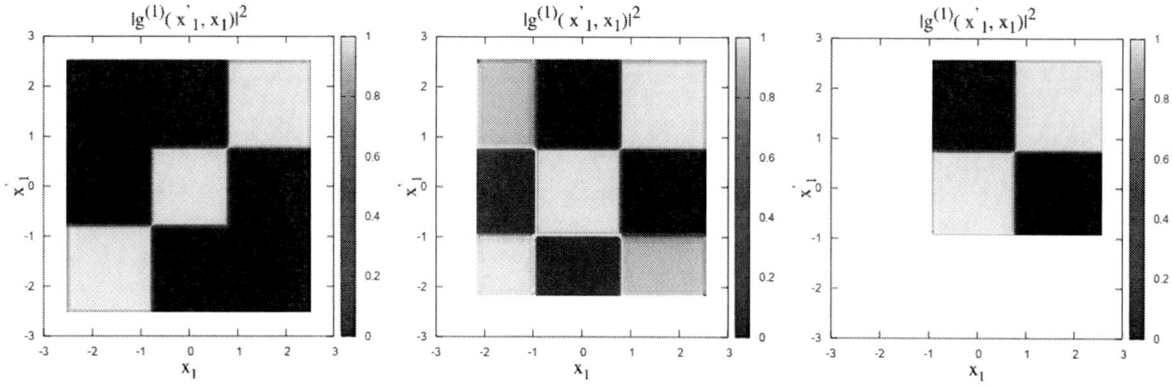

(*a*) (*b*) (*c*)

Figure 3. Modulus squared of the normalized first-order correlation function $|g^{(1)}(x'_1, x_1)|^2$ for different tilting angles (α) is shown here. (*a*) Corresponds to $\alpha = 0$, (*b*) for $\alpha = 0.9$ and (*c*) for $\alpha = 2$.

3.3. Coherence Properties

To get a deeper understanding about the properties of the condensate, the modulus squared of the normalized first-order correlation function $|g^{(1)}(x'_1, x_1)|^2$ at different tilting parameter is studied. Fig.3(a) shows $|g^{(1)}(x'_1, x_1)|^2$ for $\alpha = 0$. It is observed that there is no coherence between the wells and bosons reside in individual wells. Similar scenario is observed till $\alpha = 0.8$. For $\alpha > 0.8$, we observe correlation between the bosons of the third and first well and vice-versa apart from the correlation of bosons in the same well (Fig.3(b)). But for this value of α, there is no correlation between the bosons of the middle well with the other two wells. Of course, there is correlation among the bosons in the same well. With further increase of the tilting angle α, this observed coherence between different wells is destroyed gradually (Fig.3(c)). For higher values of α ($\alpha = 5$), the observed coherence between different wells totally destroyed and the third well becomes fully populated.

4. Conclusions

We have studied the static properties such as the ground state energy and occupation density of the natural orbitals of a system of 90 bosons in presence of a tilted triple well potential using MCTDHB approach. In presence of a tilting potential, the ground state energy per particle decreases starting with a very high value and it falls off rapidly for higher values of α corresponding to a fixed barrier height and interparticle interaction. Further, a transition from fragmented to condensate state is observed with increase in tilting angle.

References

[1] M. H. Anderson, J. R. Ensher, M. R. Matthews, C. E. Wiemann, and E. A. Cornell, Science **269**, 198 (1995).
[2] A. I. Streltsov, O. E. Alon, and L. S. Cederbaum, Phys. Rev. Lett. **99**, 030402 (2007).
[3] O. E. Alon, A. I. Streltsov, and L. S. Cederbaum, Phys. Rev. A **77**, 033613 (2008).
[4] *Many-body Schrödinger dynamics of Bose-Einstein condensates*, Kaspar Sakmann, (Springer, 2011).
[5] S. Sachdev, K. Sengupta, and S. M. Girvin, Phys. Rev. B **66**, 075128 (2002).
[6] *Geometry of the time-dependent variational principle*, P. Kramer and M. Saracen, (Springer, Berlin, 1981).
[7] F. Dalfovo, S. Giorgini, L. P. Pitaevskii, S. Stringari, Rev. Mod. Phys. **71**, 463 (1999).
[8] A. I. Streltsov and L. S. Cederbaum, Phys. Rev. A **70**, 053607 (2004).
[9] O. Penrose and L. Onsager, Phys.Rev. **104**, 576 (1956).
[10] S. Klaiman, A. U. J. Lode, A. I. Streltsov, L. S. Cederbaum, and O. E. Alon, Phys. Rev. A, **90**, 043620 (2014).

Journal of Physics: Conference Series **759** (2016) 012037 doi:10.1088/1742-6596/759/1/012037

A DMFT+CTQMC Investigation of Strange Metallicity in Local Quantum Critical Scenario

Swagata Acharya[1], **M. S. Laad**[2], **A. Taraphder**[1,3]

[1] Department of Physics, Indian Institute of Technology Kharagpur, Kharagpur 721302, India
[2] Institute of Mathematical Sciences, Taramani, Chennai 600113, India
[3] Centre for Theoretical Studies, Indian Institute of Technology Kharagpur, Kharagpur 721302, India.

E-mail: acharya.swagata@phy.iitkgp.ernet.in

Abstract. "Strange" metallicity is now a pseudonym for a novel metallic state exhibiting anomalous infra-red (branch-cut) continuum features in one- and two-particle responses. Here, we employ dynamical mean-field theory (DMFT) using low-temperature continuous-time-quantum Monte-Carlo (CTQMC) solver for an extended periodic Anderson model (EPAM) model to investigate unusual magnetic fluctuations in the strange metal. We show how extinction of Landau quasiparticles in the orbital selective Mott phase (OSMP) leads to (i) qualitative explication of strange transport features and (ii) anomalous quantum critical magnetic fluctuations due to critical liquid-like features in dynamical spin fluctuations, in excellent accord with data in some f-electron systems.

Finding the exotic strange metal phase in cuprates and certain f-electron systems is by now a fundamental problem. Strong electronic correlations change the analytic structure of the charge- and spin-fluctuation propagators from an infra-red (Landau overdamped at certain critical wave-vectors) pole to branch-cut structure. Experimentally, for f-electron systems, proximity to antiferromagnetic quantum critical phase (AF-QCP) seems to be necessary for generating the soft electronic 'glue' that results in high T_c superconductivity (HTSC). Whether the critical electronic glue responsible for Cooper pairing that show unconventional superconductivity on the border of $T = 0$ destruction of anti-ferromagnetic order is linked to AF critical fluctuations or to dualistic fluctuations accompanying Kondo breakdown (1) is thus an open issue. Rephrased somewhat, the issue is: what is the microscopic origin of the anomalous fluctuation spectra in strange metals? How are we to understand their links to equally anomalous fermiology and transport in the same picture? One often sees unique signatures of strange metallicity with fractional exponent and ω/T-scaling in inelastic neutron scattering (INS) (2), almost T-independent spin relaxation rate in nuclear magnetic resonance (NMR) (3), quasilinear-in-T resistivity without saturation and unusual low-energy optical response, $\simeq \omega^{-\eta}$ with $0.7 \leq \eta < 1.23$ (4) in cuprates as well as the 122-Fe arsenides. Theoretical rationalizations of these findings contrast strongly, ranging from local Kondo destruction, to quantum phase transitions (QPTs) associated with proximity to various ($T = 0$) ordered or momentum-selective Mott phases. These may not be mutually exclusive scenarios, but it must be emphasized that specific local quantum critical signatures mentioned above, which *require* negligible momentum dependence of the self-energy and two-particle vertices, do not easily fit into traditional itinerant Hertz-Moriya-Millis views of quantum criticality.

These issues motivate our work: we show how extinction of Landau Fermi Liquid (LFL) quasiparticles in an OSMP also enables a natural understanding of the unique branch-cut continuum form of spin- and charge fluctuations solely as a consequence of the dual (itinerant-localized) character of carriers. An OSMP is required, both in the FL^* theory (5) as well as in the Kondo destruction view, both of which describe the same physical picture of a correlated LFL phase separated from the ordered AF phase by a spin liquid region, in which the local moments decouple from the itinerant carriers. An underlying OSM transition in a microscopic fermionic theory can naturally lead to such a situation. We begin with the EPAM (1; 6) $H = H_{band}+H_{int}+H_{hyb}$, where $H_{band} = \sum_{k,\sigma} \epsilon_k c^\dagger_{k,\sigma} c_{k,\sigma} + \sum_{<i,j>} t_{ff} f^\dagger_{i,\sigma} f_{j,\sigma}$, $H_{int} = U_{ff} \sum_i n_{if\uparrow} n_{if\downarrow} + U_{cc} \sum_i n_{ic\uparrow} n_{ic\downarrow} + \sum_{i,\sigma,\sigma'} U_{fc\sigma\sigma'} n_{if} n_{ic}$ and $H_{hyb} = \sum_{k,\sigma} V_{fc}(k)(f^\dagger_{k,\sigma} c_{k,\sigma} + h.c)$.

Here ϵ_k, t_{ff} are kinetic energies, U_{cc}, U_{ff}, U_{fc} are local Hubbard interactions (intra and inter orbital) and V_{fc} is inter-orbital hybridization terms for c and f electrons respectively. Such a model is a natural one for describing mixed valent f-electron systems such as Yb- and Ce-based intermetallics. Theoretically, fluctuating valence and the possibility of associated soft valence fluctuations, along with its interplay with spin fluctuations are germane issues for study, but related aspects have received less attention in literature. It is natural to work in the rotated

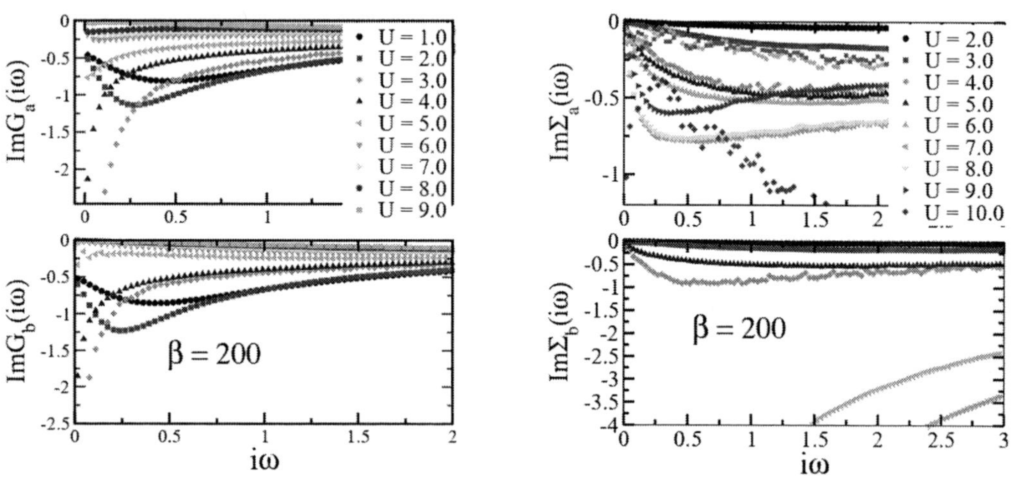

(a) (b)

Figure 1. $ImG_{a,b}(\omega)$ and $\text{Im}\Sigma_{a,b}(i\omega_n)$ as a function of $i\omega$ for a two-orbital square lattice with inter orbital hybridization t_{ab} (witha d-wave form factor) are shown as a function of U.

a, b fermion basis (linear combinations of c, f fermionic operators) which diagonalizes the non-interacting part of H above. For $U_{fc} < U^{(c)}_{fc}$ a stable strongly correlated LFL phase gives way, for $U_{fc} > U^{(c)}_{fc}$, to a selective-Mott metal with an infra-red branch-cut form of the one-electron propagator as a consequence of the strong scattering induced "inverse" orthogonality catastrophe (OC) (6), as found earlier using the multi-band iterated perturbation theory as an impurity solver in the DMFT context. It is obviously of interest to ask whether these appealing features survive upon use of a more exact impurity solver. Here, we have solved H (Eq.1) within DMFT using the continuous-time quantum Monte Carlo (CT-QMC) hybridization expansion solver (7). A very positive feature of our study is in extending CTQMC results to much lower $T = t_a/500$ (upto β=500=23K$\sim$ $2meV$) than hitherto done. Now, local one- and two-particle dynamical responses can be reliably computed, in contrast to diagrammatic solvers like IPT, where two-particle susceptibilities need reliable knowledge of the fully dynamical but local irreducible vertex. This is presently a demanding task, unless they can be argued to be irrelevant, as in large-N approaches to DMFT (5).

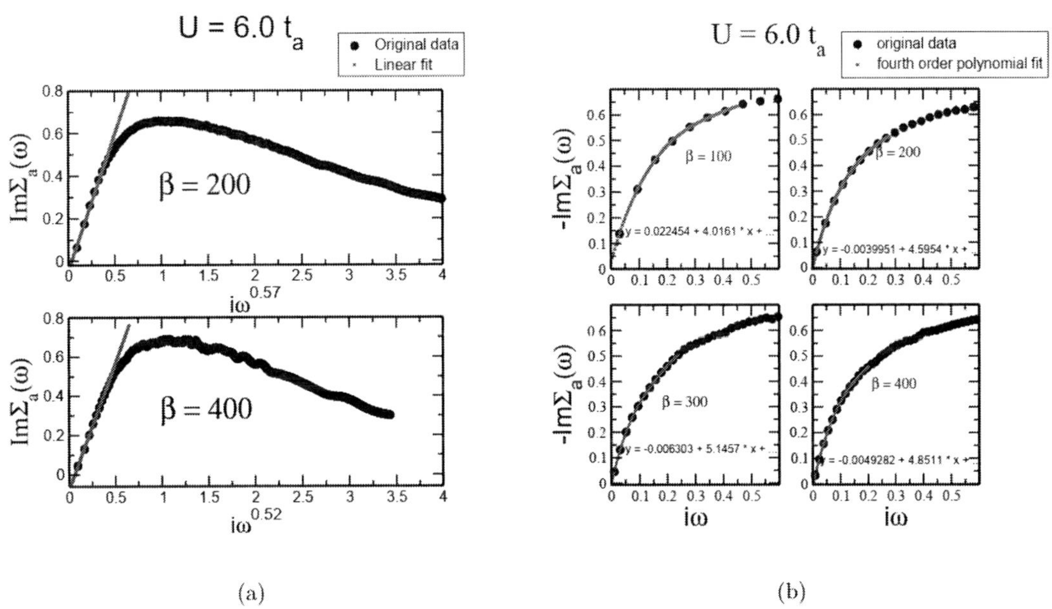

Figure 2. (a) $Im\Sigma_{a,b}(i\omega)$ as a function of $i\omega$ for a two-orbital square lattice with inter orbital hybridization t_{ab} and $U = 6.0$ at low T. Clear power-law behavior in $\text{Im}\Sigma_a(i\omega) = C(i\omega)^{1-\eta}$ with $\eta = 0.43, 0.48$ for $\beta = (k_BT)^{-1} = 200, 400$ up to high energy $O(0.5)$ eV, is visible. (b) Fourth order ploynomial fit to the low energy sector shows the evolution of ω=0 intercept and coefficient of linear term ($\sim$ mass enhancement) with temperature.

In the present work, we obtain the full vertex-corrected two-particle dynamical responses for the EPAM using the hybridization expansion-based CTQMC method as an impurity solver. Thereafter, we explicitly show that the spin and charge susceptibilities possess a long-time (in imaginary-time) quantum-critical form, and use a trick inspired by conformal symmetry to extract an effective analytic form in *real* frequencies. Such a trick has been widely employed in other contexts (5). This will allow us to discuss transport and magnetic fluctuations in the strange metal phase, and to discuss underlying microscopic features leading to the specific anomalies, in detail.

In Fig. 1, we show the imaginary parts of the one-fermion Green functions for the a, b orbital states, along with the corresponding self-energies. It is clear thereby that as U and $U'(= U_{ab} = U/3)$ increase, the semimetallic behavior for $U = 0$ (due to the $\mathbf{k}$-dependent hybridization) changes into a normal, albeit correlated LFL metallicity at moderate U. Beyond $U = 4.0$, however, an orbital-selective differentiation of electronic states clearly reveals itself in both $\text{Im}G_{a,b}(i\omega_n)$ and $\text{Im}\Sigma_{a,b}(i\omega_n)$: the a-states are in an incoherent metal state, while the b-states are in the Mott insulating regime. This dualistic feature exists in the whole parameter region $4.0 < U < 9.0$ for the given set of one-electron parameters in the EPAM, and thus this is an orbital-selective Mott *phase*, rather than a QC point. For $U > 10.0$, we find that a full Mott insulator phase obtains.

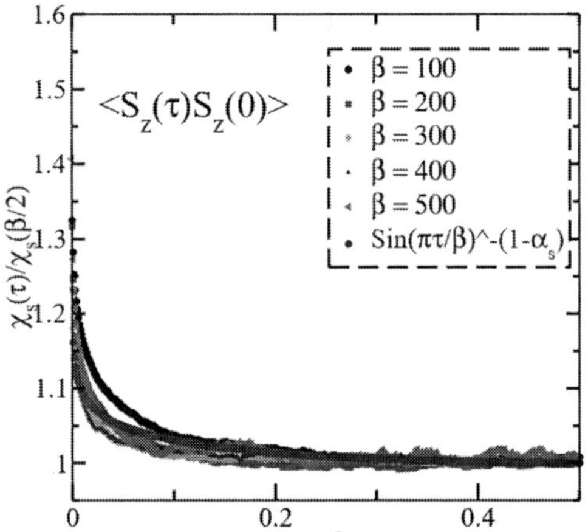

(a)

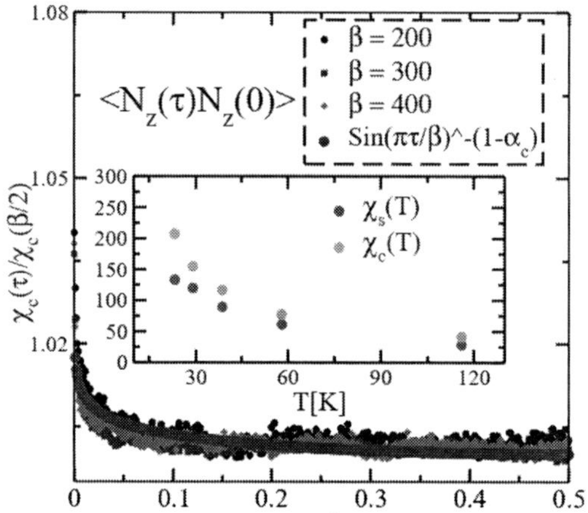

(b)

Figure 3. $Im\chi_{s,c}(\tau)$ as a function of τ/β. Both, the dynamical spin (s) and charge (c) susceptibilities show quantum critical scaling, with $\chi_{s,c}(\tau)/\chi_{s,c}(\beta/2) \simeq [\sin(\pi\tau/\beta)]^{-}(1-\alpha_{s,c})$, with $\alpha_s(0.9) \neq \alpha_c(0.98)$. Within our numerical accuracy, this constitutes a high-D realization of "spin-charge separation". The inset shows the static local susceptibilities where $\chi_c(T)$ is more singular than $\chi_s(T)$ at low temperatures.

Focussing on the OSMP, we now exhibit the a, b-fermion self-energy, $\text{Im}\Sigma_{a,b}(i\omega_n)$, and the full *local* spin susceptibility, $\text{Im}\chi_s(i\omega_n)$, for the orbital-selective Mott phase Fig. 1, 2. Interestingly, we find clear evidence of anomalous fractional power-law exponents in both. In Fig. 2, we see that $Im\Sigma_a(i\omega_n) \simeq (i\omega_n)^{1-\eta}$, with $\eta = 0.43(\beta = 200)$ and $\eta = 0.48(\beta = 400)$. In fact, this behavior extends up to rather high energies $O(0.5)$ eV, which is characteristic of a multi-particle electronic continuum expected to extend to high energies in a quantum critical regime. Remarkably, the *local* part of the dynamical spin susceptibility also exhibits infra-red singular and fractional power-law scaling behavior, characteristic of the strange metal. We see this as follows. Using the DMFT(CTQMC) results, we see that all $\chi_s(\tau)$ curves for $U > 4.0$ fall on each other and, upon a careful fitting with a scaling function characteristic for quantum criticality, we find that $\chi_s(\tau)/\chi_s(\beta/2)$ scales as $[\sin(\pi\tau/\beta)]^{-}(1-\alpha_s)$ with $\eta = 0.9$ Fig. 3. This is precisely the scaling form dictated by conformal invariance in the quantum critical region. It exactly corresponds to the scaling form

$$\chi_s''(\omega) \simeq \omega^{-\eta} F_s(\omega/T) \tag{1}$$

with $F_s(x) = x^{\eta}|\Gamma(\frac{1-\eta}{2} + i\frac{x}{2\pi})|^2 \sinh(x/2)$, whence we immediately read off that $T^{-\eta}\chi_s''(\omega)$ is a universal scaling function of ω/T, with a fractional-power-law dependence. Such a unique form for the dynamical spin susceptibility near magnetic QCPs is experimentally seen in more than a few systems by now, specifically in (not obviously proximate to magnetic order), $CeCu_{6-x}Au_x$ (2), among others. On a theoretical level, extinction of Landau fermionic quasiparticles in $G_b(k, \omega)$ in the OSMP directly manifests in the emergence of a critical branch-cut continuum in (one-spin-flip) spin-fluctuations. Even more interestingly, the dynamical *charge* susceptibility also exhibits similar scaling form, but with an exponent $\alpha_c \neq \alpha_s$, implying that spin and charge fluctuations propagate with distinct velocities. This is our central result, and we show that it leads to very good accord with the unusual normal-state magnetic fluctuation spectra *and* transport in strange metals. Physically, these features emerge as a direct manifestation of emergent, critical "pseudoparticles" driving the extinction of stable Landau-damped FL-like collective modes in the strange metal. This is simply because charge and spin fluctuations are themselves constructed from the now incoherent multiparticle continuum, rather than usual Landau quasiparticles.

What is the underlying physical origin of these emergent anomalous features? As in FL* or Kondo breakdown theories, emergence of critical liquid-like features do *not* need a one-to-one association with proximity to $T = 0$ magnetic order, since DMFT only accesses strong *local* dynamical correlations. In the OSMP, $V_{fc}(k)$ plays an especially distinct role. Explicitly, the tendency of V_{fc} to transfer a a-fermion into an b-fermion band is dynamically blocked in the OSMT. This is because the lower-Hubbard band now corresponds to all singly occupied b-states, so action of V_{fc} must create a doubly occupied (two opposite-spin electrons in the b-orbital) intermediate state. However, this lies in the *upper* Hubbard band in the b-sector and thus the resulting term now has the form $V'_{fc}(n_{i,b,-\sigma}b^{\dagger}_{i,\sigma}a_{j,\sigma} + h.c)$, which couples the a-fermion to a Gutzwiller-projected b-fermion, and thus has no interpretation in terms of a coherent one-electron-like state any more. In this sector, this is a high-energy state, and is asymptotically projected out from the low-energy Hilbert space (space of states with energy less than the selective-Mott gap). It is this emergent projective aspect that is at the root of irrelevance of $V_{fc}(k)$ at one-electron level (thus, this is a Kondo destruction (1)) and emergence of strange metal features we find above. Thus, we find a mechanism for the destruction of the LFL picture to be very closely related to the "hidden-FL" view of Anderson (8). In absence of a lattice Kondo scale, the hopping of an b-electron from site i to the bath or vice-versa (and similarly for the a-electron) thus creates a "suddenly switched-on" local potential for the b-electrons. A direct upshot is that the inter-band spin fluctuations (now local triplet excitons of the type $(a^{\dagger}_{i\sigma}b_{i\sigma'}+h.c)$ with $\sigma' = -\sigma$) thus also experience a local potential that is (incoherently) switched on and off as a function of time. The resulting problem is precisely the "inverse" of the Anderson-Nozieres-de Dominicis X-ray edge problem in the local limit of DMFT, and is a central feature of Anderson's hidden-FL theory (8). At two-particle level, this reflects itself in a divergent number of soft, *local* spin fluctuation modes, manifesting itself as an infra-red singularity along with "local quantum critical" ω/T scaling and anomalous exponents in the spin fluctuation spectrum. This is completely borne out by our CTQMC results (Fig. 2) The resulting spin polarizability, $\Pi(\omega) \simeq \chi^{-1}_{zz}(\omega) \simeq A|\omega|^{1-\alpha}$ in the QC metal, in stark contrast to $\Pi(\mathbf{q},\omega) \simeq -ia\omega$, known in the heavy LFL metal. The fully renormalized $\chi(\mathbf{q},\omega)$ now reads

$$\chi(\mathbf{q},\omega) = \frac{1}{\Pi(\omega) + J(\mathbf{q})} \tag{2}$$

where $J(\mathbf{q}) \simeq -c + (\mathbf{q} - \mathbf{Q})^2 + ...$ (e.g, $Q = (\pi,\pi)$ for Neel AFM). At finite T, $\Pi(\omega,T) = AT^{\alpha}f(\omega/T)$.

Physical Response In The Strange Metal

Remarkably, it turns out that our results afford a consistent and surprisingly good quantitative description of both transport and neutron results in the strange metal. Let us consider the consequences of our numerical findings in more detail in this section, with a specific focus on f-electron metallic systems.

(i) The dynamical spin susceptibility shows an explicit ω/T-scaling with anomalous fractional exponent α_s. This form is well-documented as the characteristic behavior in real systems where quantum critical features cannot be rationalized within a conventional picture based on a quantum Landau-Ginzburg theory. It is qualitatively consistent with observations in

$CeCu_{6-x}Au_x$ (2), and may also be applicable to other f-electron systems where there is independent evidence of the relevance of strong, quasilocal liquid-like critical fluctuations near destruction of magnetic order. From the form of the dynamical spin susceptibility, we can also immediately read off that the spin relaxation rate in nuclear-magnetic resonance (NMR) studies will vary very weakly with T, as $1/T_1 \simeq T^{0.1}$ or that $1/T_1T \simeq T^{-0.9}$. This is quite close to the measured $1/T_1T \simeq T^{-1}$ in near-optimally doped cuprates (3), and it will be of interest to study NMR relaxation rates near the QCPs in f-electron systems to see whether these local critical features are more generic.
(*ii*) The fact that $\mathrm{Im}\Sigma_{a,b}(i\omega) \simeq (i\omega)^{1-\eta}$ with $(1-\eta) = 0.52, 0.57$ for $\beta = 400, 200$ allows study of *dc* and *ac* conductivities without any further approximation within DMFT. This is because irreducible vertex corrections to the current correlation functions for the conductivities in the Bethe-Salpeter equation rigorously vanish in this limit. Transport is now entirely determined by the "simple" bubble diagram composed of the full DMFT one-electron propagators. Explicitly, the *dc* resistivity is now $\rho_{dc}(T) \simeq T^{2(1-\eta)} = T^{1.04}$ for $\beta = 400$ and $T^{1.14}$ for $\beta = 200$. Such an unusual resistivity, $\simeq T$, has long been known as one of the defining features of the strange metal, not only in cuprates, but also in $D = 3$ f-electron systems. In $YbRh_2Si_2$ (YRS), for example, the *dc* resistivity does vary like T^n with n slightly higher than unity (9). The optical conductivity can also be readily estimated as $\sigma(\omega) \simeq \omega^{-2(1-\eta)} = \omega^{-1.04}, T^{-1.14}$. This is indeed consistent with data in underdoped cuprates as well as in the $D = 3$ 122-Fe arsenides (4). OSMP are known to be generic in Fe arsenides by now (10), and our work now explicitly links the anomalous optical response, at least in 122-Fe arsenides, to an underlying OSMP. It would be very interesting to search for such manifestations of anomalous metallicity in f-electron systems, in particular in YRS, where limited evidence of a quasi-linear-in-ω scattering rate and anomalous increase in optical effective mass at low energy indeed exists. Another promising avenue is that charge fluctuations seem more singular than the spin fluctuations (inset of Fig. 3), so one would expect a valence transition to occur prior to onset of magnetism. We leave these issues for future study.

To conclude, we have presented strong numerical evidence for anomalous spin and charge fluctuation spectra emerging in an orbital-selective Mott phase in an extended periodic Anderson model. This is the basic model for a wide variety of mixed-valent f-electron systems, and is also suited to describe multi-orbital transition-metal oxides in a different parameter space. We detail how extinction of the Landau quasiparticle pole structure in the one-electron propagators in the OSMP leads to such a critical fluctuation spectrum, and point out that these features are qualitatively consistent with characteristic signatures of "strange" metallicity in mixed-valent rare-earth systems. We expect these results to have broader applicability.

References

[1] C. Pepin, Phys. Rev. B 77, 245129 (2008).

[2] A. Schröder *et al.*, Nature 407, 351-355 (21 September 2000)

[3] G. Aeppli *et al.*, Science, **278**, 1432-1435 (1997).

[4] C. C. Homes et al., Nature Scientific Reports 3, Article number: 3446 (2013).

[5] T. Senthil et al., Phys. Rev. Lett. 90, 216403 (2003).

[6] M. S. Laad et al., Journal of Physics: Condensed Matter 24, 232204 (2012).

[7] B. Bauer et al., Journal of Statistical Mechanics: Theory and Experiment 2011 (2011).

[8] Philip W Anderson and Philip A Casey 2010 J. Phys.: Condens. Matter 22 164201., P. W. Anderson, Nature Physics 2, 626 (2006).

[9] E. Abrahams *et al.*, Phys Rev B **90**, 045105 (2014), and references therein.

[10] L de Medici *et al.*, Phys. Rev. Lett. **102**, 126401 (2009); S. D. Das *et al.*, Phys Rev B **92**, 155112 (2015) and references therein.

Electronic properties of aluminium and silicon doped (2, 2) graphyne nanotube

Jyotirmoy Deb, Barnali Bhattacharya and Utpal Sarkar*
Department of Physics,
Assam University, Silchar-788011, India

*E-mail: utpalchemiitkgp@yahoo.com

Abstract. With the help of density functional theory (DFT) approach the electronic properties of Al and Si doped (2, 2) graphyne nanotube (GNT) has been investigated. We have chosen two doping positions. In one situation we have replaced one carbon atom by Al/Si atom in the chain position of the GNT and in other case we have substituted the carbon atom in the hexagonal ring by Al/Si atom. The result of the cohesive energy indicates that pristine graphyne nanotube shows highest stability whereas, Al doped at chain site of the GNT exhibits lowest stability. The density of states above and below the Fermi level for all these system is mainly contributed by carbon atom. Band structure analysis shows that the band gap can be tuned by functionalization of GNT with Al and Si atoms; which makes the possibility of using them in nano-electronic devices.

1. Introduction

A huge amount of research work has been carried out on carbon based nanomaterials such as fullerene [1], carbon nanotube (CNT) [2], graphene [3] etc. by various research groups in the previous decades owing to their unique physical and chemical properties. In 1987 another allotropic form of carbon was proposed by Baughman and co-workers. They termed it as graphyne [4, 5] which become an emerging field for researchers [6-10]. Graphyne has a fascinating structure compare to other carbon based compounds. In graphene all the carbon atoms are sp^2 hybridized, whereas in case of graphyne both sp and sp^2 bonded carbon atoms are present in a crystal lattice. In graphyne there are large numbers of acetylene linkage present between the two nearest hexagonal ring as a result of which it shows enhance performance in optical and electronic applications. Theoretically it was found that graphyne exhibits some outstanding properties such as good electrical conductivity, high mechanical and chemical stability [10] which might be useful for designing optoelectronic and nanoelectronic devices, solar cell application, energy storage, gas sensor [6-9, 12, 13].

The structure of graphyne nanotubes (GNTs) are formed by rolling the graphyne sheet into a cylindrical form. The diameter and chirality of the nanotube influences the electronic property of GNT. The electronic and optical properties of graphyne sheet and tube can also be tuned by substitutional doping by various atoms [5-9]. In our present work we have studied the effect on electronic properties of GNT due to doping by Al and Si atom. We are going to investigate how the conducting nature changes due to the interaction of Al and Si atom with GNT.

On the basis of cohesive energy, the stability of above mentioned systems are determined. We have also calculated the electric and magnetic dipole moment and band gap of the considered systems and find the possible applications of Al and Si doped GNT in nano-electronics.

2. Computational details

We have performed all the calculations using density functional theory (DFT) methodology as suggested in siesta 3.2 package [14]. Perdew–Burke–Ernzerhof (PBE) form has been used to represent the exchange-correlation functional of the generalized gradient approximation (GGA) in the calculation [15]. The Double zeta polarized numerical atomic-orbital and Troullier-Martin type norm-conserving pseudo potentials [16] have been used here. The mesh cut-off value of 300 Ryd has been taken and a mixing rate for self consistent calculations was fixed at 0.05 for the grid integration. Relaxation process continues until the magnitude of the forces acting on each atom is less or equal to 0.01 eV/Å. To avoid the interactions between different layers of graphyne system a vacuum of 20 Å has been kept along X and Y directions. 1×1×18 k points was used for sampling of Brillouin zone. To study the most stable configuration of these systems, the cohesive energy is calculated using the following relation,

$$\textbf{Cohesive energy} = \frac{\mathrm{E(system)} - [\mathrm{nE(C)} + \mathrm{mE(D)}]}{\mathrm{n+m}}$$

where E(system), E(C) and E(D) are energies of total system, carbon and doped atoms (D=Si, Al) and n, m are number of carbon and doped atoms present in the system respectively.

For the calculation of dipole moment (μ) of Al and Si doped GNT we have used the following relation;

$$\mu = (\mu_X^2 + \mu_Y^2 + \mu_Z^2)^{\frac{1}{2}}$$

where μ_X, μ_Y, and μ_Z are components of dipole moment along X, Y, and Z direction respectively.

3. Results and discussion

Our present work emphasises on the study of electronic properties of (2, 2) GNT due to presence of Al and Si atoms at different doping positions. We have replaced sp and sp^2 hybridized carbon atom by Al and Si atoms. The optimized geometries are presented in figure 1.

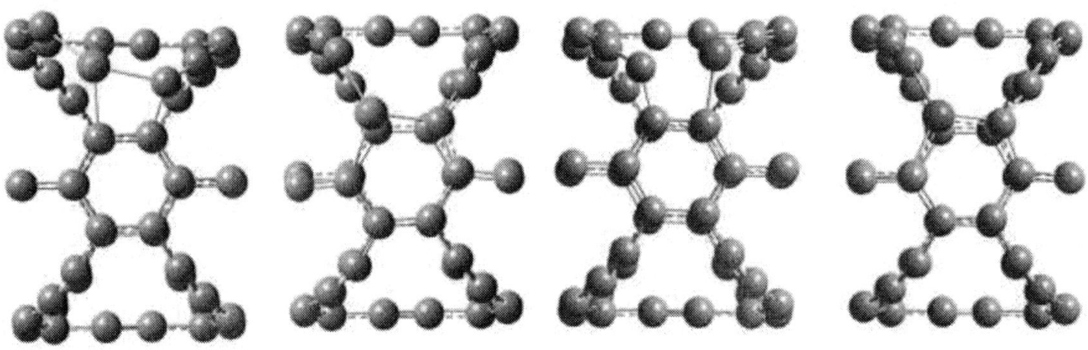

(a) **(b)** **(c)** **(d)**

Figure 1. Optimized geometries of (a) Al doped GNT at chain position; (b) Al doped GNT at ring position; (c) Si doped GNT at chain position; (d) Si doped GNT at ring position.

The lattice constant values for all the above mentioned systems are presented in Table 1. The lattice constant of pristine systems is 6.9155 **Å**, however substitutional doping by Al and Si atom slightly increases it. The increasing trend follows the order GNT < 'Si doped GNT at chain position' < 'Al doped GNT at chain position' < 'Si doped GNT at ring position' < 'Al doped GNT at ring position'.

Table 1 The lattice constant, cohesive energy, energy gap, electric and magnetic dipole moment of Al and Si doped graphyne nanotube

System	Lattice Constant (Å)	Cohesive energy (eV)	Energy gap (eV)		Dipole moment (Debye)	Magnetic moment (μ_B)
(2, 2) GNT	**6.9155**	**-9.3071**	**0.7520**		**0.0000**	**0.0000**
(2, 2) GNT + Si (ring)	**6.9657**	**-9.2025**	**0.4181**		**0.9811**	**0.0000**
(2, 2) GNT + Si (chain)	**6.9181**	**-9.2015**	**0.2212**		**0.7939**	**0.0000**
(2, 2) GNT + Al (ring)	**6.9677**	**-9.1641**	**↑0.4558**	**↓0.4969**	**0.6691**	**0.9707**
(2, 2) GNT + Al (chain)	**6.9270**	**-9.1586**	**↑0.3172**	**↓0.1668**	**0.2539**	**1.0018**

3.1 Cohesive energy

The results of cohesive energy enable us to determine the stable configuration of the system. The more and more negative cohesive energy physically signifies the greater stability of the system. The pristine graphyne nanotube has highest stability. Doping by Si and Al atom decreases the stability of the system. The stability due to doping by Al and Si atoms at two different doping positions is in the order GNT > 'Si doped GNT at ring position' > 'Si doped GNT at chain position' > 'Al doped GNT at ring position' > 'Al doped GNT at chain position'.

3.2 Electric and magnetic dipole moment

Spin polarized calculation reveals that splitting of spin does not take place in case of (2, 2) pristine GNT and Si doped GNT. Thus these systems do not possess magnetic character. Whereas, Al doped at chain and ring position of GNT initiates the spin splitting. This may occur due to redistribution of charge carriers between Al atom and carbon atom of the GNT. Thus the Al doped GNT systems are showing magnetic characteristics. The magnitudes of magnetic moment are 0.9707 μ_B and 1.0018 μ_B for Al doped at ring and chain position respectively. This considerable change in magnetic moment can be measured with the help of a SQUID magnetometer [17] and this may enable us to design a chemical or gas sensor based on Al doped GNT.

There is no induced electric dipole moment in case of intrinsic GNT system but doping by Si and Al atom originates electric dipole moment in the GNT system. The trend of dipole moment of the above mentioned systems follow the order 'Si doped GNT at ring position' > 'Si doped GNT at chain position' > 'Al doped GNT at ring position' > 'Al doped GNT at chain position'.

3.3 Band structure analysis

The spin polarized band structure is represented in Fig. 2. The result indicates that doping by Al and Si atom at different doping position significantly decreases the band gap of GNT. The pristine GNT has a band gap of 0.752 eV [18] and showing semiconducting behaviour. Al and Si doped at chain and ring position of GNT exhibits same nature of conduction as that of pristine GNT. In case of Al atom at chain position of GNT, the majority spin channel has a band gap of 0.3172 eV whereas minority spin channel possess a band gap of 0.1668 eV and the same atom when placed at ring position of GNT, both the spin channel have a band gap of 0.4558 eV for majority spin and 0.4969 eV for minority spin. When one carbon atom at chain site is substituted by Si atom, the band gap of pristine GNT system decreases to 0.2212 eV and when Si atom replaces the carbon atom at ring position the band gap of the system is 0.4181 eV. We have also noticed that the valence band maximum (VBM) and conduction band minimum (CBM) for the above mentioned systems are located at the X- point of the rectangular Brillouin zone and due to this feature all these systems are direct band gap semiconductors.

The modulation of band gap is possible due to doping by Al and Si atom. The reasonable tuning of band gaps depends not only on dopant atoms but also on doping positions. Adjustment of band gap is one of the major requirements for a system to be used in nano-electronics devices and here we have fulfilled this criterion by doping with Si and Al atom at chain and ring position respectively.

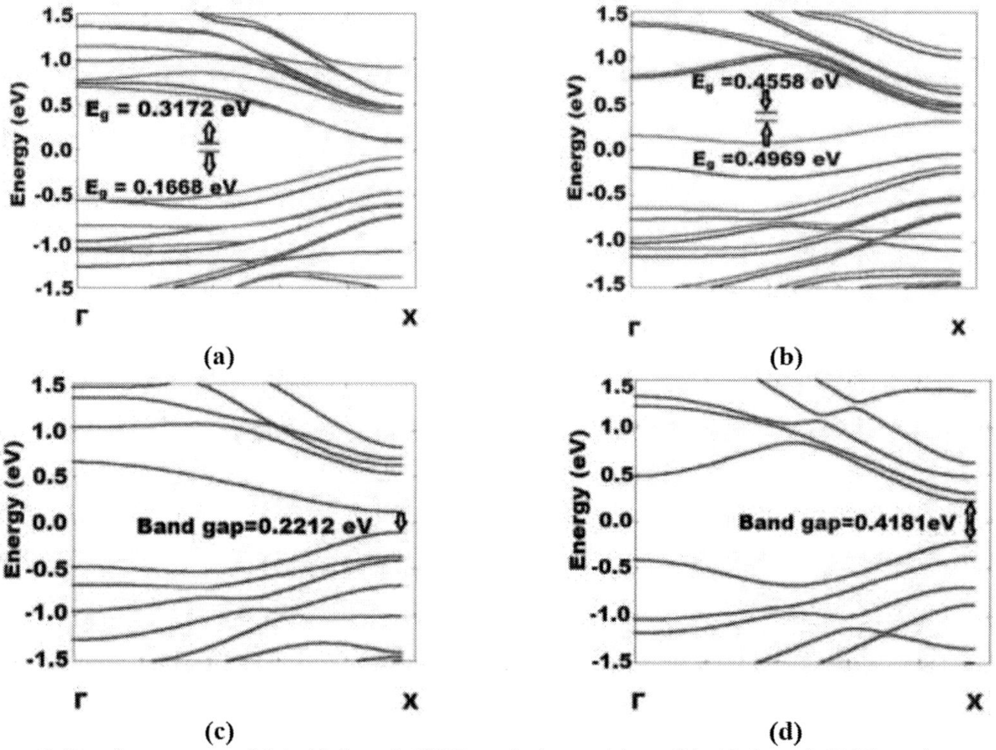

Figure 2. Band structure of (a) Al doped GNT at chain position; (b) Al doped GNT at ring position; (c) Si doped GNT at chain position; (d) Si doped GNT at ring position.

3.4 Density of states

We have presented projected density of states (PDOS) in **Figure 3**. Positive and negative values of PDOS represent up spin and down spin respectively. We mainly focus on the electronic states near the Fermi level, to study the changes in the electronic properties of GNT due to the presence of dopant atoms. In case of Al doped at chain and ring position of GNT, the contribution of carbon atom both for majority and minority spin channel is predominant in the valence band (VB) compare to that of conduction band (CB) (Fig 3a), whereas Al atom mainly participates in the CB. The contribution of aluminium atom is more significant when Al is situated at ring position of GNT (Fig 3b). PDOS plot for Al at chain and ring position of GNT also reveals that density of states is not symmetrical (upper panel of figure 3a and 3b) for both the spin channels but in case of Si at chain and ring position of GNT, the density of states for up and down spin is symmetrical (upper panel of figure 3c and 3d) in nature as a result no magnetisation is induced when Si is doped at chain and ring site of GNT. In case of Si doped at chain and ring position of GNT, the contribution of carbon atom again dominates in VB compare to CB, whereas in case of Si doped at chain position of GNT (Fig 3c), Si atom mainly participates in VB but when Si is at ring site of GNT (Fig 3d) the same atom mostly contributes in CB.

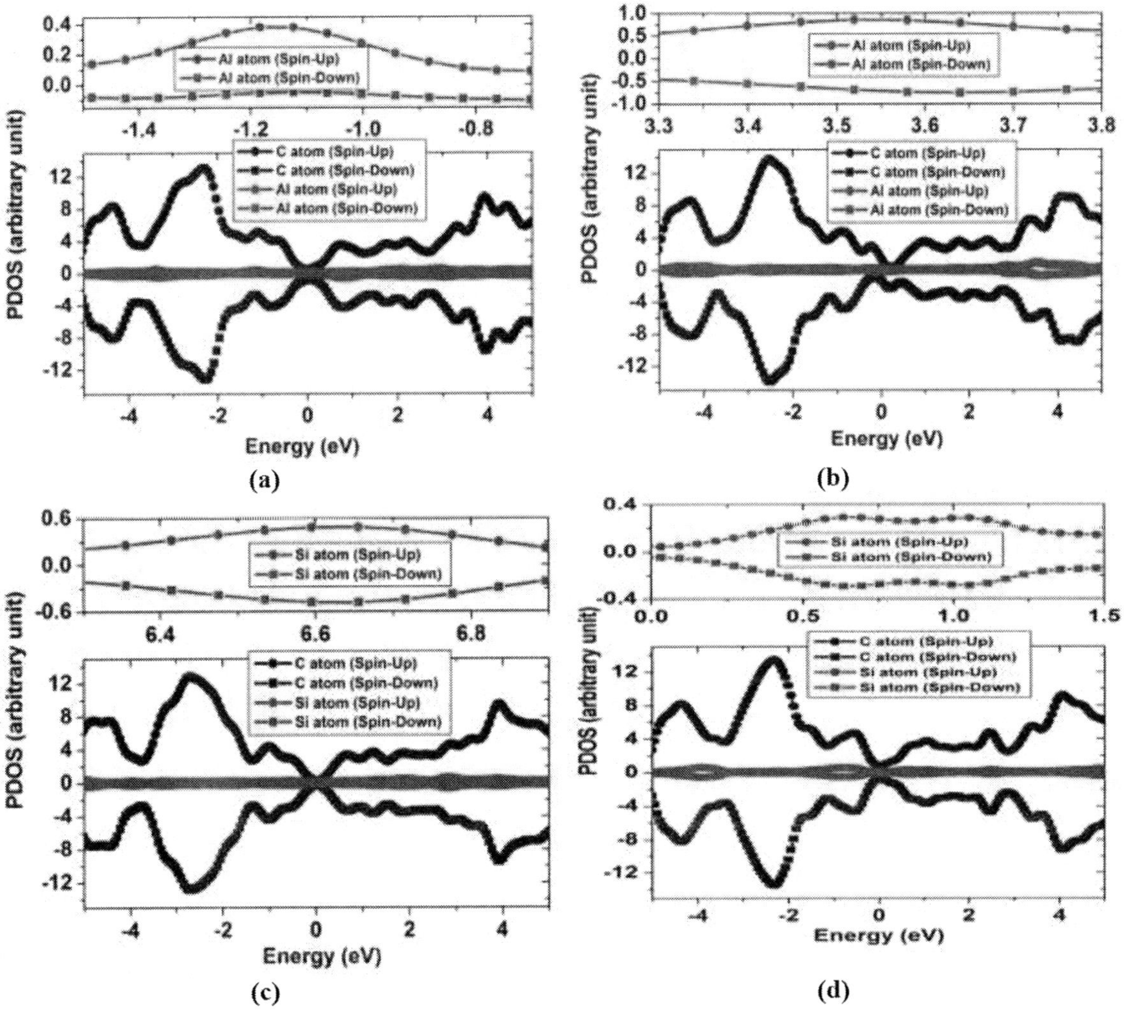

Figure 3. PDOS plot for (a) Al doped GNT at chain position; (b) Al doped GNT at ring position; (c) Si doped GNT at chain position; (d) Si doped GNT at ring position.

4. Conclusion

We have analysed the effect in electronic properties of GNT due to functionalization by Al and Si atoms using first principle calculations. On the basis of cohesive energy we can infer that stability of the GNT system decreases due to replacement of carbon atom by Al and Si atom. Among all the structures Al doped at chain position of GNT shows lowest stability. Band structure analysis confirms that all the structures are direct band gap semiconductor and doping by Al and Si atom decreases the band gap significantly. This property enables us to use these systems in technological applications where tuning of band gap is required. Al doped GNT exhibits magnetic character. Carbon atom mainly participates in the density of state for all these system. Substitutional doping by Al and Si atom influences the electronic properties of GNT; which makes the possibility of using them in nano-electronic applications.

Acknowledgement

Dr. U. Sarkar thanks International Centre for Theoretical Physics, Trieste, Italy for hosting him as a regular associate.

Journal of Physics: Conference Series **759** (2016) 012038 doi:10.1088/1742-6596/759/1/012038

References

[1] Kroto HW, Heath JR, O'Brien SC, Curl RF and Smalley RE 1985 *Nature* **318** 162

[2] Iijima S 1991 *Nature* **354** 56

[3] Novoselov KS, Geim AK, Morozov SV, Jiang D, Zhang Y, Dubonos SV, Grigorieva IV and Firsov A A 2004 *Science* **306** 666

[4] Baughman RH, Eckhardt H and Kertesz M 1987 *J Chem Phys* **87** 6687

[5] Michio Inagaki and Feiyu Kang 2014 *J. Mater. Chem. A* DOI: 10.1039/c4ta01183

[6] Bhattacharya B, Singh N B, Mondal R and Sarkar U 2015 *Phys Chem Chem Phys* **17** 19325

[7] Bhattacharya B, Singh N B and Sarkar U 2015 *Int J Quantum Chem* **115** 820

[8] Sarkar U, Bhattacharya B and Seriani N 2015 *Chemical Physics* **461** 74

[9] Singh N B, Bhattacharya B and Sarkar U 2014 *Struct Chem* **25** 1695

[10] Bhattacharya B, Singh N B and Sarkar U 2014 *Journal of Physics: Conference Series* **566** 012014

[11] Yang Y and Xu X 2012 *Comput. Mater. Sci.* **61** 83

[12] Kang B, Liu H and Lee JY 2014 *Phys Chem Chem Phys* **16** 974

[13] Turabekova MA, Dinadayalane TC, Leszczynska D and Leszczynski J 2012 *J Phys Chem C* **116** 6012

[14] Deb J, Bhattacharya B, Singh N B and Sarkar U 2016 *Struct Chem* **27** 1221

[15] Ordej´on P, Artacho E and Soler JM 1996 *Phys Rev B* **53** R10441

[16] Perdew JP, Burke K and Ernzerhof M 1996 *Phys Rev Lett* **77** 3865

[17] Troullier N and Martins J 1990 *Solid State Commun* **74** 613

[18] Lee J W, Kim H S, Lee J Y and Kang J K 2006 *Appl. Phys. Lett.* **88** 143126

[19] Deb J, Bhattacharya B and Sarkar U 2016 *AIP Conference Proceedings* **1731** 050081

Modeling of absorption and scattering properties of core -shell nanoparticles for application as nanoantenna in optical domain

Jutika Devi[1], Rashmi Saikia[2] and Pranayee Datta[1]
[1]Department of Electronics & Communication Technology, Gauhati University, Guwahati, Assam-781014, India
[2]Department of Electronics & Communication Engineering, Tezpur University, Napaam, Assam-784028, India

E-mail: jutika.electronics@gmail.com

Abstract. The present paper describes the study of core-shell nanoparticles for application as nanoantenna in the optical domain. To obtain the absorption and extinction efficiencies as well as the angular distribution of the far field radiation pattern and the resonance wavelengths for these metal-dielectric, dielectric-metal and metal-metal core-shell nanoparticles in optical domain, we have used Finite Element Method based COMSOL Multiphysics Software and Mie Theory. From the comparative study of the extinction efficiencies of core-shell nanoparticles of different materials, it is found that for silica - gold core - shell nanoparticles, the resonant wavelength is greater than that of the gold – silver, silver-gold and gold-silica core - shell nanoparticles and also the radiation pattern of the silica-gold core-shell nanoparticle is the most suitable one from the point of view of directivity. The dielectric functions of the core and shell material as well as of the embedded matrix are extremely important and plays a very major role to tune the directivity and resonance wavelength. Such highly controllable parameters of the dielectric - metal core - shell nanoparticles make them suitable for efficient coupling of optical radiation into nanoscale structures for a broad range of applications in the field of communications.

1. Introduction

The concept of nanoantennas at optical frequency has recently opened up new fields of experimental and theoretical research in nanotechnology and antenna science. The growing interest in optical nanoantennas can be attributed to the ability of metal nanoparticles to support plasmon resonances.The strongly enhanced surface plasmon resonance of noble metal nanoparticles at optical frequencies makes them excellent absorbers as well as scatterers. It is well known that the plasmon resonance wavelength of metal nanoparticles is sensitive to the nanoparticle's size, shape and the surrounding medium. The possibility of tuning the localized surface plasmon resonance wavelength, through the visible to near infrared region, makes metal nanoparticles very important and promising for the potential applications in the field of communication, quantum optics, information processing and in several linear and nonlinear optical processes. In addition to pure metal nanoparticles, in recent years, core - shell nanoparticles, in which nanoparticles of one material(dielectric/metal) are coated with a thin layer of another material (metal/dielectric), have received particular attention due to their many unique properties[1].Oldenburg et al [2], from Mie scattering theory, obtained that for silica-gold core-shell nanoparticles of core radius (60±6)nm and shell thickness (11±4)nm, dipole plasmon resonance occurs at 830nm, whereas for core radius (135±6 nm) and shell thickness (15±4)nm, the dipole resonance peak shifts to 1350nm.Dipole resonance peak having the highest scattering intensity out of all the other modes (quadrupole, octopole etc) is of great significance from the application point of view.

In the present study, metal-dielectric, dielectric-metal and metal-metal core-shell nanoparticles (eg. gold-silica, silica-gold, gold-silver, silver –gold) are modelled and simulation is done using COMSOL Multiphysics software to obtain the absorption and extinction efficiencies and the resonance wavelengths in the optical domain. The far field radiation patterns for the nanoantennas modeled by core-shell nanoparticles of different core to shell size ratios have also been studied using Mie theory [2]. Keeping in view the significance of dipole resonance and availability of characterization facilities in our lab for extending the present work to application domain, core size - shell thickness variation, for simulation, is limited within the range (66-30) nm - (5-41) nm.

2. Modeling and Simulation

2.1 Theoretical Background: The physical understanding of light scattering by small particles begins with the electric dipole concept. The polarization (i.e. the induced dipole) of materials exposed to electromagnetic fields is determined by dielectric function.

2.1.1 The dielectric function for metal particles are decomposed into two parts [3, 4].One component is the Drude free electron term and the other part is the substantial contribution of the inter band electrons. The complex permittivity of bulk metal particles can be estimated by Lorentz-Drude model and is expressed as:

$$\varepsilon_{bulk}=1-\frac{\omega_{pb}^2}{\omega^2+i\omega\gamma_0}+\frac{\omega_{pb}^2}{\omega_0^2-\omega^2-i\omega\gamma_b} \qquad (1)$$

where ω_{pb} is bound electron plasma frequency, ω_0 is bound electron resonant frequency, $\frac{1}{\gamma_b}=T_b$ is bound electron decay time, ω is the angular frequency of the incident radiation and $\frac{1}{\gamma_0}=T_0$ is the free electron scattering time.

For metal nanoparticles of size smaller than the mean free path, the decay time (γ_f) has been proved to be size dependent. The size dependency can be realized by the relation

$$\gamma_f=\frac{1}{T_0}+\frac{2v_f}{d} \qquad (2)$$

v_f is the Fermi velocity and d is the diameter of the particle.
Thus equation (1) can be modified to

$$\varepsilon_{(d,\omega)}=1-\frac{\omega_{pf}^2}{\omega^2+i\omega\gamma_f}+\frac{\omega_{pb}^2}{\omega_0^2-\omega^2-i\omega\gamma_b} \qquad (3)$$

This modified mathematical expression for decay time can be used to calculate the complex dielectric function of metal nanoparticles of different sizes at different frequencies.

For estimation of complex effective dielectric function of core-shell nanoparticles, the Effective Medium Approximation models are generally used. Effective Medium Theories are physical models based on constitutive properties of individual components and their volume fractions in the composite. There are many Effective Medium Theories and each theory is more or less accurate under different conditions. In most of the works at the nanoscale till date, Maxwell-Garnett model, Bruggeman model and Loyenga model [5] below the percolation threshold are used. But in all these models, the coating of shell materials on the core materials are not taken into account and the core-shell nanoparticles embedded in a host matrix are considered as two component systems considering the dielectric function of core material only. In our study, we have used Hashin-Shtrikman model [5] to calculate the complex dielectric function of core-shell nanoparticle first and then the core-shell nanoparticles & the host matrix are considered as two component system thus making the system more realistic. In the present work, Drude-Lorentz model is used for calculation of optical properties of metal nanoparticles first, and then the Hashin-Shtrikman bound for core-shell nanoparticles are applied, and based on the literature available on basic formulation of Rayleigh model [6], the Rayleigh model is finally used to calculate the effective complex dielectric function of the core-shell nanoparticle based nanocomposites.

2.1.2 The cross-sections for scattering (C_{sc}) and absorption (C_{ab}) are obtained from the scattering field radiated by this dipole, which is induced by the incident plane wave. The resulting expressions are

$$C_{sc}=\sigma_{geom}Q_{sc}\ ,\quad Q_{sc}=\frac{8}{3}q^4\left|\frac{\varepsilon_d-1}{\varepsilon_d+2}\right|^2, C_{abs}=\sigma_{geom}Q_{abs} \text{ and } Q_{abs}=4qIm\left|\frac{\varepsilon_d-1}{\varepsilon_d+2}\right| \quad (4)$$

Where $\sigma_{geom}=\pi a^2$ is the geometrical crosssection, a is the radius of the nanoparticle, $q=ka$ is the dimensionless size and ε_d *is* the relative dielectric function.

The extinction cross section is $C_{ext}=C_{sc}+C_{abs}$ (5)

2.2 Method: Modelling and simulation of the optical properties of core-shell nanoparticles are of great importance to achieve insight of the surface Plasmon phenomena. In this study, an analytical model based on an approximate Mie theory has been designed using Finite Element Method based COMSOL Multiphysics software. The simulation domain comprises of a core-shell nanoparticle with an air domain truncated by a perfectly matched layer (PML). The Maxwell's equations for the scattering of electromagnetic radiation by spherical particles have been solved with the model. The model geometry is shown in figure1. In this model, the wavelength dependent optical properties such as scattering cross section and extinction cross section have been simulated. The far field calculations are done on the inner boundary of the PML. The model is nonuniformly meshed ensuring approximately 10 mesh points between the core-shell nanoparticles, the scattering and far field domain.

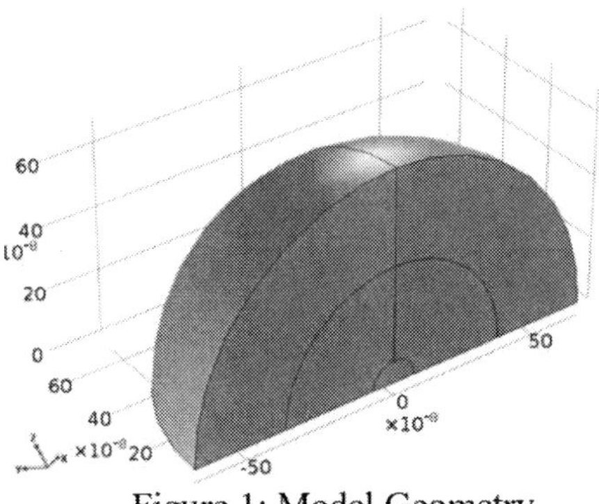

Figure 1: Model Geometry

The symmetry of the system is maintained by simulating half of the system. The simulation calculates the local electromagnetic field at every mesh point with an incident plane wave propagating in x direction with the electric field polarized along the z-axis. This imposes the following boundary conditions on the symmetry planes:

Symmetry in electric field:

$$nxH\big|_{y=0}=0 \qquad (PMC) \qquad (6)$$

Symmetry in magnetic field:

$$nxE\big|_{z=0}=0 \qquad (PEC) \qquad (7)$$

For the scattered field formulation, the background incident plane wave is defined as:

$$E_{inc}=\{0,0,E_0e^{-jk_0x}\} \qquad (8)$$

where E_0 is the plane wave amplitude, $k_0=\frac{\omega}{c}$ is the wave vector in the air, ω is the circular frequency and c is speed of light in air.

Journal of Physics: Conference Series **759** (2016) 012039 doi:10.1088/1742-6596/759/1/012039

If a plane wave is incident on an object, then the wave amplitude from the scattering object, at a large distance, is given by:

$$\psi(r) \approx e^{ikx} + f(\theta)\frac{e^{ikr}}{r} \quad (9)$$

where $f(\theta)$ is the scattering amplitude which is a function of scattering angle θ. COMSOL Multiphysics includes a built –in far-field computation node that calculates the far-field variables defined as:

$$E_{far} = \lim_{r\to\infty} r\, E_{sca} \quad (10)$$

where E_{sca} represents the scattered electromagnetic wave.

3. Results and Discussions

We have modelled gold-silica, silica-gold, silver –gold, gold-silver core-shell nanoparticles with different core to shell size ratio as well as gold nanoparticles of varying size. The surrounding medium is considered as water. The resonance wavelength of the simulated extinction spectrum for single gold nanoparticle and for various core-shell materials are listed in table 1(a) to 1(e).Table 1(a) shows that for gold nanoparticles in water, resonance wavelength decreases with decreasing particle size. This theoretical finding is supported by experimental observations [7].

Table 1: Resonance Wavelength for different sizes (a) Gold (b) Silica Core - Gold Shell (c) Gold Core – Silica Shell (d) Silver Core – Gold Shell (e) Gold Core – Silver Shell nanoparticles.

Table 1(a): Gold Nanoparticle

Core Size (nm)	Resonance Wavelength (nm)
66	556
60	552
50	537
30	520

Table 1(b): Silica Core – Gold Shell

Core Size (nm)	Shell size (nm)	Resonance Wavelength (nm)
66	5	980
60	11	720
50	21	550
30	41	650

Table 1(c): Gold Core - Silica Shell

Core Size (nm)	Shell size (nm)	Resonance Wavelength (nm)
66	5	680
60	11	500
50	21	720
30	41	200

Table1 (d): Silver Core – Gold Shell

Core Size (nm)	Shell Size (nm)	Resonance Wavelength (nm)
66	5	600
60	11	650
50	21	620
30	41	700

Table1 (e): Gold Core – Silver Shell

Core Size (nm)	Shell Size (nm)	Resonance Wavelength (nm)
66	5	780
60	11	620
50	21	610
30	41	580

Table1(a)-(e) reveal that for gold nanoparticles, variation in resonance wavelength is 36 nm for size variation of 36 nm, whereas, for core-shell nanoparticles, shell thickness variation of only 6 nm, keeping total size fixed at 71 nm, causes resonance wavelength variation of (50-260) nm. Thus, for core – shell nanoparticles, resonance wavelength is found to be very sensitive to shell thickness [8]. Moreover, it is found that out of all the core-shell configurations of Table 1, silica - gold core -shell nanoparticles of core size-shell thickness 66nm-5nm & 60nm-11nm show more shift in resonance wavelength with respect to corresponding gold nanoparticle. For silica-gold nanoparticles of core size 66nm-shell thickness 5 nm, resonance wavelength is found to be the highest (980nm), which falls in NIR region. For silica-gold core-shell nanoparticles, varying core size, keeping shell thickness fixed, it is found that for each case, highest resonance wavelength occurs for core size at ~ 66nm. Keeping core size fixed at 66nm and varying shell thickness, highest resonance wavelength is found to be at shell size ~ 5nm (not shown in tabular form). It is worth- mentioning that materials with resonance wavelength in the IR & NIR region are of great technological importance for a wide range of applications that include optical communications, medical imaging, treating tumours ,sensing, microscopy, spectroscopy, remote monitoring of industrial equipment etc.[9,10].The normalized extinction efficiency for 66nm-5nm core-shell nanoparticles of table 1 is shown in figure 2. For gold – silver core-shell nanoparticle, extinction efficiency is slightly higher than that of silica-gold core-shell nanoparticle but the resonance wavelength is much lower.

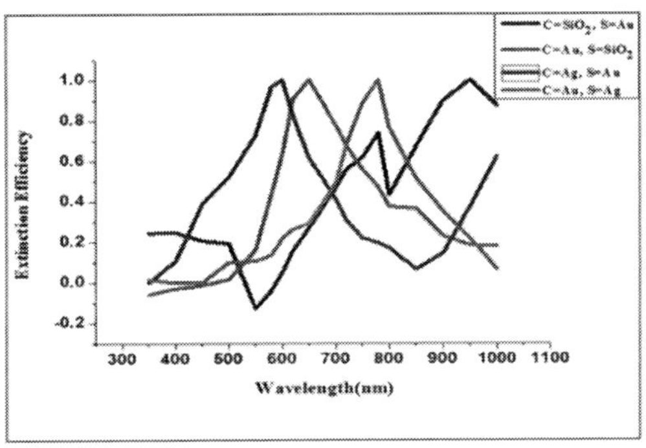

Figure 2: The extinction efficiency vs. wavelength plot for different combinations of core-shell materials embedded in water when the size of the core is 66nm and shell thickness is of 5 nm.

Thus, getting some encouraging results with 66 nm-5 nm silica-gold core-shell nanoparticles, we have simulated the extinction spectrum for silica core - gold shell nanoparticles with varying core to shell size ratio, keeping total size fixed at 71 nm, some of which are shown in figure 3. Result of figure 3 is an added support to the usefulness of 66 nm – 5 nm silica-gold core-shell nanoparticles, from application point of view.

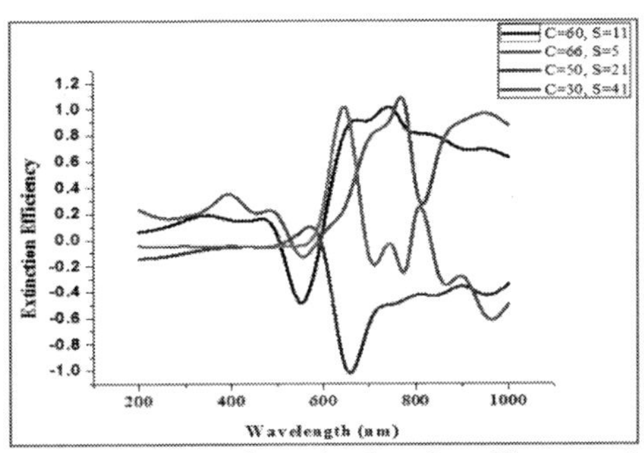

Figure 3: The extinction efficiency vs. wavelength plot for silica - gold core - shell nanocomposites for different core to shell size ratio.

The spatial distribution of scattered light intensity for gold & core-shell nanoparticles is also studied in terms of far field radiation pattern at resonance wavelengths specified in table 1(a) to 1(e) and two of them are shown in figure 4(a) and 4 (b). Out of all the samples, directivity of silica-gold core-shell nanoparticles with 66nm core and 5nm shell size is found to be highest.

Thus, it is found from the present study that by adjusting the core size and shell thickness precisely as well as by choosing proper material combination for core and shell, one can tune the directional pattern at the specified resonance wavelengths.

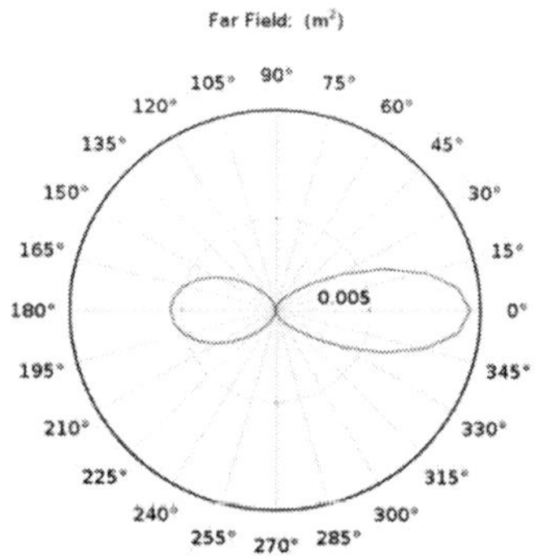

4(a)

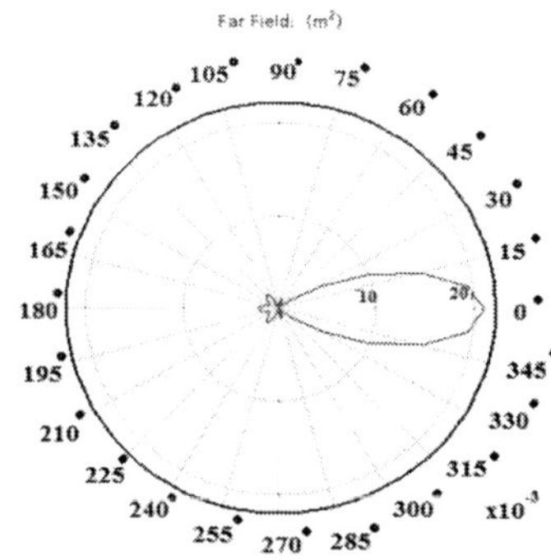

4(b)

Figure 4: Far field radiation pattern for (a) single gold nanoparticle of size 66 nm (b) silica – gold core – shell nanoparticles with core size 66 nm and shell thickness 5 nm.

4. Conclusion

From this study, it can be concluded that by choosing different materials for core & shell as well as by varying the core to shell size ratio, one can obtain optimized size ratio, for the maximum shift in resonance wavelength with respect to single component nanoparticle. Again, from the comparative study of the extinction efficiencies of core-shell nanoparticles of different materials, it is found that for silica - gold core - shell nanoparticles, the resonant wavelength is greater than that of the gold –silver, silver – gold and gold - silica core - shell nanoparticles and also the radiation pattern of the silica – gold core – shell nanoparticle is the most suitable one from the point of view of directivity. The dielectric functions of the core and shell material as well as of the embedded matrix are extremely important and play a very major role to tune the particular directivity and resonance wavelength. Such highly controllable parameters of the dielectric - metal core - shell nanoparticles make them suitable for efficient coupling of optical radiation into nanoscale structures for a broad range of applications in the field of communications as well as in medical science.

References

[1] Oldenburg et al. 1999 *Applied Physics Letters* **75**(8).
[2] Dregely D et al. 2011 *Nature Communications* **1268**(doi:10.1038/ncomm1268).
[3] Novotany L 2007 *Physical Review Letters* **98(26)** 266802.
[4] Ordal M. A. et al. 1983 *Applied Optics* **22** 7.
[5] Sadeghi H,Zolanvar A.,Ranjgar A. 2014 *American Journal of Physics* 72.
[6] Akhtar M J et al. 2014 *Frequenz- Journal of RF Engineering and Telecommunications* **68** 12.
[7] S. Link and M.A. El-Sayed 1999 *J. Phys. Chem.B* **103** 4212.
[8] Kalele S et al. 2006 *Current Science* **91** 8.
[9] Xiaoyan Y. Z. Xiong et al. 2015 *Nature* **6** 18872.
[10] Jingjing Li, Alessandro Salandrino, and Nader Engheta 2009 *Physical Review B* **79**, 195104.

First principles calculation of thermoelectric parameters of Monolayer- and Bilayer-Graphene and Heterostructures of Graphene and *h*-BN

Ransell D'Souza[a] **and Sugata Mukherjee**[b,*] [1]

S.N. Bose National Centre for Basic Sciences, Block JD, Sector III, Salt Lake, Kolkata 700098, India

Abstract. Electrical conductivity and Seebeck coefficient were calculated using Boltzmann transport theory based on first-principles bandstructure calculations for the monolayer- and bilayer-graphene. We also present calculations of the thermoelectric parameters of Graphene/*h*-BN/Graphene sandwiched heterostructures. These results will be discussed in the light of existing experimental data.

1. Introduction

Study of the thermoelectric properties of nanomaterials has become a topic of great interest in the recent past [1, 2, 3]. Electrical conductivity (σ) and Seebeck coefficient (S) of monolayer graphene (MLG) was measured in great detail by Kim and coworkers [4, 5] and others [6]. The conductivity measurements established its dependence with the square root of the carrier density, $\sigma \sim \sqrt{n}$, and the resistivity (ρ) showed the Bloch-Grüneisen type of behaviour with temperature [7]. Similar results were obtained also for the bilayer graphene (BLG) [8], where S showed slightly different behaviour than in MLG.

On the other hand, multilayers of graphene (G) and hexagonal boron nitride (*h*-BN), which is semiconducting with band gap depending on the degree of mixing of its constituents, have attracted a lot of attention for device application. Recently thermoelectric parameters of multiple layers of *h*-BN sandwiched between graphene were measured in a series of experiments by Chen et al [9, 10, 11]. Thermal conductivity (κ) of G/*h*-BN/G heterostructure and that of multilayer and bulk *h*-BN was reported and based on these measurements, the Figure of merit, $ZT = S^2\sigma T/\kappa$ was estimated.

In this paper we report theoretical calculations of thermoelectric parameters of MLG, BLG and in G/*h*-BN/G heterostructure. Our calculations of electrical transport were carried out using Boltzmann transport theory applied to band electrons, with band structure calculated from first-principles DFT based plane wave method. We have also compared our DFT based calculations with a tight-binding bandstructure based calculation of σ and S in MLG and BLG. The thermal transport in G/*h*-BN/G and in multilayer and bulk *h*-BN was carried out using equilibrium Green-Kubo theory based MD simulations performed at constant temperature and

[1] [a]ransell.d@gmail.com; ransell.dsouza@bose.res.in. [b]sugata@bose.res.in; sugatamukh@gmail.com;
*Corresponding Author

finally Z was calculated. Our results show very good agreement with available experimental data.

2. Method of Calculation

2.1. First-principles Bandstructure Calculation

Electronic structure calculations were carried out using density functional theory (DFT) with plane-wave pseudopotential based method as implemented in Quantum Espresso code [12]. Hexagonal unit cells were used for both MLG and BLG calculations and orthogonal unit cells for G/h-BN/G heterostructures. Ultrasoft pseudopotentials [13] were used to describe the core electrons and the generalized gradient approximation (GGA) [14] for the exchange-correlation potential. Monkhorst-Pack k-point grid [15] of $150 \times 150 \times 1$ and $70 \times 70 \times 4$ were used for MLG and BLG in the self-consistent calculations, while $48 \times 48 \times 2$ k-point grid was used for the G/h-BN/G heterostructures; respectively. All self-consistent calculations were performed with a 40Ry kinetic energy cutoff and a 160Ry charge density energy cutoff. For two-dimensional samples, we have kept the vacuum spacing in the unit cell to be 22Å along the z-direction which is large enough to avoid interaction between the cells. As it has been shown that the AB stacking is the most stable [16], the two monolayers in BLG and h-BN layers in G/h-BN/G heterostructures were arranged in the AB stacking, with bond lengths calculated self-consistently. We have

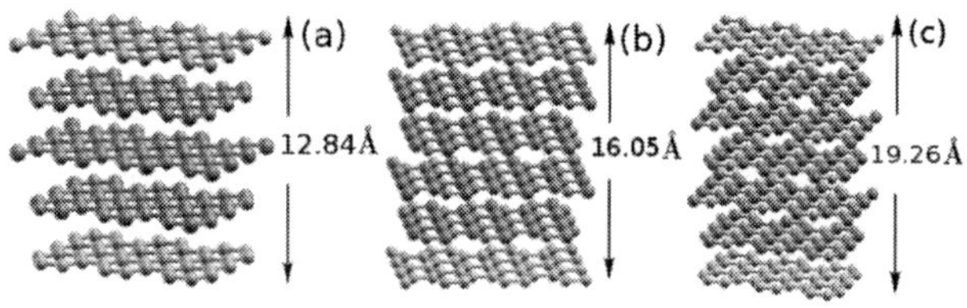

Figure 1. Ball-stick model of G/h-BN/G heterostructures with (a) three-, (b) four- and (c) five h-BN layers sandwiched between Graphene layers. Thickness of the heterostructures are indicated.

included the van der Waals interactions [17] for systems involving multiple layers.

2.2. Boltzmann Transport Theory: Electrical Transport Calculations

To calculate the transport properties, we have used the semi-classical Boltzmann transport theory applied to band electrons as implemented in the Boltztrap code [18]. The transport parameters are obtained from the group velocity, $v_\alpha(i,\mathbf{k}) = \frac{1}{\hbar}\frac{\partial \varepsilon_{i,k}}{\partial k_\alpha}$, of the band electrons, referring to the i-th energy band and α-th component of the wavevector $\mathbf{k}$.

The electrical conductivity tensor is then written, in the relaxation time approximation, as,

$$\frac{\sigma_{\alpha\beta}(T,\mu)}{\tau} = \frac{1}{V}\int e^2 v_\alpha(i,\mathbf{k})\, v_\beta(i,\mathbf{k})[\frac{-\partial f_\mu(T,\epsilon)}{\partial \epsilon}]d\epsilon \quad (1)$$

Here, f_μ is the Fermi-Dirac distribution function, V is the sample volume and τ is the electron relaxation time, which depends on the electron-electron interaction, and e is the electronic charge. Detailed studies [19, 20] have shown that τ does not depend on both the band index and $\mathbf{k}$ and is isotropic to a good approximation. Above relation was used to calculate temperature dependent resistivity of two-dimensional CBN nanomaterials recently [21]. The Seebeck coefficient tensor is expressed as,

$$S_{\alpha\beta}(T,\mu) = \frac{1}{eT}\frac{\int v_\alpha(i,\mathbf{k})v_\beta(i,\mathbf{k})(\epsilon-\mu)[\frac{-\partial f_\mu(T,\epsilon)}{\partial\epsilon}]d\epsilon}{\int v_\alpha(i,\mathbf{k})v_\beta(i,\mathbf{k})[\frac{-\partial f_\mu(T,\epsilon)}{\partial\epsilon}]d\epsilon} \quad (2)$$

ML- and BL- graphene are both gapless while calculated bandstructure of G/h-BN/G heterostructure [22] exhibits a small band gap $\sim$ 0.05 eV which changes very little for three-, four- and five-layers of h-BN in the sample. The electrical transport around the Fermi energy can be expressed in the Mott formula [23] as,

$$S = -\frac{\pi^2 k_B^2 T}{3e}\frac{d}{dE}[ln\,\sigma(E)]. \tag{3}$$

Here k_B is the Boltzmann constant. The Power factor is defined as, $P = S^2\sigma$ and the Figure of merit is defined as, $ZT = PT/\kappa$ where κ is the temperature dependent total thermal conductivity.

2.3. Thermal Transport Calculations

The thermal conductivity is calculated from the correlation function of the heat current ($\mathbf{J}$) using equilibrium molecular dynamics simulations. One can calculate the thermal conductivity κ by applying the Green-Kubo method [24, 25] or the Einstein relation [26]. There have been detailed calculations on two-dimensional stripes of Graphene and h-BN [27, 28, 29, 30, 31] using the latter method. We adopt the equilibrium Green-Kubo method as implemented in the code LAMMPS [32].Thermal conductivity is expressed as,

$$\kappa = \frac{1}{3Vk_BT^2}\int_0^\infty \langle \mathbf{J}(0)\mathbf{J}(t)\rangle dt, \tag{4}$$

where V and T are the volume and equilibrated temperature. The factor 3 accounts for averaging over the 3 dimensions and the angular brackets refers to the average over time. The macroscopic heat current is given by,

$$\mathbf{J}(t) = \sum_i \mathbf{v}_i\varepsilon_i + \frac{1}{2}\sum_{i<j}\mathbf{r}_{ij}(\mathbf{F}_{ij}\cdot(\mathbf{v}_i+\mathbf{v}_j)), \tag{5}$$

where $\mathbf{v}_i$ and ε_i are the velocity and site energy of particle i. $\mathbf{F}_{ij}$ is force on the atom i due to its neighbor j calculated from the interatomic potential.

Molecular dynamics (MD) simulations were performed first in microcanonical ensemble (NVE) and then in the Nosé-Hoover (NVT) ensemble at constant T. The constant temperature NVT ensemble requires an additional frictional term [33] to be introduced in Eq 5. As MD simulations requires an ensemble of initial velocities chosen using random number generator with a specified seed, we have used five different seeds for different sets of MD runs and the value of κ was obtained after the system was equlibriated at constant T by taking average over these runs with the standard deviation as the error bar. In all MD runs we used a large simulation cell containing around 40000 atoms to have a small error bar and a time step of 1fs. The Tersoff potential based force field [34] was used in our calculations, parameters of which were taken, for h-BN from Sevik et al. [28], for graphene from Lindsay et al. [35] and those for the B-C and N-C bonds from Kinaci et al [36], respectively. The calculated κ for two-dimensional striped Graphene and h-BN with different strip-widths [22] were found in excellent agreement with earlier simulation results [36].

3. Results and Discussions

3.1. σ and S of MLG and BLG:

The conductivity (σ), calculated from the semiclassical Boltzmann transport theory by Eq. 1 is plotted in Fig. 2(c & d) as a function of the chemical potential (μ) for temperature T=300K. In

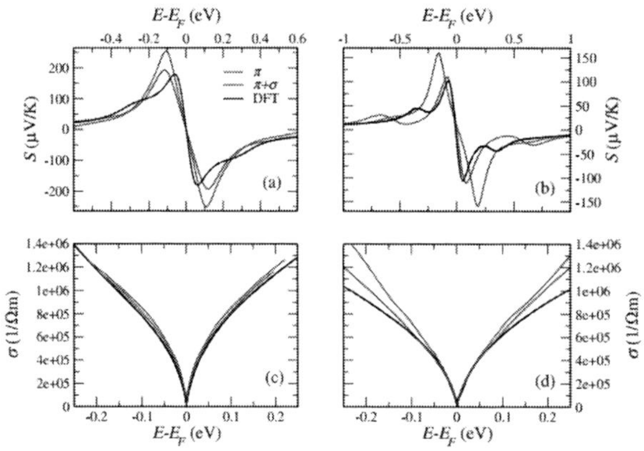

Figure 2. Calculated Seebeck coefficient (a,b) and electrical conductivity (c,d) of ML- and BL-graphene as a function of μ at $T = 300$K, respectively. Results from DFT based bandstructure are shown as black line, those from the tight-binding bandstructure with only π bands as green, and with both π and σ bands as red lines, respectively.

Fig. 2(c) we observe σ of ML graphene to be proportional to $\sqrt{n}$ where n is the charge carrier density. This $\sqrt{n}$ behavior can be explained using a tight binding model with an expansion of energy around the k point to get $\epsilon_F = \hbar v_F k_F$, where $\hbar$ is the reduced Planck constant, v_F is the Fermi velocity and k_F is the Fermi wavenumber. For a 2-dimensional sample,

$$n = \frac{k_F^2}{\pi} = \frac{\epsilon_F^2}{\pi(\hbar v_F)^2} \tag{6}$$

Therefore the MLG's density of states $D(\epsilon_F)$,

$$D(\epsilon_F) = \frac{dn}{d\epsilon} = \frac{2\sqrt{n}}{\hbar v_F \sqrt{\pi}} \tag{7}$$

Substituting this in the Einstein relation $\sigma \propto D(\epsilon_F)$, we obtain $\sigma \propto \sqrt{n}$. This behavior has been experimentally confirmed for pristine graphene [4].

The Seebeck coefficient was calculated for both ML and BL graphene using the Mott relation Eq 3 numerically and shown in Fig. 2(a & b) as a function of chemical potential μ. Our calculated S is in very good agreement with Zuev et al's [5] measurement of Seebeck coefficient for ML graphene. Using a back-gated field effect transistor, the Fermi energy can be tuned by adjusting the gate voltage V_g. Zuev et al [5] have used this method for mesoscopic graphene sample using the relation,

$$S = -\frac{\pi^2 k_B^2 T}{3e} \frac{1}{\sigma} \frac{d\sigma}{dV_g} \frac{dV_g}{dE}|_{E=E_f}. \tag{8}$$

It can be easily seen that this expression is exactly same as eq. 3. Therefore there is a direct correspondence between the gate voltage and chemical potential.

We find the resistivity of ML and BL graphene when plotted as a function of carrier density shows ρ is maximum at $n = 0$ and rapidly decreases for larger values $|n|$, which can be understood from the $\sqrt{n}$ behavior of conductivity. This has been experimentally confirmed [8, 4]. In Fig 3 we plot ρ as a function of T. This behavior of ρ at certain values of μ having the same form has

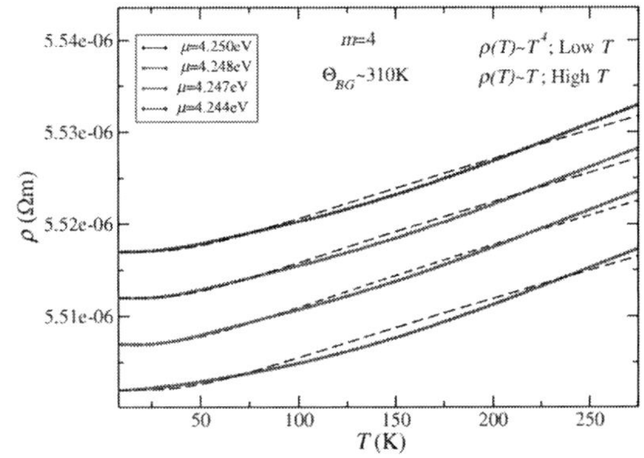

Figure 3. Calculated resistivity (ρ) of MLG as a function of T. The dashed black lines show the best fit to the Bloch-Grüneisen formula for different μ, Eq (9). Points represent results of ρ calculated from the Boltzmann Transport theory for different chemical potential μ. For clarity, the data corresponding to $\mu = 2.247$ eV, 2.248 eV and 2.25 eV are shifted by 5×10^{-9}, 10×10^{-9} and 15×10^{-9} in the y-axis, respectively.

been observed experimentally [7, 4]. In order to understand the behavior, we fit our calculated values of ρ to the Bloch-Grüneisen formula,

$$\rho(T) = \rho(0) + A\Big(\frac{T}{\Theta_{BG}}\Big)^m \int_0^{\frac{\Theta_{BG}}{T}} \frac{x^m}{(e^x - 1)(1 - e^{-x})} dx, \tag{9}$$

with m and Θ_{BG} as fitting parameters. Our best fit resulted in $m = 4$ and $\Theta_{BG} \approx 310K$ as plotted in Fig 3 as dashed lines. This suggests that at low temperatures, resistivity scales as T^4 and smoothly scales to linear T behavior at higher temperatures. This behavior has been theoretically predicted by Hwang [37] and experimentally observed by Efetov et al [7]. Our best fitted value of Θ_{BG} is in the same order of magnitude as estimated from the experimental measurements [7] and much lower than the Debye temperature $\Theta_D \approx$2100K for MLG [38].

3.2. Electrical Conductance and Thermopower of G/h-BN/G

Our calculations of electrical conductance were carried out in the relaxation time approximation assuming τ to be isotropic and independent of directions in the reciprocal lattice space. We have assumed $\tau \sim 10^{-14}$s, which is reasonable. In Fig. 4 results of the Seebeck coefficient (S) of G/h-BN/G heterostructure having 5 BN layers are shown.

As h-BN has a gap between the conduction and valence bands, the electric conduction is essentially controlled by Graphene layers at the top and bottom of the heterostructure film. Due to this, the features in S is essentially similar to that of Graphene, except at the low temperatures when the effect of the presence of a small band gap becomes evident. For $T \sim$ 40K - 300K the conductivity decreases as T increases, thereby increasing S with T as seen in the inset of Fig 4, where we plot S as a function of T at constant chemical potential. The linear dependence of S on T suggests that the mechanism for thermoelectricity is diffusive [39].

In order to compare our calculations with experimental measurements, which were performed at a constant gate voltage, we fix the chemical potential (μ) to obtain the experimental Seebeck coefficient, as indicated by the horizontal line in Fig 4. We then use the same μ in the plot of the Power factor ($S^2\sigma$) and obtain this parameter for G/h-BN/G heterostructure [22]. Our estimated value of S^2G, where G refers to the electrical conductance depending on the thickness of the sample, compares very well with the experimental data, presented in the final section.

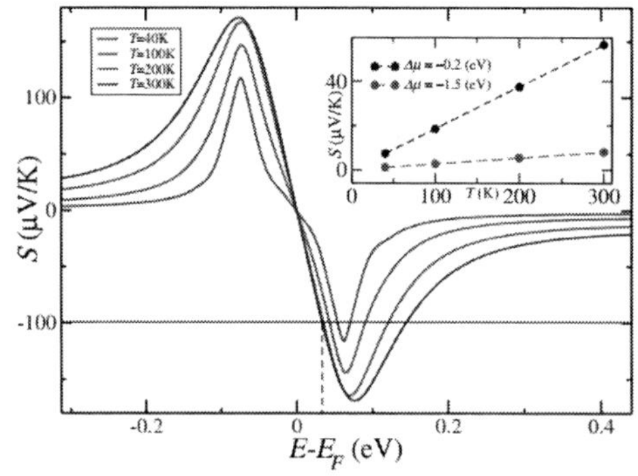

Figure 4. Calculated Seebeck coefficient of G/h-BN/G plotted against energy at various temperatures. The black horizontal line refers to the experimental value at T=300K [9]. The inset shows the Seebeck coefficient plotted against Temperature at a constant chemical potential. Taken from [22].

3.3. Thermal Conductivity of Bulk and multilayers of h-BN

The calculated κ of multilayer and bulk h-BN using the equilibrium Green-Kubo theory based MD simulation, performed at constant temperature in the Nosé-Hoover thermostat, is shown in Fig 5 for $T \sim$ 25K-400K together with the experimental data of Jo et al [11]. For each simulation run, the system was thermalized at the desired temperature first for each set of initial uniform distribution of velocities. κ was calculated for five different sets of initial velocities and the error bar was estimated from the standard deviation.

We find that κ increases with T and tends to saturate at $T \sim$ 220K for each sample of h-BN, with results for 11-layer tending towards the bulk value. For each of the three samples of h-BN multilayered films, κ shows maxima in the temperature region of 200-250K. We observed an overall agreement of our MD simulation results with recent experimental measurements [11]. For 11-layer and bulk h-BN samples the agreement between the simulation and experimental data is better in temperature range 25-300K, whereas for the 5-layer sample the MD results seem to agree only in the temperature range 100-250K.

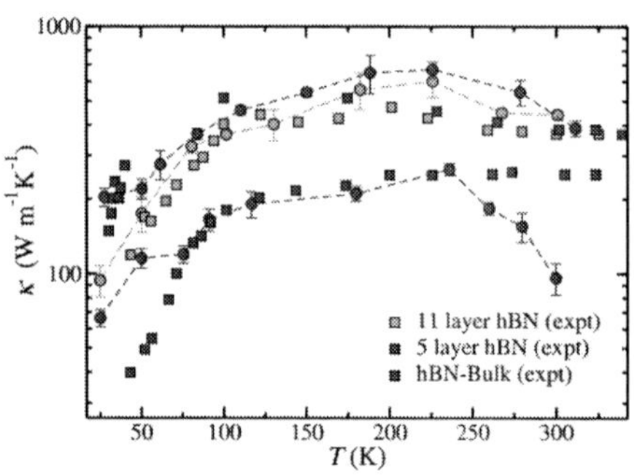

Figure 5. Calculated temperature dependence of the thermal conductivity of multilayer h-BN of different thickness and that of bulk h-BN. Experimental results [11] are shown as squares and present calculations as circles. Taken from [22].

We feel more detailed first-principles phonon based transport calculations are necessary for making a quantitative comparison with experimental data and other available calculations [40, 41, 42]. Phonon-phonon scattering and the lifetime of such processes play in an important role in thermal conductance and these can be addressed in such models.

3.4. K, S^2G and ZT of G/h-BN/G Heterostructures

In Fig 6, we show our calculated thermal conductance (K), Power Factor (S^2G) and the Figure-of-merit (ZT) of G/*h*-BN/G heterostructures with three-, four- and five-layers of *h*-BN. K and G were obtained by multiplying the corresponding κ and σ by the thickness of the heterostructures, respectively. The chemical potential (μ) was kept fixed at the experimentally observed value -99.3 μV/K of S as shown in Fig 4. With increasing thickness of *h*-BN layers, K decreases whereas S^2G increases with T, thereby increasing ZT with temperature. Our calculated values of these thermoelectric parameters agree well with the experimental data [9] for three-, four- and five-layer *h*-BN sandwiched between the Graphene layers.

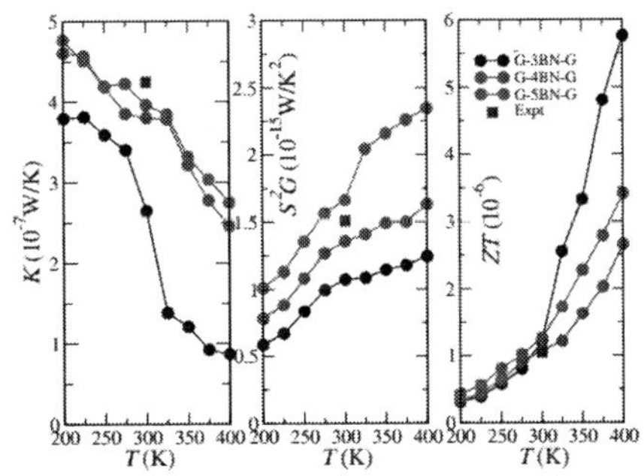

Figure 6. Calculated temperature dependence of Thermal Conductance (K), Power Factor (S^2G) and the Figure-of-merit (ZT) of G/*h*-BN/G heterostructure of different thickness shown in the left, middle and right panels; respectively. The available experimental data [9] in each panel at 300K is also indicated. Taken from [22].

We would like to stress on the fact that our calculations involve electrical transport not strictly along the vertical direction, because the Boltzmann transport theory yields smaller contributions to the electron group velocity and thereby electrical conductivity along that direction compared to those along horizontal x- and y-directions. We have calculated the z-component of the Seebeck coefficient (S_z) [22] and found this to be finite and comparable to S_x and S_y close to the Fermi energy. The total Seebeck coefficient S shown in Fig 4, however, shows a quantitative agreement with the experimental data [9]. Thus, we conclude that in the thermoelectric measurements [9] the electrical transport may not be strictly along the z-direction as the upper and lower Graphene contacts with multilayer *h*-BN would allow transport channels involving components along x- and y-directions also. A good quantitative agreement with experimental data also supports above conclusion.

In summary, we have presented a detailed computational study of the calculation of the thermoelectric parameters of monalayer- and bilayer-graphene and also of G/*h*-BN/G heterostructures with three-, four- and five-layers of *h*-BN. Our results show quantitative agreement with the available experimental data.

One of us (RD) acknowledges support from S.N. Bose National Centre in the form of a Senior Research Fellowship. All calculations were carried out in the High Performance Cluster Computer platform at S.N. Bose National Centre. The work presented in this paper is based on two separate talks given by the authors at the 28th IUPAP Conference on Computational Physics (CCP 2015), held at IIT Guwahati during 2-5 December 2015.

References

[1] Majumdar A 2004 *Science* **303** 777

[2] Zebarjadi M, Esfarjani K, Dresselhaus M S, Ren Z F and Chen G 2012 *Energy Environ. Sci.* **5** 5147
[3] Cahill D G, Braun P V, Chen G, Clarke D R, Fan S, Goodson K E, Keblinski P, King W P, Mahan G D, Majumdar A, Maris H J, Phillpot S R, Pop E and Shi L 2014 *Applied Physics Reviews* **1** 011305
[4] Bolotin K I, Sikes K J, Hone J, Stormer H L and Kim P 2008 *Physical Review Letters* **101** 096802
[5] Zuev Y, Chang W and Kim P 2009 *Physical Review Letters* **102** 096807
[6] Checkelsky J G and Ong N P 2009 *Physics Review B* **80** 081413
[7] Efetov D K and Kim P 2010 *Phys. Rev. Lett.* **105** 256805
[8] Morozov S V, Novoselov K S, Katsnelson M I, Schedin F, Elias D C, Jaszczak J A and Geim A K 2008 *Phys. Rev. Lett.* **100** 016602
[9] Chen C C, Li Z, Shi L and Cronin S B 2015 *Nano Research* **8** 666
[10] Chen C C, Li Z, Shi L and Cronin S B 2014 *Appl. Phy. Lett.* **104** 081908
[11] Jo I, Pettes M T, Kim J, Watanabe K, Taniguchi T, Yao Z and Shi L 2013 *Nano Research* **13** 550
[12] Giannozzi P and et al 2009 *J. Phys. Condens. Matter* **21** 395502
[13] Vanderbilt D 1990 *Phys. Rev. B* **41** 7892
[14] Perdew J P, Burke K and Ernzerhof M 1996 *Phys. Rev. Lett.* **77** 3865
[15] Monkhorst H J and Pack J D 1976 *Phys. Rev. B* **13** 5188
[16] Kaloni T and Mukherjee S 2011 *Modern Physics Letters B* **25** 1855
[17] Grimme S, Antony J, Ehrlich S and Krieg S 2010 *J. Chem. Phys.* **132** 154104
[18] Madsen G K H and Singh D J 2006 *Computer Physics Communication* **175** 67
[19] Schulz W W, Allen P and Trivedi N 1992 *Physical Review B* **45** 10886
[20] Allen P, Pickett W and Krakauer H 1993 *Physical Review B* **37** 14434
[21] D'Souza R and Mukherjee S 2015 *Physica E* **69** 138
[22] D'Souza R and Mukherjee S 2016 *Physica E* **81** 96
[23] Cutler M and Mott N F 1969 *Physical Review* **181** 1336
[24] Green M S 1954 *J. Chem. Phys* **22** 398
[25] Kubo R 1957 *J. Phys. Soc. Jpn.* **12** 570
[26] Zwanzig R 1965 *Annu. Rev. Phys. Chem* **16** 67
[27] Haskins J, Kinaci A, Sevik C, Sevincli H, Cuniberti G and Cagin T 2011 *ACS Nano* **5** 3779
[28] Sevik C, Kinaci A, Haskins J B and Cagin T 2011 *Phys. Rev. B* **84** 085409
[29] Sevik C, Kinaci A, Haskins J B and Cagin T 2012 *Phys. Rev. B* **86** 075403
[30] Kinaci A, Haskins J B and Cagin T 2012 *J. Chem. Phys.* **137** 014106
[31] Sevik C, Sevincli H, Cuniberti G and Cagin T 2011 *Nano Lett.* **11** 4971
[32] Plimpton S 1995 *J Comp Phys* **117** 1–19 we have used the latest release of the code LAMMPS (March 2015 version) in our MD simulations.
[33] Berber S, Kwon Y K and Tománek D 2000 *Physical Review Letters* **84** 4613
[34] Tersoff J 1989 *Phys. Rev. B* **39** 5566
[35] Lindsay L and Broido D A 2010 *Phys. Rev. B* **81** 205441
[36] Kinaci A, Haskins J B, Sevik C and Cagin T 2012 *Phys. Rev. B* **86** 115410
[37] Hwang E and Sarma S D 2008 *Phys. Rev. B* **77** 115449
[38] Pop E, Varshney V and Roy A K 2012 *MRS Bulletin* **37** 1273
[39] Scarola V W and Mahan G D 2002 *Phys. Rev. B* **66** 205405
[40] Lindsay L and Broido D A 2011 *Phys. Rev. B* **84** 155421
[41] Mortazavi B, Pereira L F C, Jiang J W and Rabczuk T 2015 *Scientific Reports* **5** 13228
[42] Ouyang T, Chen Y, Xie Y, Yang K, Bao Z and Zhong J 2010 *Nanotechnology* **21** 245701

Capturing local atomic environment dependence of activation barriers in metals using cluster expansion models

Nimish Kulkarni and Abhijit Chatterjee*

Department of Chemical Engineering, Indian Institute of Technology Bombay, Mumbai 400076 India

*Email: abhijit@che.iitb.ac.in

Abstract. It is well known that surface diffusion in metals can proceed via multiple mechanisms, such as hop, exchange and other types of concerted moves. However, the manner in which kinetic rates associated with a mechanism can depend sensitively on local atomic environment is relatively less understood. We describe recent attempts in our research group to capture the atomic environment dependence using the cluster expansion model (CEM). In particular, we focus on hop and exchange moves at the (001) surface in homoepitaxy, and show that while CEM can work remarkably well in most cases, it can sometimes provide inaccurate predictions for concerted moves.

1. Introduction

Far from the idealized picture that atoms at crystalline surfaces remain frozen at locations corresponding to the minimum energy configuration, one finds that surfaces tend to be dynamic. Understanding how surfaces can evolve is important to applications like catalysis and electrochemistry as well as materials processing. For instance, the surface density of reactive sites for a catalyst particle, or the nature of the surface roughness during corrosion can evolve with time. While most dynamical models for surface phenomena contain parameters that capture the role of atomic scale processes in a coarse-grained manner, inclusion of detailed kinetic rates could provide unprecedented insights into the surface phenomena. Unfortunately, the situation is incredibly complex for 3D structures and multicomponent materials where a plethora of kinetic mechanisms and atomic environments are encountered, i.e., the problem lies in tackling the combinatorial complexity. We report some of the preliminary advances from our group that can be used to predict kinetic rates from the local atomic environment in metal homoepitaxy. These developments indicate that one can be cautiously optimistic that our approach could pave the way for tackling more complex environments.

Metal-on-metal surface diffusion is essential to a number of applications like thin film growth [1], catalysis [2], and corrosion [3]. Fig. 1 shows an example of a metal alloy surface undergoing selective dissolution of the electrochemically active metal species in the presence of an electrolyte solution. Although the starting material was atomically flat, the dealloying process results in formation of 3D porous structures over a period of several minutes to hours. Interestingly, one finds a large number of atomic arrangements at the surface. While a large number of experimental and computational studies have identified important principles of surface diffusive processes a general understanding of how the diffusion barrier will change with the environment or admetals/substrate species is lacking. A conventional view is that surface diffusion proceeds mainly via the hopping mechanism where an adsorbed atom (adatom) hops from one lattice site to a nearest neighbor vacant site [4]. Contrary to the view that concerted pathways are unimportant, recently, we have shown that in some situations concerted pathways can play an important role in the long term evolution of the material [5]. Kinetic rates associated with the pathways have been calculated for simple cases such as a single adatom on a surface [4]. The rates can change dramatically as adatoms are introduced in the vicinity of a moving atom, i.e., its local atomic environment [6]. Predicting the rates for all possible relevant mechanisms and atomic environments that can be encountered in the dynamics presents a major challenge.

Figure 1. Large number of atomic environments can be encountered in the dynamics as a bimetallic alloy undergoes corrosion.

A promising approach to capture the local atomic environment involves the use of a cluster expansion model (CEM) [7]. CEM have been employed extensively for predicting thermodynamics properties [8], and only recently for predicting kinetic properties [6]. We describe the CEM in Sec. 2. The CEM parameters can be trained using a database of kinetic rates collected for a large number of atomic environments. However, this raises several questions, such as what is the size of the atomic environment associated with metal systems and how well can the CEM predict the environment dependent rates. The answer to these questions are provided in Sec. 2.

2. Cluster expansion model for metal homoepitaxy

Assuming that the lattice approximation is valid for the crystalline material, one can define a local environment ε consisting of N sites that can affect the kinetic rate. The value of the occupancy σ_α, which may be 0 or 1, denotes the absence or presence of an atom at the site α, respectively. Environment selected for the hop and exchange mechanism are shown in Fig. 2. Assuming that the Arrhenius law is obeyed, i.e.,

$$k(\varepsilon) = \nu(\varepsilon)\exp(-E_a(\varepsilon)/k_B T), \quad (1)$$

a CEM can be separately generated for the pre-exponential factor ν and the activation barrier E_a. Here k_B is the Boltzmann constant and T is the temperature.

We employ the embedded atom method (EAM) potential of Ref. [9] to describe the interactions between the metal atoms. Consider the example of a single Ag adatom hop. The activation barrier for a hop move is 0.5 eV. The barrier remains unchanged when one adatom is introduced into the environment at either σ_9 or σ_{19} of Fig. 3a. However, presence of adatoms at both sites σ_9 or σ_{19} results in an increase in the barrier by 0.14 eV – a synergistic effect that cannot be described by the bond-counting method. This also shows that the barriers depend on the environment in a non-intuitive manner. Fig. 2c shows that the pre-exponential factor can vary over two-orders of magnitude, whereas the activation barrier, which varies between 0.06 eV to about 1 eV depending on the environment, results in a 14 order-of-magnitude variation in the rates at 300 K. Similar behavior is observed for Al, Cu, Pd, Pt and Ni [6]. For this reason, we focus our attention on the activation barrier.

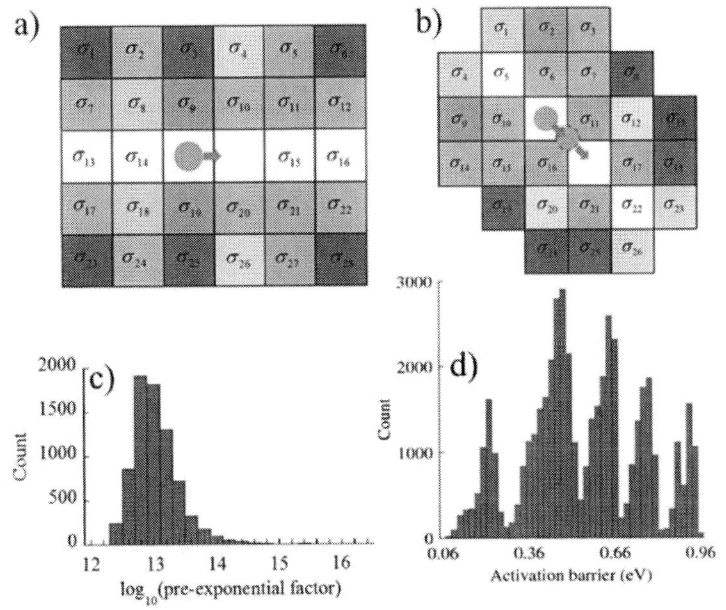

Figure 2. Atomic environments selected for a) hop and b) exchange mechanism. Only adatom locations are included in the environment. Color denotes symmetrically equivalent sites. Variation in pre-exponential factor and barrier for more than 7500 randomly selected environments is shown in panels c and d.

The CEM is written as a sum of contributions from clusters of sites of different sizes, e.g., single, pair, etc., weighted by an interaction parameter such that

$$E_a(\varepsilon) = e_0 + \sum_{\alpha} e_\alpha \sigma_\alpha + \sum_{\alpha \neq \beta} e_{\alpha\beta} \sigma_\alpha \sigma_\beta + \ldots \quad (2)$$

Here e denotes the cluster interaction (CI) parameter of a particular cluster, and e_0 is the barrier only due to interactions with substrate atoms when the environment atoms are absent. CIs are estimated using a training set containing barriers for a large number of environments. Barriers are obtained by locating saddle points on the multidimensional energy landscape using the nudged elastic band method [10]. Next, we describe some of the key findings from our analysis.

2.1. *Hop mechanism*

CEM including terms up to triplet CIs in Eq. (2) were created for Ag, Al, Pt, Pd, Ni and Cu [6]. For the environment shown in Fig. 1a, the total number of CIs that need to be estimated are 3683. This number could have been lowered by arbitrarily choosing a smaller environment at the risk for sacrificing accuracy or by exploiting the symmetry in the lattice positions. After training the CEM, most CIs are found to be negligible small and in fact less than 40 CIs are found to be sufficient to capture the effect of the atomic environment with a correlation coefficient exceeding 0.99. Thus, CEM can reliable model the local environment effects. The relevant CIs include singlet, pair and triplet interactions.

Fig. 3a-d shows that most sites that give rise to the relevant CIs are common to all metals studied. Nearest neighbor (NN) interactions play a key role, however, the interactions are anisotropic. For e.g., sites such as σ_9 and σ_{19} are found to be more important than σ_{13}. Similar behavior is observed with next NN atoms. Minor differences observed in the relevant environments shown in Fig. 3 arise because of the strength of the CIs. Interactions weaker than 0.03 eV in magnitude are considered to be too weak and are therefore ignored. Since the CIs for Pt and Pd are approximately two times larger than the ones for Ag, one finds that the relevant environment of Pt and Pd is also larger. The implication is that atoms distant from the hopping adatom can alter the rate constant by a significant amount. For instance, see locations σ_{11} and σ_{21} in Fig. 3d.

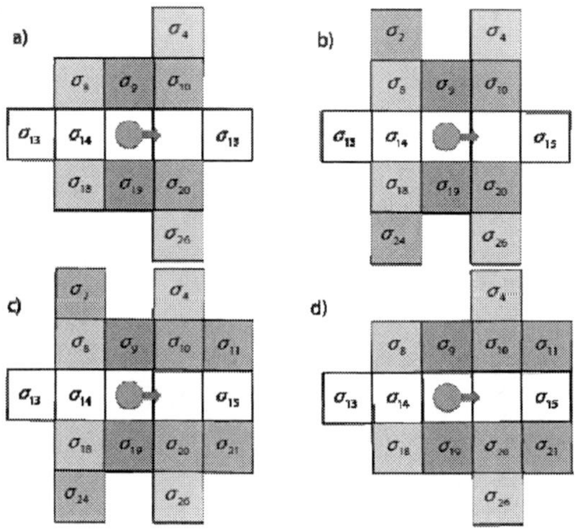

Figure 3. Environment atoms that influence the activation barrier are shown for a) Ag and Al, b) Cu, c) Ni and d) Pt and Pd.

2.2. *Exchange mechanism*

The ability of the CEM to predict the environment dependence of the hopping rates motivates its extension to concerted moves such as the exchange mechanism. In metals such as Pt, the exchange move involving a single adatom is preferred over a hop move. CEM of concerted moves have not been developed in the past. Here we describe some aspects of the model, while more details will be provided in an upcoming publication.

The first question that arises is how well the CEM is able to predict the activation barriers. A database of barriers was constructed using more than 10000 randomly selected environments for each metal from Fig. 1b. Most data (70%) was used for training the 352 CEM parameters, while remaining 10% and 20% was used for testing and validation, respectively. Triplet and larger cluster interactions were ignored. Finally, the number of interactions with magnitude greater than 0.01 eV was found to be 39, 81, 59, 70 and 169 for Ag, Pt, Pd, Cu and Ni. The agreement between the trained CEM and the NEB is excellent for Ag, Pt and Pd as shown in Fig. 4a-c. However, the agreement is not spectacular for Cu and Ni (Fig. 4d-e).

The poor agreement observed for Cu and Ni may point to the need for triplet interactions. The number of CEM parameters to be trained increases to 2952 when triplet interactions are included. Unfortunately, the dataset was insufficient for training such a large number of parameters, i.e., more barriers should be included. Moreover, the computational time required to estimate the parameters can be few days even using efficient optimization methods like the L-BFGS method. For this reason, we restrict ourselves to the pair interactions.

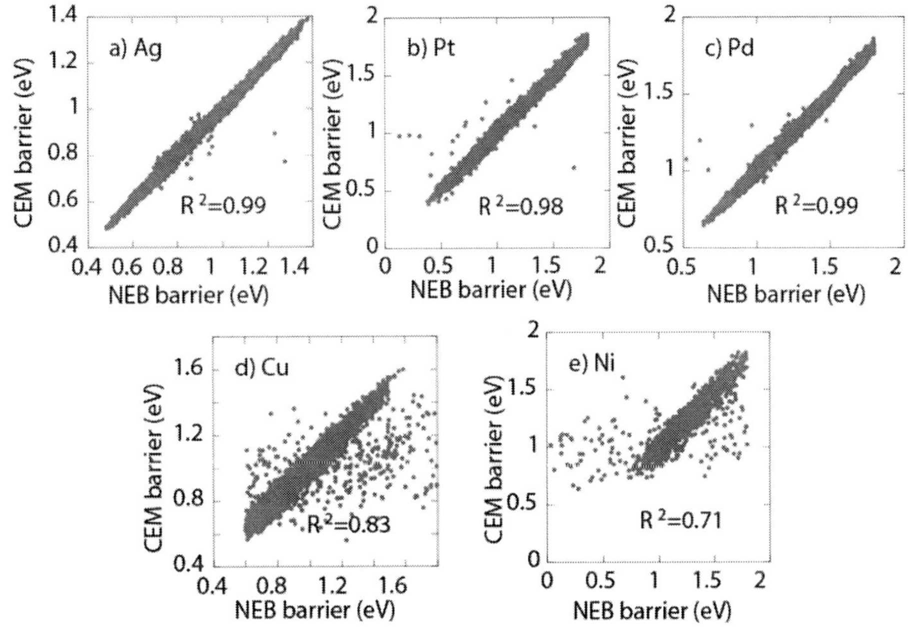

Figure 4. Parity plot for barriers calculated using CEM and NEB. While good agreement is observed for a) Ag , b) Pt and c)Pd, the agreement is poor for d) Cu and e) Ni.

The CEM reveals the relevant local environment (see Fig. 5) that influences the barrier for the exchange move. A relatively small local environment is associated with Ag and Pd, while Cu and Ni have large environments associated, i.e., longer-range interactions are present in Cu and Ni. This may result in large number of CIs for the latter, and explain the source of the difficulties encountered while training the CEM for Cu and Ni.

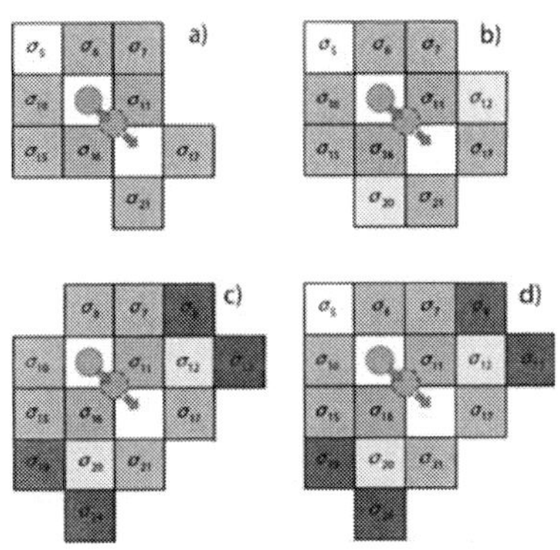

Figure 5. Environment atoms that can influence the activation barrier are shown for a) Ag and Pd, b) Pt, c) Ni and d) Cu.

Several common features are also observed. Presence of adatoms at sites labelled σ_6 and σ_{10} results in an increase in barrier by 0.27-0.56 eV depending on the metal. However, the interaction $e_{6,10}$ denoting the simultaneous presence of adatoms at σ_6 and σ_{10} is negligible. Similarly, interactions e_7 and e_{15} are also important as they raise the barrier. However, the next NN sites σ_{11} and σ_{16} are generally unimportant. Presence of an adatom at σ_{17} and σ_{21} results in a decrease in the barrier. The remaining sites participate through the pair interactions.

3. Conclusions

The cluster expansion model (CEM) has been employed for predicting the atomic environment dependence of the hop and exchange moves in homoepitaxies of Ag, Pt, Pd, Cu and Ni. The CEM containing up to triplet interactions can capture the dependence reasonably well for hop moves for all metals studied, and exchange moves in Ag, Pt and Pd. However, the model is inappropriate for Cu and Ni presumably the long-range cluster interactions present in these metals.

References

[1] Gilmer G.H., Huang H. and Roland C. 1998 *Comput. Mater. Sci.* vol 12 p 354
[2] Ziff R.M. Gulari E. and Barshad Y. 1986 *Phys. Rev. Lett.* vol 56 p 2553
[3] Erlebacher, E. et al. 2001 *Nature* vol. 410 p. 450–453
[4] Surface Diffusion: Atomistic and Collective Processes, NATO ASI Series ed Tringides, M.C.
[5] Haldar P and Chatterjee A 2015 *Modelling Simul. Mater. Sci. Eng.* vol. 23, p 025002
[6] Verma S, Rehman T and Chatterjee A 2013 *Surface Science* vol 613, p 114-125
Rehman T, Jaipal M and Chatterjee A 2013 *Journal of Computational Physics* v 243, p 244-259
[7] Sanchez J.M., Ducastelle F. and Gratias D., 1984 *Physica A* 128 p 334
[8] Curtarolo S. et al, 2013 *Nature Materials* 12 191
[9] Zhou X.W., Johnson R.A., Wadley H.N.G. 2004 *Phys. Rev. B* vol 69 p 144113
[10] Henkelman G., Uberuaga B.P. and Jónsson H. 2000 *J. Chem. Phys.* vol 113 p 9901

A hybrid density functional study of silicon and phosphorus doped hexagonal boron nitride monolayer

R E Mapasha[1], E Igumbor[1], N Chetty[1,2]

[1] Department of Physics, University of Pretoria, Hatfield campus, Pretoria 0002, RSA.

[2] National Institute for Theoretical Physics, Johannesburg 2000, RSA.

E-mail: edwin.mapasha@up.ac.za

Abstract. We present a hybrid density functional study of silicon (Si) and phosphorus (P) doped hexagonal boron nitride (h-BN). The local geometry, electronic structure and thermodynamic stability of Si_B, Si_N, P_B and P_N are examined using hybrid Heyd-Scuseria-Ernzerhof (HSE) functional. The defect induced buckling and the local bond distances around the defect are sensitive to charge state modulation q = -2, -1, 0, +1 and +2. The +1 charge state is found to be the most energetically stable state and significantly reduces the buckling. Based on the charge state thermodynamic transition levels, we noted that the Si_B, Si_N and P_B defects are too deep to be ionized, and can alter the optical properties of h-BN material.

1. Introduction

A recently synthesized single layer of hexagonal boron nitride (h-BN) has received great research attention due to its excellent mechanical properties, large exciton binding energies and high thermal conductivity [1, 2, 3, 4, 5]. The h-BN monolayer has a similar structural arrangement to graphene [6, 7] but differ in electronic properties. The h-BN monolayer has a wide band gap of around 6 eV [1, 5, 8], while graphene is a zero band gap material [6, 7]. These properties make the h-BN monolayer a suitable candidate for various advanced technological applications, including the optoelectronic applications in the ultraviolet energy range [1, 5, 8]. Furthermore, the h-BN monolayer is thermodynamically stable at room temperature [9].

During the synthesis of h-BN films, various defects are usually observed [5, 10, 11, 12]. For instance, an electron beam irradiation mostly leads to the formation of point defects [5, 10]. The presence of defects in h-BN alters the various known properties and also imposes new valuable properties. Defects such as vacancies, antisites and carbon (C) impurities in h-BN are in a high spin state [13, 14, 15, 16]. Since the defect spin states can be easily affected by the modulation of electronic chemical potential, the studies of charged defects received great research attention [17, 18]. The charged defect structures, defect formation energies and defect transition energy levels within the band gap are usually interesting properties to be inverstigated [17, 18].

The Si atom, an element that is below C on the periodic table, is well known for its technological importance. Therefore, it is expected that the creation of Si defects could make the properties of the h-BN more fascinating. Previous experimental and theoretical studies reported that the incorporation of Si impurity in the boron site is exothermic whereas in the nitrogen site is endothermic [13, 19, 20, 21, 22, 23, 24]. Previous studies using the generalized gradient approximations with the Perdew, Burke and Ernzerhof functional (GGA-PBE) [25]

Journal of Physics: Conference Series **759** (2016) 012042 doi:10.1088/1742-6596/759/1/012042

within the framework of density functional theory (DFT) also reported that Si introduces the impurity mid gap states [13, 19]. Considering the deficiency of GGA-PBE exchange correlation functional on the prediction of the band gap, hybrid functionals are required [26]. Phosphorus (P) which is the nearest neighbour of Si and an element below nitrogen on the periodic table can be an important defect in the *h*-BN monolayer. Although, there is already study of P doped *h*-BN [27], an understanding of the charged defect structures, defect formation energies and the defect transition energy levels is still scarce, more especially when using the hybrid functionals.

In this paper, the relative stability, structural and electronic properties of Si and P doped *h*-BN structures are studied using the hybrid density functional theory. The impurity defects are created in the following manner; one B atom from *h*-BN monolayer is removed and replaced by Si or P atom, and denoted as Si_B and P_B. Afterwards, one N atom is also removed and replaced by Si or P atom denoted as Si_N and P_N. These defects have shown to have an influence on the structural properties of the *h*-BN monolayer by creating a large buckling. The induced buckling and the local bond distances around the defect are sensitive to charge state modulation. Based on the thermodynamic transition levels, we noted that the Si_B, Si_N and P_B defects are deep and therefore can have a great influence on the luminescence of the *h*-BN monolayer.

2. Computational details

Our first principles calculations were performed on the framework of density functional theory (DFT) within the Vienna *ab initio* simulation packages (VASP) code. [28] For the exchange-correlation interactions, the hybrid Heyd-Scuseria-Ernzerhof (HSE) functional [26] was used. This functional predicts an excellent band gap and correct position of the charge state transition level [29]. The projector augmented wave (PAW) method [30] was used for pseudopotential construction. The plane wave energy cutoff was set to 500 eV. For the Brillouin zone integration, the Monkhorst-Pack scheme [31] was used to generate the k-points. In this study, a $4 \times 4 \times 1$ k-point mesh was employed. The atomic coordinates were allowed to relax until the residual atomic forces reduce to smaller than 0.01 eVÅ^{-1} and the total energy was allowed to converge within the 10^{-7} eV. To avoid spurious interlayer interactions, a 15 Å vacuum spacing was set along the z-axis.

3. Results and discussion

The structural relaxation of the following defects in a *h*-BN monolayer was carried out; a B or N atom get replaced by Si or P impurity to have Si_B, Si_N, P_B and P_N defect structures. Fig. 1 presents the relaxed geometries for (a) Si_B and (b) Si_N in a 6×6 supercell. The relaxed geometries for P_B and P_N defects exhibit the same buckling as for the Si_B and Si_N defects respectively, therefore are not shown.

To examine the relative stability of these defects in the neutral charge state, their formation energies are calculated using the following equation;

$$E_f[X]^0 = E_{tot}[X]^0 - E_{tot}[h\text{-BN}] + \sum_i n_i \mu_i. \qquad (1)$$

$E_{tot}[X]^0$ is the total energy of the relaxed *h*-BN supercell with Si or P defect and $E_{tot}[h\text{-BN}]$ is the total energy of pristine *h*-BN supercell. The integer n_i in Eqn.1 indicates the total number of B or N replaced (provided n_i is negative) or impurity atoms added into the system creating the defects (provided n_i is positive). The symbol μ_i represents the atomic chemical potentials for B, N, Si and P atoms. The N chemical potential is obtained from the converged energy of nitrogen gas (N_2) and for B, the energy is taken from the α-rhombohedral boron. The chemical potentials for μ_{Si} and μ_P are taken from the converged energies of Si in diamond-like structure and P in the body centered cube (BCC) structure respectively. The formation energies of the identified impurity defects Si_B, Si_N, P_B and P_N are constrained to vary within the $\Delta\mu$ values [32] as

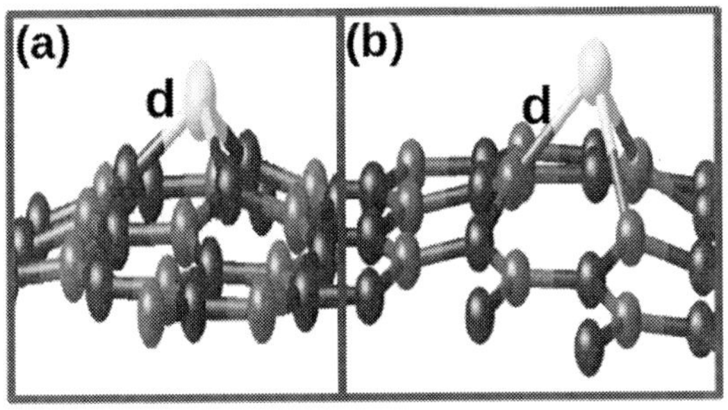

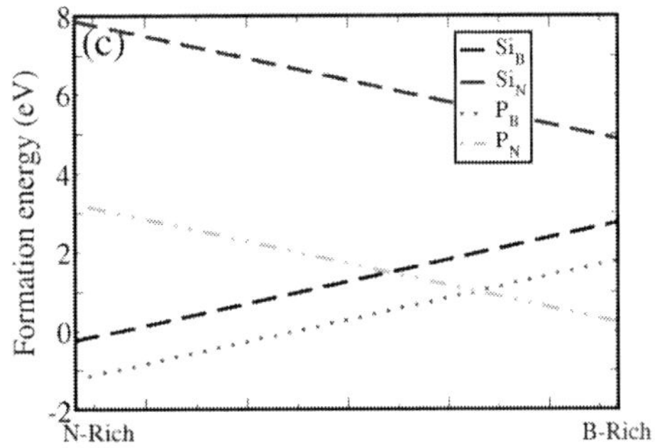

Figure 1. The optimized structures for (a) Si_B and (b) Si_N defects. The optimized geometries for P_B and P_N defects show the same buckling as for Si_B and Si_N defects. The purple, red and pink atoms represent the B, N and Si respectively. The letter d is the distance between impurity and nearest neighbor host atom. (c) The dependence of the formation energies of Si_B, Si_N, P_B and P_N impurity defects on the change in chemical potential $\Delta\mu$.

shown on Fig. 1(c). The minimum and maximum $\Delta\mu$ values of -2.64 eV and 2.64 eV (in good agreement with the previous studies [33, 34]) correspond to N-rich and B-rich extreme conditions respectively. This is to determine the growth condition in which these defects will prefer to form at a very reasonable energies.

Fig. 1(c) shows that the Si_B defect is stable under the N-rich conditions having the formation energy of -0.24 eV. This is evidence that the formation of Si-N bond is likely to occur exothermically during the synthesis of this system, in agreement with the previous studies [22, 23]. Although Si_N defect prefers the B-rich environment, it possesses the highest formation energy of about 6.00 eV. Fig. 1(c) also shows that the P_B defect is more stable in the N-rich condition. This suggests that P_B defect can easily bond with N atom. The P_N defect shows an opposing trend. These results imply that the stability of these defects greatly depends on the environment condition that corresponds to the highest amount of atomic hosts that surround the defect. Previous studies [17, 18, 33] reported that, in most cases, it cost less energy to create vacancies, antisites and impurities on the B site than on the N site in a h-BN monolayer.

We further investigate the effect of charge state q on the formation energies and structural properties of the defects mentioned above. The formation energy in the charge state q is written by extending Eq. (1) to include the information of q as follows,

$$E_f[X]^q = E_{tot}[X]^q - E_{tot}[h\text{-BN}] + \sum_i n_i\mu_i + q(\mu_e - E_v). \quad (2)$$

The variables μ_e and E_v are the electronic chemical potential and the position of the valence band maximum (VBM) [29]. Our formation energies are allowed to vary with respect to the electronic potential μ_e (Fermi level position) that ranges within the h-BN band gap of 5.73 eV obtained from our HSE calculations as shown on Fig. 2. The correction term calculated by adopting the method developed by Komsa *et al.*[35] was also added into Eq. (2) to significantly reduce the effects of spurious electrostatic interactions of the charge with its periodic images.

Table 1 presents the formation energies of the charged defects calculated at zero electronic

Table 1. The calculated formation energies in eV (at Fermi level equal to zero) of the defects Si_B, Si_N, P_B and P_N in different charge states calculated in their favourable rich conditions.

Defects	-2	-1	0	+1	+2
Si_B	9.25	3.85	-0.24	-2.73	-1.73
Si_N	11.13	6.28	4.26	3.29	3.65
P_B	9.99	4.54	-1.27	-2.63	-2.60
P_N	11.83	5.83	0.22	0.11	1.28

chemical potential. Considering charged Si_B and P_B, it is noted that the creation of holes (+1 and +2 charge states) is exothermic and the addition of electrons (-1 and -2 charge states) is endothermic. The addition of electrons can require a high amount of formation energy penalty of around 9.00 eV in the -2 charge state. In the case of charged Si_N and P_N defects, both positive and negative charge states injections give positive formation energies. Fig. 1 shows that the defected structures have a significant buckling around the impurity defect. We also find that the uncharged Si impurity on the N site (Si_N) has more buckling ($h = 1.50$ Å) than Si impurity on the B site (Si_B) which has $h = 0.90$ Å. The same trend is also seen in the case of P_B and P_N defects, where P_N has a higher buckling height than P_N. We also observe the relatively large bond length distances d around the Si defect in Si_N, than in Si_B. The large bond is the consequences of Si or P atom trying to preserve its sp^3 bonding when interacting with either B or N atom and imposing a tetrahedral-like geometry in the h-BN sheet.

Table 2. The dependence of the defect-induced structural properties on the various charge states. The Δ_h (Å) is the buckling height of defect impurity X from the h-BN monolayer.

		-2	-1	0	+1	+2
Si_B	d	1.81	1.80	1.71	1.62	1.62
	Δ_h	1.29	1.29	0.90	0.30	0.30
Si_N	d	1.96	1.96	1.94	1.96	1.99
	Δ_h	1.61	1.61	1.50	1.65	1.67
P_B	d	1.71	1.71	1.70	1.64	1.67
	Δ_h	0.96	0.96	0.90	0.31	0.35
P_N	d	1.90	1.91	1.88	1.86	1.86
	Δ_h	1.55	1.54	1.48	1.51	1.51

We further investigate the effect of charge doping on the structural properties shown on Table 2. We note that an addition of a -1 charge into Si_B and P_B defects further increases the distance d and buckling height Δ_h. This might be the result of the Coulomb repulsion between the induced electron around the defect and valence electrons in the nitrogen orbitals. It is also noted that the creation of holes in Si_B and P_B defects significantly reduces the distances d and the buckling height Δ_h. This reveals that the removal of an electron from the defect orbital reduces the possible Coulomb repulsion between the impurity atom and its nearest neighbor atoms. Interestingly, this reduction is energetically favourable. The addition of both -1 and +1 charge into Si_N and P_N defects increases the buckling slightly. This structural strain is noted to be energetically unfavourable.

Lastly, the transition energy levels ε (q/q′) are deduced from the formation energies for Si_B, Si_N, P_B and P_N impurity defects using the expression [32];

$$\varepsilon(q/q') = \frac{E_f[X]^q - E_f[X]^{q'}}{q' - q}, \tag{3}$$

where $E_f[Si]^q$ and $E_f[Si]^{q'}$ denote the formation energies at the VBM for the charge states q and q' respectively.

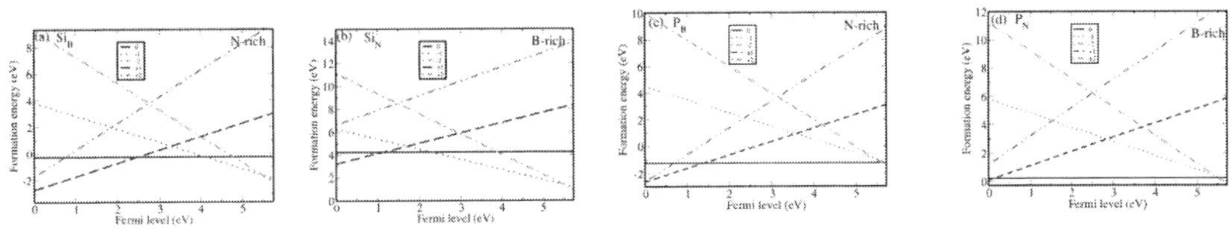

Figure 2. Calculated defect formation energies for (a) Si_B, (b) Si_N, (c) P_B and (d) P_N impurity defects in different charge states as a function of the Fermi level (electronic potential μ_e) showing the possible transition levels within the band gap of pristine h-BN monolayer.

Fig. 2 presents the variations in the formation energies for Si_B, Si_N, P_B and P_N impurity defects in different charge states as a function of the Fermi level (electronic potential μ_e) and also shows the possible transition levels within the band gap of pristine h-BN monolayer. In the case of Si_B defect, Fig. 2 (a) shows that the +1 charge state is stable from the top of VBM to the mid gap. The transition level of +1/0 is 2.52 eV above the VBM. The -1 charge state is stable from μ_e = 4.09 eV to 5.39 eV with reference to the VBM. Showing the transition level of 0/-1 at 1.64 eV below the conduction band minimum (CBM). This indicates that the defect levels +1/0 and 0/-1 in Si_B are possibly too deep to be ionized. Fig 2 (a) also shows the transition level of -1/-2 at 0.43 eV. In the case of Si_N defect, Fig. 2 (b) shows that the +1 charge state is stable from the top VBM to the μ_e = 1.26 eV (which is equal to +1/0 transition level). The 0 charge state is preferred in a relatively narrow range of μ_e = 1.10 eV. The -1 charge state is stable at a large range of μ_e = 3.29 eV, exhibiting the transition level of 0/-1 at 2.25 eV above the VBM. This indicates that this defect level is relatively deep. The results reveal that the Si impurity defect might have impact on the luminescence of h-BN monolayer.

Fig. 2 (c) shows that an addition of +1 into P_B is energetically favourable from the top of the VBM to μ_e = 1.50 eV. The neutral charge state ranges from μ_e = 1.50 eV to 5.62 eV. The defect transition level of +1/0 appears to be deep at 1.50 eV, whereas the transition level of 0/-2 is noted at 0.13 eV below the CBM. In the case of P_N, Fig. 2 (d) shows that the neutral charge state is the most energetically favourable for a wide range of band gap (about 5.35 eV), but when the μ_e is close to the VBM or CBM the +1 or -2 charge state is the most energetically favourable, respectively. The corresponding transition levels of +1/0 and 0/-2 are 0.16+VBM and 0.22-CBM respectively. This indicates that the P_N defect can relatively be easily ionized and could not have significant influence on the luminescence of h-BN monolayer.

4. Conclusions

Using hybrid density functional theory, we inverstigated the effect of Si and P impurity defects (Si_B, Si_N, P_B and P_N) in the h-BN monolayer. The local geometry, electronic structure and thermodynamic stability are examined using hybrid Heyd-Scuseria-Ernzerhof (HSE) functional. The response of these properties on the charge state modulation q = -2, -1, 0, +1 and +2 was also examined. These impurity defects cause a significant buckling in the h-BN monolayer. This structural distortion is mainly due to the effect of Si and P impurity trying to impose their sp^3 nature of bonding in the sp^2 bonded system (h-BN monolayer). The addition of electrons further increases the bond distances and the buckling heights around these Si_B, Si_N, P_B and P_N defects and is also endothermic. However, the creation of holes in Si_B and P_B defects significantly decreases the buckling height and is exothermic. Based on the charge state thermodynamic transition levels, we noted that the Si_B, Si_N and P_B are too deep to be ionized and therefore can affect the luminescence of *h*-BN material.

5. Acknowledgments

This work was supported by the University of Pretoria and National Research Foundation (NRF) of South Africa (Grant number 88246). REM thanks Dr. M P Molepo for fruitful discussions. The authors are grateful to Prof. J Pretorius and Prof. W Meyer for technical support.

6. References

[1] K Watanabe, T Taniguchi, and H Kanda, Nat. Mater. **3**, 404 (2004).
[2] D Pacile, J C Meyer, C O Girit, and A Zettl, Appl. Phys. Lett. **92**, 133107 (2008).
[3] N Alem, R Erni, C Kisielowski, M D Rossell, W Gannett, and A Zettl, Phys. Rev. B **80**, 155425 (2009).
[4] R V Gorbachev, I Riaz, R R Nair, R Jalil, L Britnell, B D Belle, E W Hill, K S Novoselov, K Watanabe, T Taniguchi, A K Geim, and P Blake, Small. **7**, 465 (2011).
[5] J C Meyer, A Chuvilin, G Algara-Siller, J Biskupek, and U Kaiser, Nano Lett. **9**, 2683 (2009).
[6] P R Wallace, Phys. Rev. **71**, 622 (1947).
[7] K S Novoselov, A K Geim, S V Morozov, D Jiang, Y Zhang, S V Dubonos, I V Grigorieva, and A A Firsov, Science **306**, 666 (2004).
[8] Y Kubota, K Watanabe, O Tsuda, and T Taniguchi, Scie. **317**, 932 (2007).
[9] Synthesis and Properties of Boron Nitride, edited by J J Pouch and S A Alteroviz (Trans Tech, Switzerland, 1990).
[10] C Jin, F. Lin, K Suenaga, and S Iijima, Phys. Rev. Lett. **102**, 195505 (2009).
[11] O L Krivanek, M F Chisholm, V Nicolosi, T J Pennycook, G J Corbin, N Dellby, M F Murfitt, C S Own, Z S Szilagyi, M P Oxley, S T Pantelides, and S J Pennycook, Nature (London) 464, 571 (2010).
[12] L Song, L Ci, H Lu, P B Sorokin, C Jin, J Ni, A G Kvashnin, D G Kvashnin, J Lou, B I Yakobson, and P M Ajayan, Nano Lett. **10**, 3209 (2010).
[13] R F Liu, and C Cheng, Phys. Rev. B **76**, 014405 (2007).
[14] M S Si, and D S Xue, Phys. Rev. B **75**, 193409 (2007).
[15] J M Pruneda, Phys. Rev. B **85**, 045422 (2012).
[16] X Wei, M Wang, Y Bando, and D Golberg, J. Am.Chem. Soc. **132**, 13592 (2010).
[17] A M Ukpong, and N Chetty, Phys. Rev. B **86**, 195409 (2012).
[18] N Berseneva, A V Krasheninnikov, and R M Nieminen, Phys. Rev Lett. **107**, 035501 (2011).
[19] Y Liu, B Gao, D Xu, H Wang, and J Zhao, Phys. Lett. A **378** 2989-2994 (2014).
[20] W Q Han, W Mickelson, and J Cumings, Appl.Phys, **93**, 1110-1114 (2008).
[21] B He, M Qiu, M F Yuen, and W J Zhang, Appl. Phys. Lett, **05**, 012104 (2014).
[22] H Murata, T Taniguchi, S Hishita, T Yamamoto, F Oba, and I Tanaka Journal of applied physics, **114**, 233502 (2013).
[23] S Majety, T C Doan, J Li, J Y Lin, and H X Jiang, AIP. Advances **3**, 122116 (2013).
[24] S Lin, X Ye, and J Huang, Phys. Chem. Chem. Phys, **17**,888-95 (2015).
[25] J P Perdew, K Burke, and M Ernzerhof, Phys. Rev. Lett. **77**, 3865 (1996).
[26] J Heyd, G E Scuseria, and M Ernzerhof, J. Chem. Phys. 118, 8207 (2003).
[27] H Kokten, and S Erkoc, Turk. J. Phys. **38** 369 - 374 (2014).
[28] G Kresse, and J Hafner Phys. Rev. B 47, 558 (1993).
[29] P Spiewak, K Sueoka, J Vanhellemont, K J Kurzydlowski, K Mlynarczyk, P Wabinski, and I Romandic Phys. B **401** 205-209 (2007).
[30] P E Blochl, and Phys. Rev. B **50**, 17953 (1994).
[31] H J Monkhorst, and J D Pack, Phys. Rev. B **13**, 5188 (1976).
[32] C G Van de Walle, and J Neugebauer, J. Appl. Phys. **95**, 3851 (2004).
[33] T B Ngwenya, A M Ukpong, and N Chetty, Phys. Rev. B **78**, 245425 (2011).
[34] S S Wise, J L Margrave, H M Feder, and W N Hurbbard, J. Phys. Chem. **70** 7 (1966).
[35] H -P Komsa, N Berseneva, A V Krasheninnikov, and R M Nieminen, Phys. Rev. B **4**, 031044 (2014).

Modelling of Palladium Gold alloy- dielectric stratified loaded plasmonic waveguide for hydrogen detection at room temperature

Sanchali Mitra, Rik Chattopadhyay, Shyamal Kumar Bhadra*

Fibre Optics & Photonics Division, CSIR- Central Glass and Ceramic Research Institute, 196 Raja S C Mullick Road, Kolkata-700032, West Bengal, India

*skbhadra@cgcri.res.in

Abstract. A plasmonic waveguide loaded with metal - dielectric stratified layer is reported for room temperature hydrogen sensing. Pd-Au alloy with atomic ratio 3:1 (Pd/Au) is used as the metal and air is used as dielectric layer in the sensor. When hydrogen is loaded in the waveguide the alloy absorbs hydrogen and the effective permittivity of the stratified medium changes. Density functional theory and Bruggeman's effective medium theory are used to calculate the permittivity of bare and hydrogenated alloy. At a specific wavelength the change in the absorption loss of the device on hydrogen absorption is used as the measurement for hydrogen detection. At 633 nm the loss decreases almost linearly with increasing hydrogen concentration. For a device length of 10 μm the change of loss is achieved about 0.03 dB for 1 % change of hydrogen concentration.

1. Introduction

Hydrogen is highly flammable gas and can burn at very low concentration of about 4% in the air. Safe and easy detection of very low concentration of hydrogen gas is a fascinating area of study. Pd is an important material for hydrogen sensing application [1-5]. But there are certain limitations of using Pd. During hydrogen absorption desorption cycle Pd membrane suffers from embrittlement and crack due to α to β phase transformation as β phase contributes more lattice expansion compared to α phase [1,2]. Alloying Pd with other metals like Au, Ag reduces volume expansion thereby reducing the probability of crack formation hence producing more stable structure. The rate of hydrogen permeation in Pd-Au and Pd-Ag alloy is higher than that of pure Pd [6, 7]. Two times higher hydrogen permeability than pure Pd is achieved using $Pd_{0.8}Au_{0.2}$ alloy [8]. Pd-Au alloy is also more resistant to sulfur poisoning w.r.t. other alloys [9]. Additionally inclusion of Au in Pd can overcome some limitations of Pd as plasmonic material.

Surface plasmon resonance based hydrogen sensor is growing area of research due to safe, hazardless, fast detection and high sensitivity. In this work we have designed stratified metal dielectric composite loaded four layer plasmonic waveguide for room temperature hydrogen sensing. Pd-Au alloy with atomic concentrations of 75% and 25% for Pd and Au is used as metal layer in the stratified media. To calculate the response of designed waveguide the permittivity of the alloy should be evaluated. Density Functional Theory (DFT) is a useful method to solve Schrödinger equation for many electron systems and can be used to evaluate various properties of any material. So we here employ DFT to calculate the complex permittivity of bare and hydrogen loaded alloy. We have used metal dielectric stratified layer instead of single metal layer to compensate the huge loss of the waveguide caused by metal. When Pd_3Au alloy absorbs hydrogen the permittivity of the alloy gets changed accordingly the loss of the waveguide for a specific wavelength changes. At wavelength 633 nm we have calculated the absorption loss that decreases almost linearly with increasing hydrogen concentration. For a device length of 10μm the decrease of absorption loss is 0.03db for 1 % change of hydrogen concentration.

2. Theoretical calculations

Plane wave Density Functional Theory (DFT) [10, 11] is used to calculate dielectric function of bare and hydrogen loaded Pd_3Au alloy. Vienna Ab Initio Simulation Package (VASP) [14] using projector-augmented wave (PAW) method [17] is used to carry out DFT calculation. Electron exchange correlation effects are described using GGA (generalized gradient approximation) with PBE (Perdew, Burke and Ernzerhof) functional [15, 16]. Permittivity is calculated by evaluating direct electronic transitions from occupied to higher energy unoccupied electronic states.

For metallic system the complex permittivity (ε(ω)) consists of interband and intraband terms. Imaginary interband part is calculated from band diagram by taking real transitions from valence band to conduction band as shown in Equation 1 [18].

$$\mathrm{Im}\left[\varepsilon_{\alpha,\beta}^{inter}(\omega)\right]=\frac{4\pi^2e^2}{V}\lim_{q\to 0}\frac{1}{q^2}\sum_{a,b}2f_{bK}\left\langle u_{aK+e_\alpha q}\middle|u_{bK}\right\rangle\left\langle u_{bK}\middle|u_{aK+e_\beta q}\right\rangle\times \left[\delta\left(\xi_{aK}-\xi_{bK}-\omega\right)-\delta\left(\xi_{aK}-\xi_{bK}+\omega\right)\right] \quad (1)$$

Here, α, β represent different Cartesian directions, δ function generate peaks at transition energies $\omega=\xi_{aK}-\xi_{bK}$ where a and b refer to two different energy bands, q is the wave vector for a given K point in first Brillouin zone, transition probability is given by $\left\langle u_{aK+e_\alpha q}\middle|u_{bK}\right\rangle\left\langle u_{bK}\middle|u_{aK+e_\beta q}\right\rangle$ where u represents the component of Bloch wave function having periodicity of lattice and V is the volume of unit cell.

Real part of the dielectric function is calculated from imaginary part using Kramers-Kronig transform as given in Equation 2.

$$\mathrm{Re}\left(\varepsilon^{inter}(\omega)\right)=\frac{1}{\pi}\rho\int_{-\infty}^{+\infty}\frac{\mathrm{Im}\left[\varepsilon^{inter}(\omega')\right]}{\omega'-\omega}d\omega' \quad (2)$$

The intraband contribution to the optical properties affects mainly the low energy infrared part of the spectra. It can be described via Drude term given in Equation 3,

$$\varepsilon^{intra}(\omega)=1-\frac{\omega_p^2}{\omega^2+i\gamma_d\omega} \quad (3)$$

Where, ω_P is plasma frequency and γ_d is the damping parameter.

ω_P represents the intraband free electron plasma frequency tensor as given in Equation 4,

$$\omega_p^2(\alpha,\beta)=-\frac{4\pi e^2}{V}\sum_{b,K}2\frac{\partial f(\xi_b)}{\partial\xi_b}\left(\hat{e}_\alpha\frac{\partial\xi_b(K)}{\partial K}\right)\times\left(\hat{e}_\beta\frac{\partial\xi_b(K)}{\partial K}\right) \quad (4)$$

Fully occupied bands do not contribute to the plasma frequency.

Damping factor γ_d is calculated from plasma frequency ω_p and conductivity σ using Equation 5 [20].

$$\gamma_d=\frac{\omega_p^2\varepsilon_0}{\sigma} \quad (5)$$

This intraband contribution is summed up with interband part to get the total dielectric function of the bare and hydrogenated alloy.

The Monkhorst-Pack mesh with 8x8x8 K grid is used for the energy calculation. Plane wave energy cut off is set to 400 eV to ensure the convergence of the total energy. For optical dielectric function calculations 16x16x16 K point grid is chosen. Geometric relaxations are done using conjugate gradient (CG) algorithm until the forces on the atoms are lower than 0.02 eV/Å. For the integration of the Brillouin zone the first order Methfessel-Paxton method with a value of 0.2 eV is used.

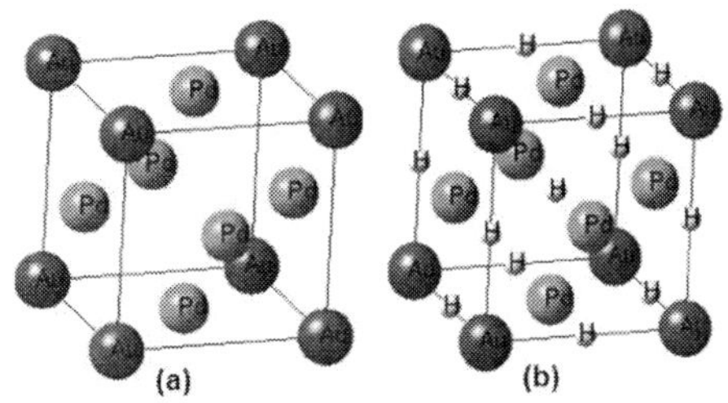

Figure 1 Atomic configurations of (a) Pd_3Au (b) Pd_3AuH_4.

For DFT calculation of Pd_3Au alloy we have used $L1_2$ structure in which Au atoms occupy corner positions whereas Pd atoms are placed in face positions [Fig. 1a]. This crystal structure is reported the most stable structure for Pd_3Au [12, 13]. For hydrogen loaded alloy hydrogen atoms are placed in octahedral interstitial sites of the Pd_3Au lattice shown in Fig. 1b. Here we consider inhomogenous distribution of hydrogen atoms in the alloy film in which both pure crystalline Pd-Au (α) phase and hydrogenated Pd-Au-H (β) phase are present together. Bruggeman's effective medium approximation [19] is used to calculate the effective dielectric function of the hydrogenated alloy having both the phases [Equation 6]. For α and β phases dielectric functions of pure Pd_3Au [Fig. 1a] and Pd_3AuH_4 [Fig. 1b] are considered.

$$f_\beta \cdot \frac{\varepsilon_\beta - \varepsilon_{eff}}{\varepsilon_\beta + 2\varepsilon_{eff}} + \left(1 - f_\beta\right) \frac{\varepsilon_\alpha - \varepsilon_{eff}}{\varepsilon_\alpha + 2\varepsilon_{eff}} = 0 \qquad (6)$$

Where ε_α, ε_β are dielectric functions of α and β phases respectively, f_β is the fraction of β phase.

For room temperature hydrogen sensing we have designed a 4 layered plasmonic waveguide shown in Figure 2b. The waveguide consists of silica substrate, higher index silica-titania film act as guiding layer, metal dielectric stratified layer on top of the guiding layer, air as the cover region. The stratified medium consists of periodic layers of Pd_3Au alloy film and air [Figure 2a]. From effective medium theory the permittivity of the stratified medium is calculated using equation 7.

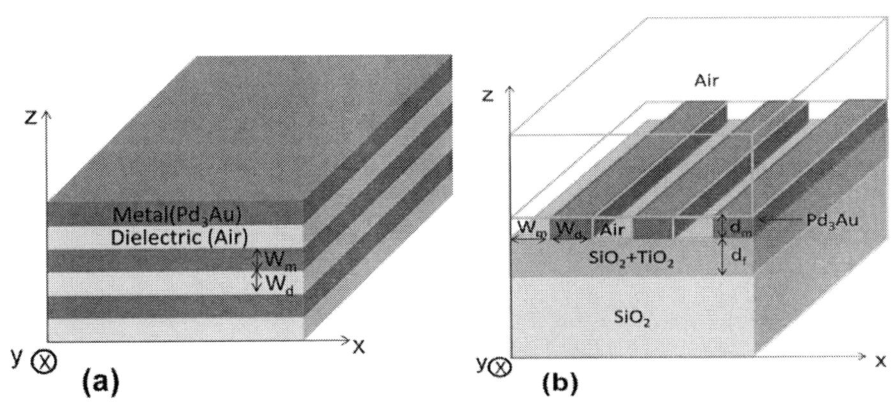

Figure 2. Schematic of (a) stratified medium (b) four layer waveguide loaded with the stratified layer. The waveguide parameters are optimized to achieve higher sensitivity. The values of the parameters are w_m=50nm, w_d= 8nm, d_f=580nm, d_m=15 nm. The refractive index of substrate (SiO_2) is 1.4573 and that of guided layer (SiO_2+TiO_2) is 1.7264.

$$\varepsilon_x = \varepsilon_y = \frac{w_m \varepsilon_m + w_d \varepsilon_d}{w_m + w_d} \qquad \text{And} \qquad \varepsilon_z = \frac{w_m + w_d}{\frac{w_m}{\varepsilon_d} + \frac{w_d}{\varepsilon_m}} \tag{7}$$

Where w_m, w_d are thicknesses and ε_m and ε_d are permittivities of the alloy layer and air respectively. ε_x, ε_y are designated as parallel orientation permittivity and ε_z as perpendicular orientation permittivity.

The plot of ε_x and ε_z are shown in Figure 3. As ε_x has negative real part in a wide wavelength region we choose parallel orientation for the metal dielectric stratified layer. The response of the waveguide is studied analytically.

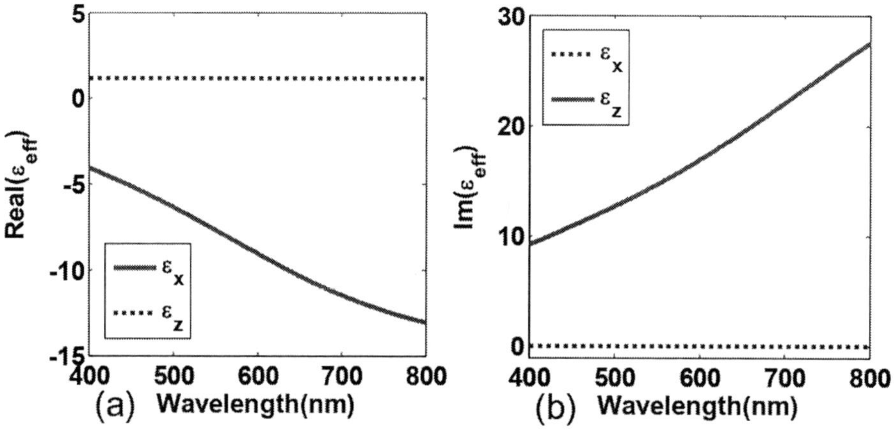

Figure 3. (a) Real part and (b) Imaginary part of effective permittivity of stratified medium for parallel (ε_x) and perpendicular (ε_z) orientation.

3. Results & Discussions

When the waveguide comes in contact with hydrogen gas the alloy absorbs hydrogen accordingly the permittivity of the alloy changes. As described in section 2 Bruggeman's effective medium theory is used to calculate permittivity of hydrogen loaded alloy. In Figure 4 we have plotted the variation of permittivity of the stratified layer on hydrogen absorption. It shows that with increasing hydrogen concentration both the real and imaginary part of complex permittivity decrease.

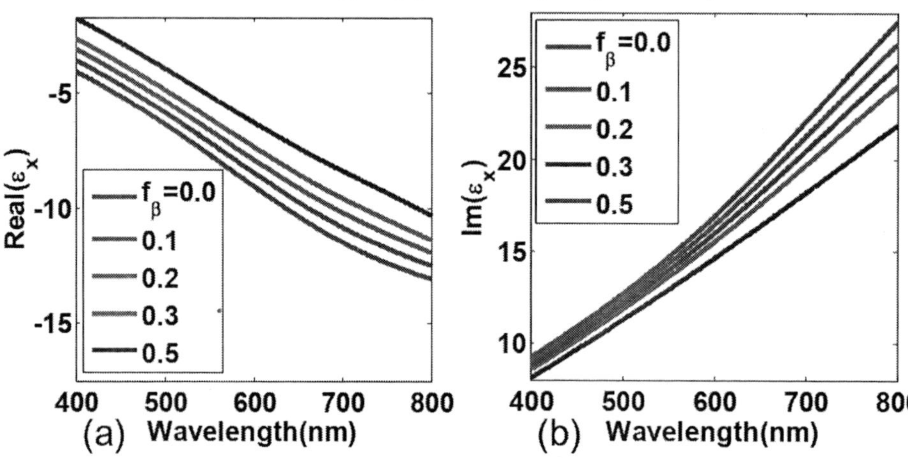

Figure 4. (a) Real part and (b) Imaginary part of permittivity of the stratified medium for different hydrogen concentration. Here f_β denotes the fraction of hydrogenated β phase to pure crystalline α phase of the alloy. With increasing hydrogen concentration f_β will increase.

We have analytically calculated the effective index of fundamental TM mode of the waveguide [21]. The mode field for fundamental TM mode at 633 nm is shown in Figure 5. The change of absorption loss of the waveguide at 633 nm on hydrogen loading is calculated. The waveguide parameters are optimized to achieve high value of sensitivity.

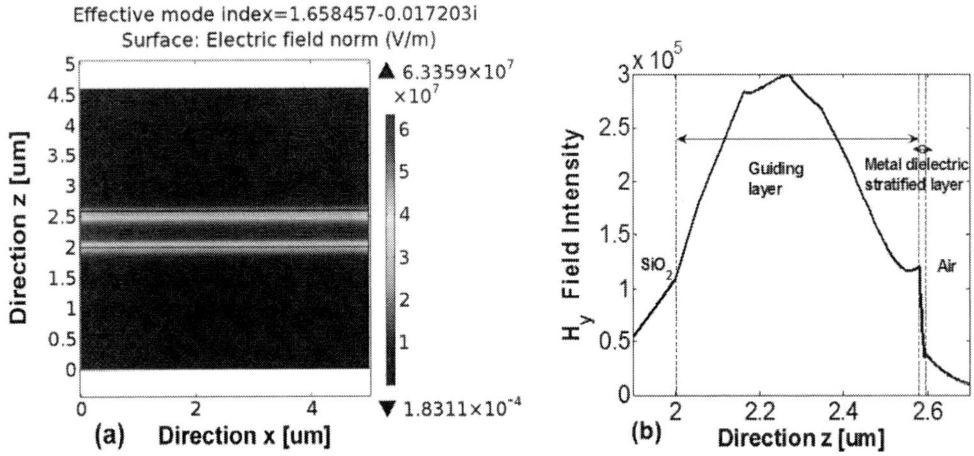

Figure 5. (a) Mode field of fundamental TM mode at 633 nm of the 4 layer stratified metal dielectric composite loaded plasmonic waveguide, (b) Intensity profile of normal component of magnetic field (H_y). The waveguide parameters are given in Figure 2.

The plot of absorption loss with variation of hydrogenated phase [Figure 6] shows the loss decreases almost linearly with increasing hydrogen concentration. The use of stratified layer instead of pure metal layer compensates the loss incurred by metal. In case of using single metal layer the absorption loss is too high to detect the transmitted power at output of the device. The slop of the graph is 0.3dB/0.1 fractional change of f_β.

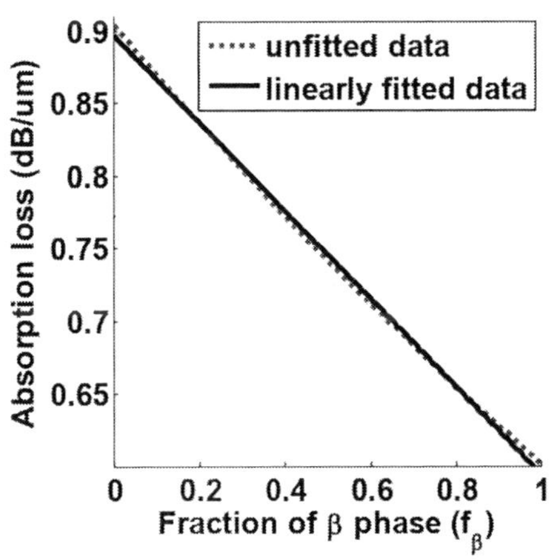

Figure 6. Plot of absorption loss of the waveguide with fraction of β phase to α phase for fundamental TM mode at 633 nm.

4. Conclusions

In this work we have reported four layer plasmonic waveguide consists of metal dielectric stratified layer for detection of hydrogen gas at room temperature. Pd_3Au alloy is used as metal layer due to its ability of absorbing hydrogen. The optical permittivity of the alloy is calculated using DFT. Bruggeman's effective medium theory is used to calculate the effective permittivity of hydrogenated alloy considering inhomogenous distribution of hydrogen atoms. During hydrogen absorption in the alloy the effective index of stratified medium changes and the overall response of the waveguide varies accordingly. The absorption loss of the waveguide for fundamental TM mode is calculated analytically. The loss at 633 nm decreases almost linearly with increasing hydrogen concentration. For 1% change of hydrogen concentration the loss will change about 0.03 dB for a device length of 10μm. The use of stratified layer instead of pure metal layer reduces the huge loss of metal so that output power can be detected properly.

Acknowledgement

The authors would like to thank Dr. K. Muraleedharan, Director, CSIR-CGCRI for support and encouragement to carry out the work and all the staff members of fiber optics and photonics division. The work was supported by Council of Scientific and Industrial Research (CSIR) 12th five year plan project on GLASSFIB (ESC 0202).

References

[1] Lewis F A 1967 *Platinum Metals Rev.* **5**(1) 21
[2] Lewis F A 1967 *The Palladium Hydrogen System* (Academic New York)
[3] Poyli M A *et al* 2012 *J. Phys. Chem. Lett.* **3**(18) 2556
[4] Maeland A J and Flanagan T B 1966 *Platinum Metals Rev.* **10**(1) 20
[5] Silkin V M *et al* 2007 *Phys. Rev. B.* **76,** 245105
[6] Sonwane C G *et al* 2006 *J. Chem. Phys.* **125**(18) 184714
[7] Flanagan T B and Wang D 2011 *J. Phys. Chem. C.* **115** 11618
[8] Gryaznov V 2000 *Separation & Purification Rev.* **29**(2) 171
[9] Chen C H and Ma Y H 2010 *J. of Membrane Sc.* **362** 535
[10] Kohn W and Sham L J 1965 *Phy. Rev.* **140** (4A) A1133
[11] Hohenberg P and Kohn W 1964 *Phys. Rev.* **136** (3B) B864
[12] Sluiter M H F *et al* 2006 *Phys. Rev. B.* **73** 174204
[13] Okamoto H and Massalski T B 1985 *J. Phase Equilibria* **6** (3) 229
[14] Kresse G and Furthmuller J 1996 *Phys. rev. B Condens. Matter* **54**(16) 11169
[15] Perdew J P *et al* 1996 *Phys. Rev. Lett.* **77** (18) 3865
[16] Perdew J P *et al* 1997 *Phys. Rev. Lett.* **78** (7) 1396
[17] Blochl P E 1994 *Phys. Rev. B* **50** 17953
[18] Harl J 2008 *The linear response function in density functional theory: optical spectra and improved description of the electron correlation, Ph. d Dissertation submitted to Universitat Wien*
[19] Bruggeman D A G 1935 *Ann. Physik (Leipzig)* **24** 636
[20] Ashcroft N W and Mermin N D 1976 *Solid state physics* (Saunders) pp 7
[21] Roy R D *et al* 2013 *Photon. Res.* **1** (4) 164

Magnetic Metamaterials: A comparative study of resonator geometry and metal conductivity

Shashank Rangu, Kamireddy Sreekar, Ravinithesh Reddy Annapureddy , Kausik Basak, Murtaza Bohra and Dibakar Roy Chowdhury*

Mahindra Ecole Centrale, Jeedimetla, Hyderabad, 500043, India
dibakar.roychowdhury@mechyd.ac.in

Abstract. In this work, split ring resonators based metamaterials are studied for microwave, terahertz and infrared frequency regimes. Two different geometries, circular and rectangular split ring resonators based metamaterials are investigated numerically for different frequency regimes. Our study indicates that the effect of metal conductivity and resonator geometry shows very little impact on the fundamental resonance mode. However the higher order modes go through significant frequency tuning because of the change in resonator geometry. We have further shown that the metal conductivity is an important parameter for the metamaterials employed in infrared domains.

1. Introduction

Metamaterials (MM) are artificial composite materials with user defined electromagnetic (EM) properties, typically made from dielectrics and arrays of structured metallic particles/features much smaller than the wavelength of the propagating electromagnetic waves [**1**]. These structured particles are generally termed as split ring resonators (SRRs). Metamaterials can strongly interact with EM waves and capable of controlling EM waves unlike any natural materials. Practically MMs can be employed to control waves with unprecedented flexibility. This has been indicated in the large number of MM based unique device/ physical concepts including electromagnetic cloaking [**2, 3**], perfect imaging [**4, 5**], optical magnetism [**6**], near field coupling [**7 - 10**], sensing [**11, 12**], electromagnetic wave modulations [**13 - 17**], polarization rotation [**18 - 20**] etc. Metamaterials are also scalable in size making its resonances tuneable for use at different wavelengths of the entire electromagnetic spectrum [1]. In this work, we have investigated the effect of split ring resonator geometry on the different resonance modes for microwave (MW), Terahertz (THz) and Infrared (IR) domain metamaterials (MMs). Particularly rectangular and circular split ring resonators are studied extensively through numerical simulations. Additionally we have included effects of using different metals with versatile conductivity in our study.

For both the geometries, we have purposefully kept resonator size along with the split gap dimension constant. These two conditions allowed us to keep the intrinsic inductance and capacitance of the individual resonator fixed. Our investigations reveal that the first order or inductive-capacitive (LC) resonance remains almost unperturbed with respect to variable resonator geometry as well as metal conductivity. This result is somewhat expected since the SRR inductance and capacitance played the major role for the first order resonance and in agreement with previous work [**14**]. However higher order resonances, such as, 2nd order and 3rd order resonances go through large frequency shift. We attribute this resonance tuning to the relative variation of the resonator arm length for the SRRs with different geometry. Additionally we have observed a reduction in resonance strength particularly for the IR MMs. Such reduction in resonance strength is attributed to the increasing loss due to the electron scattering effect at the higher operating frequencies.

2. Simulations and Discussions

In this work we have studied two geometrical configurations of metamaterials, one circular ring based MM and the other square ring based MM. In both cases resonators are composed of gold, Silver or bad silver. Bad silver is defined as metal with reduced conductivity compared to Silver. Additionally we have studied these samples at the different regions of the electromagnetic wavelengths, namely, Microwaves, Terahertz and Infrared. We have intentionally kept the split gap geometry constant for both the configurations for particular frequency regime. Lossless silicon is used as substrate for all the studied MM. Please see the figure below for detailed geometry of the structures.

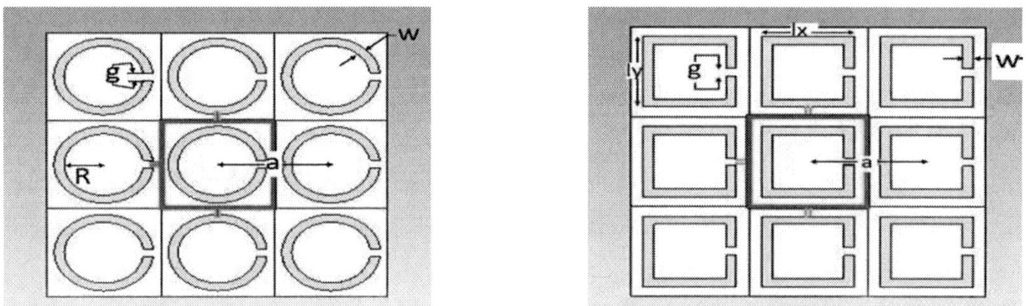

Figure 1. The periodicity of the unit cell (a) is 4500um (microwave frequency), 60 um (THz frequency) and 1.35 um (Infrared frequency). Similarly, the length of the pattern (lx) is 4500 um, 60 um, 1.35 um; the breadth of the pattern (ly) is 1515 um, 20 um, 0.496 um; the radius of the circular ring (r) is 1709.5 um, 22.57 um, 0.56 um; gap in the pattern (g) is 500 um, 4 um, 0.1 um; thickness of the pattern (t) is 3000 um, 25 um, 10 um; the thickness of the substrate (tsub) is 3000um, 25um, 10um; the width of the pattern (w) is 400um, 4um, 0.1um; respectively for microwave, terahertz and infrared regions. The frequency ranges used are: 0.5GHz to 20Hz (microwave region), 300GHz to 1800GHz (THz region) and 30THz to 80THz (Infrared region).

2.1. Simulation procedure

In this work, all the metamaterials are simulated using finite difference frequency domain (FDFD) solver using numerical computations. The pattern along with the substrate is defined as a unit cell, see figure 1 for elaboration. This is periodically repeated to form the metamaterial. In simulations we have employed electric field either parallel or perpendicular to the split gap of the resonator. Simulation results for the metamaterials for different frequency ranges are shown in the figure below.

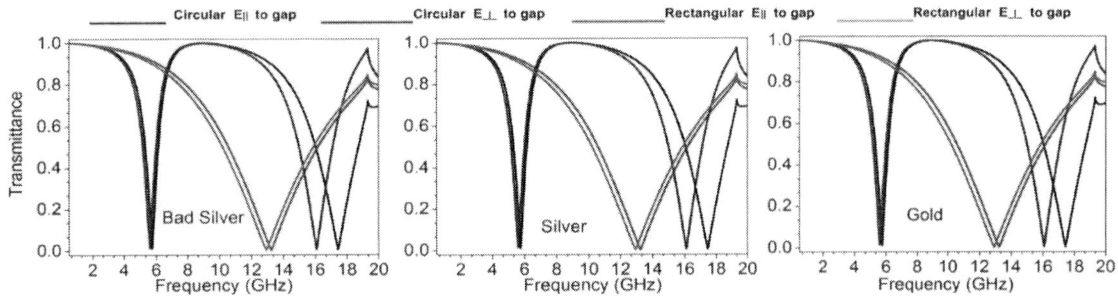

Figure 2. Simulation of metamaterials for bad silver, Silver and gold for Microwave frequency range.

As evident from figure 2, the fundamental resonance remains almost unperturbed whereas the third order resonance goes through maximum tuning due to the change in resonator shape. Fundamental resonance broadly depends on inductance and capacitance of the individual resonators [**21, 22**]. The inductance and capacitance of the metamaterials for the particular frequency regime does not alter much because of the two guiding rules we have followed in this work. First the split gap is kept constant and second the total area of the split resonator is also kept constant. Because of the constant geometry of the split gap the capacitance remains same for both the structures (rectangular and circular). Similarly size of the resonators (circular or rectangular) for the particular frequency domain is also kept constant which effectively makes the inductance of the resonators constant. Therefore these two constant parameters do not allow the fundamental resonance to shift considerably. Note that resonance frequency for the fundamental mode depends strongly on these two geometrical parameters of the resonators [**21, 22**].

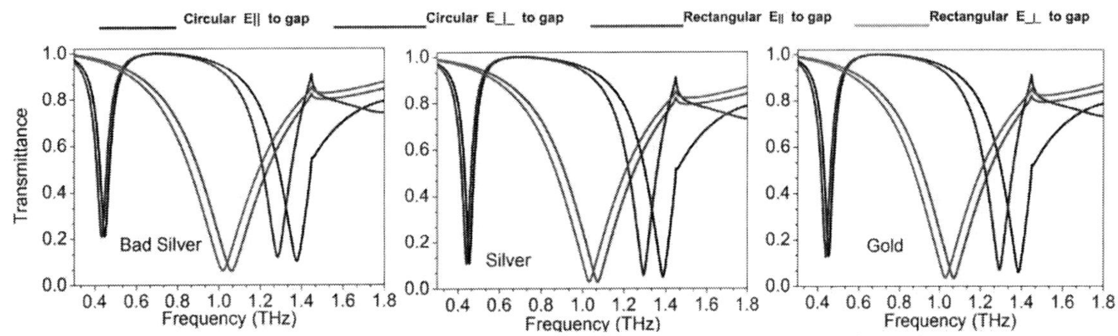

Figure. 3. Simulation of metamaterials for bad silver, Silver and gold for Terahertz frequency range.

However, the higher order resonances are more dependent on arm lengths [**23**]. This effectively results into shifting of the higher order resonances significantly, see Fig 2. For microwave frequency range, the resonance strengths remain almost same for all the resonance modes. This is because at this relatively low frequency range the metal conductivity is significantly high and therefore does not affect the resonance positions significantly, hence the resonances are not affected significantly even though different metals are employed (Figure 2). We have further studied the similar circular and rectangular resonators based metamaterials for THz frequency regime.

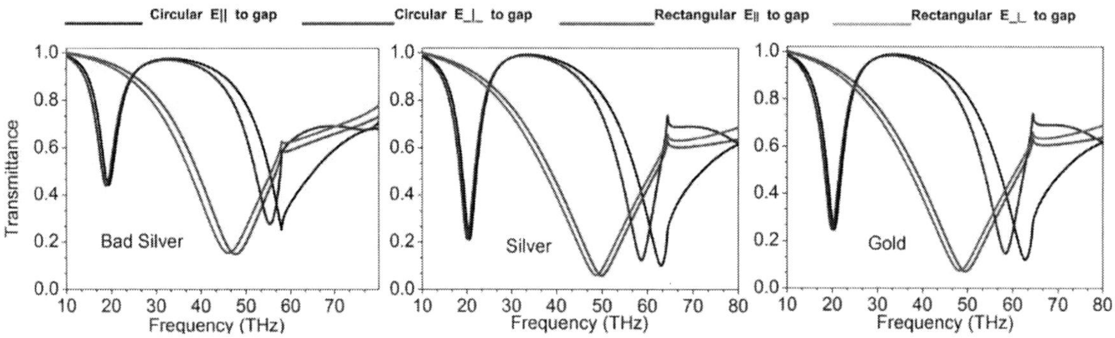

Figure 4. Simulation of metamaterials for bad silver, Silver and Gold for Infrared frequency range.

The simulation outcomes are shown in figure 3 for THz metamaterials. Because of the similar reasons as explained in the previous paragraph, the fundamental resonances remain almost intact. However the second order and third order resonances shift with change in resonator geometry. Resonance shifts toward the high frequency side in case of the circular resonator based MM because of the shorter resonator arm in the circular resonators compared to the rectangular resonator. However both the resonators are equal in size. This effect is even more prominent in case of third order resonance. The similar trend is demonstrated by the Infrared (IR) metamaterials as well as shown in figure 4. However the resonance strength reduces in a constant manner when we move from microwave to IR metamaterials through THz frequency regimes as demonstrated in figures 2, 3 and 4. The reduction in resonance strength is because of the electron-electron scattering losses. Such loss is actually increases as operating frequency increases. Our extensive numerical study comprising of circular and rectangular resonator based metamaterials for three important frequency regime clearly demonstrate that the metal conductivity does not play a significant role as long as it is higher than a critical value, for example silver conductivity. However, resonator shape can lead to significant shift in the higher order resonance modes while keeping the fundamental resonance position almost unperturbed.

Table 1. Relative shift of different resonance modes for circular and rectangular resonators as extracted from the individual simulations and depicted in figures 2, 3 and 4. Bad silver is modeled as conductor with conductivity poorer than silver and of the order of 1×10^7 S/m.

	Gold			Bad Silver			Silver		
	Mode 1	*Mode 2*	*Mode 3*	*Mode 1*	*Mode 2*	*Mode 3*	*Mode 1*	*Mode 2*	*Mode 3*
MW (GHz)	0.156	0.312	1.3455	0.156	0.312	1.326	0.156	0.312	1.3455
THz (GHz)	13.5	40.5	93	15	40.5	93	13.5	40.5	91.5
IR (THz)	0.49	1.26	4.41	0.56	1.61	2.52	0.49	1.19	4.34

The above table demonstrates the relative shift in resonance frequency after extracting from the spectrums of the metamaterials studied numerically. The 1st order mode or fundamental mode resonance frequency remains almost unperturbed when compared for different resonator geometry. Precisely the frequency shift is less than even 2 percent with respect to the individual resonances. However the second and third order resonance shifts are relatively more prominent of the order of 4 percent and 7 percent compared to the corresponding individual resonances, respectively.

3. Conclusions

In this work, we have studied the influence of different geometry and constitutive metals on split ring resonators based magnetic metamaterials through extensive numerical simulations. We have extended our study to metamaterials operating at different wavelength regimes of the electromagnetic spectrum, termed as microwave, terahertz and infrared. Our investigations show that resonator geometry does not have much impact on the fundamental resonances irrespective of the regime of operations. However resonator shape has strong influence on the higher order resonances. We attribute this resonance shift to the change in the metal arm length. For the metamaterials studied in the infrared domains, the resonances lose strength when compared to the microwave and terahertz metamaterials for the similar modes. We attribute such loss in resonance strength to electron scattering which is dominant at higher frequencies. Overall, this study of metamaterials with respect to different geometry, metals and operational wavelength regimes can be useful for designing MM for a particular application in different operational frequency regimes.

Acknowledgements:

Authors would like to acknowledge the support from Mahindra Ecole Centrale to conduct this research work.

References

[1] J.B. Pendry et al. IEEE Trans. Microwave Theory Tech **47**, 2075 (1999)
[2] D Schurig, J Mock, B J Justice, S A Cummer, J B Pendry, A F Starr, D R Smith, *Science* **314**, 977 (2006)
[3] S A Cummer, B I Popa, D Schurig, D R Smith, J Pendry, *Physical Review E,* 74, 036621 (2006)
[4] J. B. Pendry, Phys. Rev. Lett. **85**, 3966 (2000)
[5] D. R. Smith, J. B. Pendry, M. C. K. Wiltshire, *Science,* **305**, 788 (2004)
[6] A I Kuznetsov, A E Miroshnichenko, Y Hsing Fu, J B Zhang, B Luk'yanchuk, Scientific Report, **2**, 492 (2012)
[7] Na Liu, Hui Liu, Shining Zhu & Harald Giessen, Nature Photonics, **3**, 157 (2012)
[8] D Roy Chowdhury, J F O'Hara, A J Taylor, and A K Azad, *Applied Physics Letters* **104**, 101105 (2014)
[9] D Roy Chowdhury, N Xu, W Zhang, R Singh, Journal of Applied Physics **118**, 023104 (2015)
[10] D A Powell, M Lapine, M V Gorkunov, I V Shadrivov, Y S Kivshar, *Physical Review B*, **82**, 155128 (2010)
[11] J F O'Hara, R Singh, I Brener, E Smirnova, J Han, A J Taylor, W Zhang Optics Express **16**, 1786 (2008)
[12] W Withayachumnankul, H Lin, K Serita, C M Shah, S Sriram, M Bhaskaran, M Tonouchi, C Fumeaux, and D Abbott **20**, 3345 2012()
[13] HT Chen, WJ Padilla, JMO Zide, AC Gossard, AJ Taylor, RD Averitt, Nature **444**, 597 (2006)
[14] D Roy Chowdhury, R Singh, A J Taylor, H T Chen, and A K Azad, Applied Physics Letters **102**, 011122 (2013)
[15] M Liu, H Y Hwang, H Tao, A C Strikwerda, K Fan, G R Keiser, A J Sternbach, K G West, S Kittiwatanakul, J Lu, S A Wolf, F G Omenetto, X Zhang, K A Nelson, and R D Averitt, Nature, **487**, 345 (2012)
[16] D Roy Chowdhury, A K Azad, W Zhang, and R Singh, *IEEE Transactions on Terahertz Science and Technology*, **3**, 783 (2013)
[17] W Withayachumnankul, and D Abbott, *IEEE Photonics Journal*, **1**, 99 (2009)
[18] D-H Kwon, P. L. Werner, and D. H. Werner, *Optics Express*, **16**, 11802 (2008)
[19] N. K. Grady, J. E. Heyes, D. Roy Chowdhury, Y. Zeng, M. T. Reiten, A. K. Azad, A. J. Taylor, D. Dalvit, and H.-T. Chen *Science*, **340**, 1304 (2013)

[20] L Cong, W Cao, X Zhang, Z Tian, J Gu, R Singh, J Han, W Zhang, *Applied Physics Letters*, **103**, 171107 (2013)
[21] C Rockstuhl, F Lederer, C Etrich, T Zentgraf, J Kuhl, H Giessen, Optics express **14**, 8827 (2006)
[22] D Roy Chowdhury, R Singh, J F O'Hara, H T Chen, A J Taylor, A K Azad, Applied Physics Letters **99**, 231101 (2011)
[23] A K Azad, A J Taylor, E Smirnova, J F O'Hara, *Applied Physics Letters,* **92**, 011119 (2008)

Computational Study of Au_mSi_n (m+n=2-6) Nanoalloy Clusters Invoking Density Functional Based Descriptors

P. Ranjan [1], **A. Kumar** [1], **T. Chakraborty** [2]

[1] Department of Mechatronics Engineering, Manipal University Jaipur, Dehmikalan, Jaipur-303007

[2] Department of Chemistry, Manipal University Jaipur, Dehmikalan, Jaipur-303007

Email: tanmoy.chakraborty@jaipur.manipal.edu, tanmoychem@gmail.com

Abstract. In this study, electronic and optical properties of Au_mSi_n (m+n=2-6) nanoalloy clusters are systematically investigated in terms of the Density Functional Theory (DFT) with the generalized gradient approximation (GGA). Conceptual DFT based global theoretical descriptors have been used to reveal experimental properties qualitatively. In this venture, experimental properties of Au_mSi_n (m+n=2-6) nanoalloy clusters are correlated in terms of DFT based descriptors viz. HOMO-LUMO gap, Global Hardness (η), Global Softness (S), Electronegativity (χ) and Electrophilicity Index (ω). Our computed bond length of this silicon-gold cluster exhibits a close agreement with experimental bond length. Regression analysis has been done in terms of correlation between our computed descriptors and their experimental counterpart.

1. Introduction

In the recent decades, nanomaterials have deeply integrated into everyday's life. Since last few years, nanomaterials and nanotechnology have emerged as frontier research domain of science and technology [1]. The classification of nanoparticles is done in terms of size range of 1-100 nm. That particular size range exists between the levels of atomic / molecular and bulk material [1-5]. Due to existence of a large number of quantum mechanical and electronic effects, nanoparticles possess various unique physico-chemical properties [2-4]. But, there are still some instances of nonlinear transition of certain physical properties, which may vary depending on their size, shape and composition [6, 7]. A large number of experimental and theoretical reports are available for describing the effects of size and structure to change the optical, electronic, magnetic, chemical and other physico-chemical properties of nanoparticles [1,3,4]. A deep insight into the research of nanoparticles with well –defined size and structure may lead to some other alternatives for better performance [8]. The nanoparticles, due to its various applications in the areas of biological labeling, photochemistry, catalysis, information storage, magnetic device, optics, sensors, photonics, optoelectronics, nanoelectronics etc. have got immense importance [1,3,9-11].

Due to unique electronic, catalytic, optical and magnetic properties gold nanoclusters are very much popular and a potential candidate for fabrication, catalysis, bio-physics and medicine [12-28]. A number of reports are available for describing that impurity atoms can enhance the above mentioned properties of doped gold clusters that are sensitive towards the nature of dopant atom [28-

30]. Now a days, different compositions of nano alloys are being utilized for advancement of methodologies and characterization techniques [29-31]. A deep study of core-shell structure of nano compounds is very much popular as because its properties can be tuned through the proper control of other structural and chemical parameters. Silicon doped metal nanoclusters have been focus of extensive theoretical and experimental investigation due to its high importance in the field of microelectronics and semiconductor industry [28, 32]. Pal et al. [33] have reported clusters of $Au_{16}M^-$ (M=Si, Ge, Sn) and observed that $SiAu_5^-$ cluster have a tetrahedral 3D structure but MAu_x^- (x=6, 7; M=Si, Ge, Sn) have quasi-planar structures. It reflects that Au-Au interactions and Au-M interactions in the MAu_x^- clusters have a dangling unit of Au-Si, resembling that of larger Si-doped gold cluster $SiAu_{16}^-$. Majumder et al. [34] studied clusters of Au_5M (M= Na, Mg, Al, Si, P and S) and found that impurities with p electrons (Al,Si,P) exhibits nonplanar geometries while with s electrons (Na, Mg) shows planar geometries, the exception occurs for Au_5S. Sun et al. [35] reported the cage structure of $Au_{16}Si$ and observed that endohedral configuration is metastable and the silicon atom prefers to be on the exterior surface of the cage. Though, a number of experimental and theoretical studies have been done on this particular type of compounds, a theoretical analysis invoking Density Functional Theory (DFT) is still unexplored.

DFT is one of the most successful approaches of quantum mechanics to explore the electronic properties of materials in terms of quantitative descriptors. As for the larger systems electron density is more manageable as compared to wave function, DFT is very much popular to study the many- body systems [8]. Super conductivity of metal based alloys [36], magnetic properties of nano alloy clusters [37, 38] quantum fluid dynamics [39], molecular dynamics [40], nuclear physics [41, 42] can be extensively studied by DFT methodology. Recently we have established the importance of DFT based descriptors in the domain of drug designing and nano-engineering materials [43-47]. The study of density functional theory is broadly classified into three sub categories viz. theoretical, conceptual, and computational [48-51]. The conceptual density functional theory is highlighted following Parr's dictum "Accurate calculation is not synonymous with useful interpretation. To calculate a molecule is not to understand it" [52].

In this venture, we have studied Au_mSi_n, (m+n=2-6) nano-clusters in terms of DFT based conceptual descriptors. An attempt has been made to correlate the computed descriptors of the compounds with their experimental counterparts.

2. Computational details

In this study, we have made a computational analysis on the nano alloy clusters of Au_mSi_n; where m+n = 2-6. 3d modeling and structural optimization of all the compounds have been performed using Gaussian 03 [53] within Density Functional Theory framework. For optimization purpose, Generalized Gradient Approximation (GGA) exchange correlation with basis set LanL2dz has been adopted. Invoking Koopmans' approximation [54], we have calculated ionization energy (I) and electron affinity (A) of all the nano alloys using the following ansatz-

$$I = -\varepsilon_{HOMO} \quad (1)$$

$$A = -\varepsilon_{LUMO} \quad (2)$$

Thereafter, using I and A, the conceptual DFT based descriptors viz. electronegativity (χ), global hardness (η), molecular softness (S) and electrophilicity index (ω) have been computed. The equations used for such calculations are as follows-

$$\chi = -\mu = \frac{I+A}{2} \quad (3)$$

Where, μ represents the chemical potential of the system.

$$\eta = \frac{I-A}{2} \quad (4)$$

$$S = \frac{1}{2\eta} \quad (5)$$

$$\omega = \frac{\mu^2}{2\eta} \quad (6)$$

3. Results and discussions

In this study, a detail theoretical analysis of Au_mSi_n (m+n=2-6) nanoalloy clusters has been performed interms of electronic structure theory. The HOMO (Highest Occupied Molecular Orbital)-LUMO (Lowest Unoccupied Molecular Orbital) energy gap is a remarkable parameter for examining the kinetic stability of nano alloy clusters. The electronic energy gap indicates the energy required for an electron to move from the HOMO to the LUMO. The orbital energies in form of HOMO -LUMO energy gap along with computed DFT based descriptors, namely molecular electronegativity, global hardness, global softness, and electrophilicity index have been reported in the Table 1. The result from Table 1 indicates that HOMO-LUMO gap of Au_mSi_n (m+n=2-6) is maintaining a direct relationship with their evaluated global hardness values. As the frontier orbital energy gap increases, the hardness value increases. From experimental point of view, it is an expected trend. As the molecule possesses the highest HOMO-LUMO gap, it will be the least prone to response against any external perturbation. From Table 1, it is clear that Au_4Si has highest HOMO-LUMO gap which establishes magical stability in the series of nano alloy clusters. Though there is no such available quantitative data of optical properties of aforesaid clusters, we can assume that there must be a direct qualitative relationship between optical properties of Au-Si nano clusters with their computed HOMO-LUMO gap. The assumption is based on the fact that optical properties of materials are interrelated with flow of electrons within the systems which in turn depend on the difference between the distance of valence and conduction band. There is a direct linear relationship between HOMO-LUMO gap with the difference in the energy of valence-conduction band [55]. On that basis, we may conclude that optical properties of instant nano alloy clusters increases with increase of their hardness values. Similarly softness data exhibits an inverse relationship towards the experimental optical properties. The linear correlation between HOMO-LUMO gap along with their computed global hardness is lucidly plotted in the Figure 1.The high value of correlation coefficient (R^2=0.996) observed in the Figure 1, validates our predicted model. From the obtained correlation coefficients of several DFT based descriptors and their HOMO-LUMO gaps, it can be concluded that the best linear relationship is observed in case of hardness and the least one for electronegativity (R^2=0.071) of these nano clusters.

A comparative analysis has been made between experimental bond length and our computed data of the species namely Au_2, Si_2 and Au-Si. The same is reported in the Table 2. An excellent agreement between our computed bond length with reported experimental values [28, 32, 56-58] established faith in the accuracy and reliability of the applied method used in this work which supports and validates our computational analysis.

Table 1. Computed DFT based descriptors of Au_mSi_n (m+n=2-6) nanoalloy clusters

Species	HOMO-LUMO Gap (eV)	Electronegativity (eV)	Hardness (eV)	Softness (eV)	Electrophilicity Index (eV)
AuSi	1.632	4.680	0.816	0.612	13.416
$AuSi_2$	2.748	4.367	1.374	0.363	6.939
$AuSi_3$	2.394	4.707	1.197	0.417	9.254
$AuSi_4$	2.639	4.748	1.319	0.378	8.541
$AuSi_5$	2.775	4.870	1.387	0.360	8.547
Au_2Si	2.911	5.183	1.456	0.343	9.228
Au_3Si	1.687	4.952	0.843	0.592	14.537
Au_4Si	3.237	4.897	1.687	0.296	7.532
Au_5Si	1.877	4.367	0.938	0.532	10.158
Au_2Si_2	1.714	4.748	0.857	0.583	13.151
Au_2Si_3	2.775	4.870	1.387	0.360	8.547
Au_2Si_4	2.667	4.680	1.333	0.375	8.214
Au_3Si_2	2.394	5.006	1.197	0.417	10.468
Au_3Si_3	2.557	4.652	1.278	0.390	8.464
Au_4Si_2	2.639	4.965	1.319	0.378	9.342

Table 2. The calculated bond length (Å) of Au_2, Si_2 and AuSi species

Species	Computed Bond Length	Experimental Bond Length
Au_2	2.48	2.47[56]
Si_2	2.36	2.28[32]
AuSi	2.36	2.26[28,32, 57,58]

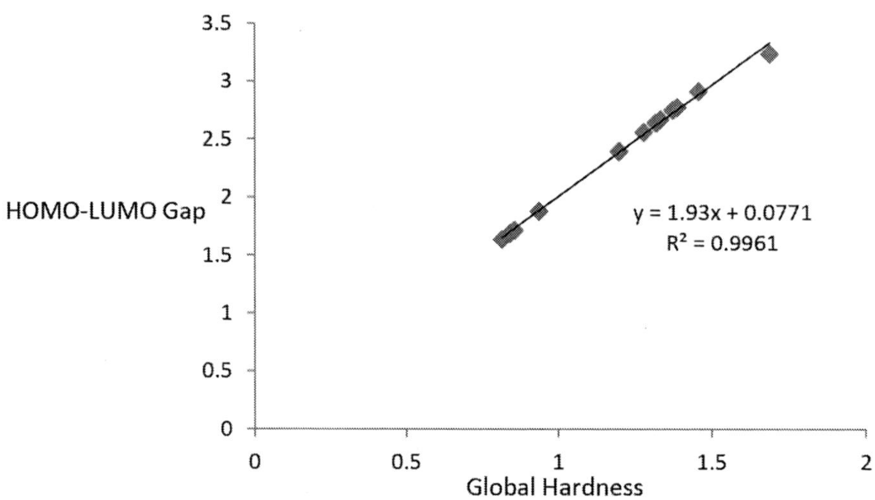

Figure 1. A linear correlation between Global Hardness Vs HOMO-LUMO Gap

4. Conclusion

In this report, an investigation has been done on electronic and optical properties of the mixed gold-silicon nano-alloy clusters in terms of conceptual DFT based descriptors viz. global hardness, HOMO-LUMO gap, electronegativity, softness, and electrophilicity index. Our computed result reveals that HOMO-LUMO energy gap runs hand in hand along with the molecular hardness of the compounds. Considering the direct relationship between optical properties of Au-Si nanoalloy clusters with their computed HOMO-LUMO gap, a qualitative relationship has been established between optical property of the mentioned compounds and their global hardness values. This trend is consistent with other experimental facts also. The high value of regression coefficient between HOMO-LUMO gap and hardness supports our approach pertaining to modeling. Our computed bond lengths for the species Au_2, Si_2 and Au-Si are numerically very close to the reported experimental values.

References

[1] Khosousi A Z and Dhirani A A 2008 *Chem. Rev.* **108** 4072.
[2] Daniel M C and Astruc D 2004 *Chem. Rev.* **104** 293.
[3] Ghosh S K and Pal T 2007 *Chem. Rev.* **107** 4797.
[4] Chaudhuri R G and Paria S 2012 *Chem. Rev.* **112** 2373.
[5] Alivisatos A P 1996 *Science* **271**, 933.
[6] Kastner M A 1993 *Phys. Today* **46**, 24
[7] Haruta M 2002 *Cattech* **6** 102.
[8] Ismail R 2012 Theoretical studies of free and supported nanoalloy clusters Ph.D. Thesis University of Birmingham UK.
[9] Roucoux A, Schulz J and Patin H 2002 *Chem. Rev.* **102**, 3757.
[10] Flores B M M, Kharisov B I, Perez V M J, Martinez P E and Lopez S T 2011 *Ind. Eng. Chem. Res.* **50** 7705
[11] Murray R W 2008 *Chem. Rev.* **108** 2688.
[12] Huang X, Sayed I E, Qian W and Sayed M E 2006 *J. Am. Chem. Soc.* **128** 2115.
[13] Hainfeld J, Slatkin D and Smilowitz H 2004 *Phys. Med. Biol.* **49** N309.
[14] Wu D, Zhang X, Liu P, Zhang L, Fan F and Guo F M 2011 *Curr. Nanosci.* **7** 110.
[15] Zhang X, Guo M, Wu D, Liu P, Sun Y, Zhang L, She Y, Liu Q and Fan F 2011 *Int. J. Mol. Sci.* **12** 2972.
[16] Chen J, Wang D, Xi J, Au L, Siekkinen A, Warsen A, Li Z, Zhang H, Xia Y and Li X 2007 *Nano. Lett.* **7** 1318.
[17] Robinson J M, Takizawa T and Vandre D D 2000 *J. Microsc.* **199** 163.
[18] Whetten R L, Khoury J T, Alvarez M M, Murthy S, Vezmar I, Wang Z L, Stephens P W, Cleveland C L, Luedtke W D and Landman U 1996 *Adv. Mater* **5** 428.
[19] Andres R P, Bein T, Dorogi M, Feng S, Henderson J I, Kubiak C P, Mahoney W, Osifchin R G and Reifenberger R 1996 *Science* **272** 1323.
[20] Mirkin C A, Letsinger R L, Mucic R C and Storhoff J J 1996 *Nature* **382** 607.
[21] Alivisatos A P, Johnsson K P, Peng X G, Wilson T E, Loweth C J, Bruchez M P and Schultz P G 1996 *Nature* **382** 609.
[22] Andres R P, Bielefeld J D, Henderson J I, Janes D B, Kolagunta V R, Kubiak C P, Mahoney W J Osifchin R G 1996 *Science* **273** 1690.
[23] Garzon I L, Michaelian K., Beltran M R, Amarillas A P, Ordejon P, Artacho E, Portal D S, Soler J M 1998 *Phys. Rev. Lett.* **81** 1600.
[24] Garzon I L, Rovira C, Michaelian K, Beltran M R, Ordejon P, Junquera J, Portal D S, Artacho E and Soler J M 2000 *Phys. Rev. Lett.* **85** 5250.
[25] Michaelian K, Rendon N and Garzon I L 1999 *Phys. Rev. B* **60** 2000.
[26] Schaaff T G and Whetten R L 2000 *J. Phys. Chem. B* **104** 2630.
[27] Priyanka and Dharamvir K 2013 *J. Phys. Chem. Chem. Phys.* **15** 12340.

[28] Gautam S, Goel N and Dharamvir K 2014 *RSC Adv.* **4** 13927.
[29] Ferrando R, Jellinek J and Johnston R L 2008 *Chem. Rev.* **108** 845.
[30] Liu H B, Pal U, Medina A, Maldonado C and Ascencio J A 2005 *Phys. Rev. B* **71** 075403.
[31] Baletto F and Ferrando R 2005 *Rev. Mod. Phys.* **77** 371.
[32] Chuang F, Hsu C, Hsieh Y and Albao M A 2010 *Chin. J. Phys.* **48** 82.
[33] Pal R, Wang L M, Huang W, Wang L S and Zeng X C 2009 *J. Am. Chem. Soc.* **131** 3396.
[34] Majumder C, Kandalam A K and Jena P 2006 *Phys. Rev. B* **74** 205437.
[35] Sun Q, Wang Q, Chen G and Jena P 2007 *J. Chem. Phys.* **127** 214706.
[36] Wacker O J, Kümmel R and Gross E K U 1994 *Phys. Rev. Lett.* **73** 2915.
[37] Illas F and Martin R L 1998 *J. Chem. Phys.* **108** 2519.
[38] Johnson D D, Nicholson D M, Pinski F J, Gyorffy B L and Stocks G M 1990 *Phys. Rev. B* **41** 9701.
[39] Kümmel S and Brack M 2001 *Phys. Rev. A* **64** 022506.
[40] Car R and Parrinello M 1985 *Phys. Rev. Lett.* **55** 2471.
[41] Koskinen M, Lipas P and Manninen M 1995 *Nucl. Phys. A* **591** 421.
[42] Schmid R N, Engel E and Dreizler R M 1995 *Phys. Rev. C* **52** 164.
[43] Ranjan P, Dhail S, Venigalla S, Kumar A, Ledwani L and Chakraborty T 2015 *Mater. Sci.-Pol.* **33** 719.
[44] Chakraborty T and Ghosh D C 2012 *Int. J. Chem. Model.* **4** 413.
[45] Ranjan P, Kumar A and Chakraborty T 2014 Computational Study of Nanomaterials Invoking DFT based Descriptors *Environmental Sustainability: Concepts, Principles, Evidences and Innovations* ed G C Mishra (New Delhi: Excellent Publishing House) chapter 33 pp 239-242.
[46] Ranjan P, Venigalla S, Kumar A and Chakraborty T 2015 A Theoretical Analysis of Bi-metallic $AgAu_n$, (n=1-7) Nano-alloy Clusters Invoking DFT based Descriptors *Research Methodology in Chemical Sciences: Experimental and Theoretical Approaches* ed T. Chakraborty and L. Ledwani (USA: Apple Academic Press & CRC Press).
[47] Ranjan P, Venigalla S, Kumar A and Chakraborty T 2014 *New Front. Chem.* **23** 111.
[48] Parr R G and Yang W 1995 *Annu. Rev. Phy. Chem.* **46** 701.
[49] Kohn W, Becke A D and Parr R G 1996 *J. Phys. Chem.* **100** 12974.
[50] Liu S and Parr R G 1997 *J. Chem. Phys.* **106** 5578.
[51] Ziegler T 1991 *Chem.Rev.* **91** 651.
[52] Geerlings P and Proft F D 2002 *Int. J. Mol. Sci.* **3** 276.
[53] Gaussian 03, Revision C.02, Frisch M J, Trucks G W et al., Gaussian, Inc. Wallingford CT 2004.
[54] Parr R G and Yang W 1989 *Density Functional Theory of Atoms and Molecules* (Oxford University Press, Oxford).
[55] Xiao H, Kheli J T and Goddard W A III 2011 *J. Phys. Chem. Lett.* **2** 212.
[56] Huber K P and Herzberg G 1979 *Constraints of Diatomic Molecules* (Van Nostrand Reinhold Company New York).
[57] Scherer J J, Paul J B, Collier C P and Saykally R J 1995 *J. Chem. Phys.* **102**, 5190.
[58] Scherer J J, Paul J B, Collier C P, O'Keefe A and Saykally R J 1995 *J. Chem. Phys.* **103**, 9187.

Journal of Physics: Conference Series **759** (2016) 012046 doi:10.1088/1742-6596/759/1/012046

Study of lattice dynamics of Fe_2SiO_4- and Mg_2SiO_4-spinels

M M Sinha[1] and Harleen Kaur
Department of Physics
Sant Longowal Institute of Engineering and Technology,
Longowal, Sangrur-148106 Punjab, India
[1]Email: mm_sinha@rediffmail.com

Abstract. Fe_2SiO_4 is an end member of Mg-rich (Mg, Fe)$_2SiO_4$, which is believed to be a major mineral of the Earth's transition zone [1, 2]. The presence of iron has a pronounced effect on elastic and thermodynamic properties of rock-forming minerals. These properties play crucial role in the interpretation of the geophysical data and thus have a large influence on our knowledge of the earth's interior. An understanding of the elastic properties of silicate will be further helpful to the interpretation of seismological data, in particular the variation in the depth range of transition zone of earth's interior. The spinel form of magnesium-iron orthosilicate, (Mg, Fe)$_2SiO_4$, is believed to be one of the most abundant minerals in the mantle's transition zone and is found to be stable in ambient conditions and therefore, a detailed study of zone centre phonons of this stable phase of orthosilicates (Mg, Fe)$_2SiO_4$ is of high interest. Hence, in the present study, the zone centre phonons of antiferromagnetic Fe_2SiO_4-spinel and Mg_2SiO_4-spinel have been studied by using short range force constant model involving interatomic interactions upto first three neighbours. The zone centre phonons of Fe_2SiO_4 is compared with that of Mg_2SiO_4 in order to study the effect of the cation exchange on the dynamic and thermodynamic properties of (Mg, Fe)$_2SiO_4$-spinel. The calculated results are compared and analyzed with exiting experimental results.

1. Introduction

The spinel form of magnesium-iron orthosilicate, (Mg, Fe)$_2SiO_4$ is the most abundant mineral in the Earth upper mantle and is believed to have dominant influence on crucial geophysical processes in this part of the Earth [1,2]. The olivine-spinel phase transition and crystal structure of silicate spinel have been extensively investigated for understanding possible phase transformations and crystal chemistry of the orthosilicate. Detailed characterization of this mineral and its end members, Mg_2SiO_4 and Fe_2SiO_4, is, therefore, crucial for future understanding of the main Earth processes. (Mg,Fe)$_2SiO_4$ is known to exists in three phases:(i) low-pressure α orthorhombic (olivine), (ii) moderate-pressure β orthorhombic (wadslayite), and (iii) high-pressure γ cubic (ringwoodite) [3] So far, several experimental [4] and theoretical [5,6] studies have been reported concerning the structural, dynamical, and thermodynamical properties of Mg_2SiO_4, crystallizing in all three $\{\alpha,\beta,\gamma\}$ phases. Whereas, Fe_2SiO_4 is known to exist only in two stable phases (α fayalite and γ spinel) represents much more challenging material as it belongs to the group of transition metal orthosilicates, additionally possessing nonvanishing magnetic moment [7].

The knowledge of the properties of all phases of (Mg,Fe)$_2SiO_4$ is of great importance for earth sciences, since it determines crucial geophysical properties of the earth's interior such as possible phase transitions and thermodynamics of the rock forming minerals. These properties play crucial role in the interpretation of the geophysical data, and thus have a large influence on our knowledge of the earth's interior. However, it is very difficult and expensive to reproduce the pressure conditions of the deep regions of the earth in the laboratory and it is more difficult to perform precise measurements in such conditions. The fundamental link between microscopic (atomistic) behaviour of minerals and macroscopic properties can be made through the lattice dynamics of the minerals which is determined by interatomic interactions, and this in turn determine the basic thermodynamic properties of the minerals. In comparison to the Fe_2SiO_4 fayalite, its high-pressure-spinel phase is a much less studied system. Therefore, in the present investigation the interatomic interaction and zone center phonons of high-pressure-spinel phase of Fe_2SiO_4 has been studied by using an angular force constant model [8]. We compare them with those obtained for Mg_2SiO_4-spinel in order to understand the influence of cation exchange on lattice dynamics and thermodynamics in the ringwoodite $(Fe,Mg)_2SiO_4$.

2. Crystal structure

Spinel phase of $(Mg,Fe)_2SiO_4$ is cubic and exists in a γ-spinel structure that belongs to space group Fd3m (O_h^7) [9]. This cubic structure consists of isolated SiO_4 tetrahedra, with Mg/Fe atoms occupying the interstitial sites between SiO_4 groups. The primitive cell has 2 formula units including 14 atoms (figure 1), so there are 42 normal modes for each point in the Brillouin zone, among which 3 are acoustic and 39 are optical modes. The optical modes at the BZ centre may be divided by symmetry as

$$\Gamma = A_{1g} + 2A_{2u} + E_g + 2E_u + T_{1g} + 4T_{1u} + 2T_{2u} + 3T_{2g}.$$

Here subscripts g and u denote symmetric and antisymmetric modes with respect to the centre of inversion, while R and IR represent Raman and infrared active modes. All E modes are doubly degenerate, and all T modes are ternary degenerate. A_{1g}, E_g, and all T_{2g} modes are Raman active, while all T_{1u} modes are infrared active.

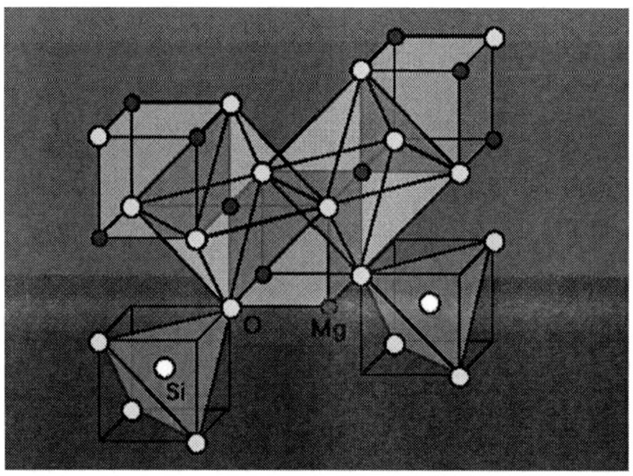

Figure 1. Crystal structure of Mg_2SiO_4-spinel

3. Interatomic interaction

The determination of the strengths of the interatomic interaction in theses orthosilicates is important from the point of view of their thermo dynamical properties. One of the few procedures suitable are the interatomic force constant investigations. Therefore, in the present investigation a de Launey angular force constant (DAF) model [8] has been applied for the calculation of the interatomic interaction of (γ-spinel). In this model, the relative displacement of the reference atom and one of its neighbours is considered. The restoring force on the reference atom is taken to be proportional to the component of the relative displacement perpendicular to a line joining the two atoms. In the present analysis , the short range force constants between the first three neighbours α_1, α_2, α_3 (central) for the interatomic interactions for Si-O, Mg(Fe)-O , and Mg(Fe)-Mg(Fe) are evaluated by fitting the measured zone-center phonon frequencies [5,7,10-11] to the corresponding analytical expressions for A_{1g}, E_g, and ΣT_{1u} obtained in Ref. 12 by solving a dynamical matrix of order (42x42) at k = 0 for a compound with an ideal spinel structure. The force constants, thus calculated are listed in Table 1. It is obvious from the table 1 that the short range Si-O forces of the SiO_4 tetrahedra are larger than other interatomic forces in both orthosilicates. The interatomic interactions between Si-O is covalent in nature may be the reason of dominance over the other interactions. The rattling of cations is the reason for this kind of interactions in minerals [13]. There is almost negligible change (0.9%) in force constant involving Si-O atoms whereas 10% and 7% changes in the values of force constants involving (Fe/Mg-O) and (Fe/Mg-Fe/Mg) respectively.

Table 1. Short-range force constants (in K dyne cm^{-1})

Force constant	Mg_2SiO_4	Fe_2SiO_4
α_1 [Si-O]	10.05	9.96
α_2 [Mg(Fe)-O]	2.14	2.35
α_3 [Mg(Fe)-Mg(Fe)]	0.57	0.48

4. Zone centre phonons

The short range force constants thus calculated are taken as input parameters to compute zone-centre (ZC) phonon frequencies for $(Mg/Fe)_2SiO_4$. The calculated ZC phonons along with other results [5,7,10-11] are given in table 2. In the case of the spinel with a cubic Fd3m symmetry, there are three independent displacements, one for each non-equivalent atom (Mg/Fe, Si, and O).

Table 2. Calculated and measured and other calculated [7,10-12] zone-center (ZC) phonon frequencies (in cm^{-1}) for Mg_2SiO_4 and Fe_2SiO_4

Species	Mg_2SiO_4			Fe_2SiO_4	
	Present calculation	Piekarz et al.[10]	Observed [11,12]	Present calculation	M. Derzsi et al. [7]
A_{1g}	133	128	133	130	130
E_g	59	55	59	59	59
$T_{2g}(1)$	145	124	126	144	127
$T_{2g}(2)$	82	91	95	78	85
$T_{2g}(3)$	37	45	48	37	44
$T_{1u}(1)$	146	143	132	145	131
$T_{1u}(2)$	74	89	87	66	76
$T_{1u}(3)$	66	71	71	50	50
$T_{1u}(4)$	45	56	56	28	23
$A_{2u}(1)$	133	126		131	126
$A_{2u}(2)$	78	89		48	29
$E_u(1)$	84	83		84	79
$E_u(2)$	61	56		32	29
T_{1g}	59	55		59	59
$T_{2u}(1)$	74	65		73	63
$T_{2u}(2)$	47	35		23	26

The eigen vector calculations suggest that Oxygen and silicon ions contribute to majority of the phonon modes. Oxygen ions contribute to the majority of high energetic modes. Table 2 suggests that Raman-active modes of A_{1g} symmetry and E_g symmetry are in excellent agreement with observed results [7,10-11] in these silicates. The eigenvector calculations predict that A_{1g} and E_g vibrations involve only oxide ion displacements. The $F_{2g(2)}$ and $F_{2g(3)}$ phonons are characterized by large oxygen and very small silicon displacements, but the lowest energy $F_{2g(1)}$ phonons derive predominantly from a vibration of the Si sublattice. The contribution of silicon atoms to other than Si-O stretching modes is very small. The phonon modes with the largest energy (T_{1u}) involve predominantly (Mg/Fe) vibrations. Comparing the zone center phonon energies of both spinel structures (table 2), it is clear that substitution of all Mg atoms by Fe atoms causes a lowering of all phonon modes. It is found that Raman modes are almost same in two spinels because these modes are due to either Si or O ion vibrations. But there is mark able decrease in magnitudes of infrared active modes [$A_{2u}(2)$, $E_u(2)$ and $T_{2u}(2)$] of Fe_2SiO_4 than that of Mg_2SiO_4. The most of these modes is due to the vibrations of (Fe/Mg). It is due to larger mass of iron compare to magnesium and is in agreement with other results [7]. The calculated Raman modes are in agreement with experiment [10] whereas, there is a deviation in infrared active modes. The consideration of more interatomic interaction may improve the results.

5. Conclusions

The interatomic interaction (Si-O) is found to be dominating over other interatomic interactions. The rattling of cations may be the cause for this dominance. A comparison of ZC phonons with theory and other calculations shows the agreement of the results. The substitution of magnesium (Mg) by iron (Fe) is mainly influencing some of the infrared active modes. The lowering of phonon modes caused by the replacement of Mg by Fe may lead to change in bulk modulus and hence significant enhancement of entropy and heat capacity.

6. Refrences

[1] Ringwood A E and Major A , 1970 *Phys. Earth Planet. Inter.* **3** 89
[2] Dziewonski A M and Anderson D L 1981 *Phys. Earth Planet. Inter.* **25** 297
[3] Ringwood A E and Major A 1966 *Earth Planet. Sci. Lett.* **1** 241
[4] Ohtani E and Kumazawa M 1981*Phys. Earth Planet. Inter.* **27** 32; Meng Y, Fei Y, Weidner D J, Gwanmesia G D and Hu J 1994 *Phys. Chem. Miner.* **21** 407; Chopelas 2000 *A Am. Mineral.* **85** 270
[5] Piekarz P, Jochym P T, Parlinski K and Łażewski J 2002 *J. Chem.Phys.* **117** 3340; Yu Y G and Wentzcovitch R M 2006 *J. Geophys. Res.* **111**, B12202
[6] Zhou J, Zhang H, Chen Yue, Shong Jun, Chen Zhuo, Yang Juan, Zheng Zhou and Wang Feng 2014 *Physica B* **449** 95
[7] Derzsi M, Piekarz P, Tokar K, Jochym P T, Lazewski J, Sternik M and Parlinski K 2011 *J. Phys.: Condens. Matter* **23** 105401
[8] Launey J de 1956 *Solid State Physics Series* **2** 219
[9] Hazen R M, Downs R T, Finger L W and Ko J 1993 *Am. Mineral.* **78** 1320
[10] Chopelas A, Boehler R, and Ko T 1994 *Phys. Chem. Mineral.* **21** 351
[11] Akaogi M, Ross N L, McMillan P and Navrotsky A 1984 *Am. Mineral.* **69** 499
[12] Gupta H C, Prasher A, Gupta V B and Tripathi B B 1990 *Phys. Stat. Sol. B* **160**, K19
[13] Sinha M M, Singh N P, Prasad K and Gupta H C 1995 *Phys. Stat. Sol. B* **190** K27

Journal of Physics: Conference Series **759** (2016) 012047 doi:10.1088/1742-6596/759/1/012047

A three-dimensional fully coupled thermo-mechanical model for Self-reacting Friction Stir Welding of Aluminium AA6061 sheets

Piyush Singh, Pankaj Biswas and Sachin D. Kore

Department of Mechanical Engineering, Indian Institute of Technology Guwahati, Guwahati, Assam, PIN-781039, India

E-mail: s.piyush@iitg.ernet.in, pankaj.biswas@iitg.ernet.in, sdk@iitg.ernet.in

Abstract. In the present work a three dimensional model of self-reacting friction stir welding in aluminium alloy AA6061 has been developed based on the Computational Fluid Dynamics (CFD) approach using COMSOL Multiphysics software. The temperature dependent material properties have been incorporated in the model from available literature. A slip-stick contact between the workpiece and tool surface has been considered with the slip factor varying linearly with distance. The methodology adopted has been validated with experimental results available in the literature. The temperature distribution observed has been found to be asymmetric about the weld centre line. The maximum temperature has been observed on the advancing side of the weld. However, the temperature distribution across the thickness has been found to be almost symmetric about the mid thickness plane. An hourglass shaped temperature distribution has been observed across the cross-section of the weld. The material flow velocity distribution shows that the deformation zone is limited to a very small region around the tool.

1. Introduction

Friction Stir Welding (FSW) is a solid state welding process which was invented in 1991 at The Welding Institute, UK [1]. The present work is based on a variant of FSW known as Self-Reacting Friction Stir Welding (SRFSW) also known as Bobbin Tool Friction Stir Welding (BTFSW). The schematic of the process and the different stages are as shown in figure 1. The most important difference of this variant is the absence of the backing plate due to a modification in tool design which is inevitable in FSW. SRFSW is a full penetration welding technique and thus eliminates any chances of root defect which is quite common in FSW. It also promises to increase mobility of the process and bring about more homogeneity in the joints.

3D CFD models have been developed for studying BTFSW for AA2024-T3 assuming the material as a highly viscous shear thinning non-Newtonian fluid [2, 3]. In order to understand the material flow in BTFSW Chen et al. [4] carried out computer simulations and material tracer tests. Hilgert et al. [5] developed a moving geometry model in COMSOL and using some custom written scripts in MATLAB to account for the tool motion. The present work aims to develop a multiphysics FEM model of the SRFSW process for Aluminium alloy AA6061-T6 using its temperature dependent properties and study the temperature distribution in the workpiece and the bobbin tool. Also study the material flow in the vicinity of the tool based on computational fluid dynamics approach.

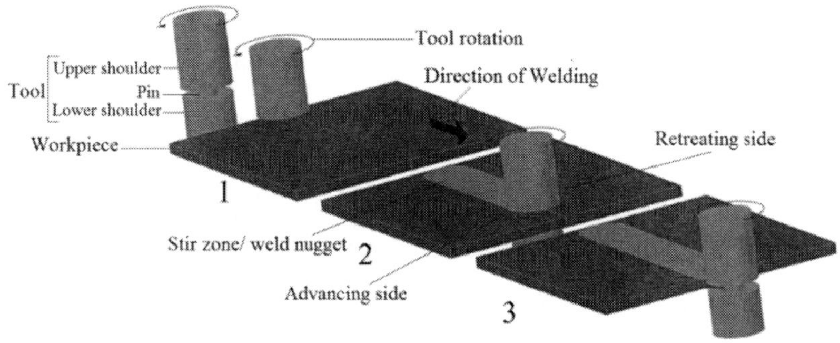

1: Tool approach and dwell 2: Tool advance and welding 3: Tool withdrawal

Figure 1. Schematic of the SRFSW process showing the different stages

2. Methodology

An attempt has been made to develop a three dimensional fully coupled thermo-mechanical model based on computational fluid dynamics (CFD) approach for which a small region of the workpiece has been assumed to behave as a fluid. COMSOL Multiphysics software has been used in the present work. The temperature dependent properties of the workpiece and tool material have been incorporated into the model. The detailed description of the methodology has been presented in the following subsections.

2.1. Model description

The workpiece with a thickness of 4 mm and tool which has a simple cylindrical geometry with 8 mm diameter pin and 18 mm diameter shoulders have been used in the model. The workpiece and tool geometry is shown in figure 3. A small region along the weld line with a width of 30 mm has been considered to behave as a fluid and the remaining as solid. The fixture holding the workpiece in its position during welding has not been considered in this study.

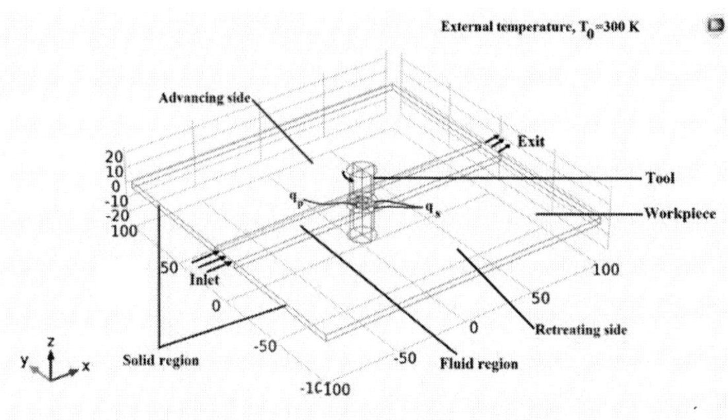

Figure 3. The workpiece and the tool geometry as used in the developed model along with the boundary conditions

Table 1. Composition of AA6061-T6 in weight percentage [6].

	Al	Cu	Si	Fe	Mg	Mn	Cr	Zn	Ti	Others
Weight %	95.85-98.56	0.15-0.4	0.4-0.8	0.7 (max)	0.8-1.2	0.15 (max)	0.04-0.35	0.25 (max)	0.15 (max)	0.05 each (max), total 0.15

2.2. Material properties

The workpiece is made of Aluminium alloy AA6061-T6 with composition as shown in table 1 [6]. The tool material is stainless steel. Temperature dependent properties of both workpiece and tool materials have been incorporated in to the model. Figure 4 shows the variation of specific heat and thermal conductivity respectively of the two materials with respect to temperature [7][8][9].

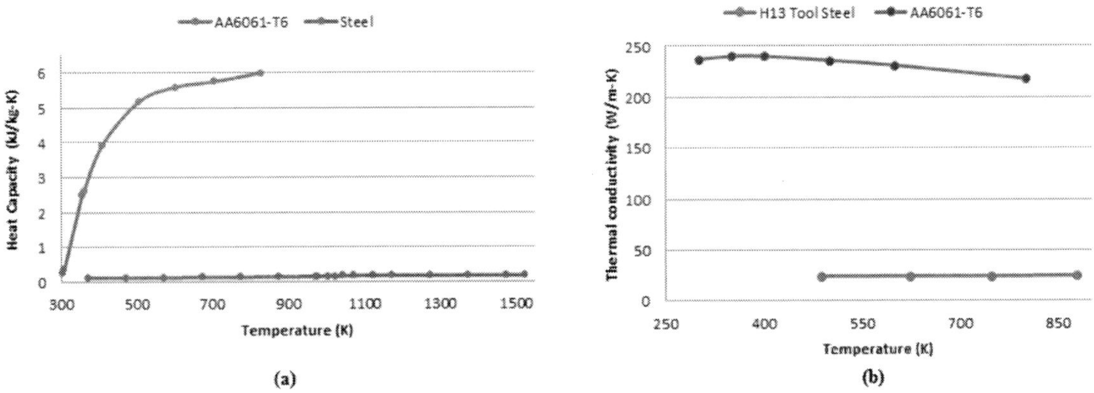

Figure 4. Variation of properties of aluminium with temperature: (a) Specific heat and (b) Conductivity [7] [8] [9]

2.3. Boundary conditions

The contact between the tool and workpiece is not exactly known. Some researchers have assumed fully sticking condition but it eliminates heat input due to friction. In the present work the contact condition has been considered to be partially slip and stick and hence a slip factor (δ) has been introduced as given by eq. 1, where x, y are the coordinates from the origin at the centre of the tool and R_s is the shoulder radius.

$$\delta = \frac{\sqrt{x^2 + y^2}}{R_s} \quad \ldots\ldots\ldots\ldots\ldots (1)$$

The coefficient of friction between two surfaces varies with the relative velocity and the actual coefficient of friction at each position is given by the relation in eq. 2 [10], where μ and μ_0=0.4 are the actual and nominal coefficient of friction at various locations respectively, ω the angular velocity and λ=1 s/m a constant to make the expression dimensionless.

$$\mu = \mu_0 \times e^{-\lambda \times \delta \times \omega \times \sqrt{x^2+y^2}} \quad \ldots\ldots\ldots\ldots (2)$$

The strain rate has not been considered constant at each position but has been calculated using the relation in eq. 3 [10], where u, v and w are the velocity components along x, y and z respectively.

$$\dot{\varepsilon} = \sqrt{\left(\left(\frac{2}{3}\right) \times \left(\left(\frac{du}{dx}\right)^2 + \left(\frac{dv}{dy}\right)^2 + \left(\frac{dw}{dz}\right)^2 + 0.5 \times \left(\frac{du}{dz} + \frac{dw}{dx}\right)^2 + 0.5 \times \left(\frac{dw}{dy} + z\right)^2 + 0.5 \times \left(\frac{du}{dy} + \frac{dv}{dx}\right)^2\right)\right)} \ldots\ldots\ldots\ldots\ldots (3)$$

The flow stress at the different locations is calculated using the eq. 4 where Z is the Zener-Holloman parameter calculated using eq. 5 [11, 12]. Q (134158.4 J mol^{-1}) is the activation energy, R is the universal gas constant, T is the absolute temperature, A (1.26×10^8 s^{-1}), α (0.03055MPa^{-1}) and n (3.24644) are material constants for AA6061 [12].

$$\sigma = \frac{1}{\alpha} \times \sinh^{-1}\left(\left(\frac{Z}{A}\right)^{\left(\frac{1}{n}\right)}\right) \quad \ldots\ldots\ldots\ldots\ldots (4)$$

$$Z = \dot{\varepsilon} \times e^{\left(\frac{Q}{R \times T}\right)} \quad \ldots\ldots\ldots\ldots\ldots (5)$$

The dynamic viscosity in the fluid region has been calculated using eq. 6 [11, 12].

$$\eta = \frac{\sigma}{(3 \times \dot{\varepsilon})} \ldots\ldots\ldots\ldots\ldots (6)$$

The heat input due to the shoulder and the pin are calculated using eq. 7 and 8 respectively [11]. No external load has been applied and the pressure (P) developed in the process has been used to calculate the heat developed. A step function has been used in the relations to prevent heat input if the temperature in the workpiece at a location reaches solidus temperature (T_m). r_p is the pin radius and T is the actual temperature at any location.

$$q_s = \left(\left((1-\delta) \times \frac{\sigma}{\sqrt{3}} \right) + \delta \times \mu \times P \right) \times \sqrt{u^2 + v^2 + w^2} \times step(T_m - T) \ldots\ldots\ldots\ldots\ldots (7)$$

$$q_p = \left(\left((1-\delta) \times \sqrt{u^2 + v^2 + w^2} \times \frac{\sigma}{\sqrt{3}} \right) + \delta \times \frac{\mu}{\sqrt{3 \times (1+\mu^2)}} \times r_p \times \omega \times \sigma \right) \times step(T_m - T) \ldots\ldots\ldots\ldots\ldots (8)$$

The ambient temperature has been assumed to be 300 K. The inlet surface of the workpiece as shown in figure 3 has been assumed to remain at the ambient temperature. Heat flow due to radiation has been neglected. However, heat dissipation due to convection has been considered from the upper and lower exposed surfaces. A physics controlled mesh has been developed using the software as shown in figure 5. The entire mesh contained 151922 elements with a minimum element quality of 0.01781 and an average element quality of 0.6945.

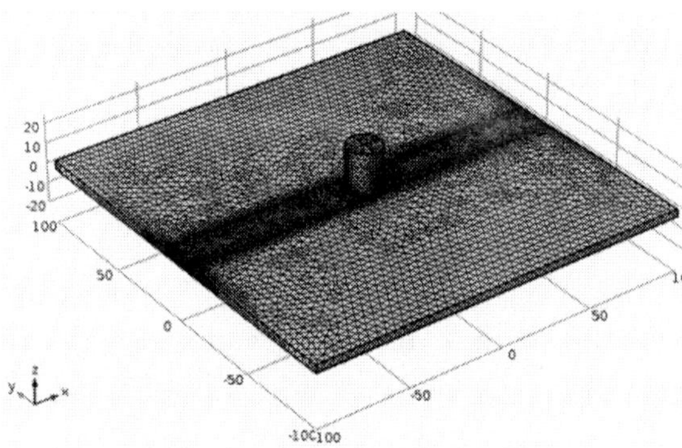

Figure 5. Mesh used in the study

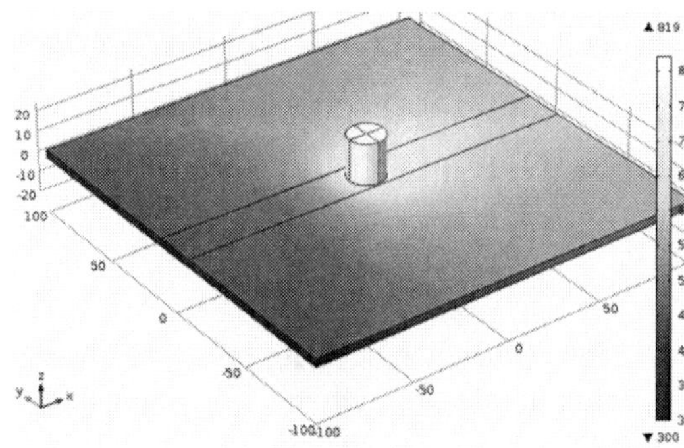

Figure 6. Surface temperature (K) distribution

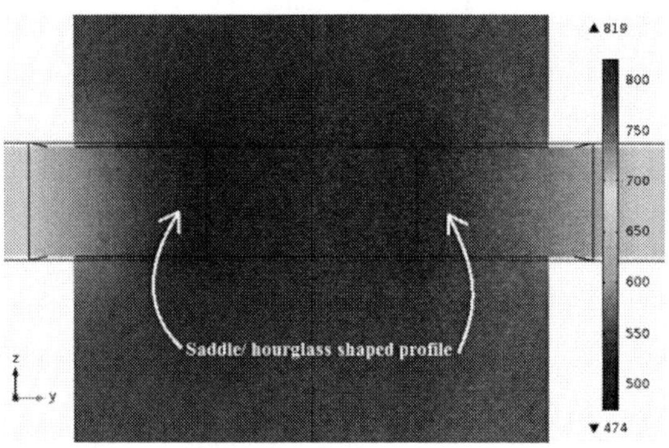

Figure 7. Temperature (K) profile in yz plane at x=0

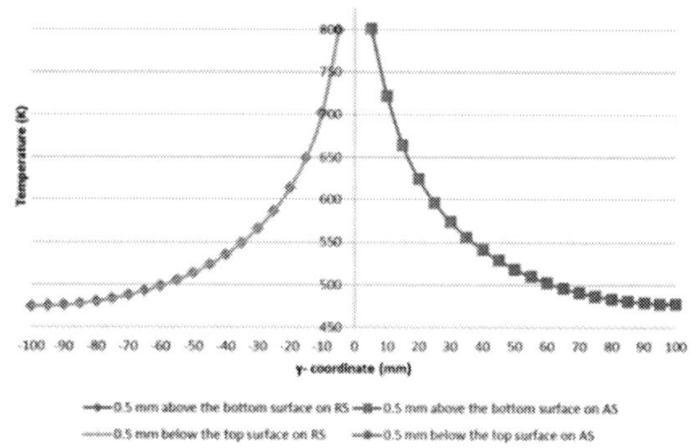

Figure 8. Variation of temperature in the workpiece along the y-coordinate at x=0

3. Results and discussion

The surface temperature distribution obtained from the simulation is shown in figure 6. The temperature slice at x=0 shown in figure 7 establishes that due to heat input from two shoulders and the pin a saddle shaped temperature profile is formed in SRFSW unlike triangular profile formed in Conventional FSW. The temperature variation along the y direction (perpendicular to the welding direction) is shown in figure 9 and in the region closer to the tool is shown in figure 10. A higher temperature has been observed on the corresponding locations of the advancing side (AS) of the weld in comparison with the retreating side (RS). However, the variation in the temperature along the top and bottom surfaces of the workpiece is almost same. From the variation of temperature along the z-direction at 1 mm from the tool pin-workpiece interface as shown in figure 11 it has been observed that the temperature distribution is almost symmetric about the mid thickness which can be attributed to the uniform heat input from the two shoulders and the absence of backing plate unlike CFSW. The advancing side has been observed to be at a higher temperature. The flow can be visualized using the velocity streamline shown in figure 12. The velocity profile in the fluid region at x=0 is shown in figure 13. The profile is similar to experimental observations of a saddle shaped weld nugget zone. Between the tool shoulders some material has been observed rotating with the tool. Figure 14 shows the pressure distribution in the model.

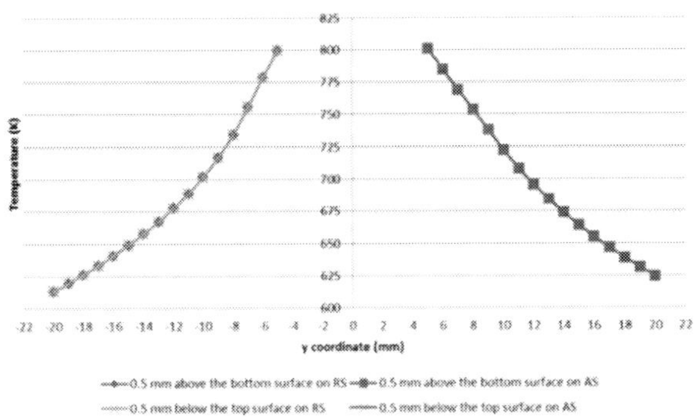

Figure 9. Variation of temperature in the workpiece below the tool shoulder along the y-coordinate at x=0

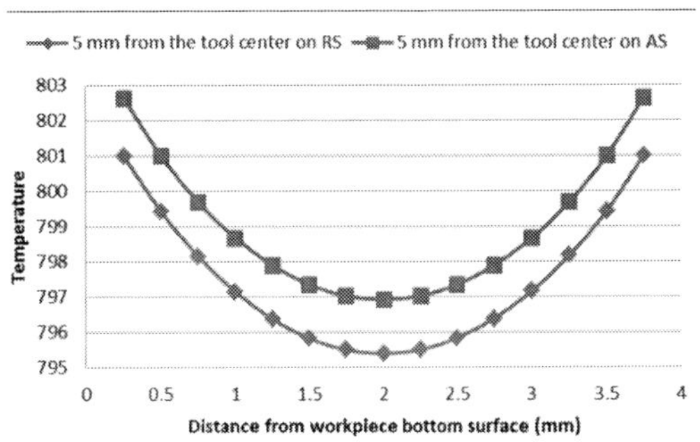

Figure 10. Variation of temperature in the workpiece along z direction

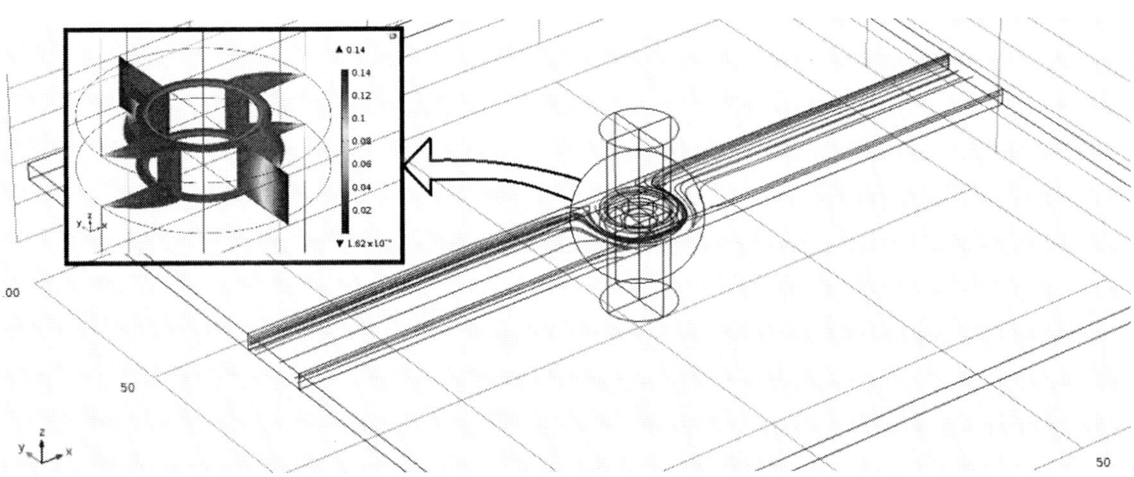

Figure 11. Velocity streamline in the model showing the material flow. The velocity (m/s) slice at x=0 and velocity vectors are shown in the inset

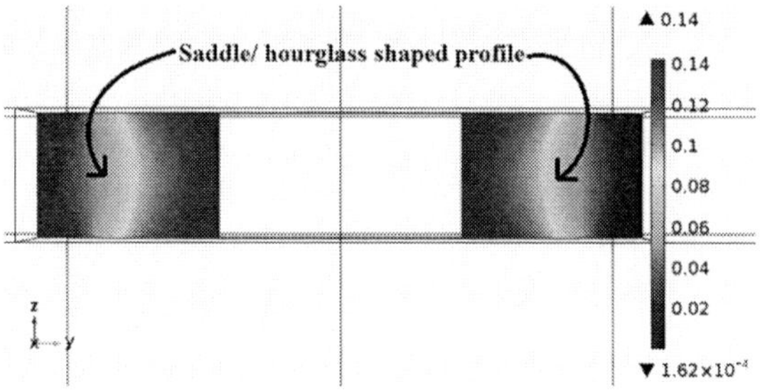

Figure 12. Slice at x=0 showing the velocity (m/s) profile in the fluid region

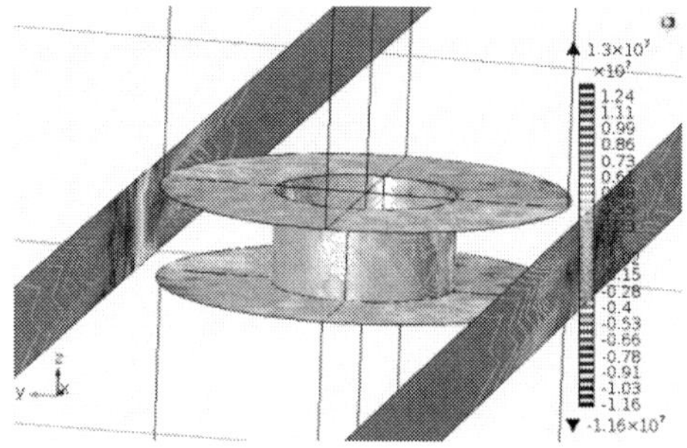

Figure 13. Pressure distribution calculated by the model

4. Conclusions

Saddle shaped temperature and velocity profiles are observed which are similar to the weld nugget zone observed in experiments. The saddle shaped profile provides more homogeneity to the weld than the triangular shaped profile in CFSW which results in non-uniform mechanical properties across the thickness. A higher temperature has been observed at the corresponding locations of the advancing side which is similar to the observations in CFSW due higher frictional heat input as a result of higher relative velocity. The results indicate that some material keeps moving with the tool during welding due to the sticking condition. A slip-stick contact using a linearly varying slip factor has been used in the model and found to provide satisfactory results. The pin plays an important role in material mixing as indicated by the velocity profile and streamlines.

References

[1] Thomas W, Nicholas E, Needham J, Murch M, Templesmith P and Dawes C, Friction stir welding Patent Application No. PCT/GB92102203 and Great Britain Patent Application No. 9125978.8, (1991)

[2] Hilgert J, Huetsch L L, Santos J F dos and Huber N 2010 *Excerpt from the Proceedings of the COMSOL Conference Paris*

[3] Hilgert J, Santos J F dos and Huber N 2011 *Friction Stir Welding and Proceesing VI Edited by: Rajiv Mishra, Murray W. Mahoney, Yutaka Sato, Yuri Hovanski, and Ravi Verma TMS (The Minerals, Metals & Materials Society)*

[4] Chen S J, Lu A L, Yang D L, Lu S, Dong D H, Dong C L 2013 *Proc. of the 1st Intl Joint Sym on Joining and Welding, Osaka, Japan, 6-8 November 2013,* **ISBN 978-1-78242-164-1**

[5] Hilgert J, Schmidt H, Santos J F dos 2009 *Excerpt from the Proceedings of the COMSOL Conference Milan*

[6] ASM Material Data Sheet Al6061-T6, http://asm.matweb.com/search/SpecificMaterial.asp?bassnum=MA6061t6, dated 30.11.2015

[7] Compton A H 1915 *Phys. Rev.* **6**, 377

[8] Specific heats of some solids, https://www.bnl.gov/magnets/staff/gupta/cryogenic-data-handbook/Section8.pdf, dated 16-11-2015

[9] AISI Type H13 Hot Work Tool Steel, http://www.matweb.com/search/datasheet_print.aspx?matguid=e30d1d1038164808a85cf7ba6aa87ef7, dated 30.11.2015

[10] Nandan R, Roy G G, Lienert T J, Debroy T 2007 *Acta Materialia* **55** 883–895

[11] Nandan R, Roy G G and Debroy T 2006 *Metall Mater Trans A, Volume* **37A** 1247-1259

[12] Hamilton C, Kopyscianski M, Senkov O and Dymek S 2013 *Metall Mater Trans A, Volume* **44A**, 1730-1740

Young's modulus of defective graphene sheet from intrinsic thermal vibrations

Siby Thomas[1], M S Mrudul[1], K M Ajith[1*] and M C Valsakumar[2]

[1]*Computational Physics Lab, Department of Physics, National Institute of Technology Karnataka (NITK), Surathkal, Mangalore - 575025, India.*

[2]*Department of Physics, Indian Institute of Technology Palakkad, Ahalia Campus, Kozhipara, Palakkad, Kerala, India - 678557*

E-mail: `ajith@nitk.ac.in`

Abstract. Classical molecular dynamics simulations have been performed to establish a relation between thermally excited ripples and Young's modulus of defective graphene sheet within a range of temperatures. The presence of the out-of-plane intrinsic ripples stabilizes the graphene membranes and the mechanical stability is analyzed by means of thermal mean square vibration amplitude in the long wavelength regime. We observed that the presence of vacancy and Stone-Wales (SW) defects reduces the Young's modulus of graphene sheets. Graphene sheet with vacancy defects possess superior Young's modulus to that of a sheet with Stone-Wales defects. The obtained room temperature Young's modulus of pristine and defective graphene sheet is $\sim$ 1 TPa, which is comparable to the results of earlier experimental and atomistic simulation studies.

Keywords: Classical molecular dynamics, Young's modulus, defects, graphene

(Some figures may appear in colour only in the online version)

1. Introduction

Graphene, the one atom thick honey-comb arrangement of carbon atoms has attracted the attention of the scientific community due to its exceptional properties. Because of its unique electronic properties, monolayer graphene has become a good candidate for various device applications [1]. Graphene shows extra ordinary mechanical properties and the Young's modulus of graphene and its derivatives like carbon nanotubes (CNTs) are well studied using different techniques which include experiment, theory and computation [2]. Free standing graphene is the hardest known material having an elastic modulus of $\sim$ 1 TPa. The highly robust nature of graphene could be useful to adjust its mechanical properties according to the demand. Lee *et al* measured the intrinsic breakdown strength and elastic properties of free standing monolayer graphene using atomic force microscope (AFM) and nano indentation techniques [3]. The superior strength and light weight nature of graphene is highly useful in flexible electronic devices.

The stability of two dimensional (2D) materials is a long standing topic of debate in condensed matter physics. A consequence of the famous Mermin-Wagner theorem is that a system of particles interacting with a sufficiently short range potential cannot exist in a perfectly ordered state at any nonzero temperature in two or lower dimensions [4, 5]. But, later in 2007, Fasolino *et al* reported that the ripples play a crucial role in the existence of graphene membranes. Due to the presence of these out-of-plane intrinsic fluctuations, graphene sheet acquires its stability [6]. The structural defects in materials have been receiving a lot of attention from the past few years. In this work, we have considered the commonly occurring vacancy and and Stone-Wales (SW) [7] defects in graphene sheet to investigate the Young's modulus. Mono vacancy defect, which results from the missing of a carbon atom in the graphene lattice is the simplest defect that can be observed in the graphene sheet. Stone-Wales defect is a topological defect in which the adjacent four hexagons in graphene are transformed into two pentagons and two heptagons by rotating one of the C-C bonds by 90° and which forms a non-hexagonal ring in the lattice without altering the connectivity of the network. Young's modulus is one of the most important properties in engineering design, which helps to measure of the stiffness of an elastic material. In the current study, we investigate the temperature dependence on the Young's modulus of defective graphene by means of classical Molecular Dynamics simulation (MD) using reactive empirical bond order (REBO) potential [8].

2. Theoretical aspects to obtain the Young's modulus of graphene

Out-of-plane height fluctuations are present in graphene even when it is in equilibrium configuration. Graphene has both optical and acoustic vibration modes present and aligned in the out-of-plane (z) direction. Here, the acoustic vibration mode is a flexural mode with low frequency and hence it will be fully excited even at very low temperatures also. The main contribution to the thermal mean-square vibration amplitude (TMSVA) of graphene is from the flexural mode in the z direction. In this work, we consider the phonon modes at temperature T, closely following Krishnan *et al* who experimentally investigated the stiffness of single-walled carbon nanotubes (CNTs) by observing their out-of-plane fluctuations at room temperature using transmission electron microscope and obtained the Young's modulus as 1.25 TPa [9]. According to the plate theory of Landau and Lifshitz [10], the equation of oscillation of a graphene membrane in the z direction is given as,

$$\rho \frac{\partial^2 z}{\partial t^2} + \frac{D}{h} \Delta^2 z = 0, \tag{1}$$

where, $D = \frac{1}{12} Y h^3 (1 - \mu^2)$ and Δ is the two-dimensional Laplacian. Y and μ are the Young's modulus and Poisson's ratio of the system. ρ and h are the density and thickness of the graphene sheet respectively. In graphene, all the modes are independent at thermal equilibrium condition, then the TMSVA at a particular temperature is given by,

$$< \sigma^2 >= 0.31 \frac{(1 - \mu^2 S)}{h^3} \frac{k_B T}{Y} \tag{2}$$

where, Y is the Young's modulus, S is the area of the sheet, k_B is the Boltzmann constant and T is the temperature. The inter layer spacing of graphene is taken as 3.33 Å [3] and the out-of-plane height fluctuations $< \sigma^2 >$, due to the thermal energy, is taken to be the spatial average of TMSVA over the area. Then the equation for Young's modulus of graphene is,

$$Y = 0.3 \frac{S}{h^3} \frac{k_B T}{< \sigma^2 >} \tag{3}$$

3. Details of Molecular Dynamics simulations

In this work, Classical Molecular Dynamics (MD) simulations based on the standard Velocity-Verlet time stepping algorithm has been carried out using Large scale Atomic/Molecular Massively Parallel Simulator (LAMMPS) software package [11]. To solve the equations of motion of the pristine and defective graphene sheet, an integration time step of 0.1 fs is used. We have used 100,000 MD steps for the analysis of the effect of vacancy and Stone-Wales defects on the Young's modulus of graphene. We used 20000 carbon atoms for the simulation and employed fixed boundary in the X direction, periodic boundary condition in the Y and Z directions and keep a vacuum separation of 15 Å in the Z direction. We also optimized the equilibrium lattice constant by minimizing the total energy of the graphene sheet using the REBO potential. This bond order potential has been shown to describe fairly well most of the structural and thermodynamic properties of different phase of carbon. The reason for the particular interest of this bond order potential is that, it correctly reproduces the best results of graphene with earlier atomistic studies [6]. Bond order potential correlate the bond length, bond strength and co-ordination, which help to changes the energetics of the conjugated bonds with desired accuracy. To control the system temperature, a Nose-Hoover type chain thermostat incorporated with canonical ensemble is used for the simulation. The mean squared height fluctuations ($< \sigma^2 >$) has been deduced from the MD simulation and investigated the values of Young's modulus using equation (3). Figure 1 (a) and (b) shows that graphene sheets containing randomly generated mono vacancy and SW defect at room temperature.

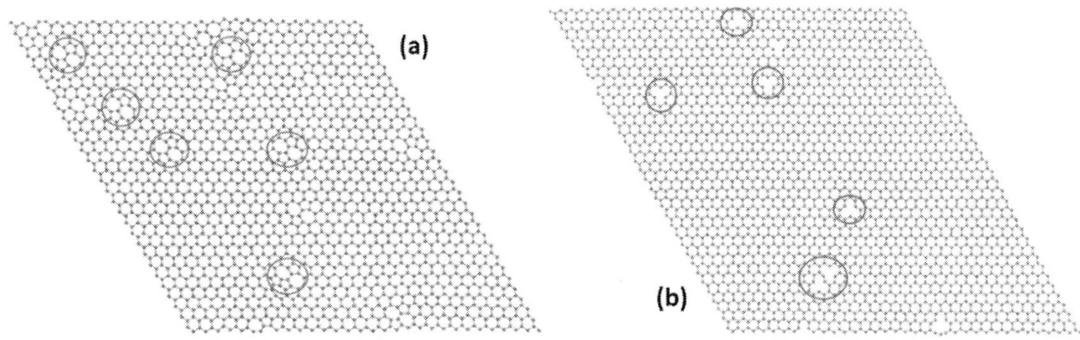

Figure 1. Figure shows graphene sheets containing randomly generated (a) Stone-Wales (SW) and (b) mono vacancy defects at room temperature.

4. Results and discussions

Study of defects is very important as they can drastically modify properties of materials. Incorporating defects for the analysis of mechanical properties helps to understand how the material behaves at different physical situations. Experimentally synthesized samples always contain different types of defects and the computational study of various types of possible defects gives a more realistic picture prior to experiment. The presence of defects may lead to reduction of, or complete loss of, symmetry of the perfect honeycomb lattice of graphene. Usually vacancies, topological defects, grain boundaries *etc* are the commonly occurring defects in low dimensional materials. Stone-Wales (SW) defect is one of the prominent topological defects observed in graphene in addition to vacancies. Figure 2 (a) shows the change in Young's modulus of pristine graphene with temperature. We observed that, at high temperatures also, graphene sheet has an Young's modulus $\sim$ 1 TPa. As the temperature increases, the out-of-plane fluctuations also increases, which leads to change the Young's modulus with temperature. Figure 2 (b) shows the change in Young's modulus with temperature of graphene with vacancy and SW defect. The Young's modulus of graphene with Vacancy defects is seen to be superior as compared to graphene with SW defect. At high temperature, graphene sheet with vacancy and SW defect show an Young's modulus of $\sim$ 1.15 TPa, which is comparable to the Young's modulus of pristine graphene sheet and this result is consistent with that from earlier studies.

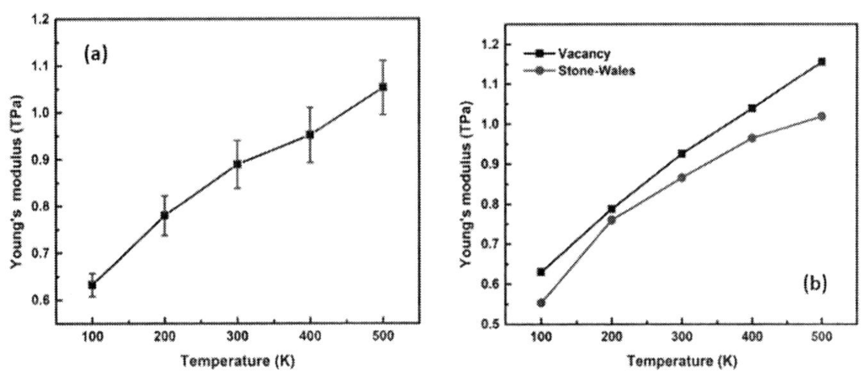

Figure 2. Figure shows the variation in Young's modulus with temperature of (a) pristine graphene sheet and (b) sheet containing randomly generated vacancy and SW defects.

The well known Mermin-Wagner theorem implies that the presence of long-wavelength fluctuations in a 2D crystal will destroy its long-range order. 2D membranes or sheets embedded in a 3D space have a tendency to show height fluctuations. These height fluctuations in membranes can be suppressed by the anharmonic coupling between bending and stretching modes. That is, the existence of a 2D membrane results in large height fluctuations on its surface [4, 6]. In our study, we also observed the height fluctuations in graphene sheet, which shows a resemblance to our earlier work of hexagonal boron nitride (h-BN), where we extracted the bending rigidity of the pristine and defective h-BN sheet in the harmonic regime [12]. Here, we found that, as the thermal energy increases, the atoms start vibrating from their mean position and the material reaches an equilibrium state after the simulation carried out a longer time. This leads to an increase of the out-of-plane height fluctuations with temperature and these fluctuations considerably change with defect concentration as shown in figure 3 (a) and (b). Krishnan *et al* [9] deduced the room temperature elastic modulii of CNT to be 1.25 TPa. The result obtained from our investigation on the Young's modulus of graphene (1 ± 0.25) using the TMSVA is comparable to the recent experimental [3] as well as atomistic simulation methods [2].

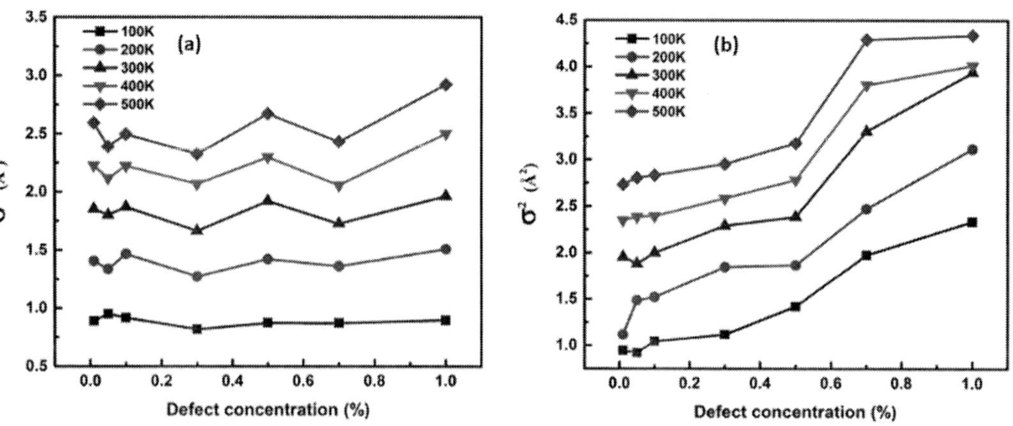

Figure 3. Figure shows variation in height fluctuations with defect concentration at various temperature of graphene sheet with randomly generated (a) vacancy and (b) Stone-Wales defects.

To analyze the variation of Young's modulus of graphene with temperature and defect concentration, we varied the temperatures from 100 K to 500 K and also the mono-vacancy and Stone-Wales defect concentration from 0.01% to 1.0%. We noticed that the Young's modulus decreases with increase of vacancy and SW defect concentration as shown in figure 4 (a) and (b). We also observed that graphene with vacancy defect shows a superior Young's modulus as compared to that with SW defects. We observe that the total number of atoms in graphene sheet is unaltered by the presence of SW defects, and hence preserve the sp^2 C-C bonding by keeping two heptagons and two pentagons corresponding to a single SW defect. On the other hand, the number of atoms reduces with addition of vacancies. Here, the missing carbon atoms breaks the perfect bonding in graphene and which results in the formation of dangling bonds, leading to a sharp decrease in the Young's modulus. In the case of vacancy, as the defect concentration increases, the bond breaking happens and the interatomic interactions become weaker, resulting in a decrease in the Young's modulus. It has been reported that the interaction between dangling bonds or the chemical instabilities due to the missing atoms in graphene sheet affects its mechanical properties [13]. At a higher temperature (1000 K), we noticed a significant decrease of Young's modulus. This is due to the collective excitation of optical phonon mode along with the flexural mode at high temperature and the equation (3) cannot be valid in that condition.

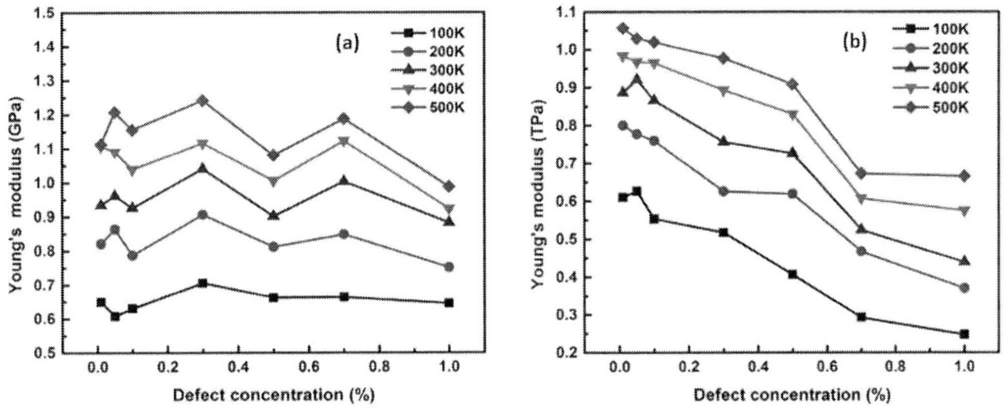

Figure 4. Figure shows the variation in Young's modulus with defect concentration at various temperature of graphene sheet with randomly generated (a) vacancy and (b) Stone-Wales defects.

5. Conclusions

Classical MD simulations have been performed to investigate the effect of defects on the Young's modulus of free standing monolayer graphene sheet within a range of temperatures using the method of thermal mean square vibration amplitude. Our results shows that the presence of vacancy and Stone-Wales defects reduces the Young's modulus of graphene sheet and the Young's modulus decreases with increasing defect concentration. The calculated room temperature Young's modulus of graphene is $\sim$ 1 TPa which is comparable with the results of the earlier experimental and atomistic simulation studies. Due to the collective excitation of optical phonon mode along with the flexural mode at high temperature, the Young's modulus shows a decease at high temperatures. Further study and analysis is desirable to engineer the mechanical properties of graphene according to the demand.

References

[1] Geim A K and Novoselov K S. 2007. *Nature Mater.*, **6**, 183.
[2] Jiang J W, Wang J S and Li B. 2009. *Phys. Rev B.*, **80**, 113405.
[3] Lee C G, Wei X D, Kysar J W and Hone J. 2008. *Science*, **321**, 385.
[4] Mermin N D. Crystalline order in two dimensions. 1968. *Phys. Rev.*, **176**, 250-254.
[5] Nelson D, Weinberg S and Piran T (Eds). Statistical mechanics of membranes and surfaces. World Scientific, Singapore (2004).
[6] Fasolino A, Los J H and Katsnelson M I. 2007. Intrinsic ripples in graphene. *Nat. Mater.*, **6**, 858-861.
[7] Stone A J and Wales D J. 1986. *Chem. Phys. Lett.*, **128**, 501-503.
[8] Donald W B, Olga A S, Judith A H, Steven J S, Boris N and Susan B S. 2002. *J. Phys. Condens. Matter*, **14**, 783-802.
[9] Krishnan A, E. Dujardin, Ebbesen T W, Yianilos P N and Treat M M J. 1998. *Phys. Rev. B.*, **58**, 14013.
[10] Landau L D and Lifshitz E M. Theory of Elasticity (Pergamon, Oxford, 1995).
[11] Plimpton S J. 1995. *Comput. Phys.*, **117**, 1-19. (http://lammps.sandia.gov)
[12] Thomas S, Ajith K M, Chandra S and Valsakumar M C. 2015. *J. Phys.: Condens. Matter*, **27** 315302.
[13] Jing et al. 2012. Effect of defects on Young's modulus of graphene sheet: A molecular dynamics simulation. *RSC Advances*, **2**, 9124-9129.

Numerical analysis of terahertz surface plasmon polaritons propagating in a parallel plate configuration

Sukhdeep Kaur[1,*], Sarabjeet Kaur[1], Manpreet Kaur[1] and Gagan Kumar[2]

[1]Department of Physics, Guru Nanak Dev University, Amritsar, Punjab-143005, India
[2]Department of Physics, Indian Institute of Technology Guwahati, Assam -781039, India

Email: *[1]sukhdeep.iitd@gmail.com

Abstract: In this paper, we present numerical analysis of terahertz (THz) surface plasmon polaritons propagating in a parallel plate configuration. A planar metal surface has the ability to support loosely bound surface waves, called Zenneck waves at terahertz frequencies. Two parallel metal plates separated by a narrow vacuum region can lead to the highly confined terahertz surface plasmon modes at specific frequencies. The frequency of the terahertz mode can be changed with a change in the gap between the plates. Further, we observed that the carrier concentration of plates strongly effect the frequency of the terahertz mode as well as its properties. We have examined in detail the dispersion properties of the terahertz surface plasmons supported by the parallel plate configuration with a varying gap width as well as carrier density of the plates.

1. Introduction

Terahertz radiations cover a narrow range of frequencies between optics and microwave region in the electromagnetic spectrum. In recent years, terahertz related research and technology development has acquired a significant interest of the scientific community all around the globe. This is because of the various applications that terahertz radiations can advance [1-5].This include high speed communication [6], sub wavelength imaging [7], slow light devices [8], bio sensing [9], spectroscopy etc. The development of the waveguides at terahertz frequencies has been one of the primary focus of research in this fast growing area of science and technology. Although, several waveguide geometries have been examined both experimentally and numerically in last more than a decade, however parallel metal configuration has been one of the most promising and widely explored techniques as it has proven its potential in applications such as imaging, sensors and strong field confinement. In this direction, Zhan *et al* [10] have experimentally demonstrated two dimensional confinement of THz waves in tapered waveguide. It is accomplished by coupling of plasmonic edge modes across the air gap. By optimizing the width and separation, THz waves have been confined down to a size of 10 μm (≈λ/260). Mueckstein and Mitrofanov [11] have studied THz surface plasmon waves excited on a gold surface by a focused beam. Jeon and Grischkowsky [12] have experimentally studied the propagation of THz Zenneck surface wave on an aluminium sheet, which are basically THz surface plasmons. They obtained weak guided THz surface plasmon propagation with high attenuation rate due to metal surface than the one predicted by Zenneck theory. Grischkowsky and co-workers have also developed a comprehensive understanding of the terahertz modes supported by the parallel metal configuration and experimentally demonstrated the ability of this configuration in sensing with greater sensitivity.

The excitation of surface plasmons and their propagation in metals at terahertz frequencies is very challenging as metal exhibit very high conductivity. Zenneck had reported the lossely bound terahertz surface waves in metal, calling them Zenneck waves which are confirmed by Grischkowsky and his co-workers as mentioned above. By placing another sheet of metal separated by a thin air gap can reduce the radiative losses. In this regard parallel metal configurations have been developed and examined for various applications as discussed above. However, an analytical description and a comprehensive understanding in this direction is not well established. Furthermore, a change in the metals with other materials such as semiconductors can greatly influence the terahertz modes propagating in the gap. Therefore, a comprehensive understanding of the terahertz modes and their propagation properties depending upon the carrier concentration of the side plates is greatly required.

In this paper, we examine the propagation properties of the terahertz (THz) surface plasmon polaritons in a parallel plate configuration. We examine the dispersion relations of the terahertz surface modes with different widths of the air film surrounded by the two parallel plates. We developed a theoretical model to obtain dispersion relations of terahertz modes. Next, we have examined the dispersion relations with a change in the carrier concentration of the plates. The dispersion relations are plotted for different carrier concentrations which are chosen keeping in mind silicon substrates under different carrier dopant level.

2. Dispersion relation of terahertz surface plasmon wave in a parallel plate configuration

We consider two parallel metal half spaces $z < -b/2$ and $z > b/2$ separated by a thin vacuum region ($-b/2 < z < b/2$) (Figure 1). Surface plasmon waves are coherent delocalized electron waves that exist at the interface between two materials where the real part of the dielectric function changes sign across the interface (e.g. a metal-dielectric interface such as metal sheet in air).

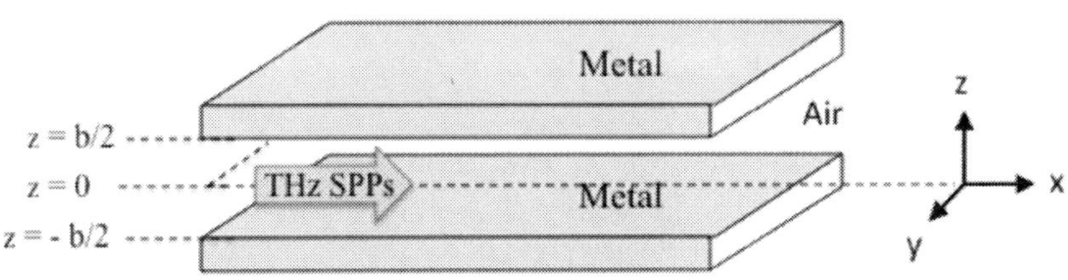

Figure 1. Schematic of a parallel plate configuration with a thin vacuum layer sandwiched between them.

The effective permittivity of metal at frequency ω is,

$$\epsilon_m = \epsilon_r - \frac{\omega_p^2}{\omega^2(1+i\nu/\omega)} \tag{1}$$

where ϵ_r is the lattice dielectric constant of the metal, $\omega_p = (n_0 e^2/m\epsilon_0)^{1/2}$, n_0, $-e$ and m are the density, charge and effective mass of the free electron, ϵ_0 is free space permittivity, c is the speed of light in vacuum and ν are the collisions. Further, we may write it in terms of real and imaginary parts i.e.

$$\epsilon_m \approx \epsilon_r - \frac{\omega_p^2}{\omega^2} + i\frac{\nu}{\omega}\frac{\omega_p^2}{\omega^2}$$

The field distribution for the surface plasmon mode for the said configuration with t, x variation is given as exp[-i (ωt-k_x x)]. The wave equation governing electric field in three different media is given by,

$$\frac{\partial^2 E_x}{\partial z^2} - \left(k_x^2 - \frac{\omega^2}{c^2}\epsilon_m\right) E_x = 0, \tag{2}$$

where $k_x = \frac{\omega}{c}\left(\frac{\epsilon_m}{1+\epsilon_m}\right)^{1/2}$, using equation (1) we may get real and imaginary parts as $k_x = k_{xr} + ik_{xi}$ and k_{xi} will govern the decay loss. Within parallel plate configuration, propagation of wave is in x-direction and mode of propagation of surface plasmon polariton is basically a TM mode. In zx plane, component of electric field obtained from the waveguide equations are given by

$$E_z = \frac{i}{(\frac{\omega^2}{c^2}\epsilon_m - k^2)}\left\{k\frac{\partial E_x}{\partial y} + \omega\frac{\partial B_x}{\partial z}\right\}, \tag{3}$$

$$E_y = \frac{i}{(\frac{\omega^2}{c^2}\epsilon_m - k^2)}\left\{k\frac{\partial E_x}{\partial y} - \omega\frac{\partial B_x}{\partial z}\right\}, \tag{4}$$

Using Gauss law $\vec{\nabla}.\vec{E} = 0$, we obtain the wave equation governing E_x in the three media, where $\epsilon_m = 1$ for $-b/2 < z < b/2$ and ϵ_m for $z > b/2$ and $z <- b/2$.

In **region I**, $z > b/2$ the wave equation is given as,

$$\frac{\partial^2 E_x}{\partial z^2} - \left(k_x^2 - \frac{\omega^2}{c^2}\epsilon_m\right) E_x = 0, \tag{5}$$

The solution of equation (5) will be,

$$E_x = B_3 e^{\beta_1 z} + B_0\, e^{-\beta_2 z} \tag{6}$$

For $z > b/2$, at $z \to \infty$, $B_3 = 0$, one may write $E_x = B_0\, e^{-\beta_1 z}$ and $E_z = \frac{ik_x}{\beta_1} E_x$ where $\beta_1 = \left(k_x^2 - \frac{\omega^2}{c^2}\epsilon_m\right)^{1/2}$. The total electric field in region $z > b/2$ will be, $\vec{E} = E_x\hat{x} + E_z\hat{z}$, i.e.,

$$\vec{E}_I = \left[\hat{x} + \frac{ik_x}{\beta_1}\hat{z}\right] B_0\, e^{-\beta_1 z} \tag{7}$$

In **region II**, $-b/2 < z < b/2$, $\epsilon = 1$ (vacuum region), the component of electric field is,

$$\frac{\partial^2 E_x}{\partial z^2} - \left(k_x^2 - \frac{\omega^2}{c^2}\right) E_x = 0, \tag{8}$$

Following Liu and Tripathi [13], one may obtain solution in component form from above equation, $E_x = B_2 e^{-\beta_2 z} + B_1 e^{\beta_2 z}$ and $E_z = \frac{ik_x}{\beta_2} E_x$, where $\beta_2 = \left(k_x^2 - \frac{\omega^2}{c^2}\right)^{1/2}$. The total electric field in this region is given by, $\vec{E} = E_x\hat{x} + E_z\hat{z}$, gives

$$\vec{E}_{II} = \left[\hat{x} - \hat{z}\frac{ik_x}{\beta_2}\right] B_1 e^{\beta_2 z} + \left[\hat{x} + \hat{z}\frac{ik_x}{\beta_2}\right] B_2 e^{-\beta_2 z} \tag{9}$$

In **region III**, $z <- b/2$, the wave equation will be,

$$\frac{\partial^2 E_x}{\partial z^2} - \left(k_x^2 - \frac{\omega^2}{c^2}\epsilon_m\right) E_x = 0 \tag{10}$$

Solution of equation (10) will be given by, $E_x = B_3 e^{\beta_1 z} + B_0\, e^{-\beta_2 z}$. For $z <- b/2$, at $z \to -\infty$, $B_0 = 0$, therefore, $E_x = B_3 e^{\beta_1 z}$ and $E_z = \frac{ik_x}{\beta_1} E_x$ will give the complete solution as $\vec{E} = E_x \hat{x} + E_z \hat{z}$, i.e.,

$$\vec{E}_{III} = \left[\hat{x} - \hat{z}\frac{ik_x}{\beta_1}\right] B_3 e^{\beta_1 z} \tag{11}$$

Thus, in each region we have,

$$\vec{E}_I = \left[\hat{x} + \frac{ik_x}{\beta_1}\hat{z}\right] B_0\, e^{-\beta_1 z}\ , \qquad z > b/2 \tag{12}$$

$$\vec{E}_{II} = \left[\hat{x} - \hat{z}\frac{ik_x}{\beta_2}\right] B_1 e^{\beta_2 z} + \left[\hat{x} + \hat{z}\frac{ik_x}{\beta_2}\right] B_2 e^{-\beta_2 z}\ , \qquad -b/2 < z < b/2 \tag{13}$$

$$\vec{E}_{III} = \left[\hat{x} - \hat{z}\frac{ik_x}{\beta_1}\right] B_3 e^{\beta_1 z}\ , \qquad z <- b/2 \tag{14}$$

Applying boundary conditions, continuity of E_x and $\epsilon_m E_x$ at $z = b/2$ and $z = -b/2$, i.e. we obtain the following equations,

$$B_1 e^{\beta_2 {}^{b}/_2} + B_2 e^{-\beta_2 {}^{b}/_2} = B_0 e^{-\beta_1 {}^{b}/_2} \tag{15}$$

$$B_1 e^{\beta_2 {}^{b}/_2} - B_2 e^{-\beta_2 {}^{b}/_2} = -\left(\epsilon_m \frac{\beta_2}{\beta_1}\right) B_0 e^{-\beta_2 {}^{b}/_2}, \tag{16}$$

$$B_1 e^{-\beta_2 {}^{b}/_2} + B_2 e^{\beta_2 {}^{b}/_2} = B_3 e^{-\beta_1 {}^{b}/_2}, \tag{17}$$

$$B_1 e^{-\beta_2 {}^{b}/_2} - B_2 e^{\beta_2 {}^{b}/_2} = \left(\epsilon_m \frac{\beta_2}{\beta_1}\right) B_3 e^{-\beta_1 {}^{b}/_2}, \tag{18}$$

Solving these equations, we obtain the dispersion relation,

$$\frac{{\beta_2}^2}{{\beta_1}^2} = \left|\frac{1-e^{\beta_2 b}}{1+e^{\beta_1 b}}\right|^2 \frac{1}{|\epsilon_m|^2}, \tag{19}$$

Simplifying further we obtain,

$$(4k_x c\omega^2 - 2k_x c\, \omega_p^2 + \omega^2\omega_p)\exp\left(-\frac{\omega^2 b}{2k_x c^2}\right) = (2k_x c\omega_p^2 + \omega^2\omega_p)\exp(-k_x b), \tag{20}$$

In order to estimate the decay loss, we can write the dispersion relation as following,

$$F = \frac{{\beta_2}^2}{{\beta_1}^2} - \left|\frac{1-e^{\beta_2 b}}{1+e^{\beta_1 b}}\right|^2 \frac{1}{|\epsilon_m|^2},$$

One may separate it in real and imaginary parts as $F(k_{xr} + ik_{xi}, \epsilon_{mr} + i\epsilon_{mi}) = 0$. Upon simplification, the one may obtain the absorption coefficient, ${k_{xi}}^{-1}$ of the field in the vacuum region given by,

$$k_{xi} = -\epsilon_{mi}\frac{\partial F/\partial \epsilon_m}{\partial F/\partial k_x} \cong k_{xr}\frac{\epsilon_{mi}}{2}\left(1 - \frac{1}{1+\epsilon_{mr}}\right), \tag{21}$$

One may use equation (21) to calculate the propagation length of the plasmonic mode in vacuum region sandwiched by metal regions. Since metal behaves like conductors at terahertz frequency, field penetration inside the metal for terahertz is negligible e.g. skin depth inside Au metal is ~ 80 nm at 1 THz. Therefore, one can neglect the loss of terahertz field inside the metal regions.

We have solved the dispersion relations given by equation (20) for different values of air gap and carrier densities of the plates. Dispersion relation is examined for two different values of air gap viz. $b = 1mm\ and\ 500\ \mu m$ for the carrier densities $n = 10^{18}/cm^3$, and $n = 10^{20}/cm^3$ of the plates. Figure 2 shows the dispersion relations for parallel plate configuration when b = 500 μm for plate carrier densities $n = 10^{18}/cm^3$, and $n = 10^{20}/cm^3$ respectively. In both the plots, one may note that as the wave number increases, initially frequency monotonically increases and then saturates at higher wave number. For higher value of wave number k_x, the dispersion curves saturate. The plasma

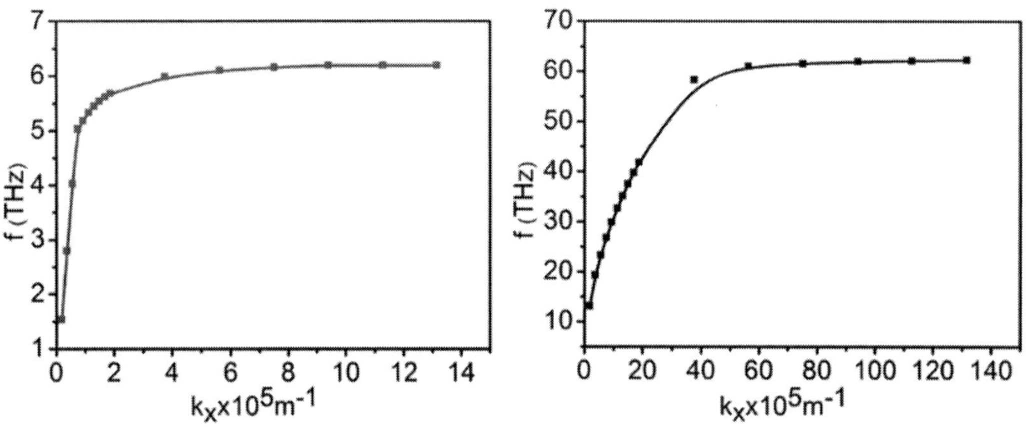

(a) (b)

Figure 2. Dispersion relation of terahertz surface plasmon polariton propagating inside the parallel plate configuration for two different values of carrier concentration of the plates (a) for $n = 10^{18}/cm^3$ and (b) $n = 10^{20}/cm^3$. The width of the air gap is same for the plots i.e. b = 500 μm.

frequency gets changed with density of the plates. In case of $n = 10^{18}/cm^3$, it is $\omega_p = 56.4x10^{12}\ rad/s$ and for $n = 10^{20}/cm^3$, it is $\omega_p = 56.4x10^{13}\ rad/s$. It may be noted that as the carrier density of the plates in increased, the saturation value of the dispersion relation decreases signifying a lower cut-off of the terahertz modes supported by the parallel plate configuration. We also examine the dispersion relation of surface plasmon mode while changing the width of the air gap for two different carrier concentrations of the metal plates. Figure 3 shows dispersion relations of the terahertz surface plasmon polaritons propagating inside the parallel plate configuration with gap width b = 1 mm for two differnt values of carrier densities of the plates, (a) $n = 10^{18}/cm^3$ and (b) $n = 10^{20}/cm^3$. It is important to highlight that the terahertz field in the vacuum region sandwiched by

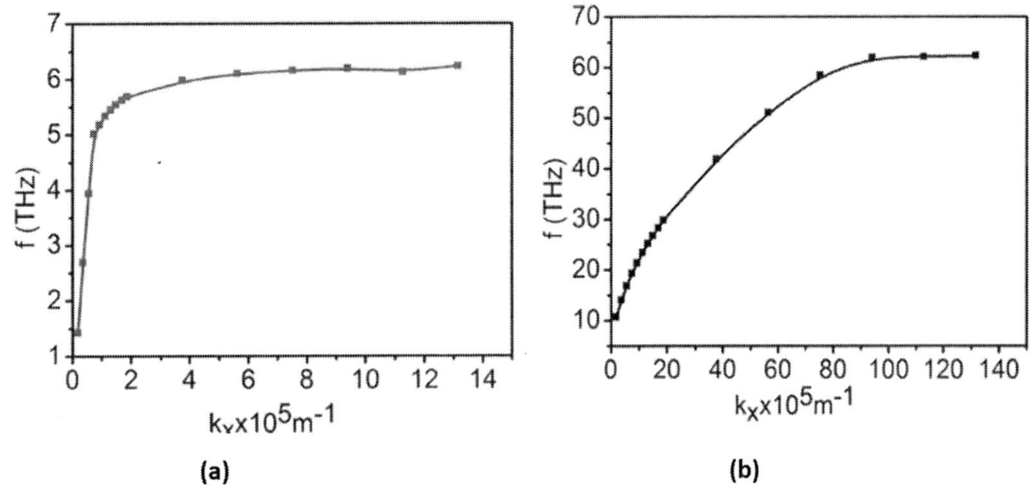

Figure 3. Dispersion relation of terahertz surface plasmon polaritons propagating inside the parallel plate configuration for two different values of carrier densities of the plates (a) for $n = 10^{18}/cm^3$ and (b) $n = 10^{20}/cm^3$. The width of the air gap is same for the plots i.e. $b = 1mm$.

two metal plates could be significant for variety of application such as sensing, acceleration etc. Therefore it is important to analyse the field profile of the plasmon mode as it propagates through the

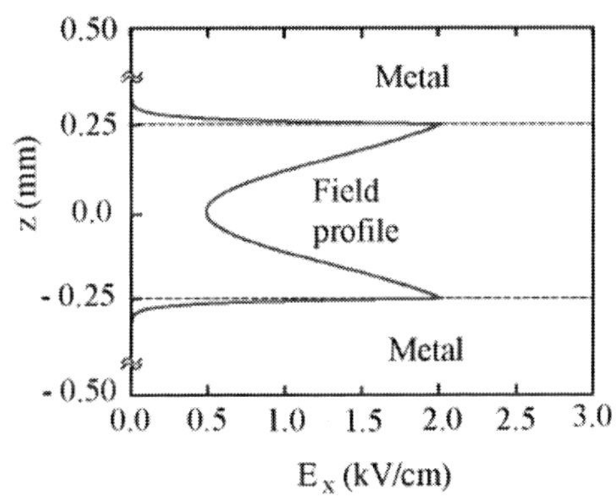

gap. In order to have the perceptivity of fields in different regions of the parallel plate configuration,

Figure 4.THz field profile in parallel plate configuration comprising two metallic layers. The field is maximum at the metal air interface, however decays exponentially in both the media with different decay constants. The field profile is calculated for the parameters: $a = 500\,\mu m$, $\omega_p = 30.4\,THz$, $B_o = 2\,kV/cm$, $\omega = 9.64\,THz$.

in Figure 4, we have plotted terahertz field profile as it travels in the waveguide. Equations (7)-(11) are used to calculate the fields of THz in different regions for parameters: $a = 500\,\mu m$, $\omega_p = 30.4\,THz$, $B_o = 2\,kV/cm$, $\omega = 9.64\,THz$, $k_x = 0.19\times10^5\,m^{-1}$. It may be noted that the field is maximum at the surface of the metals and minimum in the middle. Further, the THz field penetration inside the plates is very low because of their high conductivity. We calculated $1/e$ penetration depth ~ 3 μm for the above given parameters. The field in E_y direction is negligible for the surface plasmon mode.

In summary, we carried out analysis for the propagation of terahertz surface plasmon modes supported by the parallel plate configuration. The dispersion properties of the modes are examined for different widths of the air gaps as well as carrier densities of the plates. The dispersion relation saturates at higher frequency when carrier density of the plate is increased for all the values of air gaps. The results indicate that one may design the waveguide with different parameters such as air gap and parallel plate properties to excite modes at specific frequencies. The above analysis could be significant in designing parallel plate THz plasmonic waveguides with materials such as semiconductors replacing conventional wave guiding techniques comprising metals.

Acknowledgements

The authors would like to acknowledge the funding from the project (Project No. PHY/P/GK/02) for the financial assistance.

References

[1] Liu S, Zhang C, Hu M, Chen X, Zhang P, Gong S, Zhao T & Zhong R 2014. Coherent and tunable terahertz radiation from graphene surface plasmon polaritons excited by an electron beam. *App. Phys. Lett.*, **104** 201104.
[2] Ramakrishnan G, Kumar N, Planken P C M, Tanaka D and Kajikawa K 2012. Surface enhanced terahertz emission from a hemicyanine self assembled monolayer. *Opt. Express* **20** 4067.
[3] Chan W L, Deibel J and Mittleman D M 2007. Imaging with terahertz radiation. *Rep. Prog. Phys.* **70** 1325-79.
[4] Welsh G H, Hunt N T and Wynne K 2007. Terahertz pulse emission through laser excitation of surface plasmon in a metal grating. *Phys. Rev. Lett.* **98** 026803.
[5] Kumar G 2015. Controlling terahertz surface plasmon properties on a periodically structured silicon surface. *J. Spectro* **2015** 543985.
[6] Alferness R C 1981. Guided wave devices for optical communication. *IEEE J. Quantum Electron.***17** 946.
[7] Zhang X and Liu Z 2008. Superlenses to overcome the diffraction limit. *Nat. Mater.* **7** 435.
[8] Gan Q, Fu Z, Ding Y J and Bartoli F J 2008. Ultrawide bandwidth slow light system based on THz plasmonic graded metallic grating structures, *Phys. Rev. Lett.* **100** 256803.
[9] Anker J N, Hall W P, Lyandres O, Shah N C, Zhao J and Duyne R P V 2008. Biosensing with plasmonic nonsensors. *Nat. Mater.* **7** 442-53.
[10] Zhan H, Mendis R and Mittleman D M 2010. Superfocusing terahertz waves below λ/250 using plasmonic parallel plate waveguide. *Opt. Express* **18** 9643-50.
[11] Mueckstein R and Mitrofanov O 2011. Imaging of terahertz surface plasmon waves excited on a gold surface by a focused beam. *Opt. Express* **19** 3212-17.
[12] Jeon T I and Grischkowsky D, 2006. Terahertz Zenneck surface wave (THz Surface Plasmon) propagation on a metal sheet, *Appl. Phys. Lett.* **88** 061113.
[13] Liu C S and Tripathi V K 2007 *Electromagnetic theory for Telecommunication* (India: Cambridge University Press)

Ab - initio non-adiabatic couplings among three lowest singlet states of H_3^+: Construction of multisheeted diabatic potential energy surfaces

Bijit Mukherjee, Saikat Mukherjee and Satrajit Adhikari*

Department of Physical Chemistry
Indian Association for the Cultivation of Science
Jadavpur, Kolkata -700 032, INDIA

E-mail: pcsa@iacs.res.in

Abstract. We calculate the adiabatic potential energy surfaces and non-adiabatic interactions among the three lowest singlet states ($1^1A'$, $2^1A'$ and $3^1A'$) of H_3^+ in hyperspherical coordinates for a fixed hyperradius, $\rho = 9$ bohr as functions of hyperangles, θ ($0 \leq \theta \leq 90°$) and ϕ ($0 \leq \phi \leq 360°$). All *ab initio* calculations are performed using MRCI level of methodology implemented in quantum chemistry package, MOLPRO. The ground ($1^1A'$) and the first excited ($2^1A'$) states exhibit several conical intersections as functions of ϕ for $\theta \geq 70°$. Subsequently, we carry out adiabatic to diabatic transformation (ADT) to obtain ADT angles for constructing single-valued, continuous, smooth and symmetric 3×3 diabatic potential energy matrix to perform accurate scattering calculations.

1. Introduction

The Born - Oppenheimer (BO) treatment, one of the most fundamental theories in quantum chemistry, separates the motions of fast moving electrons from the slow moving nuclei.[1,2] This results into two important quantities, *viz.*, the adiabatic potential energy surfaces (PESs) and the non-adiabatic coupling terms (NACTs) among the electronic states. It is the NACTs, which dictate the contributions of the excited electronic states into a molecular process even if it is occurring on the ground electronic state. In order to calculate correct transition probabilities for various processes or study radiationless decay and scattering processes, one has to take into account the NACTs between the electronic states. These coupling terms become sharp functions of nuclear coordinates close to conical intersections (CIs) in the nuclear configuration space (CS), those which occur at the avoided crossings or degenerate points between the adiabatic PESs. Thus, in order to perform stable numerical calculations, one has to carry out a unitary transformation to obtain the so called diabatic representation, where the kinetic couplings among the electronic states become smooth functions of nuclear coordinates. The transformation from adiabatic to diabatic representation of Schrödinger Equation (SE) for a given sub-Hilbert space is guaranteed to produce "correct" and "accurate" diabatic PESs under the following conditions: (a) The non-removable components of NACTs are approximately zero[3]; (b) The vector fields created by the NACTs satisfy the so called Curl Conditions.[4,5] Baer[6] and his co-workers[7,8] carried out the beyond BO treatment for two state sub-Hilbert space and formulated the diabatic Hamiltonian matrix in terms of adiabatic PESs and adiabatic to diabatic transformation (ADT) angles.[9–11] Sarkar *et al.* generalized[12–15] the BO treatment for any three/four coupled electronic states, where the explicit form of NACTs, Curl - Divergence equation and diabatic Hamiltonian matrix are derived in terms of ADT angles. This approach paves[16,17] a practical way to handle the NACTs with singularities at any point(s) or along a seam in the CS to construct continuous, single-valued, symmetric and smooth diabatic PESs to carry out scattering processes.

The molecular ion H_3^+ being an important hydrogenic compound, exists in substantial amount in interstellar space. This species, first spectroscopically observed by Oka,[18] soon gained much theoretical and experimental attention owing to its important role being played in astrophysics. This seemingly simple ion only with two electrons possesses a rich electronic structure making it a target for a large number of *ab initio* calculations. Until now many accurate PESs near the ground state equilibrium geometry have been reported.[19,20] Meyer *et al.*[19] carried out calculations using full configuration interaction and the obtained local PESs were fitted using 69 carefully selected points with energies up to 25000 cm^{-1}, and then, achieved near spectroscopic accuracy. Röhse *et al.*[20] applied CISD-R12 method to achieve higher accuracy and Cencek *et al.*[21] included both adiabatic and relativistic corrections. In spite of the large amount of work on H_3^+, very less attention has been paid on its excited states. Schaad and Hicks[22] performed a systematic study of excited electronic states of H_3^+ which included the detailed study of the ground triplet state only. Later on, a complete PES has been reported,[23–25] including vibrational[24–27] and rovibrational[23,28] calculations. Several diabatic PESs have been made using the singlet states of H_3^+, few of which are: the Preston and Tully surface,[29] the Kamisaka-Bian-Nobusada-Nakamura[30] surface, the double many-body expansion potential surface constructed by Varandas*et al.*[31] using dressed DIM method. Recently, *ab initio* calculated NACTs and adiabatic PESs are employed to construct "first principle based and accurate" diabatic PESs by Mukherjee *et al.*[32] In this article, we explore the non-adiabaticities among the lowest three singlet states of H_3^+ by calculating the adiabatic PESs and NACTs for the states $1^1A'$, $2^1A'$, $3^1A'$ in the hyperspherical coordinates using the *ab initio* quantum chemistry package MOLPRO.[33] We employ beyond BO theory for three state sub-Hilbert space, calculate ADT angles by plugging NACTs in ADT equations, and construct diabatic Hamiltonian matrix for a fixed ρ as a function of θ and ϕ using adiabatic PESs and ADT angles.

2. Theory

Beyond Born-Oppenheimer theory for three state sub-Hilbert space

We assume a three - state electronic sub - Hilbert space (i.e., a complete space for the present case) with CIs among the states anywhere in the nuclear CS. The BO expansion[1,2] of the molecular wavefunction for this subspace of the Hilbert space and the total electron - nuclei Hamiltonian in the adiabatic representation are:

$$\begin{aligned} |\Psi(s_e, s_n)\rangle &= \sum_{i=1}^{3} \psi_i(s_n)|\xi_i(s_e|s_n)\rangle, \\ \hat{\mathbf{H}} &= \hat{\mathbf{T}}_{s_n} + \hat{\mathbf{H}}_e(s_e|s_n), \\ \hat{\mathbf{T}}_{s_n} &= -\frac{\hbar^2}{2}\sum_i \left(\frac{\nabla^2_{s_n,i}}{m_i}\right), \\ \hat{\mathbf{H}}_e(s_e|s_n)|\xi_i(s_e|s_n)\rangle &= u_i(s_n)|\xi_i(s_e|s_n)\rangle, \end{aligned} \tag{1}$$

where the eigenfunctions $[|\xi_i(s_e|s_n)\rangle]$ of the electronic Hamiltonian, $\hat{\mathbf{H}}_e(s_e|s_n)$, are defined by the sets of nuclear (s_n) and electronic (s_e) coordinates with nuclear coordinate dependent eigenvalues, $u_i(s_n)$. $\hat{\mathbf{T}}_{s_n}$ is the nuclear kinetic energy (KE) operator and the expansion coefficient, $\psi_i(s_n)$, are actually the nuclear wavefunctions. When the total electron - nuclear

Hamiltonian and the BO expansion of the molecular wavefunction [Eq. (1)] for a sub - Hilbert space are substituted in the time independent SE $[\hat{\mathbf{H}}\Psi(s_e, s_n) = E\Psi(s_e, s_n)]$, we can arrive at the following compact form of kinetically coupled nuclear equation:

$$-\frac{\hbar^2}{2m}\left(\vec{\nabla}+\vec{\tau}\right)^2\psi + (\mathbf{u}-E)\,\psi = 0, \tag{2}$$

where the non - adiabatic coupling and adiabatic PES matrices have the following form:

$$\vec{\tau} = \begin{pmatrix} 0 & \vec{\tau}_{12} & \vec{\tau}_{13} \\ -\vec{\tau}_{12} & 0 & \vec{\tau}_{23} \\ -\vec{\tau}_{13} & -\vec{\tau}_{23} & 0 \end{pmatrix} \quad \text{and} \quad \mathbf{u} = \begin{pmatrix} u_1 & 0 & 0 \\ 0 & u_2 & 0 \\ 0 & 0 & u_3 \end{pmatrix}, \tag{3}$$

respectively. The non - adiabatic coupling matrix elements are defined as:

$$\vec{\tau}_{ij} \;=\; \langle \xi_i(s_e|s_n)|\vec{\nabla}\xi_j(s_e|s_n)\rangle. \tag{4}$$

If the three states constitute a sub - Hilbert space, it is possible to transform ($\psi = \mathbf{A}\psi^d$) the adiabatic nuclear SE [Eq. (2)] to the diabatic one:

$$-\frac{\hbar^2}{2m}\boldsymbol{\nabla}^2\psi^d + (\mathbf{W}-E)\psi^d = 0, \qquad \mathbf{W} = \mathbf{A}^\dagger\mathbf{u}\mathbf{A} \tag{5}$$

under the condition:

$$\vec{\boldsymbol{\nabla}}\mathbf{A} + \vec{\boldsymbol{\tau}}\mathbf{A} = 0. \tag{6}$$

This equation is known as the ADT condition.[6] As the model form of (3×3) **A** has to be an orthogonal matrix and the ortho - normality conditions demand the fulfillment of six relations, three independent variables, *viz.*, Euler like angles of rotation $[\Theta_{12}(s_n)$, $\Theta_{23}(s_n)$ and $\Theta_{13}(s_n)]$, commonly called ADT angles, are the natural requirement to construct the three - state **A** matrix by taking the product of three (3×3) rotation matrices[9], $\mathbf{A}_{12}(\Theta_{12})$, $\mathbf{A}_{23}(\Theta_{23})$, and $\mathbf{A}_{13}(\Theta_{13})$ as shown below:

$$\mathbf{A}(\Theta_{12},\Theta_{23},\Theta_{13}) \;=\; \mathbf{A}_{12}(\Theta_{12})\cdot\mathbf{A}_{23}(\Theta_{23})\cdot\mathbf{A}_{13}(\Theta_{13}). \tag{7}$$

The above model form of **A** matrix [Eq. (7)] and the anti - symmetric form of $\vec{\boldsymbol{\tau}}$ matrix [Eq. (3)] are substituted in Eq. (6), where the simple manipulation as performed by Baer *et al.*[9,10] leads to the following equations for ADT angles:

$$\vec{\nabla}\Theta_{12} \;=\; -\vec{\tau}^{12} + \tan\Theta_{23}(\vec{\tau}^{13}\cos\Theta_{12} - \vec{\tau}^{23}\sin\Theta_{12}), \tag{8a}$$

$$\vec{\nabla}\Theta_{23} \;=\; -(\vec{\tau}^{13}\sin\Theta_{12} + \vec{\tau}^{23}\cos\Theta_{12}), \tag{8b}$$

$$\vec{\nabla}\Theta_{13} \;=\; -\frac{1}{\cos\Theta_{23}}(\vec{\tau}^{13}\cos\Theta_{12} - \vec{\tau}^{23}\sin\Theta_{12}), \tag{8c}$$

which in turn brings the explicit form[12] of $\vec{\boldsymbol{\tau}}$ matrix elements in terms of ADT angles:

$$\vec{\tau}_{12} \;=\; -\vec{\nabla}\Theta_{12} - \sin\Theta_{23}\vec{\nabla}\Theta_{13}, \tag{9a}$$

$$\vec{\tau}_{23} \;=\; \sin\Theta_{12}\cos\Theta_{23}\vec{\nabla}\Theta_{13} - \cos\Theta_{12}\vec{\nabla}\Theta_{23}, \tag{9b}$$

$$\vec{\tau}_{13} \;=\; -\cos\Theta_{12}\cos\Theta_{23}\vec{\nabla}\Theta_{13} - \sin\Theta_{12}\vec{\nabla}\Theta_{23}. \tag{9c}$$

The components of *ab initio* calculated NACTs are substituted in Eq. (8) to obtain the ADT angles. The diabatic potential matrix elements can be calculated by using those ADT angles and adiabatic PESs given in Eq. (5).

3. *Ab initio* calculation for PESs and NACTs:

The calculations of the PESs and NACTs for the H_3^+ system are carried out using the hyperspherical coordinates, namely, the hyperradius, ρ and the two hyperangles, θ and ϕ. The domains for these hyperangles are $0 \leq \theta \leq \pi/2$ and $0 \leq \phi \leq 2\pi$. The hyperradius ρ solely determines the size of the triangle formed by the three particles, whereas the two hyperangles describe the shape of the triangle. The *ab initio* calculations are performed in terms of the inter-particle distances of the three nuclei, which are related to the hyperspherical coordinates[34] as given below:

$$R_1 = \frac{\rho}{\sqrt{2}} d_3 \left[1 + \sin\theta \cos(\phi + \epsilon_3)\right]^{1/2}, \quad (10a)$$

$$R_2 = \frac{\rho}{\sqrt{2}} d_1 \left(1 + \sin\theta \cos\phi\right)^{1/2}, \quad (10b)$$

$$R_3 = \frac{\rho}{\sqrt{2}} d_2 \left[1 + \sin\theta \cos(\phi - \epsilon_2)\right]^{1/2}, \quad (10c)$$

where $d_i = \sqrt{m_i(m_j + m_k)/\mu M}$, $M = \sum_i^3 m_i$, $\mu = \sqrt{(m_1 m_2 m_3/M)}$, $\epsilon_2 = 2\tan^{-1}(m_3/\mu)$ and $\epsilon_3 = 2\tan^{-1}(m_2/\mu)$. For H_3^+, $m_1 = m_2 = m_3 = 1.008$ amu, $M = 3.024$ amu, $\mu = 0.582$ amu, $d_1 = d_2 = d_3 = 1.075$ and $\epsilon_2 = \epsilon_3 = 120.0$.

In our present work, we have calculated the three lower adiabatic PESs ($1^1A'$, $2^1A'$ and $3^1A'$) and the NACTs, $\tau_k^{12}, \tau_k^{13}, \tau_k^{23} [k = \theta, \phi]$ for a fixed value of ρ (= 9.0 bohr) by varying θ and ϕ in increments of $\Delta\theta = 1°$ and $\Delta\phi = 2°$ within the aforesaid ranges. As the location of CIs is close to $\rho = 9.0$ bohr, the amplitudes of the NACTs are quite prominent in its corresponding θ-ϕ CS. The *ab initio* data points for the three PESs are obtained in C_s symmetry with the MOLPRO package[33] using Dunnings[35] correlation consistent *cc-pV5Z* basis set. We performed MCSCF calculations involving the first six (6) electronic states with an active space comprising two (2) electrons in ten (10) active orbitals generating fifty six (56) configuration state functions (CSFs). Taking the reference for the first three (3) singlet states, we subsequently carried out the internally contracted MRCI calculations. The calculated minimum energy of the ground singlet state ($1^1A'$) of H_3^+ at $\rho = 9$ bohrs is -1.175 hartree, appears at $\theta = 74°$ and $\phi = \{55°, 65°, 145°, 155°, 295°, 305°\}$. These configurations correspond to the inter-particle distances as: $R_1, R_2, R_3 = \{1.41, 8.52, 8.11\}$ bohrs. The ground state ($1^1A'$) energy of singlet H_3^+ appears as -1.0120 hartree at $\theta = 0°$ in the equilateral arrangement with side lengths 6.84 bohrs. At the same geometry the first and second excited states remain degenerate ($^1E'$) with energy -0.9968 hartree.

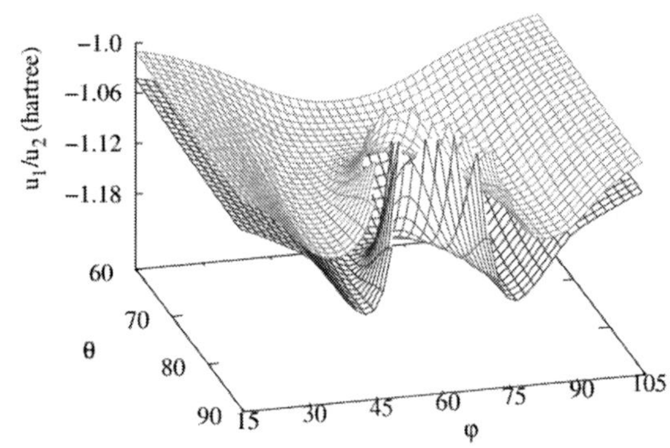

Figure 1 The CI seam between the ground ($1^1A'$) and first excited ($2^1A'$) states.

The ground ($1^1A'$) and the first excited ($2^1A'$) states exhibit three CIs at $\theta = 70°$ with $\phi = \{60°, 180°, 300°\}$ corresponding to isosceles geometries, whereas for $\theta \geq 70°$, three semi-circular CI seams along $\phi = (30° - 90°), (150° - 210°)$ and $(270° - 330°)$ come into exist. The three CIs formed at $\theta = 70°$ are due to the accidental degeneracy occurring at the three equivalent C_{2v} geometries while the six CI seams start at close to $\theta = 90°$ corresponding to a particular scalene triangular geometry and end again at close to $\theta = 90°$ corresponding to the equivalent scalene geometry traversing through the C_{2v} isosceles geometries. In addition to that, the pseudo Jahn-Teller (PJT) interaction comes into play between the upper two states ($2^1A'$ and $3^1A'$) at $\phi = \{0°, 120°, 240°\}$. We depict one of the three CI seams between the two lowest singlet adiabatic PESs of H_3^+ system along $\phi = 15° - 105°$ in Figure 1. Since there is three-fold symmetry in the PESs, all the three CI seams are equivalent. On the other hand, the three lowest singlet adiabatic PESs of H_3^+ are depicted in Figure 2 for the fixed value of $\rho = 9$ bohrs over the entire range of θ and ϕ.

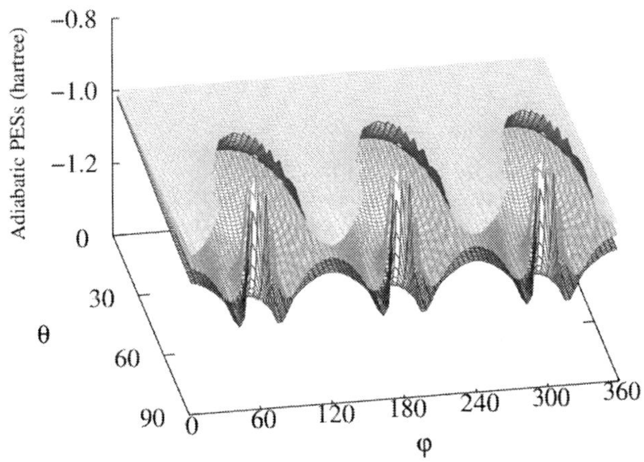

Figure 2 Adiabatic PESs of the lowest three singlet A' states of H_3^+ cluster.

We have employed the so called numerical finite difference method DDR in MOLPRO to calculate the NACTs ($\tau_k^{ij}[k = \theta, \phi], i = j = 1, 2, 3, i < j, i \neq j$) among the states. The ϕ component of the NACT, τ_ϕ^{12} takes up maximum magnitude along the three semi-circular CI seams for $\theta = 70°$ to $90°$ and at ϕs having values at $\{30°, 90°, 150°, 210°, 270°, 330°\}$. The corresponding NACT for the θ component (τ_θ^{12}) also blows up along the three CI seams in the same ranges of θ and ϕ as that of τ_ϕ^{12}, touching θ at $\theta \approx 70°$ but unlike τ_ϕ^{12}, it reaches maximum at $\phi = \{60°, 180°, 300°\}$ for $\theta \approx 70°$. The maximum magnitudes of τ_k^{13} and τ_k^{23} $[k = \theta, \phi]$ are quite smaller in comparison to τ_k^{12} $[k = \theta, \phi]$. In case of τ_k^{23} $[k = \theta, \phi]$, they exhibit pseudo-Jahn Teller (PJT) effect around $\theta = 90°$. In Figure 3, the elements of the NACTs between the ground and the first excited state for (a) ϕ and (b) θ components are displayed as functions of the stereographic projections β and γ, which are defined as: $\beta = \sin\theta\cos\phi$ and $\gamma = \sin\theta\sin\phi$.

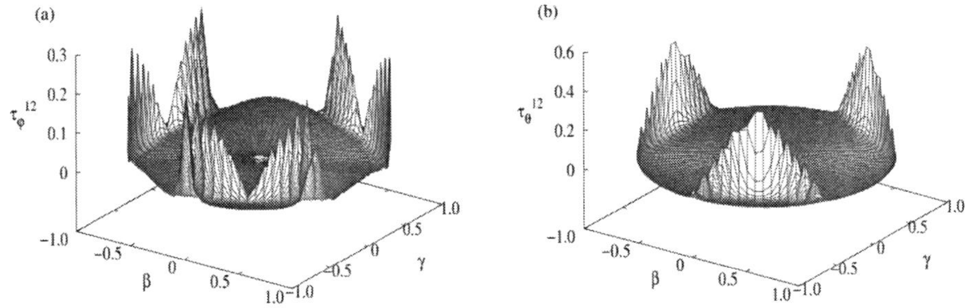

Figure 3 Functional forms of the non-adiabatic coupling (NAC) matrix elements (a) τ_ϕ^{12} and (b) τ_θ^{12}, between the ground and the first excited states as functions of β and γ. The NACTs blow up along the three CI seams.

4. ADT and diabatic potential matrix

The *ab initio* calculated NACTs are substituted in ADT equations (Eqs. 8) and those stiff differential equations are solved by using Backward Differentiation Formula over a 2D grid of geometries represented by 89 × 180 grid points in hyperspherical coordinates (θ, ϕ) ranging from $\theta = 1°$ to $89°$ and $\phi = 1°$ to $359°$. It is possible to integrate the two sets of coupled (θ and ϕ) differential equations (Eqs. 8) by choosing infinitely different paths and each path will generate different sets of θ-ϕ dependent ADT angles $(\Theta_{12}, \Theta_{23}, \Theta_{13})$. Though the various sets of ADT angles obtained due to different choice of paths are expected to show gauge invariance, each set of such angles produces different ADT matrix and such matrices will be related through orthogonal transformation.[14] At present case, we consider a rectangular path, where the differential equations for the θ grid are integrated with positive increment from $\theta = 1°$ to $89°$ for each positive step integration of the differential equations for the ϕ grid from $1°$ to $359°$. The obtained ADT angles are then used to calculate the elements of the ADT matrix. Figure 4 (a) and (b) represent 1D cuts of ADT angles (Θ_{12}) between the ground and the first excited state and ADT matrix elements (A_{11}/A_{22}) along the ϕ coordinate at various θs $\{= 50°, 70°, 89°\}$, respectively. Figure 4(a) shows that the ADT angle Θ_{12} approaches a magnitude of π for $\theta \geq 70°$ whereas for the same values of θ, the A_{11}/A_{22} elements show six sign flips within the full ϕ range (see Figure 4(b)) which asserts the presence of six CIs. Finally, the elements of the diabatic potential matrix are displayed in Figure 5 which appeared as single-valued, continuous, smooth and symmetric. Only ϕ range $(0° - 120°)$ is used in the Figure 5 due to the overall three-fold symmetry in the CS. In Figure 6 we have depicted the two diagonal elements of the diabatic PESs crossing each other properly along the CI seam at $\phi = 30° - 90°$ and $\theta \geq 70°$.

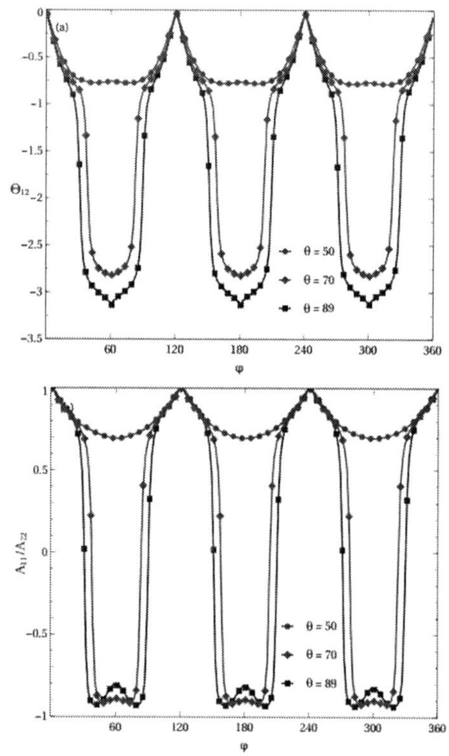

Figure 4 1D cuts of (a) Θ_{12}, (b) A_{11}/A_{22} elements of the ADT matrix as function of ϕ at various θ values.

5. Summary

We have calculated the first three singlet adiabatic PESs of H_3^+ and the NACTs between those states by MOLPRO quantum chemistry package in hyperspherical coordinates for a fixed hyperradius ($\rho = 9$ bohr). The ground $(1^1A')$ and the first excited $(2^1A')$ states exhibit three CIs as function of ϕ at $\theta = 70°$ and six CIs for $\theta \geq 70°$. Moreover, at $\phi = \{0°, 120°, 240°\}$, the states $2^1A'$ and $3^1A'$ exhibit strong PJT interactions, which also contributes to the overall non-adiabaticity of the system. Since the three singlet states ($1^1A'$, $2^1A'$ and $3^1A'$) constitute a sub-Hilbert space over the range of the selected CS, we guarantee the validity of the curl conditions, and thus, construct the diabatic Hamiltonian. The elements of the diabatic PES matrix are found to be single-valued, continuous, symmetric and smooth functions of the hyperangles.

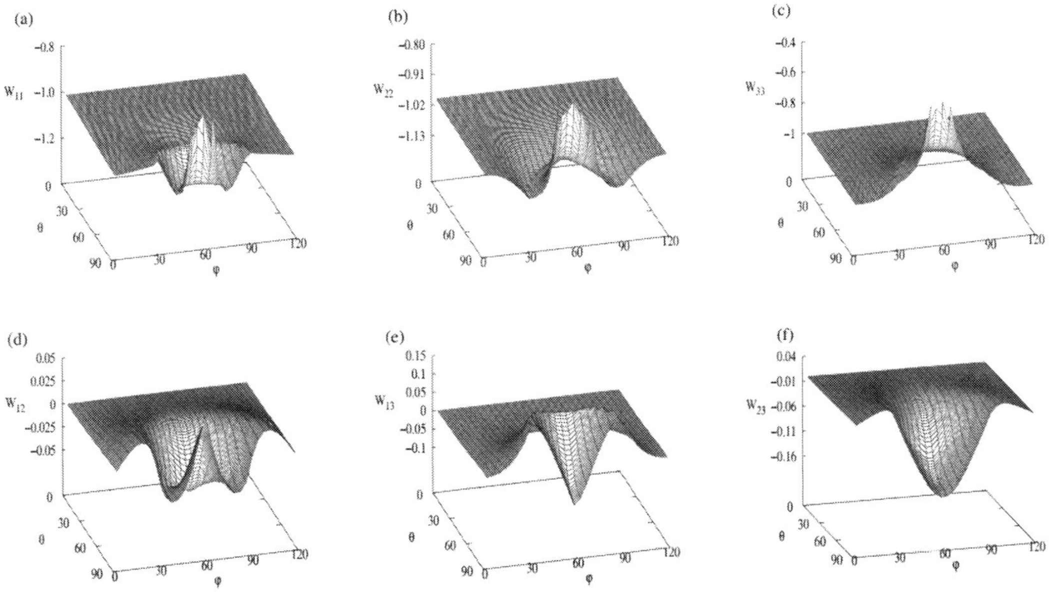

Figure 5 The diabatic PESs (panel (a)-(c)) and the coupling elements (panel (d)-(f)) as functions of θ and ϕ.

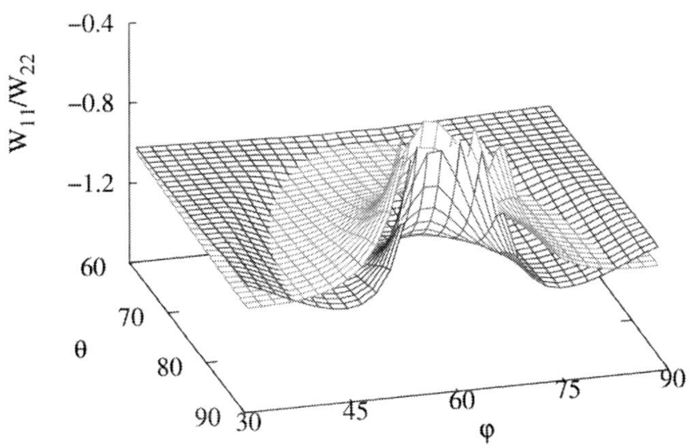

Figure 6 The crossing between the two diagonal elements of diabatic potential energy matrix as functions of θ and ϕ.

6. Acknowledgments

BM (File No: 09/080(0960)/2014-EMR-I) and SM (IF110026) acknowledge CSIR India and DST INSPIRE program, respectively for research fellowship and SA acknowledges IACS for CRAY super computing facility.

References

[1] Born M and Oppenheimer J R 1927 *Ann. Phys. (Leipzig)* **84** 457–484

[2] Born M and Huang K 1954 *Dynamical Theory of Crystal Lattices* (Oxford University Press, Oxford)

[3] Matsunaga N and Yarkony D R 1998 *Mol. Phys.* **93** 79

[4] Baer M 2002 *Phys. Rep.* **358** 75

[5] Baer M 2006 *Beyond Born - Oppenheimer: Conical intersections and Electronic nonadiabatic coupling terms* (Wiley - Interscience)

[6] Baer M 1975 *Chem. Phys. Lett.* **35** 112

[7] Baer R, Charutz D, Kosloff R and Baer M 1996 *J. Chem. Phys.* **105** 9141

[8] Charutz D M, Baer R and Baer M 1997 *Chem. Phys. Letts.* **265** 629

[9] Top Z H and Baer M 1977 *J. Chem. Phys.* **66** 1363

[10] Alijah A and Baer M 2000 *J. Phys. Chem. A* **104** 389

[11] Vértesi T, Vibók A, Halász G J and Baer M 2004 *J. Chem. Phys.* **121** 4000

[12] Sarkar B and Adhikari S 2006 *J. Chem. Phys.* **124** 074101

[13] Sarkar B and Adhikari S 2008 *J. Phys. Chem. A* **112** 9868

[14] Mukherjee S, Bandyopadhyay S, Paul A K and Adhikari S 2013 *J. Phys. Chem. A* **117** 3475

[15] Mukherjee S and Adhikari S 2014 *Chem. Phys.* **440** 106

[16] Paul A K, Sardar S, Sarkar B and Adhikari S 2009 *J. Chem. Phys.* **131** 124312

[17] Paul A K, Ray S, Mukhopadhyay D and Adhikari S 2011 *J. Chem. Phys.* **135** 034107

[18] Geballe T R and Oka T 2006 *Science* **312** 1610

[19] Meyer W, Botschwina P and Burton P R 1986 *J. Chem. Phys.* **84** 891

[20] Röhse R, Kutzelnigg W, Jaquet R and Klopper W 1994 *J. Chem. Phys.* **45** 753

[21] Cencek W, Rychlewski J, Jaquet R and Kutzelnigg W 1998 *J. Chem. Phys.* **108** 2831

[22] Schaad L J and Hicks W V 1974 *J. Chem. Phys.* **61** 1934

[23] Friedrich O, Alijah A, Xu Z R and Varandas A J C 2001 *Phys. Rev. Lett.* **86** 1183

[24] Sanz C, Roncero O, Tablero C, Aguado A and Paniagua M 2001 *J. Chem. Phys.* **114** 2182

[25] Varandas A J C, Alijah A and Cernei M 2005 *Chem. Phys.* **308** 285

[26] Cuervo-Reyes E, Rubayo-Soneria J, Aguado A, Paniagua M, Tablero C, Sanz C and Roncero O 2002 *Phys. Chem. Chem. Phys.* **4** 6012

[27] Viegas L P, Alijah A and Varandas A J C 2005 *J. Phys. Chem. A* **109** 3307

[28] Alijah A, Viegas L P, Cernei M and Varandas A J C 2003 *J. Mol. Spectrosc.* **221** 163

[29] Preston R K and Tully J C 1971 *J. Chem. Phys.* **54** 4297

[30] Kamisaka H, Bian W, Nobusada K and Nakamura H 2002 *J. Chem. Phys.* **116** 654

[31] Viegas L P, Alijah A and Varandas A J C 2007 *J. Chem. Phys.* **126** 074309

[32] Mukherjee S, Mukhopadhyay D and Adhikari S 2014 *J. Chem. Phys.* **141** 204306

[33] Werner H J, Knowles P J, Knizia G, Manby F R, Schütz M *et al.* 2010 Molpro, version 2010.1, a package of ab initio programs see http://www.molpro.net

[34] Billing G D and Muckerman J T 1989 *J. Chem. Phys.* **91** 6830

[35] Dunning T H 1989 *J. Chem. Phys.* **90** 1007

Effect of oxide barrier height in spin dependent tunneling in MTJ of FeO-MgO multilayer structure

M Bhattacharjee[1], H Nemade[13] and D Bandyopadhyay[12]

[1]Centre for Nanotechnology, Indian Institute of Technology Guwahati, India
[2]Department of Chemical Engineering, Indian Institute of Technology Guwahati, India
[3]Department of Electronics & Electrical Engineering, Indian Institute of Technology Guwahati, India
Email: mitradip@iitg.ernet.in, dipban@iitg.ernet.in

Abstract. We study the spin dependent tunneling current properties through oxide multilayers in a magnetic tunnel junction (MTJ). For this purpose, nonequilibrium Green's function approach along-with the density-functional theory have been applied. We employed three structural models of FeO-MgO-FeO multi-layer with three different width of FeO and MgO layer. An atomistic model is considered to describe the effect of oxide multilayers of different heights. Spin dependent study for tunneling reveals that the parallel spin shows higher tunneling current whereas anti-parallel spin conducts very less. Further, the lowest tunneling current is obtained for the case where the FeO and MgO each has 3 atomic layers of height whereas the tunneling current is highest in 4 atomic layers of FeO/1 atomic layers of MgO/4 atomic layers of MgO multilayer structure. Importantly, when the MgO or FeO layers are increased or decreased from this level, the tunneling current decreases significantly. The study reveals that the layer height in the tunneling domain can be important factor for tuning and adjusting tunneling current in the nanoscale regime of oxide layer thickness.

Keywords: Tunneling, MTJ, barrier height, spin dependent current.

1. Introduction

Magnetic tunnel junction (MTJ) with MgO and Al_2O_3 barrier is one of the very active areas of research because of their high tunnel magnetoresistance (TMR) alongside their applications related to magnetic random access memories (MRAM) [1]. Theoretically, it has been shown that Fe/MgO/Fe layers in MTJ can exhibit TMR up to 1000% whereas many groups have experimentally fabricated Fe/MgO/Fe multilayer MTJ, which shows a TMR of maximum ~200% at room temperature [2,3]. The same experimental TMR value for MgO barrier is increased up to 604% when the ferromagnetic electrodes were replaced by CoFeB. The deviation in the result of theoretical and experimental value is attributed to the interfacial properties of the barrier and electrode. It also has been reported that the spin dependent tunneling depends highly on the different properties of the interface [4]. Therefore, the study of interfacial properties in MTJs is important for the improvement in tunneling properties. Thus far, the first principal calculation for Fe/MgO interface has been performed by many research groups [5]. It has also been showed that

Journal of Physics: Conference Series **759** (2016) 012051 doi:10.1088/1742-6596/759/1/012051

the growth of FeO layer at the barrier electrode interface gives rise the structural coherency, which leads to high TMR values [6]. Theoretical investigations of interfaces in Fe/MgO/Fe based MTJs also revealed that the interface roughness can play an important role in TMR [7]. Experimentally, the morphologies, diffusion, and reactions at the interfaces of multilayers are now investigated using spectroscopic techniques using third generation synchrotron radiation sources, which highlights that increase in structural order due to annealing gives rise to lower TMR value [8,9]. MgO can be grown on Fe (001) using molecular beam epitaxy (MBE), pulsed laser deposition (PLD), RF-sputtering etc. However, at the time of growth it is predicted and investigated that there is a possibility of growth of thin FeO layer at the interfaces, which affects the spin dependent tunneling current and changes the TMR value significantly. In this direction, the present study focuses to a theoretical discussion on the height of FeO and MgO layer over spin dependent tunneling properties and the effects of different layer heights on tunneling current.

2. Theoretical Formulation and Results

In this paper, investigation has been carried out with oxide multilayer effect of FeO-MgO with different thickness in which the overall barrier height is taken to be constant.

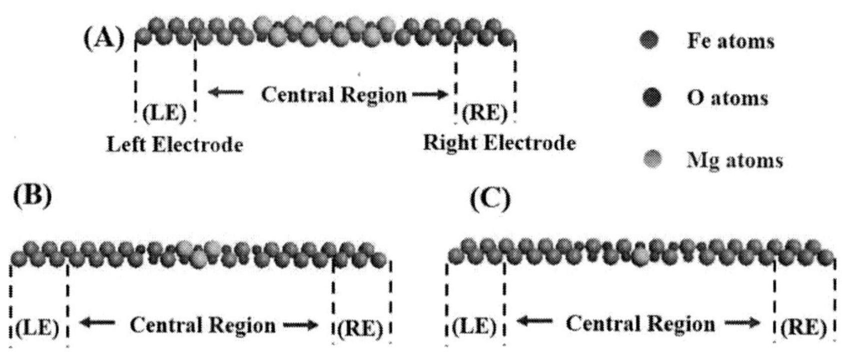

Figure 1: Schematic diagram of the atomistic device where (a) consists of 9 atomic layers of MgO, (b) consists of 3 atomic layers of FeO/ 3 atomic layers of MgO/ 3 atomic layers of MgO and (c) consists of 4 atomic layers of FeO/ 1 atomic layers of MgO/ 4 atomic layers of MgO.

The device configurations used for the simulation are shown in Figure 1. We employed three structural models of FeO-MgO-FeO multi-layer with three different width of FeO and MgO layer as described in the later. The device configuration is simple where oxide layers are put in between two Fe electrodes in order to measure the spin dependent tunneling current through the oxide multilayer. The oxide region taken here is 9 atomic layers thick and both the electrodes are 3 atomic layers thick. The lattice constant of Iron used for the simulation is 2.866 Å. Total 42 numbers of atoms are there in the device. Such small number of atoms is taken because the simulation time increases drastically once the number of atoms in the device is larger. The left and right electrodes are of 3 atomic layers each. The simulations have been performed in an iron/insulator/iron system. Three sets of simulation has been carried out by using different combination of oxides like (a) nine atomic layer MgO, (b) three atomic layer FeO/ three atomic layer MgO/ three atomic layer FeO, (c) four atomic layer FeO/ one atomic layer MgO/ four atomic

layer FeO. In all the simulations, the oxide multilayer height has taken the same as nine atomic layers. Total numbers of atom used in all the simulations are 42 out of that 6 atoms are used for both the electrodes. The calculation is performed to calculate the I-V characteristics after constructing the device. The calculator used here is "ATK-DFT (Device)" with 300 K electron temperature. SGGA (Spin dependent General Gradient Approximation) has been implemented as exchange correlation. The numbers of Monkhorst-Pack grid points are taken as 1, 1, and 100 which is used to resolve the Brillouin Zone along A, B and C directions respectively. The voltage bias between left and right electrode is applied from 0.25 Volt to 1 Volt. The final I-V characteristics for -1 to 1 bias voltage are obtained after symmetrizing each of the plots.
Here we study the spin dependent tunneling current properties through oxide multilayers in a magnetic tunnel junction (MTJ) as described earlier. For that purpose nonequilibrium Green's function approach along-with the density-functional theory have been applied. Three different structural models of FeO-MgO-FeO multi-layer with different width of FeO and MgO layer are employed.

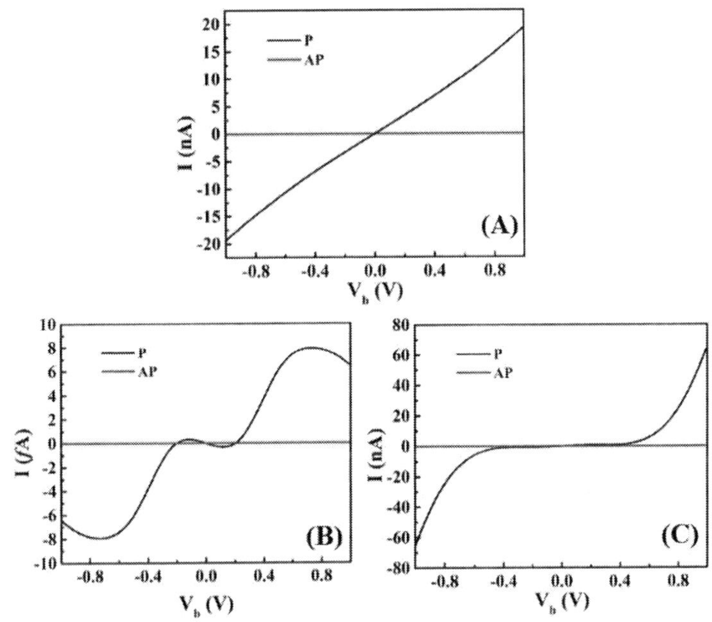

Figure 2: The spin dependent tunneling current in parallel (P) and anti-parallel (AP) situation for (A) 9 atomic layer MgO barrier (B) 3 atomic layers of FeO/ 3 atomic layers of MgO/ 3 atomic layers of MgO (C) 4 atomic layers of FeO/ 1 atomic layers of MgO/ 4 atomic layers of MgO. The current in case of 3 atomic layers of FeO/ 3 atomic layers of MgO/ 3 atomic layers of MgO is in femtoampere (*f*A) but in other two cases the current is in nanoampere (nA).

Figure 2 shows the tunneling current through the oxide multilayer in 3 different combinations of FeO and MgO layer and in each case the current in anti-parallel condition is negligible compared to parallel condition. The electrons in the case of anti-parallel magnetization scatter more that leads to a less current through the device. Moreover the tunneling current changes at different three combinations as shown in Figure 2.

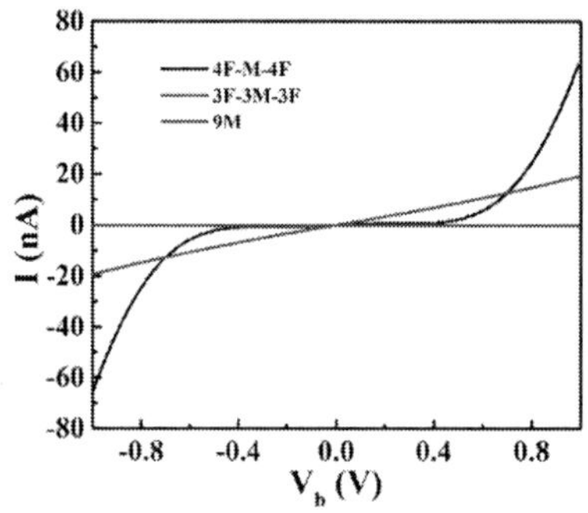

Figure 3: Tunneling current through (a) 9 atomic layer MgO barrier (9M), (b) 3 atomic layers of FeO/ 3 atomic layers of MgO/ 3 atomic layers of MgO barrier (3F-3M-3F), (c) 4 atomic layers of FeO/ 1 atomic layers of MgO/ 4 atomic layers of MgO (4F-M-4F) barrier in parallel magnetization condition.

The current through the MTJ in all 3 mentioned configurations in parallel condition are described in Figure 3 and it is clear that the tunneling current is highest in case of structure (c). However, the tunneling current shows almost linear characteristics in case of structure (a). So the growth of FeO layer may create an obstacle in tunneling and thus the growth of FeO is an important issue in fabricating Fe/MgO/Fe MTJ. Moreover, the structure (c) shows an interesting characteristics where the current start rising after a certain bias voltage. This characteristics is very similar to a diode characteristics where the mentioned critical bias voltage can be correlated with the threshold voltage.

3. Conclusions

In summary, this study of tunneling current through a multi-layer oxide barrier reveals the effect of height of oxide layer in MTJs to improve the tunneling current through it. Moreover electron spin shows a significant effect in electron transport not only in single layer but in the multi-layer structure as well. In each case of the present study suggests that the tunneling current is higher in case of parallel magnetization condition. The highest value of tunneling current is achieved in the third case of the study with 4 atomic layers of FeO/ 1 atomic layers of MgO/ 4 atomic layers of MgO (4F-M-4F) barrier.

Acknowledgments

We thank DeitY grant no. 5(9)/2012-NANO, and FIST-grant no. SR/FST/ETII-028/2010, Government of India, for financial aids. We also thank the support from Centre for Nanotechnology, IIT Guwahati for providing required facilities.

References

[1] Zhu J.-G. and Park C 2006 *Materials Today* **9** 11
[2] Butler W H, Zhang S –G and Schulthess T C 2001 *Physical Review B* **63** 5
[3] Parkin S S P, Kaiser C, Panchula A, Ricel P M, Hughes B, Samant M and Yang S -H 2004 *Nature Materials* **3** 12
[4] Aktsipetrov O A, Kim E M, Kapra R V and Murzina T V 2006 *Physical Review B* **73** 14

[5] Ravan B A, Shokri A A and Yazdani A 2010 *Solid State Communications* **150** 3–4
[6] Velev J P, Belashchenko K D, Jaswal S S and Tsymbal E Y 2007 *Applied Physics Letters* **90** 7
[7] Tusche C, Meyerheim H L, Jedrecy N, Renaud G, Ernst A, Henk J, Bruno P and Kirschner J 2005 *Physical Review Letters* **95** 17
[8] Djayaprawira D D, Tsunekawa K, Nagai M, Maehara H, Yamagata S, Watanabe N, Yuasa S, Suzuki Y and Ando K 2005 *Applied Physics Letters* **86** 9
[9] Han Y, Han J, Choi H J, Shinc H –J and Hong J 2011 *Journal of Materials Chemistry* **21** 38

Enhanced terahertz transmission through a periodic array of tapered rectangular apertures

Koijam Monika Devi *, **Amarendra K. Sarma** and **Gagan Kumar**

Department of Physics, Indian Institute of Technology Guwahati, Guwahati-781039, India.

*koijam@iitg.ernet.in

Abstract: We numerically analyze extraordinary terahertz transmission properties of an array of rectangular shaped apertures perforated periodically on a thin metal film. The apertures are tapered at different angles to achieve higher field concentration at the tapered end. The periodic sub-wavelength scale apertures ensure plasmonic behavior giving rise to the enhanced transmission of a specific frequency mode decided by the periodicity. We compare results of transmission with the rectangular shaped apertures of same parameters and observe a significant increase in the transmission. We have compared results of our numerical simulations with theory and have found them consistent.

1. Introduction

THz radiation covers a very narrow range of frequencies (0.3-3 THz) and lies between the microwave and infrared regions of the electromagnetic spectrum. It has been explored for its applications and significance in various fields of science and technology including sensing, imaging, medicine and communication [1-4] . One of the most promising application of THz radiations is THz imaging as it can penetrate the opaque materials such as clothing, paper, cardboard, plastics etc. Also because of the non-ionizing character, they are considered harmless biologically. These features have led to the terahertz imaging applications as a replacement to the X-ray or infrared imaging. Furthermore, terahertz radiations have the ability to provide physical insight of a material that is not accessible using X-ray or infrared radiation [5]. This increases the significance of these radiations.

THz optics has entered a new phase of research in recent years. A lot of research has been reported to realize the THz devices and improve their performance. In this connection, theoretical investigations on the importance of Spoof plasmons in structured medium for the enhancement of THz transmission has been widely reported [6-8]. Spoof plasmons (SPs) are localized electromagnetic surface wave modes, which can provide an easy way to confine light in sub-wavelength scales structures. Structures supporting SPs include a metal dielectric interface, metal with corrugations or holes, hybrid planar waveguides etc. [9-12] . Using the concept of SPs, interesting phenomenon such as extraordinary transmission has been demonstrated experimentally for an array of apertures fabricated in a metal film in the sub-wavelength regime[13]. Extraordinary transmission is the phenomenon of greatly enhanced transmission for an array of periodic apertures because of the generation of SPs in the periodic structure. Experimental as well as theoretical investigations suggesting the importance of geometry of a structure in the transmission properties has been reported and studied in detail [14-16]. Researchers have examined various shapes and dimensions of the structure including circular, rectangular and square apertures on the terahertz transmission properties. These structures enhance the interaction between the electromagnetic waves and the metal [17, 18]. A 2D array of tapered apertures has been found to efficiently concentrate THz radiation. In 2010, an experimental demonstration by Nguyen et. al reported the transmission properties of a conical tapered aperture fabricated on a metal film and observed an increased concentration of THz radiation at the tapered end. Further enhancement in the concentration was observed by placing periodic apertures in close proximity to the tapered aperture [19, 20].

The ability to focus and concentrate THz radiation would mean a better implementation of it for various applications including near field imaging, biological sensing and development of optoelectronic devices. Inspired with earlier studies, we report here a new geometry of rectangular tapered apertures patterned in 2-dimensions and study extraordinary terahertz transmission. These structures can be easily fabricated especially in doped silicon making it an experimentally feasible and realizable phenomenon. In this paper, we study the variation of terahertz transmission and concentration by varying the tapering of apertures. The apertures are tapered at different angles and transmission properties are examined in details. As the size of the exit plane is decreased, we calculated a corresponding increase in the transmission efficiency and the concentration magnitude.

2. Proposed geometry and simulation

In our numerical analysis, we approximated metal with perfect electrical conductor as it behaves like a perfect conductor at terahertz frequencies. Simulations were performed for an array of rectangular apertures perforated on a thin metal film. We performed simulations for three different sizes of the apertures i.e. length's' and width 'w' = 500 μm, 323 μm and 232 μm. The thickness (t=500 μm) and the periodicity of the apertures (p=1mm) was kept fixed throughout the study. The arrays were tapered at different angle 'θ' and the transmission properties were studied for each case. Figure 1(a) and 1(b) shows the schematic diagram of the 2D array of tapered apertures. Figure 1(c) shows the side view of a single tapered rectangular aperture, where θ is the angle by which holes are tapered and D_1, D_2 represents the size of apertures at the input and exit planes. One can calculate D_2 for given values of D_1 and the tapering angle. For larger values of θ, the size of apertures at the exit plane decreases for fixed width of the substrate.

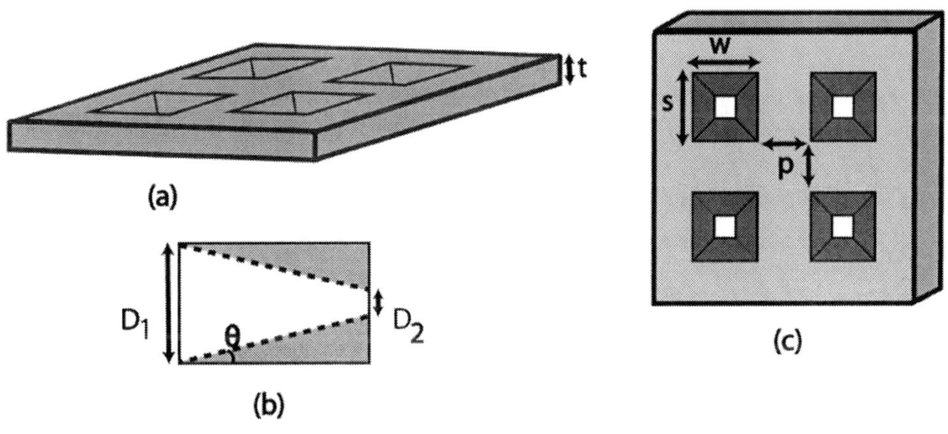

Figure 1: (a) Schematic representation of the tapered apertures arranged periodically in 2-dimension. (b) Front view of the array of tapered apertures. (c) Side view of a single aperture tapered at angle θ w.r.t normal at the input plane.

When the THz radiation is incident on the structure, the periodic nature of the array excites SPs on both interfaces of the structure. Although THz radiation cannot penetrate through metal, the geometry of the structure enables the excited SPs to strongly couple through the apertures. This coupling causes an enhancement in the transmission. This enhancement in transmission can be seen at a particular frequency where the SPs are excited and this frequency is decided by the periodicity of the structure. Moreover, as the THz radiation is incident on the tapered structure, a fraction of the incident radiation gets reflected and a fraction of it gets transmitted. Because of the scattering inside

Journal of Physics: Conference Series **759** (2016) 012052 doi:10.1088/1742-6596/759/1/012052

the apertures, the transmitted light gets slowed down and is accumulated at the tapered end causing an enhancement in the concentration of field at the tapered end. The numerical simulations were performed using the technique of finite element method. The whole geometry was meshed with grids of size λ/10 indicating a sub-wavelength regime.

3. Results and Discussions

We used frequency domain solver to obtain the transmission properties of the tapered apertures arranged periodically. Figure 2(a) and 2(b) shows the transmission amplitude for an array of tapered apertures in comparison with the reference apertures. In figure 2(a), blue traces shows the transmission amplitude for the tapered apertures with tapering angle $\theta = 10^{\circ}$, having D_1=500 μm, D_2= 323 μm, and the red traces represents the transmission amplitude for an array of rectangular non-tapered apertures having length 's' and width 'w' both 323 μm which is considered as reference. In figure 2(b), we obtained transmission amplitude for another tapering angle $\theta = 15^{\circ}$ with D_1 unchanged. In this case if one calculate D_2 for the same thickness, then it turns out to be 232 μm. The blue and red traces represents tapered and non-tapered array of apertures respectively. In both cases, it is evident that the transmission is greater for the tapered aperture as compared to the non-tapered reference aperture. We calculated more than 20 times enhancement in the transmission amplitude for tapering angle 15° compared with the rectangular apertures.

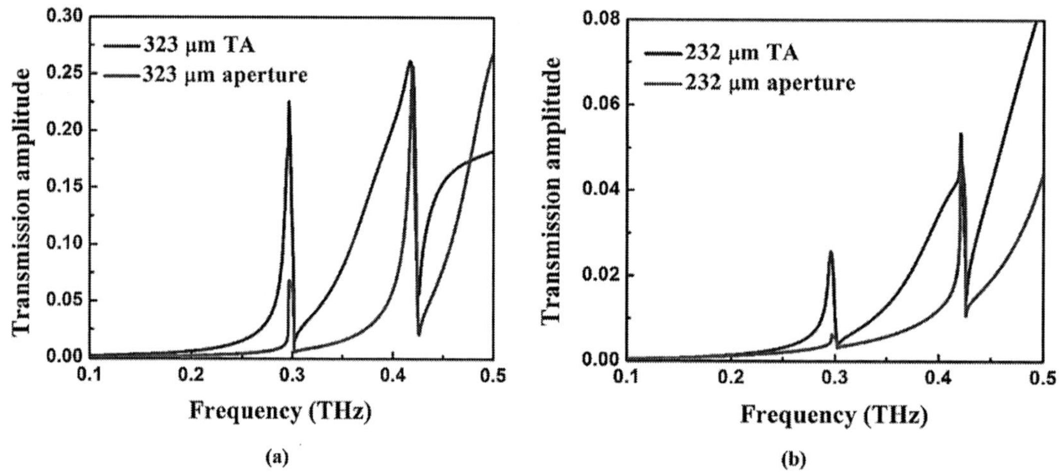

Figure 2: Transmission amplitude for (a) a tapered aperture with parameters: D_1=500 μm, D_2= 323 μm and $\theta = 10^{\circ}$. The reference signal corresponds to the rectangular apertures of size 323 μm. (b) Transmission amplitude for periodic tapered apertures corresponding to parameters: D_1=500μm, D_2= 232 μm and $\theta = 15^{\circ}$. The reference signal corresponds to the rectangular apertures of size 232 μm.

Figure 3 shows the transmission efficiency of the tapered apertures in comparison with the reference apertures. The transmission efficiency for an aperture is calculated using the formula $t_x^E(v) = t_x^E(v)/t_{500\mu m}^E(v)$, where x is the dimension for the tapered aperture or the reference aperture. The quantity x ranges from 232 μm to 323 μm for different tapering angles. Here, the quantity $t_x^E(v)$ is the relative spectral transmission efficiency of any aperture relative to the input. The value $t_{500\mu m}^E(v)$ gives the transmission amplitude for an array of rectangular apertures having length and width= 500 μm. Figure 3(a) shows comparison for tapering angle $\theta = 10^{\circ}$ having D_1=500 μm, D_2= 323 μm, and the reference aperture. In the figure, blue traces signifies the transmission efficiency for a tapered aperture, while the red one represents the same for reference apertures. Similarly, figure 3(b) shows the transmission efficiency comparison for tapered apertures with tapering angle $\theta = 15^{\circ}$ having D_1=500 μm, D_2= 232 μm and the reference aperture. Again the blue and red traces represents transmission efficiency for tapered and non-tapered apertures respectively. One may note that there is a significant enhancement in the efficiency as we increase the tapering angle. In our analysis, we calculated about 25 times enhancement in the transmission efficiency for tapering angle of 15°compared with the rectangular apertures.

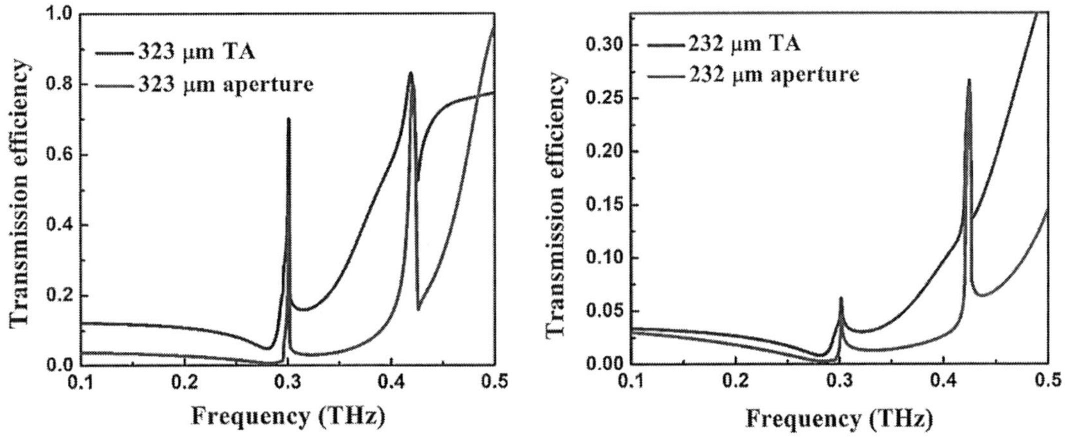

(a) (b)

Figure 3: Transmission efficiency for (a) a tapered aperture (TA) with parameters: D_1=500 μm, D_2= 323 μm and $\theta = 10^{\circ}$ and the reference aperture, (b) Transmission efficiency for periodic tapered apertures corresponding to parameters: D_1=500μm, D_2= 232 μm and $\theta = 15^{\circ}$and the reference aperture.

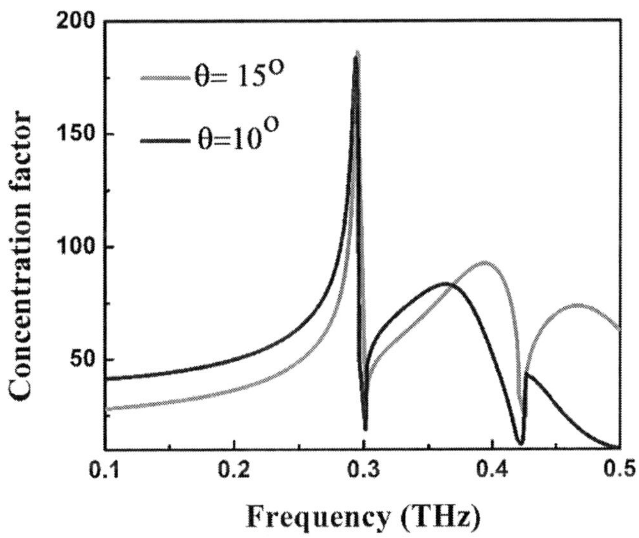

Figure 4: The amplitude concentration factor for tapered apertures for different tapering angle, θ.

Next, we calculate the concentration factor of the terahertz field at the tapered end. The amplitude concentration factor is given by $f_{\mathrm{E}}(v) = \mathrm{t}^{\mathrm{E}}_{\mathrm{X,TA}}(v)/\mathrm{t}^{\mathrm{E}}_{x}(v)$, where x is the dimension of the aperture. The results are shown in figure 4 for different tapering angles. The blue traces represents concentration factor for the tapering angle of 10° with corresponding dimensions as D_1=500 μm and D_2= 323 μm and the green traces represents the amplitude concentration factor for tapering angle of 15° with D_1=500 μm, D_2= 232 μm. It can be noticed that the amplitude concentration factor is greatly enhanced when the tapering angle is increased. In our analysis, we calculated an enhancement of eight times for the tapering angle 15° compared with the 10°. It is evident from the above results that one can control the transmission efficiency and the concentration factor by changing the angle through which the apertures are tapered. The analysis could be significant for the application where strong terahertz field concentration is required such as near field imaging, sensing etc. It will further help the researcher to design the experiments in this direction.

4. Conclusion

In conclusion, we numerically examined the terahertz transmission properties of the periodically arranged apertures tapered at different angles and compare results with the rectangular apertures of same parameters. We observed a significant increase in the transmission amplitude as well as transmission efficiency for the tapered apertures. A subsequent increase in the transmission efficiency as well as concentration of field was noticed as we decreased the size of the aperture at the exit plane. For an apertures tapered at 15°, we calculated a 20 fold increase in transmission and 25 time enhancement in the efficiency compared to the rectangular apertures of same size at the input plane. The terahertz transmission properties can be controlled with the tapering angle as well as the other parameters of the apertures. Our study could be significant for near field imaging technology and in the development of devices operating in the THz domain.

Acknowledgement

The authors would like to thank the Board of Research in Nuclear Sciences (BRNS), Project no. PHY1P1GK102 for their financial assistance.

5. References

1. Hu, B. and M. Nuss, *Imaging with terahertz waves.* Optics letters, 1995. **20**(16): p. 1716-1718.
2. Federici, J.F., et al., *THz imaging and sensing for security applications—explosives, weapons and drugs.* Semiconductor Science and Technology, 2005. **20**(7): p. S266.
3. Federici, J. and L. Moeller, *Review of terahertz and subterahertz wireless communications.* Journal of Applied Physics, 2010. **107**(11): p. 111101.
4. Krügener, K., et al., *Terahertz meets sculptural and architectural art: Evaluation and conservation of stone objects with T-ray technology.* Scientific reports, 2015. **5**.
5. Jepsen, P.U., D.G. Cooke, and M. Koch, *Terahertz spectroscopy and imaging–Modern techniques and applications.* Laser & Photonics Reviews, 2011. **5**(1): p. 124-166.
6. Garcia-Vidal, F., L. Martin-Moreno, and J. Pendry, *Surfaces with holes in them: new plasmonic metamaterials.* Journal of optics A: Pure and applied optics, 2005. **7**(2): p. S97.
7. Yu, Z., et al., *Terahertz spoof plasmonic coaxial microcavity.* Applied optics, 2014. **53**(6): p. 1118-1123.
8. Yin, S., et al., *Spoof surface plasmon polaritons in terahertz transmission through subwavelength hole arrays analyzed by coupled oscillator model.* Scientific reports, 2015. **5**.
9. Offerhaus, H., et al., *Creating focused plasmons by noncollinear phasematching on functional gratings.* Nano letters, 2005. **5**(11): p. 2144-2148.
10. Pendry, J., L. Martin-Moreno, and F. Garcia-Vidal, *Mimicking surface plasmons with structured surfaces.* Science, 2004. **305**(5685): p. 847-848.
11. Maier, S.A., *Plasmonics: fundamentals and applications.* 2007: Springer Science & Business Media.
12. Kumar, G., et al., *Planar plasmonic terahertz waveguides based on periodically corrugated metal films.* New Journal of Physics, 2011. **13**(3): p. 033024.
13. Ebbesen, T.W., et al., *Extraordinary optical transmission through sub-wavelength hole arrays.* Nature, 1998. **391**(6668): p. 667-669.
14. Koerkamp, K.K., et al., *Strong influence of hole shape on extraordinary transmission through periodic arrays of subwavelength holes.* Physical Review Letters, 2004. **92**(18): p. 183901.
15. Ghaemi, H., et al., *Surface plasmons enhance optical transmission through subwavelength holes.* Physical review B, 1998. **58**(11): p. 6779.

16. Zhang, X., et al., *Effects of Compound Rectangular Subwavelength Hole Arrays on Enhancing Optical Transmission.* Photonics Journal, IEEE, 2015. **7**(1): p. 1-8.
17. Ruan, Z. and M. Qiu, *Enhanced transmission through periodic arrays of subwavelength holes: the role of localized waveguide resonances.* Physical review letters, 2006. **96**(23): p. 233901.
18. Lee, J., et al., *Terahertz electromagnetic wave transmission through random arrays of single rectangular holes and slits in thin metallic sheets.* Physical review letters, 2007. **99**(13): p. 137401.
19. Diwekar, M., S. Blair, and M. Davis, *Increased light gathering capacity of sub-wavelength conical metallic apertures.* Journal of Nanophotonics, 2010. **4**(1): p. 043504-043504-11.
20. Nguyen, T.D., Z.V. Vardeny, and A. Nahata, *Concentration of terahertz radiation through a conically tapered aperture.* Optics express, 2010. **18**(24): p. 25441-25448.

Terahertz surface plasmon polaritons propagation in slanted pillar geometries

Maidul Islam and Gagan Kumar*

Department of Physics, Indian Institute of Technology Guwahati, Guwahati-781039, Assam, India

*gk@iitg.ernet.in

Abstract: In this paper, we present detailed numerical study of periodically corrugated surfaces to establish the directional dependence of terahertz (THz) surface plasmon polaritons propagation. A metal surface approximated as perfectly electrical conductor at terahertz frequencies, is periodically patterned with slanted sub-wavelength scale pillars forming a plasmonic waveguide. The slanted pillar geometries are found to support highly confined terahertz modes at specific frequencies which can be tailored with the slanted pillar parameters and their angle of tilt with respect to the normal. The terahertz mode propagation with different angles through which pillars are tilted w.r.t. the normal is examined in detail. From these results we observed that the slanted geometries are more efficient in guiding terahertz in a particular direction.

Introduction:

In recent years, there has been significant interest in developing devices and networks for high speed communication. THz radiations covering a frequency range of 0.3 to 3THz has been examined in last more than a decade for various applications including high speed communication **[1-4]**. THz waves can promise data transfer at much higher speed because of their ability to transfer data signals at tera bits per second. However to realize devices operating at terahertz one need to develop efficient waveguides exhibiting low loss and high directional confinement to transfer signals **[5-7]**. In this direction, a lot of research has been reported in last more than a decade both experimentally and theoretically **[8-12]**. Williams *et al.* first time experimentally reported the highly confined surface plasmon polaritons propagation by structuring a dielectric and depositing a thick copper layer on it **[13]**. Around the same time, Zhu *et al.* designed and fabricated one dimensional array of apertures in a thin stainless steel foil and thereby forming a planar plasmonic waveguide and experimentally demonstrated straight and passive guided wave devices **[14]**. In an attempt to reduce propagation loss Kumar *et al.* experimentally examined the ability to guide THz radiation using 1D arrays of periodically spaced rectangular blind holes in metal films **[15]**. The blind hole pattern led to the propagation of THz surface plasmon only on one side of the waveguide thereby reducing loss. The waveguiding approaches in guiding terahertz on a planar surface as discussed above have been inspired by the seminal work by sir Pendry and co-workers in 2004 **[16]**.

In last decade, there have been several theoretical studies reported as well on THz waveguides. In this respect, Wood *et al.* investigated dispersion properties of surface plasmon polaritons in slanted dimpled surface patterns and examined their field confinement **[17]**. Rusina *et al.* has developed fundamental understanding in designing the optimum and efficient metal /dielectric nanowaveguides suitable for the THz nanofocusing **[18]**. They reported nanoconcentration of THz radiation by considering a parallel plate waveguide that consists of a dielectric slab sand-witched between two thick metal plates. Making further advancement in the field of THz plasmonic based waveguides, Kumar *et al.* showed propagation of THz beam on a heavily doped silicon surface patterned with pillars **[6]**. They observed that patterned silicon surface acts as an effective medium for highly confined terahertz surface mode propagation. Although these developments on terahertz waveguides have been able to overcome several hurdles in the field of THz science and but active and passive components in this narrow range of frequency are still lacking. One of the challenge has been to selectively guide a terahertz modes in a particular direction than the other in an efficient way.

The guiding of terahertz in a particular direction under the low divergence can lead to variety of applications such as long distance sensing, identification **[19]** and high speed transmission of information. In THz planar plasmonic waveguides, we discover than one can bend or tilt the corrugated structure to transfer terahertz signal more efficiently in a particular direction than the other. We have addressed this ability by designing slanted pattern on a thin metal sheet approximated as perfectly electrical conductor at terahertz frequencies.

In this paper, we design terahertz planar plasmonic waveguides by periodically corrugating a thin metal surface with slanted pillars. The sub-wavelength scale pillars tilted as different angles ensure the plasmonic response in our numerical analysis. We performed both time domain and frequency domain numerical simulations to precisely analyse the mode propagation in plasmonic waveguides comprising pillars at different angles. While placing the sources at the centre of the waveguide and detectors at both the ends, we examined terahertz waveguide transmission with different tilting angles of the pillars. Finally we summarize the results in the conclusion section.

Waveguide geometry and Dispersion properties:

We designed and numerically examined several waveguide geometries with varying angle of slanted pillars. The schematic of the designed planar THz waveguide is shown in Figure 1. Figure **1(a)** shows a 3 dimensional view of the waveguide, whereas **(b)** depicts a side view of the waveguide. Throughout our study, length (l), width (w), height (h), and periodicity (p) of the tilted pillars are kept constant. These parameters are: *length (l) = 100μm, width (w) = 500μm, height (h) = 100μm, periodicity (p) = 400μm*. In order to examine the effect of angles through which the pillars are tilted w.r.t. normal on the terahertz surface mode propagation, we considered pillars with different tilt angles of 5^0, 10^0, 20^0, 40^0 and 60^0.

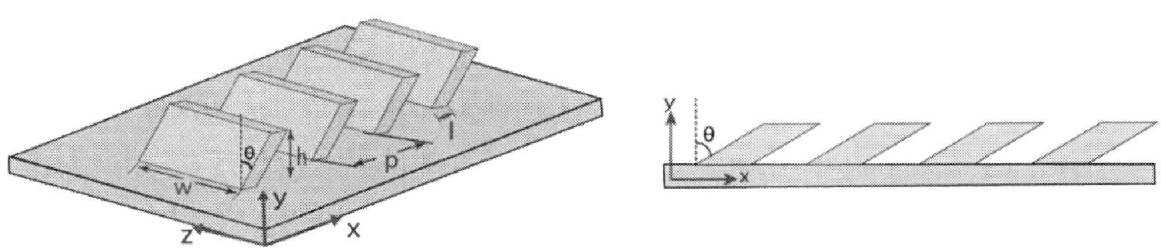

(a) **(b)**

Figure 1: Schematic of the planar THz waveguide. (a) 3-Dimensional view of the waveguide comprising one dimensional array of pillars tilted with respect to the surface. The other constant parameters labelled as 'l', 'w', 'h' and 'p' are length, width, height, and periodicity respectively. (b) Depicts the side view of pillar patterned waveguide with pillars making angle θ with the normal.

In order to examine THz mode propagation in slanted pillar geometries, we examined dispersion relations of the pillar waveguides using a Finite Element Eigen mode solver, with the periodic boundary condition along propagation direction and absorbing layers along the transverse direction. The dispersion relations are calculated for two waveguides with pillars tilted at 10^0 and 60^0. The dispersion relation was calculated by numerically modelling one unit cell of the periodic structure. Perfect electrical conducting (PEC) boundary condition was assumed for the pillar waveguide owing to the high conductivity of metal at terahertz frequencies. Figure 2(a) shows dispersion relations of slanted waveguide pattern. The blue and red curves correspond to surface electromagnetic modes supported by the pillars tilted at angle 10^0 and 60^0 respectively Here, initially we get monotonic increment of frequency and corresponding phase i.e. wavenumber. Up to a certain phase, there is a constant group velocity since the dispersion relation exhibit linear relationship between the frequency and wavenumber and after that we get a saturation in the dispersion i.e. zero group velocity at frequency 0.31THz for tilt angle 10^0 and at frequency 0.22THz for tilt angle 60^0. But just before the saturation, there will be a constant phase velocity. In these slanted waveguide, we observed that dispersion relation shifting away from the light line when these pillars are more tilted with respect to surface.

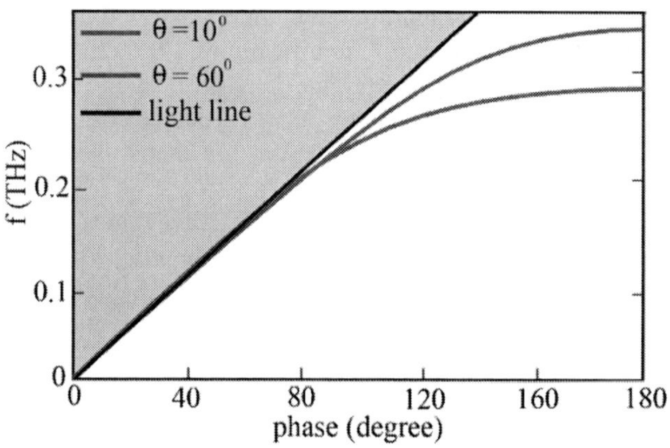

Figure 2: (a) Numerically calculated dispersion properties for the pillar patterned waveguides with pillar tilted at angles 10^0 and 60^0 with respect to normal.

Result and discussion:

We use modified technique of terahertz time domain spectroscopy to examine the designed slanted pillar patterns. We numerically obtain the time domain signal of the terahertz waveform at the other end of the waveguide using a waveguide port and convert it to frequency domain signal using Fast Fourier Transform. The waveguide transmission spectra are obtained using finite element time domain method. First, we examine the 3 cm long planar waveguide with pillars tilted at 5^0. The results in the form of frequency domain spectra are shown in Figure 3. The inset shows the corresponding time domain signal. One may notice a resonant behaviour in the spectra with anti-resonant frequency at 0.31 THz in the present case. In our numerical analysis, we observed a very little change in the anti-resonant frequency with a change in the angle of pillars because of the slight change in the height of the pillars when they are titled. However, one may observe a significant change in the resonant frequency when height of the pillars is changed.

Next, we examine the terahertz mode propagation of the terahertz modes in the forward and backward directions of the pillar patterned waveguides and compare the results. We place the terahertz source in the middle of the 4 cm long waveguide and examine terahertz modes at 2 cm distance from the source in both the directions. The pillars of the waveguide are tilted in one direction. The frequency domain spectra indicates a difference in output waveguide amplitude calculated at both the ends. The results are shown in Figure 3(b). The blue and red traces represent signal in the forward and backward directions respectively for the pillars tilted at $\mathbf{60^0}$. We observe a higher waveguide transmission in the forward direction i.e. the direction in which pillars are tilted. We get an output amplitude of ~ 95 % in the forward direction compared with the 54% signal amplitude in the backward direction for the wave propagation of same distance in both directions. Therefore the proposed waveguides, allow us to propagation the signal more efficiently in one direction by changing the angle through which pillars are tilted. One may also control the THz waveguide transmission depending of the angle.

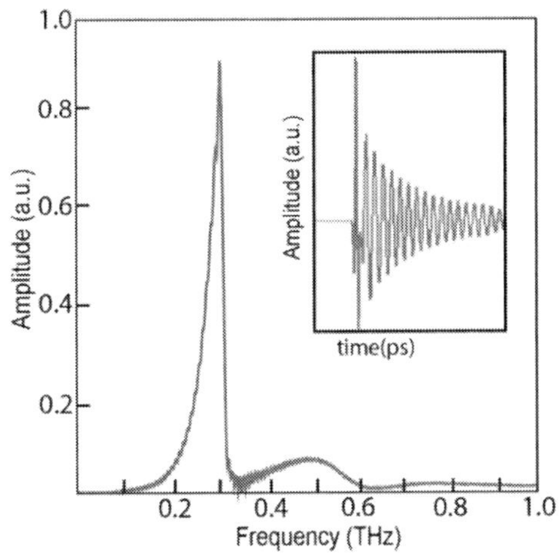

(a)

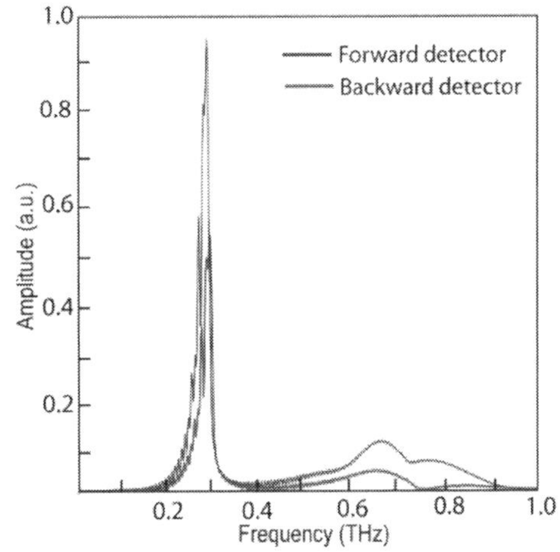

(b)

Figure 3: (a) Numerically simulated waveguide transmission spectrum on a PEC substrate patterned with slanted pillars at an angle 5^0 w.r.t the surface. The other parameters are: length 'l' = 100μm, width 'w' = 500μm, height 'h' = 100μm and periodicity 'p' = 400μm. The inset shows the corresponding time-domain spectrum. (b) Frequency domain spectra of the pillar patterned waveguides with pillar angle 60^0 w.r.t. normal in forward and backward directions. The detectors are placed at a distance of 2 cm from the centre from both the ends.

In Figure 4, we examine the field profile of the terahertz mode supported by a particular waveguide with slanted pillars for the parameters: *length (l) = 100μm, width (w) = 500μm, height (h) = 100μm, periodicity (p) = 400μm* and tilt angle of 60^0 with respect to normal. The field profiles are examined at the ON-resonance (0.31 THz) and OFF-resonant frequencies (0.5THz). One may note that at resonant frequency, the field is highly confined to the pillars geometries, however it is leaking away from the surface at other frequencies not at resonance. It is to be mentioned that in our numerical study, we have assumed tetrahedral shaped grids of size ~ λ/10 indicating sub wavelength regime.

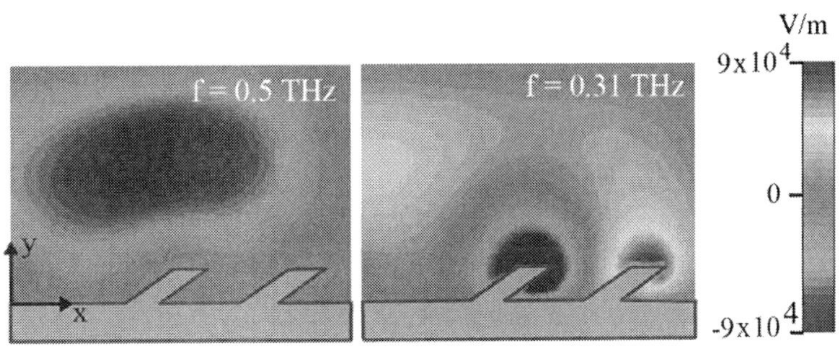

Figure 4: (a) Field profiles in the x-y plane of the waveguide at ON-resonance (0.31THz) and OFF-resonance (0.5THz) frequency.

We next consider one essential property of the periodic slanted pillars waveguide pattern, which is confinement and that is relevant for waveguide application. Firstly, we analyse confinement along vertical direction. We have numerically measured the vertical confinement by finite-difference time-domain simulation for the slanted pillars at angle 40^0 with respect to normal, by translating the detectors away from the surface. Figure 5(a) plots the measured peak spectral amplitude as a function of distance above the surface, along with an exponential fit which brings out a 1\e length of 0.25mm approximately. After that, we considered the confinement of the terahertz modes in both the transverse directions by placing detectors at different position in order to have a comprehensive understanding of the THz mode propagation. The horizontal confinement of the resonant mode was calculated also by finite-difference time-domain simulation. Figure 5(b) plots the calculated field amplitude as a function of the transverse coordinate z also for slanted pillars pattern at angle 40^0 with respect to normal, which shows a full-width at half-maximum (FWHM) of approximately 500μm FROM THE Gaussian fit of the results. The information on the waveguide transmission properties could be useful in fabrication and further demonstration of such waveguide experimentally. It may be noted that although our study on slanted geometries is based on numerical simulations, but one can fabricate such waveguides on semiconductors. One may explore deep reactive ion etching for fabrication of slanted structures with suitable arrangement of the substrate in tilted configuration.

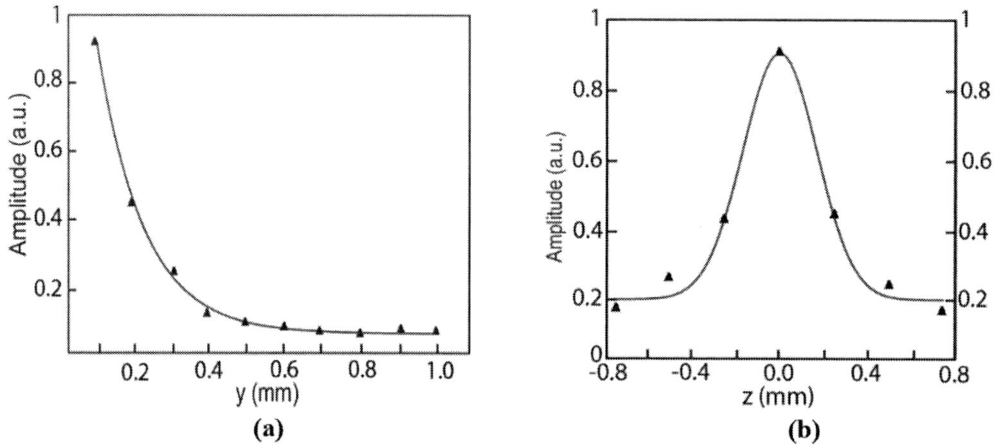

Figure 5: (a) THz waveguide transmission of the THz signal probed at different positions above the surface. The data is exponentially fitted to obtain 1\e propagation length. (b) A plot of THz field measurements in lateral direction to estimate the confinement of propagating wave in the lateral direction.

Conclusion:
Briefly, we have examined in details the planar plasmonic terahertz waveguides comprising sub-wavelength scale pillars tilted at different angles. We have numerically calculated the dispersion relations of the fundamental Eigen modes supported by slanted pillar waveguides. It is observed that a change in the angle of tilt results in a change of dispersion properties. We confirmed the existence of various surface modes in dispersion relations independently through the time domain and frequency domain solver. A change in the angle of tilt changes the cut-off frequency of the terahertz supported by the pillar waveguide. Further, we analyse the terahertz transmission amplitude at two opposite ends of the waveguide by placing source at the centre with pillars tilted in a particular directions. We found that transmission is more than 40% in the direction of tilted pillars. This clearly indicated that one can guide terahertz in a particular direction more efficiently with suitable designing parameters. We have also calculated the mode confinement above the surface (1\e propagation length ~ 0.25mm) and on the surface along transverse directions (FWHM ~ 0.5mm). The field profiles of the modes indicates a strong confinement at the resonant frequencies. The terahertz waveguides discussed here could play a crucial role in designing active devices and networks at terahertz frequencies.

Acknowledgements:

The authors would like to acknowledge the funding from the project (Project No. PHY/P/GK/01) for the financial assistance.

References:

1. Grady, N.K., et al., *Terahertz metamaterials for linear polarization conversion and anomalous refraction.* Science, 2013. **340**(6138): p. 1304-1307.
2. Otsuji, T., et al., *Graphene-based devices in terahertz science and technology.* Journal of Physics D: Applied Physics, 2012. **45**(30): p. 303001.
3. Knap, W., et al., *Nanometer size field effect transistors for terahertz detectors.* Nanotechnology, 2013. **24**(21): p. 214002.
4. Docherty, C.J. and M.B. Johnston, *Terahertz properties of graphene.* Journal of Infrared, Millimeter, and Terahertz Waves, 2012. **33**(8): p. 797-815.
5. Li, S., et al., *Terahertz surface plasmon polaritons on a semiconductor surface structured with periodic V-grooves.* Optics express, 2013. **21**(6): p. 7041-7049.
6. Kumar, G., et al., *Terahertz surface plasmon waveguide based on a one-dimensional array of silicon pillars.* New Journal of Physics, 2013. **15**(8): p. 085031.
7. Nguyen, T., et al., *Terahertz plasmonic properties of highly oriented pyrolytic graphite.* Applied Physics Letters, 2013. **102**(17): p. 171107.
8. Kumar, G., et al., *Planar terahertz waveguides based on complementary split ring resonators.* Optics express, 2011. **19**(2): p. 1072-1080.
9. Lin, C., et al. *A 2× 40 Gbps wireless communication system using 0.14 THz band oritho-mode transducer.* in *Infrared, Millimeter, and Terahertz waves (IRMMW-THz), 2015 40th International Conference on*. 2015. IEEE.
10. Nagel, M., A. Marchewka, and H. Kurz, *Low-index discontinuity terahertz waveguides.* Optics express, 2006. **14**(21): p. 9944-9954.
11. Quan, X., S. Zhang, and H. Li. *High Directivity Horn Antenna of Metamaterial in Terahertz*. in *2015 International Power, Electronics and Materials Engineering Conference*. 2015. Atlantis Press.
12. Song, Z., et al., *Teeter-totter effect of terahertz dual modes in C-shaped complementary split-ring resonators.* Journal of Applied Physics, 2015. **118**(4): p. 043108.
13. Williams, C.R., et al., *Highly confined guiding of terahertz surface plasmon polaritons on structured metal surfaces.* Nature Photonics, 2008. **2**(3): p. 175-179.
14. Zhu, W., et al., *Engineering the propagation properties of planar plasmonic terahertz waveguides.* Selected Topics in Quantum Electronics, IEEE Journal of, 2011. **17**(1): p. 146-153.
15. Kumar, G., et al., *Planar plasmonic terahertz waveguides based on periodically corrugated metal films.* New Journal of Physics, 2011. **13**(3): p. 033024.
16. Pendry, J., L. Martin-Moreno, and F. Garcia-Vidal, *Mimicking surface plasmons with structured surfaces.* Science, 2004. **305**(5685): p. 847-848.
17. Wood, J., et al., *Spoof plasmon polaritons in slanted geometries.* Physical Review B, 2012. **85**(7): p. 075441.
18. Rusina, A., et al., *Nanoconcentration of terahertz radiation in plasmonic waveguides.* Optics express, 2008. **16**(23): p. 18576-18589.
19. Liu, H.-B., et al., *Detection and identification of explosive RDX by THz diffuse reflection spectroscopy.* Optics Express, 2006. **14**(1): p. 415-423.

Few-body model approach to the bound states of helium-like exotic three-body systems

Md A Khan

Department of Physics, Aliah University, IIA/27, Newtown, Kolkata-700156, India

E-mail: drakhan.rsm.phys@gmail.com, drakhan.phys@aliah.ac.in

Abstract. In this paper, calculated energies of the lowest bound S-state of Coulomb three-body systems containing an electron (e^-), a negatively charged muon (μ^-) and a nucleus (N^{Z+}) of charge number Z are reported. The 3-body relative wave function in the resulting Schrödinger equation is expanded in the complete set of hyperspherical harmonics (HH). Use of the orthonormality of HH leads to an infinite set of coupled differential equations (CDE) which are solved numerically to get the energy E.

PACS numbers: 02.70.-c, 31.15.-Ar,31.15.Ja, 36.10.Ee.

Keywords: Hyperspherical Harmonics (HH), Raynal-Revai Coefficient (RRC), Renormalized Numerov Method (RNM), Exotic Ions.

1. Introduction

Atoms and ions containing exotic particles like muon, kaon, taon, baryon and their antimatters are of immense have become an interesting research topic in many branches of physics including atomic, nuclear and elementary particle physics, plasma and astrophysics, experimental physics [1, 2]. Apart from being the simplest exotic atom formed by replacing an orbital electron of neutral helium, the muonic helium atom ($^4\mathrm{He}^{2+}\mu^-e^-$) is an important topic of investigation since its first formation and detection by Souder et al [3]. Muon being about 207 times heavier than an electron, the size of the muonic helium atom is smaller by a factor of about $1/400$ of the ordinary electronic helium atom. Muonic helium atoms are the unusual pure atomic three body systems without any restriction due to Pauli exclusion principle for electron and muon being non-identical fermions. These are formed as by-products of the process of muon catalyzed fusion, hence are useful to understand the fusion reactions properly [4, 5]. The electromagnetic interaction between the electron and negatively charged muon can be better understood by this simplest muonic system by precise measurements of hyperfine structure [6, 7].

As exotic particles are mostly unstable, their parent atoms (or ions) are also very short lived. These exotic short-lived atoms or ions can be formed by trapping the accelerated exotic particles inside matter and replacing one or more electron(s) in an ordinary atom by exotic particle(s). The absorbed exotic particle revolves round the nucleus of the target atom in orbit of radius equal to that of the electron before its ejection from the atom, which subsequently cascades down the ladder of resulting exotic atomic states by the emission of X-rays and Auger transitions before being lost on its way to the nucleus. If the absorbed exotic particle is a negatively charged muon, it passes through various intermediate atmospheres before being trapped in the vicinity of the atomic nucleus [8]. In the course of its journey inside the matter, it scatters from atom to atom as free electron and gradually loses its energy until it is captured into an atomic orbit. In the lowest energy level (1S), it experiences only Coulomb interaction with nuclear protons while it experiences weak interaction with the rest of the nucleons. As discussed above, exotic muonic atoms (or ions) are produced by replacing one or more electron(s) of neutral atoms by one or more exotic particle(s) like muon, pion, kaon, anti-proton having an electric charge equal to that of the electron [9]. The most studied exotic few-body Coulomb system are the muonic atoms (or muonic ions) which are formed by removing one or more orbital electron(s) by one or more negatively charged muon(s). However the present communication we shall consider only those atoms or ions in which the positively charged nucleus is being orbited by one electron and one negatively charged muon.

These atoms have been under theoretical scanner of several authors [10, 11, 12, 13, 14, 15, 16, 17, 18, 19, 20, 21, 22, 23, 24, 25, 26] for being investigated. In this communication, we adopt hyperspherical harmonics expansion (HHE) approach for a systematic study of the ground state of atoms/ ions containing an orbital electron plus a negatively charged muon revolving round the positively charged nucleus forming a three-body system. In our model, we assume that the electromagnetic interaction of the valence particles with the nucleus is sufficiently weak to influence the internal structure of the nucleus. Again, the fact that the muon is much lighter than the nucleus allows us to regard the nucleus to remain a static source of electrostatic interaction. However, a hydrogen-like two-body model, consisting a quasi-nucleus ($\mu^- N^{(Z-2)+}$, formed by the muon and the nucleus) plus an orbital electron can also be tested due to much smaller size of muonic orbital than that for an electronic orbital.

In HHE formalism for a general three-body system of three unequal mass particles the choice of Jacobi coordinates correspond to three different partitions and in the ith partition, particle labeled by 'i' remains as a spectator while the remaining two particles labeled 'j' and 'k' form the interaction pair. So the total potential contains three binary interaction terms (i.e. $V = V_{jk}(r_{jk}) + V_{ki}(r_{ki}) + V_{ij}(r_{ij})$) and for computation of matrix element of V(r_{ij}), potential of the (ij) pair, the chosen HH is expanded in the set of HH corresponding to the partition in which $\vec{r_{ij}}$ is proportional to the first Jacobi vector [27] by the use of Raynal-Revai coefficients (RRC) [28]. The energies obtained for the lowest bound S-state is compared with the ones of the literature.

In Section II, we briefly introduce the hyperspherical coordinates and the scheme of the transformation of HH belonging to two different partitions. Results of calculation and discussions will be presented in Section III and finally we shall draw our conclusion in section IV.

2. HHE Method

The choice of Jacobi coordinates for systems of three particles of mass m_i, m_j, m_k is shown in Fig.1.

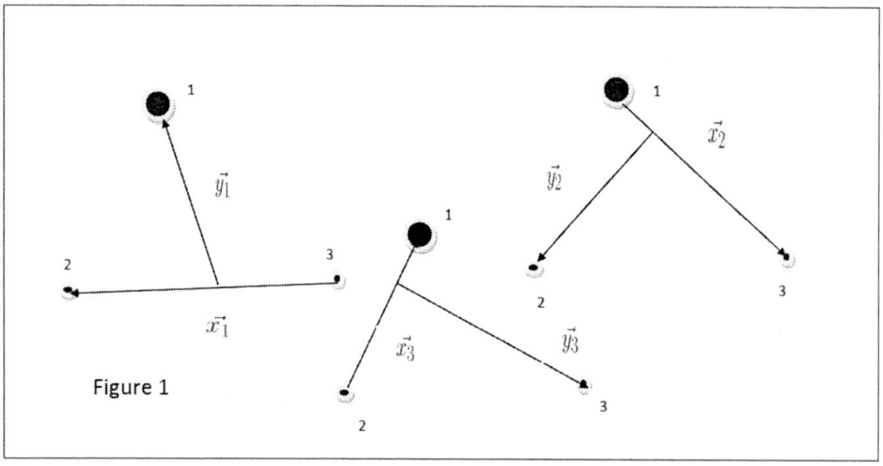

Figure 1. Jacobi coordinates in different partitions of a three-body system.

The Jacobi coordinates [29] in the ith partition can be defined as:

$$\left.\begin{array}{rcl} \vec{x_i} & = & \left[\frac{m_j m_k M}{m_i(m_j+m_k)^2}\right]^{\frac{1}{4}} (\vec{r_j} - \vec{r_k}) \\ \vec{y_i} & = & \left[\frac{m_i(m_j+m_k)^2}{m_j m_k M}\right]^{\frac{1}{4}} \left(\vec{r_i} - \frac{m_j\vec{r_j}+m_k\vec{r_k}}{m_j+m_k}\right) \\ \vec{R} & = & (m_i\vec{r_i} + m_j\vec{r_j} + m_k\vec{r_k})/M \end{array}\right\} \qquad (1)$$

where $M = m_i + m_j + m_k$ and the sign of $\vec{x_i}$ is determined by the condition that (i, j, k) should form a cyclic permutation of (1, 2, 3).
The Jacobi coordinates are connected to the hyperspherical coordinates [30] as

$$\left.\begin{array}{rclcrcl} x_i & = & \rho\cos\phi_i & ; & y_i & = & \rho\sin\phi_i \\ \rho & = & \sqrt{x_i^2 + y_i^2} & ; & \phi_i & = & tan^{-1}(\eta_i/\Psi_i) \end{array}\right\} \qquad (2)$$

The relative three-body Schrődinger's equation in hyperspherical coordinates can be written as

$$\left[-\frac{\hbar^2}{2\mu}\left\{\frac{\partial^2}{\partial\rho^2} + \frac{5}{\rho}\frac{\partial}{\partial\rho} + \frac{\hat{K}^2(\Omega_i)}{\rho^2}\right\} + V(\rho,\Omega_i) - E\right]\Psi(\rho,\Omega_i) = 0 \qquad (3)$$

where $\Omega_i \rightarrow \{\phi_i, \theta_{x_i}, \phi_{x_i}, \theta_{y_i}, \phi_{y_i}\}$, effective mass $\mu = \left[\frac{m_i m_j m_k}{M}\right]^{\frac{1}{2}}$, potential $V(\rho,\Omega_i)$ $= V_{jk} + V_{ki} + V_{ij}$. The square of hyper angular momentum operator $\hat{K}^2(\Omega_i)$ satisfies the eigenvalue equation [30]

$$\hat{K}^2(\Omega_i)\mathcal{Y}_{K\alpha_i}(\Omega_i) = K(K+4)\mathcal{Y}_{K\alpha_i}(\Omega_i) \qquad (4)$$

where the eigen function $\mathcal{Y}_{K\alpha_i}(\Omega_i)$ is the hyperspherical harmonics (HH). An explicit expression for the HH with specified grand orbital angular momentum $L(=|\vec{l_{x_i}} + \vec{l_{y_i}}|)$ and its projection M is given by

$$\begin{array}{rcl} \mathcal{Y}_{K\alpha_i}(\Omega_i) & \equiv & \mathcal{Y}_{Kl_{x_i}l_{y_i}LM}(\phi_i, \theta_{x_i}, \phi_{x_i}, \theta_{y_i}, \phi_{y_i}) \\ & \equiv & {}^{(2)}P_K^{l_{x_i}l_{y_i}}(\phi_i)\left[Y_{l_{x_i}}^{m_{x_i}}(\theta_{x_i}, \phi_{x_i})Y_{l_{y_i}}^{m_{y_i}}(\theta_{y_i}, \phi_{y_i})\right]_{LM} \end{array} \qquad (5)$$

with $\alpha_i \equiv \{l_{x_i}, l_{y_i}, L, M\}$ and $[]_{LM}$ denoting angular momentum coupling. The hyper-angular momentum quantum number $K(= 2n_i + l_{x_i} + l_{y_i};\ n_i \rightarrow$ a non-negative integer) is not a conserved quantity for the three-body system. In a given partition (say partition "i"), the wave-function $\Psi(\rho, \Omega_i)$ is expanded in the complete set of HH

$$\Psi(\rho, \Omega_i) = \sum_{K\alpha_i} \rho^{-5/2} U_{K\alpha_i}(\rho) \mathcal{Y}_{K\alpha_i}(\Omega_i) \tag{6}$$

Substitution of Eq. (6) in Eq. (3) and use of ortho-normality of HH, leads to a set of coupled differential equations (CDE) in ρ

$$\begin{aligned} &\left[-\frac{\hbar^2}{2\mu}\frac{d^2}{d\rho^2} + \frac{(K+3/2)(K+5/2)\hbar^2}{2\mu\rho^2} - E\right] U_{K\alpha_i}(\rho) \\ &+ \sum_{K'\alpha'_i} < K\alpha_i \mid V(\rho, \Omega_i) \mid K'\alpha'_i > U_{K'\alpha_i'}(\rho) = 0. \end{aligned} \tag{7}$$

where

$$< K\alpha_i | V(\rho, \Omega_i) | K', \alpha'_i >= \int \mathcal{Y}^*_{K\alpha_i}(\Omega_i) V(\rho, \Omega_i) \mathcal{Y}_{K'\alpha_i'}(\Omega_i) d\Omega_i \tag{8}$$

3. Results and discussions

For the present calculation, we assign the label 'i' to the nucleus of mass $m_i = m_N$ (and charge +Ze), the label 'j' to the negatively charged muon of mass $m_j = m_\mu$ (and charge -e) and the label 'k' to the electron of mass $m_k = m_e$ (and charge -e). Hence, for this particular choice of masses, Jacobi coordinates of Eq. (1) in the partition "i" become

$$\left.\begin{aligned} \vec{x_i} &= \left[\frac{m_\mu m_e (m_N + m_\mu + m_e)}{m_N (m_\mu + m_e)^2}\right]^{\frac{1}{4}} (\vec{r_j} - \vec{r_k}) \\ \vec{y_i} &= \left[\frac{m_\mu m_e (m_N + m_\mu + m_e)}{m_N (m_\mu + m_e)^2}\right]^{-\frac{1}{4}} \left(\vec{r_i} - \frac{m_\mu \vec{r_j} + m_e \vec{r_k}}{m_\mu + m_e}\right) \end{aligned}\right\} \tag{9}$$

and the corresponding Schrödinger equation (Eq. (7)) is

$$\begin{aligned} &\left[-\frac{\hbar^2}{2\mu}\left\{\frac{d^2}{d\rho^2} - \frac{(K+3/2)(K+5/2)}{\rho^2}\right\} - E\right] U_{K\alpha_i}(\rho) \\ &+ \sum_{K'\alpha_i'} < K\alpha_i \mid \frac{\beta_i}{\rho cos\phi_i} - \frac{Z}{\rho\left|\beta_i sin\phi_i\ \hat{y_i} - \frac{1}{2\beta_i} cos\phi_i \hat{x_i}\right|} \\ &- \frac{Z}{\rho\left|\beta_i sin\phi_i \hat{y_i} + \frac{1}{2\beta_i} cos\phi_i \hat{x_i}\right|} \mid K'\alpha_i' > U_{K'\alpha_i'}(\rho) = 0 \end{aligned} \tag{10}$$

where $\beta_i = \left[\frac{m_\mu m_e (m_N + m_\mu + m_e)}{m_N (m_\mu + m_e)^2}\right]^{\frac{1}{4}}$ and $\mu = \left(\frac{m_N m_\mu m_e}{m_N + m_\mu + m_e}\right)^{\frac{1}{2}}$ is the effective mass of the system. In atomic units we take $\hbar^2 = m_e = m = e^2 = 1$. Masses of the particles involved in this work are partly taken from [30, 31, 32, 33].

In the ground state of electron-muon three-body system, the total orbital angular momentum, L=0 and there is no restriction (on lx_i) due to Pauli exclusion principle as electron and muon are non-identical fermions. Since $L = 0$, $l_{x_i} = l_{y_i}$, and the set of quantum numbers represented by α_i is $\{l_{x_i}, l_{x_i}, 0, 0\}$. Hence, the quantum numbers $\{K\alpha_i\}$ can be represented by $\{Kl_{x_i}\}$ only. Eq. (10) is solved following the method described in our previous work [30] to get the ground state energy E.

One of the major drawbacks of HH expansion method is its slow rate of convergence for Coulomb-type long range interaction potentials, unlike for the Yukawa-type short-range potentials for which the convergence is reasonably fast [29, 34]. Hence, to achieve the desired degree of convergence, sufficiently large K_m value has to be included in the calculation. But, if all K values up to a maximum of K_m are

Journal of Physics: Conference Series **759** (2016) 012054 doi:10.1088/1742-6596/759/1/012054

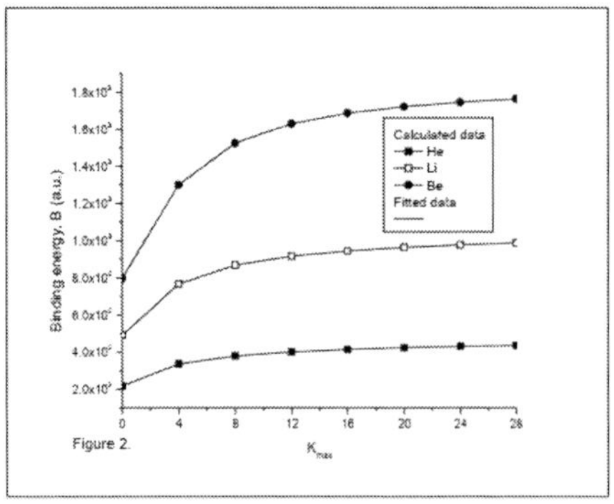

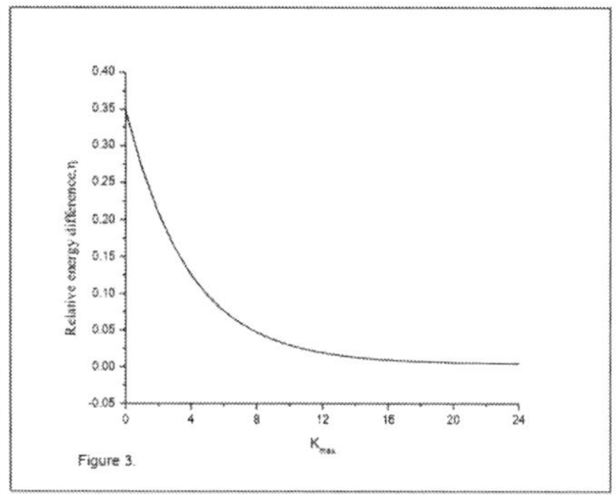

Figure 2. Pattern of dependence of the ground-state energy (B) of muonic atom/ions on the increase in K_{max}.

Figure 3. Pattern dependence of the ground-state relative energy difference $\eta = \frac{B_{K_m+4} - B_{K_m}}{B_{K_m+4}}$ of muonic helium ($^{\infty}He^{2+}\mu^- e^-$) on the increase in K_{max}.

included in the HH expansion then the number of the basis states can be determined by relation

$$N_{K_m} = \frac{(K_m + 2)(K_m + 4)}{8} \tag{11}$$

It follows from Eq. (11) that number of basis states and hence the size of coupled differential equations (CDE) (Eq. (7)) increases rapidly with increase in K_m. For the available computer facilities, we are allowed to solve up to $K_m = 28$ reliably. The calculated ground state energies (B_{K_m}) with increasing K_m for muonic helium ($^{\infty}He^{2+}\mu^- e^-$), muonic lithium ($^{\infty}Li^{3+}\mu^- e^-$) and muonic beryllium ($^{\infty}Be^{4+}\mu^- e^-$) are presented in columns 2, 4 and 6 of Table 1. Energies for a number of muonic atom/ions of different atomic number (Z) at $K_m = 28$ are presented in column 4 of Table 2. The pattern of convergence of the energy of the lowest bound S-state with respect to increasing K_m can be checked by gradually increasing K_m values in suitable steps (dK) and comparing the relative energy difference $\eta = \frac{B_{K+dK} - B_K}{B_{K+dK}}$ with that found in the previous step. From Table 1, it can be seen that at $K_m = 28$, the energy of the lowest bound S-state of $e^- \mu^-\ ^{\infty}He^{2+}$ converges up to 3rd decimal places and similar convergence trends are observed in the remaining cases. The pattern of increase in binding energy (B) with respect to increasing K_{max} is shown in Figure 3 for few representative cases. In Figure 4 the relative energy difference η is plotted against K_{max} to demonstrate the relative convergence trend in energy. The calculated ground state energies muonic three-body systems of different nuclear charge Z (and of infinite nuclear mass), have been plotted against Z as shown in Figure 4 to study the dependence of the bound state energies on the strength of the nuclear charge using data from Table 2. The curve of Figure 4 shows a gradual increase in energy with the increase in the strength nuclear charge Z approximately following the empirical equation

$$\begin{aligned} B(Z) = & -154.64851 + 4.47088Z + 131.84786Z^2 \\ & - 2.79311Z^3 + 0.02174Z^4 \end{aligned} \tag{12}$$

Eq. (12) may be used to estimate the ground state energy of muonic atom/ions of given Z assuming infinite nuclear core. Finally, in Table 2, energies of the lowest bound

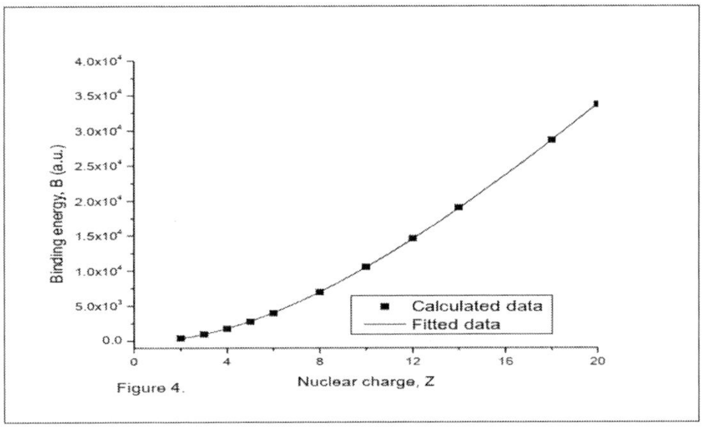

Figure 4. Pattern of dependence of the ground-state energy (B) of muonic atom/ions on the increase in nuclear charge Z.

S-state of several muonic three-body systems obtained by numerical solution of the coupled differential equations by the renormalized Numerov method [35] have been compared with the ones of the literature wherever available. Since reference values are not available for systems having nuclear charge $Z > 3$, we made a crude estimation of the ground-state $(1s_e 1s_\mu)$ energies following the relation

$$B_{est}^{3B} = \frac{Am_N Z^2}{2}\left[\frac{1}{1+Am_N} + \frac{m_\mu}{m_\mu + Am_N}\right] \tag{13}$$

where $m_N = 1836$, muon mass $m_\mu = 206.762828$ (in atomic unit) and A is mass number of the nucleus. Here we assumed two hydrogen-like subsystems for the muonic atom/ions. This estimate can further be improved by assuming a compact $(A(Z), \mu^-)_{1s}$ positive muonic ion and an electron in the 1s state, "feeling" a (Z-1) charge and an $Am_N + m_\mu$ mass. For this case the estimation formula (Eq. (13)) can be replaced by

$$B_{est}^{2B} = \frac{1}{2}\left[\frac{Am_N + m_\mu}{1 + Am_N + m_\mu}(Z-1)^2 + \frac{Am_N m_\mu}{m_\mu + Am_N} Z^2\right] \tag{14}$$

4. Conclusion

In conclusion, we note that the calculated ground-state energy of muonic helium and muonic lithium at $K_m = 28$ listed in column 4 of Table 2 are greater than the corresponding reference values listed in column 5 of Table 2. This discrepancies may have arose due to the CUSP conditions applicable to Coulomb systems which have not been accommodated in the present calculations. It can be seen that the estimated energies (B_{est}) in columns 2 & 3 of Table 2 are less than the corresponding calculated energies (B_{calc}) in column 4 of Table 2 for systems having $Z \leq 10$ while the same for $Z > 10$ becomes larger than B_{calc}. This may happen due to weaker correlation between electron-muon pair in systems having nuclear charge $Z < 10$ while a stronger correlation in systems having $Z > 10$. It may also be noted that estimated energy listed in column 3 of Table 2 agrees fairly with the reference values

Table 1. Energy (B) of the lowest bound S-state of electron-muon three-body systems at different K_{max} along with the corresponding relative energy difference η.

	Binding energies (B) and corresponding relative energy difference (η)					
System	$^{\infty}He^{2+}\mu^-e^-$		$^{\infty}Li^{3+}\mu^-e^-$		$^{\infty}Be^{4+}\mu^-e^-$	
K_{max}	B	η	B	η	B	η
0	217.78577	0.352207	490.78306	0.360783	798.35396	0.385798
4	336.19645	0.110907	767.78745	0.116652	1299.82240	0.146813
8	378.13435	0.055125	869.17847	0.052009	1523.49147	0.065150
12	400.19520	0.033222	916.86372	0.029708	1629.66350	0.033758
16	413.94723	0.022228	944.93547	0.019627	1686.59904	0.020329
20	423.35753	0.015913	963.85254	0.014046	1721.59813	0.013800
24	430.20330	0.011949	977.58415	0.010573	1745.68917	0.010153
28	435.40598	0.006945	988.03084	0.005984	1763.59496	0.006227

Table 2. Energy (B) of the lowest bound S-state of electron-muon-nucleus three-body systems.

	Binding energies expressed in atomic unit (a.u.)			
	Estimated		Calculated	Other Results
System	$B^{3B}_{est}[Eq.(13)]$	$B^{2B}_{est}[Eq.(14)]$	$B_{K_m=28}$	
$e^-\mu^{-3}He^{2+}$	400.574	399.064	420.424	399.042[a], 399.043[b]
$e^-\mu^{-4}He^{2+}$	404.212	402.702	424.017	402.637[c], 402.641[d]
$e^-\mu^{-\infty}He^{2+}$	415.537	414.026	435.406	414.036[e], 414.037[f]
$e^-\mu^{-6}Li^{3+}$	917.814	915.291	970.737	915.231[e], 915.231[g]
$e^-\mu^{-7}Li^{3+}$	920.224	917.701	973.166	917.649[e], 917.650[g]
$e^-\mu^{-\infty}Li^{3+}$	934.957	932.433	988.031	932.457[e]
$e^-\mu^{-9}Be^{4+}$	1641.703	1638.161	1745.087	
$e^-\mu^{-\infty}Be^{4+}$	1662.146	1658.603	1763.595	
$e^-\mu^{-10}B^{5+}$	2568.319	2563.753	2732.374	
$e^-\mu^{-\infty}B^{5+}$	2597.104	2592.535	2761.200	
$e^-\mu^{-12}C^{6+}$	3705.224	3699.628	3937.535	
$e^-\mu^{-\infty}C^{6+}$	3739.829	3734.231	3971.528	
$e^-\mu^{-16}O^{8+}$	6602.337	6594.666	6907.068	
$e^-\mu^{-\infty}O^{8+}$	6648.585	6640.910	6949.141	
$e^-\mu^{-20}Ne^{10+}$	10330.524	10320.754	10486.654	
$e^-\mu^{-\infty}Ne^{10+}$	10388.414	10378.641	10534.362	
$e^-\mu^{-24}Mg^{12+}$	14889.783	14877.894	14539.329	
$e^-\mu^{-\infty}Mg^{12+}$	14959.316	14947.424	14590.826	
$e^-\mu^{-28}Si^{14+}$	20280.115	20266.085	18956.238	
$e^-\mu^{-\infty}Si^{14+}$	20361.291	20347.257	19010.171	
$e^-\mu^{-32}S^{16+}$	26501.520	26485.328	23653.644	
$e^-\mu^{-\infty}S^{16+}$	26594.340	26578.142	23709.055	
$e^-\mu^{-40}Ar^{18+}$	33564.416	33546.038	28574.033	
$e^-\mu^{-\infty}Ar^{18+}$	33658.461	33640.078	28624.646	
$e^-\mu^{-40}Ar^{18+}$	33564.416	335546.038	28574.033	
$e^-\mu^{-\infty}Ar^{18+}$	33658.461	33640.078	28624.646	
$e^-\mu^{-40}Ca^{20+}$	41437.550	41416.966	33152.978	
$e^-\mu^{-\infty}Ca^{20+}$	41553.656	41533.066	33709.888	

[a]Ref[2, 10, 11, 37], [b]Ref[38], [c]Ref[2, 10, 11, 38, 39], [b]Ref[1, 20, 40], [e]Ref[2], [f]Ref[11], [g]Ref[18]

for for Z=2,3. Finally, it can also be added that in the cases of highly charged muonic ions relativistic correction together with proper inclusion of Kato's cusp conditions [36] (in the limit $r_{jk} \rightarrow 0$) is important for obtaining improved results.

The author acknowledges Aliah University for providing computer facilities.

References

[1] Huang K N 1977 *Phys. Rev.* A **15** (5) 1832.
[2] Ancarani L U, Rodriguez K V and Gasaneo G 2010 *In: EPJ Web of Conferences* **3** 02009
[3] Souder P A *et al* 1975 *Phys. Rev. Lett.* **34** 1417
[4] Eskandari M R *et al* 2002 *Inter. J. Modern Phys.* C **13** 265
[5] Eskandari M R*et al* 2002 *Can. J. Phys.* **80** 1099
[6] Orth H *et al* 1982 *Phys. Rev. Lett.* **45** 1483
[7] Gardner C J *et al* 1982 *Phys. Rev. Lett.* **48** 1168.
[8] Borie M 1982 *Rev. Mod. Phys.* **54** 67
[9] Richard I L 1982 *Am. J. Phys.* **50** 562
[10] Frolov A M 2002 *Phys. Rev.* A **65** 024701
[11] Frolov A M 2000 *Phys. Rev.* A **61** 022509
[12] Sultanov R A and Guster D 2003 *Jour. of Computational Physics* **192** (1) 231
[13] Yalcin Z and Simsek M 2002 *Int. J. Quantum Chem.* **88** (6) 735
[14] Dodd L R 1974 *Phys. Rev.*A **9** (4) 637
[15] Rodriguez K V *et al* 2010 *Int. J. Quantum Chem.* **110** (10) 1820
[16] Smith Jr V H and Frolov A M 1995 *J. Phys.* B **28** (7) 1357
[17] Frolov A M 2005 *Phys. Lett.*A **345** 173
[18] Frolov A M 2006 *Phys. Lett.*A **353** 60
[19] Frolov A M and Smith Jr V H 2004 *J. Phys.*B **37** 2917
[20] Frolov A M and Wardlaw D M 2011 *Eur. Phys. J.* D **63** (3) 339
[21] Halpern A 1964 *Phys. Rev. Lett.* **13** 660
[22] Carter B P 1966 *Phys. Rev.* **141** 863
[23] Delves L M and Kalotas T 1968 *Australian J. Phys.* **21** 1
[24] Kamimura M 1988 *Phys. Rev.* A **38** 621
[25] Vinitsky S I *et al* 1980 *Zh. Eksp. Teor. Fiz.* **79** 698; [1980 *Sov. Phys. JETP* **52** 353]
[26] Krivec R and Mandelzweig V B 1997 *Phys. Rev.* A **56** 3614
[27] Khan Md A *et al* 1999 *FIZIKA B (Zagreb)***8** (4) 469
[28] Raynal J and Revai J 1970 *Il Nuo. Cim.* A**68** (4) 612
[29] Beiner M and Fabre de la Ripelle M 1971 *Nuvo Cim.* **1** (14) 584
[30] Khan Md A 2012 *Eur. Phys. Jour.* D **66** 83
[31] Cohen E R and Taylor B N 1998 *Phys. Today***51** (8) BG9
[32] Cohen E R and Taylor B N 2000 *Phys. Today* **53** (8) BG11
[33] Mohr P J and Taylor B N 2002 *Phys. Today* **55**(8) BG6
[34] Ballot J L and Fabre de la Ripelle M 1980 *Ann. Phys. (N. Y.)* **127** 62
[35] Johnson B R 1978 *J. Chem. Phys.* **69** 4678
[36] Kato T 1957 *Commn. Pure. Appl. Math.* **10** 151
[37] Krivec R and Mandelzweig V B 1998 *Phys. Rev.* A **57** 4976
[38] Eskandari M R and Rezaie B *Phys. Rev.* A **72** (1) 012505
[39] Pachucki K 2001 *Phys. Rev.* A **63** 032508
[40] Chen M K 1989 *Phys. Rev.* A**40** (10) 5520

Finite element modelling and analysis of the effect of frequency on the electromagnetic compression of tubes

Ramesh Kumar*, Sachin D Kore

Indian Institute of Technology Guwahati, Guwahati-781039 India
*ramesh.mech@iitg.ernet.in

Abstract. Electromagnetic compression of the tube is a high strain rate forming process, in this process formability of the material can be increased by reducing the spring back effect developed in the process. High electrical conductivity materials like copper, aluminum are generally used for electromagnetic compression. In electromagnetic compression, frequency of the current pulse is an important parameter to be analyzed. In this work the effect of current pulse frequency on the electromagnetic compression of the tube has been studied. The tube and coil were modelled and analyzed using finite element software LS-DYNA. It has been found that the deformation obtained during tube compression were increases first with the increase in frequency, obtained a maximum value and then decreases with the further increase in the frequency. The change in frequency greatly influences the deformation obtained during the electromagnetic compression of tubes.

1. Introduction

Electromagnetic (EM) compression is high energy and high strain rate forming process in which the amount of plastic deformation obtained is more than that of in the conventional forming process. It has many features like very rapid process, less spring back, high precision forming, and large forming limit [1]. EM compression can be used for different processes of the EM manufacturing such as cladding, welding, crimping. Typically, this technique was used for manufacturing of bimetallic tubes, joining of thin tubes of similar or dissimilar materials [2], internal metallic mandrel [3]. The peculiar advantages of EM compression technology are in assembly and joining of lightweight metals such as aluminum and magnesium, because it improves the formability, strain distribution and reduces wrinkles, spring-back and distortion [3].

In EM compression magnetic pressure is the driving force which causes the compression of the tube. When the current discharges from the capacitor bank an eddy current will produce in the tube. Because of the two opposing magnetic field, either by primary and eddy current magnetic pressure will generate in the tube [4]. The magnetic pressure generate in the tube causes the compression of the tube. In the EM module of LS-DYNA, magnetic field, electric field and eddy current were calculated by solving the Maxwell's equations [5].

The current frequency is one of the key parameters of the electromagnetic compression process. The dimensions, shape and material properties of the forming components such as coil, workpiece and die will affect the frequency for a particular EM machine. The capacitance of the machine and inductance of the machine, coil and tube will determine the frequency of the current [3].

The components of EM compression of the tube and the process are shown in figure 1. It consists of power supply, capacitor bank to store the energy, high voltage switch, coil tool to generate magnetic fields, Rogowski coil to measure the discharge current, R-C Integrator will help while measuring the discharge current in the tube.

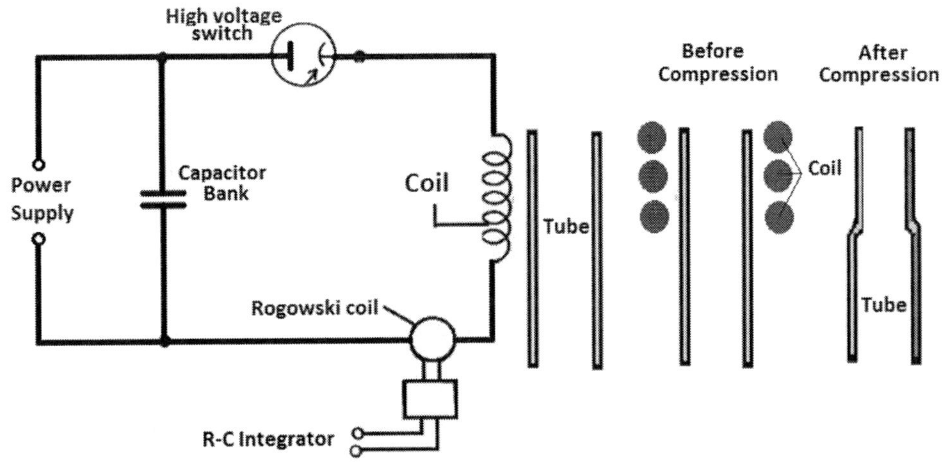

Figure 1. Different components and the process of EM compression of tubes.

1.1. Johnson-Cook materials model

In the finite element simulation the Johnson-Cook constitutive equation was used to model the behaviour of tubes. The Johnson-Cook equations which is combination of plastic strain and plastic strain rate can be described as [6]:

$$\sigma=\left(A+B\varepsilon^{n}\right)\left(1+C\ln\dot{\varepsilon}_{p}\right)\left(1-\left(\frac{T-T_{r}}{T_{m}-T_{r}}\right)^{m}\right) \quad (1)$$

Where σ is flow stress of the material, ε is equivalent plastic strain, $\dot{\varepsilon}_p$ is plastic strain rate, T is absolute temperature, T_r is the room temperature, T_m is the melting temperature of the material and A, B, C, n, m are constants. Constants in this equation are obtained from simple mechanical tests such as isothermal tension test and torsion tests, which is given in table 1. For EM simulations the model was simulated for different values of discharge voltages, ranging from 3.0 kV to 5.4 kV. In the simulation the frequency of the current was also varied from 1 kHz to 35 kHz.

Table 1. Values of Johnson-Cook material constant parameters [7].

Materials	A (MPa)	B(MPa)	n	C	T_m (K)	m
1050A-H24	110	150	0.36	0.014	918	1

Material properties of Al 1050 used for the simulation in the EM module of the finite element software LS-DYNA is listed in table 2.

Table 2. Material properties for tube Al 1050 [8].

Density (ρ)	Modulus of rigidity (G)	Modulus of Elasticity (E)	Poisson's Ratio (γ)	Heat Capacity (C)	Thermal Conductivity (k)	Electrical resistivity (ρ)
2705 Kg m^{-3}	26 GPa	69 GPa	0.33	896 J $Kg^{-1}K^{-1}$	227 W $m^{-1}K^{-1}$	2.9×10^{-8} Ω m

Journal of Physics: Conference Series **759** (2016) 012055 doi:10.1088/1742-6596/759/1/012055

2. Results and Discussion

The deformation patterns obtained during the experiment at various energy levels are shown in figure 2. In the experiment the discharge voltages were varied from 3 kV to 5.4 kV and it was found that at 5.4 kV of discharge voltage the compression obtains in the tube was maximized because further increase in voltage leads to penetration of tube edges between them.

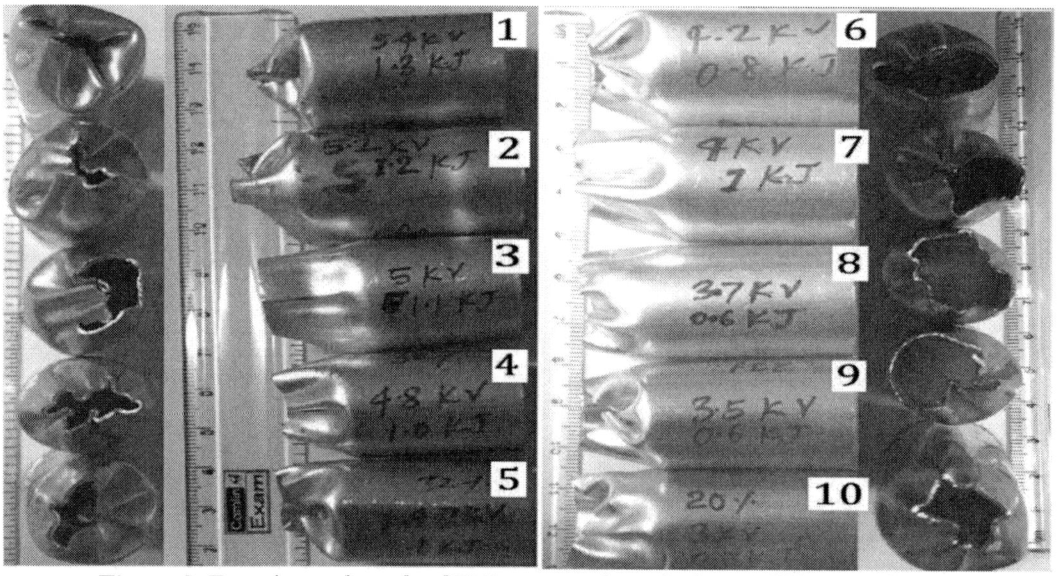

Figure 2. Experimental result of EM compression of tubes at different voltages.

Deformation patterns obtained during the experiment at different discharge voltages and from simulation at same discharge voltages are shown in figure 3. It was observed that the deformation patterns obtained from the simulation have a good agreement with that of the experimental deformation pattern at the same discharge voltages.

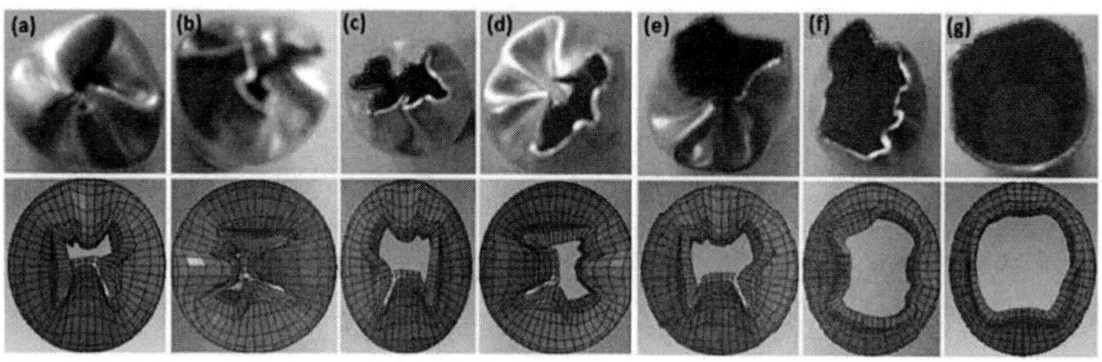

Figure 3. Deformation pattern of tubes obtained in experiment and simulations at different voltages (a) 5.4 kV, (b) 5.2 kV, (c) 5.0 kV, (d) 4.8 kV, (e) 4.2 kV, (f) 3.5 kV, (g) 3.0 kV.

After an experimental validation of the model based on the deformation patterns, obtained at different levels of discharge voltages the frequency of the model were varied by changing the capacitance value, and keeping constant inductance equal to 1.2 μH. The values of the capacitance used for variation in the frequency are listed in table 3.

Table 3. Values of capacitance and frequency at different voltages.

Voltage (kV)	3	3.5	3.7	4	4.2	4.8	5	5.2	5.4
Frequency (kHz)	1	5	10	15	20	25	30	35	40
Capacitance (μ F)	21108	844	211	93	52	33	23	17	13

2.1. Variation in velocity with frequency and discharge voltage

The velocity of the tube was found to be increasing as the frequency was increased from 1 kHz to 15 kHz, and further increase in frequency from 15 kHz to 35 kHz leads to decrease in velocity. The optimum value of frequency was found to be 15 kHz. The effect of frequency on velocity vs time of tube is shown in figure 4. The velocity of the tube was increasing as the discharge voltage was increased, and the corresponding variation is shown in figure 5. At the maximum voltage of 5.4 kV for the given configuration the maximum velocity was found 158 m/s.

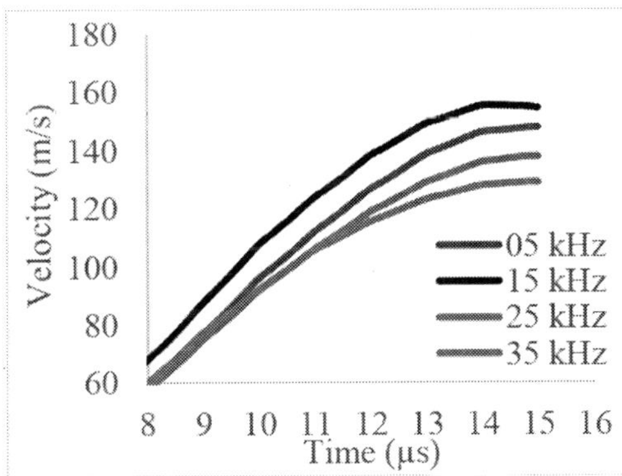

Figure 4. Variation in the velocity with time at different frequencies at 5.4 kV.

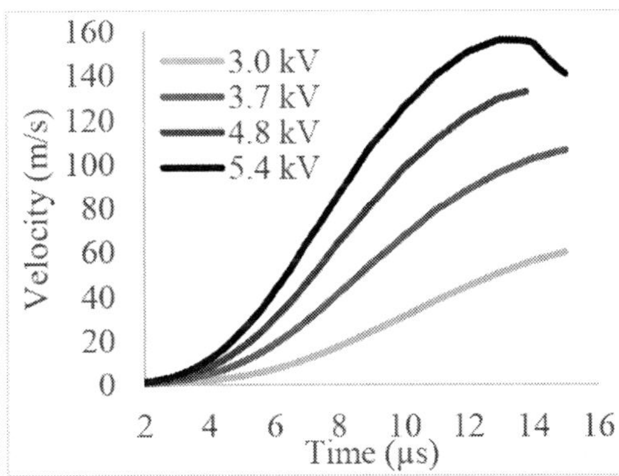

Figure 5. Variation in velocity with time at different discharge voltages at 15 kHz.

2.2. Variation in stress with frequency and discharge voltage

The stress developed in the tube was found to be increasing as the frequency was increased from 1 kHz to 15 kHz, and further increase in frequency from 15 kHz to 35 kHz leads to decrease in stress. The optimum value of frequency was found to be 15 kHz. The effect of frequency of stress vs time is shown in figure 6. Stress developed in the tube was increasing as the discharge voltage was increased, and the corresponding variation is shown in figure 7. At the maximum voltage of 5.4 kV for the given configuration the maximum stress was 125 MPa.

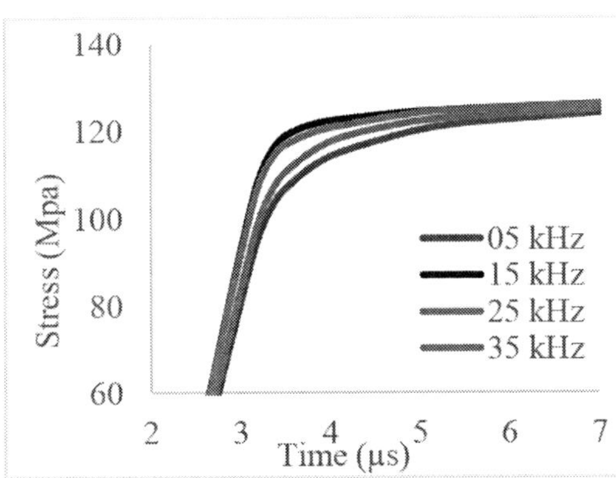

Figure 6. Variation in the stress with time at different frequencies at 5.4 kV.

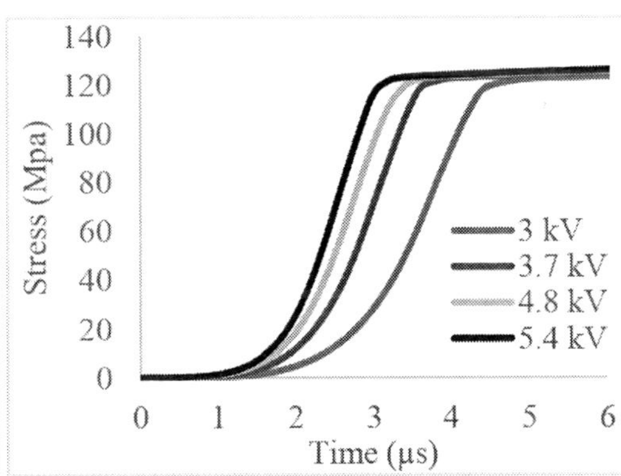

Figure 7. Variation in stress with time at different discharge voltages at 15 kHz.

2.3. *Variation in plastic strain with frequency and discharge voltage*

The plastic strain developed in the tube was found to be increasing as the frequency was increased, and further increase in frequency leads to decrease in plastic strain. The optimum value of frequency was found to be 15 kHz. The effect of frequency of plastic strain is shown in figure 8. Plastic strain value was increasing as the discharge voltage was increased, and the corresponding variation is shown in figure 9. At the maximum voltage of 5.4 kV for the given configuration the maximum plastic strain was 1.3.

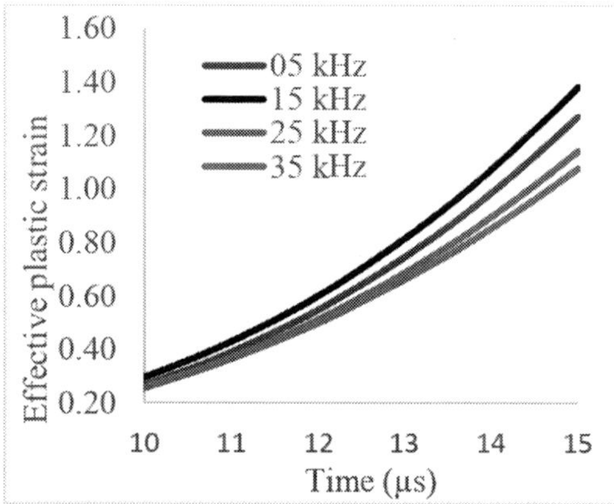

Figure 8. Variation in the effective plastic strain with time at different frequencies at 5.4 kV.

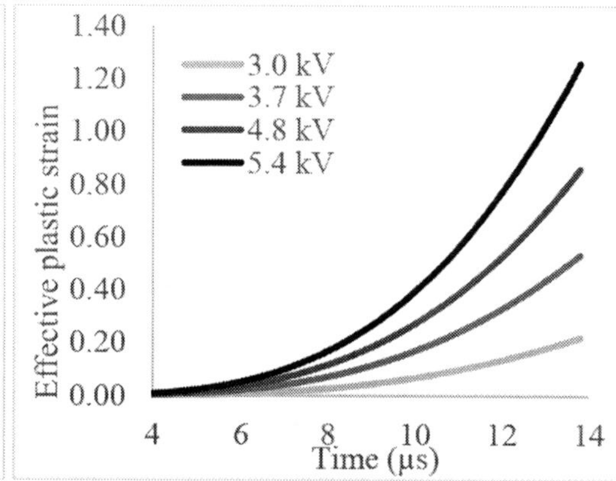

Figure 9. Variation in effective plastic strain with time at different discharge voltages at 15 kHz.

2.4. *Effect of frequency on skin depth*

Due to primary discharge current in the circuit, an eddy current is induced in flyer tube which is alternating in nature and alternating electric current flow mainly at the skin of the conductor, at an average depth called the skin depth [9] and it is measured as by equation (2). For different values of frequency and discharge voltage the values of skin depth are listed in table 4.

Table 4. Variation in the skin depth with frequency at different voltages.

Voltage (kV)	3.0	3.5	3.7	4.0	4.2	4.8	5.0	5.2	5.4
Frequency (kHz)	1	5	10	15	20	25	30	35	40
Skin depth (mm)	2.670	1.190	0.845	0.690	0.598	0.534	0.488	0.452	0.422

$$\delta = \left(\frac{\rho}{\pi \mu_0 \mu_r f} \right)^{1/2} \tag{2}$$

Where ρ is the electrical resistivity of the workpiece, μ_o is the permeability of the free space ($\mu_o = 4\pi \times 10^{-7}$ Hm^{-1}), μ_r is the relative permeability of the material and f is the frequency of the discharge current and it is given by equation (3). The variation in the skin depth with frequency is shown in figure 10.

$$f = \frac{1}{2\pi\sqrt{LC}} \tag{3}$$

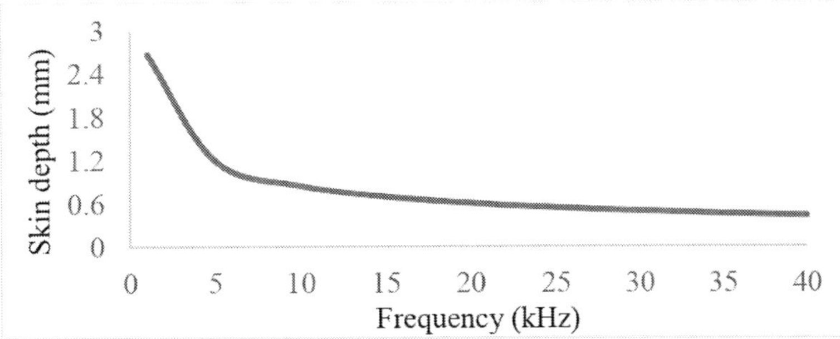

Figure 10. Variation in the skin depth with frequency.

3. Conclusions

Comparison between the numerical and experimental results were carried out based on the deformation patterns and showed good agreement between them. Increase in frequency first increases the velocity, strain and stress and then decrease. Strain rate hardening has the largest effect on EM compression. In the simulation result, the optimum frequency value was found 15 kHz for Al 1050 alloy. In the simulation maximum resultant velocity of the tube was found 258 m/s, the maximum stress was 125 MPa and the maximum plastic strain was 1.3 at 5.4 kV discharge voltage and 15 kHz frequency.

References

[1]. G. K. Fenton and G. S. Daehn, 1998 Modeling of Electromagnetically Formed Sheet Metal, *Journal of Materials Processing Technology,* vol. 75, pp. 6-16.

[2]. H. Hokari, T. Sato, K. Kawauchi and A. Muto, 2009 Magnetic impulse welding of aluminium tube and copper tube with various core materials, *Welding International,* vol. 12, no. 8, pp. 619-626.

[3]. Y. Haiping and L. Chunfeng, 2009 Effects of current frequency on electromagnetic tube compression, *Journal of Materials Processing Technology,* vol. 209, pp. 1053-1059.

[4]. S. H. Lee and D. N. Lee, 1994 A Finite Element Analysis of Electromagnetic Forming for Tube Expansion, *Journal of Engineering Materials and Technology,* vol. 116, pp. 250-254.

[5]. P. L. Eplattenier, G. Cook, C. Ashcraft, M. Burger, J. Imbert and M. Worswick, 2009 Introduction of an Electromagnetism Module in Ls-Dyna for Couple Mechanical-Thermal-Electromagnetic Simulations, *Steel Research Int,* vol. 80, no. 5.

[6]. G. Johnson and W. Cook, 1983 A constitutive model and data for metals subjected to large strain, high strain rates and high temperatures," *Proceedings Seventh International Symposium on Ballistic,* pp. 541-547.

[7]. S. E. Hilde and A. M. Emil, 2013 Blast loaded aluminium plates," Norwegian University of Science and Technology, Norwegian.

[8]. [Online]. Available: http://www.matweb.com/search/datasheet.aspx?matguid=b1aa69c0528a40729478403542a8c94a&n=1.

[9]. S. D. Kore, J. Imbert, Y. Zhou and M. Worswick, 2011 Magnetic Pulse Welding, *ASM Handbook,* vol. 06A, pp. 704-709.

Efficient shortcut techniques in evanescently coupled waveguides

Koushik Paul and Amarendra K. Sarma*

Department of Physics, Indian Institute of Technology Guwahati, Guwahati, Assam, India

*Email: *aksarma@iitg.ernet.in*

Abstract. Shortcut to Adiabatic Passage (SHAPE) technique, in the context of coherent control of atomic systems has gained considerable attention in last few years. It is primarily because of its ability to manipulate population among the quantum states infinitely fast compared to the adiabatic processes. Two methods in this regard have been explored rigorously, namely the transitionless quantum driving and the Lewis-Riesenfeld invariant approach. We have applied these two methods to realize SHAPE in adiabatic waveguide coupler. Waveguide couplers are integral components of photonic circuits, primarily used as switching devices. Our study shows that with appropriate engineering of the coupling coefficient and propagation constants of the coupler it is possible to achieve efficient and complete power switching. We also observed that the coupler length could be reduced significantly without affecting the coupling efficiency of the system.

1. Introduction

In recent years the methods of shortcut to adiabatic passage (SHAPE) have gained considerable attention owing to its many possible applications in atomic and optical physics [1-8]. Shortcut methods originated due to the requirement of speeding up the adiabatic processes in atomic and quantum physics [9]. Transitionless quantum driving (TQD) and Lewis-Riesenfeld Invariant (LRI) methods are most widely used in this regard. The TQD approach drives a quantum system in a way such that adiabatic states become stationary enabling infinitely fast population transfer among the diabatic states [1,10]. On the other hand, the LRI approach exploits the property of dynamical invariants of the system to inverse engineer the conditions for instantaneous population exchange between the states. These methods have already been used in various contexts such as: creation of entanglement between atoms in cavity [11], atom cooling in harmonic traps [12] and control of spin in quantum dot [13], to mention a few. It is interesting to note that, drawing inspiration from many analogies with quantum physics, these methods have been applied even in wave-optics [14-16]. In this regard, coupled waveguides in integrated optics are particularly interesting due to its tremendous practical applications [17,18]. In general, the function of a waveguide coupler is to split coherently an optical field incident on one of the input ports and direct the two parts to the output ports. Adiabatic passage technique has been exploited successfully in two and three waveguide couplers in order to study the eigenmode evolution of optical power [19-23]. For a sufficiently long coupler, where adiabaticity is satisfied, the system follows its initial eigenmode, causing power transfer from one waveguide to the other. Large device length causes higher transmission loss and makes designing practical devices difficult. However, there are significant opportunities to make couplers more efficient and small in dimension using shortcut methods. Several new studies in this regard have been reported recently [24,25].

In this work we choose a waveguide coupler of length $\boldsymbol{L}$ consists of two single mode waveguides with $\boldsymbol{z}$ being the direction of propagation. The schematic of the system is shown in Fig.1. Waveguides are tapered in nature so that propagation constants $\boldsymbol{\beta_1}$ and $\boldsymbol{\beta_2}$ varies along $\boldsymbol{z}$. Separation between the waveguides is not constant and the evanescent coupling coefficient between the waveguides ($\boldsymbol{\kappa}$) also varies along $\boldsymbol{z}$ and is maximum at length $\boldsymbol{L/2}$ where the separation is minimal.

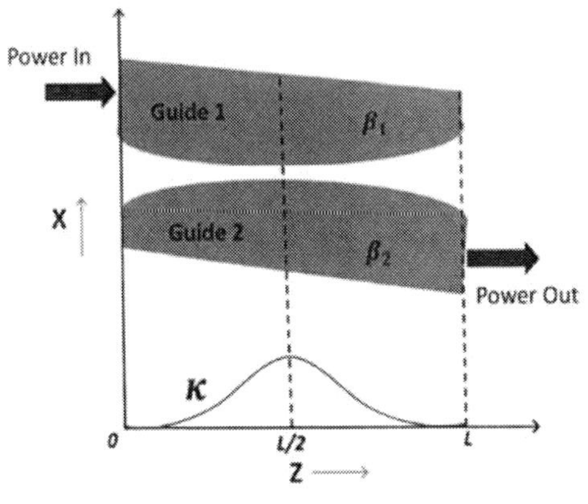

Figure 1. Schematic diagram for tapered wave-guide coupler of length L. Propagation direction is along z direction. β_1 and β_2 are the propagation constants. Separation between the waveguides varies along z and the coupling coefficient κ attains a maximum at $z = L/2$.

2. Shortcut Methods

To illustrate the basic principles, let us start from the coupled mode theory for coupled optical waveguides. Coupled mode equations are expressed as $d/dz[a_1(z), a_2(z)]^T = -i\, H(z)[a_1(z), a_2(z)]^T$. We recognize $H(z)$ as the Hamiltonian and is given by:

$$H(z) = \begin{pmatrix} \Delta(z) & \kappa(z) \\ \kappa(z) & -\Delta(z) \end{pmatrix} \tag{1}$$

Here a_1 and a_2 are the modal amplitudes of the respective waveguides. Δ is the mismatch parameter, defined as $\Delta = (\beta_2 - \beta_1)/2$ and z is the direction of propagation. When the rate of variation of $\Delta\,(z)$ and $\kappa\,(z)$ is small enough with respect to z, the system undergoes adiabatic evolution along z. Adiabatic condition for such system is $\kappa_0 L \gg 1$, where κ_0 is the coupling coefficient at $L/2$. This clearly suggests the requirement for large length of the coupler. By using shortcut methods, we will show how to circumvent this issue, even when adiabatic condition is violated.

2.1 Lewis-Riesenfeld Invariant approach

The Hamiltonian (1) can be written in the following form:

$$H(z) = \kappa(z)\sigma_x + 0\,\sigma_y + \Delta(z)\sigma_z \tag{2}$$

Here σ_x, σ_y and σ_z are the well-known Pauli matrices for spin $\frac{1}{2}$ particles. These operators satisfy Lie algebra: $[\sigma_i\,, \sigma_j] = 2i\epsilon_{ijk}\sigma_k$. The Hamiltonian satisfies SU(2) symmetry and hence there exist an invariant which would satisfy the usual invariant equation: $dI(z)/dz = \partial I(z)/\partial z - i[I(z), H(z)]$. According to Lewis-Riesenfeld theory [26], this invariant can be written as follows:

$$I(z) = \frac{\Omega}{2}\begin{pmatrix} \cos\gamma(z) & \sin\gamma(z)\, e^{i\beta(z)} \\ \sin\gamma(z)\, e^{-i\beta(z)} & -\cos\gamma(z) \end{pmatrix} \tag{3}$$

Here Ω is an arbitrary constant which has the dimension of $\kappa(z)$. $\gamma(z)$and $\beta(z)$ are the parameters which characterizes $I(z)$ and satisfies the following conditions:

$$\dot{\gamma}(z) = 2\kappa(z)\sin\beta(z) \tag{4a}$$
$$\left(2\Delta(z) + \dot{\beta}(z)\right)\sin\gamma(z) = 2\kappa(z)\cos\gamma(z)\cos\beta(z) \tag{4b}$$

where overdot represents derivative with respect to z. It is important to note that $I(z)$ and $H(z)$ does not commute normally. To make $I(z)$ and $H(z)$ commute at the boundaries so that the eigenstates exactly match at the ends of the coupler, we impose $[H(0), I(0)] = 0$ and $[H(L), I(L)] = 0$. With straightforward calculations we obtain the following constraints:

$$[\kappa(z)\sin\gamma(z)\sin(z)]_{\ z=0,L} = 0 \tag{5a}$$
$$\left[\sqrt{2}\kappa(z)\cos\gamma(z) - \Delta(z)\sin\gamma(z)\, e^{\pm i\beta(z)}\right]_{z=0,L} = 0 \tag{5b}$$

These constraints help us to determine the required boundary conditions for $\beta(z), \gamma(z)$ and $\kappa(z)$. We find that, the boundary conditions can be satisfied only if $\kappa(0) = \kappa(L) = 0$ and $\sin\gamma(0) = \sin\gamma(L) = 0$. $\beta(z)$ helps us to configure $\kappa(z)$ and $\Delta(z)$. It is to be noted that β cannot be chosen to be zero as $\kappa(z)$ is finite. We chose $\beta(z)$ such that $\kappa(z)$ keeps its amplitude minimal. With the above considerations, we set the boundary conditions as follows:

$$\gamma(0) = \pi;\quad \gamma(L) = 0;\quad \dot{\gamma}(0) = 0;\quad \dot{\gamma}(L) = 0 \tag{6}$$
$$\beta(0) = -\frac{\pi}{2};\quad \beta(L) = -\frac{\pi}{2};\quad \dot{\beta}(0) = \frac{3\pi}{2L};\quad \dot{\beta}(L) = -\frac{3\pi}{2L} \tag{7}$$

Using the above boundary conditions one can easily construct the parameters γ and β, and thereby all the other necessary parameters to study the evolution of optical power within the waveguides.

2.2 Transitionless Quantum Driving algorithm

According to Berry's Transitionless driving algorithm [10], when the adiabatic conditions are not satisfied (for relatively small lengths), we need to find an additional interaction term in order to avoid non-adiabatic effects. To construct such interaction term we use unitary transformation to go to the adiabatic basis, $\{A_j\}$, using: $[A_1, A_2]^T = U_0^{-1}\,[a_1, a_2]^T$. The driving Hamiltonian is expressed as: $H_a = i\,\Sigma_j|\partial_z A_j\rangle\langle A_j|$, which when written in explicit form looks like:

$$H_a(z) = \begin{pmatrix} 0 & -i\dot{\theta}/2 \\ i\dot{\theta}/2 & 0 \end{pmatrix} \tag{8}$$

Here $\theta(z) = \tan^{-1}\kappa(z)/\Delta(z)$ is the angle of mixing. $\dot{\theta}$ represents additional coupling between those waveguides. We add $H_a(z)$ back to the original Hamiltonian to find an effective coupling that can be realized through appropriate engineering of the coupler. Using the unitary transformation, $U_1 = e^{-i\phi/2}|a_1\rangle\langle a_1| + e^{i\phi/2}|a_2\rangle\langle a_2|$ the effective Hamiltonian can be expressed as follows:

$$H_{eff}(z) = \begin{pmatrix} \Delta_{eff}(z) & \kappa_{eff}(z) \\ \kappa_{eff}(z) & -\Delta_{eff}(z) \end{pmatrix} \tag{9}$$

Here $\kappa_{eff} = \sqrt{\kappa^2 + \dot{\theta}^2/4}$ and $\Delta_{eff} = \Delta - \dot{\phi}/2$. The unitary transformations U_0 and U_1 are connected via the parameters θ and ϕ. These parameters need to be adjusted such that the adiabatic bases of the unitary transformations are equivalent at the boundary. This leads to the following boundary conditions: $\dot{\theta}(0) = \dot{\theta}(L) = 0$ and $\theta(0) = \theta(L) = 2\pi$. However in our case ϕ can be chosen arbitrarily as it has no effect on power evolution.

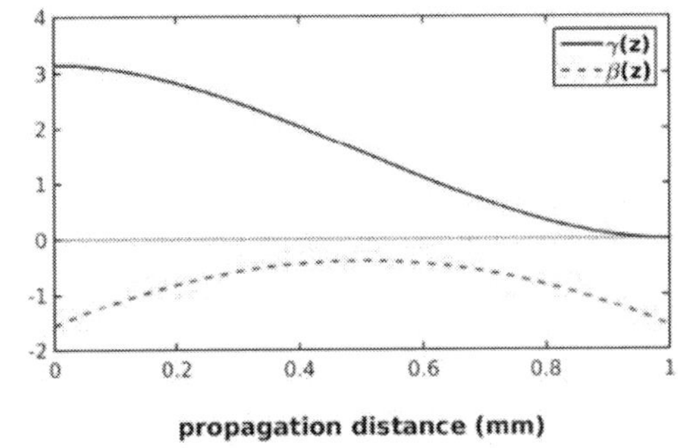

Figure 2. (color online) profile of $\beta(z)$ (dashed) and $\gamma(z)$ (solid) determined through polynomial ansatz with the coefficients determined using the boundary conditions.

3. Results and Discussion

Using the conditions in Eq. (6) and (7), $\gamma(z)$ and $\beta(z)$ can be constructed. We followed the polynomial ansatz to interpolate them at the intermediate points. We choose them as $\gamma(z) = \sum_{i=0}^{3} g_i z^j$ and $\beta(z) = \sum_{i=0}^{3} b_i z^j$, where g_js and b_js are determined using boundary conditions. Fig. (2) shows the spatial profile of γ and β. The mismatch parameter Δ and coupling coefficient , as determined through the invariant method, are shown in Fig. (3a). The profile of Δ_{eff} and κ_{eff}, determined from the transitionless driving algorithm is shown in Fig. (3b), where Δ and κ are chosen as per the famous Allen-Eberly scheme [27]:

$$\kappa(z) = \kappa_0 \operatorname{sech}[(2\pi(z - z_0))/L], \qquad \Delta(z) = \Delta_0 \tanh[2\pi(z - z_0)/L] \tag{10}$$

We have taken: $\kappa_0, \Delta_0 = 1\, mm^{-1}$ and $z_0 = L/2$. Similarity between both the methods are evident from Fig. (3a) and (3b). The strength of the couplings determined from both the methods is almost same, however the mismatch is much larger in case of the invariant method.

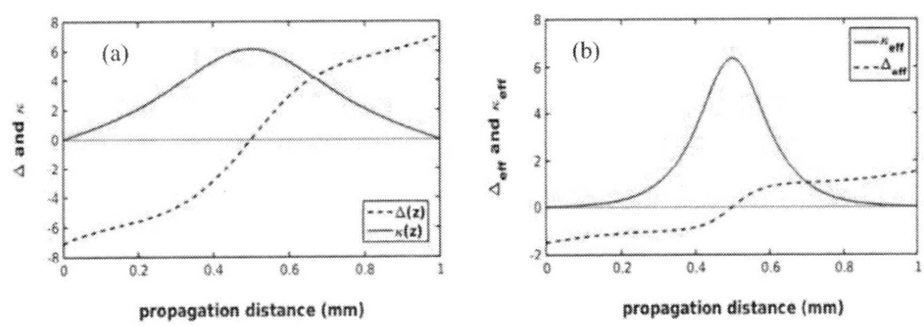

Figure 3. (color online). Spatial profile of coupling and mismatch. (a) $\Delta(z)$ (dashed) and $\kappa(z)$ (solid) determined through Invariant based method, (b) $\Delta_{eff}(z)$ (dashed) and $\kappa_{eff}(z)$ (solid) determined from transitionless driving method.

We have numerically solved the Master equation for density matrix to study the power evolution inside the waveguides [24]. Fig. (4) shows fractional power evolution, defined as $P_2(z)/P_1(0)$, within

Journal of Physics: Conference Series **759** (2016) 012056 doi:10.1088/1742-6596/759/1/012056

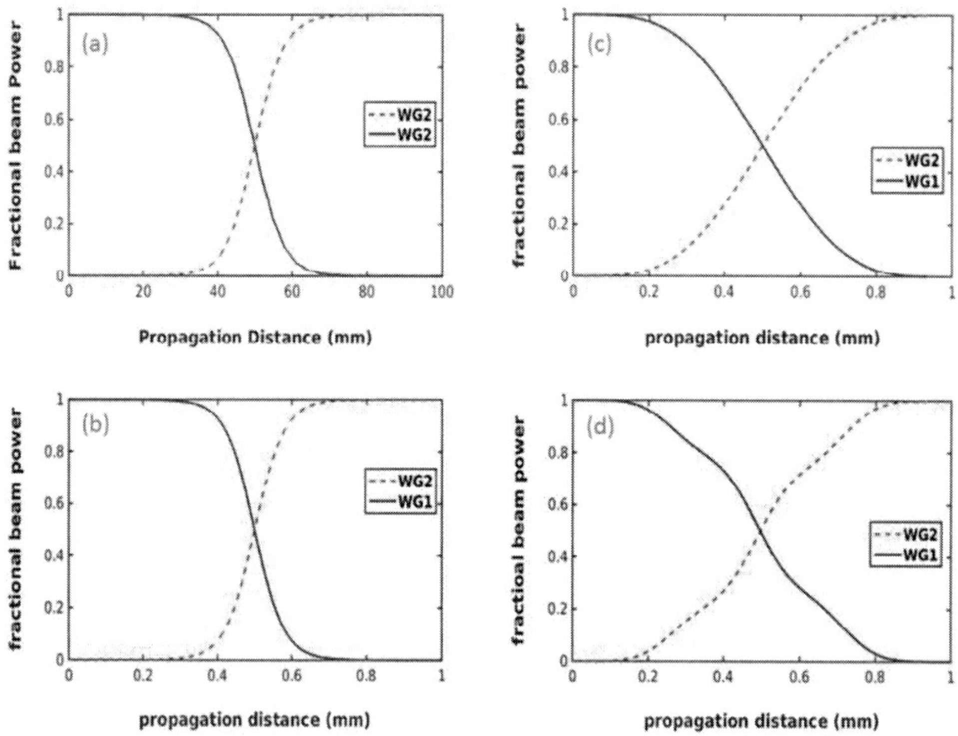

Figure 4. (color online) Spatial evolution of fractional beam power with respect to z for (a) adiabatic case with $L = 100mm$, (b) Transitionless driving where $L = 1mm$, (c) & (d) Invariant based approach using the Hamiltonian and the invariant respectively, $L = 1mm$.

the coupler. Optical power is launched in the first waveguide while the other one is kept empty. For adiabatic regime, in Fig. (4a) we have chosen a 100 mm coupler device, which is sufficiently large so that the adiabatic condition is being satisfied. Complete power transfer is observed. Fig. (4b) shows the power evolution through the transitionless driving method which follows the adiabatic path exactly but within a coupler size of merely 1mm. In Fig. (4c), $\kappa(z)$ and $\Delta(z)$ are taken according to Eq. (4) and used in the original adiabatic Hamiltonian. It exactly shows the adiabatic nature of the evolution but again only 1mm long coupler is enough to complete the power transfer. In Fig. (4d), we have used the Lewis-Reisenfeld invariant instead of the actual Hamiltonian, which precisely matches with Fig. (4c) at the boundaries but does not follow the adiabatic path.

4. Conclusions

In conclusion we have described two very efficient methods for power transfer in waveguide couplers using the analogy between quantum formalism and coupled mode theory for waveguides. We have shown that with judicious choice of coupling coefficient and mismatch profile it is possible to construct a coupler, which is considerably small in size compared to adiabatic couplers. With recent development of fabrication techniques, separation between the waveguides and profile of taper can be controlled accurately and hence designing κ and Δ, as predicted by these theories, is achievable. The proposed methods may be used to design couplers suitable for photonic circuits, of which couplers are integral components.

Acknowledgement

K. Paul would like to thank MHRD, Government of India for a research fellowship.

References

[1] Chen X, Ruschhaupt A, Lizuain I, Guéry-Odelin D and Muga J G, 2011 *Phys. Rev. Lett.* **105** 123003
[2] Chen X and Muga J G, 2012 *Phys. Rev. A.* **86** 033405
[3] Giannelli L and Arimondo E, 2014 *Phys. Rev. A.* **89** 033419
[4] Ibanez S, Martinez-Garaot S, Chen X, Torrontegui and Muga J G, 2011 *Phys. Rev. A.* **84** 023415
[5] Torosov B T, Valle G D and Longhi S, 2013 *Phys. Rev. A.* **87** 052502
[6] Torosov B T, Valle G D and Longhi S, 2014 *Phys. Rev. A.* **89** 063412
[7] Muga J G, Chen X, Ibanez S and Gue´ry-Odelin D, 2009, *J. Phys. B: At. Mol. Opt. Phys.* **42** 241001
[8] Schaff J F, Capuzzi P, Labeyrie G and Vignolo P, 2011 *New J. Phys.* **13** 113017
[9] Torrontegui E, Ibáñez S, Martínez-Garaot S, Modugno M, del Campo A, Guéry-Odelin D, Ruschhaupt A, Chen X and Muga J G 2013 *Adv. At. Mol. Opt. Phys.* **62** 117
[10] Berry M V, 2009 *J. Phys. A: Math. Theor.* **42** 365303
[11] Lu M, Xia Y, Shen L T, Song J and Ba An N, 2014 *Phys. Rev. A* **89** 012326
[12] Chen X, Ruschhaupt A, Schmidt S, del Campo A, Guéry-Odelin D and Muga J G 2010 *Phys. Rev. Lett.* **104** 063002
[13] Ban Y and Chen X 2014 *Sci. Rep.* **4** 6258
[14] Tseng S Y, *Opt. Express* 2013 **21** 21224
[15] Tseng S Y, Wen R D, Chiu Y F, and Chen X, 2014 *Opt. Express* **22** 1884
[16] Tseng S Y, 2014 *Opt. Lett.* **39** 6600
[17] Nikolopoulos G M, 2008 *Phys. Rev. Lett.* **101** 200502
[18] G. P. Agrawal, 2007 *Applications of Nonlinear Fiber Optics*, 2nd Ed. (San Diego: Academic Press)
[19] Longhi S, 2006 *Phys. Rev. E.* **73** 026607
[20] Sun X, Liu H C and yariv A, 2009 *Opt. Lett.* **34** 280
[21] Dreisow F, Szameit A, Heinrich M, Keil R, Nolte S, Tunnermann A and Longhi S, 2009 *Opt. Lett.* **34** 2405
[22] Huang W, Rangelov A A and Kyoseva E, 2014 *Phys. Rev. A.* **90** 053837
[23] Longhi S, Della Valle D, Ornigotti M and Laporta P, 2007 *Phys. Rev. B.* **76** 201101
[24] Paul K and Sarma A K, 2015 *Phys. Rev. A.* **91** 053406
[25] Tseng S Y and Jhang Y W, 2013 *IEEE Photon. Technol. Lett.* **25** 2478
[26] Lewis H R and Riesenfeld W B, 1969 *J. Math. Phys.* **10** 1458
[27] Allen L and Eberly J H, 1975 *Optical resonance and two-level atoms* (New York: Wiley)

Journal of Physics: Conference Series **759** (2016) 012057 doi:10.1088/1742-6596/759/1/012057

Experimental and theoretical study of 3-methyl-4-nitrobenzoic acid using DFT and IVP methods

J Prashanth[1], Jai Kishan Ojha[2], B Venkatram Reddy[1*] and G Ramana Rao[1]

[1] Department of Physics, Kakatiya University, Warangal 506009, Telangana, India
[2] Department of Physics, Govt. Degree & P.G. College, Mancherial-504208, Adilabad, Telangana, India

*E-mail: bvreddy67@yahoo.com

Abstract. The Fourier Transform Infrared (FTIR) and FT-Raman spectra of 3-methyl-4-nitrobenzoic acid have been recorded in the range 4000-400 cm^{-1} and 3500-50 cm^{-1}, respectively. The optimized geometry of the molecule, its vibrational frequencies have been computed using the Density Functional Theory (DFT) employing B3LYP/6-311++G basis set. The scaled values of harmonic vibrational frequencies so obtained have been compared with their experimental counter parts. The scaling factors have been refined to reproduce the frequencies with an RMS error of 9.26 cm^{-1} between the experimental and computed frequencies. The theoretically predicted FTIR and FT-Raman spectra agree satisfactorily with those of experimental spectra. A 89-parameter modified valence force field was evaluated by solving inverse vibrational problem (IVP) using Wilson's GF matrix method. The force constants were refined using 44 experimental frequencies of this molecule in overlay least-squares technique. The average error between observed and computed frequencies was found 11.61cm^{-1}. PED and eigen vectors computed in the process were used to make unambiguous vibrational assignments of all the fundamental vibrations of this molecule. The values of dipole moment and hyperpolarizability were determined to study the NLO behaviour of this molecule. The HOMO and LUMO energies were also evaluated for this molecule.

1. Introduction

Vibrational assignment for complex systems can be proposed based on frequency agreement between the computed and observed fundamental frequencies. The spectroscopists have been investigating to develop 'a priori' method of predicting vibrational frequencies of a given molecule for the past two and a half decades. In this process, two different and distinct methods, namely classical experimental approach [1] and theoretical quantum mechanics approach [2-13], have been evolved.

In experimental approach, Snyder and Schachtschneider [1] proposed the overlay least-square technique wherein a set of related molecules can be treated together to reduce the number of force constants. This method has been applied successfully to some representative molecular systems as reported in our earlier work on nitrotoluenes, dimethylanilines and some substituted methylbenzenes [14]; and on other molecules as mentioned therein.

Spectroscopic investigation has been carried out in theoretical approach in our earlier work on 3,6-dichloro-4-methylpyridazine and 3,6-dichloropyridazine-4-carboxylic acid by DFT method using B3LYP functional with 6-311++G(d,p) basis set [15]; and 4-methyl-3-nitrobenzoic acid employing DFT method employing B3LYP/6-311++G level of theory [16]. We have recently reported the results of both theoretical (DFT) and experimental (inverse vibrational problem (IVP)) methods for some pyridine-dicarboxylic acids [17]. Now, we have taken up the task to test further in case of 3-methyl-4-nitrobenzoic acid (3M4NA) by DFT method and solving IVP using Wilson's GF matrix method [18].

Benzoic acid occurs widely in plants and animal tissues. It is used in miticides, contrast media in urology, cholocystographic examinations and the manufacture of pharmaceuticals. The derivatives of benzoic acid are essential components of the Vitamin B-complex [19]. Further, the herbicidal activity of a molecule containing carboxylic acid group is mainly due to the presence of the acid moiety or a group that is easily convertible to this moiety [20]. The vibrational analysis of benzoic acid and d_5-deuterated benzoic acid was made by Furic and Durig using IR and Raman spectroscopy [21]. Normal coordinate analysis was carried out for urea meta nitrobenzoic acid for the assignment of vibrational frequencies using potential energy distribution by Gunasekaran et al [22]. Sundaraganesan et al reported vibrational analysis of 5-amino-2-chlorobenzoic acid [23]; p-chlorobenzoic acid [24]; 3,4-diaminobenzoic acid [25]; 5-amino-2-nitrobenzoic acid [26]; and 2-chlorbenzoic acid [27] employing ab initio HF and DFT methods using FTIR and FT-Raman spectra. Methylnitrobenzoic acid and its derivatives are also known for their local anaesthetic action [28]. Further, it has been demonstrated that 4-Methyl-3-nitrobenzoic acid is a potent inhibitor cancer cell chemotaxis and may be developed into a novel anti-metastatis drug [29]. Hence, we have taken up experimental and theoretical investigation of 4-Methyl-3-nitrobenzoic acid using its FTIR and FT-Raman spectra and suggested vibrational analysis by Density Functional Theory (DFT) [16]. From the above, it is clear that the experimental (infrared and Raman) and theoretical investigation of 3-methyl-4-nitrobenzoic (3M4NA) acid are by no means complete. Hence, a systematic investigation of vibrational spectra of this molecule employing experimental and theoretical methods is undertaken. The purpose of this investigation is:

1. To record FTIR and FT-Raman spectra of this molecule to get complete information on its vibrational frequencies.
2. To make DFT (Density Functional Theory) calculations in order to identify the most stable rotational isomer in the ground state, calculate theoretical vibrational frequencies using scaling, obtain the values of first order hyperpolarizability and to evaluate HOMO-LUMO energies.
3. To solve inverse vibrational problem using optimized geometries and employing overlay least-squares technique.
4. To make unambiguous vibrational assignments of all the fundamentals of the molecule using PED and eigen vectors in conjunction with scaled DFT results.

2. Spectral measurements

The molecule 3M4NA was obtained from Aldrich Chemical Company, USA and used as such for the spectral measurements. The room temperature FTIR spectrum of the compound was recorded using Bruker IFS 66V spectrometer employing KBr optics in the spectral range 4000-400 cm^{-1} with a scanning speed of 30cm^{-1} min^{-1} with spectral width 2.0cm^{-1}. The FT-Raman spectrum of this molecule was recorded in the 3500-50 cm^{-1} spectral region using FRA 106 Raman module with a resolution of of ±2 cm^{-1}, using 1064 nm line of Nd:YAG laser operating at 200 mw power for excitation.

3. Methods of computations

3.1. Quantum chemical calculations

3.1.1. Molecular geometry

The starting point for calculations is to determine the most stable conformer for the molecule under investigation. The bonds around which rotation are permitted are the C-C_α and C=O of acid group. Hence, the molecule 3M4NA was subjected to a rigorous conformational analysis resulting from the rotation of these bonds using Gaussian 09w software package [30] implemented on Pentium-V (3.2GHz) workstation. The computations were established by employing the standard density functional triply-parameter hybrid model DFT/B3LYP [31, 32] using 6-311++G basis set. These computations yielded four stable rotational conformers as shown in figures 1(a), 1(b), 1(c) and 1(d), respectively. The conformer in figure 1(a) has the lowest global minimum energy at -664.61788 Hartree. Hence, this is accepted as the most stable conformer for optimizing the geometry of this molecule. Subsequent calculations were performed with this optimized structure as shown in figure 2. The same figure contains numbering of atoms also.

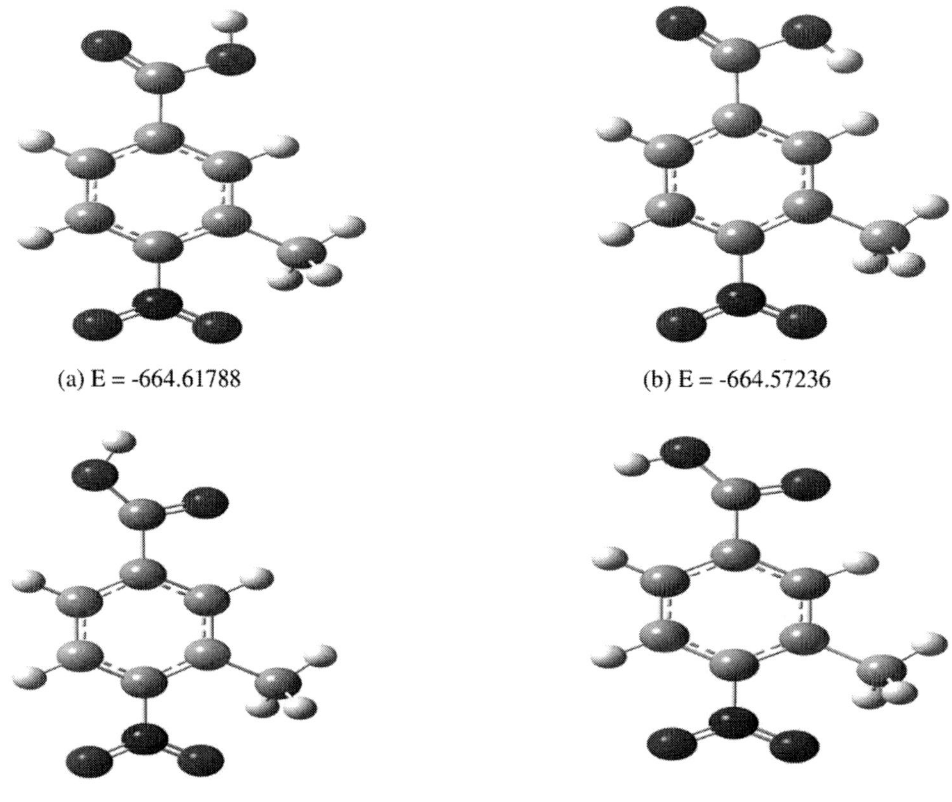

Figure 1. Conformers of 3M4NA (Energy, E in Hartree)

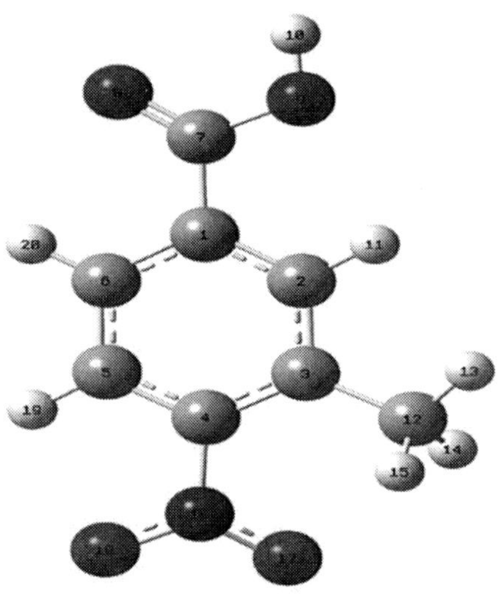

Figure 2. Optimized molecular structure

The experimental and simulated FTIR and FT-Raman spectra are presented in figures 3 and 4. For plotting simulated IR and Raman spectra, a pure Lorentzian band shape was used with a full width at half maximum (FWHM) 0f 10 cm^{-1}. The optimized structure parameters namely bond lengths, bond angles and dihedral angles of 3M4NA in its most stable conformation are presented in Table 1. Scaling of the force constants was made according to scaled quantum mechanical (SQM) procedure [33, 34] employing selective scaling in the natural coordinate representation [35, 36]. Transformation of the force field, normal coordinate analysis, least-square refinement of scale factors and calculation of potential energy distribution (PED) were established with the MOLVIB program (version 7.0) written by Suindius [37,38].

Table 1: Optimized geometrical parameters of 3M4NA

Bond	Bond length (Å)	Bond angle	Value (in ^{0})	Torsional angle	Value (in ^{0})
C1-C2	1.393	C1-C2-C3	120.11	C1-C2-C3-C4	-1.0
C2-C3	1.397	C2-C3-C4	122.54	C2-C3-C4-C5	1.0
C3-C4	1.410	C3-C4-C5	115.78	C3-C4-C5-C6	1.0
C4-C5	1.408	C4-C5-C6	122.46	C3-C2-C1-C7	-178.0
C5-C6	1.388	C6-C1-C2	118.70	O8-C7-C1-C2	157.0
C6-C1	1.406	C2-C1-C7	122.95	C2-C1-C7-O9	-22.0
C1-C7	1.490	C1-C7-O8	123.19	C1-C7-O9-H10	-10.0
O7-C8	1.226	C1-C7-O9	117.79	C6-C1-C2-H11	178.0
C7-O9	1.384	C7-C9-H10	114.92	C4-C5-C6-H20	178.0
O9-H10	0.974	C1-C2-H11	122.46	C2-C3-N12-O13	20.0
C2-H11	1.079	C2-C3-N12	115.68	C2-C3-N12-O14	-159.0
C3-N12	1.466	C3-N12-O13	117.94	C2-C3-C4-C15	-179.0
N12-O13	1.270	C3-N12-O14	119.00	C3-C4-C15-H16	-70.0
N12-O14	1.266	C3-C4-C15	125.24	C3-C4-C15-H17	48.0
C4-C15	1.507	C4-C15-H16	111.27	C3-C4-C15-H18	169.0
C15-H16	1.090	C4-C15-H17	112.22	C3-C4-C5-H19	-179.0
C15-H17	1.089	C4-C15-C18	109.74	C4-C5-C6-H20	179.0
C15-H18	1.089	C4-C5-H19	118.18	C4-C3-C2-C1	-1.0
C5-H19	1.081	C5-C6-H20	120.86	C5-C4-C3-C2	1.0
C6-H20	1.080	O13-N12-O14	123.04	C6-C5-C4-C2	1.0
		H16-C15-H17	106.27		
		H17-C15-H18	108.70		

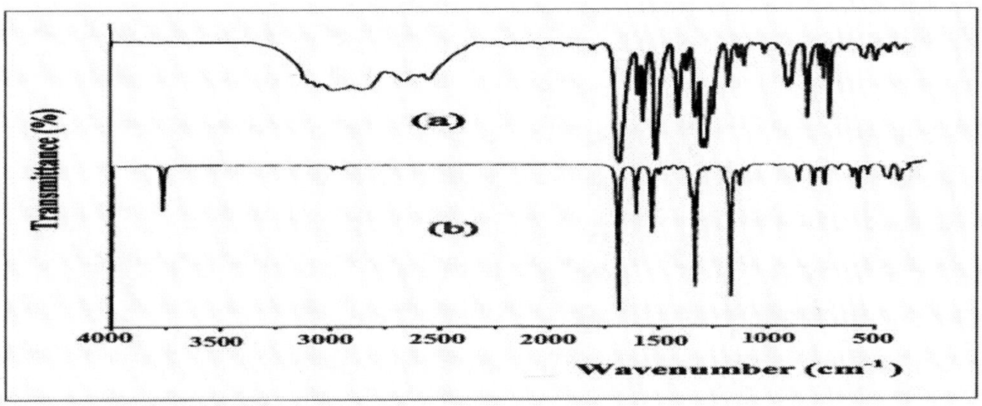

Figure 3. FTIR spectrum of 3M4NA: a) Observed, b) Simulated with DFT

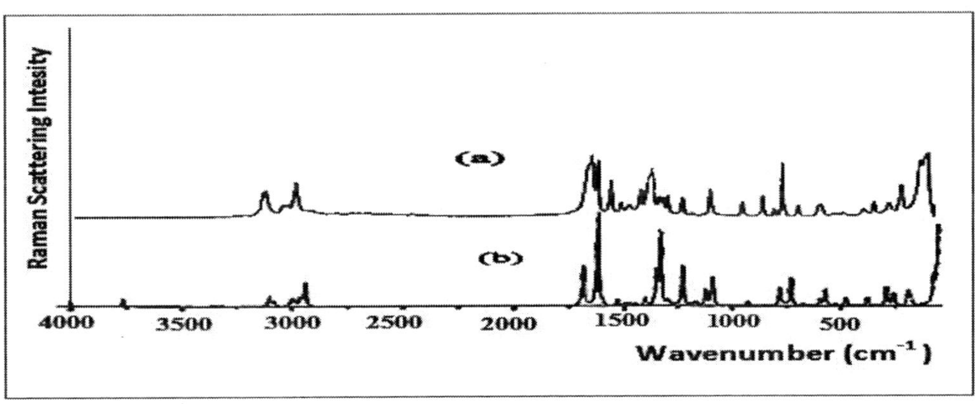

Figure 4. FT-Raman spectrum of 3M4NA: a) Observed, b) Simulated with DFT

3.1.2. Vibrational frequencies

The molecule consists of 20 atoms. Hence, it has 54 fundamentals, distributed as 36 in-plane vibrations of a′-species and 18 out-of-plane vibrations of a″-species in C_s point group symmetry for the molecule. The vibrational modes of the molecule were described using the geometrical parameters presented in Table 1. Experimental frequencies (both IR and Raman) and computed frequencies (both unscaled and scaled), as obtained from DFT/B3LYP/6-311++G level of theory, are reported in Table 2. The rms error between unscaled frequencies and observed frequencies was found to be 90.41 cm^{-1}. On using the refined scale factors, this error was reduced to 9.23cm^{-1} for this molecule.

Table 2: Observed, calculated frequencies (in cm^{-1}) with DFT/B3LYP/6-311++G and IVP; and vibrational assignment of 3M4NA

S.No	Mode[a]	Obs. freq.(cm^{-1})		Cal. freq.(cm^{-1})			Vibrational assignment[b]
		IR	Raman	DFT		IVP	
				Unscaled	Scaled		
a'-species							
1	ν(C-C)1	-	675	699	679	673	18a(32)+1(23)+7b(18)
2	ν(C-C)8a	1587	1590	1626	1603	1589	8a (65)+3(12)+6a(10)
3	ν(C-C)8b	1612	1618	1648	1623	1620	8b(76)+6b(10)
4	ν(C-C)14	1269	1269	1240	1250	1265	14(78)+18a(12)
5	ν(C-C)19a	-	1484	1496	1473	1498	19a(67)+7b(24)
6	ν(C-C)19b	-	-	1510	1498	1525	19b(51)+3(49)
7	ν(C-H)2	3102	-	3221	3102	3100	2(100)
8	ν(C-H) 20a	-	3082	3207	3082	3089	20a(97)
9	ν(C-H)20b	-	3002	3206	3002	3002	20b(95)
10	ν(C-C_α)13	1200	1204	1172	1207	1195	14(46)+13(24)+ν(C_α-OH)(13) +18a(10)
11	ν(C-N)7a	-	-	1093	1094	1075	7a(48)+6a(20)+8a(13)+19a(11)
12	ν(C-C')7b	1161	1160	1163	1166	1171	7b(40)+1(32)+12(11)+18a(11)
13	β(CH)3	1297	1288	1234	1279	1265	3(73)+14(15)
14	β(CH)18a	1127	-	1112	1122	1118	1(43)+18a(25)+ν(C_α-OH)(11)
15	β(CH)18b	-	1078	1061	1085	1090	18b(65) +19b(17)
16	β(CC_α)15	-	256	251	258	237	15(68)+9a(18)
17	β(CN)9b	304	-	300	294	314	9b(50)+γ(NO_2)(16)
18	β(CC')9a	338	-	370	380	364	9a(43)+9b(22) +γ(NO_2)(19)
19	β(CCC)6a	534	-	505	490	502	6a(35)+13(25)+δ(C_α-OH)(19)
20	β(CCC)6b	-	372	425	395	390	6b(32)+1(23)+7b(16)
21	β(CCC)12	836	837	850	852	833	12(29) +δ(C_α=O)(24) +1(22)
a''-species							
22	π(CH)5	853	-	873	890	870	5(78)
23	π(CH)11	789	785	800	783	794	11(75)+10b(10)
24	π(CH)17b	-	-	935	945	936	17b(85)
25	π(CC_α)10b	-	201	192	196	188	10b(29)+10a(21)+ω(OH)(19)
26	π(CN)17a	-	-	180	182	169	17a(75) +4(23)
27	π(CC')10a	270	-	257	264	262	10a(31)+τ(CC_α)(22) +τ(NO_2)(18)
28	τ(CCCC)4	736	-	733	733	740	4(55)+ω(C_α=O)(23) +5(13)
29	τ(CCCC)16a	586	-	574	573	572	16a(81) +ω(OH)(14)
30	τ(CCCC)16b	-	496	482	510	517	16b(81)+5(10)
Vibrations of methyl group							
31	ν_s(CH_3) a'	-	2943	3043	2943	2915	ν_s(CH_3)a' (99)
32	ν_{as}(CH_3) a'	2984	2981	3108	2984	2987	ν_{as}(CH_3)a' (99)
33	δ_s(CH_3) a'	-	-	1378	1360	1371	δ_s(CH_3)a'(99)
34	δ_{as}(CH_3) a'	1399	1393	1423	1408	1397	δ_{as}(CH_3)a'(69)+19a(16)
35	γ(CH_3) a'	1035	-	997	1028	1046	γ(CH_3)a'(69)+δ_{as}(CH_3)a'(30)
36	ν_{as}(CH_3) a"	-	-	3121	2961	2977	ν_{as}(CH_3)(92)a"
37	δ_{as}(CH_3") a"	-	1443	1514	1447	1453	δ_{as}(CH_3) a"(86)
38	γ(CH_3") a"	-	-	1026	1040	1019	γ(CH_3)a"(66) +δ_{as}(CH_3)a"(19)
39	τ(CH_3) a"	-	83	78	79	72	τ(CC')(72)+16b(13)

Cont'd on next page

Journal of Physics: Conference Series **759** (2016) 012057 doi:10.1088/1742-6596/759/1/012057

Vibrations of nitro group							
40	$\nu_s(NO_2)$ a'	1348	1348	1366	1335	1344	$\nu_s(NO_2)(37)$ +19b(19) +7a(19)
41	$\nu_{as}(NO_2)$ a'	1525	1530	1581	1530	1526	$\nu_{as}(NO_2)(69)$ +19b(16)
42	$\delta(NO_2)$ a'	764	-	756	760	790	$\delta(NO_2)(38)+\gamma NO_2(36)$ +12(14)
43	$\Upsilon(NO_2)$ a"	-	569	573	576	572	$\gamma(NO_2)(31)$+6b(28) +9b(14)
44	$\omega(NO_2)$ a"	665	-	666	667	679	$\omega(NO_2)(59)$+4(30)
45	$\tau(NO_2)$ a"	-	-	48	53	40	τ(CN)(82)
Vibrations of acid group							
46	$\nu(C_\alpha{=}O)$ a'	1689	-	1793	1687	1698	$\nu(C_\alpha{=}O)(70)+\delta(OH)(11)$
47	$\nu(C_\alpha\text{-}OH)$ a'	1315	1304	1344	1304	1306	$\nu(C_\alpha\text{-}OH)(22)$ +12(19) +18a(18)
48	$\delta(OH)$ a'	1429	-	1425	1410	1401	$\delta(OH)(71)+\nu(C_\alpha{=}O)(15)$
49	$\nu(O\text{-}H)$ a'	-	-	3768	3768	3768	$\nu(O\text{-}H)(100)$
50	$\delta(C_\alpha{=}O)$ a'	604	-	603	602	615	$\delta(C_\alpha{=}O)(25)$ +18a(20) + $\nu(C_\alpha\text{-}OH)(16)$
51	$\delta(C_\alpha\text{-}OH)$ a'	-	468	471	476	460	$\delta(C_\alpha{=}O)(27)$ +12(18)+ $\delta(C_\alpha\text{-}OH)(13)$
52	$\omega(OH)$ a"	928	929	927	928	936	$\omega(OH)(32)$ +11(31)+10a(13)
53	$\tau(C_\alpha{=}O)$ a"	-	116	159	130	144	$\omega(C_\alpha{=}O)(68)$+11(14) +$\tau(C_\alpha\text{-}O)$
54	$\tau(C_\alpha\text{-}O)$ a"	-	-	58	63	54	$\tau(CC_\alpha)(77)+\omega(OH)(11)$

[a]: Mode in Wilson's notation [39]. C_α represents Carbon atom of acid group and C' refers to Carbon atom of methyl group.
[b]: Assignment shown in this column is based in IVP. Number in the parenthesis % of PED and number before the parenthesis is vibrational mode. PED less than 10% is not shown. ν, stretching; β, in-plane bending; γ, in-plane rocking; ω, wagging; δ, deformation; π, out-of-plane bending; τ, torsion.

3.2. Computations based on classical mechanics
This section deals with the solution of inverse vibrational problem, wherein Wilson's GF-matrix method was employed for the normal coordinate analysis [39]. The computations were carried out using the structure parameters (Table 1) obtained from DFT computations, whereas the internal coordinates and the symmetry coordinates were similar to those used in our earlier work on 4-methyl-3-nitrobenzoic acid [16], which were derived following the suggestions made by Brooks and Cyvin for benzene [40] and the modifications mentioned in our earlier paper [41].

The initial force constants were transferred from our earlier work on nitrotoluenes and methylbenzenes [41] for methyl nitrobenzene part, whereas those of acid group were taken from the literature [42, 43]. The initial set of 89 force constants was subsequently refined with a damping factor of 0.10 employing overlay least-squares method using Schachtschneiders program [44]. The refinement converged in four cycles with an average error of 11.61 cm^{-1} between the observed and calculated frequencies. The PED obtained with final set of force constants, vibrational assignments along with calculated frequencies and mixing of modes are presented in Table 2 for 3M4NA.

4. Results and discussion
4.1. Vibrational assignments
Assignments of both in-plane and out-of-plane vibrations were made by DFT and IVP methods by referring to potential energy distribution (PED) and eigen vectors obtained in the computations. A considerable help from Gauss View program was derived in DFT. There are no significant differences between the PED and hence the vibrational assignments in the two methods. Hence, the vibrational assignment that arrived based on IVP is presented in Table 2. The results presented in Table 2 are self-explanatory. Hence, the discussion is unwarranted.

4.2. First order hyperpolarizability
The total molecular dipole moment (μ_t) and its components; total molecular polarizability (α_t) and its components; and first order hyperpolarizability (β_t) and its components were computed using DFT/B3LYP/6-311++G level of theory. The first order hyperpolarizability is a measure of non-linear optical (NLO) effects. Urea is a prototypical molecule used in the NLO properties of molecular systems. Hence, it was used frequently as a threshold value for the purpose of comparison. Based on

DFT computations, the calculated values of dipole moment and first order hyperpolarizability of 3M4NA are 4.0731 Debye and 1.982 x 10^{-30} cm^5/esu, whereas the corresponding values for Urea are 1.3732 Debye and 0.3728 x 10^{-30} cm^5/esu, respectively. Thus, we find that the value of μ_t and β_t for 3M4NBA is greater than the corresponding threshold values of Urea. Hence, it can be concluded, on the basis of the magnitude of the first order hyperpolarizability, that 3M4NA may be a potential applicant in the development of NLO materials.

4.3. Frontier molecular orbitals

As presented in our earlier paper [17] and references cited therein, the frontier molecular orbitals (HOMO and LUMO) and their properties such as energy are very useful for Physicists and Chemists and are very important parameters for quantum chemistry. The HOMO and LUMO computed energies at B3LYP/6-311++G level are illustrated in figure 5. The energy values of HOMO, LUMO and difference between them are found -0.34510 eV, -0.21580 eV and 0.1293 eV for 3M4NA, respectively. It is seen from the HOMO and LUMO energy values that the chemical potential of the molecule under investigation is negative and hence, the compound is stable.

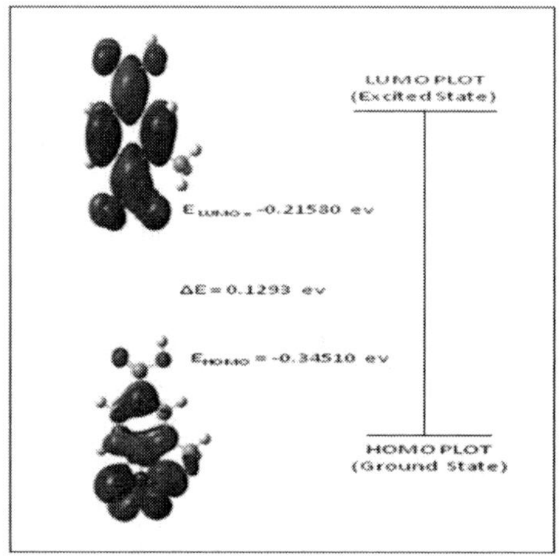

Figure 5. Frontier molecular orbitals of 3M4NA

Conclusion

The most stable conformer with the minimum energy having C_s point group symmetry has been suggested for 3-methyl-4-nitrobenzoic acid. The structure parameters evaluated in the study for nitrobenzoic acid are in agreement with the available literature values. Both DFT and IVP methods have demonstrated that the calculated frequencies are in good agreement with the corresponding experimental values. The vibrational assignments arrived at on the basis of DFT and IVP results are in good agreement. It is also observed that there is no mixing of in-plane vibrations with those of the out-of-plane vibrations in the molecule even though such mixing is not symmetry-forbidden. A 89-element modified valence force filed is evaluated for this molecule. It is also demonstrated that title compound exhibits NLO property. HOMO-LUMO energies are also determined for the molecule.

Acknowledgement

The authors sincerely acknowledge the University Grants Commission (UGC), New Delhi, India (F. No. 41-960/2012 (SR), dt.26/7/2012) for the financial support granted for this investigation. The authors are also thankful to the Sophisticated Analytical Instrumentation Facility (SAIF), IIT Madras, Chennai, India for the spectral measurements recorded.

References:

[1] Snyder R G and Schachtschneider J H 1963 Spectrochim. Acta A. 19 117.
[2] Andzelm J and Wimmer E 1992 J. Chem. Phys. 96 1280.
[3] Handy N C, Maslen P E, Amos R D and Andrews J S 1992 Chem. Phys. Lett. 197 506.
[4] Murray C W, Laming G J, Handy N C and Amos R D 1992 Chem. Phys. Lett. 199 551.
[5] Johnson B J, Gill P M W and Pople J A 1993 J. Chem. Phys. 98 5612.
[6] Florian J and Johnson B G 1994 J. Chem. Phys. 98 3681
[7] Parr R G and W.Yang 1989 Density functional Theory of Atoms and Molecules, Oxford, New York
[8] Konh W and Sham L J 1965 Phys. Rev. A 140 1133.
[9] Baker J, Scheine A and Andzelm J 1993 Chem. Phys. Lett. 216 380.
[10] Lopez Tocon I, Becucci M, Pietraperzia G, Castellucci E and Otero J C 2001 J. Mol Struct. 565 421.

[11] Pulay P, Fogarasi G, Pongor G, Boggs J E and Vargha A 1983 J. Am. Chem. Soc. 105 7037.
[12] Yoshida H, Ehara A and Matsuura H 2000 Chem. Phys. Lett. 325 477.
[13] (a) Sundius T 2002 Vib. Spectrosc. 29 89.
(b) Molvib (V 7.0): Calculation of Harmonic Force Fields and Vibrational Modes of Molecules 2002, QCPE Program No. 807.
[14] Ojha J K, Venkatram Reddy B and Ramana Rao G 2012 Spectrochim. Acta A. 96 632.
[15] Prabavathi N, Senthil Nayaki N and Venkatram Reddy B 2015 Spectrochim. Acta A 136 1134.
[16] Prashanth J, Ramesh G, Laxman Naik J, Ojha J K, Venkatram Reddy B and Ramana Rao G 2015 Optics and Photonics Journal 5 91.
[17] Naik J L, Venkatram Reddy B and Prabavathi N 2015 J. Mol. Struct. 1100 43.
[18] Wilson E B 1934 Phy. Rev. 45 706.
[19] Varasanyi G, Assignments for Vibrational Spectra of Seven Hundred Benzene Derivatives, 1974 Vol. I, Adam Hilger, London.
[20] James C, Ravikumar C, Sundius T, Krishnakumar V, Kesavamurthy R, Jayakumar V S and Hubert Joe I 2008 Vib. Spectrosc. 47 10.
[21] Furic K and Durig J R 1986 Chem. Phys. Lett. 126 92.
[22] Gunasekaran S, Natarajan R K, Syamala D and Rathikha R 2006 Indian J. Pure and Appl. Phys. 44 315.
[23] Sundaraganesan N, Dominic Joshna B, Settu K 2007 Spectrochim. Acta A 66 381.
[24] Sundaraganesan N, Anand B, Meganathan C, and Dominic Joshua B 2008 Spectrochim. Acta A 69 871.
[25] Sundaraganesan N, Dominic Joshna B, Meganathan C, Meenakshi R and Cornard J P 2008 Sectrochim. Acta A 70 376.
[26] Ramalingam M, Sundaraganesan N, Saleem H and Swaminathan J 2008 Specrochim. Acta A 71 23.
[27] Sundaraganesan N, Dominic Joshua B and Radjakoumar T 2009 Ind. J. Pure and Appl. Physics 47 248.
[28] Tyagi O.D and Yadav M 1990 Text Book of Synthetic Drugs, Fifth ed., Anmol Publications, New Delhi, pp 49.
[29] Guo H, Li M, Chen P, Blake D J, Kong X, Hao X, Niu R and Zhang N 2012 Cancer lett. 324 (2) 232.
[30] Frisch M J, Trucks G.W. et al 2009 Gaussian, Inc., Wallingford CT.
[31] Becke A D 1993 J. Chem. Phys. 98 5648.
[32] Lee C, Yang W and Parr R G 1988 Phys. Rev. B 37 785.
[33] Berces A and Ziegler T 1993 J. Chem. Phys. 98 4793.
[34] Pulay P, Fogarasi G, Pongor G, Boggs J E and Vargha A 1983 J. Am. Chem. Soc.105 7037.
[35] Fogarasi G and Pulay P in: J. R. Durig (Ed), 1985 Vibrational Spectra and structure, Elsevier, Amsterdam, Vol. 14, Chapt. 3, pp 125.
[36] Fogarasi G, Zhou X, Tayler P W and Pulay P 1992 J. Am. Chem. Soc.114 8191.
[37] Sundius T 1990 J. Mol. Struct. 218 321.
[38] (a) Sundius T 2002 Vib. Spectrosc. 29 89.
(b) Molvib (V 7.0): Calculation of Harmonic Force Fields and Vibrational Modes of Molecules 2002, QCPE Program No. 807.
[39] Wilson Jr E B 1939 J. Chem. Phys. 7 1047; 1941 9 76.
[40] Brooks W V J and Cyvin S J 1962 Spectrochim. Acta A 18 397.
[41] Ojha J K, Venkatram Reddy B and Ramana Rao G 2012 Spectrochim. Acta A. 96 632.
[42] Ashok Babu V and Ramana Rao G 1987 Ind. J. Pure Appl. Phys. 25 209.
[43] Daisy Magdaline J and Chithambarathanu T 2012 Ind. J. Pure Appl. Phys. 50 7.
[44] Schachtschneider J H 1966 Technical report No. 57-65, shell development co., Emeryville, California.

Non-adiabatic effects in the $\tilde{A}^2B_2$ and $\tilde{B}^2A_1$ states of $CH_2F_2^+$ through coupling vibrational modes

Rudraditya Sarkar

School of Chemistry, University of Hyderabad, Hyderabad-500046, India

E-mail: rudra.smgr@gmail.com

Abstract. Vibronic coupling between the two energetically close-lying excited electronic states ($\tilde{A}^2B_2$ and $\tilde{B}^2A_1$) of $CH_2F_2^+$ is studied in this article. A reduced dimensional model Hamiltonian is constructed in a diabatic representation and using standard vibronic coupling theory. A detailed topographical analysis of the $\tilde{A}$-$\tilde{B}$ coupled surfaces is presented here and nuclear dynamics study on these surfaces is also studied by using time-dependent wavepacket propagation approach.

1. Introduction

The photophysics of difluoromethane (Freon 32) received renewed attention of experimentalists [1, 2, 3, 4] over the last few decades due its applicability in refrigerant industry instead of chlorofluorocarbons. Only a few theoretical calculations of its cationic electronic states [5] are available in the literature. Recently we focussed on the vibronic dynamics of the cationic electronic states of difluoromethane [6] by using the multi-states-multi-modes (MSMM) vibronic coupling theory developed by Köppel, Domcke and Cederbaum [7]. In this article, we mainly concentrate on the vibronic interaction between the two energetically close-lying $\tilde{A}^2B_2$ and $\tilde{B}^2A_1$ excited electronic states of cationic difluoromethane ($CH_2F_2^+$). A detailed account of the effect of individual coupling vibrational modes as well as the cumulative effect of all coupling vibrational modes on the coupled $\tilde{A}-\tilde{B}$ states is discussed here. The internal conversion dynamics of the excited $\tilde{B}^2A_1$ electronic state to the energetically lower $\tilde{A}^2B_2$ electronic state of $CH_2F_2^+$ is discussed.

2. Theoretical methodology: The vibronic coupling model

The first ($\tilde{A}^2B_2$) and second ($\tilde{B}^2A_1$) excited states of difluoromethane radical cation originate from removal of one electron from the occupied b_2 and a_1 molecular orbitals (MOs), respectively, at its ground electronic state of neutral difluoromethane belonging to C_{2v} equilibrium symmetry. The selection of these two states are made because of their energetic proximity. As we are dealing with two electronic states of $CH_2F_2^+$, the constructed model diabatic vibronic Hamiltonian is represented by a 2⊗2 matrix,

$$\mathcal{H} = -\frac{1}{2} \sum_{i \in a_1, a_2, b_1, b_2} \omega_i \left(\frac{\partial^2}{\partial Q_i^2} \right) \mathbf{1} + \begin{pmatrix} \mathcal{H}_{11} & \mathcal{H}_{12} \\ \mathcal{H}_{21} & \mathcal{H}_{22} \end{pmatrix}. \quad (1)$$

Where first term of Eq. 1 represents the kinetic energy of the reference state and **1** represents a (2 × 2) unit matrix. The diagonal matrix element of the second term of Eq. 1, (H_{11}/H_{22}), represents the diabatic electronic energy of the cationic electronic states and the off-diagonal element of the second term of Eq. 1, (H_{12}/H_{21}), represents the coupling between them. All the elements (H_{ij}) of this matrix are function of the dimensionless normal coordinate (NC) of the vibrational modes of neutral reference. The dimensionless NCs are represented here, as Q_g and Q_u, for totally symmetric vibrational modes and coupling vibrational modes, respectively. The nine vibrational modes of CH_2F_2 transform according to the following IREPs of C_{2v} symmetry point group

$$\Gamma = 4a_1 + 1a_2 + 2b_1 + 2b_2. \quad (2)$$

Elementary symmetry selection rule is presented in Eq. 3, allows coupling of $\widetilde{A}$ and $\widetilde{B}$ states (in first order) through the vibrational modes of b_2 symmetry

$$B_2 \otimes b_2 \otimes A_1 \supset A_1. \quad (3)$$

Where the symmetry of electronic states and vibrational modes are denoted by the upper and lower case letters, respectively. The matrix elements of the vibronic Hamiltonian (cf. Eq. 1) are expanded in a Taylor series as

$$H_{ii} = \sum_{k\in a_1} \frac{1}{2}\omega_k Q_g^{k\,2} + \sum_{k\in b_2} \frac{1}{2}\omega_k Q_u^{k\,2} + E_i + \sum_{k\in a_1} \kappa_k^i Q_g^k + \sum_{k\in a_1} \frac{1}{2}\gamma_k^i Q_g^{k\,2} + \sum_{k\in b_1,b_2,a_2} \frac{1}{2}\gamma_k^i Q_u^{k\,2} \quad (4)$$

$$H_{ij} = \sum_{k\in b_2} \lambda_k^{i-j} Q_u^k. \quad (5)$$

The first two terms in Eq. 4 describe the harmonic potential energy of neutral molecule in its electronic ground state and corresponding ω_k values are the harmonic frequencies of the vibrational modes. The term E_i represents the vertical ionization energies of the $\widetilde{A}$ and $\widetilde{B}$ electronic states. The term κ_k corresponds to the linear intra-state coupling parameter of the totally symmetric (a_1) vibrational modes, whereas the term γ_k represents the quadratic intra-state coupling parameters for all vibrational modes. The off-diagonal coupling term, λ_k^{i-j} corresponds to the linear inter-state coupling parameter between $\widetilde{A}$ and $\widetilde{B}$ through coupling vibrational modes.

The adiabatic potential energy along a coupling vibrational mode is given by

$$V_{1,2}(Q_u^k) = \frac{1}{2}\omega_k Q_u^{k\,2} + \frac{1}{2}(\gamma_k^2 + \gamma_k^1)Q_u^{k\,2} + \frac{1}{2}(E_1 + E_2) \mp \sqrt{\left\{(E_1 - E_2) + \frac{1}{2}(\gamma_k^2 - \gamma_k^1)\right\}^2 + 4\lambda^2 Q_u^{k\,2}} \quad (6)$$

A characteristic feature of new minima is observed in lower adiabatic surface $V_1(Q_u^k)$, whereas the upper surface becomes steeper. The symmetry of the nuclear geometry at the new minima is lower than the symmetry of equilibrium geometry of the reference state, this phenomenon is known as "the breaking of molecular symmetry". It is known that the symmetry breaking is simply a consequence of repulsion of the diabatic surfaces via the vibronic coupling [7]. The value of dimensionless normal coordinate at the minimum of the lower adiabatic PES is represented by following equation (excluding the γ_k^i):

$$Q_u^k \left(\omega_k - \frac{\lambda^2}{\sqrt{\left(\frac{E_2 - E_1}{2}\right)^2 + \lambda^2 Q_u^{k\,2}}} \right) = 0. \quad (7)$$

In this equation, $\Delta = \frac{E_2 - E_1}{2}$ and $x = \frac{\lambda^2}{\omega_k \Delta}$ and x is a dimensionless quantity. The three roots of Eq. 6 have the following forms:

$$Q_u^k = 0; \quad Q_u^k = \pm \frac{\lambda}{\omega_k} \sqrt{1 - \frac{1}{x^2}}. \tag{8}$$

If the value of $x < 1$, then the second and third roots of Eq. 7 become imaginary. So the validity of second and third roots remain only when $x \geq 1$ and when $x < 1$ first root $Q_u^k = 0$ is valid. As a result, two equivalent minima form at $Q_u^k \neq 0$ in the lower adiabatic PES when $x > 1$ and the previous minimum at $Q_u^k = 0$ is converted as local maximum. The stabilization energy due to this symmetry breaking phenomenon is $E_s = \Delta\left(\frac{(1-x)^2}{2x}\right)$. No symmetry breaking occur for $x < 1$ and molecule does not get any stabilization due to this phenomenon. Only just above the threshold value of $x = 1$, the stabilization energy quadratically increases with x, whereas at the larger value of x, a linear dependency is observed.

After inclusion of M number of coupling vibrational modes in Eq. 6, the generalized formula of x becomes:

$$x = \sum_{k=1}^{M} x_k. \tag{9}$$

Where, x_k is the dimensionless x parameter for k^{th} coupling mode and $x_k = \frac{\lambda_k^2}{\omega_k \Delta}$. It is seen from Eq. 9 that due to multi-mode effect x is generated from the contribution (x_k) of each coupling vibrational mode. In this way symmetry breaking phenomenon of a molecule becomes cumulative effect of all coupling vibrational modes. So in order to give an explanation of Eq. 9, one can say that if a single coupling vibrational mode fails to introduce a minimum at $V_1(Q_u^k)$ at $Q_u^k \neq 0$, then due to the multi-mode effect of the other coupling vibrational modes, there will be a possibility to form a minimum in the Q_u^k sub-space under the condition of $x \geq 1$.

3. Results and discussion

In the recent past we established a diabatic vibronic coupling model for the first four electronic states of $CH_2F_2^+$ by performing extensive electronic structure calculations and reported the vibronic energy level structure of these electronic states [6]. In this article we consider the first ($\widetilde{A}^2B_2$) and second ($\widetilde{B}^2A_1$) excited electronic states of $CH_2F_2^+$ because of their energetic proximity (~0.40 eV). It is found that the vibronic structures of three excited states, $\widetilde{A}^2B_2$, $\widetilde{B}^2A_1$ and $\widetilde{C}^2A_2$ of $CH_2F_2^+$ are highly overlapping and they form the second photoelectron band of CH_2F_2 [2, 6]. These three states are coupled through multiple conical intersections (CIs) and are well separated from the ground electronic states ($\widetilde{X}^2B_1$) of $CH_2F_2^+$ [6].

The two vibrational modes of b_2 symmetry are represented as ν_8 and ν_9 vibrational modes in our previous paper [6]. The same representation is followed in this article. The vibrational frequencies, first-order (κ) and second-order (γ) intra-state coupling parameters of totally-symmetric modes (represented as ν_1,ν_2, ν_3 and ν_4) and the γ value of ν_8 and ν_9 vibrational modes are reproduced here and given in Table 1. The vertical ionization energies of considered states and the inter-state coupling (λ) of ν_8 and ν_9 vibrational modes are listed in Table 2. It is noted that all the coupling parameters and VIEs are calculated by CAS(14,11)SCF-MRCI level of theory with cc-pVTZ basis set. The effect of inter-state coupling through these vibrational modes ν_8 and ν_9 are presented in Figs. 1 and 2, respectively.

The individual dimensionless x_k parameter values for the coupling modes ν_8 and ν_9 are calculated by using the parameters from Tables 1 and 2 and those values are given in Table 2. It is clear from this table that both the values are lower than the threshold value of x_k =1 (cf. Sec. 2). Thus both these coupling vibrational modes unable to create double minima at the lower adiabatic surface $[V_1(Q_u^k)]$ due to their individual effect on the coupled-surfaces. The individual

Table 1. The vibrational frequencies, κ and γ values of totally symmetric (a_1) and b_2 symmetric vibrational modes are reproduced here from Ref. [6]. The κ and γ values of $\widetilde{A}^2B_2$ and $\widetilde{B}^2A_1$ (in parentheses) of $CH_2F_2^+$ are listed here.

Vibrational mode	Frequencies (eV)	κ (eV)	γ (eV)
$\nu_1(a_1)$	0.3867	0.0582 (-0.1078)	0.0061 (-0.0105)
$\nu_2(a_1)$	0.1948	0.1614 (0.2303)	0.0131 (0.0033)
$\nu_3(a_1)$	0.1421	-0.2717 (-0.0459)	0.0070 (-0.0236)
$\nu_4(a_1)$	0.0667	0.1635 (-0.1275)	-0.0263 (0.0014)
$\nu_8(b_2)$	0.1852		-0.0467 (-0.0068)
$\nu_9(b_2)$	0.1414		-0.1005 (0.0097)

vibronic coupling effect of these two coupling vibrational modes is reflected in the curvature of two PESs (cf. Figs. 1 and 2): one without considering the inter-state vibronic coupling (λ) and another with the inclusion of λ. The upper coupled-surface V_2, is steeper near the minimum at Q_u^k =0 than the upper uncoupled-surface, whereas the lower coupled-surface V_1 is relatively flat compared to the lower uncoupled-surface. This scenario is depicted in both Figs. 1 and 2, for the individual effect of ν_8 and ν_9 coupling vibrational modes, respectively. In these figures dashed lines reprsent the coupled-surfaces and the solid line represent the uncoupled surfaces. So it is established that the symmetry breaking and the formation of double minima at the coupled lower adiabatic surface is not possible due to the two-states-single-mode interaction. We mentioned in the Sec. 2. that the lowering of symmetry of the coupled lower surface also occurs due to the cumulative interaction of all participating coupling vibrational (here, ν_8 and ν_9) modes. The cumulative dimesionless x parameter value ($\sim$1.042) of these two coupling vibrational modes is just above the threshold value of x =1, which suggests a lowering of symmetry of the lower coupled surface due to the simultaneous distortion along two coupling vibrational modes. As a result of this two-states-multi-modes interaction, the lower coupled surface gets stabilization in (Q_u^8, Q_u^9) sub-space and the upper coupled surface becomes steeper.

Table 2. The vertical ionization energies of the $\widetilde{A}^2B_2$ and $\widetilde{B}^2A_1$ electronic states of $CH_2F_2^+$ and the inter-state coupling parameters between these two states are reproduced here from Ref. [6]. The dimensionless x_k parameters and excitation strength are tabulated in last two columns in the table, respectively.

Electronic state	Vertical ionization energy (eV)	Vibrational mode	Inter-state coupling parameter (λ (eV))	Dimensionless x_k parameter	Excitation strength
$\widetilde{A}^2B_2$	15.34	ν_8	0.0983	0.2609	0.14
$\widetilde{B}^2A_1$	15.74	ν_9	0.1487	0.7810	0.55

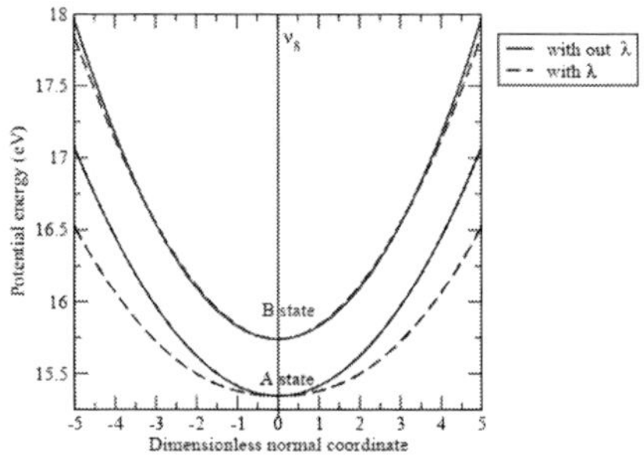

Figure 1. The effect of inter-state coupling between the $\widetilde{A}^2B_2$ and $\widetilde{B}^2A_1$ electronic states of $CH_2F_2^+$ through ν_8 vibrational mode is shown here. The solid (black and blue) lines represent the 1-D PESs in absence of inter-state coupling and the dashed (black and blue) lines represent the 1-D PESs in presence of inter-state coupling between these two states.

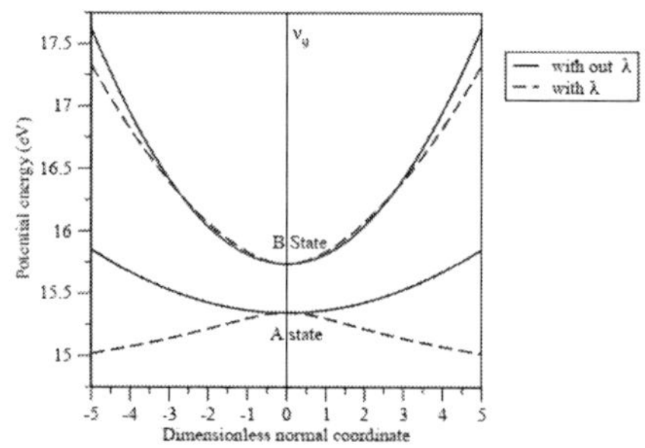

Figure 2. The effect of inter-state coupling between the $\widetilde{A}^2B_2$ and $\widetilde{B}^2A_1$ electronic states of $CH_2F_2^+$ through ν_9 vibrational mode is shown here. The solid (black and blue) lines represent the 1-D PESs in absence of inter-state coupling and the dashed (black and blue) lines represent the 1-D PESs in presence of inter-state coupling between these two states.

The internal conversion (IC) dynamics between the doublet $\widetilde{A}$ and $\widetilde{B}$ states of $CH_2F_2^+$ via $\widetilde{A}$-$\widetilde{B}$ conical intersections is studied by time-dependent wavepacket propagation approach as implemented in MCTDH programing modules [8, 9, 10, 11, 12]. The decay and growth of adiabatic (solid line) and diabatic (dashed line) electronic populations are measured by keeping the initial wavepacket in the $\widetilde{B}$ state of $CH_2F_2^+$ and is plotted in Fig. 3. The value of electronic population of $\widetilde{B}$ state is 1.0 (diabatic) and 0.67 (adiabatic) at t=0 fs and it rapidly decays to the $\widetilde{A}$ state of $CH_2F_2^+$. The indication of the latter phenomena is accounted by the electronic population growth of the latter state. From Fig. 3, it can be seen that the rate of IC via adiabatic path is faster than the diabatic path. The rate of electronic population transfer from $\widetilde{B}$ state to $\widetilde{A}$ state is found to be ~5 fs and ~14 fs, in adiabatic and diabatic framework, respectively. Since, the coupling strength (excitation strength) of ν_9 vibrational mode is ~4 times higher than the vibrational mode ν_8 (cf. Table 2), the contribution of ν_9 vibrational mode is higher in this electronic population transfer. The quasi-degeneracy between the $\widetilde{A}$-$\widetilde{B}$ CIs and the energy minimum of $\widetilde{B}$ state, is another factor for this rapid IC from the $\widetilde{B}$ state to $\widetilde{A}$ state of $CH_2F_2^+$. The calculated energy of the minimum of $\widetilde{A}$-$\widetilde{B}$ CIs is 15.47 eV [13] (within second-order approximation) and the energy minimum of $\widetilde{B}$ state is 15.46 eV [13]. The location of energy minimum of $\widetilde{B}$ state is only ~0.01 eV lower than the $\widetilde{A}$-$\widetilde{B}$ CIs, which allows a strong mixing of the lower vibrational levels of $\widetilde{B}$ state with the higher vibrational levels of $\widetilde{A}$ state. This causes a huge broadening of the spectral profile of these states. The effect of coupling vibrational modes on the $\widetilde{A}$-$\widetilde{B}$ coupled state spectral broadening is shown in Fig. 4. The effect of ν_8 and ν_9 coupling vibrational modes along with the totally symmetric vibrational modes are shown in panels a and b, respectively in the same figure. From the Fig. 4, it can be seen that the spectral broadening of panel b is higher than the panel a, which is in good accord with the higher coupling strength of ν_9 vibrational mode.

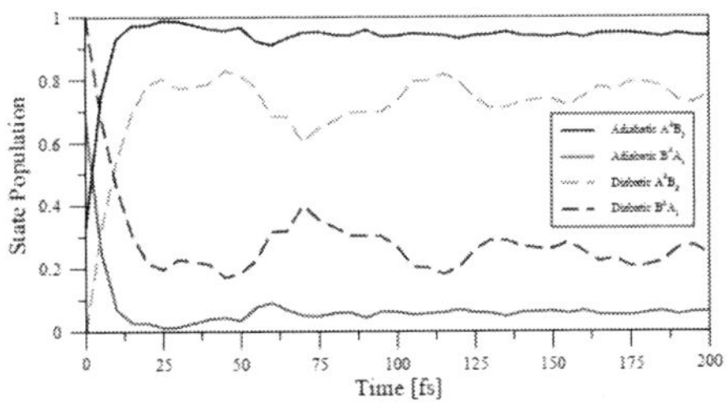

Figure 3. Adiabatic and diabatic state population analysis of the $\widetilde{B}^2A_1$ state as a function of time in femto-second, keeping the initial wavepacket in $\widetilde{B}$ state in the coupled $\widetilde{A} - \widetilde{B}$ surface of $CH_2F_2^+$.

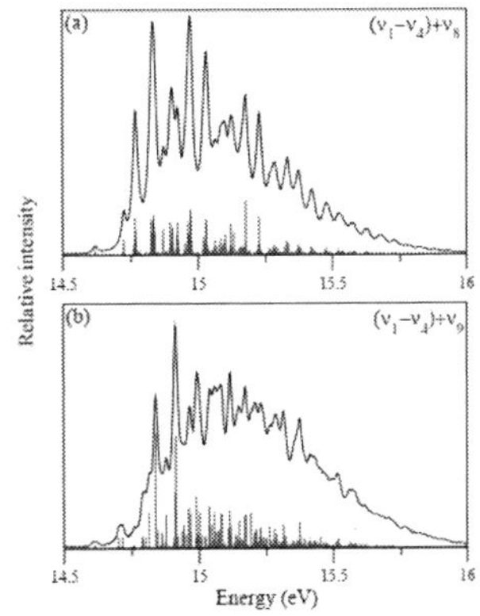

Figure 4. The effect of ν_8 and ν_9 coupling vibrational modes along with totally symmetric vibrational modes on the coupled $\widetilde{A}$-$\widetilde{B}$ state spectrum are shown in panel a and b, respectively.

4. Summary

A theoretical account of vibronic coupling between the two closely lying excited states ($\widetilde{A}^2B_2$ and $\widetilde{B}^2A_1$) of $CH_2F_2^+$ is presented in this article. A model 2⊗2 vibronic Hamiltonian is constructed for the purpose. The effect of coupling vibrational modes (ν_8 and ν_9) on the coupled $\widetilde{A}^2B_2$-$\widetilde{B}^2A_1$ surface is studied here by constructing the two-states-single-mode as well as two-states-multi-modes model Hamiltonian. The result shows that the symmetry breaking and stabilization of lower coupled adiabatic surface is not possible through single mode interaction rather it is possible via cummulative interaction of both coupling modes. The presence of moderate inter-state coupling and the quasi-degeneracy between the $\widetilde{A} - \widetilde{B}$ CIs with the minimum of $\widetilde{B}$ state facilitates the internal conversion between these two states.

Acknowledgements

Rudraditya Sarkar acknowledges UGC, New Delhi, India for his doctoral fellowship. Prof. Susanta Mahapatra, University of Hyderabad, is highly acknowledged for his guidance to shape up this article and valuable discussions.

References

[1] C. R. Brundle, M. B. Robin and H. Basch 1970 J. Chem. Phys. **53**, 2196.
[2] T. Pradeep and D. A. Shirley 1993 J. Electron Spectrosc. Relat. Phenom. **66**, 125.
[3] D. P. Seccombe, R. P. Tuckett and B. O. Fisher 2001 J. Chem. Phys. **114**, 4074.
[4] P. W. Forysinski, P. Zielke, D. Luckhaus, and R. Signorell 2010 Phys. Chem. Chem. Phys. **12**, 3121.
[5] K. Takeshita 1990 Chem. Phys. Lett. **165**, 232.
[6] R. Sarkar and S. Mahapatra 2015 Molecular Physics. **113**, 3073.
[7] H. Köppel, W. Domcke, and L. S. Cederbaum 1984 Adv. Chem. Phys. **57**, 59.
[8] L. S. Cederbaum, W. Domcke, H. Köppel, and W. von Niessen 1977 Chem. Phys. **26**, 169.
[9] H.-D. Meyer, U. Manthe and L.S. Cederbaum 1990 Chem. Phys. Lett. **165**, 73.
[10] U. Manthe, H.-D. Meyer and L.S. Cederbaum 1992 J. Chem. Phys. **97**, 3199.
[11] M.H. Beck, A. Jäckle, G.A. Worth, H.-D. Meyer 2000 Phys. Rep. **324**, 1.
[12] G. A. Worth, M. H. Beck, A. Jäckle, and H.-D. Meyer The mctdh package, Version 8.4, (2007), University of Heidelberg, Heidelberg, Germany. See: http://mctdh.uni-hd.de.
[13] R. Sarkar and S. Mahapatra 2016 J. Phys. Chem. A. 120, 3504.

Journal of Physics: Conference Series **759** (2016) 012059 doi:10.1088/1742-6596/759/1/012059

Atom assisted cavity cooling of a micromechanical oscillator in the unresolved sideband regime

Bijita Sarma and Amarendra K Sarma

Department of Physics, Indian Institute of Technology Guwahati, Guwahati-781039, Assam, India

s.bijita@iitg.ernet.in, aksarma@iitg.ernet.in

Abstract. The ground state cooling of a mechanical oscillator in an optomechanical cavity containing an ensemble of identical two-level ground-state atoms is studied in the highly unresolved-sideband regime. The system exhibits electromagnetically-induced transparency-like quantum interference effect. The mutual interaction with the cavity optical field gives rise to an indirect coupling between the atomic and mechanical modes. In presence of this interaction, the noise spectrum gets modified and leads to asymmetric cooling and heating rates. Using the quantum master equation, time evolution of the average phonon number is studied and it is observed that the average phonon occupancy in the mechanical resonator exhibits ground-state cooling.

1. Introduction

Recently there has been a surge of interest in the study of micromechanical systems in the context of cavity optomechanics. Research in this field has been primarily carried out with a view towards the application of mechanical oscillators for ultrasensitive measurements and as tools for quantum information processing. Since many highly sensitive measurements are limited by thermal noise, it is a prerequisite to go down to the ground-state of these oscillators for these possible applications [1]. Thus ground-state cooling of micromechanical resonators has become a necessary prerequisite for all such studies. In analogy to the laser cooling of ions in strong binding regime [2], conventional cavity cooling of mechanical oscillators requires the condition of resolved-sideband regime, where the cavity mode decay rate should be lower than the mechanical oscillator resonance frequency. However, this condition is hard to be fulfilled in experiments. Here, we consider the ground state cooling of a micromechanical oscillator in an optomechanical cavity that traps an ensemble of identical two-level atoms. Atomic systems have drawn much curiosity in the context of application in the field of optomechanics due to the viability of coupling atoms to cavity and their comparatively lower decay rate [3]. We carry out the analysis in unresolved-sideband regime and we show that even in the highly unresolved-sideband regime, cavity cooling is possible by coupling the mechanical oscillator to the atomic ensemble via the cavity field. It is interesting to observe that this coupling can be harnessed to obtain unequal heating and cooling rates of the mechanical oscillator, thereby reaching the ground-state cooling.

2. Theory

We consider an optomechanical cavity containing an ensemble of identical two-level ground-state atoms as shown schematically in figure 1(a). The cavity is driven by an intense pump laser of frequency ω_l. The Hamiltonian of the system reads

$$H = H_0 + H_I + H_{drive} + H_{bath} \tag{1}$$

where, $H_0 = \omega_c c^\dagger c + \omega_m b^\dagger b + \omega_a S_z$ is the free Hamiltonian of the system, where c and b denote the bosonic annihilation operators of the cavity and mechanical modes, and $S_z = \sum_i \sigma_z^i$ is the collective z- spin operator of the atoms. The second term $H_I = g_{OM} c^\dagger c(b^\dagger + b) + J_0(S_- c^\dagger + S_+ c)$ describes the optomechanical and the atom-field interactions, where g_{OM} is the single-photon optomechanical coupling strength and J_0 is the averaged atom-photon coupling strength. The coherent pumping of the optical mode is described by $H_{drive} = \varepsilon_p(c^\dagger e^{-i\omega_l t} + c e^{i\omega_l t})$ with amplitude $\varepsilon_p = \sqrt{\frac{k_c P_{in}}{\hbar\omega_l}}$. The last term, H_{bath} describes the dissipations of the system. We transform the spin algebra of the atoms to a collective bosonic operator, $a = S_-/\sqrt{N}$. For large atom number and weak atom-photon coupling, $S_z \approx -N/2 + a^\dagger a$. In the frame rotating with the input laser frequency ω_l, the Hamiltonian reads

$$H = -\Delta_c c^\dagger c - \Delta_a a^\dagger a + \omega_m b^\dagger b + g_{OM} c^\dagger c(b^\dagger + b) + J(c^\dagger a + a^\dagger c) + \varepsilon_p(c^\dagger + c) \tag{2}$$

Here, $\Delta_c = \omega_l - \omega_c$ and $\Delta_a = \omega_l - \omega_a$ are the detunings of the cavity mode and the atomic mode with respect to the input laser. $J = J_0\sqrt{N}$ is the collective atom-photon coupling strength. The time evolution of the system operators are described by nonlinear Heisenberg-Langevin equations

$$\dot{c} = \left(i\Delta_c - \frac{k_c}{2}\right)c - ig_{OM}c(b^\dagger + b) - iJa - i\varepsilon_p - \sqrt{k_c}c_{in}(t) \tag{3.1}$$

$$\dot{a} = \left(i\Delta_a - \frac{k_a}{2}\right)a - iJc - \sqrt{k_a}a_{in}(t) \tag{3.2}$$

$$\dot{b} = \left(-i\omega_m - \frac{\gamma_m}{2}\right)b - ig_{OM}c^\dagger c - \sqrt{\gamma_m}b_{in}(t) \tag{3.3}$$

where,k_c, k_a and γ_m are the decay rates of the cavity mode, atomic mode and the mechanical mode respectively. c_{in}, a_{in} and b_{in} are the corresponding input vacuum noise operators with zero mean value and nonzero correlation functions given by $\langle c_{in}(t)c_{in}^\dagger(t')\rangle = \delta(t-t')$, $\langle a_{in}(t)a_{in}^\dagger(t')\rangle = \delta(t-t')$, $\langle b_{in}(t)b_{in}^\dagger(t')\rangle = (n_{th}+1)\delta(t-t')$ and $\langle b_{in}^\dagger(t)b_{in}(t')\rangle = n_{th}\delta(t-t')$ [4]. n_{th} is the bath phonon number given by, $n_{th} = [exp\left(\frac{\hbar\omega_m}{k_B T}\right) - 1]^{-1}$, where k_B is Boltzmann constant and T is the bath temperature. For a strong input laser drive, equations (3.1)-(3.3) can be linearized around the steady state mean values as, $O = O_s + \delta O$. The Langevin equations for the fluctuation terms are given by

$$\dot{\delta c} = \left(i\Delta_c' - \frac{k_c}{2}\right)\delta c - iG(\delta b^\dagger + \delta b) - iJ\delta a - \sqrt{k_c}c_{in}(t) \tag{4.1}$$

$$\dot{\delta a} = \left(i\Delta_a - \frac{k_a}{2}\right)\delta a - iJ\delta c - \sqrt{k_a}a_{in}(t) \tag{4.2}$$

$$\dot{\delta b} = \left(-i\omega_m - \frac{\gamma_m}{2}\right)\delta b - iG(\delta c^\dagger + \delta c) - \sqrt{\gamma_m}b_{in}(t) \tag{4.3}$$

where, $G = g_{OM}c_S$ is the enhanced optomechanical coupling strength due to the driving optical field and $\Delta_c' = \Delta_c - g_{OM}(b_s + b_s^\dagger)$ is the modified detuning. Then the linearized Hamiltonian of the system is given by

$$H_L = -\Delta_c'\delta c^\dagger\delta c - \Delta_a\delta a^\dagger\delta a + \omega_m\delta b^\dagger\delta b + G(\delta c^\dagger + \delta c)(\delta b^\dagger + \delta b) + J(\delta c^\dagger\delta a + \delta a^\dagger\delta c) \tag{5}$$

The radiation pressure force on the mechanical oscillator for the hybrid system is estimated as $F = -\frac{\partial H_I}{\partial x} = -G\,[\delta c^\dagger + \delta c]/x_{ZPF}$, where, x_{ZPF} is the zero-point fluctuation of the mechanical oscillator. The spectral density, for the system is calculated using $S_{FF}(\omega) = \int dt\, e^{i\omega t}\langle F(t)F(0)\rangle$, that equals to be

$$S_{FF}(\omega) = \frac{G^2|\chi(\omega)|^2}{x_{ZPF}^2}[k_c + k_a J^2|\chi_a(\omega)|^2] \tag{6}$$

where, $\chi(\omega) = [\{\chi_c(\omega)\}^{-1} + J^2\chi_a(\omega)]^{-1}$ is the total response function of the hybrid system. Response functions of the cavity optical mode and atomic modes are given by $\chi_c(\omega) = \left[-i(\omega + \Delta_c') + \frac{k_c}{2}\right]^{-1}$ and $\chi_a(\omega) = \left[-i(\omega + \Delta_a) + \frac{k_a}{2}\right]^{-1}$ respectively.

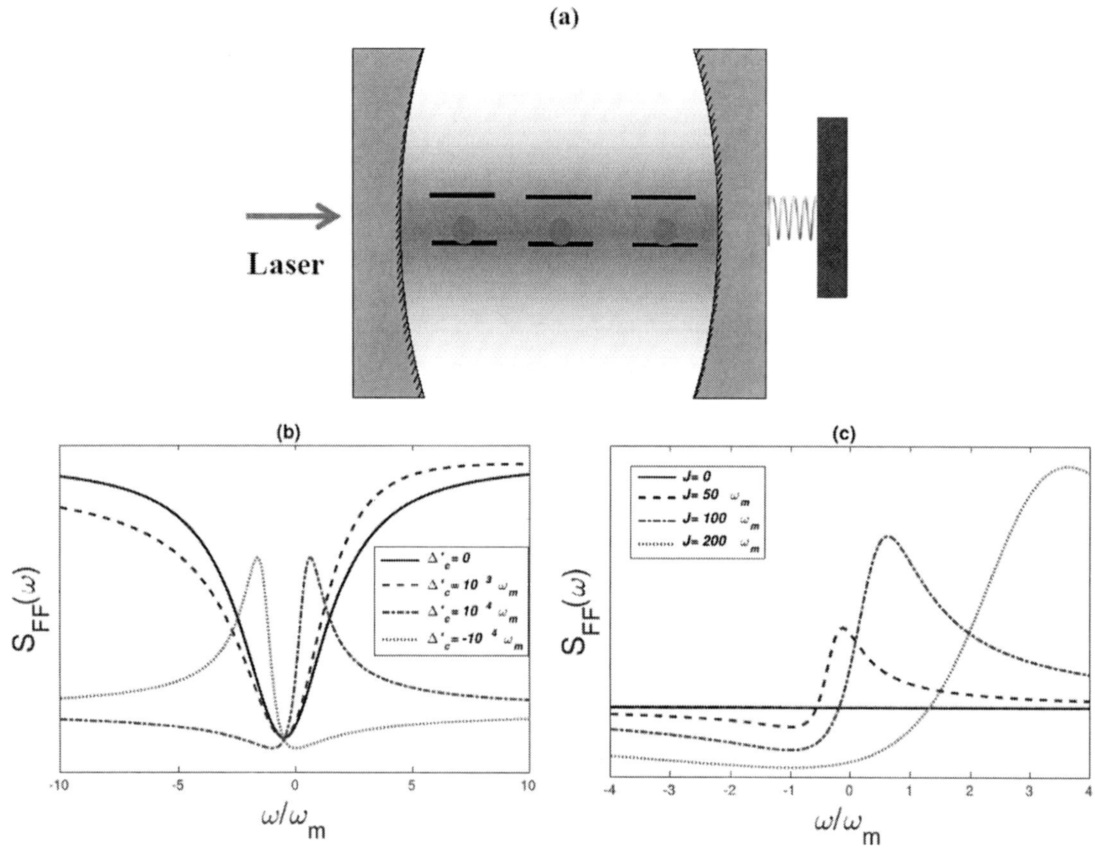

Figure 1. (a) Schematic diagram of an optomechanical cavity containing a two-level atomic ensemble; (b) plot of noise spectrum with varying Δ_c' for $J = 100\omega_m$; (c) plot of noise spectrum with varying J for $\Delta_c' = 10^4\ \omega_m$. Other parameters are $\gamma_m = 10^{-5}\omega_m$, $k_c = 10^4\omega_m$, $k_a = 0.5\ \omega_m$, $\Delta_a = 0.5\ \omega_m$, $G = 50\omega_m$ and $n_{th} = 10^3$.

3. Results

In absence of the atomic ensemble, the setup reduces to a standard optomechanical system. In that case, optimum cooling occurs in the resolved sideband regime $k_c \ll \omega_m$ at cavity detuning $\Delta'_c = -\omega_m$. Here, we are considering the coherent interaction among the optical mode, the mechanical mode and the atomic mode. The optical mode is coupled to the mechanical mode via the optomechanical interaction and in turn the atomic mode is also connected to the optical mode. Thus these modes form a Λ type three-level system. This type of systems exhibits quantum interference effect that is similar to the electromagnetically-induced transparency (EIT) in atomic systems. In figure 1(b), the noise spectrum for different values of modified cavity-field detuning in the unresolved-sideband regime is shown. For a generic optomechanical system, the noise spectrum is standard Lorentzian. However, in presence of the atomic ensemble, the spectrum is not Lorentzian. The cooling and heating rates of the mechanical resonator are given by $A_- = S_{FF}(\omega_m)x_{ZPF}^2$ and $A_+ = S_{FF}(-\omega_m)x_{ZPF}^2$ respectively. The plots clearly indicate that the cooling and heating rates are not equal. The large asymmetry between mechanical cooling and heating processes results in the possibility of enhancing the cooling rate while suppressing the heating rate. As can be seen from figure 1(b), the asymmetry in heating and cooling rates is prominent at $\Delta_c' = 10^4\ \omega_m$ and $-10^4\ \omega_m$. However, the detuning $\Delta_c' = 10^4\ \omega_m$ would give rise to net cooling effect via the enhancement of cooling rate and inhibition of heating rate; whereas

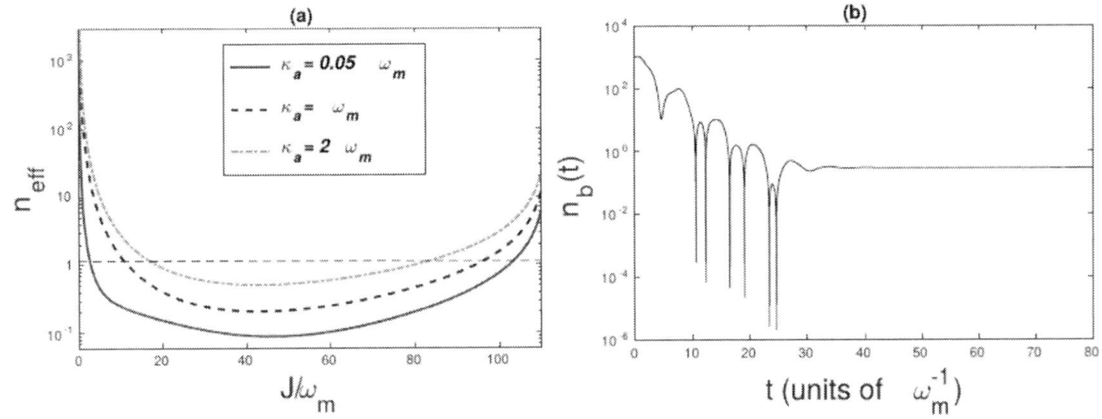

Figure 2. (a) Plot of steady-state cooling limit as function of J for varying k_a; (b) Time evolution of the mean phonon number in the mechanical resonator in presence of the atomic ensemble, starting from $n_{th} = 10^3$. Parameters considered are $k_c = 10^4\omega_m$, $k_a = 0.5\ \omega_m$, $\Delta'_c = 10^4\omega_m$, $\Delta_d = 0.5\ \omega_m$, $\gamma_m = 10^{-5}\omega_m$, $J = 100\omega_m$ and $G = 50\omega_m$.

$\Delta'_c = -10^4\ \omega_m$ would give rise to heating of the mechanical oscillator by the enhancement of heating rate and inhibition of cooling rate. Hence, by tuning the cavity-field detuning, the cooling rate of the mechanical resonator can potentially be enhanced while reducing the heating rate. For further calculations, detuning is taken to be $\Delta'_c = 10^4\ \omega_m$. In figure 1(c), the variation of noise spectrum as a function of the atom-cavity coupling strength J is shown. In the unresolved-sideband regime, in the absence of the atomic ensemble, the noise spectrum is flat as illustrated in the plot. For optimum value of J near $100\ \omega_m$, the asymmetry in cooling and heating rates is prominent.

In the highly unresolved regime, $k_c \gg \omega_m$ and for $k_c \gg (k_a, \gamma_m)$, the cavity mode can be considered as a perturbation. Under this assumption, the three-mode system can be reduced to a two-mode system and in analogy to the standard cavity optomechanical system; the steady-state cooling limit is approximated as $n_{eff} = n_{classical} + n_{quantum}$. For effective resolved sideband regime ($k_{eff} \ll \omega_m$), the classical and quantum cooling limits are given by $n_{classical} = \frac{4G_{eff}^2 + k_{eff}^2}{4G_{eff}^2 k_{eff}}\gamma_m n_{th}$ and $n_{quantum} = \frac{k_{eff}^2 + 8G_{eff}^2}{16(\omega_m^2 - 4G_{eff}^2)}$ [5]. The modified parameters for the effective atomic-mechanical mode interaction are given by: $\Delta_{eff} = \Delta_a - \eta^2\Delta_c'$, $k_{eff} = k_a + \eta^2 k_c$, $G_{eff} = \eta G$, where $\eta = J/\sqrt{\Delta_c'^2 + \left(\frac{k_c}{2}\right)^2}$. Even when the cavity is in the unresolved-sideband regime, the effective interaction of the atomic and mechanical modes can bring the system back to effective resolved-sideband regime. In figure 2(a) we show the steady-state cooling limit n_{eff} as a function of normalized atom-cavity coupling J, for different values of atomic line-width k_a. The plots show the broad range of coupling strength for optimum cooling. It is worth to be noted that for lower values of k_a, the cooling limit is lower and the range of J is broader. Furthermore, we study the time evolution of the average phonon number $n_b(t) = \langle \delta b^\dagger \delta b\ (t)\rangle$ in the mechanical resonator, using the master equation approach. The quantum master equation for the system is given by

$$\dot{\rho} = i[\rho,\ H_L] + \frac{k_c}{2}\left(2\delta c\rho\delta c^\dagger - \delta c^\dagger\delta c\rho - \rho\delta c^\dagger\delta c\right) + \frac{k_a}{2}\left(2\delta a\rho\delta a^\dagger - \delta a^\dagger\delta a\rho - \rho\delta a^\dagger\delta a\right) + \frac{\gamma_m}{2}(n_{th}+1)\left(2\delta b\rho\delta b^\dagger - \delta b^\dagger\delta b\rho - \rho\delta b^\dagger\delta b\right) + \frac{\gamma_m}{2}n_{th}(2\delta b^\dagger\rho\delta b - \delta b\delta b^\dagger\rho - \rho\delta b\delta b^\dagger) \quad (7)$$

To find out the mean phonon number $n_b(t)$, we solve a system of linear differential equations $\partial_t\langle\hat{o}_i\hat{o}_j\rangle = Tr(\dot{\rho}\hat{o}_i\hat{o}_j) = \sum_{m,n}\mu_{m,n}\langle\hat{o}_m\hat{o}_n\rangle$ where, $\hat{o}_i$, $\hat{o}_j$, $\hat{o}_m$, $\hat{o}_n$ are one of the operators:

$\delta a^\dagger$, $\delta b^\dagger$, $\delta c^\dagger$, δa, δb and δc. $\mu_{m,n}$ are the corresponding coefficients. In figure 2(b), the time evolution of the mean phonon number is shown. The bath phonon number is considered to be 10^3. Initially the average phonon number in the mechanical oscillator is equal to the bath phonon number and all other second order moments are zero. The plot shows that the average phonon occupancy in the mechanical resonator tends to zero with time indicating ground state cooling of the resonator.

4. Conclusion

We have studied the cooling of a mechanical resonator in an optomechanical cavity containing an ensemble of two-level atoms in the highly unresolved-sideband regime. The system exhibits quantum interference effect that is similar to the electromagnetically-induced transparency (EIT) in Λ-type atomic systems. Though the atomic and mechanical modes are not coupled directly, the interaction with the cavity optical field gives rise to an indirect coupling between them. The noise spectrum gets modified due to quantum interference effect and leads to asymmetric cooling and heating rates. Using the quantum master equation, time evolution of the mean phonon number is studied and it is observed that the average phonon occupancy in the mechanical resonator exhibits ground-state cooling.

References

[1] Marquardt F, Chen J P, Clerk A A and Girvin S M 2007 *Phys. Rev. Lett.* **99** 093902
[2] Wineland D J and Itano W M 1979 Phys. Rev. A **20** 1521-40
[3] Camerer S, Korppi M, Jöckel A, Hunger D, Hänsch T W and Treutlein P 2011 *Phys. Rev. Lett.* **107** 223001
[4] Gardiner C W and Zoller P 1991 *Quantum Noise* (Berlin: Springer-Verlag)
[5] Dobrindt J M, Wilson-Rae I and Kippenberg T J 2008 *Phys. Rev. Lett.* **101** 263602

Journal of Physics: Conference Series **759** (2016) 012060 doi:10.1088/1742-6596/759/1/012060

Black Hole Paradoxes

Pankaj S. Joshi
Tata Institute of Fundamental Research, Homi Bhabha Road, Colaba, Mumbai 400005, India
E-mail: psj@tifr.res.in

Ramesh Narayan
Harvard-Smithsonian Center for Astrophysics, 60 Garden Street, Cambridge, MA 02138, USA
E-mail: rnarayan@cfa.harvard.edu

Abstract. We propose here that the well-known black hole paradoxes such as the information loss and teleological nature of the event horizon are restricted to a particular idealized case, which is the homogeneous dust collapse model. In this case, the event horizon, which defines the boundary of the black hole, forms initially, and the singularity in the interior of the black hole at a later time. We show that, in contrast, gravitational collapse from physically more realistic initial conditions typically leads to the scenario in which the event horizon and space-time singularity form simultaneously. We point out that this apparently simple modification can mitigate the causality and teleological paradoxes, and also lends support to two recently suggested solutions to the information paradox, namely, the 'firewall' and 'classical chaos' proposals.

One of the most spectacular predictions of the general theory of relativity is the black hole, an object that plays a central role in modern physics [1,2,3] and astrophysics [4,5]. Black holes are, however, plagued by fundamental paradoxes that remain unresolved to this day. First, the black hole event horizon is teleological in nature [6] which means that we need to know the entire future space-time of the universe to determine the current location of the horizon. This is essentially impossible. Second, any information carried by infalling matter is lost once the material falls through the event horizon. Even though the black hole may later evaporate by emitting Hawking radiation [7], the lost information does not reappear, which has the rather serious and disturbing consequence that quantum unitarity is violated [8]. Here we propose that the above paradoxes are restricted to a particular idealized model of collapse first studied in the 1930s [9, 10] in which the event horizon, which defines the boundary of the black hole, forms initially, and the singularity in the interior of the black hole forms at a later time. In contrast, gravitational collapse from more reasonable and/or physically more realistic initial conditions often leads to models in which the event horizon and the singularity form simultaneously. We show that this apparently simple modification mitigates the causality and teleological paradoxes and at the same time lends support to two recently proposed solutions to the information paradox, namely, the "firewall" [11] and "classical chaos" [12].

A black hole is expected to form naturally in the universe whenever a massive star runs out of nuclear fuel at the end of its life and collapses under its own self-gravity. As a result, the infalling mass is compressed to a space-time singularity of infinite curvature and density. This singularity is, however, hidden from view because it is covered by an event horizon, a one way membrane through which particles, light rays and signals can enter but from which nothing can escape.

The above standard picture of black hole formation is motivated by the classic work of Oppenheimer & Snyder [9] and Datt [10], who studied the collapse of a "dust" cloud (fluid with mass density but no pressure) of uniform density. As the space-time diagram in Figure 1 shows, the key feature of this idealized collapse model is that the event horizon forms already at the space-time point A, where local conditions in the infalling dust cloud are perfectly normal, and the density and curvature are finite. The singularity, where the curvature diverges, forms at the point S, which lies in the future of A; indeed, S lies in the future for all worldlines within the dust cloud when they cross the horizon, e.g., the point B. In addition, the location of the horizon at points A and B depends on the entire future history of infalling matter, including shells such as C which fall in later and whose very existence is unknown to the gas falling through A or B. This illustrates the teleological nature of the event horizon: the location of the event horizon is determined by the entire future history of space-time, a profoundly paradoxical situation.

Another problem is that black holes run into a major conflict with quantum theory. A black hole swallows all information carried by matter falling in through the event horizon. When the hole subsequently evaporates by emitting Hawking radiation [7], the mass energy that was swallowed is returned to the external universe, but in an uncorrelated mixed form that carries no information. Thus, a black hole takes in pure quantum states and converts them to mixed states [8]. This violates quantum unitarity, which is a very disturbing prospect and is dubbed the black hole information paradox.

The above paradoxes have attracted considerable attention over the years and various solutions have been proposed. A rather radical solution was suggested recently [11] in which the event horizon is replaced by a firewall. According to this proposal, an infalling observer encounters a firewall of outgoing bolts of radiation at the horizon and is destroyed. Thus, the event horizon is replaced by a curvature singularity — the firewall — associated with outgoing radiation at the horizon. The information carried in by the observer is absorbed at the firewall and is presumably returned via pure quantum states when the singularity radiates or the hole evaporates, thus solving the information paradox. Proponents of the firewall hypothesis present it as "the most conservative resolution" [11] of the information paradox.

The firewall proposal, however, faces several objections, including the fact that CPT invariance of quantum gravity rules out the model [12]. In our view, the most serious problem is the fact that the firewall is by construction located at the event horizon, but the location of the horizon is determined teleologically. Somehow, the firewall singularity must sense the future space-time and thereby decide where it ought to be located. There is no information, either in local conditions of the collapsing matter or in signals received from the past, that provides any indication that a firewall must form.

There are also other models proposed to resolve these issues, such as the 'fuzzball' scenario (see e.g. [13], and also [14]). Yet another alternative to the firewall model was suggested by Hawking [12] proposing that gravitational collapse produces only an apparent horizon and not a true event horizon, and that therefore no information needs to be lost in the collapse. In order to avoid the horizon, Hawking suggests that the region of the collapsed object inside the event horizon develops a chaotic metric and matter fields. Such a chaotic collapsed object would radiate chaotically but deterministically, so quantum unitarity is preserved. However, the chaos model again suffers from the teleological problem. Chaos must be generated at and inside the horizon, but how does the infalling material know that it should become chaotic when local conditions are perfectly normal and when the location of the horizon must be determined by all of the future? Moreover, for near-spherically symmetric collapse models such as those considered here, any chaos is likely to be restricted to regions close to the singularity, and the causal structure of the solution does not permit signals to propagate from this region out to the

Oppenheimer–Snyder Collapse -- Teleological Paradox

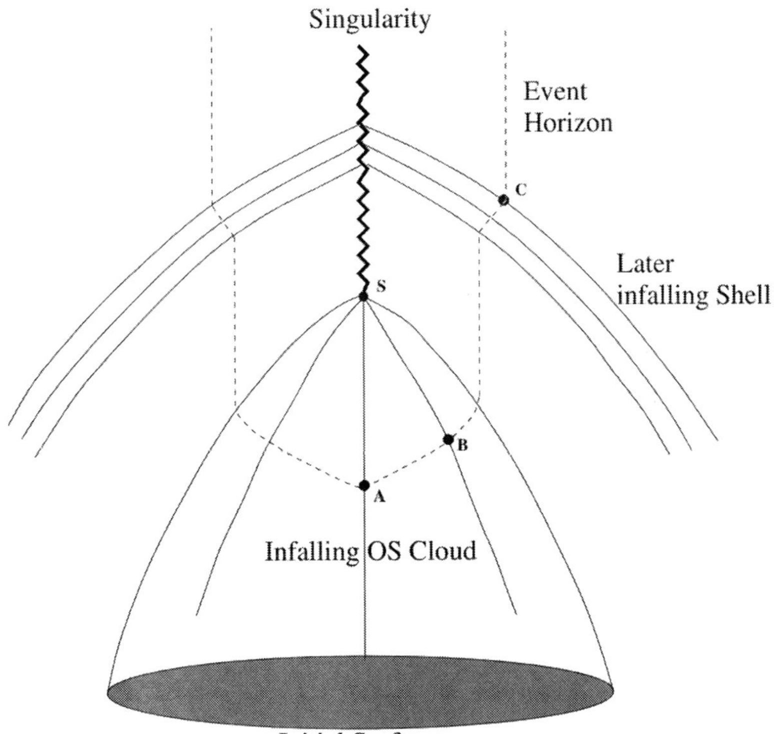

Figure 1. Gravitational collapse of a spherical pressureless matter cloud (dust cloud) of uniform density whose initial state is shown by the shaded region at the bottom. Time increases in the vertical direction, and space is represented by the horizontal directions. The thin solid lines show worldlines of infalling matter. The event horizon (dashed line) first forms on the worldline of the center of the cloud, at the space-time point A, where conditions are very non-singular. Even point B, on the worldline of matter farther out in the cloud, is at an earlier time than the epoch corresponding to S, where the singularity (thick jagged line) first forms. Although the point C, on a subsequently infalling shell of matter, is at a later time than S, no signals can reach this point from the singularity. Therefore, if any new physics, e.g. firewall or chaos, is to be triggered at A or B or C, it has to be via information traveling back from the future, either along the horizon from future infinity or along worldlines of matter backward from the singularity. This is the teleological paradox, which is caused by the fact that the event horizon is causally cutoff from the singularity, a consequence of cosmic censorship.

horizon.

The problems described above arise as a result of assuming a foundational principle of black hole theory, namely, the so-called cosmic censorship conjecture [15]. This conjecture, which is motivated by the space-time diagram shown in Figure 1, suggests that, generically, every singularity that forms as a result of collapse is hidden from view behind an event horizon. That is, the singularity is always cutoff from the external universe. This has the profound consequence that it rules out any signals or communication from the singularity (or its vicinity) to the event horizon. In other words, no worldline at the horizon ever receives any information from the past about the singularity; the only way it can find out about the existence of the singularity is by receiving information from the future (teleological nature of the horizon). We propose that this is the root cause of the black hole paradoxes that we are facing today.

A reasonable alternative emerges if we agree or assume that any solution to the information paradox, such as firewall or chaos, requires information to be received at the horizon about the existence of a singularity in the space-time, and moreover, that this information should be received via signals from the past. If such signals are able to reach the event horizon, they could carry information related to "new physics" that might emerge in the vicinity of the singularity due to the extreme nature of all physical quantities there, and this information could potentially provide a causal trigger to generate either a firewall or chaos. By this reasoning, cosmic censorship and firewalls/chaos are mutually incompatible, since cosmic censorship requires the future of a worldline to determine its present behaviour (teleological communication).

Fortunately, it is now known that the cosmic censorship conjecture is not fully correct since large classes of physically reasonable gravitational collapse models have been worked out within Einstein gravity and been shown to transcend cosmic censorship [16]. In these solutions, the event horizon either is delayed or does not form at all [17], allowing a naked space-time singularity to be visible to the external universe. Moreover, such solutions are by no means fine-tuned. They occur for a wide range of physically reasonable initial conditions.

To illustrate the last point, we note that the (OSD) collapse shown in Figure 1 corresponds to an initial density distribution of the dust cloud of the form

$$\rho(r) = \rho_0, \qquad r \leq r_c. \tag{1}$$

This model is physically rather unlikely since it requires a finite density at the cloud boundary. A homogeneous density sphere is not a good approximation for a massive star. Consider instead the following centrally-peaked model of the initial density, which has the same radius and mass as the constant density model, but is arguably more reasonable,

$$\rho(r) = \frac{5}{2}\rho_0 \left[1 - \left(\frac{r}{r_c}\right)^2\right], \qquad r \leq r_c. \tag{2}$$

The collapse of this centrally-peaked cloud, starting from dilute initial conditions ($r_c \gg GM_c/c^2$ where M_c is the total mass of the cloud), gives the causal structure shown in Figure 2 [18,19]. The key difference from Figure 1 is that the two distinct points A and S in the former have collapsed to a single point. As a result, the event horizon originates at the singularity itself, and the singularity becomes naked. Signals emitted at or near the singularity can have different histories. An infinite family of rays escapes to infinity (making the singularity at least partially visible), some rays are trapped and fall back on the singularity, and a select few rays travel along the horizon, connecting the singularity to points on the horizon. The existence of these last rays — null geodesics that emerge from the singularity and travel along the event horizon — is the key result we highlight here.

Before exploring the consequences, we note that a wide range of spherically-symmetric dust collapse models with different initial density and velocity structure have the same qualitative causal structure as that shown in Figure 2. Furthermore, models that include pressure with reasonable equations of state also behave similarly [20], as does at least one non-spherically symmetric model [21]. The general conclusion is that the causal structure shown in Figure 2 is as plausible as the standard OSD solution (Figure 1), and possibly more realizable in realistic physical situations. One caveat is that models with rotation have not so far been explored, so their collapse properties are not known.

The model illustrated in Figure 2 produces a radically different scenario compared to the OSD picture of collapse (Figure 1) which has formed the basis of all discussions so far on the

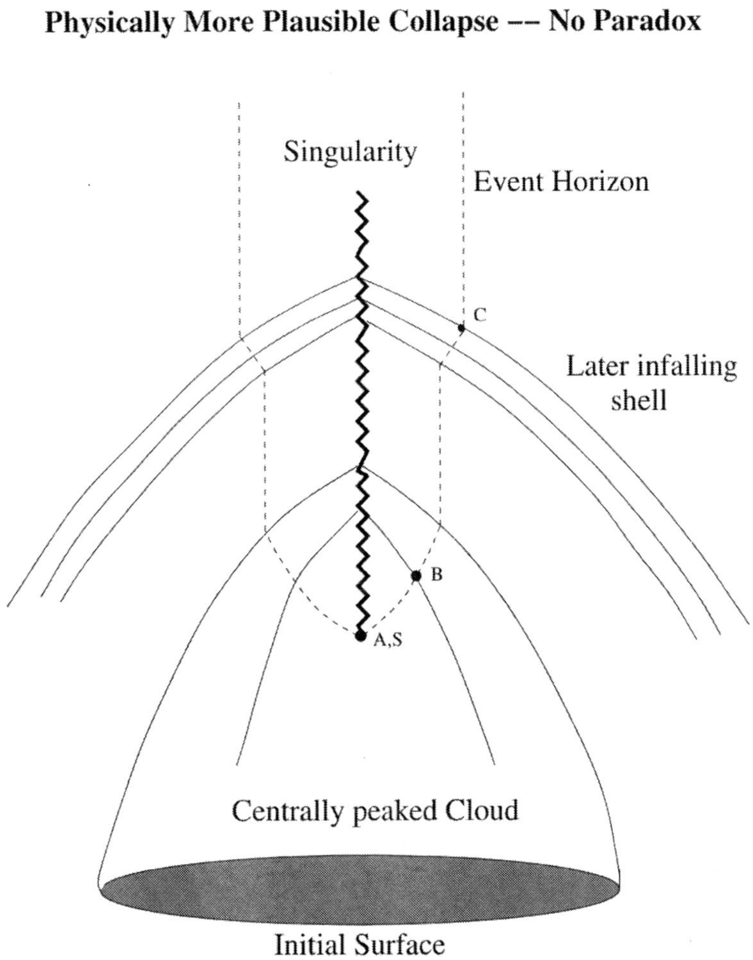

Figure 2. Gravitational collapse of a spherical dust cloud with a density maximum at the center (equation 2). A naked singularity S forms at A, and the event horizon (dashed line) also originates at A. As a result, the singularity can communicate with the entire event horizon, including points such as B and C. New physics, e.g., firewall or chaos, could thus be triggered on the horizon by signals from S, i.e., from the past. Because of the presence of a naked singularity, this model violates cosmic censorship, which is rather restrictive and is confined to special models such as in Figure 1. Correspondingly, there is no teleological paradox. A wide range of physically reasonable initial conditions of the cloud gives collapse with the causal structure shown here.

causal structure of black holes, the information paradox, cosmic censorship. etc. Fluid at the center of the cloud (initial density peak) no longer enters the horizon when it is physically very regular with modest density and curvature. Instead, by the time this matter reaches the horizon, it has already attained extremely high densities and has an arbitrarily large space-time curvature. We expect the matter to behave very much like the hot big bang in reverse, and to become arbitrarily hot and radiation-dominated. Moreover, as the curvature approaches the Planck scale (or other appropriate scale), new physical phenomena associated with quantum gravity should emerge. Most importantly, signals from this ultra-dense region in the quantum gravity regime will flow out along the horizon, conceivably modifying physics throughout the

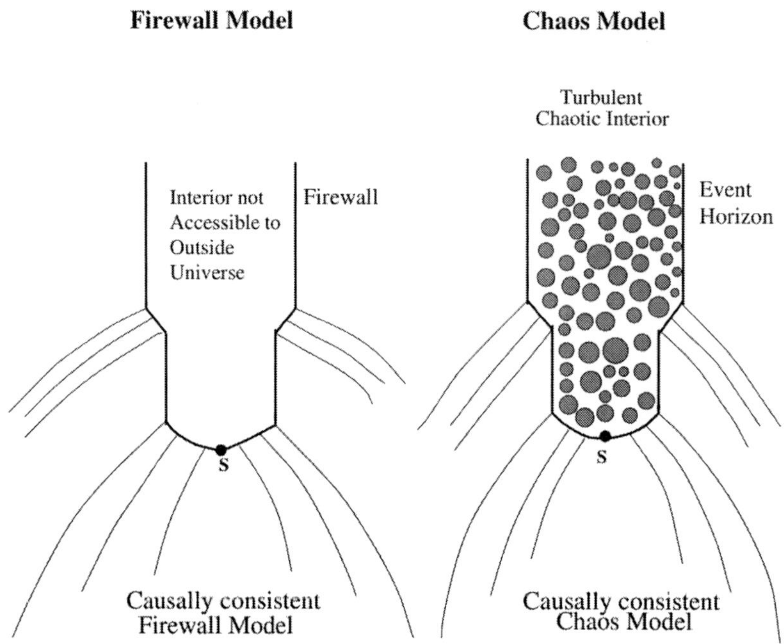

Figure 3. A causally consistent version of the firewall (left) and chaos (right) models. Both models require some exotic phenomenon — a singular firewall or chaotic dynamics — to switch on suddenly at the horizon (thick solid lines). To trigger this behaviour, a warning of some sort must reach fluid that is about to cross the horizon. Because of cosmic censorship, no such warning from the past is possible if the collapse behaves as in Figure 1. In contrast, when the event horizon connects to the singularity in the past, as in the case of the naked singularity scenario shown in Figure 2, signals from the singularity travel to all points on the horizon. Since the singularity is a region of extreme physical conditions corresponding to the quantum gravity regime, signals originating from here could, in principle, trigger the necessary behaviour for the firewall and chaos models. (No specific trigger mechanism is described here since the focus is on establishing causality.)

horizon. Two ways in which this might happen are illustrated in Figure 3.

In one scenario, the quantum matter at the singularity in Figure 2 is radiated away along outgoing rays. The maximum burst of radiation will arguably be along the event horizon because close to that surface and below the singularity the densities, pressures and all other physical quantities attain their maximum and unbounded values. The firewall could then originate at the naked singularity and propagate as a singular wall of outgoing radiation. As material farther out in the cloud approaches the firewall, e.g., point B, even though its own local properties may be quite regular, when it hits the singularity at the firewall, it will be absorbed and will add to the firewall (Figure 3 left). We do not have any explanation of how the latter might happen, but neither was an explanation offered with the original firewall hypothesis. Our contribution here is to show that it is possible to have a firewall originate at a singular point and then evolve causally, without any need for a teleological connection to the future.

Another interesting point is that the null and timelike paths in the vicinity of the naked singularity have been examined in earlier work [22,23] and found to have a complex structure that might well give rise to chaotic behaviour in the interior of the horizon. If this chaos, which will originate in the naked singularity, is able to expand and fill the interior of the horizon, then it is conceivable that the entire interior of the black hole could become chaotic, thereby resembling the classical chaos model [12] (Figure 3 right). Once again, we do not attempt to explain the detailed dynamics of such a model. What we show is that a chaos model could potentially be causally consistent without having to invoke teleological properties.

With reference to chaos from gravitational collapse, we know that the inner region of Kerr geometry is unstable and might be chaotic. If the collapse of a massive rotating star results in a configuration described by the Kerr metric on the outside, this may offer a way to produce chaos in the interior, as suggested by Hawking [12]. However, we note that the unstable region does not extend outside the inner horizon and so chaos cannot propagate all the way to the outer horizon as needed to solve the information paradox. Thus the chaos model requires some other trigger to generate the necessary turbulence. Our proposal in Figure 3 is a possible solution.

In summary, we suggest that some of the problems that have plagued black hole physics might be the result of (i) relying on the classical OSD picture of gravitational collapse of a constant density cloud (Figure 1), in which the event horizon forms much earlier than the singularity, and (ii) assuming that this model and its associated cosmic censorship describes the generic behaviour of collapse. By making use of physically more realistic gravitational collapse models, e.g., those with initial density higher at the center (Figure 2), a very different picture of black hole formation through collapse emerges in which the horizon and the singularity generically form at the same epoch. Such models violate cosmic censorship and can potentially resolve the event horizon and information paradoxes.

Finally, a number of astronomical observations indicate that astrophysical black holes have dark surfaces, and this has been used to argue that they possess event horizons [4,24,25]. In actual fact, the evidence indicates only that the objects possess apparent horizons, so the observations are in principle consistent with the chaos model [12]. The situation is less clear in the case of the firewall model [11], since there has been little discussion of the radiative properties of the firewall.

Acknowledgment: RN thanks the Tata Institute of Fundamental Research for hospitality while this work was completed, and the NSF for partial support under grant AST1312651.

REFERENCES

[1] Susskind, L. and Lindesay, J., "An introduction to black holes, information, and the string theory revolution : the holographic universe", World Scientific, (2005).

[2] Page, D. N., "Hawking radiation and black hole thermodynamics", New Journal of Physics, (2005), **7**, p.203.

[3] Sen, A., "Black hole entropy function, attractors and precision counting of microstates", GRG Journal, (2008), **40**, p. 2249.

[4] Narayan, R., "Black holes in astrophysics", New Journal of Physics, (2005), **7**, p199.

[5] Begelman, M. and Rees, M., "Gravity's Fatal Attraction", (2010), Cambridge, UK: Cambridge University Press.

[6] Hawking, S. W. and Ellis, G. F. R., "The large-scale structure of space-time.", Cambridge (UK): (1973), Cambridge University Press.

[7] Hawking, S. W., "Black hole explosions?", Nature, (1974), **248**, p30.

[8] Hawking, S. W., "Breakdown of predictability in gravitational collapse", Phys. Rev.D, (1976), **14**, p 2460.

[9] Oppenheimer, J. R. and Snyder, H., "On Continued Gravitational Contraction", Phys. Rev. (1939), **56**, p455.

[10] Datt, B., "Uber eine Klasse von Lösungen der Gravitationsgleichungen der Relativität", Zeitschrift fur Physik, (1938), **108**, p314.

[11] Almheiri, A. and Marolf, D. and Polchinski, J. and Sully, J., "Black holes: complementarity or firewalls?", Journal of High Energy Physics, **2**, p62.

[12] Hawking, S. W., "Information Preservation and Weather Forecasting for Black Holes", ArXiv e-prints 1401.5761,

[13] Mathur, S., see e.g. http://arXiv.org/abs/arXiv:1308.2785 and references therein.

[14] Kalyana Rama S., e-Print: arXiv:1211.5645 [hep-th]

[15] Penrose, R., "Gravitational Collapse: the Role of General Relativity", Nuovo Cimento Rivista Serie, (1969), **1**, p252.

[16] Joshi, P. S., "Gravitational Collapse and Spacetime Singularities", Cambridge University Press, Cambridge, UK, 2007.

[17] Joshi, P. S. and Malafarina, D. and Narayan, R., "Distinguishing black holes from naked singularities through their accretion disc properties", Classical and Quantum Gravity, (2014), **31**, p015002.

[18] Joshi, P. S. and Dwivedi, I. H., "Naked singularities in spherically symmetric inhomogeneous Tolman-Bondi dust cloud collapse", Phys. Rev. D, 1993), **47**, p5357.

[19] Deshingkar, S. S. and Jhingan, S. and Joshi, P. S., "On the Global Visibility of the Singularity in Quasi-Spherical Collapse", GRG Journal (1998), **30**, p1477.

[20] Joshi, P. S. and Malafarina, D., "Recent Developments in Gravitational Collapse and Spacetime Singularities", International Journal of Modern Physics D, (2011), **20**, p2641.

[21] Joshi, P. S. and Królak, A., "Naked strong curvature singularities in Szekeres spacetimes", Classical and Quantum Gravity, (1996), **13**, p3069.

[22] Deshingkar, S. S. and Joshi, P. S., "Structure of nonspacelike geodesics in dust collapse", Phys. Rev. D, (2001), **63**, p.024007.

[23] Deshingkar, S. S. and Joshi, P. S. and Dwivedi, I. H., "Appearance of the central singularity in spherical collapse", Phys. Rev. D, (2002), **65**, p084009.

[24] Narayan, R. and McClintock, J. E., "Advection-dominated accretion and the black hole event horizon", New Astron. Rev., (2008), **51**, p733.

[25] Broderick, A. E. and Loeb, A. and Narayan, R., "The Event Horizon of Sagittarius A*", Astrophys. J., 2009, **701**, p 1357-1366.

Journal of Physics: Conference Series **759** (2016) 012061 doi:10.1088/1742-6596/759/1/012061

Strongly correlated classical plasmas under external forcing and dissipation - an example using Molecular Dynamics

Harish Charan and Rajaraman Ganesh[1]

Institute for Plasma Research, Gandhinagr,Gujarat, INDIA

E-mail: [1]`ganesh@ipr.res.in`

Abstract. Systems with excess of average potential energy per particle than its kinetic energy develop strong correlations. It is well known that such systems are not amenable to standard procedures of BBGKY hierarchy. This results in failure of both kinetic and fluid models. Phenomenology is the normal way out. If such a strongly correlated system is further subjected to strong drive and/or dissipation, "near first principles" computational methods such as Molecular Dynamics become necessary. Formation of Rayleigh-Bénnard convection cells (RBCC), where a liquid is under the action of external gravity and external temperature gradient, is one such phenomena. We report here, the formation of RBCC in 2-dimensional strongly coupled Yukawa liquids, characterized by coupling strength Γ (ratio of average potential energy to kinetic energy per particle) and screening parameter κ (ratio of average inter particle distance to Debye length). We observe the existence of a critical external temperature difference, beyond which RBCC are seen to emerge. Beyond this critical external temperature difference, the strength of the maximum convective flow velocity is shown to exhibit a new, linear relationship with external temperature difference and with a slope independent of (Γ, κ). The time taken for the transients to settle down (τ_s) is found to be 10,000 to 20,000 ω_{pd}^{-1}, where ω_{pd} is dust plasma frequency. At very high values of Γ and/or low values of κ, RBCC are seen to get suppressed.

1. Introduction

Strongly coupled plasmas (SCP) can be created in laboratories by immersing dust particles/grains of size of few micrometers and mass of the order of 10^{-13}kg in conventional glow discharge plasmas. Dust grains can acquire charge upto $Q \approx 10^4$e, where e is the electronic charge, grain temperature $T \approx 0.01$eV and gain plasma frequency $f_d \approx$ 1-50 Hz. SCPs can exhibit very high potential energy over kinetic energy per particle. SCP are found in variety of the systems in nature, for example, in dusty (complex) plasma [1], astrophysical plasmas [2], charged colloidal suspension and ultra-cold plasma [3] to mention a few. Due to the large mass and low temperature, grain medium exhibits a slow time response than the background plasma. An average inter-grain distance "a" can be of few millimeters [4, 5]. These stretched time and long length scales of strongly coupled plasmas make it possible to study the "kinetic level" dynamics [6] while at the same time address large scale hydrodynamic phenomena.

Depending on the value of the potential energy over kinetic energy per particle, dusty (complex) plasma can exist in variety of states, for example if the potential energy is very high then dusty (complex) palsma can form "crystalline phase" [4]. In the recent years dusty palsmas have been a keen area of interest in experimental studies of the fundamental phenomena, for example, formation [4] and melting of Coulomb crystal [7], Mach cones excited by laser in Coulomb crystal [8], dynamics of 3-dimensional Yukawa balls [9], shear bending of the Yukawa liquids [10], head-on collision of the dust-acoustic solitons [11] to mention a few. There have been also very interesting simulation studies, for example, large scale Molecular Dynamics (MD) simulations of 2D strongly coupled Yukawa liquids have been performed to study "hydrodynamics" instability such as Kelvin-Helmholtz instability [12] and coherent dipole and tripole structures [13] as initial value problems.

Formation of the pattern such as Rayleigh-Bénard convection cells (RBCC) is one the fundamental non-equilibrium phenomena in a driven dissipative system. In the past simulations studies have been performed for the analysis of RBCC in the conventional fluids described by Lenard-Jones (LJ) potential, for example, formation of RBCC has been addressed by MD simulations [14, 15] and Direct Simulation Monte Carlo [16].

MD simulation of strongly coupled plasmas is a challenging problem with compounded difficulties in the presence of drive and dissipation. An MPI based molecular dynamics code in two and three dimensions has been developed to address the above said issues [12, 13, 17, 18]. The code is modular and is capable of handling a variety of model classical interaction potentials, external forcing, dissipation and a range of boundary conditions. Here we try to address the effect of strong coupling on the formation of the steady state RBCC, where dusty (complex) palsma is subjected to an external temperature gradient [19], in the presence of gravity [20] lead to steady state RBCC. A more detailed description of the work can be found in Ref. [21], where the effect of strong coupling, results for different values of aspect ratio, effect of dust-neutral drag, results for the smaller degrees of freedom etc. have also been addressed.

2. Strongly coupled (dusty/complex) plasma dynamics governed by Yukawa interaction

The inter-grain potential between two grains in a strongly coupled (dusty/complex) plasma is given by Yukawa potential $\phi(r) = [Q^2/(4\pi\epsilon_0 r)]exp(-r/\lambda_D)$ where, Q is the charge on grain, "r" is inter-grain distance, ϵ_0 is the permittivity of vacuum, λ_D is the Debye length of the background plasma. Yukawa systems can be quantified [17] by two dimensionless parameters: (i) Coupling parameter $\Gamma = Q^2/(4\pi\epsilon_0 a k_B T)$ and (ii) Screening parameter $\kappa = a/\lambda_D$. Here "a" (Wigner-Seitz radius) $= 1/\sqrt{n\pi}$, T is dust grain temperature, "m" is mass of dust grain, k_B is Boltzmann constant and n is the 2-D aerial number density. The length, time and energy are normalized with a, ω_{pd}^{-1} and $Q^2/(4\pi\epsilon_0 a)$ respectively, where ω_{pd} is the dust-plasma frequency given by $[Q^2 n/(4\pi\epsilon_0 m a)]^{1/2}$. Other quantities in MD units are given in Table 1. For typical

Table 1. Reduced units used the current study

Reduced quantity	Reduced units
Distance (r)	$\frac{r}{a}$
Time (t)	$t\omega_{pd}$
Temperature (T)	$\frac{k_B T}{(\frac{Q^2}{4\pi\epsilon_0 a})}$
Energy (E)	$\frac{E}{(\frac{Q^2}{4\pi\epsilon_0 a})}$
Acceleration due to gravity (g)	$\frac{g}{(a\omega_{pd}^2)}$

laboratory dusty plasma parameters [22], earth's gravity in our units is found to be $g_e \approx 0.063$.

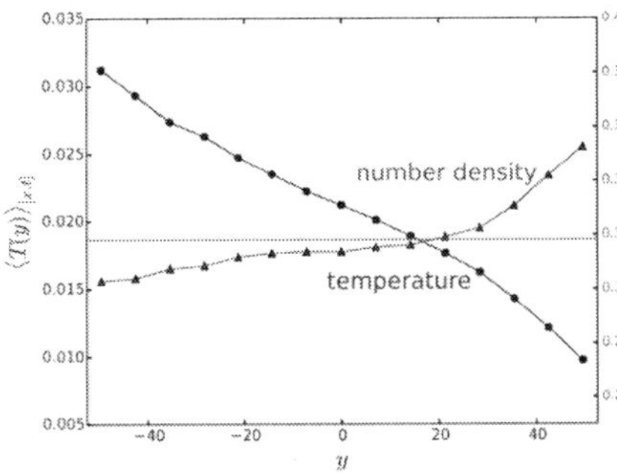

Figure 1. x-averaged density $\langle n(y)\rangle_{[x,t]}$ and temperature $\langle T(y)\rangle_{[x,t]}$ profile for $\Gamma = 50, \kappa = 4,\ g = 0.003g_e$ and $\frac{\Delta T}{T} = 1$.

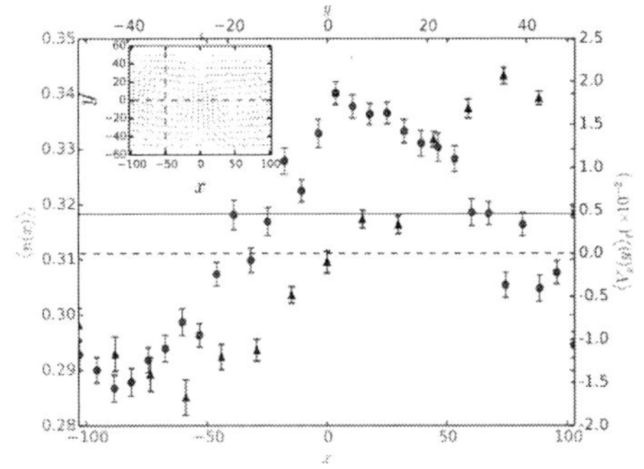

Figure 2. Averaged density $\langle n(x)\rangle_t$ variation with "x" at "y" $\approx$ 0 and horizontal component ("x") of fluid velocity $\langle V_x(y)\rangle_t$ variation with "y" at "x" $\approx$ -50. for $\Gamma = 50, \kappa = 4,\ g = 0.003g_e$ and $\frac{\Delta T}{T} = 1$.

In this work, we use $g \approx 0.0006\ g_e$ to $0.03\ g_e$ or "milli-gravity" values.

3. Simulation details

We use an upgraded version of two dimensional Multi Potential Molecular Dynamics (MPMD) [18, 20] code. We perform MD simulation of 7200 particles interacting through Yukawa potential. The $\hat{x}$ boundaries are periodic and $\hat{y}$ boundaries are specularly reflecting. The upper and lower $\hat{y}$ boundaries are maintained at low and high temperature respectively. We choose a 2-dimensional vertical cross sectional area of a 3-dimensional system, with $L_x = 212a$ and $L_y = 106a$ which gives the aspect ratio $\beta = \frac{L_x}{L_y} = 2$, average number density n = $7200/L_xL_y = 1/\pi$. As shown by Donkó et. al. [23], that for a given κ, viscosity of a strongly coupled Yukawa fluid is a convex function of Γ reaching a minimum for certain Γ. We choose a broad range of values of Γ = 1, 25, 50, 100, 200, 500, 800, 1000 and κ = 0.5, 1, 2 and 4 in the current study. These parameters are routinely attainable in laboratory conditions [4, 5]. We use Gaussian thermostat [24] in the presence of gravitational field to attain a desired bulk temperature T (=1/Γ) of the liquid. The time for thermalizing the system is 500 ω_{pd}^{-1} and then we let the system to evolve micro-canonically for another $500\omega_{pd}^{-1}$ at end of which it attains a thermal equilibrium corresponding to the desired Γ. A Leapfrog integrator with the time step size $\Delta t = 0.01$ is used such that the fluctuation in the total energy without the thermostat is $< 10^{-3}$ over an interval of $1000\omega_{pd}^{-1}$ To generate a temperature gradient in the system we follow the method used by Vannitsem and Mareshchal [25]. In the present study we use relative temperature difference $\frac{\Delta T}{\overline{T}}$ as a key parameter for deciding the onset of the convection cells, where $\overline{T}$ is the average temperature.

4. Results

The system of particles is fluidized on to a 30×15 grid to analyze the "fluid" properties of the system. In FIG.-1 we plot "x"-averaged density and temperature profile for (Γ = 50, κ = 4), $g = 0.003g_e$ and $\frac{\Delta T}{\overline{T}} = 1$. The density inversion due to the external temperature gradient is clearly seen in FIG.1, where density is seen to stay close to its mean value set at the start of the simulation. In FIG.-2 we plot the averaged horizontal "x"-fluid-velocity $\langle V_x(y)\rangle_t$ as a function of "y" at "x" $\approx$ -50 and density $\langle n(x)\rangle_t$ as a function of "x" at "y" $\approx$ 0. In FIG.-3 we plot the "fluid" velocity field and local vorticity, constructed from the particle information. It is clearly

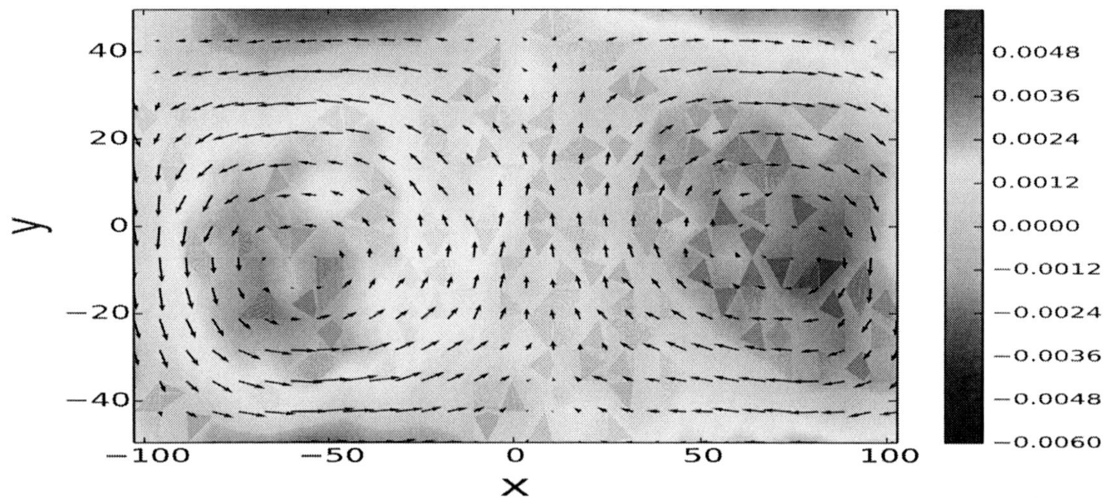

Figure 3. The snapshots show vorticity ($\vec{\omega} = \vec{\nabla} \times \vec{v}$) plot of the system. Blue and red regions correspond to negative and positive vorticity, respectively, and the color-map label shows the magnitude of local vorticity. Arrows indicate direction of the local "fluid" velocity for $\Gamma = 50, \kappa = 4$, $g = 0.005 g_e$ and $\frac{\Delta T}{T} = 1.7$. The quantities are obtained by taking the successive time averages, for the period of 20,000 ω_{pd}^{-1}. Grain velocities in the region are fluidized through a 30 × 15 grid with 16 particles per cell, to construct local vorticity and velocities.

seen that after the system has reached steady state, the system exhibits a two roll structure for $\beta = 2$. The convection cells do not emerge quickly but go through different transient stages and eventually attain a steady state of one or two rolls of convection cells depending on the value of β ($= \frac{L_x}{L_y}$). For the range of (Γ, κ) and $\frac{\Delta T}{T}$ values considered here, typical time for the transients to settle down (τ_s) to a steady state RBCC is found to be about 10,000-20,000 ω_{pd}^{-1} or 1-2 million time steps. As shown in FIG.4, relative temperature difference ($\frac{\Delta T}{T}$) is seen to scale with maximum horizontal flow velocity ($\langle V_x \rangle_t^{max}$). It is evident from FIG.-5 that RBCC are seen to emerge after the system has reached its steady state, where the total energy is not changing too much around its equilibrium values. It is observed that for a given (Γ, κ), there exists a critical value of relative temperature difference $(\frac{\Delta T}{T})_c$ beyond this value RBCC are seen to emerge and the horizontal convective flow velocity ($\langle V_x \rangle_t^{max}$) increases linearly with $\frac{\Delta T}{T}$. It is interesting to note that the slopes are independent of (Γ, κ) and appear to be "universal" for the Yukawa liquids. This novel linear dependency is demonstrated perhaps for the first time for the strongly coupled systems. The quantities in FIG.1 and 2 are obtained by taking the time average for the period of 20,000 ω_{pd}^{-1} and are obtained for $\Gamma = 50, \kappa = 4$ and $\frac{\Delta T}{T} = 1$. However for very strongly coupled systems, meaning for high Γ (and/or low κ) values the system does not form the steady state RBCC. For example at high Γ (say ≥ 500) we find that the transient phase prolongs up to $\tau_s \approx$ 100,000 ω_{pd}^{-1} and the formation of steady state RBCC is rendered difficult even at very high values of $\frac{\Delta T}{T}$ [see FIG.-6]. In FIG.-6 we plot the velocity field plot for $\Gamma = 1000$ and $\kappa = 4$, as we can see the strong coupling effect is seen to suppress the formation of steady state RBCC. We believe this could be due to a complex dependence of the non-equilibrium transport coefficients at high Γ. We also observe formation of steady state RBCC for a range of β values for Yukawa fluid [not shown here, for details see Figure 12 and

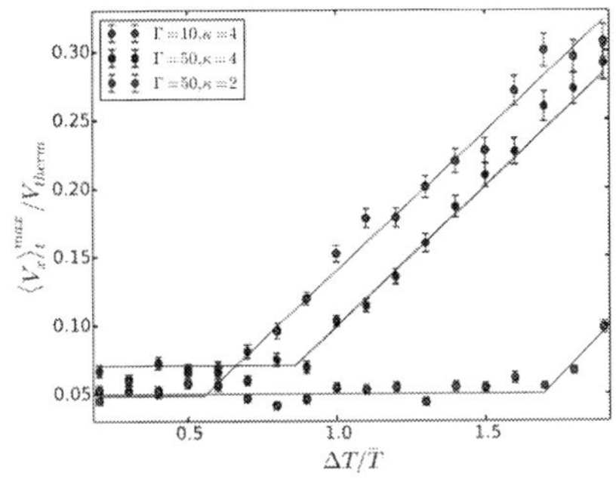

Figure 4. The averaged maximum horizontal convective flow velocity $\langle V_x \rangle_t^{max}/V_{therm}$ as a function of relative temperature difference $(\Delta T/\overline{T})$ for different values of (Γ, κ).

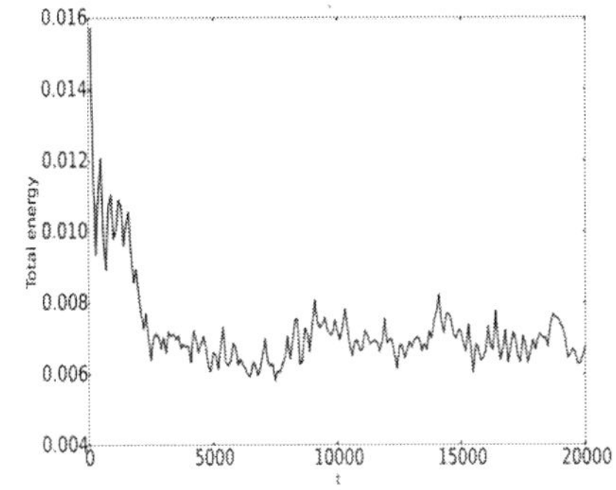

Figure 5. Long time evolution of total energy for $\Gamma = 50, \kappa = 4$ (transient phase) and $\frac{\Delta T}{T} = 1.7$.

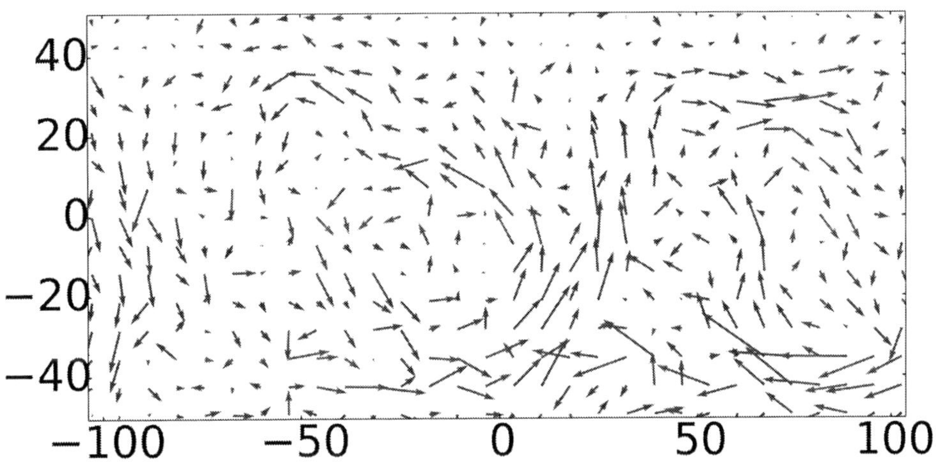

Figure 6. Fluid velocity field for $\Gamma = 1000, \kappa = 4$, $g = 0.00025g_c$ (transient phase) and $\frac{\Delta T}{T} = 1.8$. The velocities shown here are averaged over the period (100,000 ω_{pd}^{-1}).

13 of [21]].

5. Summary

The phenomena of Rayleigh-Bénard instability in strongly coupled Yukawa liquids using molecular dynamics simulations has been addressed. The convection cells take a time period called, transient period τ_s, for the full development. The time taken for the steady state to form is found to be $\tau_s \approx 10{,}000$ - $20{,}000\ \omega_{pd}^{-1}$. A linear relationship between average maximum convective horizontal velocity and relative temperature is found for strongly coupled Yukawa liquids and slopes are found to be independent of (Γ, κ) values. The strong coupling effects are seen to suppress the formation of steady state convective cells and the transient phases are seen to prolong for up to 100,000 ω_{pd}^{-1} or more. Steady state RBCC is also seen for a range of values of aspect ratio β.

In future many unanswered observations will be addressed, for example, physical reason behind origin of linear relationship between average maximum convective horizontal velocity and relative temperature difference and universality of slope etc.

6. Acknowledgments

The work has been performed on Udbhav cluster at Institute for Plasma Research.

References

[1] Shukla P K 2001 *Physics of Plasmas (1994-present)* **8**
[2] Merlino R L and Goree J A 2004 *Phys. Today* **57(7)** 32
[3] Rolston S L 2008 *Physics* **1** 2
[4] Thomas H, Morfill G E, Demmel V, Goree J, Feuerbacher B and Möhlmann D 1994 *Phys. Rev. Lett.* **73**(5) 652–655
[5] Lai Y J and I L 2002 *Phys. Rev. Lett.* **89**(15) 155002
[6] Morfill G, Rubin-Zuzic M, Rothermel H, Ivlev A, Klumov B, Thomas H, Konopka U and Steinberg V 2004 *Phys. Rev. Lett.* **92**(17) 175004
[7] Thomas H M and Morfill G E 1996 *Nature (London)* **379**(6568) 806–809
[8] Melzer A, Nunomura S, Samsonov D, Ma Z W and Goree J 2000 *Phys. Rev. E* **62**(3) 4162–4176
[9] Ivanov Y and Melzer A 2009 *Phys. Rev. E* **79**(3) 036402
[10] Chan C L, Woon W Y and I L 2004 *Phys. Rev. Lett.* **93**(22) 220602
[11] Sharma S K, Boruah A and Bailung H 2014 *Phys. Rev. E* **89**(1) 013110
[12] Joy A and Ganesh R 2010 *Phys. Rev. Lett.* **104**(21) 215003
[13] Joy A and Ganesh R 2011 *Phys. Rev. Lett.* **106**(13) 135001
[14] Rapaport D C 1988 *Phys. Rev. Lett.* **60**(24) 2480–2483
[15] Mareschal M, Malek Mansour M, Puhl A and Kestemont E 1988 *Phys. Rev. Lett.* **61**(22) 2550–2553
[16] Watanabe T and Kaburaki H 1997 *Phys. Rev. E* **56**(1) 1218–1221
[17] Hartmann P, Kalman G J, Donkó Z and Kutasi K 2005 *Phys. Rev. E* **72**(2) 026409
[18] Joy A and Ganesh R 2009 *Phys. Rev. E* **80**(5) 056408
[19] Rothermel H, Hagl T, Morfill G E, Thoma M H and Thomas H M 2002 *Phys. Rev. Lett.* **89**(17) 175001
[20] Charan H, Ganesh R and Joy A 2014 *Physics of Plasmas (1994-present)* **21** 043702
[21] Charan H and Ganesh R 2015 *Physics of Plasmas* **22** 083702
[22] Fortov V, Ivlev A, Khrapak S, Khrapak A and Morfill G 2005 *Physics Reports* **421** 1 – 103 ISSN 0370-1573
[23] Donkó Z and Hartmann P 2008 *Phys. Rev. E* **78**(2) 026408
[24] Evans D J, Hoover W G, Failor B H, Moran B and Ladd A J C 1983 *Phys. Rev. A* **28**(2) 1016–1021
[25] Vannitsem S and Mareschal M 1995 *Phys. Rev. E* **51**(6) 5564–5570

Bayesian Cosmological inference beyond statistical isotropy

Tarun Souradeep[1], Santanu Das[1] and Benjamin Wandelt[2]

1. IUCAA, Post Bag 4 Ganeshkhind, Pune 411007 India
2. IAP, Lagrange Institute, Sorbonne University, Paris

Abstract. With advent of rich data sets, computationally challenge of inference in cosmology has relied on stochastic sampling method. First, I review the widely used MCMC approach used to infer cosmological parameters and present a adaptive improved implementation SCoPE developed by our group. Next, I present a general method for Bayesian inference of the underlying covariance structure of random fields on a sphere. We employ the Bipolar Spherical Harmonic (BipoSH) representation of general covariance structure on the sphere. We illustrate the efficacy of the method with a principled approach to assess violation of statistical isotropy (SI) in the sky maps of Cosmic Microwave Background (CMB) fluctuations. The general, principled, approach to a Bayesian inference of the covariance structure in a random field on a sphere presented here has huge potential for application to other many aspects of cosmology and astronomy, as well as, more distant areas of research like geosciences and climate modelling.

1. Introduction

With the different experimental data the cosmology has become a play room for the data analysts. In recent past we get data from WMAP, Planck, BICEP, BOSS etc and there are many new experiments coming up. We need to analyse the high volume of data and draw inferences from the data. For many data analysis process we require advanced numerical techniques. In this proceeding we discuss two cosmological parameter estimation methods in which we used some advanced numerical methods.

First, we demonstrate a cosmological parameter estimation for different theoretical model by comparing the C_l with the observed C_l. We develop a MCMC code, named as SCoPE [1, 2, 3, 4, 5], based on global Metropolis Hastings algorithm [6, 7, 8, 9]. In general the Metropolis Hastings algorithm is pretty slow. Therefore, we have developed few novel technique that makes the MCMC method faster. In the first section we briefly discuss about these techniques that make SCoPE faster then standard Monte Carlo methods.

In the second part of the proceeding we have discussed about a Monte Carlo method that we use for Bayesian estimation of the isotropy violation in the CMB sky. For the first time, we develop an algorithm for fast basyan estimation of the isotropy violation on the CMB sky. We use Hamiltonion Monte Carlo for the estimation process. The stability of the algorithm was a big issue there. Therefore, we have applied different integration method for testing the stability of the method. We have discussed that in the second part of the proceeding.

2. Metropolis Hastings

For cosmological parameter estimation we use Metropolis Hastings (MH) algorithm. The Metropolis-Hastings (MH) is one of the most widely used MCMC sampler, in which the posterior i.e. $\pi(\theta)$ is sampled using a random walk. A standard Markov Chain at each step i, randomly chooses a candidate value θ_{i+1} from the proposal distribution $q(.|\theta_i)$. The candidate value only depends on the current data point θ_i. The new data point is then accepted with probability $\alpha = \min(1, \pi(\theta_{i+1})/\pi(\theta_i))$. If the new data point rejected, the previous point is replicated by increasing its weight by +1. The chain of data-points thus generated, approximate the target posterior distribution $\pi(\theta)$. The proposal distribution is generally taken to be a Gaussian distribution i.e. $q(\theta_{i+1}|\theta_i) = N\exp(-\frac{s}{2}u_i[C_{ij}]^{-1}u_j)$, where $u_i = \theta_{i+1} - \theta_i$ and C_{ij} is the covariance matrix. s is the step size. Theoretical optimum step size for an ideal distribution that provide the best acceptance rate is $s = 2.4/\sqrt{n}$ for a n dimensional MCMC sampler. The covariance matrix is provided as an input to the program. As the exact covariance matrix is unknown before the analysis, in practice an approximate covariance matrix, often based on some previous analysis, is provided. If no prior information is available about the covariance between parameters then some approximate diagonal matrix is also often used. However, in such cases the acceptance rate of the sampler may reduce drastically and can be ensured to remain reasonable only by trial and error. Therefore a prior knowledge of the covariance is required. Parallelization of MH sampler is generally done by running multiple chains. Whether it is better to run a longer chain than running multiple short chains has been addressed and argued by many authors [10, 11]. But, for running multiple parallel chains proper mixing between the chains has to be ensured. Therefore, each of the multiple chains has to be long enough so that the it can represent an unbiased sample of the population. Gelman-Rubin "R" statistics is generally used for testing the mixing of chains. For convergence, the chains have to be long enough such that R is very close to unity[12]. For practical purposes it is taken as $R < 1.2$. However, this criterion though necessary is often not sufficient for ensuring proper sampling. It is desirable to devise an implementation of MH algorithm that allows the individual chains to run in parallel and increase the acceptance rate of the models of a chain. Apart from that the mixing of the chains are also necessary. Therefore in SCoPE we have modified standard MH algorithm to accomplish effective parallelization through prefetching together with all other above mentioned features, namely, enhanced acceptance, regular covariance update from samples.

2.1. Embellishing the standard Metropolis-Hastings algorithm

2.1.1. Prefetching In MCMC it is extremely useful to speed up the generation of a single chain, through parallelization rather than using multiple chains. In our work we make the individual chains parallel by precomputing several draws from the posterior distribution ahead of time via multiple evolution of models simultaneously in parallel and then use only the values that are needed. Prefetching is a draw level parallelization in a single chain. The method can be explained by taking the binary tree of a Metropolis algorithm as shown in Fig. 1. In a kth level binary tree there are total $2^k - 1$ nodes, each of which represents a possible future state of a metropolis algorithm. The branches at the left child of any node represent the accepted steps and the right child represents the rejected states. If we have enough computational resources then all $2^k - 1$ nodes can be evaluated simultaneously and k steps of a MCMC chain can be carried out in parallel simultaneously. Though the method of prefetching allows to parallelize a single chain, it only uses k steps out of $2^k - 1$ computations. The rest of the computations are not utilized. Therefore, many argue against computing all the nodes of the binary tree. Rather if we know the acceptance rate at a point of time from the previously accepted data points, we can statistically identify and precompute only the most probable chain and hence avoid any unnecessary wastage of computation power. It is easy to see that if the acceptance probability at any point of time is less than 0.5 then the extreme right chain (1-3-7-15-..) of Fig. 1 will

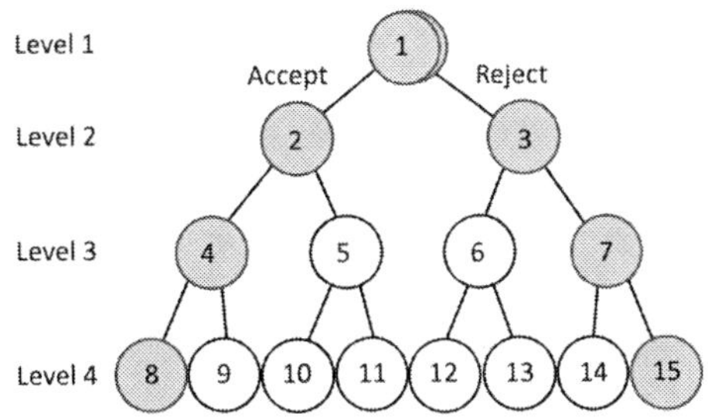

Figure 1. Prefetching scheme is explained in the text with the help of above figure

be most probable chain. In a similar manner if the acceptance rate is more than 0.5 then the extreme left chain (1-2-4-8-..) will be the most probable chain [13]. Therefore, by pre-evoluting only the most probable chain, we parallelize the code and at the same time we can manage the computational resources in a better way.

2.1.2. Delayed rejection One of the major problems with the MCMC method is the choice of the step size for the proposal distribution. It will be better idea if the rejected sample from one step can be used to determine the proposal distribution for the next sample by varying stepsize accordingly. This increases acceptance rate but at a cost of violation of the Markovian property. But if we can find some method that can change the acceptance probability of the sample point to compensate the step size variation then that will be useful. The concept of delayed rejection can briefly be explained as follows. Suppose at some step i, the position of a chain is $\theta_i = x$. Suppose at this time a candidate y_1 is accepted from $q_1(x, y_1)$ and accepted with probability $\alpha_1(x, y_1) = \min(1, \pi(y_1)q_1(y_1, x)/\pi(x)q_1(x, y_1))$ as in the standard MH algorithm. For a Markov chain, $q_1(y_1, x)$ is time symmetric, i.e. $q_1(y_1, x) = q_1(x, y_1)$. Therefore, the acceptance ratio only depends on the posterior. A rejection at any step suggests that there is a local bad fit of the correct proposal and a better one, $q_2(x, y_1, y_2)$, can be constructed in light of this. But, in order to maintain the same stationary distribution the acceptance probability of the new candidate, y_2, has to be properly computed. A possible way to reach this goal is to impose detailed balance separately at each stage and derive the acceptance probability that preserves it. It can be done by changing the acceptance probability as [14, 15, 16, 17, 18, 19]

$$\alpha_2(x, y_1, y_2) = \min\left(1, \frac{\pi(y_2)q_1(y_2, y_1)q_2(y_2, y_1, x)\left[1 - \alpha_1(y_2, y_1)\right]}{\pi(x)q_1(x, y_1)q_2(x, y_1, y_2)\left[1 - \alpha_1(x, y_1)\right]}\right). \quad (1)$$

The Markovian property of the chain will not get destroyed, but still the sample choice can be made dependent on the previously accepted data point. We implement this in our algorithm.

2.1.3. Inter-chain covariance adaptation The practical problem in implementing MH is the tuning problem of the proposal distribution q so that the sampling is efficient. One of the recent improvements in the MCMC efficiency is to introduce adaptive samplers. The adaptive MCMC uses the sample history and automatically tune the proposal distribution in the sampling process [20, 21, 22, 23]. In adaptive metropolis algorithm, the covariance matrix from the samples obtained so far is used as the covariance of a Gaussian proposal. The most common parallelization scheme of MCMC method is to run parallel chains instead of running a single one. If in each chains proposal distribution is adapted using the local covariance matrix then the acceptance probability a chain may improve, however, the inter-chain mixing will not improve. If some chain stuck at some local minima then the local covariance matrix corresponding to that chain will be erroneous. So, in case of a local peak the local covariance update will give covariance corresponding to the local peak. In that case, the mixing of chains will slow down and sampling may not be proper. Therefore, we have adapted the concept of the inter-chain covariance update in the adaptation technique. We run several parallel chains, and randomly update the covariance matrix taking the data points accepted till then from all the chains. This means we have used the covariance after ith accepted step as $\text{Cov}(\theta_{11}.., \theta_{i1}, \theta_{12}, .., \theta_{i2}....\theta_{1n}, ..., \theta_{in})$, where, n is the number of chains. This speeds up the mixing of the chains and covers the sample space faster. The value of the covariance matrix will freeze after few adaptations and hence we will be using same Gaussian proposal after few steps, which is important to guarantee proper sampling. However, if the adaptive covariance is not frozen, it may give rise to unfair sampling as the Gaussian proposal will vary between steps. Therefore, the process of the adaptive covariance calculation is only used for the initial burn in process and after that the adaptation is stopped, during which the covariance calculation attains partial convergence.

3. Estimation of Isotrypy violation with Hamiltonion Monte Carlo

The previous section we briefly demonstrate how the standard MH algorithm can be modified to make a faster parameter estimation algorithm. However, MH algorithm is mainly applicable when the number of parameters is small, of order of 10's. However, if the parameter space is high of the order of 10^6-10^7 then MH is not at all efficient. To explore such a big parameter space we need to use Gibbs sampling or Hamiltonian Monte Carlo (HMC) etc. In the problem of estimating isotropy violation in CMB we use HMC [24, 25, 26, 27].

In CMB the observed sky map is a convolution of the real sky temperature with the instrumental beam with an addition of instrumental noise. Therefore, $\tilde{T}(\gamma)$ the actual temperature signal of the CMB sky along the direction γ is linearly related to the observed sky temperature, $\tilde{d}(\gamma)$,as

$$\tilde{d}(\gamma) = \hat{B}(\gamma, \gamma_0)\tilde{T}(\gamma_0)d\Omega\gamma_0 + \tilde{N}(\gamma) \tag{2}$$

where $B(\gamma, \gamma_0)$ is the instrumental beam profile and$\tilde{N}(\gamma)$ is the instrumental noise. For a perfectly circular beam profile, $B(\gamma, \gamma_0) \equiv B(\gamma \cdot \gamma_0)$, assumed in this work, it is easy to deconvolve the effect of the beam after inferring the power spectra. However, if the beam is not circular symmetric then the effect of the beam depends on the full scan pattern of the experiment and its deconvolution may be non-trivial [28, 29, 30, 31, 32]. For data defined on a sphere, it is convenient to work in the spherical harmonic space. The CMB signal, $\tilde{T}(\gamma)$, then can be expanded in terms of spherical harmonics as

$$\tilde{T}(\gamma) = \sum_{l=0}^{\infty} \sum_{m=-l}^{l} a_{lm} Y_{lm}(\gamma)\,, \tag{3}$$

where $Y_{lm}(\gamma)$ are the spherical harmonic functions and a_{lm} are the coefficients in the spherical harmonic basis. Similarly, the observed data, $\tilde{d}(\gamma)$, can also be expanded in spherical harmonics with coefficients, d_{lm}. For a perfectly statistically isotropic sky, the two point correlation function on sky can be expressed in terms of the angular power spectrum, C_l , alone as

$$\langle a_{lm} a^*_{l'm'} \rangle = S_{lml'm'} = C_l \delta_{ll'} \delta_{mm'} \tag{4}$$

Here $\langle \dots \rangle$ denotes the ensemble average. However, in presence of SI violation, C_l does not provide a full description of the covariance matrix. A general covariance matrix can be expanded in the BipoSH representation as

$$\langle a_{lm} a^*_{l'm'} \rangle = \sum_{m,m'} (-1)^{m'} A^{LM}_{ll'} C^{LM}_{lml'-m'}\,, \tag{5}$$

where $C^{LM}_{lml'-m'}$ are the Clebsch Gordon coefficients and $A^{LM}_{ll'}$ are the BipoSH spectra that provide a natural generalisation of the angular power spectrum.

The observed sky map contains the noise with the CMB signal and our goal is to calculate the posterior distribution of $A^{LM}_{ll'}$ from the observed sky map, i.e. $P(A^{LM}_{ll'}|d_{lm})$ or $P(S_{lml'm'}|d_{lm})$. However, its difficult to calculate this Probability directly as there are no direct method to find out the a_{lm}s. Therefore, in practice instead of finding this distribution, people find a joint probability distribution of the $S_{lml'm'}$ and a_{lm}, i.e. $P(S_{lml'm'}, a_{lm}|d_{lm})$ and then marginalize over a_{lm}.. This can be obtained directly by using the Bayesian Technique

$$\begin{aligned} P(S_{lml'm'}, a_{lm}|d_{lm}) &= P(d_{lm}|a_{lm})P(a_{lm}|S_{lml'm'})P(S_{lml'm'}) \\ &= \frac{1}{\sqrt{|N_l|}}\exp\left[-\frac{1}{2}\left(d^*_{lm}-a^*_{lm}\right)^T N_l^{-1}\left(d_{lm}-a_{lm}\right)\right] \\ &\quad \times\frac{1}{\sqrt{|S_{lml'm'}|}}\exp\left(-\frac{1}{2}a^{*T}_{lm}S^{-1}_{lml'm'}a_{lm}\right)P(S_{lml'm'}) \end{aligned} \tag{6}$$

$P(S_{lml'm'})$ is the prior on $S_{lml'm'}$. Here we use a flat prior on $S_{lml'm'}$ i.e. $P(S_{lml'm'}) = 1$. We also consider a isotropic noise and N_l is the noise power spectrum.

We tried to sample a_{lm} amd $A^{LM}_{ll'}$ simultaneously from a given map d_{lm} and noise covariance matrix N_l. Therefore we have of the order of 10^7 variables.

3.1. Calculating the equations of motion

Hamiltonian Monte Carlo (HMC) technique based on the Classical Hamiltonian Mechanics relies on the fact that the density of a group of particles with random momenta placed in a potential will trace the potential given that all of them start from random momentum drawn from a normal distribution with mean 0 and co-variance M, where M is a positive definite matrix called the mass matrix and can be chosen independently. It is known that HMC method can sample the distribution more effectively even in very high dimensional space in comparison to other conventional MCMC methods. In a Hamiltonian Monte Carlo algorithm we need to define a conjugate momentum and a mass corresponding to each of its parameters. We consider the conjugate momentum to a_{lm} and $A^{LM}_{ll'}$ are $p_{a_{lm}}$ and $p_{A^{LM}_{ll'}}$ and a corresponding mass $m_{a_{lm}}$, $m_{A^{LM}_{l_1l_2}}$ respectively. The mass matrices are the positive definite quantity by their definition. Thus the Hamiltonian for the motion of this ensemble of particles is

$$H = \sum_{lm}\frac{p^2_{a_{lm}}}{2m_{a_{lm}}} + \sum_{LMll'}\frac{p^2_{A^{LM}_{ll'}}}{2m_{A^{LM}_{ll'}}} - \ln(P(S_{lml'm'}, a_{lm}|d_{lm}))\,. \tag{7}$$

Using Hamiltonian mechanics

$$\dot{a}_{lm} = p_{a_{lm}}/m_{a_{lm}} \tag{8}$$

and

$$\dot{p}_{a_{lm}} = -\frac{\partial H}{\partial a_{lm}} = N_l^{-1}\left(d^*_{lm}-a^*_{lm}\right) - S^{-1}_{lml'm'}a^*_{lm}, \tag{9}$$

Similarly, the equation of motion for $A^{LM}_{ll'}$ will be

$$\dot{A}^{LM}_{ll'} = p_{A^{LM}_{ll'}}/m_{A^{LM}_{ll'}} \tag{10}$$

and

$$\dot{p}_{A_{ll'}^{LM}} = -\frac{\partial H}{\partial A_{ll'}^{LM}} = -\frac{1}{2}\partial_{A_{ll'}^{LM}} \ln|S| + \partial_{A_{ll'}^{LM}} \left(a_{lm}^{*} \left(S^{-1}\right)_{lml'm'} a_{l'm'}\right) \tag{11}$$

HMC is performed in two steps. In the first step, values of the momentum variables are chosen from the Gaussian distribution of mean 0 and variance m_x, where $x \in (a_{lm}, A_{ll'}^{LM})$. In the next step a Matropolis update is performed from the state $(p_{a_{lm}}, p_{A_{ll'}^{LM}}, a_{lm}, A_{ll'}^{LM})$ to a new state $(p_{a_{lm}}^1, p_{A_{ll'}^{LM}}^1, a_{lm}^1, A_{ll'}^{LM1})$ by integrating the equation equation of motions through a trajectory of length Δt. The Hamiltonian is computed in this new state and the state is accepted with probability $\min(1, \exp(-\Delta H))$, where ΔH is the change in the Hamiltonian between these two states. If the new state is accepted then similar operation is performed considering $(a_{lm}^1, A_{ll'}^{LM1})$ as the new position variable, otherwise the position is not updated from $(a_{lm}, A_{ll'}^{LM})$. Choice of $m_{a_{lm}}$ and $m_{A_{ll'}^{LM}}$ decides the stability of the integration process. HMC algorithm in general uses the Leapfrog integration algorithm due to its time reversal symmetry and almost symplectic nature. However, in our calculation we have used a fourth order Forest and Ruth integrator , which is a symplectic integrator that involves three Leapfrog steps, works better then a simple Leapfrog.

3.2. Computational implementation

3.2.1. Overcoming the time complexity There are practical limitations arising from carrying out the computationally challenging analysis in reasonable time. The first practical issue that we face is related to inverting the covariance matrix. A brute force inversion of the matrix is computationally prohibitive. The inversion of $\left(S^{-1}\right)_{lml'm'} a_{l'm'}$ can be done by using Gauss Seidel method. Brute force inversion of $S_{lml'm'}$ is difficult and time consuming except for small l_{max} . However, using the fact that, in case of CMB signal the off-diagonal components of the matrix $S_{lml'm'}$ are expected to be much smaller than the diagonal components dominated by A_{ll}^{00} , we can use Taylor series expansion to invert the matrix. Taylor series up to the first order gives us considerably good results.

3.2.2. Stability of numerical integration Another computational issue is the choice of the numerical integration method and the mass matrix. In Hamiltonian integrators, though Leapfrog integrator is common because the integrator preserves the Hamiltonian in phase space (symplectic), the propagation error being huge we have to use a fourth order symplectic integrator, namely Forest and Ruth integrator [25, 33], which performs better and the propagational errors are contained at a manageable level. Forest-Ruth algorithm is a combination of three Leapfrog steps. It can be shown that for the absolute stability of the integration process we need to take the mass matrix of the particles to the inverse of the covariance matric of the corresponding quantities [24].

4. Conclusion

In this proceeding we discuss two MCMC techniques that we use for cosmological parameter estimation and discuss the techniques which we use to make the algorithm faster. In our modified global Metropolis Hastings algorithm, SCoPE, the individual chains can run in parallel and a rejected sample can be used to locally modify the proposal distribution without violating the Markovian property. The latter increases the acceptance probability of the samples in chains. The prefetching algorithm allows us to increase the acceptance probability as much as required, provided requisite number of multiple cores are available in the computer. Apart from these, due to the introduction inter-chain covariance update the code can start without specifying any input covariance matrix. The mixing of the chains is also faster in SCoPE.

In the second part of the proceeding we discuss about a general method to infer the underlying covariance structure of a CMB sky using a completely Bayesian technique. We employ the Bipolar Spherical Harmonic representation of the general covariance matrix underlying random fields on a sphere and outline the method for a fully Bayesian inference of the angular power spectrum and the BipoSH coefficients simultaneously from a single observed map. We use Hamiltonian Monte Carlo for sampling the posterior distributions of the BipoSH parametrization of the covariance. We also discuss the time complexity and choice of the mass matrix and integration method for the HMC algorithm that can provide stable numerical integration.

References

[1] S. Das and T.Souradeep, SCoPE: An efficient method of Cosmological Parameter Estimation JCAP, 2014, 1407, 018

[2] S. Das, S. Mukherjee and T. Souradeep, Revised cosmological parameters after BICEP 2 and BOSS JCAP, 2015, 1502, 016

[3] S. Mukherjee, P. K. Aluri, S. Das, S. Shaikh and T. Souradeep, Direction dependence of cosmological parameters due to cosmic hemispherical asymmetry 2015

[4] S. Das, S. Mitra and S. T. Paulson, Effect of noncircularity of experimental beam on CMB parameter estimation JCAP, 2015, 1503, 048

[5] S. Mukherjee, S. Das, M. Joy and T. Souradeep, Estimation of Inflation parameters for Perturbed Power Law model using recent CMB measurements JCAP, 2015, 1501, 043

[6] W. K. Hastings, Monte Carlo Sampling Methods Using Markov Chains and their Applications, Biometrika, 1970, 57, 97-109

[7] A. Lewis, S. Bridle, Cosmological parameters from CMB and other data: a Monte-Carlo approach, Phys. Rev. D66:103511, 2002

[8] N. Metropolis, A. Rosenbluth, M. Rosenbluth, A. Teller and E. Teller, Equations of State Calculations by Fast Computing Machines, Journal of Chemical Physics, 1953, 21, 1087-1092

[9] A. Lewis, Efficient sampling of fast and slow cosmological parameters, Phys. Rev., 2013, D87, 103529

[10] C. J. Geyer, Practical Markov Chain Monte Carlo, Statistical Science, 1992, 7(4): 473-483

[11] A. Gelfand and A. Smith, Sampling-Based Approaches to Calculating Marginal Densities, Journal of the American Statistical Association, 1990, 85: 398-409

[12] A. Gelman and D. B. Rubin, Inference from iterative simulation using multiple sequences, Statist. Sci. 7 (1992) 457-511

[13] J. Rosenthal, Parallel computing and Monte Carlo algorithms, Far East J. Theor. Stat., 4: 207-236, 2000

[14] A. Mira, On Metropolis-Hastings algorithms with delayed rejection, Metron, Vol. LIX, No. 3-4, pp. 231-241 (2001)

[15] L. Tierney and A. Mira Some, Adaptive Monte Carlo methods for Bayesian inference, Statist. Med. 18 2507, 1999

[16] P. J. Green and A. Mira, Delayed rejection in reversible jump Metropolis-Hastings, Biometrika, 2001 88 1035-1053

[17] http://helios.fmi.fi/~lainema/dram/

[18] H. Haario, M. Laine, A. Mira and E. Saksman, DRAM: Ecient adaptive MCMC, Statistics and Computing, 2006, 16(3): 339-354

[19] M. Trias, A. Vecchio, J. Veitch, Delayed rejection schemes for efficient Markov-chain Monte-Carlo sampling of multimodal distributions, 2009, arXiv:0904.2207

[20] W. Gilks, G. Roberts and S. Sahu, Adaptive Markov chain Monte Carlo through regeneration, Journal of the American Statistical Association, 1998, 93(443): 1045-1054

[21] H. Haario, E. Saksman and J. Tamminen, An adaptive Metropolis algorithm, Bernoulli, 2001, 7(2): 223-242

[22] G. O. Roberts and J. S. Rosenthal, Coupling and Ergodicity of Adaptive MCMC, Journal of Applied Probability, 2007, 44(2): 458-475

[23] C. Andrieu and E. Moulines, On the ergodicity properties of some adaptive MCMC algorithms, Annals of Applied Probability, 16 (2006), no. 3, 1462-1505

[24] S .Das, B. D.Wandelt and T. Souradeep, Bayesian inference on the sphere beyond statistical isotropy JCAP, 2015, 1510, 050

[25] J. Taylor, M. Ashdown, and M. Hobson, *Fast optimal CMB power spectrum estimation with Hamiltonian sampling*, *Mon.Not.Roy.Astron.Soc.* **389** (2008) 1284, [arXiv:0708.2989].

[26] A. Hajian, *Efficient Cosmological Parameter Estimation with Hamiltonian Monte Carlo*, *Phys.Rev.* **D75** (2007) 083525, [astro-ph/0608679].

[27] S. Duane, A. Kennedy, B. J. Pendleton, and D. Roweth, *Hybrid monte carlo*, *Physics Letters B* **195** (1987), no. 2 216 – 222.

[28] S. Das, S. Mitra, A. Rotti, N. Pant, and T. Souradeep, *Statistical isotropy violation in WMAP CMB maps due to non-circular beams*, arXiv:1401.7757.

[29] N. Joshi, S. Das, A. Rotti, S. Mitra, and T. Souradeep, *Revealing Non-circular beam effect in WMAP-7 CMB maps with BipoSH measures of Statistical Isotropy*, arXiv:1210.7318.

[30] C. Armitage and B. D. Wandelt, *Deconvolution map-making for cosmic microwave background observations*, *Phys. Rev.* **D70** (2004) 123007, [astro-ph/0410092].

[31] E. Keihänen and M. Reinecke, *ArtDeco: a beam-deconvolution code for absolute cosmic microwave background measurements*, *Astronomy & Astrophysics* **548** (Dec., 2012) A110, [arXiv:1208.1399].

[32] E. Keihänen, K. Kiiveri, V. Lindholm, M. Reinecke, and A.-S. Suur-Uski, *The impact of beam deconvolution on noise properties in CMB measurements: Application to Planck LFI*, *ArXiv e-prints* (June, 2015) [arXiv:1506.0855].

[33] R. K. J. M. K. Jain, R. K. Iyengar, *Numerical Methods for Scientific and Engineering Computation*. New age international (P) limited, publishers, 2005.

Solar Transients Disturbing the Mid Latitude Ionosphere during the High Solar Activity

Shivangi Bhardwaj[1], Parvaiz A Khan[2], Roshni Atulkar[1], Bhupendra Malvi[3], Azad Ahmad Mansoori[3] and P K Purohit[1]

[1]*National Institute of Technical Teachers' Training & Research, Bhopal-462002, India*
[2]*Department of Electronics and Communication Engineering, Islamic University of Science and Technology, Pulwama-192122, J & K, India.*
[3]*Department of Electronics, Barkatullah University, Bhopal-462026, India*

E-mail: sbshivangi@gmail.com

Abstract. We investigate the effect of solar transients on the mid latitude ionosphere during the high solar activity period of solar cycle 23 i.e 2003 and 2004. A mid latitude station, Guangzhou (23.1N, 113.4E) was selected to carry out the investigation. The ionospheric behaviour at the selected station is characterized by considering the critical frequency of F2 layer (foF2) obtained by using the ground based Ionosonde observations. Then we selected two types of solar transients viz. solar flares and Coronal Mass Ejections (CMEs). To quantify the effect of solar flares we have considered the X-ray flux (1–8Å) and EUV flux (26–34nm). Similarly to quantify the effect of CMEs, we have considered the geomagnetic storms, because during high solar activity the geomagnetic storms are caused by CMEs. From our analysis we conclude that during the geomagnetic storms the value of foF2 decreases as compared to quiet days thereby showing a negative effect. On the contrary we found that during solar flares there is sudden and intense increase in foF2. We also performed a correlation analysis to access the magnitude of association between changes in flux values and peak values of Dst during flares and storms with the corresponding changes and peak values of foF2. We found that a strong correlation exists between the enhancements/decrements in foF2 and enhancements in flux values and Dst .We conclude, while geomagnetic activity suppresses ionospheric activity the flares enhance the same.

1. Introduction

Ionosphere, the upper part of earth's atmosphere, is highly vulnerable to solar disturbances. During solar disturbance like solar flares and Coronal Mass Ejections huge amount of energy is released from the sun in very short duration of time. This energy eventually reaches the geospace and is transferred to the magnetosphere where it causes several changes in the magnetosphere. Since the magnetosphere and ionosphere of earth form a coupled system therefore the state of ionosphere also gets disturbed largely due to such disturbances. The solar flares and CMEs affect the ionosphere differently.

During solar flares the release of energy occurs in the form of radiations. The increase in the radiation flux at specific wavelengths like X-ray and EUV is considerably large. Consequently, as the amount of X-ray and EUV flux impinging on the ionosphere increases, the ionization process in the ionosphere shoots up leading to sudden and intense increase in the electron density of ionosphere.

Therefore, flares abruptly change the structure and state of the ionosphere in the sunlit hemisphere, causing several types of disturbances commonly known as sudden ionospheric disturbances [1]. The ionospheric response to solar flares has been studied widely [2, 3, 4, 5]. It has been found that solar flares affect the ionosphere in many ways like increase of critical frequency of the F2 region [2], increase in electron density in the E and F1 region [6], sudden increase of the Total Electron Content (SISTEC) in the upper ionosphere.

Although, the solar flares influences of ionosphere are quite straight forward the geomagnetic influences of ionosphere are highly complex. The most important and prominent feature that has been observed during geomagnetic storms is the depression in foF2 and the Total Electron Content (TEC) [7, 8, 9]. The significant changes have also been reported in other ionospheric parameters, such as the virtual height of the F-layer (h'F), the peak height of the F2-layer (hpF2), the maximum electron density of the F2-layer (NmF2) [9, 10, 11]. The negative storm effects are thought to be caused due to the alteration in the global theromospheric winds and neutral composition. The change in neutral composition (increase in N_2/O) is responsible for the dramatic reduction in electron density [12, 13, 14]. These changes are transported to mid-latitudes by the storm-time global circulation. However, the positive F2-region effects at mid-latitudes are believed to be produced by travelling atmospheric disturbances (TADs).

2. Event Selection

In order to study the response of ionosphere to the two types of solar transients viz solar flares and Coronal Mass Ejections, we have selected the high activity phase of solar cycle 23 i.e. year 2003 and 2004. In order to study their effect on ionosphere we have selected a mid latitude station, Guangzhou (23.1N, 113.4E). After selecting the station and period of study we then proceed to the selection of solar flare and CME events. We have selected eight X-class solar flares on the basis of their visibility at the selected station and the full availability of data sets. The selection of CMEs was made by selecting geomagnetic storms produced by CMEs. The complete list of selected flares and storms is provided in table 1 and table 2.

Table 1. The complete catalog of all the selected solar flare events along with ionospheric parameters.

S. No	Event Date	Start Time	Peak Time	End Time	GOES Class	Active Region	Location	ΔfoF2	ΔX-ray	ΔEUV
1	27-May-03	22:56	23:07	23:17	X1.3	10365	S06W20	4.15	8.67E-05	3.39E+09
2	28-May-03	00:17	00:27	00:39	X3.6			4.55	1.33E-04	3.23E+09
3	29-May-03	00:51	01:05	01:12	X1.2	10365	S07W46	4.45	1.32E-04	3.59E+09
4	10-Jun-03	23:19	00:02	00:12	X1.3	10375	N12W60	4.9	2.23E-04	5.48E+09
5	23-Oct-03	08:19	08:35	08:49	X5.4	486	S16E67	4.35	8.23E-05	2.17E+09
6	26-Oct-03	05:57	06:54	07:33	X1.2	486	S15E32	3.9	3.64E-05	3.11E+09
7	26-Feb-04	01:50	02:03	02:10	X1.1	564	N14,W28	7.4	4.98E-04	2.13E+10
8	16-Jul-04	01:43	02:06	02:12	X1.3	649	S10,E26	3.6	7.98E-05	3.98E+09

3. Data and method of analysis

In order to study and compare the effect of solar flares and geomagnetic storms on mid latitude ionosphere, we used several parameters describing these phenomena. We know flares are characterized by a dramatic increase in X-ray and EUV flux. The solar X-ray flux is monitored by GOES Satellite series. For our study we have taken the solar X-ray flux data from the National Geophysical Data Center (NGDC) server at http://www.ngdc.noaa.gov/stp/GOES/. We have taken the X-ray flux in the range 1Å to 8Å with 5 min resolution. Similarly, the EUV flux is monitored by an EUV spectrometer onboard Solar and Heliophysical Observatory (SOHO) spacecraft. The SEM experiment measures EUV flux, integrated in the two wavelength bands 26-34nm and 0.1-50nm and can be accessed at http://www.usc.edu/dept/space_science/semdatafolder/. In our analysis we have used the SEM EUV flux in the wavelength range 26-34nm band with 5 min temporal resolution.

Table 2. The complete catalog of all the selected geomagnetic storms.

S.No	Event Date	Peak Dst	Peak foF2	ΔfoF
1	29 May 2003	-144	9.5	-3.87
2	20 November 2003	-422	14	-1.30
3	11 February 2004	-93	14	-1.11
4	30 August 2004	-129	9.1	-2.38
5	08 November 2004	-373	12.8	-1.64
6	10 March 2004	-78	14.2	-3.36
7	17 July 2004	-76	11.8	-0.63
8	04 April 2004	-117	14.1	-1.11
9	30 October 2003	-383	14.5	-0.49
10	12 July 2003	-105	13.9	0.60

The intensity and occurrence of geomagnetic storms is monitored by defining a storm intensity index commonly known as Disturbed Storm Time (Dst) index which is defined as an average of horizontal components of the ground magnetic field from selected observatories located around equator. The data for Dst indices with 1h resolution for each storm were obtained from the web server http://wdc.kugi.kyoto-u.ac.jp/.

In our present investigation we have used the ground based ionosonde observation to characterize the changes in the ionosphere as result of flares and storms. We have used the critical frequency of F2 layer, symbolized as foF2. The data of foF2 was taken from the database of NGDC. The NGDC maintains a huge database of more than 250 stations and can downloaded from its server at http://spidr.ngdc.noaa.gov/spidr/. We have used the 15 min values of foF2 over the Guangzhou station.

We took the peak values of X-ray flux, EUV flux, Dst index and foF2 to investigate the relation between the peak values. We also calculated the enhancement/decrements in the foF2 as well X-ray flux, EUV flux and Dst index to quantify the amount of changes. The enhancements in the flux values were calculated from the background flux and are designated as ΔX-ray and ΔEUV flux. However, the change in the foF2 was calculated in two ways separately for flares and storms. To investigate the flare effect the change was calculated from the median of the day. While for storm effects the change was calculated from the quietest day of the month and represented as ΔfoF2.

4. Results and discussion

After obtaining the required data sets and doing the all the processing and analysis the final data sheets were prepared and graphs were constructed to obtain the results and finally arrive at general conclusions. The study was conducted in two parts. In the first part the effect of solar flares on the ionosphere was studied and in the second part the effect of geomagnetic storms was studied.

4.1. Effect of solar flares on the ionospheric foF2

We know during solar flares a dramatic increase occurs in the solar X-ray flux and EUV flux. Consequently, the ionization process in the ionosphere will be increased significantly which will be reflected in the foF2 parameter. First, we constructed the time profile foF2 and the X-ray flux. The time profile X-ray flux with the corresponding profile of foF2 for four of the selected flares is shown in figure 1. In each panel of the figure 1 we can clearly notice a sharp peak in the X-ray flux curve which corresponds to a solar flare. However, the foF2 follows a diurnal pattern i.e. increasing from morning achieving a normal diurnal peak in the afternoon and then decreasing back to minimum in the evening. Such type of behavior is characteristic of mid latitude ionosphere. But whence the X-ray flux peaks the foF2 curve undergoes a sudden and sharp jump instantaneously and then follows the normal diurnal pattern. In otherwords, during the time of solar flare the value of foF2 rises over the normal background value indicating an increase in foF2. This feature can be observed in all the four panels of the figure 1 with magnitude of increase being different for different flares. Therefore, we conclude from figure 1 that during solar flares the value of foF2 is increased due to increase in the X-ray flux. The increase is almost instantaneous, intense and short lived.

Along with increase in X-ray flux, the flares are also characterized by significant increase in the EUV flux. The figure 2 show the temporal changes in the EUV flux along with corresponding changes in the foF2 during four of the selected solar flare events. Here we find a sharp peak in the wine colored curve indicating a solar flare. So we find during flare the flux of EUV radiation is significantly increased. The effect of this increase in EUV flux on the foF2 can be observed as a sudden increase in the foF2 over the normal diurnal pattern. Similar behaviour was observed for the other six flare events also. Figure 1 and 2 also shows a very good agreement between the radiation flux enhancements and the positive changes in ionospheric foF2. As the flux start increasing from their background levels due to a solar flare corresponding to it the foF2 also increases.

To access the magnitude of association between the solar radiations and the foF2 we took the peak values of both types of parameters and constructed the scatter plots and derived the corresponding correlation coefficients. We found that peak values of foF2 do not follow a very good correlation with the peak values of X-ray flux (R=0.17) and EUV flux (R=0.33). But on the otherhand when we examined the correlation of ΔfoF2 with ΔX-ray flux and ΔEUV flux, we found a very strong correlation between the two. The correlation coefficients were 0.97 and 0.93 respectively with ΔX-ray flux and ΔEUV flux as depicted in figure 3.

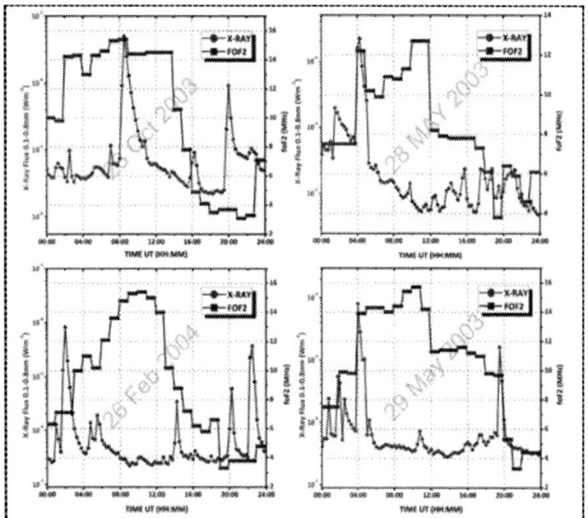

Figure 1. The time profile of foF2 along with the X-ray flux (1-8Å) profile during four of the selected flare events.

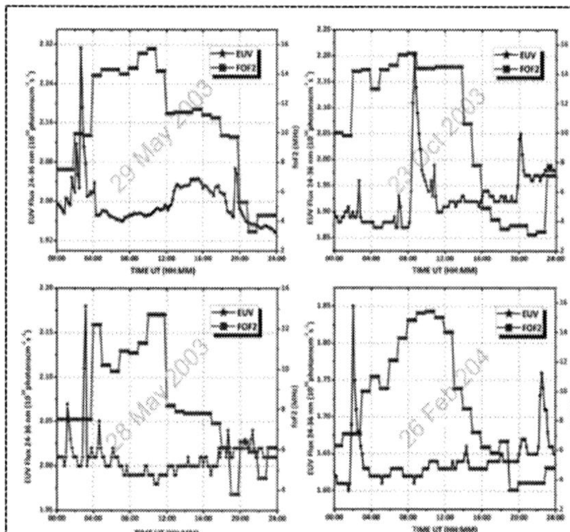

Figure 2. The time profile of foF2 along with the EUV flux (26-34nm) profile during four of the selected flare events.

4.2. Effect of geomagnetic storms on the ionospheric foF2

We now present the results describing the effect of geomagnetic storms on the foF2. The figure 4 shows the temporal variation of foF2 on the storm day, quietest day of the month and two control days. The nine panels represent the nine events. In all of events we find that the values of foF2 on the storm day are smaller than the values of foF2 on the quietest day of month. In some event the value of foF2 on the storm day is also smaller than the values on the two control days. However, in two cases (29 Aug. 2004 and 29 May 2003) the value of foF2 on the storm day is higher than the corresponding values on the control days. Therefore, it can be easily observed from the figure 4 that the value of foF2 is decreased due to a geomagnetic storm. Hence, it can be concluded that the geomagnetic storms produce a negative effect on the ionosphere. The negative effect of storms can be much clearly noticed in figure 5 where we have plotted the foF2 along with Dst index during one of the selected events.

From this figure we notice that the Dst index achieves the peak on the 30-08-2004, indicating the main phase of geomagnetic storm. In the second panel of figure 5 we can clearly notice that on that day the values of foF2 is significantly lower than the corresponding values on the other days. The peak of foF2 is significantly lower than the peak values on the other days. Therefore, figure 5 clearly depicts that geomagnetic storms have a negative effect on the ionospheric foF2 at mid latitudes. In order to quantify the magnitude of association between the two, we derived the correlation between the peak values of foF2 and the peak values of Dst, which is shown in top panels of figure 3. From the figure we notice a strong correlation between peak values of foF2 and Dst as well as between ΔfoF2 and Dst. The corresponding correlation coefficients are 0.63 and 0.94 showing that ΔfoF2 exhibits a very strong correlation with Dst than does peak foF2.

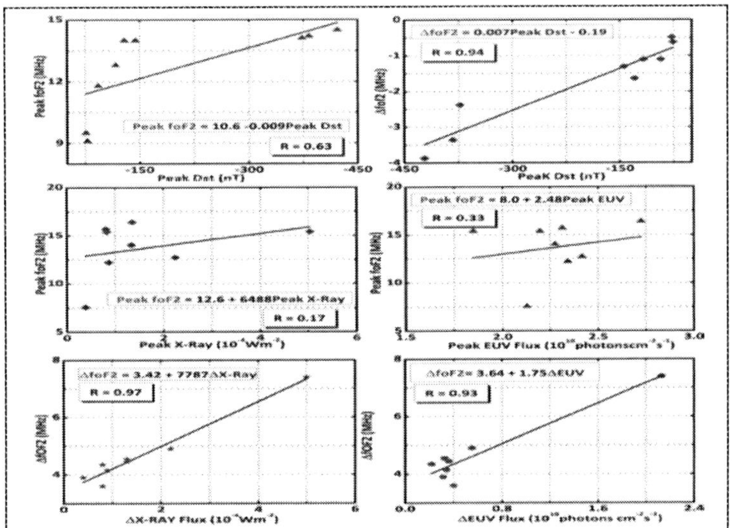

Figure 3. The scatter plot and correlation of foF2 and its changes with the X-ray flux, EUV flux, Dst and their changes.

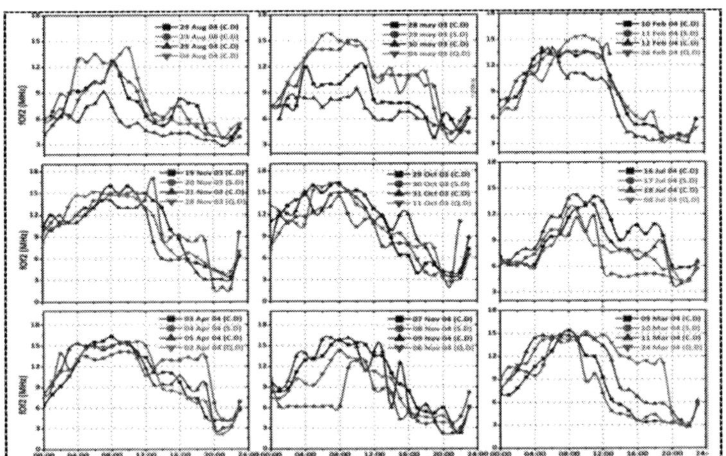

Figure 4. The time profile of foF2 during the storm day, quitest day of the month and two control days.

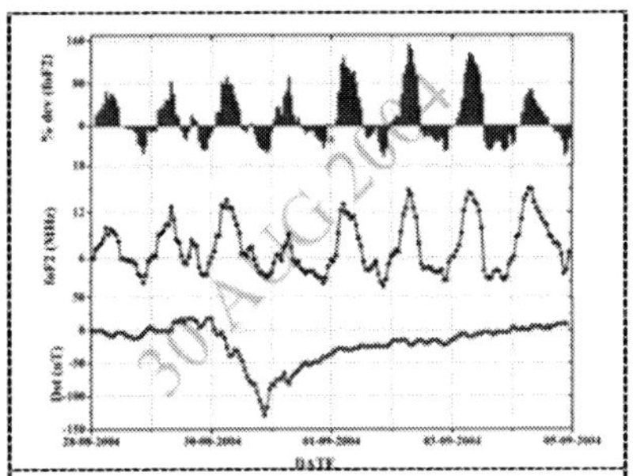

Figure 5. The time profile of foF2 along with profile of Dst during the 30-08-2004 storm event.

5. Conclusions

We studied and compared the response of ionosphere to solar flares and geomagnetic storms. From our analysis we draw the following conclusions:

- The value of foF2 undergoes an enhancement during the solar flares while during geomagnetic storms the value of foF2 suffers a decrease.
- The peak values of foF2 do not follow a very good correlation with the corresponding peak values of Xray flux, EUV flux and Dst. However, the enhancements/decrements of the foF2 follow a very strong correlation with the enhancements of X-ray flux, EUV flux and Dst.

6. Acknowledgements

The authors are thankful to various online data server like National Geophysical Data Center (NGDC), WDC Kyoto, for the data used in the study.

7. References

[1] Davies K 1990 *Ionospheric Radio* (Peter Peregrinus, London)
[2] Knecht R W and Davies K 1961 Solar flare effects in the F region of the ionospheric *Nature* **190** 797–798
[3] Zhang D H and Xiao Z 2005 Study of ionospheric response to the 4B flare on 28 October 2003 using International GPS Service network data *J. Geophys. Res.* **110** A03307 doi:10.1029/2004JA010738.
[4] Tsurutani B T et al 2005 The October 28, 2003 extreme EUV solar flare and resultant extreme ionospheric effects: Comparison to other Halloween events and Bastille Day event *Geophys. Res. Lett.* **32**, L03S09 doi: 10.1029/2004GL021475.
[5] Liu J Y et al 2006 Solar flare signatures of the ionospheric GPS total electron content *J. Geophys. Res.* **111** A05308 doi: 10.1029/2005JA011306.
[6] Thome G D and Wagner L S 1971 Electron density enhancements in the E and F regions of the ionosphere during solar flares *J. Geophys. Res.* **76** 6883–6895 doi: 10.1029/JA076i028p06883.
[7] Lakshmi D R, Rao B N C, Jain A R, Goel M K and Reddy B M 1991 Response of equatorial and low latitude F region to the great magnetic storm of 13 March 1989 *Ann. Geophys.* **9** 286–290.
[8] Lakshmi D R, Veenadhari B, Dabas R S and Reddy B M 1997 Sudden post midnight decrease in equatorial F-region electron densities associated with severe magnetic storms *Ann. Geophys* **15** 306–313.
[9] Batista I S, de Paula E R, Abdu M A and Trivedi N B 1991 Ionospheric effects of the March 13, 1989, magnetic storm at low and equatorial latitudes *J. Geophys. Res.* **96** 13,943–13,952.
[10] Afraimovich E L et al 2002 Simultaneous radio and optical observations of the midlatitude atmospheric response to a major geomagnetic storm of 6–8 April 2000 *J. Atmos. Sol.-Terr. Phys.* **64** 1943–1955.
[11] Yeh K C, Lin K H and Conkright R O 1992 The global behavior of the March 1989 ionospheric storm *Canadian Journal Physics* **70** 532–543.
[12] Chandra S and Spencer N W 1976 Thermospheric storms and related ionospheric effects *J. Geophys. Res.* **81** 5018–5026.
[13] Forbes J M 1989 Evidence for equatorial penetration of electric fields, winds, and compositional effects in the Asia/Pacific sector during the September 17–24, 1984 ETS interval *J. Geophys. Res.* **94** 16999–17008.
[14] Fuller-Rowell T J, Codrescu M V and Moffett R J 1994 Response of the thermosphere and ionosphere to geomagnetic storms *J. Geophys. Res.* **99** 3893–4011.

Effect of non zero cosmological constant on the motion of light ray

Sarani Chakraborty, A. K. Sen

Department Physics, Assam University, Silchar-788011, India

E-mail: sarani.chakraborty.phy@gmail.com, asokesen@yahoo.com

Abstract.

Cosmological constant, the value of the energy density of the vacuum of space, was introduced by Einstein as an extension of general relativity. Recent studies show that cosmological constant has a noticeable effect on the shape of space-time i.e. it will influence the path of light ray. In this study the effect of non-zero cosmological constant on the path of light ray in three different types of space-time geometry namely Schwarzschild de Sitter, Kerr de Sitter and Kerr- Newman de Sitter has been discussed.

1. Introduction

Einstein explained gravitation as a demonstration of geometry of space-time and gave the final formulation in 1915 [1]. General relativity predicted that the universe must either expand or contract. Einstein was a believer of static universe and to make the universe static he introduced an anti-gravity force by adding the cosmological constant. Later when Hubble gave the proof of expanding universe, he dropped that idea and called it as the biggest mistake of his life. Up to 1990, scientists thought that the rate of expansion of the universe slows over time for gravitational pull of matter of the universe. But observation indicates that the universal expansion is speeding up. So there must be force acting against gravity and cosmological constant is an example of such type of force. Hence the idea of cosmological constant came back.

One of the most important predictions of general relativity is the deflection of light ray in presence of gravitational mass. The first order contribution of mass on the path of light ray was calculated by Einstein himself. After Einstein, higher order contribution of mass towards light deflection angle was calculated by Keeton and Petters [2]. Recently, Iyer and Petters [3] calculated it for strong field and found that under weak field approximation their expression matches with that of Keeton and Petters [2].

The light deflection angle for a rotating mass (Kerr mass) was calculated by Iyer and Hansen [4]. Bozza [5] obtained the lensing formula and calculated all other components related to lensing. Azzami et. al [6, 7] calculated the two individual components (parallel and perpendicular to equatorial plane) of light deflection angle in quasi-equatorial regime. Chakraborty and Sen [8] have recently obtained the light deflection angle for a charged, rotating body in the equatorial plane and showed how deflection angle changes with charge. The off equatorial light deflection for Kerr geometry was also studied by Chakraborty and Sen [9].

On the other hand, some authors have used the material medium approach, where the gravitational effect on the light ray was calculated by assuming some effective refractive index assigned to the medium through which light is propagating [8]. Sen [10] and Roy and Sen [11] used this method to calculate the gravitational deflection of light without any weak field approximation for Schwarzschild and Kerr geometry. Nandi and Islam [12] used the optical-mechanical analogy of general relativity (based on Fermat's principle) and obtained the refractive index.

All these above mentioned work were done without considering the effect of cosmological constant and as we already discussed, it is more realistic to study the trajectory of light ray by considering non zero cosmological constant (de-Sitter universe).

Here we will **review** and discuss the effect of non-zero cosmological constant on the path of light ray in three different types of space-time geometry namely Schwarzschild de Sitter, Kerr de Sitter and Kerr Newman de Sitter. **We will also make a comparison of effect of not zero cosmological constant among the three above mentioned line elements and draw a set of conclusions.**

2. Schwarzschild de Sitter line element.

Schwarzschild de Sitter space-time represents the exterior of a static mass under de-Sitter background. The metric for the Schwarzschild de Sitter space-time is given by (using units such that G =c =1) [13],

$$ds^2 = -f_r(r)dt^2 + \frac{dr^2}{f_r(r)} + r^2(d\vartheta^2 + \sin^2 d\varphi^2)$$

where, $f_r = 1 - \frac{2m}{r} - \frac{\Lambda r^2}{3}$ and Λ is the Cosmological constant with m being the mass of the lens object.

3. Kerr de Sitter line element.

In general relativity the geometry of space time around rotating massive body can be described by Kerr metric. According to this metric, frame dragging, an unusual prediction of the general relativity is exhibited by such rotating bodies. The prediction of this effect is that all objects coming close to a rotating mass would be entrained to participate in its rotation, not because of any applied force or torque that can be felt, but because of the curvature of space time associated with the body [14]. At close enough distance all object even light must rotate with the body. According to General relativity, rotating bodies drag space time around them, a phenomenon known as frame dragging. The deflection produced in presence of a rotating black hole explicitly depends on the direction of light ray. Compared to the zero spin Schwarzschild case, the bending angle is greater for direct orbit and smaller for retrograde orbits [4].

Here we will discuss the Kerr space time under de-Sitter background i.e. Kerr de-Sitter space time. The Kerr de-Sitter line element is given by [15],

$$ds^2 = \rho^2\left(\frac{dr^2}{\Delta} + \frac{d\vartheta^2}{\Delta_\Lambda}\right) + \sin^2\vartheta\frac{\Delta_\Lambda}{\rho^2}\left(\frac{adt - (r^2+a^2)d\varphi}{1+\frac{\Delta_\Lambda}{3}a^2}\right)^2 - \frac{\Delta}{\rho^2}\left(\frac{dt - a\sin^2\vartheta d\varphi}{1+\frac{\Delta_\Lambda}{3}a^2}\right)^2$$

Where,

$$\rho^2 = r^2 + a^2\cos^2\vartheta$$

$$\Delta = r^2 - 2mr + a^2 - \frac{\Lambda r^2}{3}(r^2 + a^2)$$

$$\Delta_\Lambda = 1 + \frac{\Lambda a^2 \cos^2\vartheta}{3}$$

a=J/m, J is the angular momentum of the gravitating body.

4. Kerr-Newman de Sitter line element

In the general theory of relativity, the Kerr–Newman line element represents the most generalized form of the space time curvature, where all three factors (mass distribution, rotation, charge) have their contribution. Here we will discuss the Kerr–Newman line element under de-Sitter back ground i.e. Kerr-Newman de-Sitter line element [16],

$$ds^2 = \frac{\Delta_r^{KN}}{\Sigma^2\rho^2}(dt^2 - a\sin^2\vartheta d\varphi)^2 - \frac{\rho^2}{\Delta_r^{KN}}dr^2 - \frac{\rho^2}{\Delta_\vartheta}d\vartheta^2$$
$$-\frac{\Delta_\vartheta \sin^2\vartheta}{\Sigma^2\rho^2}(adt - (r^2+a^2)d\varphi)^2$$

Where,

$$\Delta_\vartheta = 1 + \frac{a^2\Lambda}{3}\cos^2\vartheta$$
$$\Sigma = 1 + \frac{a^2\Lambda}{3}$$
$$\Delta_r^{KN} = (1 - \frac{\Lambda}{3}r^2)(r^2+a^2) - 2mr + e^2$$

And e is the static charge.

5. Bending angle due to Schwarzschild de Sitter line element.

The concept of measurement of the effect of the cosmological constant on the light deflection angle was first introduced by Rindler and Ishak [17]. They derived the formula for light deflection angle under de Sitter back ground. According to their work a positive cosmological constant diminishes the bending angle.

Bhadra et al [13] obtained the light deflection angle in the Schwarzschild de Sitter (SDS) geometry. In their paper they computed the bending angle in the SDS space-time, taking the proper cosmological constant involved in the solution of the light trajectory. They showed that when the light path from the reference source is taken into consideration, the resultant bending in the SDS geometry will appear to increase rather than decrease due to the cosmological constant effect. They also discussed the possibility of the detection of the effect from the bending angle measurement. Their expression for bending angle is given by,

$$\alpha = \frac{4m}{r_0} - 2mr_0(\frac{1}{d_{LS}} + \frac{1}{d_{OL}}) - \frac{\Lambda r_0}{6}(d_{OL} + d_{LS}) + \frac{\Lambda r_0^3}{6}(\frac{1}{d_{LS}} + \frac{1}{d_{OL}})$$

Here r_0 is the closest approach of the light ray towards the gravitating body. d_{LS} and d_{OL} are the distance between lensing object and source and lensing object and observer. Their work shows that cosmological constant does effect the gravitational deflection of light.

6. Bending angle due to Kerr de Sitter line element.

Kraniotis [18] derived the frame dragging and deflection angle in terms of the Appell and Lauricella hypergeometric functions and the Weierstrass modular form for a Kerr and Kerr–(anti) de Sitter space times.

Sultana [15] in his paper used the method of Rindler and Ishak to obtain a weak field approximation for the bending angle in the Kerr–de Sitter space time. He derived the null geodesic equations for the Kerr–de Sitter metric and obtained the bending angle which includes contributions from the cosmological constant in the equatorial plane,

$$\alpha = \frac{4m}{b} - \frac{4am}{b^2} + \frac{15\pi m^2}{4b^2} - \frac{\Lambda b^3}{6m} - (\frac{75\pi^2}{512} + \frac{101}{96})bm\Lambda + \frac{5\pi\Lambda b^2}{32} + \frac{5\pi\Lambda ab}{16} - \frac{ab^2\Lambda}{6m}$$

Here b is the impact parameter.

7. Bending angle due to Kerr-Newman de Sitter line element.

Kraniotis [16] obtained the solution of null geodesics that describe photon orbits in the space time of a rotating electrically charged black hole (Kerr-Newman) including the contribution from the cosmological constant.

He derived the analytic solutions of the lens equations in terms of Appell and Lauricella hypergeometric functions and the Weierstrass modular form. He used the analytical solution to obtain the exact expression of the deflection angle of an equatorial light ray in the gravitational field of a Kerr-Newman de Sitter black hole.

8. Conclusion.

From the above discussion we can conclude the following points,

1. Cosmological constant has a noticeable effect on the path of the light ray. When compared with Schwarzschild expression for bending, we find that there are some extra terms in the expression for deflection which occur due to the presence of cosmological constant. If we set the cosmological constant to zero, deflection angle will reduce to that of Schwarzschild, Kerr or Kerr-Newman deflection angle.
2. The positive value of cosmological constant reduces the amount of deflection [17]. However when the light path from the reference source is taken into consideration, the resultant bending angle in the SDS geometry will appear to increase rather than decrease due to the effect of cosmological constant [13].
3. One should do the correction for cosmological constant to the expression of light deflection angle to learn the exact geometry of space time curvature.

References

[1] Misner C W, Thorne K S and Wheeler J A 1973 Gravitation W H Freeman and Company.
[2] Keeton C R and Petters A O 2005 Phys. Rev. D 72 104006.
[3] Iyer S V and Petters A O 2007 Gen. Relativ. Gravit. 39 1563–82.
[4] Iyer S V and Hansen E C 2009 Phys. Rev. D 80 124023.
[5] Bozza V 2003 Phys. Rev. D 67 103006.
[6] Aazami A B, Keeton C R and Petters A O 2011 J. Math. Phys. 52 092502.
[7] Aazami A B, Keeton C R and Petters A O 2011 J. Math. Phys. 52 102501.
[8] Chakraborty S and A K Sen 2015 Class. Quantum Grav. 32 115011.
[9] Chakraborty S and A K Sen 2015 arXiv:1504.03124 [gr-qc].
[10] Sen A K 2010 Astrofizika 53 4.
[11] Roy S and Sen A K 2015 Astrophys Space Sci. 360 23.
[12] Nandi K K and Islam A 1995 Am. J. Phys. 63, 251.
[13] Bhadra A, Biswas S and Sarkar K 2010 Phys. Rev. D 88 063003.
[14] Chakraborty S and A K Sen 2014 JPCS 481 012008.
[15] Sultana J 2013 Phys. Rev. D 88 042003.
[16] Kraniotis G V 2014 Gen. Relativ. Gravit. 46 11.
[17] Rindler W and Ishak M 2007 Phys. Rev. D 76 043006.
[18] Kraniotis G V 2005 Class. Quantum Grav. 22 4391–4424.

Frame-dragging from charged rotating body

Anuj Kumar Dubey and A. K. Sen

Department of Physics, Assam University, Silchar-788011, Assam, India

E-mail: danuj67@gmail.com

Abstract. In the present paper, we have considered the three parameters: mass, charge and rotation to discuss their combined effect on frame dragging for a charged rotating body. If we consider the ray of light which is emitted radially outward from a rotating body then the frame dragging shows a periodic nature with respect to coordinate ϕ (azimuthal angle). It has been found that the value of frame dragging obtains a maximum at, $\phi = \frac{\pi}{2}$ and a minimum at $\phi = \frac{3\pi}{2}$.

1. Introduction

Considerations on rotation are important in general relativity and frame dragging is one of its examples. Frame-dragging is a general relativistic feature of all solutions to the Einstein field equations associated with the rotating masses. General relativity is dealing with the precession phenomena such as perihelion advancement precession, geodetic precession, Lense-Thirring precession [1] and Pugh-Schiff precession [2].

In 1960, L. I. Schiff [2] showed that an ideal gyroscope in orbit around the Earth or other massive body would undergo two relativistic precessions with respect to a distant undisturbed inertial frame: (1) a geodetic precession in the orbit plane due to the local curvature of space time; (2) a frame-dragging precession due to the Earths rotation.

After Pugh-Schiff precession, a proposal came for Gravity Probe-B (GP-B) in 1959 and finally launched in 2004, as an experimental confirmation of frame-dragging. The objective of Gravity Probe-B experiment was to design, develop, conduct, and analyze the data of a flight experiment to test the predictions of Albert Einstein's general relativity.

In GP-B experiment, Schiffs formula has been used for the combined gyroscope precession: $\Omega = \frac{3GM}{2c^2R^3}(\mathbf{R} \times \mathbf{V}) + \frac{GI}{c^2R^3}[\frac{3R}{R^2}(\omega.\mathbf{R}) - \omega]$. Here the first and second terms are the geodetic, and the frame-dragging effect. G and c are the gravitational constant and the velocity of light respectively. M, I and ω are the mass, moment of inertia, and angular velocity of the earth respectively. R and v are the instantaneous distance and velocity of the gyroscope respectively.

Everitt et al. in 2008 [3] had confirmed that the geodetic effect (caused by the earth's mass; the amount by which the earth warps the local space-time in which it resides) and the frame-dragging effect (caused by the rotation of the earth; the amount by which the rotating earth drags its local space-time around with it). In the 642 km orbit of GP-B, the predicted geodetic and frame-dragging precessions are 6606 marcs/yr and 39 marcs/yr respectively [3].

Collas and Klein in 2004 [4] had showed that a prototype example of frame-dragging arises in the Kerr metric. A test Particle with zero angular momentum released from a non-rotating frame, far from the source of the Kerr metric, accumulates non-zero angular velocity in the same angular direction as the source of the metric, as the test particle plunges toward the origin. This dragging of inertial frames or frame-dragging, is due to the influence of gravity alone, and has no counterpart in Newtonian physics. Frame-dragging is a general relativistic feature, not only of the exterior Kerr solution, but of all solutions to the Einstein field equations associated with rotating sources. They also showed that surprising frame-dragging anomalies can occur in certain situations. They had considered the examples of axially symmetric solutions to the field equations in which zero angular momentum test particles, with respect to non-rotating coordinate systems, acquire angular velocities in the opposite direction of rotation from the sources of the metrics. They refer to this phenomenon as negative frame-dragging.

Mei and Jiu in 2006 [5] had showed that the angular velocity of frame-dragging is negative in the situation where non zero components of the metric $g_{\phi\phi}$ is positive. Under this situation, the zero angular momentum test particle may acquire an angular velocity in the opposite direction of rotation from the source of the metric. Such negative frame-dragging is also called frame-dragging anomaly in general relativity. Its occurrence is dependent on the coefficient of the metric in the ϕ coordinate. They also showed that negative frame-dragging occurs in non-rotating reference frames for the metrics of all solutions of the Kerr family which are associated with rotating sources.

Epstein in 2008 [6] had discussed that a Hamiltonian approach makes the phenomenon of frame-dragging from the appearance of the drag velocity in the Hamiltonian of a test particle in an arbitrary metric. A distinction is made between manifest frame-dragging such as that in the Kerr metric, and the hidden frame-dragging that can be made manifest by a coordinate transformation such as that applied to the Robertson-Walker metric. They also showed that a zone of repulsive gravity is found in the extreme Kerr metric. They also explained the frame-dragging in special relativity as a manifestation of the equivalence principle in accelerated frames.

Dubey and Sen in 2015 [7] had obtained the expression of angular velocity of frame-dragging in Kerr Geometry by considering the ray of light emitted radially outward from the surface of a rotating compact object. They obtained the general expression of frame-dragging as a function of coordinates, ϕ (azimuthal) and θ (latitude). Dubey and Sen in 2015 [8] had also obtained the general expression of angular velocity of frame-dragging in Kerr-Newman Geometry by considering a charged rotating body having electric charge (Q) and magnetic charge (P).

2. Geometry of charged rotating body

As a result of work carried out by R. H. Price, B. Carter, W. Israel, D. C. Robinson, and S. W. Hawking, black holes, from the point of view of the outside observer, can possess only three distinguishing characteristics: mass, charge and angular momentum.

When rotation is taken into consideration spherical symmetry is lost and off-diagonal terms appear in the metric and the most useful form of the solution of Kerr family (Kerr metric and Kerr-Newman metric) is given in terms of t, r, θ and ϕ, where t, and r are Boyer-Lindquist coordinates running from - ∞ to + ∞, θ and ϕ, are ordinary spherical coordinates in which ϕ is periodic with period of 2π and θ runs from 0 to π. Covariant form of metric tensor for Kerr family (Kerr 1963 [9], Newman et al. 1965 [10]) in terms of Boyer-Lindquist coordinates with signature (+,-,-,-) is expressed as:

$$ds^2 = g_{tt}c^2dt^2 + g_{rr}dr^2 + g_{\theta\theta}d\theta^2 + g_{\phi\phi}d\phi^2 + 2g_{t\phi}cdtd\phi \quad (1)$$

where g_{ij}'s are non-zero components of Kerr family.

As detailed in our previous work Dubey and Sen 2015 [8], if we consider the three parameters: mass (M), rotation parameter (a) (defined as angular momentum per unit mass setting speed of light unity), and charge (electric (Q) and / or magnetic (P)). Then it is easy to include charge in the non-zero components of g_{ij} of Kerr metric, simply by replacing $(r_g r)$ with $(r_g r - Q^2 - P^2)$. Here r and $r_g = (2GM/c^2)$ are the radial coordinate and Schwarzschild radius respectively.

Journal of Physics: Conference Series **759** (2016) 012065 doi:10.1088/1742-6596/759/1/012065

Non-zero components of g_{ij} of Kerr-Newman metric are given as follows (page 261-262 of Carroll 2004 [11]; Dubey and Sen 2015 [8]):

$$g_{tt} = (1 - \frac{r_g r - Q^2 - P^2}{\rho^2}) \tag{2}$$

$$g_{rr} = -\frac{\rho^2}{\Delta} \tag{3}$$

$$g_{\theta\theta} = -\rho^2 \tag{4}$$

$$g_{\phi\phi} = -[r^2 + a^2 + \frac{(r_g r - Q^2 - P^2)a^2 sin^2\theta}{\rho^2}]sin^2\theta \tag{5}$$

$$g_{t\phi} = \frac{a sin^2\theta(r_g r - Q^2 - P^2)}{\rho^2} \tag{6}$$

with

$$\rho^2 = r^2 + a^2 cos^2\theta \tag{7}$$

and

$$\Delta = r^2 + a^2 - r_g r + Q^2 + P^2 \tag{8}$$

If we replace $(r_g r - Q^2 - P^2)$ by $(r_g r - Q^2)$ and further if we put rotation parameter of the source (a) equal to zero, then it reduces to Reissner-Nordstrom metric. Also if we replace $(r_g r - Q^2 - P^2)$ by $(r_g r)$ then the Kerr-Newman metric reduces to Kerr metric and further if we put rotation parameter of the source (a) equal to zero then it reduces to Schwarzschild metric.

3. Frame-dragging from charged rotating body

As detailed in our previous work Dubey and Sen 2015 [8], the Lagrangian $\pounds$ of the test particle can be expressed as:

$$\pounds = \frac{g_{ij}\dot{x}^i\dot{x}^j}{2} \tag{9}$$

From the Lagrangian of the test particle, we can obtain the momentum of the test particle as:

$$P_i = \frac{\partial \pounds}{\partial \dot{x}^i} = g_{ij}\dot{x}^j \tag{10}$$

Thus corresponding momentum in the coordinates of t, r, θ, and ϕ are given by

$$P_t \equiv -E = c^2 g_{tt}\dot{t} + c g_{t\phi}\dot{\phi} \tag{11}$$

$$P_r = g_{rr}\dot{r} \tag{12}$$

$$P_\theta = g_{\theta\theta}\dot{\theta} \tag{13}$$

$$P_\phi \equiv L = c g_{t\phi}\dot{t} + g_{\phi\phi}\dot{\phi} \tag{14}$$

Since the space time of the Kerr family is stationary and axially symmetric, the momenta P_t and P_ϕ are conserved along the geodesics. So we obtain two constants of motion: one is corresponding to the conservation of energy (E) and the other is the angular momentum (L) about the symmetry axis.
Using equations (11) and (14) we can write:

$$c\dot{t} = \frac{g_{\phi\phi}E + c g_{t\phi}L}{c g_{t\phi}^2 - c g_{tt} g_{\phi\phi}} \tag{15}$$

and

$$\dot{\phi} = -\frac{g_{t\phi}E + cg_{tt}L}{cg_{t\phi}^2 - cg_{tt}g_{\phi\phi}} \tag{16}$$

Using equations (15) and (16), we can obtain the angular velocity of frame-dragging ($\frac{d\phi}{cdt}$) as (similar expression was given by Collas and Klein 2004 [4]; Mei and Jiu 2006 [5]; Dubey and Sen 2015 [7]):

$$\frac{\dot{\phi}}{c\dot{t}} = \frac{u^{\phi}}{u^{t}} = \frac{d\phi}{cdt} = -\frac{g_{t\phi}E + cg_{tt}L}{g_{\phi\phi}E + cg_{t\phi}L} \tag{17}$$

Considering L=0, Collas and Klein 2004 [4]; Mei and Jiu 2006 [5], had obtained angular velocity of frame-dragging of a zero angular momentum test particle as:

$$\frac{d\phi}{cdt} = -\frac{g_{t\phi}E}{g_{\phi\phi}E} \tag{18}$$

Considering the ray of light emitted radially outward from the surface of a compact object (from a rotating body with radius 'R'), the general expression of angular velocity of frame-dragging ($\frac{d\phi}{cdt}$) in Kerr Geometry was given as (Eqn. (46) of Dubey and Sen 2015 [7]):

$$\frac{d\phi}{cdt}(\phi,\theta) = \frac{\frac{Rsin\phi sin\theta}{sin^2\theta}(1 - \frac{r_g r}{\rho^2}) + \frac{r_g ra}{\rho^2}}{-\frac{r_g ra}{\rho^2}(Rsin\phi sin\theta) + (r^2 + a^2 + \frac{r_g ra^2}{\rho^2}sin^2\theta)} \tag{19}$$

Considering a rotating body having electric charge (Q) and magnetic charge (P), the general expression for angular velocity of frame-dragging ($\frac{d\phi}{cdt}(\phi,\theta)_{KN}$) for a charged rotating body in Kerr-Newman Geometry was given as (Eqn. (29) of Dubey and Sen 2015 [8]):

$$\frac{d\phi}{cdt}(\phi,\theta)_{KN} = \frac{\frac{Rsin\phi sin\theta}{sin^2\theta}(1 - \frac{(r_g r - Q^2 - P^2)}{r^2 + a^2cos^2\theta}) + \frac{(r_g r - Q^2 - P^2)a}{r^2 + a^2cos^2\theta}}{-\frac{(r_g r - Q^2 - P^2)a}{r^2 + a^2cos^2\theta}(Rsin\phi sin\theta) + (r^2 + a^2 + \frac{(r_g r - Q^2 - P^2)a^2}{r^2 + a^2cos^2\theta}sin^2\theta)} \tag{20}$$

In the above expression of frame-dragging given by equation (20):

- If we substitute P=0 and a=0 then we can obtain the coresponding expression of frame-dragging in Reissner-Nordstrom Geometry.
- If we substitute Q=P=0, then we can obtain the coresponding expression of frame-dragging in Kerr Geometry.
- If we substitute Q=P=0 and a=0, then we can obtain the coresponding expression of frame-dragging in Schwarzschild Geometry, which can be considered as Geodetic effect.

If we substitute $\theta = \frac{\pi}{2}$ (equator) in equation (20). We can obtain the expression of frame-dragging ($\frac{d\phi}{cdt}$) for the equatorial plane in the Kerr-Newman Geometry as:

$$\frac{d\phi}{cdt}(\phi,\theta = \frac{\pi}{2}) = \frac{[1 - (\frac{r_g}{r} - \frac{Q^2+P^2}{r^2})](Rsin\phi) + a(\frac{r_g}{r} - \frac{Q^2+P^2}{r^2})}{-a(\frac{r_g}{r} - \frac{Q^2+P^2}{r^2})(Rsin\phi) + (r^2 + a^2 + a^2(\frac{r_g}{r} - \frac{Q^2+P^2}{r^2}))} \tag{21}$$

If we substitute $\theta = \frac{\pi}{2}$ (equator) and charge Q=P=0 in equation (20). We can obtain the expression of frame-dragging ($\frac{d\phi}{cdt}$) for the equatorial plane in the Kerr Geometry as:

$$\frac{d\phi}{cdt}(\phi,\theta = \frac{\pi}{2}) = \frac{(1 - \frac{r_g}{r})(Rsin\phi) + \frac{r_g a}{r}}{-\frac{r_g a}{r}(Rsin\phi) + (r^2 + a^2 + \frac{r_g a^2}{r})} \tag{22}$$

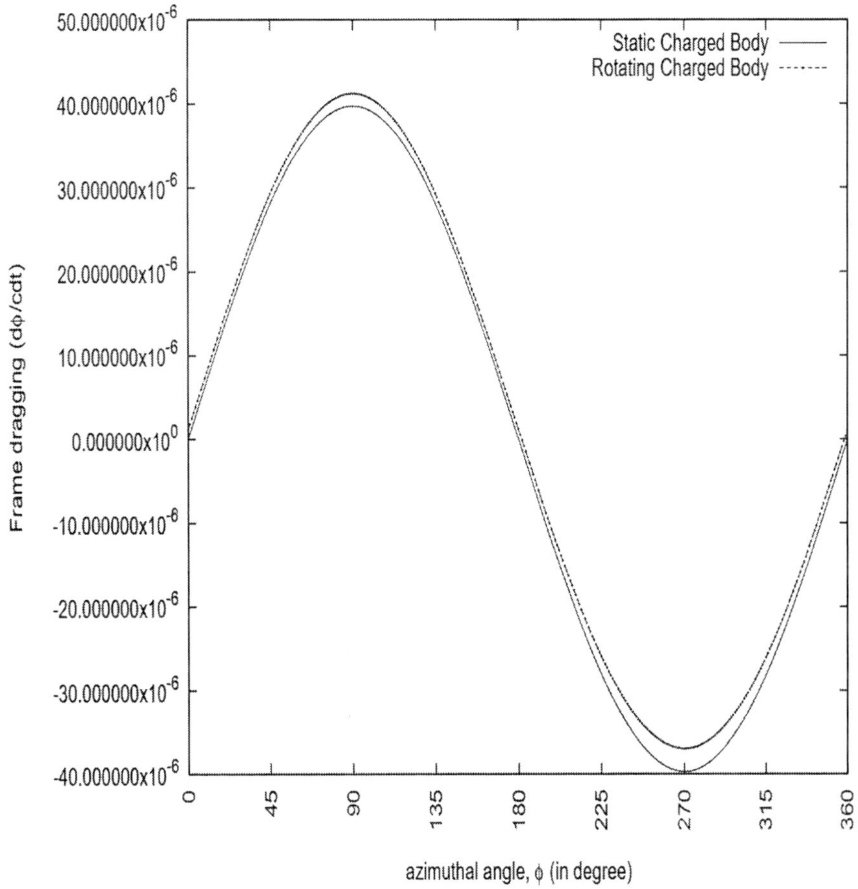

Figure 1. Shows the variation of frame-dragging $\frac{d\phi}{cdt}(\phi, \theta = \pi/2)$ versus ϕ for pulsar PSR J1748-2446 ad, from $\phi = 0$ to 2π, at fixed $\theta = \pi$ and at fixed value of charge.

For doing numerical calculations, we have considered a rotating star such as pulsar PSR J 1748-2446ad (Hessels et al. 2006 [12]), having mass (M= 1.35 Solar Mass), physical radius (R= 20.10 km), Schwarzschild radius (r_g= 4.05 km), angular velocity of rotation ($\omega = 4.50 \times 10^3$ rad/s), rotation parameter (a = 2.42 km) and a fixed value of charge.

Using equation (21), we made a plot (Figure-1) which shows the variation of frame-dragging $(\frac{d\phi}{cdt}(\phi, \theta = \pi/2)$ versus azimuthal angle (ϕ) for PSR J 1748-2446ad from ϕ= 0 to 2π, at fixed $\theta(= \pi/2)$ and fixed value of charge. Here in the plot, solid line represents the frame-dragging for charged static body while dashed line for charged rotating body of same mass.

From Figure-1, it is clearly seen that the frame-dragging from a rotating star (as pulsars) shows a periodic nature with respect to coordinate ϕ (azimuthal angle). The value of frame-dragging is maximum at $\phi = \pi/2$ and minimum at $\phi = 3\pi/2$. If we compare the frame-dragging for static body and rotating body of same mass and charge, then we notice that the effect of rotation is to increase the value of frame-dragging. It may be also noted that, the effect of charge on frame-dragging is not very significant.

Using the concept of frame-dragging, the general expression of gravitational redshift in Kerr and Kerr-Newman geometry have been discussed in details in our previous works Dubey and Sen 2015 [7]; Dubey and Sen 2015 [8].

4. Conclusions

We can conclude from the present work that,

(i) Frame-dragging is a function of the azimuthal angle (ϕ) and latitude ($\pi/2 - \theta$).

(ii) Frame-dragging is mainly affected by the rotation of the central body.

(iii) Frame-dragging is also affected by the electric and magnetic charge present on the rotating body.

(iv) The effect of charge on frame-dragging is not very significant, in comparison to the rotation effect on frame-dragging.

(v) If we consider the ray of light which is emitted radially outward from a rotating star then the frame-dragging shows a periodic nature with respect to coordinate ϕ (azimuthal angle).

(vi) The value of frame-dragging obtains a maximum at, $\phi = \frac{\pi}{2}$ and a minimum at $\phi = \frac{3\pi}{2}$.

5. Acknowledgement

We wish to thank Dr. Atri Deshmukhya, Department of Physics, Assam University, Silchar, India for useful suggestions and inspiring discussions. Finally we are thankful to the anonymous referee of this paper, for very useful comments.

References

[1] Lense J and Thirring H 2008 *Zeitschrift fur Physik* **19** 156–163
[2] Schiff L I 2008 *Phys. Rev. Lett.* **4** 215–217
[3] Everitt C W F, Adams M, Bencze W, Buchman S, Clark B, Conklin J, DeBra D B, Dolphin M, Heifetz M, Hipkins D, Holmes T, Keiser G M, Kolodziejczak J, Li J, Lipa J, Lockhart J M, Muhlfelder B, Ohshima Y, Parkinson B W, Salomon M, Silbergleit A, Solomonik V, Stahl K, Taber M, Turneaure J P, Wang S and Worden Jr P W 2008 *The Gravity Probe B Experiment, Testing Einstein's Universe. Gravity Probe B Data Analysis-Status and Potential for Improved Accuracy of Scientific Results* 5 (Prepared by Robert Kahn, Stanford University, Science Results NASA Final Report December 2008)
[4] Collas P and Klein D 2004 *General Relativity and Gravitation* **36** 1197–1206
[5] Mei H T and Jiu W Y 2006 *Chinese Physics* **15** 232–234
[6] Epstein K J 2008 *General Relativity and Gravitation* **40** 1367–1378
[7] Dubey A K and Sen A K 2015 *International Journal of Theoretical Physics* **54** 2398–2417
[8] Dubey A K and Sen A K 2015 *Astrophysics and Space Science* **360**(29) 1–7
[9] Kerr R P 1963 *Phys. Rev. Lett.* **11**(5) 237–238
[10] Newman E T, Couch E, Chinnapared K, Exton A, Prakash A and Torrence R 1965 *Journal of Mathematical Physics* **6** 918–919
[11] Carroll S M 2004 *Spacetime and geometry: An introduction to general relativity* vol 1 (Pearson, Addison Wesley)
[12] Hessels J W, Ransom S M, Stairs I H, Freire P C, Kaspi V M and Camilo F 2006 *Science* **311** 1901–1904

Journal of Physics: Conference Series **759** (2016) 012066 doi:10.1088/1742-6596/759/1/012066

Hybrid computer modelling in plasma physics

J Hromadka, T Ibehej and R Hrach

Department of Surface and Plasma Science, Charles University in Prague, V Holesovickach 2, 180 00 Prague 8, Czech Republic

E-mail: hromadka.jakub@seznam.cz

Abstract. Our contribution is devoted to development of hybrid modelling techniques. We investigate sheath structures in the vicinity of solids immersed in low temperature argon plasma of different pressures by means of particle and fluid computer models. We discuss the differences in results obtained by these methods and try to propose a way to improve the results of fluid models in the low pressure area. There is a possibility to employ Chapman-Enskog method to find appropriate closure relations of fluid equations in a case when particle distribution function is not Maxwellian. We try to follow this way to enhance fluid model and to use it in hybrid plasma model further.

1. Introduction

Computer modelling techniques [1] are well established tools in plasma physics research. There are two main kinds of them - particle and fluid computer models. Whereas particle models are more precise and give us more detailed information about systems of interacting particles, fluid models usually do not require so high computer performance. It was proven [2] that combination of both techniques, so called hybrid models, can bring valuable results for low temperature plasma at medium pressures where drift diffusion approximation can be used in the fluid part of the model. Our contribution wants to identify differences between particle and fluid model results and discuss possible ways to extend applicability of hybrid modelling techniques to other pressure regimes.

2. Conventional modelling techniques in plasma physics and their limits

2.1. Particle models

Particle models are based on solution of equations of motion of very large number of particles in the modelled ensemble. Thus, this modelling technique provides the most precise results, especially in non-equilibrium cases. Common particle modelling technique in plasma physics, which is implemented also in our model, is so-called PIC/MCC method (= Particle-In-Cell/Monte Carlo Collisions).

Solution process of this method consists of several steps. First of all, computational domain is initialized by charged particles of different species and with Maxwell velocity distribution. After that, iterative cycle starts. Cloud-In-Cell algorithm assigns charge density to all nodes of computational mesh and Poisson equation is solved (e.g. by finite difference method) to obtain potential distribution. Consequently, particle motion in this electrostatic field is resolved by velocity Verlet algorithm. Effects of scattering processes are solved by null-collision method;

our model uses its improved version [3]. Iterative cycle ends when convergence of solution is reached.

2.2. Fluid models

Fluid model equations are derived from Boltzmann equation as its velocity moments. Following set of fluid equations is very often used in low temperature plasma physics:

$$\frac{\partial n_{e,i}}{\partial t} + \nabla \cdot \boldsymbol{\Gamma}_{e,i} = 0, \tag{1}$$

$$\boldsymbol{\Gamma}_{e,i} = \pm \mu_{e,i} n_{e,i} \boldsymbol{E} - D_{e,i} \nabla n_{e,i}, \tag{2}$$

$$\Delta \phi = -\frac{e}{\epsilon_0} \left(n_i - n_e \right), \tag{3}$$

where n marks number density, $\boldsymbol{\Gamma}$ particle flux, μ mobility coefficient, D diffusivity coefficient and ϕ is potential of electric field $\boldsymbol{E}$.

Equation 2 is called drift-diffusion approximation and it is derived from general equation of momentum conservation:

$$mn\frac{\mathrm{d}\boldsymbol{v}}{\mathrm{d}t} = \pm en\boldsymbol{E} - \nabla \cdot \boldsymbol{P} + \sum \boldsymbol{F_s}, \tag{4}$$

where stress tensor $\boldsymbol{P}$ and collisional frictional force $\boldsymbol{F_s}$ are defined as following:

$$\boldsymbol{P} = \int m\boldsymbol{w}\boldsymbol{w} f(\boldsymbol{r}, \boldsymbol{v}, t) \mathrm{d}^3\boldsymbol{v}, \qquad \boldsymbol{F_s} = \int m\boldsymbol{v} \left[\frac{\delta f}{\delta t}\right]_s \mathrm{d}^3\boldsymbol{v}, \tag{5}$$

where $f(\boldsymbol{r}, \boldsymbol{v}, t)$ marks particle distribution function and $\left[\frac{\delta f}{\delta t}\right]_s$ is collisional term describing collisions of one particle species with particles of species s. Velocity $\boldsymbol{w}$ is defined as $\boldsymbol{w} = \boldsymbol{u} - \boldsymbol{v}$, where $\boldsymbol{u} = \int \boldsymbol{v} f(\boldsymbol{r}, \boldsymbol{v}, t) \mathrm{d}^3\boldsymbol{v}$.

Derivation of drift-diffusion approximation from general momentum equation takes into account several approximations: 1. Inertial term in full momentum equation is neglected, 2. steady-state is considered only, 3. plasma is treated as ideal gas (particles interact only during collisions), 4. special case of velocity moment of the collisional term in Boltzmann equation: $\boldsymbol{F_s} = -mn\boldsymbol{v}\nu$, where ν is frequency of elastic collisions. These considerations limit application of drift-diffusion model in plasma physics.

3. Comparison between particle and fluid models

2D particle and fluid models were used to investigate sheath structure and IV characteristics of a cylindrical probe with $2 \cdot 10^{-3}$ m radius to compare their results. We have considered electropositive argon plasma where temperature of electrons was 23 600 K and that of Ar^+ ions was 300 K. Computational domain had dimensions of $(4.0 \times 4.0) \cdot 10^{-2}$ m. Regular rectangular mesh with 400×400 cells was used to solve Poisson equation by finite difference method in particle model. Scattering processes that were considered in particle model; $Ar + e^-$ collisions: elastic scatter, excitation and ionization [4]; $Ar + Ar^+$ collisions: elastic scatter and charge transfer [5]. Equations of fluid model were solved by finite element method on triangular mesh with approximate number $4 \cdot 10^5$ of triangles. Open source project FENICS [6] was used to build the fluid model. Our study was done for different values of neutral gas pressure and table 1 presents corresponding plasma densities. Exact pressure dependence of plasma density was neglected, square root dependence was considered.

Table 1. Plasma density for different values of neutral gas pressure

Neutral gas pressure [Pa]	1.33e-1	1.33e0	1.33e1	1.33e2
Plasma density [m^{-3}]	5.03e13	1.59e14	5.03e14	1.59e15

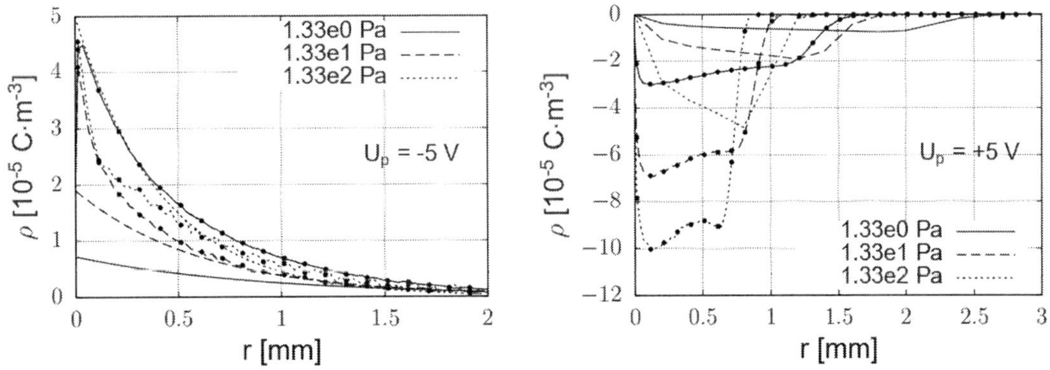

Figure 1. Net charge density in the vicinity of biased cylindrical probe for different pressures of neutral gas on the background computed by 2D particle (lines with markers) and fluid model (without markers). Two probe biases were considered, $U_p = \pm 5\,\mathrm{V}$.

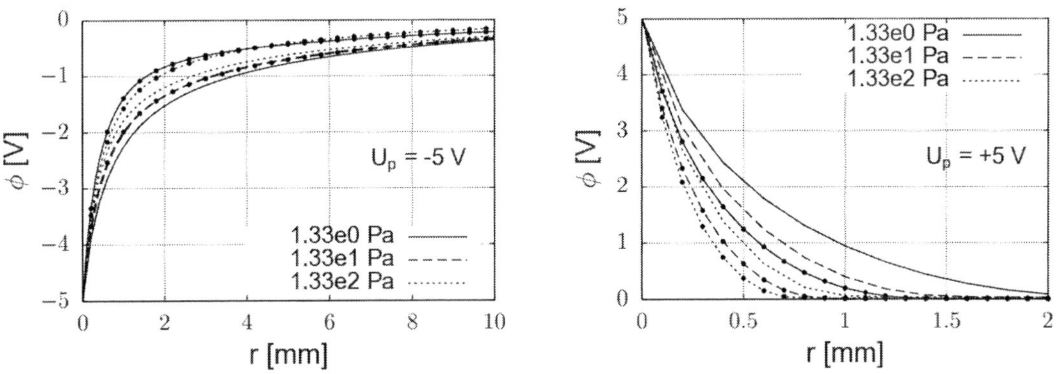

Figure 2. Potential profile in the vicinity of biased cylindrical probe for different pressures of neutral gas on the background computed by 2D particle (lines with markers) and fluid model (without markers). Two probe biases were considered, $U_p = \pm 5\,\mathrm{V}$.

3.1. Plasma sheath around cylindrical Langmuir probe

Figures 1 and 2 show sheath structure around cylindrical probe in terms of net charge density and potential profile for different pressures. It can be seen that fluid model is able to capture formation of sheath in some level of accuracy. Still fluid model results differ from that obtained by particle model.

For almost all investigated cases net charge density in the near surroundings of a probe obtained by fluid model is lower than that obtained by particle model. Fluid solution tend to

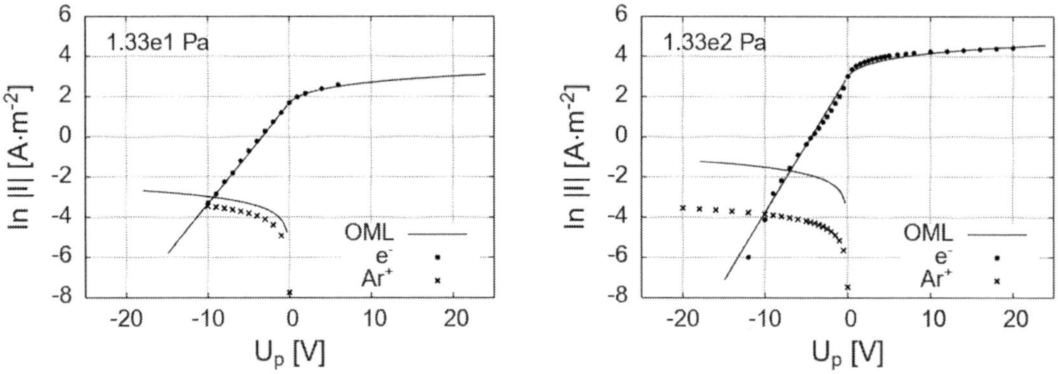

Figure 3. Comparison of IV characteristics of cylindrical Langmuir probe computed by 2D particle model with OML theory for two different pressures of neutral gas.

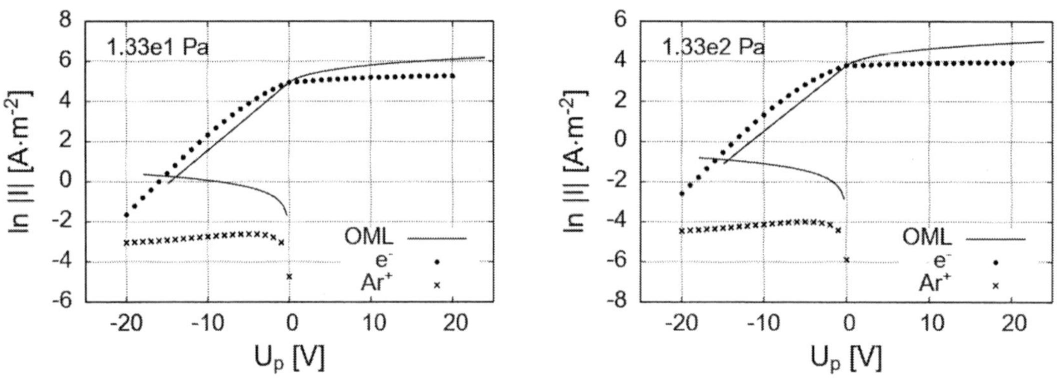

Figure 4. Comparison of IV characteristics of cylindrical Langmuir probe computed by 2D fluid model with OML theory for two different pressures of neutral gas.

preserve diffusion profile which is logaritmic in 2D case. As a result we can see that potential is shielded out on larger distance in fluid case.

It can be stated that differences between particle and fluid model results are greater for lower pressures. We also expect greater differences between fluid and particle model results in a case the average kinetic energy per particle increases. If a particle has higher kinetic energy it is more probable that it will undergo one of the inelastic scattering processes (excitation, ionization etc.) and it will affect particle's velocity distribution function in a way that it will become much more disctint from Maxwellian distribution (which is an assumption of fluid plasma models).

3.2. IV characteristic of cylindrical probe

Differences between particle and fluid models were also investigated for the case of IV characteristics of cylindrical Langmuir probe, results are presented on figures 3 and 4.

Particle model results are in very good agreement with OML theory for 13.3 Pa pressure. For higher pressure we can see disagreement caused mainly by scattering processes that are not considered by OML theory. Disagreement is observed especially for ion current.

Fluid model provides electron current that is qualitatively in agreement with OML theory. The agreement is better for lower pressure. At low pressure fluid model is able to capture increasing electron current for positive probe bias. However, fluid model is not reliable in computation of ion current at all.

4. Concept of hybrid models

Main idea which is behind hybrid models is to combine advantages of both particle and fluid models so that the resultant model would be sufficiently precise and fast enough to solve large 3D problems. Results obtained by hybrid modelling approach have already been published, e. g. [7]. There are many ways how the particle and fluid models can be combined. It was proven [2] that so-called iterative hybrid model can provide reliable results in medium pressure area. Since this model uses drift diffusion approximation it can not be used for low pressure problems.

To extend usability of hybrid models to the low pressure regime Chapman Enskog method [8] can be used. It is a method of solution of Boltzmann kinetic equation with prescribed collisional term and force law between interacting particles. Particle distribution function $f(\boldsymbol{r}, \boldsymbol{v}, t)$ is expanded around equilibrium Maxwell distribution according to small parameter ϵ which can be e. g. mean free path or Larmor radius:

$$f(\boldsymbol{r}, \boldsymbol{v}, t) = f_0 + \epsilon f_1 + \epsilon^2 f_2 + \dots \tag{6}$$

Usually, only first two terms are considered and inserted into Boltzmann kinetic equation. Consequently, this integral equation is solved for f_1 using e. g. Sonine polynomials. Obtained distribution function is used in equations 5 to derive expressions for stress tensor $\boldsymbol{P}$ and collisional frictional force $\boldsymbol{F_s}$.

5. Conclusions

In our contribution we compared results of two well established computer modelling techniques used in plasma physics - particle and fluid models. The differences between these techniques were discussed on sheath structure around cylindrical Langmuir probe and its IV characteristic. We identified that differences are greater for low pressure where drift diffusion approximation fails. We also discussed hybrid modelling techniques and possibilities of their usage in low pressure area.

Acknowledgments

The study was supported by the Charles University in Prague, project GA UK No. 220215.

References

[1] Birdsall C K and Langdon A B 1991 *Plasma physics via computer modelling* (Bristol: IOP Publishing)
[2] Hruby V and Hrach R 2010 *IEEE Trans. Plasma Sci.* **38** 2328
[3] Roucka S and Hrach R 2011 *IEEE Trans. Plasma Sci.* **39**, 3244
[4] Bogaerts A, Gijbels R and Goedheer W 1999 *Jpn. J. Appl. Phys.* **38**, 4404
[5] Phelps A 2013 *Compilation of electron cross sections* http://jila.colorado.edu/avp/collision data/
[6] Logg A, Mardal K A, Wells G N et al. 2012 *Automated Solution of Differential Equations by the Finite Element Method* (Berlin: Springer)
[7] Bogaerts A, Gijbels R 1999 *IEEE Trans. Plasma Sci.* **27**, 1406
[8] Chapman S and Cowling T G 1990 *The mathematical theory of non-uniform gases: an account of the kinetic theory of viscosity, thermal conduction, and diffusion in gases* (Cambridge University Press)

Journal of Physics: Conference Series **759** (2016) 012067 doi:10.1088/1742-6596/759/1/012067

Morphological Investigation of Disturbed Ionosphere during Intense Geomagnetic Storms

Bhupendra Malvi[1], Prateek S. Srivastav[1], Azad A. Mansoori[1], Roshni Atulkar[2], Shivangi Bhardwaj[2] and P. K. Purohit[2]

[1]*Department of Electronics, Barkatullah University, Bhopal-462026, India*
[2]*National Institute of Technical Teachers' Training and Research, Bhopal-462002, India*

Email: bhup1201@gmail.com

Abstract. Geomagnetic Storms are the disturbed magnetic conditions, influenced and induced by Interplanetary Magnetic Field and the Charged Particle's motion around the Earth, respectively, in Geospace. As the ionosphere is woven by the earth's magnetic field it responds to the change in that. During the geomagnetic storms the filled-in plasma between the magnetic field lines, geomagnetic flux tubes, redistributes itself in effect of the magnetic field forcing. In the present study we have done the investigation of the morphology of the ionosphere over the mid and high latitude regions during intense Geomagnetic Storms. We got fairly convincing results; in three cases decrease of the critical frequency of F2 layer (foF2) and in one case enhancement of the critical frequency of F2 layer (foF2) at mid and high latitudes.

1. Introduction

Solar transients; Solar Flares, Coronal Mass Ejections (CMEs), Solar Energetic Particles (SEPs) are the drivers of the Space Weather Effect in Geo-Space. When the gigantic cloud of plasma released through solar transient phenomena interacts with the Earth's magnetic environment it leads to the geomagnetic storms [1, 2]. Geomagnetic storms can be characterized by a depression in the H component of geomagnetic field for tens of hours. This depression in H component of earth's magnetic field is caused by the Ring Current encircling the Earth in a westward direction [3, 4]. Earth's ionosphere responds to varying solar and magnetospheric conditions. The ionospheric electron density over an altitude and location depends variably on the solar EUV Flux, X-ray Flux and the dynamic effects of neutral winds and electric fields. During geomagnetic storm due to the compression of earth's magnetosphere by solar wind electric fields have been observed along the geomagnetic field lines to the high latitude ionosphere. Sometimes this electric field penetrates to low latitudes and at high latitudes they lead to the rapid convection of plasma which may drive even the neutral winds via collisions and energetic particles precipitate into the lower thermosphere and below, increasing ionospheric conductivity and expanding the auroral zone. These intense electric currents are responsible for the coupling of high latitude ionosphere with magnetosphere and the enhanced energy input leads to considerable heating of the ionized and neutral gases. In this way the uneven expansion of the thermosphere generates pressure gradients which drive strong neutral winds. This disturbed thermospheric circulation changes the neutral composition and moves the plasma up and down along magnetic field lines, changing rates of production and recombination of ionized gas. During this phenomenon disturbed neutral winds generate polarization electric fields by dynamo effect by the collision with the plasma in the presence of Earth's magnetic field. There are two types of effects, in time scale, on the Earth produced by solar transients; prompt and delayed. Geomagnetic Storm effects are delayed effects due to cloud of particles ejected from Sun. Geomagnetic storm effect on the earth's atmosphere in the E region has widely been studied at high latitudes [5]. In the high latitude E region where particle precipitation in cusp and aurora are a source of substantial ionization gets affected by geomagnetic storms. Most of the ionospheric electron content is found in the F2 layer of ionosphere which tends to lie near a constant pressure level in the thermosphere [6]. Solar winds can raise or lower hmF2, characteristic height of F2 layer, above or below this level. Equator ward winds can lead to a rise in the F2 peak height up to regions of reduced loss which also produce increase in electron density of F2 region (NmF2). A drop in hmF2 due to pole ward winds decreases NmF2. Due to local time variation of winds and changes in neutral composition at middle latitudes, negative ionospheric storm effects are found to be observed very often in morning and positive storm effects in the afternoon and evening.

2. Event Selection Criterion

To Investigate the variability of ionosphere during Intense Geomagnetic Storms we have used the digital ionosonde observations carried out at two latitude station of northern hemisphere viz. mid latitude station Sofia, SQ143 (42.7° N, 23.4° E) and high latitude station College, CO764 (64.9° N, 147.8° W). For this study we have selected the four geomagnetic storms that occurred during 1999 to 2001. Only the four most intense (Dst ≤ -200 nT) geomagnetic storms were considered for the present study. The availability of various data sets was also checked and events with bad or missing data were removed from the analysis.

3. Data Sets and Sources

To accomplish this study we have we have used two sets of data: (1) Storm Intensity Index (Dst) and (2) critical frequency of F2 layer (foF2) of ionosphere measured by means of ground based radio sounding system, Ionosonde.

The data of Dst Index is available at various sites for downloading. However, for our study we have downloaded the data of Dst index from Space Physics Data Facility OMNI website (http://omniweb.gsfc.nasa.gov/). We have used the hourly values of Dst for the investigation.

For the present investigation we have used the manually scaled hourly averaged values of foF2 (critical frequency of F2 layer) over mid latitude station Sofia, SQ143 (42.7° N, 23.4° E) and high latitude station College, CO764 (64.9° N, 147.8° W) from (http://spidr.ngdc.noaa.gov/spidr/) SPIDR data services.

4. Results and Discussion

The disturbed and adverse geomagnetic conditions have a significant impact on the ionosphere. To investigate the variability of ionospheric foF2 during disturbed geomagnetic conditions we have selected four intense geomagnetic storms occurred during the year 1999 to 2001. The geomagnetic storms of 22 October 1999, 17 September 2000, 31 March 2001, and 11 April 2001 were selected to study the behavior of ionospheric delay foF2 during these geomagnetic storms. All the four cases are discussed serially as follows;

4.1. Case 1: 22 October 1999

One of the intense geomagnetic storms was observed on 22 October 1999. The storm had a sudden commencement phase followed by the initial and main phase. The storm intensity index-Dst had the minimum or peak value of -237 nT. The behavior of ionospheric foF2 over mid latitude station, Sofia and high latitude station, College before, after and on the storm day along with Dst index are shown in figure 1. From the Figure we find that the value of foF2 is decreased significantly over mid latitude as well as high latitude during the geomagnetic storm main phase i.e. 22 October 1999. However the effect of geomagnetic storm over high latitude is more. Thus the impact of geomagnetic storm is clearly seen on the mid and high latitude ionospheric foF2 as a negative effect.

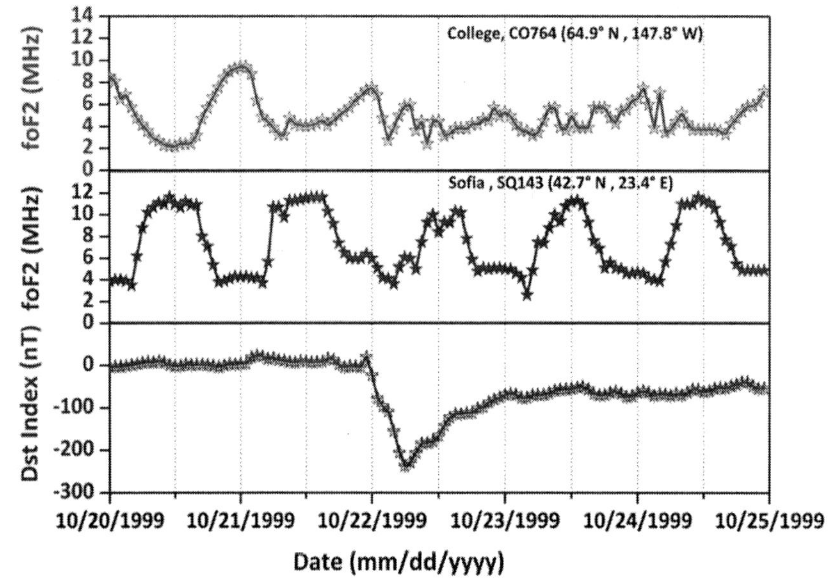

Figure 1. The temporal evolution of foF2 over mid latitude station, Sofia and high latitude station, College with Dst during the geomagnetic storm of 22 October 1999.

4.2. Case 2: 17 September 2000

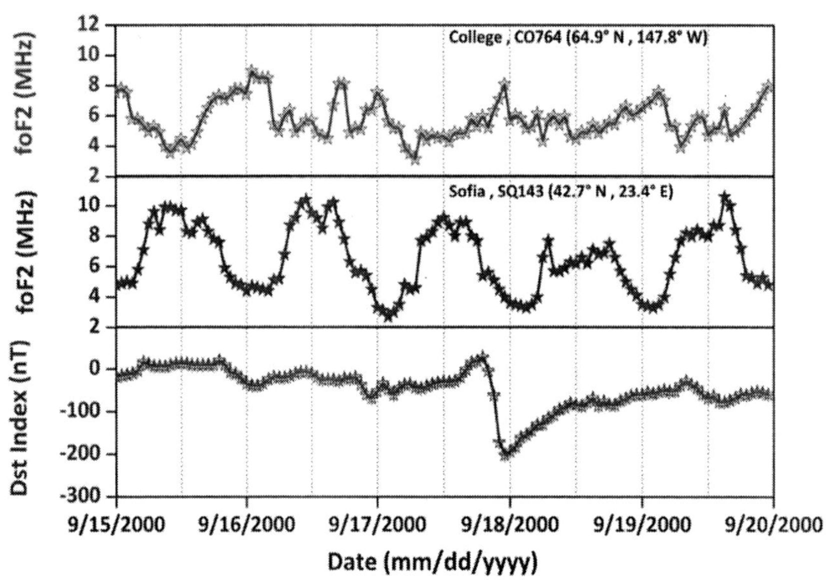

Figure 2. The temporal evolution of foF2 over mid latitude station, Sofia and high latitude station, College with Dst during the geomagnetic storm of 17 September 2000.

Another intense storm was observed on 17 September 2000. The minimum Dst recorded during this storm was -201 nT. The effect of this geomagnetic storm on the ionospheric foF2 is shown in figure 2 along with changes in storm intensity index Dst. The main phase of the geomagnetic storm was observed on 17 September 2000. From the figure we find that during the main phase of geomagnetic storm the value of foF2 decreased over mid latitude as well as high latitude. However the effect of geomagnetic storm over mid latitude is more significant. Hence during this storm also we observed a negative effect over both latitude stations, mid and high.

4.3. Case 3: 31 March 2001

The effect of geomagnetic storm of 31 March 2001 on the ionospheric foF2 is shown in figure 3. The geomagnetic storm that occurred on 31 March 2001 was an intense geomagnetic storm with peak or minimum Dst -387 nT. From the figure we clearly see the decrease in Ionospheric foF2 over both mid and high latitude stations during the day of storm. However the effect of geomagnetic storm over high latitude is clearer than mid latitude. Hence this storm also shows a negative effect over mid latitude station as well as high latitude station.

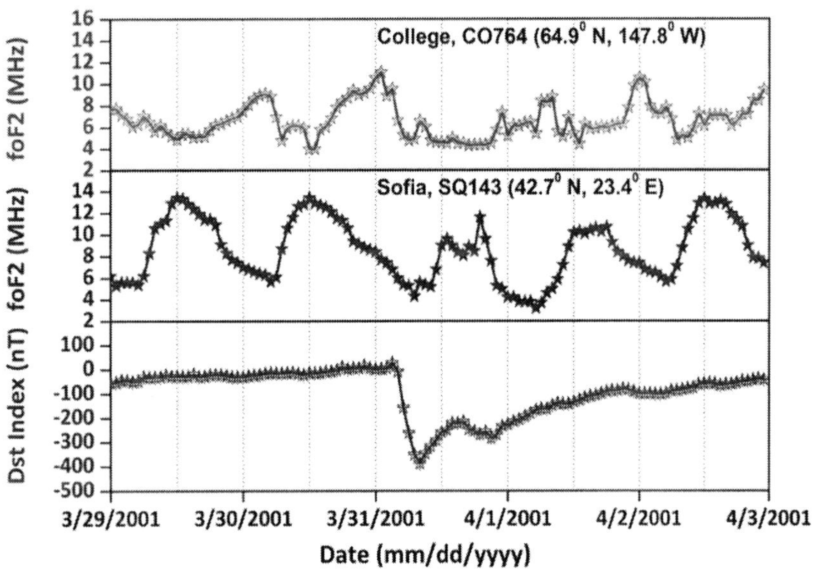

Figure 3. The temporal evolution of foF2 over mid latitude station, Sofia and high latitude station, College with Dst during the geomagnetic storm of 31 March 2001.

4.4. Case 4: 11 April 2001

The effect of geomagnetic storm of 11 April 2001 on the ionospheric foF2 along with changes in storm intensity index Dst is shown in figure 4. The minimum Dst observed for this geomagnetic storm was -271 nT. From the figure we find that Ionospheric foF2 shows a clear enhancement over mid and high latitude stations. Thus the impact of geomagnetic storm is clearly seen on the mid and high latitude ionospheric foF2 as a positive effect.

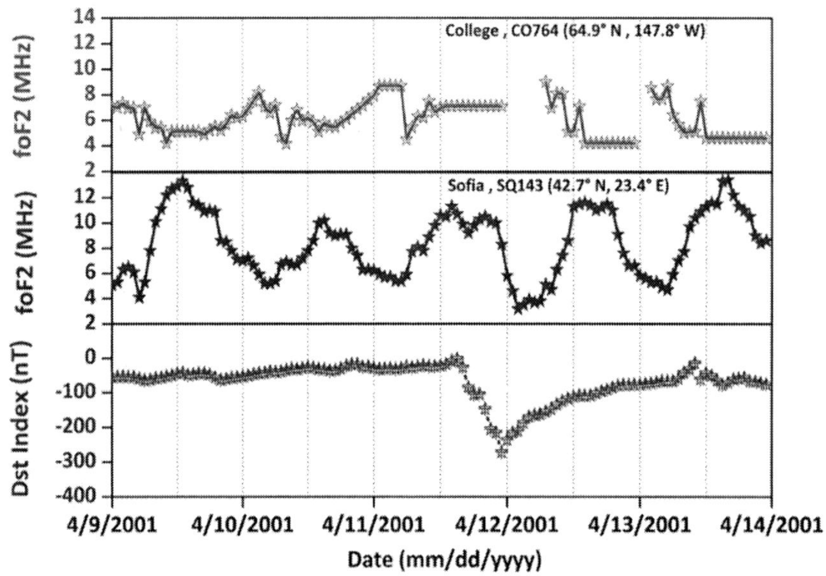

Figure 4. The temporal evolution of foF2 over mid latitude station, Sofia and high latitude station, College with Dst during the geomagnetic storm of 11 April 2001.

After compiling the above results we see that most of the times intense geomagnetic storm produce negative effect on ionospheric foF2 at mid latitudes as well as at high latitudes in northern hemisphere. The cause of simultaneous decrease of foF2 at high and middle latitudes as a revelation of depletion in F2 region plasma density are thought to be due to the changes in the neutral wind produced predominantly be the Joule heating in the auroral zone. In the cases where we observed the positive effect in midlatitude station is due to the magnetospheric electric field finding a way into the midlatitude has been found playing a major role in the ion density hence the foF2 during the main phase of the storm. East ward component of this magnetospheric electric field increase the local production on the dayside which reflects in the storm time density increase. Midlatitude F-region is has seasonal dependence as well as the neutral wind at F-region heights have so.

5. Conclusions

We summarize our main findings as follows;

- Most of the times intense geomagnetic storm produce negative effect on ionospheric foF2 at mid latitudes as well as at high latitudes in northern hemisphere. We have taken four intense geomagnetic storms for investigating the effect of intense geomagnetic storms on ionosphere and we get negative effect on mid and high latitude ionosphere of northern hemisphere during three cases out of four cases. However in one case out of four we got positive effect also.
- We also find that a particular geomagnetic storm produces same effect at all latitudes of a particular hemisphere. As in our study we got same effect on mid and high latitude ionosphere of northern hemisphere during each case.

- We also find that most of the times high latitude ionosphere shows more significant effect to the adverse geomagnetic conditions compare to mid latitude ionosphere in northern hemisphere, whatever the effect is negative or positive.

6. References

[1] Gonzalez W D and Tsurutani B T 1987 Criteria of interplanetary parameters causing intense magnetic storms (Dst < 100 nT) *Planet Space Sci.* **35** 1101–1109.

[2] Sugiura M and Chapman S 1960 The average morphology of geomagnetic storms with sudden comencement, Abandl. Akad. Wiss. *Gottingen Math. Phys.* **K1(4)**.

[3] Daglis I A et al 1999 The terrestrial ring current: Origin, formation, and decay *Rev. Geophys.* **37** 407– 438.

[4] Kamide Y et al 1998 Current understanding of magnetic storms' Storm-substorm relationships *J. Geophys. Res.* **103** 17,705– 17,728.

[5] Bauske R and Prölss G W 1997 Modeling the ionospheric response to traveling atmospheric disturbances *J. Geophys. Res.* **102**, 14,555–14,562.

[6] Rishbeth H and Edwards R 1989 The isobaric F2-layer *J. Atmospheric Terrest. Phys.* **51** 321–338.

Study of trapped particle nonlinearity in Ion Acoustic solitary wave using Vlasov simulation

Debraj Mandal, Devendra Sharma
Institute for Plasma Research, Bhat, Gandhinagar, 382428, India
E-mail: debraj@ipr.res.in

Abstract. Collective processes in a system of many interacting particles like plasma involves solving the kinetic Vlasov equation in a multiple species phase-space. One of the challenges in simulation of trapped particle equilibrium in collision-less plasmas is simultaneous resolution of multiple time scales for the evolution of the species distribution functions (f). The high resolution Vlasov simulation developed to perform such analysis resolves the trapped particle populations that are unaccounted for in the linear dispersion analysis of the plasmas. The parallel version of the flux balance Vlasov simulation code is implemented using a combination of MPI and open-MP computation technique. The simulations shed light on the nonlinear effects in regime of small amplitude limit that produce coherent structures known as Solitary electron holes (SEH). The SEH develop in ion acoustic wave (IAW) branch of plasma wave in presence of electron current and are analyzed to show their distinct behavior in the kinetic, particle trapping regime.

1. Introduction

In the kinetic description of plasma in collision less regime, the equation governing the evolution of collective processes is the Vlasov equation. In small amplitude regime obtained by linearization of Vlasov equation considering the potential as a small perturbation, the linear dispersive waves are recovered that interact with resonant particles resulting in the damping or growth of the perturbation. Exact nonlinear solutions are possible by acknowledging the existence of particles trapped by the perturbations, for example those resulting in Bernstein-Greene-Kruskal (BGK) mode solutions[1]. While this approach treats density of the trapped particles as a free parameter, there exist other approaches where trapped particles have an equilibrium distribution with finite temperature [2]. The latter produces trapped particle structures where the nonlinearity introduced by the amplitude dependent density of trapped particles influences the local potential. This nonlinearity substitutes the more conventional sources of nonlinearities in the plasma usually absent in small amplitude limit.

The present work analyzes the coherent trapped particle equailibria known as Solitary Electron Holes (SEH) using the Vlasov simulations where solutions involve resolving multiple time scales associated with electrons and ion species. For example, the collective modes produced by thermal electrons operates at a frequency different from the bounce frequency ($\omega_{be} = \sqrt{e\phi k^2/2m_e}$) of the electrons in ion acoustic waves that evolve at a yet slower ion response time. In the present work the "Flux Balance"[3] method is used to solve the Vlasov equation numerically for both electrons and ion phase-space distributions in a current caring plasma. We compare our result to the analytical model of trapped particle modes while limiting the study to the small amplitude limit where the the conventional fluid nonlinearities have a negligible role.

2. Numerical Technique for Vlasov Simulation

The flux balance method used in the present simulations is based on splitting the analytical form of the Valsov equation along the configuration and velocity dimensions in the phase space and affecting a flux corrected time evolution as discussed below.

2.1. Splitted time evolution

In electrostatic case the Vlasov equation can be written as,

$$\frac{\partial f}{\partial t} + v\frac{\partial f}{\partial x} + E(x,t)\frac{\partial f}{\partial v} = 0 \tag{1}$$

where, f is the distribution function and $E(x,t)$ is the electric field which can be calculated using the Poisson equation,

$$\frac{\partial E}{\partial x} = \int_{-\infty}^{+\infty} f_i(x,v)dv - \int_{-\infty}^{+\infty} f_i(x,v)dv \tag{2}$$

Using time splitting method one can split the Eq. (1) in two similar differential equation as [3],

$$\frac{\partial f^*}{\partial t} + v\frac{\partial f^*}{\partial x} = 0; \quad f^*(x,v,t=n\Delta t) = f^{**}(x,v,t=(n-1)\Delta t), \tag{3}$$

$$\frac{\partial f^{**}}{\partial t} + E(x,t=n\Delta t)\frac{\partial f^{**}}{\partial x} = 0; \quad f^{**}(x,v,t=n\Delta t) = f^*(x,v,t=n\Delta t), \tag{4}$$

which are solved simultaneously, the time evolved f from one of the equations being used as the initial condition for the other. Electric field is calculated by solving the Poisson Eq. (2). We use a leap-frog type scheme for splitting the time, which is second order and symmetrical in time.

2.2. Flux Balance method

Since both the differential equation Eq. (3) and (4) are identical, each of them can be solved by flux balance technique. In the Fig. (1) we explain the Flux Balance technique using three adjacent cells, $(i-1)th$, ith and $(i+1)th$ of width Δx each with X as the corresponding position. Assuming that the distribution function f is smooth in each elementary cell and velocity is positive, in the Fig. 1 the lower grid represents f at time t and the upper grid at time $t+\Delta t$, where Δt is the time step. For Δt sufficiently small, the fluid or the distribution function $f(t)$ with in the length $v\Delta t$ from one cell moves to the next cell at time $t+\Delta t$. In the fig. (1) the region shaded with horizontal lines represents fluid loss, F^i_{loss}, from ith cell to $(i+1)th$ cell and the region shaded with vertical lines represents fluid gain, F^i_{gain}, from $(i-1)th$ cell to ith cell. The new value of f at the ith cell at time $t+\Delta t$ is therefore $f^i(t+\Delta t) = f^i(t) + F^i_{gain} - F^i_{loss}$. Values F^i_{loss} and F^i_{gain} can be obtained from the Eq. (3) as,

$$F^i_{loss} = \frac{1}{\Delta x}\int_{x_i+\Delta x/2-v\Delta t}^{x_i+\Delta x/2} f(h-x_i)dh; \qquad F^i_{gain} = \frac{1}{\Delta x}\int_{x_{i-1}+\Delta x/2-v\Delta t}^{x_{i-1}+\Delta x/2} f(h-x_{i-1})dh;$$

where, h is the integration variable. Similarly expressions of F^i_{loss} and F^i_{gain} are obtainable for the negative velocities and are applicable to both parts of the splitted equations.

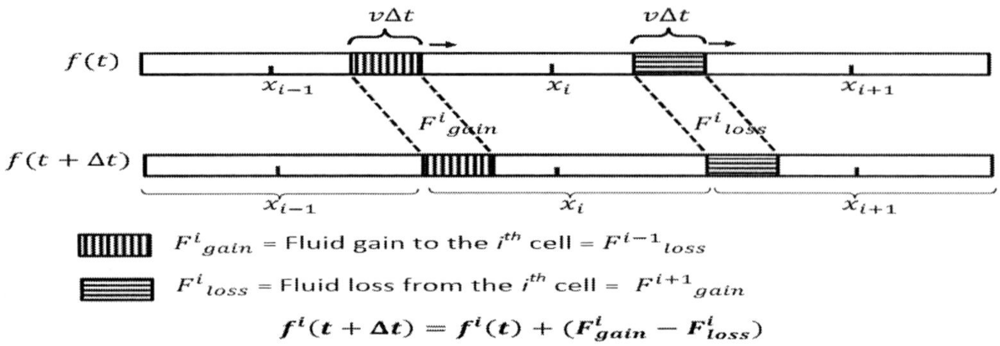

Figure 1. Block diagram of the Flux Balance Technique.

In order to preserve the positivity of the distribution function, the velocity range of the species phase space needs to be chosen carefully in combination of the time step. The required temporal and spatial resolution determine the limiting values of the species velocity space these are therefore estimated based on the phase space dimensions relevant to the problem to ensure the positivity of f. The fluctuations of 1st norm of distribution function $L^1(t) = \int\int f(x,v)dxdv$ is additionally observed and maintained below a required tolerance through out the simulation.

2.3. Parallelization using MPI and Open-MP

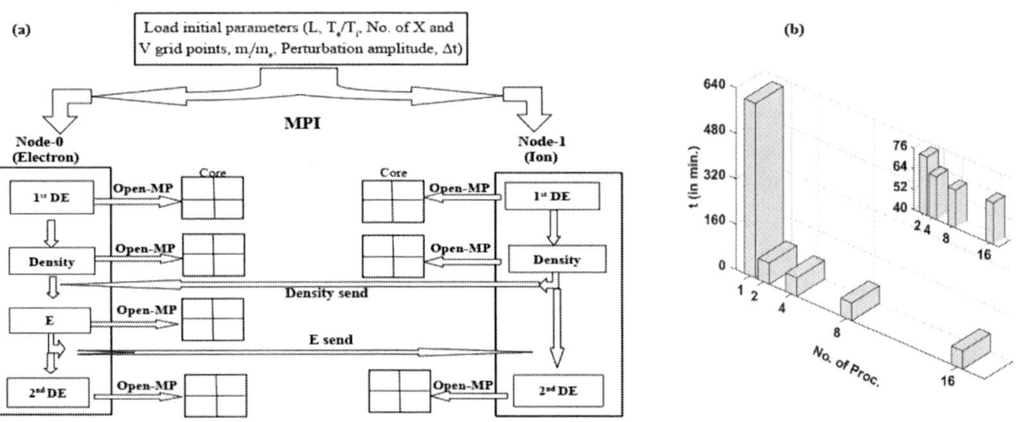

Figure 2. (a) Block diagram of the parallelization scheme. (b) Variation of time taken for 1000 time steps in (4096×8192) grid with different no. of Processors. Number of processor 1 for the serial code and 2 for only 2 MPI nodes, 4 (2×2) for 2 MPI nodes and 2 Open-MP cores in each MPI node, 8 (2×4) for 2 MPI nodes and 4 OpenMP cores in each MPI node. The subplot shows the time taken for 2,4,8 and 16 processors.

To gain computational efficiency and speed both the MPI and Open-MP schemes are adopted. Fig. (2)(a) presents the block diagram of parallelization technique. Equations (3) and (4) are needed to be solved for both electrons and ions. In order to to achieve a speed up the above solution procedure is implemented on two different nodes for electrons and the essential data sharing is done using Message Passing Interface (MPI) library. The internal computations associated with both the species are distributed on four cores using the OpenMP scheme. Fig. (2)(b) presents the computation time for the 1000 time step in a (4096×8192) dual mesh grid with increasing number of processor showing that a computational speed up of about 8 times was achievable up on an optimization and implementation of MPI to use minimum two MPI nodes as compared to the time taken for a single processor (serial) execution represented by the bar at number of processors = 1. The bars at even larger number of processors (4 and more) represent only the effect of adding more OpenMP cores while using only two MPI nodes. This parallel scheme also helpful for small number of grid points also. The efficiency of the parallel scheme compared to a serial one increases with the increase of number of grid points. Here we study evolution of solitary structures of sufficiently long spatial scales that are well resolved by a grid resolution of 512×512. Higher resolution grids are required for studying time evolution of small structures.

3. Solitary Wave Generation by Trapping Nonlinearity

3.1. Numerical Result

In order to simulate a current carrying plasma in 1D electrostatic simulations a stationary Maxwellian velocity distribution function is used for ions while a shifted Maxwellian for electron distribution, as presented in Fig. (3)(A)

$$f_j(x, v_j, t = 0) = f_{0j}(x) \exp\left[-\frac{m_j(v_j - u_j)^2}{2T_j}\right] \tag{5}$$

where the subscript $j = e$ and i correspond to electron and ion, m_j, T_j, v_j and u_j are the mass, temperature, velocity co-ordinate and drift velocity of the species j respectively. The quantity f_{0j} is the normalization constant. The normalizations used here are, potential $\tilde{\phi} = e\phi/T_e$, length $\tilde{x} = x/\lambda_{De}$, time $\tilde{t} = t\omega_{pe}$, velocity $\tilde{v} = v/v_{the}$, electric field $\tilde{E} = eE/T_e\lambda_{De}$ and temperature $\tilde{T}_j = T_j/T_e$, where λ_{De}, v_{the} and ω_{pe} are the plasma Debye length, electron thermal velocity and electron plasma oscillation frequency, respectively. Initially, a cosine density perturbation is applied to both electron and ion such that total density can be written as, $n_i = n_e = n_0 + n_1 \cos(2\pi x/L)$, where n_1 and n_0 are the amplitude of the density perturbation and the background density, respectively. In present simulation a 512×512 dual-mesh grid in phase-space and periodic boundary condition at the end of simulation zone L in x is used for both electron and ion for solving vlasov equation for the initial parameter n_1=0.008, T_i =0.05, $m_i/m_e = 1836$, $v_{0i} = 0$. Three different combinations of u_e and L, corresponding to Case- I (0.8 v_{th}, 15 λ_{De}), Case-II (0.8 v_{th}, 30 λ_{De}) and Case-III (1.2 v_{th}, 15 λ_{De}) are used.

The time evolution of the potential amplitude ψ for all the three cases is presented in Fig. (3)(B). Due to high Landau damping rate of electron plasma wave (EPW) with large wave number in Case-I and Case-III EPW are damped quickly and a linear IAW is excited, whose growth rate confirms with linear Landau growth rate (up to ∼15 ω_{pe}^{-1}). This is followed by a phase where the electron traps in the IAW becomes nonlinear showing a nonlinear growth rate and finally reaching an almost saturated state. A residual growth rate is however present due to the presence of electron current. Since, case-I (dark blue line) and II (faint violet line) have same amount of electron current, which is lesser than the Case-III (dark red line), the residual growth rate is higher for Case-III. In Case-II, due to a small Landau damping rate of EPW both the EPW and the IAW coexist. The time evolution of the electron phase space (left) nearer to the IAW phase velocity location ($C_s = 0.023v_{th}$) and the potential (right) for Case-II are presented in Fig. (3)(C). Initially there is no phase space structure close to velocity $C_s = \sqrt{T_e/m_i}$. Because of the presence of a positive slope in the electron distribution function at C_s the IAW starts to grow via the wave particle interaction. At a future time ($t > 25\omega_{pe}$) electrons are trapped in IAW and form a electron phase-space hole. The IAW become nonlinear

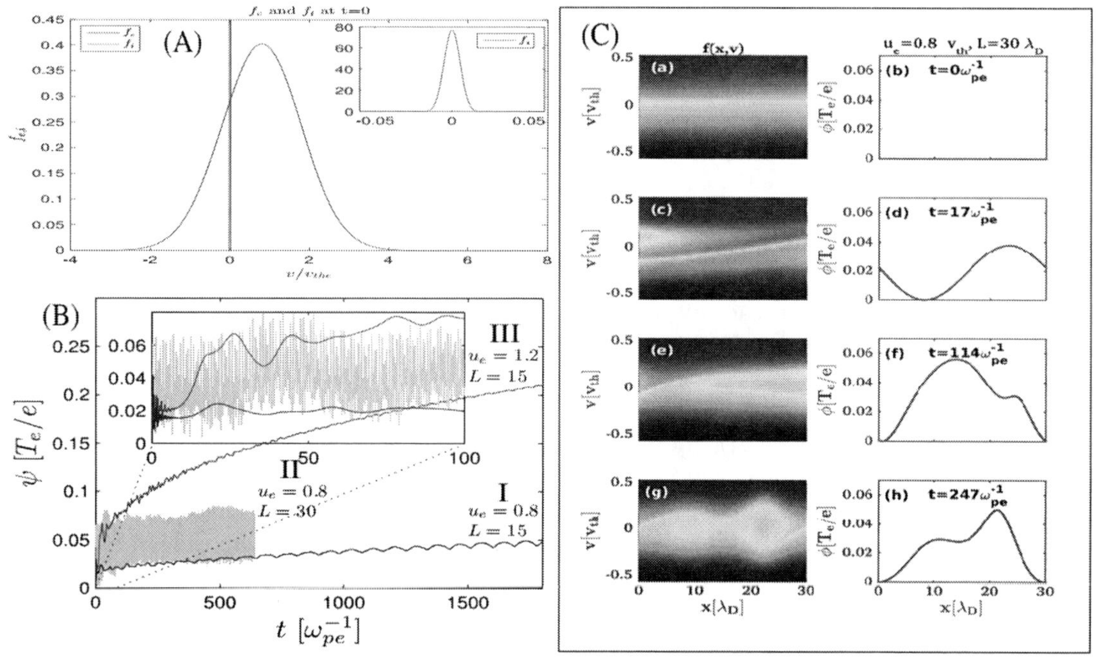

Figure 3. (A) Initial electron (blue solid line) and ion (red solid line) distribution function for Case-I ($u_j = 0.08v_{th}$). The subplot shows the full ion distribution function. (B) Temporal evolution of the wave amplitude ψ for the Case-I (dark blue solid line) ($u_e = 0.8v_{the}$), II (faint violet solid line) ($u_e = 0.8v_{the}$), III (dark red solid line) ($u_e = 1.2v_{the}$) respectively. The subplot shows evolution up to $t = 100\omega_{pe}$ in detail. (C) Temporal evolution of of the electron distribution function near to the IAW phase velocity location (left) and the potential $\phi(x,t)$ (right) for Case-II)

and finally it converts to a solitary wave. There is a second structure where the trapped particles are not thermalized, and therefore unable to produce a solitary wave like structure.

3.2. Analytical SEH generation

We now compare the electron phase-space holes created in our simulations with the SEH structure which represent the valid solutions of the Vlasov-Poison system in presence of trapped particles. The analytic expression of electron distribution function for SEH solutions in presence of electron current having trapping of electron in corresponding potential structures is given as [2]

$$f_{theory}(x,v) = \frac{1+K}{\sqrt{2\pi}} \begin{cases} \exp\left[-\frac{1}{2}\left(\sigma_e\sqrt{2\epsilon_e} - \tilde{v}_D\right)^2\right], & \epsilon_e > 0 \\ \exp\left(-\tilde{v}_D^2/2\right)\exp\left(-\beta\epsilon_e\right), & \epsilon_e \leq 0; \end{cases} \tag{6}$$

Where, $\epsilon_e = v^2/2 - \phi(x)$, x,v and ϕ are normalized quantity. $K = k_0^2\psi/2$. v_0 is the phase velocity of the wave in the electron frame. $\tilde{v}_D = v_D - v_0$, where, v_D is the drift velocity of the electron, determining the electron current in the system. Variations with $k_0^2 > 0$ and $k_0^2 = 0$ correspond to a sinusoidal wave and a solitary wave solution, respectively. β is the trapping parameter. For the solitary wave solutions phase velocity of the SEH structure can be determined via the

Nonlinear Dispersion Relation (NDR),

$$-\frac{1}{2}Z_r'\left(\tilde{v}_D/\sqrt{2}\right)-\frac{\theta}{2}Z_r'\left(u_0/\sqrt{2}\right)=\frac{16}{15}\left[\frac{3}{2}b(\beta,\tilde{v}_D)\right]\psi^{1/2} \tag{7}$$

where, $b(\beta,\tilde{v}_D)=\frac{1}{\sqrt{\pi}}\left(1-\beta-u_0^2\right)\exp(-\tilde{v}_D^2/2)$. The solution of solitary potential wave in term of β, ψ and v_0 is, $\phi(x)=\psi\,\mathrm{sech}^4\left[\sqrt{\frac{b(\beta,\tilde{v}_D)}{15}}\sqrt{\psi}x\right]$. The Fig. (4)(A) and (B) presents the analytical

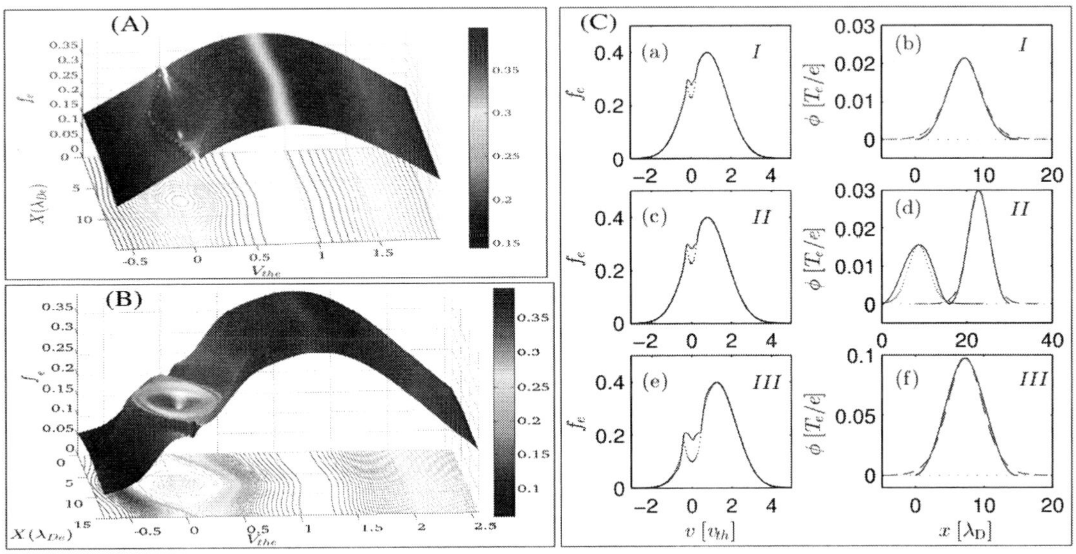

Figure 4. (A) Analytical electron distribution function given by Eq. (6). (B) Simulated electron distribution at a time when electron are trapped in the IAW for Case-III. (C)(Left) Comparison of the simulated electron distribution function $f_e(v)$ (solid line) to the analytic distribution function f_{fit} (dotted line) Eq. (6) for Case-I, II and III in the subplots C(a), C(c) and C(e) respectively. (Right) comparison of the simulated potential $\phi(x)$ (solid line) to the analytical solitary wave solution (dashed/dotted line) for Case-I, II and III in the subplots C(b), C(d) and C(f) respectively.

electron distribution function $f_{theory}(x,v)$ for SEH and simulated distribution function with trapping in Case-III respectively. One of the remarkable differences in the two phase-space plots is the existence of an ergodic separatrix region in the simulated f as compare to a well defined and sharp separatrix used in the analytic expression representing a rather ideal case. The ergodicity in simulations is attributed to a self-consistent evolution of the hole structure from a simplistic amplitude perturbation in the initial f. Fig. (4)(C) presents the further comparison of analytical and simulated solutions. The parameters required to construct the analytic solutions, v_0 and ψ are taken from the simulations and the parameter β is calculated using the NDR. Value of β have used for three different cases are β = -9.57, -8.90, -7.07 for Case-I, II and III respectively. As seen from the distribution function, despite a considerably ergodic separatrix region the simulated distribution function shows a good agreement with the analytical SEH solution. In the potential structures (right column in (C)) also the simulated results show a reasonable agreement with the analytical profiles at the hole locations [4]. In presence of electron current due to trapped particle instability a nonlinear growth rate is exist. But since the bounce frequency of the electron are higher than the growth rate, the trapped particle are thermalized and produce a new solitary structure with different ψ and β, which can be calculated by NDR Eq. (7). The potential structure follow the analytical expression for $\phi(x)$ at any time with good agreement. It is additionally observed that for the SEH where a relatively large ergodicity is present at the separatrix and trapped particle equilibrium is very week, the agreement remains poor, as in Case-II where the smaller one of the two solitons generated is recovered to have a relatively larger width.

4. Summary and Conclusion

We have used the general flux balance technique for solving Vlasov equation numerically including both electron and ion are mobilities. The procedure is parallelized using both MPI and OpenMp schemes to achieve the computational speed up. We have presented simulations of self-consistently generated ion acoustic solitary structures in a Vlasov plasma produced from a nonlinearity originating from the kinetic effects of trapped electron population. These coherent nonlinear solitary wave structure are shown to follow the analytical solutions, which consider only trapped particle nonlinearity. The presented simulations resolve a large range of time scales including fast electron plasma waves, an intermediate electron bounce frequency and a much smaller ion acoustic frequency with exact mass ratio between electrons and ions, highlighting nonlinear effects produced by trapping of resonant electrons in slower ion acoustic waves at very small amplitudes.

References

[1] I. B. Bernstein, J. M. Green and M. D. Kruskal 1957 *Phys. Rev.* **108**, 546.
[2] H. Schamel 2000 *Phys. Plasmas* **7**, 4831.
[3] E. Fijalkow 1999 *Comput. Phys. Commun.* **116**, 319.
[4] D. Mandal and D. Sharma 2014 *Phys. Plasma* **21**, 102107.

Evaluation of long term solar activity effects on GPS derived TEC

Azad A Mansoori[1], Parvaiz A Khan[2], Rafi Ahmad[3], Roshni Atulkar[3], Aslam A M[1], Shivangi Bhardwaj[3], Bhupendra Malvi[1], P K Purohit[3] and A K Gwal[1]

[1]*Space Science Laboratory, Department of Physics & Electronics, Barkatullah University, Bhopal – 462026, MP, India*
[2]*Department of Electronics and Communication Engineering, Islamic University of Science and Technology, Pulwama-192122, J & K, India*
[3]*National Institute of Technical Teachers' Training and Research, Bhopal – 462002, MP, India*

Email: azadahmad199@gmail.com

Abstract. The solar activity hence the solar radiance follows a long term periodic variability with eleven years periodicity, known as solar cycle. This drives the long term variability of the ionosphere. In the present problem we investigate the long term behaviour of the ionosphere with the eleven year cyclic solar activity. Under the present study we characterize the ionospheric variability by Total Electron Content (TEC) using measurements made by Global Positioning System (GPS) and solar cycle variability by various solar activity indices. We make use of five solar activity indices viz. sunspot number (Rz), solar radio Flux (F10.7 cm), EUV Flux (26-34 nm), flare index and CME occurrences. The long term variability of these solar activity indices were then compared and correlated with the variability of ionospheric TEC, at a mid latitude station, Usuda (36.13N, 138.36E), of Japan, during the solar cycle 23 and ascending phase of cycle 24. From our study, we found that long term changes in the ionospheric TEC vary synchronously with corresponding changes in the solar activity indices. The correlation analysis shows that all the solar activity indices exhibit a very strong correlation with TEC (R =0.76 –0.99). Moreover the correlation between the two is stronger in the descending phase of the solar cycle. The correlation is found to be remarkably strongest during the deep minimum of the solar cycle 24 i.e. between 2007- 2009. Also we noticed a hysteresis effect exists with solar radio flux (F10.7 cm) and solar EUV flux (26-34 nm). This effect is absent with other parameters.

1. Introduction

The solar activity is characterised with a variation pattern of eleven years called solar cycle. During which there occurs periodic variation in the emission of radiation from the Sun. It has been now established that the primary source for the formation of the Earth's ionosphere are the solar extreme ultraviolet radiations (EUV) and solar X-rays [1]. The solar EUV and X-ray radiation can vary by more than a factor of 2 from solar minimum to solar maximum and by as much as 50% during a solar rotation i.e. these solar radiations follow the solar cycle variation [2, 3, 4, 5]. Therefore the regular or solar cycle variations in solar EUV and X-ray radiations will affect the variability and dynamics of the Earth's ionosphere. Since the variation of solar X-ray and extreme ultraviolet (EUV) radiations are the main cause for changes in the ionosphere, in the absence of continuous long term records of these fluxes, solar proxies have been used to represent solar emissions. Sunspot number (Rz), solar 10.7 cm radio flux (F10.7) are the generally used as solar proxies [6]. Ionosphere is highly variable in space and time (diurnal, seasonal and solar cycle), with geographical location (low-latitudes, equatorial regions, high-latitudes and mid-latitudes), and with solar and geomagnetic disturbances [7, 8, 9]. Several studies have revealed the solar cycle effects of ionosphere [10, 11, 12] In the studies conducted in past, the solar cycle effects on the ionosphere have been realized by using different ionospheric parameters derived from ground based and satellite based measurements, such as Total Electron Content (TEC), critical frequency (foF2), peak electron density (NmF2) and peak height (hmF2) of the F2 layer [1, 3, 4, 6, 13, 14, 15, 16, 17, 18, 19, 20, 21], as well as various solar activity indices like sunspot number, solar 10.7 cm radio flux, EUV Flux etc.

These studies have revealed several interesting and important features of the solar cycle variation of ionospheric parameters. These studies revealed that ionosphere parameters such as critical frequency (foF2), the peak electron density (NmF2) and the Total Electron Content (TEC) are strongly controlled by solar activity in a rather complicated manner. It is found that the relationship of the ionospheric parameters : TEC, foF2, NmF2 and hmF2 with Sunspot Number, F 10.7 cm, EUV flux and X-ray flux may be roughly linear [16, 22], two segmented linear pattern [11], quadratic or higher order polynomial[17, 23].

Recent works show the ionospheric parameters approximately increase linearly with solar proxies during low and medium solar activity levels; however they tend to saturate during high solar activity level [13, 14, 16, 20, 21, 24, 25]. The true manifestation of this saturation effect is still not fully understood. In some studies a hysteresis effect has also been reported which is small at low and high latitudes but substantial at mid latitudes [3, 22].

Ionospheric research gets a significant attraction from the GPS community because ionosphere has practical importance in GPS applications as it influences the trans-ionospheric radio wave propagation. The ionospheric parameter which produces most of the effects on radio signals is Total Electron Content (TEC). By modelling TEC, the evaluation of the ionospheric error and corrections of these errors can be done for differential GPS. GPS signals experiences time delay when passing through the ionosphere. This delay of signals is directly proportional to integrated free electron density (TEC) along the ray path from the GPS satellite to a receiver. The magnitude of TEC is highly variable in space and time and it is a function of geomagnetic latitude, local time, season and 11 year sunspot cycle [26, 27, 28, 29].

The starting of the cycle 24 or the ending of cycle 23 was quite unique. During this period solar activity remained remarkably low. The sun went spotless for more than 70% of the days of the years 2007-2008. This represents the deep minimum of the cycle 24. It has been found that during this deep minimum the relationship of ionospheric parameters with solar indices like F10.7, deviated considerably from the past behaviour [30] as well as IRI over estimated the values for this period [31]. Therefore this period is interesting to study the behaviour of ionosphere under such a low solar activity.

2. Data Sets and Methodology

To investigate the long term variability of the ionospheric TEC with solar cycle variability we have considered the solar cycle 23 and initial phase of solar cycle 24 i.e. 1998 – 2011 as our period of study. For the purpose a mid latitude station of Japan namely Usuda (36.13N, 138.36E) is been selected.

To characterize the long term solar activity we have taken five solar activity indices viz Sunspot Number (Rz), solar radio Flux (F 10.7 cm), EUV Flux (26-34 nm), Flare Index and CME Occurrences. The data for Sunspot Number, Solar radio flux and Flare Index were taken from National Geophysical Data Center (NGDC) URL: http://www.ngdc.noaa.gov/. The CME occurrences were calculated from the CME catalog provided at SOHO URL: http://cdaw.gsfc.nasa.gov/CME_list/. The EUV flux (26-34nm) was taken from the URL: http://www.usc.edu/dept/space_science/sem_data/sem_data.html. Then we constructed the long term averages of these data sets and prepared the monthly and yearly averaged values of these data and then investigated their relationship with ionospheric TEC.

A network of GPS receivers is spread almost over the entire globe to accomplish the regular monitoring of the ionosphere. A huge database of the data collected at the International GPS Service

(IGS) stations is being maintained. The data is free to users. For our study, we have selected a mid latitude station of Japan, Usuda (36.13N, 138.36E). The TEC data were taken from the URL: http://sopac.ucsd.edu/dataArchive/. The raw data taken is in RINEX format, which is then processed to get the required Total Electron Content (TEC). The temporal resolution of the data is usually 30s. We then prepared the monthly and annual mean values of all the solar indices and the ionospheric parameter TEC to investigate the long term variability of TEC with selected solar indices that characterize long term cyclic variability of solar activity.

3. Results and Discussion

Primarily we compare the monthly changes of each five selected solar indices with the corresponding changes in the TEC. Figure 1 shows the changes in monthly averaged values of Sunspot Number, F 10.7, CME Occurrence, EUV flux (24 -36 nm) and Flare Index with the monthly averaged values of Total Electron Content (TEC) at Usuda. We find that the TEC follows a synchronous variation with all the solar indices. The variation of TEC with Rz, F 10.7 cm and Solar EUV flux has more close agreement than with Flare Index and CME occurrence. The result is quite obvious since the primary source of ionospheric ionization is solar radiation flux. Moreover we also notice the solar indices and TEC follow an extraordinary agreement during the deep solar minimum of the cycle 24.

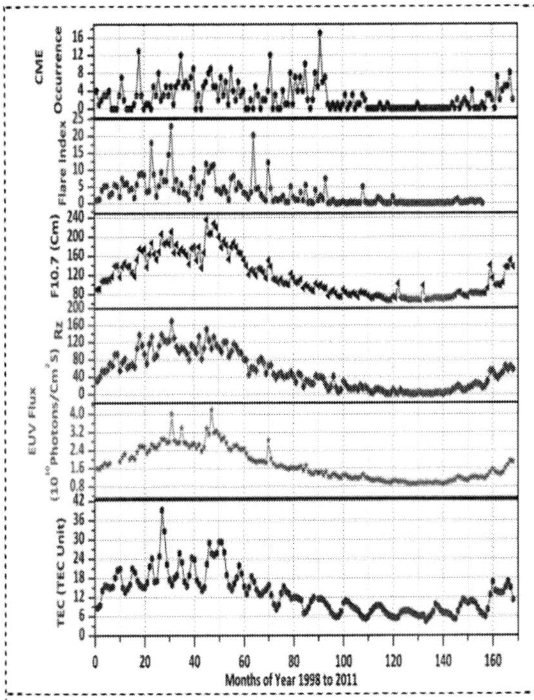

Figure 1. Monthly variation of TEC with sunspot number, F 10.7, CME occurrences, EUV flux and flare index during solar cycle 23 and 24.

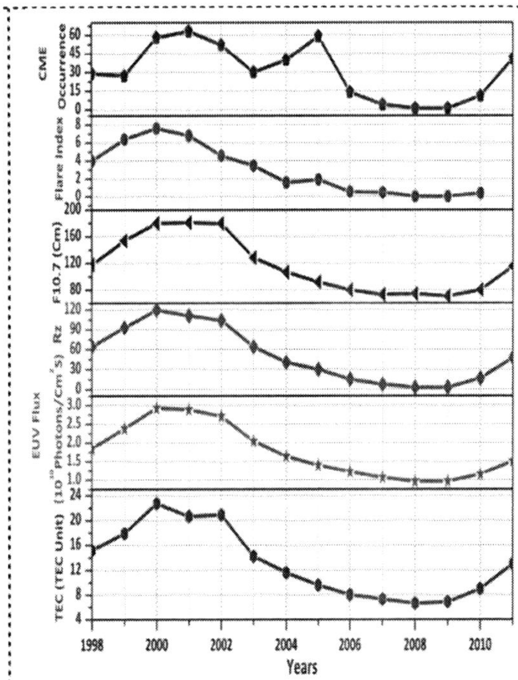

Figure 2. Annual variation of TEC with sunspot number, F 10.7, CME occurrences, EUV Flux and flare index during solar cycle 23 and 24.

We further study the relationship of TEC with the solar activity indices using the yearly averages to get a smoothened pattern of variation between the two. The annual variation of TEC with Sunspot Number, Solar Radio Flux, CME occurrence, EUV Flux, and Flare Index is shown in Figure 2. The Figure 2 reflects the similar pattern as is observed in the Figure 1; however, the variation is smooth. The annual changes in the TEC at Usuda follow an extraordinary synchronization with the annual changes in solar activity. It shows that the long term variability of the solar activity is reflected in the

Journal of Physics: Conference Series **759** (2016) 012069 doi:10.1088/1742-6596/759/1/012069

ionosphere very clearly. However, the variation of TEC is very much consistent with the annual changes in solar radiation flux than with other indices like CME occurrence etc.

To quantify the association between long term solar activity indices and the TEC, we constructed the scatter plots and derived the correlation between the annual averages of solar indices and TEC. Figure 3 shows the scatter plot and correlation analysis between the yearly averaged values of TEC with Sunspot Number, F 10.7, CME occurrence, EUV flux and Flare Index during the solar cycle 23 and 24. It is quite evident from the Figure 3 that all the solar indices exhibit a strong correlation with the TEC at Usuda. We also calculated the correlation coefficients between all the solar indices and the TEC. The correlation coefficients of Sunspot Number, F 10.7, CME Occurrence, EUV flux (24 -36 nm), and Flare Index with TEC are 0.99, 0.99, 0.76, 0.98 and 0.96 respectively. It shows that TEC almost follows one to one correlation with sunspot number, F 10.7, EUV flux (24 -36nm) and flare index. However, the correlation of TEC with CME occurrences is comparatively lesser.

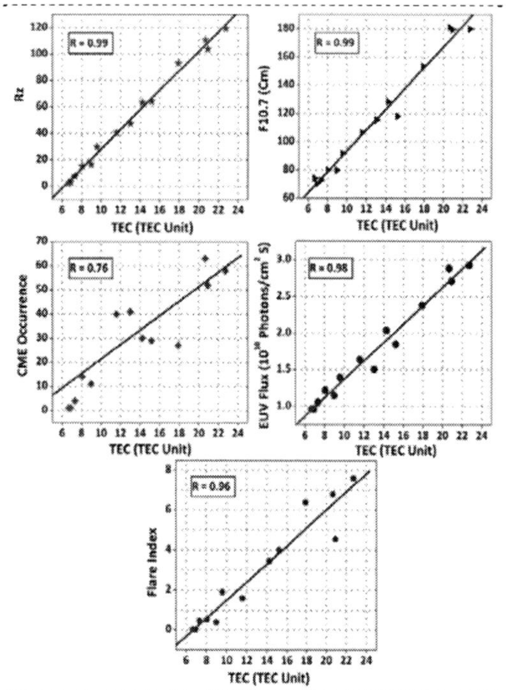

Figure 3. Scatter and correlation of TEC with sunspot number, F 10.7 cm, CME occurrences, EUV flux (24-36 nm) and flare index during solar cycle 23 and 24.

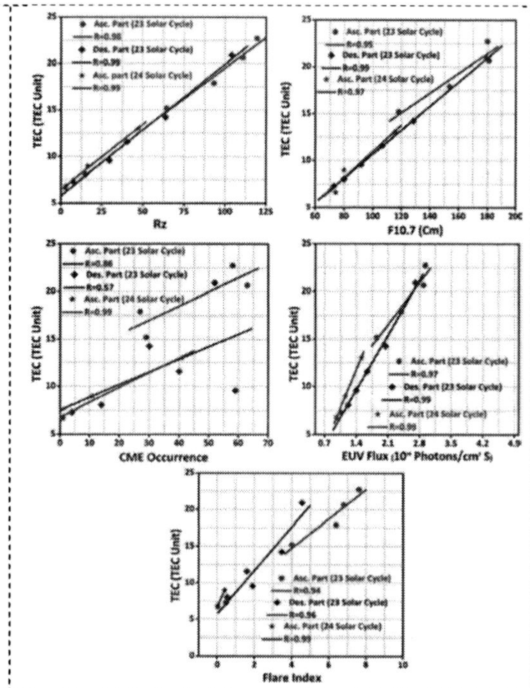

Figure 4. Correlation of TEC with sunspot number, F 10.7cm, CME occurrences, EUV flux and flare index during ascending and descending phase of solar cycle 23 and deep minimum of solar cycle 24.

Then we divide the solar cycle into two parts namely ascending and descending phases. We also take the deep minimum of cycle 24 as ascending part of solar cycle 24. We obtained the correlation of TEC with each solar index during the two branches of solar cycle as well as during the deep minimum of cycle 24 separately. The correlation of each solar index with TEC for each case is shown is Figure 4. From the Figure we found that TEC exhibit a very strong correlation with all the five solar indices during both the phase of the solar cycle. The correlation between the two is also remarkable during the deep minimum of the cycle 24. The correlation coefficients of TEC with sunspot number, F 10.7, EUV flux (24- 36 nm), flare index and CME occurrence are 0.98, 0.95, 0.97, 0.94 and 0.86 respectively during the ascending phase of the cycle 23. While during the descending phase of solar cycle 23 the correlation coefficients are 0.99, 0.99, 0.99, 0.96 and 0.57 respectively with sunspot number, F 10.7, EUV flux, flare index and CME occurrence. Here we notice that during the descending phase of the

cycle 23 the correlation with CME occurrence is weak. This is due to fact that during descending phase of the solar cycle the solar activity decreases due to which less number of CME are released from Sun and hence frequency of geomagnetic storms also decreases. Moreover the ionosphere behavior depends more on radiation flux than on geomagnetic storms. The correlation of TEC with sunspot number, F 10.7, EUV flux (24- 36 nm), flare index and CME occurence are 0.99, 0.97, 0.99, 0.99 and 0.99 respectively. It is interesting to note that TEC follow a remarkable correlation with solar indices during the deep minimum of cycle 24. Another interesting feature we observe in the analysis is the hysteresis effect (Figure 4) which exists only with solar radio flux F 10.7 cm and EUV flux (24 – 36 nm). The hysteresis effect is completely absent with other solar indices like sunspot number, flare index or CME occurrence.
The high degree of correlations obtained for these solar activity indices suggest that the production and ionization of ionosphere and it dynamics is completely controlled by the level of solar radiance that is received by the ionosphere and the ionosphere closely vary with it. Therefore, the ionospheric variability will also follow the cyclic variability of solar activity. The least correlation achieved with CME occurrence indicate that it does not involve the radiations but the magnetic field.

4. Conclusions
From our study on the solar cycle variations of the ionospheric TEC we conclude our results as:

- The ionospheric TEC has a synchronized variation with the solar cycle. The TEC exhibits an excellent association with the solar radio flux F10.7 cm, Sunspot number, and Solar EUV flux indicating that which is more responsive to them than to Flare Index and CME occurrence.
- The variations of the TEC during the deep minimum of the cycle 24 i.e during 2007-2009 follow exceptionally high correlation with Solar radio flux and Solar EUV flux.
- The correlation of TEC with solar indices is stronger during the descending phase of the solar cycle than in the ascending phase.
- The TEC shows a hysteresis effect with solar radio flux F 10.7 cm and solar EUV flux (24 -36 nm) which is absent with other solar indices.

5. References
[1] Adler N O, Elias A G and Manzano J R 1997 Solar cycle length variation: Its relation with ionospheric parameters *J. Atmos. Sol. Terr. Phys.* **59(2)** 159– 162.
[2] Klobuchar J A 1991 Ionospheric Effects on GPS *GPS World* **2 (4)** 48-51.
[3] Kane R P 1992 Sunspots, solar radio noise, solar EUV and ionospheric foF2 *J. Atmos, Terr. Phys.* **54** 463– 466.
[4] Kane R P 2003 Solar EUV and ionospheric parameters: A brief assessment *Adv. Space Res.* **32(9)** 1713–1718.
[5] Lean J L et al 2001 Variability of a composite chromospheric irradiance index during the 11-year activity cycle and over longer time periods *J. Geophys. Res.* **106** 10,645– 10,658.
[6] Bilitza D 2000 The importance of EUV indices for the International Reference Ionosphere *Phys. Chem. Earth, Part C* **25(5– 6)** 515– 521.
[7] Rishbeth H and Garriott O K 1969 Introduction to Ionospheric Physics, *Elsevier*, New York.
[8] Kawamura S et al 2002 Annual and semiannual variations of the midlatitude ionosphere under low solar activity *J. Geophys. Res.* **107(A8)** 1166, doi:10.1029/2001JA000267.
[9] Yu T et al 2004 Global scale annual and semiannual variations of daytime NmF2 in the high solar activity years *J. Atmos. Sol. Terr. Phys.* **66** 1691– 1701.
[10] Balan N, Bailey G J and Jayachandran B 1993 Ionospheric evidence for a nonlinear relationship between the solar e.u.v. and 10.7 cm fluxes during an intense solar cycle *Planet. Space Sci.* **41(2)** 141 – 145.
[11] Liu J Y, Chen Y I and Lin J S 2003 Statistical investigation of the saturation effect in the ionospheric foF2 versus sunspot, solar radio noise, and solar EUV radiation *J. Geophys. Res.* **108(A2)** 1067, doi:10.1029/ 2001JA007543.

[12] Afraimovich E L et al 2008 Global electron content: A new conception to track solar activity *Ann. Geophys.* **26** 335– 344.
[13] Balan N et al 1994a Variations of ionospheric ionization and related solar fluxes during an intense solar cycle *J. Geophys. Res.* **99** 2243–2253.
[14] Balan N, Bailey G J and Su Y Z 1996 Variations of the ionosphere and related solar fluxes during solar cycles 21 and 22 *Adv. Space Res.* **18(3)** 11 –14.
[15] Kouris S S, Bradley P A and Dominici P 1998 Solar-cycle variation of the daily foF2 and M (3000)F2 *Ann. Geophys.* **16** 1039–1042.
[16] Lei J et al 2005 Variations of electron density based on long-term incoherent scatter radar and ionosonde measurements over Millstone Hill *Radio Sci.* **40** RS2008, doi:10.1029/ 2004RS003106.
[17] Liu L et al 2007 The dependence of plasma density in the topside ionosphere on solar activity level *Ann. Geophys.* **25(6)** 1337–1343.
[18] Mikhailov A V and Mikhailov V V 1995 Solar cycle variations of annual mean noon foF2 *Adv. Space Res.* **15(2)** 79– 82.
[19] Rao M S J G and Rao R S 1969 The hysteresis variation in F2-layer parameters *J. Atmos. Terr. Phys.* **31** 1119– 1125.
[20] Richards P G 2001 Seasonal and solar cycle variations of the ionospheric peak electron density: Comparison of measurement and models *J. Geophys. Res.* **106** 12,803–12,819.
[21] Sethi N K, Goel M K and Mahajan K K 2002 Solar cycle variations of foF2 from IGY to 1990 *Ann. Geophys.* **20** 1677–1685.
[22] Chakraborty S K and Hajra R 2008 Solar control of ambient ionization of the ionosphere near the crest of the equatorial anomaly in the Indian zone *Ann. Geophys.* **26** 47–57.
[23] Chen Y, Liu L and Le H 2008 Solar activity variations of night time ionospheric peak electron density *J. Geophys. Res.* **113** A1130 doi:10.1029/2008JA013114.
[24] Balan N, Bailey G J and Moffett R J 1994b Modeling studies of ionospheric variations during an intense solar cycle *J. Geophys. Res.* **99** 17,467– 17,475.
[25] Gupta J K and Singh L 2001 Long term ionospheric electron content variations over Delhi *Ann. Geophys.* **18** 1635– 1644.
[26] Soicher H 1988 Traveling ionospheric disturbances (TIDs) at mid-latitudes: solar cycle phase dependence *Radio Sci.* **23** 283–291.
[27] Van Velthoven P J 1990 Medium-scale irregularities in the ionospheric electron content Ph.D. Thesis, Technische Universiteit Eindhoven.
[28] Jakowski N, Schlüter S and Sardo´n E 1999 Total electron content of the ionosphere during the geomagnetic storm on 10 January 1997 *J. Atmos. Sol. Terr. Phys.* **61** 299–307.
[29] Tsurutani B T et al 2004 Global dayside ionospheric upliftment and enhancement associated with interplanetary electric fields *J. Geophys. Res.* **109** A08302.
[30] Chen Y, Liu L and Wan W 2011 Does the F 10.7 index correctly describe solar EUV flux during the deep solar minimum of 2007-2009 *J. Geophys. Res.* **116** A 04304,doi:10.1029/2010JA016301.
[31] Luhr H and Xiong C 2010 IRI-2007 model overestimates electron density during the 23/24 solar minimum *Geophys. Res. Lett.* **37** L23101, doi:10.1029/2010GL045430.

Journal of Physics: Conference Series **759** (2016) 012070 doi:10.1088/1742-6596/759/1/012070

Lattice QCD simulations using the OpenACC platform

Pushan Majumdar

Department of Theoretical Physics, Indian Association for the Cultivation of Science, 2A & 2B Raja S.C. Mullick Road, Jadavpur, Kolkata 700032

E-mail: tppm@iacs.res.in

Abstract. In this article we will explore the OpenACC platform for programming Graphics Processing Units (GPUs). The OpenACC platform offers a directive based programming model for GPUs which avoids the detailed data flow control and memory management necessary in a CUDA programming environment. In the OpenACC model, programs can be written in high level languages with OpenMP like directives. We present some examples of QCD simulation codes using OpenACC and discuss their performance on the Fermi and Kepler GPUs.

1. Introduction

Lattice Quantum Chromodynamics (Lattice QCD) is a framework for computing properties of strong interactions through numerical simulations. In this formulation the continuum of space-time is replaced by a Euclidean lattice of the same dimension. In this formulation, matter fields such as quarks are placed on the lattice sites while the gauge fields are placed on the links which connect the sites. Space-time symmetries are sacrificed in this formulation, but gauge symmetry is maintained. The finite lattice spacing introduces an ultra-violet cutoff and the finite volume of the lattice provides an infra-red cutoff. Nevertheless to compare results obtained from the lattice with experiments one must extrapolate to zero lattice spacing and this is a very resource intensive process which goes as a^{-7} [1] where a is the lattice spacing.

A finite lattice means that the underlying path-integral has finite degrees of freedom and is amenable to evaluation using numerical simulations on computers. However the data size for such a problem is moderately large. Since we want to run our programs on GPUs which have very limited memory it is useful to make an estimate of the memory requirements. A rough estimate can be obtained as follows. For strong interactions we want to simulate a box with sides of 4 fermi (10^{-15}m) and a between 0.1 and 0.01 fermi. With a lattice spacing of 0.1 fm we need $40^4 = 2.56 \times 10^6$ sites and 10.24×10^6 links. In terms of memory requirement this means about 1 GB for matter fields and 1.5 GB for gauge fields. Making the lattice spacing finer to $a = 0.01$ fm would increase the memory requirements by about 4 orders of magnitude to 10 TB for matter fields and 15 TB for gauge fields. Current state of the art simulations use $a \sim 0.05$ fermi.

With an ever-increasing need for higher accuracy for numerical results, increase in data volume coupled with saturation of single core performance, efficient parallelization has become more and more important for lattice QCD computations. Fortunately lattice QCD codes can be parallelized relatively easily as for a large fraction of the simulation time they perform repeated matrix vector operations with vector sizes of about 10^7 elements. While parallelization between different nodes using MPI like platforms have been around for some time, the current hardware trend is to have many cores on a single chip or card. Our concern in this article is exploring parallelization on GPUs which have about 500 cores on a single card. Chip architectures which fall in this category are multicore CPUs $\sim$ 18 cores on a chip with about 2GB of memory per core, Xeon-PHIs with 60 cores on a chip with about 250 MB of memory per core and GPUs with 500 or 1000 cores on a chip with only about 12 MB per core.

The same OpenMP code can run on multi-core CPUs and the Xeon-PHI in its native mode although different organizations of the data is required for efficient operations as the memory available per core differs by about an order of magnitude. Running programs on GPUs require more effort. GPUs are programmed primarily with CUDA which is a C like language. It takes a reasonable effort to port existing codes to CUDA and run them efficiently. First attempts to run lattice simulations on GPUs was reported in [2]. Currently a repository of CUDA codes for lattice QCD simulations is maintained at `https://github.com/lattice/quda` [3]. Recently pure gauge theory codes have also been ported to GPUs [4]. Nevertheless a lot of existing lattice QCD codes have been written, tested and optimized in FORTRAN. CUDA FORTRAN exists but is not very useful. It only uses wrappers around the CUDA C functions for interfacing with FORTRAN routines. For individuals or small groups a much more viable option is to use a directive based programming similar to OpenMP. OpenACC provides one such option.

2. QCD simulation algorithm and the GPU architecture

Computations in QCD boil down to estimating integrals over probability distributions. The expectation value of an operator $\mathcal{O}$ is

$$\langle \mathcal{O} \rangle = \int \mathcal{D}\bar{\psi}\mathcal{D}\psi\mathcal{D}U \;\; \mathcal{O}[\bar{\psi}, \psi, U] \, e^{-S[\psi, U]} \tag{1}$$

where $e^{-S[\psi,U]}$ is the probability for the configuration ψ and U. Here $\bar{\psi}$ and ψ are grassman variables and S has the structure $\bar{\psi}\mathcal{M}\psi + \mathcal{F}[U]$ where $\mathcal{M}$ is a lattice Dirac operator and $\mathcal{F}[U]$ is the lattice gauge action. Since there are no simple representations for grassman variables on the computer, they are integrated over and the resulting determinant re-exponentiated using auxiliary bosonic fields to give

$$\langle \mathcal{O} \rangle = \int \mathcal{D}\phi^\dagger \mathcal{D}\phi\mathcal{D}U \;\; \tilde{\mathcal{O}}[\phi^\dagger, \phi, U] \, e^{-(\phi^\dagger \mathcal{M}^{-1}\phi + \mathcal{F}[U])}. \tag{2}$$

This is the form in which the expectation values are actually computed.

The algorithm of choice for performing these simulations is known as a Hybrid Monte Carlo (HMC) algorithm. The algorithm has three main components. The first step is the introduction of a Langevin noise by choosing random momentum variables from a gaussian distribution and a heat bath step in choosing random bosonic auxiliary fields. In the second step Molecular Dynamics (MD) equations are used to evolve the system through phase space for a given set of momenta and bosonic auxiliary fields. The third step is a Metropolis accept/reject step depending on the values of the Hamiltonian before and after the second step. This algorithm is exact in the sense that it has no systematic bias and satisfies the detailed balance condition.

The Langevin step requires between 10^6 and 10^{10} random numbers at each step and one simulation requires about 10^4 steps. Thus lattice QCD simulations require very high quality random number generators. The molecular dynamics step involves calculating the action of the inverse of a matrix on a vector and this is usually implemented through a conjugate gradient routine as $\mathcal{M}$ has positive definite eigenvalues. This is the most expensive step and takes up 85–95% of simulation time. Thus the conjugate gradient routine is the main target for parallelization. The accept/reject step is of course intrinsically serial.

The basic lattice QCD matrix vector operation requires 66 Flops and 120 bytes of data movement assuming single precision arithmetic. In our calculations, as in most others, double precision arithmetic is used. Thus lattice QCD computations are bandwidth limited. The GPU sits in a PCI slot on the motherboard. Thus the main bottle-neck for lattice QCD programs is the slow data movement between the CPU and the GPU. The maximum speed is about 5 GB/s or 8 GB/s depending on whether it is PCI 2.0 or 3.0. One possibility would be to avoid the CPU completely, but that is impossible as input-output and if-then clauses are evaluated on the CPU. Also with the Fermi GPU, BLAS functions and MPI calls are launched from the CPU. The main challenge in GPU programming for lattice QCD codes is how to get around these bottle-necks. To get any reasonable speed-up, at least the entire conjugate gradient must be on the GPU. Data must be copied from CPU to GPU at the start of the routine and the result copied back at the end of the routine.

The organization of the data on the GPU must be different from that of the CPU as the GPU memory is organized differently. As the CPU to main memory bandwidth is around 25 GB/s, the maximum performance without reusing data is about 13 GFLOPS. A similar calculation for the GPU would yield figures around 90 GFLOPS for a X2090 or 144 GFLOPS for a K40M. Therefore the main optimization would be to reuse data efficiently. Another big difference is that the caches on the GPU are much smaller than the CPU while there are a lot of registers. In fact the total memory offered by the registers is more than the cache. The bandwidth per core between the GPU main memory and the processor is also much smaller than that of the CPU. So it is much more beneficial to put variables directly into the register on the GPU. The CUDA compiler puts arrays into the main memory while scalar variables are put in registers. But our analysis shows that for lattice QCD calculations it would be more useful to put as many array elements into the registers as possible. One possible way to do that is to use code generators. Code generators can automatically convert arrays into lots of scalar variables. This can increase the performance of a lattice QCD CUDA code by a factor of several [5].

3. The OpenACC standard and code examples

OpenACC is a programming standard for parallel computing developed by Cray, CAPS, NVIDIA and PGI. The standard is designed to simplify parallel programming of heterogeneous CPU/GPU systems [6].

The OpenACC Application Program Interface describes a collection of compiler directives to specify loops and regions of code in standard C, C++ and Fortran to be offloaded from a host CPU to an attached accelerator, providing portability across operating systems, host CPUs and accelerators. The directives and programming model defined in this document allow programmers to create high-level host+accelerator programs without the need to explicitly initialize the accelerator, manage data or program transfers between the host and accelerator, or initiate accelerator startup and shutdown [7].

The OpenACC codes look very similar to OpenMP codes in their directive structure. One major difference though is the creation of a data region which tells the compiler when to move certain data items to the GPU or bring them back to the CPU. We illustrate the various constructs below using the conjugate gradient routine as an example. To highlight the similarity to OpenMP we keep the commented out OpenMP directives along with the OpenACC directives.

The part of the code to be off-loaded to the accelerator begins with the `!$ACC data` command and ends with the `!$ACC end data` command. The `!$ACC data` accepts lists of variables with attributes such as `copy, copyin, copyout` and `create`. The attribute `copy` implies that the values of the variables must be copied from the CPU memory to the GPU memory at the start of the data region and copied back to the CPU memory once the data region ends. The attribute `copyin` implies that the variable must be copied into the GPU memory but the result need not be copied back. Similarly `copyout` means that the variables need not be copied into the

GPU, but the results must be copied back. Finally the attribute `create` means those variables are local to the GPU they do not require transfer between the CPU and the GPU memory.

```
subroutine congrad(nitcg)
... All kinds of definitions and declarations ...

!$ACC  data copy(nitcg,alpha,betad,betan) copyin(nx,iup,idn,u,r)
!$ACC+ copyout(x,y) create(ud,ap,atap,p)
*
      call linkc_acc
!!$OMP parallel do default(shared)
!$ACC  parallel loop collapse(2) reduction(+:betan) present(p,r,x)
      do l = 1, mvd2
      do ic=1,nc
         p(l,ic) = r(l,ic) ; x(l,ic) = (0.,0.)
         betan=betan+conjg(r(l,ic))*r(l,ic)
      end do
      end do
!      betan=real(zdotc(mv3d2,r,1,r,1))
!$ACC  update host(betan)
      if (betan.lt.delit) go to 30
!$ACC  parallel present(beta,betan,betad,alphan)
       beta=betan/betad ; betad=betan ; alphan=betan
!$ACC  end parallel
       do nx = 1, nitrc !              Main loop of conjugate gradient begins
       nitcg ← nitcg+1    ;    ap = 0

       call fmv(0,mvd2,ap,p)  !                  Matrix-vector multiplication

       alphad=⟨ap,ap⟩ + ⟨p,p⟩  ;  alpha=alphan/alphad
       atap ← p   ;   x ← x + alpha * p

       call fmtv(atap,ap)  !                     Matrix-vector multiplication

       r ← r - alpha * atap
       betan=⟨r,r⟩
!$ACC  update host(betan)       !               Exit condition evaluated on CPU
       if (betan .lt. delit) go to 30
       beta=betan/betad ; betad=betan ; alphan=betan
       p ← r + beta * p
       end do       !                  Main loop of conjugate gradient ends
30     continue
*
       y = 0      !                     Solution on the second half lattice
       call fmv(mvd2,mv,y,x)  !                  Matrix-vector multiplication
*
!$ACC  end data
       return
```

As one can see from the above example, the code requires no change from the original CPU code apart from the insertion of various directives. The nature of the directives is quite close to that of OpenMP with a few extra attributes. One such attribute is `present` followed by a list of variables. This tells the compiler that the variables are already present in the GPU memory and the compiler need not copy it from the CPU. Another directive special to the accelerators is `update host()`. This tells the compiler to copy the value of the variable at that point in execution to the CPU from inside a data region. This is often used for evaluating if-then conditions on the CPU when the variable on which the condition depends is evaluated on the GPU. The directives `parallel loop` and `parallel` work in the same way as in OpenMP.

In the Fermi GPU, the MPI calls and BLAS functions are launched from the CPU. So we found it was faster to implement the vector operations using parallel loops rather than make BLAS calls. MPI calls unfortunately still need data transfer between the CPU and the GPU and that requires one more construct `update device()` which works like the `update host()` but in the reverse direction. We illustrate a basic MPI call below.

```
                  ⋮
              ap_loc ← 0
!$ACC  parallel loop present(u,ud,ap_loc,p,iup,idn)
      do  l = base+1, base+nvd2
        v1 = ap_loc(1,l-base)
                  ⋮
Lines identical to single node version
                  ⋮
        ap_loc(3,l-base) = v3
      enddo
!$ACC update host(ap_loc)
      call MPI_ALLGATHER(ap_loc,3*nvd2,MPI_DOUBLE_COMPLEX,
     +           ap,3*nvd2,MPI_DOUBLE_COMPLEX,MPI_COMM_WORLD,ierr)
!$ACC  update device (ap)
```

In this example the local array `ap_loc` is combined to form the global array `ap` which is made available to all processes. For multi-GPU applications one must worry about synchronizing the data between the CPU and the GPU after each communication call. Usually in OpenACC this can be handled by compiler options.

4. Results

In this section we describe the performance we obtained for our programs. These are either a single node two flavour ($N_f = 2$) RHMC code for staggered or a single node two flavour Wilson fermion lattice QCD simulation program with OpenMP and OpenACC directives, on different kinds of hardware - compiler combinations. All calculations were performed in double precision.

In figure 1 we present the performance in GFLOPS [9] of a single node $N_f = 2$ RHMC code for staggered fermion lattice QCD simulation on three different hardware - software combinations. While the Ivybridge node with the Cray compiler is faster due to its higher clock speeds upto two threads, beyond that the Haswell node with the Intel compiler is faster. The Cray node with the K20X GPU turns out 3 to 4 times faster in this case compared to the node only with CPUs. For the GPU code almost all the parallel parts of the program have been pushed onto the GPU with very little data transfer required between the CPU and GPU. This results in the flat behaviour for the GPU code almost independent of the number of CPU threads.

The conjugate gradient routine achieves a higher performance and we present its performance in table 1. The performance of the conjugate gradient routine had to be estimated from a slightly slower version of the program as in the final version, most of the molecular dynamics trajectory

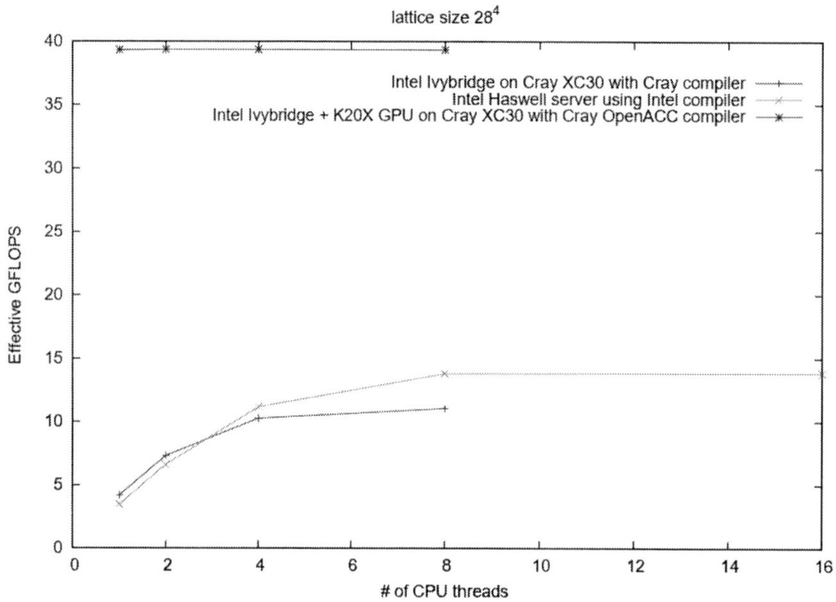

Figure 1. Performance of a single node $N_f = 2$ RHMC code for staggered fermion lattice QCD simulation on three different hardware - software combinations. The + corresponds to one Intel 10 core Ivybridge CPU at 2.8 GHz with 8 cores for computing on a Cray XC30 using the Cray compiler, the $\times$ corresponds to a server with two 10 core Haswell CPUs at 2.3 GHz and the $*$ corresponds to one node of a Cray XC30 with one CPU as in + and a K20X GPU using the Cray OpenACC compiler.

is inside a single ACC data region and timing estimates of a single routine without slowing down the program is difficult to get. Our best estimate is that on the K20X the conjugate gradient achieves about 55 GFLOPS.

# of threads	1	2	4	8	16
Ivybridge (2.8 GHz)	3.870	6.753	9.204	9.609	–
Haswell (2.3 GHz)	3.292	6.236	10.388	12.513	12.244

Table 1. Performance (in GFLOPS [9]) of the conjugate gradient routine in a two flavour RHMC code for simulation of staggered fermions for the Ivybridge and Haswell CPUs.

Gustafson's law states that better parallelization can be achieved for larger data sizes. Figure 2 is an illustration of this. With the lattice size changing from 16^4 to 24^4, better parallelization is achieved. For lattice size 28^4, a position independent code has to be built and the program has to be linked dynamically. This is slower than the default statically linked executable on the Cray and very likely this results in the drop in performance.

Global memory access in GPU is slow. Therefore an optimal ordering of array indices is

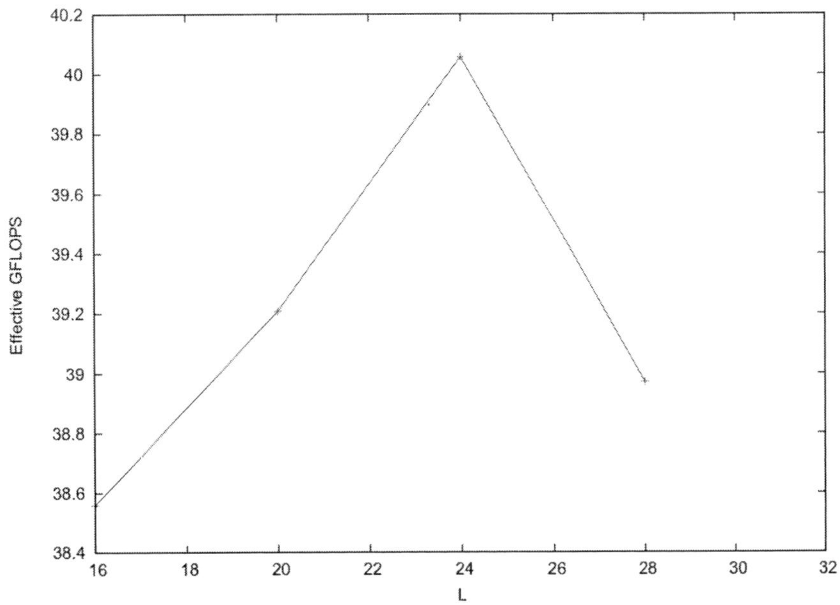

Figure 2. Performance of the Cray OpenACC compiler with varying data size. The x-axis denotes the length of one side so that the volume of the lattice is L^4

# of threads	1	2	4	8	16
Order A	673.6	623.1	522.7	474.5	447.7
Order B	651.7	505.9	325.2	243.2	194.7

Table 2. Timing in seconds of the conjugate gradient routine in a Wilson fermion simulation program for two different storage schemes of the vector on Fermi X2090 GPU. Order A corresponds to array(3,4,nsite) while Order B corresponds to array(nsite,4,3) with nsite $= 20^4$.

essential for arrays being declared on the GPU. In table 2 we present the timings of the conjugate gradient routine for the Wilson fermion simulation program for two different storage schemes of the vectors. There is a 3.5 times difference in speed depending how the array is defined. Order A is faster on the CPU while order B is faster on the GPU (on FORTRAN compilers). Thus while the conjugate gradient speeded up another routine got slowed down by a factor 2. Nevertheless since the conjugate gradient takes up most of the computation time, there was an overall speed-up by more than a factor 2 for storage order B.

5. Discussion

- In terms of both cost and power consumption, GPUs provide affordable supercomputing resources.
- OpenACC seems to be a comparatively easy path to port ones codes from the CPU to the GPU as it avoids a lot of the accelerator memory management issues and other associated

issues such as initializing the accelerator and shutting it down or generating the individual accelerator kernels.

- The OpenACC coding effort is only marginally higher than OpenMP. Almost each OpenMP directive can be replaced with a OpenACC directive. Only additional directives are creation of a data region to demarcate the portion of the code to be off-loaded to the accelerator and updating the host or the device from within a data region.
- Real speedup is possible only when a comparatively large chunk of the computation (compared to the amount of data transferred between the CPU and the GPU) such as the whole conjugate gradient is performed on the GPU.
- Performance of single GPU is roughly equivalent to 128 cores of cluster with QDR infiniband interconnect [8].
- For the programs analyzed, the advantage of the GPU increases with increasing data size.
- There can be a huge penalty in performance (several times) for data access on the GPU if the data is not optimally organized in the GPU memory.
- Comparing between an optimized pure CPU MPI code and an OpenACC code on a single Cray node, the MPI code is faster for lattice sizes of about 20^4. Beyond that the OpenACC code is faster.
- Further performance gains can be obtained by using mixed-precision routines and improved storage schemes.

Acknowledgments
We would like to acknowledge ILGTI-TIFR for funding the GPU portion of the Cray on which these studies were carried out. IACS for funding the rest of the Cray without which the GPU portion wouldn't run as well as the Cray India team for help at various stages during the development of the OpenACC codes.

References
[1] Ukawa A for CP-PACS and JLQCD Collaborations 2002 *Nucl. Phys. Proc. Suppl.* **106** 195
[2] Egri G I, Fodor Z, Hoelbling C, Katz S D, Nogradi D and Szabo K K 2007 *Comput. Phys. Commun.* **177** 631
[3] Clark M A, Babich R, Barros K, Brower R and Rebbi C 2010 *Comput. Phys. Commun.* **181** 1517
[4] Cardoso N and Bicudo P 2013 *Comput. Phys. Commun.* **184** 509
[5] Chakrabarty A and Majumdar P 2012 Hybrid Monte Carlo with Wilson Dirac operator on the Fermi GPU *Preprint* arXiv:1207.2223 [hep-lat]
[6] http://en.wikipedia.org/wiki/OpenACC
[7] http://www.openacc-standard.org/
[8] Majumdar P 2014 Lattice Simulations using OpenACC compilers *PoS* **LATTICE2013** 031
[9] All FLOP measurements were done using the craypat utility on a Cray XC30. The precise command used was `pat_build -w -T function-name executable`. This measurement was done for a single thread version of the program and the effective GFLOPS were calculated by dividing the number of FLOP thus obtained by the run-time of the OpenMP or OpenACC program.

Detailed 3D Simulation of the GEM-based detector

P Bhattacharya[1], S Biswas[2], B Mohanty[1], N Majumdar[3] and S Mukhopadhyay[3]

[1] School of Physical Science, National Institute of Science Education and Research, Jatni, Khurda, Odisha, India 752050

[2] Physics Department, Bose Institute, Block - EN, Saltlake, Kolkata, India 700091

[3] Applied Nuclear Physics Department, Saha Institute of Nuclear Phyics, 1/AF, Bidhannagar, Kolkata, India 700064

E-mail: purba.bhattacharya85@gmail.com

Abstract. The operation of Micro Pattern Gaseous Detectors (MPGDs) has often suffered from effects such as distortion of the electric field due to space charge, despite their widespread use in particle-physics and nuclear-physics experiments, astro-particle research, medical imaging, material science etc. To keep distortions due to space-charge at a manageable level, a lower ion feedback is required while maintaining substantial detector gain and good resolution. Thus, a proper optimization of the detector geometry, field configuration and gas mixtures are required to have a higher electron transparency and lower ion backflow. In our work, Garfield simulation framework has been adopted as a tool to evaluate the fundamental features of Gas Electron Multiplier (GEM). Our study begins with the computation of electrostatic field and its variation with different geometrical and electrical parameters using the neBEM toolkit. Different efficient algorithms have been implemented to increase the computational efficiency of the field solver. Finally, ion backflow and electron transparency of single and quadruple GEMs with different geometry and field configurations suitable for the ALICE-TPC, have been studied.

1. Introduction

ALICE at the LHC is planning a major upgrade of its detector systems, including the main tracking device, Time Projection Chamber, to cope with an increase of the LHC luminosity after 2018 [1]. This implies replacement of the present Multi Wire Proportional Counter (MWPC) based readout system of the TPC by continuously operated amplification scheme. The Micro-Pattern Gaseous Detectors (MPGDs) [2] are among the possible options as tracking and triggering detectors in highly luminous environments. The ALICE-TPC collaboration are currently working on the adoption of the Gas Electron Multiplier (GEM) [3] as the readout system of the TPC [4]. The new readout chambers will employ stacks of four GEM foils for gas amplification and anode pad readout. However, several questions related to the operation of the GEM have to be answered before it can be finally chosen as the possible solution. The major challenge is to have a low ion feedback in the drift volume to keep distortions due to space-charge at a manageable level [5]. At the same time, to retain the specific energy loss for a good particle identification, the fraction of primary electrons that partcipate in the avalanche process, should be large. With a proper optimization of the detector geometry, field configuration and gas mixtures, a high electron transparency, as well as, low ion backflow can be obtained.

Journal of Physics: Conference Series **759** (2016) 012071 doi:10.1088/1742-6596/759/1/012071

Keeping the requirement in mind, in the present work, the electron transparency and ion backflow fractions of GEM-based detectors have been numerically estimated. We initiated with the computation of electrostatic field and its variation with different geometrical and electrical parameters. Parallel processing and other efficient algorithms have been implemented to increase the computational efficiency of the field solver. Detailed studies have been carried out in order to optimize various aspects of the simulation framework so that complex physics problems can be approached keeping the demand of computational resources at an acceptable level. The effects of detector geometry, electric and magnetic field on electron transparency and ion backflow fraction have been carried out extensively. To begin with, single GEM configurations have been studied in detail. A good understanding of the complex physics process in this device has allowed us to deal with relative ease the quadruple GEM configurations, which is being considered as a possible option for the ALICE TPC. In this manner, it has been possible for us to achieve a understanding of the likely optimum configurations of the quadruple GEMs.

2. Simulation Tools

Garfield [6] simulation framework which provides interface to other software packages, such as neBEM (nearly exact Boundary Element Method) [7], HEED [8] and Magboltz [9], has been used for the present work. neBEM is the toolkit to compute 3D electrostatic field whereas, HEED has been used for primary ionization calculation and Magboltz for computing electron and ion transport properties in gas.

3. Numerical Models

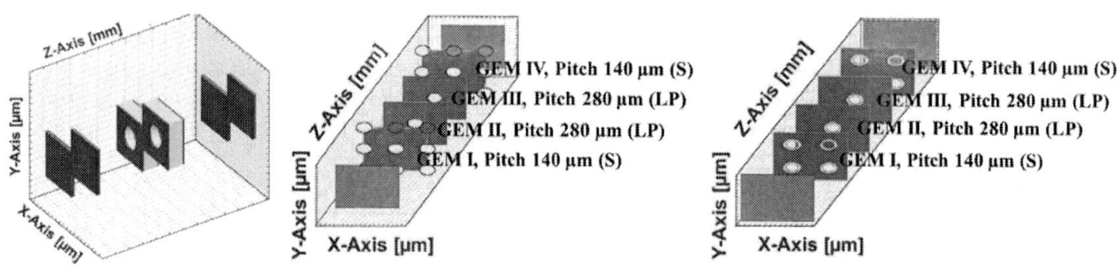

(a) (b) (c)

Figure 1: Model for (a) single and quadruple GEM with (b) aligned and (c) misalgned holes

The geometrical parameters of GEM-based detectors, used in the numerical work, are listed in table 1. The model of a basic GEM cell built using Garfield, is shown in figure 1(a). It represents a GEM foil, having two bi-conical shaped holes placed in a staggered manner along with a readout anode and a drift plane on either sides of the foil. The distance between top surface of the GEM and the drift plane is called the drift gap whereas that between the lower surface and the readout plate is named induction gap. In comparison to single GEM, in case of multi GEM detector, several GEM foils are placed in between the drift and the read-out plane. The naming scheme used in this work, numbers the foils in the increasing order towards the anode plate. The top most GEM is called GEM I and the others are GEM II, GEM III and so on. The gap in between GEM I and II is called transfer gap I and that between GEM II and III is called transfer gap II etc. For example, the simulation models of two different quadruple GEM devices are shown in figure 1. Among the four foils, GEM I and GEM IV, have the pitch of 140 μm (denoted as S), whereas the middle two foils have a larger pitch of 280 μm (denoted as LP). This arrangement, denoted S-LP-LP-S, employs asymmetric transfer fields and foils with low optical transparency and, thus, allows to block ions efficiently. Two different geometrical variation in case of four GEM foils have been considered. In the first case (QGemI), the central hole of the basic unit from all the four GEM foils are perfectly aligned (figure 1(b)). In the other case (QGemII), as shown in figure 1(c), the first and the last foils (S) are aligned with each other whereas the second and third foils (LP) are misaligned with them. The basic cell structure then has been repeated along both positive and negative X and Y-Axes to build the actual dimension of the real detector. With the help of these models, the field configuration of the detectors have been simulated using appropriate voltage settings. These are followed by the simulation of electron transmission and ion backflow fraction in Ne : CO_2 : N_2 (90:10:5) gas mixture.

Table 1: GEM design parameters.

Polymer substrate	50 μm
Copper thickness	5 μm
Outer hole diameter	70 μm
Inner hole diameter	50 μm
Hole to hole pitch	140 / 280 μm
Drift Gap	3 mm
Transfer gap	2 mm
Induction gap	2 mm

Table 2: Voltage of Quadruple GEM.

Drift Field	0.4 kV/cm
ΔV_{GEMI}	275 V
Transfer Field I	4 kV/cm
ΔV_{GEMII}	235 V
Transfer Field II	2 kV/cm
ΔV_{GEMIII}	284 V
Transfer Field III	0.1 kV/cm
ΔV_{GEMIV}	345 V
Induction Field	4 kV/cm

For the calculation of electron transmission, 10,000 electrons have been injected in the drift gap in random positions. These electrons are made to drift towards the GEM foil using the Microscopic tracking routine. In this procedure, a electron drift path goes through millions of collisions such as elastic or inelastic collision, excitation, ionization, attachment etc. The electron transmission has been estimated as the ratio between the number of electrons that reach the anode plate to that created in the drift volume.

The electrons during their drift produce avalanche inside the GEM foil. The ions created during the avalanche process, as well as the primary ions, have been considered for the calculation of ion backflow. The fraction of total ions that drift back to drift volume, gives the desired estimation.

4. Results

4.1. Field Solver

A commonly accepted approach to solve electrostatic problems for arbitrary three-dimensional geometries is the Boundary Element Method (BEM). A novel formulation of the BEM, the nearly exact BEM (neBEM) solver was developed to resolve some of the major drawbacks of usual BEM. The neBEM uses exact integration of the Greens function and its derivative in its formulation. In this approach, these integrations for rectangular and triangular elements having uniform charge density have been obtained as a closed-form analytical expressions using symbolic mathematics. Thus, these foundation expressions account for truly distributed nature of charge density on a given element. The major advantage achieved through the use of the proposed closed-form expressions is that the accuracy is enhanced throughout the physical domain including near-field region without using any special formulation in any part of the domain. During the course of this work, the neBEM toolkit has been improved significantly. The major challenge in these developments has been to increase the efficiency of the solver, while maintaining its precision.

Code parallelization: Recently, we have successfully implemented Open Multi-Processing (OpenMP) for the neBEM field solver. It is an Application Programming Interface (API) for multi-platform shared memory multiprocessor programming. We have tested these implementations on 2, 4 and 8 cores. The observed reduction in the computational time has been found to be significant while the precision of the solution has been found to be preserved. In the following table 3, we present a comparison of the time taken to solve charge densities for typical problems of single GEM detector involving 1126 number of elements and 116 repetitions of the basic structure.

Adaptive Modelling: In adaptive meshing the solution is usually attempted at a given spatial discretization and the solver is expected to increase or decrease the meshing to meet the desired accuracy specifications. For neBEM, we have presently implemented an algorithm which allows us to ignore the finer variations of charge densities on a primitive provided (i) it is not on the base device and (ii) it is at a far enough location so that the influence of the average charge density on the primitive is equivalent to the influence that is estimated preserving the real charge density variation on the primitive. The effect of using different values of primAfter on the same problem of GEM detector has been listed in table 4. In the present case, it has been noted that setting primAfter = 5 has negligible effect on the evaluated potential and field for this device.

Fast Volume: As is expected, the time to estimate potential and field for a complex device is significant. Reduction of time taken to estimate the electrostatic properties becomes increasingly important when complex processes such as Avalanche, Monte-Carlo tracking and Micro-Tracking are being modelled. In order to model these phenomena within a reasonable span of time, we have implemented the concept of using pre-computed values of potential and field at large number of nodal points in a set of suitable volumes. These volumes are chosen such that they can be repeated to represent any region of a given device and simple trilinear interpolation is used to find the properties at non-nodal points. The associated volume is named as the Fast Volume and in the table 5, in case of GEM detector, the potential and fields estimated by direct evaluation and those using FastVol have been compared.

Table 3: Effect of OMP.

Core 1	20.67 min
Core 2	11.32 min
Core 4	6.02 min
Core 8	4.77 min

Table 4: Effect of AM.

PrimAfter 0	1.47 min
PrimAfter 10	0.97 min
PrimAfter 5	0.60 min
PrimAfter 2	0.58 min

Table 5: Effect of FastVol.

Problem	Without	With
Charge density	1 min	32 min
Field Map	41 hr	4 min
Avalanche	2 days	19 sec

4.2. Electron Transmission

Electron transmission can be presented as a function of two mechanisms: electron focusing and transverse diffusion. Electron focusing depends not only on the field ratio, but also on different geometrical parameters. On the other hand, the transverse diffusion is mainly affected by the electric field and the gas composition. Besides that, the electron attachment coefficient, can also influence transmission. For a single GEM detector, the total electron transmission (ϵ_{tot}) can be identified as the multiplication of two efficiencies, the collection efficiency (ϵ_{coll}) and the extraction efficiency (ϵ_{ext}). These two efficiencies are defined as follows:

$$\epsilon_{coll} = \frac{\text{Electrons reached inside the GEM foil}}{\text{Electrons created in drift volume}}; \ \epsilon_{ext} = \frac{\text{Electrons reached the readout plane}}{\text{Electrons present inside the GEM foil}} \quad (1)$$

The variations of ϵ_{coll}, ϵ_{ext} and ϵ_{tot} under different field configurations have been plotted in figure 2. For a fixed V_{GEM} and $E_{Induction}$, ϵ_{coll} decrease with the increase of the drift field,

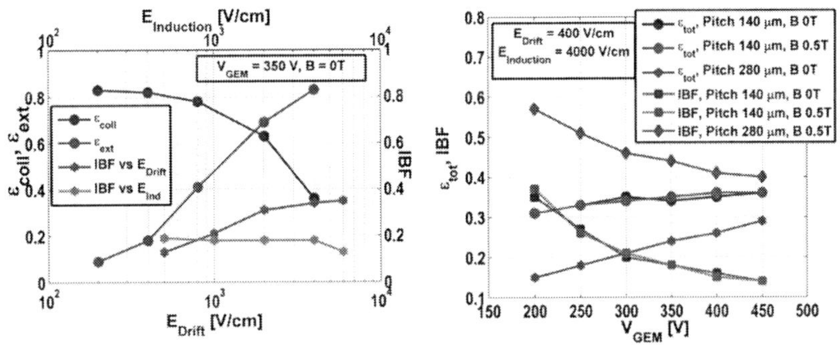

(a) (b)

Figure 2: Variation of (a) ϵ_{coll}, ϵ_{ext} and IBF with E_{Drift} and (b) $E_{Induction}$; (b) ϵ_{tot} and IBF with V_{GEM}. Effect of pitch and magnetic field are shown in (b).

Table 6: ϵ_{coll} and ϵ_{ext} of quadruple GEMs.

Geometry	B (T)	ϵ_{collI}	ϵ_{extI}	ϵ_{collII}	ϵ_{extII}	$\epsilon_{collIII}$	ϵ_{extIII}	ϵ_{collIV}	ϵ_{extIV}
QGemI	0	0.989	0.349	0.067	0.310	0.151	0.135	0.929	0.372
QGemI	0.5	0.993	0.354	0.071	0.305	0.142	0.130	0.930	0.374
QGemII	0.5	0.893	0.391	0.069	0.313	0.175	0.128	0.928	0.360

whereas at a fixed V_{GEM} and E_{Drift}, the increase of induction field, increases ϵ_{ext} as shown in figure 2(a). The change of V_{GEM} only has effect on ϵ_{coll} and thus ϵ_{tot} (figure 2(b)). It is also seen from figure 2(b), for the same voltage configuration, the smaller pitch GEM foils is better in terms of higher electron transmission, whereas 0.5 T magnetic field plays no significant effect on it.

The voltage configuration, used for quadruple GEM detectors, are listed in table 2. For the multi-GEM detectors, the electron transmission can be also identified as the multiplication of collection and extraction efficiencies of the individual GEM foils. ϵ_{coll} and ϵ_{ext} of individual GEM foils for a quadruple GEM detectors (QGemI) have been listed in table 6. No significant effect of the magnetic field on ϵ_{tot} has been observed till now.

4.3. Ion Backflow Fraction

The backflow fraction mainly depends on the field ratio and the transverse spread of the electron avalanche. Thus, a proper optimization of the field in the drift volume, GEM hole and induction regions is necessary to prevent those ions from entering the drift volume. The ion backflow of a single GEM can be reduced by decreasing E_{Drift} (figure 2(a) because less number of field lines will get out of the hole into the drift volume. At higher E_{Drift}, the ratio between E_{Drift} and E_{GEM} is large resulting in the drift of more number of ions into the drift volumm. No significant effect of $E_{Induction}$ has been observed except at the higher $E_{Induction}$ (figure 2(a)). At higher E_{GEM}, the ratio between E_{Drift} and E_{GEM} is small and thus a large fraction of ions is collected at the top surface of the GEM foil. From figure 2(b), it is also seen that the GEM foil with standard pitch of 140 μm gives less backflow fraction, whereas no significant effect of 0.5T magnetic field on backflow has been observed.

A better suppression of the backflow can be achieved by using multiple GEM structures. The ion collection efficiencies of four foils for the two different geometry have been listed in table 7.

Table 7: ion collection of quadruple GEM detectors.

Geometry	B (T)	GEMI	GEMII	GEMIII	GEMIV	Drift
QGemI	0	0.025	0.004	0.013	0.932	0.027
QGemII	0.5	0.023	0.004	0.013	0.930	0.028
QGemII	0.5	0.059	0.005	0.012	0.923	0.001

The backflow for the first case is $\sim$ 2.7%, whereas the misaligned holes decrease its value to $\sim$ 0.1%. The effect of magnetic field has been studied in conjuction with the case when there is no magnetic field. No effect of 0.5T field on overall backflow fraction has been observed.

5. Conclusion

In the present work, the electron transmission and ion backflow fraction for different GEM-based detectors have been evaluated numerically. Several modifications have been made in the simulation framework to make it significantly efficient and, as a result, suitable to study the complex physics problems of such devices. Efficiency of the field-solver has been enhanced to a great extent. Detailed studies have been carried out in order to optimize various aspects of the simulation framework so that complex physics problems can be approached keeping the demand of computational resources at an acceptable level, without losing on the precision front. Study of single GEM detectors shows that higher electron transmission and lower backflow fraction can be obtained with higher GEM voltage, lower drift field and higher induction field. GEM foils with larger pitch gives better electron transmission, as well as less backflow fraction. Multi-GEM devices are found to be better in terms of lower backflow fraction though the electron transmission is affected adversely. Several studies on quadruple GEM detectors with various geometry and field configuration have been performed. No significant effect of 0.5T magnetic field has been observed. Further work is necessary to achieve a comprehensive understanding as well as to find out an optimal geometry and field configuration in the context of the ALICE-TPC upgrade scenario.

Acknowledgments

We thank the members of the RD51 and ALICE-TPC colloboration for their help and valuable suggestions. We also acknowledge our respective Institutions for supporting us with the necessary facilities.

References

[1] Gasik P 2013 *J. Instrum.: Conf. Series* **9** C04035
[2] Hoch M 2004 *Nucl. Instrum. Method A* **535** 1
[3] Sauli F 1997 *Nucl. Instrum. Meth. A* **386** 531
[4] ALICE Collaboration 2014 *CERN-LHCC-2015-002 /ALICE-TDR-106-ADD-1*
[5] Ball M, Eckstein and Gunji T 2013 *J. Instrum.: Conf. Series* **9** C04025
[6] http://cern.ch/garfield.
[7] http://cern.ch/neBEM.
[8] http://cern.ch/heed.
[9] http://cern.ch/magboltz.

Journal of Physics: Conference Series **759** (2016) 012072 doi:10.1088/1742-6596/759/1/012072

Effect of surface roughness on the electrostatic field of an RPC

A Jash[1,3], N Majumdar[1], S Mukhopadhyay[1] and S Chattopadhyay[2]

[1] Applied Nuclear Physics Division, Saha Institute of Nuclear Physics, Kolkata 700064, India

[2] Experimental High Energy Physics Division, Variable Energy Cyclotron Centre, Kolkata 700064, India

[3] Experimental High Energy Physics Division, Homi Bhabha National Institute, Mumbai 400085, India

E-mail: abhik.jash@saha.ac.in

Abstract. The inner surfaces of the electrodes encompassing the gas volume of a Resistive Plate Chamber (RPC) have been found to exhibit asperities with three kind of feature grossly. The desired uniform electric field within the gas volume of RPC is expected to be affected due to presence of these asperities, which will eventually affect the final response from the detector. In this work, an attempt has been made to model the highly complex roughness of the electrode surfaces and compute its effect on the electrostatic field within RPC gas chamber. The different aspects of surface structures have been modeled and their effect on the electrostatic field has been calculated. The calculations have been performed numerically using Finite Element Method (FEM) and Boundary Element Method (BEM) and the two methods have been compared in this context.

1. Introduction

The Iron CALorimeter (ICAL) setup at India-based Neutrino Observatory [1] will be able to track the muons produced by the interaction of neutrinos with the iron, used as the target material, and produce information necessary for studying different parameters of neutrino oscillation and mass hierarchy. This will be achieved by deploying about 29,000 Resistive Plate Chambers (RPC) of dimension 2 m $\times$ 2 m as its active detector elements, the response from which will provide information about the position and the timing of the passing muons. The requirement of large area coverage combined with muon detection with good position ($<$ 1 cm) and timing resolution ($<$ 1 ns) disapproves non-uniform response from any part of the detector which may affect the overall performance of the setup. Long term stability in the detector performance is another important requirement in general. All these call for a detailed investigation on the detector response and understanding its dependence on factors like device geometry, material grade, gas mixture etc. including environmental parameters.

An extensive study of the electric field of the detector can pave a way to evaluate the response as it depends critically on the field configuration for a given gas mixture and environmental conditions. As the device geometry and the material are the fundamental configuring parameters of the field for a given potential, the imperfections or artifacts in the design and the material are likely to distort the electric field locally depending on their nature. Glass or bakelite is commonly used for making the resistive plates in an RPC and it has been found to develop

Journal of Physics: Conference Series **759** (2016) 012072 doi:10.1088/1742-6596/759/1/012072

some asperities (typically $< 10\mu m$) on its surface in the process of production and handling. The long term operation of the RPCs may also contribute by affecting the surface through various chemical processes. Depending on their shape and dimension, these asperities may give rise to spark, dark current etc. and thus lead to gradual degradation of the detector. In many cases, several methods of surface treatment have been adopted to cure the roughness or a better grade (better surface planarity) has been used instead. So, a detailed study on the effect of rough surface on the field configuration deems necessary in deciding the acceptable value of roughness as well as prediction and interpretation of the experimental data. In the present work, numerical simulation has been used to estimate this in case of a bakelite-RPC. The RPC has been modeled following the design parameters of a prototype fabricated for INO R&D work. The asperities on the inner surface of the bakelite electrode have been configured on the basis of the measurement of surface profile of the same bakelite grade used for building the prototype. The calculation has been performed following two numerical methods, namely, Finite Element Method (FEM) and Boundary Element Method (BEM).

In the following, section 2 describes the details of the modeling of the surface roughness based on the measurements and their implementation in the software. The results are discussed in Section 3 and the summary of the work can be found in Section 4.

2. Surface Morphology & Numerical Modeling

An RPC [2] is basically a parallel plate configuration comprising of two resistive plates encompassing a gas volume. A potential difference across the gas volume is applied through a conductive coating used on the outer surface of the plates. Two sets of mutually orthogonal readout strips are placed on either sides of the parallel plate configuration to collect the signal induced by the passage of muons through the gas volume. The design of the bakelite-RPC prototype has been discussed in detail in a previous work where the effects of various geometrical components on the electric field have been evaluated [3]. In the following, the measurements of the surface morphology of the P-120 grade and the numerical modeling of the asperities will be discussed.

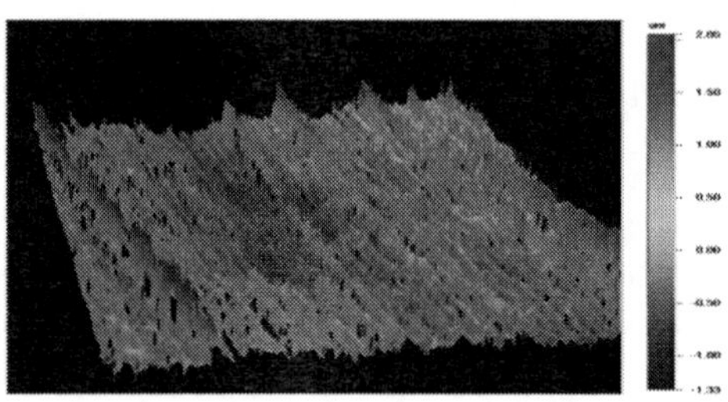

Figure 1. Surface roughness : $R_a = 158.15$ nm, $R_t = 1.73$ μm.

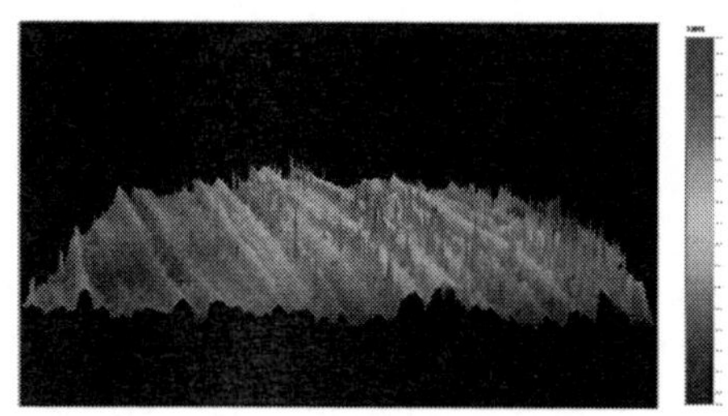

Figure 2. Surface roughness : $R_a = 282.04$ nm, $R_t = 3.37$ μm.

2.1. Surface Morphology

The imaging of a P-120 bakelite sample has been done using BRUKER ContourGT-K 3D optical microscope. A few sampling has been made at several locations with scanning area of dimension of 640 μm × 480 μm (shown in figures 1 and 2, for example). It is evident from the data that the average roughness of the surface, represented as R_a, varies between 150 - 300 nm while the range, R_t (distance between the peak and the valley) lies between 1 - 5 μm. By close inspection, following three gross structures have been found as asperities on the surface: (a) spike, (b) ridge and (c) wave. In the present work, only the effect of their shapes have been analyzed without considering their spatial distribution.

2.2. Numerical Modeling

To use the computational resources economically, a simplified model of RPC with dimension 5 cm × 5 cm, consisting of only the bakelite plates and the conductive coats have been considered while the said asperities have been distributed on the inner surface of the upper bakelite plate following a definite pattern. A voltage difference of 12 kV has been applied between the coats on either sides. The three features of the rough surface along with their modeling scheme have been discussed below:

(i) Spike - The randomly distributed spikes have been modeled using boxes of different heights and widths although with a periodic distribution. The smallest box has a cross-section of 1 μm × 1 μm and height 250 nm. Boxes with height in multiples of that of the smallest box have been placed along X-direction while those with width in multiples of the same of the smallest one have been arranged along Y. Both the height and the width of the boxes have been increased to a maximum limit of 1 μm and 4 μm, respectively, and then reduced again to reach the smallest values along the respective directions. The space between the edges of any two consecutive boxes has been made 8 μm for computational convenience which will be discussed later. For another series of boxes, placed along the diagonal, both the height and width have been increased in multiples of the smallest box and reduced back to the smallest values. The reason behind the scheme (shown in figure 3) has been to study the individual as well as the combined effect of the variation in height and width of the box in a systematic way.

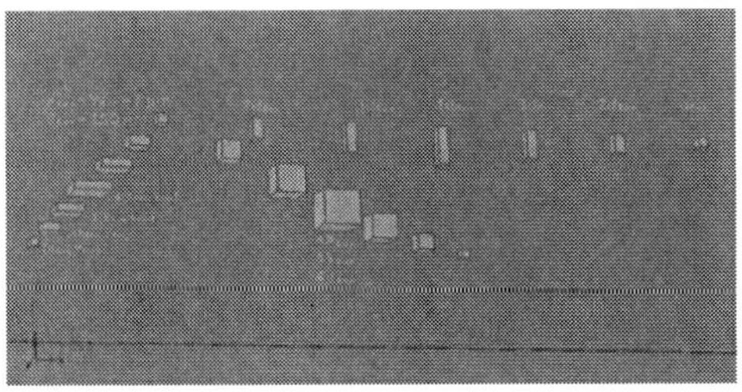

Figure 3. Surface roughness of spike-like structure modeled using a distribution of boxes of different heights and widths.

(ii) Ridge - Presence of some ridge-like structures with sharp heads can be seen in figure 1. This kind of structure has been modeled using five triangular prism-like blocks placed on the inner surface of the upper plate with their sharp edges facing the gas volume. The ridges have been placed parallel to each other along the X-direction. The height, base cross-section and pitch of the ridges are 5 μm, 25 μm × 650 μm and 50 μm respectively. The model is shown in figure 4.

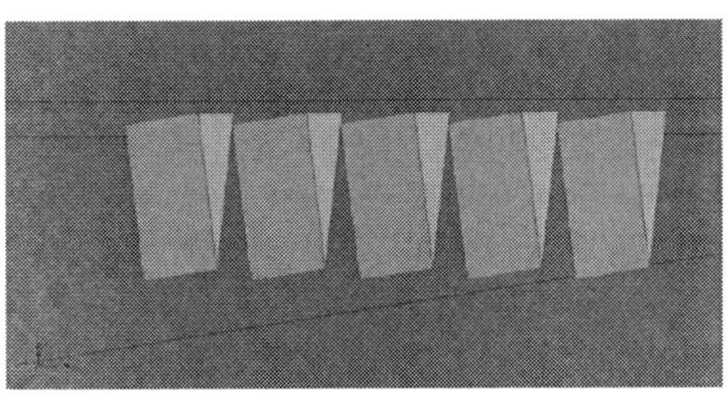

Figure 4. Surface roughness of ridge-like structures modeled using a series of triangular prism-like blocks.

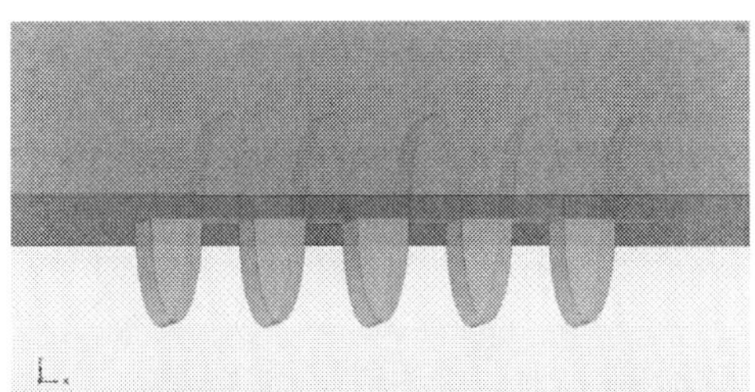

Figure 5. Surface roughness of wave-like structure modeled using a sinusoidal wave.

(iii) Wave - The surface profile measurement has also exhibited a wave-like feature as can be seen from figure 2. It has been modeled in the shape of a sinusoidal wave having amplitude of 10 μm and a wavelength of 40 μm as shown in figure 5. The width of the wave is 20 μm. The roughness feature has been incorporated on the upper bakelite plate as mentioned in earlier case. Obviously, the trenches of depth 10 μm have been implanted on the plate material.

2.3. Numerical implementation

Two different toolkits have been used to compute the field map, COMSOL Multiphysics® [4] v5.2 and neBEM [5] v1.8.16 which calculate the field following FEM and BEM respectively. In COMSOL®, the whole geometry has been meshed using free tetrahedral elements except the conductive coating region which has required a special treatment due to its very small thickness [3]. Different sizes of the elements have been used to mesh different parts of the geometry depending on their dimensions. In neBEM, the surfaces of the geometry have been discretized

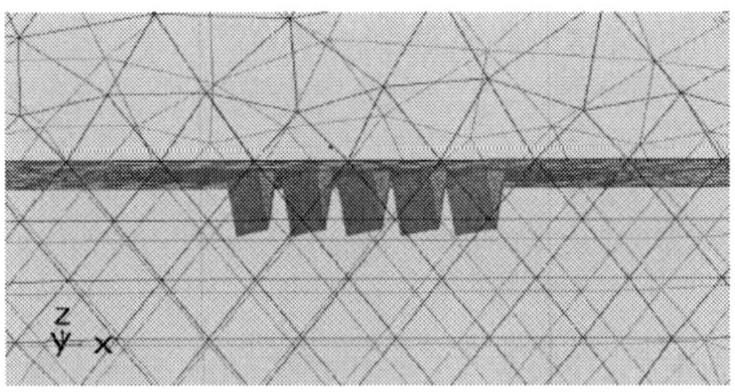

Figure 6. The ridge-like structure meshed in COMSOL®.

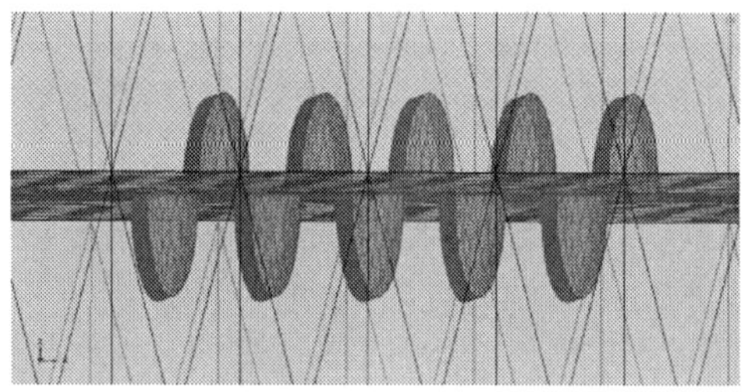

Figure 7. The wave shaped structure meshed in COMSOL®.

using rectangular or triangular elements according to the shape of the geometrical components. The geometry has been built in Garfield [6] using Constructive Solid Geometry (CSG) approach. The charge distribution on these smaller elements has been calculated using Green's function technique from the supplied boundary conditions. The potential and field at any point have been calculated directly from this charge distribution using the same Green's function technique.

Closer placement of rough structures has required a finer mesh to resolve the geometry which in turn has increased the consumption of computational resources. In modeling the spike-like structures, the boxes have been placed 8 μm away from each other as an optimum value of separation. Another difficulty in solving the models arises due to the presence of sharp edges, whose effect can be observed only by using infinitesimally small mesh. Figures 6 and 7 have shown the three dimensional meshing scheme in COMSOL® for the ridge and the wave-like structures respectively. Figure 8 has shown the two dimensional scheme of discretization, used by neBEM to solve the ridge-like structure.

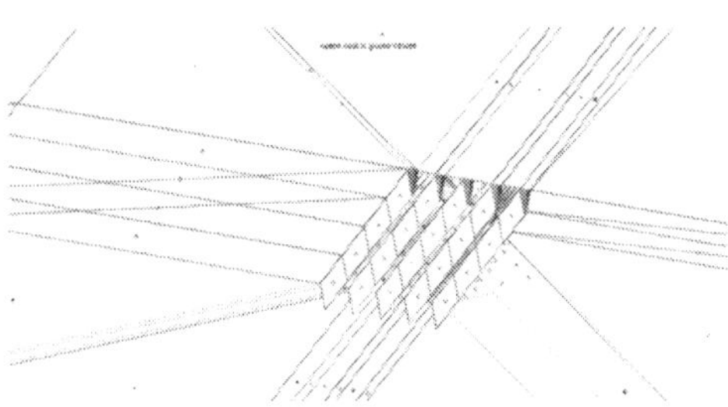

Figure 8. The ridge-like structure discretized using triangular and rectangular elements in neBEM.

3. Result

Here the results of the electric field computed by COMSOL® and neBEM have been discussed. All the plots have shown the variation in the value of Z-component of the electric field (E_z) for different features of the surface roughness as this is the major component of electric field which can affect the RPC response.

3.1. Spike

The effect of the box structures having different width on the electric field can be seen from figure 9 where the variation of E_z along the Y-direction has been plotted. The calculation has been performed at several locations away from the box and compared in the figure. The following

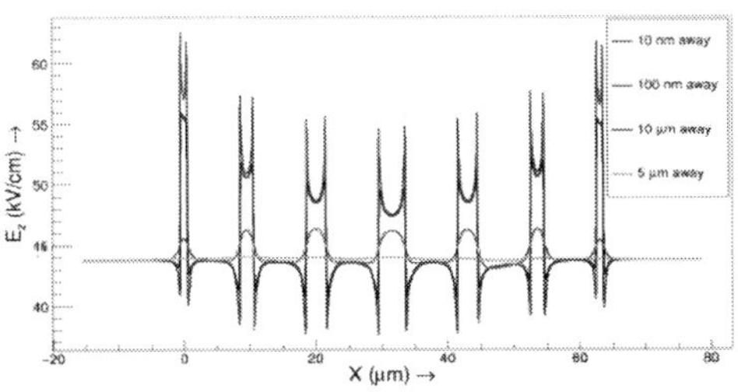

Figure 9. E_z Vs Y plot from neBEM for different distances from the smallest box.

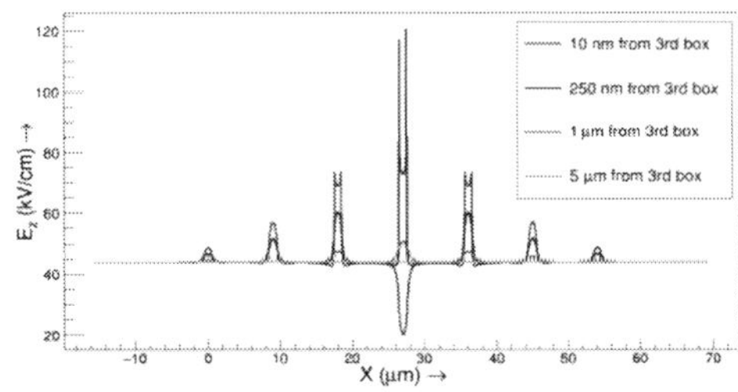

Figure 10. E_z Vs X plot from neBEM for different distances from the third box.

observations can be made from the plot: (a) The field value is enhanced at the positions of the boxes, (b) The magnitude is larger when the box is narrower in width, (c) A sharp rise occurs along the edges of the boxes, (d) All these effects get more pronounced in close vicinity of the boxes. The field is found to increase by 30.5% maximum at 10 nm away and can rise upto 42.8% at the edges while it falls close to the regular value (within 0.6%) around 5 μm. The effect of height of the box structures can be analyzed from figure 10 where the variation of field values calculated along the X-direction and at several positions away from the third box have been shown. As evident from the figure, the observations are similar to that made in earlier case. It is to be noted that in case of the highest box (at the center of the distribution), a dip on the line of 10 nm has occurred as it cuts through the box. From these two plots it can be concluded that the taller and narrower spikes will affect the field most, though the affected region will stretch with the increase in width.

3.2. Ridge

The effect of the ridge on the electric field as calculated at different distances away from the head of the ridge has been shown in figure 11. The enhancement in field value reduces as one goes away from the ridge structures as is evident from the figure. The field value can change by 46.2% to 0.6% in a range from 10 nm to 50 μm away from the structures. The contour plot of E_z in the vicinity of these ridge-like structures can be seen from the figure 12 as calculated using COMSOL®.

3.3. Wave

The change in the value of E_z along lines at different distances from the wave peaks has been shown in figure 13, calculated using COMSOL®. The effect of this wavy structure extends upto 50 μm away from the peaks where the relative deviation in the field with respect to its normal or regular value reduces to 1.9% only while it is 59.2% within 10 nm of the peak and - 21.2% at

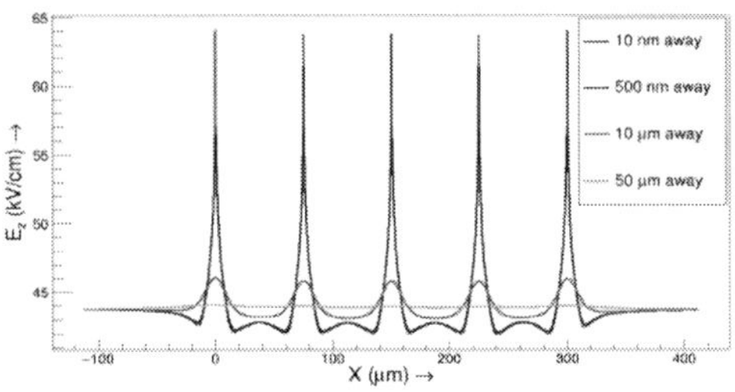

Figure 11. E_z Vs X plot along the lines at different distances from the edges of the ridges from neBEM.

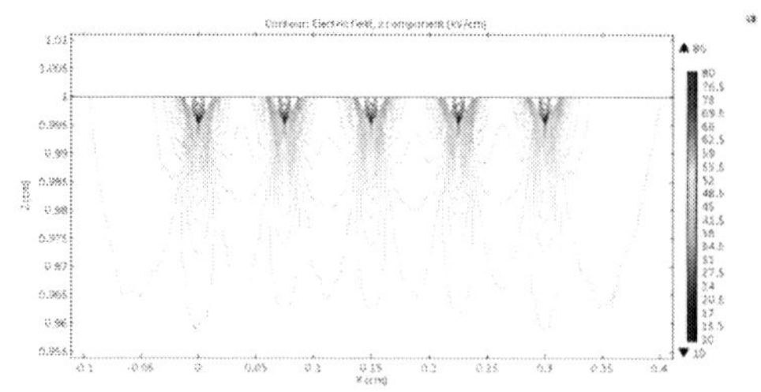

Figure 12. Contour plot of E_z in X-Z plane from COMSOL®.

the troughs. Figure 14 shows the surface contour plot of the E_z values 100 nm away from the peaks as has been computed with the COMSOL®.

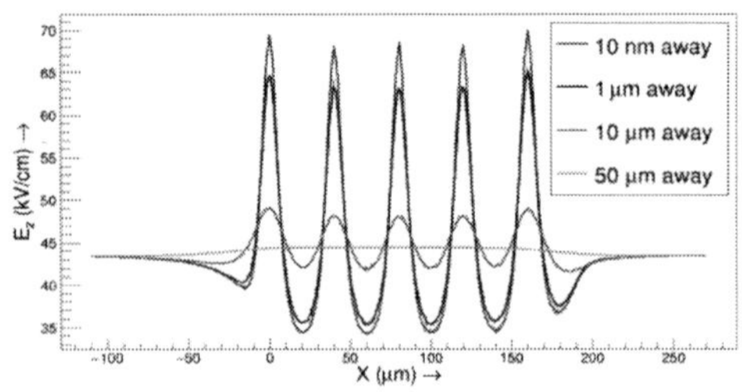

Figure 13. E_z Vs X plot from COMSOL® at different distances from the peaks of the wave.

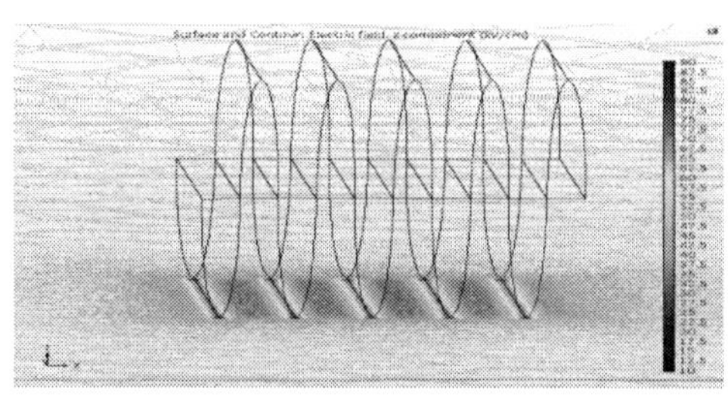

Figure 14. Contour plot of E_z from COMSOL® on the surface, 100 nm away from the peaks of the wave-shaped profile.

3.4. Comparison of FEM and BEM

The computation of E_z has been performed using two solvers and compared. Table 1 shows a comparison between these two methods in terms of different parameters to solve the case of spike-like structures (see figure 3) as an example. The second column of the table shows the number of total elements used for meshing the geometry. The third and fourth columns show the time taken to solve the problem and the system memory used during that process respectively. In case of COMSOL®, the quoted values are the ones that have been taken by the solver while using the optimized meshing scheme. The process of finding out the optimum meshing scheme itself may be a time-consuming affair for COMSOL® while for neBEM, it is significantly simpler. The fifth column of the table shows the convergence criterion for COMSOL® which is the relative error (R.E.) in the calculated values with respect to the set limit (10^{-5} here) for the last two iterations. COMSOL® has required to iterate 190 times to achieve this. For neBEM, the error has been estimated by finding out the deviation in the calculated values at the collocation points with respect to the given values as obtained from the boundary conditions. The maximum error has been found to be 10^{-4} on one of the elements in this case.

4. Summary

The electrostatic field map for different features of surface roughness of bakelite plate in an RPC has been computed using two solvers following different methodology. The value of the

Table 1. Comparison of FEM and BEM using different parameters.

Solver	Elements	Time	Memory	Error/Convergence
COMSOL®	2×10^6	15 mins	~ 4 GB	R.E. $= 10^{-5}$ in 190 iterations
neBEM	17×10^3	2 hours	~ 8.8 GB	Maximum $= 10^{-4}$

electric field near the asperities increases to 60% at the most of the regular value depending on the shape and size of the asperities. The distortion in the field is found to be confined within a small region (10 - 200 μm). This distorted field map is likely to affect the gas transport properties and production of avalanche close to the resistive plates and alter the detector behavior. Considering the overall distribution of asperities on the surface, the detector response near the plates may be significantly altered which in turn may affect the physics performance of INO ICAL detector. The roughness properties of the RPC plates need to be investigated. The plates should be made to undergo surface treatment, depending on the results of the surface study.

References

[1] ICAL Collaboration S. Ahmed et al. 2015 *arXiv* : 1505.07380
[2] Santonico R and Cardarelli R 1981 *NIM* **187** p377
[3] Jash A, Majumdar N, Mukhopadhyay S and Chattopadhyay S 2015 *JINST* **10** P11009
[4] COMSOL: a multiphysics simulation tool `www.comsol.co.in`
[5] Majumdar N and Mukhopadhyay S 2007 *JINST* **2** P09006
[6] Garfield: Rob Veenhof `http://garfield.web.cern.ch/garfield`

Exact solutions to model surface and volume charge distributions

S Mukhopadhyay[1], N Majumdar[1], P Bhattacharya[1], A. Jash[1] and D.S.Bhattacharya[1]

[1] Applied Nuclear Physics Department, Saha Institute of Nuclear Phyics, 1/AF, Bidhannagar, Kolkata, India 700064

E-mail: `supratik.mukhopadhyay@saha.ac.in`

Abstract. Many important problems in several branches of science and technology deal with charges distributed along a line, over a surface and within a volume. Recently, we have made use of new exact analytic solutions of surface charge distributions to develop the nearly exact Boundary Element Method (neBEM) toolkit. This 3D solver has been successful in removing some of the major drawbacks of the otherwise elegant Green's function approach and has been found to be very accurate throughout the computational domain, including near- and far-field regions. Use of truly distributed singularities (in contrast to nodally concentrated ones) on rectangular and right-triangular elements used for discretizing any three-dimensional geometry has essentially removed many of the numerical and physical singularities associated with the conventional BEM. In this work, we will present this toolkit and the development of several numerical models of space charge based on exact closed-form expressions. In one of the models, Particles on Surface (ParSur), the space charge inside a small elemental volume of any arbitrary shape is represented as being smeared on several surfaces representing the volume. From the studies, it can be concluded that the ParSur model is successful in getting the estimates close to those obtained using the first-principles, especially close to and within the cell. In the paper, we will show initial applications of ParSur and other models in problems related to high energy physics.

1. Introduction

Gaseous ionization detectors exhibit a complex array of physics processes. Numerical simulation is an important tool to understand these detectors at the device level. Extensive simulation is used to analyze the performance and characteristics of existing detectors and to improve their design. With the advent of Micro-Pattern Gas Detectors (MPGD), the importance of simulation has grown further. These detectors have intricate patterns of micron-order dimensions that makes their geometries of an essentially three-dimensional nature. While this design aspect serves the purpose for which they are used, namely excellent rate capability, spatial, temporal and energy resolutions, analysis of these detectors has become difficult due to the mentioned features. In addition, for almost all the gaseous ionization detectors, there are problems related to space charge accumulation that are extremely difficult to handle computationally. For example, even geometries as simple as the resistive plate chambers (RPC) in their various design variations, that are also the detector of choice for a large variety of experiments, can be significantly affected by space charge. Wire chambers, such as drift tubes, or multi-wire proportional chambers, all suffer from this problem that severely limits their usage in present high luminosity experiments, their upgrades and new experiments being planned for the future.

During recent times, a simulation framework for gaseous detectors have been developed by the RD51 collaboration in which Garfield [1] has played a central role. Garfield simulates transport, amplification and signal induction in ionization detectors, and provides links to codes that compute parameters necessary for this computation. For example, interface is provided to Heed for calculating primary ionization and cluster formation [2], to Finite Element Method (FEM) codes such as Ansys and Boundary Element Method (BEM) codes such as neBEM for computing the electromagnetic field [3, 4], and to Magboltz for estimating Townsend, attachment coefficients and transport properties [5]. Among these tools, neBEM is a recently formulated BEM toolkit [6, 7] that has the distinct advantage of being able to truly represent charge distributions on elements of rectangular and triangular shapes and incorporate their effects while solving the electromagnetic field problems involving 3D geometries of arbitrary complexity. In this brief report, we will restrict ourselves to demonstrating the accuracy of neBEM while computing electric field for an MPGD with complex geometry. In addition, preliminary use of this approach to solve problems related to volume distribution of charges will also be presented.

2. The neBEM toolkit and its applications

Some of the features expected in a field-solver handling gaseous detectors in general, and MPGDs, in particular, may be listed as its ability to handle large length scale variations (meter to microns), to provide potential and field at arbitrary locations within a volume of interest, to model devices with arbitrary complexity in terms of geometrical shape and material components. The neBEM approach was developed with these aims and it has been reasonably successful in achieving its goals [8]. In what follows, we will briefly discuss the approach and discuss its ability to solve electromagnetic field in one of the MPGDs.

The potential can be obtained by solving the Poisson's equation adopting the BEM approach, among other possibilities.

$$\nabla^2\phi(\vec{r}) = -\rho(\vec{r})/\epsilon_0$$

The solution yields the distribution of charges which is obtained through the satisfaction of a given potential configuration. The potential $\phi(\vec{r})$, for a point charge q at $\vec{r}'$ is known to be

$$\phi(\vec{r}) = \frac{q}{4\pi\epsilon_0|\vec{r} - \vec{r'}|}$$

Superposition holds for a general charge distribution and leads to

$$\phi(\vec{r}) = \int \frac{\rho(\vec{r'})dv'}{4\pi\epsilon_0|\vec{r} - \vec{r'}|} = \int G(\vec{r}, \vec{r'})\rho(\vec{r'})dv' \tag{1}$$

where, the free space Green's function for the Laplace operator in three dimensions is given as

$$G(\vec{r}, \vec{r'}) = \frac{1}{4\pi\epsilon_0|\vec{r} - \vec{r'}|}$$

Similarly, a general charge distribution yields a field

$$\vec{E}(\vec{r}) = -\nabla\Phi$$

or,

$$\vec{E}(\vec{r}) = -\nabla\left(\int G(\vec{r}, \vec{r'})\rho(\vec{r'})dv'\right)$$

Thus,

$$\vec{E}(\vec{r}) = \int \frac{\rho(\vec{r})(\vec{r} - \vec{r'})dv'}{4\pi\epsilon_0|\vec{r} - \vec{r'}|^3} \tag{2}$$

The boundary conditions are specified in a form suitable for a given problem. The conditions can be Dirichlet (known potential values), Neumann (known field values, or other known conditions normal to an interface) or Robin / Mixed (mixture of Dirichlet and Neumann) on the material boundaries / interfaces. The following integral equation of the first kind can be easily set up if we consider only the Dirichlet problem at present, for ease of discussion,

$$\phi(\vec{r}) = \int_{vol} G(\vec{r}, \vec{r'})\rho(\vec{r'})dv' \tag{3}$$

Here, the potential at a point $\vec{r}$ is $\phi(\vec{r})$, the charge density at an infinitesimally small volume dv' is $\rho(\vec{r}')$, while the permittivity of free space is ϵ_0. In order to solve the problem, we need to find $\rho(\vec{r})$ as a function of space. This solution needs to be capable of yielding the known distribution of $\phi(\vec{r})$. Potential and field at any point in the computational domain can be obtained as soon as the charge distribution on the boundaries and all the interfaces has been solved.

In order to apply the BEM technique, the first step is to discretize the boundaries and interfaces of a given problem. The shape of the discretized elements can be arbitrary, but the most common are rectangular or triangular. Between these two, the triangular elements offer more flexibility and can be used to model geometries of any variety for the both FEM and BEM. The charge distribution on the discretized elements that satisfies eq. 3 are solved for, once meshing is over. A large number of solution approaches exist, among which the constant element collocation approach is one of the most popular. In this zero-th order formulation, the charge distribution on each element is assumed to be uniform and equivalent to point charges located at the centroid of the elements. This approach provides a good optimization between accuracy and computational complexity.

This approximation, although popular due to reasons mentioned above, lead to numerical and physical singularities that can lead to grossly inaccurate results. As a result, a very large number of BEM formulations have been developed to deal with issues related to specific problems. The neBEM approach was developed to provide a formulation that could deal with many of the issues together. This was achieved by the use of symbolic integration to arrive at exact close-form analytical expression for potential and field due to singularities distributed on rectangular and triangular elements [6, 7]. While this cannot be elaborated here, we will demonstrate efficacy of the approach by comparing its results with that of FEM for an MPGD with a complex shape.

Micro Wire Detector (MWD), one among the MPGDs, is a lithographically produced cathode mesh of square holes on one side of a kapton foil and anode strips on the other, running along the middle axis of cathode mesh holes. The kapton is used so that it provides a mechanical joint between the cathode mesh of square holes and the anode strips. The detailed structure of this detector has been described in [9]. For the calculations using neBEM, the same geometry has been used here. However, the shape of kapton foil has been considered rectangular for the present studies, instead of trapezoidal in [9]. A drift plane with a voltage 1.11kV has been considered which is 785 mm away from anode strip following [9].

As an example of extremely difficult situation, field along a line around the anode just 1 micron away from the surface is considered here 1(b) [9]. Sampling of field values for neBEM has been considered to be as small as 0.1 micron. In addition, it should be noted that no smoothing or filtering has been applied as post-processing. From fig. 2(a), sharp rise in the field values is observed at all the four edges while smooth variation of field is observed on each of the four surfaces. Field values are found to decrease sharply once the points are beyond anode surfaces. FEM computation is significantly more jagged and is clearly unable to produce correct results near and at the edges.

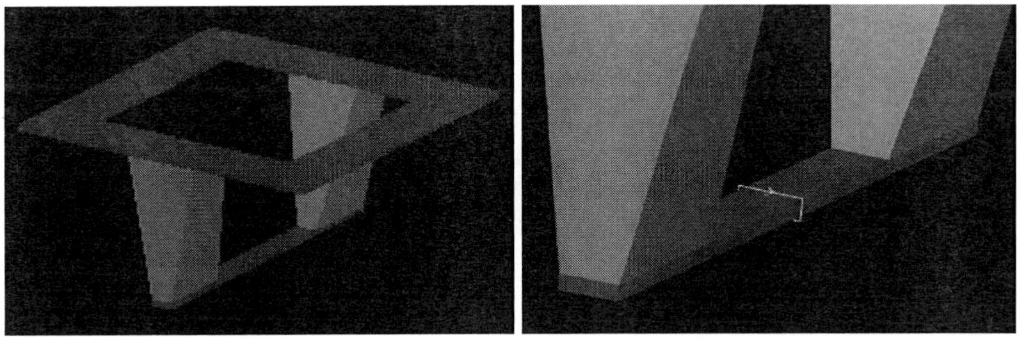

(a) (b)

Figure 1: Micro Wire Detector (a) FEM model used in [9] (b) line around anode along which electric field has been computed

The effect of discretization has also been studied in relation to this problem. In the earlier computation, we had used 20 elements to represent the top surface and 10 elements on the side surface. The elements were made successively smaller towards the edges. In order to study effect of using coarse discretization, we also used larger elements of fixed size, only 3 elements each to represent both top and side surfaces. Although there is significant difference between the results, the overall trend is represented well by the larger elements, as shown in fig. 2(b). It is important to note that there is no jaggedness (at 0.1 micron sampling) despite the use of unreasonably large elements.

(a) (b)

Figure 2: Electric field around the anode (a) comparison between neBEM and FEM [9] results (b) Effect of discretization on the estimate

There may be several reasons that lead to the difference between the results obtained using neBEM and FEM. In order to impose periodicity on the geometry, we have considered a 33 matrix of microwire units for neBEM, instead of invoking symmetry. The Neumann boundary condition has been used on the side boundaries of the FEM computational domainIn [9] in order to fulfill the same requirement. The potential and field at any arbitrary point using neBEM have been obtained using the closed form analytic influence coefficients have been used. This is the principal reason leading to the smooth solution obtained using neBEM.

3. Particles on Surface - ParSur

Algorithms used for the evolution of a charge cluster in three dimensions involves huge computational expenditure. Drastic simplifying assumptions are made in order to reduce the magnitude of the expenses. It has been found that numerical simulation of reduced dimensions (1.5D or 2D) can reproduce most of the essential features of the evolution of a cluster. However, for even more realistic simulations, we need to develop more accurate, but efficient models and tools. In recent times, space charge problems in 3D have been successfully analyzed using the popular Particle-in-Cell (PIC) model. This model has been applied to a wide variety of fields, including plasma physics, solid and fluid mechanics and several other areas of science and technology [?]. It has its own set of problem which includes computational expenditure, particle discretization errors, fixed-grid formalism. Thus, we need to look for a better model in order to study the problem of cluster evolution in a gas detector.

As has been discussed above, it is easy for neBEM to represent distributed surface charges. So, the present approach involves the representation of a given volume by a set of surfaces. The evolving volume charge distribution is smeared on the various surfaces depending on their proximity to them. This leads to the name of the approach, Particles-on-Surface (ParSur). Once the smearing is over, it is easy to estimate precisely the effect of charge distributed on any given rectangular or triangular surface using the analytic expressions of neBEM. Using ParSur, the effect of space charge is expected to be simulated accurately. Below, we have presented results from a brief study on very simple implementations of both PIC and ParSur and made a brief comparison.

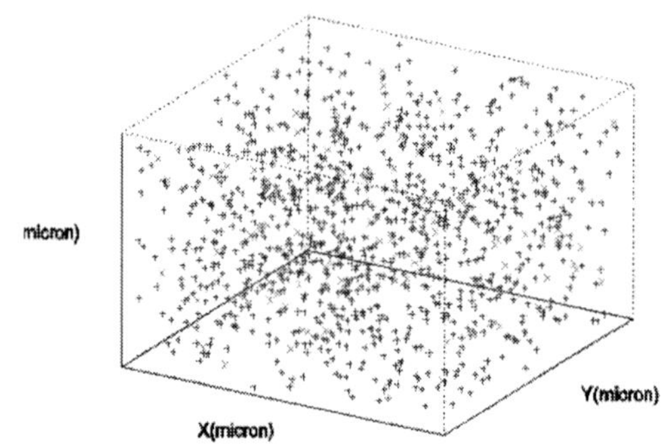

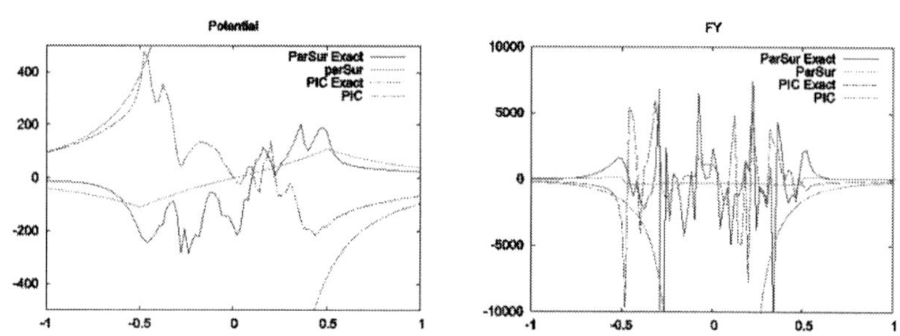

(a) (b)

Figure 3: Effect of volume distribution of ions and electrons (a) Potential (b) Field

To begin with, we have considered a simple rectangular volume filled with ions and electrons. The number of charged particles is large and they are randomly placed within the volume as shown in fig. ??. The potential and field of this volume distribution of charges have been estimated using the approaches discussed above and by the exact method, i.e., by summing up the influence of each and every charge within the volume. The comparison in fig. 3(a) and fig. 3(b) shows that the ParSur approach is significantly closer to the exact values in both the cases.

Finally, we have considered a volume filled with charges of only one sign - either positive or negative. In this case, unlike in the earlier one, the potential and field does not get cancelled on an average. In figs. 4(a) and 4(b), it is once again clear that the estimates made by ParSur follows the trend shown by the exact curve.

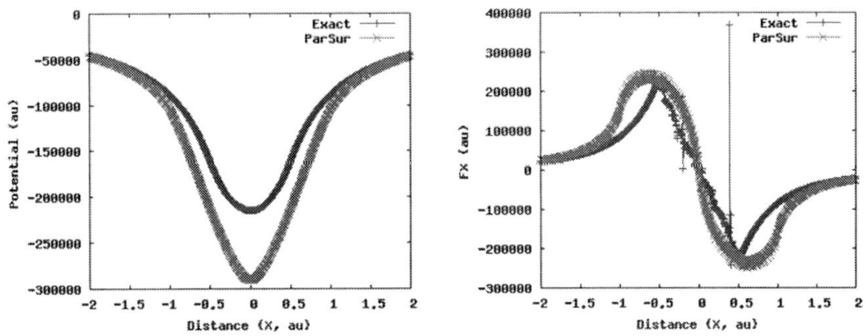

(a) (b)

Figure 4: Effect of volume distribution of electrons only (a) Potential (b) Field

4. Conclusion

We have introduced the neBEM approach for solving problems related to charge distributions. Based on close-form exact solutions, this approach has been developed as a toolkit and has been used to solve many problems related to nuclear or particle physics detectors. One such example involving micro-wire detectors has been presented and compared with FEM solutions to illustrate the efficacy of the approach. Finally, simple problems related to volume charge distributions have been solved using several approaches, among which results from ParSur, a model based on the neBEM approach, has been presented. It has been found that the new model is quite capable in estimating potential and field for these simple volume distributions. In future, ParSur will be developed further to address problems related to detector space charge accumulation problems.

References

[1] http://cern.ch/garfield.
[2] http://cern.ch/heed.
[3] http://cern.ch/neBEM.
[4] http://www.ansys.com
[5] http://cern.ch/magboltz.
[6] Mukhopadhyay S and Majumdar N, 2006 *Engg. Analysis Boun. Elem.* **30** 687
[7] Mukhopadhyay S and Majumdar N, 2006 *Engg. Analysis Boun. Elem.* **33** 105
[8] Majumdar N and Mukhopadhyay S, 2007 *J. Instrum.* **2** P09006
[9] http://consult.cern.ch/writeup/garfield/examples/micropattern/microwire/index.htm

Journal of Physics: Conference Series **759** (2016) 012074 doi:10.1088/1742-6596/759/1/012074

Simulation of π^0-γ Separation Study for Proposed CMS Forward Electromagnetic Calorimeter

Ashim Roy[1], Shilpi Jain[2], Sunanda Banerjee[3], Satyaki Bhattacharya[1], Gobinda Majumder[4]

[1]Saha Institute of Nuclear Physics, Kolkata, India
[2]National Central University, Taiwan
[3]Fermi National Accelerator Lab, United States
[4]Tata Institute of Fundamental Research, India

E-mail: ashim.roy@saha.ac.in

Abstract. The Forward Electromagnetic Calorimeter of the CMS detector is going to be upgraded in the high luminosity running as the energy of the present Electromagnetic Calorimeter (PbWO4) will degrade in the high luminosity (luminosity $10^{34} cm^{-2} s^{-1}$) running due to extensive radiation (hadron flux $10^{13} neutrons\ cm^{-2}$). Shashlik Electromagnetic Calorimeter which consists of alternate layers of 1.5 mm LYSO(Ce) crystal plates and 2.5 mm Tungsten absorbers, was a proposal for high luminosity running. One of the performance points for any electromagnetic calorimeter is the ability to separate π^0s from true photons, since final states with photons are a clean and one of the most important final states in proton-proton collisions at the LHC. The objective of this project is to study the possibility of π^0 and γ separation in the Shashlik detector using Multivariate Analysis (MVA) technique.

1. Introduction

The performance of the Electromagnetic Calorimeter (ECAL), particularly the endcap component, will suffer due to the increased hadron flux (10^{13} neutrons cm^{-2}) in a high luminosity environment (luminosity 10^{34} $\text{cm}^{-2}\text{s}^{-1}$). The transparency of the ECAL endcap will be degraded and as a result, its energy resolution will worsen. For integrated luminosity of 500 fb^{-1}, the transmission loss of light in the $PbWO_4$ crystals will be reduced by a factor larger than 25[1] in the forward region. It will be hard to recover the performance of the calorimeter through calibration. Thus it becomes important to replace the ECAL endcap with a new one which can survive in a high luminosity scenario and can give stable resolution as a function of integrated luminosity.

There were several proposals for upgrading the calorimeter system (ECAL + HCAL) for high luminosity run. For ECAL endcap, Shashlik ECAL, with alternate layers of scintillators and absorbers was one of the proposals for high luminosity running. Many materials were considered for the scintillator. Out of these inorganic scintillators, LYSO (cerium doped lutherium yittrium silicate) seems to be the best candidate based on the following considerations:

(i) radiation hardness,

(ii) fast response and also high light output,

(iii) short radiation length (1.14 cm) leading to a smaller detector size.

Figure 1 shows the layout for the Shashlik detector.

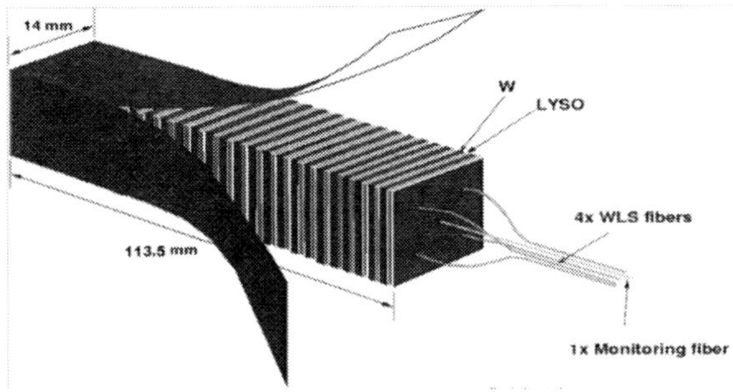

Figure 1. The layout of a Shashlik cell showing alternate layers of scintillator and absorber with 4 light collection fibers and one monitoring fiber.

2. Simulation

A stand-alone detector setup consisting of alternative layers of absorbers (2.5 mm thick tungsten) and scintillators (2 mm thick LYSO) was defined in the framework of GEANT4 [2]. GEANT4

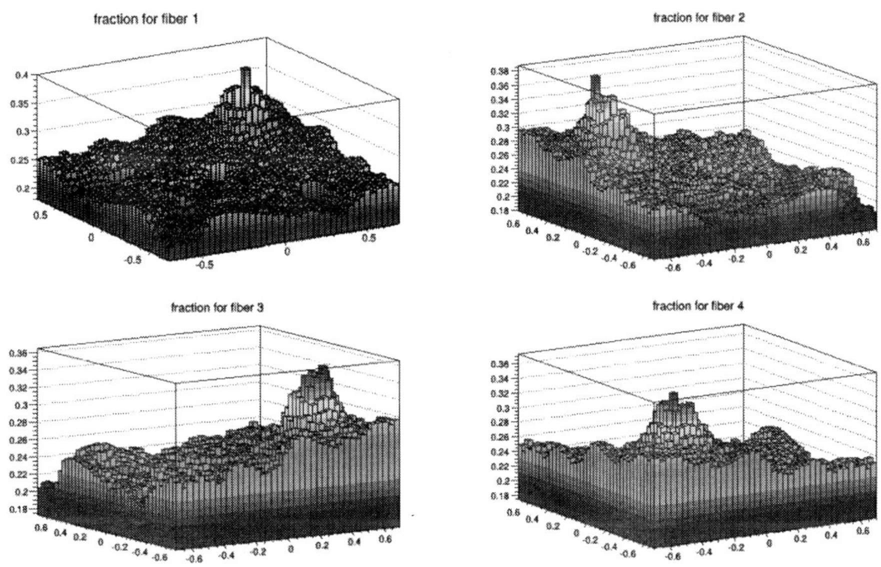

Figure 2. Figure on top-left shows the distribution of probability as a function of X,Y position for fiber 1, on the top-right for fiber 2, on the bottom-left for fiber 3 and on the bottom right for fiber 4. These four distributions are symmetric since the four fibers are symmetrically placed.

version 9.6.p02 was used with the physics list QGSP_FTFP_BERT. Light saturation effect was introduced through the use of Birk's law:

$$w \;=\; 1.0 - k_1 \cdot \ln\left(k_0 \cdot \frac{dE}{dx}\right)$$

with $k_0 = 0.0333\ \mathrm{MeV}^{-1}\cdot\mathrm{g}\cdot\mathrm{mm}^{-2}$ and $k_1 = 0.254$[3]. The weight factor w was restricted in the range 0.1:1.0. For π^0-γ separation studies, 28 layers of absorber and 29 layers of scintillators in a 11×11 matrix was defined. Transverse size of each cell in each layer was chosen to be 14 mm. Five fiber paths are defined of which the central fiber was for calibration and the other four fibers were at positions (±3.5 mm, ±3.5 mm) with respect to the center. They can be read out individually or to a combined output. The fibers were of diameter 1.6 mm and were inserted in holes of diameter 1.6 mm.

Energy deposit at a given point in the scintillator plate will be sampled differently by the four fibers. The probability of distribution of this light among the fibers depends on the transmission coefficient of the scintillator and hence on integrated luminosity. This has been estimated from a separate study using SLitrani[4].

3. π^0/γ Separation

Any search which has a final state of photons, has a background contribution from jets which fake real photons. This is because π^0's in jets decay to 2γ's almost 99.9% of the times. For high energy π^0's the angle between the two photons becomes so small that it becomes very difficult to separate photons coming from the decay of π^0, and the real photons which are coming either from the interaction vertex or from radiation off the leptons. With a transverse size of 14 mm for the detector layer, it becomes a tough task to separate out photons coming from the decay of π^0 and the real photons of energy more than 50 GeV.

In this study, an attempt has been made to explore the idea of exploiting the information from the four fibers for π^0 and γ separation. The idea behind using information from all the four fibers individually is that a larger fraction of the deposited energy from a single photon will be collected by the fiber which is closest to the impact point, while π^0, decaying to a pair of photons will have two impact points on the Shashlik detector and the sharing of light among the fibers will significantly increase.

3.1. Shower Shapes

In general, the lateral shower profile tends to be broader for photons coming from π^0 compared to that of prompt photons. This holds true for lower energy π^0's (less than 100 GeV). These shower shape variables are useful for discriminating between π^0's and photons. For all shower shape variables, the tower with maximum energy deposit is first looked for and then the shape parameters are formed around it. The following shape variables are considered:

$S1/S9$: This ratio makes use of $S1$, the energy deposited in the maximum hit tower, and $S9$, the energy deposited in 3×3 array around the maximum hit tower. The left plots of figure 3 shows the distribution of $S1/S9$ for 50 GeV, 70 GeV and 150 GeV photons and π^0's.

$S1/S4$: This ratio uses $S4$, the energy deposited in 2×2 array including the maximum energy tower. Four possible 2×2 arrays are possible which include the maximum energy tower. The combination which corresponds to the largest sum total energy is used in determining the ratio. The middle plots of figure 3 shows the distributions of $S1/S4$ for 50, 70 and 150 GeV photons and π^0's.

2-D distribution of F_{16} VS F_9: The variables, F_9 and F_{16}, are defined through equation 1.

$$F_9 = \frac{S9 - S1}{S9}, F_{16} = \frac{S16 - S4}{S16} \tag{1}$$

where $S16$ is the energy deposited in the 4×4 array of towers which includes that 2×2 array of towers which has maximum energy in the four possible combinations as explained above. The right plots of figure 3 shows the 2-D distribution of F_{16} and F_9 for 50, 70 and 150 GeV photons and π^0's.

The performance of the variables $S1/S9$, $S1/S4$ and F_{16} vs F_9 are summarized below:

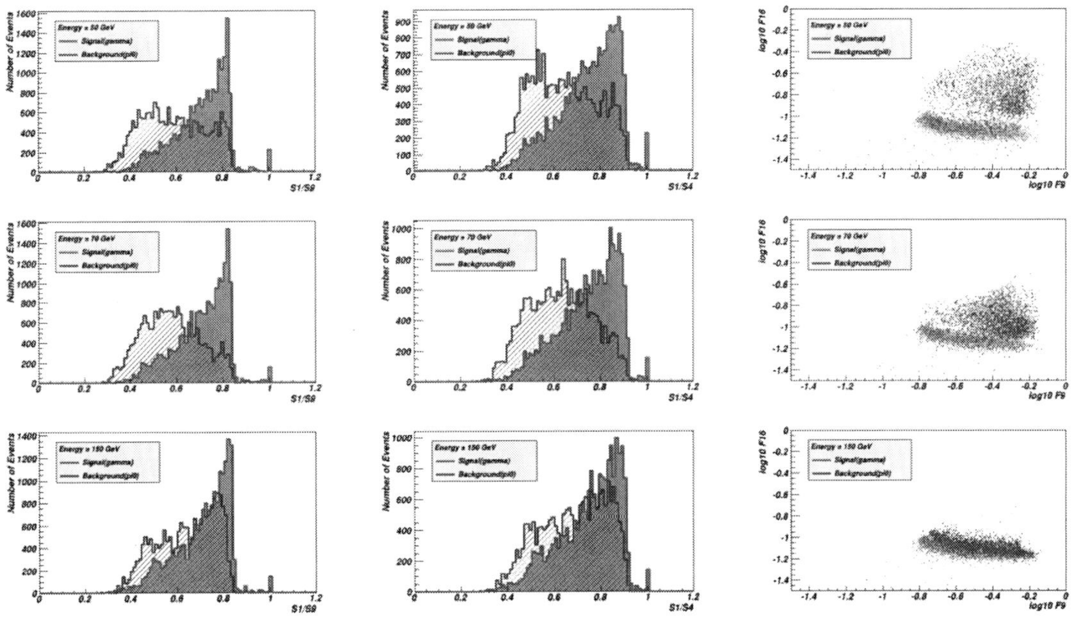

Figure 3. The distribution of S1/S9, S1/S4, F9F16 for 50 GeV, 70 GeV and 150 GeV photons and π^0's. The blue hatched histogram is for photons and the red hatched histogram is for π^0's.

- Shower shape variables $S1/S9$ and $S1/S4$ lose the sensitivity for γ/π^0 separation at energies above 70 GeV;
- The 2-D distribution of F_{16} vs F_9 performs better for 70 GeV γ/π^0 discrimination compared to $S1/S9$ and $S1/S4$. But again it loses power for discrimination at energies above 70 GeV.

3.2. Study using Multivariate Analysis (MVA)

As has been seen from the previous Sections, 3.1, that the discriminating power is reduced significantly for π^0's of energy above 100 GeV. An analysis has been carried out exploring the discriminating power gained by employing multivariate techniques to the problem. The following MVA classifiers were examined in this analysis:

(i) Boosted Decision Tree (BDT)
(ii) Gradient Boosted Decision Trees (BDTG)
(iii) Artificial Neural Network (ANN)

In this analysis, the TMVA[6] package within ROOT[5] was used. All the MVA techniques, mentioned above, are implemented in it. TMVA was trained with a sample of photons and π^0's. Energy from each individual tower in 3×3 array, or energy from each fiber in the 3×3 array was fed into TMVA. This analysis was done using photons and π^0's at 200 GeV. Two types of samples were produced:

Fixed gun sample: These were produced with the gun position fixed at (0 mm, 4 cm) in (x, y), whereas, z being 3.195 m.

Random gun sample: In this case the gun position was uniformly distributed in X and Y directions between 7 mm and + 7 mm i.e. within the central crystal.

Following two sets of training variables were used separately to train TMVA:

Coarse grain information: Input to TMVA is the ratio of energy from each tower in 3×3 array to total energy in 3×3 array.

Fine grain information: Input to TMVA is the ratio of energy from each individual fiber in 3×3 array to total energy in 3×3 array.

These energies are scaled to the total energy collected in the 3×3 array.

3.2.1. MVA using fixed gun samples In this case the TMVA was trained and tested independently using 10000 events from fixed gun samples of 200 GeV photons and π^0's. Two sets of training variables, coarse grain and fine grain information were used separately to train the TMVA. Figure 4 shows the background rejection versus signal efficiency curve for the case of both coarse grain and fine grain information.

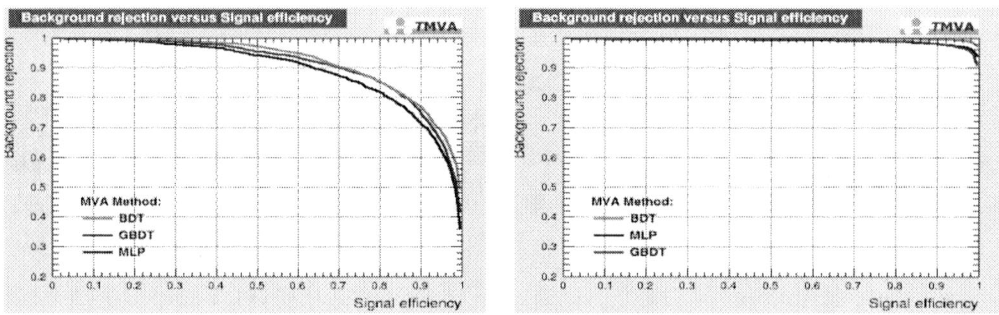

Figure 4. he background rejection versus signal efficiency curve of different MVA methods for fixed gun sample. The figure on the left are for coarse grain information and that on the right refers to fine grain information.

3.2.2. MVA using random gun samples Here TMVA was trained and tested separately using 10000 events from random gun samples of 200 GeV photons and π^0's. Two sets of training variables, coarse grain and fine grain information were used separately to train the TMVA. Figure 5 shows the background rejection versus signal efficiency curve for the case of both

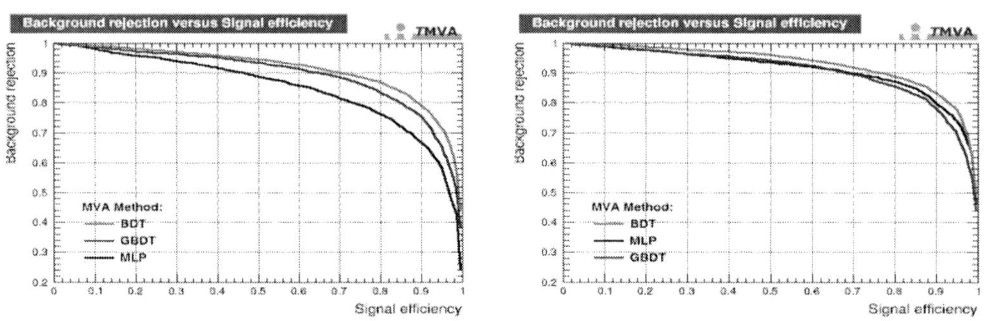

Figure 5. The background rejection versus signal efficiency curve of different MVA methods for random gun sample. The figure on the left are for coarse grain information and that on the right refer to fine grain information.

coarse grain and fine grain information. It can be seen from the figure 4 and 5, that the fine grain information is better for discriminating signal from from background.

3.3. Application of random gun samples in real application

In this case the TMVA was trained using 20000 events from random gun samples of 200 GeV photons and π^0's. The ratio of energy from each individual fiber in the 3×3 array to total energy were used as training variables for the TMVA. Two different samples were used to train the TMVA:

Unbinned random sample: The random gun sample was produced over all regions of the central crystal.

Binned random sample: The central crystal was divided into 7×7 bins. Random samples were produced in each bin separately.

For both the cases, the TMVA was tested using 20000 events from random gun samples of 200 GeV photons and π^0's produced over all regions of the central crystal. Figure 6 shows background rejection versus signal efficiency plots for binned as well as unbinned random gun sample.

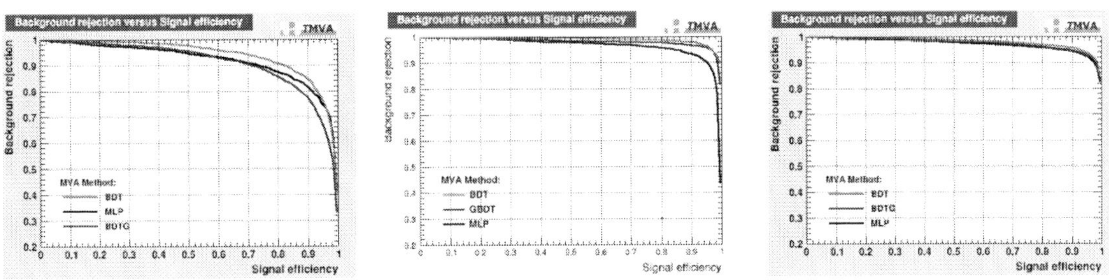

Figure 6. Background rejection versus signal efficiency for binned and unbinned random samples. The left figure is for unbinned random sample, the middle figure is for a bin where the sample is randomized over a region from −1 mm to +1 mm around centre of the central crystal in both X and Y direction and the bottom right figure is for a region from −1 mm to +1 mm in the X direction and +3 mm to +5 mm in the Y direction.

3.4. Comparison of various methods

A comparison in performance is made among all the methods described in the previous sections. This comparison is shown for 200 GeV photons and π^0's. If a method shows good separation

Variable	Background rejection (%)		
	ϵ_{signal}=80%	ϵ_{signal}=85%	ϵ_{signal}=90%
S1/S9	33.2	27.7	21.2
S1/S4	32.7	26.8	20.1
MVA(BDTG): coarse grain (fixed gun sample)	86.3	82	76.5
MVA(BDTG): fine grain (fixed gun sample)	99.1	99	98.5
MVA(BDTG): fine grain (unbinned random gun sample)	86.3	83.3	78.9
MVA(BDTG): fine grain (binned random gun sample)	92.3	91	89.3

Table 1. Table showing the background rejection for signal efficiencies of 80%, 85% and 90% for various methods. This is shown for energy point of 200 GeV.

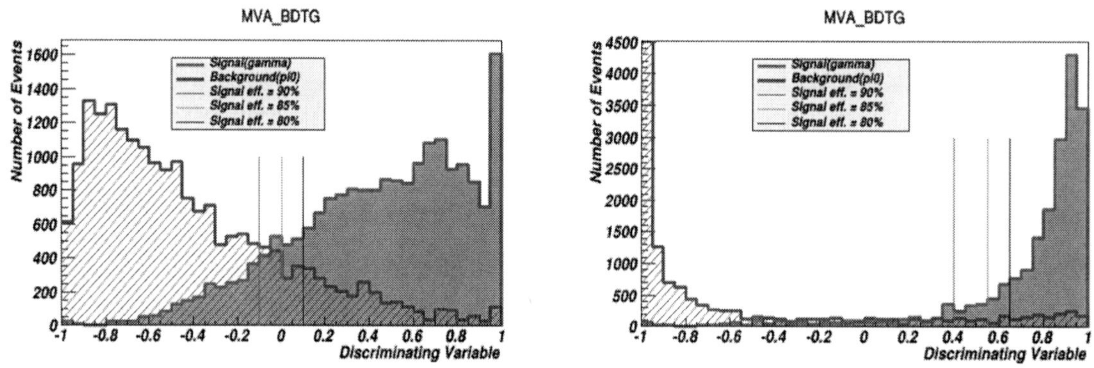

Figure 7. Comparison of BDTG MVA method for unbinned and binned random gun sample. The black, green and pink lines are the point where the cut is applied to achieve 80%, 85% and 90% signal efficiency respectively.

power for this high energy point, then it is good for lower energy points as well. BDT gives the best response for the case of both coarse grain and fine grain information as can be seen from figures 4 and 5. The response of BDTG is more or less same like BDT. Here the comparison is made using the BDTG. Table 1 shows the background rejection for all the various methods for a signal efficiencies of 80%, 85% and 90%.

4. Summary

A simulation study of $\pi^0 - \gamma$ separation of the Shashlik detector, proposed as a candidate of CMS endcap electromagnetic calorimeter in the Phase II upgrade, has been presented in this paper. This study shows that the fine grain information of the shower profile collected by individual fibers is useful for separation between π^0 and γ at high energies. With the MVA technique a background rejection efficiency of 90% with signal efficiancy 90% is achieved, which is approximately three times better than the best background rejection that could be achieved by cut-based methods. We conclude that the $\pi^0 - \gamma$ separation power of the Shashlik calorimeter can be improved significantly by emplyoying an MVA based method with fine grain information as input and this approach should work with other types of calorimeters.

5. References

[1] Technical Proposal for the Phase-II Upgrade of the CMS Detector 2015 *CERNLHCC* **010**, *LHCC-P* **008**.
[2] S. Agostinelli et al., GEANT4: A Simulation toolkit 2003 *Nucl. Instrum. Meth. A* **506** 250
[3] Y. Koba et al., Scintillation Efficiency of Inorganic Scintillators for Intermediate Energy Charged Particles 2011 *Progress in Nuclear Science and Technology* **1** 218
[4] F. X. Gentit, SLitrani *http://gentitfx.fr/SLitrani/*
[5] ROOT *http://root.cern.ch/drupal/*
[6] A. Hoecker et al. TMVA Toolkit for Multivariate Data Analysis with ROOT 2013 *http://tmva.sourceforge.net/*

Track distortion in the Large prototype of a Time Projection Chamber for the International Linear Collider

Deb Sankar Bhattacharya[1,3,4], Purba Bhattacharya[2], Supratik Mukhopadhyay[1], Nayana Majumdar[1], Sudeb Bhattacharya[5], Sandip Sarkar[1], Paul Colas[3], David Attié[3], Serguei Ganjour[3] and Aparajita Bhattacharya[4] .

(1) Applied Nuclear Physics Division, Saha Institute Nuclear Physics , Kolkata - 700064 , West Bengal, India. (2) School of Physical Sciences, National Institute Science Education and Research, Jatni, Khurda, Odisha - 752050, India. (3) DSM/IRFU/SPP, CEA Saclay, 91191 Gif Sur Yvette, France. (4) Department of Physics, Jadavpur University, Jadavpur, Kolkata 700032, West Bengal, India. (5) Retired Senior Professor, Applied Nuclear Physics Division, Saha Institute of Nuclear Physics, Kolkata - 700064, India

E-mail: dsb.physics@gmail.com

Abstract. A Micromegas (MM) based Time Projection Chamber (TPC) can meet the ILC requirements of continuous 3-D tracking, excellent spatial resolution and efficient pattern recognition. Seven MM modules which are commissioned on the end-plate of a Large Prototype TPC (LPTPC) at DESY, have been tested with a 5 GeV electron beam, under a 1 T magnetic field. Due to the grounded peripheral frame of the MM modules, at short drift, the electric field near the detector edge remain no longer parallel to the TPC axis. This causes signal loss along the boundaries of the MM modules as well as distortion in the reconstructed track. In presence of magnetic field, the distorted electric field introduces E×B effect. A detailed numerical study has been accomplished to understand the features of this distortion. Four Micromegas modules have been simulated resembling the experimental setup. The field lines, drift line of electrons considering diffusion in gas, nature of track distortion, residuals are numerically calculated in presence and in absence of magnetic field. The E×B effect has been simulated as well. Simulated results follow the experimental observations.

1. Introduction

The International Linear Collider (ILC) [1] is proposed to be a discovery and precision measurement machine where e^- and e^+ will collide initially up to 500 GeV at the electron center of mass frame. It provides opportunities to measure Higgs couplings at different energies and Higgs self-coupling at the upgraded 1 TeV energy scale. The International Large Detector [2] is one of the two detector concepts at the ILC. A Time Projection Chamber [3] has been foreseen to be installed just beyond the vertex detectors to accomplish continuous 3-D tracking for the ILD. Within a high luminosity collider, the TPC is supposed to register events at a very high rate and the physics measurements demand the position resolution of the detectors around 100 μm. Micro-pattern Gaseous Detectors (MPGD) [4] at the TPC anode plane can meet such challenges. A Large Prototype of the Time Projection Chamber (LPTPC) has been installed

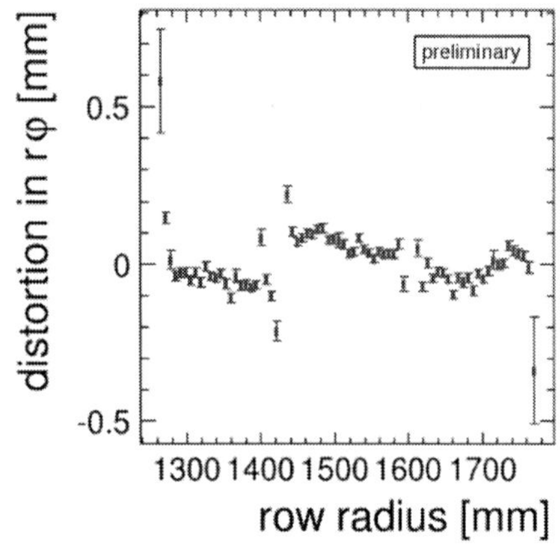

Figure 1: After the alignment correction, the observed distortion at B = 0 T [6].

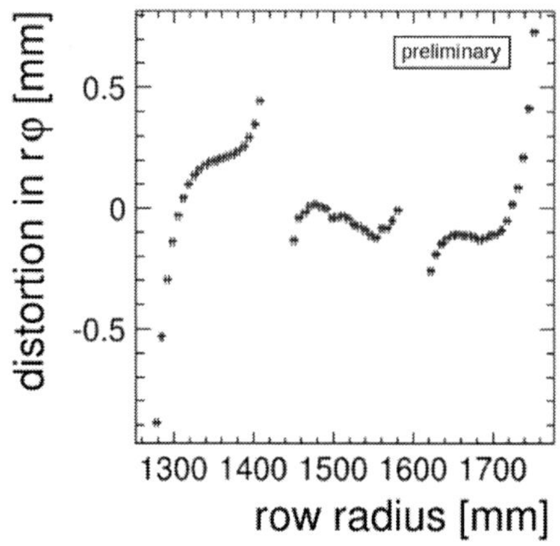

Figure 2: Observed distortion at B = 1 T. The E×B effect can be seen near the edges [6].

[5] at DESY since 2008 and different MPGDs have been tested on the LPTPC end-plate [6]. Upto seven resistive Micromegas (MM) [7] modules have been studied at the LPTPC end-plate from 2009 through 2015 and they have shown promising performance [6, 8] as required for the ILD-TPC. The keystone-shaped modules have identical size of 17 cm × 22 cm so as to fit in the end-plate. The amplification gap and the micro-mesh wire pitch have been chosen to be 128 μm and 63 μm respectively. Different types of resistive layers with surface resistance around 3 - 5 MΩ/sqr have been tested [9] at the LPTPC with an electron beam of energy ranging from 1 - 6 GeV, under a magnetic field of 1 T. The anode read out of a Micromegas module is segmented in 1726 pads of size 3 mm × 7 mm and arranged in 24 rows. The gas composition of the LPTPC is an admixture of Ar:CF_4:Isobutane (95:3:2).

In case of Micromegas, a supporting thin frame that holds the detector along its boundaries to the metal structure of the module, lies in ground potential. During analysis, it is found that the residual of the pad hits (explained in section 3.3) on the extreme rows of the MM modules have larger magnitude with respect to the other rows. Figure 1 shows the row-wise distribution of the residuals (an alignment correction is performed) when the magnetic field is not applied and in figure 2 the row-wise distribution of the residuals are plotted when a magnetic field of 1 T is applied. Near the boundaries of the modules, in case of applied magnetic field, the residuals take a 'S'-like shape [6].

In the present report, an attempt is taken to investigate the track distortion for the Micromegas modules by means of numerical calculations. In a previous report [10], we have addressed the simulation of three MM modules to explain the observed track distortion as an effect of distorted electric field (E×B effect) near the edge of the detectors. In this report, we have further broadened the scope by investigating the effect on four MM modules which represents the experimental set-up [6] more closely.

2. Numerical Modeling

The work has been accomplished within the Garfield [11] simulation frame work. The field solver used for this calculation is the *nearly exact Boundary Elementary Method* (neBEM) [12]. HEED [13] is used to calculate primary ionization and Magboltz [14] is used for computing drift, diffusion, Townsend and attachment coefficient. The Garfield framework provides suitable interfaces to the software components neBEM, HEED and Magboltz among others.

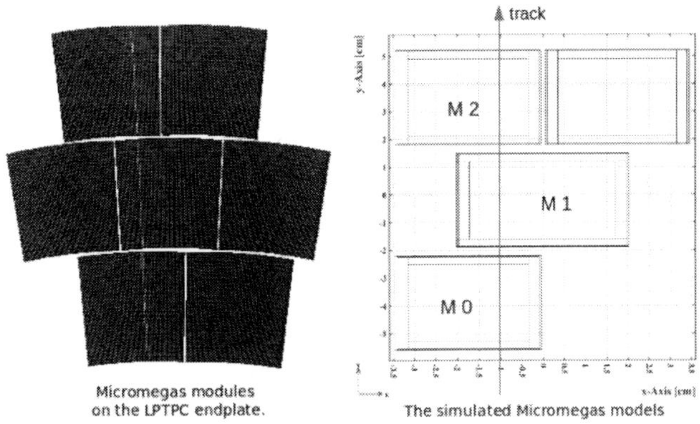

Figure 3: The simulated Micromegas models are arranged in a manner similar to the MM modules on the LPTPC end-plate.

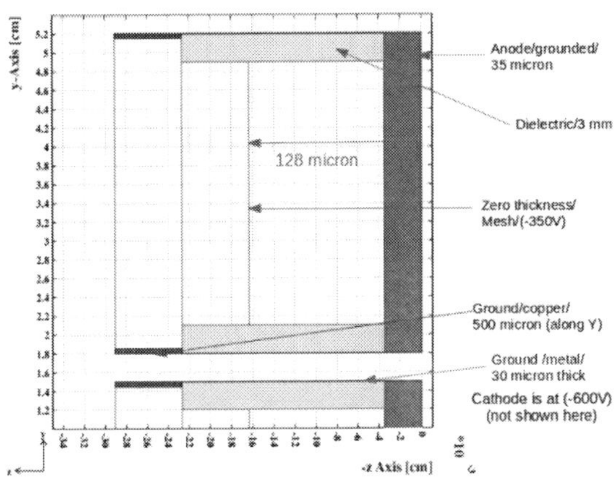

Figure 4: Detailed geometry of the simulated Micromegas models (a section in the Y-Z plane).

2.1. Geometry

The model geometry used for this calculation is simpler than that of real Micromegas on the LPTPC end-plate mainly to reduce computational complexity. Four standard Micromegas are taken instead of the seven resistive Micromegas. The shapes of the models are 3.4 cm × 3.4 cm squares instead of the 17 cm × 22 cm keystone-shaped Micromegas. However, few critical parameters are taken unchanged. The amplification gap is taken 128 μm. The photoresist support for the mesh is taken to be 3 mm thick. The vertical copper frame, which is grounded, is taken to be 30 μm thick (figure 3). The gaps between two Micromegas models are taken as 3 mm. Zero thickness model of the micro-mesh has been used. The anode lies in ground potential. The micro-mesh and the cathode (drift plane) are biased to -350 V and -600 V respectively. The cathode rests 1 cm above the micro-mesh. A challenging part of this computation is to deal with such geometry that is extended to a few centimeters in one axis and only few micro meter in the perpendicular direction.

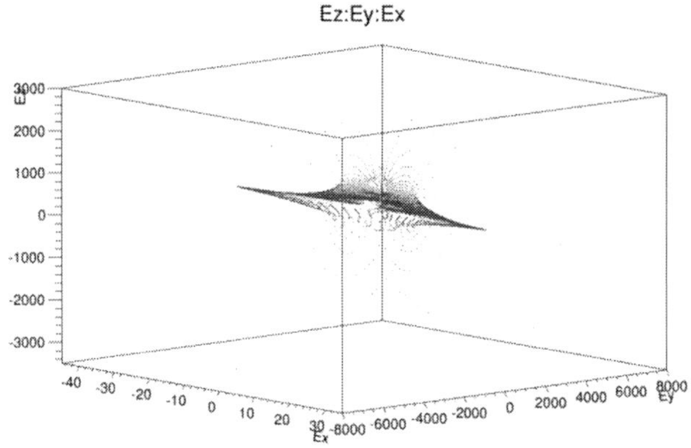

Figure 5: E_X,E_Y and E_Z (in V/cm) on the gap between the models are plotted. The E_X and E_Y components has significant magnitude that cause distortion.

3. Results

3.1. Electric Field

The electric field between the cathode and the micro-mesh near the middle of the detector is perpendicular to the anode plane, i.e., parallel to the TPC axis. Because of the grounded frame and the dielectric support on the boundaries of the detector, the electric field at the vicinity of the detector edge does not remain parallel to the TPC axis. There exist transverse components of the field (in X and Y direction) near the gap between the modules. Because of these transverse components of the field, the electrons near the edge of the detector follow a transverse path and mostly hit the dielectric. This causes signal loss for the pads on the edge. This also leads to increase of the residual for the hits near the edges. In figure 4 the X, Y and Z components of electric fields on a plane perpendicular to the anode at X = -1 cm are plotted. A large value of the Y component of the field can be seen.

3.2. Drift of electrons

A track is defined out of the electrons on 456 equidistant points along X = -1 cm (figure 2) and Z = 0.5 cm (the drift distance). The same gas as used in the LPTPC is considered for the gas transport calculations. As the field solver has computed the field values for the entire model, the track drifts towards the anode accordingly. 100 such tracks have been considered from the same position to generate statistics. The drift lines and their end points (figure 5) clearly show that due to the distortion in the electric field, a significant number of electrons do not hit the anode pad near the edge and are lost on the dielectric. Magnetic field is applied to reduce the transverse diffusion as well.

3.3. Distribution of residuals

The residual of a pad hit is defined [15] as, $\Delta X = X_{hit} - X_{track}$, where, X_{hit} is the measured position of the hit and X_{track} is the position of the hit according to the track fitting. In present calculation, the track has the true coordinates and that is why reconstruction and track fitting is not necessary. The coordinates of the individual electrons that finally hit the anode are obtained from the calculation. From the initial positions of the electrons on the track (X_{track}) and from the final positions (X_{hit}), the residuals of individual hits are obtained. The residuals are then averaged over the number of events (tracks). When plotted, the distribution of the residuals along the direction of the track length shows (figure 6) the effect of electric field distortion near the edges of the detectors. The magnitude of the residuals are high near the module gaps.

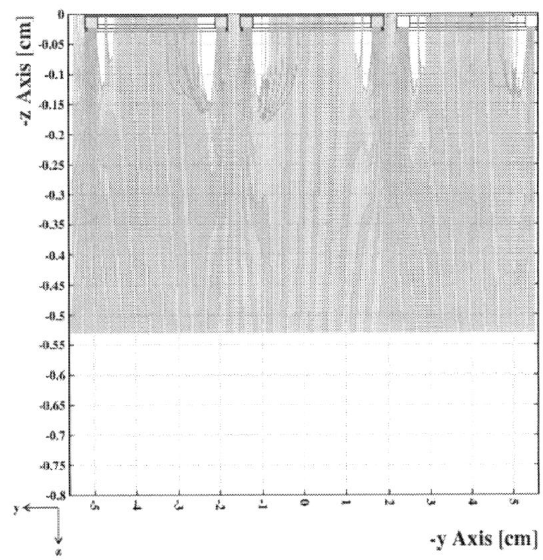

Figure 6: Drift of the electrons is shown at B=1T. A small area adjacent to the dielectric is devoid of hits while significant signal loss can be seen on the dielectric.

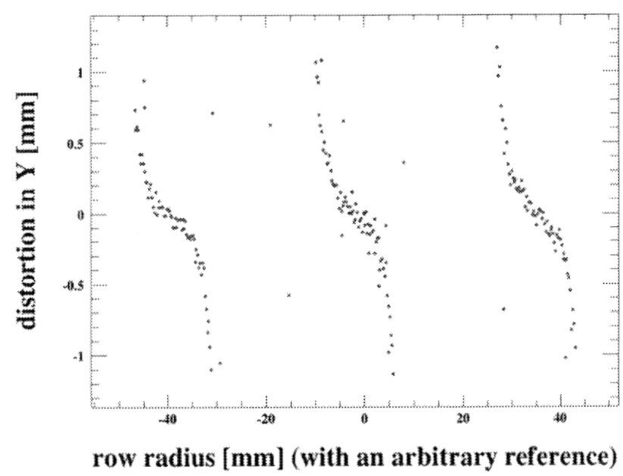

Figure 7: Simulated distortion at B=0T. The residuals are plotted against the row radius (along the direction of the track). Drift, z = 0.5 cm.

On application of a magnetic field of 1 T, the signs of the residuals change (figure 7). The reason of such change is because of the Lorentz force that comes to play once the magnetic field is applied. Near the gap, the magnetic field together with the transverse components of the electric field introduce the Lorentz force. Therefore, the electrons, when the magnetic field is applied, are moved in a different direction by the Lorentz force. And hence the change in sign occurs.

4. Conclusion

The entire simulation is done with a number of simplifications in the geometry of the detectors. Hence, a direct comparison with the experimental results is possibly not applicable. However, the intrinsic parameters, like amplification gap, thickness of the dielectric and the ground frame, the inter-modular distances and the TPC gas are maintained true to the experimental values in

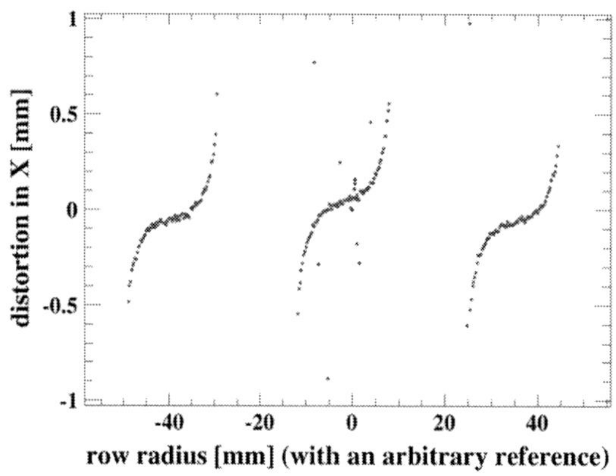

Figure 8: Simulated distortion at B=1T. The residuals are plotted against the row radius (along the direction of the track). Drift, z = 0.5 cm. The sign of the residual changes with respect to the case of B=0T.

this calculation. The miniaturization of the detector modules in X-Y dimensions are made only to avoid computational delay and complexities. The nature and the magnitude of the distortion closely match the results as seen in the Micromegas based LPTPC. The prime motivation of the study have been to understand the distortion and to explore possibilities for improvements. Further study will be continued in more details.

5. Acknowledgment

We are thankful to the members of LCTPC collaboration and RD51 collaboration for their helpful suggestions. We sincerely acknowledge IFCPAR/CEFIPRA (Project No. 4304-1) for partial financial support. We would like to thank the respective institutes for their supports.

References

[1] T.Behnke, (ed.) et al., *International Linear Collider Technical Design Report: Executive Summary* **1** (2013).
[2] T.Behnke, (ed.) et al., *International Linear Collider Technical Design Report: Detectors* **4** (2013).
[3] H.J.Hilke, *Time projection chamber, Reports on Progress in Physics* **73** (2010) 36pp.
[4] A. Oed, *Micro pattern structures for gas detectors, Nucl. Instr. and Meth. A* **471** (2001) 109-114.
[5] Thesis of Peter Schade : 'Development and Construction of a Large TPC Prototype for the ILC and Study of τ Polarisation in $\tilde{\tau}$ Decays with the ILD Detector', *UniversitätHamburg*, 2009.
[6] D.S. Bhattacharya et al, *Test of MPGD modules with a large Time Projection Chamber, Proceedings of The European Physical Society Conference on High Energy Physics* (2015).
[7] Y. Giomataris et al, *MICROMEGAS: a high-granularity position-sensitive gaseous detector for high particle-flux environments, Nucl. Instrum. Methods A.* **29** (1996) 376.
[8] Thesis of Wenxin Wang: 'A Large Area Micromegas TPC for Tracking at the ILC', *University Paris – Sud*, 2013.
[9] D. Attie, *Beam tests of Micromegas LC-TPC large prototype, JINST* **6** (2011) C01007.
[10] P. Bhattacharya et al, *Numerical Study of Electrostatic Field Distortion on LPTPC End-Plates based on Bulk Micromegas Modules, Proceedings of The 4th International Conference on Micro Pattern Gaseous Detectors* (2015).
[11] http://cern.ch/garfield.
[12] http://cern.ch/neBEM.
[13] http://cern.ch/heed.
[14] http://cern.ch/magboltz.
[15] D.C. Arogancia et al, *Study in a beam test of the resolution of a Micromegas TPC with standard readout pads, Nucl. Instrum. Methods A.* **602** (2009) 403-414.

Journal of Physics: Conference Series **759** (2016) 012076 doi:10.1088/1742-6596/759/1/012076

Excited State Mass Spectra of Ω_c^0 Baryon

Z Shah[1], K Thakkar[2], A K Rai[1] and P C Vinodkumar[3]

[1]Department of Applied Physics, SVNIT, Surat, Gujarat-395007, India
[2]Department of Applied Sciences & Humanities, GIDC Degree Engineering College, Abrama, Navsari, Gujarat-396406, India
[3]Department of Physics, Sardar Patel University, V.V. Nagar, Anand, Gujarat-388120, India

E-mail: zalak.physics@gmail.com

Abstract. We have calculated the radial and orbital excited states of singly charmed baryon Ω_c using the Hypercentral Constituent Quark Model (hCQM). The confinement potential is assumed as coulomb plus power potential (CPP$_\nu$). The ground state and excited state masses are determined with and with out first order correction to the potential. Furthermore, we plot graph between Mass(M) $\rightarrow$ Potential Index(ν). Our calculated results are in good agreement with experimental and other theoretical predictions.

1. Introduction

Baryon spectroscopy is broadly studied by various theoretical approaches and many experimental groups [1]. We know that baryons are strongly interacting fermions and made up of three quarks. Baryons consisting with u and d quarks are called nucleons ($I_3 = \frac{1}{2}$) or $\Delta(I_3 = \frac{3}{2})$ resonances. Combination of u, d and s quark are called hyperons. The singly charmed baryon is composed of a charm quark and two light quarks belong to SU(4) multiplets. SU(4) group includes all of the baryons containing zero, one, two or three charmed quarks. The multiplet numerology of the tensor product of three fundamental rendition is: $4\times4\times4 = 20+20_1^{'}+20_2^{'}+\bar{4}$. Representation shows totally symmetric 20-plet, the mixed symmetric $20^{'}$-plet and the total anti symmetric $\bar{4}$ multiplet [2].

One of the single charmed baryon, Ω_c^0 (ssc) belongs to a sextet of flavor symmetric states. It is a combination of two light quark (strange quark) and heavy quark (charm quark). The WA62 collaboration reported the first ground state of Ω_c^0 (Mass = 2740 $\pm$ 20 MeV) [3]. Since then, resonances observed simultaneously by a large variety of experiments like Belle, BaBar, CLEO etc.[1, 4, 5, 6]. Ω_c^0 belongs to a sextet but decays weakly [7]. PDG-2014 listed experimental results of ground states $\Omega_c^0(\frac{1}{2}^+)$ and $\Omega_c^0(\frac{3}{2}^+)$ but excited states measurements are still not reported.

In the past decades, various phenomenological Constituent quark models have been used and all models reproduce the baryon spectrum. The non-relativistic as well as relativistic approach like Isgur-Karl model [8], the Capstick-Isgur model [9], the interacting quark-diquark model [2], the Hypercentral Model [10],etc. had been used to calculate the ground state mass spectroscopy of charmed baryons. D. ebert et al. [11] calculated radial and orbital excited states by relativistic quark-diquark model. Moreover, lattice QCD results show the excited state mass spectroscopy of singly as well as doubly and triply charmed baryons [12, 13, 14].

In this paper, we calculate the radial and orbital excited states of Ω_c^0 baryon in Hypercentral Constituent Quark Model (hCQM). This model has been applied for the description of ground state properties of the various baryons [15, 16, 17, 18, 19, 20, 21] previously. The paper organized as follows: after the brief introduction, we give the details of model. Finally we tabulate the results and discuss the outcome of the model.

2. The Model

Baryon is made up of three quarks. The Jacobi Co-ordinates ($\vec{\rho}$ and $\vec{\lambda}$) are used to describe baryon as a bound state of three constituent quarks [15, 16, 17, 18, 19, 20, 21],

$$\vec{\rho} = \frac{1}{\sqrt{2}}(\vec{r_1} - \vec{r_2}) \qquad \vec{\lambda} = \frac{1}{\sqrt{6}}(\vec{r_1} + \vec{r_2} - 2\vec{r_3}) \tag{1}$$

such that

$$m_\rho = \frac{2m_1 m_2}{m_1 + m_2} \qquad m_\lambda = \frac{3m_3(m_1 + m_2)}{2(m_1 + m_2 + m_3)} \tag{2}$$

Where $m_1 = 0.500\ GeV$, $m_2 = 0.500\ GeV$ and $m_3 = 1.275\ GeV$ are three different constituent quark masses.

The angle of the Hyper spherical coordinates are given by $\Omega_\rho = (\theta_\rho, \phi_\rho)$ and $\Omega_\lambda = (\theta_\lambda, \phi_\lambda)$. We define hyper radius x and hyper angle ξ by,

$$x = \sqrt{\rho^2 + \lambda^2} \qquad \xi = arctan\left(\frac{\rho}{\lambda}\right) \tag{3}$$

The hypercentral potential contains three-body effects. The Hamiltonian of three body baryonic system for the hCQM is,

$$H = \frac{P_\rho^2}{2m_\rho} + \frac{P_\lambda^2}{2m_\lambda} + V(\rho, \lambda) = \frac{P_x^2}{2m} + V(x) \tag{4}$$

Where, $m = \frac{2m_\rho m_\lambda}{m_\rho + m_\lambda}$, is reduced mass. The hyperradial Schrodinger equation corresponds to the Hamiltonian can be written as,

$$\left[\frac{d^2}{dx^2} + \frac{5}{x}\frac{d}{dx} - \frac{\gamma(\gamma+4)}{x^2}\right]\Psi_\gamma(x) = -2m[E - V(x)]\Psi_\gamma(x) \tag{5}$$

where Ψ_γ(x) is the hypercentral wave function and γ is the grand angular quantum number. We consider transformation, $\phi_\gamma(x) = x^{\frac{5}{2}}\Psi_\gamma(x)$. Thus, the hyperradial Schrodinger equation reduces to,

$$\left[\frac{-1}{2m}\frac{d^2}{dx^2} + \frac{\frac{15}{4} + \gamma(\gamma+4)}{2mx^2} + V(x)\right]\phi_\gamma(x) = E\phi_\gamma(x) \tag{6}$$

For the present study we consider the hypercentral potential V(x) as the color coulomb plus power potential with relativistic correction [22, 23, 24] written as,

$$V(x) = V^0(x) + \left(\frac{1}{m_\rho} + \frac{1}{m_\lambda}\right)V^{(1)}(x) + V_{spin} \tag{7}$$

$$V^{(0)}(x) = \frac{\tau}{x} + \beta x^{\nu} \tag{8}$$

$$V^{(1)}(x) = -C_F C_A \frac{\alpha_s^2}{4x^2} \tag{9}$$

Table 1. Masses of radial excited states of Ω_c^0 (in GeV).

(ν)	A	B	Others	A	B	Others
	$1^2S_{\frac{1}{2}}$			$1^4S_{\frac{3}{2}}$		
0.5	2.695	2.695	2.695 ±0.017 [1]	2.748	2.754	2.766 ± 0.0020 [1]
0.7	2.695	2.695	2.718[2]	2.742	2.752	2.776[2]
0.9	2.695	2.695	2.698[11]	2.748	2.750	2.768[11]
1.0	2.695	2.695	2.681[14]	2.740	2.745	2.764[14]
2.0	2.695	2.695	2.699[26]	2.751	2.757	2.767[26]
	$2^2S_{\frac{1}{2}}$			$2^4S_{\frac{3}{2}}$		
0.5	3.075	3.089		3.113	3.128	
0.7	3.104	3.123	3.088[11]	3.137	3.160	3.123[11]
0.9	3.140	3.154		3.176	3.190	
1.0	3.147	3.164		3.178	3.197	
2.0	3.279	3.306		3.320	3.349	
	$3^2S_{\frac{1}{2}}$			$3^4S_{\frac{3}{2}}$		
0.5	3.349	3.371		3.371	3.394	
0.7	3.421	3.454	3.489[11]	3.441	3.476	3.510[11]
0.9	3.505	3.531		3.527	3.552	
1.0	3.529	3.561		3.548	3.580	
2.0	3.866	3.921		3.892	3.948	
	$4^2S_{\frac{1}{2}}$			$4^4S_{\frac{3}{2}}$		
0.5	3.602	3.633		3.617	3.647	
0.7	3.724	3.769	3.814[11]	3.737	3.783	3.83[11]
0.9	3.863	3.900		3.877	3.914	
1.0	3.907	3.953		3.920	3.966	
2.0	4.491	4.576		4.510	4.595	
	$5^2S_{\frac{1}{2}}$			$5^4S_{\frac{3}{2}}$		
0.5	3.843	3.880		3.853	3.890	
0.7	4.017	4.074	4.102[11]	4.026	4.084	4.114[11]
0.9	4.215	4.263		4.225	4.273	
1.0	4.283	4.343		4.292	4.352	
2.0	5.150	5.267		5.164	5.281	

A= without first order correction. B= with first order correction.

A hyper-Coulomb, $\tau = -\frac{2}{3}\ \alpha_s$ related to strong coupling constant; $\frac{2}{3}$ is the color factor for baryon, β corresponds to the string tension of the confinement. Potential index, ν varies from

0.5 to 2.0. $C_F = \frac{4}{3}$ and C_A =3 are the Casimir charges of the fundamental and adjoint representation. Coupling constant α_s is given by,

$$\alpha_s = \frac{\alpha_s(\mu_0)}{1 + \frac{33-2n_f}{2\pi}\alpha_s(\mu_0)ln\left(\frac{\mu}{\mu_0}\right)} \tag{10}$$

The spin-dependent part of the three-body interaction potential of eq. (7) is taken as

$$V_{spin}(x) = -\frac{1}{4}A\alpha_s\frac{e^{-x/x_0}}{xx_0^2}\sum_{i>j}\frac{\vec{\sigma_i}\vec{\sigma_j}}{6m_im_j}\vec{\lambda_i}\vec{\lambda_j} \tag{11}$$

where x_0 is the hyperfine parameter of the model. The spin-dependent part, $V_{SD}(x)$ contains three types of the interaction terms, such as the spin-spin term $V_{SS}(x)$, the spin-orbit term $V_{\Gamma S}(x)$ and tensor term $V_T(x)$ given by [25],

$$V_{SD}(x) = V_{SS}(x)\left[S(S+1) - s_\rho(s_\rho+1) - \frac{3}{4}\right] + V_{\Gamma S}(x)(\vec{\Gamma}\cdot\vec{S}) + V_T(x)\left[S(S+1) - \frac{3(\vec{S}\cdot\vec{x})(\vec{S}\cdot\vec{x})}{x^2}\right] \tag{12}$$

The Spin-orbit and the tensor term describe the fine structure of the states, while the spin-spin term gives the spin singlet triplet splittings. The coefficient of these spin-dependent terms of (12) can be written in terms of the vector, $V_V(x)$ and scalar, $V_S(x)$ parts of the static potential as

$$V_{\Gamma S}(x) = \frac{1}{2m_\rho m_\lambda x}\left(3\frac{dV_V}{dx} - \frac{dV_S}{dx}\right) \tag{13}$$

$$V_T(x) = \frac{1}{6m_\rho m_\lambda}\left(3\frac{d^2V_V}{dx^2} - \frac{1}{x}\frac{dV_V}{dx}\right) \tag{14}$$

$$V_{SS}(x) = \frac{1}{3m_\rho m_\lambda}\nabla^2 V_V \tag{15}$$

The baryon spin average mass in this hypercentral model is given by

$$M_B = \sum_{i=1} m_i + BE + \langle V_{SD}(x)\rangle \tag{16}$$

We take $A = A_0/(\omega+\gamma+\frac{3}{2})^2$, where A_0 is arbitary constant. We fix potential parameter β and hyperfine parameter τ for each choices of ν using ground state experimental masses of $J^P = \frac{1}{2}^+$ and $J^P = \frac{3}{2}^+$ single charmed flavor baryons (Ω_c^0) . For radial excited states, potential parameter β is found to vary as $\beta = \beta_0\sqrt{\omega+\gamma+\frac{3}{2}}$ and for orbital excited states, potential parameter $\beta = \beta_0\sqrt{\omega+\gamma+1}$. where β_0 correspond to ground state.

Table 2. Masses of Orbital excited states of Ω_c^0 (in GeV).

Ω_c^0	0.5		0.7		0.9		1.0		2.0			
	A	B	A	B	A	B	A	B	A	B	[11]	[2]
$(1^2P_{1/2})$	2.985	3.000	2.997	3.019	3.022	3.037	3.022	3.041	3.164	3.221	3.055	2.977
$(1^2P_{3/2})$	2.982	2.996	2.991	3.012	3.010	3.024	3.007	3.024	3.067	3.110	3.054	2.986
$(1^4P_{1/2})$	2.987	3.001	3.000	3.023	3.028	3.043	3.030	3.050	3.212	3.277	2.966	2.977
$(1^4P_{3/2})$	2.984	2.998	2.994	3.015	3.016	3.030	3.014	3.033	3.115	3.165	3.029	2.959
$(1^4P_{5/2})$	2.979	2.993	2.986	3.006	3.000	3.013	2.994	3.010	2.986	3.017	3.051	3.014
$(2^2P_{1/2})$	3.254	3.277	3.308	3.342	3.377	3.402	3.394	3.427	3.740	3.839	3.435	
$(2^2P_{3/2})$	3.251	3.273	3.301	3.334	3.364	3.389	3.377	3.408	3.622	3.703	3.433	
$(1^2P_{1/2})$	3.256	3.278	3.312	3.346	3.383	3.408	3.403	3.436	3.800	3.907	3.384	
$(2^4P_{3/2})$	3.253	3.275	3.305	3.338	3.370	3.396	3.386	3.417	3.681	3.771	3.415	
$(2^4P_{5/2})$	3.248	3.270	3.295	3.328	3.353	3.378	3.363	3.393	3.523	3.589	3.427	

A= without first order correction. B= with first order correction.

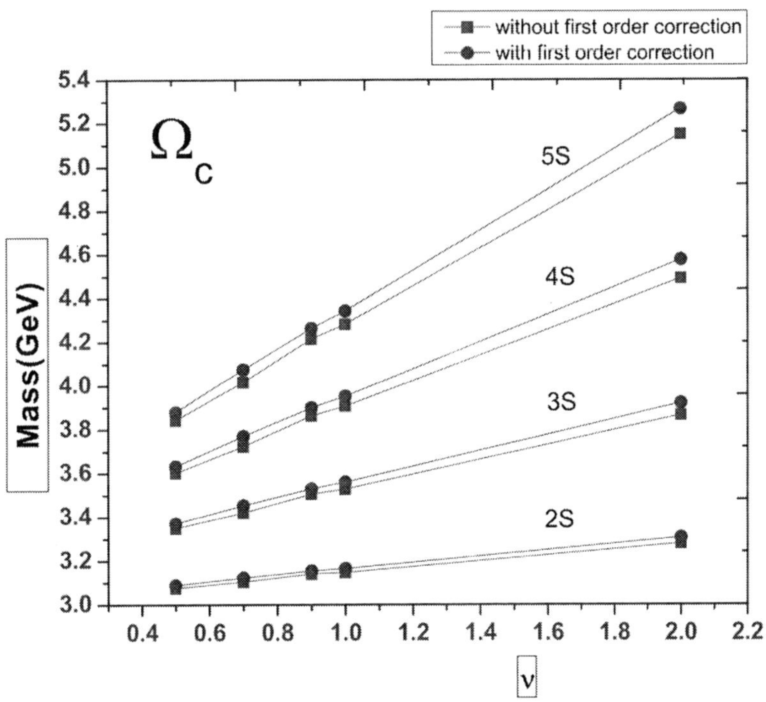

Figure 1. Variation of mass with potential index in S states

3. Result and Discussion

Using Hypercentral Constituent Quark Model(hCQM), we have calculated the mass spectra of the single charmed baryon Ω_c^0. We calculate the radial excited states upto 5S states and orbital excited states upto 2P states. Here, we have incorporated the first order correction in the hypercentral power potential, for potential index 0.5 to 2. Our results are close to the D. Ebert et al. (quark-diquark model) [11] results at the potential index 0.7 for S states, but for P-states the results are close at potential index 1.0 (see Tables (1-2)). The result obtained using the first order corrections are higher (few MeV) than without correction. As we are moving from lower excited states to higher excited states these correction effects increasing.

We have shown the variation of mass with potential index in Fig-1. We have plotted the masses of S states without correction to the potential and with first order correction to potential ($J^P = \frac{1}{2}^+$ only). One can see the result obtained from with correction lies over the without correction. We have observed same trend in the $J^P = \frac{3}{2}^+$. We will apply our model to calculate doubly as well as triply charmed Ω's in future.

Acknowledgments
One of the author, Zalak Shah is very much thankful to Dr. Nilamani Mathur for giving his valuable suggestions during the poster session in CCP-2015. A. K. Rai acknowledges the financial support extended by DST, India under SERB fast track scheme SR/FTP /PS-152/2012 and also to SVNIT (Inst. Res. Grant (Dean (*R&C*)/1488/2013-14)).

References

[1] Olive K A 2014 *Chin. Phys.C* **38** 090001
[2] Roberts W and Pervin M 2008 *Int. J. Mod. Phys. A* **23** 2817
[3] Biagi S F et al. 1985 *Z. Phys. C* **28** 175
[4] Aubert B et al. [BABAR Collaboration] 2007 *Phys. Rev. Lett.* **99** 062001
[5] Solovieva E et al. [Belle Collaboration] 2009 *Phys. Lett. B* **672** 1
[6] Cronin-Hennessy D et al. [CLEO Collaboration] 2001 *Phys. Rev. Lett.* **86** 3730
[7] Hai-Yang Cheng 2015 *arXiv:1508.07233v2*
[8] Isgur N and Karl G, 1978 *Phys. Rev. D* **18** 4187; 1979 *Phys. Rev. D* **19** 2653
[9] Capstick S and Isgur N 1986 *Phys. Rev. D* **34** 2809
[10] Giannini M M and Santopinto E, 2015 *Chin. J.Phys.* **53** 020301
[11] Ebert D, Faustov R N, Galkin V O 2011 *Phys. Rev. D* **84**, 014025
[12] Padmanath M and Mathur N 2015 *arXiv:1508.07168v1 [hep-lat]*
[13] Padmanath M, Mathur N Edwards, R and Peardon M 2014 *arXiv:1410.8791v1 [hep-lat]*
[14] Briceno R A, Lin H W and Bolton D R 2012 *Phys. Rev. D* **86** 094504
[15] Patel B, Rai A K and Vinodkumar P C 2008 *Pramana J. Phys.* **70** 797; 2008 *J.Phys.G* **35** 065001
[16] Thakkar K, Majethiya A and Vinodkumar P C 2011 *Chin. Phys. C* **36** 385
[17] Thakkar K, Patel B, Majethiya A and Vinodkumar P C *Pramana J. Phys.* **77**, 1053 (2011)
[18] Thakkar K and Vinodkumar P C 2013 *AIP Conf. Proc.* **1536** 1097
[19] Majethiya A, Vinodkumar P C 2009 *Eur. Phys. J. A***42** 213
[20] Shah Z, Thakkar K, Rai A K, Vinodkumar P C 2015 *Proceedings of the DAE-BRNS Symp. on Nucl. Phys.* **60** 688
[21] Shah Z, Thakkar K, Rai A K, Vinodkumar P C 2016 *AIP Conf. Proc.*(Accepted)
[22] Koma Y, Koma M and Wittig H 2006 *Phys. Rev. Lett.* **97** 122003
[23] Devlani N, Kher V and Rai A K 2014 *Eur. Phys. J. A* **50** 154
[24] Rai A K and Rathaud D 2015 *Eur. Phys. J. C* **75** 462
[25] Voloshin M B 2008 *Prog. Part. Nucl. Phys.***51** 455
[26] Valcarce A, Garcilazo H and Vijande J 2008 *Eur. Phys. J. A* **37** 217

Implication of θ_{13} on Majorana Neutrino mass matrices with One Texture Zero and One Vanishing Eigenvalue

Madan Singh[a*], **Radha Raman Gautam**[b], **Manmohan Gupta**[a]

[a] Department of physics, Centre of Advanced Study, P.U., Chandigarh, India.
[b] Department of physics, Himachal Pradesh University, Shimla 171005, India.

**email id: singhmadan179@gmail.com*

Abstract. The latest neutrino mixing data has established the relatively large value of reactor mixing angle, $\theta_{13} \sim 9^0$. In view of that and assuming Majorana nature of neutrinos, we have re-examined the texture one–zero neutrino mass matrices with one vanishing neutrino mass in the flavor basis. We observe that for $m_1 = 0$ (Normal mass ordering), all the six one zero classes are now phenomenologically ruled out and for $m_3 = 0$ (Inverted mass ordering), only four classes out of total six are viable with the latest neutrino oscillation data at 3σ confidence level.

1. Introduction

In recent times, experiments like Double Chooz, Daya Bay and RENO Collaborations [1-3] have established a non-zero and relatively large value of the reactor mixing angle θ_{13}. The result has motivated the experimentalists to pin down the long-standing problem of CP violation and mass ordering in neutrino sector. During past one year, several interesting experimental features have been observed, which though at low CL (<2σCL), trigger optimism in lepton sector. For instance, according to recent global fit results of neutrino oscillation [4, 5], best fit value of Dirac CP-violating phase (δ) lies near 270^0.

In the absence of any convincing theory, several phenomenological schemes [6, 7, 8] have been adopted in the literature to reduce the number of free parameters in neutrino mass matrix, however texture zeros [6, 7, 8] have been the most successful among them. In flavor basis, the texture three zero was found to be incompatible with experimental data [6], whereas the analysis of two-zero textures as propounded by Frampton, Glashow, Marfatia (FGM) [6], restrict the number of experimentally compatible classes to seven. In case of texture one zero, all the six possible classes are experimentally viable [9]. The condition of one texture zero in the neutrino mass matrix is less restrictive (and hence less predictive) than the condition of two texture zeros. However, one can further reduce the number of free parameters of one texture zero Majorana neutrino mass matrices by assuming one of the neutrino masses to be zero, which is still an experimentally viable situation.

The Majorana neutrino mass matrix contains nine parameters which include three neutrino masses (m_1, m_2, m_3), three mixing angles $(\theta_{12}, \theta_{23}, \theta_{13})$ and three CP violating phases (δ, ρ, σ). The condition of one vanishing neutrino mass allows reducing the number of free parameters of neutrino mass matrix from nine to seven. In Ref. [9, 10], a particular attention has been paid on texture one zero with one vanishing neutrino mass. In the light of the precise measurement of reactor mixing angle θ_{13}[1-3], we have re-investigated the texture one zero with one vanishing neutrino mass.
Assuming Majorana nature of neutrinos, we consider neutrino mass matrix having one zero element with one vanishing lightest neutrino mass ($m_1 = 0$ or $m_3 = 0$). In the flavor basis (i.e. charged lepton mass matrix is diagonal), the Majorana mass matrix, being complex symmetric, contains six independent entries. If one of the elements is assumed to be zero, then we obtain six possible one- zero classes:

$$P_1 = \begin{pmatrix} 0 & \times & \times \\ \times & \times & \times \\ \times & \times & \times \end{pmatrix}, \quad P_2 = \begin{pmatrix} \times & \times & \times \\ \times & 0 & \times \\ \times & \times & \times \end{pmatrix}, \quad P_3 = \begin{pmatrix} \times & \times & \times \\ \times & \times & \times \\ \times & \times & 0 \end{pmatrix};$$
$$P_4 = \begin{pmatrix} \times & \times & \times \\ \times & \times & 0 \\ \times & 0 & \times \end{pmatrix}, \quad P_5 = \begin{pmatrix} \times & 0 & \times \\ 0 & \times & \times \\ \times & \times & \times \end{pmatrix}, \quad P_6 = \begin{pmatrix} \times & \times & 0 \\ \times & \times & \times \\ 0 & \times & \times \end{pmatrix}. \tag{1}$$

where ' $\times$' denotes non-zero element.
The rest of the paper is structured as follows: In section 2, we discuss the methodology used to reconstruct the Majorana mass matrix and hence to obtain the constraints on mixing angles $(\theta_{12}, \theta_{23}, \theta_{13})$ and CP violating phases (δ, ρ, σ), by imposing the condition of texture one zero and vanishing mass eigen value. In section 3, we present the numerical analysis. In section 4, we summarize our work.

2. Methodology

In the flavor basis, the Majorana neutrino mass matrix M_ν, depending on three neutrino masses (m_1, m_2, m_3) and the flavor mixing matrix U can be expressed as [6,7],

$$M_\nu = U \begin{pmatrix} \lambda_1 & 0 & 0 \\ 0 & \lambda_2 & 0 \\ 0 & 0 & \lambda_3 \end{pmatrix} U^T, \tag{2}$$

where $\lambda_1 = m_1 e^{2i\rho}$, $\lambda_2 = m_2 e^{2i\sigma}$, $\lambda_3 = m_3$ and ρ, σ are two Majorana phases. The flavor mixing matrix U can be parameterized [6,7] as

$$U = \begin{pmatrix} c_{12}c_{13} & s_{12}c_{13} & s_{13} \\ -c_{12}s_{23}s_{13} - s_{12}c_{23}e^{-i\delta} & -s_{12}s_{23}s_{13} + c_{12}c_{23}e^{-i\delta} & s_{23}c_{13} \\ -c_{12}c_{23}s_{13} + s_{12}s_{23}e^{-i\delta} & -s_{12}c_{23}s_{13} - c_{12}s_{23}e^{-i\delta} & c_{23}c_{13} \end{pmatrix}, \tag{3}$$

where $c_{ij} = \cos\theta_{ij}$, $s_{ij} = \sin\theta_{ij}$ (i, j=1, 2, 3).
If one of the element of M_ν is considered zero, i.e. $M_{lm} = 0$, we obtain the following constraint equation

$$\sum_{i=1}^{3} U_{li} U_{mi} \lambda_i = 0 \cdot \tag{4}$$

where l, m run over e, μ and τ. The one vanishing neutrino mass condition give rise to two possibilities either m_1=0 or m_3=0, corresponding to normal or inverted mass ordering.

Case I: $m_1 = 0$ (normal mass ordering)

Using Eq. (4), one can deduce the relation for neutrino mass ratio $\frac{m_2}{m_3}$ and Majorana phase (σ) as

$$\frac{m_2}{m_3} = \frac{|U_{l3}U_{m3}|}{|U_{l2}U_{m2}|}, \qquad \sigma = \frac{1}{2}\arg\left[-\frac{m_3}{m_2}\times\frac{U_{l3}U_{m3}}{U_{l2}U_{m2}}\right]. \tag{5}$$

Since m_1 is zero, therefore Majorana phase ρ become irrelevant in this case and can be dropped out. The neutrino masses (m_1, m_2, m_3) can be expressed in terms of experimentally known mass squared differences $\left(\delta m^2 = |m_2^2 - m_1^2|,\ \Delta m^2 = |m_3^2 - m_2^2|\right)$ as

$$m_1 = 0, \quad m_2 = \sqrt{\delta m^2}, \quad m_3 = \sqrt{\Delta m^2 + \delta m^2}. \tag{6}$$

Using Eqs. (5) and (6), we can constrain the ratio of solar and atmospheric neutrinos mass squared differences, R_ν ($=\delta m^2/\ \Delta m^2$) in terms of mixing angles ($\theta_{12}, \theta_{23}, \theta_{13}$) and Dirac CP violating phase (δ)

$$R_\nu = \left[-\left(1-\frac{|U_{l2}U_{m2}|^2}{|U_{l3}U_{m3}|^2}\right)\right]^{-1}. \tag{7}$$

Case II: $m_3 = 0$ (Inverted mass ordering)

Using Eq. (4), one can derive the relations for neutrino mass ratio $\frac{m_2}{m_1}$ and Majorana phase difference $(\rho - \sigma)$ as

$$\frac{m_2}{m_1} = \frac{|U_{l1}U_{m1}|}{|U_{l2}U_{m2}|}, \qquad \rho - \sigma = \frac{1}{2}\arg\left[-\frac{m_2}{m_1}\times\frac{U_{l2}U_{m2}}{U_{l1}U_{m1}}\right]. \tag{8}$$

Here, $\rho - \sigma$ is only Majorana physical phase. From Eq. (8), it is clear that Majorana phases (ρ, σ) are linearly co-related. The neutrino mass spectrum is given as

$$m_1 = \sqrt{\Delta m^2 - \delta m^2}, \quad m_2 = \sqrt{\Delta m^2}, \quad m_3 = 0. \tag{9}$$

From Eqs. (8) and (9), the relation for R_ν can be obtained as

$$R_\nu = 1 - \left|\frac{U_{l2}U_{m2}}{U_{l1}U_{m1}}\right|^2. \tag{10}$$

The expression for Jarlskog rephasing parameter J_{CP}, which is a measure of CP violation

$$J_{CP} = s_{12}c_{12}s_{23}c_{23}s_{13}c_{13}^2 \sin\delta. \tag{11}$$

The effective neutrino mass term for neutrinoless double beta decay ($0\nu\beta\beta$) is given as

$$\langle m \rangle_{ee} = \left|m_1 c_{12}^2 c_{13}^2 e^{2i\rho} + m_2 s_{12}^2 c_{13}^2 e^{2i\sigma} + m_3 s_{13}^2\right|. \tag{12}$$

The observation of neutrinoless double beta decay would imply lepton number violation and the Majorana nature of neutrinos.

3. Numerical analysis

In this section, we discuss the numerical procedure for our calculations based on the approach given in the section 2. For singular patterns, we consider either m_1=0 or m_3=0 corresponding to normal and inverted mass ordering, respectively. To begin with, we span the parameter space of neutrino oscillation parameters (δm^2 , Δm^2, θ_{12}, θ_{23}, θ_{13}, δ) lying in 3σ range by randomly generated points of the order of 10^7. Since the Dirac CP-violating phase (δ) is unconstrained at 3σ level, the full range of δ [0^0, 360^0] is considered. Taking into account the five oscillation parameters, namely $\theta_{12}, \theta_{23}, \theta_{13}, \delta m^2, \Delta m^2$ as inputs along with the constraint of one texture zero and a vanishing neutrino mass and, the ranges of other experimentally unknown parameters can be constrained. Eqs. (7) and (10) incorporate the important constraints of a vanishing mass eigenvalue and one texture zero. In the analysis, we consider a conservative limit on sum of neutrino masses (Σ) (i.e. Σ <1eV) at 3σ C.L. In addition, for effective neutrino mass term for $0\nu\beta\beta$ decay, we have chosen relatively large value of upper bound at 3σ C.L. i.e. $\langle m\rangle_{ee}$< 0.5eV.
In our numerical analysis, the main result is that for texture one- zero model with $m_1 = 0$ (normal mass ordering), all the six classes are now found to be incompatible with latest neutrino oscillation data at 3σ level. This is due to the relatively large value of precisely measured reactor mixing angle θ_{13}. On the other hand, only classes P_2, P_3, P_5, P_6 with $m_3 = 0$ are found to be compatible with latest neutrino mixing data at 3σ C.L [4]. The numerical results for all these classes (P_2, P_3, P_5, P_6) are found to be approximately similar to the analysis given by Lashin et al. [9]. However, we have provided a significant discussion in order to distinguish these classes from each other. The correlation plots of these classes are given in Figures 1 and 2. For Class P_1 with m_3=0, with the help of Eq. (10), we find the exact expression of mass-squared difference ratio (R_ν) as $R_\nu = 1 - \tan^2\theta_{12}$, showing dependence on mixing angle θ_{12}. Using 3σ range of θ_{12} [4], we obtain $0.40 \le R_\nu \le 0.61$, which is in conflict with the experimental range of R_ν. Therefore, class P_1 is ruled out with latest mixing data. Similarly, one can check that class P_4 also remains ruled out with current experimental data as also predicted in Ref. [9]. The two independently viable classes (P_2, P_5) have been discussed below:

Class P_2 with m_3=0: Using best fit value of current experimental data [5], we obtain, $|\rho - \sigma| \approx -0.1^0$, $\delta \approx 87.61^0$. Therefore, $\rho \approx \sigma$, as also shown in Figure 1 (a). The best fit value of effective mass term, $\langle m\rangle_{ee} \approx 0.0487$eV, is found to be O ($10^{-2}$), which lies well within the sensitivity limit of future neutrinoless double beta decay experiment. The phenomenological implications of class P_3 is similar to class P_2 due to permutation symmetry [9].

Class P_5 with m_3=0: The correlation plot [Figure 2 (b)] predicts $\delta \approx 90^0, 270^0$. Hence, this class allows approximate maximal CP violation. The preliminary hint given by latest global fits data of neutrino oscillation [5, 6], on the best fit value of δ coincidently supports our numerical result. From Figure 2(d), it is clear that the parameter space of Jarlskog rephasing parameter (J_{CP}) excludes the possibility of J_{CP}=0. On the other hand, Majorana phases (ρ, σ) remain unrestricted with latest experimental data, allowing the full range [0^0, 180^0] [Figure 2(a)]. The phenomenological implications of class P_6 is similar to class P_5 due to permutation symmetry. All these features discussed above are quite helpful in distinguishing the two independent experimentally viable classes (P_2, P_5).
Interestingly, out of four viable classes (P_2, P_3, P_5, P_6) at 3σ C.L, classes (P_2, P_3) are found to be ruled out at 1σ C.L, while remaining viable classes (P_5, P_6) predict $\delta \approx 270^0$ at same confidence level.

Journal of Physics: Conference Series **759** (2016) 012077 doi:10.1088/1742-6596/759/1/012077

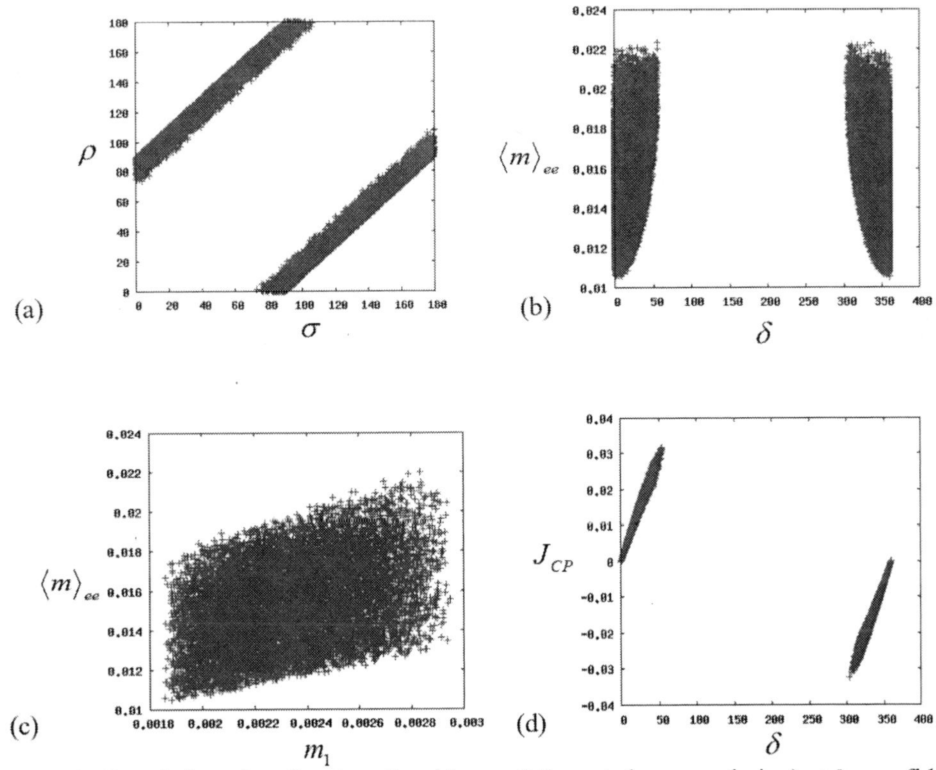

Figure 1. Correlation plots for class P_2 with $m_3=0$ (inverted mass ordering) at 3σ confidence level.

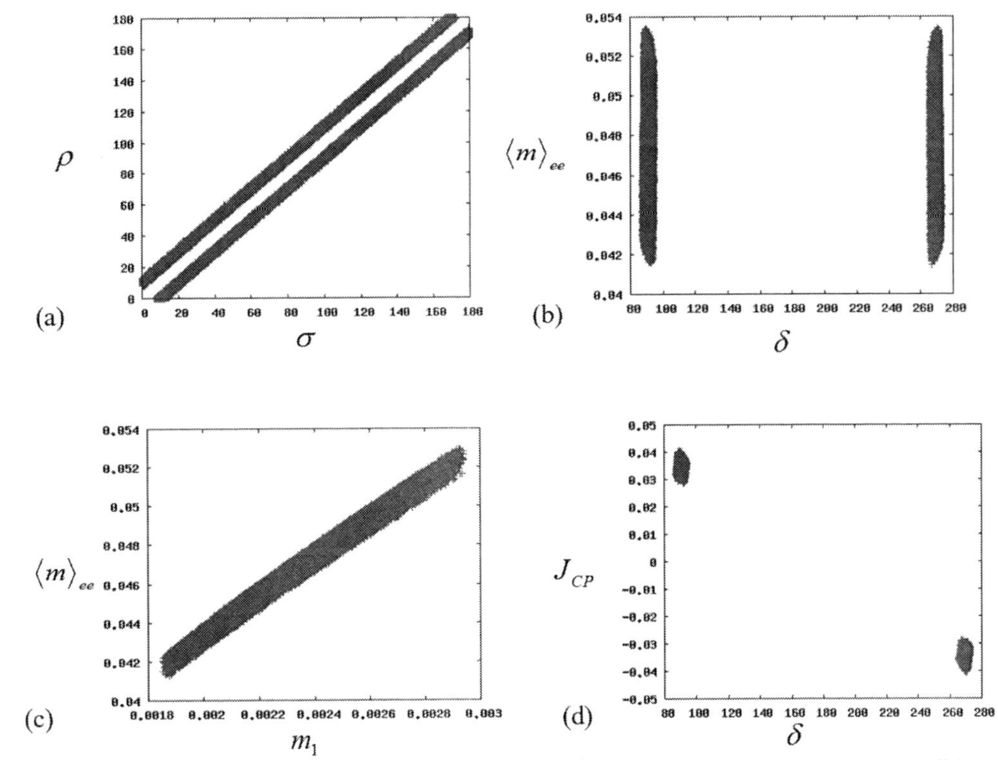

Figure 2. Correlation plots for class P_5 with $m_3=0$ (inverted mass ordering) at 3σ confidence level.

4. Summary and conclusions

We have done a systematic study of texture one-zero classes with one vanishing neutrino mass using latest global fits for neutrino oscillation parameters. We observe that the all the six classes with normal mass ordering (NO) are now ruled out at 3σ confidence level, whereas in case of inverted mass ordering, only four classes (P_2, P_3, P_5, P_6) are found to be viable with latest experimental data [4]. For these classes, the CP violating Majorana phases (ρ, σ) remain unconstrained; however, they exhibit strong linear correlation. In particular, for classes P_5 and P_6, the Dirac CP-violating phase (δ) is observed to be very close to maximal CP-violation i.e. $\delta \approx 90^0, 270^0$ ($\delta \approx 270^0$) at 3σ (1σ) C.L. This prediction is significant considering the fact that latest hints on best fit value of CP-violating phase (δ) lies in proximity of 270^0. Since no experiment has been able to fix absolute neutrino mass scale, therefore, the possibility of vanishing lowest neutrino mass cannot be ruled out at present. The future experiments will test the survivability of singular texture one zero classes along with their predictions on neutrino masses and mixings.

References

[1] Y. Abe et al., [Double Chooz collaboration], *Phys. Rev. Lett.* **108,** 131801 (2012), arXiv: 1112.6353 [hep-ex].

[2] F. P. An et al., [Daya Bay collaboration], *Phys. Rev. Lett.* **108**, 171803 (2012), arXiv: 1203.1669 [hep-ex].

[3] Soo-Bong Kim, for RENO collaboration, *Phys. Rev. Lett.* **108**, 191802 (2012) , arXiv:1204.0626[hep-ex].

[4] D. V. Forero, M. Trtola, J. W. F. Valle, *Phys. Rev.* **D 90**, 093006 (2014), arXiv:1405.7540 [hep-ph].

[5] M. C. Gonzalez-Garcia, M. Maltoni, J. Salvado and T. Schwetz, *JHEP* **1411**, 052 (2014) , arXiv:1409.5439 [hep-ph].

[6] Paul H. Frampton, Sheldon L. Glashow and Danny Marfatia, *Phys. Lett.* **B 536**, 79 (2002), hep-ph/0201008.

[7] Zhi-zhong Xing, *Phys. Lett.* **B 530**, 159 (2002), hep-ph/0201151.

[8] P. O. Ludl, S. Morisi, E. Peinado, *Nucl. Phys.* **B 857**, 411 (2012); D. Meloni, G. Blankenburg, *Nucl. Phys.* **B**, **867**, 749 (2013); W. Grimus, P. O. Ludl, *J. Phys. G,* **40**, 055003 (2013); X.Liu, S. Zhou, *Int. J. Mod. Phys.* **A,** 28 (2013) 1350040; J.Liao, D. Marfatia and K. Whisnant, *JHEP* **09**, 013 (2014); L. M. Cebola, D. E. Costa, R. G. Felipe, arXiv: 1504.06594 [hep-ph], H. Fritzsch, Zhi-zhong Xing, S. Zhou, *JHEP* **1109**, 083 (2011), arXiv: 1108.4534 [hep-ph].

[9] E. I. Lashin and N. Chamoun, *Phys. Rev.* **D85**, 113011 (2012), arXiv:1108.4010 [hep ph].

[10] Z.Z. Xing; *Phys. Rev.* **D 69**, 013006 (2004).

Journal of Physics: Conference Series **759** (2016) 012078 doi:10.1088/1742-6596/759/1/012078

Multi-scale Modeling of Radiation Damage: Large Scale Data Analysis

M Warrier[1] and U Bhardwaj[1] and S Bukkuru[2]

[1]Computational Analysis Division, Bhabha Atomic Research Center, Visakhapatnam, Andhra Pradesh, India

[2]Nuclear Physics Department, Andhra University, Visakhapatnam, Andhra Pradesh, India

E-mail: Manoj.Warrier@gmail.com

Abstract. Modification of materials in nuclear reactors due to neutron irradiation is a multi-scale problem. These neutrons pass through materials creating several energetic primary knock-on atoms (PKA) which cause localized collision cascades creating damage tracks, defects (interstitials and vacancies) and defect clusters depending on the energy of the PKA. These defects diffuse and recombine throughout the whole duration of operation of the reactor, thereby changing the micro-structure of the material and its properties. It is therefore desirable to develop predictive computational tools to simulate the micro-structural changes of irradiated materials. In this paper we describe how statistical averages of the collision cascades from thousands of MD simulations are used to provide inputs to Kinetic Monte Carlo (KMC) simulations which can handle larger sizes, more defects and longer time durations. Use of unsupervised learning and graph optimization in handling and analyzing large scale MD data will be highlighted.

1. Introduction

Materials in nuclear reactors are subjected to neutron fluxes. The neutrons pass through the material losing energy by elastic and inelastic collisions with the nucleus of the atoms constituting the material [1]. This creates energetic primary knock-on atoms (PKAs) which cause local collision cascades in the material. This results in atoms getting displaced from their crystal positions, creating defects (interstitials and vacancies). The timescale of these collision cascades are of the order of a few pico-seconds [2]. Depending on the energy of the PKA, the extent of the cascades can be from less than a nanometer to several hundred nanometers. The created defects can then diffuse throughout the material. As they diffuse they can recombine and reduce damage, or form defect clusters which change the material micro-structure and cause degradation in its properties: viz. radiation induced amorphization, radiation induced hardening, thermal and electrical conductivity degradation, radiation induced segregation, high temperature embrittlement and dimensional instabilities like rrradiation growth, creep and swelling ([3] and references therein).

The changes in micro–structure happen throughout the material for the duration of reactor operation. Therefore Multi-scale simulations are necessary to model the change in the material micro-structure [4, 5, 6]. The sequence of steps in such a study are:

(i) Creation of PKAs by neutrons. The energy spectrum of the PKA is obtained from neutron total scattering cross-sections [1].

(ii) Creation of interstitials and vacancies by the energetic PKA, thier in-cascade clustering and distribution. The creation of interstitials and vacancies can be modelled by Binary collision Approximation (BCA) Monte Carlo methods [7, 8] or by Molecular Dynamics (MD) [9, 10, 11].

(iii) The number of Frenkel pairs, their configuration and spatial distribution from the cascade simulations will be an input to lattice and/or object Kinetic Monte Carlo (KMC) simulations [6, 12] of diffusive–recombination of the interstitials, interstitial clusters and vacancies to form voids and precipitates. Note that the diffusion and precipitation can occur within the grains or along grain boundaries.

The micro-structural and compositional changes due to irradiation obtained by the above sequence of simulations can be inputs to dislocation dynamics [13] and phase field simulations [14] which can be used to obtain the changes in material properties.

Binary collision approximation Monte Carlo codes (BCA-MC) have been widely used for sputtering, range and back-scattering calculations. However better models are needed in modeling the creation of interstitials and vacancies [15]. We have used MD to study the directional effects of PKA bombardment on the number of displaced atoms in a collision cascade in crystal Cu, W and Fe [15, 16] for modest PKA energies of 1-5 keV. We have carried out several thousand MD simulations with the PKA directed along random crystal directions and shown that typically a hundred random directions in the crystal have to be explored before the average number of displaced atoms become constant. We have obtained the direction averaged (i) number of displacements, (ii) distributions of interstitials and vacancies and (iii) in-cascade cluster size distributions of these defects in [15]. The requirement for large scale post processing of several tera-bytes of data from the MD runs is described in Section.2.

In order to obtain the diffusion co-efficient of the defects, MD simulations for interstitialcy diffusion were carried out [17, 18]. In interstitialcy diffusion the interstitial atom is same as the host material atom and the interstitial displaces a lattice atom and the lattice atom becomes the new interstitial. Therefore the identity of the atom keeps changing. In order to identify the interstitial and create a random walk from which the diffusion can be parametrized, we use unsupervised learning [19] and the graph-tree data structure [17, 20]. This is described in Section.3. We illustrate the multi-scale model by using the average values gleaned from the MD simulations in a kinetic Monte Carlo model for diffusive-recombinations of defects. This is presented in Section.4. Finally we present our conclusions.

2. Large scale data analysis: MD studies of directional effects

We use the Large-scale Atomic Molecular Massively Parallel Simulator (LAMMPS) [21] for our MD simulations of collision cascades. MD outputs the atomic positions as a function of time at specified intervals and this data has to be post-processed in order to identify defects and classify them as belonging to clusters or not. This classification is further complicated by existence of "crowdion configuration" in which several neighbouring interstitials and vacancies crowd around and the number of unpaired defects have to be identified [2]. A typical MD simulation of a collision cascade requires that the simulation box size be large enough to accommodate the cascade. Since the PKA is launched in random direction and secondary cascades can take any random direction, we take the simulation box size to be around 2.5 times the expected range of the PKA. The expected range is found by initial scoping runs. The collision cascade has a time scale of a few pico–secs. Typically, a $50 \times 50 \times 50$ unit cell simulation of bcc W is sufficient to simulate collision cascades of PKA of energy 5 keV. If the atomic identities and positions of such a cascade simulation of 10 pico–secs are output every 100 MD time steps, it will lead to a 1.25 Giga-byte file. If 200 to 1000 PKA directions have to be explored, this means a 250 Giga-byte to 1 Tera-byte of data for post-processing at one PKA energy.

The defects were identified by an algorithm similar to the method of spheres [2] and is described in [16, 15]. In this method, a interstitial–vacancy pair is said to have been created if the interstitial is away from its lattice position by a value r_{disp}. We have used an unsupervised clustering algorithm (called Max-space clustering) to identify the value of r_{disp} such that the displaced interstitials could be separated from the natural vibrational amplitudes of the atoms [17, 18]. This was done as follows:

- Find the displacements (hitherto called offsets) of all atoms from their original lattice positions.
- Sort the atoms based on their offsets.
- Find the difference between offsets between consecutive atoms from the sorted list and therefore the pair of atoms (say *i* and *i+1*) having maximum difference in their offsets.
- Set r_{disp} = average value of the offsets of atoms *i* and *i+1*.

It is shown that this method of choosing offsets is better than existing methods [18].

Using the r_{disp} obtained from this method, in-situ identification of the displaced atoms was carried out by adding two new "computes" in LAMMPS, one for outputting the number of displacements, the maximum range, the range of the PKA, etc and the other to dump the positions of the identified displaced atoms for post-processing the in-cascade clustering of defects. This results in only around 500 MB of data output for post-processing 1000 random directions of a PKA at a specified PKA energy compared to the earlier 1TB data output. It also did not lead to extra computational time because the extra time for in-situ data processing was offset by the input/output (I/O) time required to write the several Giga–bytes of data files to disk. Typical direction averaged distributions of interstitials and vacancies for PKA energies of 1-5 keV in Cu is shown in Fig.1 and Fig.2 respectively. The averaging was done for 200 MD simulations of randomly directed PKA at each energy. It was seen that 200 random directions is sufficient for the variances in the number of defects produced to stabilize [15]. This can now be a realistic input for initializing defect positions in a KMC code (Section.4).

3. Interstitialcy diffusion from MD

The defects obtained from the above described collision cascades thermalize with the material atoms at the end of the cascade and diffuse. The main channel of diffusion of the interstitial atoms is by *interstitialcy diffusion*, in which, the interstitial displaces a lattice atom and occupies its position. The displaced lattice atom is the new interstitial. In an MD simulation, in order to identify the random walk of the interstitial, the changing identity of the interstitial atom also has to be tracked. The interstitial can be identified from each time frame output by the Max-space clustering algorithm described in the previous section. However, many a time, the interstitial is part of a cluster of atoms in a crowdion or dumbbell configuration (hitherto referred to as ICA - Interstitial Configuration Atoms). In order to identify the interstitial from the ICA we use the graph-tree data structure [20]. All ICA in a time-step are represented by nodes in a level (fig.3). Nodes across levels, i.e. ICA at the previous and next time-step, are connected by an edge with weight w equal to the displacement between the positions. All the nodes at level l (parent-nodes) are connected to all the nodes at level l + 1. We link every node to a parent node which has least weighing edge to it. Specifically, for every node nl of the level l, we calculate the cost function c (nl) = min(w (ni (l1) , nl)) for i [1,m], where m is the

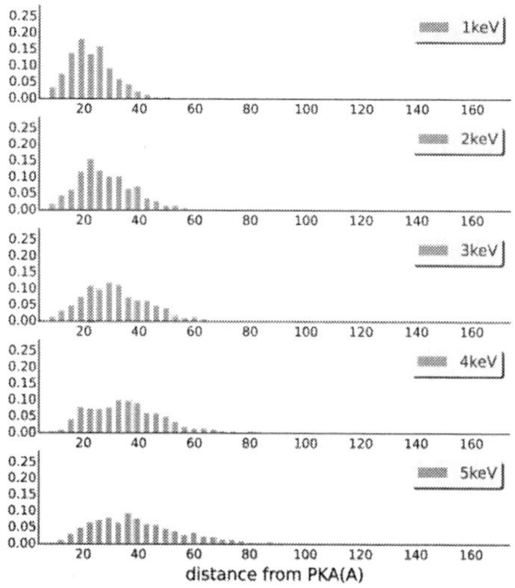

Figure 1. Distribution of the interstitials as a function of distance from the origin of the PKA for energies 1-5 keV in Cu

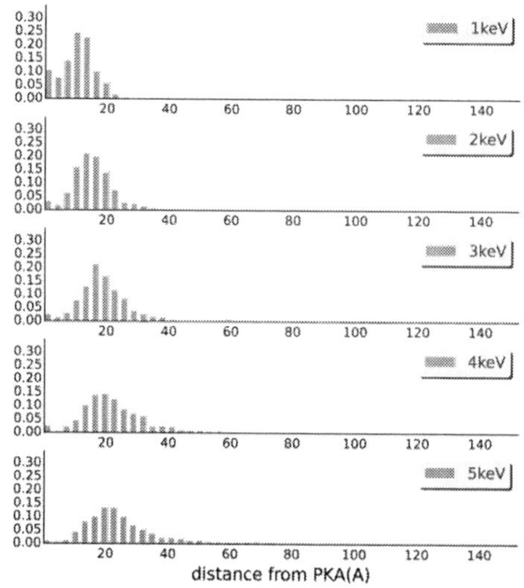

Figure 2. Distribution of the vacancies as a function of distance from the origin of the PKA for energies 1-5 keV in Cu

number of parent nodes, and mark a link to the parent node which yields minimum cost. Details of the graph-tree algorithm and various cost functions explored are presented elsewhere [17].

We have carried out MD simulations to determine (i) r_{disp}, (ii) the migration energy E_m and (iii) the diffusion coefficient for interstitialcy diffusion [18, 17]. A $10 \times 10 \times 10$ unit cell simulation box of various fcc and bcc metallic crystals of interest with a randomly placed single interstitial of the same specie was first equilibrated using a NPT MD simulation at various temperatures below the melting point of the crystal. NVE simulations were then carried out for nano–seconds with a time step of a femto–sec. Atomic positions were output every 100 MD time steps. Post-processing of the data to identify the ICA and the diffusion trajectory was carried out. Fig.4 shows a typical trajectory of Cu interstitial. The diffusion coefficient of the interstitial at each temperature was then obtained from the diffusion trajectory. Assuming an Arrhenius relation for the diffusion with temperature, the migration energy was determined. Fig.5 shows the variation of the interstitialcy diffusion coefficient of Cu with temperature. From the plot the E_m for interstitialcy diffusion of Cu is calculated to be 0.08 eV which compares well with published results of 0.12 eV [22]. The pre–factor for the diffusion is 7.9×10^{-8}. It was also seen that almost all the jumps in the random walk are 1st nearest neighbor (1NN) jumps.These diffusion parameters can and characteristics can also be used as inputs to a KMC code (Section.4).

4. Scaling up with Kinetic Monte Carlo

We have developed a simulation code for Object KMC (OKMC) simulations based on the residence time algorithm [23]. The code was validated by reproducing the interstitialcy diffusion obtained from the MD simulations in Section.3. We simulate cascade evolution of a uniform PKA energy spectrum in the energy range 1-5 keV, uniformly distributed over a $100\times100\times100$ unit cell W crystal lattice. The interstitials and vacancies are distributed from the distributions obtained

Journal of Physics: Conference Series **759** (2016) 012078 doi:10.1088/1742-6596/759/1/012078

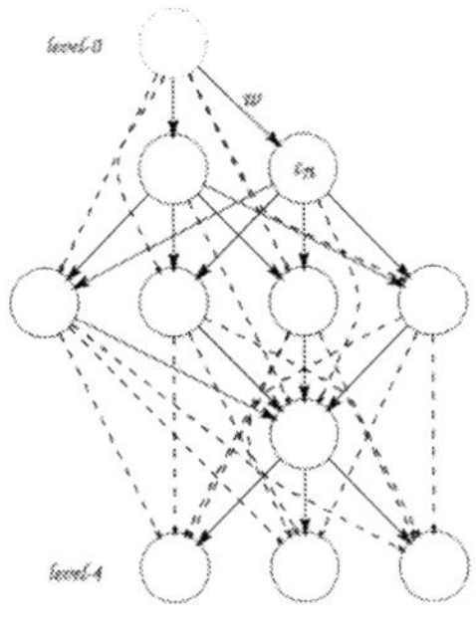

Figure 3. Representing ICA as a Graph Tree.

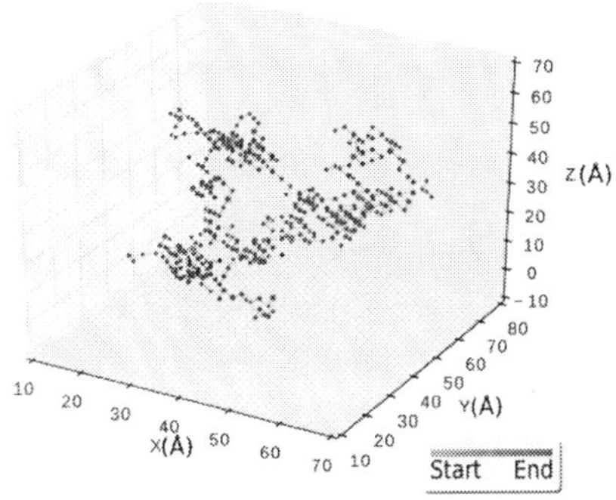

Figure 4. Interstitialcy diffusion trajectory of Cu at 900 K

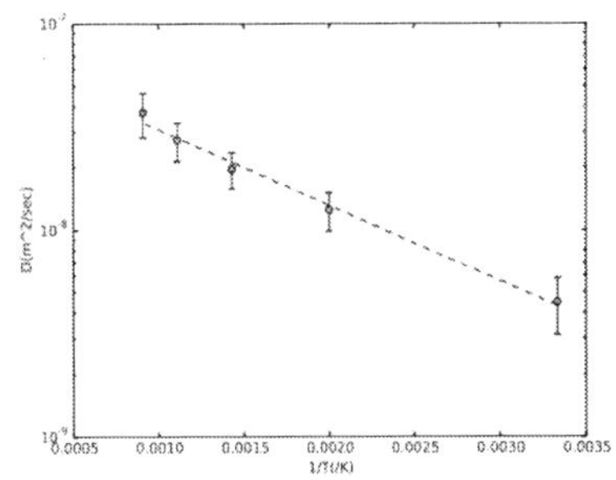

Figure 5. Interstitialcy diffusion coefficient of Cu as a function of inverse temperature

from the MD cascades described in Section.2, but for W from [15]. The migration energies and the jump sizes were specified from similar MD simulations as described in Section.3, but for W [17]. New defects were introduced in the simulation volume at the rate of 0.1/micro-sec for 10 micro-secs. Clusters of defects of size lesser than four are allowed to diffuse uni-directionally since rotation of a cluster is very rare [24]. The larger clusters are not allowed to diffuse in these simulations. The simulation was carried out till there were no more particles that could diffuse. The evolution of the clustering of interstitials and vacancies at 10 micro-secs, 28 milli-secs and at the end of the simulation (5.6 hours) are shown in Fig.6, Fig.7 and Fig.8 respectively. The results demonstrate that basic diffusion-agglomeration can be carried out by the KMC code. Work is going on to further validate it with published results.

Figure 6. Interstitial and Vacancy clustering at 10 micro-secs from OKMC Simulations

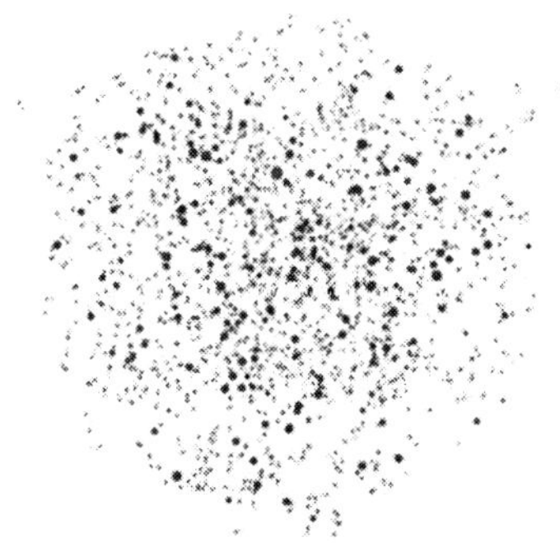

Figure 7. Interstitial and Vacancy clustering at 28 milli-secs from OKMC simulations

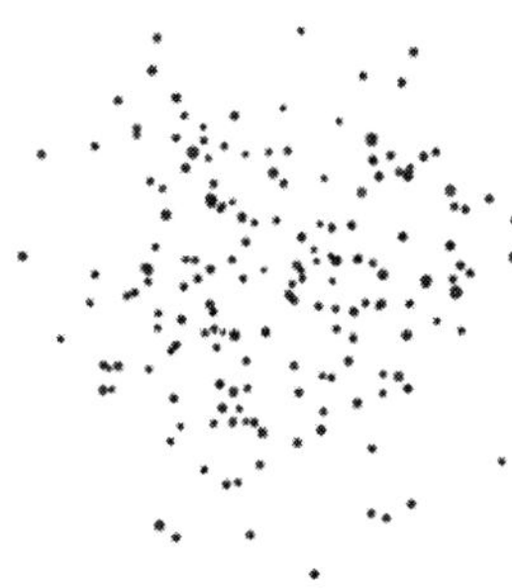

Figure 8. Interstitial and Vacancy clustering at 5.6 Hours from OKMC simulations

5. Conclusions

We have described MD simulations to obtain (i) directional averaged values of the number of defects, their distribution around the PKA origin and their in–cascade clustering during a collision cascade in Section.2 and (ii) diffusion parameters and characteristics of single interstitials in Section.3. The data gleaned from the MD simulations are passed as inputs to a KMC simulation which simulates the diffusive-recombination of the defects to form clusters as described in Section.4. Initial results of the clustering for a randomly distributed, uniform spectrum of 1-5 keV PKA is presented thereby making the link from atomistic calculations to defect clustering at higher spatial and time scales, thereby demonstrating a multi-scale model for radiation damage.

A Max-space algorithm was used to identify the defects and a graph-tree data structure was used to obtain the interstitial from a group of interstitial cluster atoms so that the random walk of the interstitial can be obtained. One of the themes of this conference was on Computational Physics Education. The methods for data analysis used in this paper are not normally part of a Computational Physics curriculum. We therefore feel the need to emphasize that perhaps data structures and Artificial Intelligence (AI) algorithms need to be taught in advanced Computational Physics courses because they give a physicist structured ways to handle and analyze large data.

Acknowledgments

We would like to acknowledge the support lent to us by N. Shaktivel, H. Hemani and others maintaining the Computer Clusters at BARC-Visakhapatnam. The large scale MD simulations would not have been possible without them. One of the authors (S. Bukkuru) acknowledges a BRNS grant (No:2013/36/32-BRNS) which helped him carry out this work.

References

[1] Greenwood L R, Smither L K 1985 SPECTER: Neutron Damage Calculations for Materials Irradiations, *Argonne National Laboratory Report* **ANL/FPP/TM-197**
[2] Stoller R E 2012 *Comprehensive Nuclear Materials* **Vol.1** Konings R J M (Ed.) Elsivier, p 293
[3] Zinkle S J 2012 *Comprehensive Nuclear Materials* **Vol.1** Konings R J M (Ed.) Elsivier, p 65
[4] Stoller R E, Mansur L K 2005 *ORNL report* **ORNL/TM-2005/506**
[5] Björkas C, Nordlund K, Caturla M J 2012 *Phys. Rev. B* **85** 024105
[6] Monasterio P R, Wirth B D and Odette G R 2007 *J. Nucl. Mater.* **361** p127
[7] Eckstein W 1991 *Computer Simulations of Ion-Solid Interactions, Springer series in Material Science* **10** Springer Verlag
[8] Robinson M T, Torrens I M 1974 *Phys. Rev. B* **9** p 5008
[9] Stoller R E, Calder A F 2000 *J. Nucl. Mater* **283-287** (5) p746
[10] Malerba L 2006 *J. Nucl. Mater.* **351** p 28
[11] Björkas C, Nordlund K, Malerba L, Terentyev D, Olsson P 2008 *J. Nucl. Mater.* **372** p 312
[12] Domain L M C, Becquart C S, Malerba L 2004 *J. Nucl. Mater.* **335** p 121
[13] Amodeo R J, Ghoneim N M 1990 *Phys. Rev. B* **41** (10) p 6958
[14] Rokkam S, El-Azab A, Millett P, Wolf D 2009 *Modelling Simul. Mater. Sci. Eng.* **17** p 064002
[15] Warrier M, Bhardwaj U, Hemani H, Schneider R, Mutzke A, Valsakumar M C 2015 *J. Nucl. Mater.* **467** p 457
[16] Warrier M, Valsakumar M C 2014 *Fusion Sci. Tech.* **65** (2) p 222
[17] Bhardwaj U, Bukkuru S, Warrier M 2016 *J. Comp. Phys.* **305** p 263
[18] Bukkuru S, Bhardwaj U, Rao A D P, Warrier M, Valsakumar M C 2015 *Identifying Interstitials in Molecular Dynamics simulations - Max Space Clustering Method, XXVII IUPAP Conf. on Comp. Phys., 2-5 Dec 2015, IIT-Guwahati, India (these proceedings)*
[19] Grygorash O, Zhou Y, Jorgensen Z, 2006 *Minimum spanning tree based clustering algorithms, in: 18th IEEE Int. Conf. on Tools with Artificial Intelligence, ICTAI06* pp. 73
[20] Puntambekar A A *Data Structures* Technical Publications, Pune, India
[21] Plimpton S J 1995 *J. Comp. Phys.* **117** p 1
[22] Corbett J W, Smith R B, Walker R M 1959 *Phys. Rev.* **114** (6) p 1452
[23] Bortz A B, Kalos M H, Lebowitz J L 1975 *J. Comp. Phys.* **17** p 10
[24] Domain L M C, Becquart C S 2004 *J. Nuc. Mater.* **335** p 121

Non-equilibrium Thermodynamics of Rayleigh-Taylor instability

Tapan K Sengupta[1], **Aditi Sengupta**[2], **K S Shruti**[1], **Soumyo Sengupta**[3] **and Ashish Bhole**[1]

[1] High Performance Computing Laboratory, Department of Aerospace Engineering, I.I.T. Kanpur, Kanpur, U.P., India
[2] Department of Engineering, University of Cambridge, Cambridge, U.K.
[3] Department of Mechanical and Aerospace Engineering, Ohio State University, Ohio, U.S.A.

E-mail: tksen@iitk.ac.in

Abstract. Rayleigh-Taylor instability (RTI) has been studied here as a non-equilibrium thermodynamics problem. Air masses with temperature difference of 70K, initially with heavier air resting on lighter air isolated by a partition, are allowed to mix by impulsively removing the partition. This results in interface instabilities, which are traced here by solving two dimensional (2D) compressible Navier-Stokes equation (NSE), without using Boussinesq approximation (BA henceforth). The non-periodic isolated system is studied by solving NSE by high accuracy, dispersion relation preserving (DRP) numerical methods described in Sengupta T.K.: High Accuracy Computing Method (Camb. Univ. Press, USA, 2013). The instability onset is due to misaligned pressure and density gradients and is evident via creation and evolution of spikes and bubbles (when lighter fluid penetrates heavier fluid and vice versa, associated with pressure waves). Assumptions inherent in compressible formulation are: (i) Stokes' hypothesis that uses zero bulk viscosity assumption and (ii) the equation of state for perfect gas which is a consequence of equilibrium thermodynamics. Present computations for a non-equilibrium thermodynamic process do not show monotonic rise of entropy with time, as one expects from equilibrium thermodynamics. This is investigated with respect to the thought-experiment. First, we replace Stokes' hypothesis, with another approach where non-zero bulk viscosity of air is taken from an experiment. Entropy of the isolated system is traced, with and without the use of Stokes' hypothesis. Without Stokes' hypothesis, one notes the rate of increase in entropy to be higher as compared to results with Stokes' hypothesis. We show this using the total entropy production for the thermodynamically isolated system. The entropy increase from the zero datum is due to mixing in general; punctuated by fluctuating entropy due to creation of compression and rarefaction fronts originating at the interface and reflecting from the walls.

Introduction

RTI arises whenever fluids of different densities at rest are in an unstable arrangement. This results in distortion of the interface by baroclinic vorticity generation due to misaligned density and pressure gradients, aided by gravitational acceleration [2, 3, 4, 5, 6, 27, 28]. RTI plays a major role in various contexts, as in inertial confinement fusion, combustion problems, supernovae explosions and geophysics. RTI constitutes an important modeling and simulation test case for codes involving various characteristics and aspects of turbulence, transport and diffusion in flows governed by non-uniform density gradients [7]. In rapid compression and expansion processes, the system ceases to be in thermodynamic equilibrium [30]. RTI is a non-equilibrium thermodynamic problem wherein the expansion and rarefaction of waves can be captured only by implementing the compressible flow formulation without using BA. For variations in temperature not exceeding 10 degrees, variation in density is at most 1% and this small variation can be ignored [4]. Atwood number (A) considered here is 0.1, where $A = \frac{\rho_u - \rho_l}{\rho_u + \rho_l}$ and ρ_u is the density of the upper fluid and ρ_l that of the lower, as shown in the schematic of the problem in figure 1 with a temperature difference $\Delta T = T_l - T_u$ of $70K$ between two compartments, where T_l is the temperature of the lower fluid and T_u that of the upper. Mikaelian [43] states that "*...direct numerical simulations are the best way to assess the validity of BA*" in his investigation of BA in RT and Richtmyer-Meshkov (RM) instabilities. Wei and Livescu [25] note that there are higher order effects which are obscured by BA for lower A values.

Experimental studies have been carried out on different aspects of RTI. Investigation of the growth of instability and turbulent mixing at the interface of two fluids, with different density ratios and experimental tank cross-sections [13], displayed increase in the width of the mixing region due to stretching of large scale motions. Similar experiment was performed, with and without tilting the 2D experimental tank, to study RTI [14]. Mixing efficiency in RTI by instantaneous removal of a plate between the fluid layers has also been investigated [15]. Measured velocity field imposed at the onset of RTI experiment was used to initialize three-dimensional (3D) simulations in [16]. RTI leading to turbulent flow was experimented with variable accelerations in [18]. Recently, an experimental investigation on the effects of forced small-wavelength, finite-bandwidth initial perturbations on miscibility of RTI leading to turbulence has been carried out [17].

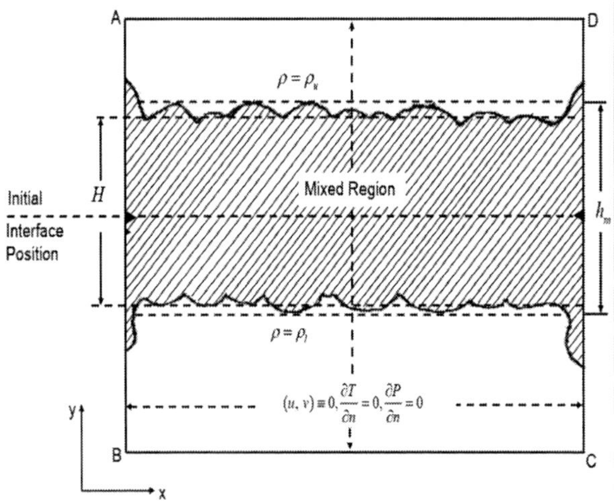

Figure 1. Schematic of RTI in a box. The mixing region, boundary conditions and the initial interface position have been depicted.

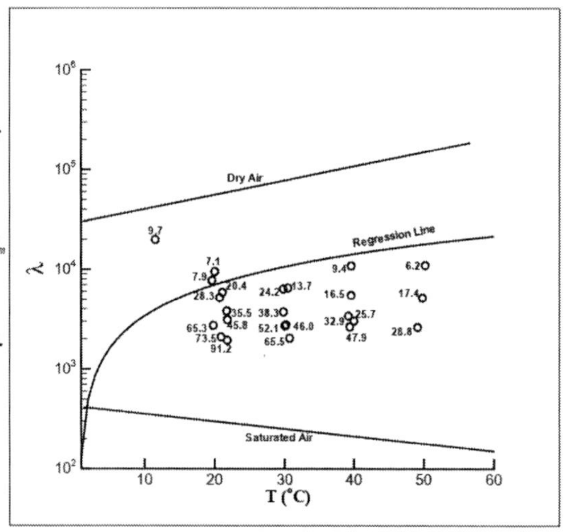

Figure 2. Variation of non-dimensional λ with T and relative humidity from [42], along with the variation obtained by regression analysis of these data.

Youngs carried out 2D and 3D numerical simulations of turbulent mixing of miscible fluids by RTI [19, 20]. Transition stages of 3D RTI between two incompressible, miscible fluids were presented using DNS in [7]. In [21], LES was used to simulate RTI and a relation between the rate of growth of the mixing layer and the net mass flux through the plane related with the initial position of the interface was obtained. Monotone implicit LES (MILES) using finite-volume technique was implemented for the numerical study of the influence of initial perturbation on turbulent RTI in [22]; Cabot and Cook [23] simulated RTI for a Reynolds number (Re) up to

32,000, which was supposed to explain type-Ia supernova. Detailed evolution of single-mode RTI was studied in [24]; the growth stages with high and low perturbation Re (defined with respect to A, kinematic viscosity ν, gravitational acceleration and perturbation wavelength) was studied in [25]; RTI driven by stabilizing and destabilizing accelerations in intermediate stages of simulation was investigated in [26]. Total entropy of RTI in an isolated system has been recently studied in a non-equilibrium framework [6, 27] using high-resolution DRP schemes [8] with compressible flow formulation.

Formulation of the problem and numerical implementation

In the present study two cases are investigated. Case 1 represents a rectangular domain with dimension $0 \leq x \leq 60$ and $0 \leq y \leq 30$ and Case 2 represents a square, with dimension $0 \leq x \leq 60$ and $0 \leq y \leq 60$. In Case 1, the domain is discretized with uniformly spaced 3000 points in the x-direction and 1800 points in the y-direction and in Case 2, 3000 points in the x-direction and 3600 points in the y-direction. No-slip, adiabatic conditions are imposed on the walls (AB, BC, CD, AD) as shown in figure 1. The governing equations and the problem formulation are as given in [6, 27]. The system of equations can be expressed in terms of fluxes, given in Hoffman & Chiang [12] as follows:

$$\frac{\partial \hat{Q}}{\partial t} + \frac{\partial \hat{E}}{\partial x} + \frac{\partial \hat{F}}{\partial y} = \frac{\partial \hat{E}_v}{\partial x} + \frac{\partial \hat{F}_v}{\partial y} \tag{1}$$

where the conserved variables, inviscid and viscous fluxes are, respectively, given by, $\hat{Q} = \Big[\rho, \rho u, \rho v, \rho e_t\Big]^T$; $\hat{E} = \Big[\rho u, \rho u^2 + p, \rho uv, (\rho e_t + p)u\Big]^T$; $\hat{F} = \Big[\rho v, \rho uv, \rho v^2 + p + (\rho - \rho_s)gy, (\rho e_t + p)v\Big]^T$; $\hat{E}_v = \Big[0, \tau_{xx}, \tau_{xy}, u\tau_{xx} + v\tau_{xy} - q_x\Big]^T$ and $\hat{F}_v = \Big[0, \tau_{yx}, \tau_{yy}, u\tau_{yx} + v\tau_{yy} - q_y\Big]^T$. In these expressions, ρ, p, u, v, T and e_t represent the density, the fluid pressure, Cartesian components of fluid velocity, the absolute temperature and the specific internal energy of the fluid, respectively, and τ_{xx}, τ_{xy}, τ_{yx}, τ_{yy} are the components of the symmetric viscous stress tensor and are related to the gradients of velocity as

$\tau_{xx} = \Big(2\mu\frac{\partial u}{\partial x} + \lambda\nabla - \cdot\vec{V}\Big)$; $\tau_{yy} = \Big(2\mu\frac{\partial v}{\partial y} + \lambda\nabla \cdot \vec{V}\Big)$ and $\tau_{xy} = \tau_{yx} = \mu\Big(\frac{\partial u}{\partial y} + \frac{\partial v}{\partial x}\Big)$

Physical properties, viz. dynamic viscosity (μ), second coefficient of viscosity (λ), specific heat capacity (C_v), and thermal conductivity (κ) are regarded as constants in this formulation. The heat conduction terms, q_x and q_y are given by, $q_x = -\kappa\frac{\partial T}{\partial x}$ and $q_y = -\kappa\frac{\partial T}{\partial y}$. The closure of system of equations is provided by the ideal gas equation and is used to define energy as, $e_t = C_vT + \frac{(u^2+v^2)}{2}$.

Stokes hypothesis

The Lame coefficients λ and μ appearing in the stress-strain constitutive relation (with μ as shear viscosity and λ as second coefficient of viscosity or dilatational viscosity [32]) is related to compressibility of the flow [30]. Stokes' hypothesis states that bulk viscosity ($\kappa = \lambda + \frac{2}{3}\mu$) is set to zero for any flow. This is irrelevant with incompressibility assumption, which in itself is an approximation, even for a very low speed flow. It implies that the thermodynamic pressure is equal to the mechanical pressure [33, 34, 35, 36]. There have been critiques [32, 33, 34, 35, 36, 41] on Stokes' hypothesis [37]. It is weakly justified for mono-atomic rigid molecule gases [33, 36]. Compilations by Liebermann [32], Karim and Rosenhead [38] and Rosenhead [39] state that λ is independent of μ and can be orders of magnitude higher in amplitude, while the sign is reversed. Emanuel [33] and Gad-el-Hak [34] questioned the hypothesis and its applicability for flows such as re-entry into planetary atmosphere. Cramer [40] showed that many common fluids including diatomic gases can have bulk viscosities thousand times larger than λ. Rajagopal [41] suggested a need for a new approach in obtaining NSE for all fluids.

Without Stokes' hypothesis, λ is given as, $\lambda = (-\frac{2}{3} + \frac{\kappa}{\mu})\mu$, where $\kappa = -7.383 \times 10^{-4} + 3.381 \times 10^{-4}\, T$ (with T in K), which is obtained by a regression analysis of experimental data in [42], shown in figure 2. Thus, the viscous normal stress components in the present formulation are given by

$$\tau_{xx} = \frac{1}{Re}\left[\left(\frac{4}{3} + \frac{\kappa}{\mu}\right)\frac{\partial u}{\partial x} + \left(-\frac{2}{3} + \frac{\kappa}{\mu}\right)\frac{\partial v}{\partial y}\right] \tag{2}$$

$$\tau_{yy} = \frac{1}{Re}\left[\left(\frac{4}{3} + \frac{\kappa}{\mu}\right)\frac{\partial v}{\partial y} + \left(-\frac{2}{3} + \frac{\kappa}{\mu}\right)\frac{\partial u}{\partial x}\right] \tag{3}$$

The experimental λ variation with temperature is shown for dry and saturated air in figure 2, as discrete points and the curved line is obtained by using regression analysis of these data. We specifically focus on the role of Stokes' hypothesis on non-equilibrium thermodynamics related to RTI and its computation in the present study. We do not require any initial perturbations here, as strong rotationality appears at the junctures of the interface with the side-walls [6, 27]. This study is similar in scope to the experiments reported in [13] and [14]. In the present computations, we have used $Re = 10^7$, $Pr = 0.7$, $\Delta T = 70K$, $Ga = 0.23489$, $Ec = 0.01$ and $Fr = 12.2586$ [27, 6] and such large values are indicative of the ability to perform calculations very accurately using the proposed formulation and numerical methods to capture the thermodynamic process which is far from equilibrium.

Computations of the model thought-experiment are performed using compressible NSE described in [6, 11, 12]. Non-periodic boundary conditions have been implemented in both vertical (y) and horizontal (x) directions. High-resolution DRP schemes [8] are implemented to carry out the simulations. The numerical and physical dispersion relations are almost the same over a large range of space and time scales. Spatial discretization of convective fluxes are performed using high resolution OUCS3 scheme [1, 9]. RTI results have been validated with Read's experiment [13] as shown in [6, 27] where the general features and primary stages of RTI have also been described.

Results and Discussion

Entropy of the non-equilibrium thermodynamic system without Stokes hypothesis: The present study is for a non-equilibrium thermodynamic process involving air as the working substance driven by large thermal gradient, even though the flow speeds are negligibly small. Here, the entropy of the system at any point (x_m, y_n) in the system is given by

$$\Delta\bar{s}_{mn}(t + \Delta t) = c_p \ln\frac{T(x_m, y_n, t + \Delta t)}{T(x_m, y_n, t)} - R\ln\frac{p(x_m, y_n, t + \Delta t)}{p(x_m, y_n, t)} \tag{4}$$

We will refer to the condition at the onset as the zero-level datum for the entropy. One can compute the entropy of the entire system by using equation (4) for each computational cell and then summing over the entire domain. The resultant entropy of the system as a function of time is shown in figure 3 for the rectangular domain (Case 1), with and without Stokes' hypothesis. Here, the entropy change with bulk viscosity is shown for three sub-cases, confirming higher rise when Stokes' hypothesis is not used. Identical grid and time step has been used for computing the three sub-cases, varying κ in figure 3. It can be noted that consistently the entropy is higher without Stokes' hypothesis and with an additional 10% increase in κ in Case 1, as one would expect due to the presence of higher viscous normal stress. The figure reveals that the time-variation of entropy is identical for the computations using with and without Stokes' hypothesis for up to $t \approx 4$, when the convection in the box is negligibly small. Beyond this time, one

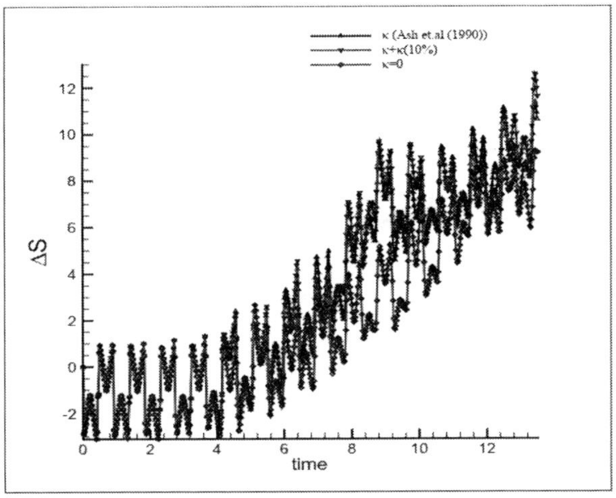

Figure 3. Total entropy for rectangular domain (Case 1) plotted as function of time; △ represents results without Stokes' hypothesis and ◇ represents results with Stokes' hypothesis and symbol ▽ represents results with an increased value (1.1κ).

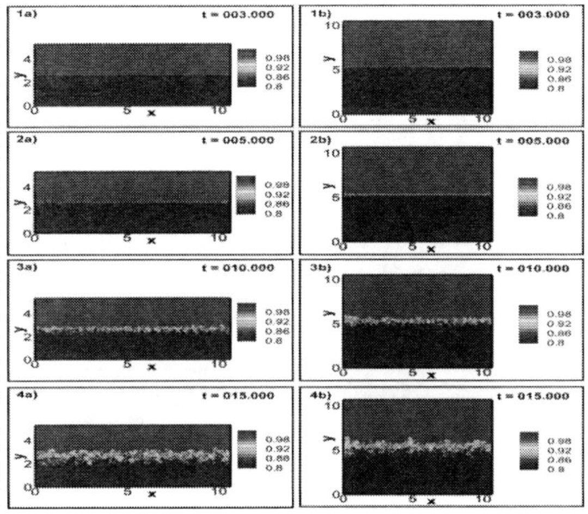

Figure 4. Total entropy for rectangular and square domain plotted as function of time. Symbol △ represents results without Stokes' hypothesis for Case 1; □ represents Case 2 without Stokes' hypothesis and ◇ for Case 2 with the hypothesis.

notices that the computed entropy is higher without the restriction of Stokes' hypothesis. This is due to the additional dissipative term $\nabla(\kappa.\nabla u)$ appearing in the momentum equation and $\kappa(\nabla u)^2$ in the energy equation. The dissipative structures arise in the systems (control volume or computational cells) which are able to exchange energy or mass with their neighbours or corresponding cells to establish a macroscopic internal order. The dissipation arising due to these structures or fronts correspond to a low entropy value visible in figures 3 and 4 and may indeed be a source of order [31]. In figure 4, entropy change is compared between rectangular and square domains. Also in this figure, the variation of entropy for Case 2 is compared with and without Stokes' hypothesis. The relation between entropy and vorticity is given by Crocco's theorem [29], as discussed in [27]. The rate of mean increase of total entropy is also greater and approximately double in Case 2 compared to Case 1, which is due to the fact that mixing occurs early and at smaller scales in the square box compared to the rectangular box as will be explained later with respect to figure 8. Interface instability causes spikes and bubble formation in the presence of mixing. The increase in the atypical M structures in the total entropy plot is due to pressure waves being created at the interface, which propagate in the y-direction and reflect from the top and bottom walls while the background mixing due to diffusion shows up of continuously increasing entropy. For the square box, pressure pulses take a longer time to reach the top of the box and reflect from the wall as rarefaction waves.

Pressure Waves:
Apart from the mean entropy increasing for computed NSE (with and without Stokes' hypothesis), there is also atypical sharp variation of entropy, as shown in figures 3 and 4. This is due to the creation of multiple pressure pulses, as the working medium (air) allows formation of pressure waves starting from the interface. This is shown in figures 5 and 6 for Case 1, with and without Stokes' hypothesis by solid and dashed lines respectively. Figures 5 and 6 show the pressure across y at $x = 30$ with the former dominated by pressure pulses for Case 1 at the onset of RTI, while figure 6 shows events at a later time when mixing is predominant. Effects

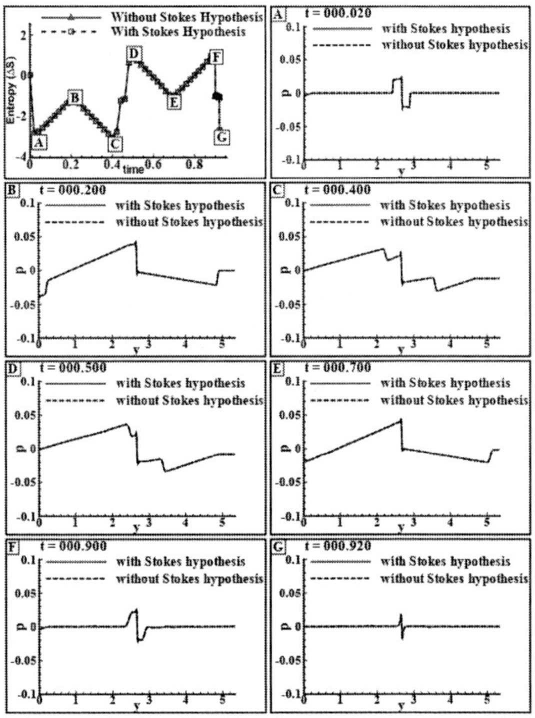

Figure 5. Variation of pressure along y for $x = 30$ at $(A)\ t = 0.02$, $(B)\ t = 0.2$, $(C)\ t = 0.4$, $(D)\ t = 0.5$, $(E)\ t = 0.7$, $(F)\ t = 0.9$, $(G)\ t = 0.92$ showing the cycle during onset of RTI. Reflection of pressure waves from the walls and interface is seen at different instances, as illustrated in top left frame.

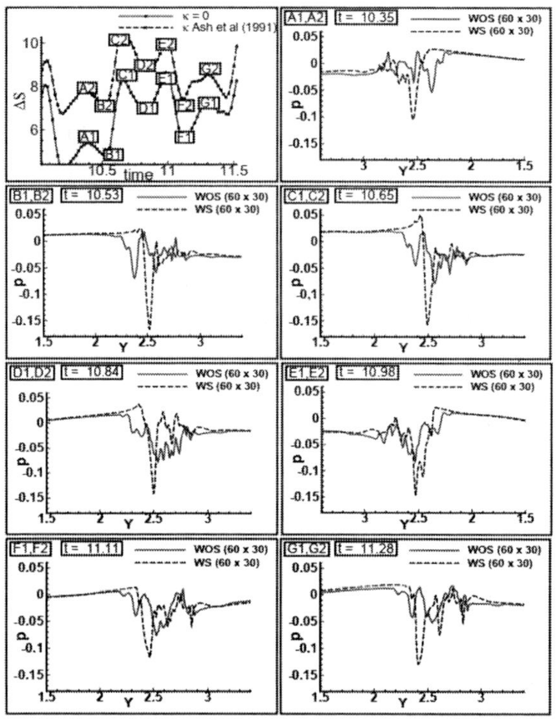

Figure 6. Variation of pressure along y for $x = 30$, with and without Stokes' hypothesis at $(A)\ t = 10.35$, $(B)\ t = 10.53, (C)\ t = 10.65$, $(D)\ t = 10.84$, $(E)\ t = 10.98$, $(F)\ t = 11.11$, $(G)\ t = 11.28$ showing one cycle which is determined by both mixing and pressure waves.

of bulk viscosity are visible at the interface ($y = 2.4 - 3$) shown in figure 6, during $t = 10.35$ to 11.25. The formation of pressure pulses from the interface towards the top and bottom of the box is associated with the fluid dynamic transient response. These pulses upon reflection from solid wall changes from compression to rarefaction pulses and vice versa and move back towards the original interface. These compression and rarefaction of the medium give rise to entropy changes, as given by equation (4), and has one is to one correspondence for the variation of entropy, with compression giving rise to increase in entropy, while rarefaction reduces entropy of the system.

Reckinger *et al* [10] in their compressible N-S, energy and species mass fraction formulation, used periodicity in x- and y-directions, using non-reflecting boundary condition along with an additional buffer layer of increased viscosity to assist damping of the pressure wave. This is to ensure that they simulate *"an infinite domain such that any pressure wave approaching the numerical boundaries is not reflected and thus does not interact with and disturb the growth of the instability"*. In the present simulations, we instead capture the pressure waves by implementing non-periodic boundary conditions in x- and y-directions to report the significance of atypical entropy variation for the thermodynamic system for both rectangular and square domain in figures 5 and 6.

Evolution of density for Cases 1 and 2 are shown in figure 7 for the complete domain. Here, the

Journal of Physics: Conference Series **759** (2016) 012079 doi:10.1088/1742-6596/759/1/012079

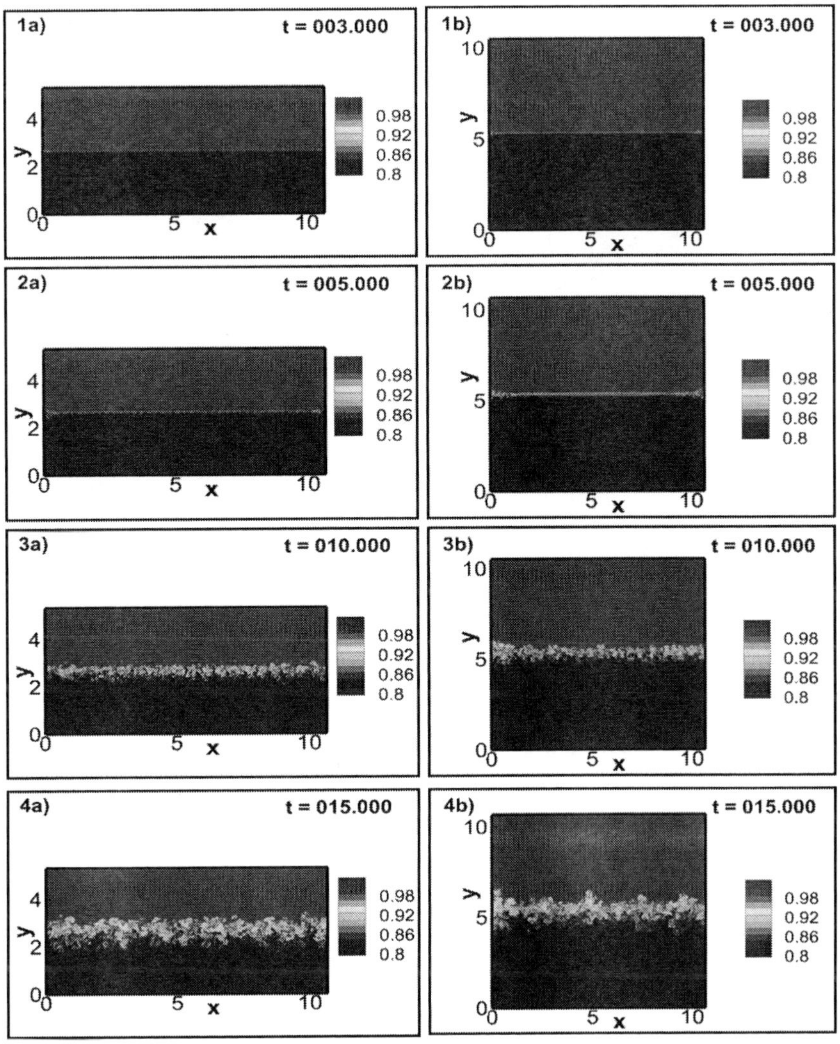

Figure 7. Density contours for the present simulations for rectangular (left) and square boxes (right) shown in $(1a, 1b)$ $t = 3.0$, $(2a, 2b)$ $t = 5.0$, $(3a, 3b)$ $t = 10.0$, $(4a, 4b)$ $t = 15.0$.

flow is not periodic in x-direction and the points of interface meeting the side-walls are the major sources of initial perturbations, as observed in frames $2a$) and $2b$) at $t = 5$. These disturbances originating at the side-walls dictate the flow evolution during RTI evident in frames $3a$) and $3b$) at $t = 10$. Although there are no visible differences in figure 7, indeed there are significant differences right from the onset (in terms of vortical structures), as seen in the zoomed pictures of the interface in figure 8. For example, in frame $2a$), one can notice distinct vortical structure at $t = 5$, while for the square box, flow evolves more due to delayed pressure reflections from top and bottom wall, which is responsible for delayed vortex structure formations. Quite contrary to classical fluid flow description, here the flow evolves from small scales, which propagate rapidly along the interface from both the ends (the triple points at the junction of the side-walls and the interface).

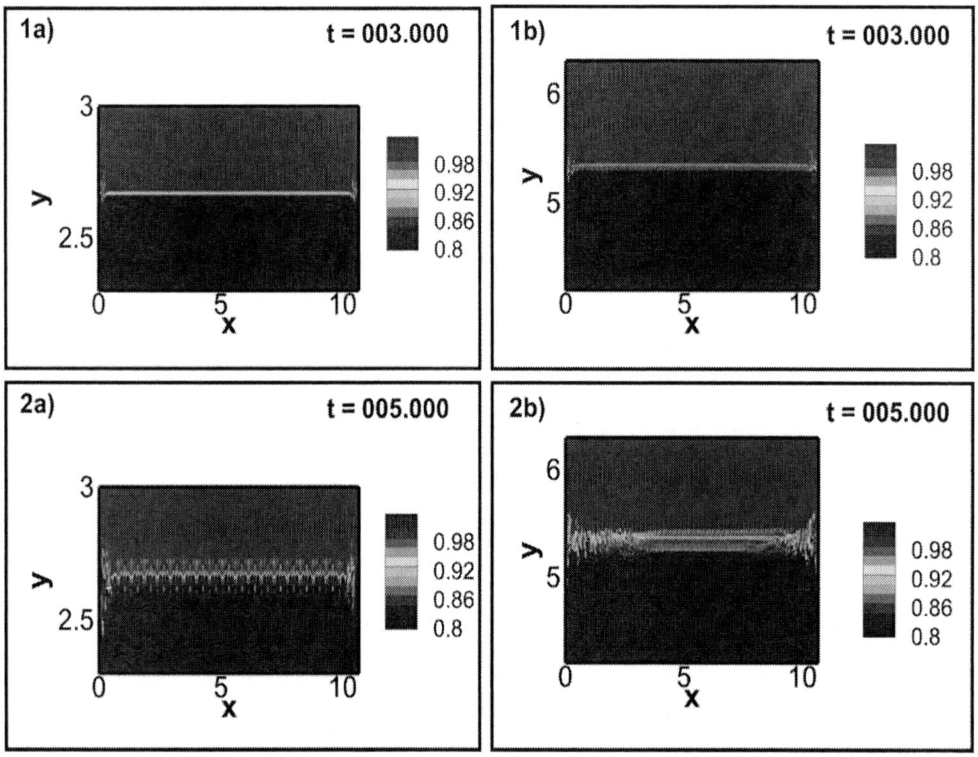

Figure 8. Zoomed density contours for the present simulations for rectangular (left) and square boxes (right) shown at: $(1a, 1b)$ $t = 3.0$, and $(2a, 2b)$ $t = 5.0$ Twisting and stretching at the interface can be visualized in frame $2b$) which is absent in frames $1a$) and $2a$).

Summary

In the present analysis, Ash *et al.* [42]'s experimental data for air is used for a new model of second coefficient of viscosity, so that the bulk viscosity (κ) is not zero. In classical approach using Stokes' hypothesis, one uses $\kappa = 0$ for compressible NSE. These results are compared in figures 3 to 6, from the onset of RTI till the time when mixing is complete ($t = 14$). It is clearly evident that the Stokes' hypothesis promotes dissipation, while the calculations without the same hypothesis show sharper entropy variation. Initial growth is determined by formation of pressure waves, which is dependent on the value of bulk viscosity and this is evident in figures 3 and 4. The present research helps one obtain an alternative to using traditional Stokes' hypothesis, which results in a more accurate representation of entropy variation that exhibits the non-equilibrium thermodynamics during RTI in an inhomogeneous system. In the near future, we will extend the present study of RTI to a 3D finite non-periodic box.

References

[1] Sengupta, T K 2013 *High Accuracy Computing Methods: Fluid Flows and Wave Phenomena* (New York: Cambridge University Press)
[2] Rayleigh Lord 1883 *Proc. London Math. Soc.* **14** pp 170-77
[3] Taylor G I 1950 *I. In Proc. R. Soc. London A* bf 201 pp 192-6
[4] Chandrasekhar S 1961 *Hydrodynamics and Hydrodynamic Stability* (Clarendon: Oxford)
[5] Sharp D H 1984 *Phys. D* **12** pp 3-18
[6] Bhole A, Sengupta S, Sengupta A, Shruti K S and Sharma N 2015 *In Proc. of IUTAM Symp. Advances in Computation, Modeling and Control of Transitional and Turbulent Flows* (eds. T K Sengupta *et al.*) World Sci. Publ. Co. Singapore pp 471-81

[7] Cook A W and Dimotakis P E 2001 *J. Fluid Mech.* **443** pp 69-99
[8] Sengupta T K, Rajpoot M K and Bhumkar Y G 2011 *Comp. & Fluids* **47** pp 144-54
[9] Sengupta T K, Ganeriwal G and De S 2003 *J. Comp. Phys.* **192** pp 677-94
[10] Reckinger S J, Livescu D and Vasilyev O V 2010 *Phys. Scr.* **T142** pp 014064(1-6)
[11] Sengupta T K, Bhole A and Sreejith N A 2013 *Comp. & Fluids* **88** pp 19-37
[12] Hoffmann K A and Chiang S T 2011 *Computational Fluid Dynamics* **2** (Kansas: Engineering Education Systems)
[13] Read K I 1984 *Phys. D* **12** pp 45-58
[14] Andrews M J and Spalding D B 1990 *Phys. Fluids A* **2** pp 922-27
[15] Linden P F and Redondo J M 1991 *Phys. Fluids A* **3** pp 1269-77
[16] Dalziel S B, Linden P F and Youngs D L 1999 *J. Fluid Mech.* **399** pp 1-48
[17] Roberts M S and Jacobs J W 2016 *J. Fluid Mech.* **787** pp 50-83
[18] Dimonte G and Schneider M 1996 *Phys. Rev. E* **54** pp 3740-3
[19] Youngs D L 1984 *Phys. D* **12** pp 32-44
[20] Youngs D L 1991 *Phys. Fluids A* **3** pp 1312-20
[21] Cook A W, Cabot W and Miller P L 2004 *J. Fluid Mech.* **511** pp 333-62
[22] Ramaprabhu P, Dimonte G and Andrews M J 2005 *J. Fluid Mech.* **536** pp 285-319
[23] Cabot W H and Cook A W 2006 *Nature Phys.* **2** pp 562-68
[24] Ramaprabhu P, Dimonte G, Woodward P, Fryer C, Rockefeller G, Muthuraman K, Lin P H and Jayaraj J 2012 *Phys. Fluids* **24** pp 074107(1-21)
[25] Wei T and Livescu D 2012 *Phys. Rev. E* **86** pp 046405(1-14)
[26] Ramaprabhu P, Karkhanis V and Lawrie A G W 2013 *Phys. Fluids* **25** pp 115104(1-33)
[27] Sengupta T K, Sengupta A, Sengupta S, Bhole A, Shruti K S 2016 *Int. J. of Thermophysics* **37:36** pp 1-25
[28] Lawrie A G W 2009 *Ph.D. thesis* (DAMPT, University of Cambridge, UK)
[29] Shapiro A H 1954 *The Dynamics and Thermodynamics of Compressible Fluid Flow* (USA: Ronald Press Co.)
[30] Landau L D and Lifshitz E M 1987 *Fluid mechanics* (USA: Elsevier)
[31] Glansdorff P and Prigogine I 1971 *Thermodynamic Theory of Structure, Stability, and Fluctuations* (London: Wiley-Interscience)
[32] Liebermann L N 1949 *Phys. Rev.* **75** pp 1415-25
[33] Emanuel G 1990 *Phys. Fluids A* **2** pp 2252-55.
[34] Gad-el-Hak M 1995 *J. Fluids Engg.* **117** pp 3-5
[35] Graves R E and Argrow B M 1999 *J. of Thermophysics and Heat Transfer* **13** pp 337-42
[36] Buresti G 2015 *Acta Mechanica* **226** pp 3555-9
[37] Stokes G G 1845 *Tran. Cambridge Philo. Soc.* **8** pp 287-305
[38] Karim S M and Rosenhead L 1952 *Rev. of Modern Phys.* **24** pp 108-16
[39] Rosenhead L 1954 *In Proc. R. Soc. London A* **226** pp 1-6
[40] Cramer M S 2012 *Phys. Fluids* **24** pp 066102(1-23)
[41] Rajagopal K R 2013 *Int. J. of Non-Linear Mech.* **50** pp 141-51
[42] Ash R L, Zuckerwar A J and Zheng Z 1991 *NASA CR-187783*
[43] Mikaelian K O 2014 *Phys. Fluids* **26** pp 054103(1-16)

Influence of nonlinear chemical reactions on the transport coefficients in oscillatory Couette flow

Swarup Barik and D. C. Dalal
Department of Mathematics, IIT Guwahati, Guwahati, India
E-mail: s.barik@iitg.ernet.in, durga@iitg.ernet.in

Abstract. A multiple-scale method of averaging is applied to the study of transport of a chemical species in oscillatory Couette flow where the species may undergoes a reversible phase exchange with the boundary wall and nonlinear chemical reactions both within the fluid and at the boundary wall. Analytical expressions are obtained for transport coefficients. The results shows how the transport coefficients are influenced by the reversible phase exchange reaction kinetics and the rate and degree of the nonlinear decay chemical reaction.

1. Introduction

Transport of a solute in a fluid flow has several important applications in diverse fields such as biology, chemistry, chromatography and environmental fluid mechanics. Taylor [9] first studied theoretically as well as experimentally the transport of a solute dissolved in a fluid flowing through a circular pipe of constricted diameter. Since then, this mechanism has been termed as "Taylor dispersion". In the system, reaction may take place in the bulk flow (mobile phase) and at the sorbet boundaries (immobile phases). It may be reversible or irreversible in nature and it may also be classified as linear or nonlinear.

Dispersion in oscillatory Couette flow without reaction has been studied by Mazumder and co-workers [1, 5] using a modified moments method. Solute transport in oscillatory Couette flow with linear phase exchange reaction has been analysed by Ng and Bai [4]. Revelli and Ridolfi [7, 8], Paul and Mazumder [6] have shown the effects of nonlinear chemical reactions on solute transport. In spite of its environmental and industrial applications, less attention has been paid to the study of nonlinear reactions on solute transport.

The main purpose of this study is to explore the effects of nonlinear chemical decay phenomena on transport coefficients in an oscillatory Couette flow. A 2-D mathematical model is formulated by considering the effects of diffusion, advection, decay and a phase exchange kinetics between mobile and immobile phases. In the present work, nonlinear decay reaction terms are included in both in mobile and immobile phases and this is a generalization of the work of Ng and Bai [4]. This study may help to model shear-driven flows encountered in micro motors, micro channels and other micro fluidic systems.

2. Problem under consideration

A laminar, one-dimensional plane Couette flow between two infinite parallel flat walls with a separation width h is considered. The lower boundary at $y = 0$ is assumed to be stationary whereas the upper boundary at $y = h$ oscillates in its own plane with a prescribed velocity

$$u(h,t) = U[1 + \phi Re(e^{-i\omega t})],$$

where U is the steady component of the velocity of the upper boundary, ϕ is a factor such that ϕU is the amplitude of the oscillatory velocity component, ω is the angular frequency of the plate oscillation, t is time and Re stands for real part. The velocity profile of the flow can be taken as (Ng and Bai [4])

$$u(y,t) = u_s + Re(u_w e^{-i\omega t}),$$

where $u_s(y) = Uy/h$ and $u_w(y) = U\phi\frac{\sin\sigma y}{\sin\sigma h}$ are the steady and oscillatory components of the velocity field respectively. Here $\sigma^2 = i\omega/\nu$ or $\sigma = (1+i)/\delta$ and $\delta = \sqrt{2\nu/\omega}$, where ν is the kinematic viscosity of the fluid.

The chemical considered in this study is assumed to be completely miscible in the fluid and would undergo a reversible sorptive phase exchange across the fluid-lining interface. It is possible that in the lining the chemical exists in a form different from that in the flowing fluid. To distinguish between the forms of the chemical that exist in the fluid and in the lining, we referred to them as the mobile and the immobile phases, respectively. Because of reversible phase exchange reaction, the chemical sticks into immobile phase from mobile phase and reversely release into mobile phase from immobile phase. It is also assumed that the chemical undergoes nonlinear decay reaction both in mobile and the immobile phases and at equilibrium stage, these two phases will have their concentrations in a fixed ratio

$$C_s/C = \alpha,$$

where C is the concentration (mass of chemical per bulk volume of fluid) of the mobile phase, C_s is the concentration (mass of chemical per surface area of wall) of the immobile phase and α the partition coefficient having the dimension of length.

If the equilibrium is not attained, the phase exchange kinetics along with nonlinear decay chemical reaction (defined by Δ_s) generally takes place either in forward or in backward direction (Paul and Mazumder [6]) according to

$$\frac{\partial C_s}{\partial t} = k(\alpha C - C_s) + \Delta_s,$$

where k is the reversible reaction rate constant.

The problem for the transport of the chemical can be formulated as follows:

$$\frac{\partial C}{\partial t} + u\frac{\partial C}{\partial x} = D\frac{\partial^2 C}{\partial x^2} + D\frac{\partial^2 C}{\partial y^2} + \Delta, \quad 0 < y < h, \tag{1}$$

where D is the molecular diffusivity of the chemical in the fluid assumed to be constant and Δ represents the effect of nonlinear chemical reaction occurring in the bulk flow.

The corresponding reaction rates, Δ and Δ_s are prescribed according to the power laws,

$$\Delta = -\lambda C^n, \quad \Delta_{si} = -\lambda_{si} C_{si}^m, \quad (i = 1, 2),$$

where λ and λ_s are the reaction rate constants for the mobile and immobile phases respectively and n and m are the corresponding degrees of nonlinearity.

The boundary conditions for (1) are taken as

$$D\frac{\partial C}{\partial y} = \frac{\partial C_{s1}}{\partial t} = k_1(\alpha_1 C - C_{s1}) - \lambda_{s1} C_{s1}^m, \quad y = 0, \tag{2}$$

$$-D\frac{\partial C}{\partial y} = \frac{\partial C_{s2}}{\partial t} = k_2(\alpha_2 C - C_{s2}) - \lambda_{s2} C_{s2}^m, \quad y = h. \tag{3}$$

The subscripts 1 and 2 are used to denote quantities associated with the lower and upper boundaries.

Journal of Physics: Conference Series **759** (2016) 012080 doi:10.1088/1742-6596/759/1/012080

3. Assumptions for asymptotic analysis

Homogenization technique proposed by Mei *et al.* [2] has been applied in this study. It is assumed that a sufficiently long time has passed since the discharge of the chemical into the flow so that the length scale for the axial spreading of the chemical cloud is much greater than the channel depth, and $\epsilon = h/L (\ll 1)$ used as the perturbation parameter. Along with two different length scales h and L, three distinct time scales have also been considered for lateral diffusion, advection/reaction and axial diffusion (Ng [3]). Their ratios are

$$T_0 : T_1 : T_2 = 1 : \frac{1}{\epsilon} : \frac{1}{\epsilon^2}.$$

Based on these time scales, fast, medium and slow time variables have been introduced accordingly as follows

$$t_0 = t, \quad t_1 = \epsilon t, \quad t_2 = \epsilon^2 t.$$

The original time derivative becomes, according to the chain rule:

$$\frac{\partial}{\partial t} = \frac{\partial}{\partial t_0} + \epsilon \frac{\partial}{\partial t_1} + \epsilon^2 \frac{\partial}{\partial t_2}.$$

Dependent variables are expanded in power series to carry on perturbation analysis:

$$C = C^{(0)} + \epsilon C^{(1)} + \epsilon^2 C^{(2)} + O(\epsilon^3),$$

$$C_{si} = C_{si}^{(0)} + \epsilon C_{si}^{(1)} + \epsilon^2 C_{si}^{(2)} + O(\epsilon^3), \quad (i = 1, 2).$$

By perturbation analysis, 1-D transport equation has been obtained that describes the averaged behavior of the 2-D phenomenon.

4. Transport coefficients

Let us introduce the following normalised variables

$$\widehat{C}^{(0)} = \frac{C^{(0)}}{C_0}, \widehat{\alpha}_1 = \frac{\alpha_1}{h}, \widehat{\alpha}_2 = \frac{\alpha_2}{h}, \widehat{\beta}_1 = \frac{\beta_1 h}{D}, \widehat{\beta}_2 = \frac{\beta_2 h}{D}, Da_1 = \frac{k_1 h^2}{D}, Da_2 = \frac{k_2 h^2}{D}, \widehat{\lambda} = \frac{\lambda}{k_*} C_0^{n-1},$$
$$\widehat{\lambda}_{s1} = \frac{\lambda_{s1}}{k_1} h^{m-1} C_0^{m-1}, \widehat{\lambda}_{s2} = \frac{\lambda_{s2}}{k_2} h^{m-1} C_0^{m-1}, \widehat{\sigma} = \sigma h, \widehat{\eta} = \eta h, \widehat{\delta} = \frac{\delta}{h},$$

where $\beta_j = -\frac{i\omega k_j \alpha_j}{k_j - i\omega}$, ($j = 1, 2$ and $i = \sqrt{-1}$). The Damkohler number Da is the ratio of phase exchange rate to the diffusion rate. Here in the expression of $\widehat{\lambda}$, it is considered that $k_* = k_1$ only when the lower boundary is sorbing (Case 1), $k_* = k_2$ only when the upper boundary is sorbing (Case 2) and without loss of generality $k_* = k_1 = k_2$ when both boundaries are sorbing (Case 3). Using this normalised variables, the effective transport equation can be obtained by perturbation analysis of (1) - (3) as

$$\frac{\partial \widehat{C}^{(0)}}{\partial t} - \left[\frac{D}{R} + \widehat{D}_{Ts} \frac{U^2 h^2}{D} + \widehat{D}_{Tw} \frac{U^2 h^2}{D}\right] \frac{\partial^2 \widehat{C}^{(0)}}{\partial x^2} + \langle \bar{u} \rangle \widehat{\zeta}_1 \frac{\partial \widehat{C}^{(0)^n}}{\partial x} + \langle \bar{u} \rangle \widehat{\zeta}_2 \frac{\partial \widehat{C}^{(0)^m}}{\partial x} + \langle \bar{u} \rangle \widehat{\zeta}_3 \frac{\partial \widehat{C}^{(0)}}{\partial x}$$
$$+ k_* \widehat{\chi}_1 \widehat{C}^{(0)^{2n-1}} + k_* \widehat{\chi}_2 \widehat{C}^{(0)^{n+m-1}} + k_* \widehat{\chi}_3 \widehat{C}^{(0)^n} = 0, \quad (4)$$

where $\widehat{D}_{Ts}$ and $\widehat{D}_{Tw}$ are respectively the dispersion coefficients due to the steady and oscillatory flows and $R = 1 + \widehat{\alpha}_1 + \widehat{\alpha}_2$. The overbar denotes time average and angle brackets denote section average. The terms $\widehat{\zeta}_i$'s and $\widehat{\chi}_i$'s ($i = 1, 2, 3$) are the effective coefficients for the advection and reaction respectively. Analytical expressions for the dispersion coefficients for the steady and oscillatory flows *i.e.* $\widehat{D}_{Ts}$ and $\widehat{D}_{Tw}$ have been derived but are not given here. Graphical variations for $\widehat{D}_{Tw}$ can be seen from figure (1).

Analytical expressions for all the advection and reaction coefficients can be obtained as

$$\begin{aligned}
\widehat{\zeta}_1 &= -\frac{\widehat{\lambda} Da_*}{6R^3}\left[5\widehat{\alpha}_1^2 + 3\widehat{\alpha}_2^2 - 4\widehat{\alpha}_1\widehat{\alpha}_2 + \widehat{\alpha}_1 - \widehat{\alpha}_2 + 12\left(\frac{\widehat{\alpha}_1}{Da_1} + \frac{\widehat{\alpha}_2}{Da_2}\right)\right],\\
\widehat{\zeta}_2 &= \frac{1}{R^2}\left[\widehat{\lambda}_{s1}\widehat{\alpha}_1^m + \widehat{\lambda}_{s2}\widehat{\alpha}_2^m\right], \qquad \widehat{\zeta}_3 = \frac{1}{R},\\
\widehat{\chi}_1 &= -\frac{n\widehat{\lambda}^2 Da_*}{3R^3}\left[\widehat{\alpha}_1^2 + \widehat{\alpha}_2^2 - \widehat{\alpha}_1\widehat{\alpha}_2 + 3\left(\frac{\widehat{\alpha}_1}{Da_1} + \frac{\widehat{\alpha}_2}{Da_2}\right)\right],\\
\widehat{\chi}_2 &= \frac{m\widehat{\lambda}}{R^2}\left[\widehat{\lambda}_{s1}\widehat{\alpha}_1^m + \widehat{\lambda}_{s2}\widehat{\alpha}_2^m\right], \qquad \widehat{\chi}_3 = \frac{\widehat{\lambda}}{R}.
\end{aligned}$$

Here in the expression of $\widehat{\zeta}_1$ and $\widehat{\chi}_1$, one has to take $Da_* = Da_1$ for Case 1, $Da_* = Da_2$ for Case 2 and without loss of generality, it is assumed that $Da_* = Da_1 = Da_2$ for Case 3.

If there is no nonlinear reaction in the bulk flow and in the boundary, the chemical undergoes only reversible phase exchange with the flowing fluid i.e., $\widehat{\lambda} = \widehat{\lambda}_{s1} = \widehat{\lambda}_{s2} = 0$.

So, the equation (4) is reduced to

$$\frac{\partial \widehat{C}^{(0)}}{\partial t} - \left[\frac{D}{R} + \widehat{D}_{Ts}\frac{U^2h^2}{D} + \widehat{D}_{Tw}\frac{U^2h^2}{D}\right]\frac{\partial^2 \widehat{C}^{(0)}}{\partial x^2} + \langle\bar{u}\rangle\widehat{\zeta}_3\frac{\partial \widehat{C}^{(0)}}{\partial x} = 0, \tag{5}$$

and this equation coincides with the one obtained by Ng and Bai [4] in their work. This shows that, in the absence of nonlinearity the results are in good agreement with the existing literature.

5. Results and discussions

In this problem, effects of different parameters and chemical reaction on some transport coefficients have been studied.

The influence of $\widehat{D}_{Tw}$ with $\widehat{\delta}$ is shown in figure (1), where $\widehat{\delta}$ is the oscillation parameter. The value of $\widehat{\delta}$ will be smaller for higher angular frequency. The dispersion coefficient $\widehat{D}_{Tw}$ increases monotonically with $\widehat{\delta}$ for all sorbing cases. In other words, for a lower frequency the dispersion coefficient value will be higher. The enhancement effect gradually decreases as $\widehat{\delta}$ increases and beyond unity no remarkable effect on $\widehat{D}_{Tw}$ is observed. Again, among all three sorbing cases, Case 1 gives largest dispersion coefficient and Case 3 gives the least. Also, when both boundaries are inert, $\widehat{D}_{Tw}$ does not change in the plotted range of $\widehat{\delta}$.

Graphs of the reaction and advection coefficients $\widehat{\chi}_1$ and $\widehat{\zeta}_2$ against $\widehat{\alpha}_1$ and $\widehat{\alpha}_2$ are shown in figures (2a) and (2b) respectively for different degrees of nonlinearity. For a fixed value of n, initially the graph of $\widehat{\chi}_1$ decreases sharply to a minimum with increase of $\widehat{\alpha}_1$ and $\widehat{\alpha}_2$ and then it asymptotically approaches to zero. Large degree of nonlinearity for the decay reaction in the bulk flow reduces the value of the coefficient $\widehat{\chi}_1$. Figure also shows that advection coefficient $\widehat{\zeta}_2$ increases for $m > 2$ and remain stationary for $m \leq 2$ as $\widehat{\alpha}_1$ and $\widehat{\alpha}_2$ increase. Large degree of nonlinearity for the decay reaction at the boundaries shows the large value of the coefficient $\widehat{\zeta}_2$.

The variations of transport coefficients $\widehat{\chi}_1$ and $\widehat{\zeta}_1$ with $\widehat{\lambda}$ can be seen from figure (3) for different values of retention parameters $\widehat{\alpha}_1$ and $\widehat{\alpha}_2$. It is clear from the figure that the coefficients $\widehat{\chi}_1$ and $\widehat{\zeta}_1$ are found to decrease. The changes are more noticeable for $\widehat{\chi}_1$ than those for $\widehat{\zeta}_1$ and the trends are undisturbed by any variations in $\widehat{\alpha}_1$ and $\widehat{\alpha}_2$.

Figure (4) shows the behavior of the advection coefficient $\widehat{\zeta}_2$. It is seen from the figure that for all values of $\widehat{\alpha}_1$ and $\widehat{\alpha}_2$, $\widehat{\zeta}_2$ increases with the increase of $\widehat{\lambda}_{s1}$ and $\widehat{\lambda}_{s2}$, moreover, larger $\widehat{\alpha}_1$ and $\widehat{\alpha}_2$ increase the rate of enhancement of $\widehat{\zeta}_2$.

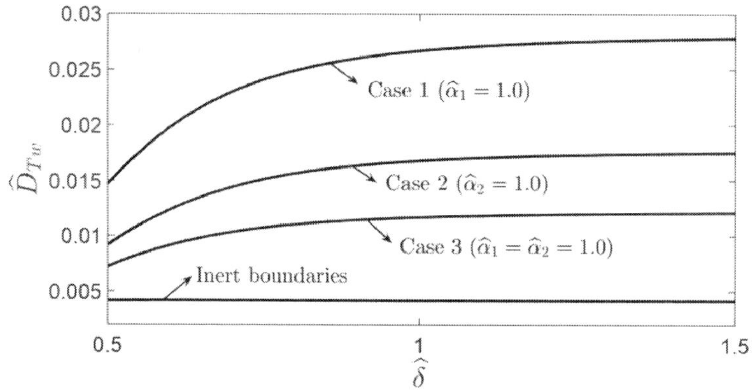

Figure 1. Variations of oscillatory dispersion coefficient $\widehat{D}_{Tw}$ with oscillation parameter $\widehat{\delta}$, where $\phi = 1.0$, $Sc = 0.1$ and $Da_1 = Da_2 = 1.0$.

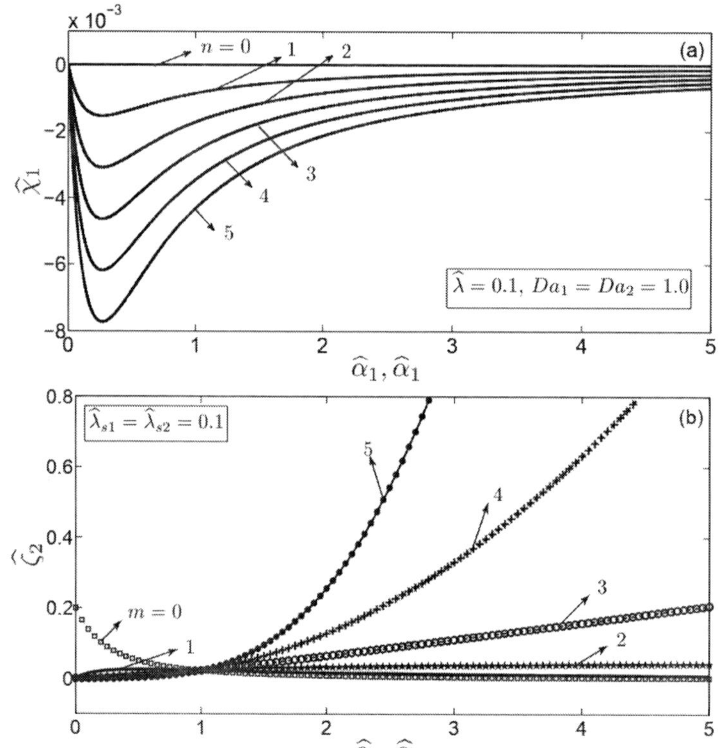

Figure 2. The transport coefficients $\widehat{\chi}_1$,$\widehat{\zeta}_2$ as a function of $\widehat{\alpha}$ for different degrees of nonlinearity.

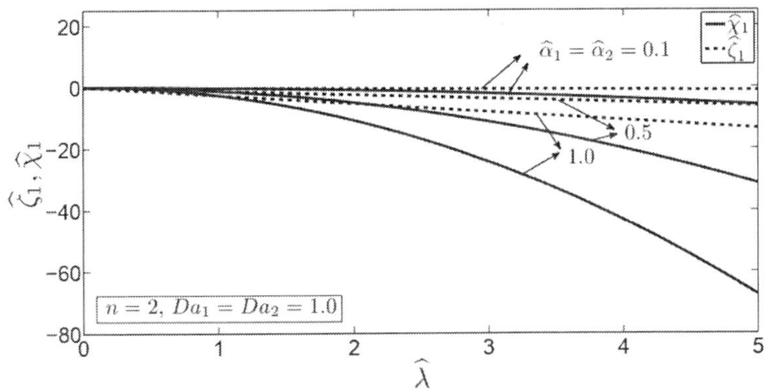

Figure 3. The transport coefficients $\widehat{\chi}_1$,$\widehat{\zeta}_1$ as a function of $\widehat{\lambda}$ for different values of $\widehat{\alpha}$.

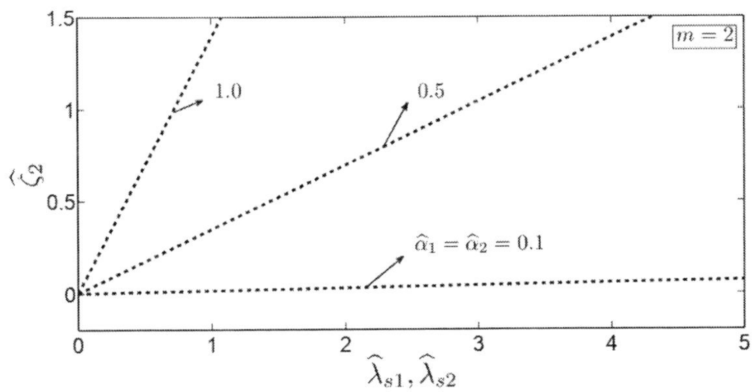

Figure 4. The advection coefficient $\widehat{\zeta}_2$ as a function of $\widehat{\lambda_s}$ for different values of $\widehat{\alpha}$.

6. Conclusions

Effects of different parameters and nonlinear reactions both within the bulk flow and at the wall on some transport coefficients are accounted explicitly. Some important conclusions are

(i) Lower angular frequency will give larger dispersion coefficients and coefficient is highly influence by wall sorption than inert boundary walls.

(ii) If the nonlinearity of the mobile phase reaction increases then its corresponding coefficient decreases whereas trend is opposite for the immobile phase reaction.

(iii) If the decay reaction rate in the bulk flow increases then corresponding coefficient decreases while result is opposite for decay reaction at the boundaries.

References

[1] Bandyopadhyay S and Mazumder BS 1999 *Int. J. Eng. Sci.* **37** 1407
[2] Mei CC, Auriault JL and Ng CO 1996 *Adv. Appl. Mech.* **32** 278
[3] Ng CO 2006 *Proc. R. Soc. A* **462** 481
[4] Ng CO and Bai YC 2005 *Acta. Mech.* **178** 65
[5] Paul S and Mazumder BS 2008 *Int. J. Eng. Sci.* **46** 1203
[6] Paul S and Mazumder BS 2011 *Int. J. Heat Mass Tran.* **54** 75
[7] Revelli R and Ridolfi L 2002 *J. Fluid Mech.* **472** 307
[8] Revelli R and Ridolfi L 2003 *Adv. Water Resour.* **26** 815
[9] Taylor GI 1953 *Proc. R. Soc. A* **219** 186

Journal of Physics: Conference Series **759** (2016) 012081 doi:10.1088/1742-6596/759/1/012081

Moffatt vortices in the lid-driven cavity flow

Sougata Biswas†, **Jiten C. Kalita**‡

Department of Mathematics, Indian Institute of Technology Guwahati, PIN 781039, India

E-mail: †b.sougata@iitg.ernet.in, ‡jiten@iitg.ernet.in

Abstract. In incompressible viscous flows in a confined domain, vortices are known to form at the corners and in the vicinity of separation points. The existence of a sequence of vortices (known as Moffatt vortices) at the corner with diminishing size and rapidly decreasing intensity has been indicated by physical experiments as well as mathematical asymptotics. In this work, we establish the existence of Moffatt vortices for the flow in the famous Lid-driven square cavity at moderate Reynolds numbers by using an efficient Navier-Stokes solver on non-uniform space grids. We establish that Moffatt vortices in succession follow fixed geometric ratios in size and intensities for a particular Reynolds number. In order to eliminate the possibility of spurious solutions, we confirm the physical presence of the small scales by pressure gradient computation along the walls.

1. Introduction

In Stokes flow near a corner between two intersecting solid boundaries or a solid boundary and a free surface, the existence of a sequence of counter-rotating vortices was first established theoretically by H. K. Moffatt [14, 15]. A flow visualization experiment by Taneda [19] in a V-notch endorsed the existence of these vortices. They are known as "Moffatt vortices", named aptly after H. K. Moffatt. Such vortices are formed due to rotation of a cylinder or external stirring force close to the solid boundaries. A careful undermining into the existing literature reveals that the study of Moffatt vortices in Stokes flow on different geometries [1, 11, 12, 13, 14, 15, 16] has so far been established mostly through theoretical studies. There have been very few numerical studies [4, 17] on this topic available in the existing literature. In this work, we study the existence of Moffatt eddies at the bottom corners for the flow in Lid-driven cavity at moderate Reynolds numbers by using a recently developed Higher-Order-Compact (HOC) scheme by Kalita *et al.* [8] on non-uniform space grids. We further establish that the Moffatt eddies in succession follow geometric ratios of the size and intensities which again depend on Reynolds numbers. By grid independence study and computing pressure gradients along the walls of the cavity, we confirm that the smallest scales captured by us are not numerical artifacts.

This paper is organized in four sections. The problem is briefly described in Section 2, while Section 3 deals with numerical procedures and other related issues, Section 4 with results and discussions and finally, in Section 5 we summarize the whole work.

2. Problem Description

We have considered the classical Lid-driven square cavity problem (see Figure 1) which is one of the most extensively solved problems in CFD ([3, 5, 6, 8, 9, 18]) as it displays almost all fluid mechanical phenomena for incompressible viscous flows. The governing equations for the flow

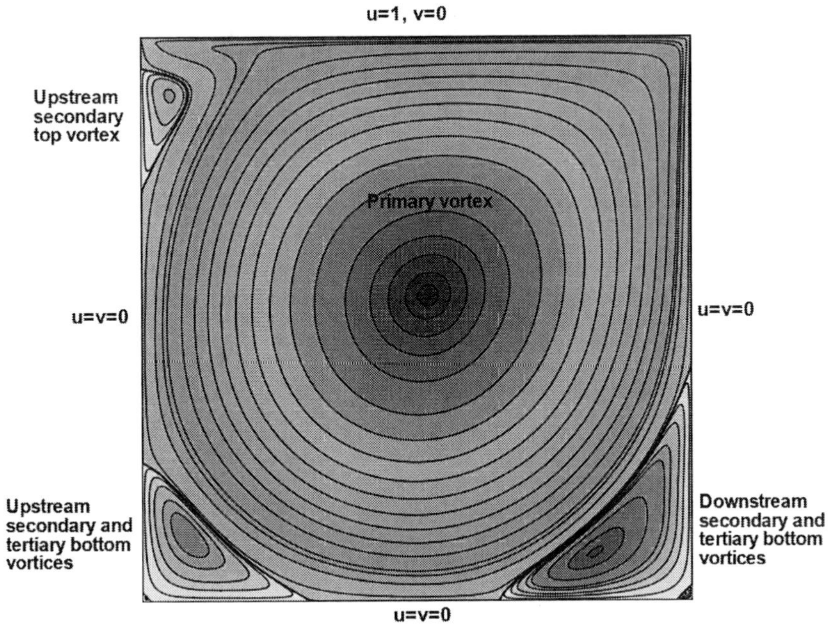

Figure 1. Schematic of the lid-driven square cavity flow along with the streamlines for $Re = 3200$.

in cavity are the two-dimensional (2D) incompressible Navier-Stokes (N-S) equations which in the non-dimensional primitive-variables form are given by

$$\frac{\partial u}{\partial x} + \frac{\partial v}{\partial y} = 0, \tag{1}$$

$$\frac{\partial u}{\partial t} + u\frac{\partial u}{\partial x} + v\frac{\partial u}{\partial y} = -\frac{\partial p}{\partial x} + \frac{1}{Re}\nabla^2 u, \tag{2}$$

$$\frac{\partial v}{\partial t} + u\frac{\partial v}{\partial x} + v\frac{\partial v}{\partial y} = -\frac{\partial p}{\partial y} + \frac{1}{Re}\nabla^2 v, \tag{3}$$

Here u and v are the velocities along the x- and y- directions, t is the time, p is the pressure and $Re = \frac{UL}{\nu}$ is the Reynolds number where U is the characteristic velocity (velocity of the lid), L is the characteristic length (side of the square cavity) and ν is the kinematic viscosity.

Introducing streamfunction ψ and vorticity ω, the above equations (1)-(3) transform to

$$\omega_t - \frac{1}{Re}(\omega_{xx} + \omega_{yy}) + (u\omega_x + v\omega_y) = 0, \tag{4}$$

$$\psi_{xx} + \psi_{yy} = -\omega(x, y), \tag{5}$$

$$\text{with} \quad u = \psi_y,\ v = -\psi_x \quad \text{and} \quad \omega = v_x - u_y. \tag{6}$$

3. Numerical Procedure and Other Related Issues

The unsteady convection-diffusion equation in 2D for a flow variable ϕ can be written as

$$b\frac{\partial \phi}{\partial t} - \nabla^2\phi + c(x,y,t)\frac{\partial \phi}{\partial x} + d(x,y,t)\frac{\partial \phi}{\partial y} = f(x,y,t), \tag{7}$$

where $b > 0$ is a constant, c and d are the convection coefficients in the x- and y-directions, respectively, and f is a forcing function. The HOC scheme by Kalita *et al.* [8] for this equation on non-uniform space grid is given by:

$$\begin{aligned}
&b[1 + (H_1 + H_2c)\delta_x + (K_1 + K_2d)\delta_y + \{H_2 - 0.5(x_f - x_b)(H_1 + H_2c)\}\delta_x^2 \\
&+\{K_2 - 0.5(y_f - y_b)(K_1 + K_2d)\}\delta_y^2]\delta_t^+\phi_{ij}^n \\
&+[-A_{ij}\delta_x^2 - B_{ij}\delta_y^2 + C_{ij}\delta_x + D_{ij}\delta_y + G_{ij}\delta_x\delta_y - H_{ij}\delta_x\delta_y^2 - K_{ij}\delta_x^2\delta_y - L_{ij}\delta_x^2\delta_y^2]\phi_{ij}^n \\
&= F_{ij},
\end{aligned} \tag{8}$$

where δ_t^+ denotes the forward difference operator for time with uniform time step Δt and n represents the time level; the coefficients A_{ij}, B_{ij}, C_{ij}, D_{ij}, G_{ij}, H_{ij}, K_{ij}, L_{ij} and F_{ij} can be found in Kalita *et al.* [8].

In order to capture Moffatt vortices on relatively coarser grids, it is essential that the regions in the neighborhood of the solid boundaries are filled with cluster grids. To generate a centro-symmetric grid with clustering near the walls, we use the stretching function [9]

$$x_i = \frac{i}{i_{max}} - \frac{\lambda}{2\pi}\sin\left(\frac{2\pi i}{i_{max}}\right),\ 0 < \lambda \leq 1 \tag{9}$$

in both x- and y-directions.

The boundary conditions for velocity on the top wall are given by $u = U = 1$, $v = 0$. On other walls of the cavity, the velocities are zero i.e, $u = 0$, $v = 0$. For streamfunction vorticity formulation, streamfunction values are zero along the four walls i.e, $\psi = 0$. For vorticity a transient HOC wall boundary approximation [8] has been used.

In order to solve the algebraic systems associated with the finite difference approximation (8) the biconjugate gradient stabilized method (BiCGStab) [8, 10] has been used because non-uniform grid invariably leads to non-symmetric matrices.

4. Results and Discussions

For the problem under consideration, computations were carried out in a time marching manner till steady state was reached for $100 \leq Re \leq 3200$ on grids of sizes 81×81, 161×161 and 321×321. The Reynolds numbers considered are moderate enough to negate the effect of three dimensionality occurring above a certain critical Reynolds number Re_C [2].

4.1. Evidence of Moffatt vortices in the cavity

It is a well known fact that even the best available computational resources are incapable of capturing all the members of the so called infinite sequence of Moffatt vortices. With the computational resource available at our hand, we have been able to capture the first few members in that sequence. We now present those members in terms of streamfunction contours: in Figure 1, a global view of them can be seen for $Re = 3200$, while in Figure 2-3, one can see their local (magnified) visualization for $Re = 100$. Because of the extremely small scales of the tertiary and quaternary vortices, they cannot be depicted in the global view (eg., see Figure 1). For these vortices, the following nomenclature has been adopted. For the vortices at the left corner, BL1 denotes the secondary vortex which is the first one to appear in the sequence. Likewise, BL2, BL3, BL4, $\cdots$ denote the tertiary, quaternary and post-quaternary vortices respectively in that sequence in the same corner. In a similar way, for the Moffatt vortices in the right corner, the vortices in the same sequence are denoted by BR1, BR2, BR3, BR4, $\cdots$ etc.

Journal of Physics: Conference Series **759** (2016) 012081 doi:10.1088/1742-6596/759/1/012081

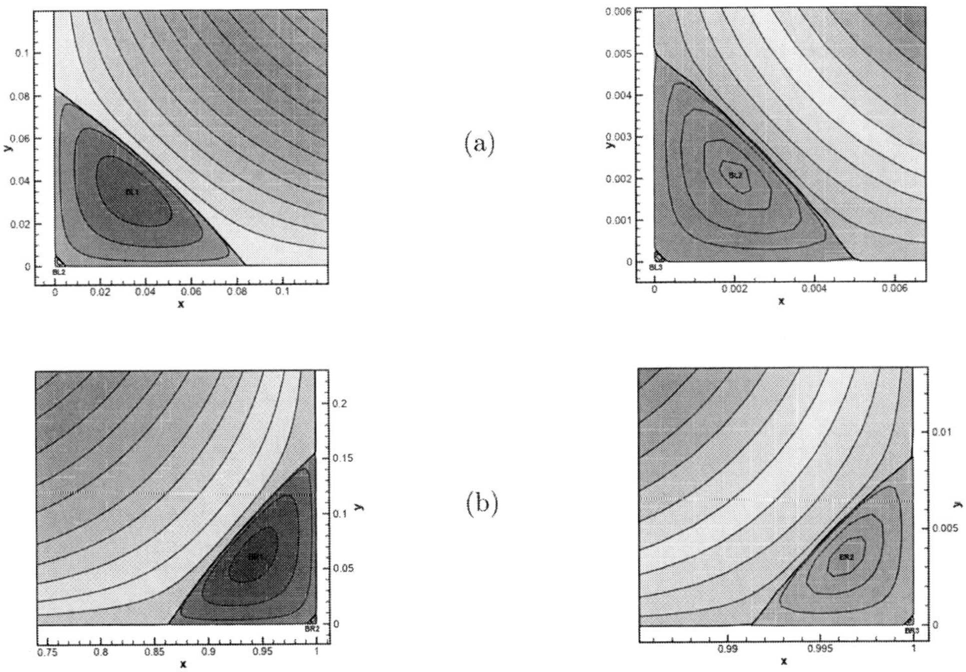

Figure 2. Moffatt vortices for $Re = 100$ on grid size 161×161: (a) BL1, BL2, BL3 at bottom left corner and (b) BR1, BR2, BR3 at bottom right corner.

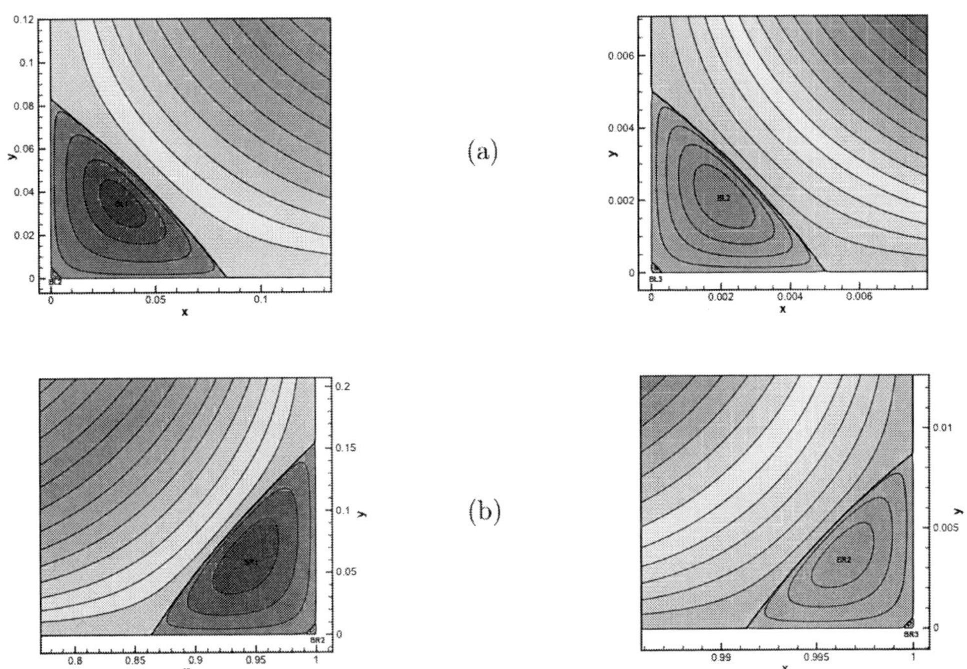

Figure 3. Moffatt vortices for $Re = 100$ on grid size 321×321: (a) BL1, BL2, BL3 at bottom left corner and (b) BR1, BR2, BR3 at bottom right corner.

4.2. Qualitative description of Moffatt vortices

In Table 1, we provide properties of Moffatt vortices in terms of their intensities and sizes along with the location of their centers. A comparison of our finest grid computations with the

Table 1. Properties of Moffatt vortices for $Re = 100$.

Vortex	Grid size	Intensity (ψ)	Center location (x, y)	Size
BL1	81×81	1.527017e-6	(0.03469, 0.03469)	0.0490696
	161×161	1.799623e-6	(0.03482, 0.03480)	0.0492360
	321×321	1.754731e-6	(0.03470, 0.03471)	0.0490840
BL2	81×81	-4.076319e-11	(0.00220, 0.00221)	0.0031273
	161×161	-4.815543e-11	(0.00222, 0.00221)	0.0031389
	321×321	-4.800864e-11	(0.00220, 0.00220)	0.0031167
BL3	81×81	—	—	—
	161×161	1.262750e-15	(0.00013, 0.00013)	0.0001971
	321×321	1.212033e-15	(0.00013, 0.00013)	0.0001958
BR1	81×81	1.057025e-5	(0.94104, 0.05955)	0.0838020
	161×161	1.260218e-5	(0.94110, 0.06456)	0.0873908
	321×321	1.239697e-5	(0.94321, 0.06179)	0.0839218
BR2	81×81	-2.961589e-10	(0.99655, 0.00349)	0.0049061
	161×161	-3.469230e-10	(0.99654, 0.00348)	0.0049096
	321×321	-3.383937e-10	(0.99653, 0.00347)	0.0049033
BR3	81×81	—	—	—
	161×161	9.422597e-15	(0.99976, 0.00023)	0.0003274
	321×321	9.047695e-15	(0.99977, 0.00023)	0.0003259

Table 2. Comparison of secondary and tertiary vortex data with benchmark results.

Re	Vortex	Ref	Grid size	Intensity (ψ)	Center location (x, y)	Size
100	BL1	Present	321 × 321	1.754731e-6	(0.03470, 0.03471)	0.0490840
		Ghia *et al.*[6]	129 × 129	1.74877e-6	(0.0313, 0.0391)	0.0500849
	BL2	Present	321 × 321	-4.800864e-11	(0.00220, 0.00220)	0.0031167
		Ghia *et al.*[6]	129 × 129	–	–	–
	BL3	Present	321 × 321	1.212033e-15	(0.00013, 0.00013)	0.0001958
		Ghia *et al.*[6]	129 × 129	–	–	–
	BR1	Present	321 × 321	1.239697e-5	(0.94321, 0.06179)	0.0839218
		Ghia *et al.*[6]	129 × 129	1.25374e-5	(0.9453, 0.0625)	0.0830562
	BR2	Present	321 × 321	-3.383937e-10	(0.99653, 0.00347)	0.0049033
		Ghia *et al.*[6]	129 × 129	–	–	–
	BR3	Present	321 × 321	9.047695e-15	(0.99977, 0.00023)	0.0003259
		Ghia *et al.*[6]	129 × 129	–	–	–
1000	BL1	Present	321 × 321	2.331454e-4	(0.08333, 0.07757)	0.1138463
		Ghia *et al.*[6]	129 × 129	2.31129e-4	(0.0859, 0.0781)	0.1160966
		Erturk *et al.*[5]	601 × 601	2.3261e-4	(0.0833, 0.0783)	0.1143231
		Botella *et al.*[3]	160 × 160	2.334528e-4	(0.0833, 0.0781)	0.1141862
	BL2	Present	321 × 321	-6.135485e-9	(0.00470, 0.00470)	0.0066468
		Ghia *et al.*[6]	129 × 129	–	–	–
		Erturk *et al.*[5]	601 × 601	-8.4221e-9	(0.0050, 0.0050)	0.0070710
		Botella *et al.*[3]	160 × 160	-6.39800e-9	(0.00484, 0.00484)	0.0068447
	BL3	Present	321 × 321	1.577689e-13	(0.00028, 0.00028)	0.0003959
		Ghia *et al.*[6]	129 × 129	–	–	–
		Erturk *et al.*[5]	601 × 601	–	–	–
		Botella *et al.*[3]	160 × 160	–	–	–
	BR1	Present	321 × 321	1.731098e-3	(0.86570,0.11281)	0.1753926
		Ghia *et al.*[6]	129 × 129	1.75102e-3	(0.8594, 0.1094)	0.1781480
		Erturk *et al.*[5]	601 × 601	1.7281e-3	(0.8633, 0.1117)	0.1765326
		Botella *et al.*[3]	160 × 160	1.729717e-3	(0.8640, 0.1118)	0.1760546
	BR2	Present	321 × 321	-4.87726e-8	(0.99261,0.00739)	0.0104510
		Ghia *et al.*[6]	129 × 129	-9.31929e-8	(0.9922, 0.0078)	0.0110308
		Erturk *et al.*[5]	601 × 601	-5.4962e-8	(0.9917, 0.0067)	0.0106667
		Botella *et al.*[3]	160 × 160	-5.03944e-8	(0.99232, 0.00765)	0.0108399
	BR3	Present	321 × 321	1.333030e-12	(0.99957, 0.00042)	0.0006010
		Ghia *et al.*[6]	129 × 129	–	–	–
		Erturk *et al.*[5]	601 × 601	–	–	–
		Botella *et al.*[3]	160 × 160	–	–	–

benchmark results [3, 5, 6] are presented in Table 2. One can see that our results are extremely close to them. Note that, extreme clustering enabled us to capture the quaternary vortices, which was not visible in [3, 5, 6]. The intensity of a vortex is defined by the streamfunction value at the center of the vortex. Following the work of H. K. Moffatt [14, 15], we measure the size of a vortex in terms of the Euclidean distance between the center of the vortex and the corner of the cavity.

We observe that the intensity as well as size of these vortices decrease rapidly, thus corroborating our findings with the theory. For a fixed Reynolds number, as grid size increases, one can observe the extreme closeness of the values of intensity and size in smaller scales also.

In Table 3, we present the size and intensity ratio of two vortices in succession for step sizes $\frac{1}{81}, \frac{1}{161}, \frac{1}{321}, \frac{1}{641}$ and for zero-grid-step limit ($h \to 0$). The ratios (both size and intensity) for step sizes $\frac{1}{641}$ and zero-grid-step limit are obtained by the Richardson's extrapolation and Lagrange's interpolation [7] of the computed data on coarser grids respectively. From Table 1 and Table 3,

Table 3. Intensity ratio and size ratio between two consecutive vortices for $Re = 100$.

		Step Size (h)				
Eddy Ratio	**Properties**	1/81	1/161	1/321	1/641	$h \to 0$
BL1:BL2	Intensity	0.3746e5	0.3737e5	0.3655e5	0.3649e5	0.3667e5
	Size	15.69063	15.68534	15.74826	15.75246	15.73847
BL2:BL3	Intensity	—	0.3813e5	0.3961e5	0.3970e5	0.3938e5
	Size	—	15.92053	15.91181	15.91123	15.91317
BR1:BR2	Intensity	0.3569e5	0.3632e5	0.3663e5	0.3665e5	0.3658e5
	Size	17.08097	17.79980	17.11507	17.06942	17.22158
BR2:BR3	Intensity	—	0.3681e5	0.3740e5	0.3743e5	0.3731e5
	Size	—	14.99573	15.04196	15.04504	15.03477

we observe that the intensity as well as the size for a particular grid-size go down in geometric progression with a common ratio which establishes H. K. Moffatt's theory [14, 15] accurately.

4.3. Small scales resolution

By computing pressure gradients along the left and right walls of the cavity we conclude that our solution is not a spurious one. In Figure 4, we present changes in sign in pressure gradient in

(a)

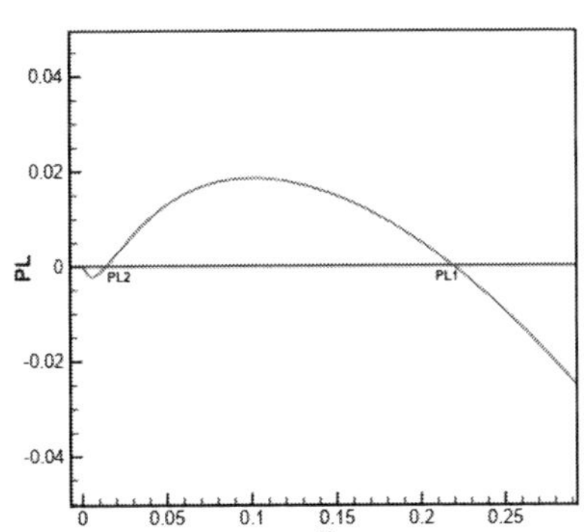

(b)

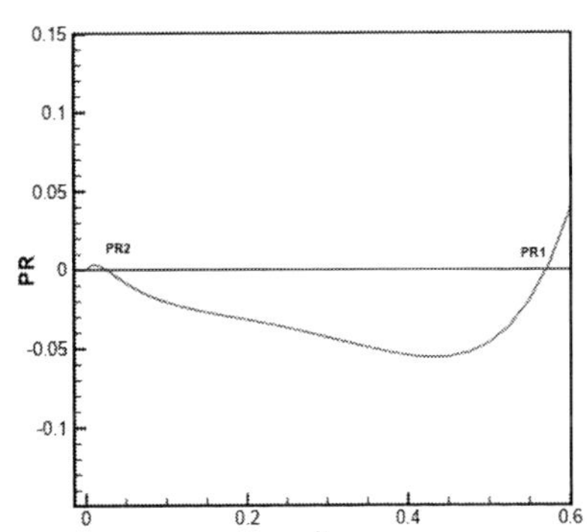

Figure 4. Pressure gradients along (a) left wall and (b) right wall for $Re = 100$ on grid size 81×81.

which PL1, PR2 denotes sign change from negative to positive and PL2, PR1 denotes positive to negative. It is well known that when pressure gradient changes sign along a wall, flow separation takes place and paves the way for the creation of a vortex. As such, the total number of vortices is equal to the total number of changes in sign in pressure gradient which establishes that our solution is accurate even in smaller scales and is free from numerical artifacts. Similar facts can be observed for the grid sizes 161×161 and 321×321 and other Reynolds numbers considered here as well.

5. Conclusion

We have established the existence of Moffatt vortices in the Lid-driven cavity flow by utilizing an HOC scheme at moderate Reynolds numbers, which, to the best of our knowledge has not been carried out before. The outcome of the computation of the ratio of size and intensity of these vortices strongly corroborates with the prediction of the theory of H. K. Moffatt. These has been further strengthened by the extrapolation and interpolation of the critical data.

References

[1] Anderson D M and Davis S H 1993 *J. Fluid Mech.* **18** 1-31
[2] Auteri F, Parolini N and Quartapelle L 2002 *J. Comput. Phys.* **183** 1-25.
[3] Botella O and Peyret R 1998 *Comput. Fluids* **27** 421-433
[4] Collins W M and Dennis S C R 1976 *J. Fluid Mech.* **6** 417-432
[5] Erturk E, Corke T C and Gökcöl 2005 *Int. J. Numer. Meth. Fluids* **48** 747-774
[6] Ghia U, Ghia K N and Shin C T 1982 *J. Comput. Phys.* **48** 387-411.
[7] Hoffman J D 2001 *Numerical Methods for Engineers and Scientists* (New York: Marcel Dekker)
[8] Kalita J C, Dass A K and Nidhi N 2008 *J. Comput. Appl. Math.* **214** 148-162
[9] Kalita J C 2007 *Eng. Appl. Comput. Fluid Mech.* **1** 36-48
[10] Kelley C T 1995 *Iterative Methods for Linear and Nonlinear Equations* (Philadelphia: SIAM)
[11] Kirkinis E and Davis S H 2014 *J. Fluid Mech.* **746** R3
[12] Malhotra C P, Weidman P D and Davis A M J 2005 *J. Fluid Mech.* **522** 117-139
[13] Malyuga V S 2005 *J. Fluid Mech.* **522** 101-116
[14] Moffatt H K 1964 *J. Fluid Mech.* **18** 1-18
[15] Moffatt H K 1964 *Arch. Mech. Stosowanej* **2** 365-372
[16] Shankar P N 2005 *J. Fluid Mech.* **539** 113-135
[17] Shankar P N 1993 *J. Fluid Mech.* **250** 371-383
[18] Shankar P N and Deshpande M D 2000 *Annu. Rev. Fluid Mech.* **32** 93-136.
[19] Taneda Sadatoshi 1979 *J. Phys. Soc. Japan* **46** 1935-42

Flexible flapping wings can exhibit quasi-periodic motion!

Chandan Bose[1] **& Sunetra Sarkar**[2]

[1] Department of Applied Mechanics, Indian Institute of Technology Madras, Chennai 600036, India

[2] Department of Aerospace Engineering, Indian Institute of Technology Madras, Chennai 600036, India

E-mail: cb.ju.1991@gmail.com, sunetra.sarkar@gmail.com

Abstract.
The dynamics of a flexible flapping wing is investigated by modelling it as a coupled nonlinear fluid-structure interaction (FSI) system in the low Reynolds number flow regime in accordance to the flight of flapping wing micro air vehicles (MAVs). A bifurcation analysis, by varying the free-stream wind velocity (U_∞) as the control parameter, revealed the presence of a new dynamics in the form of a quasi-periodic attractor in the flapping wing motion. The structural and aerodynamic nonlinearities present in the system cause a supercritical Hopf bifurcation, where stable limit cycle oscillation emerges from fixed point response beyond a critical value of the free-stream velocity. Further increasing the control parameter, another bifurcation named Neimark-Sacker bifurcation takes place and as a result, the flapping wing exhibits quasi-periodic motion. The presence of Neimark-Sacker bifurcation in the flapping flow-field dynamics is an interesting find and the present work focuses on it's associated dynamical behaviour. Various dynamical system tools like frequency spectra, phase space, Poincaré section, first return map have been implemented successfully to confirm the presence of quasi-periodicity.

1. Introduction

The amazing flying capability of insects has fascinated the researchers from the beginning of the twentieth century to investigate the unsteady aerodynamics of the flapping flight. The need for designing advanced flapping wing Micro Air Vehicles (MAVs) has boosted the research efforts in this direction. The combination of small dimensional size and slow speeds result in a low Reynolds number flight regime (10, 000 to 30, 000) which is well below the flight of conventional aircrafts. Hence at low Re, the viscous boundary layer is thicker and can result in a high drag and also there is deterioration of performance due to laminar boundary layer separation [1]. MAVs also experience highly unsteady flows in their environment and there is a need to overcome the additional drag and loss of lift without an increase in the size [2]. Thus, the natural flyers exploit the coupling between the unsteady flow and the wing deformation to enhance their aerodynamic performance in these adverse conditions.

With the change of kinematic, structural or aerodynamic parameters in a flapping flight, the wing-wake interaction can lead to an abrupt transition from periodicity to aperiodicity in the wake topology as well as in the aerodynamic loads which can indeed lead to an unpredictable body motion due to very low body inertia of extremely light weight flapping wing MAVs. This underlines the need for specialized control algorithms. The aperiodic wake topology has been reported in recent literature in the flow past a rigid airfoil with forced flapping [3, 4, 5]. But unfortunately, there is no conclusive depiction of the transition from periodic to aperiodic regime from a dynamical systems point of view. Ashraf *et. al.* [5] has mentioned about the quasi-periodic transition to aperiodicity in the flow past a plunging airfoil based on the frequency spectra of the thrust coefficients. Although they have observed multiple peaks in the frequency spectra, no evidence in support of the existence of incommensurate frequencies was presented. Moreover, they have neglected the effect of flexibility and the subsequent fluid structure interaction effects, which is not realistic, especially when very light weight MAVs are involved.

As most of the earlier reported works discussing the dynamics of flapping wing MAVs such as Ashraf *et al.* [5] assumed the wing to be rigid, there is a crucial need to understand the role of structural flexibility on the overall dynamical performance. Further, most of the earlier works are either based on inviscid potential flows without considering any separations [6] or simplified point vortex models for leading edge vortices [7]. As such simplified models limit the understanding, we propose to use a full fidelity N-S solver in the present work. To summarize, the present work investigates the dynamics of flexible flapping systems in terms of the wing motion for comparable fluid-solid mass values (light weight MAVs) using a high fidelity flow solver.

The present FSI framework is comprised of a high fidelity Navier-Stokes based viscous aerodynamic solver and a structural model with cubic nonlinearity in the pitching motion, coupled by a weak coupling method. Bifurcation analysis considering Free stream velocity (U_∞) as the control parameter brings a new dynamics into light. It reveals that the flapping wing undergoes quasi-periodic motion in the presence of two incommensurate frequencies beyond a critical threshold of the control parameter. Nonlinear time series analysis has been carried out to categorize the dynamics.

This paper is organized as follows: Section 2 provides computational methodology, section 3 discusses the bifurcation analysis and the nonlinear time series analysis and finally this paper ends with the concluding remarks in section 4.

2. Computational methodology

2.1. Governing Equations and the Flow solver

The flow is governed by the incompressible Navier-Stokes (N-S) equation as follows,

$$\nabla.\vec{u} = 0, \ \frac{\partial \vec{u}}{\partial t} + [\vec{u}.\nabla]\vec{u} = -\nabla p/\rho + \nu\nabla^2\vec{u}. \tag{1}$$

Here, $\vec{u}$ is the velocity of the flow, p is the pressure, ν is the kinematic viscosity, ρ is the fluid density, $\nabla = \frac{\partial}{\partial x}\vec{i} + \frac{\partial}{\partial y}\vec{j}$, is the two dimensional gradient vector and t is time. The simulations are performed in OpenFOAM®, an open source finite-volume based CFD solver. The arbitrary Lagrangian Eulerian (ALE) formulation has been used for discretizing the N-S equation on a deformed flow mesh using radial basis function (RBF) interpolation scheme. The pressure-velocity coupling is achieved through PISO algorithm [9] with three PISO corrector loops. The algorithms used here for solving the pressure, velocity and mesh motion have been discussed in detail by Bos [10]. A rectangular computational domain shown in Fig.1 was used for the analysis. The computational domain is discretized using a combination of structured and unstructured grids as shown in Fig. 2. The grid independent study and the flow solver validation have been discussed in details in [11].

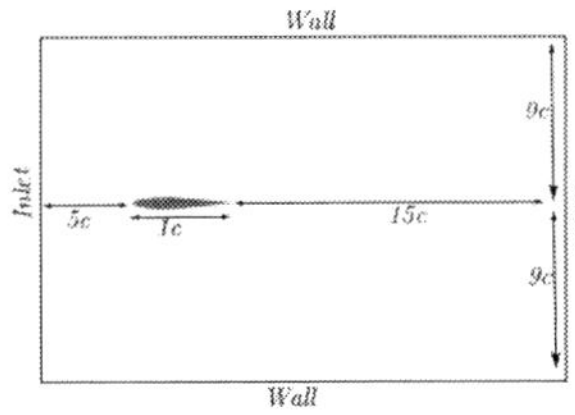

Figure 1. Computational domain

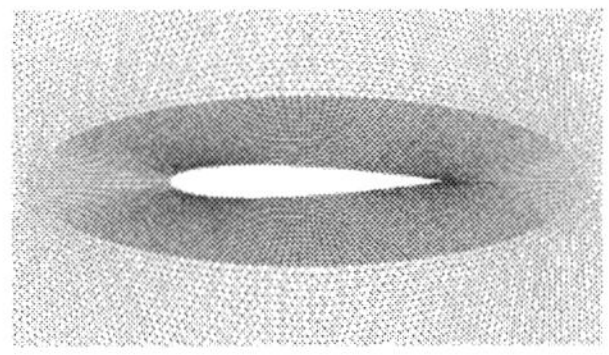

Figure 2. Close-up view of the mesh

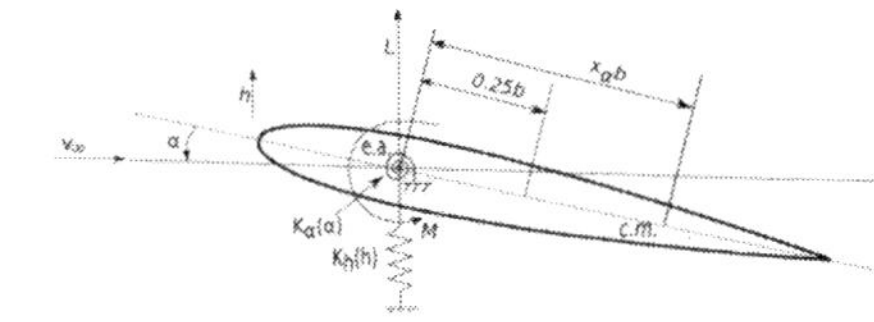

Figure 3. Schematic of a symmetric airfoil in pitch and plunge degrees of freedom

2.2. Structural solver

The structural part is comprised of an airfoil with pitch-plunge degrees of freedom as shown in Fig. 3. The airfoil is restrained by non-linear springs. The directions of positive lift and moment are indicated in the figure. The equations of motion are given by [6],

$$m\ddot{h} + S\ddot{\alpha} + C_h\dot{h} + K_h(h) = L(t), \; K_h(h) = K_{h1}h + K_{h2}h^3; \tag{2}$$

$$S\ddot{h} + I\ddot{\alpha} + C_\alpha\dot{h} + K_\alpha(\alpha) = M(t), \; K_\alpha(\alpha) = K_{\alpha1}\alpha + K_{\alpha2}\alpha^3. \tag{3}$$

In the above equations, h and α are the plunge and pitch displacements respectively, m is the mass of the airfoil, S is the first moment of inertia, I is the second moment of inertia, K_{h1}, K_{h2}, $K_{\alpha1}$ and $K_{\alpha2}$ are the spring stiffness coefficients, C_h and C_α are the damping coefficients, $L(t)$ is the unsteady lift and $M(t)$ is the moment about the quarter chord. Lift and moment are calculated by the flow solver at each time step. The structural response (Eq. 2-3) is solved using an explicit fourth-order Runge-Kutta method. The time step for integration is equal to that of the flow solver. A partitioned approach based weak coupling method has been used in the present investigation to couple the flow solver and the structural solver. In the partitioned approach, two separate solvers for fluid and structure are used jointly by exchanging information from one time step to another in an alternating fashion explicitly [12]. The coupling methodology has been discussed in details in [11].

3. Results and discussions

3.1. Bifurcation Analysis

Bifurcation analysis helps us to investigate the qualitative changes in the system behaviour by changing a control parameter. It is essential for the design of flexible flapping devices to define a stable operation regime. The flexible flapping wing studied here is inherently a coupled nonlinear dynamical system. In the present study, the flapping motion is limited to two degrees of freedom with pitch and plunge motion. The non-dimensionalised free-stream velocity ($U^* = \frac{U_\infty}{b\,\omega_\alpha}$) has been chosen as the bifurcation parameter to investigate its effect on the system dynamics. A schematic plot of the bifurcation diagram is presented in Fig. 4.

A summary of the dynamics of the FSI system in response to the change of the bifurcation parameter, U^* can be seen in Fig. 5. The value of the solid to fluid mass ratio is kept low at 5, in order to keep the focus on the light weight MAV wings. As we gradually increase the control parameter, a rich dynamics is revealed and multiple bifurcations take place with the dramatic qualitative change in the system behaviour. In order to investigate those changes, the pitch response is chosen to plot the bifurcation diagram. Corresponding to every U^*, the amplitudes of the local maxima in the computed time series of pitch response is plotted to get the upper branch of the bifurcation diagram. The number of local maxima, at a given parameter, gives the period of oscillations: a single local maximum indicates a limit cycle oscillation, two local maxima values suggest period two oscillations and so on.

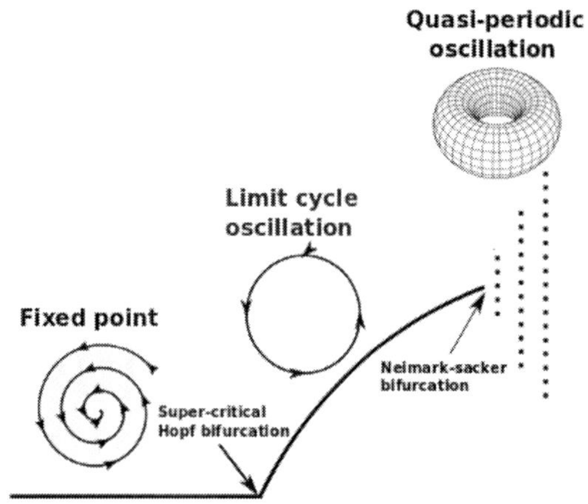

Figure 4. Schematic of the bifurcation diagram

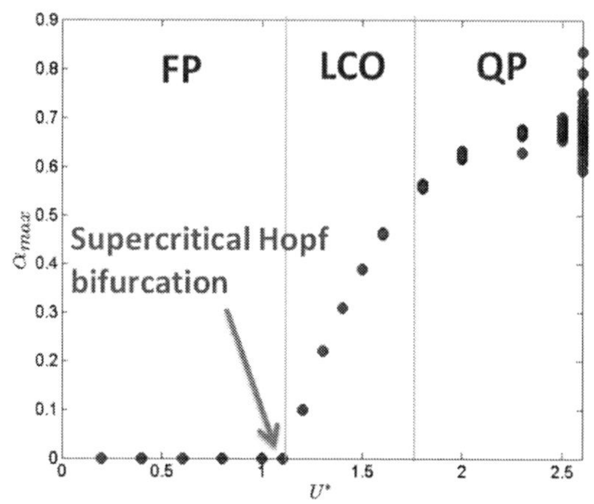

Figure 5. Bifurcation diagram of the FSI system

The first qualitative change takes place in the system at $U^* = 1.1$ where a supercritical Hopf-bifurcation takes place and as a result, a stable fixed point response becomes unstable and gives way to a stable self-excited limit cycle oscillation. The point at which the bifurcation occurs is called the "Hopf point". The amplitude of the single frequency LCO increases with the increase of bifurcation parameter. Further increasing the control parameter, beyond $U^* = 1.7$, another change in the dynamics is observed: a second Hopf bifurcation of the periodic solution occurs, introducing a new frequency which is incommensurate with the first one in the bifurcating solution. This kind of bifurcation is called a secondary Hopf or Neimark-Sacker bifurcation [13]. It occurs when two complex conjugate eigenvalues exit the unit circle away from the real axis. As a result, the system behaviour changes from limit cycle oscillation to quasi-periodic oscillation with two incommensurate frequencies. The local maxima in the oscillations no longer have a constant amplitude as additional frequencies appear in the response which are not harmonics of the earlier existing fundamental frequency. The regions of fixed point response, limit cycle oscillations and quasi-periodic oscillations are shown as the region of 'FP', 'LCO' and 'QP' respectively in Fig. 5.

3.2. Nonlinear time series analysis

For a better understanding of the dynamics and to confirm the type of bifurcations involved, nonlinear time series analysis tools such as time histories, state-space plots, Poincaré sections, first return maps and the frequency spectra are implemented in the present study. Nonlinear time series analysis techniques provide tools for systematic analysis and identification of specific signatures of time series data generated by a complex nonlinear system.

The time histories of the pitch response at different values of U^* are shown in Fig. 6 and the frequency spectra of the pitch response at $U^* = 2.6$ are presented in Fig. 7. The time history prior to the supercritical Hopf bifurcation shows a damped response where the response settles down to a stable equilibrium point beyond the transient motion. The time history of limit cycle oscillations has the appearance of a uniform trace as shown in Fig. 6(b) and the corresponding frequency spectra have one fundamental frequency in the form of a dominant peak. The envelope of time history for quasi-periodic response given in Fig. 6(c) looks different from that of Fig. 6(b) and shows uneven peaks. The corresponding frequency spectra of a two frequency quasi-periodic time history contain two fundamental peaks with an irrational

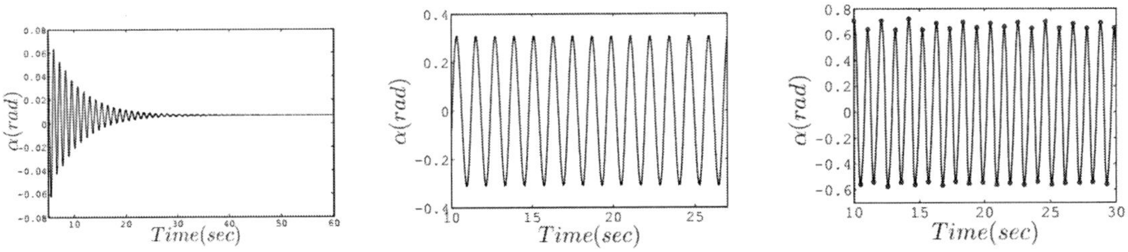

(a) $U^* = 1.0$ (b) $U^* = 1.3$ (c) $U^* = 2.6$

Figure 6. Time histories pitch response

frequency ratio attributing to two incommensurate frequencies. The other non-harmonic peaks occur as linear combinations of the two fundamental incommensurate frequencies. In case of $U^* = 2.6$, the frequency spectra clearly shows two incommensurate frequencies '$f1$' and '$f2$' and the other frequencies are '$f1+f2$', '$2f1+f2$', '$2f1$', '$3f1$', '$4f1$' and '$5f1$' which clearly indicate the linear combination of the mentioned incommensurate frequencies, $f1$ & $f2$ as shown in Fig. 7.

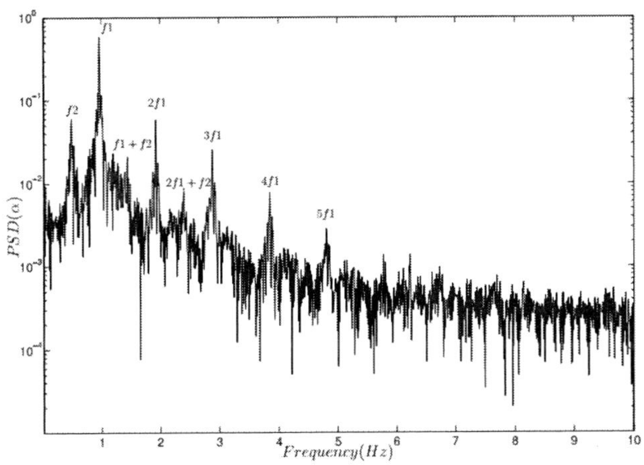

Figure 7. FFT of pitch response at $U^* = 2.6$

The space defined by the independent coordinates required to describe a motion is called a state space and the independent coordinates are called state variables [13]. In the present work, the physical state-space, projected in two dimensions, is constructed by the state variables (α, $\dot{\alpha}$). Phase space provides us a qualitative picture of the evolution of the trajectories and presence of the attractor of the system. The phase portraits for the pitch response at different U^* are presented in Fig. 8(a), 8(d) & 8(g). The phase portrait corresponding to the damped response of the wing is a stable spiral in the phase space as shown in Fig. 8(a), whereas the phase portrait of the limit cycle oscillation represents a unique closed one-dimensional attractor which depicts the periodic nature of the flapping motion as shown in Fig. 8(d). However, with the introduction of a new incommensurate frequency due to Neimark-Sacker bifurcation, the single loop turns into a dense toroidal structure. A toroidal structure in the phase space is an indication of quasi-periodic oscillations. Due to the presence of the incommensurate frequencies, the phase space trajectory evolves on the surface of a torus, never closing itself [14]. Among the three, the phase portrait corresponding to $U^* = 2.6$ is quasi-periodic and represents a shape of T^2 torus as shown in Fig. 8(g). The deformed shape of the attractor can be attributed to the fact that the quasi-periodic attractor is a superposition of two incommensurate fundamental frequencies and their harmonics [15].

A Poincaré section converts a continuous time evolution into a discrete time mapping. Thus, it is a very useful tool to characterize the time series whose phase space dimension is high (> 3). The reduced attractor is not identical to the original attractor but it preserves the topological properties of the original attractor [16]. In this present work, the Poincaré sections have been constructed in a stroboscopic manner with the known time period of the limit cycle oscillation. Hence, the Poincaré section acts like a stroboscope, freezing the components of the motion commensurate with the limit cycle time period (say T). So, we will get a single point corresponding to the limit cycle oscillation, whereas if we get a collection of k (say) discrete points on the Poincaré section, the corresponding motion is periodic with the period kT. When the ratio of the added frequency and the limit cycle frequency is not a rational number, it denotes two frequency quasi-periodic motion and the points on the corresponding Poincaré section will fill up a closed smooth curve densely [13]. The Poincaré sections for the pitch response have been plotted for $U^* = 1, 1.3, 2.6$ in Fig. 8(b), 8(e) & 8(h) respectively. As discussed above, they correspond to the damped, periodic and quasi-periodic dynamics.

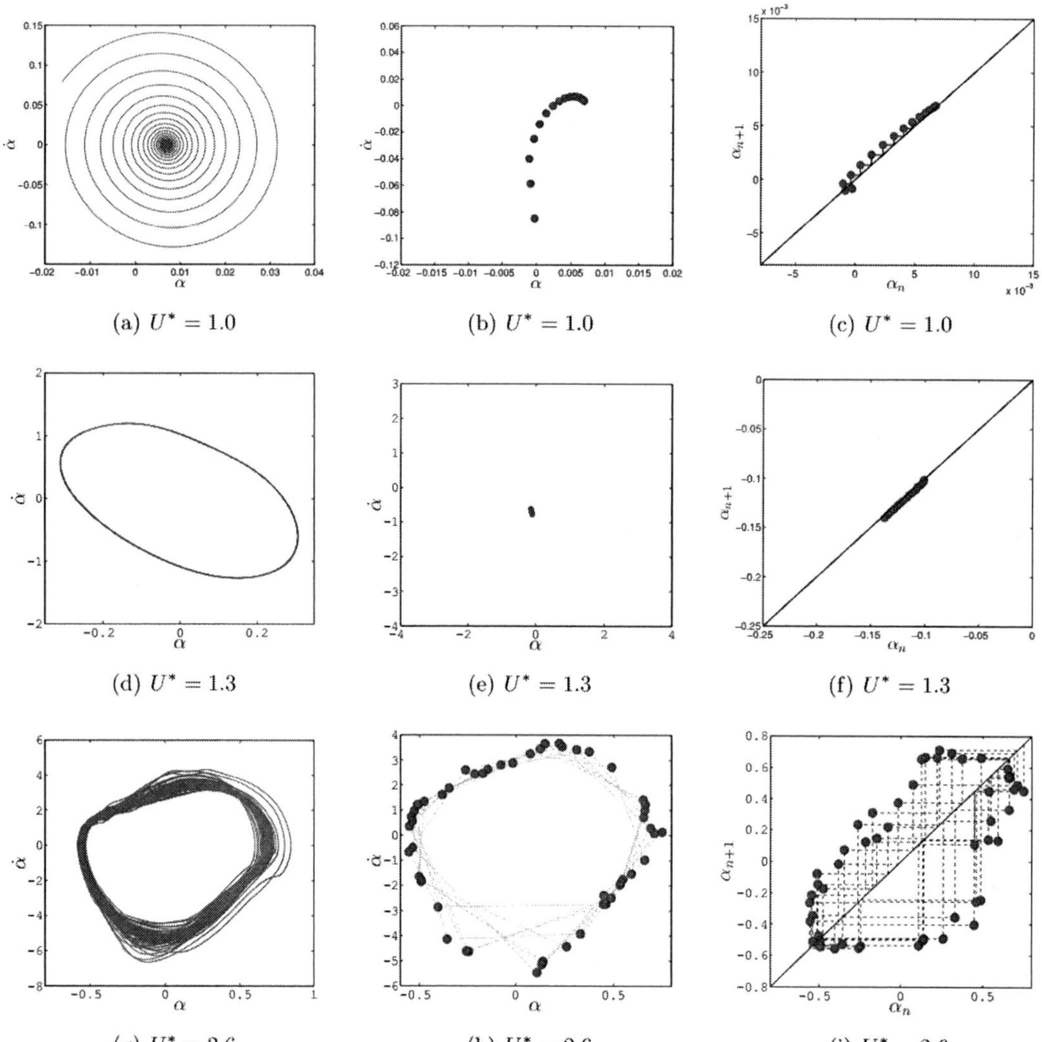

(a) $U^* = 1.0$ (b) $U^* = 1.0$ (c) $U^* = 1.0$

(d) $U^* = 1.3$ (e) $U^* = 1.3$ (f) $U^* = 1.3$

(g) $U^* = 2.6$ (h) $U^* = 2.6$ (i) $U^* = 2.6$

Figure 8. Phase portraits (left), Poincaré sections (middle) & first return maps (right)

To analyze the dynamics further, we have implemented 1D mapping technique of first return map which is essentially constructed from the Poincaré sections as discussed before. Using the first return map, we investigate the evolution of the discretized version of the system derived from the continuous evolution. In the first return map, the points on the phase space frozen by the stroboscopic Poincaré section (x_n) are collected and they are plotted against the next iterates (x_{n+1}) that occur and a 2D plot (x_n) vs (x_{n+1}) is obtained. For a periodic attractor, the successive iterates are equal, hence mapping lies on the $x = y$ diagonal, else it moves away from the diagonal. Thus, a first return map can be utilized to explore the evolution of the phase space trajectories in a lower dimensional space preserving the same properties. Fig. 8(c), Fig. 8(f) & Fig. 8(i) present first return maps for damped response, limit cycle oscillation and quasi-periodic attractor respectively. The first return map for a limit cycle oscillation is represented by a diagonal line along $x = y$ line, whereas it is a loop in the case of a quasi-periodic state. Further, the cobweb diagrams are constructed to understand the evolution of the iterates of the return map.

4. Conclusion

A new dynamics of quasi-periodicity is observed in the motion of a flexible flapping wing in the bifurcation analysis considering the non-dimensional free-stream velocity as the control parameter. It has been observed that with the increase of the control parameter, at first, a supercritical Hopf bifurcation takes place at a critical value of the parameter due to which the flapping motion undergoes a self-sustained LCO. Thereafter, further increasing the control parameter, a Neimark-Sacker bifurcation occurs over the LCO making the flapping motion quasi-periodic. The time series of pitch response has been categorized using nonlinear time series analysis tools. The existence of the incommensurate frequencies in the frequency spectra of the pitch response at $U^* = 2.6$ is evidently shown in the present study. Further, the toroidal phase space along with the closed Poincaré section and a loop structure in the first return map confirm the quasi-periodic nature of the pitch response at $U^* = 2.6$. For the future work, the authors are working to understand the dynamics of the flapping wing at the higher control parameter values beyond quasi-periodic dynamics.

References

[1] Mueller T J & DeLaurier J D 2003 *Annual Review of Fluid Mechanics* **35(1)** 89-111
[2] Young J 2005 *Ph.D. Thesis, University of New South Wales, Australian Defence Force Academy, School of Aerospace, Civil and Mechanical Engineering.*
[3] Blondeaux P, Guglielmini L & Triantafyllou M S 2005 *AIAA Journal* **43** 918-922
[4] Lentink D, Heijst G F V, Muijres F T & Leeuwen J L V 2010 *Biology Letters, rsbl20090806*
[5] Ashraf M A, Young J & Lai J C S 2012 *AIAA Journal* **50(11)** 23082324
[6] Lee B H K, Jiang L Y & Wong Y S 1999 *Journal of Fluids and Structures* **13 (1)** 75-101
[7] Ramesh K, Murua J & Gopalarathnam A 2015 *Journal of Fluids and Structures.* **55** 84-105
[8] Bos F M, Van Oudheusden B W, & Bijl H 2013 *Computers and Fluids* **79** 167-177.
[9] Ferziger J & Peric M 2002 *Computational Methods for Fluid Dynamics,3rd edition, Springer-Verlag, Berlin.*
[10] Bos F M, Van Oudheusden B W, & Bijl H 2013 *Journal of Fluids and Structures* **42** 130-151.
[11] Bose C, Badrinath S, Gupta S, & Sarkar S 2016 *Procedia Engineering* **144** 883-890.
[12] Michler C, Hulshoff S, Van Brummelen E & De Borst R 2004 *Computers & fluids* **33** 839-348.
[13] Nayfeh A H, & Balachandran B 2008 *Applied Nonlinear Dynamics: Analytical, Computational and Experimental Methods. John Wiley & Sons.*
[14] Kabiraj L, Saurabh A, Wahi P, & Sujith R I 2012 *Chaos: An Interdisciplinary Journal of Nonlinear Science* **22(2)** 023129
[15] Saha A K, Muralidhar K & Biswas G 2000 *Journal of Engineering Mechanics* **126(5)** 523-532
[16] Berge P, Pomeau Y, & Vidal C 1984 *Order within Chaos. Wiley and Sons.*

Journal of Physics: Conference Series **759** (2016) 012083 doi:10.1088/1742-6596/759/1/012083

Computational Study of Fluid Flow Characteristics past Circular Cylinder due to Confining Walls with Local Waviness

R Deepakkumar[1], S Jayavel[1] and Shaligram Tiwari[2]

[1]IIITD&M Kancheepuram, Chennai-600127, Tamilnadu, India
[2]Department of Mechanical Engineering, IIT Madras, Chennai-600036, India

sjv@iiitdm.ac.in

Abstract. Flow past circular cylinder confined by walls with local waviness near the cylinder has been computed and the vortex shedding characteristics are studied at Reynolds number 200. The governing equations are solved using finite volume based CFD solver, 'Ansys Fluent 15.0'. In the present study, flow fields have been computed with waviness placed at four different locations. The effect of the location of waviness on vortex formation and its shedding frequency has been closely captured and presented. The flow parameters such as drag, lift and Strouhal number are calculated using the field data obtained from computations. It has been observed that the vortex shedding is significantly controlled by changing the location of local waviness. Among the four different locations considered in the present study, the waviness located near the wake region effectively suppresses the shedding. Moreover, the inward projection of the waviness plays a vital role in vortex suppression. It is also observed this method of vortex suppression leads to increase in drag.

1. Introduction

Flow physics in the downstream of flow past an unconfined circular cylinder is an interesting phenomenon. Flows over civil structures and heat exchanger tubes are such examples. Confining walls in the proximity of the cylinder significantly influence the alternate vortex shedding. Further, with high confinement, the shedding gets completely suppressed (Tiwari *et al.* 2006). Confining walls and protrusions in the confining boundaries such as vortex generators can also be effectively utilized to control the flow characteristics (Jayavel and Tiwari 2008; Jayavel and Tiwari 2009). For example, cylinder confined by rigid walls effectively controls the shedding characteristics, Chen *et al.* (1995), Sahin *et al.* (2004), Tiwari *et al.* (2006) and Singha*et al.* (2010) have studied the effect of wall confinement on vortex shedding and reported increased drag on the cylinder. From the work of Dipankar and Sengupta (2005), it is observed that when the cylinder is exposed to an oncoming shear flow by placing near to a solid wall, that weaken the shed vorticity from the cylinder at Reynolds number 1200. Also they have explained the effect of plane wall on flow around the cylinder in terms of vorticity dynamics and vortex induced instability mechanism. Strykowski and Sreenivasan (1990), Mittal and Raghuvanshi (2001) have used a secondary small control cylinder in the wake of the main cylinder and effectively controlled the shedding. The presence of splitter plate in the wake also effectively controls the shedding, Kwon and Choi (1996); Jayavel and Tiwari (2010); Dehkordi and Jafari (2010) have identified the location and length of the splitter plate to reduce drag and lift acting on the cylinder. From the available literature, it is clear that the effect of wall confinement on vortex shedding characteristics and especially waviness near the cylinder would be of more interest. The present work is aimed to study the physical phenomena of flow past circular cylinder confined between parallel walls with waviness in selected region close to the cylinder.

2. Problem statement

This numerical study is aimed at the control of vortex shedding on a cross flow past circular cylinder confined between two parallel rigid walls with waviness in the walls near to the cylinder as shown in Figure 1. The Reynolds number taken for the study is 200 (Sahin *et al.* 2004), which ensures the shedding in presence of plain wall confinement. Waviness has been located at different locations in the plain wall as shown in Figure 2 and the shedding characteristics are monitored. The Figure 1 shows the various dimensions of the computational domain using cylinder diameter (D) as the characteristic length scale. The cylinder confinement is $2D$, upstream and downstream lengths are $6D$ and $24D$ respectively. The wavelength and amplitude of waviness are $2D$ and $0.15D$ respectively. The location of waviness is specified in terms of horizontal distance between the cylinder center and the wave center. The four different locations considered in the present study are shown in Figure 2.

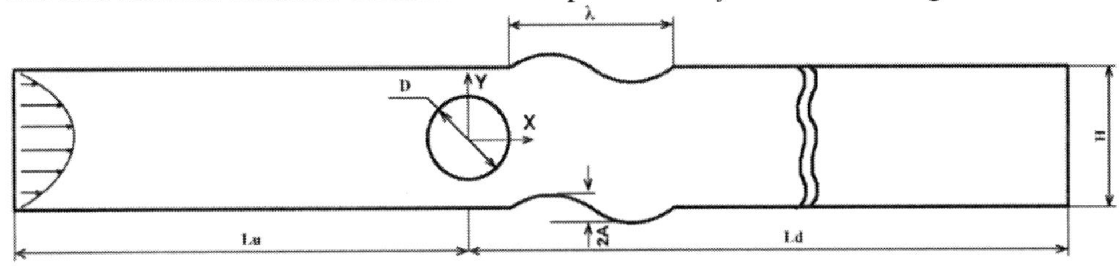

Figure 1. Computational domain.

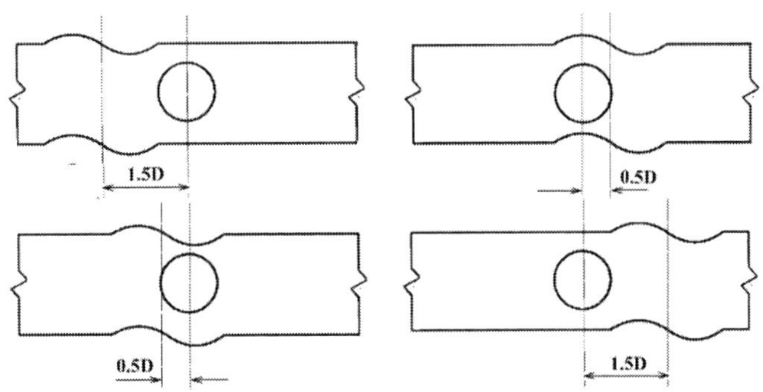

Figure 2. Different location of waviness.

3. Numerical solution methodology

In order to obtain the flow fields in the computational domain, the entire domain has been discretized into finite number of small control volumes as shown in Figure 3 using the Ansys ICEMCFD 15.0. The block structured grid methodology has been used to generate body fitted structured grid, so as to improve convergence of the solution. To capture the steep gradients near the cylinder, 'O' grid has been generated around the cylinder with very fine mesh size. The governing equations are solved using the solver Ansys Fluent 15.0 with appropriate boundary conditions. At the inlet, fully developed flow condition has been imposed using user defined function. Outflow condition with zero-gradient for variables has been assumed at the domain exit. The no-slip boundary condition is used for rigid solid walls such as side walls and cylinder surface. As the Reynolds number (Re=200) considered in the present study ensures shedding with plain wall confinement, the entire simulations of wavy wall confinement have been conducted using unsteady solver. The time step taken for the study is 0.002 based on Courant–Friedrichs–Lewy (CFL) condition and the computations are carried out for 5000 time-steps. Second order implicit method is chosen for transient formulation. The numerical scheme, COUPLED algorithm is employed for coupling pressure and velocity while solving the momentum and pressure-based continuity equations. As the grids are aligned with flow direction, the convective terms are discretized using QUICK scheme to improve the solution accuracy. Second order interpolation scheme is used for calculating the pressures at cell face. The gradients of solution variables at cell centers are determined using least-square cell-based approach. The convergence criterion for the residuals of continuity, X-Velocity and Y-Velocity components are taken as 1×10^{-6}.

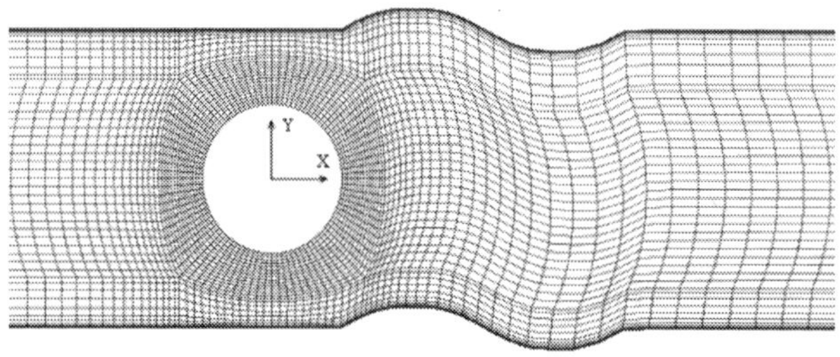

Figure 3. Schematic representation of mesh used in the present study.

4. Results and discussion

The following sections details the grid independent results, validation of the numerical methodology, the importance of waviness in the side wall on vortex shedding and its control. The effects of location of waviness on the shedding characteristics are also presented and discussed.

4.1. Grid independence and validation

In order to obtain an optimum grid size based on reduced computational time and improved solution accuracy, a thorough grid independence study has been conducted with various grid sizes as listed in Table-1 and the study has been extended for various Reynolds numbers. The grids are generated using block structured method, which facilitates the grid refinement around the cylinder and the confining walls. The percentage variation in C_D between (G_1 & G_2) and (G_2 & G_3) are 0.25 and 0.17, respectively at Re = 20, whereas at Re = 200 the variations are 0.17, 0.07. On the other hand, the percentage variation of C_{pb} between (G_1 & G_2) and (G_2 & G_3) are 6 and 1.76, respectively at Re = 20. It is noticed that for a fixed Reynolds number, the percentage variation with change in grid size is more for base pressure co-efficient (C_{pb}) as compared to drag coefficient (C_D). The value of C_{pb} is monitored from a nodal point, whereas C_D is calculated using the force values averaged over the cylinder surface, this averaging process make the value of C_D less sensitive to grid size. Hence a nodal point observation is better choice to conduct grid independence tests. In the present study, the maximum percentage variation of C_{pb} between (G_2 & G_3) is 3.96 at Reynolds number 200. Therefore, the grid size G_2 has been chosen as the optimum grid size for further computations. Moreover, the present computational methodology has been validated with the results of Sahin *et al.* (2004). The drag computed from the present study and those available in the literature shows good agreement with each other and the maximum deviation between the results is 1.82% at Re = 10. The results of grid independence study and validation of the results are not presented due to space constraint.

Table 1. Number of nodes and cells used in various grid sizes.

Grid	Total nodes	Total cells
G_1	26053	25550
G_2	43364	42782
G_3	139679	138528

4.2. Significance of local waviness

In order to study the significance of waviness on vortex shedding and its control, simulation has been carried out on trough only confined cylinder and corresponding flow characteristics such as drag and lift-signals are observed and shown in Figures 4(a) & (b), respectively. The drag and lift-signal confirms the presence of shedding, which is also observed from the streamline plots as shown in Figure 5. Hence it is clear that the trough only confinement is not enough to suppress shedding.

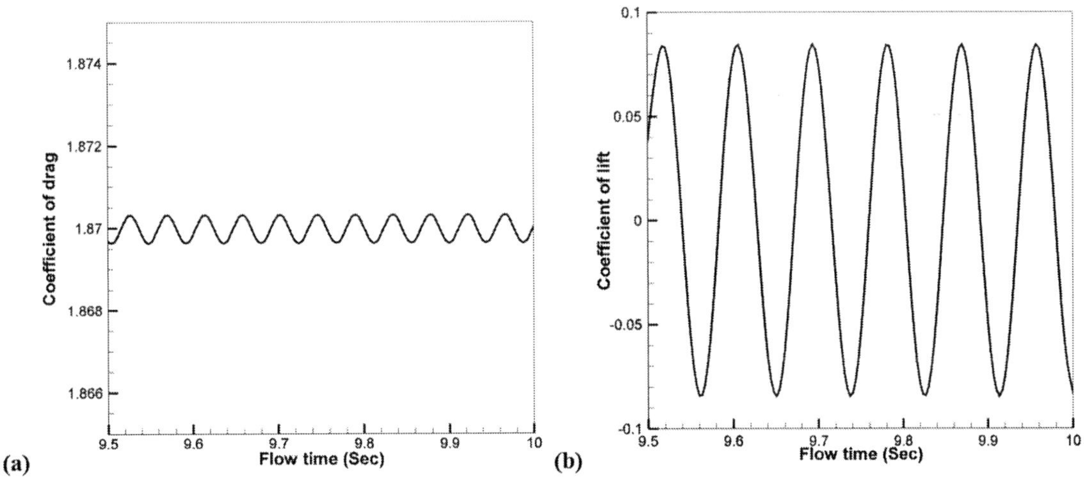

Figure 4. Effect of trough only confinement at Re = 200 (a) drag signal and (b) lift signal

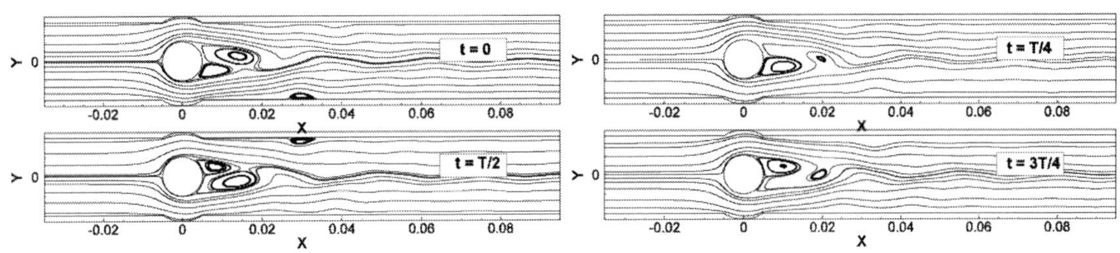

Figure 5. Streamlines at various flow time with trough only confinement at Re = 200

4.3. Effect of local waviness on shedding characteristics

The Figures 6(a) & (b) are plotted to observe the variation of C_D and C_L with flow time for various locations of waviness for a time period of 0.5 second and compared with the similar results corresponding to Pw confinement (reference case). The drag force on the cylinder is higher with waviness in all cases except Case-1. In case-1, the approaching flow striking the cylinder is slightly inclined with respect to the main flow direction. Thus, the total force on the cylinder is reinforced by lift force, hence drag reduced. With Case-1, the time-averaged flow past the cylinder is one-sided *i.e.*, increased flow rate in the gap between bottom wall and cylinder creates more pressure and leads to the shift of fluctuating lift-signal towards positive y-direction as shown in Figure 6(b) as compared to alternating lift of Pw configuration. The increased drag in case-2 and case-3 is due to the presence of wavy-crest near the cylinder which reduces flow passage and creates local acceleration of flow near the cylinder surface. The suppression of shedding and no fluctuation in lift is observed for the Cases 3 and 4 (Figure 6b) but offers more drag on the cylinder as compared to Pw configuration. In case-3, the maximum drag experienced is due to partially attached flow, that leads to increase in separation angle thus more shear on the cylinder surface. On the other hand, the more flow passage between cylinder and confining wall the drag is lesser in case-4 as compared with case-3. In presence of wavy confinement, the percentage increase in drag force is 17.28 and 2.64 for the Cases 3 and 4, respectively. It is also identified that with Cases 1 and case 2, fluctuating lift-signal is shifted either towards positive or negative y-direction depending on the direction of the approaching flow. With Case-1, drag is slightly reduced (0.7%) due to inclined approaching flow. But, with Case-2 drag increases (12.21%) due to increase in local confinement of cylinder. The percentage increase in maximum lift is 32.18 and 205.5%, respectively with Cases 1 & 2. This drastic rise in lift force with Case-2 is due to unequal confinement of the cylinder and approaching flow direction. It is also noticed that with Case-3, the crest of the waviness suppresses the shedding effectively and results in no fluctuation in lift. The quantitative results such as coefficient of drag, coefficient of lift and Strouhal number are computed for all the four cases are presented in the Table-2. The streamlines as shown in Figure 7 indicates the presence of vortex shedding for the Cases 1 & 2, and complete suppression of vortex shedding for the Cases 3 & 4. Finally, it is observed that the suppression of shedding is achieved by some penalty in increased drag or lift force on the cylinder. Hence, to reduce or nullify this effect the study can be extended to impose the waviness in out-phase configuration (OPC). With this OPC configuration the reduced flow passage between confining wall and cylinder surface can be avoided.

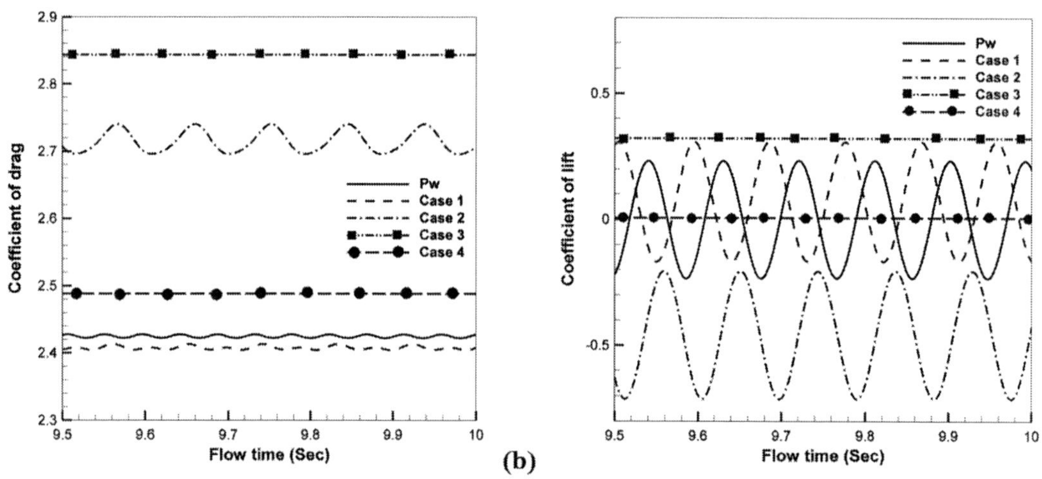

(a) (b)

Figure 6. Effect of location of waviness on (a) C_D and (b) C_L

Table 2. Effect of placement of waviness on C_D, C_L and St

Config.	**Mean C_D**	**Peak to Peak C_L**	**RMS C_L**	**Mean C_L**	**St**
Pw	2.424	-0.233 to 0.233	0.165	0	0.355
Case 1	2.407	-0.167 to 0.308	0.181	0.032	0.347
Case 2	2.72	-0.205 to -0.712	0.493	-0.458	0.339
Case 3	2.843	No fluctuation	0.322	0.322	No shedding
Case 4	2.488	No fluctuation	0.005	0.005	No shedding

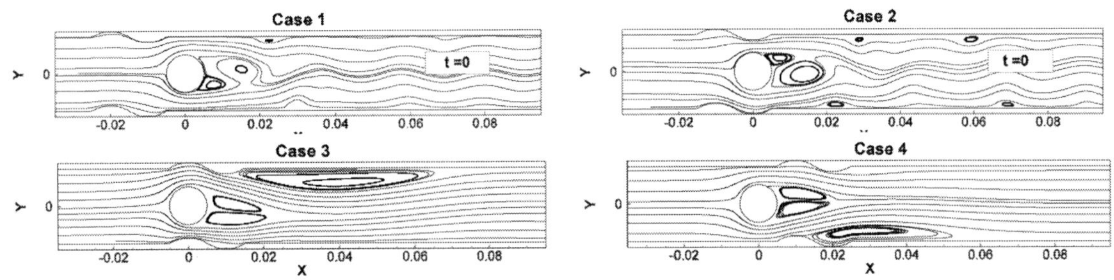

Figure 7. Streamline plots for various cases

5. Conclusion

The significance of local waviness in the confining side walls on vortex shedding from the cylinder surface has been studied using two dimensional numerical simulations at Reynolds number 200. Local waviness in the confining walls has been placed at four different locations with reference to cylinder center. In general, waviness located in upstream controls the shedding. Whereas, waviness in downstream completely suppresses the shedding. The flow direction approaching the cylinder, the gaps between the cylinder and the confining walls and the location of the crest region are the three major parameters that control the shedding characteristics. In the present work, the local waviness in the confining walls are in same phase, the study could be extended with waviness in different phase.

References

[1] Dehkordi, B.G., Jafari, H.H., 2010. On the Suppression of Vortex Shedding From Circular Cylinders Using Detached Short Splitter-Plates. J. Fluids Engg. 132, 044501, 1-4.
[2] Dipankar, A., Sengupta, T.K., 2005. Flow past a circular cylinder in the vicinity of a plane wall. J. Fluids Struct. 20, 403–423.
[3] Jayavel, S., Tiwari S., 2008. Numerical study of flow and heat transfer for flow past inline circular tubes built-in a rectangular channel in presence of vortex generators. Numer. Heat Transfer - Part A 54, 777-797.
[4] Jayavel, S., Tiwari S., 2009. Numerical investigation of incompressible flow past circular tubes in confined channel. CFD Letters 1, 1-14.
[5] Jayavel, S., Tiwari S., 2010. Effect of vortex generators and integral splitter plate on heat transfer and pressure drop for laminar flow past channel-confined tube banks. Heat Transfer Engg. 31, 383-394.
[6] Kwon, K., Choi,H., 1995. Control of laminar vortex shedding behind a circular cylinder using splitter plates. Phys. Fluids 8, 479-486.
[7] Mittal, S., Raghuvanshi, A., 2001.Control of Vortex Shedding Behind Circular Cylinder for Flows at Low Reynolds Numbers. Int. J.Numer. Meth. Fluids, 35(4), 421-447.
[8] Sahin, M., Robert, G., Owens, A., 2004. Numerical investigation of wall effects up to high blockage ratios on two-dimensional flow past a confined circular cylinder. Phys. Fluids 16, 1305-1320.
[9] Tiwari, S., Pratish P Patil., Jayavel, S., Biswas, G., 2006. Flow field characteristics near first transition for flow past a circular tube confined in a narrow channel. NCFMFP-2006, 1132.
[10] Singha, S., Sinhamahapatra, K.P., 2010. Flow past a circular cylinder between parallel walls at low Reynolds numbers. Ocean Engineering 37, 757-769.
[11] Strykowski, P.J., Sreenivasan, R., 1990. On the formation and suppression of vortex shedding at low Reynolds number.J. Fluid Mech. 218, 71-107.

Journal of Physics: Conference Series **759** (2016) 012084 doi:10.1088/1742-6596/759/1/012084

Effect of Rayleigh Number and Density Stability Ratio on characteristics of Double-Diffusive Salt Fingers

Manoj Dhiman, Faria Rehman, O.P Singh

School of Engineering, Indian Institute of Technology Mandi, India

E-mail: manoj_d@students.iitmandi.ac.in,rehmanfaria@gmail.com,om@iitmandi.ac.in

Abstract. In this paper, we investigate the finger evolution pattern and physics behind the layer formation in a two layer model using numerical simulations. A three-dimensional (3D) numerical code has been developed to solve Navier-Stokes equations along with energy and concentration conservation equations. Simulations were carried out to study the effect of Rayleigh number and density ratio on the characters of salt finger. Instabilities in form of rolls grow from side walls in all simulated cases. Square planform is found to be stable only at low Ra_T. Individual components flux is found to be larger in 3D case compared to 2D, but the flux ratio is found to be less in 3D.

1. Introduction

Salt finger convection is a buoyancy driven flow in which the density of fluid depends upon two scalar quantities with different diffusion rate and distributed such that fluid is initially gravitationally stable. The presence of faster diffusing component makes the fluid lighter and slower diffusing increases the heaviness. The increase in density due to presence of slower diffusing component in top layer is balanced by faster diffusing component. Initially, for system to be stable the higher diffusing component in top layer has to be higher than the slower diffusing component. Salt fingers are found in many natural and engineering environments, a few examples are oceanography, water bodies [1], metallurgy, geology, and stellar dynamics [2–5]. The difference in diffusion rate of components make the system gravitationally unstable, locally, near the interface, which makes it conducive for perturbations to develop and grow. The perturbations develop in forms of cells which grows slowly inside the density inverted zone and at much faster rate after crossing this zone. While doing so, they exchange more of higher diffusing components horizontally which makes the system run for longer time.

Since last decade, researchers have started reporting results from 3D simulations, which are more close to reality and provide and extra dimensions giving useful insights into the system With advancements made in computer and electronics the size of computational problem doesn't pose a big challenge any more. If in 2D system model, fingers evolve from line interface as alternating rising and falling structures, in 3D they rise from a plane and the pattern they form, on the interface plane, is still an unresolved topic.

The linear stability analysis of governing equations gives degenerated solutions for wave number selection and the suitable combination of variables chosen by system is attributed to the non-linear interactions between the fingers.

$$k^2 + l^2 = k_{max}^2 \tag{1}$$

Table 1: Basic Notations

H	Height of domain	u,v,w	components of velocity vector
α	coefficient of thermal expansion	k_t	coefficient of thermal diffusivity
β	coefficient of salinity expansion	k_s	coefficient of salt diffusivity
ΔT	Initial temp difference between two layers	ν	coefficient of salt diffusivity
ΔS	Initial salinity difference between two layers	Ra_T	$\frac{g\beta_T \Delta T H^3}{\nu k_t}$(Thermal Rayleigh no.)
Ra_s	$\frac{g\beta_S \Delta S H^3}{\nu k_s}$(Thermal Rayleigh no.)	R_ρ	$\frac{Ra_T}{Ra_S}$

where, k_{max}^2 corresponds to the maximum growth rate wave number. k and l are the in-plane (horizontal) wave numbers. Strauss [6] made first attempt to resolve this mystery, by proposing that rolls are more stable but this was contradicted by experiment conducted by Shirtcliffe and Turner [7] in which they showed square patterns in a triangular tank. Procter and Holyer [8] showed mathematically using weak non-linear analysis that equilibrated fingers prefer rolls over square cells. Simulations were run on a three dimensional in-house code to solve N-S equations, and parallelized it using OpenMP. The importance of Rayleigh number on salt finger has been shown by [9–12] The system at different values of R_ρ and Rayleigh numbers, Ra_S and Ra_T is explored by us. In this paper, observations are reported from a limited range of dimensionless parameters for a thermohaline sysem, which are, Rayleigh number (Ra_T) = 10^4, 10^5,10^8 for density stability ratio (R_ρ) = 1, 1.1 and 1.5.

2. Mathematical Formulation

Mathematical model of double diffusive convection involves incompressible continuity equation, Navier-Strokes equations for momentum and energy, and equation for concentration transport. The boundary conditions are slip, and adiabatic for mass and heat. The mathematical expressions are as follows:

Continuity:

$$\frac{\partial u}{\partial x} + \frac{\partial v}{\partial y} + \frac{\partial w}{\partial z} = 0, \tag{2}$$

X-momentum:

$$\frac{\partial u}{\partial t} + u\frac{\partial u}{\partial x} + v\frac{\partial u}{\partial y} + w\frac{\partial u}{\partial z} = -\frac{\partial p}{\partial x} + \mathbf{Pr}\left(\frac{\partial^2 u}{\partial x^2} + \frac{\partial^2 u}{\partial y^2} + \frac{\partial^2 u}{\partial z^2}\right) \tag{3}$$

Y-momentum:

$$\frac{\partial v}{\partial t} + u\frac{\partial v}{\partial x} + v\frac{\partial v}{\partial y} + w\frac{\partial v}{\partial z} = -\frac{\partial p}{\partial y} + \mathbf{Pr}\left(\frac{\partial^2 v}{\partial x^2} + \frac{\partial^2 v}{\partial y^2} + \frac{\partial^2 v}{\partial z^2}\right) \tag{4}$$

Z-momentum:

$$\frac{\partial w}{\partial t} + u\frac{\partial w}{\partial x} + v\frac{\partial w}{\partial y} + w\frac{\partial w}{\partial z} = -\frac{\partial p}{\partial z} + \mathbf{Pr}\left(\frac{\partial^2 w}{\partial x^2} + \frac{\partial^2 w}{\partial y^2} + \frac{\partial^2 w}{\partial z^2}\right) - \mathbf{Ra}_S\mathbf{Pr}_S + \mathbf{Ra}_T\mathbf{Pr}T \tag{5}$$

Energy:

$$\frac{\partial T}{\partial t} + u\frac{\partial T}{\partial x} + v\frac{\partial T}{\partial y} + w\frac{\partial T}{\partial z} = \left(\frac{\partial^2 T}{\partial x^2} + \frac{\partial^2 T}{\partial y^2} + \frac{\partial^2 T}{\partial z^2}\right) \tag{6}$$

Concentration:

$$\frac{\partial S}{\partial t} + u\frac{\partial S}{\partial x} + v\frac{\partial S}{\partial y} + w\frac{\partial S}{\partial z} = \frac{\mathbf{Pr}}{\mathbf{Sc}}\left(\frac{\partial^2 S}{\partial x^2} + \frac{\partial^2 S}{\partial y^2} + \frac{\partial^2 S}{\partial z^2}\right) \tag{7}$$

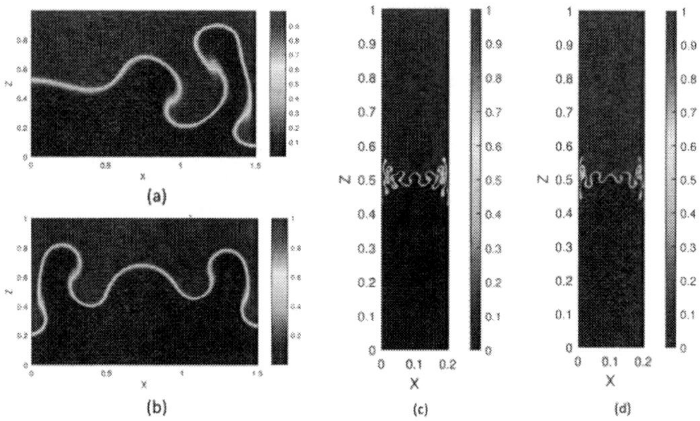

Figure 1: Comparison of 2D and 3D concentration field at $R_\rho = 1.5$, $\tau = 100$ (a) 2D, $Ra_T = 7E4$, time 2 hours, (b) 3D, $Ra_T = 7E4$, at time = 1.5 hours, (c) 2D, $Ra_T = 7E8$ at time= 25 sec, and (d) 3D, $Ra_T = 7E8$ at time = 20 sec

3. Planforms in 3-Dimensions

As shown in figure 1a and 1b, the wave number is a strong function of Ra_T. From figure 2 One thing which is common to all results is that the perturbation starts with rolls from each side wall and crosses where they meet, because they may be the result of T-points (where boundary and two layers of fluid meets, which is a highly unstable region). The widening of these rolls as they propagate towards centre can be ascribed to the changes brought by near wall fluxes. The roll which are thin and high in energy due to density inversion is adjacent to the wall is always a trough, due to the density inversion, they are thin and high in energy. Each trough undergoes constructive interference with the perpendicular roll which produces high velocity jets. The second roll, which is a crest, interferes constructively with second perpendicular roll, but it takes time to invert the flow of previously generated thin high energy roll. This roll eventually produces patterns of up and down-going cells developed near the corners.

At fixed $R_\rho = 1.5$, for $Ra_T = 7 \times 10^8$ when these rolls meet at the centre of interface, a round spot in vertical velocity field was observed. Outer rolls wither away to give a symmetric pattern which lasts only for 1∼2 minutes but, the inner roll is still there due to its large width. At centre the round spot converts slowly to cross, as the rolls emerge from bottom and at one time they completely disrupts the previous symmetry to divide the whole plane into four symmetric quadrants. The convection at these parameters is violent at the interface and after 15 minutes all the symmetry is gone from the plane and patterns observed was similar to what Radko and Stern [13] showed in their paper. Merging, splitting and generation of new fingers might be the cause for randomness which appear in plane-form.

Vertical velocities in rolls looks like to be dependent upon their width; thinner near the walls having large velocity, and wider towards the centre with less velocity. In the corners large vertical velocities are found due to constructive interference of thin rolls. At $Ra_T = 7 \times 10^8$ with $R_\rho = 1.1$ (figure 2h, i), the effect of high Ra_S is disruptive on fingers which makes system highly unstable and a proper interference of rolls at the centre was not observed, however a very random breakup was present at central zone. At $Ra_T = 7 \times 10^4$ and $R_\rho = 1.5$, the most stable system simulated, adopts widely reported structure of squares, which continues till the fingers reaches mid height of the domain. For $R_\rho = 1.5$ and $Ra_T = 7 \times 10^8$ at coarse grids of 97^2X241, with aspect ratio of 0.4 : 0.4 : 1.0, worm like patterns on mid-section are observed as shown in figure 2g. They stays like that with very subtle changes over large time period.

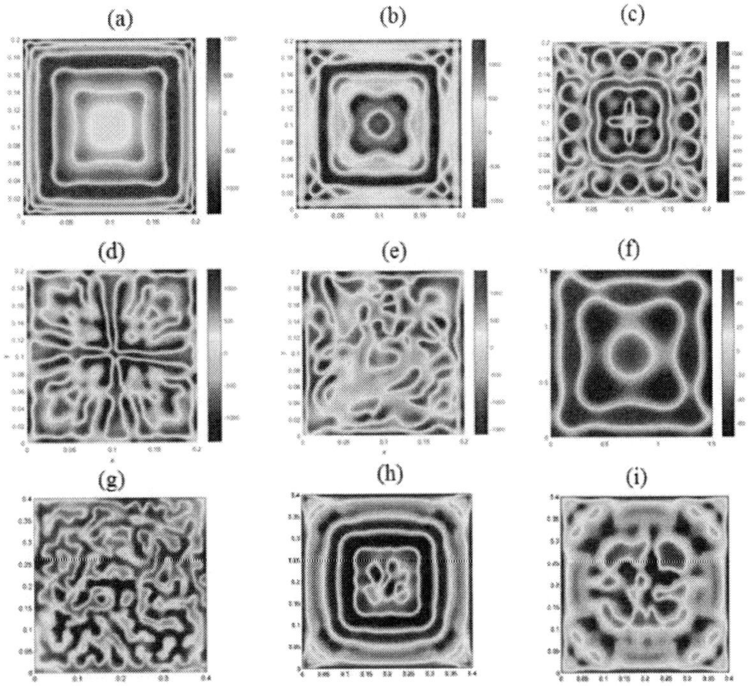

Figure 2: (a) to (h) shows the changes in planform with time at 13, 20, 32, 60, and 107 seconds, for $R_\rho = 1.5$, $Ra_t = 7 \times 10^8$. Figure (h) shows almost square planform at same R_ρ but at $Ra_t = 7 \times 10^4$. Figure (g) for $R_\rho = 1.1$ and $Ra_t = 7 \times 10^8$ at equilibrium state. Figure (h) and (i) for $R_\rho = 1.1$ and $Ra_t = 7 \times 10^8$ and at time = 20 and 25 seconds, respectively.

4. Staircases formation

After the system has passed through its strong convection period and left with small kinetic energy, and steps in concentration field (as shown in figure 4) sandwiching small magnitude vertical fingers were observed. Based upon the observed rolls, the reason might be side heating phenomenon. In this phenomenon, fluid is stably stratified with concentration of salt and heat decreasing from top to bottom and is heated from side vertical wall. A fluid parcel adjacent to the side wall rises due to heat, owing to buoyancy. During its rise salt diffuses very slowly out of it and its vertical motion is halted when its density becomes equal to that of surrounding. This parcel is forced to take slightly tilted path from horizontal towards the opposite wall and turns downward due to heat lost in this journey. The trailing fluid comes from its originating region. This completes the whole circulation and stacks of them has been observed in experiments.

Just before staircases starts to happen, full domain divides into two horizontal halves based on concentration of heat and salt in the fluid. Figure 5 shows the slice of concentration and temperature when staircase begin to form. More heat and salt concentrates in one half and less in the other half maintaining zero horizontal gradient in density, and clearly two separate zones are observed all through the height of system. Heat diffuses from high concentration zone and sets in many layers of horizontally downward tilted flow towards the other half, which is seen from the velocity components in the horizontal plane. What remains unexplained by analogising it with side wall heating phenomena is the existence of vertical cells within two layers.

5. Flux Ratio

Convective fluxes of individual components are higher in 3D system as compared to its counterpart in 2D. The reason owes to high kinetic energy which gives very less time for

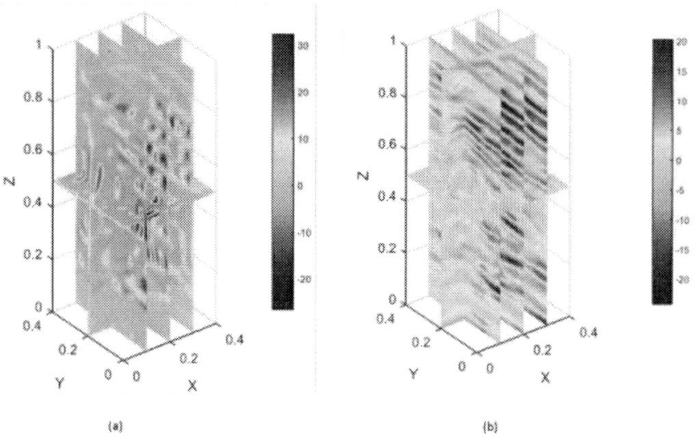

Figure 3: (a) Shows the vertical velocity field and (b) the horizontal velocity field. These figures are the evidence of vertical fingers between parallel flows. Slices are taken at x = 0.1, 0.2, 0.3; y = 0.2 and z = 0.5. the magnitude of vertical an horizontal flow is almost equal

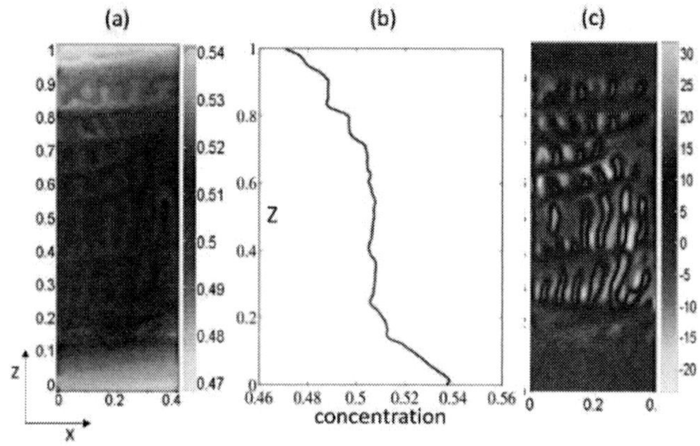

Figure 4: Slice at y = 0.2 (a) is concentration field, (b) horizontal mean of concentration and (c) shows the vertical velocity field.

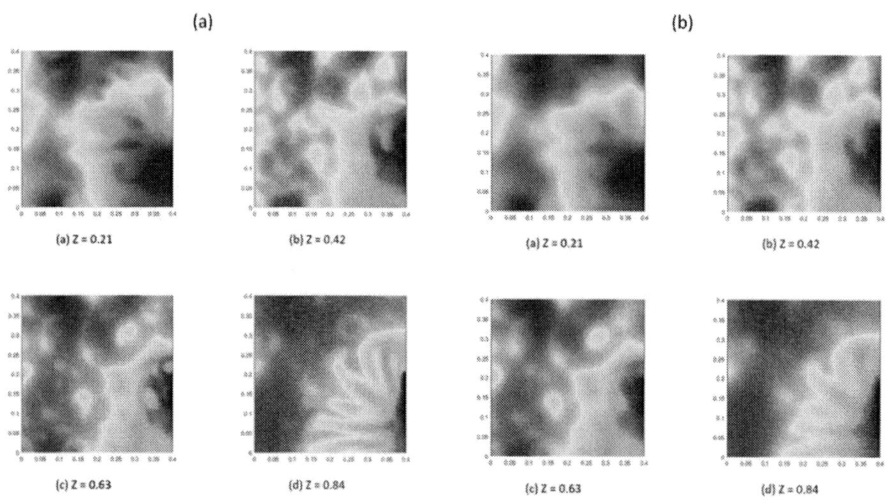

Figure 5: (a) Temperature field and, (b) Concentration field at value of Z = 0.21, 0.42, 0.63 and 0.84

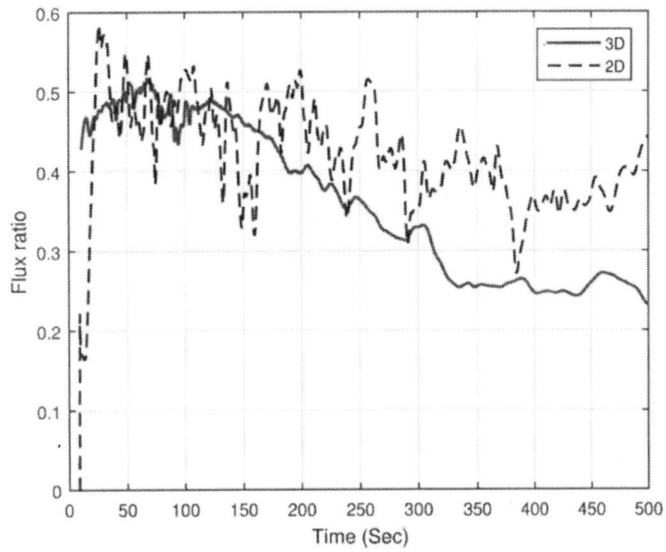

Figure 6: Comparison between 2D and 3D flux ratio at $R_\rho = 1.5$,$Ra_T = 7E8and\tau = 100$

horizontal diffusion to happen. This small time is too less for salt to diffuse and is a cause for large difference in individual components flux. Due to this large difference in individual component's flux, flux ratio is lower in 3D as compared to 2D fingers as shown in figure 6. Also flux ratio in 3D have less fluctuations compared to its counterpart in 2D.

6. Conclusion

In this paper we addressed the salt fingering in 2D and 3D. We shows density ratio alone is not sufficient and any one Rayleigh number is necessary to describe the system uniquely. The new observation of staircase in 3D is advantage over 2D model. From numerical simulation we observed that perturbations in 3D starts with rolls, however it is only observable at low values of Rayleigh number (Ra_T) at fixed density ratio that square planform is stable. We further notice that individual fluxes of heat and salt remains high in 3D in comparision to 2D, however flux ratio have less values.

7. References

[1] Schmid M, Busbridge M *et al.* 2010 *Limnology and Oceanography* **55** 225–238
[2] Schmitt R W 1983 *Physics of Fluids* **26** 2373
[3] Singh O P, Ranjan D, Srinivasan J and Sreenivas K 2011 *Journal of Geography and Geology* **3** 42
[4] Traxler A, Stellmach S, Garaud P, Radko T and Brummell N 2010 **677** 23 (*Preprint* 1008.1807)
[5] Stellmach S, Traxler a, Garaud P, Brummell N and Radko T 2010 **677** 18 (*Preprint* 1008.1808)
[6] Straus J M 1972 *Journal of Fluid Mechanics* **56** 353–374
[7] Shirtcliffe T and Turner J 1970 *Journal of Fluid Mechanics* **41** 707–719
[8] Proctor M R and Holyer J Y 1986 *Journal of fluid mechanics* **168** 241–53
[9] Shen C Y and Veronis G 1997 *Journal of Geophysical Research: Oceans* **102** 23131–23143
[10] Sreenivas K, Singh O and Srinivasan J 2009 *Physics of Fluids (1994-present)* **21** 026601
[11] Hage E and Tilgner A 2010 *Physics of Fluids (1994-present)* **22** 076603
[12] Singh O and Srinivasan J 2014 *Physics of Fluids (1994-present)* **26** 062104
[13] Radko T 2008 *Journal of Fluid Mechanics* **609** 59–85

Journal of Physics: Conference Series **759** (2016) 012085 doi:10.1088/1742-6596/759/1/012085

Gap Filling of Precipitation Data by SSA - Singular Spectrum Analysis

A S F Filho, G A R Lima[1]
Departamento de Engenharia Sanitária e Ambiental, Universidade Federal de Mato Grosso Av. Fernando Corrêa da Costa, nº 2367, 78060-900, Cuiabá, MT, Brazil
E-mail: garlima@ufmt.br

Abstract. From the macroscopic standpoint, the precipitation time series is obtained from observation of natural systems rather than in the laboratory. These time series are often full of gaps (missing values) due to the conditions under which the measurements are made. Missing values give rise to various problems in spectral estimation, inhibit statistical analysis and in specifying boundary conditions for numerical models. Hence, gap filling is necessary in environmental science. The aim of this study is to highlight the application of the SSA forecasting algorithm to fill in missing values to real-life time series. It was applied to several monthly precipitation time series recorded over a large savannah area in Brazil. The results are promising and the accuracy and reliability depend on the pattern and relative length of the gaps with respect to the total length of the time series and presence of noise.

1. Introduction

Although, precipitation long varying trend analyses on different spatial and temporal scales has been of great concern because of the attention given to global climate change [1], the worldwide number of active monitoring points is still quite low [2]. Furthermore, time series are affected by non-climatic factors caused by malfunctioning or maintenance of measuring equipment, changes in location of monitoring points, systematic errors and data transcription. As a consequence, time series generally contain many missing values. Missing values impose gradual bias and are the major source of problems in the analysis and modelling of spatial and temporal variability. In other words, missing values impose a limit on the quality of information that can be extracted from the observed values (records). Incomplete time series reveal only part of the reality. Examining the full extent of the precipitation time series, with the missing data filled in, allows for better significance testing in the spectral analysis because all the information contained in the time series can be extracted and the information quality maximized. In this sense, SSA is a widely used method of time series analysis [3-5] and filling gaps [6-10]. State-of-the-art gap filling algorithms based on the SSA exploit the information contained in periodic temporal patterns to fill in the observations. Today there are three SSA based approaches for filling in missing values. The first approach was designed to deal with stationary time series only and preserved only pairs of valid vector components in the vectors' inner products [6]. In the second approach the missing values are estimated iteratively by successive application of SSA and the complete time series are composed by imputation of estimated missing values. But the other values are taken from the initial time series [7]. The third approach is an extension of SSA forecasting algorithms known as Caterpillar-SSA where the main methodological

[1] To whom any correspondence should be addressed.

concept is the components' separability. In this approach, neither a parametric model nor stationarity conditions have to be assumed for the time series. [8]. Following the last approach, in this paper we highlight the application of the algorithm to fill gaps in precipitation time series recorded at wide savannah area in Brazil.

2. Methodology - SSA Algorithm

The SSA algorithm decomposes the initial time series into a sum of a small number of interpretable components such as a slowly varying trend, seasonal, periodic or oscillating components, and also structureless components of noises. It is based on the singular value decomposition (SVD) of a trajectory matrix constructed upon the initial time series. The algorithm consists of two complementary stages: decomposition and reconstruction. Each stage, comprises two steps: Embedding and SVD are the steps of the first stage, grouping and diagonal averaging are the steps of the second stage. The extension of the SSA algorithm for the analysis of time series with missing values has the same structure, but the steps are somewhat different [11]. A brief description of the algorithm is presented below. A detailed discussion can be found in [11,12] and references therewith.

Let $X_N = (x_1, ..., x_N)$ be a time series (input) consisting of N data points, some part of which is unknown (missing values).

2.1 *First stage - Decomposition*

Step one – Embedding. The embedding procedure transforms the initial time series into the sequence of L-dimensional lagged vectors $\{X_i\}_{i=1}^{K}$ where $K = N - L + 1$. Some of the lagged vectors contain missing components. The window length L is the only parameter in the method. Let us fix the L, $1 < L < N$. Selection of the proper L depends on the problem in hand and on preliminarily information about the time series to be analized. Let C be the set of indices such that the lagged vectors X_i, with $i \in C$ are complete. Let us collect all complete lagged vectors $X_i, i \in C$, into the matrix $\tilde{\mathbf{X}}$. If there are no missing values, then the matrix $\tilde{\mathbf{X}}$ coincides with the trajectory matrix of the series $\mathbf{X}_N$.

Step two – SVD. Let $\tilde{\mathbf{S}} = \tilde{\mathbf{X}}\tilde{\mathbf{X}}^{\mathrm{T}}$. Denote by $\lambda_1 \geq ... \geq \lambda_L \geq 0$ the ordered eigenvalues of the matrix $\tilde{\mathbf{S}}$ and by $U_1,...,U_L$ the orthonormal system of the eigenvectors of the matrix $\tilde{\mathbf{S}}$ corresponding to these eigenvalues, $d = max\{i : \lambda_i > 0\}$.

2.2 *Second stage - Reconstruction*

Step three a - Reconstruction stage. This stage demands the grouping to make subgroups of the decomposed trajectory matrices and diagonal averaging to reconstruct the new time series from the subgroups choosing the subspace and projection of the complete lagged vectors. Let a set of indices $I_r = \{i_1, ..., i_r\} \subset \{1, ..., d\}$ be chosen and the subspace $M_r = span(U_{i_1}, ..., U_{i_r})$ be formed. The choice of the eigenvectors (i.e., their indices) corresponding to X_N^I is the same as in Basic SSA. The complete lagged vectors can be projected onto the subspace M_r in the usual way:

$$\hat{X}_i = \sum_{k \in I_r} (X_i, U_k) U_k, \quad i \in C$$

Step three b - Projection of the incomplete lagged vectors. For each **Q**-incomplete lagged vector with missing components in the positions from the set **Q**, the given step consists of two parts: (a) calculation of $\hat{X}_i\big|_{J_L/Q}, i \notin C$, and (b) calculation of $\hat{X}_i\big|_{\mathbf{Q}}, i \notin C$. Since adjacent lagged vectors have common information which enables processing of empty vectors with $Q=J_L=\{1,...,L\}$. Note that Step three b may change the vectors $\hat{X}_i, i \in C$. The result of Steps 3a and 3b is the matrix

$\hat{\mathbf{X}}=[\hat{X}_1 : ... : \hat{X}_K]$ which serves as an approximation to the trajectory matrix of the series $\mathbf{X}_N^{I}$, under the proper choice of the set I_r.

Step four - Diagonal averaging. In the last step of the algorithm, the matrix $\hat{\mathbf{X}}$ is transformed into the new series $Y_N^{(1)}$ (the reconstructed time series) by means of the diagonal averaging with a simultaneous filling in of the missing data.

The algorithm was applied to the longest available precipitation time series in order to fill gaps and identify slowly varying trends and periodic components from 31 monitoring points located at Alto Médio Cuiabá River Basin as is illustrated in figure 1. The source of data is available in the Hidroweb geodatabase [13].

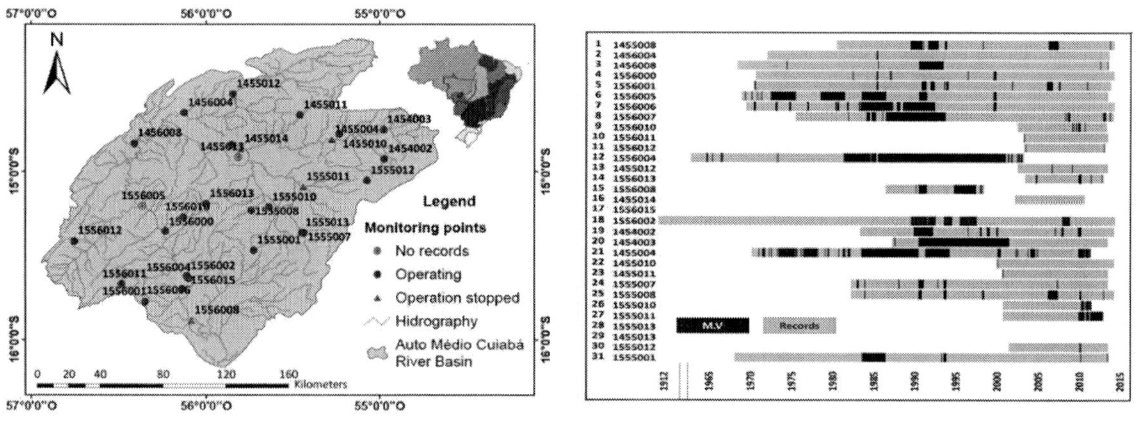

Figure 1. Spatial and temporal distribution of monitoring points at Alto-Médio-Cuiabá River Basin

2.3 ***Gap-filling procedures*** - As noise is produced by error of measurements or uncertainties, in accordance with [9] when the distribution of gaps follows a seasonal pattern, it generates spurious, intermittent peaks in the reconstructed time series. In the case of precipitation records, during the reconstruction stage, the spectral method also gives rise to negative values for imputation of some missing values in the dry season. To avoid this, the idea is thus to fill in missing values of small time series formed by breaking the initial time series into 12 sub-time series by means of a delay (or lag with k=12) according to each month of the calendar year. We consider reconstruction using signal plus noise components by the successive application of the SSA algorithm for each month. Then all predicted values are inserted in the initial time series where the values are missing. Thus a complete time series is obtained as seen in [7]. As imputation means to substitute missing values with plausible values for a stated purpose and all the predicted missing values fall in the range of minimum and maximum records for each hydrologic year of each time series, the ANOVA-LR (Analysis of Variance of Linear Regression Analysis) between complete and reconstructed time series was selected as a measure of "goodness of fit" between the models and time series (confidence interval 95%). Finally, the complete time series was again analyzed to illustrate the capability of the algorithm to extract a long varying trend, and periodic and noise components. All of the results were obtained by means of *CatMV1.0* and *Caterpillar-SSA 3.30* software. The window lengths were assigned by following the theoretical recommendation [14].

3. Results

Although the study was carried on for 31 time series, the results are shown for only four. These four time series have the highest number of missing values and all types of gaps uniformly distributed (figure 2 (a)), seasonal (figures 2 (b) and (c) and prolonged gaps (figure 2 (d)). These are also found in the other 27 times series. Figure 2 shows the summarized results of the decomposition stage for the initial time series. Visual inspection shows that independently of the type of gaps, series length and noise quantity, the algorithm is capable of extracting the slowing varying trend and seasonality, which correspond to the start and the end of each hydrological year.

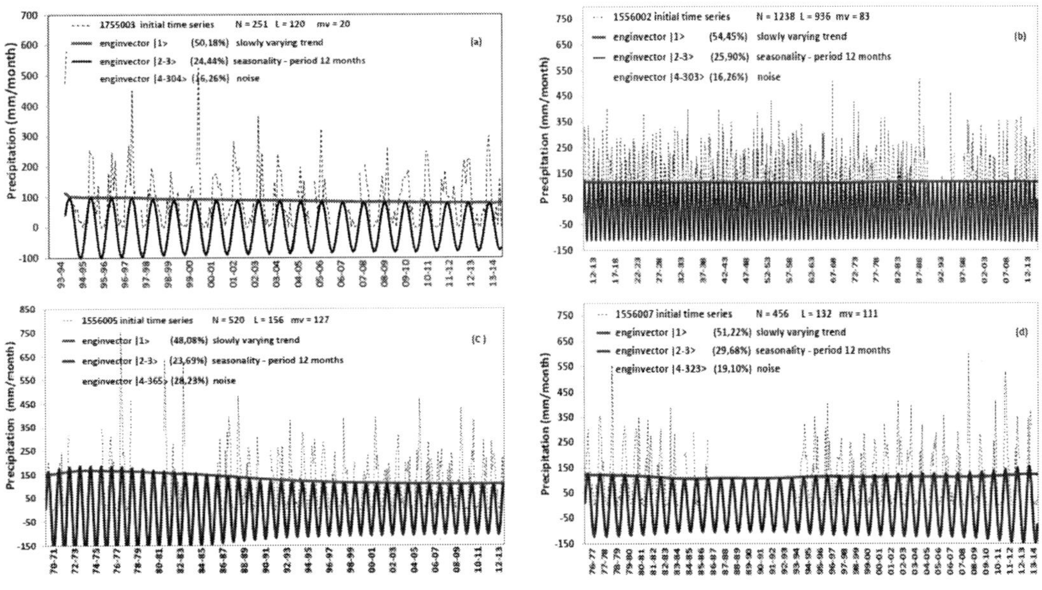

Figure 2. Initial precipitation time series and decomposition after SVD

Figure 3 shows the results of the decomposition stage for the complete time series after the imputation of missing values. Visual inspection of figures 3 (b) and (c) shows new extracted periodic components |4-5⟩, which correspond to the start and the end of the rainy and dry seasons respectively.

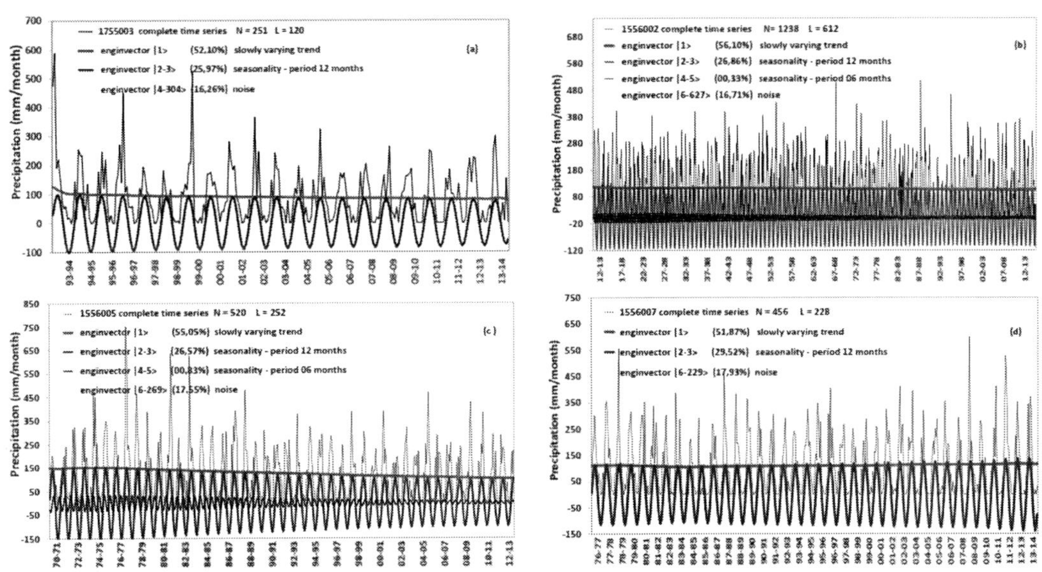

Figure 3. Complete precipitation time series and its decomposition after SVD

Complete time series tend to have more information about oscillations and noise reduction. Figure 4 shows the *w*-correlation matrix for the first 14 reconstructed components of the complete time series in a 20-grade scale from white to black corresponding to the absolute values of correlations from 0 to 1. It is possible to see the approximate separability of groups such as the slowly varying trend represented by eigentriple |1⟩, seasonal components by eigentripe |2-3⟩ and eigentripe |4-5⟩ and noise by others eigentriples |6 .... *k*⟩. This means that the selection of the proper window length allows extraction (separation) of all the components of the series and the filling in of missing values at the reconstruction stage.

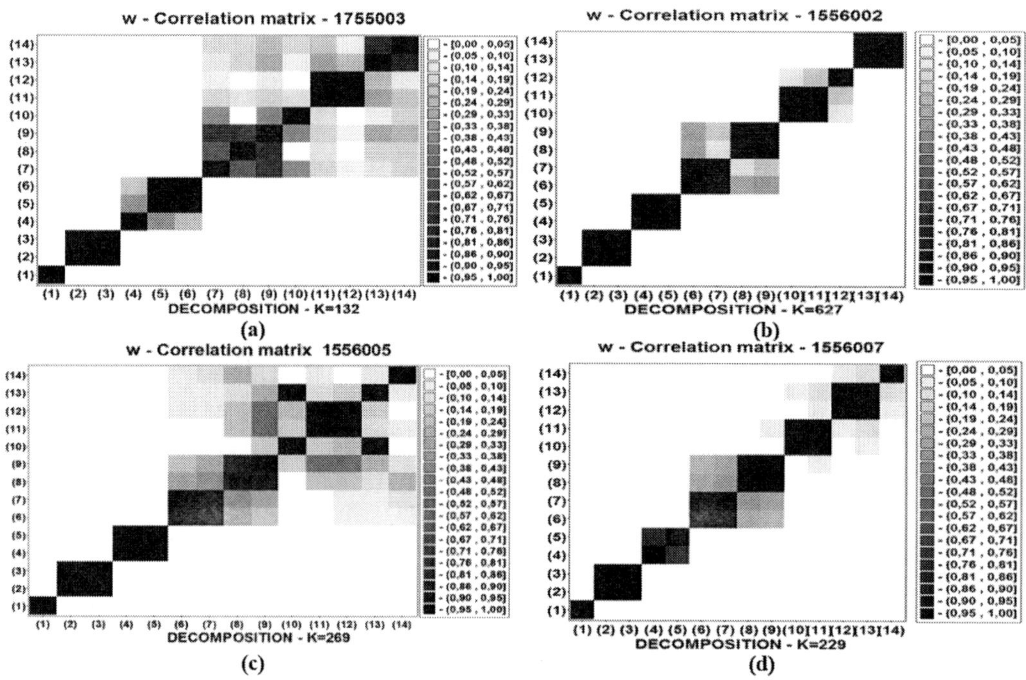

Figure 4. Matrix of w-correlations for the first 14 reconstructed components

A relative measure of goodness fit is shown in Table 1. The correlation coefficient *R* measures the degree of linear association between complete and reconstructed time series. The determination of the coefficient R^2 of the regression line indicates the percentage of those records in complete time series that can be explained by the records predicted during the reconstruction stage. In other words, more than 90% of the precipitation records can be explained by the linear regression and consequent gap-filling strategy.

Table 1 – Summary statistical of regression - ANOVA

	N	*R*	R^2	*SD*	*F*	*p-value*
	1755003					
P(complete) = - 6.16 *P*(reconstructed) +0,55	251	0.986	0.9723	43.34	17839.45	0.000
	1556002					
P(complete) = - 9.26 *P*(reconstructed) +1,05	1238	0.991	0.9820	59.94	19634.34	0.000
	1556005					
P(complete) = - 8.39 *P*(reconstructed) + 1,04	520	0.925	0.8556	44.62	3061.90	0.001
	1556007					
P(complete) = - 4.21 *P*(reconstructed) +1,02	456	0.989	0.9781	16.32	20820.42	0.000

In Table 1, the column *SD* presents the standard deviation of the data in relation to the regression line (unexplainable variation) in the records. The *F* represents the ratio of explained by unexplained variance and determines the overall statistical significance of the regression equation. As the *p*-value is less than the confidence level 5%, the regression is not rejected. According to LRA results, all time series were reconstructed successfully.

4. Conclusions

One of the conclusions is that the SSA caterpillar algorithm can deal with the non-stationary nature inherent in precipitation records by filling in missing values and in the simultaneous extraction of long varying trends and periodic components. The gap-filling strategy adopted while utilizing a conceptually simple and computationally efficient algorithm represents an adaptation of existing techniques to create an operational method that is applicable to precipitation time-series. In all cases, the adopted strategy prevented negative values and spurious, intermittent peaks in the reconstruction stage and was particularly helpful for situations with continuously missing data, which quickly become prohibitive, without breaking the initial time series by means of a delay (or lag with k=12). The results are promising but further analysis is required in order to address real-world limitations associated with sensitivity of types of gaps and the impact of noise on the liability of predicted missing values.

While the study was restricted to precipitation, the strategy could be readily adapted to a wide variety of environmental time-series and combined with the SSA caterpillar algorithm to have a very wide range of applicability.

Acknowledgements

A S F Filho thanks Fundação de Amparo à Pesquisa do Estado de Mato Grosso (FAPEMAT) for a graduate fellowship and to John C. Carpenter for feedback on grammar and writing.

5. References

[1] Longobardi A and Villani P 2009 *Int. J. Climatol.* DOI: 10.1002/joc.2001
[2] Hulme M 1995 *Weather* **50**, 34-42
[3] Vautard M Ghil M 1989 *Physica D*, **35**, 395-424
[4] Allen M and Smith L 1996 *J. Clim.*, **9**, 3373-3404
[5] Vautard R Yiou P and Ghil M 1992 *Physica D*, **58**, 95-126
[6] Schoellhamer D 2001 *Geophys. Res. Lett.*, **28**, 16, 3187-3190
[7] Kondrashov D and Ghil M 2006 *Nonlin. Processes Geophys.* **13**, 2, 151-159
[8] Golyandina N and Osipov E 2007 *J. Stat. Plan. Infer* 137, **8**, 2642-2653
[9] Musial J P Vertraete M M Gobron N 2011 Atmos. Chem.Phys., **11**, 7905-7923
[10] Buttla J V Zscheischler J and Mahecha M D 2014 *Nonlin. Processes Geophys.* **21**, 203-215
[11] Golyandina N and Zhigljavsky A 2013 *Singular Spectrum Analysis for Time Series* Springer
[12] Golyandina N Nekrutkin V and Zhigljavsky A 2001 *Analysis of Time Series Structure: SSA and related techniques*, Chapman & Hall-CRC, New York - London
[13] http:// http://www.snirh.gov.br/hidroweb/, August 2015
[14] Golyandina N 2010 Statistics and Its Interface, **3** (3), 259–279

Journal of Physics: Conference Series **759** (2016) 012086 doi:10.1088/1742-6596/759/1/012086

An accurate predictor-corrector HOC solver for the two dimensional Riemann problem of gas dynamics

Bidyut B. Gogoi

Department of Mechanical Engineering, Indian Institute of Technology Guwahati, Guwahati - 781039, INDIA

E-mail: b.gogoi@iitg.ernet.in

Abstract. The work in the present manuscript is concerned with the simulation of two-dimensional (2D) Riemann problem of gas dynamics. We extend our recently developed higher order compact (HOC) method from one-dimensional (1D) to 2D solver and simulate the problem on a square geometry with different initial conditions. The method is fourth order accurate in space and second order accurate in time. We then compare our results with the available benchmark results. The comparison shows an excellent agreement of our results with the existing ones in the literature. Being a finite difference solver, it is quite straight-forward and simple.

1. Introduction

Over the past few decades, simulation of the Riemann problem of gas dynamics has been an interesting and challenging task amongst the researchers and there have been several attempts on developing numerical schemes for the Euler equations of gas dynamics. A close look at the literature reveal that there are plenty of numerical methods for the one-dimensional (1D) Riemann problem of gas dynamics. Most of these numerical methods are lower order accurate in space, for example, the AUSM, HLLC solver, Godunov, Rusanov, Lax-Friedrich, Lax-Wendroff, MacCormack methods etc. but recently Euler/Navier-Stokes equations have also been solved by compact schemes with near spectral accuracy [1, 2]. Very recently, we have developed a class of fourth order compact schemes [3] for the one-dimensional euler equations of gas dynamics. These schemes were based on finite difference formulation and were spatially fourth order accurate and temporally second order accurate or first order accurate depending on a weighted average parameter. These schemes were seen to capture shock fronts and the contact discontinuities with very correct locations and highly satisfactory sharpness. Unlike the earlier proposed HOC schemes, the numerical results obtained by the present HOC schemes show almost no smearing across all the discontinuities. In the present work, we have extended this 1D solver to a 2D solver which is then used to simulate the 2D Riemann problem of gas dynamics. Several researchers [4–9] worked on the 2-D Riemann problem with varied numerical approaches. However, in majority of these methods, upwind schemes have been devised, which use Riemann problems [5–9] to construct numerical flux in the finite volume setting. Higher order version of these schemes give spurious wiggles at discontinuities like shocks [8, 10]. Compared to the existing methods for the 2D Riemann problem, our proposed method is quite straight-forward and simple.

The paper has been arranged in four sections. Section 2 deals with the problem description, basic formulation and numerical procedure, Section 3 the numerical test cases and finally, Section 4 summarizes the whole work.

2. The problem and discretization

Fig. 1 shows the geometrical set-up of the 2D Riemann problem. The four quadrant has four different initial conditions. At time $t = 0$, the four separations are removed and gas is allowed to flow.

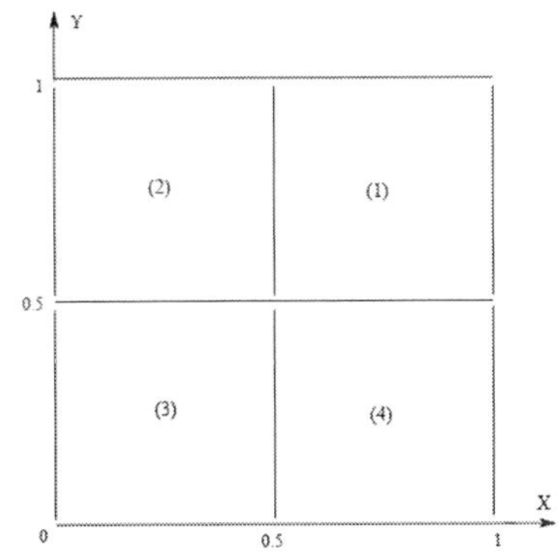

Figure 1. Problem Geometry

The flow is governed by the 2D Euler equations, which can be written as

$$U_t + F(U)_x + G(U)_y = 0, \tag{1}$$

where

$$U = \begin{bmatrix} \rho \\ \rho u \\ \rho v \\ E \end{bmatrix}, F(U) = \begin{bmatrix} \rho u \\ \rho u^2 + p \\ \rho uv \\ u(E+p) \end{bmatrix} G(U) = \begin{bmatrix} \rho v \\ \rho uv \\ \rho v^2 + p \\ v(E+p) \end{bmatrix} \tag{2}$$

Here ρ is the density, u and v are the x- and y-component of velocities, p is the pressure and E is the total energy. The system is closed by specifying an equation of state

$$E = \frac{p}{(\gamma - 1)} + \frac{1}{2}\rho(u^2 + v^2), \tag{3}$$

The above set of equations together with the following four initial conditions

$$(p, \rho, u, v)(t = 0) = \begin{cases} (p, \rho, u, v)(1) & \text{if x} > 0.5 \text{ and } \text{y} > 0.5 \\ (p, \rho, u, v)(2) & \text{if x} < 0.5 \text{ and } \text{y} > 0.5 \\ (p, \rho, u, v)(3) & \text{if x} < 0.5 \text{ and } \text{y} < 0.5 \\ (p, \rho, u, v)(4) & \text{if x} > 0.5 \text{ and } \text{y} < 0.5 \end{cases}$$

is the Riemann problem of gas dynamics in two space dimension.

In order to develop the higher order compact (HOC) scheme for the 2D Riemann problem (extending our recently developed 1D scheme [3]), let us consider the unsteady 2-D purely convection equation for a transport variable ϕ in some continuous domain with suitable boundary conditions

$$\frac{\partial \phi}{\partial t} + c(x,y,t)\frac{\partial \phi}{\partial x} + d(x,y,t)\frac{\partial \phi}{\partial y} = f(x,y,t) \tag{4}$$

where c and d are convection coefficients and f is the forcing function.

The steady-state form of equation (4) may be wriiten as

$$c(x,y)\frac{\partial \phi}{\partial x} + d(x,y)\frac{\partial \phi}{\partial y} = f(x,y) \tag{5}$$

Considering a uniform mesh of step sizes h and k along x and y directions respectively, a standard central difference approximation to equation (5) at the (i,j)-th node is given by

$$c\delta_x\phi_{i,j} + d\delta_y\phi_{i,j} - \tau_{i,j} = f_{i,j} \tag{6}$$

where $\phi_{i,j}$ denotes $\phi(x_i, y_j)$; while δ_x and δ_y denotes first order central difference operators along x- and y-directions respectively.

The truncation error $\tau_{i,j}$ in (6) is of the form

$$\tau_{i,j} = \left[c\frac{h^2}{6}\frac{\partial^3\phi}{\partial x^3} + d\frac{k^2}{6}\frac{\partial^3\phi}{\partial y^3}\right]_{i,j} + O(h^4, k^4) \tag{7}$$

Now, to derive a fourth order compact finite difference formulation of (5), we differentiate it twice w.r.t. x and then w.r.t. y, which results in the following equations

$$c\frac{\partial^3\phi}{\partial x^3} = \frac{\partial^2 f}{\partial x^2} - \frac{\partial^2 c}{\partial x^2}\frac{\partial \phi}{\partial x} - 2\frac{\partial c}{\partial x}\frac{\partial^2\phi}{\partial x^2} - \frac{\partial^2 d}{\partial x^2}\frac{\partial \phi}{\partial y} - 2\frac{\partial d}{\partial x}\frac{\partial^2\phi}{\partial x\partial y} - d\frac{\partial^3\phi}{\partial x^2\partial y} \tag{8}$$

$$d\frac{\partial^3\phi}{\partial y^3} = \frac{\partial^2 f}{\partial y^2} - \frac{\partial^2 d}{\partial y^2}\frac{\partial \phi}{\partial y} - 2\frac{\partial d}{\partial y}\frac{\partial^2\phi}{\partial y^2} - \frac{\partial^2 c}{\partial y^2}\frac{\partial \phi}{\partial x} - 2\frac{\partial c}{\partial y}\frac{\partial^2\phi}{\partial x\partial y} - c\frac{\partial^3\phi}{\partial x\partial y^2} \tag{9}$$

Putting these expressions for higher derivatives in (7) and inturn putting them back in (6) results in the following HOC scheme for (5).

$$\left[\alpha_{i,j}\delta_x^2 + \beta_{i,j}\delta_y^2 + A_{i,j}\delta_x + B_{i,j}\delta_y + C_{i,j}\delta_x\delta_y + D_{i,j}\delta_x\delta_y^2 + E_{i,j}\delta_x^2\delta_y\right]\phi_{i,j} = F_{i,j} \tag{10}$$

The coefficients appearing in (10) are given by:

$$\alpha_{i,j} = \frac{h^2}{3}\delta_x c_{i,j};\ \beta_{i,j} = \frac{k^2}{3}\delta_y d_{i,j};\ A_{i,j} = \left[1 + \frac{h^2}{6}\delta_x^2 + \frac{k^2}{6}\delta_y^2\right]c_{i,j};$$

$$B_{i,j} = \left[1 + \frac{h^2}{6}\delta_x^2 + \frac{k^2}{6}\delta_y^2\right]d_{i,j};\ C_{i,j} = \frac{1}{3}\left(h^2\delta_x d_{i,j} + k^2\delta_y c_{i,j}\right);$$

$$D_{i,j} = \frac{c_{i,j}k^2}{6};\ E_{i,j} = \frac{d_{i,j}h^2}{6};\ F_{i,j} = \left[1 + \frac{h^2}{6}\delta_x^2 + \frac{k^2}{6}\delta_y^2\right]f_{i,j}$$

For the unsteady Equation (4), HOC formulation is similar to (10), but coefficients c and d are now functions of x, y and t and the R.H.S. is of the form $f(x,y,t) - (\partial\phi)/(\partial t)$. Taking Δt as the time-step and using (10) and applying forward differencing for $(\partial\phi)/(\partial t)$ we have the following HOC approximation for the unsteady Equation (4).

$$\left[1 + \frac{h^2}{6}\delta_x^2 + \frac{k^2}{6}\delta_y^2\right]\delta_t^+\phi_{i,j}^n + \left[\alpha_{i,j}\delta_x^2 + \beta_{i,j}\delta_y^2 + A_{i,j}\delta_x + B_{i,j}\delta_y + C_{i,j}\delta_x\delta_y + D_{i,j}\delta_x\delta_y^2 + E_{i,j}\delta_x^2\delta_y\right]\phi_{i,j}^n = F_{i,j} \tag{11}$$

where δ_t^+ is the forward difference operator and the superscript n represents the n-th time level.

To simulate the 2D Riemann problem, we convert each equations in (1) to purely convection equations as

$$\frac{\partial \rho}{\partial t} + u\frac{\partial \rho}{\partial x} + v\frac{\partial \rho}{\partial y} = f_1 \tag{12}$$

$$\frac{\partial m}{\partial t} + u\frac{\partial m}{\partial x} + v\frac{\partial m}{\partial y} = f_2 \tag{13}$$

$$\frac{\partial n}{\partial t} + u\frac{\partial n}{\partial x} + v\frac{\partial n}{\partial y} = f_3 \tag{14}$$

$$\frac{\partial E}{\partial t} + u\frac{\partial E}{\partial x} + v\frac{\partial E}{\partial y} = f_4 \tag{15}$$

where

$$m = \rho u, \qquad n = \rho v,$$
$$f_1 = -\rho\left(\frac{\partial u}{\partial x} + \frac{\partial v}{\partial y}\right),$$
$$f_2 = -\frac{\partial p}{\partial x} - m\left(\frac{\partial u}{\partial x} + \frac{\partial v}{\partial y}\right)$$
$$f_3 = -\frac{\partial p}{\partial y} - n\left(\frac{\partial u}{\partial x} + \frac{\partial v}{\partial y}\right)$$
$$f_4 = -u\frac{\partial p}{\partial x} - p\frac{\partial u}{\partial x} - v\frac{\partial p}{\partial y} - p\frac{\partial v}{\partial y} - E\left(\frac{\partial u}{\partial x} + \frac{\partial v}{\partial y}\right).$$

Then we use the following predictor-corrector strategy as summarized below

1. First solve (12) for ρ using the developed HOC.
2. Replace f_1 by $f_1 = f_1 + \delta(\frac{\partial^2 \rho}{\partial x^2} + \frac{\partial^2 \rho}{\partial y^2})$, and again solve (12) but with the updated ρ.
3. Solve equation (13), with ρ obtained in 2, by putting $m = \rho u$.
4. Replace f_2 by $f_2 = f_2 + \delta(\frac{\partial^2 m}{\partial x^2} + \frac{\partial^2 m}{\partial y^2})$, and again solve (13) but with the new m obtained in step 3 and new u obtained from new m as $u = \dfrac{m}{\rho}$, where ρ is the one obtained in step 2.
5. Solve now equation (14), with new u (i.e. again do $u = \dfrac{m}{\rho}$ after step 4) and using $n = \rho v$, where ρ is the one obtained in step 2.
6. Replace f_3 by $f_3 = f_3 + \delta(\frac{\partial^2 n}{\partial x^2} + \frac{\partial^2 n}{\partial y^2})$, and again solve (14) but with the new n obtained in step 5 and new v obtained from new n as $v = \dfrac{n}{\rho}$, where ρ and u are the ones obtained in step 2 and 4.
7. With latest values of ρ, u and v (here again do $v = \dfrac{n}{\rho}$ after step 6) write energy E as $E = \dfrac{p}{(\gamma - 1)} + \dfrac{1}{2}\rho(u^2 + v^2)$. Solve equation (15) with all the new values.
8. Lastly, replace f_4 by $f_4 = f_4 + \delta(\frac{\partial^2 E}{\partial x^2} + \frac{\partial^2 E}{\partial y^2})$, and again solve (15) with these new values along with the updated $p = (\gamma - 1)(E - \frac{1}{2}\rho(u^2 + v^2))$, obtained from step 7.
9. Repeat steps 1-8 until final time T is reached. Note that δ should be a very small entity close to zero. We take it 0.01. Since it adopts a two step strategy where it predicts one value at one

step and again solve for it, we termed it as a *predictor-corrector* HOC.

Parameters used: CFL = 0.475 and grid size 401 × 401. $C_i = \sqrt{(\gamma * p_i/\rho_i)}$, where $i = 1, 2, 3, 4$ and $\gamma = 1.4$ and C = $\max(C_i)$ and $\Delta t = (CFL * min(h, k))/C$.

3. Results

We simulate 4 different configurations and compare our results with those of Kurganov and Tadmor [4], which gives us a good comparison with theirs.

We consider the following four initial datas (taken from Kurganov and Tadmor [4]):

Configuration 3:

$(p, \rho, u, v)(1) = (1.500, 1.5000, 0.000, 0.000)$, $(p, \rho, u, v)(2) = (0.300, 0.5323, 1.206, 0.000)$
$(p, \rho, u, v)(3) = (0.029, 0.1380, 1.206, 1.206)$, $(p, \rho, u, v)(4) = (0.300, 0.5323, 0.000, 1.206)$.

Configuration 5:

$(p, \rho, u, v)(1) = (1.0000, 1.0000, -0.7500, -0.5000)$, $(p, \rho, u, v)(2) = (1.0000, 2.0000, -0.7500, 0.5000)$
$(p, \rho, u, v)(3) = (1.0000, 1.0000, 0.7500, 0.5000)$, $(p, \rho, u, v)(4) = (1.0000, 3.0000, 0.7500, -0.5000)$.

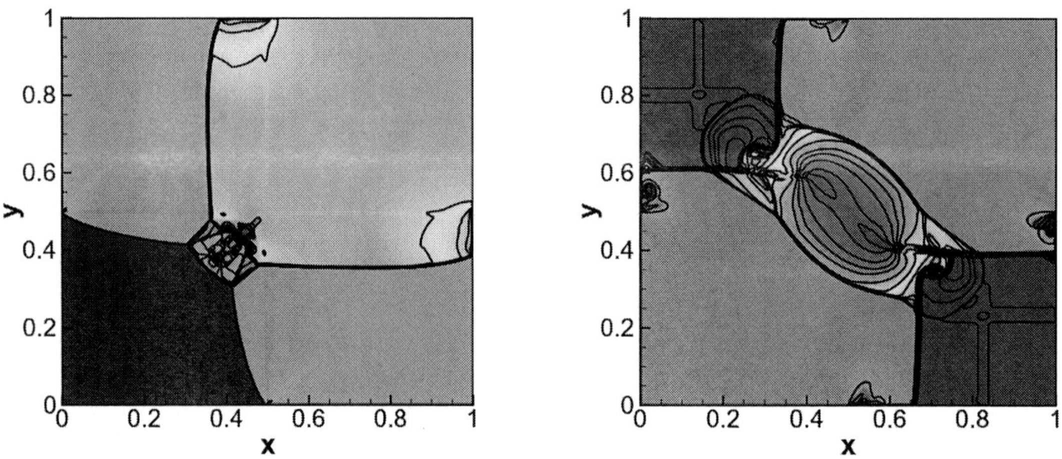

Figure 2. *2-D Riemann problem: Configuration 3 at $t = 0.23$ (left), Configuration 5 at $t = 0.3$ (right).*

Configuration 11:

$(p, \rho, u, v)(1) = (1.000, 1.0000, 0.1000, 0.0000)$, $(p, \rho, u, v)(2) = (0.400, 0.5313, 0.8276, 0.0000)$
$(p, \rho, u, v)(3) = (0.400, 0.8000, 0.1000, 0.0000)$, $(p, \rho, u, v)(4) = (0.400, 0.5313, 0.1000, 0.7276)$.

Configuration 12:

$(p, \rho, u, v)(1) = (0.400, 0.5313, 0.0000, 0.0000)$, $(p, \rho, u, v)(2) = (1.000, 1.0000, 0.7276, 0.0000)$
$(p, \rho, u, v)(3) = (1.000, 0.8000, 0.0000, 0.0000)$, $(p, \rho, u, v)(4) = (1.000, 1.0000, 0.0000, 0.7276)$.

4. Conclusion

A new higher order compact (HOC) finite difference solver has been proposed for the two-dimensional Riemann problem of gas dynamics. The scheme is fourth order accurate in space.

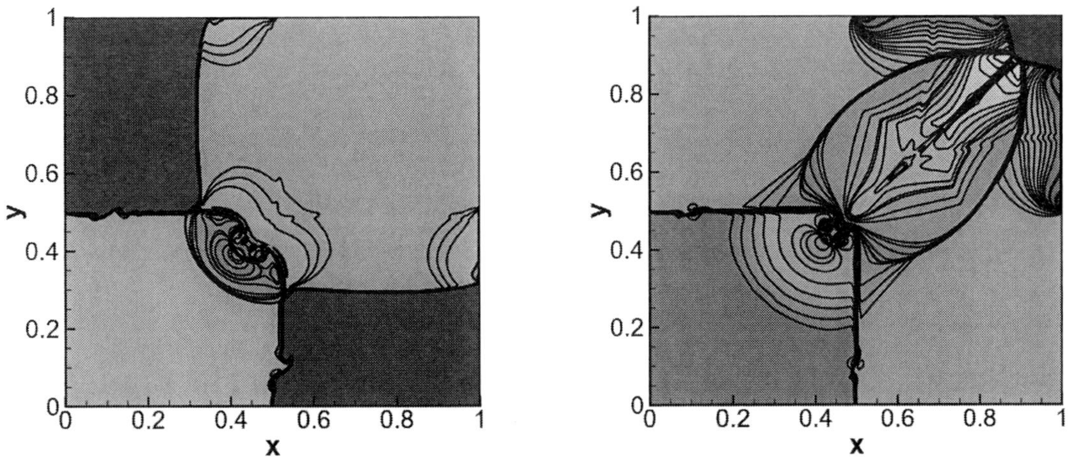

Figure 3. *2-D Riemann problem: Configuration 11 at* $t = 0.3$ *(left), Configuration 12 at* $t = 0.25$ *(right).*

The robustness of the scheme has been illustrated by its application to some of the test cases (with different initial conditions) of this problem on a 401×401 grid. In all the numerical cases, our computed numerical solutions are found to be in excellent match with those of the established results in the literature. Being simple and straight-forward, this method is quite easier to understand and implement, Moreover, the implicitness of the scheme allows us to choose a very larger time-step in contrast to the earlier existing finite difference methods for the Riemann problem. The proposed scheme thus has a very good potential for implementation in more intriguing problems of gas dynamics.

References

[1] Sengupta T K 2013. *High Accuracy Compact Schemes: Fluid Flows and Wave Phenomena* (Cambridge University Press, USA)
[2] Sengupta T K, Bhole A, Sreejith N A 2013 *Computers and Fluids* **88**, 19
[3] Gogoi B B 2015 *Applied Mechanics and Materials* **813-814**, 643
[4] Kurganov A, Tadmor E 2002 *Numerical Methods for Partial Differential Equations* **18**, 584
[5] Schulz-Rinne C W, Collins J P, Glaz H M 1993 *SIAM Journal of Scientific Computing* **14** 1394
[6] Liska R, Wendroff B 2003 *SIAM Journal of Scientific Computing* **25** 995
[7] Chang T, Chen G Q, Yang S 1995 *Discrete and Continous Dynamical Systems* **1** 555
[8] Zaghi S 2014 *Computer Physics Communications* **185** 2151
[9] Hirsch C 2002. *Numerical Computation of Internal and External Flows: The Fundamentals of Computational Fluid Dynamics* (Springer-Verlag, Springer Publishing Group, Heidelberg, New York, USA)
[10] Sengupta T K, Ganeriwal G, Dipankar A 2004 *Journal of Scientific Computing* **21(3)** 253

Journal of Physics: Conference Series **759** (2016) 012087 doi:10.1088/1742-6596/759/1/012087

Computational studies on small wind turbine performance characteristics

N Karthikeyan and T Suthakar
Department of Mechanical Engineering,
National Institute of Technology-Tiruchirappalli, Tamil Nadu-620015, India.

Abstract: To optimize the selection of suitable airfoils for small wind turbine applications, computational investigation on aerodynamic characteristics of low Re airfoils MID321a, MID321d, SG6040, SG6041, SG6042 and SG6043 are carried out for the Reynolds number range of $(0.5\text{-}\ 2)\times10^5$. The BEM method is used to determine the power coefficient of the rotor from the airfoil characteristics; in addition, the blade parameters like chord and twist are also determined. The newly designed MID321a airfoil shows better aerodynamic performance and maximum power coefficient as compared with other investigated airfoils for wider operating ranges.

1. Introduction

Wind is a cost effective, efficient source of energy and is abundant in nature and roughly 10 million MW of power can be harnessed from it. Wind turbines convert the energy in the flowing air to mechanical energy without the use of fossil fuels. Compared to other renewable energy extracting devices, wind turbine efficiency is significantly higher. Large wind turbines are installed at places where there are favorable wind condition, is referred as commercial wind turbines. Such turbines attain power coefficient of 0.45, whereas small wind turbines have power coefficient of around 0.25 and are generally installed any places irrespective of wind conditions. These turbines are classified into three types namely micro (1 kW and >1.5m), midrange (5 kW and > 2.5m) and mini (20+, > 5m) [1] based on the power and rotor radius.

Small wind turbines usually operate at Reynolds number less than 5×10^5. In this Reynolds number range, laminar flow gets separated at the upper surface of the airfoil and is reattached to the surface as turbulent causing laminar separation bubble, which increases the drag of the airfoil. The aerodynamic behavior of the airfoil is affected by laminar separation bubble significantly. The wind tunnel study was conducted on 15 low Reynolds number airfoils for free and fixed conditions to establish the database for small wind turbine applications [2]. In a continuing effort to improve the aerodynamic properties of airfoils at these Reynolds number range four airfoils were developed, one root and three primary airfoils namely SG6040 and SG6041, SG6042, and SG6043[3].

Further studies have been conducted [4] on these airfoils and the results were validated against previous airfoils. Thus SG6043 airfoil was combined with GOE15, E422, Eppler560 and S1223, the performance of SG6043_Eppler422 was found to be better than the other combinations at Reynolds number of 0.38×10^5, $1.28\times10^5$and 2.05×10^5. To improve the roughness sensitivity of airfoil at the tip section of blade, USPT2 airfoil was designed with implication of multi-objective genetic algorithm and it reached a maximum lift to drag ratio of 72 at Re of 2×10^5, modified airfoil was better than its base SG6043 airfoil [5]. AF300 airfoil was designed by modifying the trailing edge of high lift airfoils S1210 and S1223, which are taken as base airfoils and new airfoil attained the C_{lmax} of 1.72, 1.81 and 1.86 at Re 0.7×10^5, $1.28\times10^5$and 2.05×10^5respectively for angle of attack of 14° [6].

This investigation focuses on the performance prediction of low Reynolds number airfoils for small wind turbine applications and the blade parameters are determined using BEM approach. The literature studies are explained in Introduction, determination of Reynolds number and aerodynamic characteristics are described in Aerodynamic design section and the subsequent sections deal with the Computational methodology and the outcome of numerical simulations.

2. Aerodynamic Design

Many parameters play vital role on the design of wind turbine blade that includes aerodynamic behavior, generator characteristics, blade strength, rigidity and noise levels. The main objectives of small wind turbine design are maximizing its energy extraction and improving its starting characteristics at low wind speed; rotor aerodynamics plays a crucial role in the minimization of cost of energy generation. However, the total energy produced depends on both the behavior of the power output and wind's probability distribution [7, 8]. The diameter is determined by the following equation (1),

$$P = c_P \frac{1}{2} \rho \eta_{all} \left(\frac{\pi}{4} d^2 \right) V_0^3 \qquad (1)$$

where, P-power, c_p-power coefficient, ρ-density of air, η_{all}-overall efficiency of generator, V_o-free stream velocity of air and d -the diameter of rotor.

Most of the wind turbine analyses are determined by the common design and analysis method, which is known as blade element momentum method (BEM) and it was proposed by Glauert [7, 8].
The outline of BEM is given below.

BEM is the combination of momentum and blade element theory. In this theory some assumptions are made such that the blade is divided into strips along the radial location and no interaction between the strips is considered. The momentum theory estimates the thrust and torque at different locations of the rotor. The blade element theory determines the forces on the blade elements by using sectional lift and drag data from the performance of airfoils corresponding to the angle of attack. The BEM relationship was developed by balancing the local blade forces with momentum forces and relations. The BEM is usually implemented in a convergence scheme, starting with assumed values of a and a'. By summing the contributions of each annulus, the rotor thrust and torque are found through the iterative process.

2.1 Iterative method summary

The following procedure can be used to predict the power coefficient of rotor for specified tip speed ratio by iterative process [8]:

1. Initialize a and a', typically set $a = a' = 0$.
2. Calculate the relative wind angle φ with the equation (2), and angle of attack α with equation (3),

$$\phi = \tan^{-1}\left[\frac{1-a}{1+a'}\right] \qquad (2)$$

$$\phi = \theta + \alpha \qquad (3)$$

3. Get C_l and C_d at the appropriate and Reynolds number.
4. Recalculate a and á with equations (4) and (5),

$$a = \frac{1}{\frac{4F\sin^2\phi}{\sigma C_l\cos\theta}+1} \quad (4)$$

$$a' = \frac{1}{\frac{4F\sin\phi\cos\phi}{\sigma C_l}-1} \quad (5)$$

5. Repeat steps 2-4 until a and a' converge to stable values. Repeat process for each element, saving the final from each element.
6. Calculate c_p using equations (6).

$$c_p = \frac{8}{\lambda^2}\int \lambda_r^3 F a'(1-a)\left[1-\frac{C_d}{C_l}\tan\phi\right] d\lambda_r \quad (6)$$

where, a -Axial reduction factor, a' -Angular reduction factor, σ -Solidity ratio, ϕ - Relative wind flow angle, λ -Tip speed ratio, θ -Pitch angle, C_l -Lift coefficient, C_d -Drag co-efficient , λ_r -local tip speed ratio and F -Tip loss factor.

2.2 Airfoil selection and Reynolds number prediction

Airfoils are the basic elements (shapes) in the wind turbine blade cross sections, which convert the kinetic energy in wind to mechanical energy effectively. Aerodynamic behavior such as lift and drag characteristics changes due to angle of attack and Reynolds number. Appropriate selection of Reynolds number would improve the annual energy production.

$$Re = V_0 c / v \quad (7)$$

Where, Re is Reynolds number, V_0 is free stream velocity, c is chord and v is kinematic viscosity of air. Range of velocity is taken from 3 to 14m/s, chord values ranging from 0.25 to 0.35m and v is taken as $=16.5\times10^{-6}$ m^2/s at Standard atmospheric condition.

Based on the atmospheric conditions, predicted Reynolds number comes under the range of 6×10^4-2×10^5. Selected airfoils for an investigation are given in table 1.

Table 1. Airfoil name and geometrical characteristics

Airfoil Name	Thickness (t/c %)	Camber(%c)
MID321a	0.08	0.031
MID 321d	0.14	0.035
SG6040	0.16	0.025
SG6041	0.10	0.020
SG6042	0.10	0.038
SG6043	0.10	0.055

3. Computational study

In this study, aerodynamics performance of the airfoil is predicted through the Xfoil tool, which was developed by Mark Derla in 1986 at the Massachusetts Institute of Technology (MIT). This tool is specially tailored for subsonic airfoil design and analysis purpose. It is built with panel code algorithm which is the combination of e^n laminar to turbulent transition method and the set of integral boundary layer formulations.

The airfoil co-ordinates are obtained from the airfoil database website [9, 10]. The basic geometrical requirements of the airfoil for carrying simulation in this tool are panel number and specific inflow condition (Ncrt) values [11]. Thus the minimum paneling parameter and Ncrt values are maintained at 160 and 9 respectively.

The iterative procedure is used to solve the BEM relationship by using Matlab tool. Before initializing the iteration process some parameters needed to be assumed. The blade is divided into ten strips and aerodynamic loads are calculated for each strip along the span of rotor. Then the overall blade power coefficient is predicted from the individual strips of blade using trapezium formula of Numerical Integration. Finally blade parameters such as chord and twiste angle have been calculated.

4. Result and discussion

The proposed design of small wind turbine configurations are 3 bladed, 2.25m radius, TSR 7, fixed pitch-variable speed with design wind speed of 7 m /s for the power of 1kW. These values were given as an input to the BEM calculations.

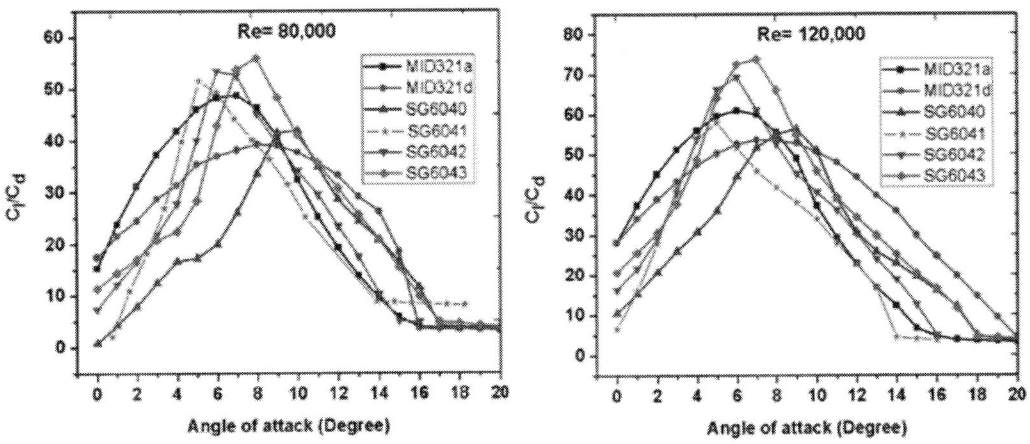

Figure 1. C_l/C_d versus Angle of attack for Reynolds number of 0.8×10^5 and 1.2×10^5.

The aerodynamic behavior of the airfoils at the Reynolds number of 0.8×10^5 and 1.2×10^5 with angle of attack ranging from 0 to 20° is shown in figure 1. At the Reynolds number of 800,000, MID321a and MID 321d airfoils shows a better performance for wider range of angle of attack compared to the other four airfoils whereas SG6040, SG6041, SG6042 and SG6043 airfoils performed at narrow angle of attack with maximum lift to drag ratio of 40,45, 55and 58 respectively. At Reynolds number of 120,000, MID321a and MID321d produces lift to drag ratio of 60 and 55, while SG6040, SG6041, SG6042 and SG6043 airfoils attain the maximum lift to drag ratio of 56, 59, 70 and 75 those are operated at narrow angle of attack.

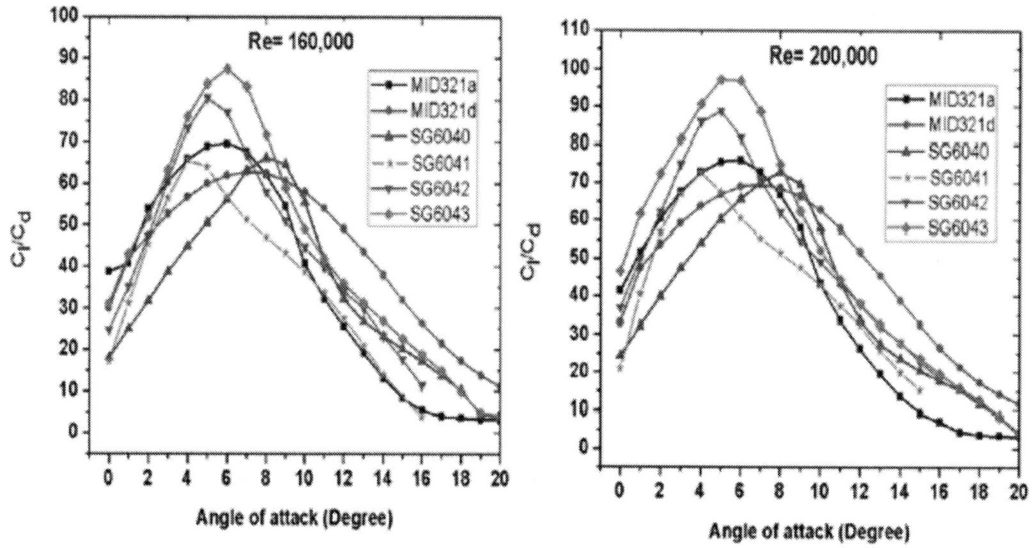

Figure 2. C_l/C_d versus Angle of attack for Reynolds number of 1.6×10^5 and 2×10^5.

The aerodynamic behavior of the airfoils at the Reynolds number of 1.6×10^5 and 2×10^5 with angle of attack varying from 0-20° is shown in figure 2. At Reynolds number 1.6×10^5, MID321a and MID321d

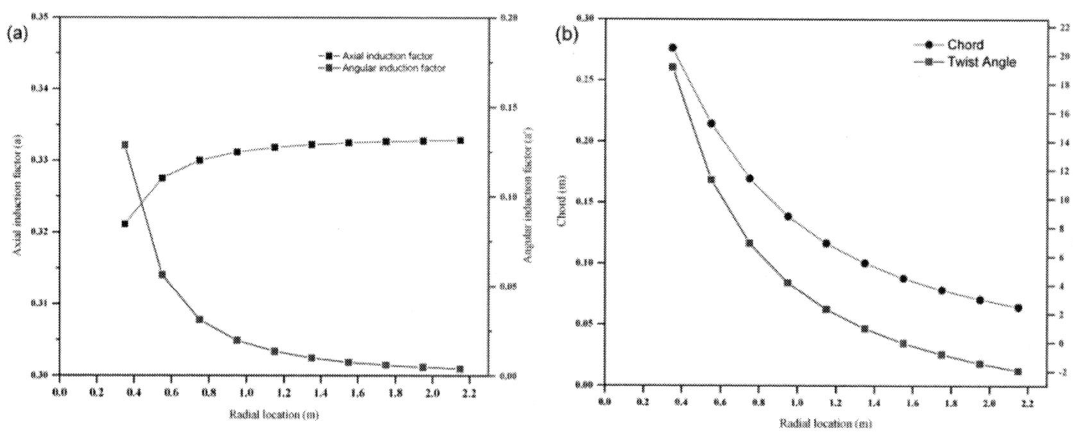

Figure 3. (a) Axial, Angular inductions factors, (b) Chord and Twist values of MID321a airfoil at $Re=0.8\times10^5$.

produced limited glide ratio (C_l/C_d) compared to the remaining airfoils. SG6042 and SG6043 airfoils attained lift to drag ratio of 80 and 90 at short range of angle of attack. At higher Reynolds number of 2×10^5, MID321a and MID321d airfoils produced very limited glide ratio, while SG6043 airfoil attained a glide ratio of 100.

From the above results, MID321a and MID321d airfoils generate quite better lift to drag ratio for wider angle of attack compared to other four airfoils. The aerodynamic properties of Mid321a airfoil lift to drag ratio at 0.8×10^5 Reynolds number is selected. Thus it is used to BEM approach; the maximum achievable power coefficient is determined through the iterative procedure. For the maximum power coefficient, chord and twist values are derived as shown in figure 3. The axial and angular induction factors are plotted in the figure 3(a), which is attained from the BEM method. The chord and twist values are higher at root section and lower at tip section is shown in figure 3(b).

5. Conclusion

The operating characteristics of a small wind turbine having 3 blades and of 2.5 m diameter are estimated and aerodynamic properties of the concerned airfoils are determined using Xfoil tool. Subsequently the airfoil that shows the best performance is subjected to BEM calculations. Based on the obtained results, MID321a and MID321d airfoils are showed quite better lift to drag ratio at wider range of angle of attack for entire Re range whereas SG6042 and SG6043 airfoils are attained maximum glide ratio with narrow angle of attack. Optimal blade geometry parameters like chord and twist are calculated using aerodynamic characteristics of MID321a airfoil at 0.8×10^5 Reynolds number. The blade parameters such as chord values are varied from 276mm to 64mm and twist angle are varied from 19 ° to -1.95°. The maximum power coefficient value of 0.4 is attained by MID321a airfoil for Re 0.80×10^5.

References

1. Karthikeyan, N., Kalidasa Murugavel, K., Arun Kumar, S., & Rajakumar, S. (2015). Review of aerodynamic developments on small horizontal axis wind turbine blade. *Renewable and Sustainable Energy Reviews*, **42**, 801–822.
2. Giguere, P., & Selig, M. S. (1997). Low Reynolds number airfoils for small horizontal axis wind turbines. *Wind Engineering, **21**(**6**)*, 367-380.
3. Giguère, P., & Selig, M. S. (1998). New Airfoils for Small Horizontal Axis Wind Turbines. *Journal of Solar Energy Engineering*, ***120*(2)**, 108.
4. Wata, J., Faizal, M., Talu, B., Vanawalu, L., Sotia, P., & Ahmed, M. R. (2011). Studies on a low Reynolds number airfoil for small wind turbine applications. *Science China Technological Sciences*, ***54*(7)**, 1684–1688.
5. Ram, K. R., Lal, S., & Rafiuddin Ahmed, M. (2013). Low Reynolds number airfoil optimization for wind turbine applications using genetic algorithm. *Journal of Renewable and Sustainable Energy*, **5(5).**
6. Singh, R. K., Ahmed, M. R., Zullah, M. A., & Lee, Y.-H. (2012). Design of a low Reynolds number airfoil for small horizontal axis wind turbines. *Renewable Energy*, **42**, 66–76.
7. Wood, D., (2011*) Small Wind turbines, Analysis, Design, and Applications*. Green Energy and Technology, Springer.
8. Manwell, JF, McGowan, JG, Rogers, Al. (2010) *Wind Energy Explained: theory, Design and Application*, Wiley.
9. Derla, M., and Youngen, H. (2006) XFOIL -6.96 User Guide, Massachusetts Institute of Technology, http://web.mit.edu/drela/Public/web/xfoil/.
10. http://tracfoil.free.fr
11. Derla, M.,(1989) Xfoil: an analysis and design system for low Reynolds number airfoils, In: *Low Reynolds Number Aerodynamics*: Proceedings of the Conference Notre Dame, Ind, USA, 5-7 June,**54, 1-12**, Springer, Berlin, Germany.

Laminar convective heat transfer characteristic of Al_2O_3/water nanofluid in a circular microchannel

K Trinavee[1, 3], T K Gogoi[1] and M Pandey[2]
[1]Department of Mechanical Engineering, Tezpur University, Tezpur, Assam, India
[2]Department of Mechanical Engineering, IIT Guwahati, Assam, India

[3]kumari.trinavee779@gmail.com

Abstract. In this study, laminar convective heat transfer characteristics Al_2O_3/water nanofluid in a circular microchannel is investigated using a two-phase (discrete phase) model. The computational fluid dynamic code FLUENT (ANSYS) is employed to solve the coupled momentum and energy equations with the boundary conditions of uniform wall heat flux and velocity at channel inlet. Detail analysis is done showing variation of wall temperature, fluid bulk mean temperature, heat transfer coefficient, Nusselt number, shear stress, friction, pressure drop, entropy generation etc. along the microchannel at two particle volume concentrations (1% and 4%) of the nanofluid. Comparison of results is provided between base and nanofluid and also for two cases, one with constant property and the other with variable temperature thermal conductivity and viscosity. Results show that heat transfer is enhanced in case of the nanofluid with low entropy generation and the heat transfer parameters increase with increase in nanoparticle volume concentration and Reynolds number. However, use of nanofluid also causes increase in pressure drop and shear stress. A comparison of the constant and variable property model showed that heat transfer is further enhanced; entropy, shear stress and pressure drop further decrease when temperature dependent properties of the nanofluid are considered instead of constant properties.

1. Introduction

Use of nanofluids in heat exchangers and cooling devices has received significant research interest in the recent past. Many experimental studies have been carried out to evaluate convective heat transfer characteristics of nanofluids in tubes [1-3], microsystem with single microchannel [4], multiple parallel rectangular micro-channels [5, 6] and wide rectangular microchannel [7]. Al_2O_3/water in various volume concentrations were considered in the above studies under laminar flow condition except in in Hozzat et al. [3] where the study was done under turbulent flow conditions considering Al_2O_3, CuO, and TiO_2 nanoparticles and carboxymethyl cellulose (non-Newtonian base fluid). The above studies have reported heat transfer enhancement with nanofluids with higher concentration in the base fluid. It was also observed that the heat transfer enhancement is more in the entrance region compared to the fully developed region [1, 5].

Simultaneously numerical analyses have also been done to predict heat transfer behavior of nanofluids under laminar [7-16] and turbulent flow conditions [17-19]. Some studies [8, 9, 17] used single phase method while some others [7, 12, 14, 19] employed two-phase method for modeling nanofluids. The single phase method, where the nanofluid is considered as homogeneous mixture of nanoparticles, is simple and computationally efficient. In the two-phase method, the nanoparticles and the base fluid are considered as solid and liquid with different velocity and temperatures. Bianco et al. [12] used both single-phase and two-phase methods considering constant and temperature dependent properties of the Al_2O_3/water nanofluid. Kalteh et al. [14] investigated laminar forced convection heat transfer of Cu/water nanofluid inside an isothermally heated parallel plate microchannel using single and two phase models. There is another method based on Boltzmann theory that considers nanoconvection due to Brownian motion and particle collision [10, 11, 20].

The numerical studies discussed above were basically conducted on circular tubes [8, 12, 17-19], rectangular micro channels [9, 10], trapezoidal channel [11], parallel disc [13], parallel plate microchannel [14, 15] and wide rectangular channel [7] etc. The heat transfer performance of a system varies with the shape and size of the coolant flow channel. Theoretical analyses on a circular microchannel heat sink using nanofluids are not many. Sohel et al. [21] made a comparative analysis in a circular microchannel using Al_2O_3/water, TiO_2/water and CuO/water nanofluids under laminar flow condition. Sohel et al. [22] in another study considered circular microchannels and minichannels to analyze entropy generation rate under turbulent flow condition using Cu, Al_2O_3 nanoparticles and water, ethylene glycol base fluids. Zhang et al. [23] presented an experimental study using Al_2O_3/water nanofluid with 0.25%, 0.51% and 0.77% volume concentrations through a circular microchannel of 0.5 mm inner diameter. In this study, laminar convection of Al_2O_3/water nanofluid in a circular microchannel with 1% and 4% particle volume concentrations is numerically investigated. Ansys (Fluent) 14.0 version is used to solve the problem. Both constant and temperature dependent thermophysical properties of nanofluid are taken into consideration and a comparative analysis is provided with complete details showing variation of heat transfer, entropy generation and flow parameters such pressure drop, friction and shear stress etc.

2. Mathematical modeling

The geometry of the circular microchannel is shown in Figure 1. It is assumed that Al_2O_3/water nanofluid is incompressible and its flow through the microchannel is developing, steady and laminar. Al_2O_3 particles are spherical with 38 nm diameter. Due to symmetry, only the upper half of the channel is considered to save the computational time. A two phase model [12] is employed to analyze the flow and heat transfer problem of nanofluid in the circular microchannel.

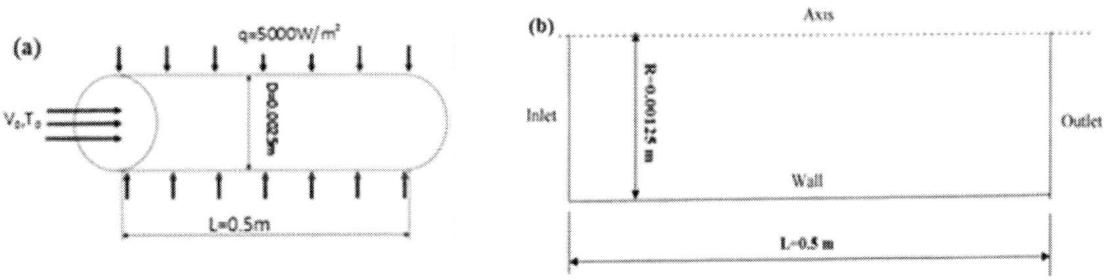

Figure 1. (a) Geometry of the circular microchannel (b) computational domain

2.1. Governing equations

The steady-state governing equations for the Eulerian-Lagrangian discrete two phase model are expressed as follows [12].

Continuity equation: $$\nabla.\left(\rho\vec{V}\right)=0 \tag{1}$$

Momentum equation: $$\nabla.\left(\rho\vec{V}\vec{V}\right)=-\nabla P+\nabla.\left(\mu\nabla\vec{V}\right)+S_m \tag{2}$$

Energy equation: $$\nabla.\left(\rho\vec{V}C_pT\right)=\nabla.\left(\lambda\nabla T\right)+S_e \tag{3}$$

Particle momentum and energy equations in Lagrangian form are as follows.

$$\frac{d\vec{V}_p}{dt} = F_D\left(\vec{V} - \vec{V}_p\right) + \frac{\vec{g}\left(\rho_p - \rho\right)}{\rho_p} + \vec{F} \tag{4}$$

$$\rho_p C_p \frac{dT_p}{dt} = \frac{6h}{d_p}\left(T - T_p\right) \tag{5}$$

$\vec{V}_p$ and T_p in equations (4) and (5) are the nanoparticle velocity and temperature respectively. ρ_p is the density and d_p is the diameter of the nanoparticles. Eq. (4) is the force balance equation for a particle immersed in the fluid and $\vec{F}$ in Eq. (4) accounts for additional forces due to rotation of reference frame, thermophoretic force, Brownian force etc. The gravity term in Eq. (4) and additional forces '$\vec{F}$' are however neglected in this analysis. F_D is the drag coefficient defined according to Stokes' law. Details about the calculation procedure of F_D is highlighted in Ref. [12]. In Eq. (5), h is the heat transfer coefficient, ρ and T refer to fluid density and temperature respectively. For the details including the source/sink terms (S_m and S_e) in Eq. (2) and Eq. (3), one can refer to the article in Ref. [12].

2.2. Boundary conditions

At the channel inlet, uniform axial velocity, V_0, and temperature, $T_0 = 293$ K, are assumed. Pressure outlet condition is specified at the channel exit. On the channel wall, the no-slip condition and uniform heat flux of 5000 W/m^2 are imposed.

2.3. Thermophysical Properties of the Nanofluids

The thermophysical properties of Al_2O_3 particles and base fluid are as follows [2, 12].

$\rho_p = 3880\ \text{kg/m}^3;\ Cp_p = 733\ \text{J/kgK};\ k_p = 36\ \text{W/mK}$

$\rho_{bf} = 998.2\ \text{kg/m}^3;\ Cp_{bf} = 4182\ \text{J/kgK};\ k_{bf} = 0.597\ \text{W/mK};\ \mu_{bf} = 9.93\text{x}10^{-4}\ \text{kg/ms}$

Case1: Constant property model

Density: $\rho_{nf} = (1-\phi)\rho_{bf} + \phi\rho_p$ (6)

Specific heat: $Cp_{nf} = (1-\phi)Cp_{bf} + \phi Cp_p$ (7)

In this case, viscosity and thermal conductivity of the nanofluid is a function of T_0 only.

Dynamic viscosity: $\phi = 1\% \rightarrow \mu_{nf} = 3.4\times10^{-2} - 1.975\times10^{-4}T_0 + 2.912\times10^{-7}T_0^2$ (8)

$\phi = 4\% \rightarrow \mu_{nf} = 4.051\times10^{-2} - 2.353\times10^{-4}T_0 + 3.475\times10^{-7}T_0^2$ (9)

Thermal conductivity: $\phi = 1\% \rightarrow k_{nf} = 0.003352T_0 - 0.3708$ (10)

$\phi = 4\% \rightarrow k_{nf} = 0.004961T_0 - 0.8078$ (11)

Case 2: Temperature dependent property model

Density and specific heats are same with constant property model. However, viscosity and thermal conductivity vary with temperature in this case. Temperature dependent viscosity and thermal conductivity of nanofluid with different concentration are expressed as follows [12, 15].

Dynamic viscosity: $\phi = 1\% \rightarrow \mu_{nf} = 3.4\times10^{-2} - 1.975\times10^{-4}T + 2.912\times10^{-7}T^2$ (12)

$\phi = 4\% \rightarrow \mu_{nf} = 4.051\times10^{-2} - 2.353\times10^{-4}T + 3.475\times10^{-7}T^2$ (13)

Thermal conductivity: $\phi = 1\% \rightarrow k_{nf} = 0.003352T - 0.3708$ (14)

$\phi = 4\% \rightarrow k_{nf} = 0.004961T - 0.8078$ (15)

The viscosity and thermal conductivity of base fluid are also considered temperature dependent in the second case [12, 15].

$$\mu_{bf} = 7.57\times10^{-2} - 6.37\times10^{-4}T + 1.80\times10^{-6}T^2 - 1.73\times10^{-9}T^3 \tag{16}$$

$$k_{bf} = -1.13 + 9.71\times10^{-3}T - 1.31\times10^{-3}T^2 \quad (17)$$

2.4 Numerical method

Ansys fluent [24] version 14.0 was used to solve the present problem using a control volume approach. A second order upwind method was used for solving the momentum and energy equations. Pressure and velocity were coupled using Semi Implicit Method for Pressure Linked Equations (SIMPLE) algorithm. The convergence criteria was set to 10^{-6}. Three different non-uniform grids were considered to check grid independency of the numerical results. The non-uniform grid 500x50 was found acceptable. Grid independency was achieved as the wall temperature obtained with this grid size was almost the same with the non-uniform grid 700x60 (Refer Figure 2).

3. Results and discussion

The study was carried out for the base fluid (water) and nanofluid with ϕ = 1% and ϕ =4% considering both constant and temperature dependent properties to determine the heat transfer characteristics at three different Reynolds number viz. Re=500, 1000 and 1900. Figure 3 shows variation of temperature at the channel wall (T_w) for the nanofluid with ϕ =1% and ϕ =4% for both constant and variable property models at Re=500. Bulk mean temperature (T_m) variation of the nanofluid is also shown for variable property model. It was seen that both T_w and T_m increase along the channel and in case of the nanofluid with ϕ =4%, T_w and T_m were found comparatively less. Moreover, T_w also reduced slightly when variable properties were considered instead of fixed.

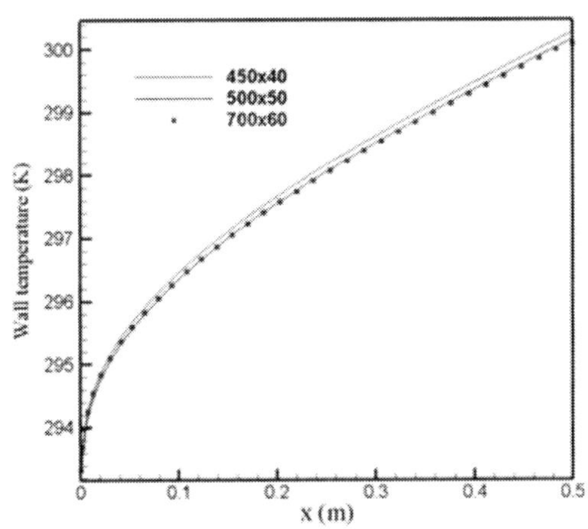

Figure 2. Wall Temperature at various grid size

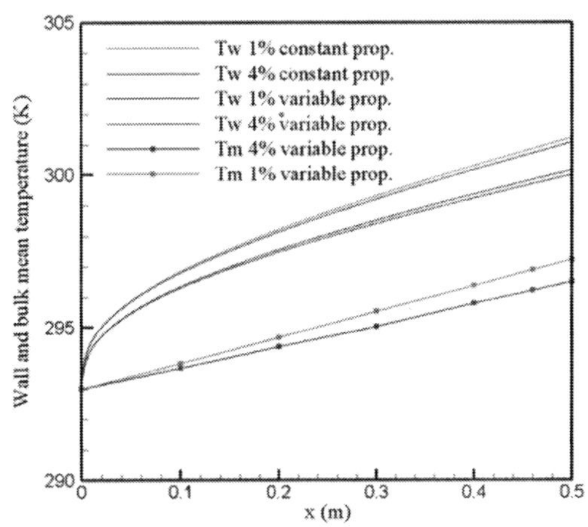

Figure 3. Wall temperature for constant and variable property models at Re=500 (grid size 500x50) and bulk mean temperature for variable property model with ϕ =1% and ϕ =4%

Variation of heat transfer coefficient along the microchannel is shown in Figure 4(a) for the base and nanofluid for both constant and variable property models. Heat transfer coefficient increased significantly in case of the nanofluid and it was also more for the nanofluid with ϕ =4% compared to that with ϕ =1%. Moreover, at a given ϕ, higher heat transfer coefficient was obtained with the variable property model. Figure 4(b) shows variation of heat transfer coefficient along the microchannel at various Reynolds numbers for the nanofluid with ϕ =4% (variable property). It was seen that heat transfer coefficient was more at higher Reynolds number. Heat transfer enhancement with nanofluids at higher particle concentration was reported in Refs. [1-7]. Thermal conductivity of the nanofluid is more compared to the base fluid which increases due to dispersion of nanosize solid particles in the base fluid.

Journal of Physics: Conference Series **759** (2016) 012088 doi:10.1088/1742-6596/759/1/012088

This is depicted in Figure 5(a) which shows that thermal conductivity is more for the nanofluids and it is the highest for the nanofluid with ϕ=4%. Figure 5(b) shows that viscosity also increases significantly in case of the nanofluid and it was the maximum for the nanofluid with ϕ=4%. Further, it was seen that the viscosity of the nanofluid decreases along the channel while in case of the base fluid it shows an increasing trend.

Effect of higher viscosity of the nanofluid can be seen in Figures 6 (a) and 6 (b). It was found that the shear stress and skin friction coefficient was higher for the nanofluid compared to those of the base fluid and more so in case of the nanofluid with higher ϕ. Further, lower values of shear stress and skin friction coefficient were obtained in case of the variable property model. Bianco et al. [12] also found higher heat transfer coefficient and lower shear stress with the temperature dependent model. Figure 6 (c) shows that for the same outlet pressure of 1 atm., the pressure drop is more in case of the nanofluid and it increases with ϕ. However, the pressure drop was comparatively less for temperature dependent model. Heat transfer with finite temperature difference is an irreversible process, therefore, it is expected that maximum heat transfer should occur with minimum entropy generation. Figure 6 (d) shows that less entropy is generated when nanofluid is used and entropy decreases with increase in ϕ. Also, entropy generation is less in case of the variable property model.

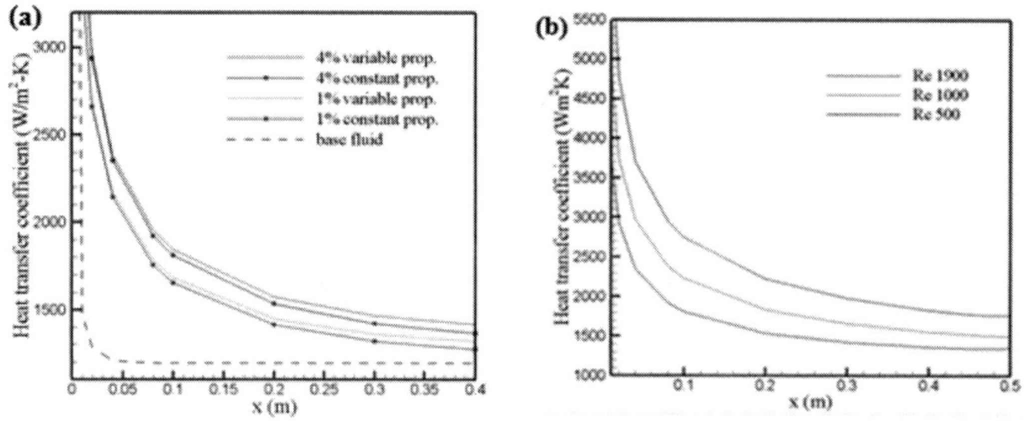

Figure 4. Heat transfer coefficient variation along the tube for (a) base and nanofluid with ϕ=1% and ϕ=4% at Re = 500 for both constant and variable property models (b) nanofluid with ϕ=4% at various Reynolds Numbers only for variable property model.

Variation of Nusselt number (Nu) along the microchannel for the base and naofluid with ϕ =1% and ϕ=4% is shown in Figure 7(a) for both the constant and variable property models. It was found that Nu decreases rapidly along the channel and assumes almost a constant value in case of the base fluid. In case of the nanofluid, however, it changes gradually and Nu value is more than that of the base fluid at all positions. Moreover, Nu was more for the nanofluid with ϕ=4% compared to that with ϕ=1%. Further, higher value of Nu was obtained with the variable property model. It was also observed that variation of heat transfer coefficient and Nu follow almost the same trend along the microchannel. Figure 7 (b) shows that Nu also increases with Re for both the base and nanofluid.

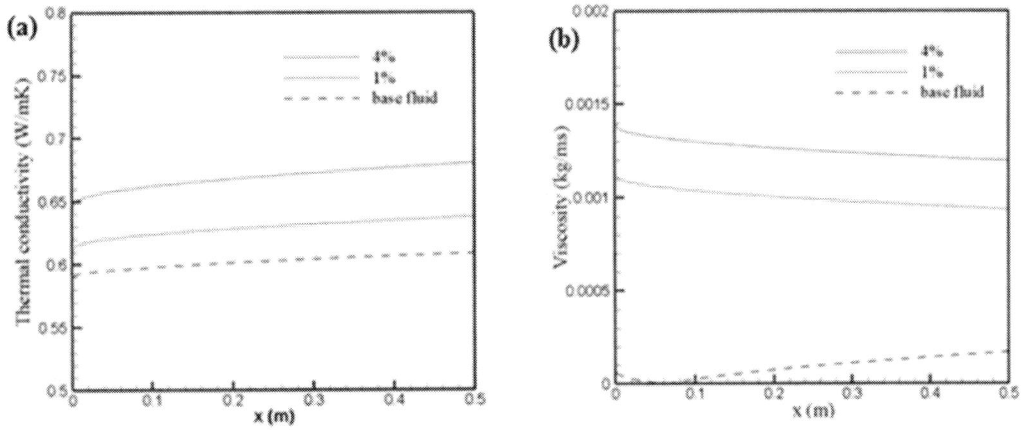

Figure 5. Variation of (a) thermal conductivity (b) viscosity for nanofluid with ϕ=1% and ϕ=4% along the microchannel wall at Re=500 for the variable property model

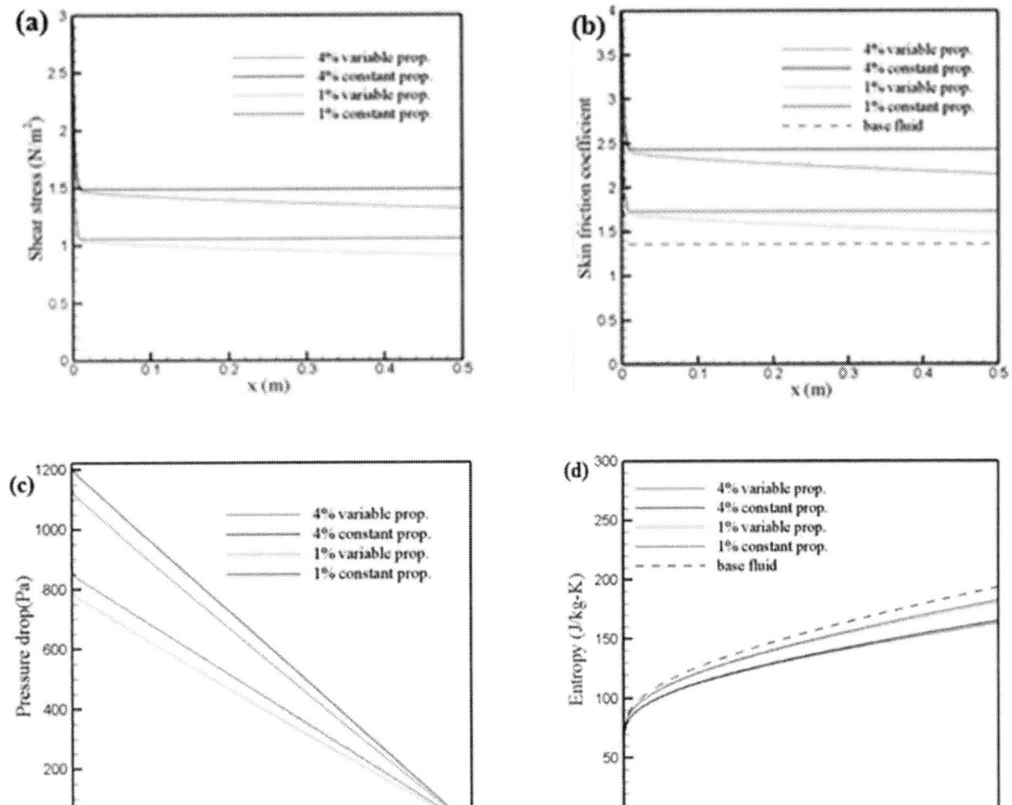

Figure 6. Variation of (a) entropy (b) pressure (c) shear stress (d) friction coefficient along the channel wall for base and nanofluid with ϕ=1% and ϕ=4% for both constant and variable property models at Re=500

4. Conclusions

In this study, laminar convective heat transfer characteristics of Al_2O_3 nanofluid in a circular microchannel with 1% and 4% volumetric concentrations is presented and compared with those of the base fluid (water). Ansys Fluent is used to solve the problem using a discrete two phase model. Full details of heat transfer capability of nanofluid is provided along with comparison of results between base and nanofluid considering both constant and temperature dependent properties of nanofluid. Based on the results, the following conclusions were drawn.

1. Compared to the base fluid, heat transfer coefficient and Nusselt number were more for the nanofluid and these two parameters increased with ϕ and Re.
2. T_w and T_m were less in case of nanofluid and these were minimum for the nanofluid with ϕ=4%. T_w and T_m further reduced when temperature dependent model was considered.
3. Thermal conductivity of the nanofluid was more than the base fluid and it increased with ϕ, however viscosity also increased simultaneously.
4. Due to higher viscosity, the shear stress, friction and pressure drop were more in case of the nanofluid and values of these parameters increased with ϕ. However, the variable property model gave slightly better results.
5. Enhancement in heat transfer was achieved with the nanofluid with low entropy generation.

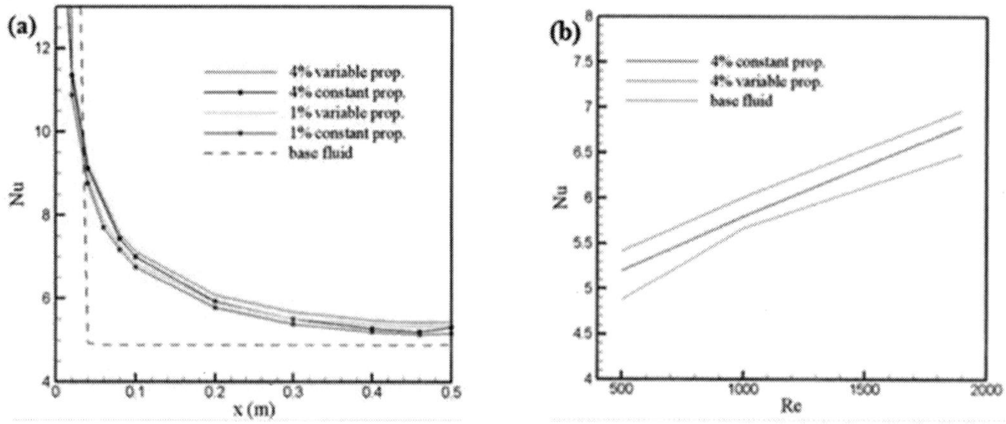

Figure 7. Nusselt number variation with (a) channel length for base and nanofluid with ϕ=1% and ϕ = 4% at Re=500 (b) Reynolds number for base and nanofluid with ϕ=4% for both constant and variable property models at the central location of the microchannel

REFERENCES:

[1] Wen D and Ding Y 2004 *Intl. J. of Heat and Mass Transfer* **47**(24) 5181-88
[2] Heris S Z, Esfahany M N and Etemad S G 2007 *Intl. J. of Heat and Fluid Flow* **28** 203-210
[3] Hojjat M, Etemad S G, Bagheri R, Thibault J 2011 *Intl. J. of Thermal Sciences* **50** 525-531
[4] Jung J Y, Oh H S, Kwak H Y 2009 *Intl. J. of Heat and Mass Transfer* **52** 466–472
[5] Lee J and Mudawar I 2007 *Intl. J. of Heat and Mass Transfer* **50**(3) 452 – 463
[6] Ho C J, Wei L C and Li Z W 2010 *Applied Thermal Engineering* **30** 96–103
[7] Kalteh M, Abbassi A, Saffar-Avval M, Frijns A, Darhuberd A and Harting 2012 *Applied Thermal Engg* **36** 260–268
[8] Maiga S E B, Palm S J, Nguyen C T, Roy G, Galanis N 2005 *Intl. J. of Heat and Fluid Flow* **26** (4) 530–546
[9] Chein R, Huang G 2005 *Applied Thermal Engineering* **25** 3104–14
[10] Jang S P, Choi S U S 2006 *Applied Thermal Engineering* **26** 2457–63
[11] Li J, Kleinstreuer C 2008 *Intl. J. of Heat and Fluid Flow* **29** 1221–32
[12] Bianco V, Chiacchio F, Manca O and Nardini S 2009 *Applied Thermal Engineering* **29**(17)3632–42
[13] Feng Y and Kleinstreuer C 2010 Nanofluid *Intl. J. of Heat and Mass Transfer* **53** 4619–28
[14] Kalteh M, Abbassi A, Saffar-Avval M, Harting J 2011 *Intl. J. of Heat and Fluid Flow* **32** 107–116

[15] Mashaei P R, Hosseinalipour S M. and Bahiraei M 2012 M. *J. of Applied Mathematics* Article ID 259284 1-18

[16] Hatami M and Ganji D D 2014 *Energy Conversion and Management* **78** 347–35

[17] Maiga S E B , Cong Tam N, Galanis N , Roy G, Mare T and Coqueux M 2006 *Intl. J. of Numerical Methods in Heat and Fluid Flow* **16** (3) 275–292

[18] Namburu P K, Das D K, Tanguturi K M and Vajjha R S 2009 Intl. J. of Thermal Sciences **48** 290–302

[19] Bianco V, Manca O and Nardini S 2011 *Intl.J. of Thermal Sciences* **50** 341-349

[20] Mohammeda H A, Bhaskarana G, Shuaiba N H and Saidurb R 2011 *Renewable and Sustainable Energy Reviews* **15** 1502–12

[21] Sohel M R, Saidur R, Faizul M, Sabri M , Kamalisarvestani M, Elias M M and Ijam 2013 A *Intl Communications in Heat and Mass Transfer* **42** 75–81

[22] Sohel M R, Saidur R, Hassan N H, Elias M M, Khaleduzzaman S S, Mahbubul I M 2013 *Intl. Communications in Heat and Mass Transfer* **46** 85–91

[23] Zhang H, Shao S, Xu H and Tian C 2013 *Applied Thermal Engineering* **61** 86-92

[24] ANSYS FLUENT 14.0 User's Guide, November 2011

Journal of Physics: Conference Series **759** (2016) 012089 doi:10.1088/1742-6596/759/1/012089

Open Source Computing in Physics Education

Ananda Dasgupta

Indian Institute of Science Education and Research Kolkata, Mohanpur Campus, India - 741252

E-mail: `adg@iiserkol.ac.in`

1. Introduction

No one can deny that computers are important tools for a physicist. The aspect that we are most familiar with is that they are great for solving research problems! What is not so widely appreciated is that they are also great as teaching tools. They are grossly underused in this aspect. In this talk, I will try to demonstrate how an instructor can make use of the computer to make her course both more attractive and effective . The focus of the lecture was on three aspects

- writing your own animations for physics teaching.
- teaching students to explore physics through computing.
- using open source software for simple experiments.

For the sake of brevity, I will discuss only the first one in this article. In this, I will assume a reasonable amount of computational skill on the part of the instructor. However, one does not have to be an expert in computational physics in order to make use of the methods described here.

2. Animations as a teaching tool

All teachers will agree that animations can be very useful for improving student comprehension of complicated topics. They can serve as complements for traditional laboratory experiments, as well as aids for thought experiments allowing students to explore parameter regimes that might to too tedious or too expensive otherwise. Given their potential, it is natural that a lot of work has already been done on this topic. A huge repository of animations dealing with various aspects of physics are already available. This includes java applets, shockwave animations, vpython animations and a lot, lot more. Some of these are commercial, while a lot is available for free on the internet. It needs to be stressed that some of the free software is of extremely high quality - often they are superior to their commercial counterparts.

Despite the easy availability of commercial and non-commercial software there are several stumbling blocks on the road to adapting the available ones for the classroom. Even if you manage to find one that deals with the subject at hand, the available software may not be exactly suitable for the purpose at hand. At the very least, you may want to use other values for the parameters which are hard-coded. What's even more likely is that the software may not deal with the exact aspect of the problem that you want to explore. The best solution to this problem is to write your own animations!

This solution comes with its own set of problems though. After all, it takes time and effort to learn a programming language! Given the busy schedule most educators have it would be too much to expect this from them. Even if one is used to programming for solving research problems it takes a lot of time and effort to write animations. A way forward that I propose here is to make use of the animation capabilities built into **gnuplot**.

3. Gnuplot

There are several reason why we recommend gnuplot. To begin with, the best reason is that it is open source. It is available free of cost - this may be an important consideration for many of us. Given the wide variety of operating systems that are in use, another plus point is that versions of gnuplot is available for most of them. The most important consideration is that it is relatively easy to create high quality plots using this software and what's even more important for the present discussion - animations are also easy to create!

3.1. Plotting with gnuplot

By default, gnuplot is used for creating 2D and 3D plots. For example, the command

```
plot sin(x)
```

produces the following graph

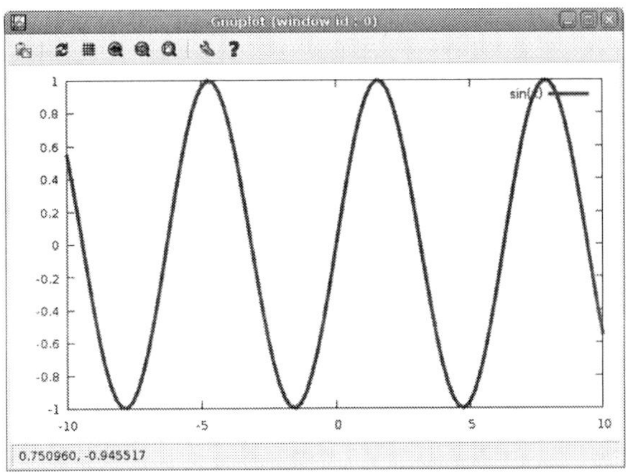

As you can see, by default gnuplot takes the independent variable to run from -10 to +10. This and many other aspects can be customized. For details, please refer to the gnuplot documentation.

We will be more interested in gnuplot scripts. At the most basic level, a script is nothing but a sequence of gnuplot commands written in a file. for example, consider a file with the following lines

```
set sample 1000
plot sin(x)
pause 3
plot sin(x) + sin(3*x)/3
pause 3
plot sin(x) + sin(3*x)/3 + sin(5*x)/5
pause -1
```

will display the graphs shown below

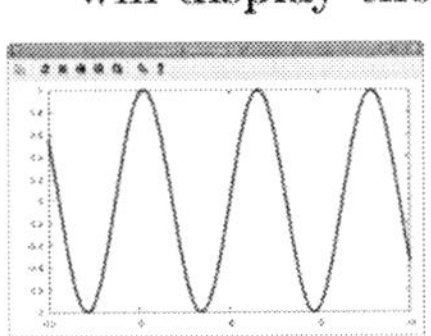

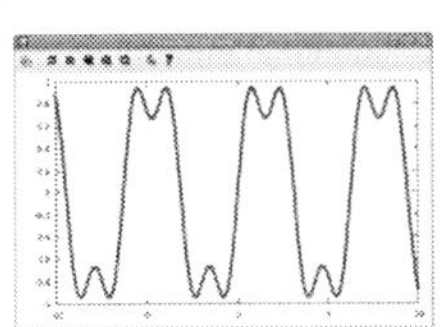

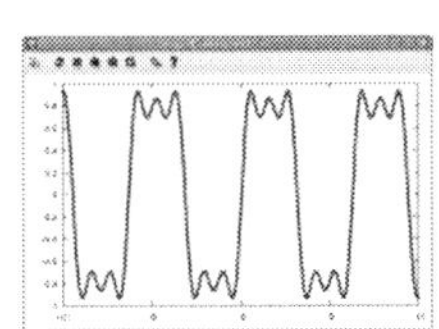

after intervals of 3 seconds (that's what the 'pause 3' are for). By default gnuplot plots a curve by plotting 100 equally spaced points. The first line in the script changes this number of points to 1000 - leading to a smoother curve. The final pause -1 keeps the pauses the program indefinitely (that is until the user hits a key) so that the picture persists on the screen.

3.2. The reread command

When gnuplot encounters the **reread** command in a script it loads the script again. This is the basic step that allows us to write animations in gnuplot. To see how this works take a look at the script 'easyWave.plt' containing the following

```
# file 'easyWave.plt'
# Plots a running wave

plot sin(x-ct)
pause sp
ct = ct + sp
if (ct<5) reread
```

To run this script open the gnuplot interpreter and enter the commands

ct = 0
sp = 0.05
load 'easyWave.plt'

and you will see a wave running across your screen.

If you want to run the script directly without opening the interpreter, you must put the three lines above in another script **initWave.plt** . Running this script by typing **gnuplot initWave.plt** at the command line (for linux) or double-clicking the icon (in Windows) will give you the running wave.

3.3. More complex examples - Fourier series again

The Fourier series for the square wave

$$\mathrm{sq}(x) = \sum_{i=0}^{\infty} \frac{\sin[(2i+1)x]}{2i+1}$$

leads to the partial sums

$$\mathrm{sq}_n(x) = \sum_{i=0}^{n} \frac{\sin[(2i+1)x]}{2i+1}$$

which obey

$$\mathrm{sq}_n(x) = \mathrm{sq}_{n-1}(x) + \frac{\sin[(2n+1)x]}{2n+1}$$

We can use this recursively, along with $\mathrm{sq}_0(x) = \sin(x)$, to calculate $\mathrm{sq}_n(x)$

To create an animation that will show the partial fourier sums for the square wave in succession, we need the initialization script **initFourier.plt** :

```
# initFourier.plt
# Fourier series for the square wave

sq(x,n) = n == 0? sin(x):sin((2*n+1)*x)/(2*n+1)+sq(x,n-1)
```

```
set sample 1000

n = 0
load 'Fourier.plt'
```

and the main script **Fourier.plt**

```
t = sprintf('Fourier Series summed to %d terms',n+1)
set title t font 'Helvetica,16'

p [-10:10] [-1.5:1.5] 4/pi*sq(x,n) t 'square'

n = n+1

pause -1 "Hit enter to continue"

if (n<10) reread
pause -1 "Hit enter to quit"}
```

Running this script by typing

`gnuplot initFourier.plt` will result in the display of the first ten partial sums of the Fourier expansion of a square wave, with gnuplot waiting each time before moving on to the next graph for the user to strike a key.

4. Conclusion

In conclusion let me emphasize that it is quite easy to master the art of creating high quality, flexible animations using gnuplot, which can go a long way towards aiding the teaching-learning process.

[1] The gnuplot website *http://www.gnuplot.info.*

Journal of Physics: Conference Series **759** (2016) 012090 doi:10.1088/1742-6596/759/1/012090

Non-Boltzmann Ensembles and Monte Carlo Simulations

K. P. N. Murthy

School of Physics, University of Hyderabad, Hyderabad 500 046 Telengana India

and

Manipal Centre for Natural Sciences, Manipal University, Manipal 576 104, Karnataka, India

E-mail: `kpnmsp@uohyd.ac.in`

Abstract. Boltzmann sampling based on Metropolis algorithm has been extensively used for simulating a canonical ensemble and for calculating macroscopic properties of a closed system at desired temperatures. An estimate of a mechanical property, like energy, of an equilibrium system, is made by averaging over a large number microstates generated by Boltzmann Monte Carlo methods. This is possible because we can assign a numerical value for energy to each microstate. However, a thermal property like entropy, is not easily accessible to these methods. The reason is simple. We can not assign a numerical value for entropy, to a microstate. Entropy is not a property associated with any single microstate. It is a collective property of all the microstates. Toward calculating entropy and other thermal properties, a non-Boltzmann Monte Carlo technique called Umbrella sampling was proposed some forty years ago. Umbrella sampling has since undergone several metamorphoses and we have now, multi-canonical Monte Carlo, entropic sampling, flat histogram methods, Wang-Landau algorithm *etc.* This class of methods generates non-Boltzmann ensembles which are un-physical. However, physical quantities can be calculated as follows. First un-weight a microstates of the entropic ensemble; then re-weight it to the desired physical ensemble. Carry out weighted average over the entropic ensemble to estimate physical quantities. In this talk I shall tell you of the most recent non-Boltzmann Monte Carlo method and show how to calculate free energy for a few systems. We first consider estimation of free energy as a function of energy at different temperatures to characterize phase transition in an hairpin DNA in the presence of an unzipping force. Next we consider free energy as a function of order parameter and to this end we estimate density of states $g(E, M)$, as a function of both energy E, and order parameter M. This is carried out in two stages. We estimate $g(E)$ in the first stage. Employing $g(E)$, we generate an entropic ensemble. In the second stage, we estimate $g(E, M)$ by averaging $g(E)$ over partial entropic ensemble for various values of the order parameter M. We present results on Ising spin system which exhibits second order phase transition and on a double strand DNA, which exhibits first order phase transition.

1. Introduction

Corresponding to a thermodynamic property, we have, in statistical mechanics, a random variable. The average of the random variable over a suitable statistical ensemble, *e.g.* microcanonical for an isolated system, canonical for a closed system, and grand canonical for an open system, gives the value of the corresponding thermodynamic property. For example, consider internal energy U. Corresponding to this, in statistical mechanics, we have energy E -

the kinetic energy and interaction energy of the atoms and molecules of the macroscopic object. A numerical value of E can be assigned to each microstate. The value of E fluctuates when an equilibrium system switches from one microstate to another. These fluctuations are an integral part of an equilibrium description. The average of E over a canonical ensemble at a given temperature equals U. The computation of average energy is then straight forward. Generate a canonical ensemble employing Monte Carlo method. A simple arithmetic average of energy over a Monte Carlo sample gives the required answer. The statistical error associated with the estimated average is also computed from the same sample. Such a neat computational scheme is possible because a numerical value for energy can be assigned to each microstate.

How does one calculate entropy ? We notice that we can not assign a value of entropy to any single microstate. Entropy is a property that belongs collectively to all the microstates. While energy is a " private " property, entropy is a " social " or " public " property, see below.

Let $\{C_i\ :\ i = 1,\ 2,\ \cdots\}$ denote the microstates of an equilibrium system and $\{p(C_i)\ :\ i = 1,\ 2,\ \cdots\}$, the corresponding probabilities. The Boltzmann-Gibbs-Shannon entropy is given by $S = -k_B \sum_i p(C_i) \ln p(C_i)$. For an isolated system, $p(C_i) = 1/\widehat{\Omega}(E, V, N)$, where $\widehat{\Omega}$ denotes the number of microstates. For a simple fluid $\widehat{\Omega}$ depends on energy E, volume V, and number of particles N. We have $S = k_B \ln \widehat{\Omega}(E, V, N)$. For a closed system at temperature $T = 1/[k_B\beta]$, we have, $p(C_i) = Q^{-1} \exp[-\beta E(C_i)]$, where $Q(T, V, N) = \sum_i \exp[-\beta E(C_i)]$ is the canonical partition function.

2. Metropolis Algorithm

Consider a system characterized by probabilities $\{p(C_\nu)\ :\ \nu = 1, 2, \cdots\}$. Our aim is to generate a large number of microstates of the system consistent with the given probabilities. To this end, start with an arbitrary initial microstate C_0 and generate a Markov chain, $C_0 \to C_1 \to \cdots C_i \to C_{i+1} \cdots$, employing Metropolis rejection algorithm[1], see below.

Let C_i be the current microstate and $p_i = p(C_i)$, its probability; we make a change in the current microstate and construct a trial microstate C_t. For example, if we are simulating Ising spin system, select a spin randomly from the current spin configuration, and flip it to get a trial microstate. Let $p_t = p(C_t)$. Calculate $p = \min.(1, p_t/p_i)$. Generate a random number ξ uniformly and independently distributed between zero and unity. If $\xi \le p$ accept the trial state and advance the Markov chain to $C_{i+1} = C_t$. If not, reject the trial state and advance the Markov chain to $C_{i+1} = C_i$. Repeat the process on the microstate C_{i+1} and proceed to generate a long Markov chain. The asymptotic part of the chain shall contain microstates belonging to an ensemble characterized by $\{p(C_\nu)\ :\ \nu = 1, 2, \cdots\}$.

3. Boltzmann Sampling

When the probabilities $\{p(C_\nu)\ :\ \nu = 1, 2. \cdots\}$ describe a physical ensemble *e.g.* canonical ensemble or grand canonical ensemble, we call it Boltzmann sampling. Notice Metropolis algorithm requires only the ratio of probabilities; hence we need to know $p(C)$ only upto a normalization constant. It is precisely because of this we can simulate a closed system, since we need to know only the Boltzmann weight, $\exp[-\beta E(C)]$, and not the canonical partition function. Also the algorithm obeys detailed balance and hence generates a reversible Markov chain, see *e.g.* [2], appropriate for describing an equilibrium system.

Generate a Markov chain until it equilibrates; continue the Markov chain and collect a set of large number of microstates $\{C_i\ : i = 1, 2, \cdots M\}$. Let O be a property of interest and $O(C)$, its value, when the system in a microstate C. Then,

$$\langle O \rangle \ = \ \operatorname*{Limit}_{M\to\infty} \frac{1}{M} \sum_{i=1}^{M} O(C_i). \tag{1}$$

To calculate the statistical error associated with our estimate of the mean, we proceed as follows. We calculate the second moment

$$\langle O^2 \rangle \quad = \quad \mathop{Limit}_{M\to\infty} \frac{1}{M} \sum_{i=1}^{M} O^2(C_i), \tag{2}$$

and the variance $\sigma^2 = \langle O^2 \rangle - \langle O \rangle^2$. The statistical error is then given by $\sigma/\sqrt{M}$. Notice that the Boltzmann sampling described above, depends crucially on our ability to assign a numerical value of the property O to each microstate of the system.

Consider estimating a property like entropy. We can not assign a numerical value for entropy to a microstate. Entropy is a collective property of all the microstates of an equilibrium system. Hence entropy can not be easily obtained from Boltzmann sampling. We need non-Boltzmann sampling techniques and to this, we turn our attention below.

4. Non-Boltzmann Sampling

Torrie and Valleau[3] were, perhaps, the first to propose a non-Boltzmann algorithm to calculate thermal properties. Their method, called Umbrella sampling, has since undergone a series of metamorphoses. We have multi-canonical Monte Carlo algorithm of Berg and Neuhaus[4], entropic sampling of Lee[5] and the algorithm of Wang and Landau[6]. We describe below the algorithm due to Wang and Landau[6].

Let $\widehat{\Omega}(E)$ denote the density of states of the system under simulation. The microcanonical entropy is given by logarithm of the density of states : $S(E) = \ln \widehat{\Omega}(E)$. Let C be a microstate and $E_C = E(C)$, its energy. We define an ensemble characterized by the probabilities $p(C) \propto 1/\widehat{\Omega}(E_C)$, defined for all the microstates of the system. We employ Metropolis rejection technique to generate microstates based on these probabilities. The probability of acceptance of a trial microstate is given by, $p = \min.(1, p_t/p_i) = \min.[1, \widehat{\Omega}(E_i)/\widehat{\Omega}(E_t)]$. Note, if the trial state belongs to a low entropy region, it gets accepted with unit probability; if not, its acceptance probability is less than unity. Thus, in the rejection step, low entropy regions are preferred, statistically. This preference cancels statistically exactly, the natural tendency of random sampling trial states from regions of high entropy. As a result, the ensemble generated by the algorithm will have equal number of microstates in equal regions of energy. In other words, the histogram of energy of the microstates of the ensemble shall be flat. But a crucial point remains : we do not know the density of states $\widehat{\Omega}$, as yet.

Wang and Landau[6] proposed to estimate the density of states in an initial learning run. We define a function $g(E)$ and set it to unity for all E. We also define a histogram of energy $H(E)$ and set it to zero for all E. Start with an initial microstate C_0. Let $E_0 = E(C_0)$ be its energy. Update $g(E_0)$ to $\alpha \times g(E_0)$ where α is called the Wang-Landau factor, set to $\alpha_0 = e$, in the first iteration. Also update $H(E_0)$ to $H(E_0) + 1$. Construct a Markov chain of microstates as per Metropolis rejection technique taking the probabilities proportional to $1/g(E(C))$. Every time you advance the Markov chain, update g and H. The Markov chain shall be inhomogeneous, since the transition probabilities depend on the time. Carry out the simulation of the Markov chain, until the histogram is flat over a range of energy. Then set $\alpha = \alpha_1 = \sqrt{\alpha_0}$, reset $H(E) = 0 \ \forall \ E$ and proceed with the second iteration of the learning run. The value of α tends to unity upon further iterations. After some twenty five iterations, $\alpha = \alpha_{25} \approx 1 + 3 \times 10^{-7}$. The histogram would be flat over the range of energy of interest. Flatter the histogram, closer is $g(E)$ to the true density of states $\widehat{\Omega}(E)$. We take $g(E)$ at the end of last iteration of the learning run, that leads to a reasonably flat histogram, as an estimate of $\widehat{\Omega}(E)$. We can define suitable criteria for flatness of the histogram. For example we can say the histogram is flat if the smallest and largest entries in the histogram do not differ from each other by more than say 10%. Note we calculate the density of states only upto a multiplicative constant. The microcanonical entropy is given by $S(E) = \ln g(E)$, upto an additive constant. We can then employ the machinery of thermodynamics to calculate all the thermal properties, starting from the microcanonical entropy. Also, the converged density of states can be employed in the production run and a large number of microstates generated. These microstates belong to an ensemble which we call as entropic ensemble or Wang-Landau ensemble. The Wang-Landau ensemble is un-physical. Nevertheless, physical quantities can be estimated by un-weighting and re-weighting, see below.

We first attach a statistical weight of unity to each microstate of the Wang-Landau ensemble. Then we divide the statistical weight by $p(C) = 1/g(E(C))$. This is called un-weighting. Upon un-weighting, the ensemble of weighted microstates becomes microcanonical. In other words, weighted averaging over the microstates of the (unphysical) Wang-Landau ensemble is equivalent to averaging over the (physical) microcanonical ensemble. We can further re-weight to the desired ensemble. For example, to calculate the average over canonical ensemble, we multiply by the Boltzmann weight, see below.

Let $\{C_i : i = 1, 2, \cdots, M\}$ denote the microstates of the Wang-Landau ensemble. We carry out weighted average after un-weighting and re-weighting,

$$\langle O \rangle = \underset{M\to\infty}{\text{Limit}} \frac{\sum_{i=1}^{M} O(C_i)W(C_i)}{\sum_{i=1}^{M} W(C_i)} \tag{3}$$

where the weight factor is given by,

$$W(C_i) = g(E(C_i)) \exp(-\beta E(C_i)) \tag{4}$$

$\langle O \rangle$ is the desired average over a canonical ensemble.

Thus, employing Wang-Landau algorithm, mechanical as well as thermal properties can be calculated. So far so good. There are a few outstanding issues with the Wang-Landau algorithm in particular and Non-Boltzmann Monte Carlo methods in general. The principal amongst them concerns the slow convergence of the density of states, for systems with continuous Hamiltonians, see *e.g.* [7] for a discussion on this issue. Another issue is that the algorithm does not obey detailed balance. There are also problems associated with calculation of statistical error bars. We shall not address these and related problems in this talk. Instead in what follows, we shall describe how does one calculate a thermal property like (Helmholtz) free energy as a function of energy (or order parameter), for temperatures close to transition point, employing Wang-Landau algorithm. We shall also illustrate the technique on a few examples.

5. Calculation of Free Energy

The free energy of a closed system at temperature T is given by, $F(T, V, N) = -k_B T \ln Q(T, V, N)$, where $Q(T, V, N)$ is the canonical partition function. We can also define microcanonical free energy for an isolated system as,

$$F(U, V, N) = U(S, V, N) - \left(\frac{\partial U}{\partial S}\right)_{V,N} S \,, \tag{5}$$

where $S(U, V, N)$ is the microcanonical entropy given by $S = k_B \ln \widehat{\Omega}(E, V, N)$.

For an equilibrium system, free energy can not be simultaneously a function of both energy and temperature : an isolated system with a given energy has an unique temperature; a closed system at a given temperature has a unique (average) energy. However we may be interested in estimating the penalty in terms of excess free energy required to keep a closed system in a non-equilibrium state with its energy E different from the equilibrium energy $U(T)$. To this end, following Joon Chang Lee[8], we define a phenomenological free energy F_L, which is simultaneously a function of both energy and temperature. Let $F(T)$ denote the free energy of the closed system at temperature T. We have $F_L(E,T) \geq F(T)$ for all energies, and equality obtains when $E = U(T)$. The excess free energy is given by $\Delta F = F_L(E,T) - F(T)$, where,

$$F_L(E,T) = -k_B T \ln \sum_C g(E(C)) \ \exp\left[-\beta E(C)\right] \ \delta\left(E(C) - E\right). \tag{6}$$

The sum runs over all the microstates belonging to the entropic ensemble collected during the production run of the Wang-Landau Monte Carlo simulation.

5.1. Phase Transition in hpDNA : Free Energy versus Energy

For purpose of illustration, we present the results on free energy profiles of a DNA-hairpin (hp-DNA) subjected to a constant unzipping force. DNA has two strands of equal length twisted around a common axis to form a double helix. Each strand can be considered as a chain of monomers or bases. When complementary bases at the two ends of a strand pair up, hairpin structure occurs, with a loop and stem. The structure can be closed or open. The transition from closed to open state is called denaturation or unbinding. If denaturation occurs by virtue of temperature, it is called melting. Force induced melting has been extensively studied, see [9] and the references cited therein. We present results on chain of length 110 units. We determine the transition temperature from the peak of the specific heat profile. We take two temperatures in the neighbourhood and on either side of the transition point, to calculate free energy profiles. We have considered four site occupation, bond fluctuation model[10], on a two dimensional square lattice. For details of the simulation see [9].

We first calculate heat capacity and determine the temperature at which it is maximum. We take this as transition temperature T_C. Then we estimate the free energy *versus* energy profiles and these are depicted in Figures 1 - 3, for $T = T_C$, $T < T_C$, and $T > T_C$ respectively. It is clear the transition is first order.

6. Free Energy versus Order Parameter

Landau free energy is usually expressed as a function of order parameter M. Landau, Tsai, and Exler[11] defined a joint density of states $g(E,M)$ and employed it in the Wang-Landau algorithm. Let E_i and M_i be the energy and order parameter of the current microstate; let E_t and M_t denote the same quantities for the trial microstate. The acceptance probability in the Metropolis algorithm is taken as, $p = \min.[1, g(E_i,M_i)/g(E_t,M_t)]$. The joint density of states is updated at every step. We also monitor the histogram $H(E,M)$. The histogram now is a two dimensional surface. When the surface becomes flat it indicates that $g(E,M)$ has converged to the true two dimensional density of states, $\widehat{\Omega}(E,M)$. Obtaining a flat histogram in two dimensional space requires considerable computing time[11]. A good strategy[12] shall involve estimating the two dimensional density states in the production run, see below. The learning run is reserved, as usual, for obtaining a converged density of states in energy space only.

We assemble, in the production run, an entropic ensemble. We first un-weight the microstate and then re-weight it to a microcanonical ensemble. We estimate the joint density of states as a weighted average over the entropic ensemble. This is equivalent to averaging over a microcanonical ensemble. Thus,

$$\langle g(E_i, M_j)\rangle = \frac{\sum_C \delta(E(C) - E_i) \ \delta(M(C) - M_j) \ g(E(C))}{\sum_C \delta(E(C) - E_i) \ g(E(C))} \quad \forall \ E_i, \ M_j. \tag{7}$$

In the above, the sum runs over all the microstates of the entropic ensemble. Having estimated the joint density of states, we calculate the Landau free energy as

$$F_L(M_j, T) = -k_B T \ln \sum_i \langle g(E_i, M_j)\rangle \exp(-\beta E_i) \tag{8}$$

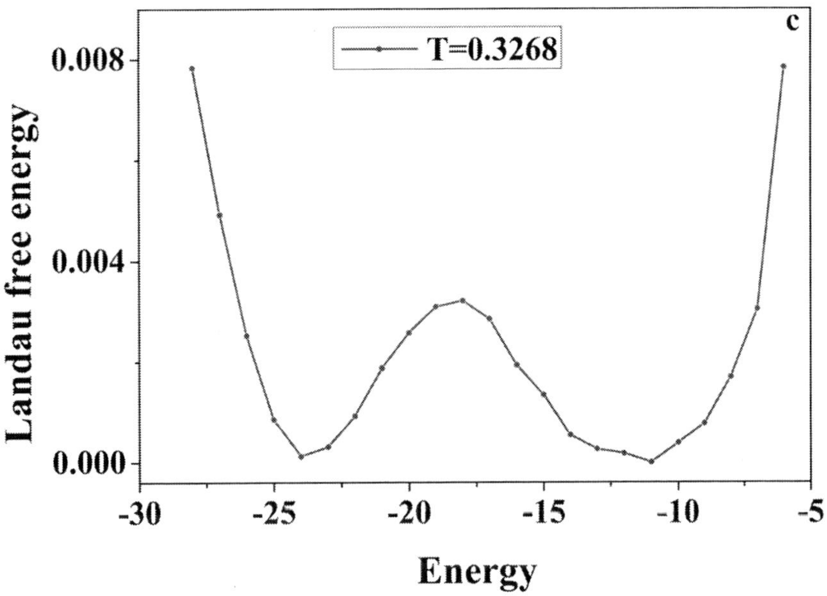

Figure 1. Variation of free energy with energy at $T = T_C$ and unzipping force $f = 0.04$. Temperature at which the heat capacity peaks is taken as T_C. The chain length is 110; reproduced from [9]

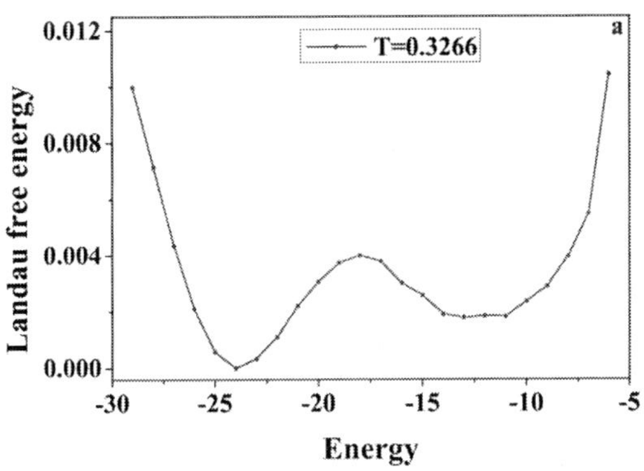

Figure 2. Same as Figure 1, except $T < T_C$; reproduced from [9]

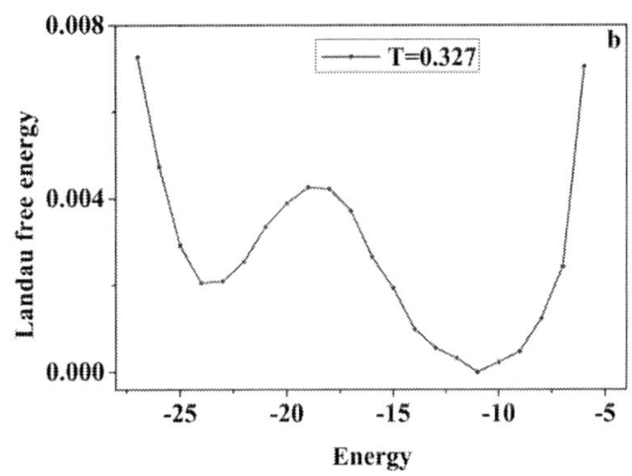

Figure 3. Same as Figure 1, except $T > T_C$; reproduced from [9]

where we take T to be very close to the transition temperature.

6.1. Ising Spins, dsDNA and Order Parameter

Figure 4 depicts the results on Landau free energy, $F_L(M,T)$ versus magnetization, M, for a two dimensional Ising spin system. The spins are located on the vertices of a 32×32 square lattice. We have estimated the joint density of states, $g(E,M)$, in the production run of Wang-Landau simulation. For details see [13]. Free energies for $T < T_C$ (bottom curve), $T = T_C$ (middle curve), and $T > T_C$ (top curve) are shown. The transition temperature T_C was located from the specific heat profiles.

Figure 5, reproduced from [13], depicts results on variation of heat capacity with temperature, for double strand DNA. We have calculated free energy employing both the methods : (i) standard Wang-Landau algorithm in which $g(E,M)$ is obtained in the learning run and (ii) two-stage method in which $g(E,M)$ is estimated in the production run by un-weighting and

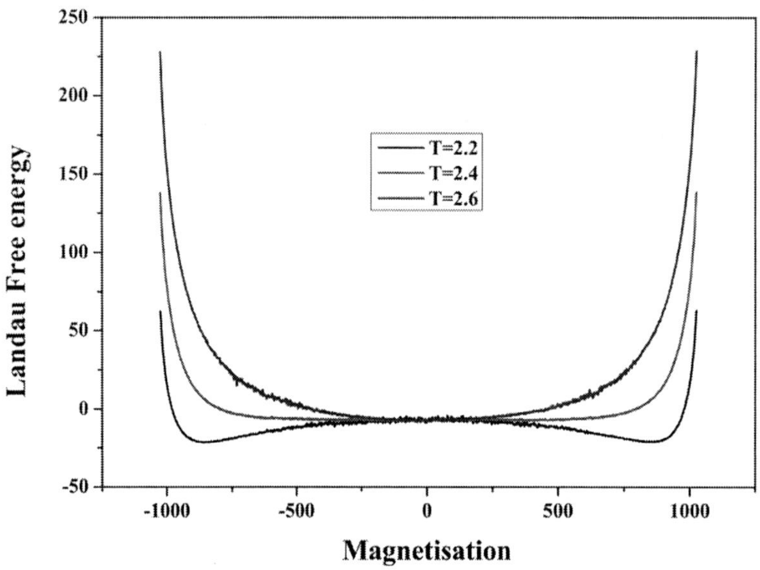

Figure 4. Free energy *versus* magnetization for zero field Ising spin system; joint density of states has been calculated employing the two stage method. Two dimensional square lattice with $L = 32$. The top curve is for $T > T_C$; the middle curve is for $T = T_C$; and the bottom curve is for $T < T_C$; reproduced from [13]

re-weighting procedures. The two results agree with each other, demonstrating that the two-stage method is as good as the standard Wang-Landau sampling toward predicting mechanical properties. I must mention that the two-stage method took considerably much less computer time.

Figure 6 reproduced from [13], shows results on Landau free energy as a function of the order parameter for dsDNA. The end to end distance is taken as order parameter and is denoted by the symbol ξ. We have presented results at $T = T_C$, $T < T_C$ and at $T > T_C$. The transition is first order.

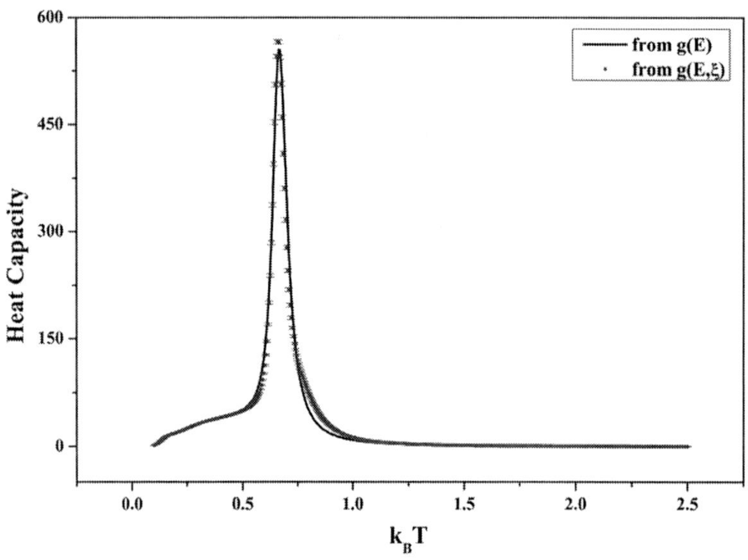

Figure 5. Heat capacity of dsDNA as a function of T calculated from $g(E)$, estimated by standard Wang-Landau algorithm and from $g(E, \xi)$, obtained from the two stage method; reproduced from [13]

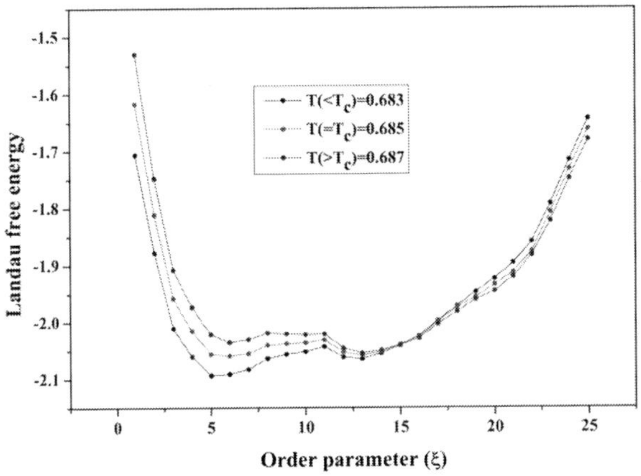

Figure 6. Free energy *versus* order parameter for a double strand DNA - dsDNA of length 80 in the presence of unzipping force $f = 0.02$. The end to end distance is taken as order parameter. Joint density of states, $g(E, \xi)$, has been calculated employing the two stage method. Top curve is for $T < T_C$; the middle curve is for $T = T_C$; the bottom curve is for T > T_C; reproduced from [13]

7. Summary

I have introduced briefly the basic principles of non-Boltzmann Monte Carlo methods. The density of states is estimated through an iterative process. Logarithm of density of states gives entropy; all other thermal properties can be estimated from entropy, employing the machinery of thermodynamics. Though a non-Boltzmann ensemble is unphysical, we can extract averages of mechanical properties, over desired physical ensembles by employing un-weighting and re-weighting techniques. Thus both thermal and mechanical properties can be estimated employing non-Boltzmann Monte Carlo simulation methods. I have illustrated the usefulness of non-Boltzmann ensemble on three systems : an hairpin DNA, Ising spins, and double stranded DNA. Results on free energy profiles at temperatures close to transition point were presented. I believe it is time we take a Non-Boltzmann Monte Carlo, as a method of first choice for simulating macroscopic systems.

Acknowledgements

This talk is based on Ref. [9] and [13] which report work carried out in collaboration with V S S Sastry, M Suman Kalyan, and R Bharath; and to them I owe my thanks. Computations were carried out at CMSD, University of Hyderabad, Hyderabad, and the Shakti cluster, MCNS, Manipal University, Manipal.

References

[1] N. Metropolis, A. W. Rosenbluth, M. N. Rosenbluth, A. H. Teller, and E. Teller 1953 J. Chem. Phys. **21** 1087; see also G. Bhanot 1988 Rep. Prog. Phys. **51** 429; K. P. N. Murthy 2004, *Monte Carlo Methods in Statistical Physics* (India : Universities Press)
[2] K. P. N. Murthy, and V. S. S. Sastry 2005 *Markov Chain Monte Carlo Methods in Statistical Physics* PoS(SMPRI2005) 018
[3] G. M. Torrie, and J. P. Valleau 1977 J. Comp. Phys. **23** 187
[4] B. A. Berg, and T. Neuhaus 1992 Phy. Rev. Lett. **68** 9
[5] J Lee 1993 Phy. Rev. Lett. **71** 211; Erratum 1993 **71** 2353
[6] F. Wang, and D. P. Landau 2001 Phy. Rev. Lett. **86** 2050
[7] D Jayasri, V. S. S. Sastry, and K. P. N. Murthy 2005 Phys. Rev. E **72** 36702; P Poulin, F Calvo, R Antoine, M Broyer, and P Dugord 2006 Phys. Rev. E **73** 56704.
[8] Joon Chang Lee 2002 *Thermal Physics : Entropy and Free Energies* (Singapore : World Scientific)
[9] M. Suman Kalyan, and K. P. N. Murthy 2015 Physica **A 428** 38
[10] I. Carmesin, and K. Kremer 1988 Macromolecules **21** 2819
[11] D. P. Landau, Shan-Ho Tsai, and M. Exler 2004 Am. J. Phys. **72** 1294
[12] C. Geravais, T. Wüst, D. P. Landau, and Y. Xu 2009 J. Chem. Phys. **130** 215106
[13] M. Suman Kalyan, R. Bharath, V. S. S. Sastry, and K. P. N. Murthy 2016 J. Stat. Phys. **163** 197

Research Activity in Computational Physics utilizing High Performance Computing: Co-authorship Network Analysis

Sul-Ah Ahn and Youngim Jung*

Korea Institute of Science and Technology Information, 245 Daehak-ro, Yuseong-gu, Daejeon 34141, Republic of Korea

acorn@kisti.re.kr

Abstract. The research activities of the computational physicists utilizing high performance computing are analyzed by bibliometirc approaches. This study aims at providing the computational physicists utilizing high-performance computing and policy planners with useful bibliometric results for an assessment of research activities. In order to achieve this purpose, we carried out a co-authorship network analysis of journal articles to assess the research activities of researchers for high-performance computational physics as a case study. For this study, we used journal articles of the Scopus database from Elsevier covering the time period of 2004-2013. We extracted the author rank in the physics field utilizing high-performance computing by the number of papers published during ten years from 2004. Finally, we drew the co-authorship network for 45 top-authors and their coauthors, and described some features of the co-authorship network in relation to the author rank. Suggestions for further studies are discussed.

1. Introduction

The purpose of bibliometrics is to map the literary production in a given field and to determine the structure of a field of study. Bibliometric analyses in various research fields have been conducted during decades. Lately, we also conducted a bibliometric analysis in the high-performance computing field [1]. For the computational physics field utilizing high-performance computing (ComPhy-HPC), it is of interest that bibliometric analyses are known to be very rare in ComPhy-HPC and related fields, because the high-performance computing is an emerging field of computational research at the intersection of science and computation [2]. By use of bibliometric approaches, we analyzed the research activities of the computational physicists utilizing high-performance computing. Our work includes the results of a bibliometric study of the computational physics utilizing high-performance computing, featuring in co-authorship network analysis of journal articles. In our knowledge, this is the first time that the co-authorship analysis is performed on journal publication data in the computational physics filed utilizing high-performance computing.

Earlier, a concept of scientific collaboration network is firstly introduced by Newman [3]. He took note on networks of scientists in which two scientists are considered connected if they have jointly coauthored a paper. After this, the affiliation networks of scientist, or co-authorship network, in which a link between two scientists is established by their co-authorship of one or more scientific papers are constructed [4]. It is assumed in the further study that co-authorship indicates a level of scientific collaboration [5]. Recently, Morel *et. al.* utilized the co-authorship network analysis as a powerful tool in order to plan strategically research, development and capacity building programs on neglected diseases [6]. More recently, Ding studied scientific collaboration and endorsement by using network analysis of co-authorship and citation networks [7].

2. Our Work

We carried out a co-authorship network analysis of journal articles to assess the research activities of computational physicists utilizing high-performance computing as a case study. The data sources used for this study are journal articles of the Scopus database from Elsevier covering the time period of 2004-2013 for journal articles. We extracted author rank in the high performance computing field by the number of papers published during ten years from 2004. We calculated the weight of co-authorship and draw the co-authorship network for top-45 authors and their coworker.

2.1. Data

Computational physics utilizing High Performance Computing (ComPhy-HPC) was selected as the test area. Papers were collected from the Scopus database covering the time period of 2004-2013. Based on a set of search terms related to ComPhy-HPC, the following queries were formed: SUPERCOMPUT* PHYSICS, HIGH PERFORMANCE COMPUT* PHYSICS, HIGH THROUGHPUT COMPUT* PHYSICS, CLUSTER COMPUT* PHYSICS, GRID COMPUT* PHYSICS

2.2. Methodology for Analyzing Co-author Network

Analysis of co-authorship network reveals collaboration using published papers. Refinement of the source data was necessary because the first names of the authors are abbreviated and the authors who have multiple affiliations are duplicated in the Scopus database.

We generated a pathfinder network using symmetric square matrix whose elements are correlation value between authors. Pathfinder network algorithm is helpful to describe the structure of the network, because it simplifies the complex network in matrix representation [8]. For the visualization of the generated coauthor network, NodeXL has been utilized.

3. Results and Discussions

3.1. Author Ranks

In total, 4,403 papers were gathered. The corresponding authors of all articles have not been identified. Only 4,197 articles among 4,403 articles, the corresponding author of each article are identified. The number of authors involving in authoring those articles is 11,460. More than 78.7% authors, tallying 9,022 authors in the author list have authored only once. Only 46 scholars in this field have authored at least ten articles during this period. ComPhy-HPC is emerging and novel technology in physics field and the limited number of scholars have studied related with this topic. Figure 1 illustrates the number of articles published and indexed in the Scopus database for the last ten years.

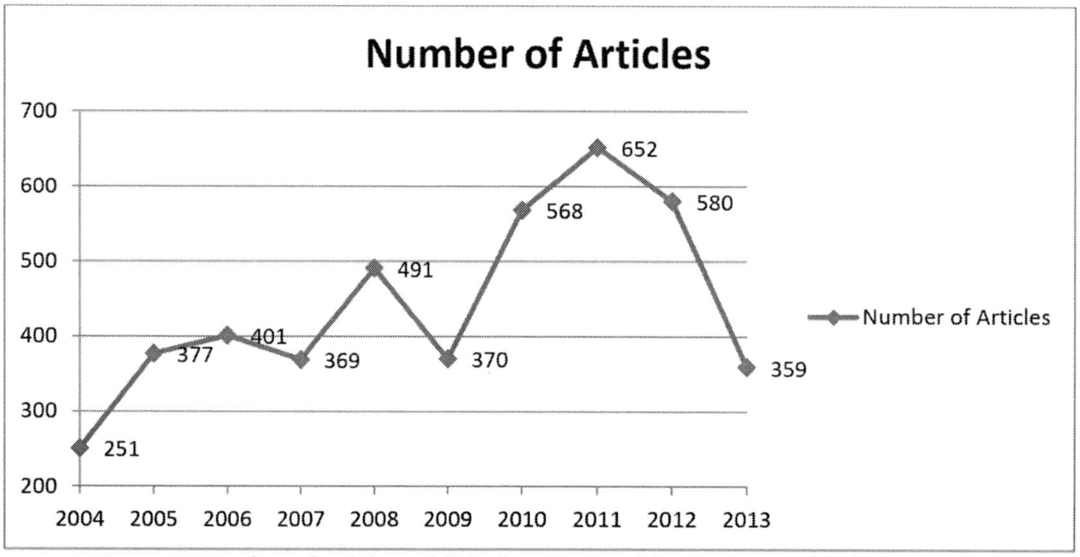

Figure 1. Number of articles published and indexed in the Scopus database by Year

Table 1 presents the author rank in the target field by the number of papers published from 2004 to 2013.

Table 1. ComPhy-HPC author ranks in 2004-2013 (Top 10)

Rank	Author Name	Number of papers	Percentage over publication by top 45 authors
1	Gauss Jürgen	26	8.7
2	Schaefer III Henry F.	22	7.4
3	Maeno T	18	6.0
3	Sfiligoi Igor	17	5.7
5	Graciani R	14	4.7
5	Saiz P	14	4.7
7	Wenaus T	13	4.3
7	Bunčič P	13	4.3
7	Bartlett Rodney J	13	4.3
10	Tsaregorodtsev A	12	4.0
10	Szalewicz Krzysztof	12	4.0
10	Garzoglio G	12	4.0
10	Grigoras C	12	4.0

Ranked as the top 1 author, Gauss, Jürgen published 26 papers in ComPhy-HPC field during decade. His affiliation is the Johannes Gutenberg University of Mainz, well known as the top tier of high performance computing. His work is focused on the application of the quantum chemical calculation and the development of new quantum chemical methods.

Maeno T is ranked as the 3rd best publishing author, with 18 papers published. His affiliation is the Brookhaven National Laboratory, well-known for nuclear and high-energy physics. The main topics in his publication includes the design, development and commissioning of innovative new distributed computing software critical to the ATLAS experiment.

3.2. Co-authorship and Co-authorship Network in ComPhy-HPC

In the co-authorship network, nodes represent authors and an edge shows that linked authors (nodes) have worked together at least for publishing one paper. The width of the edge represents the weight of the co-authorship between the two authors. The weight of the co-authorship is calculated by the cosine coefficient [9] of the two authors who authored articles published from 2004 to 2013 together. Only 45 top authors and their coauthors who involved in authoring at least ten publication among 11,460 authors have been selected for calculating the cosine coefficient. Table 2 shows the cosine coefficient, or weight of co-authorship among 45 top authors and their coauthors in ComPhy-HPC.

Table 2. Co-authoring rank among 45 top authors and their coauthors in ComPhy-HPC (2004-2013)

Co-authoring Rank	Author Name	Author Name	Weight of Co-Authorship (Cosine Coeff.)
1	Negri A	Vercesi V	0.9987
2	Luminari L	Vercesi V	0.99422
2	Luminari L	Werner P	0.99422
4	George S	Vercesi V	0.99065
4	George S	Werner P	0.99065
6	Buncic P	Harutyunyan A	0.98705
7	Grigoras C	Betev L	0.98644
8	Sciaba A	Flix J	0.89658
9	Saiz P	Grigoras C	0.80007
10	Maeno T	Vercesi V	0.79691
10	Maeno T	Werner P	0.79691
12	Schaefer III Henry F	Allen Wesley D	0.78716
13	Graciani R	Tsaregorodtsev A	0.74148
14	Wenaus T	Maeno T	0.71652
15	Furano F	Carminati F	0.67777
16	Stewart G A	Wenaus T	0.6049
17	Andreeva J	Saiz P	0.55235
18	Sfiligoi Igor	Bonacorsi D	0.20799
19	Krylov Anna I	Head Gordon Martin	0.16717
20	Bartlett Rodney J	Hirata So	0.14447
21	Gauss Jürgen	Schaefer III Henry F	0.14207
22	Garzoglio G	Sfiligoi Igor	0.11282

23	Szalewicz Krzysztof	Bartlett Rodney J	-0.02445

Negri A and Vercesi V are ranked as the top-1 co-author couple, who have the strongest correlation for the co-authorship among 45 top authors and their coauthors in HPC.

Figure 2, the co-authorship network among 45 top authors and their coauthors in ComPhy-HPC is represented using NodeXL.

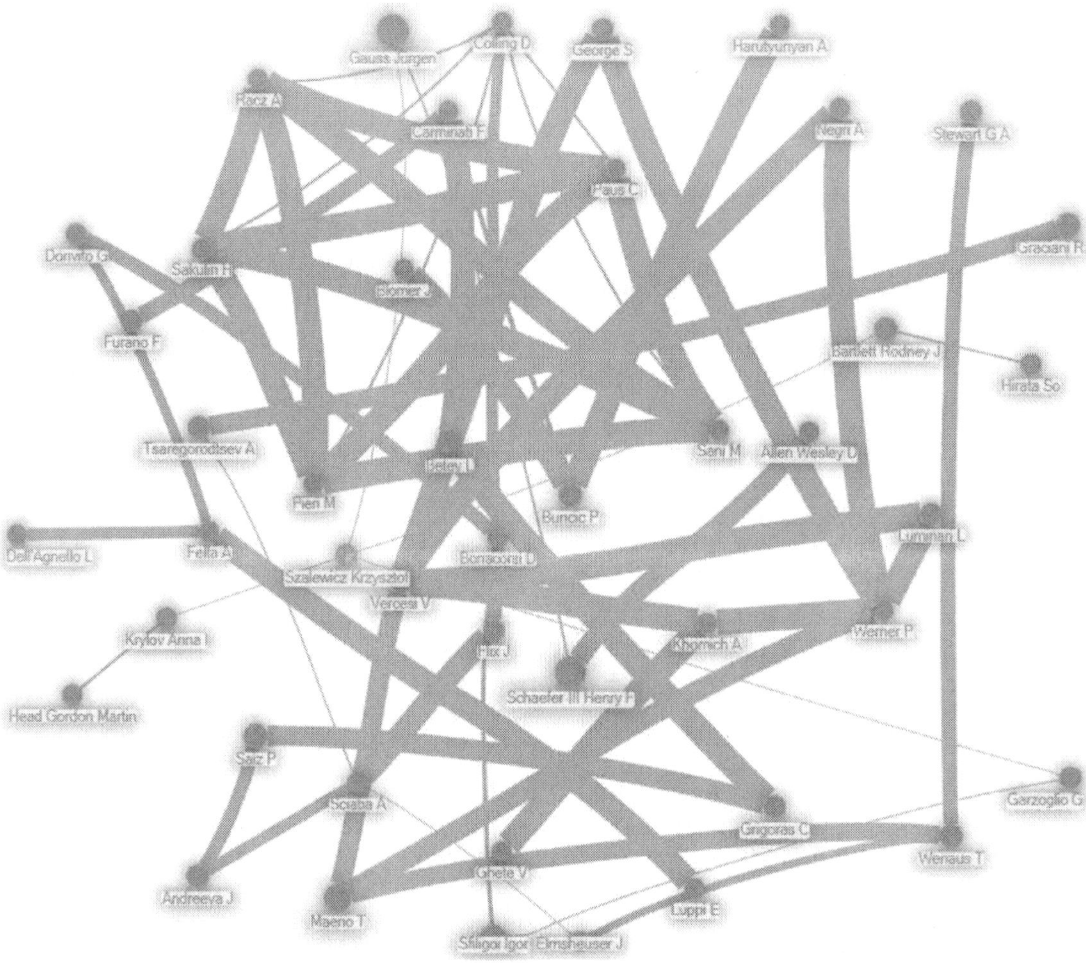

Figure 2. Co-authorship network of 45 top authors and their coauthors in ComPhy-HPC (2004-2013)

Gauss Jürgen and Schaefer III Henry F, the top-1 and top-2 author according to number of publishing, seem to have relatively weak correlation for the co-authorship among 45 top authors and their coauthors in HPC. George S and Werner P are not listed in the top 10 list in the author rank, but they show relatively strong correlation for the co-authorship (top 4 in table 2). As shown in figure 2, Werner P is also connected to Maeno T (top 3 in table 1), which reveals collaboration between authors may easily propagate through dominant authors.

4. Summary

We performed a co-authorship network analysis of journal articles to evaluate the research activities in the computational physics field utilizing high-performance computing (ComPhy-HPC). The data sources used for this study are journal articles of the Scopus database from Elsevier covering the time period of 2004-2013 for journal articles. We extracted author rank in the ComPhy-HPC field by the number of publication during ten years. We calculated the weight of co-authorship and draw the co-authorship network for 45 top-authors and their coauthors. High ranked authors by publication are not also strong in co-authorship among 45 top-authors and their coauthors. Some authors, not much high ranked authors by publication, show relatively more strong correlation for the co-authorship. In order to provide significant bibliometric results for an evaluation of research activities to computational physicists utilizing high-performance computing and policy planners, further studies using co-authorship network analysis or citation network [10] are suggested to be done for more than 45 top-authors in the ComPhy-HPC field.

References

[1] Ahn S-A and Jung Y 2015 *ICIC Express Letters, Part B: Applications* **6** 1437
[2] 2013 *High-Performance Computing Using FPGAs,* ed W Vanderbauwhede and K Benkrid (New York: Springer-Verlag)
[3] Newman M E J 2001 *Proc. Natl. Acad. Sci. USA* **98** 404
[4] Newman M E J 2001 *Physical Review E* **64** 016131.
[5] Newman M E J 2004 *Proc. Natl. Acad. Sci. USA* **101**(suppl. 1) 5200.
[6] Morel C M, Serruya S J, Penna G O and Guimariaes R 2009 *PLOS/Neglected Tropical Diseases* **3** e501
[7] Ding Y 2011 *J. Informetr.* **5** 187
[8] Lee J 2006 *Journal of the Korean Society for Library and Information Science* **40** 333
[9] Egghe L and Leydesdorff L 2009 *J. Am. Soc. Inf. Sci. Technol.***60** 1027
[10] Erman N and Todorovski L 2009 Mapping the e-Government Research with Social Network Analysis *EGOV 2009 (LNCS vol 5693)* ed M A Wimmer *et al* (Heidelberg: Springer) pp 13-25

Identifying Interstitials in MD Simulations – Max Space Clustering Method

S Bukkuru [1], U Bhardwaj[2], ADP Rao[1], M Warrier[2]
[1]Dept. of Nuclear Physics, Andhra University, Visakhapatnam, India
[2]Computational Analysis Division, BARC, Visakhapatnam, India

E-mail: `srinivasaraobukkuru@gmail.com`

Abstract. Molecular Dynamics (MD) simulations are suitable to study the details of atomistic processes. Identification of interstitials is one of the main difficulties in the analysis of defect dynamics. To identify the interstitials, different methods use different criteria. All the methods, except the Wigner-Seitz method, assume some input parameters which may lead to an erroneous number of interstitials. We describe an unsupervised clustering algorithm, called Max Space clustering method, to identify interstitials without assuming any input parameters. We show that the algorithm is suited for identifying interstitials at high temperatures when amplitude of atomic oscillations become larger, a fact ignored by other models.

1. Introduction

Safe and successful operation of any nuclear energy system greatly relies on the development of suitable component materials. Changes in the material micro-structure due to high neutron doses will lead to material property changes. This has implications for the material lifetime in reactors. This has led to increased interest in both the experiments and simulations to understand the changes in micro-structure due to radiation damage [1].

Multi-scale methods are necessary to span the length and time scales involved in radiation damage due to neutron irradiation [2, 3, 4]. Neutrons create energetic primary knock-on atoms (PKAs) that can cause localized collision cascades. Interstitials and vacancies formed in the collision cascades diffuse and recombine to either neutralize each other or to form interstitial and vacancy clusters. This diffusive-recombination can occur throughout the operational lifetime of the reactor material, leading to significant changes in the physical and mechanical properties of materials.

In contrast to the time and length scales required for the observable radiation induced mechanical property changes, the primary radiation damage event that initiates these changes is smaller by many orders of magnitude [3]. Molecular dynamics (MD) is an atomistic simulation method that can provide details of atomistic processes generally upto a few nanoseconds [5, 6, 7]. Techniques such as Kinetic Monte Carlo (KMC) [8, 9] or Mean field rate theory [10] are required to bridge the gap between MD and experimentally observable time and length scales. For example, the diffusion of interstitials, thier migration energy and details of thier diffusion can be obtained from MD and used in KMC simulations at the higherscales [8, 9, 11].

Interstitialcy diffusion – wherein an interstitial displaces a lattice atom, thereby making the lattice atom an interstitial – plays a prominent role in the diffusive recombination of Frenkel pairs created during the neutron irradiation. The identity of the diffusing atom changes continuously. Moreover, the interstitial can be part of dumbbell and crowdion configurations [3, 7]. Therefore, identifying the changing interstitial and its trajectory from MD simulations is a complicated task. We have developed a robust and computationally efficient algorithm called Max-Space clustering method to identify the probable interstitials from MD simulations. Then we use graph optimization technique to identify the interstitial from dumbbell and crowdion configurations thereby creating a diffusion trajectory for the interstitial. In Section 2, we describe existing methods to identify interstitials and thier drawbacks. In Section 3, we describe our MD simulations of Cu,Fe and W interstitialcy diffusion. In Section.4, we describe the Max-Space clustering method and the graph optimization technique. We present our results and carry out a comparative study of existing methods of identifying interstitials with our Max-Space Clustering (MSC) method.

2. Identification of interstitials–Existing methods

MD has become a powerful tool to study the atomistic processes. Though numerous methods are present for identification of interstitials, we perform a comparative study of few commonly used methods.

2.1. Existing Methods

In the course of simulation, usually one has to follow any one of the following available criteria to identify the point defects for visualization and analysis: (i) Potential energy, (ii) Atomic Bonding, (iii) Geometric Methods: (a) Wigner-Seitz cell method (b)Lindemann's sphere (LS) method (c) The nearest neighbour sphere (NNS) method (d) The effective sphere (ES) method.

At very low temperatures, perhaps the simplest way of identifying the interstitials is to consider the atoms with a potential energy significantly higher (typically 2eV) than the equilibrium value as interstitials [12, 13]. Though this method is very simple to implement, it can not provide details about the defect structures. When atomic bonding is considered, if an atom does not have correct coordination number of bonds, it is recognized as an interstitial [14].

Using the geometric methods, defect analysis is done by the comparison of atomic positions with respect to a geometric structure such as Wigner-Seitz cell[15], or LS [16, 17, 18], or NNS [19] or ES[3]. These geometric structures are centered on ideal lattice sites. Wigner-Seitz cell is a Voronoy polyhedron centered on an ideal lattice site. The LS method uses a sphere with a radius equal to atomic mean squared displacement at the melting point of the crystal being considered. The NNS method proposed by M. J. Caturla et al. uses a sphere with radius equal to half of the nearest neighbour distance r_{NN}, whereas the ES method proposed by R. E. Stoller employs a sphere with radius equal to 30% of the inter-atomic separation of the crystal being studied [3]. Stoller, in his recent paper, also used a sphere with radius equal to 0.15 times the lattice parameter before ballistic phase and 0.25 times the lattice parameter after the ballistic phase to study the fragmentation of cascades into sub-cascades [20].

In geometric methods, if the geometric structure contains no atoms, the lattice postion is labelled as a vacancy and either if it contains an additional atom, or if an atom is not associated with any geometric structure, then it is labelled as an interstitial. Nowadays, perhaps ES and NNS methods are the most commonly used methods. Hence we compare our results with these in section 5.

2.2. Drawbacks of existing Methods

Methods using potential energy, atomic bonding and LS overestimate number of interstitials even at lower temperatures. They do not provide any details of defect structure [6]. Wigner-Seitz cell method is computationally expensive despite of its advantage that it does not involve any arbitrary cut-off size as an input parameter. Though LS criteria overestimates the number of interstitials even at low temperatures, it could be the motivation behind the widely used NNS and ES methods. Both NNS and ES methods give a rough number of interstitials. Though ES method is the most successful method among the others and computationally faster than Wigner-Seitz cell method, it overestimates the number of interstitials at higher temperatures and also it can not correct the number of interstitials for dumbbell and crowdion configurations.

While using the any one of the above mentioned criteria, results are affected by the thermal vibration of atoms in the simulation cell. To eliminate the thermal vibrations several techniques can be used, [21, 22, 23]. But they are very computer intensive and cannot be employed to each configuration obtained from the simulation.

To overcome the above mentioned difficulties, we present a rigorous unsupervised clustering algorithm, called Max-Space Clustering (MSC) [24, 25]. Apart from the MSC algorithm, we also use graph optimization technique to correct the number of interstitials for dumbbell and crowdion configurations and to get a smooth trajectory. These are described in Section 4.

3. MD Simulations

We use Large-scale Atomic/Molecular Massively Parallel Simulator (LAMMPS) code [26] for the classical MD study of interstitialcy diffusion in Cu, Fe and W single crystals. A cubic simulation box of $10 \times 10 \times 10$ unit cells is used for each element. A single atom is introduced close to the center of the crystal. It causes an abrupt increase in pressure and leads to the deformation of the crystal. To overcome this,energy minimization of the system is performed till the potential energy of the crystal becomes minimum. This is done by iteratively adjusting the atom coordinates. The EAM potential by Foiles et al. [27] was used to simulate Cu. The EAM/alloy potential by Zhou et al. was used to simulate W [28]. For Fe, the concentration dependent EAM potential by Stukowski et al. [29] was used.

An NPT ensemble was used to carry out MD simulations at 0 *bar* pressure with periodic boundary conditions in all the three directions till the system equilibrates. After the equilibration, an NVE ensemble is used to study interstitialcy diffusion for $10ns$. MD simulations were carried out at intervals of 100 K in the temperature range 300K–1100K and 300K–1200K for Cu and Fe respectively and at the intervals of 200 K in the temperature range 300K–2900K for W. The chosen temperature range is so as to have a valid potential well before the melting point. A time step of 1 femto-second was used for both, NPT and NVE ensembles. The atomic co-ordinates were output for every 100 time steps during the NVE simulations to study the interstitial transport. Hitherto, each output for every 100 time-steps is referred as a *frame* and the displacement of an atom from its nearest lattice position is referred to as *offset*.

4. Obtaining the Interstitial Diffusion Trajectory

4.1. Max-Space Clustering -Algorithm

Identification of interstitials in the analysis of defect dynamics is an issue of critical importance [6, 22, 23]. In this section, we describe Max-Space clustering method, which can identify interstitials without assuming any input cut-off values. The only required input is the positions of all atoms in a given time-step. At any given time-step, the interstitial exists as a single interstitial atom (SIA) or as part of a dumbbell or crowdion configuration. It has to be identified in each successive MD frame to produce a smooth trajectory. Dumbbell and crowdion configurations are shown in Fig.1 and Fig.2. In dumbbell configuration, two atoms share one lattice position. In crowdion configuration, three or more atoms share two or more lattice positions such that n atoms share $n-1$ lattice positions. The MSC algorithm helps us to seperate these probable interstitials from the vibrating lattice atoms from low temperatures to the temperatures close to the melting point.

The MSC algorithm to identify the probable interstitials in a given MD time-step is as follows:

- Find out the offsets of each atom in the frame.

Figure 1: Dumbbell configuration of interstitial

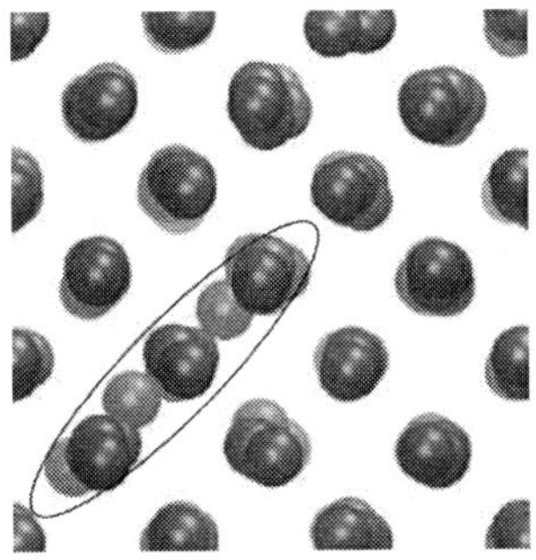

Figure 2: Crowdion configuration of interstitial

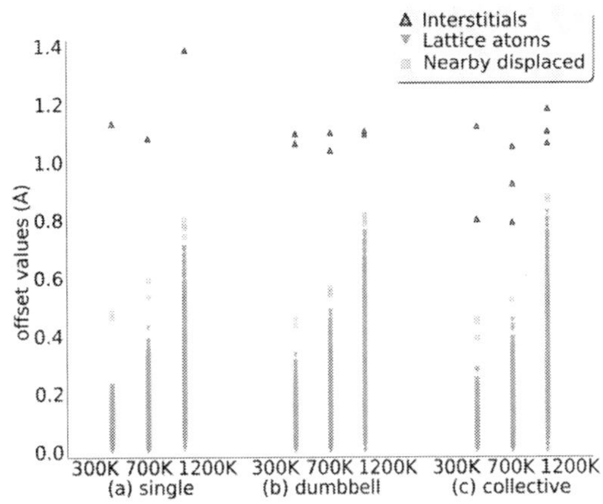

Figure 3: Max-Space clustering of SIA, dumbbell and crowdion configuration offsets and lattice atom offsets–Cu,Fe and W respectively

- Sort the atoms by their offsets in descending order.
- Find the difference in offsets between successive members of the sorted list of atoms.
- Find mid-point of successive offsets with the maximum difference. This is hitherto referred to as "threshold offset".
- Group the atoms above and below the mid point of successive offsets with the maximum difference.

Two groups of atoms are created by the above algorithm as shown in Fig3. One group having one or more number of atoms with large offsets (probable interstitials) and the other group of atoms having smaller offsets (lattice atoms). The second group of atoms with smaller amplitudes also includes another sub-group of atoms with slightly higher amplitudes. This sub-group of atoms are the strongly displaced atoms nearby the interstitials and can be seen in Figs.1 & 2 and they are indicated by squares in Fig.3. The atoms in the group having larger offsets are all probable candidate interstitials.

4.2. Identifying the true interstitial
The graph optimization algorithm to identify the interstitialcy diffusion path is as follows:

- The positions of probable interstitials obtaned from MSC algorithm, in each MD frame, are considered to be nodes of a graph structure [30].
- All the nodes across successive MD frames are connected thereby creating possible diffusion pathways.
- A cost function is constructed for the different diffusion pathways such that it minimizes the distance between interstitials across consecutive frames and it maximizes the offset of interstitial.
- The diffusion path having the lowest value for the cost function is chosen.

This technique is described more in detail in [24]

5. Results and Conclusions

At a given temperature, the average threshold offset of all the MD frames is calculated. A least square fit of the average threshold offset at different temperatures for Cu, Fe and W is carried out. A staight line (MSC) is obtained for Cu, Fe and W as shown in Figs.4–6. Apart from these, two more lines are plotted corresponding to the NNS and ES thresholds in these figures. Notethat the offsets shown in these figures are of single frames. The straight line obtained by each method separates the interstitials from lattice atoms. Atoms of the offsets, appearing above any line are labelled as interstitials, and appearing below that line are labelled as lattice atoms in a given time-step in each method.

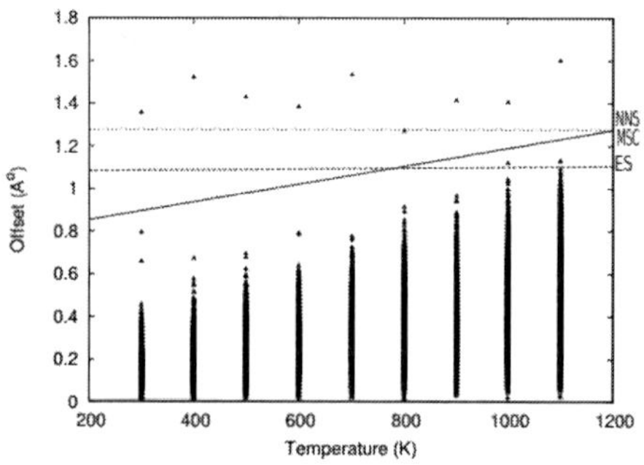

Figure 4: Interstitial selection in SIA configuration – Cu

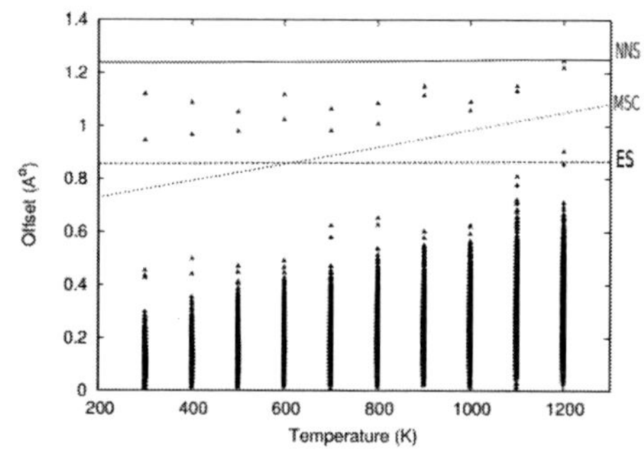

Figure 5: Interstitial selection in dumbbell configuration – Fe

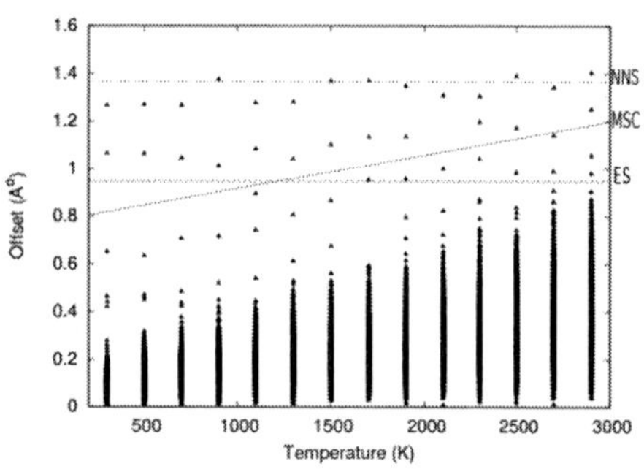

Figure 6: Interstitial selection in crowdion configuration – W

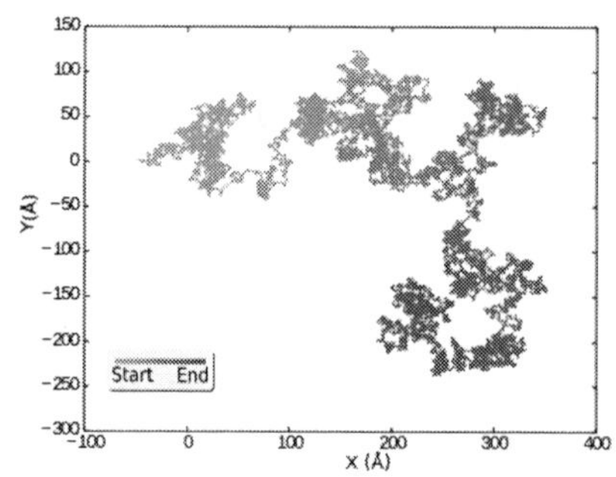

Figure 7: Cu interstitialcy diffusion trajectory at 900K using MSC and graph optimization (colour-online)

Figs.4–6 show that NNS method underestimates the number of interstitials and in some frames, it can not detect any interstitial even though it exists. ES method is ambiguous in choosing the threshold offset such as "15% or 25% or 30% of lattice parameter" [3, 20] and does not provide any justification for the selected threshold offset. In this article we use "30% of lattice parameter" criterion of ES method for comparision. With the other criteria, number of interstitials predicted by the ES method further increases. Note that the ES method is the best amongst the "sphere methods". It however overestimates the number of interstitials as the temperature increases and it can not correct for dumbbell and crowdion configurations.

Though it is not a good idea to have a single threshold offset value for all the frames even at a given temperature, for a simple and most efficient way of finding interstitials, threshold offset separation values at 300K and the slope of the separating lines for Cu, Fe and W are given in table 1. Using these two parameters, threshold offsets at any temperature can be calculated.

Table 1: Threshold offsets

Element	Threshold Offset at 300K(Å)	Slope(Å/K)
Cu	0.89	4.38×10^{-4}
Fe	0.77	2.92×10^{-4}
W	0.82	1.38×10^{-4}

Fig.7 shows the interstitialcy diffusion trajectory of Cu using the MSC and the graph optimization algorithm. These methods are being used to obtain interstitialcy diffusion parameters from MD simulations and are being used in KMC simulations [11]. We conclude by stating that these methods can be used for multiple interstitial diffusion tracking in collision cascades.

Acknowledgments
One of the authors (S Bukkuru) would like to acknowledge a BRNS grant which made this study possible.

References

[1] S.J.Zinkel, J.T. Busby, Mater. Today 12 (2009) 12-19
[2] M. Warrier, M. C. Valsa Kumar, Fusion Science and Technology, (2014) DOI: 10.13182/FST13-657
[3] R.E. Stoller, Primary Radiation Damage Formation, in R.J.M. Konings (Ed.), Comprehensive Nuclear Materials, Vol. 1, Elsevier, 1 (2012) 293-332.
[4] C.S.Becquart, C. Domain, L. Malerba, M. Hou, Nucl. Instrum. and Meth. B 228 (2005) 181-186.
[5] Wei Cai, Ju Li and Sidney Yip, 1.09 - Molecular Dynamics, Vol-1, Comp. Nuc. Mat., Elsevier, (2012) 249-265.
[6] K. Nordlund, M. Ghaly, R.S. Averback, M.J. Caturla, et al., Phys.Rev. B 57 (1998) 7556.
[7] M. Warrier, U. Bhardwaj, H. Hemani, R. Schneider, A. Mutzke, M.C. Valsakumar, Jnl. Nuc. Mat., 467(2015)457-464.
[8] P.R. Monasterio,B.D. Wirth,G.R. Odette, Jnl. Nuc. Mat.,Volume 361, Issues 2–3, (2007), Pages 127–140
[9] C. Domain, C.S. Becquart, B.D.Wirth, Jnl. Nuc. Mat., Volume 335, Issue 1, (2004) Pages 121–145
[10] R. Stoller, S. Golubov, C. Domain, C. Becquart, Jnl. Nucl. Mater. 382 (2008) 77–90.
[11] Warrier M, Utkarsh B, Bukkuru S, Multi-scale Modeling of Radiation Damage: Large Scale Data Analysis, XXVII IUPAP Conf. on Comp. Phys., 2-5 Dec 2015, IIT-Guwahati, India (these proceedings)
[12] M. Ghaly, R.S. Averback, Phys. Rev. Lett. 72 (1994) 364.; T. Diaz de la Rubia, G.H. Gilmer, Phys. Rev. Lett. 74 (1995) 2507.
[13] T. Diaz de la Rubia, G.H. Gilmer, Phys. Rev. Lett. 74 (1995) 2507.
[14] M. Tang, L. Colombo, J. Zhu, T. Diaz de la Rubia, Phys. Rev. B 55 (1997)14279.
[15] J.B. Gibson, A.N. Goland, M. Milgram, G.H. Vineyard, Phys. Rev. 120 (1960) 1229.
[16] F. A. Lindemann, Physik. Z. 1910, 11, 609.
[17] A. Magnus, F. A. Lindemann, Zeitschr. fu r Elektrochemie, 8 (1910) 269.
[18] V. A. Borodin, Nuc. Inst. and Methods in Phys. Res. B 282 (2012) 33–37
[19] M. J. Caturla, T. Diaz de la Rubia, L.A. Marqués, G.H. Gilmer, Phys. Rev. B 54 (1996)16683.
[20] E Antoshchenkova, L. Luneville, D. Simeone, R.E. Stoller, M. Hayoun, Jnl. Nuc. Mat., 458 (2015) 168–175.
[21] L.A. Marques, L. Pelaz, P. Castrillo, J. Barbolla, Phys. Rev. B 71 (2005) 085204
[22] Ivan Santos et al., Materials Science and Engineering B 124–125 (2005) 372–375.
[23] Pedro López, Iván Santos, María Aboy, Luis A. Marqués and Lourdes Pelaz, Nuclear Instruments and Methods in Physics Research B 352 (2015) 156–159
[24] Bhardwaj U, Bukkuru S, Warrier M 2016 J. Comp. Phys. 305 p 263
[25] U. Bhardwaj, M. Tech. Thesis, Homi Bhabha National Institute (2015)
[26] S.J. Plimpton, J. Comp. Phys., 117 (1995) 1–19.
[27] S. M. Foiles, M. I. Baskes, and M. S. Daw, Phys. Rev. B, 33(12) (2007) 7983–7991.
[28] X.W. Zhou, H.N.G. Wadley, R.A. Johnson, D.J. Larson, and et al, Acta Mater., 49(19) (2001) 4005–4015.
[29] A. Stukowski, B. Sadigh, P. Erhart, and A. Caro. Modeling Simul. Mater. Sci. Eng. 89(34)(2009) 3371–3391.
[30] Thomas H. Cormen; Charles E. Leiserson, Ronald L. Rivest, Clifford Stein (2001),Introduction to Algorithms (2nd ed.). MIT Press and McGrawHill.

A study on modeling and simulation of Multiple-Gate MOSFETs

Mandeep Singh, Gaurav Kumar, Sushanta Bordoloi and Gaurav Trivedi

Dept. of EEE, Indian Institute of Technology Guwahati, Assam-781039, India

E-mail: s.mandeep@iitg.ernet.in

Abstract. Endless scaling of planar MOSFET over the past four decades has delivered proliferating transistor density and performance to integrated circuits (ICs) at the cost of increase in short channel effects (SCEs). As a result of narrow channel lengths in deeply scaled MOSFETs, off-state leakage current happens to increase the power requirement of device by forcing drain potential to lose its leverage over the electrostatics of channel. Multigate devices (Double Gate FET, FinFET) promise better immunity to SCEs by concealing the issues caused by scaling of planar technology, and also exhibits better scalability with increased level of integration. In this paper, a study on modeling and simulation of multiple-gate MOSFETs is presented. This paper describes the development in semiconductor technology from planar to non-planar devices, benefits of multigate MOSFETs over previous technologies. Effects of varying different parameters (such as Fin thickness, Fin height, use of different gate material etc.) in multigate (specially FinFET) devices is also described. Improvement in device physics models along with technology, and importance of computational physics for efficient modeling of nanoscale devices is presented.

1. Introduction

As the semiconductor technology moves towards nanoscale length, complexity to design and fabricate devices has increased to achieve desired accuracy and results. Endless scaling of bulk CMOS resulted in SCEs, leakage current, power consumption and threshold voltage roll-off. Several alternatives to CMOS scaling were used to move forward with scaling, are as follows: i)use of strained silicon to increase mobility whereby enhancing drive current, ii) high-K materials are being used to achieve same level of results without compromising drive current, iii) Silicon on Insulator (SOI) emerged as one of the best alternative prior to multigate technology to reduce power consumption, leakage, with better isolation, high reliability and speed. Multigate transistors emerged as the new hope for semiconductor technology, extending development beyond nanometer length with better control over channel, eliminating SCEs and threshold voltage roll-off. Among multigate devices FinFET shows promising results with variation in various parameters such as Fin thickness, Fin height and use of different gate materials.

Even with recent advancement in technology, it is hard to model and simulate nanoscale devices because of the level of physics involved at this structural complexity. Device modeling has moved from classical to quantum models in order to incorporate physical phenomenon involved in short channel devices. Complexity in device modeling and simulation further increases as we include quantum models for nano length devices. Parallel processing of simulation environment with graphical processing units and other accelerators have shown improvement in accuracy and speed for device simulation.

This paper is organized as follows. In Section 2, shift from planar to non-planar technology, multigate technology and its benefits over previous technology is described along with variations in different parameters of FinFET device and models used for modeling of multigate devices are described. In Section 3, improvement in device physics models along with technology and importance of computational physics for better study of nanoscale devices is described. Section IV concludes the paper along with discussion of the work presented in this paper.

2. Beyond Planar CMOS Technology

Planar CMOS scaling was limited by SCEs, leakage current, power consumption and threshold voltage roll-off. Several advanced techniques are discussed here to reduce the drawbacks of CMOS scaling.

2.1. *Strained Si*

As the device size shrinks, mobility decreases, in order to enhance mobility silicon is strained, resulting in higher drive current (I_{ON}) under the same conditions i.e. with lower supply voltage and thick gate oxide. Trade-off can be achieved among current drive, short channel effects and power consumption. In strained silicon, atoms are stretched beyond their atomic distance as shown in the figure 1, which is achieved by developing a thin layer of Si epitaxially on material with higher lattice constant. Strained Si have significant disadvantage in the form of self-heating effects due to lower thermal conductivity and dislocation effects in epitaxial layer [1]. Self-heating effects can be overcome by reducing the thickness of epitaxial layer in the device structure. The remarkable advantage of strained si is improved mobility even at high effective fields, resulting in approximate 70 % increment in mobility with 30 % of speed improvement [2].

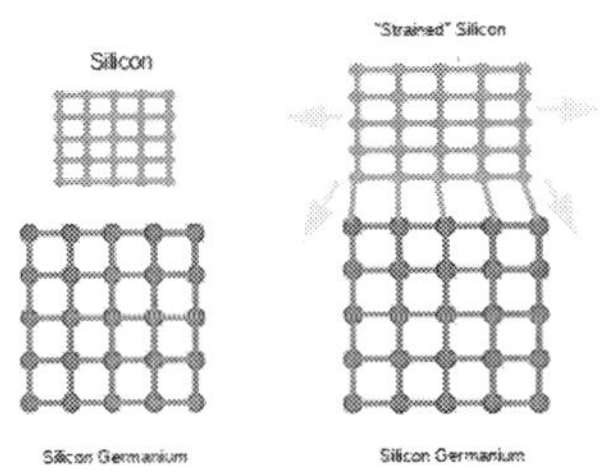

Figure 1. Strained Silicon.

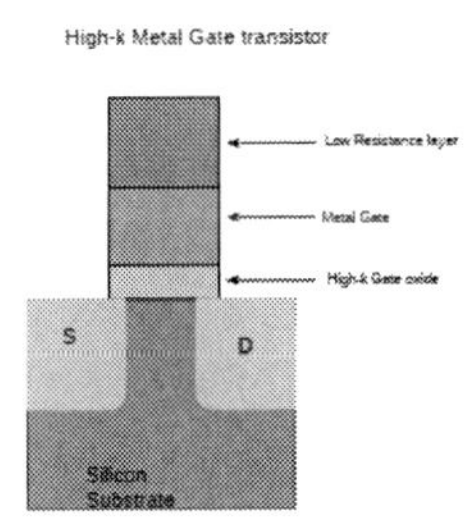

Figure 2. High K Metal Gate transistor.

2.2. *High K material*

With constant scaling of device size into nanoscale technology, scaled gate oxide results in exponential increase of gate leakage current. With the number of transistors on a single chip growing exponentially, power dissipation becomes a big problem. To reduce the gate leakage current in ultra-small devices without adjusting the ON state current of MOS device, reduced electrical oxide thickness (EOT) of gate dielectric are required. High K material in case of fully depleted silicon on insulator (FD-SOI) or thin film (e.g. FinFET) has impoved short channel effects and thus loosens up the requirements for EOT scaling. Direct interface between high-K dielectric and poly-silicon gate has some limitation such as photon scattering which degrades the carrier mobility, thermal instability and a direct shoot up in threshold voltage value. To eliminate these limitations, metal gate electrode is used with the high-K gate insulator as shown in figure 2.

2.3. Silicon on Insulator (SOI)

SOI structure is similar to the conventional bulk structure except the presence of buried layer which is grown by the oxygen ions in the Si as shown in figure 3, which provides full dielectric isolation to active Si layer from the main substrate with the help of buried oxide layer, that allows the use of relatively high resistivity main substrate, causing a significant reduction in parasitic [3] and leakage current in substrate. Contrary to bulk Si transistor, SOI technology has become attractive way for the fabrication of advanced CMOS ICs because of low supply voltage, minimum power consumption features [4], higher drive current, excellent device isolated structure, high reliability and high speed due to elimination of vertical as well as sidewall capacitances. SOI CMOS transistors suffers from self heating effect which appears only when the transistor is in conduction mode. But this effect certainly will not eliminate the global usage of SOI for CMOS ICs.

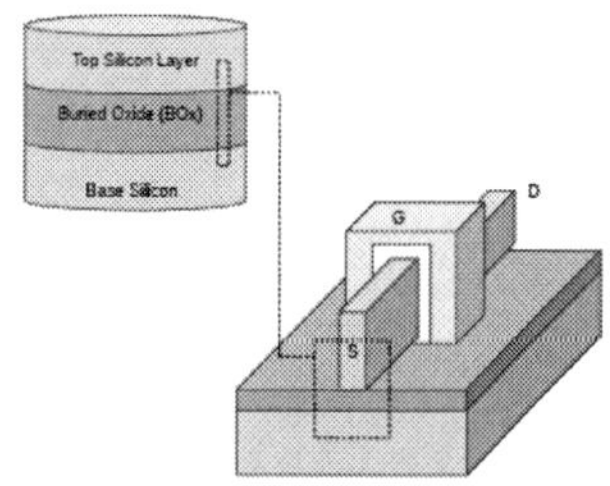

Figure 3. Silicon on insulator.

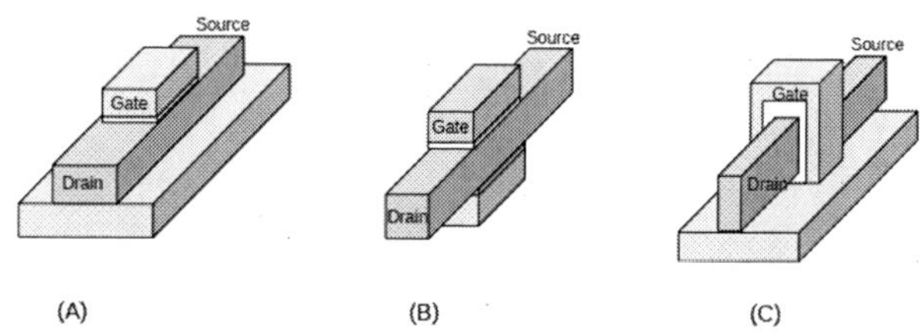

Figure 4. (A) Bulk MOSFET (B) Double-gate MOSFET (C.) FinFET (Trigate MOSFET)

2.4. Multigate transistors

With so many alternative after the bulk MOS transistor, a more promising device structure or alternative was required to extend the growing semiconductor technology to overcome short channel effects such as drain induced barrier lowering (DIBL), threshold voltage roll off and to get the precise control over channel current. Multigate transistor emerged as the promising candidate for enhancing the performance and scaling properties of device, among multigate transistors dual-gate MOSFET (DG-MOSFET) as shown in figure 4 with aligned planar gate structure provides improved channel control but its fabrication process is more challenging. With ease of fabrication, FinFET and Trigate FETs have emerged as the dominant structure. FinFET has an advantage to be fabricated on bulk as well as on SOI, obtained by etching thin silicon strip called fins with gate electrodes occupying the sidewalls of silicon strips. SOI FinFETs have better isolation as they are isolated by very shallow trench isolation than bulk FinFETs which have common substrate for isolation.

Thickness and height of Fin are the two important parameter which are used to control short channel effects [5]. Thinner Fin has much better control over short channel effects [6] than that of thick Fin, as in later case channel barrier is lowered by drain electric field due to the reduction in source/Fin and drain/Fin capacitances. Fin thickness plays an important role to overpower short channel effects, as DIBL is increased with the increment in Fin thickness. Drive current capability of FinFET is mainly controlled by Fin height, as noticed in[7] increment in Fin height degrades off-state leakage current. In case of tri-gate MOSFET, use of different gate material shows significantly improved performance[8] in terms of surface potential, electric field and carrier velocity distribution over double or single material gate.

Modelling of multigate device started with double gate MOSFET. Two models are discussed here which have the major contribution among multigate device modelling.

- **Taur Models :** Taur [9] solved the Poisson equation for the case of intrinsic doping concentration as follows, neglecting fixed charges:

$$\frac{d^2\varphi}{dx^2} = \frac{qn_i}{\epsilon_s} e^{\frac{\varphi - V}{\phi T}} \tag{1}$$

where :

φ changes across the thickness of the film and V changes along the channel.
q is elctrostatic charge, n_i is intrinsic carrier density, T is temperature in Kelvin and ϵ_s is dielectric permitivity.

After two integration, the surface potential φ_s along the channel is obtained as function of a constant intermediate calculation parameter β :

$$\varphi_s(x) = V - 2\varphi_t \ln\left[\frac{t_s}{2\beta L_{Di}} \cos\left(\frac{2\beta}{t_s} x\right)\right] \tag{2}$$

where t_s is the silicon layer thickness, L_{Di} is the intrinsic Debye length and φ_T is the thermal voltage kT/q.

- **BSIM-MG Model :** In 2012 the BSIM-MG model [10] was adopted as industry standard. Two cases were considered mainly to formaulate models: symmetric or common gate (CMG) and asymmetric gates (IMG). This model already considers the silicon layer doping concentration and short-channel effects. In the case of symmetric gates and doped silicon layer, Poisson's equation is solved using the perturbation method approach.
The general equation has two terms: the first term attributed to inversion charges only and second term attributed to body doping Na, where φ_f is the Fermi potential.

$$\frac{d^2\varphi}{dx^2} = \frac{qn_i}{\epsilon_s} e^{\frac{\varphi - V}{\phi T}} + \frac{qN_a}{\epsilon_s} \tag{3}$$

This solution has two terms, φ_1 considering only the first term, and φ_1 solving the second term as a perturbation. After defining a function of β , the sequence of calculation is the following:

$$f(\beta) = \ln(\beta) - \ln(\cos(\beta)) - \frac{V_{gfb} - V}{2\varphi_T} - \ln\left[\frac{t_s}{2}\sqrt{\frac{qn_i{}^2}{2\epsilon_s \varphi_T N_a}} + M(\beta) = 0\right] \tag{4}$$

where :

$$M(\beta) = \frac{2C_s}{C_{ox}} \sqrt{\beta^2 \left[\frac{e^{\frac{\varphi_p}{\varphi_T}}}{\cos\beta^2} - 1\right] + \frac{\varphi_p}{\varphi_T{}^2}(\varphi_p - 2\varphi_T \ln[\cos\beta])} \tag{5}$$

and

$$V_{gfb} = V_{gs} - V_{fb} - V \tag{6}$$

Solution of equation 4 can be obtained by: 1) Newton-Rhaphson method; 2) table look approach; 3) analytical approximation with good initial solution β_0 and corrections of 3rd order h.

3. Development in device Modeling

As the device size shrinks, physics involved in device and its understanding became very important and so does the physical phenomenon used to describe device characteristic. Earlier for long channel devices, Drift Diffusion model was used to define the device physics [11], since it could not deal with hot carrier effects and it was good for gate length greater than 500nm. To overcome the hot carrier effects for shorter channel devices (upto 100nm of gate length) Hydrodynamic models [12] were used to describe the device physics. Hydrodynamic models have become a practice in industry for describing device physics with certain variation in different parameters (e.g. impact ionization coefficient, mobility, etc.). However for ultra-small devices these models do not have exact predictive power, for that some of the approximations [13] need to be relaxed and Quantum transport models should be considered with inclusion of quantum effects, small-geometry effects and SCEs to overcome quantum mechanical restraints along channel, as shown in the hierarchical map [14] of transport models.

Approximate ↑ / Exact ↓	Model	Improvements	Easy, Fast ↑ / Difficult ↓
Approximate	**Model**	**Improvements**	Easy, Fast
Semiclassical approaches	Compact models	Appropriate for circuit design	
Semiclassical approaches	Drift-Diffusion equations	Good for device down to 0.5 μm	
Semiclassical approaches	Hydrodynamics equations	Velocity overshoot effect can be treated properly	
Semiclassical approaches	Boltzmann transport equation Monte Carlo/CA methods	Accurate up to classical limits	
Quantum approaches	Quantum Hydrodynamics	Keep all classical hydrodynamics features + quantum corrections	
Quantum approaches	Quantum Monte Carlo/CA methods	Keep all classical features + quantum corrections	
Quantum approaches	Quantum-Kinetic equations (Liouville, Wigner-Boltzmann)	Accurate up to single particle description	
Quantum approaches	Green's function method	Includes correlation in both space and time domain	
Exact	Direct solution of the n-body Schrodinger equation	Can be solved only for small number of particles	Difficult

Figure 5. Hierarchy in Device Modeling[14].

Even with recent advancement in technology, the present technology [15] is unable to tackle new challenges in scaling of semiconductor devices from classical down to quantum scale devices as shown in figure 5. Classical models based on drift-diffusion model equations are certainly inadequate for device length below 200nm, even typical Hydrodynamic model do not provide exact description of device physics as they neglect some important contribution in the channel region. Computational changes results in limited performance of emerging device technology as self-consistent solution of coupled transport-field equation is required and also both the transport and the Poisson's equations are needed to be solved over 3D domain, requiring highly accurate algorithms with high-end computational platform. It is very challenging to both computationally and from device physics perspective to solve for appropriate level of accuracy required to describe and calculate the device characteristic.

4. Conclusion

A study of progress in semiconductor technology is presented with improvements among various technologies. Multigate transistors have better control over channel thus gives better performance to previous technologies. Multigate devices emerged to extend semiconductor technology beyond nano meter of channel length. With variation in certain parameters of multigate device better results were found. Device modeling and simulation plays important role in scaling. To develope a new device and test its functionality, exact modeling of device is required to understand the physics involved. To get the faster and accurate results on simulation environment, parallel processing of simulation is required on graphical processing units. Several method to fasten the solving algorithms and to get accurate results are used, such as use of finite element method (FEM) for complex geometries over finite difference method (FDM). Study in this paper provides a huge scope in development of new tools for device modeling and simulation.

References

[1] J Welser, JL Hoyt, S-I Takagi, and JF Gibbons. Strain dependence of the performance enhancement in strained-si n-mosfets. In *Electron Devices Meeting, 1994. IEDM'94. Technical Digest., International*, pages 373–376. IEEE, 1994.

[2] Kern Rim, S Koester, M Hargrove, J Chu, PM Mooney, J Ott, T Kanarsky, P Ronsheim, M Ieong, A Grill, et al. Strained si nmosfets for high performance cmos technology. In *VLSI Technology, 2001. Digest of Technical Papers. 2001 Symposium on*, pages 59–60. IEEE, 2001.

[3] Ying-Che Tseng, W Margaret Huang, Clifford Hwang, and Jason Woo. Ac floating body effects in partially depleted floating body soi nmos operated at elevated temperature: an analog circuit prospective. *Electron Device Letters, IEEE*, 21(10):494–496, 2000.

[4] GK Celler and Sorin Cristoloveanu. Frontiers of silicon-on-insulator. *Journal of Applied Physics*, 93(9):4955–4978, 2003.

[5] Gaurav Saini and Ashwani K Rana. Physical scaling limits of finfet structure: A simulation study. *International Journal of VLSI design & Communication Systems (VLSICS)*, 2(1), 2011.

[6] VS Basker, T Standaert, H Kawasaki, C-C Yeh, K Maitra, T Yamashita, J Faltermeier, H Adhikari, H Jagannathan, J Wang, et al. A 0.063 μm 2 finfet sram cell demonstration with conventional lithography using a novel integration scheme with aggressively scaled fin and gate pitch. In *VLSI Technology (VLSIT), 2010 Symposium on*, pages 19–20. IEEE, 2010.

[7] A Prathima, Kiran Bailey, and KS Gurumurthy. Impact of device parameters of triple gate soi-finfet on the performance of cmos inverter at 22nm. *International Journal of VLSI Design & Communication Systems*, 3(5):79, 2012.

[8] C. Li, Y. Zhuang, and Li Zhang. Simulation study on finfet with tri-material gate. In *Electron Devices and Solid State Circuit (EDSSC), 2012 IEEE International Conference on*, pages 1–3, Dec 2012.

[9] Huaxin Lu and Yuan Taur. An analytic potential model for symmetric and asymmetric dg mosfets. *Electron Devices, IEEE Transactions on*, 53(5):1161–1168, 2006.

[10] Mohan V Dunga, Chung-Hsun Lin, Xuemei Xi, Darsen D Lu, Ali M Niknejad, and Chenming Hu. Modeling advanced fet technology in a compact model. *Electron Devices, IEEE Transactions on*, 53(9):1971–1978, 2006.

[11] Michael S Shur. Low ballistic mobility in submicron hemts. *Electron Device Letters, IEEE*, 23(9):511–513, 2002.

[12] Massimo V Fischetti and Steven E Laux. Monte carlo simulation of transport in technologically significant semiconductors of the diamond and zinc-blende structures. ii. submicrometer mosfet's. *Electron Devices, IEEE Transactions on*, 38(3):650–660, 1991.

[13] G Baccarani and MR Wordeman. An investigation of steady-state velocity overshoot in silicon. *Solid-State Electronics*, 28(4):407–416, 1985.

[14] Dragica Vasileska and Stephen M Goodnick. Computational electronics. *Synthesis Lectures on Computational Electromagnetics*, 1(1):1–216, 2005.

[15] Eric Pop. Energy dissipation and transport in nanoscale devices. *Nano Research*, 3(3):147–169, 2010.

Extensible, Reusable, and Reproducible Computing: A Case Study of PySPH

Prabhu Ramachandran[1]

[1] Department of Aerospace Engineering, IIT Bombay, Powai, Mumbai, India 400076

E-mail: prabhu@aero.iitb.ac.in

Abstract.
In this work, the Smoothed Particle Hydrodynamics (SPH) technique is considered as an example of a typical computational research area. PySPH is an open source framework for SPH computations. PySPH is designed to be easy to use. The framework allows a user to implement an entire simulation in pure Python. It is designed to make it easy for scientists to reuse their code and extend the work of others. These important features allow PySPH to facilitate reproducible computational research. Based on the experience with PySPH, general recommendations are suggested for other computational researchers.

Keywords: SPH, Reproducible Computing, Scientific Computing

1. Introduction

Reproducibility and the ability to use and extend the research of others has always been a cornerstone of scientific research. However, even a cursory sampling of most modern publications in computational research will show that this does not hold true in practice. Research papers are often hard to extend, reuse, or reproduce. Results may not be reproducible either because some parameters are mistyped or not specified at all. The authors cannot be blamed entirely as it is easy to overlook some details of a particular computation. There is little interest or incentive in making computational results reproducible. There are several strategies to overcome these but they typically require a fair bit of effort and some knowledge of software engineering. Donoho et al. [1] provide an overview of reproducible computing along with several potential benefits and objections to its adoption by the scientific computing community.

Many researchers are not experts in software engineering, yet high performance code is also desirable. It would be ideal to have an environment where the user of a tool could write high-level code that is numerically well-behaved, mathematically consistent, and representative of the physics without worrying about high-performance or requiring expertise in software engineering principles.

In this paper we consider the Smoothed Particle Hydrodynamics (SPH) technique as an example of a typical computational research area. The SPH method was first proposed by Gingold and Monaghan [2] and Lucy [3] to simulate astrophysical problems. The method employs a Lagrangian formulation for the continuum equations and discretizes field properties into particles carrying them. This results in a Lagrangian particle method that does not require a mesh. This makes the method very general purpose. Since its initial formulation in astrophysics, it has been adapted to simulate a wide variety of problems including incompressible free-surface fluid flow [4] and solid mechanics [5]. An excellent review of SPH and its various applications is provided by Monaghan [6].

While there are some open source implementations of SPH available, they are not very easy to extend and are often tailor-made for one particular domain. In the astrophysics community GADGET-2 [7]

is one such open source package. However, this is designed only for astrophysical simulations and is implemented in C. While C is a popular language, it is more difficult than a modern programming language like Python. Modifying the source code in order to extend it, requires a reasonable amount of experience. This makes extending such codes a little more difficult. SPHysics is an open source package but is designed for incompressible flows. JOSEPHINE [8] is another recent package but implemented in Fortran 90. These packages are not very easy to extend and are tailor-made for specific domains.

Over the last several years, my group has been working on an open source framework for SPH called PySPH (`http://pysph.bitbucket.org`). PySPH is implemented in Python which is a powerful, easy-to-learn programming language. PySPH automatically generates high performance code from high-level Python code. Good software development practices are used for testing the code on a regular basis on multiple platforms. Finally, simple tools are used to automate computational tasks. This can be used to facilitate easy reproduction of results by others. This requires additional effort. These practices make it easy to extend, reuse, and reproduce any results produced from the framework without sacrificing performance. Some of the general techniques and approaches that have been used to achieve this are discussed so as to make this useful to a general audience.

The following section provides a very brief introduction to the SPH method. Thereafter, some numerical implementation details are discussed. This is followed by an overview of PySPH with a discussion on how it has been made easy to use, reuse, and extend. The last section provides a set of general recommendations to other computational scientists.

2. Smoothed Particle Hydrodynamics

2.1. Mathematical details

Consider the identity,

$$f(x) = \int_{-\infty}^{\infty} \delta(x - x')f(x')dx', \tag{1}$$

where $\delta(x)$ is the Dirac delta distribution. The SPH scheme approximates the Dirac delta using a kernel $W(x)$ and this results in,

$$f(x) \approx \int_{-\infty}^{\infty} W(x - x')f(x')dx'. \tag{2}$$

The kernel, W, is typically, compact, smooth, and symmetric. The kernel is parametrized on the parameter h which determines its spatial extent. In an SPH scheme, one represents the function f using a collection of N particles that together approximate the function. Each particle located at x_j has a functional variation determined by $W(x - x_j)$ and a weight, $f(x_j)$. The function can therefore be discretized as,

$$f(x) \approx \langle f(x) \rangle = \sum_{j=1}^{N} W_h(x - x_j)f(x_j)\Delta x_j \tag{3}$$

Clearly, since W is compact, the summation does not need to include all particles but only particles in the neighborhood of x. This is written as,

$$f(x) \approx \langle f(x) \rangle = \sum_{j \in \mathcal{N}(x)} W_h(x - x_j)f(x_j)\Delta x_j, \tag{4}$$

where $\mathcal{N}(x)$ is the set of neighboring particles at x.

Given that the particles may move, Δx_j is not constant. An elegant way to estimate it, is to use the particle density. If a particle has mass, m_j and density ρ_j, then one can write the SPH approximation of the function as,

$$\langle f(x) \rangle = \sum_{j \in \mathcal{N}(x)} \frac{m_j}{\rho_j} W_h(x - x_j)f(x_j). \tag{5}$$

Given that the functions can be represented with a collection of particles, one can easily define derivatives by using the derivative of the smooth kernel. For example, in one-dimension, one could write the derivative of the function at particle i as,

$$\frac{\partial f_i}{\partial x_i} = \sum_{j \in \mathcal{N}(x)} \frac{m_j}{\rho_j}(f_j - f_i)\frac{\partial W_{ij}}{\partial x_i}. \tag{6}$$

When solving a partial differential equation using SPH, a Lagrangian formulation is typically used. For example, for the conservation of mass, the governing equation is,

$$\frac{d\rho}{dt} = -\rho \nabla \cdot \vec{v}. \tag{7}$$

With the SPH method, this equation would be written as,

$$\frac{d\rho_i}{dt} = -\rho_i \sum_{j \in \mathcal{N}(x)} \frac{m_j}{\rho_j}\vec{v}_{ji} \cdot \nabla_i W_{ij}. \tag{8}$$

Using a similar approach, the governing differential equations for a variety of problems may be solved. The partial differential equations are thus converted to ordinary differential equations that may be integrated using a standard integrator. More details are available from the review by Monaghan [6].

2.2. *Numerical details*

Given the governing differential equations, one may discretize them as discussed in the previous section. This results in a set of coupled ordinary differential equations similar to that in equation (8). Given a particular differential equation that is suitably discretized using an SPH scheme, the typical steps involved in an SPH computation are as follows:

- Discretize the field variables into particles carrying the properties of the flow.
- For each differential equation, compute the right-hand sides (for example see equation (8)) as follows:
  - For each particle i, find the neighbors of the particle.
  - Compute the acceleration from each of the neighboring particles, j, on the particle, i.
- Integrate the differential equations.

From the above it can be seen that there are a few key aspects in an SPH computation:

(i) Find nearest neighbors of a given particle.
(ii) Find the interactions between individual particles.
(iii) Integrate the differential equations.

All SPH implementations allow researchers to specify these. The differences between SPH schemes mostly lies in how the interactions between particles is specified.

3. PySPH

As discussed in the previous section, it is important to be able to easily specify the interactions between particles and also specify how the equations of motion are to be integrated. In the following sections, some details of the PySPH implementation are discussed.

The PySPH framework allows a user to specify the interactions between particles and the integration schemes, using Python. Python [9] is a powerful and easy to use programming language. It is increasingly being used as the first introductory programming language in the top US universities [10]. It is also well known for its increasing popularity with scientists [11].

The use of Python makes it very easy to define the SPH interactions. However, Python is not designed for high-performance computing and one must use a low-level language for this. PySPH generates this code automatically from the high level Python code. PySPH also makes use of OpenMP and MPI to parallelize the code automatically. The user does not need to write any code to use the parallelization. Thus, PySPH allows a user to specify the interactions easily without sacrificing performance.

The goal of PySPH is to make it very easy for anyone to use and extend it to simulate a wide variety of SPH problems. This must be done without sacrificing correctness or performance. In order to do this, good software engineering practices must be employed. The next sections focus on a few key aspects: ease of use, accuracy and testing, reusability, and reproducibility.

3.1. Ease of use

There are several things that make PySPH easy to use.

- The use of Python makes it very easy to create new simulations and extend the existing framework.
- It is distributed under a liberal BSD license which allows it to be reused by anyone without restriction.
- The sources are version controlled using Git which allows others to easily use and contribute to the development of PySPH. The sources are currently available from `http://pysph.bitbucket.org`.
- It is cross-platform and runs on Linux, OS X, and Windows.
- The documentation on how to install and use PySPH is available online at `http://pysph.readthedocs.org`.
- When using PySPH to solve specific problems, once a user writes the Python code to solve the problem, no changes need to be made to use multiple CPU cores or multiple computers.
- About 30 useful examples are provided with PySPH. These examples can be easily run and modified by users.
- A built-in 3D viewer using Mayavi [12] for the solutions generated by PySPH is also provided. Fig. 1 provides a screenshot of the viewer.

In order to solve a typical problem with PySPH a user simply has to implement the following:

(i) Generate the initial particles with suitable properties.
(ii) Specify the inter-particle interactions, which are called "Equations" in PySPH.
(iii) Specify the integrator.
(iv) Perform any post-processing to produce results.

PySPH already implements several different SPH schemes for which the particle interaction equations are already defined. These simply need to be specified. Equations may be easily defined by the user if they are not already implemented. As a simple example, consider the continuity equation discretized in equation (8). In PySPH, one could write this equation in pure Python as follows,

```
class ContinuityEquation(Equation):
    def initialize(self, d_idx, d_arho):
        d_arho[d_idx] = 0.0

    def loop(self, d_idx, d_arho, s_idx, s_m, VIJ, DWIJ):
        vijdotdwij = DWIJ[0]*VIJ[0] + DWIJ[1]*VIJ[1] + DWIJ[2]*VIJ[2]
        d_arho[d_idx] += s_m[s_idx]*vijdotdwij
```

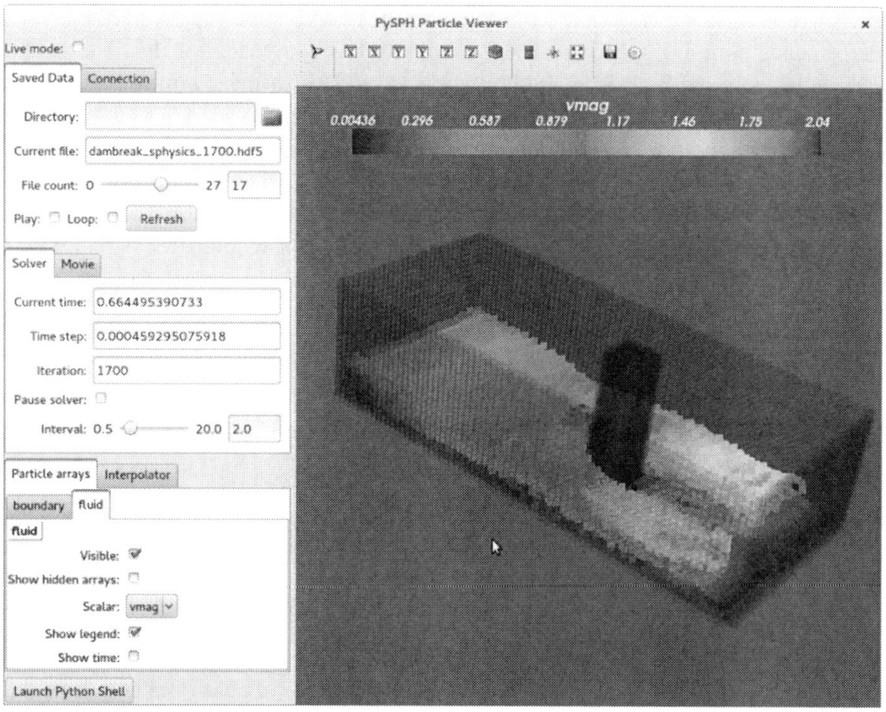

Figure 1. PySPH viewer showing a simulation of the dam-break past an obstacle.

Note that `DWIJ` is the gradient of the kernel and `VIJ = v_i - v_j`. These are automatically provided by PySPH.

Once the user has written the code to solve a problem, PySPH internally generates high-performance parallel code. This is then executed. The high-performance code is generated by the use of Cython which generates C++ code. This is automatically compiled and called from Python.

PySPH is much easier to use than a C/C++/F90 package since it does not require the user to explicitly compile anything. Users can simply run their Python scripts and the framework takes care of the rest. Extending an example is equally simple and a minimal amount of code needs to be written. We find that even undergraduate students with little experience are able to use the framework.

The approach of generating code makes it easy for users to perform SPH simulations. It also allows the PySPH developers to work on other high-performance backends transparent to the users. A GPU implementation is currently under development.

3.2. Accuracy and Testing

As an open source package that is released often, with a large number of features and an emphasis on ease of use, it is important that PySPH be free of errors and works correctly on multiple platforms. The project makes use of unit and integration tests to ensure this. There are about 250+ tests written for the project. These tests include unit tests that test the functionality of individual pieces of code as well as the entire framework. For example, one of the tests runs every example that ships with PySPH to ensure that the code successfully runs and the high-performance code is generated without error. Another test runs an example in serial and parallel, and ensures that the results are the same. The existence of these tests is extremely convenient as they immediately let the developers know if something is incorrect.

PySPH also makes use of continuous integration. This is a common software development practice used in many software projects. Every time the source code is changed and pushed to the PySPH source code repository, the entire test suite runs automatically on multiple platforms. The free services provided by `http://shippable.com` and `http://codeship.com` are used to run the tests on Linux and `http://appveyor.com` is used to run the tests on Windows. The tests are run on both Python 2.7.x and Python 3.4.x. Test result badges are displayed on the main page of the PySPH website. This makes it very easy for both the developers and users to see the current status of the software.

The combination of version control, online hosting, and continuous integration is very powerful, and allows us to develop reliable software relatively easily.

3.3. Reusability, and Reproducibility

As discussed above, by virtue of being written in Python and being freely available, PySPH enables reuse. Once a new set of equations is implemented in PySPH, it may be used by anyone.

In any area of numerical computation, there are a standard set of benchmark problems that are solved each time a new scheme is implemented. In PySPH, many standard benchmarks are implemented in such a way that a writer of a new scheme can easily reuse all the code in the existing example. For instance, if one considers the standard Lid-Driven-Cavity benchmark problem that is routinely solved (see [13] and [14]), PySPH provides a standard implementation and also includes all the post-processing to compare the results with well-established results. If a new SPH scheme is created and one wishes to compare the results with those of other implementations, one simply needs to subclass the standard example and change just the equations that are used. One does not need to re-write the initial particle distribution or the post-processing. This makes it extremely easy to compare the different schemes with minimum effort.

As can be seen, PySPH is not merely a package which is implemented in an easy to use programming language. It is also developed with the users and developers in mind, and strives to make it easier to reuse as much as possible. This is important as it allows anyone to verify the computational results produced. Furthermore, any improvements to the PySPH examples immediately benefit the entire community. This readily facilitates the comparison of schemes.

The reusability of PySPH naturally makes it easy to perform reproducible research in the area of SPH. By using PySPH, it is easy for other researchers to build on previous work. It is easier to develop new schemes as users can focus on the new scheme without having to implement the benchmark problems or worrying about performance of the scheme.

In any publication there are several computations performed. A single plot may involve tens or hundreds of simulations. In order to reproduce such results, it is imperative to make it easy to for any researcher to re-run the simulations and generate the plots. A simple way to do this is to automate the generation of the plots in a publication. The use of Python and the design of PySPH make this easy to do. For a paper that is currently under preparation, we use all the above features of PySPH. In addition, we use other libraries in Python that facilitate automation. With this we are able to automate the generation of every figure in our publication. For example, we could run the following command:

```
$ python automate.py
```

inside our publication directory and this will run simulations for a few days and produce every single plot that is used in the paper. The details of this will be discussed in a future publication.

The features discussed above are individually not very significant, however, together they facilitate reproducible research in SPH, which has not been done before. It is my belief that the general principles that are behind the PySPH project may be applied to any software developed for computational science.

4. Concluding remarks

In this paper, the SPH method has been used as a prototypical computational scheme. The PySPH framework has been discussed with a focus on its ease of use, software engineering, and reuse. It has also been mentioned that with additional effort to automate results in publications, one can easily perform reproducible computational research.

Based on the lessons learned in the development of PySPH, I have the following general recommendations to others developing software for computational science:

- Make research codes and software open source with a convenient license.
- Use a modern version control system and host the software online to facilitate easy collaboration and contribution.
- Make every effort to make the package easy to use and install.
- Generate code for high-performance.
- Test the code regularly and use continuous integration services on multiple platforms.
- The examples are just as important as the source and it is a good idea to make the examples reusable, complete, and self-contained.
- Make it easy to automate the generation of results.
- Think about reuse and reproducibility when designing and writing code.

If these general recommendations are adopted more computational research can be made reproducible.

Acknowledgements

I thank Kunal Puri and other developers who have helped develop PySPH. I thank Kadambari Devarajan and an anonymous referee whose comments have helped improve this article.

References

[1] Donoho D L, Maleki A, Rahman I U, Shahram M and Stodden V 2009 *Computing in Science and Engineering* **11** 8–18 ISSN 1521-9615
[2] Gingold R A and Monaghan J J 1977 *Monthly Notices of the Royal Astronomical Society* **181** 375–389
[3] Lucy L B 1977 *The Astronomical Journal* **82** 1013–1024
[4] Monaghan J J 1994 *Journal of Computational Physics* **110** 399–406
[5] Gray J, Monaghan J J and Swift R 2001 *Computer Methods in Applied Mechanics and Engineering* **190** 6641–6662
[6] Monaghan J J 2005 *Reports on Progress in Physics* **68** 1703–1759
[7] Springel V 2005 *Monthly Notices of the Royal Astronomical Society* **364** 1105–1134
[8] Cherfils J, Pinon G and Rivoalen E 2012 *Computer Physics Communications* **183** 1468 – 1480 ISSN 0010-4655
[9] van Rossum G *et al.* 1991– The Python programming language `http://www.python.org/`
[10] Guo P 2014 Python is now the most popular introductory teaching language at top u.s. universities `http://goo.gl/iYclqt`
[11] Perkel J M 2015 *Nature* **518** 125–126
[12] Ramachandran P and Varoquaux G 2011 *Computing in Science and Engineering* **13** 40–51
[13] Ghia U, Ghia K N and Shin C T 1982 *Journal of Computational Physics* **48** 387–411
[14] Adami S, Hu X and Adams N 2013 *Journal of Computational Physics* **241** 292–307 URL `http://linkinghub.elsevier.com/retrieve/pii/S002199911300096X`

Exascale Storage Systems the SIRIUS Way

S A Klasky[1,2,3], H Abbasi[1], M Ainsworth[4,1], J Choi[1], M Curry[5], T Kurc[6,1], Q Liu[1], J Lofstead[5], C Maltzahn[7], M Parashar[8], N Podhorszki[1], E Suchyta[1], F Wang[1], M Wolf[1,3], C S Chang[9], M Churchill[9], S Ethier[9]

[1] Oak Ridge National Laboratory, Oak Ridge, TN 37831, USA
[2] Department of Electrical Engineering & Computer Science, The University of Tennessee Knoxville, Knoxville, Tennessee 37996, USA
[3] School of Computer Science, Georgia Institute of Technology, Atlanta, GA 30332, USA
[4] Division of Applied Mathematics, Brown University, Providence, RI 02912, USA
[5] Sandia National Laboratories, Albuquerque, NM 87185, USA
[6] Department of Biomedical Informatics, Stony Brook University, Stony Brook, NY 11794, USA
[7] Computer Science Department, University of California at Santa Cruz, Santa Cruz, CA 95064, USA
[8] Rutgers Discovery Informatics Institute, Rutgers University, Piscataway, NJ 08854, USA
[9] Princeton Plasma Physics Laboratory, Princeton, NJ 08543, USA

klasky@ornl.gov

Abstract. As the exascale computing age emerges, data related issues are becoming critical factors that determine how and where we do computing. Popular approaches used by traditional I/O solution and storage libraries become increasingly bottlenecked due to their assumptions about data movement, re-organization, and storage. While, new technologies, such as "burst buffers", can help address some of the short-term performance issues, it is essential that we reexamine the underlying storage and I/O infrastructure to effectively support requirements and challenges at exascale and beyond. In this paper we present a new approach to the exascale Storage System and I/O (SSIO), which is based on allowing users to inject application knowledge into the system and leverage this knowledge to better manage, store, and access large data volumes so as to minimize the time to scientific insights. Central to our approach is the distinction between the data, metadata, and the knowledge contained therein, transferred from the user to the system by describing "utility" of data as it ages.

1. Introduction

The purpose of computing is insight, not numbers [1], a remark Richard Hamming made over 50 years ago at the dawning of the age of large scale scientific computing. The central objective of our next-generation storage and I/O system (Sirius) is to minimize the time to insight with respect to scientific computing at extreme scales. This means that we need new research into managing, storing, and retrieving the large data volumes produced by simulations, and analyzed for months afterwards.

To begin, we present an example to illustrate these challenges and to motivate the on-going work. Next we describe our guiding principles to help tackle these challenges. In the subsequent sections we describe the key applied mathematics and computer science research components of SIRIUS. Here we introduce a new approach, data refactoring using an auditor, to aid in the separation of information from data. Finally, we describe the key challenges when this approach is realized, along with the future obstacle to manage the data lifecycle at extreme scales.

1.1. Illustrative Example

Our motivating use case is a series of simulations of the ITER fusion experiments using the XGC1 [2] application. XGC1 is one of the largest applications used at the Department of Energy's (DOE's) Leadership Class Facilities with an allocation of over 300M hours in 2015. These simulations were scheduled to produce 100 PB of data over ten days of total runtime on the Titan system at the Oak Ridge Leadership Computing Facility (OLCF) and required a team of experts, including the user group and I/O and storage personnel, to help ensure that the maximum amount of information would be saved. Due to physical resource limitations, the simulation was immediately restricted to write about 10 PB. However, when the time to write and read, together with the financial cost to archive this volume of data was fully explored, the data size goal was further reduced to only 5 PB total.

This forced a careful scrutiny of the data to be generated and restriction of the output to only the most important pieces. Further, this reduced data set had to be divided into two categories: the data that would be accessed while the simulation was running or shortly thereafter and the data that would be needed after the whole campaign was finished. The former category of data would be stored on disk and the latter would be archived. Additionally, the entire team needed to figure out how to efficiently retrieve these different categories of data for post-processing analysis and visualization.

To accomplish this, the application and storage teams developed new application specific data reduction techniques and added them to the ADIOS [3] I/O middleware layer. A rudimentary discovery system was also created to track what data was on disk and what was on tape. Initially this solution was sufficient, but eventually a new problem was discovered. Although the team had reserved compute resources for the run, the entire file system was shared, resulting in very high I/O variability. Addressing this variability, caused by contention on shared resources, is an ongoing concern. Earlier efforts [4] were insufficient due to the immense data volumes, limited time available, and the number of current users.

The problems faced by this user workflow motivate our research here. The use case demonstrated that three tiers of solutions are required: application level knowledge of data and how it will be used, middleware management of data and resources, and storage system level scheduling of resources. It also provided us with new insights. First, the users and developers of the application data can, without a great loss of encoded knowledge, reduce the size of data that needs to be stored. Second, data storage must be managed with input from the users to correctly decide the target of storage operations. Third, shared resources must be managed in a way that provides sufficient performance to all applications without encumbering the applications with high variability in performance. These insights drive the SIRIUS project and its research components.

1.2. Guiding principles of SIRIUS

The exascale storage system and I/O (SSIO) community must overcome the challenges described above for both the application scientist trying to write data and those trying to read data. This must be done such that the SSIO system can fairly share the resources among the users while helping enable exascale science by prioritizing and understanding the application level data requirements. The SIRIUS project is ultimately guided by two basic principles:

Principle 1: A knowledge-centric system design that allows user knowledge to define data policies. Today, SSIO layers are written in a stovepipe fashion and quite often do not allow optimizations. We are re-designing the layers in a highly integrated fashion where users express their intentions into the system and actions will statically and dynamically optimize for both the system and for individual requests.

Principle 2: Predictable performance and data quality in the SSIO layers need to be established to maximize the information (rather than raw data) generated on the exascale systems. Without predictable performance, runs may be slowed down because of shared resource contention, which can affect key science decisions, e.g., how much data reduction should be performed. To have some confidence that sufficient output time is available, a conservative estimate, rather than nearly accurate, can be used.

ADIOS can alleviate the need for "magic" and "tricks" to optimize application I/O performance on today's file systems by placing extra annotations in the metadata to better understand the underlying storage system [3]. In SIRIUS, we are extending this by providing a systematic autonomic approach for combining intentions and other knowledge from the *user* with performance estimations and guarantees from the underlying storage. By capturing user intentions and acting upon them in the middleware, we free the user from polluting application code with system specific optimizations. These techniques will be further integrated in combing ADIOS with an object store like RADOS-Ceph [5], which is a distributed object store and file system. Ceph offers both a POSIX and object interface including features typically found in parallel file systems.

2. Data Refactoring

The classical workflow where the entire dataset is written to storage for later analysis will no longer be viable at exascale simply because the amount of generated data will be too large due to capacity and performance limitations. In the future, it will be vital to take advantage of *a priori* information (1) to gain higher performance and predictability of I/O, (2) to prioritize the most useful data for end users so that I/O can be finished under time constraints, and (3) to perform *in situ* operations and analysis before storing the information. A result of these requirements is a need for a set of techniques to reduce and restructure data, here referred to as *data refactoring*.

The fusion use case described earlier shares commonalities with many other DOE applications. *A priori* information can be provided by application scientists regarding which data should be sent to the storage system (e.g., SIRIUS) so that minimally, the most science relevant data can be available for subsequent analysis. This allows science goals to be accomplished even when the storage is busy servicing other users. As we discussed earlier, this shared resource contention is common and causes high performance variability. For our approach, data refactoring generates the data prioritization classes. This refactoring includes data re-organization and reductions. There are many data refactoring techniques and the best choice will generally be application dependent. However, our observation is that, once the choice is settled for an application, it will not typically change from run to run within an extended campaign.

One research challenge in effectively and efficiently refactoring data is understanding when the time and resources required to identify and execute the "best methods" exceeds the gains achieved. Another critical research question concerns quantifying and controlling information loss from refactoring the data and using a reduced dataset. Broadly speaking, the path from data to knowledge consists of extracting underlying models or patterns from the datasets and interpreting the resulting models. Although scientific data generally contains random components due to finite precision and measurement and calibration effects, useful scientific data is never purely random. As such, a core concern in refactoring is understanding how much *information* is present in a dataset and therefore which type(s) of refactoring will be the most effective.

Ideally, scientists would like to perform the entire analysis *in situ,* effectively circumventing the large data issue completely. The catch, of course, is that this is unlikely to be possible since, by their nature, large-scale simulations aim to discover new information often hidden in the form of higher order effects amongst the data deluge. In particular, this means if data thinning or truncation is applied haphazardly, the higher order effects may be eliminated.

Typically, entire data sets cannot be stored in easily accessible storage due to its sheer size. Yet, the data cannot be reduced prior to archiving without risking losing information. Viewed in this way, the problem would appear intractable. As noted above, though, this is not a true impediment as long as we can incorporate a user's *a priori* knowledge of models and effective refactoring techniques. The information needed to answer the scientist's particular science goals is frequently significantly smaller, and using careful information-theoretic and application-given techniques, the SSIO layer can exploit this.

Deep application knowledge means one can sometimes achieve dramatically superior data reduction compared with what one might achieve otherwise. However, even in the absence of such high level knowledge, SIRIUS must offer generic data reduction and re-organization techniques. For instance, certain basic data semantics information is needed and must be supported by the overall infrastructure. Currently we are studying three generic and one application specific refactoring methods: (1) Precision based, (2) Frequency-based, (3) Linear Auditing, and (4) Application aware histogramming. Each of these is described below.

Precision based refactoring groups together the most significant data bits of the binary representation of each object. This data generally has a higher "utility" than data with the least significant bits, and contains less entropy. This technique can involve memory intensive operations, and as such, great care must be taken to balance the CPU, memory, and storage bandwidth costs and savings. Reading this data back introduces additional challenges—hot to find the separated components of the data if full resolution is needed and how to reconstruct the "packed" data to a form the user can work with.

Multi-resolution refactoring separates a dataset using variable granularities, or resolutions. Higher granularity samples are subtracted from lower granularity ones. This technique is commonly used in streaming data using techniques such as JPEG-2000 [6]. These mechanisms support spatial random access or region of interest access at varying degrees of granularity. In relation to this project, this capability allows us to place the lowest granularity chunks in the fastest storage and the highest granularity chunks can either be written to slower tiers or not written at all. In order to fully take advantage of multi-resolution re-organization, techniques to pre-condition the data can be used to further optimize this process. This technique can be applied together with Precision-based refactoring methods as the pre-conditioner.

Linear Reconstruction is a "technique" for reducing data consists of storing every s data item and simply ignoring the rest of the data. This approach is frequently adopted as a simple, but often effective method for data reduction. The tactic assumption underlying the approach is that the discarded values could, if required, be reconstructed by interpolating the retained values. More generally, one can retain every s-th item of data, where s is a chosen stride length resulting in an s-fold reduction in the size of the data. Objections to this and related ad hoc procedures are their lossy nature and lack of a sound theoretical basis. In order to address the lossy nature, we are using the linear reconstruction procedure described above as a predictor for the discarded items of data. The difference, or deltas, between the actual value of the discarded item can be shown, under reasonable assumptions, to have a significantly lower entropy than the original data. This means that the deltas are amenable to compression using, for instance, a standard arithmetic encoder. The resulting procedure then becomes lossless. The process of reinstating the data consists of identifying the stored values at the adjacent strides s, performing linear interpolation to obtain a predicted value, using the predicted value as key to access the delta correction from the arithmetic encoder, and finally, summing the predicted and delta corrector to obtain the original data item.

The resulting procedure can be surprisingly effective. Figure 1 shows the results obtained for varying choices of stride length s for the storage of data sampled from the function $sin(x)\ exp(20x)$, $0<x<1$ at uniformly spaced points. The compression ratio is computed by comparing the number of bits that would be required to store the original, full data set with the number of bits required to store the compressed data set including the costs (and overhead) associated with the arithmetic encoder.

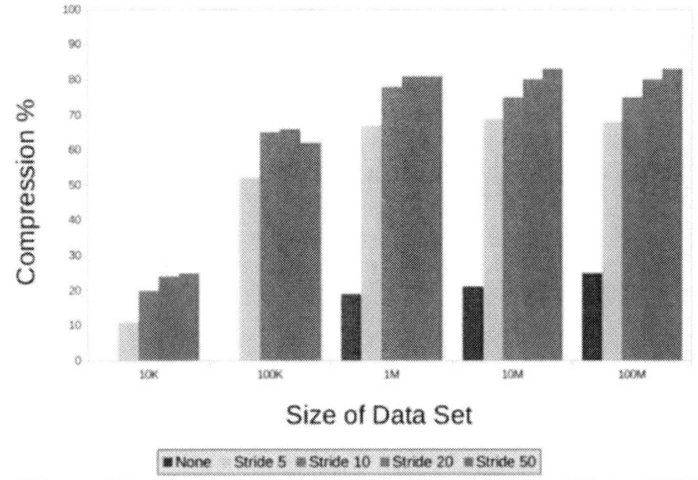

Figure 1. Compression ratio obtained using linear reconstruction for different choices of string length s.

Remarkably, the optimal choice for the stride s is considerably larger than the choice $s=10$ often used in practice despite the fact that the underlying data has a variation of several orders of magnitude across the sampling range.

Application Aware Auditor contains a class of techniques which have more domain knowledge than the generic techniques described above. In our example, physicist's understands that the particle distribution closely follows a Maxwellian distribution [7]. By using methods borrowed from perturbation theory[10] we can remove the background Maxwellian distribution and then store only the reduced non-Maxwellian piece of the distribution function. Physically, this should have a much smaller dynamic range. In our example we applied this technique using a two dimensional histogram of the particles in velocity space, and were able to get a dramatic reduction of the data to be saved using this lossy technique. This technique can be further combined with our other refactoring techniques to further reduce the potential storage cost of the information.

3. Refactoring Challenges

Refactoring large volumes of data is a challenging computing problem with three broad challenges. (1) Large volumes of data need to be refactored with *minimal impact* on the *application* and the computational *resources*. (2) The refactored data has to be created in a manner consistent with the *access patterns of the data consumer*. This requires predicting the common access patterns for data sets, identifying the most likely patterns and then correctly selecting the appropriate refactoring code path. (3) The abstraction through which data is viewed by the user needs to be *consistent in the presence of changing refactoring techniques* (in order to minimize the cognitive burden on the user), while providing sufficient transparency to enable both user and system to optimize metrics such as data layout, data accuracy and predictable performance. Next we describe these broad areas in detail and elaborate our approach for addressing these key challenges.

3.1. Resource Utilization

The generation of large volumes of data, is a costly and resource intensive apparatus. Additional work required by data refactoring can only be justified if the benefits outweigh the costs, and if the overall impact on the performance of the data pipeline is positive. Thus, any additional task to refactor,

[10] See e.g. https://en.wikipedia.org/wiki/Perturbation_theory

reorganize or compress data will need to account positively for the consumption of resources. In particular, there are three resources that are critical to scientific big data pipelines, on board system memory, additional time on the processor, and bandwidth consumed for data movement.

If we consider memory as the first constraint, data refactoring requires maintaining an original baseline state in memory, while generating the pieces to represent the refactored state. In the precision based refactoring technique, the original data is kept in memory while tiered precision arrays are generated. This requires that we double the memory requirement, and in some, very memory constrained use cases, this might render the refactoring an unviable approach. To overcome this obstacle, we have developed a method to refactor data streams by utilizing temporary windows over a stream of data. This windowed approach limits the additional memory overhead, but incurs a performance penalty when outputting data. Thus, we are studying optimal window sizes for the different refactoring techniques we have described earlier.

Similarly, for CPU and bandwidth consumption, the consumed resources will have an impact on the performance of the application. Some of this can be mitigated by utilizing asynchronous methods to compute and move the refactored data. This will require careful management of when the refactoring computation is called, and when data is moved, to minimize the overhead on the application due to resource contention. Here we are extending our past work on contention avoidance for data movement and resource sharing [8].

3.2. Refactoring Selection

There exist multiple techniques to refactor data that provide a varying set of tradeoffs between data output and access times, storage utilization, and accuracy of results. The selection of the refactoring technique becomes particularly challenging due to two factors. First, the requirements from the users evolve over periods of time, as the importance and utility of data change. Secondly, refactoring techniques can be used in combination, in any non-specific order.

Appropriately selecting the set and ordering of refactoring methods will require evaluations of the strengths of each combination within specific user constraints. The combination of user needs and data characteristics will require an autonomic controller to select the refactoring method. For example, a delta compression such as using the Linear Auditor will benefit greatly from an initial refactoring into multiple precision bins (since minor noise in the bins for higher order bits will be smoothed out). However, if the user requirement is that all data must be accessed at maximum precision, the additional work for precision based refactoring is unnecessary.

3.3. Linked Data Abstraction

Separate from the resource and performance impact of refactoring on the data intensive application, is the increased complexity of the organization and structure of the stored data. To be able to provide users with a compelling case for adopting this approach, a succinct data abstraction is required. This data abstraction must be able to provide a coherent view of discrete blocks of data, with each block having passed through multiple refactoring functions. Moreover, as the system makes decision on moving the various blocks based on importance and utility, the abstraction must be able to encapsulate the location of the blocks and allow a common API to access the data transparently. We are building this abstraction on the current ADIOS data abstraction which creates separate blocks for each process in a parallel cohort. We will extend this with deep links and attributes that describe the type and parameters of the refactoring operation. Achieving this combination of transparent data access with a performant middleware system will be one of the key challenges that will need to be solved to provide a usable data platform for exascale applications.

4. Managing Data Lifecycle

SIRIUS aims to manage the overall data life cycle, including data generation (e.g., from a simulation) or acquisition (e.g., in the case of experimental and observational data), optimized data placement, runtime data management including migration, reorganization and reduction, data consumption for knowledge discovery, and purging data from the system to optimize system operation. Key research questions addressed by SIRIUS include: (1) How can we initially place data so that it can be discovered and consumed efficiently? (2) How can the placement and migration of data across a multi-tiered storage hierarchy be optimized at runtime, both from the application and system perspective? (3) How can knowledge about the application used to better prepare the data for consumption? (4) When and how do we make the decision to purge data?

Data placement and movement: When an application outputs or accesses data, the storage and middleware layers needs to decide what data is placed where in the multi-level storage system. This placement decision can have a significant impact on data management throughout the lifetime of the data. For example, our past work on data–staging [[9],[10] on HPC systems with multi-level memory structures has shown that different output techniques targeting different layers of the storage hierarchy can have a significant impact on the overhead observed by the application for I/O operations. A key requirement is application-driven runtime mechanisms for dynamically managing data placement across the layers of the distributed storage hierarchy throughout the data lifecycle, coordinating data movement and data sharing between the components of the application workflow, with the overarching goal of maximizing the relative utility to the application as well as the system while reducing access costs. As noted before, the complexities of heterogeneous multi-level storage structures require adaptive placement policies are required to be implemented to optimally utilize storage resources vertically (across deep memory hierarchies) and horizontally (across nodes within a memory level) while accommodating dynamic application requirements and transient system states. Our approach is to increase knowledge about the data and its use within the application and leverage this knowledge to drive data placement and overall management.

Soliciting application hints: As described in the previous section, one of the main components of our proposed storage system is the ability to reorganize, refactor, and reduce data as it is generated and to reorganize and possibly regenerate the data as it is accessed. We carry this principle into the placement and movement of data by allowing applications to define hints and policies that guide what data is placed where. We will explore the use of application hints in two distinct areas. First, we will study the challenges and trade-offs of either augmenting the I/O interface with hints or allowing the addition of an external specification that defines the use case. Our experience with developing modern I/O interfaces has shown that both techniques have value [9] and we will investigate the set of hints that are embedded in the application code vs. those that are described within a non-compiled specification. Second, we will study how hints can guide data placement as data is handed off from application to storage (during a write) and from storage to application (during the read). In both cases we will study what minimal set of annotations and hints can allow the storage system to minimize data movement and optimize the resources consumed by I/O.

5. Metadata Challenges

The challenge when applying refactoring techniques, particularly application aware techniques, is how to incorporate sufficient knowledge in the storage system such that an arbitrary future client has sufficient information to recreate the desired information. Additionally, by spreading data across multiple different kinds of storage media that typically have independent namespaces, locating any particular data will be challenging.

SIRIUS will provide sufficient built-in and extensible metadata services to support efficient data access. First, detailed data metadata, such as array dimensions and other similar data, must be visible within the metadata for any effective data selection. Second, each data chunk stored within the storage hierarchy must have some way to address it. As data utility forces data migration within SIRIUS, the metadata must either dynamically track data as it moves or offer a search feature to discover where data currently resides. Which approach is superior and under what conditions still needs to be determined through further research.

6. Conclusion

Extreme scale application workflows, such as the fusion workflow described in this paper, generate very large amounts of data that need to be processed and analyzed before potential insights from the simulations can be realized. Managing, storing and retrieving these large volumes of data have become critical challenges. The central objective of the SIRIUS project is to address these challenges and minimize the *time to insight* for scientific workflows at extreme scales by enabling predictable performance across the storage system. We are creating a new understanding of how users can describe their intentions by describing the data utility, and refactor their data to achieve faster storage, retrieval, and understanding of the information contained in the data.

Acknowledgments

This project is funded by the following research grants from the Advanced Scientific Research Office at the Department of Energy, under contracts SDAV: ERKJ200/KJ0403000, RSVP: ERKJU60/KJ0402000, and Sirius: ERKJ311/KJ0402000. Sandia National Laboratories is a multi-program laboratory managed and operated by Sandia Corporation, a wholly owned subsidiary of Lockheed Martin Corporation, for the U.S. Department of Energy's National Nuclear Security Administration under contract DE-AC04-94AL85000.

References

[1] Hamming R W 1986 *Numerical Methods for Scientists and Engineers* 2nd Ed. (New York: Dover)
[2] Chang S *et al.* 2006 Integrated particle simulation of neoclassical and turbulence physics in the tokamak pedestal/edge region using XGC *FEC 2016 Proc* (Chengdu: IAEA) p 119
[3] Liu Q *et al.* 2014 Hello ADIOS: the challenges and lessons of developing leadership class I/O frameworks *Concurr Comput 26* pp 1453–1473.
[4] Lofstead J, Zheng F, Liu Q, Klasky S, Oldfield S, Kordenbrock T, Schwan K and Wolf M 2010 Managing Variability in the IO Performance of Petascale Storage Systems *SC12 Proc* (New Orleans: IEEE) pp 1-12
[5] Weil S A, Brandt S A, Miller E L, Long D D E and Maltzahn C 2006 Ceph: a scalable, high-performance distributed file system *OSDI 2006 Proc* (Berkeley: USENIX) pp 307-320
[6] Marcellin M W *et al.* 2000 An overview of JPEG-2000 *DCC 2000 Proc.* (Snowbird: IEEE)
[7] Krook M and Wu TT 1976 Formation of Maxwellian tails *PRL* **36** p 1107
[8] Abbasi H *et al.* (2010) Datastager: scalable data staging services for petascale applications *Cluster Comput* **13** pp 277-290
[9] Sun Q, Zhang F, Jin T, Bui H, Romanus M, Yu H, Kolla H, Chen J and Parashar M 2015 Adaptive Data Placement For Staging-based Coupled Scientific Workflows *SC15 Proc (Austin: IEEE)*
[10] Tong J *et al.* 2015 Exploring Data Staging Across Deep Memory Hierarchies for Coupled Data Intensive Simulation Workflows *IPDPS 2015 Proc* (Hyderabad: IEEE)

Implementation of Digital Lock-in Amplifier

Sabyasachi Bhattacharyya[1],Ragib Nasir Ahmed[2], Basab Bijoy Purkayastha[3], Kaustubh Bhattacharyya[4]

[1]Gauhati University Instititue of Science and Technology, Guwahati,India, [2,4]School of Technology, Assam Don Bosco University, Guwahati, India [3]IIT Guwahati, Guwahati, India

[1]sabya005@gmail.com,[2]ahmedragib@gmail.com, [3]basab@iitg.ernet.in, [4]kaustubh.bhattacharyya@dbuniversity.ac.in

Abstract. The recovery of signal under the presence of noise is utmost essential for proper communication. The signals corrupted due to noise can be recovered using various techniques. However the weak signals are more prone to noise and hence they can be easily degraded due to noise. In such cases, a digital lock-in amplifier becomes an essential device for recovery of such weak signals. Keeping the cost, speed and other considerations, we will present the implementation of digital lock-in amplifier and how it recovers the weak signal under extreme noisy conditions.

1. Introduction

Optimum signal recovery under noisy environments is a very significant problem that the modern day research focuses upon. Traditional amplifier designs do not provide satisfactory results in a highly degraded noise scenarios as they give very low signal power levels. The prime reason behind it is that traditional amplifier designs do not single out the specific frequency component of our interest. Such amplifiers lead to scenarios where the noise levels are in millivolt range whereas the signal levels are in microvolt range and thus leading to poor signal-to-noise ratio.

Optimum signal recovery under noisy environments is a very significant problem that the modern day research focuses upon. Traditional amplifier designs do not provide satisfactory results in a highly degraded noise scenarios as they give very low signal power levels. The prime reason behind it is that traditional amplifier designs do not single out the specific frequency component of our interest. Such amplifiers lead to scenarios where the noise levels are in millivolt range whereas the signal levels are in microvolt range and thus leading to poor signal-to-noise ratio.

In the field of communication the signals gets degraded when they are allowed passed through the channel. The signals which are weak may not reach the receiver as the noisy channel conditions degrade the original signal upto extreme case. So a digital LIA can be used for the recovery of the weak signals at the receiver.LIA is also used for the null detection in AC bridge measurement systems [3][4].The lock-in amplifier implemented in a microcontroller based DSP is inexpensive, portable and customizable [5].

The implementation of digital LIA is feasible only when the frequency of the reference signal is in the range of a few kilohertz. In that case the analog signal can be converted into digital signal by using ADC and DSPs. To covert an analog signal into digital signal the sampling frequency must be at least twice the frequency of the input signal [3]. However if frequency of the reference signal is in megahertz range then applying ADC and DSP on such signal may not be feasible due to the technical limitations of ADC and DSPs. So a combination of both analog and digital circuitry is used in the implementation of LIA [6]. Similar works has been reported in [7][8].

Lock-in amplifiers are used to detect and measure very small AC signals all the way down to nanovolt range. It works on phase-sensitive detection to single out the components of the signal at a specific reference frequency and phase rejecting all other noise frequencies. Based on this, digital phase locked (DPLL) form the basis for providing an accurate tuning to the desired frequency and phase. The performance of the lock-in amplifiers is subject to parameters such as locking accuracy, speed of locking, noise immunity, sampling frequency, integration time for hardware integration, cost etc. [9]. Similar works has been reported in [10].

The background of LIA is explained in the second section. The literature review of LIA is reported in the third section. The proposed model of digital LIA is discussed in the next section. The experimental details are explained in the subsequent section. The result and discussion is reported in the next section followed by a suitable conclusion, acknowledgement and references.

2. Lock-in Amplifier

A lock-in amplifier is a device which can extract a signal of interest at a specific frequency from an output signal which contains significant noise levels at other frequencies. The main objective of the lock-in amplifier is to increase the signal-to-noise ratio of the system by synchronizing the output signal to the input signal at a specific frequency and by filtering the other frequencies which contains the noise [11]. The essential components of LIA are shown in Figure 1.

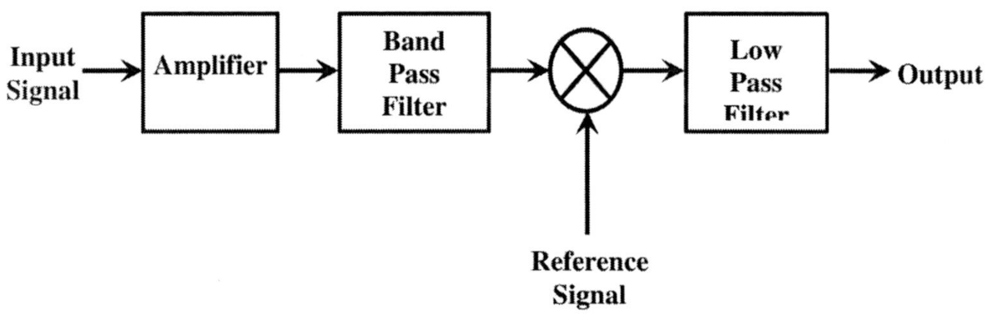

Figure 1: Block diagram of a typical lock-in amplifier [12].

The lock-in amplifier consists of an input amplifier to amplify the input signal and its output is fed to BPF and then the output of the BPF is multiplied with a reference signal and then passed through the LPF to remove the high frequency components present in the signal thereby only the DC components are present at the output of the LPF. This output signal is the desired signal of interest

3. Literature Review

Digital lock-in amplifier is widely used in the field of science and engineering. It can be used for the measurement of impedance by taking the advantage of the reduced noise enabled by the single frequency measurement [11]. The measurement of impedance can be useful for applications like semiconductor gas sensing inorganic molecules [13]. It is also used for the reduction of noise in the analysis of optical samples like infrared spectroscopy [14]. A high frequency digital LIA which works on the principle of random sampling is reported in [6]. An LIA is typically used in AC measurement system for null detection in order to measure the imbalance signal [3]. A novel digital lock in detection technique for measuring the amplitude and phase of the multiple AM signals has been reported in [15]. This detection technique requires a reference signal that is modulated at the same frequencies as their signal of interest while ignoring other frequencies [15]. The recovery of a weak signal based on the concept of zero-crossing algorithm which is implemented using digital phase locked loop is reported in [16][17].

4. Proposed Model

The block diagram of the proposed model is shown in Figure 2.

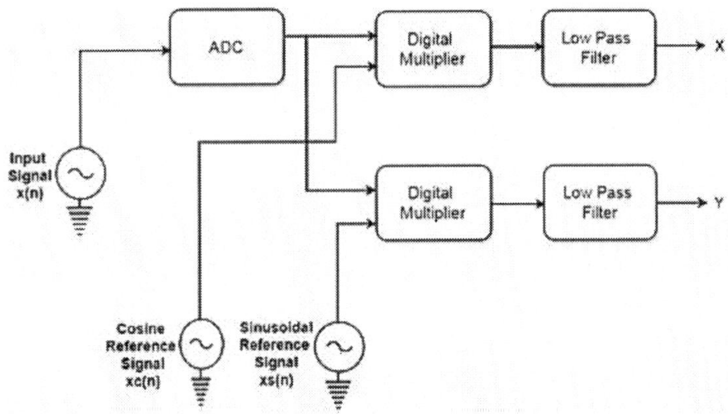

Figure 2: Block Diagram of the Proposed Model

From the block diagram show in Figure 2, we have taken two reference signals one of which has the same phase as that of the experimental signal and the other reference signal is the 90^0 phase shift with respect to the input signal. The experimental signal taken to be sinusoidal cosine waveform is multiplied with the reference signal having same phase to produce the X component or the inphase component and then passed through the LPF. The experimental signal is multiplied with the 90^0 phase shifted reference signal to produce the Y component i.e. the quadrature phase component and then passed through the LPF.

The principle of digital LIA is explained in equations (1), (2) and (3).

$$V_{in} = V_{signal}\cos(\omega_{ref}t + \theta_{signal}) \tag{1}$$

where, V_{in} is the experimental signal

ω_{ref} is the radian frequency of the experimental signal,

θ_{signal} is the phase of the experimental signal

We have taken two reference signals as described by equations (2) and (3).

$$V_{refcos} = V_{ref}\cos(\omega_{ref}t + \theta_{ref}) \tag{2}$$

$$V_{refsin} = V_{ref}\sin(\omega_{ref}t + \theta_{ref}) \tag{3}$$

Multiplying equation (1) and (2) we get $Vout_X$ as described by equation (4)

$$V_{outX} = \frac{1}{2}V_{signal}V_{ref}\cos(\theta_{signal} - \theta_{ref}) + \frac{1}{2}V_{signal}V_{ref}\cos(2\omega_{ref} + \theta_{signal} + \theta_{ref}) \tag{4}$$

Multiplying equation (1) and (3) we get the $Vout_Y$ as described by equation (5)

$$V_{outY} = \frac{1}{2}V_{signal}V_{ref}\sin(2\omega_{ref}t + \theta_{signal} + \theta_{ref}) - \frac{1}{2}V_{signal}V_{ref}\sin(\theta_{signal} - \theta_{ref}) \tag{5}$$

Equation (4) and equation (5) are passed through the low pass filter in order to eliminate the high frequency component and hence we obtain only the inphase and quadrature phase component as described by equation (6) and (7).

$$X = \frac{1}{2}V_{signal}V_{ref}\cos(\theta_{signal} - \theta_{ref}) \tag{6}$$

$$Y = \frac{1}{2}V_{signal}V_{ref}\sin(\theta_{signal} - \theta_{ref}) \tag{7}$$

The final amplitude of the lock-in amplifier is given by equation (8)

$$A = 2 \times \sqrt{X^2 + Y^2} \tag{8}$$

The phase of the lock-in amplifier is given by equation (9)

$$\text{Phase}=\tan^{-1}\frac{Y}{X} \tag{9}$$

5. Experimental Details

A heterodyne lock-in detection technique can be performed where there is a need for f_m to be larger than the possible values for f_s [15]. An experimental signal and two reference signals are created that has a time period of 0.1 seconds at a constant frequency of 5 KHz. In order to simulate the increase or decrease in the ADC sample frequency the amount of values per cycle for each signal matrix were increased or decreased which can be done while maintaining the correct frequency and same time period. The sampling frequency is set to be 70 KHz. A high sampling frequency is used so that the immunity of noise can be improved and the average error square can be reduced. The experimental signal is chosen to be a sinusoidal cosine waveform with a phase delay of $\frac{\pi}{30}$ radian and amplitude of 30 mV. The first reference signal is a sinusoidal cosine waveform with a zero phase shift. The second reference signal is sinusoidal signal waveform. The amplitude of the two reference signals are 1V each. The experimental signal is multiplied element by element with the cosine reference signal and also with the sinusoidal reference signal. The output of the two multiplied signal requires filtering which can performed by using LPF. The LPF is designed by taking the mean of the two multiplied matrices thereby producing a single DC value for each of X and Y channel outputs. The X and Y channel outputs are then used to find the amplitude of the signal and the phase difference between the signal and the reference signal. After that the noise is added to the signal to test the lock-in amplifier noise rejection performance. The noise is added using randn function in MATLAB which produces a random number with a uniform Gaussian distribution. The length of the noise matrix should be same as that of the experimental signal matrix so that they can be added to the experimental signal. This random noise matrix is multiplied with a noise level multiplication factor which is in a range of increasing incremental values to simulate the level of noise increasing in the experiment. The range of the noise multiplication factor is in the range from 0.1 to 1000 with a step size of 100 linearly spaced over the range. This range is chosen because the performance of the lock-in amplifier is not known. The SNR of the signal is used for the determination of the level of immunity of noise at which the lock-in amplifier still produces an acceptable output. The SNR of the signal is calculated as the ration of signal power to the noise power. The amplitude of the lock-in amplifier is measured for the calculated SNR for the range of noise multiplication factors. As the amplitude of the signal would increase with increase in noise and so such output is not acceptable. So an acceptable threshold must be set so that the output must not be far above the input amplitude. A value of 2 percent above the input amplitude is considered as an acceptable output. If the output amplitude is below an SNR of -40 dB then it is discarded in this case. So the range of the noise multiplicative factor is then reduced from 0.1 to 100 in order to reduce the computational time. The error between the input and output is calculated by taking the difference between each element of input and output. This is done to monitor the output error as the noise is continuously increasing. This produces only one number for each level noise in the range. This value would change each time the program will run due to the different random noise. So a set of 1000 random noise is used. The sets of 1000 random noise samples are then averaged so that a more appropriate Gaussian distribution is achieved. These average random samples are then squared to obtain an average error square for each noise level. This average error square is the square of the average error between the input amplitude and output amplitude for the set of random variables. The multiplicative noise factor is then changed to a logarithmic scale as the average error square has been plotted in the logarithmic Y axis. In order to simulate the lock-in amplifier in microprocessor signal quantization is used to simulate a range of ADC bit resolutions. A range of 2 to 16 bit resolution is used. The quantization is performed on the experimental signal after the noise is added to it. The signal matrix is converted to a sixteen bit integer and then quantized to the required bit resolution. The signal is then converted back to double perdition format. A plot of specific SNR values and their corresponding average error square value from the straight line approximation matrix is carried out in order to compare the different ADC resolutions.

6. Results and Discussion

In this section the results of the digital LIA are discussed.

An experimental cosine waveform is created having frequency of *5 KHz* with a phase delay of $\frac{\pi}{10}$ radians. The amplitude of the experimental signal is *30 mV*. The sampling frequency is *70 KHz*. It is multiplied with a cosine reference signal without phase delay and has amplitude of 1V to produce a multiplicative output signal as shown in Figure 3.

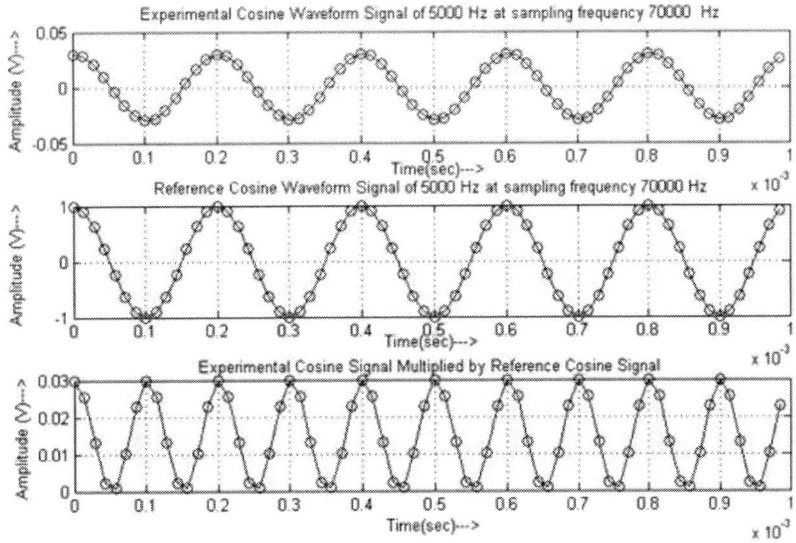

Figure 3: Experimental Cosine signal and Reference Cosine Signal and their multiplication.

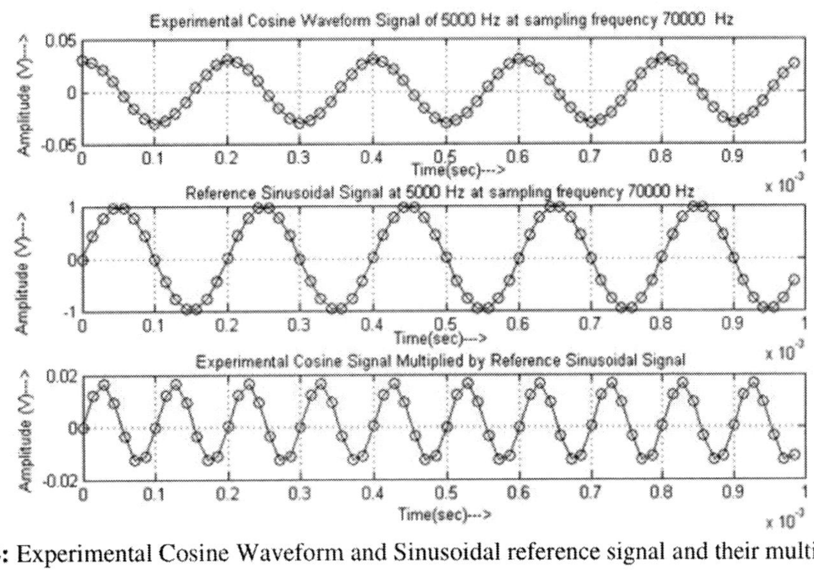

Figure 4: Experimental Cosine Waveform and Sinusoidal reference signal and their multiplication.

The experimental cosine waveform is multiplied with a sinusoidal reference signal and their plot is shown in Figure 4.

The output of the two multiplicative signals is then passed through the LPF. The LPF is designed as an averaging filter. The mean of the output of the first multiplicative signal is calculated to obtain a single DC value which is the inphase component and the mean of the second multiplicative signal is calculated to obtain a single DC value which is the quadrature phase component. The output amplitude of the signal and the phase is calculated using equation (8) and (9) respectively.

The error which occurred is then averaged and then squared to obtain the average error. The average error^2 at calculated SNR at sampling frequency 70 KHz for 1000 sets of random samples with increase in bit resolution is given in Table 1.

Table1: Average Error^2 w.r.t bit resolution and calculated SNR

SNR	Bit Resolution	Average Error
-19.2474	2 bit	0.60777
-23.4898	3 bit	0.48485
-23.4898	4 bit	0.45437
-23.4898	5 bit	0.44581
-23.4898	6 bit	0.44493
-24.0959	7 bit	0.44392
-23.4898	8 bit	0.44283
-24.0959	9 bit	0.44281
-24.0959	10 bit	0.44280
-24.0959	11 bit	0.44281
-24.0959	12 bit	0.44280
-24.0959	13 bit	0.44280
-24.0959	14 bit	0.44279
-24.0959	15 bit	0.44278
-24.0959	16 bit	0.44276

From table 1 we can conclude that the average error decreases with increase in bit resolution and then become constant at calculated SNR and at sampling frequency 70 KHz.

The results of the average error square in both logarithmic Y axis and linearly spaced scale, the quantization of the experimental signal and the output amplitude for 8 bit is shown in Figure 5.

From Figure 5, it can be seen that the average error square decreases with respect to increase in SNR in dB and finally becomes constant with further increase in SNR in dB. The average amplitude is valid at 2% error i.e. the output amplitude below SNR of -40 dB are discarded and only above SNR of above -40 dB is considered.

7. Conclusion

The digital LIA can be used for the recovery of extremely weak signal embedded in the background noise. The weak experimental signal is sampled at sampling frequency more than the Nyquist frequency and quantized. The experimental signal is converted into digital signal and an ADC bit resolution of 2 to 16 is considered. A sampling frequency of 70 KHz is considered. The average error square is calculated at calculated SNR and is seen that the average error square decreases with increase in ADC bit resolution from 2-8 bits and after that it becomes constant. Based on this 8 bit resolution can be used for the implementation of digital LIA. The output amplitude is plotted against SNR and an error of 2 percent above the input amplitude is considered as valid output.

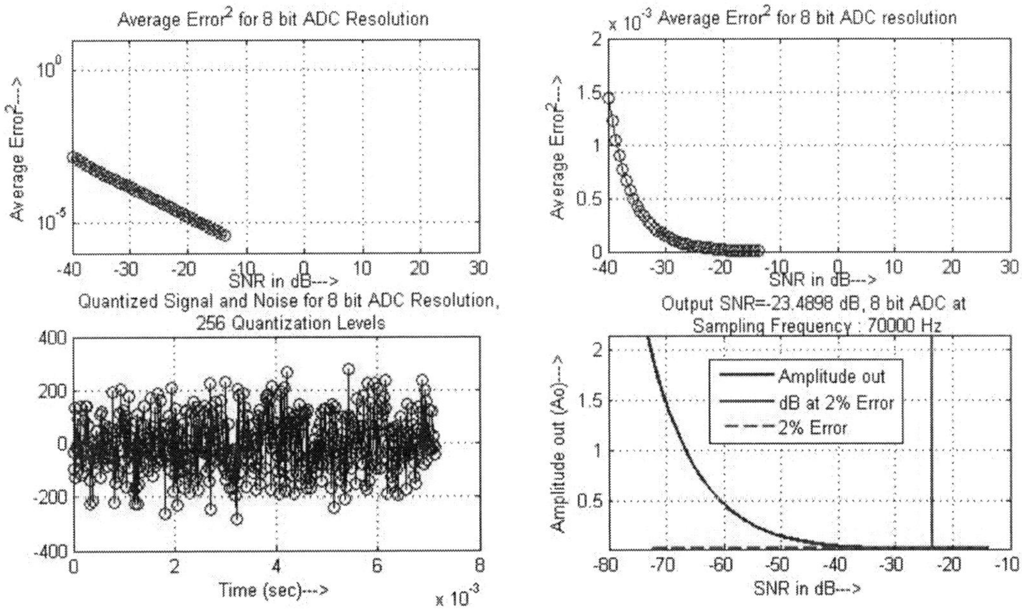

Figure 5: Result of average error^2 for 8-bit resolution and the output amplitude at sampling frequency 70 KHz against SNR in dB.

8. Future Work

Hardware FPGA implementation of digital lock-in amplifier would be chosen as a basis for future work. Further the immunity of noise can be improved by about 3 dB and the average error square can be reduced by half by increasing the sampling frequency [15]. Also by increasing the integration time the noise immunity can be improve by about 10 dB [15].

Acknowledgement
We would like to thank all the faculty members, teaching and non-teaching staff at Dept. of ECE, GUIST and Dept. of ECE, ADBU for all their help and support. We would also like to thank all our friends at GUIST and DBCET for the consistent help and support which motivated us to propose the work in a proper manner.

References

[1] Marco Zanuso, Salvatore Levantino, Alberto Puggeli, Carlo Samori, Andrea L. Lacaita, "Time-to-digital converter with 3 ps resolution and digital linearization algorithm" ISSCC Dig. Tech. Papers, pp. 340-341.

[2] Alessandro Balla, Matteo Beretta, Paolo Ciambrone, Maurizio Gattaet al, "Low Resource FPGA Based Time-to-Digital Converter", to be submitted to IEEE Transaction on Instrumentation and Measurement pp. 1-3.

[3] Zi Liu, Lixiong Zhu, Andrew Koffman, Bryan C. Waltrip and Yicheng Wang, "Digital lock-in amplifier for precision audio frequency bridge", IEEE Digital Xplore, ISBN: 978-1-4673-0442-9, pp. 586-587, 2012.

[4] D. Georgakopolous, I. Budovsky and L. Palafox, "Comparision of a lock-in amplifier and a digital sampling voltmeter as null detectors for differential measurements between sinusoidal and stepwise approximated ac signals", CPEM 2010 Conf. Digest, pp. 60, June 2010.

[5] Asutosh Kar, Pankaj Goel, Mahesh Chandra and V.K. Gupta, "A low- cost, portable alternative for a digital lock-in amplifier using TMS320C5535 DSP", IEEE Digital Xplore, ISBN: 978-1-4799-4075-2, 2014

[6] Maximiliano O. Sonnaillon, Raul Urteaga, Fabian Jose Bonetto and Martin Ordonez, "Implementation of high frequency digital lock-in amplifier", IEEE CCECE/CCGEI, ISBN: 0-7803-8886-0, pp. 1229-1232, May 2002.
[7] D. Carrica, M. Benedetti and R. Petrocelli, "Random sampling applied to the measurement of a DC signal immersed in noise", IEEE Transactions on Instrumentation and Measurement, vol. 50, no. 5, pp. 1319-1320, 2001.
[8] M.O. Sonnailon and F.J. Bonetto, "A low cost high performance DSP-based lock-in amplifier capable of measuring multiple frequency sweeps simultaneously", review of Scientific Instruments, vol. 76, 024703, 2005.
[9] Ralf Vogelgesang, "Lock-in amplifier theory", pp. 1-7, January 28, 2004.
[10] Sabyasachi Bhattachryya, Ragib Nasir Ahmed, Basab Bijoy Purkaystha and Kaustubh Bhattacharyya, "Zero crossing DPLL based phase recovery system and its application in wireless communication", in the Proceedings of IEEE International Conference on Computer Communication and Control (IC4-2015), Indore and IEEE Digital Xplore, ISBN:
[11] Andrew E. Moe, Steve R. Marx, Iqbal Bhinderwala and Denise M. Wilson, "A miniaturized lock-in amplifier design suitable for impedance measurement in cells", IEEE Digital Xplore, ISBN: 0-7803-8692-2, pp. 215-219, 2004.
[12] P.M. Maya-Hernandez, M.T. Sanz-Pascual, B. Calvo and D. Antolin, "A CMOS low power lock-in amplifier", IEEE Digital Xplore, ISBN:978-1-4577-1772-7, pp. 1-5, 2012.
[13] ShengQian, Ma, J. Schroder and P. Hauptmann, "Sensor impedance spectrum measurement interface with lock-in amplifier", Proceedings of IEEE Sensors 2002, vol. 2, pp. 1313-1316, June 2002.
[14] Andrea De Marcellis, Giuseppe Ferri, Arnaldo D' Amico and Corrado Di Natale, "A fully-analog lock-in amplifier with automatic phase alignment for accurate measurements of ppb gas concentrations", IEEE Sensors Journal, vol. 12, no. 5, pp. 1377-1380, May 2002.
[15] James M. Masciotti, Joseph M. Lasker and Andreas H. Hielscher, "Digital lock-in detection for discriminating multiple modulation frequencies with high accuracies and computational efficiency", IEEE Transactions on Instrumentation and Measurement, vol. 57, no. 1, pp. 182-184, January 2008.
[16] Sabyasachi Bhattacharyya, Ragib Nasir Ahmed, Basab Bijoy Purkayastha and Kaustubh Bhattacharyya, "Zero-crossing algorithm based phase recovery for DPLL based wireless communication", in the Proceedings of International Conference on Computing, Communication and Automation (ICCCA2015) and IEEE Digital Xplore, ISBN: 978-1-4799—88890-7, pp. 1273-1276, 2015.
[17] Sabyasachi Bhattacharyya, Ragib Nasir Ahmed, Basab Bijoy Purkayastha and Kaustubh Bhattacharyya, "Zero crossing DPLL based phase recovery system and its application in wireless communication", in the Proceedings of IEEE International Conference on Computer Communication and Control (IC4-2015), MGI, Indore, pp. 1-4,

Journal of Physics: Conference Series **759** (2016) 012097 doi:10.1088/1742-6596/759/1/012097

Designing reversible arithmetic, logic circuit to implement micro-operation in quantum computation

Gunajit Kalita and Navajit Saikia

Department of Electronics and Telecommunication Engineering, Assam Engineering College, Jalukbari, Guwahati-13, Assam. India

mailtogunajit@gmail.com

Abstract. The futuristic computing is desired to be more power full with low-power consumption. That is why quantum computing has been a key area of research for quite some time and is getting more and more attention. Quantum logic being reversible, a significant amount of contributions has been reported on reversible logic in recent times. Reversible circuits are essential parts of quantum computers, and hence their designs are of great importance.

In this paper, designs of reversible circuits are proposed using a recently proposed reversible gate for arithmetic and logic operations to implement various micro-operations (simple add and subtract, add with carry, subtract with borrow, transfer, incrementing, decrementing etc., and logic operations like XOR, XNOR, complementing etc.) in a reversible computer like quantum computer. The two new reversible designs proposed here for half adder and full adders are also used in the presented reversible circuits to implement various micro-operations. The quantum costs of these designs are comparable. Many of the implemented micro-operations are not seen in previous literatures. The performances of the proposed circuits are compared with existing designs wherever available.

1. Introduction

In recent time the emerging areas such as low-power computing, reversible logic, nanotechnology, quantum computing are gaining attention by many. These promising fields are presently explored by the researchers for better opportunities and are trying to use and address the issues related to these technologies. In digital circuits, heat emission has been a big issue which is also responsible for information loss. The researcher community is finding that the quantum technology may become means of futuristic computing. The quantum logic is reversible and so reversible logic is also getting a lot of popularity in recent times. In reversible logic circuits, inputs may be preserved in the output and feedback is avoided. And thus energy and information can be saved. The heat emission is very low and that helps circuits to last long. On the other hand, these will help to build circuits, called green circuits. Some pioneers such as IBM scientists Landauer and Bennet started investigating the issues related to heat generation in conventional electronics [4] and their literatures give idea of reversible computation [1]. Tommaso Toffoli is one of the pioneers to discuss conservative logic [9]; the very popular reversible "Toffoli" gate is named after him.

Lots of works on reversible logic have been reported that includes circuit design as well as circuit synthesis. A Reversible logic gate has bijective mapping between inputs and outputs and hence inputs are available at output. This results no information loss in the output. Thus it saves energy and information. On the other hand, traditional digital logic gate has single output for several inputs. Numbers of reversible gates have been presented by various researchers, out of these few of them [2,6,8,12] are studied to compare the designs. To discuss all of them is beyond the scope of this paper.

2. Reversible GN gate and basic circuits using GN gate

The Reversible GN gate [3] is a new gate that is proposed recently, it has three inputs and three outputs. Below, figure 1(a) shows reversible GN gate. Figure 1(b) is the quantum representation of this gate. In the quantum representation, four basic quantum gates are used and so the quantum cost [2] of the GN gate is four. The GN gate can be used to design various reversible circuits like adder, subtractor, comparator [3] etc.
Figure 2 shows a half adder using reversible GN gate, however a reversible CNOT gate is also used in the design. Figure 3 shows a full adder using two reversible GN gates, one CNOT gate and two NOT gates.

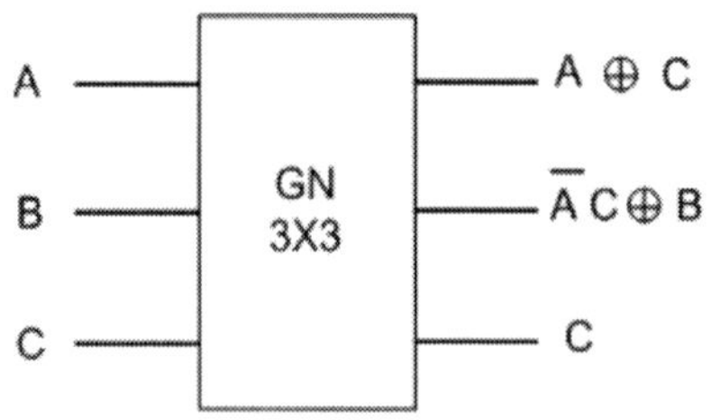

Figure 1(a): Reversible GN Gate

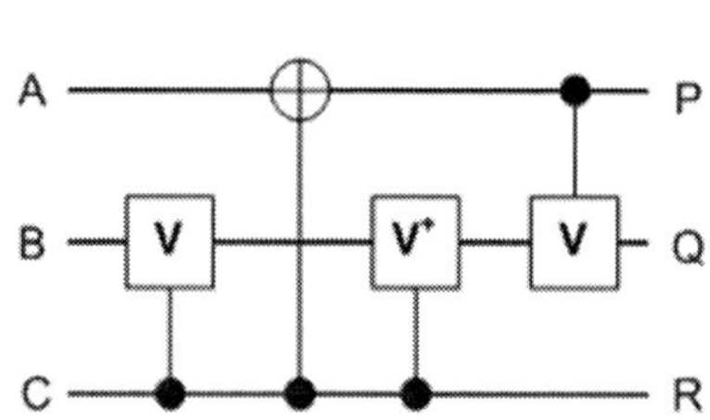

Figure 1(b): Quantum representation of GN Gate

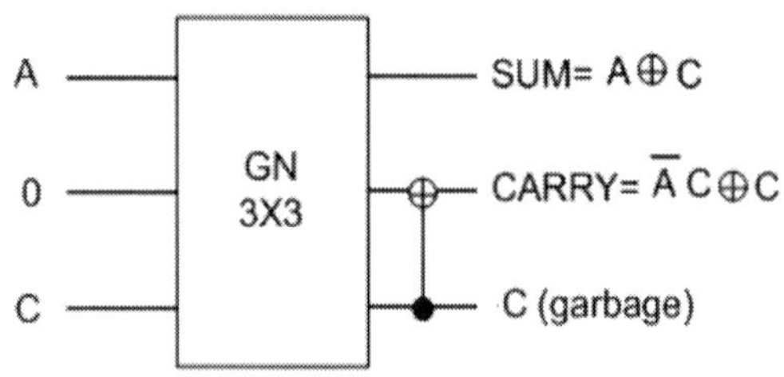

Figure 2: Half Adder Using GN Gate

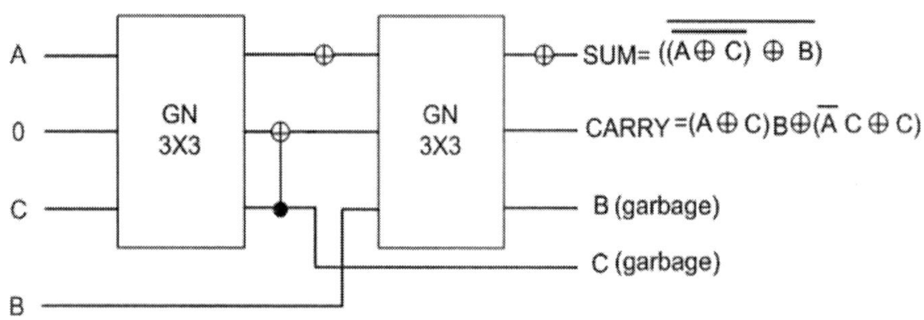

Figure 3: Full Adder circuit using GN Gate

The quantum cost of the reversible half adder shown in figure 2 is five (quantum cost of one GN gate is four and one CNOT gate is one). The quantum cost of reversible full-adder shown in figure 3 is eleven (two GN gates quantum cost is eight, one CNOT gate is one and two not gates is two).

2.1. *Truth Table of Reversible GN gate.* The Truth table of Reversible GN gate is shown below, where it has three input lines and three output lines. From the truth table it is very clear that there is bijective mapping between inputs and outputs.

Table 1. Truth table of reversible GN gate

Input			Output		
A	B	C	P	Q	R
0	0	0	0	0	0
0	0	1	1	1	1
0	1	0	0	1	0
0	1	1	1	0	1
1	0	0	1	0	0
1	0	1	0	0	1
1	1	0	1	1	0
1	1	1	0	1	1

3. Reversible arithmetic circuit

Many researchers have proposed different designs on arithmetic circuits [5,7,8,10,11]. The reversible circuit proposed here is comparable and has more number of operations. The reversible arithmetic circuit is shown in next page (scc figure 4). This circuit uses reversible multiplexer (RMUX) and reversible full adder to perform various micro-operations (shown in the table 2). In the diagram S_0 and S_1 are two select lines. A $(A_3A_2A_1A_0)$ and B $(B_3B_2B_1B_0)$ denote two 4-bit data inputs and R $(R_3R_2R_1R_0)$ denotes a 4-bit data output. These values are nothing but some binary values. The garbage outputs are marked as *'g'* and constant inputs are marked as either 0 or 1. The quantum cost of each reversible multiplexer is 6, each reversible full-adder circuit cost is 11 and not gate cost is 1. Hence total quantum cost of the arithmetic circuit = $6X4 + 11X4 + 1X4 = 72$. This cost is comparable [5,7,8,10,11] and shown better performance in terms of quantum cost. However this cost can be further optimized. And high order data bit can be used with a modified circuit and that need to be tested further. At this moment that is beyond the scope of this paper.

Eight different micro operations have been tested and shown in this paper [see table 2], besides these number of other operations can be added to this circuit. It is seen that the transfer and copy operation is similar but their operational codes are different, for example to copy it is 100 and for transfer it is 111 and hence whenever any of these operations is sought the distinct code turned to be helpful for better semantic. A reversible multiplexer used in this design has four inputs lines out of these two inputs are constant 0 and 1 ,and two select lines S_0 and S_1 and in the output side a single output and five garbage outputs lines are noticed. In this design, four such reversible multiplexers are used. Apart from these, four reversible adders (see figure 3) are also used.

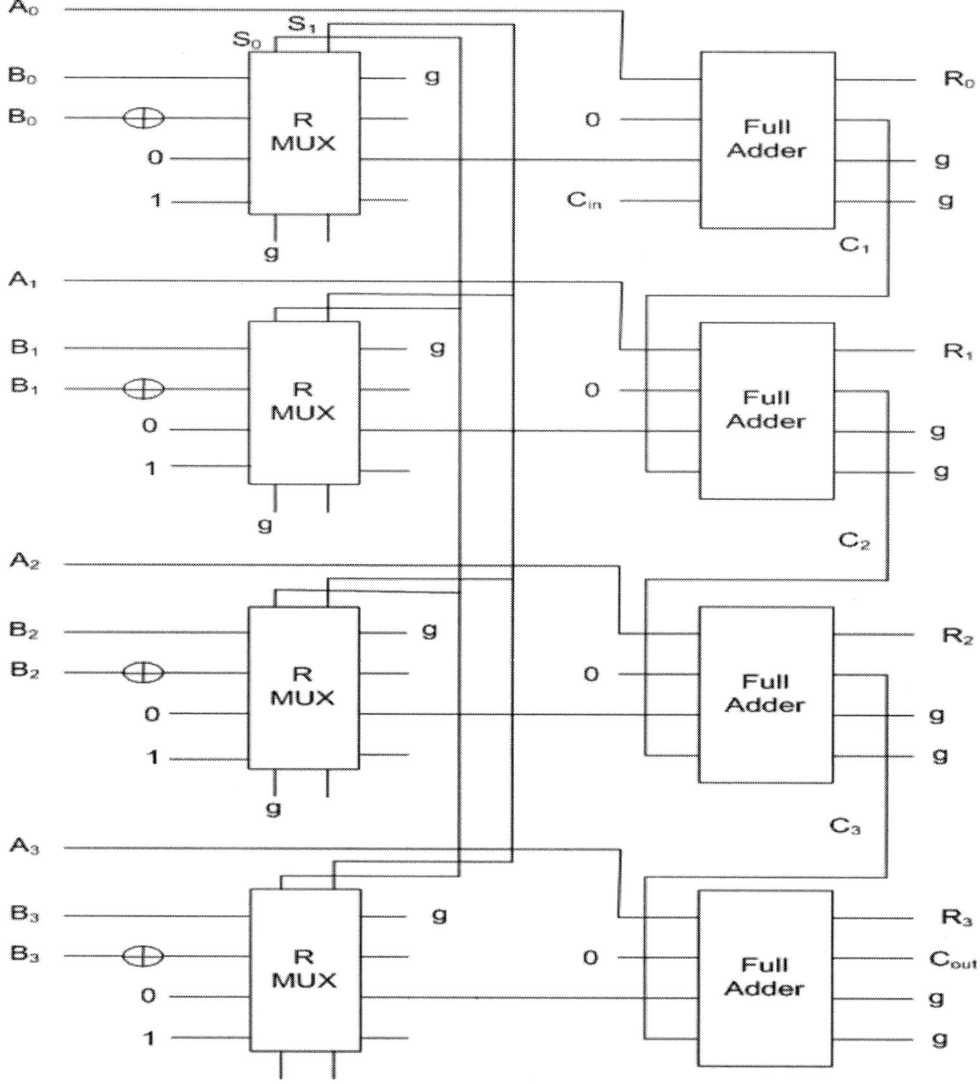

Figure 4: Reversible Arithmetic Circuit

Table 2. Arithmetic circuit function table

Select			Input	Output	Micro-operations
S_1	S_0	C_{in}		$R=A+B+C_{in}$	
0	0	0	B	R=A+B	Add
0	0	1	B	R=A+B+1	Add with carry
0	1	0	~B	R=A+~B	Subtract with borrow
0	1	1	~B	R=A+~B+1	Subtract
1	0	0	0	R=A	Transfer A
1	0	1	0	R=A+1	Increment A
1	1	0	1	R=A-1	Decrement A
1	1	1	1	R=A	Copy A

The reversible adder takes one bit directly from input, another from the output bit of the reversible multiplexer, one bit as a carry input and a constant input ‘0’. In the output side of a reversible adder, one output bit, one carry bit and two garbage outputs is seen.

4. Conclusion and future works

In this paper reversible half and full adders are designed using reversible GN gate. Besides these, some micro-operations are implemented for reversible arithmetic and logic unit, which are helpful in designing a reversible computer. The quantum costs of the design may be further optimized. More micro-operations can be added. One of the challenges is to reduce the number of garbage output in future.

References

[1] Bennet C. H. Logical reversibility of computation. IBM J. Res. Dev. 17, 525, (1973)

[2] Barenco, A et. al. Elementary gates for quantum computations. The American Physical Society. 52 , 3457- 3467(1995).

[3] Kalita, G. and Saikia, N. Reversible comparator circuit using a new reversible gate. Proc. of 6th ICCC (September) Allahabad, India , published by ACM (2015).

[4] Landauer, R. Irreversibilty and heat generation in the computing process. *IBM J. Res. Dev.* 5, 183(1961).

[5] Ravish, A et. al. Design of control unit for low power AU using reversible logic. Procedia Engineering, Elsevier 30, 631-638 (2012).

[6] Sasanian, Z et. al. Realizing reversible circuits using a new class of quantum gates. Proc. of ACM DAC, pp 36-41, June 3-7 San Francisco CA, USA (2012).

[7] Sultana, S and Radecka, K. Reversible Architecture of Computer Arithmetic. Intl. Journal of Computer Application 93(14) 6-14 (2014).

[8] Thapliyal, H and Srinivas, M.B. Novel reversible "TSG" gate and its application for designing components of primitive reversible quantum ALU. Proc. of 5th IEEE Intl. Conference on Information Communication and Signal Processing. Pp 1425-1429, Dec 6-9 Bangkok, Thailand (2005).

[9] Toffoli, T. Reversible Computing. Tech Memo MIT/LCS/TM-151. MIT Lab for Computer Science (1980).

[10] Thomsen, M.K. et. al. Reversible arithmetic logic unit for quantum arithmetic. Journal of Physics A: Mathematical and theoretical 43, 382002(10pp)(2010).

[11] Willie, R. et. al. Designing a RISC CPU in reversible logic. Proc. of 41st IEEE Intl Symposium on Multi-valued logic (ISMVL), pp 170-175, May 23-25 Tuusula (2011).

[12] Yang, G. et. al. Majority based reversible logic gates. Elsevier Journal on Theoretical Computer Science. 334 259-274 (2005).

Journal of Physics: Conference Series **759** (2016) 012098 doi:10.1088/1742-6596/759/1/012098

A Framework to Simulate Semiconductor Devices Using Parallel Computer Architecture

Gaurav Kumar[1], Mandeep Singh[1], Anand Bulusu[2], and Gaurav Trivedi[1]

[1]Dept. of EEE, Indian Institute of Technology Guwahati, Assam-781039, India
[2]Dept. of ECE, Indian Institute of Technology Roorkee, Uttarakhand-247667, India

E-mail: gkumar2012@iitg.ernet.in

Abstract. Device simulations have become an integral part of semiconductor technology to address many issues (short channel effects, narrow width effects, hot-electron effect) as it goes into nano regime, helping us to continue further with the Moore's Law. TCAD provides a simulation environment to design and develop novel devices, thus a leap forward to study their electrical behaviour in advance. In this paper, a parallel 2D simulator for semiconductor devices using Discontinuous Galerkin Finite Element Method (DG-FEM) is presented. Discontinuous Galerkin (DG) method is used to discretize essential device equations and later these equations are analyzed by using a suitable methodology to find the solution. DG method is characterized to provide more accurate solution as it efficiently conserve the flux and easily handles complex geometries. OpenMP is used to parallelize solution of device equations on manycore processors and a speed of $1.4x$ is achieved during assembly process of discretization. This study is important for more accurate analysis of novel devices (such as FinFET, GAAFET etc.) on a parallel computing platform and will help us to develop a parallel device simulator which will be able to address this issue efficiently. A case study of PN junction diode is presented to show the effectiveness of proposed approach.

1. Introduction

The recent developments of semiconductor technology is backed-up by Technology Compute Aided Designs (TCAD), which plays a vital role by providing an efficient simulation environment to create desired future devices of different geometries and materials . To approximately predict the electrical behaviour of a semiconductor device, TCAD implements various device physics, mathematical models such as Drift-Diffusion, Hydrodynamics, Quantum Corrected Boltzmann and Non-equilibrium Greens Function [1]. These models consist of a number of partial differential equations (PDEs). Approximate solution of these PDEs is computed using numerical techniques as it is impractical to get analytical solution. Various discretization methods such as finite difference (FDM), finite volume (FVM) and finite element (FEM) are available to obtain numerical solution of a PDE. The advantages and disadvantages of these methods are shown in Table 1. DG outperforms both FEM and FVM while handling complex geometries with high-order accuracy and local mass conservation [2, 3]. In 1973, Reed and Hill [4] first introduced DG method for hyperbolic equations. Later, DG method has been extended to develop different methods [2, 5, 6] for hyperbolic and nearly hyperbolic problems. As per our knowledge, DG-FEM is explained in literature but has not been implemented as of now in any of the available device simulators. In this paper, we present a semi-parallel device simulator to analyze semiconductor devices using DG-FEM as the discretization scheme. DG divides the whole geometry into smaller finite domains called "element" to find numerical solution of PDEs. DG requires substantially less information about the neighbouring elements making parallelization easier [7]. The proposed device simulator is capable of generating uniform and non-uniform mesh throughout geometry. More accurate solution is obtained using non-uniform meshing, with less computation complexity as compared to denser mesh throughout the geometry. OpenMP [8] directives are used to parallelize the solution of numerous algebraic equations on manycore processors. Parallelization in the framework is implemented during assembly process of DG-FEM.

Table 1. Comparison of different discretization schemes [2, 3].

	Complex Geometries	High-Order Accuracy	Local Mass Conservation
FDM	✗	✓	✓
FVM	✓	✗	✓
FEM	✓	✓	✗
DG-FEM	✓	✓	✓

Rest of the paper is organized as follows. In Section 2, fundamental device equations for Drift-Diffusion model are discussed. Section 3 describes discontinuous Galerkin discretization scheme. Implementation details and simulation results are discussed in Section 4. Section 5 concludes the paper.

2. Formulation of Basic Device Equations: Drift-Diffusion Model

The flowchart to build framework for a device simulator is shown in Figure 1, which explains the simulation process [9] by incorporating several device models. Simulation process is started by procuring all the data related to device geometry, material parameters, doping profile and necessary boundary conditions. Device geometry (whole domain) is further divided into smaller domains in order to solve PDEs locally to complete the process of discretization. Discretization generates a mesh throughout the whole geometry to get the solution over all nodes. Then charge is computed using an initial guess of solution. Calculated charge is then used to iteratively solve Poisson's and Continuity equations until the solution converges to a pre-defined threshold. Gummel's and Newton-Raphson algorithm [9] is used to implementation both Poisson's and continuity equation. Current is calculated at the end of simulation process for specified input parameters.

The five fundamental physics equations for Drift-Diffusion model to simulate semiconductor devices are as follows:

2.1. Poisson's equation

$$\nabla \cdot (-\epsilon \nabla \phi) = \rho \tag{1}$$

$$where \;\; \rho = q[p - n + N_d - N_a]$$

2.2. Current equations

$$J_n = qn\mu_n \nabla \phi + qD_n \nabla n \tag{2}$$

$$J_p = qp\mu_p \nabla \phi - qD_p \nabla p \tag{3}$$

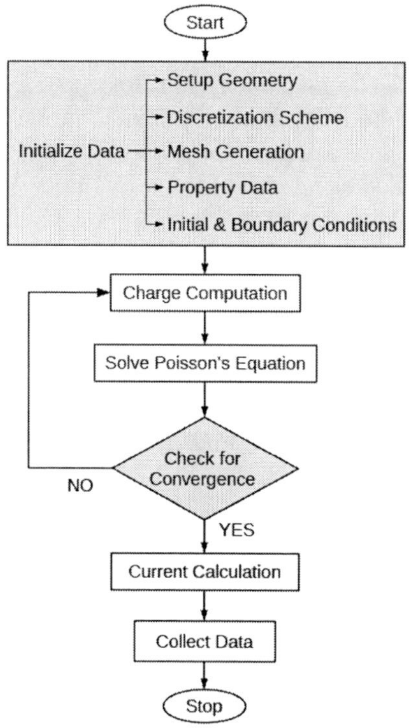

Figure 1. Process flow of a typical device simulator.

2.3. Continuity equations

$$\frac{\partial n}{\partial t} = \frac{1}{q} \nabla \cdot J_n + R_n \tag{4}$$

$$\frac{\partial p}{\partial t} = -\frac{1}{q} \nabla \cdot J_p + R_p \tag{5}$$

where ϕ is potential, μ is mobility, ρ is space charge density, D is diffusion coefficient, R is net generation and recombination rate, n & p specifies electron and hole density in conduction and valance band respectively, q is electrostatic charge, N_a is acceptor concentration, N_d is donor concentration, J is current density, ϵ is dielectric permittivity and t denotes time.

3. Discontinuous-Galerkin Discretization Scheme

DG method is a special case of FEM, in which the basis functions (used for discretization) are discontinuous piecewise polynomials. Discontinuous basis functions effectively handle interactions between element boundaries, to achieve accurate and stable results for nonlinear hyperbolic systems. Features like, high parallelizability, ease of handling complex geometries, non-linear stability and higher order accuracy makes this discretization scheme more considerable. The discontinuous basis function with degree of freedom (DOF) 1 and 2 are as shown in Figure 2 and Figure 3 respectively. The difference in basis functions for FEM and DG-FEM is as shown in Figure 4 and Figure 5 respectively. DG-FEM basis functions shows discontinuity at the node point, whereas FEM basis functions are continuous in nature.
To showcase the implementation of DG method, a generic conservation equation is considered as follows [2]:

$$\frac{\partial \tilde{u}}{\partial t} + \nabla \cdot \vec{g} = 0 \tag{6}$$

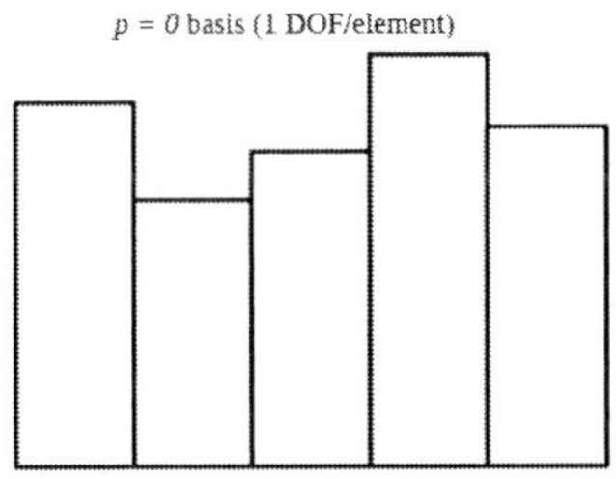

Figure 2. DG basis function with one degree of freedom per element.

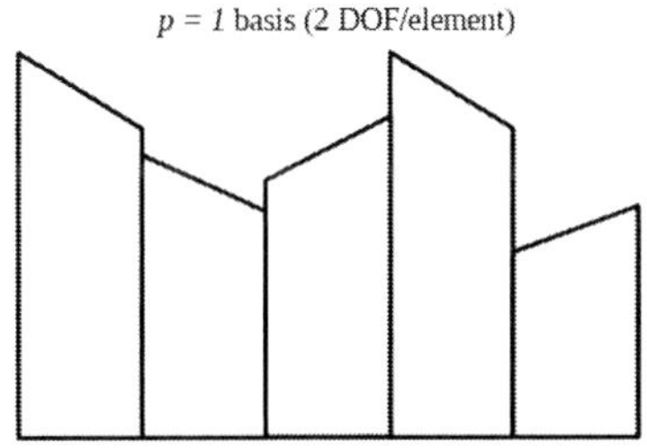

Figure 3. DG basis function with two degree of freedom per element.

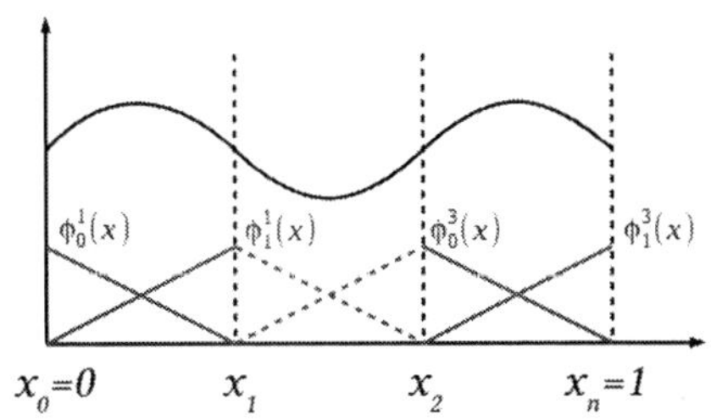

Figure 4. Basis function as used in finite element method.

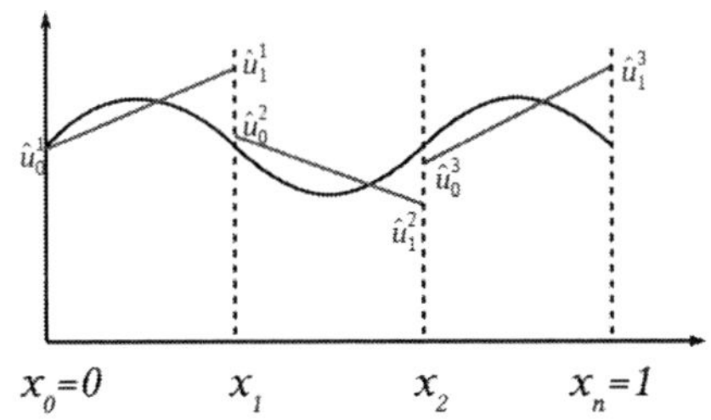

Figure 5. Basis function as used in DG finite element method.

Multiplying (6) by a basis function and integrating over the whole domain (Ω),

$$\int_\Omega v\frac{\partial u}{\partial t}dV + \int_\Omega v\,\nabla\cdot\vec{g}dV = 0, \quad \forall_v \in \Phi \tag{7}$$

$$= \sum_e \int_e v\frac{\partial u}{\partial t}dV + \sum_e\left(-\int_e \nabla v\cdot\vec{g}dV + \oint_{\partial e} v\vec{g}\cdot\vec{n}dS\right) \tag{8}$$

For the efficient conservation of flux on the boundary, two different operators (Jump and Average) are defined w.r.t. normal $\vec{n}^+ = \vec{n} = -\vec{n}^-$ ($+$ *and* $-$ denotes outflow and inflow respectively),

$$\begin{aligned} &Jump\ operator: \quad [v]_{x_k} = v_{I_k}(x_k) - v_{I_{k+1}}(x_k) \\ &Average\ operator: \quad \{v\}_{x_k} = \frac{1}{2}(v_{I_k}(x_k) + v_{I_{k+1}}(x_k)) \end{aligned} \tag{9}$$

Applying (9) in (8),

$$= \sum_e \int_e v\frac{\partial u}{\partial t}dV - \sum_e \int_e \nabla v\cdot\vec{g}dV + \sum_f \int_f [v]\{\vec{g}\} + [\vec{g}]\{v\}dS \tag{10}$$

DG discretization is then defined as,

$$\sum_e v\frac{\partial u}{\partial t} - \sum_e \int_e \nabla v\cdot\vec{g}dV + \sum_f \int_f \gamma(\tilde{u}^+, \tilde{u}^-, v^+, v^-, \vec{n})dS = 0, \quad \forall_v \in \Phi \tag{11}$$

The function parameter, γ is required to have following properties: i) stability, ii) consistent as $u^+ = u^- = \tilde{u}$, and iii) conservative as weight, $W = 1\ \forall_x \in e$, $W = 0\ \forall_x \notin e$.

The global formulation using DG-FEM for equation (6) is,

$$\sum_e \int_e v\frac{\partial u}{\partial t} - \sum_e \int_e \nabla v \cdot \vec{g} dV + \sum_f \int_f \gamma(\tilde{u}^+, \tilde{u}^-, v^+, v^-, \vec{n}) dS = 0, \quad \forall_v \in \Phi \tag{12}$$

Global solution of (6) can be written as the summation of local solution of each individual element:

$$u = \sum_e u^e \tag{13}$$

The formulation thus reduces to element-wise FEM problem, coupled by internal boundary conditions. Next section presents implementation details of framework and simulation results.

4. Implementation Details and Simulation Results

A framework to simulate semiconductor devices based on DG-FEM is presented in this paper, which is capable of simulating 2D device geometries incorporating five fundamental device equations from Drift-Diffusion model. Discretization process divides whole geometry into various small elements, which leads to the formation of a mesh like structure through out geometry in such a way that no two elements overlap each other. Proposed simulator has the flexibility to select elements of different size and shape, capable of generating both uniform and non-uniform mesh. Triangular elements have been chosen in our simulator to discretize 2D geometry, although the module used in our proposed device simulator supports linear, triangular and quadrilateral basis functions as well. The choice of triangular elements helped us to prove the effectiveness of the methods chosen to analyze semiconductor devices correctly. Faster and more accurate results have been obtained by using non-uniform mesh with respect to denser uniform mesh. Parallelization in simulator has also been achieved during the assembly process of DG-FEM while constructing a global matrix from discrete element equations. OpenMP directives have been used for parallelization and a speedup of $1.4x$ has been obtained by using Dell Precision T7610 having QuadCore processor and 16GB RAM. Future work on the parallelizing of solution of global matrix would improve the performance of our proposed device simulator. To showcase the effectiveness of proposed simulator, results for a PN junction diode are validated with TCAD device simulator Sentaurus [10] and presented as follows.

4.1. PN Junction Diode

Different parameters used to simulate a silicon based PN junction diode are shown in Table 2.

Table 2. Various device parameters used for a PN diode.

Parameters	Value
Length of P-region	30 μm
Length of N-region	70 μm
Doping Profile: N_a, N_d	10^{16} cm^{-3}
Hole Mobility: μ_p	480 $cm^2/V-s$
Electron Mobility: μ_n	1350 $cm^2/V-s$
Temperature: T	300 K

Figure 6 and Figure 7 shows the formation of uniform and non-uniform mesh for a PN junction diode respectively. Denser mesh near the junction is created in order to achieve higher accuracy. Potential profile inside a diode at thermal equilibrium is shown in Figure 8. Built-in potential of a PN junction diode is found to be approximately $0.67V$. Total current density for a forward bias voltage is as shown in Figure 9. Our simulator also supports complex geometries (FinFET, GAAFET) and the analysis of these devices with a detailed study of physical phenomenon will be presented in our future manuscripts.

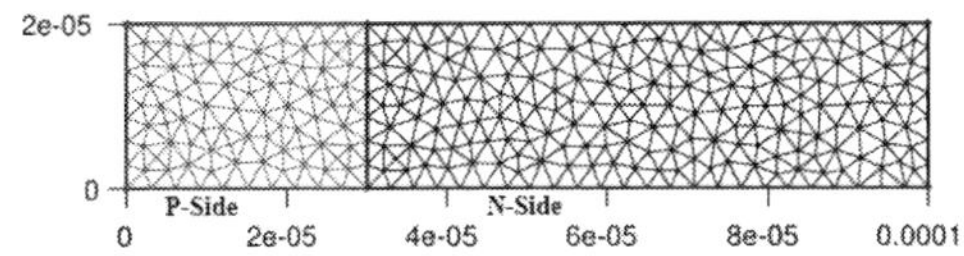

Figure 6. Uniform mesh over PN junction diode.

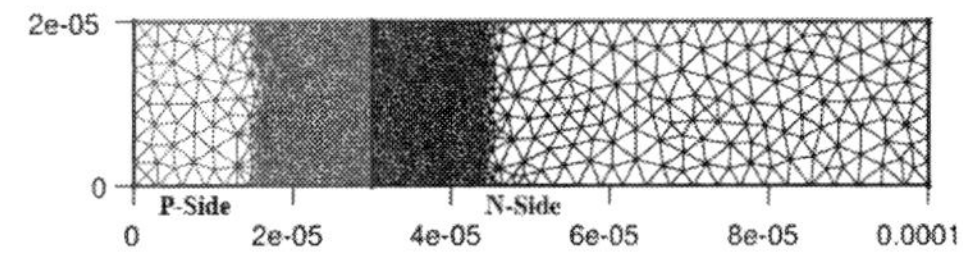

Figure 7. Non-uniform mesh over PN junction diode.

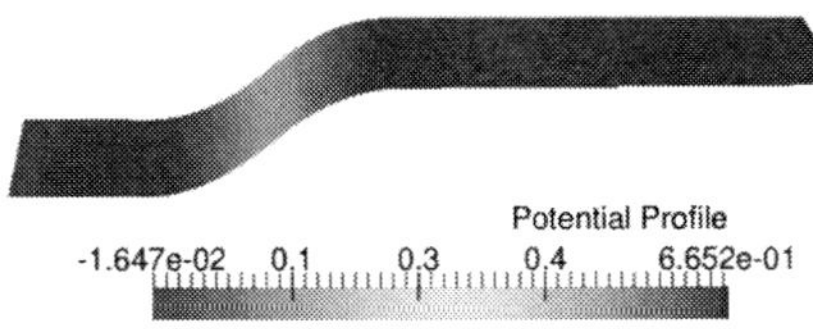

Figure 8. Potential profile of a 2D PN junction diode at equilibrium.

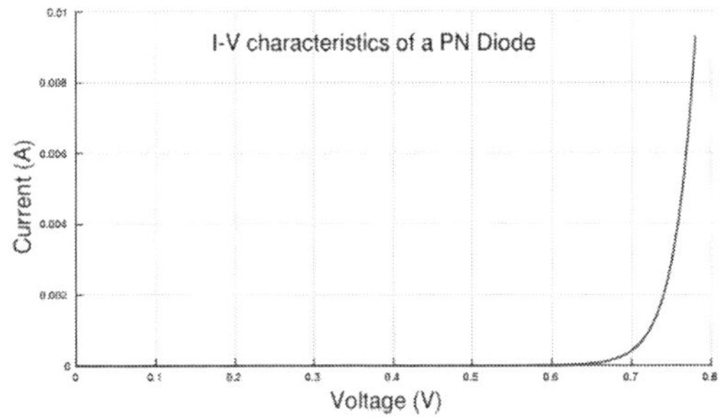

Figure 9. Forward biased current-voltage characteristics of a PN junction diode.

5. Conclusion

A parallelized and accurate DG based two-dimensional device simulator is presented in this paper to simulate semiconductor devices. Process flow of a typical device simulator and various discretization schemes to obtained the numerical solution of a device physics models has been discussed. The proposed framework uses discontinuous Galerkin finite element method to discretize PDEs of Drift-Diffusion model. DG method provides high parallelizability, easy handle of complex geometries, non-linear stability and higher order accuracy as compared to other discretization schemes. To achieve faster and accurate results, it is imperative to take advantage of current manycore computing architecture. Parallelization in the framework is achieved using OpenMP to produce $1.4x$ faster results as compared to serial computation. Simulation results for a PN diode has been presented to showcase the effectiveness of proposed framework.

References

[1] M. Pourfath, V. Sverdlov, and S. Selberherr. Transport modeling for nanoscale semiconductor devices. In *Solid-State and Integrated Circuit Technology (ICSICT), 2010 10th IEEE International Conference on*, pages 1737–1740, Nov 2010.

[2] Shu CW. Cockburn B, Karniadakis GE. *Discontinuous Galerkin Methods: Theory, Computation and Applications*. Springer-Verlag:Berlin, 2000.

[3] W. Fichner R.E. Bank, D.J. Rose. *Numerical methods for semiconductor device simulation*. ZEEE Trans. Electron. Dev., 7th edition, 1983.

[4] W.H. Reed and T.R. Hill. *Triangular mesh methods for the neutron transport equation*. Oct 1973.

[5] B. Rivire. *Discontinuous Galerkin Methods for Solving Elliptic and Parabolic Equations*. Society for Industrial and Applied Mathematics, 2008.

[6] Z. Liu and H. Chen. Discontinuous galerkin immerse finite volume element method for elliptic interface problems. In *Information Science and Technology (ICIST), 2014 4th IEEE International Conference on*, pages 115–118, April 2014.

[7] Y. Cheng, M. G. Irene, A. Majorana, and C. W. Shu. Performance of a discontinuous galerkin solver for semiconductor boltzmann equations. In *Computational Electronics (IWCE), 2010 14th International Workshop on*, pages 1–4, Oct 2010.

[8] http://openmp.org/wp/.

[9] S. Selberherr. *Analysis and Simulation of Semiconductor Devices*. Springer-Verlag Wien, 1984.

[10] Synopsys. Technology computer aided design. 1986.

Two Dimensional Numerical Simulator for Modeling NDC Region in SNDC Devices

Dheeraj Kumar Sinha, Amitabh Chatterjee and Gaurav Trivedi

Department of Electronics and Electrical Engineering, Indian Institute of Technology Guwahati, Assam, India-781039.

E-mail: d.sinha@iitg.ernet.in

Abstract. This paper deals about making a numerical simulator for observing the self-organization phenomenon like current filament formation in the negative differential conductivity (NDC) region of operation in a device using MATLAB environment. Emphasis is laid on choosing the appropriate boundary conditions in simulator to observe the NDC part of the device Current-Voltage ($I - V$) characteristics along with the appropriate isothermal physical models are included for mobility and the recombination-generation. The simulator used in this paper is designed to observe NDC region in S-shaped NDC (SNDC) devices like BJT and similar conditions can be used to observe the field domains in the N-shaped NDC (NNDC) devices with appropriate changes. Dirichlet and Neumann boundary conditions are applied to BJT and $I - V$ characteristics of the developed device simulator are compared with the commercial available TCAD device simulator *(i.e., Sentaurus)*.

1. Introduction

Detailed numerical simulation and modeling have become extremely important over the last four decades to optimize the performance and design of nano-scale semiconductor devices. It has been seen that huge number of trials are required in the experiments to optimize semiconductor device performance however it can be decimated using computer simulation for faster development cycle [1]. Numerical modeling of semiconductor devices are based on basic semiconductor equations which has been first proposed by Gummel et. al in 1964 and many papers have been published in this domain [2]. In general, the numerical simulation of the device includes a basic semiconductor physical models which describes carrier transport models in material such as drift-diffusion, hydrodynamic etc[3]. Moreover, we solve Poisson s and continuity equations for both electron and hole, to more computationally challenging physical models such as the energy balance, which exemplifies some higher moment simplification of Boltzmann transport approximation (BTE). However, the choice of appropriate physical models are dependent on level of detail is required. In particular, semiconductor device structures whose electronic properties changes on a deep nanometer scale gives an number of examples of nonlinear transport models[4, 5]. In these devices nonlinear transport mechanisms are complex because electrical or optical properties are governed by different nonlinear dynamic processes. The electronic transport properties of a semiconductor material show up most by $I - V$ characteristic form under time-independent voltage bias conditions[4]. It has been determined in a complex manner by microscopic properties of material, which specifies the current density (J) with function of local electric field (E) and by electrodes. Under low voltage or current conditions, the relation between J and E do not always exist, however it is strongly dependent in high voltage or current conditions. Moreover,

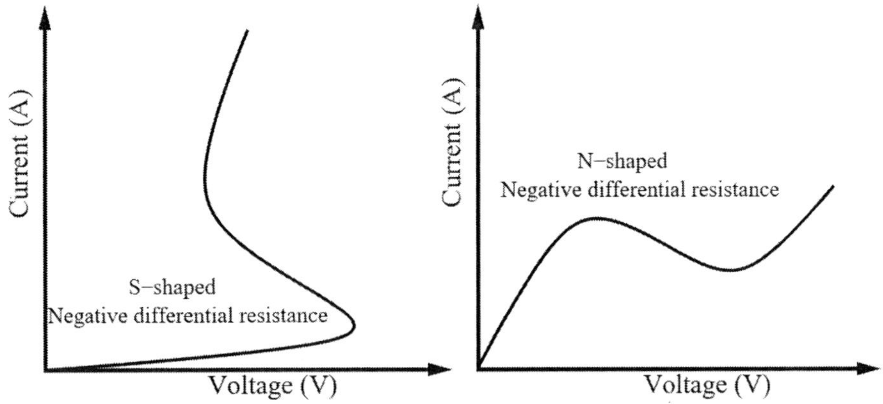

Figure 1. Typical negative differential conductivity (NDC) regions are observed in $I - V$ characteristics of the device structure under high current or voltage conditions[6].

an unstable situation has been observed when current density increases or decreases corresponds to negative differential conductivity region[6].

Two different $I - V$ characteristics of negative differential conductivity (NDC) regions have been described by S-shaped or N-shaped $J(E)$ characteristic, which have been denoted by SNDC and NNDC respectively, as shown in Figure 1. The NDC nature of the $I-V$ characteristics have been analyzed analytically and the conditions under which SNDC and NNDC curves are formed have been discussed by Scholl et al.[6]. The physics behind the formation of current filamentation in SNDC devices and Field domains in NNDC devices have also been explained[6]. There has been research work done on how to make a device simulator by solving the semiconductor equations and about the NDC operating region of the device individually but very little emphasis has been laid over the conditions to be satisfied by a simulator to observe the NDC region of the $I - V$ characteristics of a device. This paper targets at making a simulator to observe the NDC nature of the SNDC like BJT by applying appropriate boundary conditions. In this paper a generalized method is proposed to apply the appropriate boundary conditions to observe the NDC region of the $I - V$ characteristics and the same is proved by plotting the SNDC $I - V$ characteristics of a BJT. Both Neumann and Dirichlet boundary conditions are applied to the BJT and a comparison is made on choosing different boundary conditions.

The paper is organized as follows Section II briefs flow of the proposed simulator. Furthermore, two-dimensional discretization scheme is formulated in terms of basic semiconductor equations then the boundary conditions for observing the NDC curves is discussed. Finally, the results are presented and discussed for different boundary conditions.

2. Proposed Flow of Device Simulator

The basic flow of the simulator and device structure used for simulation purpose are shown in Figure 2. The structure for which the simulator is built and boundary conditions tested is a two dimensional ($2D$) NPN BJT, as shown in Figure 2(b). The structure has three contacts base, emitter and collector out of which emitter and collector contacts are heavily doped n^+ regions with $N_D = 10^{20} cm^{-3}$ and base contact is heavily doped p^+ region with $N_A = 10^{19} cm^{-3}$. The collector-base junction is formed by relatively less doped collector and base regions with $N_{Dn} = 10^{16} cm^{-3}$ and $N_{Ap} = 10^{17} cm^3$ respectively. The contact regions need to be heavily doped to have better contacts and lightly doped region near the junction helps in withstanding the practical levels of operating voltage not limited by tunneling effect. The proposed flow of device simulator is described below

(i) Using the boundary conditions as input and the matrix representation of discretized

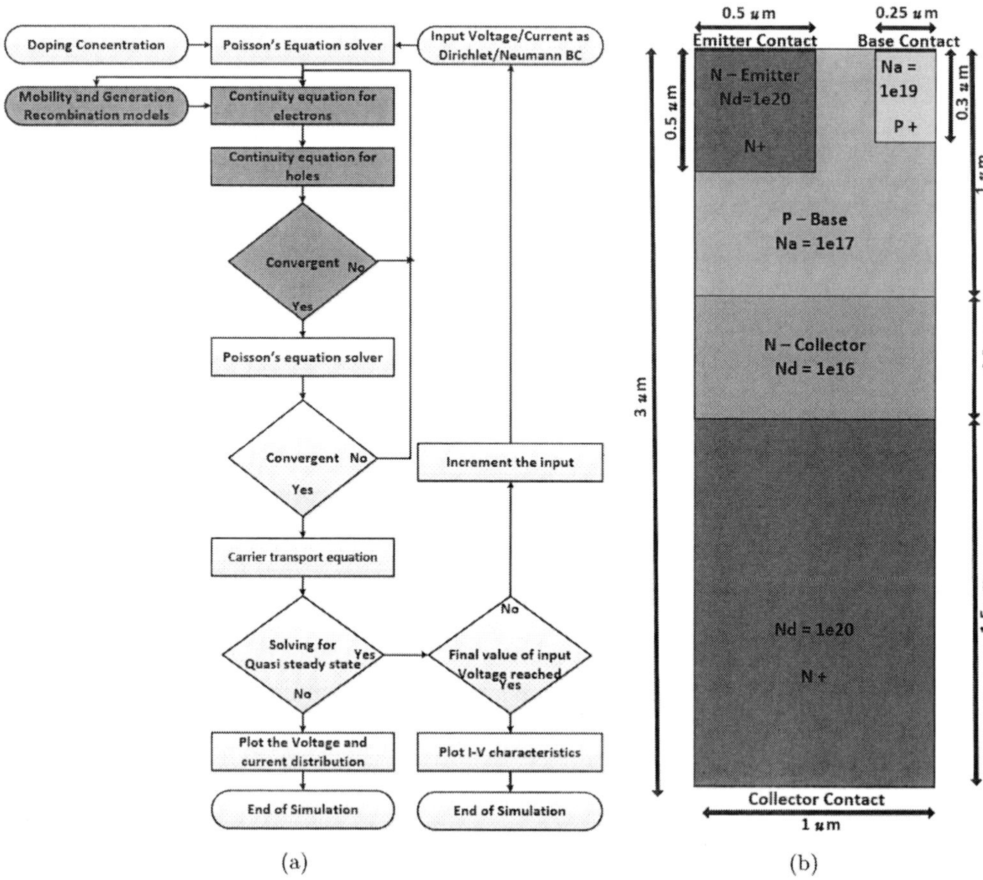

Figure 2. (a) Proposed flow of the device simulator, (b) Device under test for simulation purpose is layered BJT device structure where doping of all the regions are given.

Poisson s equation, determines the potential at every discretized node in the device structure.

(ii) From the potential obtained in Step 1, we calculate the Electric field at every point using central difference FDM method.

(iii) It can be seen that, as the mobility and net recombination rate are included into the continuity equations results non linear partial differential equation which cannot be solved directly by matrix method. Therefore, to solve these equations, we use Gummels iteration method [2], while solving the electron continuity equation consider hole concentration (p) as constant (equal to initial doping concentration for the first time) and calculate electron concentration (n).

(iv) While solving the hole continuity equation consider electron concentration as a constant and equal to that obtained in step 3 and calculate the hole concentration. Now update the hole concentration, mobility and recombination rate in step 3 and repeat these two steps till convergence is reached. At this point the electron and hole concentrations are obtained will be satisfying both the continuity equations.

(v) Substitute the electron and hole concentration obtained after convergence in step 4 into the Poissons equation and solve step 2 to step 5 until convergence is achieved. At this point the potential and carrier concentrations obtained will satisfy the Poisson s and continuity equations. Now we calculate the current density using the Drift-Diffusion equations.

(vi) Finally we plot the I-V characteristics by increasing the input and repeat step 1 to step 5 until the final value of input is reached.

2.1. Basic Semiconductor Equations

To analyze the distribution of current density and potential inside any arbitrary semiconductor device under various operating conditions it needs to be modeled mathematically using the following equations

(i) Poisson s equation
(ii) Continuity equations for electrons and holes
(iii) Carrier transport equation for electrons and holes
(iv) Mobility and Recombination-Generation rate

2.2. Discretization of Basic semiconductor equations

We need to discretized the basic semiconductor equations to solve them numerically. Though there are many ways of discretization of the Partial differential equations like finite element method (FEM), finite volume method (FVM) and finite difference method (FDM). However, we choose to discretized using five point Finite Difference Method owing to its simplicity and good accuracy. FDM Discretized basic semiconductor equations are

- Poisson s equation

$$\frac{n_d}{m_d}\psi(x+1,y)+\frac{n_d}{m_d}\psi(x-1,y)-2\left(\frac{n_d}{m_d}+\frac{m_d}{n_d}\right)+\frac{m_d}{n_d}\psi(x,y+1)+\frac{m_d}{n_d}\psi(x,y-1)= m_d.n_d\frac{q}{\epsilon}(n-p-C) \tag{1}$$

- Continuity equation for electrons

$$n(x+1,y)\left[\frac{n_d}{m_d}-\frac{n_d}{2v_t}E_x\right]+n(x-1,y)\left[\frac{n_d}{m_d}+\frac{n_d}{2v_t}E_x\right]+n(x,y+1)\left[\frac{m_d}{n_d}-\frac{m_d}{2v_t}E_y\right]+ n(x,y-1)\left[\frac{m_d}{n_d}-\frac{m_d}{2v_t}E_y\right]-n(x,y)\left[2\left(\frac{m_d}{n_d}+\frac{n_d}{m_d}\right)+temp\right]= m_dn_d\left(\frac{R(\psi,n,p)}{D_n}+\frac{1}{D_n}\frac{\partial n}{\partial t}\right) \tag{2}$$

- Continuity equation for holes

$$p(x+1,y)\left[\frac{n_d}{m_d}+\frac{n_d}{2v_t}E_x\right]+p(x-1,y)\left[\frac{n_d}{m_d}-\frac{n_d}{2v_t}E_x\right]+p(x,y+1)\left[\frac{m_d}{n_d}+\frac{m_d}{2v_t}E_y\right]+ p(x,y-1)\left[\frac{m_d}{n_d}-\frac{m_d}{2v_t}E_y\right]-p(x,y)\left[2\left(\frac{m_d}{n_d}+\frac{n_d}{m_d}\right)-temp\right]= m_dn_d\left(\frac{R(\psi,n,p)}{D_p}+\frac{1}{D_p}\frac{\partial p}{\partial t}\right) \tag{3}$$

Where

$$E_x=\frac{\partial\psi}{\partial x};\ dE_x=\frac{\partial^2\psi}{\partial x^2};\ \text{and}\ E_y=\frac{\partial\psi}{\partial y};\ dE_y=\frac{\partial^2\psi}{\partial y^2};$$

m_d/n_d = horizontal/vertical differential length and E_x, dE_x, E_y and dE_y in the continuity equation can be calculated using the central difference and five point discretization method.

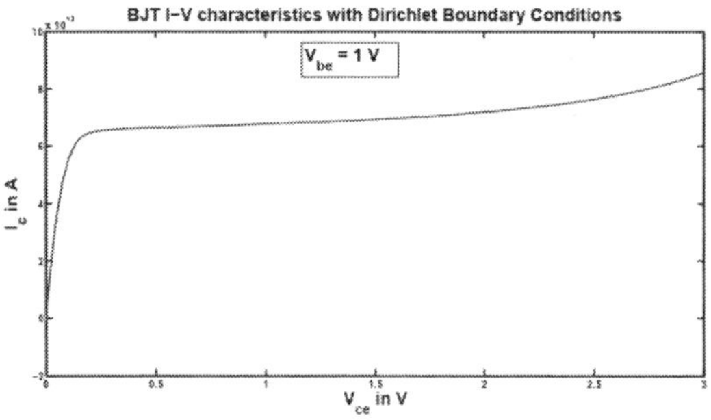

(a)

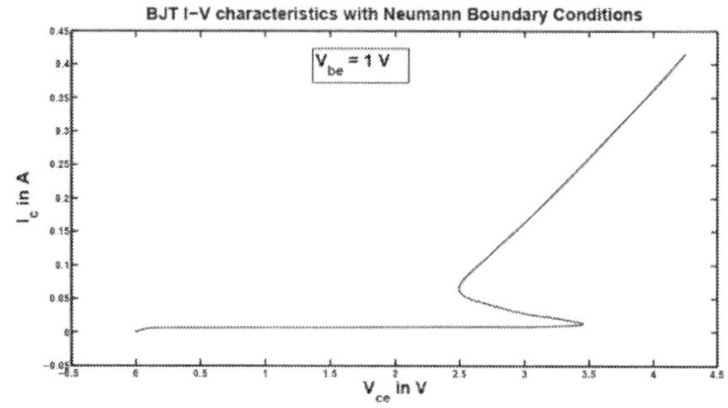

(b)

Figure 3. Proposed device simulator results when different boundary conditions are considered. (a) Dirichlet boundary conditions and voltage input are considered, (b) Neumann boundary conditions and current input are chosen.

2.3. Boundary Conditions For Observing NDC Region

Boundaries of a semiconductor device can be classified into two types such as Physical Boundaries and Artificial Boundaries. Physical boundaries include the boundaries which physically exist like contacts and interface to the insulating material while artificial boundaries are the ones which is introduced for the purpose of simulation or calculations so that it can separate one device from the neighboring devices. Boundary conditions are mathematical conditions which are specified at the boundaries to solve the partial differential equations. In general the boundary conditions are decided by the type of contacts which is dependent on the type of the input given to the device. Broadly boundary conditions are divided as

- Dirichlet boundary Dirichlet boundary condition states the value of the solution parameter so that solution of the partial differential equation needs to take at the boundaries of the domain. We solve the Poissons equation with potential as the parameter and stating Dirichlet boundary condition at the boundaries which means we are fixing the value of the potential at the boundaries which is equivalent to connecting the device to a voltage source. To plot the I-V characteristics of a device with Dirichlet boundary conditions is physically equivalent to connecting the device to a variable voltage source and checking the current flowing through the device.
- Neumann boundary Neumann boundary condition states the derivative of the of the solution parameter on the boundary of the domain that must be satisfied by the solution of the partial differential equation. Physically Neumann boundary condition is applied when you have to connect the device to a current source. When a current source is to be connected to the device it can be done in simulation terms by applying Neumann boundary conditions.

3. Results and Discussion

In the simulation it is observed that, when we apply Dirichlet boundary conditions to an SNDC device like BJT, the resulting $I-V$ characteristics are negative differential resistance as shown in Figure 3(a). But in the most cases when we use Newtons method, it doesn t converge at the first breakdown point from where the curve goes into NDC part. This is due to triggering of impact ionization in the device, which results avalanche-generated electrons and avalanche-generated holes thus current sharply increases across the terminal which results in instability in the device and does not satisfy the outer boundary conditions. However, applying Neumann boundary condition at collector contact of the BJT and Dirichlet boundary conditions at emitter and base contact, shows S-shaped negative differential resistance region as shown in Figure 3(b).

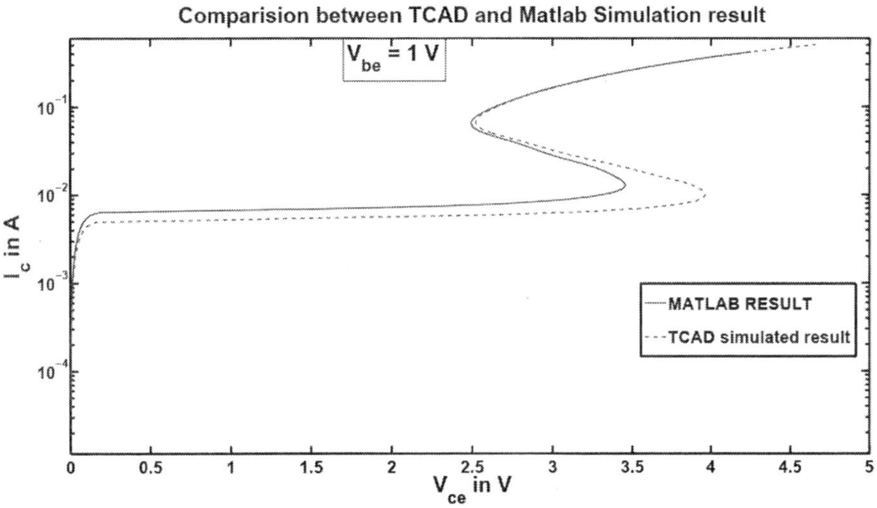

Figure 4. Comparison between commercial device simulator (Sentaurus) results and our proposed device simulator result of BJT $I - V$ characteristics with base voltage = 1 V and Neumann Boundary conditions.

To verify our results with commercial device simulator we have used Sentaurus, which is a set of TCAD tools which simulates and characterize semiconductor devices. The BJT given in Figure 4 is compared simulated results with the same base voltage and boundary conditions but using Sentaurus TCAD and the result obtained is exported and plotted in matlab and compared with the result observed from matlab. The output is showing some variation between the matlab simulated graph and TCAD simulated graph this might be because of reasons like different mobility models being used and the effects like temperature which is not used in the matlab simulator but are taken into consideration in sentaurus TCAD.

4. Conclusion

In summary, we have developed a numerical device simulator to observe negative differential conductivity region in SNDC devices such as BJT. For the simulation purpose, we have used BJT as an device structure and applied different boundary conditions across the collector terminal, It has been observed that Dirichlet boundary conditions shows NNDC behavior while applying Neumann boundary condition results SNDC nature of the device. The proposed flow of the device simulator has been discussed and further results of developed device simulator have been compared with the commercial TCAD device simulator (i.e, Sentaurus).

References

[1] P. Dubock, *"DC Numerical Model for Arbitrarily biased Bipolar Transistors in Two Dimensions"*, Electronics Letters, vol. 6, no. 3, pp. 53-55, Feb 1970.
[2] H. Gummel, *"A Self-consistent Iterative Scheme for One-dimensional Steady-state Transistor Calculations"*, IEEE Transactions on Electron Devices, vol. 11, no. 10, pp. 455-465, Oct 1964.
[3] J. Slotboom, *"Iterative Scheme for one and Two dimensional DC Transistor Simulation"*, Electronics Letters, vol. 5, no. 26, pp. 677-678, Dec 1969.
[4] D. Kennedy and R. OBrien, *"Two-dimensional Mathematical Analysis of a Planar type Junction Field-Effect Transistor"*, IBM Journal of Research and Development, vol. 13, no. 6, pp. 662-674, Nov 1969.
[5] S. Selberherr, *"Analysis and Simulation of Semiconductor Devices"*, 1st ed. Springer Science and Business Media, 2012.
[6] E. Scholl, *"Equal Areas Rules for Filamentation in SNDC Elements"*, Solid-State Electronics, vol. 29, no. 7, pp. 687-695, 1986.

High performance FDTD algorithm for GPGPU supercomputers

Andrey Zakirov, Vadim Levchenko, Anastasia Perepelkina, Yasunari Zempo

Keldysh Institute of Applied Mathematics, RAS, Miusskaya sq., 4, Moscow, 125047, Russia
Hosei University, Faculty of Computer and Information Sciences, 3-7-2 Kajino-cho, Koganei-shi, Tokyo, Japan

E-mail: mogmi@narod.ru

Abstract. An implementation of FDTD method for solution of optical and other electrodynamic problems of high computational cost is described. The implementation is based on the LRnLA algorithm DiamondTorre, which is developed specifically for GPGPU hardware. The specifics of the DiamondTorre algorithms for staggered grid (Yee cell) and many-GPU devices are shown. The algorithm is implemented in the software for real physics calculation. The software performance is estimated through algorithms parameters and computer model. The real performance is tested on one GPU device, as well as on the many-GPU cluster. The performance of up to $0.65 \cdot 10^{12}$ cell updates per second for 3D domain with $0.3 \cdot 10^{12}$ Yee cells total is achieved.

1. Introduction

The Finite Difference Time Domain (FDTD) method [1] is implemented in numerous codes for simulation of electrodynamics. The method is easily extended for simulation of other wave processes, such as acoustic and elasticity waves [2]. This paper deals with an original approach to the numerical solution of the evolutionary Cauchy problem, the Locally Recursive non-Locally Asynchronous (LRnLA) algorithms [3]. The difference between the LRnLA method and conventional methods is that the optimization of computations deals not only with layerwise computation, but traces data dependencies in a 4D space-time domain [4]. The algorithms are implemented in codes for various physics modelling [5, 6, 7, 8, 9].

The increase in performance is possible by avoiding stepwise synchronization. This is the idea of alternative algorithms for stencil computations (trapezoids, time-slicing and time-skewing [10, 11, 12, 13, 14, 15]). This approach is not wide spread, but the basics may be found in works by other authors. The so-called loop tiling and loop skewing methods result in similar algorithms for a simple domain geometry [16, 17, 18, 19]. The research on loop blocking led to creation of the cache-aware and cache-oblivious algorithms [20]. These were used for stencil computations of partial derivative equations [21, 15]. In a 1D simulation these methods lead to trapezoidal and diamond blockings of space. The generalizations to 2D and 3D are possible.

Among these approaches the LRnLA method has the following advantages. The approach takes account for the complexity of modern computers. The space-time optimization account for all parallel levels, all levels of the memory subsystem. The theory is built on the model of the available computer and allows a priori quantitative estimates of the performance of the supposed

Journal of Physics: Conference Series **759** (2016) 012100 doi:10.1088/1742-6596/759/1/012100

method implementation. The theory applies to any physics simulation with local dependencies, any amount of dimensions.

Some cache-oblivious algorithms coincide with the LRnLA algorithms for lower dimensions.

1.1. DiamondTorre algorithm and its CUDA implementation

The DiamondTorre algorithm is described as a subdivision of 4D space-time region in shapes. DiamondTorre size is defined by two parameters: the size of the base `DTS`, and its height `TH`. More detailed explanation of the algorithm, its parameters and implementation for wave equation on 1 GPU can be found in [4].

The correspondence of the DiamondTile to the components of staggered grid in X—Y projection is illustrated on fig. 2. This is defined as the basic element for DiamondTile algorithms for the FDTD scheme for Maxwell equations with the 4th order of approximation and its size is `DTS=1` by definition. It consists of the two (E-field and B-field) diamonds that are shifted along the time axis (by half of the time step) and along one coordinate axis (by 1.5 spatial step — the half width of the chosen scheme stencil). DiamondTorre consists of `TH` DiamondTile pairs, shifted in a similar manner against each other. The DiamondTorre algorithm is a process of making calculations for all points in the described 4D shape. In the GPGPU implementation, DiamondTorre is a CUDA-kernel. DiamondTorre's with the same Y-axis position are processed asynchronously by CUDA-blocks. Inside, cells with different Z-coordinates are processed by different CUDA-threads.

The DiamondTorre base size `DTS`=1 is optimal in this case. The higher `DTS` leads to higher operational intensity. But with `DTS`=2 significant performance drops are confirmed, since the data on one DiamondTile exceeds the GPU register file size, or the instruction count in one DiamondTorre kernel exceeds the instruction cache size. The register size is taken as 256 KB. If we choose the number of grid points along Z axis as $Nz = 384$ (double precision) or $Nz = 768$ (single precision), the limit of registers per CUDA-thread is estimated as $256K/(768 * 4) = 85$. This roughly corresponds to the amount of data for `DTS`=1. Also, in this case, one CUDA-kernel corresponding to DiamondTorre algorithm contains ~ 1500 instructions. One instruction takes ~ 8 B, which brings us very close to the instruction cache limit (estimated as 16 KB).

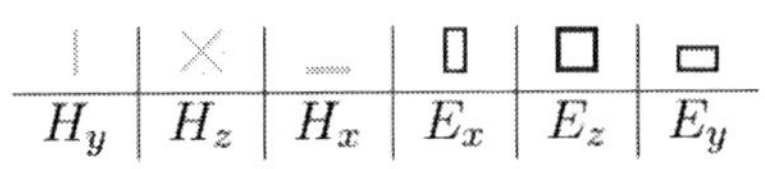

Figure 1. Field designations

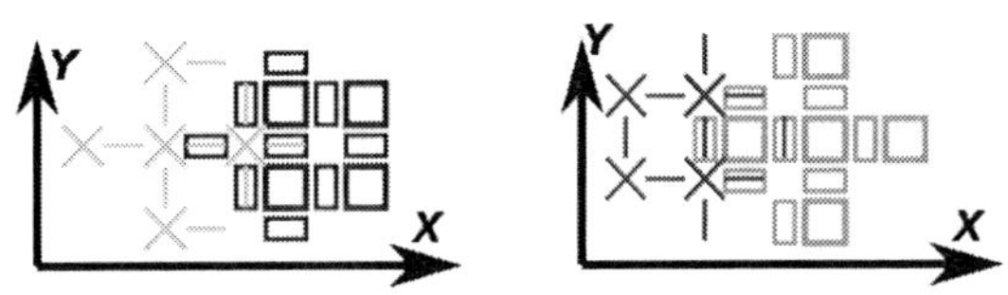

Figure 2. The field components in one DiamondTile. Two options.

If we assume that one DiamondTile is loaded into the GPU register, to calculate one B-field diamond three E-field diamonds need to be loaded. The resulting B-fields need to be saved. It concludes one half of cell updates in a Diamond, for the other half three B-field diamonds are loaded and one B-field diamond is calculated and saved. An estimate of data throughput can be made. With `TH` $\to \infty$, on average, 3 Yee cell data are to be loaded, and 1 Yee cell data should be saved per one Yee cell update, $4 \times 6 \times 8 = 192$ B total. Operation count for one cell update is about 110 Flop. Operational intensity is obtained as their ratio, 0.57 Flop/Byte.

We use roofline [22] model to estimate the efficiency of the algorithms. For any contemporary GPU our problem is memory-bound. The theoretical performance is estimated as $P/192$ Yee cell updates per second (Ys), where P is GDDR5 memory bandwidth, 192 is the necessary data throughput. On NVidia Tesla K20 ($P = 224 \cdot 10^9$ bytes per second) this equals $1.167 \cdot 10^9$ Ys.

1.2. Calculation window

The advantage of DiamondTorre algorithm is the high performance for large problems in which field data do not fit in the device memory. It is easily achieved by updating data in a "calculation window", which moves along x-axis in negative direction. Data load and save to/from the global RAM are performed asynchronously with computations. Only the data for one following group of asynchronous DiamondTorres are loaded. The data of DiamondTorres which are no longer necessary for the current `TH` update are saved and deleted from device memory.

The performance does not decline if the computation time of DiamondTorre is longer than the time that is necessary for memory copy to/from the device. The computation time increases linearly with `TH`, and the copy time is constant. So with high enough `TH` host-device transfers are concealed.

1.3. Small scale performance tests

The described algorithms are implemented in code, which features all the required methods for real physics computations: the FDTD simulation is performed in a 3D spatial domain with the 4th order accurate scheme; Perfectly Matched Layer (PML) absorbing boundary conditions, Total Field/Scattered Field (TFSF) wave source and complex material equations according to Drude, Drude-Lorenz model are used.

Small scale performance tests were conducted to find the optimal algorithm parameters. We measure performance in Yee cell updates per second (Ys). This number is usually of 10^9 order, so the main unit is GYs.

Fig. 3 shows the performance results for a problem size of $(600 \times (3 \times blocks) \times 384)$ with varied "blocks" parameter and varied `TH`. For DiamondTorres lined up along Y axis, one CUDA-block performs computation for one DiamondTorre. The tests were conducted on Tesla K20x. It has 14 streaming multiprocessors (SMs). If the number of involved SMs per device is less than 14, the performance is limited by by GDDR5 access latency. The latency is about 500 clock cycles (about 500 ns). From the Little's law [23], to cover this latency, at least 10^4 transactions are necessary.

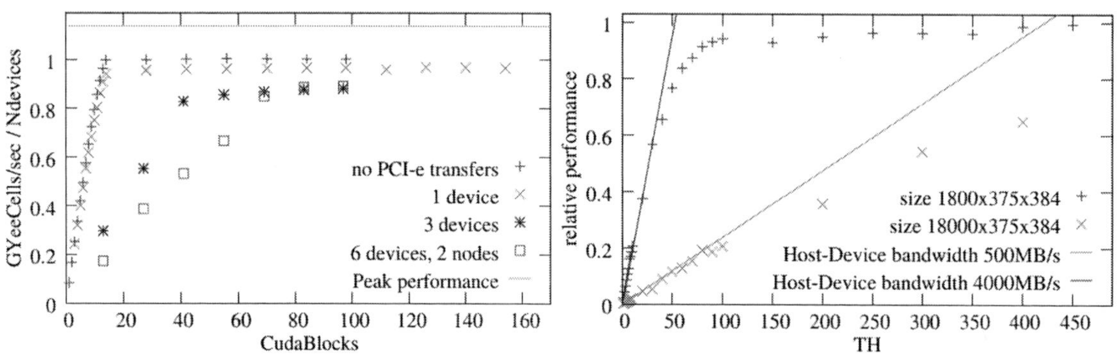

Figure 3. Performance results for varied amount of blocks. Domain size is $600 \times (3 \cdot block) \times 384$ (left). Performance results for varied TH (right). The second problem size is 10 times bigger, so it does not fit into node memory.

The register file should be enough to keep the data of all necessary diamonds. So, for the Kepler architecture, the maximum vector size (Nz, also equal to the amount of involved CUDA-threads) is equal to 384 (double precision). This is why the data throughput is not utilized completely when all 14 SMs are involved ($\sim 14 \cdot 384 \approx 6000$ transactions). It is the main reason why the performance is only 90% from the peak one. It becomes more than 10^9 GYs for sufficiently large TH. TH is inevitably smaller near region boundaries, so the performance for real problems is by few percent lower.

For low TH (fig. 3) the performance is limited by the PCI-express bandwidth. With the increase of TH the shift to limitation by GDDR5 bandwidth is confirmed. Its smoothness is explained by the cut-off of DiamondTorres at the edges of the domain. The optimal sufficient TH is 100, as can be seen on the graph. This value is used in subsequent tests, if not stated otherwise. A similar behavior of the dependency of performance rate on TH is observed when the data is stored on a disk (SSD in this case). The optimal TH becomes significantly larger (more than 500).

1.4. Multi-device and multi-node parallelism

All DiamondTorre's standing side-by-side along Y axis are asynchronous. So they may be processed by different devices inside one node, or by different nodes. Judging by the previous tests (fig. 3) the data transfers between 3 devices on one node may be concealed completely if the amount of the involved CUDA-blocks is more than 42 on each device. For the devices installed on different nodes the duration of data transfer is approximately twice bigger. In this case data transfers may be concealed completely if the amount of the involved CUDA-blocks is more than 70 on each device.

Additionally, concurrency on X axis is possible. In this case the data are subdivided in blocks in X axis and distributed between nodes. The data on adjacent nodes overlap by the calculation window size: NW points of X axis. On the i-th device the data are updated from the $(n+i\cdot TH)$-th step to the $(n+i\cdot TH{+}\mathrm{TH})$-th step, where n is an integer number (the time step on which the data of the leftmost node exists on).

2. Parallel Scaling Results on TSUBAME2.5

The parallel performance of the code has been tested on TSUBAME2.5 supercomputer. Its specifications, that are important for this study, are as follows. The amount of available node per one run is up to 300 according to the usage conditions (1408 in total). Each node has 3 NVIDIA Tesla K20x GPGPUs installed. Their total GDDR5 memory is $3\times5.625=16.875$ GB, with 208 GB data throughput each. Each node has at least 54 GB, up to 96 GB on several ones. Only a little above 40 GB from it is available for the computation. Devices are connected with PCI-e 2.0 with 4 GB/sec throughput in each direction. Each node has 120 GB SSD memory. Nodes are connected by Infiniband QDR interconnect with up to 4 GB/sec throughput.

2.1. Weak scaling

For the weak scaling the domain is scaled proportionally to the amount of nodes used. Both X and Y axis parallelism were tested. Three series were performed (fig. 4):

(i) Y axis scaling. Data transfers may be concealed (112 CUDA-blocks on each device).

(ii) Y axis scaling. Data transfers limit the performance (42 CUDA-blocks on each device).

(iii) X axis scaling. Each device contains 1890 Yee cells. All 3 devices are involved on each node, performing parallel computation in Y direction.

In the first case, the parallel efficiency is above 99% as expected. The maximum achieved performance is $0.65\cdot10^{12}$ Ys for one computation on domain with $300\cdot10^{9}$ grid points (10 TB data, 256 nodes). In the second case the performance becomes lower when the number of nodes rises from 1 to 2. This is because the Infiniband data throughput is lower than data throughput between devices on one node. The following increase in the number of nodes does not lower the performance. In the third case parallel scaling is close to ideal, but still less than the one for the first series. The main decrease occurs between 1 and 2 nodes. It is caused by a slight imbalance in node utilization.

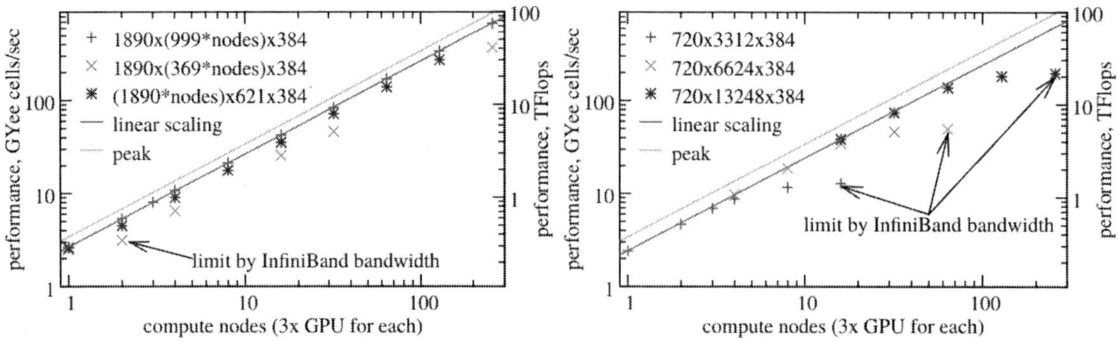

Figure 4. Weak scaling (left). Strong scaling (right)

2.2. Strong scaling

For the strong scaling we chose a fixed size domain. By increasing the amount of nodes, the domain is subdivided to more and more parts, that are to be processed concurrently. Three series were performed (fig. 4):

(i) $720 \times 3312 \times 384$ size domain is scaled on 1–32 nodes;

(ii) 4 times bigger domain ($720 \times 13248 \times 384$) is scaled on 4–64 nodes;

(iii) 16 times bigger domain ($720 \times 52992 \times 384$) is scaled on 16–256 nodes;

With the increase in amount of parallel nodes the performance decreases since the amount of CUDA-blocks per device becomes lower (see fig. 3). At the same time there is a limit on maximum Ny size per device, determined by device memory. It actually may be increased, but the Nx size and `TH` should be decreased at the same time. This would lower the general performance. By utilizing more nodes, domain size and `TH` may be increased again and performance rises.

This dependency on `TH` is shown on fig. 5. The problem size is $450 \times 62208 \times 128$ grid cells. For one node computation `TH` is equal to 15, and increases up to 150 for 8 nodes and higher. For low amount of nodes the speedup is better than linear. This is because `TH` may be optimized only when enough data is processed on each node.

Finally, strong scaling tests were performed on a problem with $38400 \times 363 \times 128$ grid points (fig. 5). On one node the performance is about 30% from the peak performance, since the size along Z axis is not optimal. It is not big enough to conceal the GDDR5 access latency, since the amount of simultaneous transactions is too low. But the acceleration is up to 40 times. Only with 128 nodes and above the data transfers are taking more time than computation, which leads to decrease in computation rate.

One node has little memory size (only about 3 times more than total device memory), and this becomes the reason for the acceleration limit. The increase in the available memory size should increase acceleration ability proportionally.

3. Conclusion

The work can be summarized as follows. The FDTD code has been developed, that allows simulation of real optical phenomena. The distinguishing feature of the code is the use of DiamondTorre LRnLA algorithm, which maximizes the performance on one device, and parallel efficiency for multi-GPU architectures. The software was tested on the TSUBAME2.5 supercomputer. The high computation rate is achieved (more than 1 billion Yee cell updates per second on one device). The problem size is not limited by device memory. The scaling for ~ 1000 devices becomes 1000 for the weak scaling, and ~ 100 for the strong scaling.

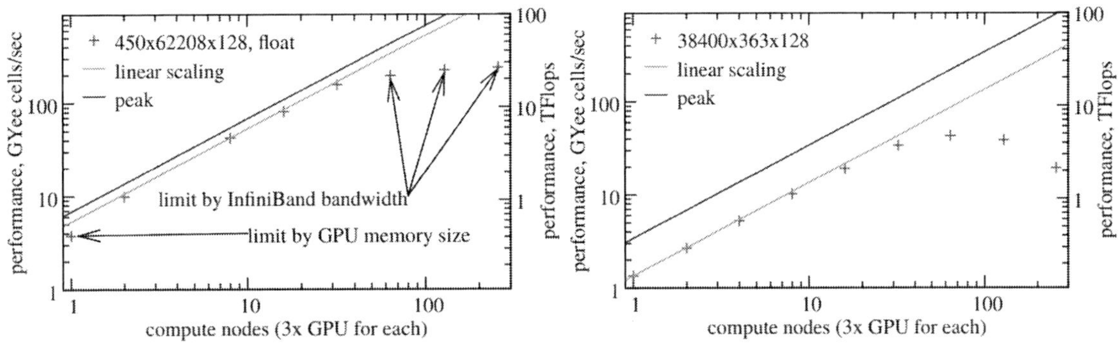

Figure 5. Strong scaling on Y axis with variable TH (left), on X axis (right)

The algorithm parameters (such as TH and problem size) allow not only qualitative, but also quantitative estimates of the performance and parallel scaling. Maximal achieved performance is $0.65 \cdot 10^{12}$ for 3D domain with $0.3 \cdot 10^{12}$ Yee cells total. For example, such size for wave optics problems corresponds to 1 cubic millimeter domain. This allows a significant breakthrough in computational nanooptics, by allowing the simulation in domains that were previously too big even for supercomputers. It may be used for simulation of complex optical devices.

The work is supported by Hosei International Fellowship grant, RFBR grant no. 14-01-31483.

References

[1] Taflove A and Hagness S C 2005 *Computational Electrodynamics: the Finite-Difference Time-Domain Method* 3rd ed (Norwood, MA: Artech House)
[2] Virieux J 1986 *Geophysics* **51** 889–901
[3] Levchenko V 2005 *J. of Inf. Tech. and Comp. Systems* **1** 68 (in Russian)
[4] Perepelkina A Y and Levchenko V D 2015 *Keldysh Institute Preprints* **18** 20
[5] Korneev B A and Levchenko V D 2015 em Procedia Computer Science **51** 1292-1302
[6] Perepelkina A Y, Goryachev I A and Levchenko V D 2013 *Journal of Physics: Conference Series* **441** 012014
[7] Perepelkina A, Goryachev I and Levchenko V 2014 *Journal of Physics: Conference Series* **510** 012042
[8] Zakirov A.V., Levchenko V.D. 2012 *Mathematical Models and Computer Simulations* **4** 2 pp 155–162.
[9] Kaplan S.A., Levtchenko V.D. et al. *Geoinformatika* 2011. **1** pp 49-55. (in Russian)
[10] Frigo M and Strumpen V 2007 *The Journal of Supercomputing* **39** 93–112
[11] Tang Y *et al.* 2011 *Proceedings of the Twenty-third Annual ACM SPAA* (NY, USA: ACM) pp 117–128
[12] Grosser T *et al.* 2013 *Proceedings of the 6th Workshop on General Purpose Processor Using Graphics Processing Units* GPGPU-6 (New York, NY, USA: ACM) pp 24–31
[13] Orozco D, Gao G 2009 Mapping the fdtd application to many-core chip architectures *Parallel Processing, 2009. ICPP '09* pp 309–316 ISSN 0190-3918
[14] McCalpin J, Wonnacott D 1999 Time skewing: A value-based approach to optimizing for memory locality Tech. rep.
[15] Strzodka R *et al.* 2011 *Proceedings of the International Conference on Parallel Processing* IEEE Computer Society p 571–581
[16] Wolf M, Lam M 1991 *Proceedings of the ACM SIGPLAN 1991* (New York, NY, USA: ACM) pp 30–44
[17] Wolfe M 1989 *Proceedings of the 1989 ACM/IEEE Conference on Supercomputing* Supercomputing '89 (New York, NY, USA: ACM) pp 655–664
[18] Wolfe M 1986 *International Journal of Parallel Programming* **15** 279–293
[19] Terrano A 1988 *Frontiers of Massively Parallel Computation, 1988. Proceedings* pp 227–229
[20] Prokop H 1999 Cache-oblivious algorithms
[21] Frigo M and Strumpen V 2005 *Proceedings of the 19th Annual International Conference on Supercomputing* pp 361–366
[22] Williams S, Waterman A and Patterson D A 2009 *Commun. ACM* **52** 65–76
[23] Little J D C 1961 *Operations Research* **9** 383–387